Animals in primary succession

The role of fauna in reclaimed lands

Edited by

J. D. MAJER

Senior Lecturer, School of Biology, Curtin University of Technology, Perth, Western Australia

The right of the University of Cambridge to print and sell all manner of books was granted by Henry VIII in 1534. The University has printed and published continuously since 1584.

CAMBRIDGE UNIVERSITY PRESS

Cambridge

New York Port Chester Melbourne Sydney

Published by the Press Syndicate of the University of Cambridge
The Pitt Building, Trumpington Street, Cambridge CB2 1RP
40 West 20th Street, New York NY 10011, USA
10 Stamford Road, Oakleigh, Melbourne 3166, Australia

First published 1989

Printed in Great Britain at the University Press, Cambridge

British Library cataloguing in publication data

Animals in primary succession
1. Reclaimed land. Animals. Ecological
aspects
I. Majer, J. D. (Jonathan D.)
591.52

Library of Congress cataloguing in publication data

Animals in primary succession.
Bibliography: 67 pp.
Includes index.
1. Reclamation of land—Environmental aspects.
2. Animal ecology. 3. Ecological succession.
I. Majer, J. D. (Jonathan David), 1949–
QH545.R38A55 1988 574.5′264 88-23699

ISBN 0 521 33400 4

UP

Contents

Contributors

Abbott, I. Department of Conservation and Land Management, P.O. Box 104, Como, Western Australia, 6152, Australia

Allen, N. T. Worsley Alumina Pty. Ltd., P.O. Box 50, Boddington, Western Australia, 6390, Australia

Butcher, J. School of Biology, Curtin University of Technology, Kent Street, Bentley, Western Australia, 6102, Australia

Dunger, W. Staatliches Museum fur Naturkunde, 89 Görlitz, Postschliessflach 425, German Democratic Republic

Eijsackers, H. Research Institute for Nature Management, Post Box 9201, 6800 HB, Arnhem, the Netherlands

Fowler, D. K. Tennessee Valley Authority, Norris, Tennessee 37828, USA

Hutson, B. R. CSIRO, Division of Soils, Private Bag P.O., Glen Osmond, South Australia, 5064, Australia

Laycock, W. A. Department of Range Management, University Station, Box 3354, Laramie, Wyoming, 82071, USA

Klimstra, W. D. Co-operative Wildlife Research Laboratory, Southern Illinois University at Carbondale, Illinois, 62901, USA

Ma, Wei-Chun Research Institute for Nature Management, Post Box 9201, 6800 HB, Arnhem, the Netherlands

Majer, J. D. School of Biology, Curtin University of Technology, Kent Street, Bentley, Western Australia, 6102, Australia

Meijer, J. Herestraat 56, 9843 AL, Grijpskerk, the Netherlands

Nawrot, J. Co-operative Wildlife Research Laboratory, Southern Illinois University at Carbondale, Illinois, 62901, USA

Nichols, O. Alcoa of Australia Limited, P.O. Box 252, Applecross, Western Australia, 6153, Australia

Parker, C. A. Department of Soil Science and Plant Nutrition,

University of Western Australia, Nedlands, Western Australia, 6009, Australia

Recher, H. Department of Ecosystem Management, University of New England, Armidale, New South Wales, 2351, Australia

Samuel, D. E. Division of Forestry, West Virginia University, P.O. Box 6125, Morgantown, West Virginia 26506-6125, USA

Unsworth, P. School of Biology, Curtin University of Technology, Kent Street, Bentley, Western Australia, 6102, Australia

Urbanek, R. Seney National Wildlife Refuge, Seney, Michigan, 49883, USA

Viert, S. R. Cedar Creek Associates Inc., P.O. Box 9557, Fort Collins, Colorado, 80525, USA

Whelan, R. J. Department of Biology, University of Wollongong, P.O. Box 1144, Wollongong, New South Wales, 2500, Australia

Wykes, B. J. School of Biology, Curtin University of Technology, Kent Street, Bentley, Western Australia, 6102, Australia

Preface

We frequently encounter land which has been grossly disturbed by mining or some other type of land-use. Until relatively recently, and still in many developing countries, this land tended to be left in its disturbed state with no form of reclamation applied to it. Following the advent of increased environmental awareness in the sixties, and the passing of various environmental protection and land reclamation acts, companies are often now required to reclaim the disturbed land. This work is sometimes performed by the company itself or by consultants or government agencies. Zoologists are seldom involved with the process.

This book brings together the information on animals in reclaimed lands and is divided into six Parts:

Part 1 is an introductory Chapter which outlines the types of land-use which are considered in this book and the various ways in which fauna can influence the reclamation of such lands. In Part 2 a series of five Chapters examines in detail the ways in which animals influence various facets of the developing ecosystem on reclaimed land. The third Part then looks at some of the more practical aspects of fauna return to reclaimed land. It considers some of the fauna-associated problems which may be encountered during the reclamation process and some of the benefits, either ecological or economic, which may accrue from a proper consideration of the part played by fauna during the reclamation process.

A selection of case studies which have been chosen to reinforce the content of the preceding Chapters is presented in the fourth Part of this book. Finally, after a synthesis Chapter in Part 5, a fully cross-referenced bibliography of fauna studies which have been performed in relation to reclaimed lands, is provided.

Coverage of the material in this book has been designed with two types of reader in mind. It is aimed at practitioners such as mine-rehabilitation officers, soil-conservation officers, urban and regional planners and other such staff. In addition, the Chapters highlight the ecological aspects of reclamation so the book should be of value to students, lecturers, research workers and other professional ecologists.

Preparation of this book involved considerable expense. I wish to thank the sponsors who contributed financially to this project, namely: the Western Australian Departments of Conservation and Environment, Conservation and Land Management and Main Roads and also Alcoa of Australia Limited, Allied Eneabba Limited, Associated Minerals Consolidated Limited, Greenbushes Tin N.L., Westralian Sands Limited and Worsley Alumina Pty. Ltd. I also wish to thank the staff of the numerous American, Australian, British and Dutch mining companies, universities or government departments who provided me with information and facilities when I visited them during the initial planning stages of this book.

I am particularly grateful to the following persons who refereed individual Chapters: Ian Abbott, Geoffrey Anderson, Barry Carbon, Lynn Day, Barry Fox, John Fox, Gordon Friend, John Greenslade, Penelope Greenslade, Jan Henry, Richard Hobbs, Barry Hutson, John Janes, Lucien Koch, Tim New, Owen Nichols, Harry Recher, James Ridsdill-Smith, David Samuel, Denis Saunders, Robert Whelan and Ken Youngson. Finally, I also wish to acknowledge the efforts of Sallie-Ann Grayson for typing the manuscript, Francoise Berlandier and Sean Kelly for drafting many of the diagrams, John van Schagen for assistance with proof reading and the editorial staff of Cambridge University Press for their help in various aspects of the preparation of this book.

Jonathan Majer
February 1988

Part 1

General introduction to the role of fauna in reclaimed lands

Introduction

J. D. MAJER

Many books have been written about land reclamation but few have considered the role of animals in this important aspect of land management. This lack of attention to fauna in the literature is also reflected on-site. Very few mining companies, or other bodies concerned with land reclamation, employ zoologists.

This is worrying because the success of most reclamation programmes is ultimately connected with the animals which colonize the land. The fertility and structure of the soil may be impeded if the right soil animals are not present and the nature of the plant succession is moulded by the activities of herbivores, pollinators and seed dispersal agents. Other animals offer new opportunities for the after-use of disturbed land. For instance gravel pits may provide valuable wildfowl refuges in what were formerly fields, new fishing and hunting areas may be created among the mines and spoil dumps of coal mining areas, and valuable agricultural grazing land may be produced on reclaimed mined areas or sea beds.

This introductory Chapter commences with an outline of the types of land disturbance which bring about the need for reclamation. It then outlines some of the ecological principles which are relevant to ecosystem development. The Chapter goes on to review the ways in which the role of fauna can be considered during the planning of reclamation and how it can influence the outcomes. It also outlines the practical value of encouraging fauna in such areas and highlights the potential contribution of reclaimed areas to fauna conservation. Finally, the history of reclamation research which has involved fauna is outlined.

The types of land reclamation covered in the book are generally those where a primary succession is involved. For instance, where plants and animals have been removed from an area by mining, where the community has been destroyed by extreme pollution, where land has been reclaimed from the sea, or where wetlands have been created in previously terrestrial habitats. It does not cover examples where land has been changed from one type of vegetation to another such as farmland to forest or native forest to plantation. However, because of the deterioration of dry-land agricultural soils in many parts of the World, a short Chapter on the role of soil biota in the rehabilitation of such lands is included.

1

Fauna studies and land reclamation technology – a review of the history and need for such studies

J. D. MAJER

Introduction

Throughout the World land is constantly being disturbed by draining, infilling, dredging, road or pipeline construction, mining, quarrying and various forms of industrial development. Following the transient use of this land attempts are often made to reclaim it to some productive or aesthetically acceptable state. The term *reclamation* is an umbrella one which covers all aspects of any process which encourages the re-use of degraded land. Included here are planning, engineering and usually, but not always, biological processes. A non-biological option may involve bitumenizing a mined out area to provide a water catchment. The biological option often involves preparation of the substrate followed by the establishment of a plant community such as a forest, crop-land, native vegetation or merely a stabilizing plant cover.

The reconstruction of ecosystems in these areas is a complicated process and the technology is, as yet, inadequate for many disturbed areas. It involves the expertize of people from a range of professions including ecologists, engineers, landscape architects and industrialists (Bradshaw, 1983). It follows that a considerable amount of research has been performed on many aspects of land reclamation.

Much of this work has entered the literature as reports, papers and conference proceedings and a number of bibliographies have been compiled on specific areas of reclamation. It is noteworthy that while

Table 1.1 *Breakdown of the number of publications concerning fauna in a selection of land reclamation bibliographies*

Name of bibliography	Number of references			
	Total	Wildlife	Fisheries	Grazing
Anon (1975). *Reclamation of Coal Mined-Land: a Bibliography with Abstracts*. Monroeville: National Coal Association.	717	17 (2.4%)	7 (1.0%)	9 (1.3%)
Anon (1980). *Western Energy and Land Use Team Publications: A Bibliography with Abstracts*. US Fish and Wildlife Service Report W/COAL-80/W28.	87	40 (46.0%)	38 (43.6%)	1 (1.1%)
Coaldrake, J. E. & Beattie, K. J. (1974). *Annotated Bibliography on the Ecology and Stabilization of Coastal Sand Dunes, Mining Spoils and other Disturbed Areas. Supplement No.* 1. Canberra: CSIRO Division of Plant Industry.	363	0 (0.0%)	0 (0.0%)	0 (0.0%)
Coaldrake, J. E., McKay, M. & Roe, P. A. (1973). *Annotated Bibliography on the Ecology and Stabilization of Coastal Sand Dunes, Mining Spoils and other Disturbed Areas*. Canberra: CSIRO Division of Plant Industry.	477	0 (0.0%)	0 (0.0%)	0 (0.0%)
Czapowskyj, M. M. (1976). *Annotated Bibliography on the Ecology and Reclamation of Drastically Disturbed Areas*. USDA Forest Service General Technical Report NE-21	591	8 (1.4%)	9 (1.5%)	0 (0.0%)
Gifford, G. F., Dwyer, D. D. & Norton, B. E. (1972). *Bibliography of Literature Pertinent to Mining Reclamation in Arid and Semi-Arid Environments*. Logan: Utah State University.	312	2 (0.6%)	4 (1.3%)	0 (0.0%)

Glenn-Lewin, D. C., Fay, G. & Cecil, S. D. (1976). *Bibliography of Strip Mine Ecology*. Iowa: Energy and Mineral Resources Research Institute.	595	23 (3.9%)	21 (3.5%)	2 (0.3%)
Lang, P. J. (1985). *Annotated Bibliography on Mining Rehabilitation: A Selected Bibliography on Australian and Overseas References completed primarily for use in South Australia, 1932–1982*. Adelaide: Department of Environment and Planning.	626	2 (0.3%)	0 (0.0%)	0 (0.0%)
Lewis, L. R., LaFevers, J. R., Perry, A. O. & Rice, Jr., W. (1976). *Integrated Mined-Area Reclamation and Land Use Planning. Volume* 4, *A Bibliography of Integrated Mined-Area Reclamation and Land Use Planning, with Annotations*. Argonne: Argonne National Laboratory.	424	6 (1.4%)	1 (0.2%)	0 (0.0%)
Smith, J. R., Carpenter, M. P. & Nawrot, R. J. (1979). *Annotated Bibliography of Illinois Mined Lands: Characterization, Utilization, Reclamation*. Carbondale: Southern Illinois University.	270	34 (12.6%)	7 (2.6%)	11 (4.1%)
Yundt, S. E. & Booth, G. D. (1978). *Bibliography, Rehabilitation of Pits, Quarries, and other Surface Mined Lands*. Ontario: Ministry of Natural Resources.	485	3 (0.6%)	1 (0.2%)	1 (0.2%)

many of the papers in these bibliographies concern the soil and vegetation in disturbed areas, few publications consider the fauna (Table 1.1). On average 6.3% of the publications make at least some reference to wildlife while 4.9% and 0.3% consider fisheries and livestock grazing respectively. This trend is reflected in the type of personnel employed by companies and agencies concerned with land reclamation. While agronomists, soil scientists and, to a lesser extent botanists abound, zoologists are seldom employed. I know of only three zoologists who are currently employed by the entire Australian mining industry.

Animals play a crucial role in the processes within mature and developing ecosystems situated on previously disturbed land. Also, in view of their potential economic or aesthetic value, they are of interest to land reclamation. For instance, mined areas may be reclaimed to provide new fisheries, productive grazing land or areas of significant wildlife conservation value.

Role of this book

There is clearly a need to bring together the current knowledge about the ways in which fauna may contribute to reclamation technology. This book was written to fill this gap in the literature. It describes how the presence of certain animals can assist, or impede, the process of ecosystem restoration and also how reclamation may be designed to cater for economically or aesthetically valuable species such as fish, cattle or wildlife. It also discusses the implications to recolonization of such problems as toxic substrates, unfavourable climatic zones or pest outbreaks.

The book integrates the information on animals with that on plants, soils and site factors. By doing so it is hoped that it will complement the already wide range of books which have been written about land reclamation. Included in this list are general reclamation texts by Bradshaw & Chadwick (1980), Cairns (1980), Chadwick & Goodman (1975), Down & Stocks (1977), Gemmell (1977), Goodman, Edwards & Lambert (1965), Hackett (1971, 1977), Holdgate & Woodman (1978), Hutnick & Davis (1973), Schaller & Sutton (1978) and University of Newcastle Upon Tyne (1972). Other relevant texts concern oil spill amelioration (Cairns & Buikema, 1986) and microbial aspects of soil reclamation (Tate & Klein, 1985). Coverage in the present volume is largely confined to terrestrial ecosystems although it does include a Chapter on wetlands created within terrestrial habitats. Those interested in the reclamation of rivers and streams are referred to the recent text by Gore (1985).

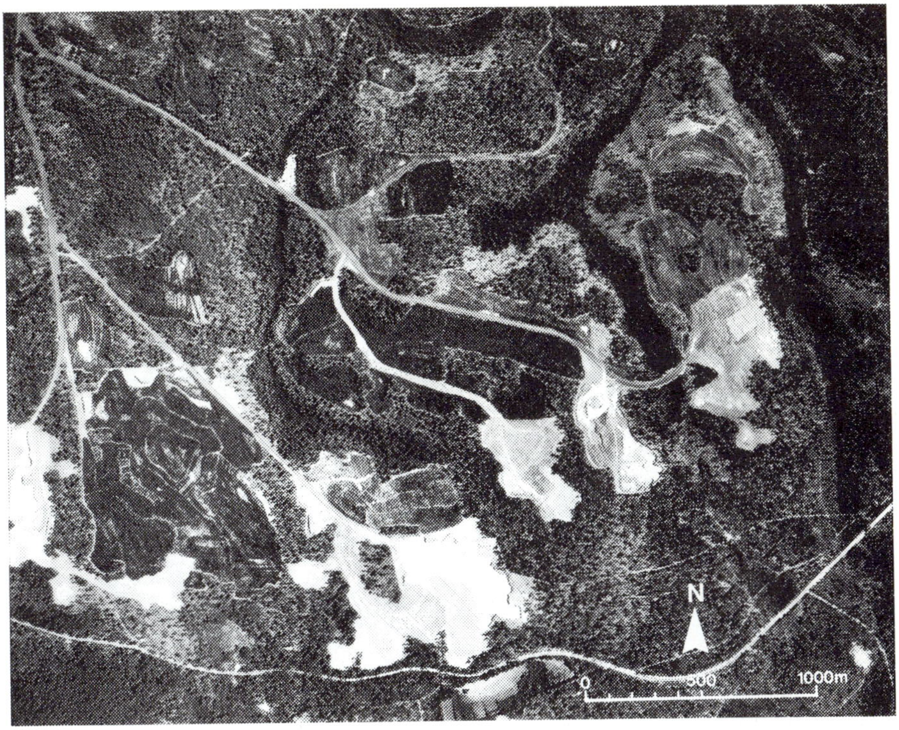

Fig. 1.1. Aerial view of a bauxite mine at Jarrahdale, Western Australia showing the distribution of mining pods within the jarrah (*Eucalyptus marginata*) forest (photograph by courtesy of Alcoa of Australia Ltd.).

Ecological disturbance

Before considering the reclamation process we should first look at what happens to the ecosystem when it is subjected to a transient land-use. This very much depends on the type of disturbance which is involved.

During surface-mining, quarrying or similar excavation work, the overlying vegetation is totally destroyed. Topsoil may, or may not, be removed for re-use and the underlying ore and overburden bodies may be taken. The overburden is often replaced and this may be followed by topsoil replacement. Most Australian bauxite mining operations follow this procedure (Tacey & Glossop, 1980). This type of operation totally destroys the vegetative parts of plants and much of the associated fauna which inhabited the soil, litter and vegetation. When the pod-type of mining is practiced, as in the south-west Australian bauxite mines, the direct impact on the ecosystem may be confined to the mined area and the area of disturbance which often surrounds it (Fig. 1.1). When extensive linear surface-mining takes place, as in the mineral sand mines

Fig. 1.2. Aerial view of coastal plain at Capel, Western Australia, showing three mineral sand surface-mining operations (V). The largest mined area is approximately 7 km long and 250 m wide (photograph by courtesy of Westralian Sands Ltd.).

of western and eastern Australia, a temporary barrier to movement of animals and plant propagules may also be formed (Fig. 1.2). This may destroy the connectivity between the pristine sites on either side of the mine (Bridgewater, 1987).

Mining and mineral processing also give rise to dumps of waste material. These may be situated in the surrounds of the mine or, in the case of mineral processing, occur at sites remote from the source of the extraction. In both cases these deposits often cover and destroy the underlying ecosystem. Overburden dumps may consist of extremely coarse material such as reject slate quarry waste (Sheldon & Bradshaw,

1976) and the sides of these dumps may be steep, with slopes of up to 40° (see review by Fox, 1984). Slime dams, slurry ponds and similar impoundments are formed when the residues from ore processing are deposited in surface depressions or artificially dammed areas. These also destroy the underlying ecosystem, frequently in areas representing localized low-land communities. The resulting areas usually possess substrates of small particle size and high moisture content. Other activities which lead to the infilling of low-lying communities include industrial waste disposal dumps such as those found in the Lower Swansea Valley (Bromley & Humphries, 1979) and urban land-fill areas (Pandit & Fotedar, 1982). The latter type of disposal tends to be carried out in wetlands – a habitat which is already becoming scarce in most parts of the World. The substrates produced by the land-uses mentioned in this paragraph may be chemically hostile to the developing ecosystem. For instance, sulphide-containing ores associated with some coal mines may oxidize on contact with the atmosphere to produce acid wastes. Other wastes are strongly alkaline or soluble salts may be present in excess. Some areas may be affected by heavy metal toxicity or, in the case of urban land-fill, a whole range of pollutants.

Another activity which leads to the formation of new land is the draining of the sea to produce land for agriculture and urban development. The extensive polders in the Netherlands are an excellent example (Ente & De Glopper, 1985). A related activity is the dredging of ports, channels and basins to maintain safe and efficient port facilities. The dredged material may be deposited in marshes, upland sites or in the water to produce new islands (Anon., 1978). Both the drained lands of the polders and the areas of dredged material create a habitat or substrate which initially lacks terrestrial organisms. They therefore provide a substrate which requires management in a similar way to those created by mining, urbanization or industry.

This review of some of the land-uses which produce new land indicates how ecosystems are, at least for a time, destroyed or drastically disturbed. The areas in question are considerable. For instance, Britain had about 340 000 ha of derelict land by 1983 (Bradshaw, 1983). In 1968, the US Bureau of Sport, Fisheries and Wildlife identified over 800 000 ha of wildlife habitat which had been damaged by mining (Spaulding & Ogden, 1968). In addition to the area directly affected by mining, there may be a multiplier effect due to wastes or pollutants entering streams or other aquatic systems. For instance, in the United States acid and heavy metals from the Blackbird mine dump have killed all aquatic life in the nearby creek. The destruction of the fishery downstream of this area has caused an estimated annual loss of US$ 1.5 million (Richardson, 1980). The problems associated with substrate toxicity and the impact of this on the fauna of reclaimed areas is discussed in Chapter 9 by Wei-Chun Ma and Herman Eijsackers.

Another multiplier effect results from the fragmentation of the original habitat by the imposition of mining or other such land-uses throughout the landscape (Fig. 1.1). The remaining fragments may effectively become habitat islands of varying shape, size and degree of isolation (Harris, 1984; Saunders *et al.*, 1987). Thus if the principles of island biogeography put forward by MacArthur & Wilson (1967) are accepted, it is likely that the conservation status of fauna in these remaining habitat islands could well be reduced.

To summarize, using Rykiel's (1979) disturbance characteristics, most of the above land-uses have the following characteristics:

Characteristic	*Impact or value*
1. Initial predominant effect	Ecosystem destroyed
2. Frequency (number of times disturbance occurs in specified time interval)	Once
3. Duration (length of time disturbance lasts)	1–> 60 years
4. Intensity (measure of severity)	Biomass eliminated
5. Reliability (probability of recurrence of the disturbance)	$P < 0.001$
6. Level (ecological echelon at which disturbance is considered)	Ecosystem
7. Scale (area which is affected)	1–> 1000 ha

The product of the disturbance is an area lacking plants and animals, often with a complete absence of soil. The 'soil' which is present often lacks stratification, organic matter, nutrients and, sometimes, adequate water. Examples of these skeletal soils include china clay wastes (Roberts *et al.*, 1982) or the red mud residues left over from the bauxite refining process (Southwell & Majer, 1982). In other areas, such as the Australian bauxite mines, the original topsoil is placed back over the mined area and only differs from its original condition in terms of underlying substrate and its physical arrangement.

In most instances it is desirable to reclaim these disturbed areas to provide some new use for the disturbed land. Before looking at the

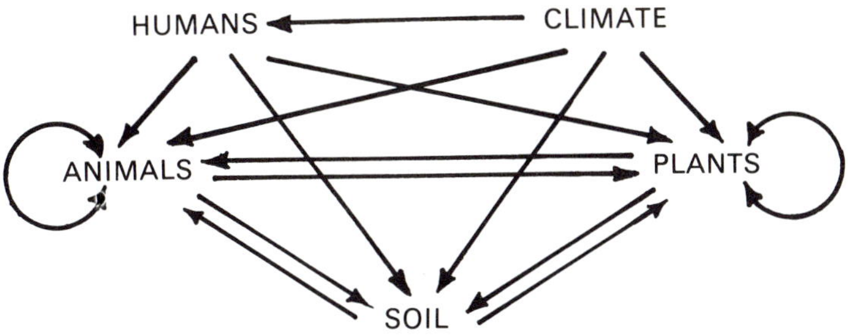

Fig. 1.3. The components of the ecosystem which should be considered during the land reclamation process (from Bradshaw, 1984).

involvement of animals in this process we should first identify exactly what is meant by reclamation.

Reclamation, restoration and rehabilitation

When the period of transient land-use has expired the land may be reclaimed to some new use. The umbrella term *reclamation* has already been defined. A biologically-based option is often preferred so *reclamation* usually involves the process of ecosystem reconstruction. Since the ecosystem consists of physical and biological components, including people (Fig. 1.3), all parts should be considered. Even in the absence of reclamation, the degraded land will often be colonized by plants and animals (Burley & Hopkins, 1984; Chapman, McGinnes & Downing, 1978). However, the resulting succession may not be compatible with, or rapid enough for, achieving the desired new land-use. In order to achieve the desired outcome the area is often rehabilitated. Langkamp (1979) defines *rehabilitation* as the return of the area to a proper (biological) state. This could mean the return of an area to long-term productive use, such as the establishment of an income earning enterprize, or for the provision of less tangible recreational or aesthetic values. He divides rehabilitation into *conditional rehabilitation*, which is effective as long as people can step in and augment or counteract natural phenomena (e.g. managing and fertilizing pasture), and *self-sustaining rehabilitation*, which is the management of the area to a state where human inputs are no longer required, nutrient cycles are closed and the components of the biota are more or less in dynamic equilibrium (e.g. a forest of native vegetation).

Bradshaw (1984) considers the process of rehabilitation by a 2-dimensional graph, with ecosystem function (biomass and nutrient content) and ecosystem structure (species and complexity) as the two

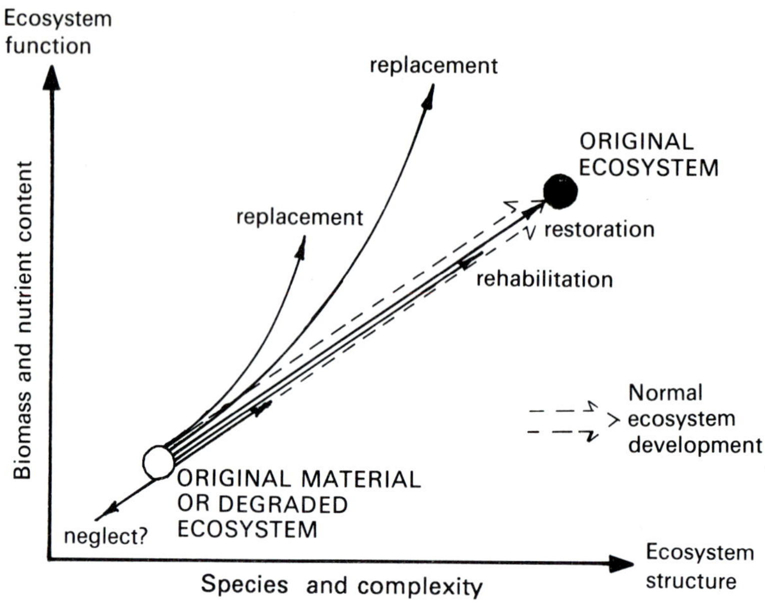

Fig. 1.4. Relationship between ecosystem structure and function during natural succession and various types of reclamation (from Bradshaw, 1984).

axes (Fig. 1.4). Natural development of the ecosystem, or succession, involves moving from a simple, low complexity state in the bottom left towards a more complex state in the top right. He places the various types of rehabilitation on the same graph and shows rehabilitation as progressing towards the original ecosystem but achieving only partial success. *Restoration* is the attainment of the ecosystem's original structure and function, but this is seldom achieved. An alternative approach is *replacement*, where an ecosystem which is different from the original one is created. The areas which are rehabilitated for grazing (discussed by William Laycock in Chapter 10) are an example of ecosystem replacement. These ecosystems tend to be high in biomass and nutrient content but low in species richness and complexity (Fig. 1.4). Bradshaw's model is well illustrated by looking at the abundance and richness of ants returning to Western Australian bauxite mines which have been rehabilitated to *Eucalyptus* monoculture or to a species-rich forest containing many components of the original native vegetation (Fig. 1.5). Finally, there is the *neglect* option in which no reclamation is carried out. Here, as mentioned above, a natural succession may proceed, or there may be further degradation due to erosion, leaching or some other wasting process (Fig. 1.4).

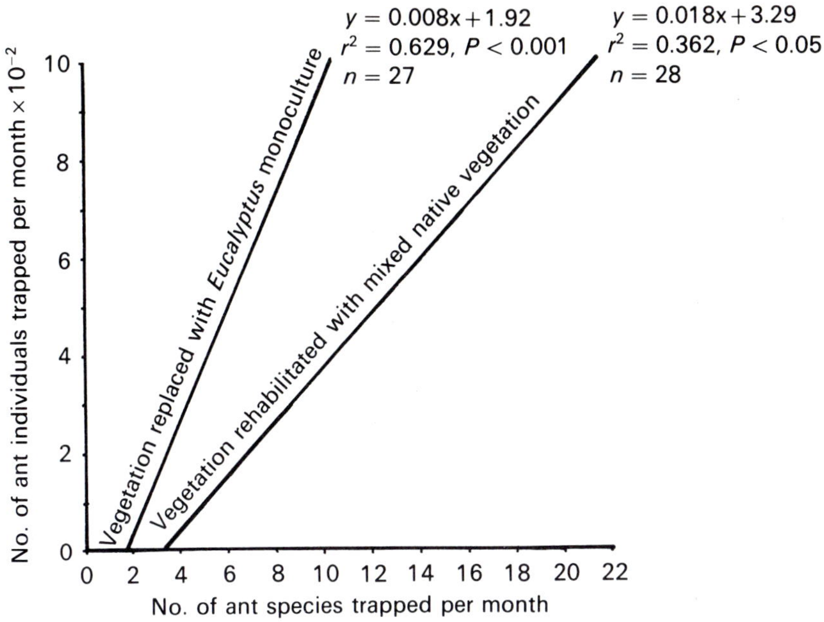

Fig. 1.5. Relationship between ant species richness and number of ants caught by pitfall trapping in Western Australian bauxite mines where the original jarrah forest was either replaced with a *Eucalyptus* monoculture or rehabilitated with mixed native vegetation (adapted from data in Majer, 1981).

Which of these various options is carried out depends on what is required by the laws of the state or country, on public opinion, financial considerations and on the physical limitations of the area. Carbon (1984) recommended that the following series of objectives be formulated in order to achieve successful rehabilitation:

1. Define the land-use capability of the degraded area
2. Define the end product we wish to achieve
3. Define the form of effective rehabilitation we can afford
4. Decide how to implement the rehabilitation
5. Implement the rehabilitation, and
6. Monitor and modify the results.

Objectives 1 and 2 may be governed by three other often-cited objectives (Morrison, 1982). These are:

1. Prevention or reduction of wind and water erosion
2. Provision of food and cover for a variety of animal species, and
3. Improvement of the visual or aesthetic quality of such sites.

Discussion of how these objectives may be achieved is outside the scope of this book. For these the reader is referred to the texts already cited, many of which contain sections on landscape architecture, engineering, soil science and botanical aspects of rehabilitation. However, to achieve a desirable plant community and soil structure, animals may be involved. The ways in which fauna may contribute to the outcome of rehabilitation are discussed in the following section.

Importance of fauna in reclamation

Soil structure and turnover

Under the general umbrella of the term 'soil structure' there are various soil attributes which are important to the well-being of the ecosystem. Included here are the distribution of pores in the soil – important for soil aeration, drainage and root penetration; soil friability – important for root penetration and burrowing by animals; and soil stratification – important for ensuring that nutrients are available at a depth where they are available to plants. Animals have an important influence on all three structural attributes. Earthworms, termites, ants and beetles all create cavities in the soil by their burrowing activities and they also increase the friability of the soil. Earthworms, termites, ants and certain vertebrates also bring soil from lower strata to the surface and hence increase nutrient availability to plants. Earthworms have generally been attributed the major role in this activity (see Satchell, 1983) although an Australian study has shown that ants can bring 841 g m^{-2} yr^{-1} of soil to the surface while earthworms turn up only 133 g m^{-2} yr^{-1} (Humphreys, 1981).

Abbott, Parker & Sills (1979) performed a study on the role of larger soil animals in agricultural soils which is relevant to the problem of land reclamation. They found that when ploughing and stocking by sheep was suspended for 7 years there was a recovery in the density of larger soil animals. This was accompanied by improvements in water permeability, soil compactability and the number of pores within the soil; all characteristics which were associated with the build-up of the fauna. The role which soil animals play in the pedogenesis of rehabilitated lands is further discussed by Ian Abbott in Chapter 2.

Litter decomposition and nutrient cycling

The soil of most degraded lands is generally deficient in the major nutrients. To overcome this a spot, or broadcast, fertilizer is usually applied. This may work in the short-term but for long-term sustainability, organisms which can recycle nutrients back from the leaf litter to the soil must colonize the area. Micro-organisms, such as fungi and bacteria play an enormous part in this process. Nevertheless, animals such as earthworms, millipedes, termites and springtails play an overriding regulatory role (Kitchell *et al.*, 1979). They break up the litter so that micro-organisms can act on it, redistribute the material to strata where nutrient release can take place more rapidly, or directly stimulate the activity of micro-organisms. It follows that the presence of these animals is necessary if nutrient cycles are to be established in the rehabilitated areas.

Southwell & Majer (1982) recently demonstrated the importance of earthworms in nutrient cycling when they introduced the worm *Eisenia fetida* into alkaline 'mud' residues from the bauxite refining process. Earthworms were fed with pelletized sheep food and, by comparison with control treatments where pellets but not earthworms were added, large increases in soil phosphorus, potassium, nitrogen and carbon occurred. The role which soil and litter fauna play in decomposition and nutrient cycling is explored in greater detail by Barry Hutson in Chapter 3.

Impact on primary productivity

The developing vegetation in rehabilitated areas is subjected to attack by a wide range of herbivores. These may feed on the roots, above-ground vegetative parts, or seeds. Furthermore, since rehabilitated areas are often characterized in their early stages by a dominance of a few plant species, some of which may not be native to the area, herbivory may be particularly intense. This is because dominant plants provide abundant resources for the build-up of herbivores and any exotic plants which are present may have poorly adapted defenses against herbivores with which they have no history of evolution. The resulting rehabilitation may therefore suffer a double onslaught – a general reduction in primary productivity due to herbivory and a possibility of pest outbreaks from time to time.

Prachar & Samuel's (1988) work on aspen (*Populus grandidentata*) suckers in a West Virginian coal mine illustrates the former phenomenon. Here browsing by deer suppressed the height increment of suckers and the overall productivity of the area was further reduced by the damaged plants becoming more susceptible to drought and other malentities. The influence of herbivores on plant productivity is discussed in Chapter 4 by Richard Urbanek. His Chapter also describes various instances of pest outbreaks in rehabilitated areas.

Impact on plant species composition

It is well known that selective grazing by mammals can increase or decrease the species diversity of vegetation. More recently it has been suggested that differential grazing by insects on particular *Eucalyptus* species may actually increase the diversity of vegetation (Morrow, 1977). The reason is that an otherwise dominant *Eucalyptus* sp. is suppressed by herbivore activity, and this in turn allows other plant species to thrive.

The implications of this are particularly important to land rehabilitation since, as mentioned, the early stages may be characterized by the abundance of a few species of plants. Herbivore activity may play an important role in reducing the dominance of these plants and thus provide opportunities for plants normally less able to compete with them. There is an almost total absence of literature on the impact of herbivores on primary successional patterns of regrowth. Consequently, much of the information on the impact of animals on plant communities in rehabilitated areas must be inferred from the studies performed in secondary successions such as those following fire. Rob Whelan takes up these issues further in Chapter 5.

Mycorrhizal associations

The nutrition and vigour of many plants is enhanced by the presence of mycorrhizal fungal associations with their roots. This is because the fungus contributes to nutrient uptake by the plant. Prior to rehabilitation the soils of a degraded area may have little or no microbial activity. There is therefore a need to build up the microbial status so that these important plant–mycorrhizal associations may be established.

There are many records of animals consuming the nutritious fruiting bodies of higher fungi. The spores are often present in the faeces so it is likely that these mycophagous animals may act as their agents of dispersal and inoculation. A recent example of this relationship involves squirrels which are found to be effective in dispersing and establishing ectomycorrhizae on *Pinus ponderosa* (Kotter & Farentinos, 1984). As a consequence, if mycorrhizae are to become established, it is desirable that the dispersing animals first colonize or move through the rehabilitated area.

Seed dispersal

The fate of seeds in rehabilitated areas is also influenced by the activities of vertebrates and invertebrates. Birds such as emus, consume plant diaspores and viable seed may pass intact through the alimentary canal. As a result, they may play an important role in seed dispersal (Noble, 1975). As well as benefiting the rehabilitation by dispersing seed within the area, such animals may enrich the area by carrying seed of new species from the adjacent vegetation.

Ants are also major vectors of seeds. Seeds of legumes and other genera which possess arils are frequently gathered by ants. These consume the aril but often discard the rest of the seed intact. By doing this the seed may be buried in the nest, away from fire and predators and protected from desiccation. Majer (1980, 1984) has demonstrated that this mutual relationship operates in mined areas which were rehabilitated some years previously. The restoration of this relationship over the period of rehabilitation is discussed in more detail in Chapter 5.

Pollination of plants

Many plants are pollinated by insects, birds or mammals. If pollen vectors are excluded, or eliminated, plants will not set seed. The ubiquity of most insects and the mobility of most bird pollinators suggests that appropriate pollinators are likely to be present in rehabilitated areas. However, if the rehabilitated area is small, vegetated by plants not present in the adjacent vegetation, and pollinated by animals not present there either, it may not support a pollen vector population.

In other instances, an area may be rehabilitated with plants having a very specific pollinator. Included here may be some of the low-growing Australian heath *Dryandra* spp. which are thought to be pollinated by honey possums (*Tarsipes rostratus*) (Hopper, 1981). These animals are unlikely to be early colonizers of disturbed land and it is therefore in such places that the long-term sustainability of certain *Dryandra* populations is likely to be a problem. The information on the breakdown of pollinator systems is, at present, largely inferential but some further thoughts on the operation of pollinator systems in rehabilitated areas are given by Richard Urbaneck in Chapter 4 and Rob Whelan in Chapter 5.

Encouragement of feral animals

The early stages of rehabilitation are often characterized by the growth of lush vegetation, particularly if grasses are used. This may provide conditions suitable for feral animals which may be unable to colonize the adjacent native vegetation. For instance, the rabbit has been introduced into Australia and become a serious pest. By colonizing a

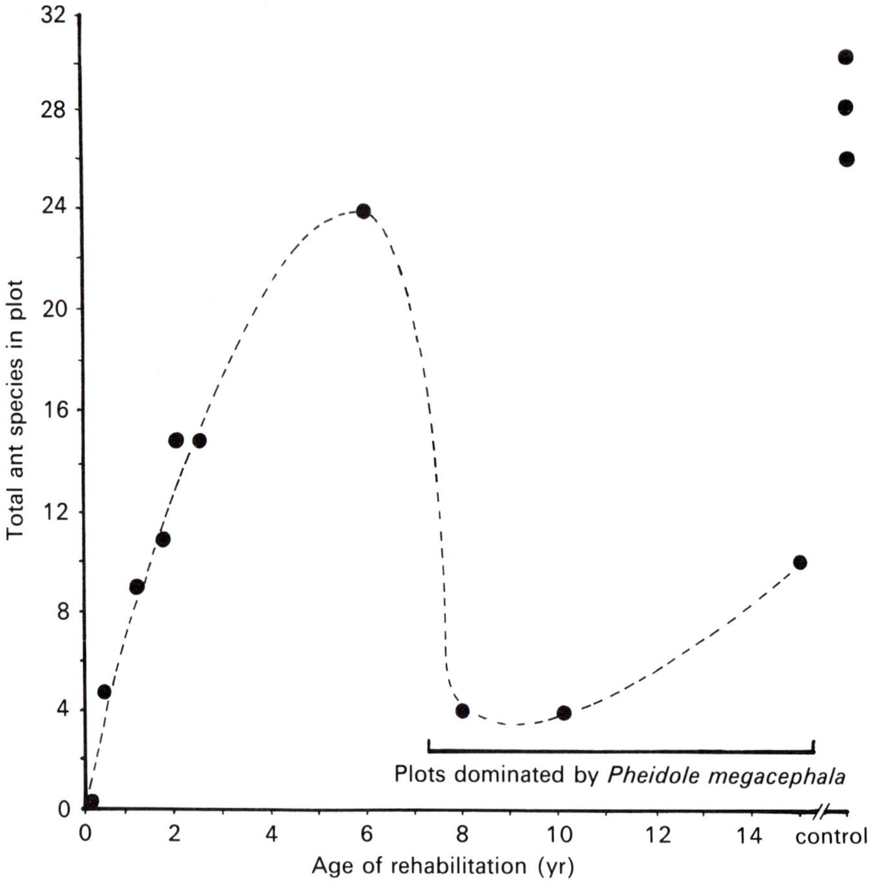

Fig. 1.6. Relationship between the return of native ant species and incidence of the introduced ant, *Pheidole megacephala*, in sand-mined areas on North Stradbroke Island, Queensland, Australia (from Majer, 1985).

rehabilitated area rabbits may forage into adjacent native vegetation, which they may not normally inhabit, and damage seedlings. Although no such cases are reported in the Australian mine-rehabilitation literature, there are known instances where rabbits have colonized agricultural enclaves within the pristine *Eucalyptus* forest where they are not normally found (King, Wheeler & Robinson, 1984).

I have recently investigated the return of ants in rehabilitated sand mines on North Stradbroke Island in eastern Australia (Majer, 1985). The return of ant species followed a linear pattern up to 7 years (Fig. 1.6), but thereafter there was a sharp decline in the number of ant species per plot. This coincided with the introduced ant pest, *Pheidole megacephala*, colonizing the mine and attaining massive densities. This ant was absent in the adjacent native vegetation and the competitive effect which it had on the native fauna is a matter for concern.

Baseline data and fauna monitoring surveys

A frequent objective of land rehabilitation is the restoration of the original biota to the previously disturbed area. This objective may become incorporated in the legislative machinery of a particular country or state. Objectives and regulations such as these clearly require the miner, and possibly other types of land user, to obtain information on the fauna present in the area which is about to be disturbed. This is one reason why the fauna needs to be surveyed prior to land disturbance. Also, since the presence of certain animals is intimately linked with the nature of the vegetation and the physical environment (Fig. 1.3), a survey of the fauna can tell us much about the vegetation, soil and other aspects of the rehabilitation which has been carried out. In other words, there is a great potential to use animals as bio-indicators of the extent, or success, of land reclamation (US Congress Office of Technology Assessment, 1986). For this to be achieved it is important that adequate pre-disturbance data on animal distribution and abundance is obtained. The importance of baseline fauna surveys in planning and monitoring reclamation is discussed in more detail in Chapter 7 by Nick Allen.

Conservation of animals

Many species of animals have become extinct, or are currently threatened with extinction. Since the year 1600 A.D. about 120 mammal and 150 bird species have become extinct. The rate of extinction has accelerated in recent times. By 1981, the United States Office of Endangered Species listed 181 species of vertebrates as endangered or threatened within the United States (Giles, 1981). Some of these extinctions resulted from natural causes such as changes in climate, or competition between different species. However, contemporary extinction primarily results from human interference with natural processes, the most important of which is the alteration or destruction of habitats.

Mining and other activities which degrade land are obvious threats to the conservation status of animals and plants. As already mentioned, these activities will, at least temporarily, eliminate or frighten animals away from the area of disturbance. This may not be important if the proportion of a particular community which is disturbed is small and isolated. However, in many cases large proportions of particular vegetation associations are affected (Fig. 1.1) and linear surface-mines may even form temporary barriers to the normal movement of animals throughout the remaining pristine vegetation (Fig. 1.2). If the conservation of wildlife is to be catered for, and this is important both for vertebrates and the lowliest invertebrates (Wells, Pyle & Collins, 1983), rehabilitation should be undertaken to minimize the threat to native fauna. In countries such as Britain, where wildlife habitat is already extremely scarce, the rehabilitation of long-term degraded land can sometimes make a positive contribution to wildlife conservation (Davis, 1979). In Chapter 12 Dave Samuel discusses the potential of rehabilitation to contribute to the conservation of fish and wildlife. The provision of wetland habitats for fauna in reclaimed lands is discussed separately in Chapter 11 by Jack Nawrot and Willard Klimstra while the design of rehabilitation methods to specifically encourage fauna recolonization is covered in Chapter 8 by Steve Viert.

Which animals are important?

Since the educational background of many rehabilitation staff includes little or no zoological training, the reader may lack the necessary information to assess the role of certain animals which may be associated with the rehabilitation. Appendix 1 provides a quick reference to the types of animals which may be found, and also to the major role they play within the ecosystem. It is a World-wide list and includes taxa which may be absent from some countries or continents. Animals which are only found in marine environments are excluded from the list although freshwater animals which might occur in water bodies within rehabilitated areas are included. The Appendix is not intended to be an aid to identification and should therefore be used in conjunction with one of the vertebrate or invertebrate texts which are mentioned at the foot of the Appendix.

Having described why animals should be considered on the reclamation agenda, the following section reviews some of the legal obligations of miners and other land users concerning fauna in their base-line and rehabilitation studies.

Legal provisions

In order to obtain a perspective on the legal provisions for reclamation it is worth looking at the situation in a range of countries. I have selected Britain, Australia and the United States since they reflect a trend in control of reclamation requirements from local government in

Britain, State government in Australia to Federal government in the United States.

In all these countries the early miners, and similar land users, generally did nothing to reclaim the land after its transient use; the land was considered to be a disposable commodity which was in plentiful supply. By the 1960s the upsurge in environmental concern was in full swing and in many countries land was no longer a plentiful commodity. This brought about not only a need to reclaim land as it became available, but also the thousands of hectares of derelict land which had accumulated since the Industrial Revolution.

Britain

Local councils in Britain have the responsibility for approving mining development and enforcing appropriate reclamation by mining companies, under the Town and Country Planning Act (1971). This arrangement leads to a tremendous variation in approach to the problem. Some local councils are extremely conscientious while others appear reluctant to insist on land reclamation as this may affect the economic viability of major local employers (Pearce, 1982). However a recent act (the Town and Country Planning (Minerals) Act (1981)) defines for mineral workings a more precise requirement for reclamation to a suitable after-use, and also establishes the need for appropriate after-use.

Superimposed on this is the British Government strategy for reclamation of existing derelict land under the Derelict Land Act (1982). The Government gives grants to local councils for reclaiming the dereliction left by coal mining and other industries. There are sometimes instances where the council may wish to reclaim a dormant coal mine but is not able to because the National Coal Board (NCB) has some hope of re-opening the mine at a later date. In this case the NCB will be responsible for the final reclamation. The Government has also altered the type of reclamation projects that qualify for a reclamation grant. In recent circulars the Government has announced that it favours plans which make the land available to new industry and other private sector money-making activities rather than to, say, wildlife.

The whole strategy is described in a recent Government publication (University of Liverpool, 1985). The reclamation requirements for mineral workings in the limestone areas of the Peak District National Park (Challenger, 1980) provide a good example of reclamation requirements at the local level. Some of the worked land is not classified as derelict under the Department of Environment's (1972) definition since it 'has blended into the landscape in the process of time to the extent that it can reasonably be considered part of the natural surroundings, or has been put to some form of acceptable use and no longer constitutes a problem, e.g. a potential nature reserve is an area of special scientific interest, a well vegetated spoil heap, quarry or pit which no longer constitutes an eyesore.' In such cases the land may already have considerable wildlife value (see Usher, 1979) and may already be managed as a nature reserve with little or no reclamation being required.

Current operators must submit planning applications for mineral operations and, in the case of the Peak District National Park, applications are examined and determined by reference to the need to conserve the Park. This is rather a specific case since we are dealing with a national park. Generally speaking local authorities have to consider various interests when deciding planning applications. Visual appearance is often of prime concern, particularly in national parks. In such cases, rehabilitation may be required so that the site matches the surrounding land or visual barriers of trees are planted. New uses such as caravan sites, playing fields or local industry may be promoted or, in some cases, a multi-use rehabilitation programme may be recommended (e.g. grazing, recreation and wildlife). Planning for wildlife may require the operator to design the mining operation in such a way that suitable habitats are created as well as to carry out a particular rehabilitation programme. Apart from these general recommendations there is no specific provision for fauna in the reclamation regulations currently in force in Britain.

Australia

The situation in Australia, where Federal and State governments operate, is quite different. All States have their own mining acts which operate on the principle that minerals are the property of the Crown (in effect the State). In giving the right to exploit a mineral the State government usually requires some undertaking by the operator to reclaim the area. This may range from a requirement to stabilize the substrate, to provide a visual barrier or to provide a vegetation cover over the degraded area. Generally speaking, the more sensitive the environment, the more stringent are the conditions imposed on the operator.

In one State, Western Australia, a system of Agreement Acts operates for major projects. Here an Agreement is drawn up between the State government and a mining company, or other major land user. Should the Agreement be approved by Parliament the developer receives the right to develop the resource in return for a series of undertakings.

The rehabilitation requirements of these Agreements are variable and depend on the sensitivity of the environment. For instance, sand mining occurs in the species-rich heath north of Perth and the minesite overlaps the boundaries of a flora reserve. In consequence, the Agreement for that minesite requires the company to prepare proposals which must be approved by the relevant minister. These must consider the environmental effects of the operation and the company must also undertake detailed research and monitoring into these effects and the resulting rehabilitation programme. Reporting to the minister must be on an annual basis and the minister must be satisfied that the rehabilitation is to a desired standard. To facilitate this the minister is advised by an interdepartmental committee of experts. They have laid down specific requirements which include rigorous quantitative standards for vegetation return. Although the Agreement does not require fauna studies, the company adopts an ecosystem approach to its rehabilitation and sponsors work on fauna in and around the rehabilitated areas.

The environmental requirements of the Agreements have also become more stringent with the passing of time. For instance, the rehabilitation requirements of early Agreements for Western Australian bauxite mines merely required that the mined areas be left tidy. The 1972 Agreement for a new bauxite minesite required the company to restore the area, reafforest and prevent soil erosion while the latest Agreement bound the company to continuous research and monitoring to gauge the effectiveness of rehabilitation techniques. The act requires the company to prepare an Environmental Review and Management Programme (see below) and this document details specific requirements like the provision of a diverse understorey of plants in the mined areas and the adoption of practices which encourage the return of wildlife, both vertebrates and invertebrates. Under the latest Agreement the company must submit annual reports on environmental investigations and research.

Rehabilitation requirements may also be placed on an operator as a result of the environmental impact statement system of appraisal. Starting from 1970 the Federal government, and more recently various State governments, have requested that environmental impact statements (or equivalent documents) be prepared for impending projects which are considered likely to have a significant impact on the environment. When these documents have been approved by the appropriate government department and minister, the operator is often requested to undertake certain rehabilitation measures and research into such procedures. In Western Australia, the obligations resulting from the appraisal of the environmental impact statement (called an Environmental Review and Management Programme in that State) become incorporated into the Agreement which is enacted for that project.

The outcome of the Australian system of enforcement of rehabilitation is that although a fairly regimented system exists, there is a variation in standards and approach between States and between projects of differing age. The final system we will look at is that in force in the United States.

United States

In many States prior to the 1960s, mining, quarrying and similar types of transient land-use proceeded with little attention being paid to rehabilitation or environmental protection. However, with the growing concern for the environment in the mid 1960s, attention was focused on the detrimental effects of surface mining, particularly for coal reserves. This was manifested by many of the States adopting regulations which required specified environmental controls and reclamation standards for the coal industry. Some States passed strong laws, others did not. Being unhappy with progress at the State level, Congress began consideration of a Federal regulating law in 1968. Following years of Congressional debate the Surface Mining Control and Reclamation Act of 1977 (SMCRA) was enacted on 3rd August, 1977 (Haynes, 1978). This act provided, at the national level, the legal basis for requiring initial environmental performance standards for achieving acceptable mining and reclamation standards. The implications of the SMCRA to fish and wildlife have been reviewed by Haynes (1978), Samuel (1979) and Nawrot and Klimstra (1983). The Act prescribes specific requirements for pre-mine planning, environmental assessment and achieving rehabilitation. It has two regulations which have important implications to wildlife (Samuel, 1979). Section 816.111 states that 'reclamation shall establish on all affected land a diverse, effective, and permanent vegetative cover of the same seasonal variety native to the area of disturbed land or species that support the approved post-mining land used'. The acceptable land-use categories defined in the SMCRA are shown in Table 1.2. Category 10 has obvious implications for fish and wildlife although fauna would probably need to be considered in categories 5–9.

Table 1.2. *Acceptable land-use categories as defined in the interim regulations of the Surface Mining Control and Reclamation Act of* 1977 (*reproduced from Haynes*, 1978).

1. *Heavy industry* – manufacturing facilities, powerplants, airports, etc.
2. *Light industry and commercial services* – office buildings, stores, parking facilities, apartment houses, motels, hotels, etc.
3. *Public services* – schools, hospitals, churches, libraries, water-treatment facilities, waste-disposal facilities, public parks and recreation facilities, transmission lines and pipelines, highways, etc.
4. *Residential* – housing (other than apartment houses) and necessary support facilities (e.g. vehicle parking and recreation facilities).
5. *Cropland* – land used primarily for growing crops, including land used in support of farming operations.
6. *Rangeland* – land used for grazing, including forest lands with an understorey vegetation suitable for grazing or browsing use.
7. *Hayland or pasture* – land used primarily for hay production, grazing of livestock, or production of livestock feed.
8. *Forest land* – exhibits at least a 25 % tree canopy or is at least 10 % stocked by trees of any size, including land formerly exhibiting such tree cover and that will be naturally or artificially reforested.
9. *Impoundments of water* – includes all impoundments for beneficial uses such as stock ponds, irrigation, fire protection, recreation or water supply.
10. *Fish and wildlife habitat and recreation lands* – includes wetlands, and all other habitats managed primarily for fish and wildlife or recreation.
11. *Combined uses* – any approved combination of land-uses, but, one land-use must be designated as the primary land-use, and one or more other land-uses are designated as secondary.

The SMCRA also requires every surface mine permit application to have a fish and wildlife plan which states 'how the plan will minimize disturbances and adverse impacts on fish and wildlife and related environmental values during surface coal mining and reclamation operations'. This plan must include details of the wildlife and their habitat, both within and adjacent to the area which is to be mined.

The SMCRA is administered through the Office of Surface Mining (OSM) within the US Department of the Interior. However, if the OSM approves, the relevant State regulating authorities can administer these regulations through their own State acts, provided these are at least as stringent as those in the Federal Act. The result is that there is relative uniformity in the standards of pre-planning, environmental appraisal and rehabilitation of surface coal mines throughout the United States. A series of handbooks have been produced to assist operators in complying with the regulations. Hinkle, Ambrose & Wenzel (1981*a*,*b*,*c*,*d*,*e*) produced a series of handbooks to accomplish fish and wildlife needs on a regional basis while Ramani & Clar (1978) developed a manual for pre-mine planning which includes provisions for wildlife.

The SMCRA applies to the coal mining industry. Other types of mining, such as phosphate in Florida, are regulated by State acts (Marion, Maehr & Frohlich, 1981) while the smaller sand and gravel quarries may be regulated at the State or local government level (Morris, 1982). In some cases these regulations make no specific requirements for wildlife although the regulations for some of the larger mines, such as the uranium mines in Wyoming, follow those of the SMCRA and therefore consider fish and wildlife to a reasonable degree.

This review of the legal provisions for reclamation in three countries has illustrated a wide diversity of approaches to the problem. If any generalizations may be made, there is probably a trend of increasing uniformity of approach to reclamation and the consideration of fauna from Britain, through Australia to the United States with the most consistent approach. Also, fish and wildlife are more likely to be considered in American rehabilitation programmes than in those operating in Australia or Britain. In the final section of this Chapter I will look at some of the trends in fauna research in rehabilitated areas and discuss how these have been influenced by the various regulations in each country.

Trends in fauna research

In Chapter 19, the books, journal articles, conference proceedings, popular articles, reports and theses which concern animals in relation to rehabilitation are listed. A total of 993 entries are listed with contributions coming from most parts of the World. Little work has been undertaken in Africa (3 entries), South America (2 entries) or Asia (4 entries).

The numbers of entries published per year in Europe (excluding Britain), Britain, Australasia and the United States are shown in Fig. 1.7. The mean number of entries per decade are given for the 1930s, 1940s and 1950s and per half decade for 1961–5 and 1966–70. Significant amounts of fauna work commenced in the 1950s in the United States, the early 1960s in Europe, the early 1970s in Britain and the mid-1970s in

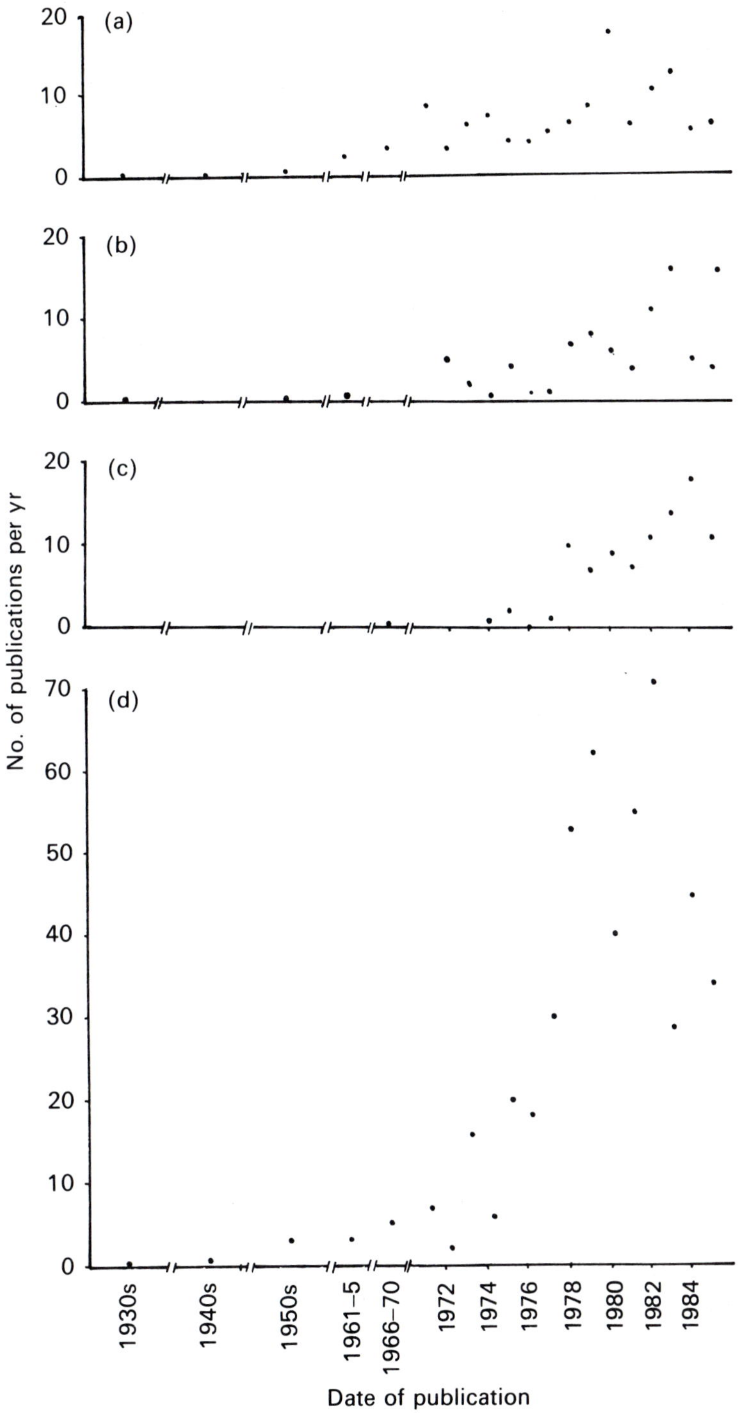

Fig. 1.7. Trends in the number of publications published per year on fauna in relation to land reclamation in (a) Europe (excluding Britain), (b) Britain, (c) Australasia, and (d) the United States. Prior to 1971 the mean number of publications per decade or half decade are shown.

Australia. Publication output increased in all four regions up until the early 1980s although the increase was most marked in the United States. Here the output of publications in 1982 was more than triple that in 1976. This upsurge is probably correlated with the increased requirements for rehabilitation research resulting from the 1977 SMCRA. There was, however, a modest increase in publication output in the other three regions during this period. This probably resulted from the increase in environmental awareness which was set in motion during the 1960s and was paralleled by the general increase in pure ecology publications over this period. There has been a decline in the output of publications in the period 1982–1983 onwards and this is probably associated with a range of reasons. First, in the United States and certain parts of Australia at least, there is a feeling that many of the answers to problems have been found and that the findings may now be applied to the operation phase. Companies and government agencies are therefore less likely to fund new projects than they were when many questions lay unanswered. Secondly, with the current decline in the economic well-being of many countries, people are more concerned with employment than with environmental matters. There is therefore less public pressure on companies and agencies to fund and carry out research into subjects such as those under consideration in this book. Thirdly, and related to this, because of the economic down-turn, funds for research are becoming increasingly difficult to obtain. This is worrying because many of the reclamation problems have yet to be solved.

The following Chapters will now focus on certain aspects of fauna in relation to rehabilitation and point out how the findings of previous studies can be applied to the rehabilitation programmes. In addition, the Chapters will also discuss the needs for further research into the role of fauna in successful rehabilitation schemes.

References

Abbot, I., Parker, C. A. & Sills, I. D. (1979). Changes in the abundance of large soil animals and physical properties of soils following cultivation. *Australian Journal of Soil Research*, **17**, 343–53.

Anon. (1978). *Handbook for Terrestrial Wildlife Habitat Development on Dredged Material.* U.S. Army Engineers Waterways Experimental Station Technical Report No. D-78-37.

Bradshaw, A. D. (1983). The reconstruction of ecosystems: presidential address to the British Ecological Society, December 1982. *Journal of Applied Ecology*, **20**, 1–17.

Bradshaw, A. D. (1984). Ecological principles and land reclamation practice. *Landscape Planning*, **11**, 35–48.

Bradshaw, A. D. & Chadwick, M. J. (1980). *The Restoration of Land: The Ecology and Reclamation of Derelict and Degraded Land.* Oxford: Blackwell.

Bridgewater, P. (1987). Connectivity: an Australian Perspective. In *Nature Conservation: The Role of Remnants of Native Vegetation*, ed. D. Saunders, G. Arnold, A. Burbidge & A. Hopkins, pp. 195–200. Sydney: Surrey Beatty & Sons.

Bromley, R. D. F. & Humphrys, G. ed. (1979). *Dealing with Dereliction. The Redevelopment of the Lower Swansea Valley*. Swansea: University College of Swansea.

Burley, J. B. & Hopkins, R. B. (1984). Potential for enhancing nongame-bird habitat values on abandoned mine lands of western North Dakota. In *Proceedings 1984 Symposium on Surface Mining, Hydrology, Sedimentology and Reclamation*, December 1984.

Cairns, Jr. J. ed. (1980). *The Recovery Process in Damaged Ecosystems*. Michigan: Ann Arbor Science.

Cairns, Jr. J. & Buikema, Jr. A. L. (1986). *Restoration of Habitats Impacted by Oil Spills*. Stoneham: Butterworths.

Carbon, B. A. (1984). Introductory comments. In *Mined Land – Is It Sustainable?* ed. B. G. Collins, pp. 4–6. Bentley: Western Australian Institute of Technology.

Chadwick, M. J. & Goodman, G. T. ed. (1975). *The Ecology of Resource Degradation and Renewal*. Oxford: Blackwell Scientific Publications.

Challenger, G. P. (1980). Reclamation for wildlife: setting the scene. *Reclamation Review*, **3**, 193–6.

Chapman, D. L., McGinnes, B. S. & Downing, R. L. (1978). Breeding bird populations in response to the natural revegetation of abandoned contour mines. In *Surface Mining and Fish/Wildlife Needs in the Eastern United States*, ed. D. E. Samuel, pp. 328–32. Morgantown: West Virginia University.

Davis, B. N. K. (1979). Chalk and limestone quarries as wildlife habitats. *Minerals and the Environment*, **1**, 48–56.

Department of Environment (1972). *Survey of Derelict and Despoiled Land: Guidance Notes*. London: H.M.S.O.

Down, C. G. & Stocks, J. (1977). *Environmental Impact of Mining*. New York: John Wiley.

Ente, P. J. & De Glopper, R. J. ed. (1985). *The Lauwerszee Fifteen Years Closed Off. Results of Research and Experience with Development and Management*. Lelystad: Rijksdienst voor de Ijsselmeerpolders.

Fox, J. E. D. (1984). Rehabilitation of mined lands. *Forestry Abstracts*, **45**, 565–600.

Gemmell, R. P. (1977). *Colonization of Industrial Wasteland*. London: Edward Arnold.

Giles, Jr. R. H. (1981). *Endangered Species: An Aid to Land-use Planning in the Coal Producing Countries of Southwestern Virginia*. Virginia Polytechnic Institute and State University Occasional Paper.

Goodman, G. T., Edwards, R. W. & Lambert, J. M. (1965). *Ecology and the Industrial Society*, Oxford: Blackwell.

Gore, J. A. ed. (1985). *The Restoration of Rivers and Streams*, Boston: Butterworth.

Hackett, B. (1971). *Landscape Planning: An Introduction to Theory and Practice*. Newcastle upon Tyne: Oriel Press.

Hackett, B. (1977). *Landscape Reclamation Practice*, Guildford: IPC Science and Technology Press.

Harris, L. D. (1984). *The Fragmented Forest. Island Biogeography Theory and the Preservation of Biotic Diversity*. Chicago: University of Chicago Press.

Haynes, R. J. (1978). The surface mining control and reclamation act of 1977 and potential impacts on fish and wildlife. In *Proceedings of the 32nd Annual Conference of the South-Eastern Association of Fish and Wildlife Agencies*, pp. 790–6.

Hinkle, C. R., Ambrose, R. E. & Wenzel, C. R. (1981*a*). *A Handbook for Meeting Fish and Wildlife Information Needs to Surface Mine Coal – OSM Region 1*. U.S. Fish and Wildlife Service Report FWS/OBS-79/48.3.1.

Hinkle, C. R., Ambrose, R. E. & Wenzel, C. R. (1981 *b*). *A Handbook for Meeting Fish and Wildlife Information Needs to Surface Mine Coal – OSM Region 2.* U.S. Fish and Wildlife Service Report FWS/OBS-79/48.3.2.

Hinkle, C. R., Ambrose, R. E. & Wenzel, C. R. (1981 *c*). *A Handbook for Meeting Fish and Wildlife Information Needs to Surface Mine Coal – OSM Region 3.* U.S. Fish and Wildlife Service Report FWS/OBS-79/48.3.3.

Hinkle, C. R., Ambrose, R. E. & Wenzel, C. R. (1981 *d*). *A Handbook for Meeting Fish and Wildlife Information Needs to Surface Mine Coal – OSM Region 4.* U.S. Fish and Wildlife Service Report FWS/OBS-79/48.3.4.

Hinkle, C. R., Ambrose, R. E. & Wenzel, C. R. (1981 *e*). *A Handbook for Meeting Fish and Wildlife Information Needs to Surface Mine Coal – OSM Region 5.* U.S. Fish and Wildlife Service Report FWS/OBS-79/48.3.5.

Holdgate, M. W. & Woodman, M. J. (1978). *The Breakdown and Restoration of Ecosystems.* New York: Plenum Press.

Hopper, S. D. (1981). Honey Possums. *S.W.A.N.S.*, **11**, 21–3.

Humphreys, G. S. (1981). The rate of ant mounding and earthworm casting near Sydney, New South Wales. *Search*, **12**, 129–31.

Hutnick, R. J. & Davis, G. ed. (1973). *Ecology and Reclamation of Devastated Land*, vol. 1. New York: Gordon and Breach.

King, D. R., Wheeler, S. H. & Robinson, M. H. (1984). Daytime locations of European rabbits at three localities in south-western Australia. *Australian Wildlife Research*, **11**, 89–92.

Kitchell, J. F., O'Neill, R. V., Webb, D., Gallop, G. W., Bartell, S. M., Koonce, J. F. & Ausmus, B. F. (1979). Consumer regulation of nutrient cycling. *Bio Science*, **29**, 28–34.

Kotter, M. M. & Farentinos, R. C. (1984). Formation of Ponderosa pine ectomycorrhizae after inoculation with faeces of tassel-eared squirrels. *Mycologia*, **76**, 758–60.

Langkamp, P. J. (1979). Reclamation terminology. *Landline*, **1**, 6–7.

MacArthur, R. H. & Wilson, E. O. (1967). *The Theory of Island Biogeography.* Princeton: Princeton University Press.

Majer, J. D. (1980). The influence of ants on broadcast and naturally spread seeds in rehabilitated bauxite mined areas. *Reclamation Review*, **3**, 3–9.

Majer, J. D. (1981). *The Role of Invertebrates in Bauxite Mine Rehabilitation.* Forests Department of Western Australia Bulletin No. 93.

Majer, J. D. (1984). The influence of ants on seeding operations in northern Australian mined areas. *Reclamation and Revegetation Research*, **2**, 299–313.

Majer, J. D. (1985). Recolonization by ants of rehabilitated mineral sand mines on North Stradbroke Island, Queensland, with particular reference to seed removal. *Australian Journal of Ecology*, **10**, 31–48.

Marion, W. R., Maehr, D. S. & Frolich, R. K. (1981). Phosphate mine reclamation and habitats for wildlife. In *Proceedings of the Surface Mining Hydrology and Reclamation Symposium*, Lexington, Kentucky, December 1981, pp. 501–6.

Morris, R. A. (1982). Regulatory land use aspects of sand and gravel mining as they affect reclamation for wildlife habitat and open space: a national perspective. In *Proceedings of Symposium on the Wildlife Values of Gravel Pits*, Minnesota, June 1982, pp. 16–23.

Morrison, D. G. (1982). Principles of revegetating mined lands. In *Proceedings of Symposium on the Wildlife Values of Gravel Pits*, Minnesota, June 1982, pp. 51–8.

Morrow, P. (1977). The significance of phytophagous insects in the *Eucalyptus* forests of Australia. In *The Role of Arthropods in Forest Ecosystems*, ed. W. J. Mattson, pp. 19–29. New York: Springer Verlag.

Nawrot, J. R. & Klimstra, W. D. (1983). Fish and wildlife permit requirements and assessment practices. In *Surface Mining Equipment Monitoring and Reclamation Handbook*, ed. L. V. A. Sendlein, H. Yazicigil & C. L. Carlson, pp. 509–33. New York: Elsevier.

Noble, J. C. (1975). The effects of emus (*Dromaius novaehollandiae* Latham) on the distribution of the nitre bush (*Nitraria billardieri* DC). *Journal of Ecology*, **63**, 979–84.

Pandit, A. K. & Fotedar, D. N. (1982). Restoring damaged wetlands for wildlife. *Journal of Environmental Management*, **14**, 359–68.

Pearce, F. (1982). The wastelands of Britain. *New Scientist*, **95**, 898–900.

Prachar, R. E. & Samuel, D. E. (1988). Deer–aspen relationship on an abandoned surface mine. *Applied Forestry* (in press).

Ramani, R. V. & Clar, M. L. (1978). *User's Manual for Premining Planning of Eastern Surface Coal Mining: Vol. 1 Executive Summary*. Report No. EPA-6000/7-78-180. Washington, D.C: U.S. Government Printing Office.

Richardson, B. Z. (1980). Reclamation of acid-producing spoils on a western surface mine. In *Proceedings of High Altitude Revegetation Workshop No. 4*, ed. C. L. Jackson & M. A. Schuster, pp. 101–12. Boulder: Colorado State University.

Roberts, R. D., Marrs, R. H., Skeffington, R. A., Bradshaw, A. D. & Owen, L. D. C. (1982). Importance of plant nutrients in the restoration of china clay and other mine wastes. *Transactions of the Institute of Mining and Metallurgy*, **91**, 42–50.

Rykiel, J. R. (1979). Ecological disturbances. In *The Mitigation Symposium: A National Workshop on Mitigating Losses of Fish and Wildlife Habitats*. Rocky Mountain Forest Range Experiment Station (Forest Service). General Technical Report RM-65, pp. 624–6.

Samuel, D. E. (1979). Surface mine reclamation for wildlife: the new technical law and you. In *Proceedings of the 1979 North-Eastern Fish and Wildlife Conference*, Providence, Rhode Island, pp. 145–50.

Satchell, J. E. ed. (1983). *Earthworm Ecology: From Darwin to Vermiculture*. London: Chapman and Hall.

Saunders, D. A., Arnold, G. W., Burbidge, A. A. & Hopkins, A. J. M. ed. (1987). *Nature Conservation: The Role of Remnants of Native Vegetation*. Sydney: Surrey Beatty and Sons Ltd.

Schaller, F. W. & Sutton, P. (1978). *Reclamation of Drastically Disturbed Lands*. Madison: American Society of Agronomy.

Sheldon, J. C. & Bradshaw, A. D. (1976). The reclamation of slate waste tips by tree planting. *Landscape Design*, **113**, 31–3.

Southwell, L. T. & Majer, J. D. (1982). The survival and growth of the earthworm *Eisenia foetida* (Lumbricidae: Oligochaeta) in alkaline residues in association with the bauxite refining process. *Pedobiologia*, **23**, 42–52.

Spaulding, W. M. & Ogden, R. D. (1968). *Effects of Surface Mining on the Fish and Wildlife Resources of the United States*. U.S. Bureau of Sport, Fish and Wildlife Publication No. 68.

Tacey, W. H. & Glossop, B. L. (1980). Assessment of topsoil handling techniques for rehabilitation of sites mined for bauxite mines within the jarrah forest of Western Australia. *Journal of Applied Ecology*, **17**, 195–201.

Tate, III, R. L. & Klein, D. A. ed. (1985). *Soil Reclamation Processes: Microbiological Analyses and Applications*. New York: Marcel Dekker.

University of Liverpool. (1985). *Transforming our Land – The Way Forward*. London: H.M.S.O.

University of Newcastle upon Tyne. (1972). *Land Reclamation*, vol. 2. Guildford: IPC Science and Technology Press.

U.S. Congress, Office of Technology Assessment. (1986). *Western Surface Mine Permitting and Reclamation*, OTA-E-279. Washington, D.C: U.S. Government Printing Office.

Usher, M. B. (1979). Natural communities of plants and animals in disused quarries. *Journal of Environmental Management*, **8**, 223–36.

Wells, S. M., Pyle, R. M. & Collins, N. M. ed. (1983). *The IUCN Invertebrate Red Data Book*. Gland: International Union for the Conservation of Nature and Natural Resources.

Part 2

Functional aspects of fauna in reclaimed land

Introduction

J. D. MAJER

The following Chapters focus in more detail on the functional aspects of fauna in reclaimed land.

Typically the soil which is replaced after mining, land-fill or dredging, or the soil which has been compacted by the activities associated with urbanization, is lacking in the structure which is necessary for aeration, water infiltration, root growth and other important soil processes. By their tunnelling and feeding activities, animals such as earthworms, springtails and mites play an important role in improving the soil structure in reclaimed land. The beneficial, and sometimes detrimental, effects of animals on soil structure are discussed in Chapter 2 by Ian Abbott.

The soil of reclaimed lands, particularly that which has been brought up from greater depths or has been subjected to some mineral extraction process, is often lacking in the nutrients which are necessary for sustainable plant growth. In the short-term the soil condition is often ameliorated by adding fertilizers. However, for long-term sustainability of the developing ecosystem it is necessary for the natural nutrient cycles to be re-established. In Chapter 3 Barry Hutson reviews the decomposition and mineralization processes within the soil and also the features of the developing ecosystem which may directly or indirectly affect the fauna and decomposition processes. The Chapter then examines the relevant data from a range of rehabilitation sites in order to develop some generalizations that will be of benefit to the reclamation procedure.

In moving from the soil to the vegetation, animals are once again found to play an important role both during, and following, the reclamation process. By reducing plant growth, reducing seed production, and causing mortality of plants, herbivorous animals may have a negative influence on primary productivity in the developing rehabilitation. Richard Urbanek reviews the impact which such animals have on the productivity of a range of legumes, grasses, composites, conifers and hardwoods which are used in various rehabilitation programmes. The control of these herbivores is discussed in general terms towards the end of the Chapter, although the actual control prescriptions are beyond the scope of this book. Instead, the interested reader is referred to additional texts which are listed in the bibliography of his Chapter.

As well as influencing primary productivity, animals may play an important role in moulding the species composition of the developing plant community. In Chapter 5, Rob Whelan outlines the many ways in which consumers can influence the development and maintenance of plant species composition. Included here are the herbivorous, seed-dispersing and pollinating activities of animals as well as the indirect influence on plants via their impact on soil condition. This Chapter culminates in a discussion of the practical implications of these activities, implications which could make management of rehabilitated land cheaper and more effective.

The re-establishment of important processes mentioned in the foregoing Chapters is reliant upon the colonization of a suitable range of animals. Unfortunately there is a dearth of long-term studies on fauna succession in reclaimed lands. In Chapter 6, I explore the likely fauna recolonization patterns in rehabilitated lands by drawing upon the results of published studies and by extrapolating from contemporary ecological theory. The likelihood of attaining the pre-disturbance animal community composition is examined and the influence of extraneous factors such as fertilizers, climate and rehabilitation technology is reviewed. In summary, this Chapter asks the important question: will the animals return?

2

The influence of fauna on soil structure

IAN ABBOTT

Introduction

The significant contribution that animals make to improving or destroying soil structure in primary successions which develop after mining, or other such disturbances, is not widely appreciated. This is because little relevant research has been carried out and, it must be confessed, much of it is badly designed and poorly analyzed. Therefore my conclusions about the influence of fauna on restoration of soil structure of disturbed areas such as mined lands must remain tentative until reclamation biologists experiment with the locally available soil fauna in relation to rehabilitation practices. Hopefully this Chapter will serve to guide such studies so that they will be more focused and rigorous than previous ones.

Important characteristics of soil structure are the geometric arrangement of soil particles, and of the pores and voids among them, the combination of these particles into aggregates, and the stability of these aggregates in water. Indeed, Lynch & Bragg (1985) state that the soil aggregates are the basic units of soil structure. Texture partly determines the proportion and diameters of pores in the soil, and organic matter contributes to aggregation of particles (Tisdall & Oades, 1982). Kubiena (1938) defined a good soil macrofabric as comprising aggregates bound together so as to form a system of interconnected voids, with the aggregates themselves internally porous. A favourable soil structure ensures proper aeration below ground, penetration by roots, adequate soil moisture content, and prevents surface crusting and erosion of topsoil.

The fauna which influences soil structure comprises three groups.

First, invertebrate animals living in the soil and litter consist of microfauna (< 200 μm length), mesofauna (200 μm–2 mm) and macrofauna (2–20 mm). Examples of each invertebrate category are: protozoans and nematodes (microfauna); mites, springtails and larvae of flies (mesofauna); and earthworms, isopods, millipedes, ants and termites (macrofauna). The second group is the vertebrate animals, such as rodents, shrews and wombats which live partly under, and partly above, the ground. The remaining animals consist of the heavier ungulates, elephants, rhinoceros, etc., vertebrates living exclusively above ground.

Of all these animals, the value of earthworms was the first to be recognized. Two centuries ago, White (1788) discussed seven functions of earthworms: 'by boring, perforating, and loosening the soil, and rendering it pervious to rains and the fibres of plants, by drawing straws and stalks of leaves and twigs into it; and most of all, by throwing up such infinite numbers of lumps of earth called worm-casts...'. These concepts have since been restated many times in scientific language, and the last 100 years of research have involved careful documentation and verification of these functions in many locations.

How does fauna influence soil structure?

Animals can have either a beneficial or detrimental effect on soil structure.

Beneficial effects

Introductory remarks Animals which play a beneficial role have a direct physical effect by burrowing into the soil and mixing it. A labyrinth of subterranean tunnels, opening occasionally to the surface, allows water to infiltrate readily, promotes aeration, and may assist deeper penetration of the soil by roots. These animals often return regularly to the surface, perhaps to feed on litter or to defaecate, and they therefore maintain permanent burrows and facilitate turnover of the soil profile. The most important invertebrates in this regard are some of the macrofauna (earthworms, ants and termites).

An indirect, chemical or microbial effect of beneficial animals involves the production of faeces and their influence on the decomposition of litter. Both meso and macrofauna are directly important, with microfauna consuming fungal and bacterial populations present in litter and soil (Seastedt, 1984). The enzyme cellulase in the guts of these animals, or the presence of certain micro-organisms which produce such enzymes, allows breakdown and fragmentation of litter, which in this state can be further decomposed by other micro-organisms, fungi and soil animals. The fragmented or decomposed litter eventually becomes mixed with the mineral component of the soil. Very importantly, the gums and other polysaccharides produced by microbial digestion of litter contribute to stability of the soil aggregates (Harris, Chesters & Allen, 1966).

Bulk density and pore space Barley (1961) stated that the importance of earthworms in modifying soil porosity (the volume of the soil not occupied by solid particles) is over-rated, but this is incorrect because different earthworm species have different habits and therefore affect the soil in different ways (Guild, 1955). For example, the percentage pore space of a soil containing a high population of surface casting species, such as *Aporrectodea longa* and *Aporrectodea nocturna*, is much greater than that of a pasture with a high population of earthworms not producing surface casts (Evans, 1948). If the dominant species defaecate underground, pore space is likely to be only negligibly affected.

Interconnected cavities, pores and holes caused by earthworms and ants may be very extensive (see photographs in Green & Askew, 1965). In mounds constructed by ants, voids 0.2–2.3 cm in diameter occupy 20% by volume, double that in the buried A horizon (Salem & Hole, 1968). The bulk density of soil in the mound was half that in the adjacent surface soil.

In contrast to earthworms and ants, when building their nests and mounds, many termites repack and cement the soil. This reduces the pore space (Lee & Wood, 1971).

Mixing Darwin (1840) was the first to record the development of a stone-free horizon beneath the soil surface, and to deduce that it was caused by earthworms preferentially ingesting small particles and translocating them towards the surface. Earthworms can also alter the distribution of coarse sand particles in the soil profile, as the more numerous small specimens tend to reject these particles when feeding. Evidence that these particles are reduced in size (comminuted) in the gizzard is reviewed by Evans (1948). Earthworm casts have been found to contain more silt and clay and fewer sand particles than topsoil (De Vleeschauer & Lal, 1981; Lal & Akinremi, 1983).

The quantity of casts deposited on the soil surface by earthworms may attain 43 t ha^{-1} yr^{-1} (Evans & Guild, 1947), implying that the upper 10 cm of soil could pass through the guts of the earthworm population in about 12 years. The activity of earthworms may be so intense that the soil horizons are destroyed (Buntley & Papendick, 1960), with the entire profile (to 90 cm depth) consisting of casts and filled worm channels in varied states of preservation.

Mounds of ants have been recorded as containing more clay than the surface soil because the ants had translocated material from the B horizon (Baxter & Hole, 1967). These mounds weighed 16 t ha^{-1} and occurred at a density of 1500 ha^{-1}. Termites act in a similar fashion. Their mounds may contain up to 63 t ha^{-1} of soil (Lee & Wood, 1971). Nye (1955) calculated that up to 0.2 t ha^{-1} of subsoil was deposited on the topsoil by termites each year. The species which were studied built mounds very rapidly but occupied them briefly. After their erosion, the soil profile had effectively been inverted.

Aeration and drainage Earthworm species which defaecate below ground can only make a space within the soil at one point at the expense of consolidating it or filling in existing air spaces elsewhere. Therefore, apart from effects on soil aggregates (see below), the only activity which can increase soil air space under field conditions is surface casting.

Large pores (1–10 mm diameter) are essential for rapid drainage, consequently earthworm burrows are very important as they shed large volumes of water quickly, preventing run-off and soil erosion. Under field conditions, the number of burrows is proportional to the density of earthworms (Tisdall, 1985).

Water-holding capacity An abundance of fine pores promotes friability and assists in maintaining enough soil moisture for plant growth (Stockdill & Cossens, 1967; Russell, 1973). The capillary spaces within the aggregates which are present in the casts of earthworms aid retention of water (Satchell, 1967; Lal & Akinremi, 1983). Termite mounds also tend to retain water more readily than adjacent soils (Lee & Wood, 1971).

Litter decomposition Certain fungi are important decomposers of plant matter because they have enzymes which are able to degrade lignin and cellulose (Burns & Martin, 1986). The mesofauna feed on these fungi (Coleman, 1986) and produce faecal pellets which are added to the soil; some members (springtails) are also important in comminuting plant residues (Hale, 1967). The macrofauna probably has the most important impact on soil structure by incorporating these residues (organic matter) into the soil. Litter is comminuted and defaecated in a form more suitable for decomposition by micro-organisms, and then mixed intimately within the mineral soil.

Many studies have shown that soils without earthworms will accumulate an undecayed mat of litter in the soil surface (Satchell, 1967; Stockdill & Cossens, 1967; Barley & Kleinig, 1964; Van Rhee, 1965; Edwards & Lofty, 1977 (plate 8); Huhta, 1979; Vimmerstedt & Finney, 1973). In terms of soil structure, this should be less desirable than if the litter had been incorporated into the soil. However, accumulation of undecayed litter on the soil surface is of more consequence to disruption of nutrient cycling (see Chapter 3).

Soil aggregates At the soil micromorphological scale, faecal pellets of invertebrates such as Enchytraeidae, Collembola and Acari can be particularly abundant. These animals re-organize humic and fine mineral material into discrete microstructural units (see Pawluk, 1985, and Rusek, 1985, for many fine photographs). These units probably contribute to soil tilth.

Rusek (1985) recognizes three stages in development of soil microstructure. The simplest type of humus is formed by faeces of Collembola, oribatid mites and certain fly larvae. The next stage consists of faecal pellets of millipedes, slaters, beetle larvae, certain fly larvae, enchytraeids and small litter-dwelling earthworms. The best humus is formed by high densities of burrowing species of earthworm. This scheme is somewhat oversimplified, as it makes no reference to the role of some mesofauna in eating and therefore disintegrating faecal pellets of certain macrofauna.

The stability of soil aggregates in water contributes to a favourable soil structure. Earthworm casts have a greater stability of aggregates than topsoil (Dutt, 1948), probably because soil particles during their passage through the gut are coated with gums and mucilages produced by bacteria (Swaby, 1950). This increases the resistance of casts to disintegration by water. De Vleeschauer & Lal (1981) found that the kinetic energy of falling raindrops required to disrupt casts was 5–54 times greater than that required to disrupt natural soil aggregates.

However, the stability of casts depends on the availability of high nitrogen plant residues, so that casts from soil under grass or lucerne are more stable than from soil under straw crops (Satchell, 1967). This is because plant matter which is high in nitrogen is decomposed more quickly by micro-organisms, resulting in more aggregating substances (Harris *et al.*, 1966).

Detrimental effects

The negative influence on soil structure which animals may have falls into two categories. Direct effects involve trampling and treading of the soil surface, causing crusting of the surface or compaction of the upper part of the profile (Witschi & Michalk, 1979). Compaction reduces the density of soil invertebrates (King & Hutchinson, 1983), although deep burrowing species of earthworm are little affected (Piearce, 1984). Dense herds or flocks of the heavier vertebrates (cattle, elephants, rhinoceros, sheep and goats) are likely to cause serious compaction problems. Kangaroos and rabbits are regarded as 'soft-footed' and hence of negligible importance. Populations of animals which indirectly degrade soil structure form the second category of detrimental fauna. By overgrazing or overbrowsing the vegetation, some animals eventually impair the production of litter onto the soil surface and this in turn reduces the density of beneficial soil and litter invertebrates. Ultimately, poor land management by humans is responsible for Worldwide degeneration in soil structure, through overstocking and too frequent cultivation.

How quickly can fauna influence soil structure?

As in the previous section it is essential to distinguish beneficial influences from adverse ones. Most of this section will concern faunal recolonization and not effects on soil structure. This is because most research has been done by soil zoologists who do not specifically measure soil structural attributes. Nevertheless, once a suitable fauna is present, some improvement in soil structure should follow within a few years.

Improvement in soil structure

The rate at which soil fauna can improve soil structure, particularly that of mined areas or other drastically disturbed land, presumably depends on how soon the soil fauna can colonize the substrate and attain high densities. This further depends upon the physical and chemical suitability of the material (see Chapter 9), on the provision of food, and on a favourable microclimate. The first aim in rehabilitation (either for amenity or productive use) should be to establish vegetative cover, not to establish a soil fauna. This is because vegetation will provide the food for the soil fauna.

Successful rehabilitation often depends on ripping open the soil to encourage rooting, adjusting pH by liming, and fertilizing. The litter produced by the vegetation provides the carbon source for the soil fauna when they colonize. One of the major roles for the soil fauna in rehabilitated land is to prevent accumulation of a thick layer of litter on the soil surface, because this would retard infiltration of water, lock up nutrients essential for plant growth, and delay the attainment of a soil structure most conducive to plant growth.

Enchytraeid worms and larval flies colonized brown coal dumps within 4 years of amelioration and planting (see Chapter 13 and Dunger, 1967). In the fifth year, earthworms colonized, and by the tenth the biomass of the macrofauna was similar to that of a nearby undisturbed soil. Neumann (1973) demonstrated that carabid beetles were pioneer colonizers of coal spoil banks but disappeared after 10 years.

Colonization of rehabilitated coal pit heaps by mites and springtails is rapid once vegetative cover has developed (Hutson, 1980). Hutson found that isotomid springtails were specially important in the first 3 years of rehabilitation because earthworms were absent. Various species of earthworms colonized coal spoils 6–13 years after rehabilitation (Dunger, 1968), and earthworms on grazed leys returned to densities similar to undisturbed land within 5 years of backfilling of the mine (Thompson, 1985). Bal (1982) recorded very dense earthworm populations on a polder 11 years after revegetation. Earthworms invaded an undisturbed podzol and completely altered it within 3 years, so that the upper 14 cm consisted entirely of their casts (Langmaid, 1964).

In contrast, earthworms took 40 years to colonize grassland established on rehabilitated peat (Cotton & Curry, 1982). They may have taken so long because of adverse pH conditions. The peat had only 15% of the total numbers of earthworms which were present in the surrounding mined soil. Obviously the occurrence of earthworms at low densities is not likely to ameliorate soil structure.

Sometimes earthworms may be absent from rehabilitated sites because of colonization difficulties, and not through unfavourable conditions of the sites. Thus, earthworms were absent from spoil on which trees had been planted 5 years earlier (Hamilton & Vimmerstedt, 1980). They had dispersed and multiplied within 1 year throughout the site after they were artificially introduced into the soil. They did not establish in contiguous non-wooded areas.

In the laboratory, Bal (1982) described how earthworms created an A0/A1 horizon 5 cm thick in a densely compacted soil over a 3-month period. In a forest soil, Nielsen & Hole (1964) removed the upper 15 cm of the A horizon and substituted B1 material. The activities of the earthworms, which were naturally present, caused a new A1 horizon to begin to develop. Extrapolation of 5 years of data suggested that a new A1 horizon would develop in 30–40 years. The critical role of a continual litter supply in promoting development of dense earthworm populations was illustrated by the same study; for 5 years all litter was removed from one plot, doubled in another, and not interfered with in a control plot. The average numbers of surface castings ha^{-1} were 135 000, 765 000 and 390 000 respectively.

Bunn (1983) identified two mound-building species of termite which occur in rehabilitated bauxite mined areas in Western Australia after 4 years and 6 years, but no data on abundance were presented (Nichols & Bunn, 1980).

My recommendation is that after appropriate rehabilitation of spoil, suitable species of earthworm (see below) should be introduced annually and their survival monitored. Repeated introduction (at least for 2–3 years) of earthworms should accelerate the amelioration of soil structure.

Deterioration in soil structure

There is not much information available on how quickly fauna can cause soil structure to deteriorate, but probably the stocking rate of sheep, goats, etc. is critical. This is illustrated by a study near Kodj Kodjin in the Western Australian wheatbelt (Abbott & Parker, unpubl. data). In September 1978, infiltration (as measured by Abbott, Parker & Sills, 1979) of water into a wheat paddock took 38 s. A small exclosure was then set up in this paddock, the wheat was harvested by December 1978 and then a typical stocking of sheep was allowed to graze the stubble and weeds. In March 1979 infiltration averaged 133 s in the trampled paddock outside the exclosure, compared with 68 s within the exclosure. By September 1979, the data were 150 and 61 s respectively. The total number of large soil animals (as defined by Abbott *et al.*, 1979) was then 7 and 301 respectively. Hence the deterioration in soil structure caused by treading of sheep took only 3 months.

If pasture is established on rehabilitated land, careful attention to the stocking rate is essential if soil compaction is to be avoided.

Discussion

Several points require amplification. This survey of soil structure in reclaimed lands revealed few instances of research on soil invertebrates other than earthworms. This is undoubtedly because ants and termites are more difficult to collect alive in large numbers, and because they have a complex social organization which is difficult to re-assemble. However, it should be recognized that in arid locations, ants and termites may be more relevant and useful than earthworms.

Researchers investigating how to establish populations of earthworms on rehabilitated land need to be aware of the species of earthworm present on nearby undisturbed land, because it is essential to evaluate species for their effectiveness in ameliorating soil structure. Thus *Eisenia fetida* requires manure or compost for its survival, *Lumbricus rubellus* is an important soil mixer, *A. longa* and *Lumbricus terrestris* are deep burrowing, surface casting species, and *Aporrectodea caliginosa*, *Aporrectodea rosea* and *Aporrectodea chlorotica* excrete their waste material underground (Van Rhee, 1965; Bal, 1982).

Despite this knowledge, it is wasted effort to inoculate rehabilitated soil with the correct species of earthworm if any adverse environmental conditions are not overcome beforehand. Clearly the ideal species of earthworm substantially improves soil structure while tolerating certain adverse soil conditions.

Finally, it is important to note an unresolved disagreement in the literature concerning the extent of contribution that the invertebrate soil fauna makes to soil structure. Barley (1959) found that in the A horizon of the soil which he studied, roots occupied more space than earthworm tunnels. Russell (1973) asserted that roots are the most important group of agents which create structural pores in undisturbed soils. In contrast, Greenland (1971) considered that growth in roots alone cannot cause the increases in porosity that occur under pasture, and that reworking of the soil into more porous aggregates was accomplished by soil animals. More recently, Tisdall & Oades (1982) have outlined a compromise between these extremes. This is that plants indirectly increase the water-stable aggregation of soil particles by providing the food that enables large populations of soil animals to build up.

Conclusion

Three fundamental activities of soil fauna are known to improve soil structure. *Burrowing* changes the porosity of the soil, thus improving infiltration of water and aeration. *Humification*, through the ingestion and fragmentation of organic matter and the production of gum-rich faecal pellets, improves the stability and friability of soil. *Mixing* involves the translocation of soil onto or near the surface, which may lead to textural inversion of the soil profile. Also, the production of faecal pellets tends to result in the incorporation of decomposed leaf litter more intimately with the mineral soil.

Provided that rehabilitation ameliorates any adverse physical and chemical conditions and that revegetation is accomplished, substantial re-establishment of invertebrate fauna can be achieved within 5 years. The resulting improvements in soil structure by invertebrates can, however, be quickly nullified if excessive treading and trampling of soil by the larger vertebrates are not mimized.

References

Abbott, I., Parker, C. A. & Sills, I. D. (1979). Changes in the abundance of large soil animals and physical properties of soils following cultivation. *Australian Journal of Soil Research*, **17**, 343–53.

Bal, L. (1982). *Zoological Ripening of Soils.* Wageningen: Centre for Agricultural Publishing and Documentation.

Barley, K. P. (1959). Earthworms and soil fertility. IV. The influence of earthworms on the physical properties of a red-brown earth. *Australian Journal of Agricultural Research*, **10**, 371–6.

Barley, K. P. (1961). The abundance of earthworms in agricultural land and their possible significance in agriculture. *Advances in Agronomy*, **13**, 249–68.

Barley, K. P. & Kleinig, C. R. (1964). The occupation of newly irrigated lands by earthworms. *Australian Journal of Science*, **26**, 290–1.

Baxter, F. B. & Hole, F. D. (1967). Ant (*Formica cinerea*) pedoturbation in a prairie soil. *Soil Science Society of America Proceedings*, **31**, 425–8.

Bunn, S. E. (1983). Termite (Isoptera) fauna of jarrah forest in the Wagerup-Willowdale region, Western Australia: relevance to the rehabilitation of bauxite minesites. *Forest Ecology and Management*, **6**, 169–77.

Buntley, G. J. & Papendick, R. I. (1960). Worm-worked soils of eastern South Dakota, their morphology and classification. *Soil Science Society of America Proceedings*, **24**, 128–32.

Burns, R. G. & Martin, J. P. (1986). Biodegradation of organic residues in soil. In *Microfloral and Faunal Interactions in Natural and Agro-ecosystems*, ed. M. J. Mitchell & J. P. Nakas, pp. 137–202. Dordrecht: Nijhoff/Junk.

Coleman, D. C. (1986). The role of microfloral and faunal interactions in affecting soil processes. In *Microfloral and Faunal Interactions in Natural and Agro-ecosystems*, ed. M. J. Mitchell & J. P. Nakas, pp. 317–48. Dordrecht: Nijhoff/Junk.

Cotton, D. C. F. & Curry, J. P. (1982). Earthworm distribution and abundance along a mineral-peat soil transect. *Soil Biology and Biochemistry*, **14**, 211–14.

Darwin, C. (1840). On the formation of mould. *Transactions of the Geological Society of London*, **5**, 505–9.

De Vleeschauer, D. & Lal, R. (1981). Properties of wormcasts under secondary tropical forest regrowth. *Soil Science*, **132**, 175–81.

Dunger, W. (1967). Die Entwicklung der Mackro- und Megafauna in rekultivierten

Haldenboden. In *Progress in Soil Biology*, ed. O. Graff & J. E. Satchall, pp. 340–52. Amsterdam: North-Holland Publishing Company.

Dunger, W. (1968). Die Entwicklung der Bodenfauna auf rekultivierten Kippen und Halden des Braunkohlentagebaues. *Abhandlungen und Berichte des Naturkundemuseums Görlitz*, **43**, 1–256.

Dutt, A. K. (1948). Earthworms and soil aggregation. *Journal of the American Society of Agronomy*, **40**, 407–10.

Edwards, C. A. & Lofty, J. R. (1977). *Biology of Earthworms*. London: Chapman & Hall.

Evans, A. C. (1948). Studies on the relationships between earthworms and soil fertility. II. Some effects of earthworms on soil structure. *Annals of Applied Biology*, **35**, 1–13.

Evans, A. C. & Guild, W. J. M. (1947). Studies on the relationship between earthworms and soil fertility. I. Biological studies in the field. *Annals of Applied Biology*, **34**, 307–30.

Green, R. D. & Askew, G. P. (1965). Observations on the biological development of macropores in soils of Romney marsh. *Journal of Soil Science*, **16**, 342–9.

Greenland, D. J. (1971). Changes in the nitrogen status and physical condition of soils under pastures, with special reference to the maintenance of the fertility of Australian soils used for growing wheat. *Soils and Fertilizers*, **34**, 237–51.

Guild, W. J. M. (1955). Earthworms and soil structure. In *Soil Zoology*, ed. D. K. McE. Kevan, pp. 83–98. London: Butterworths.

Hale, W. G. (1967). Collembola. In *Soil Biology*, ed. A. Burges & F. Raw, pp. 397–411. London: Academic Press.

Hamilton, W. E. & Vimmerstedt, J. P. (1980). Earthworms on forested spoil banks. In *Soil Biology as Related to Land Use Practices*, ed. D. L. Dindal, pp. 409–16. Washington: Environmental Protection Agency.

Harris, R. F., Chesters, G. & Allen, O. O. (1966). Dynamics of soil aggregation. *Advances in Agronomy*, **18**, 107–69.

Huhta, V. (1979). Effects of liming and deciduous litter on earthworm (Lumbricidae) populations of a spruce forest, with an inoculation experiment on *Allolobophora caliginosa*. *Pedobiologia*, **19**, 340–5.

Hutson, B. R. (1980). Colonization of industrial reclamation sites by Acari, Collembola and other invertebrates. *Journal of Applied Ecology*, **17**, 255–75.

King, K. L. & Hutchinson, K. J. (1983). The effects of sheep grazing on invertebrate numbers and biomass in unfertilized natural pastures of the New England Tablelands (NSW). *Australian Journal of Ecology*, **8**, 245–55.

Kubiena, W. L. (1938). *Micropedology*. Iowa: Collegiate Press.

Lal, R. & Akinremi, O. O. (1983). Physical properties of earthworm casts and surface soil as influenced by management. *Soil Science*, **135**, 114–22.

Langmaid, K. K. (1964). Some effects of earthworm invasion in virgin soils. *Canadian Journal of Soil Science*, **44**, 34–7.

Lee, K. E. & Wood, T. G. (1971). Physical and chemical effects on soils of some Australian termites, and their pedobiological significance. *Pedobiologia*, **11**, 376–409.

Lynch, J. M. & Bragg, E. (1985). Microorganisms and soil aggregate stability. *Advances in Soil Science*, **2**, 133–71.

Neumann, U. (1973). Succession of soil fauna in afforested spoil banks of the brown-coal mining district of Cologne. In *Ecology and Reclamation of Devastated Land*, vol. 1, ed. R. J. Hutnik & G. Davis, pp. 335–47. New York: Gordon & Breach.

Nichols, O. G. & Bunn, S. (1980). *Termite Utilisation of Rehabilitated Bauxite Mined Sites*. Alcoa of Australia Ltd Environmental Research Bulletin no. 9.

Nielsen, G. A. & Hole, F. D. (1964). Earthworms and the development of coprogenous A1 horizons in forest soils in Wisconsin. *Soil Science Society of America Proceedings*, **28**, 426–30.

Nye, P. H. (1955). Some soil-forming processes in the humid tropics. IV. The action of the soil fauna. *Journal of Soil Science*, **6**, 73–83.

Pawluk, S. (1985). Soil micromorphology and soil fauna: problems and importance. *Quaestiones Entomologicae*, **21**, 473–96.

Piearce, T. G. (1984). Earthworm populations in soils disturbed by sampling. *Biological Conservation*, **29**, 241–52.

Rusek, J. (1985). Soil microstructures – contributions on specific soil organisms. *Quaestiones Entomologicae*, **21**, 497–514.

Russell, E. W. (1973). *Soil Conditions and Plant Growth*. London: Longman.

Salem, M. Z. & Hole, F. D. (1968). Ant (*Formica exsectoides*) pedoturbation in a forest soil. *Soil Science Society of America Proceedings*, **32**, 563–7.

Satchell, J. E. (1967). Lumbricidae. In *Soil Biology*, ed. A. Burges & F. Raw, pp. 259–322. London: Academic Press.

Seastedt, T. R. (1984). The role of microarthropods in decomposition and mineralization processes. *Annual Review of Entomology*, **29**, 25–46.

Stockdill, S. M. J. & Cossens, G. G. (1967). The role of earthworms in pasture production and moisture conservation. In *Proceedings of the New Zealand Grassland Association, 1966*, pp. 168–83.

Swaby, R. J. (1950). The influence of earthworms on soil aggregation. *Journal of Soil Science*, **1**, 195–7.

Thompson, H. W. (1958). Restoration of agricultural and other soils. *Nature*, **181**, 390–1.

Tisdall, J. M. (1985). Earthworm activity in irrigated red-brown earths used for annual crops in Victoria. *Australian Journal of Soil Research*, **23**, 291–9.

Tisdall, J. M. & Oades, J. M. (1982). Organic matter and water-stable aggregates in soils. *Journal of Soil Science*, **33**, 141–63.

Van Rhee, J. A. (1965). Earthworm activity and plant growth in artificial cultures. *Plant and Soil*, **22**, 45–8.

Vimmerstedt, J. P. & Finney, J. H. (1973). Impact of earthworm introduction on litter burial and nutrient distribution in Ohio strip-mine spoil banks. *Soil Science Society of America Proceedings*, **37**, 388–91.

White, G. (1788). *The Natural History and Antiquities of Selborne in the County of Southhampton*. London: White & Son.

Witschi, P. A. & Michalk, D. L. (1979). The effect of sheep treading and grazing on pasture and soil characteristics of irrigated annual pastures. *Australian Journal of Agricultural Research*, **30**, 741–50.

3

The role of fauna in nutrient turnover

B. R. HUTSON

Introduction

There are two main benefits to be gained from encouraging the development of soil faunal activities on derelict land. Firstly, as discussed in this Chapter, their contribution to organic matter decomposition/nutrient turnover is significant. Secondly, they are important for mixing and aeration of soil, as discussed in Chapter 2. Under some circumstances, they can also be used as indicators of soil conditions (see Chapter 13).

The relative roles of various groups of soil organisms have been investigated by many workers (see Dickinson & Pugh, 1974; Swift, Heal & Anderson, 1979; Anderson, Rayner & Walton, 1984). However, ecosystems examined have usually been fully developed, and they represent areas with a very wide range of fauna. Therefore, only general conclusions can be drawn from the results of studies such as litter-bag loss trials, first used 25 years ago (Edwards & Heath, 1963). Much of the published data on the fauna of reclamation sites describes population composition, colonization and succession (see Chapters 13–16): there is far less information available on the actual functioning of the communities present, and much of that which exists relates only to earthworms (Curry & Cotton, 1983).

In this Chapter, after summarizing decomposition and mineralization processes within the soil, the fauna and their feeding habits are reviewed, along with features of the ecosystem that will directly or indirectly affect the fauna and decomposition processes. Then, data from various reclamation sites are examined and an attempt is made to draw some conclusions that will be of benefit to site management.

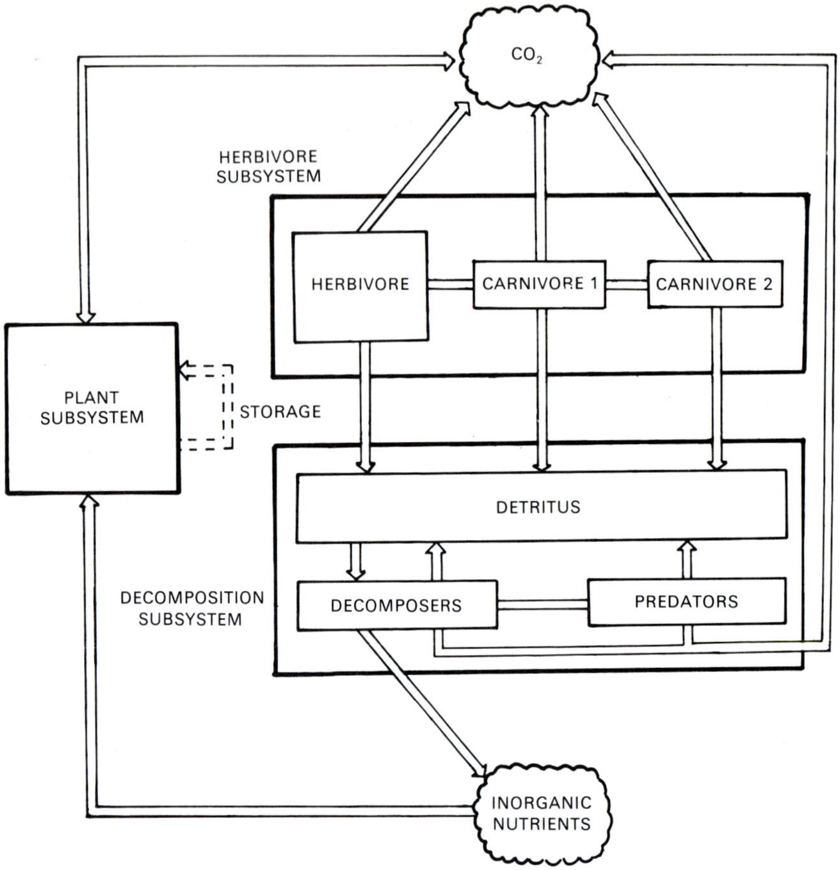

Fig. 3.1. A general model of ecosystem structure. Major pathways of matter transfer are shown by arrows, organic matter pools by rectangles, inorganic pools as clouds (based on Swift, Heal & Anderson, 1979; Fig. 1.3).

Decomposition and mineralization processes within the soil

The natural supply of organic matter to the soil is plant litter, dead roots, root exudates and sloughed off root cells. If it is to be of value for further plant growth, without depending on added components such as fertilizers, the litter must be decomposed so that the released nutrients can be re-utilized.

Within the soil, nutrients are present in two forms: inorganic and organic. Roots are not generally permeable to organic molecules and plants absorb inorganic nutrient elements from the soil solution 'pool'. Because the major fraction of each element is either insoluble or adsorbed to solid surfaces, nutrient availability is dependent more on the rate of

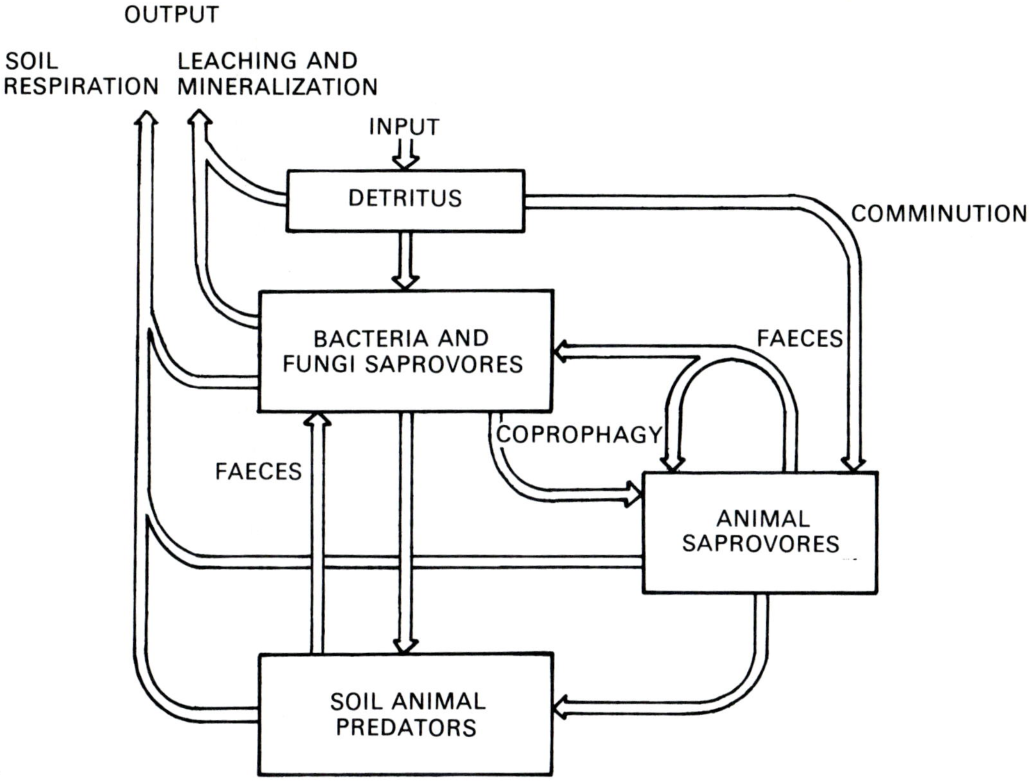

Fig. 3.2. A general model of the functioning of the organisms of the decomposition subsystem indicating feeding and respiration pathways (based on Wiegert, Coleman & Odum, 1970).

replenishment of this 'pool' than on its actual size (see Swift, Heal & Anderson, 1979). With regard to primary production, the soil mineral form is generally insignificant compared with the inorganic form derived from mineralization of organic elements by the action of decomposer organisms.

Thus, in any ecosystem, the balance between primary production and decomposition rate will determine organic matter accumulation or loss in the soil, whereas movement of essential nutrient elements through the system is an important regulator of primary production growth (Fig. 3.1).

Although the decomposers are ultimately responsible for mineralizing nutrients, the process is regulated by other factors such as: (a) the quality of the organic resource; (b) the physico-chemical characteristics of the environment; and (c) the organisms present.

Faunal contributions to decomposition will be primarily discussed here, but microorganisms cannot be disregarded due to the strong interaction between fauna and microflora (see Anderson, Rayner & Walton, 1984; Fitter *et al.*, 1985). In the absence of a fauna, the microflora immobilize nutrients within their biomass, thereby competing with plants for essential nutrients (Parker *et al.*, 1984*a*; Visser, 1985).

The effect of feeding by many of the soil fauna is to accelerate the decomposition process as a result of comminution (reduction in organic particle size) or predation (Fig. 3.2). Within the gut, food may be further comminuted and catabolic action takes place. Most primary resource feeders possess a similar basic enzyme complement with limited protease and lipase activity but strong carbohydrase activity. Most saprovores can digest starch, glycogen, sugar and hemicellulose, but the occurrence of cellulase and lignin-degrading enzymes is more limited.

Varieties of soil fauna and their feeding habits

Good general accounts of the soil fauna are given by Burges & Raw (1967) and Wallwork (1970).

General categories by size

The soil fauna can be conveniently categorized by length (van der Drift, 1951):

1. *Microfauna* (< 200 μm), including Protozoa and some small Nematoda and Acari (mites);
2. *Mesofauna* (200 μm–2 mm), including microarthropods, particularly the Collembola (springtails), Acari and Nematoda;
3. *Macrofauna* (2–20 mm), including the oligochaete Enchytraeidae (potworms), meso- and macroarthropods, i.e. most soil-inhabiting insects and spiders, Diplopoda (millipedes), Isopoda (woodlice), and smaller Mollusca (slugs and snails); and
4. *Megafauna* (> 20 mm), including oligochaete Lumbricidae and other families (earthworms), larger Mollusca and soil-inhabiting or grazing vertebrates.

Major taxonomic groupings

1. *Protozoa* – generally these are predators on bacteria and their catalytic effect may be great (see Clarholm, 1985). They can enter small pores inaccessible to larger fauna (Elliot *et al.*, 1980), feed on bacteria contained within, then become the prey of larger fauna.
2. *Nematodes* – many are bacterivores or fungivores and are not

directly involved in litter decomposition. Where they do occur, they will be consumed by the larger fauna, providing an important link in the decomposer food chain.

3. *Microarthropods* – major reviews of the overall role of this group in decomposition/nutrient turnover have been made by Harding & Stuttard (1974), Swift, Heal & Anderson (1979), and Seastedt (1984), among others. Collembola have been shown to feed on a wide range of organic material including fungal hyphae and spores, bacteria, fresh and decaying leaves, algae, faeces, exuviae, carrion and living prey. Although fresh litter may not be the preferred diet, and may not even be assimilated if eaten, it can form a significant, fragmented part of the gut contents of many Collembola (Ashraf, 1969). Of the soil-inhabiting Acari, Cryptostigmata (oribatids) are generally regarded as the most important regulators of decomposition processes (Crossley, 1977) and are fungivores and detritus feeders. However, they are generally restricted to feeding in surface litter, and not around roots or buried litter (Santos & Whitford, 1981). Mesostigmata are mainly predators, particularly on mesofauna, but some are fungal feeders. Prostigmata also include some predators but the smaller, more abundant, species are generally fungivores or graze on microflora. Astigmata feed mainly on liquids or plant detritus which is in an advanced state of decomposition. Thus, there is little evidence of mites feeding directly on undecomposed litter.
4. *Macroarthropods* – the relative importance of this group in temperate woodland litter decomposition was assessed by Edwards (1974). Diplopoda are generally important and large insect larvae (especially Diptera and Coleoptera) may also contribute significantly. In warm arid and semi-arid regions, it has been shown that termites perform similar functions to earthworms in the process of soil structure formation and organic matter turnover (Lee & Wood, 1971). They: (1) construct tunnels from nests to food sources; (2) ingest inorganic soil to provide nutrients for their symbiotic gut microflora; and (3) use faeces mixed with soil particles to construct tunnels. In primitive wood-eating termites, the hindgut contains cellulolytic protozoan symbionts whereas extensive catabolic activity by bacterial symbionts occurs in soil feeding termites.
5. *Oligochaeta* – when present, lumbricid earthworms are of much greater importance than enchytraeid potworms to litter burial,

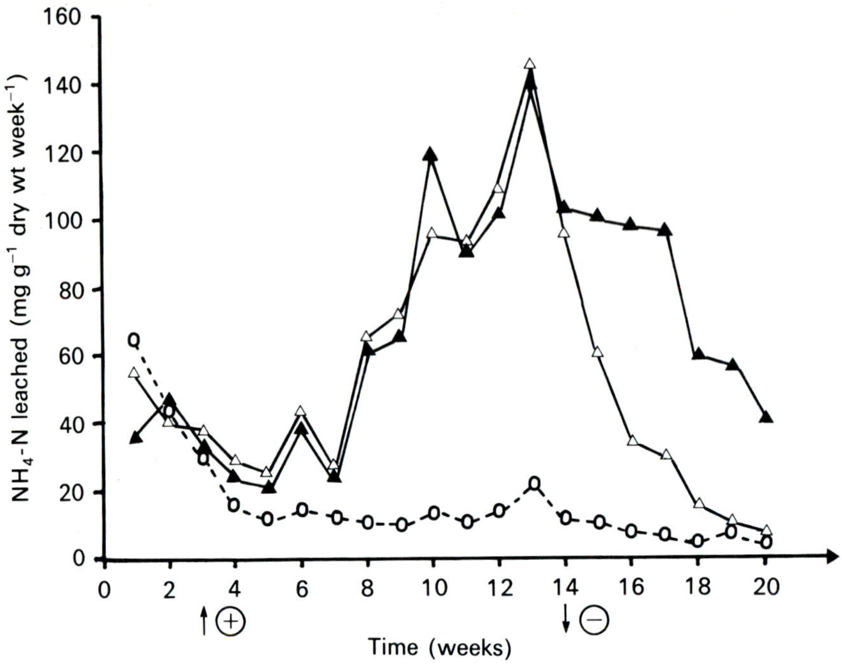

Fig. 3.3. The effects of adding and removing *Glomeris marginata* on NH_4-N release from 2 g oak litter in microcosms. o = control; ▲ = 4 animals added at week 3, not removed; △ = 4 animals added at week 3, removed at week 14 (from Anderson & Ineson, 1984).

breakdown and soil profile formation (Lofty, 1974). This is because of their role in: (1) removal of litter from the soil surface; (2) burial by casting over the litter; (3) litter fragmentation; (4) incorporation of fragmented and decomposed material throughout the soil habitat; (5) reduction of the C:N ratio in organic matter and acceleration of nitrogen re-assimilation by plants; and (6) multiplication of soil microflora within the gut.

Interactions of fauna with microorganisms

There are few quantitative data on the direct effects of faunal feeding or excretion on litter decomposition, but overall studies on energy dissipation indicate this to be only a minor factor. Studies on feeding behaviour and population dynamics have indicated a relationship between changes in faunal predator and microbial prey populations and it has been demonstrated (Fig. 3.3) that whereas microorganisms play an enormous part in decomposition processes, the fauna play an over-riding indirect regulatory role (see Coleman *et al.*, 1984; Anderson & Ineson, 1984).

The indirect effects can be divided into three separate influences on microbial activity (see Visser, 1985):

1. The comminution, mixing and channelling of soil and litter. Surface litter is thus redistributed into soil strata where nutrient release can take place more rapidly and organic material is prepared for microbial activity. Furthermore, fungal succession on decaying material in the animal's gut may be diverted from that found on uningested material of the same kind to give a competitive advantage to fast growing, short life-cycle species resulting in an enhancement of decomposition rate.
2. Direct grazing on microflora. Feeding by mesofauna on selected fungi may have a significant effect on fungal distribution patterns – it has been demonstrated that a fine balance may exist between microbial stimulation and inhibition caused by litter grazers. The degree of impact is dependent on not only invertebrate grazing pressure but also the growth rate of the grazed organisms.
3. Dispersal of microbial propagules. This may be the most important factor, but the success of the transmitted organisms in new environments remains undetermined.

In deciduous woodland soil under laboratory and field conditions, litter feeding animals were found to have a significant effect on nitrogen mineralization from leaf litter and soil organic matter, which could not be predicted from population data alone (Anderson *et al.*, 1985). They concluded that animals disrupt the normal processes of microbial mineralization and that nitrogen mineralization is enhanced at C:N ratios where nitrogen immobilization would be expected to occur.

In deserts where predatory mites feed on bacteria-feeding nematodes, which in turn consume bacteria in leaf litter, Elkins & Whitford (1982) examined decomposition of roots and buried litter. Results indicated considerably more decomposition, CO_2 evolution and nitrogen/phosphorus mineralization where mites were present compared with plots where the predatory mites were eliminated by biocides.

The influence of resource quality on decomposition

Different organic resources possess different chemical and physical attributes, decomposing at different rates even under conditions when all other factors (e.g. the physical environment) are controlled (see

Swift, Heal & Anderson, 1979). For the purpose of material comparison, the situation is further complicated by the fact that the rate of decomposition declines as organic matter decomposes.

Chemical variations

1. *Carbon and energy sources* – the bulk of carbon is stored in a variety of polymeric compounds with differing properties. Variations in their proportions and in the structure of the plant tissue components (e.g. woody tissue/green leaves, hardwoods/softwoods) will influence the effect of resource quality on decomposition and nutrient release.
2. *Nutrient sources* – nutrient concentration levels have been shown to affect microbial decomposition and may also be important selective factors for decomposer animals (see table 4.10 in Swift, Heal & Anderson, 1979).
3. *Modifiers (inhibitors/stimulators)* – most of these compounds offer plants protection against excessive herbivore grazing, or attack by pathogens, but they may temporarily inhibit activities of sensitive microorganisms in soil. There is abundant evidence for the effects on decomposers of plant polyphenolic compounds. Pesticides, fungicides and accidental pollutants may also have a profound effect (e.g. heavy metals from mining activity – see Chapter 9).

Physical variations

1. *Surface properties and 'toughness'* – it has not been clearly demonstrated that leaf palatability can be attributed to 'toughness' (e.g. heavy cuticularization) but this may be of importance.
2. *Particle size* – this determines: (1) patterns of colonization by microorganisms; (2) the ability of animals to ingest food; and (3) certain parameters of the environment such as its moisture retention and gaseous diffusion rate.

Influence of resource quality on decomposer community dynamics

Despite a high degree of variation in the succession of organisms on decomposing organic matter, four main phases can be distinguished: (1) colonization; (2) exploitation; (3) invasion; and (4) post-invasion (Swift, 1977). Phases (1) and (2) are dominated by fungi. During phase (3), animal invasion depends on conditioning of the resource and the supply of a high quality microbial tissue as a food resource, and (4) is typified by microorganisms adapted to the particulate resources formed during comminution by animals.

Influence of environmental physico-chemistry on decomposition

In the soil, most factors classed under this title have an indirect regulatory effect on decomposers, whereas some effects of climate are more direct (e.g. leaching, freezing/thawing, drying/wetting).

Moisture

Soil water is essential for the movement of most microfauna, and the tissue growth of all fauna. Some animals are protected against desiccation and can survive in relatively dry habitats. Moisture indirectly affects soil aeration, pH, etc.

Aeration

Since aeration is dependent on soil moisture, temperature and physical structure, it is difficult to separate the effects of climate and soil type from those of aeration.

pH

The activity of decomposers in litter layers is an important determinant of the pH of underlying soil. Thus, pH and its effects on microorganisms may vary locally. Meso- and macrofauna are more able to avoid these effects due to the scale of their habitat and the structure of their exoskeleton. However, pH tolerance in lumbricid earthworms has been studied in detail and found to vary considerably (e.g. Satchell, 1955).

Temperature

Experimental responses to large-scale differences are obvious, but are less easy to detect correctly at the microhabitat scale. Apart from geographical aspects of temperature differences, variations in soil may be a result of vegetation cover and fluctuations in time (seasonal and diurnal).

Leaching

Leaching may determine the initial rate of weight or nutrient loss in fresh detritus, and directly modify the resource quality. Also, reduction in particle size accelerates leaching losses (Witkamp, 1969).

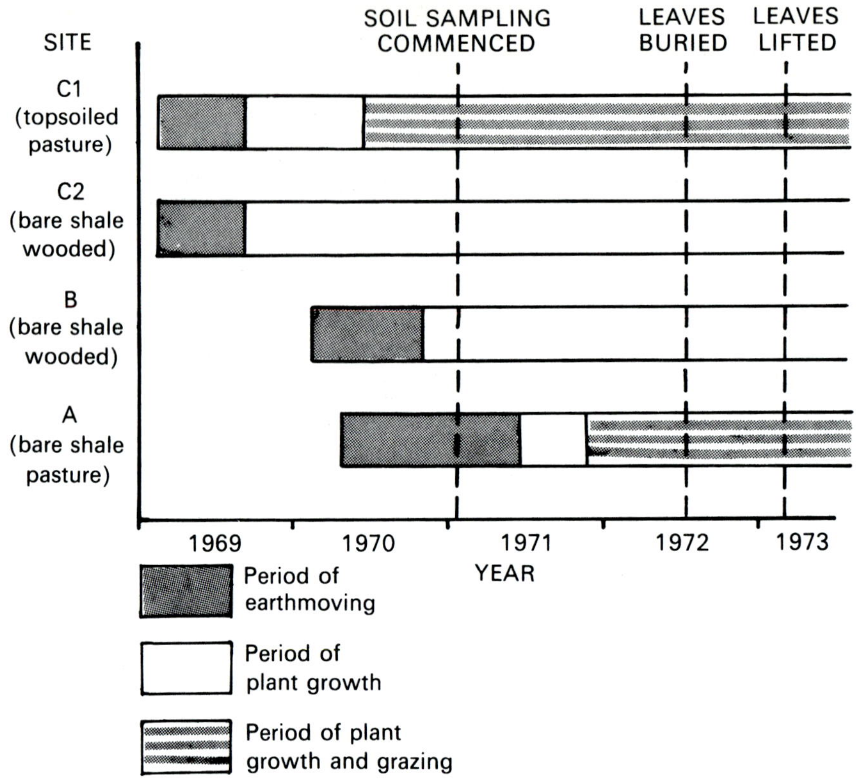

Fig. 3.4. Comparison between reclamation stages on the four pitheap sites used for litter bag loss trials (based on Hutson & Luff, 1978).

Estimates of faunal contributions to decomposition at various reclamation sites

Reclaimed coal waste (Europe)

The contribution of soil fauna to the degradation of organic matter at four sites in Northumberland, England, was compared with results from a 'normal' (control) soil by Hutson (1980*b*). When the experiment began, reclamation of the sites had been completed for between 1.5 and 3.5 years (Fig. 3.4). They were variously treated (topsoil or bare shale, woodland or permanent pasture, grazed or ungrazed), and had been rapidly colonized by Acari and Collembola (Hutson, 1980*a*).

The use of the multiple mesh-size litter bag technique also enabled estimates of the relative importance of macro-, meso- and microfauna and flora to be made. Bags of three mesh sizes, each containing three similar sized oakleaf pieces, were used. The fine mesh excluded all except the microorganisms, the medium mesh excluded only macro-

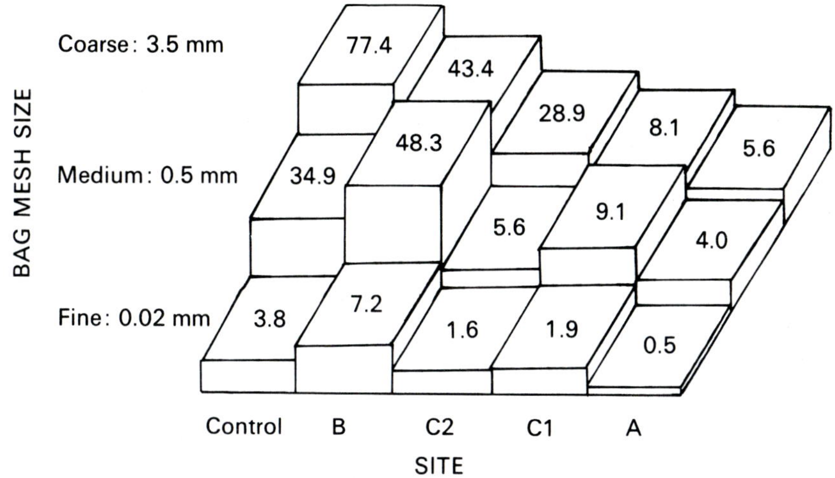

Fig. 3.5. Rates of breakdown of confined oak leaves in bags of three mesh sizes, buried in reclaimed pitheaps and a control pasture soil. Numbers on each column are the percentages of leaf material lost after eight months (from Luff & Hutson, 1977).

fauna, and the coarse mesh allowed access to all organisms. After 8 months of burial at the sites, degradation was almost complete (77 %) in coarse mesh bags at the control site, so all bags were lifted to assess leaf loss (Fig. 3.5). Significantly less degradation (35 %) had occurred in the medium mesh bags at the control site. Compared with the control medium mesh bags, greater loss had occurred from the coarse (43 %) and medium (48 %) mesh bags on the reclamation site that supported the highest Collembola density.

Unlike the control site, there were no earthworms or large saprophages (macrofauna) and few Diptera (fly) larvae or Enchytraeidae (mesofauna) on the reclaimed sites. A large number of only a relatively limited type of fauna had colonized the sites, and it was possible to determine which groups, and even suggest which species, were directly involved in leaf degradation. The results indicated a significant contribution to the initial degradation process by Collembola rather than Acari (which were mainly Prostigmata and Astigmata) and on all four sites isotomid Collembola (dominated by *Isotoma notabilis*, *Isotoma viridis* and *Cryptopygus garretti*) contributed most to collembolan biomass (see table 7 in Hutson, 1980*b*).

Both direct (litter bag loss) and indirect (metabolic activity) methods were used by Dunger (1968*a*, *b*, 1969*b*) to assess changes in litter breakdown rate as faunal colonization and succession progressed at sites

Table 3.1. *Consumption* ($kcal\ m^{-2}\ yr^{-1}$) *of litter-feeding soil animals in reclaimed brown coal dumps* (*from Dunger*, 1969*b*)

	Years after reclamation			
	1	3	7	10
Microarthropods	20	160	75	40
Enchytraeidae	5	13	31	18
Macroarthropods	1	80	70	20
Lumbricidae	1	2	300	900
% total litter consumed	0.6	4.6	27.3	76.3

in East Germany (see Chapter 13). At less-acid sites, reclaimed for only 3 years or less, his conclusion was similar to that of Luff & Hutson (1977) and Hutson (1980*b*). This was that in the absence of earthworms, microarthropods (especially Collembola, but also cryptostigmatid mites at Dunger's sites) were relatively important during the first stages of soil formation by way of their metabolic activity (see Table 13.2). Microarthropods consumed the most litter, although this was only a relatively small proportion (< 5%) of the total available (Table 3.1). After 7 years reclamation, earthworms became established and were the most significant faunal decomposers (total breakdown rose to 27%), and after 10 years, 70% of the litterfall was removed by earthworms.

Even after 15 years at Dunger's more acid sites, earthworms were limited to site edges and Collembola remained the most abundant and significant animals.

Open-cast surface mines (*USA*)

At a coal surface mine in an arid region of New Mexico, litter bags were used to measure barley-straw decomposition rate at plots variously treated with top-dressings (Elkins *et al.*, 1984). Site procedures tested were: (1) raw spoil; (2) bark-amended spoil; (3) straw-mulch-amended spoil; (4) topsoiled spoil; and (5) an adjacent, unmined control area. The aim was to evaluate the efficiency of the amendments in restarting soil-forming processes both by their effects on decomposition rate and by estimating the responses of the total soil biota in relation to the measured decomposition rates.

Throughout the study, the bark-amended site exhibited the highest litter loss rate, similar to that at the control site (Fig. 3.6). Microarthropod diversity and density were similar in bags and surrounding soil, with

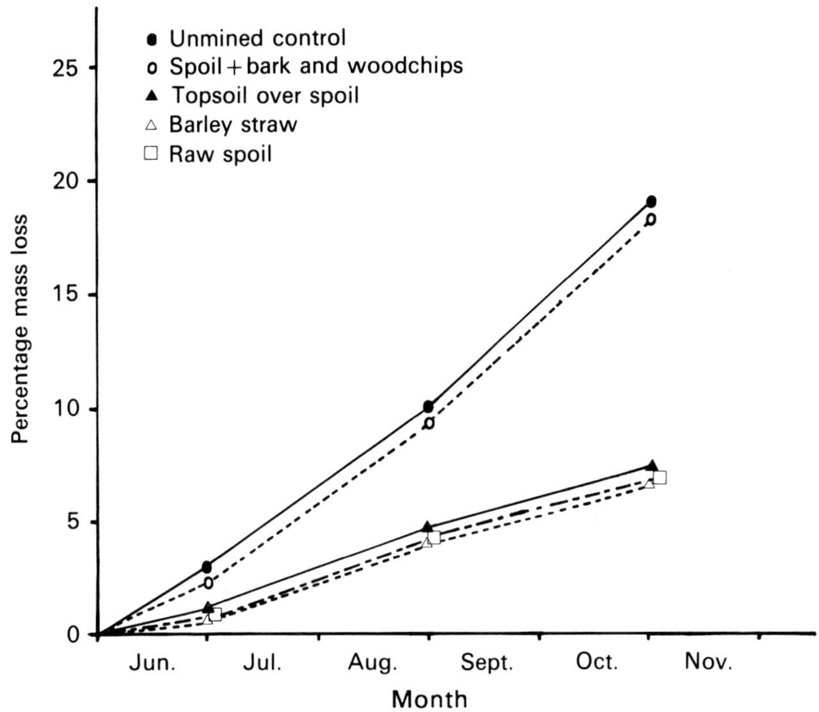

Fig. 3.6. The relationship between organic amendments mixed with raw spoils and decomposition of barley straw at surface-mine sites in New Mexico (based on Elkins *et al.*, 1984).

highest results from bark-amended and control sites. Straw-amended and topsoiled sites exhibited greatly reduced activity of microarthropods and decomposer organisms generally, indicating a lower nutrient cycling efficiency.

It is notable that in arid soils, some microarthropods have been shown to be especially important regulators of decomposition and mineralization (Santos & Whitford, 1981). Initial litter colonizers are some small fungivorous prostigmatid mites (tydeids, then tarsonemids and pyemotids), large numbers of predatory mesofauna follow, then Collembola (more general feeders) are the last group to colonize. The prostigmatid mites were well represented in the bark-amended and unmined control sites studied by Elkins *et al.* (1984). The lower level of microfloral activity recorded at the straw-amended, topsoiled and raw spoil sites was not reflected by actual microfloral numbers, thus their biomass did not appear to reflect their biological activity. Since micro- and mesofauna are known to enhance decomposition and mineralization rates, it is therefore

Table 3.2 *The effect of time after reclamation on the first-order rate constant* (k) *for barley straw decomposition in buried litter bags on mined and unmined surface-mine sites in New Mexico* (*from Parker* et al., 1984b)

Time since reclamation (years)	k(g^{-1})	r^2
unmined	1.602	0.95
4	0.307	0.94
3	8.840	0.87
2	0.562	0.90
1	0.449	0.94

Note: r^2 represents the correlation coefficient value.

apparent that attempts to evaluate the role of the microflora without data on these fauna groups, can be very misleading.

Other straw mulching and topsoil reclamation procedures have also failed to produce the environmental requirements to support a mature soil biota (Parker *et al.*, 1984*b*). After 3–4 years, the initial treatment benefit disappeared with a subsequent reduction in soil biota and decomposition rate (Table 3.2). This was not reflected immediately by above-ground production, but more long-term consequences would be likely. In a similar ecosystem, mass loss from litter bags was reported to be due primarily to removal by microarthropod activity, not to microbial mineralization and consequent CO_2 production (Parker *et al.*, 1984*a*). Evaluation of nitrogen budgets also re-emphasized the importance of microarthropods as regulators of decomposition by microbial stimulation in arid environments.

The effect of termites on soil formation from spoil in these arid areas was also examined by Whitford & Elkins (1988). Although the results indicate their importance to site reclamation, successful translocation of termite colonies to reclaimed areas has not yet been achieved.

The effects of earthworm activity on surface mine (acid shale) spoil banks in Ohio were examined using measurements of organic matter incorporation and nutrient cycling (Vimmerstedt & Finney, 1973). In the field *Lumbricus terrestris* buried leaf litter and humus (equivalent to 5 t litter ha^{-1} over 2 years), and deposited castings on the surface. In a greenhouse study using cores of litter and soil, depth of litter burial was far greater with than without earthworms (Fig. 3.7). The equivalent of 9.5 t litter ha^{-1} and 52.5 t humus ha^{-1} was buried or consumed after 174

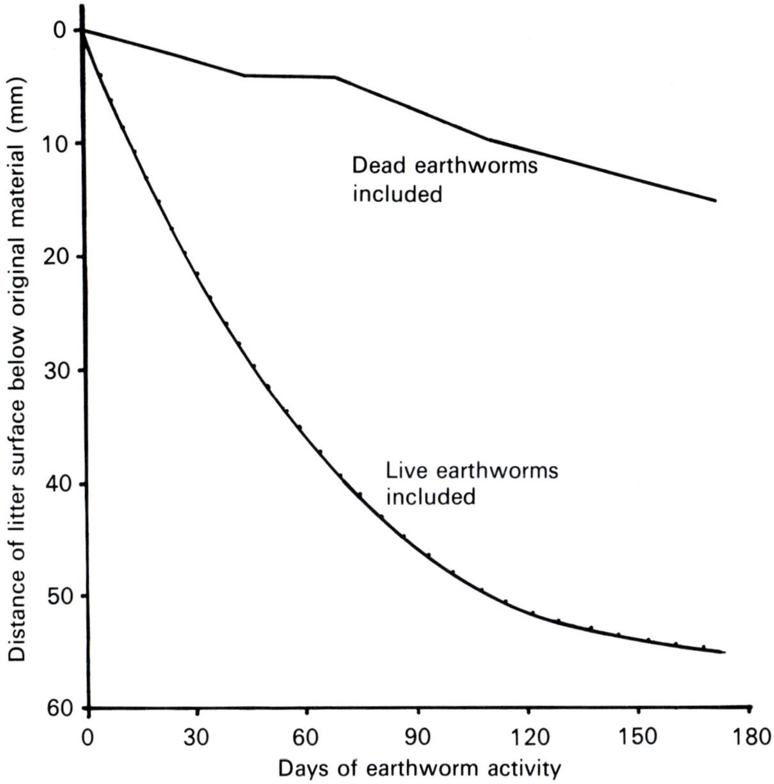

Fig. 3.7. Downward movement of litter surfaces in strip-mine spoil cores with or without live earthworms during 174 days in the greenhouse (from Vimmerstedt & Finney, 1973).

days, while 16.7 t castings ha^{-1} were deposited on the surface. There was a noticeable increase in the exchangeable cations K, Ca, Mg and available P in the mineral spoil section of the core samples.

Peat sites (Ireland)

The earthworm *L. terrestris* has been found to be a slow colonizer of reclaimed peat, but when introduced experimentally, litter disappearance was greatly accelerated (Curry, Kelly & Bolger, 1985). On the sites studied, macroinvertebrates were scarce, and consisted mainly of insect larvae and spiders, and microarthropods were by far the most abundant group. Results of faunal exclusion experiments indicated that consumption of litter by introduced worms accounted for 84% of total litter disappearance (Table 3.3). This confirmed the importance of earthworm activity in litter decomposition but indicated that on peat

Table 3.3. Salix *litter consumption by* Lumbricus terrestris *in reclaimed cutover peat, estimated by litter bag loss on sites with and without earthworms present (from Curry, Kelly & Bolger*, 1985)

Period	Litter consumption per bag (g dry wt)		Daily consumption rate (mg dry wt g^{-1} fresh wt d^{-1})
	with worms	without worms	
13 Oct. 83–8 Dec. 83	17.4±0.3	2.8±0.3	18
9 Dec. 83–2 Feb. 84	8.2±0.9	1.4±0.3	7
3 Feb. 84–29 Mar. 84	8.2±0.8	0.6±0.4	7

sites, the process is limited by their scarceness, caused by combinations of low pH, high N level and low palatability (Swift, Heal & Anderson, 1979).

Polder sites (the Netherlands)

The introduction of earthworms into 7-year-old pasture and 13-year-old orchard soils on reclaimed polder resulted in the development of mull soils, compared with mor soils in adjacent areas without worms (van Rhee, 1963). Later, increased grass yields were reported from polder soils where worms were present but there was no effect on orchard fruit yields (see Curry & Cotton, 1983).

Bauxite mine refining residues (Australia)

An investigation on the survival, growth and influence on soil fertility of the earthworm *Eisenia fetida* in soil taken from various depths of bauxite residue sites, indicated that at all sample depths within the upper 50 cm, the presence of living earthworms led to an increase in C and all measured nutrients: N, total P, available P and K (Southwell & Majer, 1982).

Conclusions for site management

With regard to litter decomposition/nutrient cycling, the benefits derived from the soil fauna have been described and the influences of the food resource quality and physico-chemical features of the environment have been outlined.

At the European coal-waste sites, litter is processed by micro-arthropods, especially Collembola and Diptera larvae in the first stages of faunal development. At wooded sites, cryptostigmatid mites and Enchytraeidae may also be important. Here, moder-like humus conditions develop, with a layer of undecomposed litter lying on the surface of the acid soil. At grassland sites, the amount of litter production is less, so that a higher proportion of that available can be assimilated by the limited fauna. After several years, earthworms may become numerous, in which case litter removal rate is greatly increased, thus altering soil-forming processes (Vimmerstedt, 1983).

Simply spreading earthworms on reclaimed industrial sites will not magically restore the soil to normal (see Curry & Cotton, 1983; and also Chapters 9 and 13). The species must be carefully selected to suit site conditions, especially pH. Liming may be required to enable earthworms to survive. *Allolobophora caliginosa* was successfully introduced into brown coal dump material but only after mulching the soil surface with composted leaf litter to provide food and stabilize fluctuations in moisture content (Dunger, 1969*a*).

A study of the total soil biota is also required to assess or monitor the success of reclamation procedures. For instance after 2 years at surface-mine sites it was noted that differences in decomposer activity had not been indicated by above-ground productivity estimates (Elkins *et al.*, 1984). Thus, reclamation success should not be assessed by only the rate of vegetation re-establishment over short periods. Further, it was concluded that topsoiling may not always be a justifiable treatment, particularly in areas where insufficient material is available.

In conclusion, management of reclaimed land should clearly aim to encourage the soil fauna in order to help develop the soil and ultimately aid plant growth with the minimum of additional fertilizers. This can be conducted in the following ways:

1. The physico-chemical conditions in the soil should be stabilized as soon as possible: extremes of moisture, temperature and pH should be minimized by appropriate treatments.
2. Mechanical cultivation of the soil should be minimized since soil fauna brought to the surface may die from desiccation, and an important food source, which may also provide an overall physical barrier, can be lost from the upper soil layers.
3. The organic content of the soil should be supplemented wherever possible in the most appropriate way: the material provides a food source encouraging faunal establishment, and subsequently provides minerals for plant growth.

4. On sites suited to their survival and development, earthworms can be introduced to enhance litter burial and decomposition: the possibility of introducing termites is uncertain at present.

References

Anderson, J. M., Huish, S. A., Ineson, P., Leonard, M. A. & Splatt, P. R. (1985). Interactions of invertebrates, micro-organisms and tree roots in nitrogen and mineral element fluxes in deciduous woodland soils. In *Ecological Interactions in Soil: Plants, Microbes and Animals*, ed. A. H. Fitter, D. Atkinson, D. J. Read & M. B. Usher, pp. 377–92. Oxford: Blackwell Scientific Publications.

Anderson, J. M. & Ineson, P. (1984). Interactions between micro-organisms and soil invertebrates in nutrient flux pathways of forest ecosystems. In *Invertebrate–Microbial Interactions*, ed. J. M. Anderson, A. D. M. Raynor & D. W. H. Walton, pp. 59–88. Cambridge: Cambridge University Press.

Anderson, J. M., Rayner, A. D. M. & Walton, D. W. H. ed. (1984). *Invertebrate–Microbial Interactions*. Cambridge: Cambridge University Press.

Ashraf, M. (1969). Studies on the biology of Collembola. *Revue d'Ecologie et Biologie du Sol*, **6**, 337–47.

Burges, A. & Raw, F. ed. (1967). *Soil Biology*. London: Academic Press.

Clarholm, M. (1985). Possible roles for roots, bacteria, protozoa and fungi in supplying nitrogen to plants. In *Ecological Interactions in Soil: Plants, Microbes and Animals*, ed. A. H. Fitter, D. Atkinson, D. J. Read & M. B. Usher, pp. 355–65. Oxford: Blackwell Scientific Publications.

Coleman, D. C., Ingham, R. E., McClellan, J. F. & Trofymow, J. A. (1984). Soil nutrient transformations in the rhizosphere via animal–microbial interactions. In *Invertebrate–Microbial Interactions*, ed. J. M. Anderson, A. D. M. Raynor & D. W. H. Walton, pp. 35–58. Cambridge: Cambridge University Press.

Crossley, Jr. D. A. (1977). The role of terrestrial saprophagous arthropods in forest soils: current status of concepts. In *The Role of Arthropods in Forest Ecosystems*, ed. W. J. Mattson, pp. 49–56. New York: Springer-Verlag.

Curry, J. P. & Cotton, D. C. F. (1983). Earthworms and land reclamation. In *Earthworm Ecology*, ed. J. E. Satchell, pp. 215–28. London: Chapman and Hall.

Curry, J. P., Kelly, M. & Bolger, T. (1985). Role of invertebrates in the decomposition of *Salix* litter in reclaimed cutover peat. In *Ecological Interactions in Soil: Plants, Microbes and Animals*, ed. A. H. Fitter, D. Atkinson, D. J. Read & M. B. Usher, pp. 393–7. Oxford: Blackwell Scientific Publications.

Dickinson, C. H. & Pugh, G. J. F. ed. (1974). *Biology of Plant Litter Decomposition* (2 vols.). London: Academic Press.

Dunger, W. (1968*a*). Die Entwicklung der Bodenfauna auf rekultivierten Kippen und Halden des Braunkohlentage-bauer. *Abhandlungen und Berichte des Naturkundemuseums, Görlitz*, **43**, 1–256.

Dunger, W. (1968*b*). Produktionsbiologische Untersuchungen an der Collembolen-Fauna gestörter Böden. *Pedobiologia*, **8**, 16–22.

Dunger, W. (1969*a*). Fragen der naturlichen und experimentellen Besiedlung kulturfeindlicher Böden durch Lumbriciden. *Pedobiologia*, **9**, 146–51.

Dunger, W. (1969*b*). Über den Anteil der Arthropoden an der Umsetzung des Bestandesabfalles in Anfangs-Bodenbildungen. *Pedobiologia*, **9**, 366–71.

Edwards, C. A. (1974). Macroarthropods. In *Biology of Plant Litter Decomposition*, vol. 2, ed. C. H. Dickinson & G. J. F. Pugh, pp. 533–54. London: Academic Press.

Edwards, C. A. & Heath, G. W. (1963). The role of soil animals in breakdown of leaf material.

In *Soil Organisms*, ed. J. Doekson & J. van der Drift, pp. 76–85. Amsterdam: North Holland.

Elkins, N.Z., Parker, L. W., Aldon, E. & Whitford, W. G. (1984). Responses of soil biota to the organic amendments in stripmine spoils in northwest New Mexico. *Journal of Environmental Quality*, **13**, 215–19.

Elkins, N. Z. & Whitford, W. G. (1982). The role of microarthropods and nematodes in a semi-arid ecosystem. *Oecologia*, **55**, 303–10.

Elliot, E. T., Anderson, R. V., Coleman, D. C. & Cole, V. (1980). Habitable pore space and microbial trophic interactions. *Oikos*, **35**, 327–35.

Fitter, A. H., Atkinson, D., Reed, D. J. & Usher, M. B. ed. (1985). *Ecological Interactions in Soil: Plants, Microbes and Animals*. Oxford: Blackwell Scientific Publications.

Harding, D. J. L. & Stuttard, R. A. (1974). Microarthropods. In *Biology of Plant Litter Decomposition*, vol. 2, ed. C. H. Dickinson & G. J. H. Pugh, pp. 489–532. London: Academic Press.

Hutson, B. R. (1980*a*). Colonization of industrial reclamation sites by Acari, Collembola and other invertebrates. *Journal of Applied Ecology*, **17**, 255–75.

Hutson, B. R. (1980*b*). The influence on soil development of the invertebrate fauna colonizing industrial reclamation sites. *Journal of Applied Ecology*, **17**, 277–86.

Hutson, B. R. & Luff, M. L. (1978). Invertebrate colonization and succession on industrial reclamation sites. In *Scientific Proceedings, Royal Dublin Society, Series A*, vol. 6, no. 11, pp. 165–74. Dublin: Royal Dublin Society.

Lee, K. E. & Wood, T. G. (1971). *Termites and Soil*. London: Academic Press.

Lofty, J. R. (1974). Oligochaetes. In *Biology of Plant Litter Decomposition*, vol. 2, ed. C. H. Dickinson & G. J. F. Pugh, pp. 467–88. London: Academic Press.

Luff, M. L. & Hutson, B. R. (1977). Soil fauna populations. In *Landscape Reclamation Practice*, ed. B. Hackett, pp. 125–47. Guildford: IPC Business Press.

Parker, L. W., Elkins, N. Z., Aldon, E. F. & Whitford, W. G. (1984*b*). Development of soil biota and some aspects of nitrogen cycling on reclaimed coal mine spoils in the arid southwest. *Journal of Environmental Quality*, **19**, 215–19.

Parker, L. W., Santos, P. F., Phillips, J. & Whitford, W. G. (1984*a*). Carbon and nitrogen dynamics during the decomposition of litter and roots of a Chihuahuan desert annual, *Lepidium lasiocarpum*. *Ecological Monographs*, **54**, 339–60.

Santos, P. F. & Whitford, W. G. (1981). The effects of microarthropods on litter decomposition in a Chihuahuan desert ecosystem. *Ecology*, **62**, 654–63.

Satchell, J. E. (1955). Some aspects of earthworm ecology. In *Soil Zoology*, ed. D. K. McE. Kevan, pp. 180–201. London: Butterworths.

Seastedt, T. R. (1984). The role of microarthropods in decomposition and mineral processes. *Annual Review of Entomology*, **29**, 25–46.

Southwell, L. T. & Majer, J. D. (1982). The survival and growth of the earthworm *Eisenia foetida* (Lumbricidae: Oligochaeta) in alkaline residues associated with the bauxite refining process. *Pedobiologia*, **23**, 42–52.

Swift, M. J. (1977). The ecology of wood decomposition. *Science Progress (Oxford)*, **64**, 179–203.

Swift, M. J., Heal, O. W. & Anderson, J. M. (1979). *Decomposition in Terrestrial Ecosystems*. Los Angeles: University of California Press.

van der Drift, J. (1951). Analysis of the animal community in a beech forest floor. *Tijdschrift voor Entomologie*, **94**, 1–168.

van Rhee, J. A. (1963). Earthworm activities and the breakdown of organic matter in agricultural soils. In *Soil Organisms*, ed. J. Doeksen & J. van der Drift, pp. 54–9. Amsterdam: North Holland.

Vimmerstedt, J. P. (1983). Earthworm ecology in reclaimed opencast coal mining sites in Ohio. In *Earthworm Ecology*, ed. J. E. Satchell, pp. 229–40. London: Chapman and Hall.

Vimmerstedt, J. P. & Finney, J. H. (1973). Impact of earthworm introduction on litter burial and nutrient distribution in Ohio strip-mine spoil banks. *Soil Science Society of America Proceedings*, **37**, 388–91.

Visser, S. (1985). Role of the soil invertebrates in determining the composition of soil microbial communities. In *Ecological Interactions in Soil: Plants, Microbes and Animals*, ed. A. H. Fitter, D. Atkinson, D. J. Read & M. B. Usher, pp. 297–317. Oxford: Blackwell Scientific Publications.

Wallwork, J. A. (1970). *Ecology of Soil Animals*. London: McGraw-Hill.

Whitford, W. G. & Elkins, N. Z. (1988). The importance of soil ecology in strip mine reclamation. In *Diverse Scientific Perspectives in Reclamation in the San Juan Basin*. New Mexico: University of New Mexico Press (in press).

Wiegert, R. G., Coleman, D. C. & Odum, E. P. (1970). Energetics of the litter–soil subsystem. In *Methods of Study in Soil Ecology. Proceedings of the Paris Symposium*, ed. J. Phillipson, pp. 93–8. Paris: UNESCO.

Witkamp, M. (1969). Environmental effects on microbial turnover of some mineral elements. I. Abiotic factors. *Soil Biology and Biochemistry*, **1**, 167–76.

4

The influence of fauna on plant productivity

RICHARD P. URBANEK

Introduction

Little information is available concerning the effects of animals upon plants in rehabilitated lands. While numerous articles on the distribution and abundance of vertebrates on these lands have been produced, the invertebrate fauna, despite its importance, has largely been neglected. In the eastern United States, most rehabilitation attempts on mined lands involve establishment of a few species of non-native, agronomic grasses and legumes or plantings of either native or non-native trees. Many grass–legume plantings are eventually used as grazing land or for mulch or hay production. In the western United States, non-native grasses and legumes, as well as some native grasses, forbs and shrubs, are often used. Grazing land again is a dominant post-mining use.

The fauna may negatively affect plant productivity in three ways: (1) reduction in growth or herbage yield; (2) reduction in seed production; and (3) mortality. Positive effects include stimulation of plant growth by feeding or soil improvement, pollination, and dispersal of seeds or symbiotic plant micro-organisms (mycorrhizae). The primary focus of this Chapter will be to review the recent literature on the effects of animals on plant productivity. Because it is impossible to do a world review, I have concentrated on the USA. Specifically, the impact of invertebrates, mainly insects, on plant species commonly used to rehabilitate mined lands in the eastern United States (Vogel, 1981) will be examined. Many of these species are used extensively in other parts of the World as well. Most insect data presented were collected on unmined areas but are also applicable to mined lands, or other types of disturbance. Unless specifically designated, information collected was from studies on unmined areas. Plant species which are important in the establishment of vegetative cover are considered; faunal effects on more specialized uses of rehabilitated lands, such as production of cash grain crops (see Table 1.2), are beyond the scope of this Chapter. Many soil organisms affect soil fertility and hence plant production, and the role of fauna in soil development has been presented in Chapters 2 and 3. Predators and parasitoids, by reducing populations of herbivorous insects, indirectly affect plant productivity. Arthropods act as vectors for numerous viruses and other pathogens which affect plant productivity. These latter two important relationships are also beyond the scope of this Chapter. Readers who are interested in these topics are referred to the excellent review in Price (1984). The limited data on the effects of vertebrates on the rehabilitation of mined lands are presented. However, effects of cattle will receive little attention, as this topic is covered in Chapter 10.

All data presented in tabular form were from studies on unmined areas or the laboratory. Where possible the losses in production relative to a control at either experimentally determined or natural field densities are indicated. Yield reductions refer to above-ground dry biomass unless indicated otherwise. The purposes of these tables are to summarize the observed impacts of some common invertebrate herbivores on plant species commonly used to rehabilitate lands in the United States and, where possible, to indicate some densities or other pertinent conditions resulting in specific effects. The reader is referred to the individual references for more specific data on experimental conditions and location of the studies.

Impacts of invertebrates on important plant species

Forage legumes

Legumes commonly used in land rehabilitation in North America are susceptible to damage by many insect herbivores (Table 4.1). Alfalfa (*Medicago sativa*) is the best studied legume species in terms of loss of productivity to insects. Larvae of the alfalfa weevil (*Hypera postica*) constitute one of its more serious pests. Hintz, Wilson & Armbrust (1976) found a linear reduction in yield over most of the larval densities observed (Fig. 4.1). Berberet, Morrison & Senst (1981) found a linear reduction related to reduced growth and delayed maturity in addition to removal of plant foliage. Damage has also been shown to be more than just insect food consumption, which accounted for only 50 % of weight

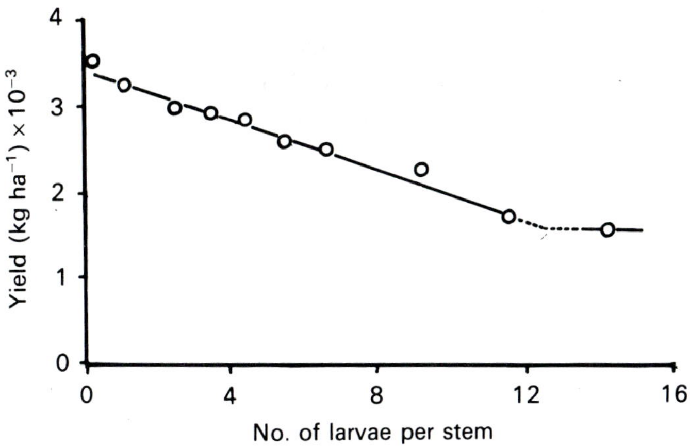

Fig. 4.1. Relationship between alfalfa yield (dry weight) and density of alfalfa weevil (*Hypera postica*) larvae (from Hintz, Wilson & Armbrust, 1976).

loss in alfalfa plants after 10 days of larval feeding (Koehler & Pimentel, 1973). The additional weight loss was thought to be due to loss of plant photosynthetic potential. First-growth alfalfa was able to compensate for 87 % of the weevil feeding, while second-growth, with less stored root energy available, could only compensate for 50 %. Adult alfalfa weevils may also reduce productivity of alfalfa. Bjork & Davis (1984) found that each adult consumed 4.5 times as much as each larva. Under field conditions new adults usually caused delays in regrowth rather than visible defoliation; therefore, injury caused by new adults was more overlooked than that of the larvae.

The most important pest of alfalfa in the eastern United States is considered to be the potato leafhopper (*Empoasca fabae*) (Cuperus *et al.*, 1983). These authors demonstrated a significant negative curvilinear relationship between leafhopper numbers and alfalfa dry-matter yield and height at harvest, with more loss per leafhopper with low than with high leafhopper numbers (Fig. 4.2). This species overwinters in the Gulf states and disperses northward in late spring; therefore, spring growth of alfalfa is not affected. Damage to plants occurs as a result of insect piercing and feeding from vascular tissue and to toxic secretions injected during the feeding process (Medler, 1941; Granovsky, 1930). Symptoms of this damage include yellowing, dwarfing and wilting. The greater the infestation rate, the less the photosynthetic and transpiration rates (Womack, 1984). Leafhopper attack is most severe in the seeding year (Faris, Baenziger & Terhune, 1981). Kouskolekas & Decker (1968) found that yield reduction declined with delayed infestation (Fig. 4.3).

Table 4.1 *Yield reductions or other losses attributable to insect feeding on legumes*

Plant species	Herbivore species	Insect density or treatment	Loss	Type of study[a]	Reference
AL	AW	peak of 2.7 larvae per stem	17% yield red.	I	Liu & Fick (1975)
AL	AW	peak of 1.1 larvae per stem	5% yield red.	I	Liu & Fick (1975)
AL	AW	1 larva	7.34 mg per larval stage	L	Koehler & Pimentel (1973)
AL	AW	peak of 7.25 larvae per stem	52% yield red.	I	Wilson, Stewart & Vail (1979)
AL	MSB	297 nymphs m^{-2}	17% yield red.	I	Wilson *et al.* (1979)
AL	FLH	0.2–2.5 per sweep	8% yield red.	I	Wilson *et al.* (1979)
AL	All insects		78% yield red.	I	Wilson *et al.* (1979)
AL	AW	1 adult	35.3 mg per adult stage	C	Bjork & Davis (1984)
AL	AW	5 adults per plant	81% red. in leaf surface	C	Bjork & Davis (1984)
AL	PLH	80 per sweep	21–28% yield red.[b]	I	Smith & Medler (1959)
AL	PLH	80 per sweep	36–48% yield red.[c]	I	Smith & Medler (1959)
AL	PLH	6 per seedling	31–46% yield red. in 6 days	L	Flinn & Hower (1984)
AL	MSB	3100 nymphs m^{-2}	20% yield red.	C	Parman & Wilson (1982)
AL	PA	1265 apterous forms per pot	44% yield red.	L	Harper & Kaldy (1982)
AL	PA	variable	23–48% yield red.	C	Kindler, Kehr & Ogden (1971)
AL	SAA	variable	19–38% yield red.	C	Kindler *et al.* (1971)
AL	PA	1 aphid	2.5 mg (fresh wt.) day^{-1}	L	Mittler & Sylvester (1961)
AL	BAA	23 aphids per stem	40% yield red.	L	Summers & Coviello (1984)
AL	ABL	1 mine per leaflet	13% yield red. in leaf wt.	—	Andaloro, Peters & Alicandro (1983)
AL	VCW	1 larva	442 mg per larval stage	L	Buntin & Pedigo (1985*a*)
RC	MSB	each 11 nymphs m^{-2}	12.5 kg ha^{-1}	I	Everly (1959)

RC, AL	MSB	—	19–36 % yield red.	I	Weaver (1950)
YSC	SCW	1 adult per 5 seedlings	40–47 % mortality[d]	C	Craig (1978)
YSC	SCW	1 adult per 3 seedlings	5–18 % mortality[e]	C	Craig (1978)
YSC	SCW	late summer defoliation	50–71 % yield red. next year	C	Craig (1978)
YSC	SCW	9 adults per 2nd-yr plant	17–18 % yield red.[e]	C	Craig (1978)
YSC	SCW	9 adults per 2nd-yr plant	34 % yield red.[d]	C	Craig (1978)
SC	SCW, CRC	larvae in roots	25 % N-fixing nodules damaged	—	Manglitz, Anderson & Gorz (1963)
LC	CRC	—	22 % yield red. in new root growth	L	Powell *et al.* (1983)
BFT	PB1	—	80 % flower bud loss	C	Guppy (1963)

Note: [a] C = cage study in field. I = insecticide trials in field. L = laboratory study.
[b] high-fertility plots; [c] low-fertility plots.
[d] Slow growth conditions; [e] good growth conditions.

Key to species codes:

AL alfalfa (*Medicago sativa*)
BFT birdsfoot trefoil (*Lotus corniculatus*)
LC Ladino clover (*Trifolium repens*)
RC red clover (*Trifolium pratense*)
SC sweet clover (*Melilotus* spp.)
YSC yellow sweet clover (*Melilotus officinalis*)

ABL alfalfa blotch leafminer (*Agromyza frontella*) DIPTERA
AW alfalfa weevil (*Hypera postica*) COLEOPTERA
BAA blue alfalfa aphid (*Acyrthosiphon kondoi*) HOMOPTERA
CRC clover root curculio (*Sitona hispidula*) COLEOPTERA
MSB meadow spittlebug (*Philaenus spumarius*) HOMOPTERA
PA pea aphid (*Acyrthosiphon pisum*) HOMOPTERA
PB1 plant bug (*Plagiognathus chrysanthemi*) HEMIPTERA
PLH potato leafhopper (*Empoasca fabae*) HOMOPTERA
SAA spotted alfalfa aphid (*Therioaphis maculata*) HOMOPTERA
SCW sweet clover weevil (*Sitona cylindricollis*) COLEOPTERA
VCW variegated cutworm (*Peridoma saucia*) LEPIDOPTERA

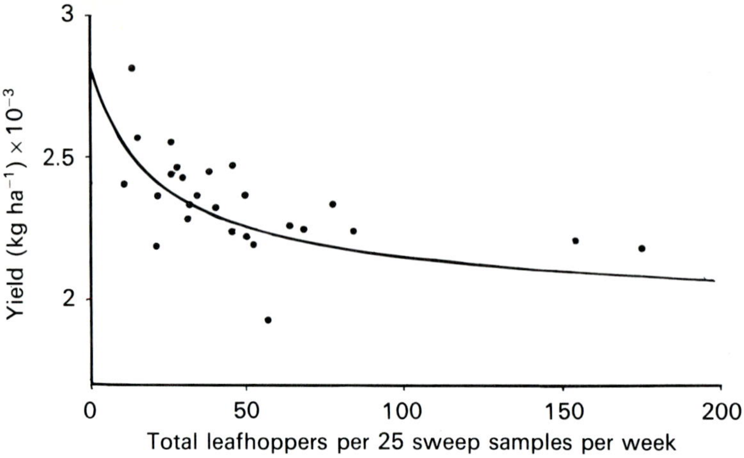

Fig. 4.2. Relationship between alfalfa yield (dry weight) and abundance of potato leafhoppers (*Empoasca fabae*) (from Cuperus *et al.*, 1983).

The reduction in alfalfa yield was directly proportional to the leafhopper density, and there was an inverse relationship with the height of the alfalfa at the time of infestation.

Another common pest of alfalfa in the eastern United States is the meadow spittlebug (*Philaenus spumarius*). Nymphs, which feed on xylem sap, are most prevalent in fields the year after seeding with a small grain companion crop; nymphal feeding shortened plant internodes, causing reduction in height and bunching of top leaves, and significantly delayed maturity (Parman & Wilson, 1982).

The variegated cutworm (*Peridoma saucia*) may also decrease productivity of alfalfa. Damage is caused primarily by consumption of leaves, petioles and axillary stems. Delays in plant development were proportional to larval density (Buntin & Pedigo, 1985*b*).

Other invertebrates responsible for damage or loss of productivity in alfalfa include several aphids (Aphididae) and the alfalfa blotch leafminer fly (*Agromyza frontella*) (Table 4.1). Other pests are the clover root curculio beetle (*Sitona hispidula*) (Cranshaw, 1985), which may also increase susceptibility of roots to fungal infection (Dickason, Leach & Gross, 1968); the slug (*Deroceras reticulatum*) and crickets (*Nemobius* spp.), which have been shown to be major defoliators of seedlings in no-tillage plots (Grant *et al.*, 1982); plant bugs (Miridae) (Walstrom, 1983); grasshoppers (Acrididae and Tettigoniidae) (Wiegert, 1965); and mites (Ridland & Halloran, 1981). In Pennsylvania, the slug (*D. reticulatum*) destroyed most of the alfalfa seedlings in oat stubble during the first

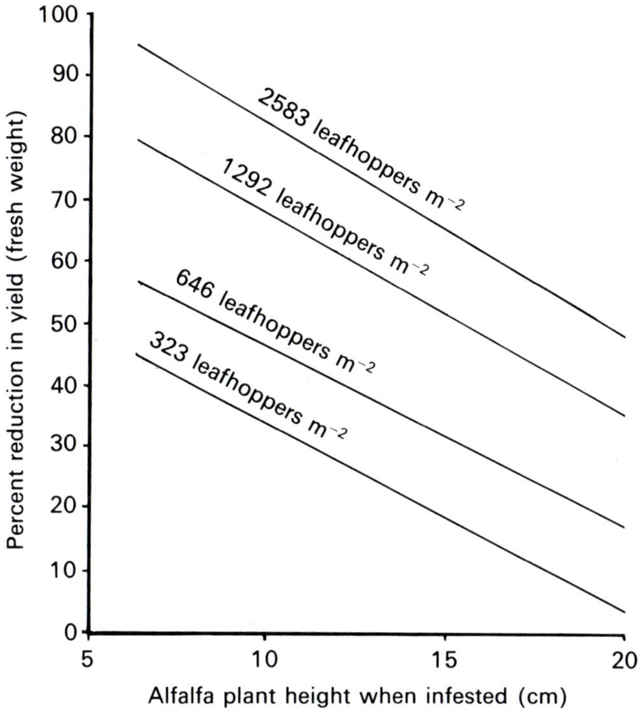

Fig. 4.3. Relationship between percent reduction in fresh-weight yield and alfalfa height at time of infestation by potato leafhoppers (*Empoasca fabae*) (from Kouskolekas & Decker, 1968).

weeks after planting (Byers, Mangan & Templeton, 1983). The most important seed losses are due to the alfalfa seed chalcid wasp (*Bruchophagus roddi*) (Brewer & Horber, 1984; Ahring, Moffett & Morrison, 1984) and plant bugs (Erdelyi *et al.*, 1981).

Red clover (*Trifolium pratense*) is attacked by many of the same insects which feed on alfalfa. In North Carolina, yield of Ladino clover (*Trifolium repens*) may decline after the first year due to root damage by clover root curculios, which predispose the roots to *Fusarium* rot (James *et al.*, 1980). In red clover, the clover root borer (*Hylastinus obscurus*) may also spread root rot pathogens (Kamm & Buttery, 1984).

Sweet clover, especially yellow sweet clover (*Melilotus officinalis*), is commonly seeded in grass–legume mixtures on rehabilitated lands in North America. A major pest is the sweet clover weevil (*Sitona cylindricollis*). Most damage is done by adults, which feed on leaves and young stems; seedlings are particularly susceptible. Effects of the root-feeding larvae of the sweet clover weevil and related clover root curculio beetle on productivity are also probably significant but little known.

Birdsfoot trefoil (*Lotus corniculatus*), another species commonly used in rehabilitation, especially in the northern United States, may incur foliage damage from meadow spittlebugs, pea aphids, potato leafhoppers, plant bugs, grasshoppers (*Melanoplus* spp.) and leaf-roller moths (Tortricidae) (Neunzig & Gyrisco, 1955). Plant bugs (*Adelphocoris lineolatus*, *Adelphocoris rapidus* and *Lygus lineolaris*) were most damaging, causing blasting (necrosis) of the floret buds. These bugs, as well as the stink bug (*Euschistus variolaris*) and crickets (*Acheta assimilus* and *Nemobius fasciatus fasciatus*), also caused seed damage by feeding on seed pods. The clover seed chalcid wasp oviposited in developing seeds.

Crown vetch (*Coronilla varia*) has been an important species used in rehabilitation in eastern North America, especially on steep slopes where erosion control is critical. Pest problems are few in comparison to those of alfalfa and other forage legumes. Wheeler (1974) noted defoliation by the differential grasshopper (*Melanoplus differentialis*), red-legged grasshopper (*Melanoplus femurrubrum*), and gelechiid moth (*Stomopteryx palpilineella*) in several highway plantings in Pennsylvania. Other abundant, or potentially damaging, species included the alfalfa plant bug (*A. lineolatus*), potato leafhopper, clover stem borer (*Languria mozardi*), nitidulid beetle (*Meligethes nigrescens*), and spider mites (*Tetranychus* spp.). While the grasshoppers and gelechiid caused defoliation in highway plantings, only small numbers were found in forage plantings. These findings agree with those from eastern Ohio, where no defoliation of extensive crown vetch plantings on surface-mined lands was noted (Urbanek, 1982); despite a substantial plant biomass on these sites, insect densities and biomass in the herbaceous layer were low relative to other grass–legume plantings.

Little information is available on herbivore-induced yield reductions in legumes on rehabilitated lands, but outbreaks and potential for such damage have been observed. On rehabilitated mined lands in eastern Ohio, larvae of the alfalfa weevil caused partial defoliation of yellow sweet clover (Urbanek, 1982). On one site, seeded the previous year to yellow sweet clover, red clover, tall fescue (*Festuca arundinacea*), and orchard grass (*Dactylis glomerata*), alfalfa weevil larvae comprised 47 % of total herbaceous layer arthropod individuals and had a density of 2483 larvae m^{-2} during the period 29 May to 10 June. This density was high, especially in view of the fact that the sweet clover did not even occur in a pure stand. Mortality of sweet clover as a result of larval infestation was not observed, probably because the outbreak did not occur until the sweet clover was near maturity.

In summary, specific data on effects of invertebrate herbivores on productivity of legumes used in rehabilitation of land in North America are lacking, but losses due to insect species on the same plant species on agricultural lands (Table 4.1) indicate that establishment of vegetation during rehabilitation could be severely hindered at certain pest species densities. Relationships between damage and herbivore density were generally linear, but were curvilinear in one study. These relationships are variable, and data are insufficient to make any broad generalizations. Additive yield reductions by different herbivore species on the same crop have been reported (Wilson, Stewart & Vail, 1979; Smith & Ellis, 1983). Loss of production is not limited to consumption of foliage. Several authors noted reduced growth or regrowth and delays in plant development and maturity. Younger plants appeared most susceptible to damage. Legumes under favourable growth conditions or on sites of higher fertility were less susceptible to damage than those on sites of low fertility or under slow growth conditions (Table 4.1). More rigorous environmental conditions on areas under rehabilitation might therefore have greater potential for pest problems.

On rehabilitated lands, legumes are particularly important for soil improvement because of their ability to fix nitrogen. Thus insect damage to these plant species could adversely affect the entire community. Aphid infestation can reduce nitrogen fixing by reducing flow of photosynthate to roots and nodules. On pea plants (*Pisum sativa*), aphid feeding resulted in fewer root nodules and a decreased acetylene reducing activity of *Rhizobium* bacteria by as much as 86% (Sirur & Barlow, 1984), and similar effects could be possible on aphid-infested legumes used in rehabilitation.

Grasses

With the exception of some groups such as grasshoppers, the effects of invertebrates on productivity of grasses are often not obvious, both because of the small size of the damaging organisms and the lack of damage to readily observable plant structures; many pests are fluid-feeders, stem-borers, or root feeders. Reductions in yield can, however, be substantial (Table 4.2).

Tall fescue (selection Kentucky-31) is the grass species most widely used for rehabilitating mined lands in the eastern United States (Vogel, 1981). Quisenberry & Yonke (1981*a*) found the leafhopper (*Forcipata loca*) caused permanent damage to fescue leaves, and the damage was proportionately related to the extent and duration of feeding. Injury was,

Table 4.2. *Yield reductions or other losses attributable to invertebrate feeding on grasses*

Plant species	Herbivore species	Insect density or treatment	Loss	Type of study[a]	Reference
SBG	LH	—	47% yield red.	I	Byers & Jung (1979)
SBG, RT, TF	LH	—	64% yield red.	I	Byers & Jung (1979)
TF	LH	—	15% yield red.	I	Kelly & Klostermeyer (1984)
TF, BS	GH1, CR	peak biomass of 927 mg m^{-2}	10% of net primary production	C	Van Hook (1971)
TF	LH1	9 per 9.5-cm diam. pot	8–22% yield red.	L	Quisenberry & Yonke (1981*b*)
TF	LH1	27 per 9.5-cm diam. pot	13–36% yield red.	L	Quisenberry & Yonke (1981*b*)
BG	LH, PH	—	3% yield red., frequent harvest	I	Hawkins *et al.* (1979)
BG	LH, PH	—	33% yield red., cut only once	I	Hawkins *et al.* (1979)
BG	FAW	97–162 sixth instar larvae m^{-2}	28% max. yield red.	C	Alvarado, Fuxa & Wilson (1983)
BG	FAW	100 3rd–6th instar larvae m^{-2}	39% max. yield red.	C	Alvarado *et al.* (1983)
BG	FAW	162 3rd–6th instar larvae m^{-2}	50% max. yield red.	C	Alvarado *et al.* (1983)
RCG	FF	—	17–24% yield red.	I	Wolf (1967)
PRG	FF	—	8–24% yield red.	I	Henderson & Clements (1977)
IRG	ASW	0.35 larvae per tiller	65% tiller mortality	—	Goldson (1982)
IRG	BC	5 adults m^{-2}	1.8 kg ha^{-1} day^{-1}	C	Blank & Olson (1981)
IRG	BC	60 adults m^{-2}	25.1 kg ha^{-1} day^{-1}	C	Blank & Olson (1981)
RG	CW	20 4th–5th instar larvae per pot	complete defoliation in 2 days	L	Kamm (1985)
Range G	GH	18 late instar nymphs or adults m^{-2}	70% yield red.[b]	—	Nerney (1958)
Range G	GH	18 late instar nymphs or adults m^{-2}	20% yield red.[c]	—	Nerney (1958)
Range G	GH	1 adult	9–91 mg day^{-1}	—	Hewitt (1977)
Range G	GH	1 adult	9–150 mg day^{-1}	—	Hewitt (1977)
Range G	GH	1 4th–5th instar or adult	43 mg day^{-1}	—	Hewitt & Onsager (1982)
Range G	WG	46.3 late instar larvae m^{-2} (roots)	88% grass cover red.	I	Ueckert (1979)

Range G	WG	46.3 late instar larvae m^{-2} (roots)	43% red. in below-ground biomass	I	Ueckert (1979)
IWG	BGB	1333 m^{-2}	13% yield red.	—	Todd & Kamm (1974)
Range G	NEM	2–6 million m^{-2}	22–37% yield red.	I	Smolik & Lewis (1982)
Range G	NEM	10 × density in untreated plots	20% of net primary prod. (roots)	I	Stanton, Allen & Campion (1981)
BGr	NE1	500 per pot, initial inoculum	13% of net primary prod. (total plant)	L	Stanton (1983)

Note: [a] C = cage study in field. I = insecticide trials in field. L = laboratory study.
[b] heavily grazed by cattle, sparse vegetation cover;
[c] moderately grazed by cattle, good vegetation cover.

Key to species codes:

BG	Bermuda grass (*Cynodon dactylon*)
BGr	blue grama (*Bouteloua gracilis*)
BS	broomsedge (*Andropogon virginicus*)
IRG	Italian ryegrass (*Lolium multiflorum*)
IWG	intermediate wheatgrass (*Agropyron intermedium*)
PRG	perennial ryegrass (*Lolium perenne*)
RCG	reed canary grass (*Phalaris arundinacea*)
Range G	range grasses, including wheatgrasses (*Agropyron* spp.), buffalo grass (*Buchloe dactyloides*), blue grama, red threeawn (*Aristida longiseta*), and hairy tridens (*Erioneuron pilosum*).
RG	ryegrass (*Lolium* spp.)
RT	redtop (*Agrostis alba*)
SBG	smooth brome grass (*Bromus inermis*)
TF	tall fescue (*Festuca arundinacea*)
ASW	Argentine stem weevil (*Listronotus bonariensis*) COLEOPTERA
BC	black cricket (*Teleogryllus commodus*) ORTHOPTERA
BGB	black grass bug (*Labops hesperius*) HEMIPTERA
CR	cricket (*Pteronemobius fasciatus*) ORTHOPTERA
CW	cutworms (*Dargida procincta* and *Pseudaletia unipuncta*) LEPIDOPTERA
FAW	fall armyworm (*Spodoptera frugiperda*) LEPIDOPTERA
FF	frit fly (*Oscinella frit*) and other Chloropidae DIPTERA
GH	grasshoppers (Acrididae) ORTHOPTERA
GH1	grasshoppers (*Melanoplus sanguinipes* and *Conocephalus fasciatus*) ORTHOPTERA
LH	leafhoppers (Cicadellidae) HOMOPTERA
LH1	leafhopper (*Forcipata loca*) HOMOPTERA
NE1	nematode (*Heliocotylenchus exallus*)
NEM	nematodes
PH	planthoppers (Delphacidae) HOMOPTERA
WG	white grub (*Phyllophaga crinita*) COLEOPTERA

however, apparently confined to the cells pierced by the stylets, as feeding did not result in necrosis, wilting, or secondary pigmentation as noted for the potato leafhopper on alfalfa.

Bermuda grass (*Cynodon dactylon*), a warm season perennial best adapted to climate in the southern United States, is also susceptible to attack by leafhoppers and planthoppers (Delphacidae). Hawkins *et al.* (1979) found yields were reduced most (33 %) in plots cut only once annually but by only 3 % in grass harvested most often (to simulate close grazing).

Stem borers are common, often undetected herbivores on many grasses. As a result of work on reed canary grass (*Phalaris arundinacea*), a cool season perennial often used on moist sites, Wolf (1967) concluded that attack by larvae of the frit fly (*Oscinella frit*), a stem-boring chloropid, may be widespread but not recognized. The frit fly has caused some seedling mortality of oats (*Avena sativa*) in Sweden, where average infestation of main shoots was 18 % (Larsson, 1984). Annual oats, like rye (*Secale cereale*) and wheat (*Triticum aestivum*), is often used as a quick-cover companion crop on newly rehabilitated lands in the eastern United States.

Another cool season species which is susceptible to infestation by stem-boring flies is perennial ryegrass (*Lolium perenne*) (Clements & Henderson, 1982). This plant species has been extensively studied in the United Kingdom, where it is the most important sown grass. An annual species, Italian ryegrass (*Lolium multiflorum*) is even more affected by stem borers. Henderson & Clements (1977) observed significant yield reductions despite no outward signs of pest attack (Table 4.2).

In the northwest United States, cutworms (Noctuidae) are important defoliators of ryegrass (Kamm, 1985). The amount of leaf foliage wasted was substantially greater than the amount actually consumed by larvae of two species studied.

The common reed (*Phragmites australis*) is a common species on shallow-flooded slurry deposits on some mines in the midwestern United States. In the Netherlands this species is attacked by the noctuid moth (*Archanara geminipuncta*); the stem-boring larvae may kill up to 80 % of the apical parts of reed shoots (Mook & Van der Toorn, 1985). The reaction of the plant, however, is formation of thinner shoots, less suitable for larval survival, in the next year. This plant response was considered partially responsible for periodic local extinction of the herbivore.

In the western United States, grasshoppers (Acrididae) are considered to be the most damaging pests on rangeland (Hewitt, 1977). Affected prairie grass species are numerous, but wheatgrasses (*Agropyron* spp.), buffalo grass (*Buchloe dactyloides*), and blue grama (*Bouteloua gracilis*) are some of the most common. The most damaging grasshopper species tend to prefer sparsely vegetated habitats and are thus favoured where range is overgrazed by cattle (Hardman & Smoliak, 1982). Grasshoppers (*Melanoplus* spp.) were the dominant herbivore species in terms of biomass during late summer on some grass–legume plantings on mined sites in the eastern United States (Urbanek, 1982), yet their effects on such areas have never been studied.

The extensive literature available on loss of plant productivity to insects would lead one to believe that this group accounts for most of the invertebrate consumption of living plant material. However, the effects of nematodes on plant productivity may be substantial, especially in grasslands where a high proportion of the biomass is below ground. Smolik & Lewis (1982) estimated plant-feeding arthropod biomass to be 0.095 and 0.085 $g\ m^{-2}$ dry weight on range in excellent and fair condition, respectively, compared to 0.54 and 0.38 $g\ m^{-2}$ of plant-feeding nematodes on the same sites. Nematodes, while not well studied, could be dominant herbivores on many sites.

Stanton & Krementz (1982) found that native shrub-steppe vegetation in Wyoming consistently supported higher densities of all nematode trophic groups than did areas disturbed by surface mining. On disturbed sites, areas with fresh rather than stockpiled topsoil had higher densities. These authors stressed the need for more work on root–micro-organism interactions and their function in rehabilitation efforts.

The effects of fertilization on grassland pest populations may be important. Moore & Clements (1984) found a higher percentage of perennial ryegrass tillers infected with stem-boring chloropid and opomyzid fly larvae in high than in low nitrogen plots, but conflicting results with other studies indicate that the effects of fertilizer on pest populations during land rehabilitation need study. The effects of harvest regime on grasses and other plant species also need further study. Because many rehabilitated areas in the United States are not managed for intensive agricultural production, results indicating greatest damage to grasses that were harvested least frequently (Hawkins *et al.*, 1979) might indicate a greater potential for damage on many rehabilitated areas.

Although a significant amount of production may be lost to herbivores, most perennial grasses tolerate herbivory well, especially on above-ground biomass. Vigorous vegetative reproduction greatly enhances growth and survival. On rehabilitated surface mines in midwestern US, insect problems on major species such as tall fescue have not been noted. On the contrary, on sites left undisturbed after rehabilitation, e.g. those not grazed, this species often forms thick cover, precluding establishment of other planted or successional species and retaining nutrients in litter or senescent biomass, thus eventually limiting its own productivity as well.

Composites

The common sunflower (*Helianthus annuus*) is a native North American annual which is sometimes used to provide autumn and winter food for wildlife on rehabilitated lands. Larvae of the sunflower stem weevil (*Cylindrocopturus adspersus*) tunnel in the cortical tissue; this activity weakens plants and may result in stalk breakage. Stalk strength is negatively correlated with number of infesting larvae (Oseto, Burr & Carlson, 1982). Injury of tissue by feeding larvae also results in the enhanced growth of several pathogenic fungi which cause stem rot (Yang, Rogers & Luciani, 1983).

While published accounts of insect damage to economically important plants are numerous, reports on native volunteer herbaceous species are not. Goldenrod (*Solidago* spp.) is one such common volunteer on rehabilitated lands in eastern North America. In Pennsylvania, Hartnett & Abrahamson (1979) found up to 38% of shoots infested with galls caused by the tephritid fly (*Eurosta solidaginis*), the gelechiid moth (*Gnorimoshema gallaesolidaginus*), or the cecidomyiid fly (*Rhopalomyia solidaginis*) and noted a 43–51% reduction in number of seeds in infested plants.

Tall goldenrod (*Solidago altissima*) is also susceptible to defoliation by the chrysomelid beetles (*Trirhabda virgata* and *Trirhabda borealis*). Densities of both adults and larvae may exceed 35 individuals per stem and cause complete defoliation. This plant is also the primary host of *Publilia concava*, a sap-feeding treehopper tended for honeydew production by ants, especially *Formica* spp. Messina (1981) examined the effect of the beetle–ant–treehopper relationship on the plant and found that stems bearing treehoppers and their attendant ants escape defoliation by the beetles. Although the treehoppers reduced growth in plant height by 25%, ants repelled adult beetles and deterred feeding by beetle larvae,which caused an 85% reduction in seed yield on plants without ants.

Conifers

Impacts of insects on productivity of some conifers commonly used in land rehabilitation in the eastern US appear in Table 4.3. Red pine (*Pinus resinosa*), native to the northern United States and Canada, has performed well on a variety of rehabilitated mined lands, especially in the northern Appalachian region. This species is susceptible to the European pine shoot moth (*Rhyacionia buoliana*) (Vogel, 1981), which injures buds and shoots and disrupts normal tree growth. In Wisconsin, Hainze & Benjamin (1984) determined the impact of the red pine shoot moth (*Dioryctria resinosella*), the larvae of which tend to attack the distal one third of shoots. The shoot tip is generally destroyed, and a variable length of the shoot is hollowed and killed (Hainze & Benjamin, 1983).

The major insect pest of eastern white pine (*Pinus strobus*) in the northeastern United States is the white pine weevil (*Pissodes strobi*). Weevil attack usually kills the terminal shoot and reduces tree growth. Wilkinson (1983) found damage was most prevalent in plantations, and repeated attacks reduced the height growth of the most heavily infested trees.

Arthropods act as vectors of several organisms pathogenic to pine. The pine sawyer (*Monochamus carolinensis*) carried a mean of 19152 nematodes per individual. These nematodes are the causative agent of pine wilt disease of Scots pine (*Pinus sylvestris*) in Missouri (Linit, Kondo & Smith, 1983). The black turpentine beetle (*Dendroctonus terebrans*) and associated blue stain fungus (*Leptographium terebrantis*) have been associated with mortality of Scots pine in Massachusetts (Highley & Tattar, 1985), and growth reductions in seedlings of this pine in Finland have tentatively been associated with a virus transmitted by the aphid (*Cinara pini*) (Holopainen & Soikkeli, 1984).

Shortleaf pine (*Pinus echinata*) and loblolly pine (*Pinus taeda*) are two species from southern United States which are commonly used in rehabilitation. Young trees, particularly of shortleaf pine, are susceptible to damage by the Nantucket pine tip moth (*Rhyacionia frustrana*) (Williston, 1958). The larvae tunnel in stems, which die back from the point of larval entry to the tip. Damage has been noted in plantations on surface-mined land in Illinois (Ashby, 1964).

Another important pest of loblolly and shortleaf pines is the southern pine beetle (*Dendroctonus frontalis*), an outbreak species that may cause tree mortality. Stressed trees such as those struck by lightning (Hodges & Pickard, 1971), or affected by drought (St. George, 1930), are particularly susceptible.

Table 4.3. *Yield reductions or other losses attributable to insect feeding on conifers*

Plant species	Herbivore species	Insect density or treatment	Loss	Reference
RP	RPSM	9-yr epidemic	38–65% height growth	Hainze & Benjamin (1984)
RP	RPSM	9-yr epidemic	16–42% radial growth	Hainze & Benjamin (1984)
RP	RPS	7-yr cycle	100% mortality	McClure (1983)
RP	PA	100+ beneath 100 bark flakes	75% crown damage	McClure (1982)
RP	WG	seedlings	20% mortality, 27% injured roots	Fowler & Wilson (1974)
JP	JPBW	—	20–99% red. in annual radial growth	Kulman, Hodson & Duncan (1963)
JP	PB	—	51–87% of conelets	Rauf, Cecich & Benjamin (1984)
LP	NPTM	7–15 yr after planting	3–17% red. in height	Young, Warren & Stephen (1979)
LP	NPTM	7–15 yr after planting	9–16% red. in diam.	Young *et al.* (1979)
SLP	NPTM	7–15 yr after planting	17–24% red. in height	Young *et al.* (1979)
SLP	NPTM	7–15 yr after planting	17–29% red. in diam.	Young *et al.* (1979)
LP	NPTM	young plants	18–26% dieback yr^{-1} for 3 yr	Lashomb, Steinhauer & Douglass (1978)
LP, SLP	SPB	100+ attacks m^{-2} of bark	mortality (threshold)	Hodges *et al.* (1979)
LP, SLP	CW, SB	—	up to 100% of seed	McLeod & Yearian (1982)
LP	CW, SB	—	86% of seed	Yates & Ebel (1978)
SLP	SB	2 nymphs per conelet for 4 days	100% conelet abortion	DeBarr & Kormanik (1975)
SLP	SB	natural, caged cones	92% conelet abortion	DeBarr & Ebel (1974)

Key to species codes:

JP	jack pine (*Pinus banksiana*)	PB	plant bug (*Platylygus luridus*) HEMIPTERA
LP	loblolly pine (*Pinus taeda*)	RPS	red pine scale (*Matsucoccus resinosae*) HOMOPTERA
RP	red pine (*Pinus resinosa*)	RPSM	red pine shoot moth (*Dioryctria resinosella*) LEPIDOPTERA
SLP	shortleaf pine (*Pinus echinata*)	SB	seedbugs (*Leptoglossus corculus* and *Tetyra bipunctata*) HEMIPTERA
CW	coneworms (*Dioryctria* spp.) LEPIDOPTERA	SPB	southern pine beetle (*Dendroctonus frontalis*) COLEOPTERA
JPBW	jack pine budworm (*Choristoneura pinus*) LEPIDOPTERA	WG	white grubs (*Phyllophaga* spp.) COLEOPTERA
NPTM	Nantucket pine tip moth (*Rhyacionia frustrana*) LEPIDOPTERA		

In addition to pines, other conifers used in rehabilitation in the eastern United States include spruces (*Picea* spp.) and larch (*Larix* spp.). They have numerous insect pests, and sawflies, especially, may cause defoliation (see Kulman, 1971). White spruce (*Picea glauca*) is attacked by the eastern spruce budworm (*Choristoneura fumiferana*) in the northern United States and Canada, where extensive outbreaks may result in severe mortality. An extensive review of the effects of defoliation on growth of trees has been presented by Kulman (1971).

Young conifers, particularly within a few years after planting, seem most susceptible to mortality, and damage may be most prevalent in plantations. These two factors cause particular concern for establishment of plantations on rehabilitated areas. Additionally, because stressed trees may be more prone to damage by herbivores, the environmental conditions on newly rehabilitated lands may enhance opportunity for growth loss or mortality. Among conifers, significant or almost complete loss of seed crops is common (Table 4.3; Lyons, 1957; Rauf & Benjamin, 1983; DeBarr, Barber & Maxwell, 1982; Ebel *et al.*,1975). In most areas of the eastern United States, reproduction of the planted conifers is not a goal of rehabilitation, but in others this loss of production could present a potential long-term problem.

Hardwoods

Because of its wide adaptability to soil conditions, rapid growth, and nitrogen-fixing capacity, black locust (*Robinia pseudoacacia*) has been the most frequently and widely used tree for surface-mine plantings in the eastern United States (Vogel, 1981). It is also susceptible to damage by the locust borer (*Megacyllene robiniae*), a major pest in rehabilitated lands and one of the few forest insects whose effects on such lands have been studied. Larvae of this cerambycid beetle tunnel in heartwood and sapwood. Damage to locust trees on mined lands has frequently been noted, but rarely quantified. In southern Illinois, Urbanek & Klimstra (1986) noted that living and dead black locust accounted for 66 and 19% of total tree density, respectively on a 25-year-old site and 30 and 28% of total tree density, respectively on a site examined 34 years after mining. High mortality was attributed primarily to locust borer. In Maryland, Harman, Van Tyne & Thompson (1985) found significant differences in infestation rates on mined and unmined land. Unmined sites had no trees with 11 or more current attacks, whereas 14% of the trees on mined sites had 11 or more attacks, and about one third of these had 21 or more attacks. The mean attack level on trees sampled by trapping of emerging adults was 8.9 on the surface-mines vs 3.8 on the unmined sites. Survival of borer larvae to adult stage was also higher on mined sites (58.6 vs 36.2%). Average height and diameter of similar-age trees were greater off mines, indicating greater stresses on mine sites. Soil pH and potash concentration were significantly lower on mined sites. Greater environmental stresses on trees on mined land may explain the increased susceptibility to locust borer attack.

Black locust is also subject to defoliation by the locust leaf mining beetle (*Odontota dorsalis*), although infestation does not usually result in tree mortality (Dominick, 1938). In Maryland, black locust is also the host of *Vanduzea arguata*, a gregarious sap-feeding treehopper, which is tended by the ant (*Formica subsericea*) (Fritz, 1983). The ants repel adults of the locust leaf miner but have little effect on the damaging larvae, which are usually hidden in leaf mines. The ants did, however, also exclude *Nabicula subcoleoptrata*, an important nabid bug predator of the leaf miner when they attended branches containing the leafhoppers. Thus ants indirectly protect beetle larvae from the predatory species. In one of two study years, defoliation by the leaf miner was significantly greater on branches containing ants. It was concluded that probably no net benefit or harm occurred to black locust as a result of harbouring the treehopper-tending ants, although a possible mechanism for harm had been identified.

On a surface-mine rehabilitation site in Pennsylvania, Loerch & Cameron (1984) studied within-tree distribution and seasonality of larvae of the bronze birch borer (*Agrilus anxius*) on young European white birch (*Betula pendula*). This buprestid species may cause dieback in crowns as well as mortality; trees growing under stress are most often attacked. Young stands of birch on surface-mine areas were often damaged.

Black alder (*Alnus glutinosa*) is a short-lived nitrogen-fixing species which is commonly used in Europe and North America. On rehabilitated sites in Denmark, Schlätzer (1973) noted partial defoliation caused by the chrysomelid beetle (*Agelastica alni*) and other species. Woolly aphids (Eriosomatidae) and various larvae have been implicated as contributors to damage in the eastern United States (Funk, 1973).

Impacts of insect herbivores on some other hardwood species are presented in Table 4.4. Hybrid poplars (*Populus* spp.) have been used in rehabilitation because of their rapid growth. They are susceptible to attack by the cottonwood leaf beetle (*Chrysomela scripta*), a chrysomelid native to North America. Its principal hosts are *Populus* and willows

Table 4.4. *Yield reductions or other losses attributable to insect feeding on hardwoods*

Plant species	Herbivore species	Insect density or treatment	Loss	Reference
HP	CWLB	—	19–48% apical tip destruction	Burkot & Benjamin (1979)
HP	TPB	—	0.5–82% of trees with split-stem lesions	Wilson & Moore (1985)
ASP	FTC	middle of outbreak	67–88% of above-ground biomass	Pollard (1972)
TP	YPW	—	1.9% of canopy	Van Hook & Dodson (1974)
TP	AB	predispose 6-yr-old trees to *Fusarium* canker	36% of trees killed back to soil surface	Anderson & Hoffard (1978)
SG	LFSB	—	13–78% seed loss	Ebel & Summerville (1983)
SLM	APH	2.25 million per 20-m tall tree	74% less annual stem wood (est.)	Dixon (1971)
BWN	AB	—	74% dieback to ground surface	Weber & McPherson (1985)
PEC	APP	266–601 per seedling after 25 days	29–41% red. in root wt.	Tedders, Wood & Snow (1982)
PEC	APP	266–601 per seedling after 25 days	34–38% red. in stem length	Tedders *et al.* (1982)
ROG	DEF	2 yr of 60–100% spring defoliation	10–100% mortality	Nichols (1968)
ROG	DEF	moderate spring defoliation	20–30% red. in radial growth	Nichols (1968)
RO	GM	90% defol. during 3 growing seasons	31% mortality	Stalter & Serrao (1983)
RO	GM	smaller (0–46 cm diam.) trees	16.5% mortality	Stalter & Serrao (1983)
RO	GM	larger trees	72.2% mortality	Stalter & Serrao (1983)

Key to species codes:

ASP aspen (*Populus tremuloides*)
BWN black walnut (*Juglans nigra*)
HP hybrid poplars (*Populus* spp.)
PEC pecan (*Carya illinoiensis*)
RO red oak (*Quercus rubra*)
ROG red oak group (several spp.)
SG sweet gum (*Liquidambar styraciflua*)
SLM sycamore-leaved maple (*Acer pseudoplatanus*)
TP tulip poplar (*Liriodendron tulipifera*)

AB ambrosia beetles (*Xylosandrus germanus* and/or *Xyleborus sayi*) COLEOPTERA
APH aphid (*Drepanosiphum platanoides*) HOMOPTERA
APP aphids (*Monelliopsis nigropunctata*, *Monellia caryella*, *Melanocallis caryaefoliae*) HOMOPTERA
CWLB cottonwood leaf beetle (*Chrysomela scripta*) COLEOPTERA
DEF defoliators, including oak leaf tier (*Croesia semipurpurana*), fall cankerworm (*Alsophila pometaria*), and forest tent caterpillar (*Malacosoma disstria*) LEPIDOPTERA
FTC forest tent caterpillar
GM gypsy moth (*Lymantria dispar*) LEPIDOPTERA
LFSB leaf-footed seedbug (*Leptoglossus oppositus*) HEMIPTERA
TPB tarnished plant bug (*Lygus lineolaris*) HEMIPTERA
YPW yellow poplar weevil (*Odontopus calceatus*) COLEOPTERA

(*Salix* spp.). Adults and larvae defoliate young trees and destroy apical tips; severe growth loss results. The tarnished plant bug (*L. lineolaris*) is also a pest of young poplars; feeding activity results in split-stem lesions up to 5 cm long. Stems and branches often break at the lesion. On mined lands in Denmark, the cerambycid beetle (*Saperda carcharias*) has caused stem mortality of hybrid poplar (Schlätzer, 1973).

Wild black cherry (*Prunus serotina*) is an important volunteer species on mined lands in the eastern United States. The eastern tent caterpillar (*Malacosoma americanum*), a pest of many species of Rosaceae, is the primary defoliator. In several infested mixed species stands in New England, 3–96% of black cherry trees contained egg masses (Waage & Bergelson, 1985). In South Carolina, tent caterpillars consumed only an estimated 3.4% of their resource base, although their potential consumption could have been much greater (Callahan, 1977). In spring, black cherry possesses foliar nectaries. In Michigan, nectar attracts predaceous ants (*Formica obscuripes*). Tilman (1978) suggested that the ant–cherry relationship is a facultative mutualism and that nectar production is timed so as to maximize the chance of successful ant predation on tent caterpillar colonies.

The American elm (*Ulmus americana*) is rarely used in plantings because of its susceptibility to Dutch elm disease, which is caused by the fungus (*Ceratocystis ulmi*) and carried in North America by the elm bark beetles (*Scolytus multistriatus* and *Hylurgopinus rufipes*). Millions of elms in Europe and North America have been killed by this disease (Strobel & Lanier, 1981). American elm is, however, a common volunteer species on many mined sites, and survives well on areas studied in southern Illinois (Urbanek, 1976).

Maples, especially silver maple (*Acer saccharinum*), but also red (*Acer rubrum*) and sugar maple (*Acer saccharum*), have been used in rehabilitation in the eastern United States. The maple leaf cutter (*Paraclemensia acerifoliella*) has caused serious damage to foliage of sugar maple in New York (Connola, 1960). In Connecticut, defoliation of sugar maple predisposed the roots to attack by the fungus (*Armillaria mellea*) and tree mortality resulted (Wargo & Houston, 1974).

Nichols (1968) summarized the causes of oak (*Quercus* spp.) mortality in Pennsylvania. In the white oak group, most mortality of white oak (*Quercus alba*) and chestnut oak (*Quercus prinus*) resulted from outbreaks of the pit-making oak scale (*Asterolecanium minus*). The appearance of mortality in oaks was coincident with girdling action by larvae of the two-lined chestnut borer (*Agrilus bilineatus*), which attacks stressed or damaged trees. As a result of examining trees defoliated by the gypsy moth (*Lymantria dispar*), Wargo (1977) indicated that mortality of oaks cannot be explained solely by one factor. He found colonization of roots of dead and dying trees by the shoe-string fungus (*Armillariella mellea*) and of the main stem by the chestnut borer. Oak mortality results from an interaction of defoliation, fungus and borer.

The gypsy moth is an outbreak species in the northeastern United States and parts of Canada. Preferred hosts are red oak (*Quercus rubra*), white oak, big-tooth aspen (*Populus grandidentata*), hop-hornbeam (*Ostrya virginiana*), serviceberry (*Amelanchier* spp.), and sugar maple, although many other hosts occur (Lechowicz & Jobin, 1983; Barbosa, 1978).

Eucalyptus spp. are commonly used in rehabilitation in Australia, where return of mined lands to pre-mining condition is a common goal. Although herbivory by insects in mature stands may be small (Ohmart, Stewart & Thomas, 1983*a*,*b*), phytophagous insects may cause high, chronic levels of damage on certain of these species, particularly on small trees or in disturbed areas. An excellent review is presented by Morrow (1977).

As with conifers, young hardwoods and young plantations such as those involved in rehabilitation may be particularly susceptible to damage by herbivores; this may result from direct feeding damage or other means such as transmission of fungal pathogens. Studies on black locust (Harman *et al.*, 1985) and white birch (Loerch & Cameron, 1984) indicated that the stresses on trees on surface-mined lands may have increased their susceptibility to herbivore attack. As with conifers many tree species involved in rehabilitation incur damage or mortality due to insect outbreaks, although the potential for such outbreaks on rehabilitated lands in the United States, with the exception of black locust, has still to be realized.

Pollination

The beneficial pollination of many plants by insects is well known. Among species used in rehabilitation, legumes are the largest group dependent on pollination by insects. In North America the most important pollinators of crops are honey bees (*Apis mellifera*) (Levin, 1983), but native bees may also be important (Free, 1970). Members of the Rosaceae are dependent on insect pollination, as are composites such as goldenrod (Gross & Werner, 1983) and many other groups. Even supposedly self-pollinating cultivars of hybrid sunflowers produced more seed in plots where bees had access (Langridge & Goodman,

1981). Study of pollen carried by syrphid flies (*Melanostoma* and *Platycheirus* spp.) suggests that even predominantly wind-pollinated plants such as cattail (*Typha* spp.), grasses, and sedges (Cyperaceae) could be pollinated by insects (Leereveld, 1982).

The relationships between insects and plants are complex; and the distinction between pollinators and phytophages may be obscure. On flatpea (*Lathyrus* spp.), for example, bruchid beetles (*Bruchus affinis*) visit flowers and transfer pollen, but the larvae feed on the seeds within flatpea pods (Labeyrie & Hossaert, 1985). Furthermore, many plant–insect pollination systems may be obligate and the absence of such pollinators may be a major factor preventing successful regeneration of plants in rehabilitated areas.

Impacts of vertebrates on vegetation of mined lands

On an abandoned coal mine in West Virginia, Prachar & Samuel (1988) found that browsing by white-tailed deer (*Odocoileus virginianus*) was most prevalent on first-year suckers of big-tooth aspen during the growing season and on second-year suckers during dormancy. Browsing suppressed height and height growth of suckers. Aspen suckers suppressed by browsing became susceptible to mortality-causing agents associated with the harsh microenvironment on the abandoned mines. On rehabilitated lands in Pennsylvania, Davidson (1970) noted browse damage to several pine species. Seedlings of jack pine, pitch pine (*Pinus rigida*), and Austrian pine (*Pinus nigra*) were browsed more heavily when within 3 m of a black locust than when farther away. Nitrogen produced by black locust may have stimulated production of more succulent tissue in the pines.

In areas of natural vegetation occurring among rehabilitated surface-mines in British Columbia, ungulates, predominantly elk (*Cervus canadensis*) and deer (*Odocoileus* spp.), caused severe hedging to redstem ceanothus (*Ceanothus sanguineus*) and willow (Stanlake, Eastman & Stanlake, 1978). Willow was also eaten in rehabilitated areas whereas planted shrubs such as black cottonwood (*Populus trichocarpa*) and trembling aspen (*Populus tremuloides*) were usually uprooted where animals had tried to browse them.

On rehabilitated lands in Denmark, Schlätzer (1973) noted browsing intensities of 52–55 % of living twigs of serviceberry (*Amelanchier canadensis*), 57–60 % of rowan (*Sorbus aucuparia*), and 13–55 % of the apple (*Malus sargenti*) by deer (*Cervus elaphus*). Effects on the plants were not presented.

Small mammals may also influence plant productivity on rehabilitated lands. In Denmark, Schlätzer (1973) noted that plantings of legumes had often failed, frequently as a result of cutting by hares (*Lepus europaeus*), followed by wind erosion and covering of plants by sand. On rehabilitated lands in Illinois, trees planted in locations where vegetation such as sweet clover was abundant suffered heavy rabbit (*Sylvilagus floridanus*) damage in winter and this usually resulted in tree mortality (Schavilje, 1941). Stanton & Krementz (1982) noted almost complete elimination, primarily by rabbits (*Sylvilagus audubonii*), of planted four-wing saltbush (*Atriplex canescens*), wheatgrasses (*Agropyron* spp.), and yellow sweet clover on rehabilitated areas in Wyoming. In Alberta, an average of 50% of young nursery-raised trees planted on rehabilitated sites were girdled by meadow voles (*Microtus pennsylvanicus*) that had become established in dense grass cover created to stabilize nearby spoil piles (Radvanyi, 1980). Tree mortality due to girdling damage ranged from 2 to 95% and was particularly heavy for deciduous species. In previous studies of maple, poplar, walnut (*Juglans nigra*) and ash (*Fraxinus* spp.) plantations in Ontario, heavy rodent damage occurred even after tree heights of 3.7–4.6 m and basal diameters of 13–15 cm had been attained.

Hingtgen & Clark (1984) examined the impact of small mammal activity on the standing crop of vegetation on areas rehabilitated after coal mining in Wyoming. Deer mice (*Peromyscus maniculatus*) comprised 85.4% of small mammals captured and occurred at average densities of 13.0 and 13.9 ha^{-1} on 2-year-old and more established rehabilitated areas, respectively. Deer mice consumed 20 g of plant matter ha^{-1} day^{-1}. They consumed 0.11% of above-ground plant standing crop during the growing season; consumption by the total small mammal community was less than 1% of peak standing crop. By contrast, the legume sainfoin (*Onobrychis viciifolia*) was significantly affected by grazing of small mammals; on 3–5-year-old sites, 13.6% of the standing crop of this species, enough to affect establishment, was consumed. The impact of selective grazing on plant community composition is explored in more detail in Chapter 5. Impacts on vegetation of species such as voles (*Microtus* spp.), which consume large amounts of plant material (McGowan, 1980), could be significant (Batzli & Pitelka, 1970), but effects of these small mammals on productivity of rehabilitated lands have been little studied.

The impact of wild ungulates on herbaceous vegetation has also received little attention. Medcraft & Clark (1986) noted heavy use of alfalfa by mule deer (*Odocoileus hemionus*) and pronghorn antelope (*Antilocapra americana*) on rehabilitated lands in Wyoming. Increased use of pastures after cattle were removed in late summer could have resulted from a positive impact of cattle grazing. The reduction in grass cover by the cattle, accompanied by the late summer and autumn greenup of alfalfa and other forbs, may have made these areas more attractive to deer and pronghorns. Under a different grazing regime on rehabilitated land in Montana, where spring and autumn cattle grazing occurred, growth of crested wheatgrass (*Agropyron desertorum*) was apparently stimulated (DePuit & Coenenberg, 1978). By the second growing season, estimated total vegetation productivity in the grazed pasture was 2029 kg ha^{-1} compared to 1206 kg ha^{-1} in the ungrazed control. This illustrates the variable effect which different grazing intensities may have; heavy grazing may reduce growth and increase mortality while light grazing may stimulate growth.

Influence of animals on seed dispersal and destruction of planted seeds

Rodents have been responsible for loss of planted seeds on some sites. White-footed mice (*Peromyscus leucopus*) were apparently responsible for red oak seeding failures on spoil banks in Pennsylvania (Bramble & Sharp, 1949). On mined lands in Indiana, this mouse was abundant and fed on a wide variety of seeds (Mumford & Bramble, 1973). In tests in Nevada, deer mice consumed or destroyed an amount of seed equal to 30–40 % of their body weight daily (Everett, Meeuwig & Stevens, 1978). The predation of broadcast seeds by birds has also been suggested (Goebel & Berry, 1976).

In Australia, seed theft by ants is common (Majer, 1980, 1984, 1985). Two different groups of ants occur on rehabilitated areas, seed harvesters which actually destroy the seed, and elaiosome collectors, which probably benefit the seed in some way. Elaiosomes are ant-attracting structures that occur on the seeds of many Australian plants. They enhance seed dispersal by ants, which store the viable seed in their nest and consume the elaiosome. Majer (1980) concluded that seed removal by ants was not a problem in the centre of rehabilitated bauxite mined areas and of only limited importance adjacent to forest. However, rapid build-up of seed-eating ants in recently rehabilitated areas resulted in seed removal rates of 100 % up to 40 m into the topsoiled area adjacent to the previously rehabilitated site; all seed removed from the more recent area was taken to the adjacent rehabilitated area (Majer, 1984,

1985). These studies also showed that seed removal varied among planted species, for example, being high for fleshy-fringed seed of *Grevillea pteridifolia*, less for *Acacia* spp., which also possess elaiosomes, and least for non-arillate seeds such as those of *Eucalyptus* spp. On sites where successful rehabilitation has been completed, seed collection and burial could be a means of maintaining a viable seed source in spite of periodic fires. The possible importance of seed dispersers in maintaining the species diversity of plant communities is discussed further in Chapter 5.

Vertebrates may also be important in seed dispersal on reclaimed lands. Fowler, Fowler & Thomas (1982) found that up to 75% of seeds of autumn olive (*Elaeagnus umbellata*) were viable after passing through digestive tracts of grey and red foxes (*Urocyon cinereoargenteus* and *Vulpes vulpes*). Autumn olive is a valuable shrub species for wildlife plantings in the eastern United States and it starts producing berries in the second or third season after planting. Mean number of autumn olive seeds per fox scat was 218 in November, and 622 seeds were found in one scat. Dispersal of seeds of many plant species by birds and other mammals is likely to be significant but remains unstudied on rehabilitated lands.

Discussion

Though both vertebrates and invertebrates are capable of reducing plant productivity, their modes of inflicting damage are different. Among invertebrates, particularly insects, there exist important defoliators and feeders on other green parts, phloem and xylem feeders, root-feeders, stem-borers, leaf-miners, flower and seed eaters, and many that transmit disease organisms during feeding. Every plant species is the host for some insect species. Due to their larger size, the feeding niches of vertebrates are more limited. While some herbaceous species, such as those in legume plantings, may be significantly affected, greatest potential for damage by wild vertebrates appears to be to small woody vegetation and occurs as a result of browsing by ungulates or girdling of young trees by rabbits or rodents.

The numerous articles on damaging effects of animals, particularly insects, upon plants reflect the general interest in fauna from an economic standpoint. These effects are often more short-term and obvious. Mattson & Addy (1975) concluded that normal insect grazing (from 5 to 30% of annual foliage crops) usually does not impair annual primary production and may actually accelerate growth. Due to compensatory plant growth, tissue destruction can rarely be translated into a simple proportional reduction in final yield (McNaughton, 1983). Accelerated plant growth as a result of grazing by cattle (Reardon, Leinweber & Merrill, 1974), grasshoppers (Dyer & Bokhari, 1976), and the hispid cotton rat (*Sigmodon hispidus*) (Howe, Grant & Folse, 1982) has been noted; the same growth was not observed after simple mechanical clipping. Insect grazers tend to ensure consistent and optimal output of plant production over the long-term. Numerous authors have commented on the role of consumers as regulators of ecosystems (Chew, 1974; Mitchell & Pfadt, 1974; Springett, 1978), although other data indicate no relationship between herbivory and regulation of plant growth rate (Lamb, 1985). An examination of plant–herbivore dynamics, as well as can be made with existing information, can be found in Crawley (1983).

Many of the presented influences of insects on plants were derived from studies related to intensive agricultural management; for example, several cuttings per year in the case of alfalfa. Most surface-mine plantings are not normally subjected to this intensive harvesting regime or level of management. Hawkins *et al.* (1979) found that the least frequently harvested grass plots had substantially greater reductions in yield attributable to insect damage than plots that were harvested more frequently. Greater potential for damage on mined lands might therefore be expected because of the more permanent vegetative cover. Rehabilitated lands, even those which have been retopsoiled, possess very different soils than those on unmined lands. By comparing fields with similar crops but different soil types, Tischler (1955) found that soil characteristics influenced composition, density, and biomass of the arthropod fauna present, including those of phytophagous forms. The influence of rehabilitated land soils on the survival, growth and reproduction of invertebrate herbivores needs to be examined.

Rehabilitation work often, but by no means always, produces simplified ecosystems. Areas rehabilitated with only a few plant species may be more susceptible to outbreaks of herbivores. Decreased efficiency of predators and parasites (Pimentel, 1961), greater success of herbivores in finding, persisting and reproducing on concentrated resources (Root, 1973), and a lack of associated resistance or interference by non-host plants that occur in diverse communities (Tahvanainen & Root, 1972; Atsatt & O'Dowd, 1976) could all be possible mechanisms promoting eruptive population growth of insect pest species. Tahvanainen (1972) and Cromartie (1975) demonstrated the importance of individualistic responses of herbivore species to specific resources present. Price (1984), in a discussion of the work of these and other investigators, stressed the need for a detailed understanding of resource availability, individualistic herbivore responses, herbivore life cycles relative to resource availability, plant habitats and associated plants in the study of communities. Although resistant plant varieties are used whenever possible in rehabilitation work, data presented in this Chapter have illustrated that productivity of most species used in the eastern United States can be limited by herbivores. Certain herbaceous species, such as alfalfa and clovers, are particularly susceptible to outbreaks of weevils, leafhoppers, spittlebugs, aphids and other insects. Monospecific tree plantations, such as black locust on mined sites, appear more susceptible to pest outbreaks than trees under other conditions. Diversity in rehabilitation planting schemes needs to be promoted, and the individual pest species associated with each plant species to be used need to be considered.

The present relatively pest-free status of some plant species currently used in rehabilitation may not be lasting. Crown vetch, for example, currently incurs little insect damage on surface mine plantings. Wheeler (1974), however, noted that crown vetch, indigenous to Europe where it is not a crop, has been of economic importance in the United States only since 1935. Alfalfa, likewise, did not begin to develop pest problems until 40 years after its introduction into the United States, but now this species is host to many damaging consumers, as previously indicated. An ultimate increase in pest populations after the establishment of a crop in monocultures in a new environment is to be expected.

Specific recommendations for control of insect and other pests are beyond the scope of this Chapter. In addition, the field of pest management changes so rapidly that chemical control measures soon become outdated. Principles of insect pest management and general control measures can be found in Metcalf & Luckmann (1982), Pfadt (1985), Davidson & Lyon (1987), and Horn (1987). In the United States specific up-to-date details and adjustments for chemical control of pests on trees and forage crops can be obtained from the Cooperative Extension Service, an organization jointly sponsored by the US Department of Agriculture, the State land-grant university, and county government, and containing an office in every county. Comparable government sponsored agencies exist in many other countries.

The Surface Mining Control and Reclamation Act of 1977 (SMCRA) made no comments on use of pesticides in rehabilitated lands in the USA; however, requirements to restore areas of prime farmland to equivalent, or higher levels, of yield as surrounding unmined prime farmland under equivalent levels of management could necessitate some chemical control measures. For other lands the SMRCA requires or permits: (1) establishment of a diverse, permanent, vegetative cover capable of self-regeneration and plant succession at least equal in extent of cover to the natural vegetation of the area; (2) use of introduced species where desirable and necessary to achieve the approved post-mining land use; or (3) an approved long-term, intensive, agricultural post-mining land use. The post-mining use of many areas as grazing land, and a responsibility period for successful rehabilitation of 5–10 years, usually means the operator can meet legal rehabilitation requirements without use of pesticides, despite loss of some plant productivity due to animals.

Although many of the dominant arthropods on rehabilitated lands in the eastern US are agricultural pests, most of these lands are not intensively used for agricultural production, and control of these arthropods by pesticides is usually not warranted. These, and co-occurring beneficial species, are integral components of community dynamics on these sites; further research is needed on their roles in nutrient recycling, energy flow and development of the system. Pesticides would also damage the litter/soil fauna, which may be instrumental in soil development and maintenance of long-term productivity (see Chapters 2 and 3).

Lands rehabilitated, or currently under rehabilitation, can represent a stress-inducing environment for plant growth and survival. Increased sensitivity of plants to the effects of animals, both vertebrate and invertebrate, might therefore be expected. While some impacts of fauna on plants have been described, much of the study of these relationships on rehabilitated lands still lies ahead. The various ways in which fauna can influence the composition of the plant community are now reviewed in the following Chapter.

References

Ahring, R. M., Moffett, J. O. & Morrison, R. D. (1984). Date of podset and chalcid fly infestation in alfalfa seed crops in the southern Great Plains. *Agronomy Journal*, **76**, 137–40.

Alvarado, E. A., Fuxa, J. R. & Wilson, B. H. (1983). Correlation of absolute population estimates of *Spodoptera frugiperda* (Lepidoptera: Noctuidae) with sweep sampling and yield in bermuda grass. *Journal of Economic Entomology*, **76**, 792–6.

Andaloro, J. T., Peters, T. M. & Alicandro, A. J. (1983). Population dynamics of the alfalfa blotch leafminer, *Agromyza frontella*, and its influence on alfalfa yield in Massachusetts. *Environmental Entomology*, **12**, 510–14.

Anderson, R. L. & Hoffard, W. H. (1978). Fusarium canker–ambrosia beetle complex on tulip poplar in Ohio. *Plant Disease Reporter*, **62**, 751.

Ashby, W. C. (1964). Vegetation development on a strip-mined area in southern Illinois. *Transactions of the Illinois State Academy of Science*, **57**, 78–83.

Atsatt, P. R. & O'Dowd, D. J. (1976). Plant defense guilds. *Science*, **193**, 24–9.

Barbosa, P. (1978). Distribution of an endemic larval gypsy moth population among various tree species. *Environmental Entomology*, **7**, 526–7.

Batzli, G. O. & Pitelka, F. A. (1970). Influence of meadow mouse populations on California grassland. *Ecology*, **51**, 1027–39.

Berberet, R. C., Morrison, R. D. & Senst, K. M. (1981). Impact of the alfalfa weevil, *Hypera postica* (Gyllenhal) (Coleoptera: Curculionidae), on forage production in non-irrigated alfalfa in the southern plains. *Journal of the Kansas Entomological Society*, **54**, 312–18.

Bjork, C. D. & Davis, D. W. (1984). Consumption of alfalfa by adult alfalfa weevils (Coleoptera: Curculionidae). *Environmental Entomology*, **13**, 432–8.

Blank, R. H. & Olson, M. H. (1981). The damage potential of the black field cricket *Teleogryllus commodus*. *New Zealand Journal of Agricultural Research*, **24**, 251–8.

Bramble, W. C. & Sharp, W. M. (1949). Rodents as a factor in direct seeding on spoil banks in central Pennsylvania. *Journal of Forestry*, **47**, 477–8.

Brewer, G. J. & Horber, E. (1984). Field infestation and alfalfa seed chalcid (Hymenoptera: Eurytomidae) development in different *Medicago* clones. *Environmental Entomology*, **13**, 1157–9.

Buntin, G. D. & Pedigo, L. P. (1985*a*). Variegated cutworm (Lepidoptera: Noctuidae) foliage consumption and larval development on alfalfa. *Journal of Economic Entomology*, **78**, 482–4.

Buntin, G. D. & Pedigo, L. P. (1985*b*). Development of economic injury levels for last-stage variegated cutworm (Lepidoptera: Noctuidae) larvae in alfalfa stubble. *Journal of Economic Entomology*, **78**, 1341–6.

Burkot, T. R. & Benjamin, D. M. (1979). The biology and ecology of the cottonwood leaf beetle, *Chrysomela scripta* (Coleoptera: Chrysomelidae), on tissue cultured hybrid *Aigeiros* (*Populus* × *Euramericana*) subclones in Wisconsin. *Canadian Entomologist*, **111**, 551–6.

Byers, R. A. & Jung, G. A. (1979). Insect populations on forage grasses: effect of nitrogen fertilizer and insecticides. *Environmental Entomology*, **8**, 11–18.

Byers, R. A., Mangan, R. L. & Templeton, Jr. W. C. (1983). Insect and slug pests in forage legume seedlings. *Journal of Soil and Water Conservation*, **38**, 224–6.

Callahan, J. T. (1977). Resource utilization by colonial Lepidoptera. In *The Role of Arthropods in Forest Ecosystems*, ed. W. J. Mattson, pp. 30–4. New York: Springer-Verlag.

Chew, R. M. (1974). Consumers as regulators of ecosystems: an alternative to energetics. *Ohio Journal of Science*, **74**, 359–70.

Clements, R. O. & Henderson, I. F. (1982). Screening of pesticides for use on established grassland to improve herbage yield. *Pesticide Science*, **13**, 617–22.

Connola, D. P. (1960). Control of maple leaf cutter, *Paraclemensia acerifoliella*, by aerial spraying. *Journal of Economic Entomology*, **53**, 957–8.

Craig, C. H. (1978). Damage potential of the sweetclover weevil, *Sitona cylindricollis* (Coleoptera: Curculionidae), in the Canadian prairies. *Canadian Entomologist*, **110**, 883–9.

Cranshaw, W. S. (1985). Clover root curculio injury and abundance in Minnesota alfalfa of different stand age. *Great Lakes Entomologist*, **18**, 93–5.

Crawley, M. J. (1983). *Herbivory: the Dynamics of Animal–Plant Interactions*. Berkeley: University of California Press.

Cromartie, W. J. (1975). The effect of stand size and vegetational background on the colonization of cruciferous plants by herbivorous insects. *Journal of Applied Ecology*, **12**, 517–33.

Cuperus, G. W., Radcliffe, E. B., Barnes, D. K. & Marten, G. C. (1983). Economic injury levels and economic thresholds for potato leafhopper (Homoptera: Cicadellidae) on alfalfa in Minnesota. *Journal of Economic Entomology*, **76**, 1341–9.

Davidson, R. H. & Lyon, W. F. (1987). *Insect Pests of Farm, Garden and Orchard*, 7th edn. New York: John Wiley & Sons.

Davidson, W. H. (1970). *Deer Prefer Pine Seedlings Growing Near Black Locust.* U.S. Department of Agriculture, Forest Service Research Note, NE-111.

DeBarr, G. L., Barber, L. R. & Maxwell, A. H. (1982). Use of carbofuran for control of eastern white pine cone and seed insects. *Forest Ecology and Management*, **4**, 1–18.

DeBarr, G. L. & Ebel, B. H. (1974). Conelet abortion and seed damage of shortleaf and loblolly pines by a seedbug, *Leptoglossus corculus*. *Forest Science*, **20**, 165–70.

DeBarr, G. L. & Kormanik, P. P. (1975). Anatomical basis for conelet abortion on *Pinus echinata* following feeding by *Leptoglossus corculus* (Hemiptera: Coreidae). *Canadian Entomologist*, **107**, 81–6.

DePuit, E. J. & Coenenberg, J. G. (1978). Plant response and forage quality for livestock grazing on coal mine spoils in Montana. In *Ecology and Coal Resource Development*, vol. 2, ed. M. K. Wali, pp. 892–8. New York: Pergamon Press.

Dickason, E. A., Leach, C. M. & Gross, A. E. (1968). Clover root curculio injury and vascular decay of alfalfa roots. *Journal of Economic Entomology*, **61**, 1163–8.

Dixon, A. F. G. (1971). The role of aphids in wood formation. I. The effect of the sycamore aphid, *Drepanosiphum platanoides* (Schr.) (Aphididae), on the growth of sycamore, *Acer pseudoplatanus* (L.). *Journal of Applied Ecology*, **8**, 165–79.

Dominick, C. B. (1938). Notes on the locust leaf miner, *Chalepus dorsalis* Thumb. *Journal of Economic Entomology*, **31**, 186–9.

Dyer, M. I. & Bokhari, U. G. (1976). Plant–animal interactions: studies of the effects of grasshopper grazing on blue grama grass. *Ecology*, **57**, 762–72.

Ebel, B. H., Flavell, T. H., Drake, L. E., Yates III, H. O. & DeBarr, G. L. (1975). *Southern Pine Seed and Cone Insects.* U.S. Department of Agriculture, Forest Service General Technical Report, SE-8.

Ebel, B. H. & Summerville, K. O. (1983). An evaluation of feeding by a seed bug, *Leptoglossus oppositus* (Say), on seed yield of sweetgum, *Liquidambar styraciflua* L. *Journal of the Georgia Entomological Society*, **18**, 316–20.

Erdelyi, C., Manninger, S., Manninger, K. & Buglos, J. (1981). Some factors affecting seed yield loss of lucerne caused by insect pests. *Acta Phytopathologica Academiae Scientarium Hungaricae*, **16**, 171–80.

Everett, R. L., Meeuwig, R. O. & Stevens, R. (1978). Deer mouse preference for seed of commonly planted species, indigenous weed seed, and sacrifice foods. *Journal of Range Management*, **31**, 70–3.

Everly, R. T. (1959). Evaluation of population estimates and the rate of loss of forage for the meadow spittlebug. *Proceedings of the Indiana Academy of Science*, **68**, 171–85.

Faris, M. A., Baenziger, H. & Terhune, R. P. (1981). Studies on potato leafhopper (*Empoasca fabae*) damage in alfalfa. *Canadian Journal of Plant Science*, **61**, 625–32.

Flinn, P. W & Hower, A. A. (1984). Effects of density, stage and sex of the potato leafhopper, *Empoasca fabae* (Homoptera: Cicadellidae), on seedling alfalfa growth. *Canadian Entomologist*, **116**, 1543–8.

Fowler, L. J., Fowler, D. K. & Thomas, J. E. (1982). Dispersal of autumn olive seeds by foxes on coal surface mines in east Tennessee. *Journal of the Tennessee Academy of Science*, **57**, 83–5.

Fowler, R. F. & Wilson, L. F. (1974). Injury to aldrin treated and untreated red pine by white grubs (Coleoptera: Scarabaeidae) and other agents during first five years after planting. *Great Lakes Entomologist*, **7**, 81–8.

Free, J. B. (1970). *Insect Pollination of Crops.* New York: Acedemic Press.

Fritz, R. S. (1983). Ant protection of a host plant's defoliator: consequence of an ant–membracid mutalism. *Ecology*, **64**, 789–97.

Funk, D. T. (1973). Growth and development of alder plantings on Ohio strip-mine banks (Discussion at end of paper). In *Ecology and Reclamation of Devastated Lands*, vol. 2. ed. R. J. Hutnik & G. Davis, pp. 483–91. New York: Gordon and Breach.

Goebel, C. J. & Berry, G. (1976). Selectivity of range grass seeds by local birds. *Journal of Range Management*, **29**, 393–5.

Goldson, S. L. (1982). An examination of the relationship between Argentine stem weevil (*Listronotus bonariensis* (Kuschell)) and several of its host grasses. *New Zealand Journal of Agricultural Research*, **25**, 395–403.

Granovsky, A. A. (1930). Differentiation of symptoms and effect of leafhopper feeding on histology of alfalfa leaves. *Phytopathology*, **20**, 121.

Grant, J. F., Yeargan, K. V., Pass, B. C. & Parr, J. C. (1982). Invertebrate organisms associated with alfalfa seedling loss in complete-tillage and no-tillage plantings. *Journal of Economic Entomology*, **75**, 822–6.

Gross, R. S. & Werner, P. A. (1983). Relationships among flowering phenology, insect visitors and seed-set of individuals: experimental studies on four co-occurring species of goldenrod (*Solidago*: Compositae). *Ecological Monographs*, **53**, 95–117.

Guppy, J. C. (1963). Observations on the biology of *Plagiognathus chrysanthemi* (Hemiptera: Miridae), a pest of birdsfoot trefoil in Ontario. *Annals of the Entomological Society of America*, **56**, 804–9.

Hainze, J. H. & Benjamin, D. M. (1983). Bionomics and pattern of attack of the red pine shoot moth, *Dioryctria resinosella* (Lepidoptera: Pyralidae), in Wisconsin. *Canadian Entomologist*, **115**, 1169–75.

Hainze, J. H. & Benjamin, D. M. (1984). Impact of the red pine shoot moth, *Dioryctria resinosella* (Lepidoptera: Pyralidae), on height and radial growth in Wisconsin red pine plantations. *Journal of Economic Entomology*, **77**, 36–42.

Hardman, J. M. & Smoliak, S. (1982). The relative impact of various grasshopper species on *Stipa–Agropyron* mixed prairie and fescue prairie in southern Alberta. *Journal of Range Management*, **35**, 171–6.

Harman, D. M., Van Tyne, M. A. & Thompson, W. A. (1985). Comparison of locust borer *Megacyllene robiniae* Forster (Coleoptera: Cerambycidae) attacks on coal strip-mined lands and lands not mined. *Annals of the Entomological Society of America*, **78**, 50–3.

Harper, A. M. & Kaldy, M. S. (1982). Effect of the pea aphid, *Acyrthosiphon pisum* (Hemiptera (Homoptera): Aphididae), on yield and quality of forage alfalfa. *Canadian Entomologist*, **114**, 485–9.

Hartnett, D. C. & Abrahamson, W. G. (1979). The effects of stem gall insects on life history patterns in *Solidago canadensis*. *Ecology*, **60**, 910–17.

Hawkins, J. A., Wilson, B. H., Mondart, C. L., Nelson, B. D., Farlow, R. A. & Schilling, P. E. (1979). Leafhoppers and planthoppers in coastal bermudagrass: effect of yield and quality and control by harvest frequency. *Journal of Economic Entomology*, **72**, 101–4.

Henderson, I. F. & Clements, R. O. (1977). Grass growth in different parts of England in relation to invertebrate numbers and pesticide treatment. *Journal of the British Grassland Society*, **32**, 89–98.

Hewitt, G. B. (1977). *Review of Forage Losses Caused by Rangeland Grasshoppers*. U.S. Department of Agriculture, Agricultural Research Service, Miscellaneous Publication Number 1348.

Hewitt, G. B. & Onsager, J. A. (1982). A method for forecasting potential losses from grasshopper feeding on northern mixed prairie forages. *Journal of Range Management*, **35**, 53–7.

Highley, L. & Tattar, T. A. (1985). *Leptographium terebrantis* and black turpentine beetles associated with blue stain and mortality of black and scots pines on Cape Cod, Massachusetts. *Plant Disease*, **69**, 528–30.

Hingtgen,T. M. & Clark, W. R. (1984). Impact of small mammals on the vegetation of reclaimed land in the northern Great Plains. *Journal of Range Management*, **37**, 438–41.

Hintz, T. R., Wilson, M. C. & Armbrust, E. J. (1976). Impact of alfalfa weevil larval feeding on the quality and yield of first cutting alfalfa. *Journal of Economic Entomology*, **69**, 749–54.

Hodges, J. D., Elam, W. W., Watson, W. F. & Nebeker, T. E. (1979). Oleoresin characteristics and susceptibility of four southern pines to southern pine beetle (Coleoptera: Scolytidae) attacks. *Canadian Entomologist*, **111**, 889–96.

Hodges, J. D. & Pickard, L. S. (1971). Lightning in the ecology of the southern pine beetle, *Dendroctonus frontalis* (Coleoptera: Scolytidae). *Canadian Entomologist*, **103**, 44–51.

Holopainen, J. K. & Soikkeli, S. (1984). Occurrence of *Cinara pini* (Homoptera, Lachnidae) in Scots pine (*Pinus sylvestris*) seedlings with disturbed growth. *Annales Entomologici Fennici*, **50**, 108–10.

Horn, D. J. (1987). *Ecological Approach to Pest Management*. New York: Guilford Press.

Howe, J. G., Grant, W. E. & Folse, L. J. (1982). Effects of grazing by *Sigmodon hispidus* on the regrowth of annual rye-grass (*Lolium perenne*). *Journal of Mammalogy*, **63**, 176–9.

James, J. R., Lucas, L. T., Chamblee, D. S. & Campbell, W. V. (1980). Influence of fungicide and insecticide applications on persistence of ladino clover. *Agronomy Journal*, **72**, 781–4.

Kamm, J. A. (1985). Cutworm defoliators of ryegrass. *Pan-Pacific Entomologist*, **61**, 68–71.

Kamm, J. A. & Buttery, R. G. (1984). Root volatile components of red clover: identification and bioassay with the clover root borer (Coleoptera: Scolytidae). *Environmental Entomology*, **13**, 1427–30.

Kelly, R. E. & Klostermeyer, L. E. (1984). Relative abundance and insecticidal control of leafhoppers (Homoptera: Cicadellidae) in a mixed tall fescue pasture in Tennessee. *Journal of the Kansas Entomological Society*, **57**, 657–61.

Kindler, S. D., Kehr, W. R. & Ogden, R. L. (1971). Influence of pea aphids and spotted alfalfa aphids on the stand, yield of dry matter, and chemical composition of resistant and susceptible varieties of alfalfa. *Journal of Economic Entomology*, **64**, 653–7.

Koehler, P. G. & Pimentel, D. (1973). Economic injury levels of the alfalfa weevil (Coleoptera: Curculionidae). *Canadian Entomologist*, **105**, 61–74.

Kouskolekas, C. & Decker, G. C. (1968). A quantitative evaluation of factors affecting alfalfa yield reduction caused by the potato leafhopper attack. *Journal of Economic Entomology*, **61**, 921–7.

Kulman, H. M. (1971). Effects of insect defoliation on growth and mortality of trees. *Annual Review of Entomology*, **16**, 289–324.

Kulman, H. M., Hodson, A. C. & Duncan, D. P. (1963). Distribution and effects of jack-pine budworm defoliation. *Forest Science*, **9**, 146–57.

Labeyrie, V. & Hossaert, M. (1985). Ambiguous relations between *Bruchus affinis* and the *Lathyrus* group. *Oikos*, **44**, 107–13.

Lamb, D. (1985). The influence of insects on nutrient cycling in eucalypt forests: a beneficial role? *Australian Journal of Ecology*, **10**, 1–5.

Langridge, D. F. & Goodman, R. D. (1981). Honeybee pollination of sunflower cultivars Hysun 30 and Sunfola. *Australian Journal of Experimental Agriculture and Animal Husbandry*, **21**, 435–8.

Larsson, H. A. (1984). Frit fly (*Oscinella frit* L.) damage in oats and its chemical control by synthetic pyrethroids. *Zeitschrift Fuer Angewandte Entomologie*, **97**, 470–80.

Lashomb, J. H., Steinhauer, A. L. & Douglass, L. (1978). Impact studies of Nantucket pine tip moth populations on loblolly pine. *Environmental Entomology*, **7**, 910–12.

Lechowicz, M. J. & Jobin, L. (1983). Estimating the susceptibility of tree species to attack by the gypsy moth, *Lymantria dispar*. *Ecological Entomology*, **8**, 171–83.

Leereveld, H. (1982). Anthecological relations between reputedly anemophilous flowers and syrphid flies. III. Worldwide survey of crop and intestine contents of certain anthophilous syrphid flies. *Tijdschrift Voor Entomologie*, **125**, 25–35.

Levin, M. D. (1983). Value of bee pollination to U.S. agriculture. *Bulletin of the Entomological Society of America*, **29**, 50–1.

Linit, M. J., Kondo, D. & Smith, M. T. (1983). Insects associated with the pinewood nematode, *Bursaphelenchus xylophilus* (Nematoda: Aphelenchoididae), in Missouri. *Environmental Entomology*, **12**, 467–70.

Liu, B. W. Y. & Fick, G. W. (1975). Yield and quality losses due to alfalfa weevil. *Agronomy Journal*, **67**, 828–32.

Loerch, C. R. & Cameron, E. A. (1984). Within-tree distributions and seasonality of immature stages of the bronze birch borer, *Agrilus anxius* (Coleoptera: Buprestidae). *Canadian Entomologist*, **116**, 147–52.

Lyons, L. A. (1957). Insects affecting seed production in red pine. IV. Recognition and extent of damage to cones. *Canadian Entomologist*, **89**, 264–71.

Majer, J. D. (1980). The influence of ants on broadcast and naturally spread seeds in rehabilitated bauxite mined areas. *Reclamation Review*, **3**, 3–9.

Majer, J. D. (1984). The influence of ants on seeding operations in northern Australian mined areas. *Reclamation and Revegetation Research*, **2**, 299–313.

Majer, J. D. (1985). Recolonization by ants of rehabilitated mineral sand mines on North Stradbroke Island, Queensland, with particular reference to seed removal. *Australian Journal of Ecology*, **10**, 31–48.

Manglitz, G. R., Anderston, D. M. & Gorz, H. J. (1963). Observations on the larval feeding habits of two species of *Sitona* (Coleoptera: Curculionidae) in sweetclover fields. *Annals of the Entomological Society of America*, **56**, 831–5.

Mattson, W. J. & Addy, N. D. (1975). Phytophagous insects as regulators of forest primary production. *Science*, **190**, 515–22.

McClure, M. S. (1982). Distribution and damage of two *Pineus* species (Homoptera: Adelgidae) on red pine in New England. *Annals of the Entomological Society of America*, **75**, 150–7.

McClure, M. S. (1983). Population dynamics of a pernicuous parasite: density-dependent vitality of red pine scale. *Ecology*, **64**, 710–18.

McGowan, K. J. (1980). *Populations of Small Mammals and their Utilization of Arthropods on Ohio Strip-mined Lands.* M.Sc. Thesis, Ohio State University, Columbus.

McLeod, P. J. & Yearian, W. C. (1982). Insect control tactics evaluated in loblolly pine seed orchards. *Arkansas Farm Research*, **31**(5), 10.

McNaughton, S. J. (1983). Compensatory plant growth as a response to herbivory. *Oikos*, **40**, 329–36.

Medcraft, J. R. & Clark, W. R. (1986). Big game habitat use and diets on a surface mine in northeastern Wyoming. *Journal of Wildlife Management*, **50**, 135–42.

Medler, J. T. (1941). The nature of injury to alfalfa caused by *Empoasca fabae* (Harris). *Annals of the Entomological Society of America*, **34**, 439–50.

Messina, F. J. (1981). Plant protection as a consequence of an ant–membracid mutualism: interactions on goldenrod (*Solidago* sp.). *Ecology*, **62**, 1433–40.

Metcalf, R. L. & Luckmann, W. H. (1982). *Introduction to Insect Pest Management*, 2nd edn. New York: John Wiley and Sons.

Mitchell, J. E. & Pfadt, R. E. (1974). A role of grasshoppers in a shortgrass prairie ecosystem. *Environmental Entomology*, **3**, 358–60.

Mittler, T. E. & Sylvester, E. S. (1961). A comparison of the injury to alfalfa by the aphids, *Therioaphis maculata* and *Macrosiphum pisi*. *Journal of Economic Entomology*, **54**, 615–22.

Mook, J. H. & Van der Toorn, J. (1985). Delayed response of common reed *Phragmites australis* to herbivory as a cause of cyclic fluctuations in the density of the moth *Archanara geminipuncta*. *Oikos*, **44**, 142–8.

Moore, D. & Clements, R. O. (1984). Stem-boring Diptera in perennial ryegrass in relation to fertilizer. I. Nitrogen level and form. *Annals of Applied Biology*, **105**, 1–6.

Morrow, P. A. (1977). The significance of phytophagous insects in the *Eucalyptus* forests of

Australia. In *The Role of Arthropods in Forest Ecosystems*, ed. W. J. Mattson, pp. 19–29. New York: Springer-Verlag.

Mumford, R. E. & Bramble, W. C. (1973). Small mammals on surface-mined land in southwestern Indiana. In *Ecology and Reclamation of Devastated Lands*, vol. 1, ed. R. J. Hutnik & G. Davis, pp. 369–76. New York: Gordon and Breach.

Nerney, N. J. (1958). Grasshopper infestations in relation to range condition. *Journal of Range Management*, **11**, 247.

Neunzig, H. H. & Gyrisco, G. G. (1955). Some insects injurious to birdsfoot trefoil in New York. *Journal of Economic Entomology*, **48**, 447–50.

Nichols, J. O. (1968). Oak mortality in Pennsylvania. *Journal of Forestry*, **66**, 681–94.

Ohmart, C. P., Stewart, L. G. & Thomas, J. R. (1983*a*). Phytophagous insect communities in the canopies of three *Eucalyptus* forest types in southeastern Australia. *Australian Journal of Ecology*, **8**, 395–403.

Ohmart, C. P., Stewart, L. G. & Thomas, J. R. (1983*b*). Leaf consumption by insects in three *Eucalyptus* forest types in southeastern Australia and their role in short-term nutrient cycling. *Oecologia*, **59**, 322–30.

Oseto, C. Y., Burr, W. F. & Carlson, R. B. (1982). Relationship of sunflower planting dates to damage by *Cylindrocopturus adspersus* (Coleoptera: Curculionidae) in North Dakota. *Journal of Economic Entomology*, **75**, 761–4.

Parman, V. R. & Wilson, M. C. (1982). Alfalfa crop responses to feeding by the meadow spittlebug (Homoptera: Cercopidae). *Journal of Economic Entomology*, **75**, 481–6.

Pfadt, R. E. (1985). *Fundamentals of Applied Entomology*, 5th edn. New York: MacMillan Publishing Co.

Pimentel, D. (1961). Species diversity and insect population outbreaks. *Annals of the Entomological Society of America*, **54**, 76–86.

Pollard, D. F. W. (1972). Estimating woody dry matter loss resulting from defoliation. *Forest Science*, **18**, 135–8.

Powell, G. S., Campbell, W. V., Cope, W. A. & Chamblee, D. S. (1983). Ladino clover resistance to the clover root curculio (Coleoptera: Curculionidae). *Journal of Economic Entomology*, **76**, 264–8.

Prachar, R. E. & Samuel, D. E. (1988). Deer–aspen relationships on an abandoned surface mine. *Journal of Applied Forestry* (in press).

Price, P. W. (1984). *Insect Ecology*, 2nd edn. New York: John Wiley and Sons.

Quisenberry, S. & Yonke, T. R. (1981*a*). Effects of *Forcipata loca* feeding on tissue of Kentucky-31 tall fescue. *Annals of the Entomological Society of America*, **74**, 521–4.

Quisenberry, S. & Yonke, T. R. (1981*b*). Responses of 'Kentucky-31' tall fescue to varying *Forcipata loca* DeLong & Caldwell infestation levels: growth chamber study (Homoptera: Cicadellidae, Typhlocybinae). *Environmental Entomology*, **10**, 550–3.

Radvanyi, A. (1980). Control of small mammal damage in the Alberta oil sands reclamation and afforestation program. *Forest Science*, **26**, 687–702.

Rauf, A. & Benjamin, D. M. (1983). Spatial distribution of jack pine cones and those attacked by insects. *Great Lakes Entomologist*, **16**, 183–5.

Rauf, A., Cecich, R. A. & Benjamin, D. M. (1984). Conelet abortion in jack pine caused by *Platylygus luridus* (Hemiptera: Miridae). *Canadian Entomologist*, **116**, 1213–18.

Reardon, P. O., Leinweber, C. L. & Merrill, L. B. (1974). Response of sideoats grama to animal saliva and thiamine. *Journal of Range Management*, **27**, 400–1.

Ridland, P. M. & Halloran, G. M. (1981). The influence of the lucerne bud mite (*Eriophyes medicaginis* Keifer) on the growth of lucerne. *Australian Journal of Agricultural Research*, **32**, 773–81.

Root, R. B. (1973). Organization of a plant–arthropod association in simple and diverse habitats: the fauna of collards (*Brassica oleracea*). *Ecological Monographs*, **43**, 95–124.

Schavilje, J. P. (1941). Reclaiming Illinois strip mined coal lands with trees. *Journal of Forestry*, **39**, 714–19.

Schlätzer, G. (1973). Some experiences with various species in Danish reclamation work. In *Ecology and Reclamation of Devastated Lands*, vol. 1, ed. R. J. Hutnik & G. Davis, pp. 33–64. New York: Gordon and Breach.

Sirur, G. M. & Barlow, C. A. (1984). Effects of pea aphids (Homoptera: Aphididae) on the nitrogen fixing activity of bacteria in the root nodules of pea plants. *Journal of Economic Entomology*, **77**, 606–11.

Smith, D. & Medler, J. T. (1959). Influence of leafhoppers on the yield and chemical composition of alfalfa hay. *Agronomy Journal*, **51**, 118–19.

Smith, S. M. & Ellis, C. R. (1983). Economic importance of insects on regrowths of established alfalfa fields in Ontario. *Canadian Entomologist*, **115**, 859–68.

Smolik, J. D. & Lewis, J. K. (1982). Effect of range condition on density and biomass of nematodes in a mixed prairie ecosystem. *Journal of Range Management*, **35**, 657–63.

Springett, B. P. (1978). On the ecological role of insects in Australian eucalypt forests. *Australian Journal of Ecology*, **3**, 129–39.

Stalter, R. & Serrao, J. (1983). The impact of defoliation by gypsy moths on the oak forest at Greenbrook Sanctuary, New Jersey. *Bulletin of the Torrey Botanical Club*, **110**, 526–9.

Stanlake, E. A., Eastman, D. S. & Stanlake, M. G. (1978). *Ungulate Use of Some Recently Reclaimed Strip Mines in Southeastern British Columbia*. British Columbia Ministry of Recreation and Conservation, Fish and Wildlife Report Number R-1.

Stanton, N. L. (1983). The effect of clipping and phytophagous nematodes on net primary production of blue grama, *Bouteloua gracilis*. *Oikos*, **40**, 249–57.

Stanton, N. L., Allen, M. & Campion, M. (1981). The effect of the pesticide carbofuran on soil organisms and root and shoot production in shortgrass prairie. *Journal of Applied Ecology*, **18**, 417–31.

Stanton, N. L. & Krementz, D. (1982). Nematode densities on reclaimed sites on a cold desert shrub-steppe. *Reclamation and Revegetation Research*, **1**, 233–41.

St. George, R. A. (1930). Drought-affected and injured trees attractive to bark beetles. *Journal of Economic Entomology*, **23**, 825–8.

Strobel, G. A. & Lanier, G. N. (1981). Dutch elm disease. *Scientific American*, **245**, 56–66.

Summers, C. G. & Coviello, R. L. (1984). Impact of *Acyrthosiphon kondoi* (Homoptera: Aphididae) on alfalfa: field and greenhouse studies. *Journal of Economic Entomology*, **77**, 1052–6.

Tahvanainen, J. O. (1972). Phenology and microhabitat selection of some flea beetles (Coleoptera: Chrysomelidae) on wild and cultivated crucifers in central New York. *Entomologica Scandinavica*, **3**, 130–8.

Tahvanainen, J. O. & Root, R. B. (1972). The influence of vegetational diversity on the population ecology of a specialized herbivore, *Phyllotreta cruciferae* (Coleoptera: Chrysomelidae). *Oecologia*, **10**, 321–46.

Tedders, W. L., Wood, B. W. & Snow, J. W. (1982). Effects of feeding by *Monelliopsis nigropunctata*, *Monellia caryella* and *Melanocallis caryaefoliae* on growth of pecan seedlings in the greenhouse. *Journal of Economic Entomology*, **75**, 287–91.

Tilman, D. (1978). Cherries, ants and tent caterpillars: timing of nectar production in relation to susceptibility of caterpillars to ant predation. *Ecology*, **59**, 686–92.

Tischler, W. (1955). Influence of soil types on the epigeic fauna of agricultural land. In *Soil Zoology*, ed. D. K. McE. Kevan, pp. 125–37. New York: Academic Press.

Todd, J. G. & Kamm, J. A. (1974). Biology and impact of a grass bug *Labops hesperius* Uhler in Oregon rangeland. *Journal of Range Management*, **27**, 453–8.

Ueckert, D. N. (1979). Impact of a white grub (*Phyllophaga crinita*) on a short grass community and evaluation of selected rehabilitation practices. *Journal of Range Management*, **32**, 445–8.

Urbanek, R. P. (1976). *Vertebrate and Floral Diversity on Strip-mined Land in Williamson and Saline Counties. Illinois.* M.A. Thesis, Southern Illinois University, Carbondale.

Urbanek, R. P. (1982). *Arthropod Community Structure on Strip-mined Lands in Ohio.* Ph.D. Thesis, Ohio State University, Columbus.

Urbanek, R. P. & Klimstra, W. D. (1986). Vertebrates and vegetation on a surface-mined area in southern Illinois. *Transactions of the Illinois State Academy of Science*, **79**, 175–87.

Van Hook, Jr. R. I. (1971). Energy and nutrient dynamics of spider and orthopteran populations in a grassland ecosystem. *Ecological Monographs*, **41**, 1–26.

Van Hook, Jr. R. I. & Dodson, G. J. (1974). Food energy budget for the yellow-poplar weevil, *Odontopus calceatus* (Say.). *Ecology*, **55**, 205–7.

Vogel, W. G. (1981). *A Guide for Revegetating Coal Minesoils in the Eastern United States.* U.S. Department of Agriculture, Forest Service General Technical Report NE-68.

Waage, J. K. & Bergelson, J. M. (1985). Differential use of pin and black cherry by the eastern tent caterpillar *Malacosoma americanum* Fab. (Lepidoptera: Lasiocampidae). *American Midland Naturalist*, **113**, 45–55.

Walstrom, R. J. (1983). Plant bug (Heteroptera: Miridae) damage to first-crop alfalfa in South Dakota. *Journal of Economic Entomology*, **76**, 1309–11.

Wargo, P. M. (1977). *Armillariella mellea* and *Agrilus bilineatus* and mortality of defoliated oak trees. *Forest Science*, **23**, 485–92.

Wargo, P. M. & Houston, D. R. (1974). Infection of defoliated sugar maple trees by *Armillaria mellea*. *Phytopathology*, **64**, 817–22.

Weaver, C. R. (1950). Improvement in hay yields resulting from control of the meadow spittlebug. *Journal of Economic Entomology*, **43**, 7–11.

Weber, B. C. & McPherson, J. E. (1985). Relation between attack by *Xylosandrus germanus* (Coleoptera: Scolytidae) and disease symptoms in black walnut. *Canadian Entomologist*, **117**, 1275–7.

Wheeler, Jr. A. G. (1974). Phytophagous arthropod fauna of crownvetch in Pennsylvania. *Canadian Entomologist*, **106**, 897–908.

Wiegert, R. G. (1965). Energy dynamics of the grasshopper populations in old field and alfalfa field ecosystems. *Oikos*, **16**, 161–76.

Wilkinson, R. C. (1983). Leader and growth characteristics of eastern white pine associated with white pine weevil attack susceptibility. *Canadian Journal of Forest Research*, **13**, 78–84.

Williston, H. L. (1958). Shortleaf versus loblolly pine in north Mississippi. *Journal of Forestry*, **56**, 761.

Wilson, L. F. & Moore, L. M. (1985). Vulnerability of hybrid *Populus* nursery stock to injury by the tarnished plant bug, *Lycus lineolaris* (Hemiptera: Miridae). *Great Lakes Entomologist*, **18**, 19–23.

Wilson, M. C., Stewart, J. K. & Vail, H. D. (1979). Full season impact of the alfalfa weevil, meadow spittlebug and potato leafhopper in an alfalfa field. *Journal of Economic Entomology*, **72**, 830–4.

Wolf, D. D. (1967). Yield reductions in reed canarygrass caused by frit fly infestation. *Crop Science*, **7**, 239–40.

Womack, C. L. (1984). Reduction in photosynthetic and transpiration rates of alfalfa caused by potato leafhopper (Homoptera: Cicadellidae) infestations. *Journal of Economic Entomology*, **77**, 508–13.

Yang, S. M., Rogers, C. E. & Luciani, N. D. (1983). Transmission of *Macrophomina phaseolina* in sunflower by *Cylindrocopturus adspersus*. *Phytopathology*, **73**, 1467–9.

Yates III, H. O. & Ebel, B. H. (1978). Impact of insect damage on loblolly pine seed production. *Journal of Economic Entomology*, **71**, 345–9.

Young, J., Warren, L. O. & Stephen, F. M. (1979). Pine growth and yield: influences of pine tip moth control. *Arkansas Farm Research*, **28**(2), 7.

5

The influence of fauna on plant species composition

R. J. WHELAN

Introduction

It is clear from the introductory Chapter of this volume that far too little is known about the impact of fauna on the vegetation structure of rehabilitated land. Only 0.3% of the published reports surveyed by Majer (see Chapter 1; p. 8) examined the impact of grazing animals and many of these reports concern the suitability of rehabilitated land as forage for domestic stock or game (see Chapters 10 and 12) rather than considering the impact of herbivores on the development or maintenance of the vegetation. Furthermore, the invasion of animals into disturbed areas is usually considered to be a passive phenomenon, following the development of appropriate vegetation, and the possible role of animals as agents of change in the developing vegetation is frequently ignored.

The need to consider herbivorous animals, and other consumers such as seed dispersers and pollinators, will depend largely upon the management objective of reclamation. With self-sustaining rehabilitation or complete restoration (see Chapter 1 for definitions) as an end-point, consumers will have an indispensable role in both the development of the vegetation after disturbance and also the maintenance of the final, equilibrium plant community. During the early stages of rehabilitation, an understanding of the facilitative functions of consumers (i.e. as in seed dispersal, nutrient cycling, etc.) may make management cheaper and more effective. Concern about the role of consumers in an ecosystem replacement program may, in fact, be confined to these facilitative functions and also to negative functions, such as suppression of planted seedlings and saplings by herbivores.

Given that land managers do need an understanding of the roles of consumers, can we define what contribution consumers make to attaining the 'species and complexity' objective of reclamation in the model proposed by Bradshaw (1983) and illustrated in Chapter 1, Fig. 1.4? Unfortunately, few studies have directly addressed the question of the importance of consumers in land rehabilitation, and there are therefore two possible approaches:

1. to wait until there are sufficient documented studies in the reclamation literature, and/or
2. to extrapolate from studies of processes operating in undisturbed communities and in communities recovering from natural disturbances.

In view of the large areas of disturbed land that have not yet been rehabilitated and the urgency with which most reclamation must now be conducted, this second approach is essential. Land managers involved in reclamation will need to base management decisions in relation to consumers and plant community structure on predictions and hypotheses drawn from a wide range of fields rather than on a body of case-studies. The aims of this Chapter are therefore as follows:

1. To examine published studies on undisturbed communities and natural disturbances to reveal general principles regarding the influence of fauna on the development and maintenance of plant species composition,
2. To integrate published studies on the importance of consumers on the development of plant communities in land reclamation programs into this framework,
3. To reveal gaps in our understanding which will be fruitful areas for further study in the field, and
4. To encourage feedback into theoretical ecology from the basic research that will need to be a part of any rehabilitation program. The development of a general theory of the role of animals in primary successions will come mostly through studies of rehabilitation of highly disturbed land (Game, Carrel & Hotrabhavandra, 1982).

Plant succession after disturbance

Introduction

The study of reclamation of a highly disturbed site is essentially the study of a primary succession (see introduction to Part 1; Game *et al.*, 1982; Bradshaw, 1983). The develoment of a community of organisms in a new habitat (e.g. new volcanic island, glacial retreat, etc.) or in a disturbed site follows a 'directional and continuous pattern of colonization and extinction by species populations' (Begon, Harper & Townsend, 1986). This process is known as a plant succession, and it often produces a sequence of readily identified plant communities, or 'seral stages', starting with a 'pioneer' community. Successions that occur on a new landform, primary successions, may take several hundred years before they reach anything recognizable as a mature, equilibrium community. Secondary successions, occurring after various degrees of disturbance to a pre-existing community (i.e. agricultural clearing, hurricanes, etc.), may be extremely rapid.

There is insufficient knowledge of the influence of animals in the development of communities through natural primary successions to permit clear predictions of the importance of animals in land reclamation (see Chapter 1), although ample opportunities exist for studies of abandoned quarries, gravel pits, disused roads etc. Predictions must therefore rest mostly on theories derived from studies on secondary successions following disturbances such as agriculture, fire, etc. The purpose of this section is to examine the various theories of plant succession and to put them in a context relevant to the land manager. The important mechanisms behind successional changes in vegetation will be highlighted and the possible roles for consumers in these mechanisms explored.

Theories of succession

Classical theory The widely accepted explanation of the mechanism of plant succession is illustrated by the following quotation from Odum (1953, p. 189):

'Thus, on a piece of bare ground, on a plowed field, or in a brand new body of water, physical conditions are extreme... the habitat has not been "mellowed" or moderated by the actions of living organisms. Therefore, only a few organisms which are especially adapted to those conditions are able to tolerate them and at the same time are able to quickly invade new areas. These organisms then develop a flourishing, though limited community, since there is little interspecific competition. Soon, however, these pioneer organisms modify the habitat by reaction on the abiotic environment, making conditions favourable for other organisms, which invade and outcompete the pioneers.'

This explanation focuses on the process of 'reaction', in which early invaders are necessary to pave-the-way for later invaders by altering the physical conditions of the site. There is little room in this explanation for a major role for consumers.

Model name and References	*Pictorial Representation*

Facilitation Model

Connell & Slatyer (1977) [also 'Classical' and 'Relay Floristics' – Clements (1916), Egler (1954)]

Initial Floristic Composition

Egler (1954)

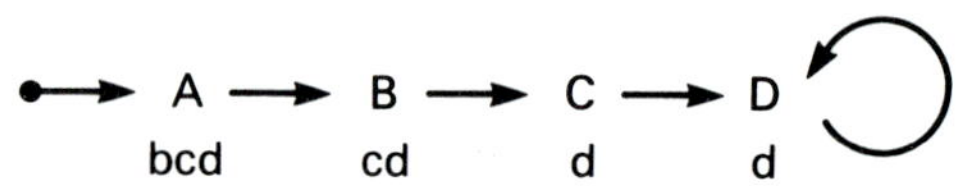

Tolerance Model

Connell & Slatyer (1977) [also 'Competitive Hierarchy' – Horn 1976]

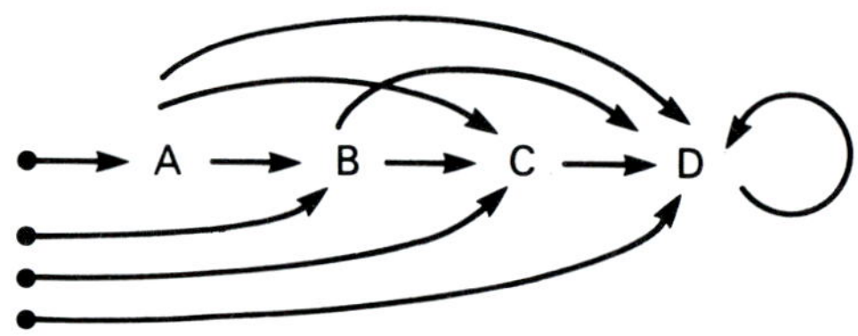

Inhibition Model

Connell & Slatyer (1977)

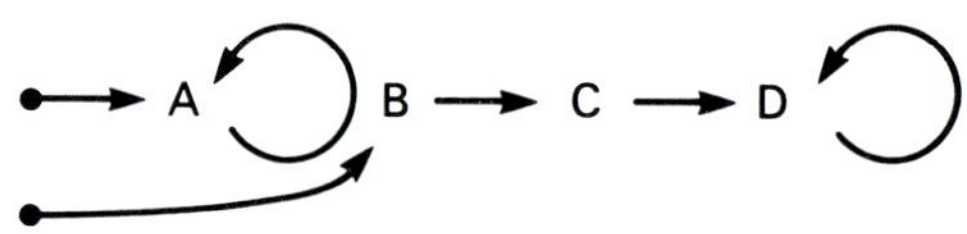

Chronic Disturbance

Horn (1976)

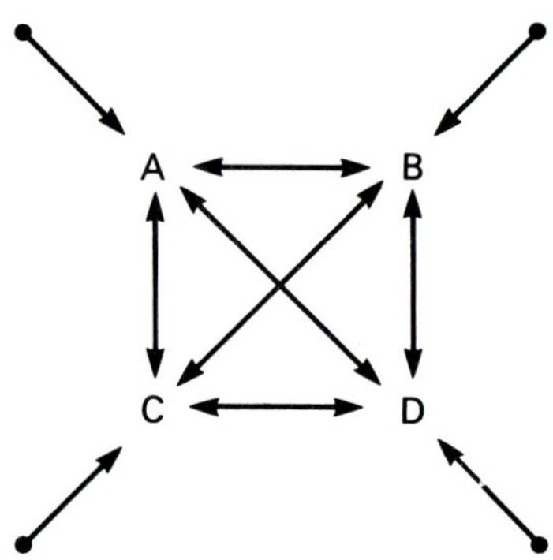

Fig. 5.1. Descriptions of various models explaining successional replacement sequences. Upper case letters represent dominant species defining a particular seral stage, lower case letters represent sub-dominants present within the seral stage. A circular arrow represents a stable, climax community. Dotted arrows represent alternative starting points for a successional sequence after disturbance (from Noble, 1981).

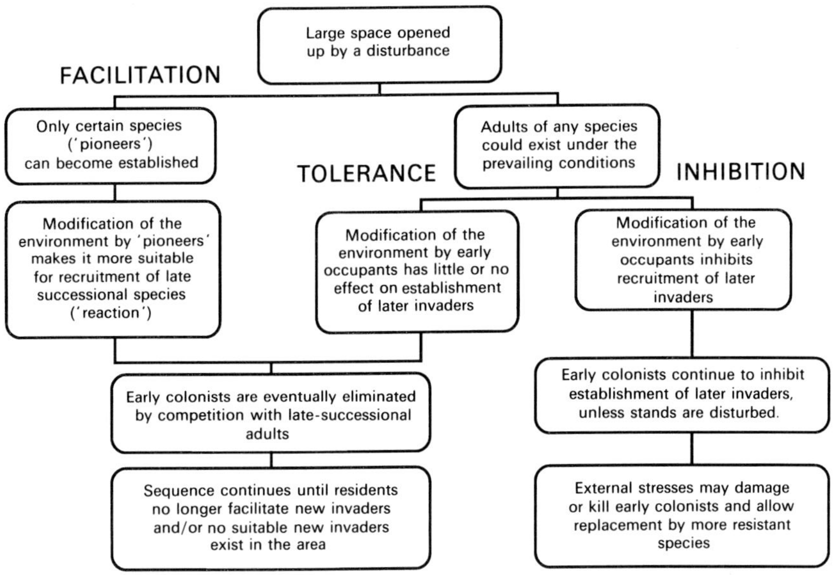

Fig. 5.2. Descriptions of the principal mechanisms that distinguish between the three main models of succession (modified from Connell & Slatyer, 1977 and Begon *et al.*, 1986).

Alternative theories There has been considerable controversy about the generality of this model of plant succession (Drury & Nisbet, 1973; Connell & Slatyer, 1977; MacIntosh, 1980) and various alternative models have been proposed. These are summarized in Figs. 5.1 and 5.2 (see also Noble, 1981).

One major area of discussion centres on the importance of the process of 'reaction', that is, the degree to which early successional plant populations 'pave-the-way' for later invaders. This is the essential feature of the Facilitation Model (see Fig. 5.2). It is perhaps generally accepted that the process of reaction is important early in primary successions, where physical conditions are so extreme that they inhibit establishment of plant propagules. However, when physical conditions are sufficiently ameliorated, either in a secondary succession, or later in a primary succession, biological processes may be predominant in determining the success of invasions of new species. Thus inhibition of potential colonists by competition may be the greatest effect of a well-developed pioneer plant population (Begon *et al.*, 1986). This view has been supported by observations of long-stable, seral stages, dominated by shrubs or herbs, which appear to be inhibiting succession to woodland or forest (Niering & Egler, 1955; Niering & Goodwin, 1974;

Clemens, Bradley & Gilbert, 1984; Buckley, Chilton & Devonald, 1981). Many experimental studies in marine intertidal systems (e.g. Dean & Hurd, 1980; Sousa, 1980) and some in old-field systems (McCormick, 1968), in which early colonists have been removed, also support this Inhibition Model and suggest that it may have widespread applicability.

Several studies of land rehabilitation also suggest that populations of pioneer plant species that invade a site soon after disturbance inhibit the colonization or growth of later invaders (Ringe, Graves & Wittwer, 1984). Clemens *et al.* (1984) concluded that woody plants should be planted in urban demolition sites as soon as possible to avoid competition from the perennial grasses and herbs that become established early in the succession.

Egler (1954) contrasted his Initial Floristic Composition Model with the Facilitation Model (which he called 'Relay Floristics'), suggesting that representatives of all seral stages may be present very soon after disturbance, and variations in life-history characteristics among groups of species explained the succession. The Tolerance Model (Connell & Slatyer, 1977) states that later invaders simply tolerate the conditions at the site and eventually outcompete the pioneers. The Chronic Disturbance Model (Horn, 1976) pertains to areas suffering chronic disturbance in small patches, proposing that any species is likely to invade an opening and locally a dominant species may therefore be replaced by any number of other species. This model applies to the phenomenon of 'patch dynamics', reviewed by Pickett & White (1985). The processes producing these kinds of changes within plant communities are considered below, under the heading 'The Regeneration Niche'. Recurrent disturbances (e.g. fires) may also occur over wide areas, rather than in small patches within an otherwise undisturbed community. In fire-prone ecosystems, recovery in a given sample area within the burned community may follow the Chronic Disturbance Model in that the species appearing in the sample are a subset of those present in the community before disturbance. It can be argued that it is inappropriate to label these patterns of community change 'succession'.

The lack of a single, generally applicable model of plant succession means that there can be no precise prediction made about the way succession will occur in a given rehabilitation site. It is therefore necessary to extract, from the various models discussed above, the relevant questions that need to be asked about any attempt at site restoration. These are as follows:

1. Is the appearance of later invaders limited by poor dispersal into the site?
2. Are later successional stages inhibited by inappropriate physical conditions?
3. Are pioneer plants required to pave-the-way for later invaders?
4. Will pioneer plants inhibit the establishment of later invaders?
5. Do the pioneers disappear because of competition from later invaders?

Effects of consumers on a succession

How might the actions of animals be involved in the answers to these question? Many authors view animal invasions into a disturbed site simply as a passive consequence of the development of appropriate vegetation stages (e.g. Bradshaw, 1983; Murdoch, Evans & Peterson, 1972; Smith, 1940). However, other authors have given animals an active role in the processes described above (Chapman, 1931; Knapp, 1974; Price, 1975; Crawley, 1983). Animals could either accelerate or retard a plant succession through their activities as herbivores, dispersers of seeds and agents of soil disturbance.

Introduction of new species: seed dispersal One effect of animals on a plant succession is the introduction of seeds of new species. Many animals disperse viable seeds over long distances (Murray, 1986), including into recently disturbed sites such as old fields (McDonnell & Stiles, 1983) and burnt areas (Whelan, 1986). Strang (1974) suggested that light grazing by mammals in African grasslands introduces propagules of tree species, thereby facilitating a succession to woodland. The same grazing animals might also create the conditions in the grassland necessary to allow the establishment of these pioneers. For example, in Africa, seeds of *Acacia nilotica*, brought from woodlands to grasslands in cattle faeces, germinate and establish in the favourable microenvironment of the dung pat (Strang, 1974). In Denmark, Jensen & Nielsen (1986) have shown that seed-eating rodents are responsible for the appearance of oaks (*Quercus robur* and *Q. petraea*) in the succession of heathland to woodland. Radioactively labelled acorns were buried by rodents in the heath, up to 34 m away from oaks.

Amelioration of soil conditions Establishment of a plant propagule is not ensured by its dispersal to a site. Physical conditions within highly disturbed sites are extreme and likely to permit germination and survival of only a few tolerant species. Bradshaw (1983) pointed to the importance of earthworms in facilitating a primary succession by generating an accumulation of fine soil on the surface of urban clearance sites in Liverpool. The activity of decomposers and burrowing animals is likely to play an important part in the amelioration of extreme soil conditions (see Chapter 2) and thereby allow colonization by later successional species.

Nutrient limitation, particularly lack of nitrogen, is a feature of the early stages of primary successions (Crocker & Major, 1955). This is also true in the rehabilitation of highly disturbed sites (Bradshaw, 1983). The questions of how herbivores contribute to cycling of nutrients early in succession and how this contribution changes during a succession are potentially very important yet they have rarely been addressed (see Chapter 3 and Lamb, 1985). Schowalter, Webb & Crossley (1981) showed that the total biomass of foliage arthropods was two times higher in undisturbed forest than in first-year regeneration after clearcutting in North Carolina. This difference was reversed in the second year after clearcutting due to an enormous increase in biomass of sucking bugs (Homoptera) in the clearcut site. The general conclusion of this study was that: 'changes in canopy arthropod community structure following disturbance alter the flow of nutrients through the arthropod community.' Blanton & Ewel (1985), in an experiment in Costa Rica, showed that the amount of leaf material cut per day by *Atta cephalotes* (leaf-cutter ants) was influenced by vegetational complexity. It was over an order of magnitude greater in a monoculture stand of cassava than in a complex forest site (Fig. 5.3). It is not at all clear, however, what impact changes such as these will have on plant community composition in a succession. Schowalter (1981) suggested that insect herbivory will facilitate succession by accelerating the decline of early successional plant species and mobilizing nutrients bound up in them.

Luxton (1982) commented that additional nutrients need to be supplied to the vegetation growing on coal shale tips because of a lack of decomposers. Nutrients accumulated in dead plant material may be lost to the system via wind and water erosion rather than being recycled (Marrs, Granlund & Bradshaw, 1980). Light grazing can increase the overall productivity of the vegetation in a rehabilitation site (Schuman *et al.*, 1984; Williamson, 1981) by reducing the litter build-up and probably by facilitating nutrient cycling. The potential importance of herbivores in the recycling of nutrients makes this a profitable area for further investigation. For example, planting legumes in a site may facilitate the

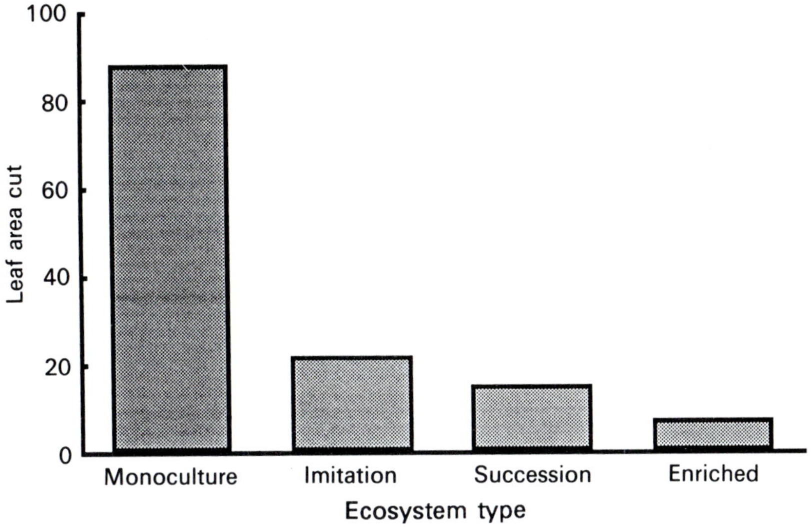

Fig. 5.3. Amount of leaf area harvested per m^2 of land surface area per day by leaf-cutting ants in four experimental ecosystems in Costa Rica. 'Monoculture' was a cassava crop, 'imitation' was a planted combination of non-indigenous species chosen to mimic the physiognomy of the natural vegetation, 'succession' was the normal successional forest developed on the unmanipulated site, and 'enriched' was the natural succession vegetation made more complex with additional input of seeds of new species to increase complexity and species diversity.

fixing of nitrogen, but introduction of herbivores may be necessary to make it available to other components of the developing community.

Retardation of succession Many studies, particularly those describing grassland–woodland boundaries in North America and Africa, have indicated that grazing animals have a central role in maintaining grassland (Daubenmire, 1968; Dyksterhuis, 1948; Strang, 1974; Smart, Hatton & Spence, 1985). These grazers are seen as producing a 'disclimax' (Clements, 1928), permanently preventing the successional development of the vegetation from grassland to shrubland and woodland. This effect of herbivores is also at work in many other successional systems. For example, old orchards, surrounded by undisturbed forest and woodland in water catchment areas near Wollongong, Australia, are currently dominated by bracken fern (*Pteridium esculentum*) and blady grass (*Imperata cylindrica*). Invasion of pioneer trees and shrubs, particularly *Acacia* and *Eucalyptus* species, is inhibited by the grazing of wallabies (*Wallabia bicolor*) and wombats (*Vombatus ursinus*) (Lambkin, 1986; Sim, 1986).

Although mammalian herbivores are more conspicuous as agents of vegetation change, recent studies have shown that insects can be just as important. In an experimental study on an old-field succession in Ontario, McBrien, Harmsen & Crowder (1983) demonstrated that herbivory by phytophagous beetles in 5-year-old fields reduced the percent cover of the relatively stable, dominant *Solidago* and permitted populations of early successional species to increase their percent cover. A study by Stinson (1983: cited in Brown, 1985) in plots of old-field succession at Silwood Park, England, involving the chemical exclusion of all insects from a site already protected from birds and mammals, produced dramatic differences in the plant community. Representatives of new plant species were accumulated more quickly in the insect-free site.

A study by Humphries (1980) illustrates this principle in operation in land reclamation. Seed predation by small rodents and grazing by rabbits limited the development of the plant community on an old quarry site in Derbyshire, England. Recent studies of the rehabilitation of the margins of newly constructed railway lines near Wollongong, New South Wales, also suggest that mammalian herbivores (rabbits, wombats and wallabies) are responsible for inhibiting tree reinvasion, keeping the plant community as a closely-cropped sward of introduced grasses and clover, initially planted to provide a rapid cover (Newland, 1986).

Facilitation of succession Crawley (1983) stated that: 'There is unlikely to be "an effect" of herbivores on succession.' In contrast to the above examples of herbivores retarding succession, Connell & Slatyer (1977) and White (1979) have highlighted a facilitatory role of herbivores in plant succession, namely the creation of gaps in a dense seral stage permitting the invasion of individuals of later-successional stages. Good support for this view comes from experimental studies in various ecosystems (e.g. Knapp, 1974; McBrien *et al.*, 1983; Lubchenco, 1983).

Various studies have shown that recruitment to populations of some plant species in Australian eucalypt forest and woodland is permitted only in the year after fire, before herbivory by reinvading herbivores becomes severe (Table 5.1). Does the grazing pressure that develops after this short period of 'release' affect the development of the post-fire plant community? The answer to this question needs long-term exclosure studies. Mature stands of *Acacia binervata* in unburnt forest on the east coast of Australia, near Wollongong, recruited densely after wildfires 19 years ago (1968) and are now declining (personal observations). Most of

Table 5.1. *Mortality due to insect feeding within a cohort of seedlings during the first and second years after fire in Western Australian woodland*

	First year			Second year		
Plant species	Eaten	Not eaten	% eaten	Eaten	Not eaten	% eaten
All species combined	68	1997	3.29	5	830	0.60
Jacksonia sternbergiana	36	256	12.33	4	113	3.41
Hardenbergia comptoniana	9	54	14.29	0	24	0
Petrophile linearis	8	364	2.15	1	214	0.46

Note: Seedlings in the 'not-eaten' category may include seedlings which died of other causes (from Whelan, 1978).

the seedlings which appear annually are grazed to the ground by swamp wallabies. Although it is not known what effect continued recruitment of *A. binervata* would have on the rest of the plant community if grazers were excluded, it is possible that this herbivory facilitates a successional shift in vegetation from a fire-prone, sclerophyllous plant community to a more mesophytic, closed forest (Whelan, 1986).

Feeding by insects has been shown to have a similar effect in conifer successions after fire in the Rocky Mountains of North America (Amman, 1977). Lodgepole pine regenerates densely from seed after wildfire. When these trees reach a suitable stage of development they become susceptible to attack by the mountain pine beetle and, in the absence of further fire, the removal of these pines permits more rapid growth of the later successional species – Douglas fir, subalpine fir and Englemann spruce.

Grazing animals can certainly create gaps in the cover of a pioneer community developing in a highly disturbed site. Game *et al.* (1982) showed that temporary grazing on the developing vegetation of abandoned surface coal mines in Missouri, U.S.A. caused a fission of coalescing patches of vegetation. It is not clear from studies such as this what effect, facilitatory or inhibitory, the creation of open patches by herbivores might have on the subsequent development of the plant community in a mine site. Longer term studies are needed. However, the process does mimic the effects of grazers in creating space in grassland swards or in marine intertidal communities and thereby maintaining high species diversity. In an experimental study, Williamson (1981) found that cattle grazing increased the species diversity of rehabilitated, surface-mined grasslands in North Dakota, perhaps by allowing establishment in the mine site of plant species that would otherwise have been excluded by competition.

Effects of plant palatability A plant community developing in a patch of regenerating vegetation will be available to a range of herbivores present in the surrounding, undisturbed community. It is therefore pertinent to ask: Are the pioneer plants in the regenerating patch more palatable than later successional plant species? Cates & Orians (1975) proposed that weedy species common as invaders in newly disturbed sites will not devote resources to anti-herbivore defences. The reasoning is that such species exploit newly disturbed sites for such a short time that they avoid a build-up of herbivores and have therefore not evolved defence mechanisms. Experimental tests of this hypothesis have yielded contradictory results (Cates & Orians, 1975; Otte, 1975); the contradiction explained by Maiorana (1978) as simply demonstrating that early successional herbivores prefer early successional plants and late successional herbivores prefer late successional plants. Nevertheless, this is an important line of investigation for future studies of vegetation rehabilitation. The appropriate question is whether the herbivores of the surrounding, undisturbed vegetation prefer the successional plant species of the regenerating disturbed site and congregate there in high densities. This does seem to be the case in patches of post-fire regeneration (Leigh & Holgate, 1979; Whelan & Main, 1979; Quinn, 1986). The possibility that post-fire vegetation is more palatable because of a pulse of nutrients made available by the fire introduces the important question of whether fertilization of rehabilitation sites has the same effect.

Related to the question of higher palatability of early successional species is the possibility, illustrated in the study of leaf-cutter ants described above (Blanton & Ewel, 1985), that plant monocultures attract higher levels of herbivory than mixed species stands (Pimentel, 1961; Root, 1973). However, there is some debate about the generality of this situation, and an alternative possibility is that the high densities of plants typical of monocultures satiate the available herbivores resulting in less damage per plant, rather than more. The answer probably lies in the characteristics of the herbivore species. Populations of specialist herbivores, especially insects, may be able to locate available host plants readily in a monoculture and will therefore increase rapidly. Similarly, generalist, mobile herbivores may recruit numbers rapidly to exploit a dense, monoculture resource. Mixed species stands, in which host plants are interspersed among non-palatable species, may give individual plants some protection from their specialist herbivores. In contrast, herbivores that are territorial or sedentary and are long-lived in relation to the plant's generation time may not be capable of building up numbers in time to exploit an abundant resource.

The magnitude of the effects of preferential grazing on a succession will depend to some extent on the importance of the preferred plant species to the development of the plant community in the site. If the Facilitation Model of succession (see above) holds in a particular site, then removal of the palatable species by herbivores will have a detrimental effect. Legumes in nutrient-poor, Australian eucalypt forests may be especially important in directing nitrogen into these systems (Shea & Kitt, 1976). Herbivory which limits recruitment of legume populations may be expected to have a detrimental effect on the plant community (Shea, McCormick & Portlock, 1979), while less severe feeding on leaves and litter will redistribute the nitrogen, and therefore be beneficial. Furthermore, the apparent antagonistic interaction between dense legume regrowth and the virulence of the dieback fungus, *Phytophthora cinnamomi* (Shea, Gillen & Kitt, 1978) point to the likely complexity of interactions between different components of the plant community. Dieback is an important pathogen favoured by bauxite mining activities in southwestern Australia (see Chapter 16 and Mulcahy, 1984). Removal of legume populations by the preferential grazing of herbivores would result in greater fungal infection of tree and other shrub species that are planted in the site or invade naturally.

Pollination The pioneer plant community developing in a disturbed patch would be expected to differ substantially from the surrounding, undisturbed vegetation. Plants will be either deliberately introduced to establish cover rapidly on the new site, or they will be well-dispersed, weedy species, including many alien species, typical of disturbed sites in the region. One consequence of this will be a very patchy distribution of plants from the point of view of a pollinator. Specialist pollinators may move between patches, exploiting nectar and transferring pollen among the plants flowering within each patch and also between patches (Feinsinger, 1976). However, generalist pollinators will forage on many appropriate plant species both within the patch and also in the surrounding undisturbed plant community. Such movements between different plant species are likely to reduce the effectiveness of pollination by moving pollen between plants of different species ('competition through interspecific pollen flow'; Waser, 1983).

Are there specialist pollinators of early successional plant species? Specialization by a pollinator seems a risky strategy when food plants might be distributed in a highly variable pattern in both space and time. In fact, several studies suggest that pioneer plant species are commonly obligate, or at least facultative, selfers. That is, they are capable of setting viable seed by self-fertilization without the aid of pollinators. This is seen to have evolved in response to the uncertainty of pollination in successional environments (Levin & Kerster, 1974). However, as plants with more specialized pollination requirements invade the post-disturbance community, seed set through outcrossing may become more problematic. Small gene pools and a lack of specialized pollinators may limit populations of outcrossing species that are not also represented in the surrounding, undisturbed community. This is a potential problem for the success of a rehabilitation program which land managers should be aware of, although I am not aware of any study which demonstrates unsuccessful establishment of a plant population due to inadequate pollination.

Consumers and the maintenance of undisturbed communities

'Ecosystem replacement' (defined in Chapter 1) may not have indigenous, mature plant communities as its final objective. Moreover, few programs aimed at complete, or self-sustaining, rehabilitation have yet reached their end-points. Nevertheless, long-term planning of such rehabilitation (see Chapter 6) must include an understanding of the complexities of ecosystem functioning (Muston, 1986), including the realization that consumers play a central role in the maintenance of the plant species composition of the ultimate community that develops on a site.

An extremely important role of consumers (i.e. herbivores, seed dispersers and pollinators) in undisturbed communities is the maintenance of species diversity. This section will therefore examine the ways in which consumers affect coexistence of species in undisturbed communities.

Herbivory

Coexistence of species Shugart (1984) argued that when trees within a community are under competitive conditions, herbivory is likely to permit coexistence. This process of predators preventing competitive exclusion among prey populations is a common feature of marine intertidal communities (Paine, 1966; Connell, 1971). Such an effect is most likely with density-dependent, generalist herbivores. This means that feeding will be concentrated on those plant populations at the highest relative abundance, creating gaps which allow rarer species to persist. When the preferred food becomes rare itself, through intense herbivory, the generalist herbivore will be able to switch to the next most abundant species. Begon *et al.* (1986 – Chapter 19) provide an excellent basic discussion of the impact of herbivores in plant communities and Crawley (1983) provides an outstanding review of the current state of knowledge, but it is clear that many of the long-presumed effects of herbivory in natural ecosystems (see Ellison, 1960) remain untested or equivocal. The impact of herbivory on the dynamics of plant communities 'involves the most difficult question in evolutionary biology, population biology and ecology' (Janzen, 1979). Nevertheless, there are several classic studies which illustrate the potentially enormous effects of herbivory on plant communities (Darwin, 1859 – Chapter 3; Tansley & Adamson, 1925; Watt, 1962, 1981; Harper, 1969). Because of the ubiquity of herbivores, their effects on plant species composition are often not recognized until enclosure or exclosure experiments are conducted (Fig. 5.4). In most instances, exclusion of herbivores has led to a marked simplification of the vegetation, with many species disappearing, probably through competitive exclusion; however, exact effects may depend upon the intensity of grazing before exclusion.

Although large herbivores such as mammals are generally thought of as the prime agents of vegetation change, insects may also be of great importance (Springett, 1978). For example, a study by Burdon & Chilvers (1974) indicated that fungi and insects on two *Eucalyptus* species were density-dependent pathogens/herbivores and therefore had the potential to prevent competitive exclusion by preferentially attacking whichever species was at the higher relative abundance. Morrow & La Marche (1978) showed that preferential grazing by insects on a faster-growing *Eucalyptus* species (*E. stellatula*) reduced its growth more than its slower-growing, potential competitor (*E. pauciflora*) (see also Morrow, 1977).

Janzen (1970) postulated a major role for specialist insect seed

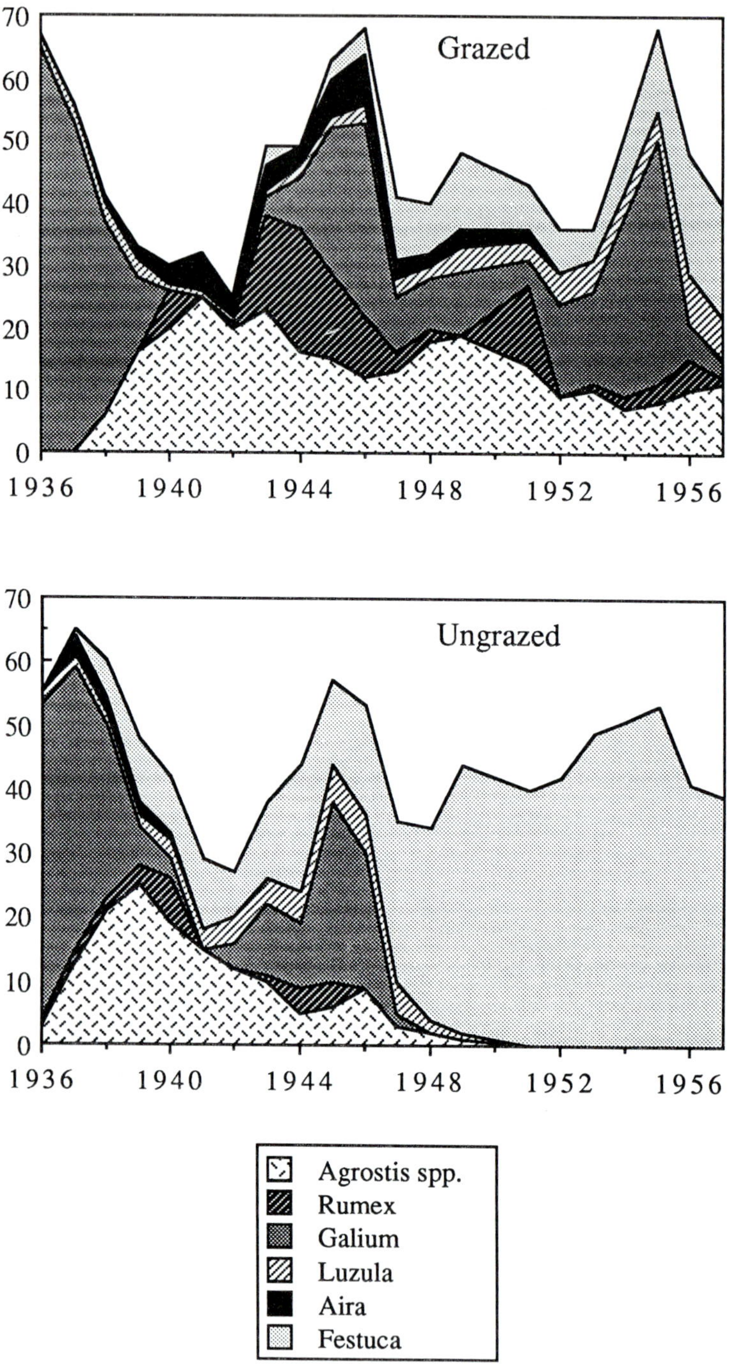

Fig. 5.4. Twenty years of changes to vegetation composition as a result of exclusion of rabbits from an English grassland (from Watt, 1960, and Miles, 1979).

predators in the maintenance of high diversity among tropical forest trees. Intense predation of seeds under and near the parent tree might permit recruitment of other species in this zone, while seedlings of the parent species are likely to recruit only at some distance. Connell (1971) argued along similar lines, but focused on seedlings rather than seeds. Although this hypothesis is appealing, experimental tests have produced various results (Hubbell, 1980; Janzen, 1972; Wilson & Janzen, 1972).

Elimination of species Intense herbivory is likely to cause a local loss of species. Many studies of overgrazing by domestic stock in rangelands demonstrate this quite clearly (Crawley, 1983). Can similar situations occur with native herbivores under natural conditions? Some herbivore-exclusion studies in both terrestrial and intertidal communities (see Begon *et al.*, 1986) suggest that, under certain conditions, herbivory can reduce species diversity by consuming preferred species totally and preventing them from re-establishing. Several studies suggest that herbivory can produce either enhanced or reduced species diversity depending on the palatabilities and physical structures of the plant species concerned and on the taxa and density of the grazers (Fig. 5.5) (During & Willems, 1985).

Frelich & Lorimer (1985) argued that poor recruitment of hemlock (*Tsuga canadensis*) in some forests in Michigan, USA, is related to browsing by deer. Areas where hemlock regeneration was apparent were characterized by low deer numbers due to hunting. The loss of *Acacia rostellifera* from many parts of Rottnest Island, Western Australia, has been attributed to intense grazing by the quokka (*Setonyx brachyurus*) when the *Acacia* was at a susceptible stage of development, namely as seedlings appearing in response to fire (Baird, 1958; Storr, 1963; Whelan, 1978). Similar effects were noted by Whelan (1978) and Leigh & Holgate (1979) where intensive grazing by marsupials killed established plants of some woody perennial shrubs that were resprouting after a fire. Caged plants regenerated successfully. Predation on seeds may also have a negative effect, causing local extinctions. For example, Pickard (1982) suggested that the introduction of seed predators (*Rattus rattus*) to Lord Howe Island is likely to prevent future recruitment to *Howea* palm populations.

It may be argued that some of these examples are special cases, in which herbivores are either introduced, or are at high abundances due to unnatural circumstances. However, land rehabilitation sites may be ideal

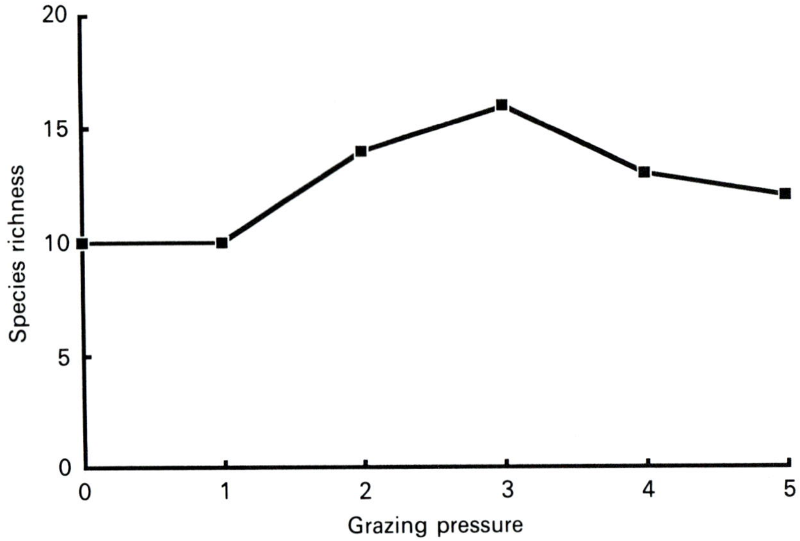

Fig. 5.5. Influence of intensity of grazing and plant species richness of communities on sand dunes. Grazing pressure increases from 0 to 5 (modified from Zeevalking & Fresco, 1977).

situations for such conditions, and this will be explored in a later section.

Nutrient cycling and plant community composition

Several studies on ecosystem dynamics have emphasized the possible importance of herbivores in the release of nutrients bound up in plant biomass in a community (Mattson & Addy, 1975; Springett, 1978; Main, 1981*a,b*). Forest trees, and presumably other plants, exhibit a substantial degree of internal nutrient cycling ('biochemical cycling'; Attiwill, 1980) in which nutrients such as phosphates are withdrawn from leaves before abscission. For example, Attiwill (1980) showed that 34 to 42% of the phosphorus requirements of hardwood trees are supplied by this internal cycling (Fig. 5.6). The effective breaking of this cycle by herbivory, and the consequent diversion of nutrients into an extra-tree cycle ('biogeochemical cycling'; Attiwill, 1980), would make these nutrients available to other components of the plant community, perhaps shorter-lived, understorey species. Hollinger (1986) showed that herbivory by the California oak moth (*Phryganidia californica*) more than doubled the flow of nitrogen and phosphorus from *Quercus agrifolia* to the forest floor compared with a non-outbreak year. Nutrients were much more rapidly mobilized from insect frass than from leaf litter.

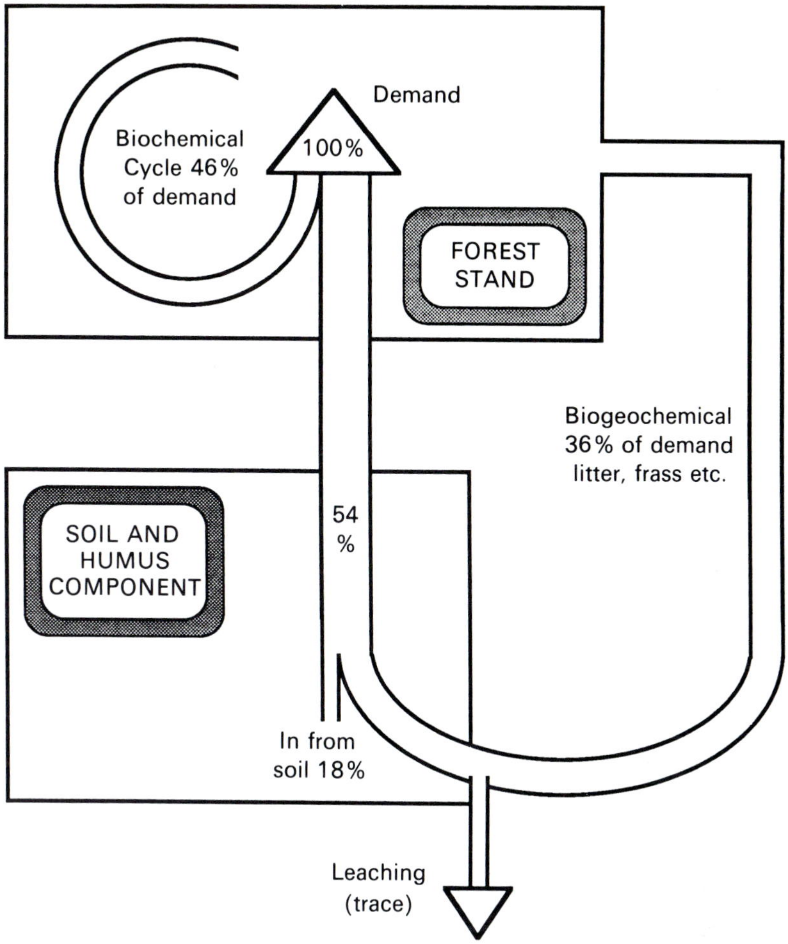

Fig. 5.6. Internal (biochemical) and external (biogeochemical) cycling of phosphorus in *Eucalyptus obliqua*. An increase in the proportion of phosphorus going into the external cycle, caused by herbivory on leaves, increases the availability of this nutrient to other components of the plant community (modified from Attiwill, 1980).

Although herbivore-mediated biogeochemical cycling of nutrients has been demonstrated, it is not clear what effect it has on plant species composition. It is reasonable to predict that in the absence of herbivore activity in a rehabilitation site, fewer plant species might be supported because of the 'locking up' of much of the nutrient pool within trees.

Chapter 3 has already examined the role of soil and litter invertebrates in nutrient cycling during land rehabilitation. In forests and woodlands of dry regions, many of the nutrients in an ecosystem are bound in dead plant material such as leaf litter, twigs, fallen logs and in standing dead boles and branches. Under these conditions, where a coarse, well-aerated litter layer and long dry seasons reduce the effectiveness of nutrient cycling by mesofauna, bacteria and protozoa, litter feeding by macrofauna may be of great importance. Availability of nutrients for plant growth may depend on the activities of organisms such as termites, millipedes, wood-boring beetles and other litter- and wood-feeders. In the dry sclerophyll forest of southwestern Australia, for example, I have observed that a widespread and abundant grasshopper genus (*Goniaea* spp.), which is cryptic against a background of *Eucalyptus* leaf litter, also eats dead leaves and may contribute significantly to nutrient cycling.

The 'regeneration niche'

Herbivory is not the only avenue through which animals can have an effect on plant community structure. Grubb (1977) pointed out that coexistence of long-lived plant species may be determined at the stage of the life-cycle during which seeds germinate and seedlings become established. Thereafter, plants of many species will be relatively tolerant of pressures such as drought, shading and competition for nutrients. If plant species in a community have different requirements for germination and establishment, then the number of species able to co-exist will be determined by the number of different 'regeneration niches' available in the site.

Animals can provide 'regeneration niches' (Grubb, 1977). Platt (1975), for example, showed that the mounds of bare soil produced by badger diggings provided an opportunity for recruitment of 'fugitive' prairie plant species which would otherwise have been unable to germinate and establish in the dense prairie sward. Harper (1977) documented various examples of recruitment to and persistence of plant populations that depended on animal-caused disturbances such as dung heaps, hoof prints, molehills etc. In Australia, numerous animals produce scratchings and mounds, including insects (ants, termites), birds (mallee fowl, lyrebirds), and monotremes and marsupials (echidnas, wombats, bettongs) (Main, 1987). However, as in other parts of the World, little is known of the importance of such diggings for plant regeneration. It is therefore impossible to assess the impact of these animals on a plant community and, consequently, the need to reproduce their effects in reclamation sites.

In some regions, ants may be especially important in allowing plant species to persist. Various authors have pointed to the abundance of myrmecochorous (i.e. seeds dispersed by ants) plant species in Australian heaths (Berg, 1981; Majer, 1982). Shea *et al.* (1979) noted that post-fire germination of legume species in *Eucalyptus marginata* (jarrah) forest in Western Australia occurred in discrete clumps. They attributed this spatial pattern to the storage of intact seeds by ants. Davidson & Morton (1981, 1984) suggested that this phenomenon may permit recruitment to populations of certain plant species, which would otherwise be out-competed. Thus these animals are both dispersing the seed and also providing the safe-site for germination and establishment. Christensen (1980) reported that similar clumps of seedlings occurring after fire in southwest Australian forest result from burial of seeds by *Bettongia penicillata* (a small macropod marsupial).

The overall effect of animals providing safe sites for regeneration seems likely to be a maintenance of high plant species diversity. Habitat heterogeneity on a fine scale, caused by animals, permits species coexistence on the scale of the whole plant community, because early-successional elements are able to coexist in the disturbed parts of a mature community (Laycock, 1958; Denslow, 1985).

Pollination and seed dispersal

Crawley (1983) pointed to the importance of pollinators in maintaining the species diversities of plant communities. Plants that are highly specialized for pollination by one or only a few species of pollinator will clearly suffer a reduction in relative abundance in the absence of the pollinator. There are few experimental tests of this logical expectation in the literature. A similar prediction could be made for the importance of seed dispersers in plant communities. The removal of seeds from rehabilitated areas has already been discussed in Chapter 4. Mention has already been made, in the present Chapter, of the potential importance of ants and small mammals in dispersing seeds within a plant community, and placing them in sites suitable for germination and establishment. Absence of seed dispersers may therefore be expected to cause a decline in abundance of those plant species depending on them for effective dispersal. In an interesting illustration of this principle, Temple (1977) suggested that the continuing decline in the abundance of the tree, *Calvaria major*, on Mauritius is due to the extinction of its specialist seed disperser, the dodo.

These lines of argument for the importance of seed dispersers and pollinators are the basis for the emphasis placed by Gilbert (1980) on animals he labelled 'mobile links' which may maintain community structure in neotropical plant communities. He argued that these mobile animals link a number of different plant species in spatially separate components of a community. A single species of seed-disperser, for example, could be the primary agent of seed dispersal for several widely separated plant species at different times of year. The loss of just one such animal species from a community would therefore have an inordinately large impact on the community. Similarly, the loss of a single plant species from a year-long series of consecutively-flowering species in a community ('keystone mutualists'; Gilbert, 1980) could lead to the local extinction of those populations of pollinators and seed dispersers which must have a food supply throughout the year. This would have feed-back repercussions on all plant species in the series.

Important considerations for reclamation studies

It is important to be able to apply the above theories and case-studies to a given rehabilitation program. However, any attempt at general application is seriously confounded by the wide variety of disturbances to be rehabilitated and the equally wide variety of end points that may be required of rehabilitation programs within a single type of disturbance. For example, land to be rehabilitated may be a patch of disturbance surrounded by undisturbed vegetation (e.g. Western Australian bauxite mine sites; Fig. 1.1) or a quarry in the midst of highly-disturbed, agricultural land. The ultimate aim of rehabilitation in either of these situations may be ecosystem replacement (silviculture, grazing land, recreation land) or complete restoration, even ecological refuges (Bradshaw, 1979; see also Chapter 1).

It is clear from the above discussion that animals are expected to have marked impacts on the development and maintenance of a plant community on a disturbed site. Furthermore, these impacts may be either detrimental or beneficial in the context of the reclamation objective for the site. It is possible to summarize the main categories of beneficial and detrimental impacts (Table 5.2) and to attempt an association of these impacts with certain characteristics of the reclamation site (Table 5.3). Valuable parallels may be drawn between the problems of management of rehabilitation sites after disturbance and the recent discussion by Main (1987) about the problems facing management of remnant patches of natural vegetation in the wheatbelt region of Western Australia.

Patch size

One of the features of disturbed communities is patchiness. From individual tree falls to old fields and burned sites, the regenerating

Table 5.2. *Summary of the various impacts of animals that may be important in ecosystem development and maintenance in rehabilitation*

Ecological Process	
Acquisition of species	Maintenance of species
(1) Beneficial effects of animals	
Herbivory redistributes nutrients bound up in tissues of pioneers.	Herbivory aids nutrient cycling within a site.
Long-distance seed dispersers bring seeds into site.	Local seed dispersers get seeds to safe sites for germination and establishment.
Herbivory and other disturbances create gaps in dense pioneer community (good if Inhibition Model of succession holds).	Herbivory prevents competitive exclusion among plant species.
Soil disturbances enhance conditions for seedling establishment.	Pollinators ensure adequate plant reproduction
	Animal diggings create 'Regeneration Niches'.
(2) Detrimental effects of animals	
Herbivores eliminate pioneer plant species (bad if Facilitation Model of succession holds).	Mobile herbivores export nutrients from site.
	Overgrazing eliminates palatable plant species.

community can be treated as an island in a sea of different vegetation. In this respect, natural disturbances are particularly good analogies to rehabilitated lands. There is no general theory regarding the impact of patch size *per se* on plant species composition (Miller, 1982), but the consequences of patch size on other processes might be predicted.

Patch size is likely to exert a strong influence on dispersal of plants into a disturbed site. Even without the activities of animals as seed dispersers, small patches will easily receive a seed rain from the surrounding plants, with a relatively complete representation of species present in the undisturbed community. Dispersal of propagules into larger patches will clearly require a dispersal vector, such as wind or animals. A large patch will therefore receive a more biased representation of species from the surrounding community, depending on the characteristics of the dispersal agent, i.e. which species of plants it feeds on and the spatial pattern of its movements to and within regenerating

Table 5.3. *Summary of the possible interactions between various site characteristics and the impacts of animals on plant species composition of rehabilitation sites*

Site characteristics	Potential problems relating to consumers
(1) Restoration	
Restoration site a patch in a 'sea' of natural vegetation; or site containing refuges within its boundaries.	Severe herbivory removes palatable pioneers. Other roles (pollination, seed dispersal, nutrient cycling, physical disturbance) readily reinstated.
Restoration site isolated from natural vegetation.	Reduction in natural seed dispersal. Pollination, nutrient cycling, decomposition reduced. Herbivory as a means of maintaining species co-existence absent.
(2) Ecosystem replacement	Herbivory on palatable new plantings, especially if fertilized. Limited pollination and seed dispersal within the site Decomposers and nutrient cycling potentially limited.

patches. Different taxa of animals will have different capacities to invade disturbed patches of a particular size (see Chapter 6).

Various authors have concluded that the strong interaction between grazing and the size of a regenerating patch will determine the magnitude of the impact of herbivory on the developing community (Whelan, 1978; Leigh & Holgate, 1979; Whelan & Main, 1979; Shea *et al.*, 1979; Quinn, 1986). This interaction exists because individual plants are particularly susceptible to grazing when regenerating after disturbance (i.e. as resprouts or seedlings) and also because herbivores congregate in small patches of regenerating vegetation. In heathlands of southeastern Australia, for example, swamp wallabies (*Macropus bicolor*) show a food preference for the lily, *Sowerbaea juncea*, especially in vegetation regenerating after fire (pers. obs.). Whelan (1986) showed that the degree of grazing on this species declined with distance from the unburned

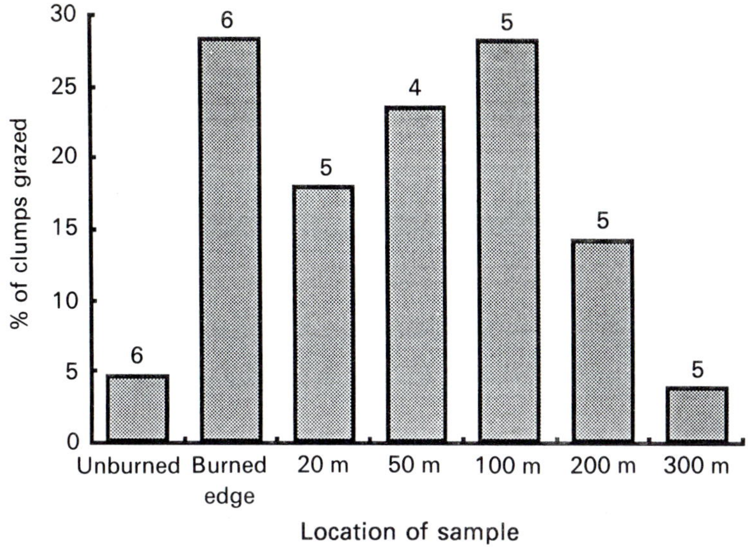

Fig. 5.7. Intensity of grazing by swamp wallabies on vanilla lily (*Sowerbaea juncea*) in relation to distance from the boundary of a fire. Clumps of the lily at each distance-class were scored for presence or absence of grazing (see Whelan, 1986).

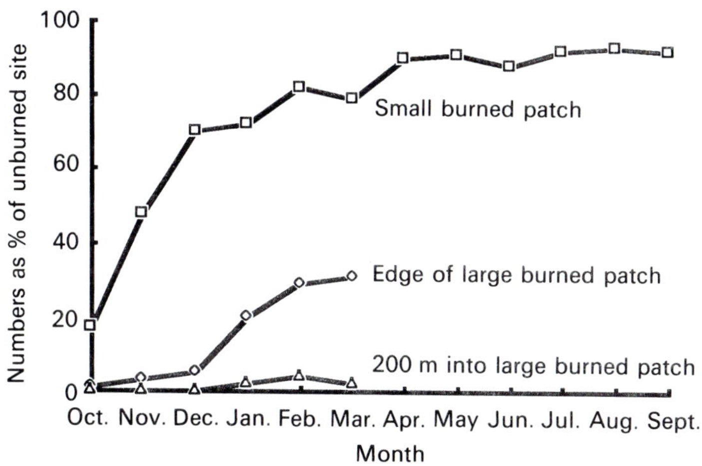

Fig. 5.8. Invasion of large and small burned areas by acridid grasshoppers in southwestern Australia. Numbers of grasshoppers at each sample date are expressed as a percentage of the numbers found in adjacent, unburnt vegetation (see Whelan & Main, 1979).

boundary (Fig. 5.7) because the wallabies shelter in the unburned vegetation during the day and move out to graze in the burnt patch by night. This pattern of grazing has also been shown to cause mortality of woody perennial shrubs (Leigh & Holgate, 1979). Smith (1940) noted that pioneer stages of old-field vegetation provided attractive food for seed-eating rodents, and the surrounding climax vegetation provided the rodent populations with cover. Because of the effects described here, future management of small patches of vegetation after prescription burning in one reserve, the University of Western Australia's Marsupial Breeding Station, near Perth, will include either culling of mammalian herbivores or fencing of the burned patches for several years (B. T. Clay, pers. comm.).

Edge effects have also been shown for insect herbivores. Whelan & Main (1979) demonstrated that herbivorous grasshoppers reinvade a small burned patch of vegetation more rapidly than a large area (Fig. 5.8), and insect herbivores were responsible for substantial mortality of seedlings appearing after fire, especially legumes. Majer (1980) showed that ants removed seeds from bauxite mines to which topsoil and seeds of native species had been added. This effect extended about 25 m into the minesite from the boundary of the adjacent forest (see Fig. 16.3).

In view of the chronically low nutrient levels and the increasing costs of artificial fertilization in rehabilitation sites, legumes provide an attractive source of nitrogen. Leguminous shrubs can be used as a nursery cover to facilitate tree establishment (Nimmo & Weatherell, 1961; cited in Bradshaw, 1983). However, herbivores which feed on these shrubs within a disturbed site but live outside the site could clearly have an inhibitory effect on succession by exporting nutrients (Main, 1981*b*). Similarly, nutrients may be removed, via insect herbivores, by predators which live outside the disturbed site. These hypotheses warrant more complete and experimental study.

Patch characteristics

Other characteristics of a patch of regenerating vegetation, as well as its size, are likely to affect the impact of animals on vegetation development. This fact has been recognized in management aimed at attracting wildlife and game animals into rehabilitated areas (e.g. Brenner, 1978). Mammalian herbivores shelter in small unburned patches of vegetation contained within a burned landscape and feed on the regenerating vegetation (Christensen & Kimber, 1975; Quinn, 1986). Herbivory by these animals would be much reduced in large, continuous burns (Leigh & Holgate, 1979).

Stanlake, Eastman & Stanlake (1978), in a study of rehabilitated mines in British Columbia, showed that ungulates were attracted to 'islands' of native vegetation in the seeded areas and that heaviest browsing was on the edges of the disturbed area adjacent to cover provided by these 'islands'. This principle was also noted by Reeders (1985) in rehabilitated bauxite mines in northern Australia. It is not clear what effect, inhibitory or facilitatory, this pattern of grazing will have on the subsequent development of the plant community, but the studies on post-fire regeneration discussed above suggest that it will be substantial. If herbivores are able to concentrate on a rehabilitated site due to its small size and/or through the provision of patches of cover within the site, attempts to establish plant species for particular purposes (e.g. forage for wildlife, legumes for nitrogen fixation) may well be unsuccessful.

The studies by Lambkin (1986) and Sim (1986) reported earlier, which demonstrated that herbivory by wallabies and wombats inhibited succession to woodland in regenerating orchards, also demonstrated the importance of an edge effect but other characteristics of the tree seedlings were also important. For example, seedlings appearing naturally were either less palatable or less easily located than hand-planted nursery stock of the same species (Lambkin, 1986). Furthermore, seedlings and saplings planted into cleared land suffered far greater herbivory than seedlings surrounded by dense bracken and blady grass. Even in sites with natural regeneration, herbivores will exhibit preferences, ignoring some species and concentrating on others.

Natural dispersal into rehabilitation sites will be strongly affected by both the size of a disturbed site and also the nature of the surrounding vegetation as a source of new plant species (Clemens *et al.*, 1984). Humphries (1980), for example, concluded that availability of plant propagules in the surrounding vegetation was a major factor determining vegetation development in English limestone quarries. Animals will be especially important as seed dispersers in large disturbed sites (Fowler, Fowler & Thomas, 1982). However, colonization of animal-dispersed species will be slow if conditions in the site do not attract the seed dispersers. The provision of perch sites for birds can facilitate seed input to a site (McDonnell & Stiles, 1983). However, when seed dispersal is predominantly by animals other than perching birds, other factors must determine plant invasions. For example, the emu (*Dromaeus novaehollandiae*) disperses the seeds of *Santalum*, *Leucopogon* and *Persoonia* in Western Australia (Main, 1981*b*). If these birds are absent, or are not attracted to a disturbed site, it is unlikely that seeds of these plant species will find their way into an area being rehabilitated. This is especially true where successful germination depends on the passage of a seed through the gut of the seed-disperser.

Karr (1968) concluded that the structural complexity of the vegetation already present in regenerating surface mines in Illinois was important in determining bird diversity. This may in turn influence the subsequent seed input to the site. Similarly, Reeders (1985) showed that bird recolonization of bauxite mines in northern Australia was associated with existing vegetation structure in the mine site. Furthermore, islands of vegetation left undisturbed within the mines facilitated bird invasions. The role of this patchiness in seed dispersal and consequent vegetation development remains to be examined.

Majer (1980) emphasized that although 'seed theft' by ants may be disadvantageous for the initial development of bauxite mine rehabilitation, seeds stored in nests are available for germination in a favourable microsite when the vegetation is eventually burned. Ants are clearly selective in the species of seeds removed and stored (Majer, 1980, 1984) and their activities would therefore influence the composition of the plant community regenerating from seed after the first burn on the site.

Surrounding vegetation

The magnitude of the effects of herbivores outlined above will depend not only upon the size of the rehabilitated site and its other characteristics, but also upon the nature of the surrounding vegetation. The surrounding vegetation is the source of plant propagules as well as herbivores, seed dispersers and pollinators. A rehabilitation site that is effectively an island within an extensive agricultural region may therefore escape intense herbivory on a planted pioneer community. On the other hand, it may lack the reserve of invertebrates important in breaking internal biochemical nutrient cycles and those involved in litter breakdown. It may also lack a continual input of seeds of desirable plant species. Main (1967) gave a good example of this principle in her description of recruitment to sandalwood (*Santalum spicatum*) populations in small island-like reserves in the wheatbelt of Western Australia.

> 'Far more serious for the persistence of these trees as a component of the bush landscape ... in settled regions where farmlands have disrupted the continuity of the bushland so that emus are precluded from farmland-enclosed bush tracts, is the absence of any natural dispersal agent of the sandalwood (and quandong) stones. Thus, even when seeds germinate now, the seedlings develop in the litter zone of the parent tree and it is unlikely that such seedlings will ever survive the shadowing tree.'

Toxicity

The effects of toxins on animals in a rehabilitation site is a question that has received very little attention. Concern has been directed at the possiblity of raising domestic stock on reclaimed land, especially if there is contamination by heavy metals (see Chapter 10 and Bradshaw, 1983). However, the beneficial functions of animals in nutrient cycling and in facilitating invasions of new plant species will be reduced in sites contaminated by heavy metals. Perhaps the search for metal-tolerant plant species for reclamation of mine spoils should be accompanied by studies on tolerance in decomposers and herbivores.

It is well known that certain plant species are associated with naturally occurring heavy metal deposits. Indeed, heavy metals have been prospected for and identified by the plants growing in the area (Cole, 1973). These plants often contain high concentrations of the metal. Most animals are very sensitive to toxins such as these, yet little is known of their effects on the activities of herbivores and decomposers. Studies of the plant and animal communities occurring around heavy metal deposits would reveal many hints about the problems likely to be encountered in reclamation of toxic sites.

Summary: future approaches

In contrast to the view expressed by Bradshaw (1983), 'from the point of view of the reconstruction of a properly functioning ecosystem, they [animals] play little part', it is clear that animals do have profound impacts in some situations. However, the foregoing discussion of the possible roles of animals in both the development and the maintenance of species composition in plant communities defies any attempt to construct a set of general principles. Some activities of animals will inhibit, and others will facilitate, a highly managed rehabilitation program. Other activities, in addition, will play a more subtle but indispensable part in producing self-sustaining ecosystems on highly disturbed sites.

At the current stage of our understanding, this review can do little more than explore the ways in which animals influence plant species composition both in mature communities and also in natural primary and secondary successions, point to the possible applications of these ideas in rehabilitation of disturbed lands, and identify areas most in need of investigation. It is quite clear that our knowledge of the impact of animals on undisturbed plant communities and natural successions is inadequate to permit accurate assessments of their likely impact on plant species composition of rehabilitated land. However, the main possibilities are summarized here:

1. Intensive herbivory is likely to reduce species diversity by eliminating populations of palatable plant species.
2. Moderate herbivory is likely to enhance species diversity for several reasons: gaps created in the cover of pioneer plants may facilitate establishment of later colonists; nutrients released from leaves and litter may become available for other species; nutrients otherwise bound in dry litter and lost through erosion would be retained within the system.
3. Seed dispersal will be important in the introduction of propagules of plant species from the surrounding vegetation. Effective seed dispersal, influenced by size and structural characteristics of the site, may accelerate a succession and increase plant species diversity within the site.
4. Pollination may be important in better developed sites, especially if species in the site require cross-fertilization and are not represented in surrounding vegetation.
5. Burrowing and trampling by animals may be important both in permitting colonization by new species and in maintaining species diversity through soil turnover and the provision of regeneration sites.
6. The importance of all the above processes will be influenced by the physical characteristics of the rehabilitation site, such as its shape, size and included 'islands' of undisturbed vegetation and also by the nature of the surrounding vegetation. Seed dispersal will be facilitated in small sites or in sites with established cover, but intensive grazing will also be permitted.

The design of sites to be rehabilitated, particularly in complete restoration programs, will need to take the above factors into account. One feature of the foregoing review is the paucity of examples coming from studies of land reclamation programs. In view of the lack of work on the mechanisms involved in natural primary successions, studies on rehabilitation are of great importance to the development of ecological theory (Bradshaw, 1983). Reclamation of severely disturbed lands offers a great opportunity for replicated, experimental studies. As stated by Game *et al.* (1982), mine sites are ecological laboratories for primary succession studies. These studies will need to be undertaken if site rehabilitation is to be successful. It is therefore important that they are designed properly, replicated where possible, the results quantified and published in the ecological literature.

Acknowledgements

I thank the following colleagues for their contributions to the development of my ideas on this topic and for providing constructive comments on an early version of this paper: Jonathan Majer, Bert and Barbara Main, Ros Muston and Richard Hobbs. The 1986 class of Bachelor of Environmental Science students at the University of Wollongong contributed by forcing me to think of practical applications and shortcomings of ecological theory. Support for the portions of my own research reported here was provided by a Postgraduate Research Studentship at the University of Western Australia, the University of Wollongong and the Australian Research Grants Scheme.

References

Amman, G. D. (1977). The role of the mountain pine beetle in lodgepole pine ecosystems: Impact on succession. In *The Role of Arthropods in Forest Ecosystems*, ed. W. J. Mattson, pp. 3–18. New York: Springer Verlag.

Attiwill, P. M. (1980). Nutrient cycling in a *Eucalyptus obliqua* (L'Herit.) forest. IV nutrient uptake and nutrient return. *Australian Journal of Botany*, **28**, 199–222.

Baird, A. M. (1958). Notes on the regeneration of vegetation of Garden Island after the 1956 fire. *Journal of the Royal Society of Western Australia*, **41**, 102–7.

Begon, M., Harper, J. L. & Townsend, C. R. (1986). *Ecology: Individuals, Populations, Communities.* Oxford: Blackwell Scientific Publications.

Berg, R. Y. (1981). The role of ants in seed dispersal in Australian lowland heathland. In *Heathlands and Related Shrublands of the World*, vol. 9B, ed. R. L. Specht, pp. 51–9. Amsterdam: Elsevier.

Blanton, C. M. & Ewel, J. J. (1985). Leaf-cutting ant herbivory in successional and agricultural tropical ecosystems. *Ecology*, **66**, 861–9.

Bradshaw, A. D. (1979). Derelict land – is the tidying up going too far? *The Planner*, **65**, 85–8.

Bradshaw, A. D. (1983). The reconstruction of ecosystems. *Journal of Applied Ecology*, **20**, 1–17.

Brenner, F. J. (1978). Food and cover evaluation of strip mine plants as related to wildlife management. In *Surface Mining and Fish/Wildlife Needs in the Eastern United States*, ed. D. E. Samuel, J. R. Stauffer, C. H. Hocutt & W. T. Mason, pp. 294–305. Morgantown: West Virginia University.

Brown, V. K. (1985). Insect herbivores and plant succession. *Oikos*, **44**, 17–22.

Buckley, G. P., Chilton, K. G. & Devonald, V. G. (1981). The influence of sward control on the establishment and early growth of ash (*Fraxinus excelsior* L.) and Norway maple (*Acer planatoides* L.). *Journal of Environmental Management*, **13**, 223–40.

Burdon, J. J. & Chilvers, G. A. (1974). Fungal and insect parasites contributing to niche differentiation in mixed species stands of eucalypt saplings. *Australian Journal of Botany*, **22**, 265–9.

Cates, R. G. & Orians, G. H. (1975). Successional status and the palatability of plants to generalist herbivores. *Ecology*, **56**, 410–18.

Chapman, R. N. (1931). *Animal Ecology.* New York: McGraw-Hill.

Christensen, P. E. S. (1980). *The Biology of* Bettongia penicillata *Gray*, 1837 *and* Macropus eugenii (*Demarest*, 1817) *in Relation to Fire.* Forests Department of Western Australia, Bulletin 91.

Christensen, P. E. S. & Kimber, P. C. (1975). Effects of prescribed burning on the flora and fauna of south-west Australian forests. *Proceedings of the Ecological Society of Australia*, **9**, 85–106.

Clemens, J., Bradley, C. & Gilbert, O. L. (1984). Early development of vegetation on urban demolition sites in Sheffield, England. *Urban Ecology*, **8**, 139–47.

Clements, F. E. (1928). *Plant Succession and Indicators.* New York: Wilson.

Cole, M. M. (1973). Geobotanical and biogeochemical investigations in the sclerophyllous woodland and shrub associations of the eastern goldfields area of Western Australia, with particular reference to the role of *Hybanthus floribunda* (Lindl.) F. Muell. as a nickel indicator and accumulator plant. *Journal of Applied Ecology*, **10**, 269–320.

Connell, J. H. (1971). On the role of natural enemies in preventing competitive exclusion in some marine animals and in rain forests. In *Dynamics of Populations*, ed. P. J. den Boer & G. R. Gradwell, pp. 298–312. Wageningen: Pudoc.

Connell, J. H. & Slatyer, R. O. (1977). Mechanisms of succession in natural communities and their role in community stability and organization. *American Naturalist*, **111**, 1119–44.

Crawley, M. J. (1983). *Herbivory.* Oxford: Blackwell Scientific Publications.

Crocker, R. I. & Major, J. (1955). Soil development in relation to vegetation and surface age at Glacier Bay, Alaska. *Journal of Ecology*, **65**, 427–48.

Darwin, C. (1859). *The Origin of Species.* London: John Murray.

Daubenmire, R. (1968). The ecology of fire in grassland. *Advances in Ecological Research*, **5**, 209–66.

Davidson, D. W. & Morton, S. R. (1981). Competition for dispersal in ant-dispersed plants. *Science*, **213**, 1259–61.

Davidson, D. W. & Morton, S. R. (1984). Dispersal adaptations of some *Acacia* species in the Australian arid zone. *Ecology*, **65**, 1038–51.

Dean, T. A. & Hurd, L. E. (1980). Development in an estuarine fouling community: the influence of early colonists on later arrivals. *Oecologia*, **46**, 295–301.

Denslow, J. S. (1985). Disturbance-mediated coexistence of species. In *The Ecology of Natural Disturbance and Patch Dynamics*, ed. S. T. A. Pickett & P. S. White, pp. 307–23. Orlando: Academic Press.

Drury, W. H. & Nisbet, I. C. T. (1973). Succession. *Journal of the Arnold Arboretum*, **54**, 331–68.

During, H. J. & Willems, J. H. (1985). Diversity models applied to a chalk grassland. In *Plant Community Ecology*, ed. R. K. Peet, pp. 321–32. The Hague: Junk.

Dyksterhuis, E. J. (1948). The vegetation of the western cross timbers. *Ecological Monographs*, **18**, 325–76.

Egler, F. E. (1954). Vegetation Science concepts: I, initial floristic composition – a factor in old field vegetation development. *Vegetatio*, **4**, 412–17.

Ellison, L. (1960). Influence of grazing on plant succession of rangelands. *The Botanical Review*, **26**, 1–66.

Feinsinger, P. (1976). Organization of a tropical guild of nectarivorous birds. *Ecological Monographs*, **46**, 257–91.

Fowler, L. J., Fowler, D. K. & Thomas, J. E. (1982). Dispersal of autumn olive seeds by foxes on coal surface mines in east Tennessee. *Journal of the Tennessee Academy of Science*, **57**, 83–5.

Frelich, L. E. & Lorimer, C. G. (1985). Current and predicted longterm effects of deer browsing in hemlock forests in Michigan, U.S.A. *Biological Conservation*, **34**, 99–120.

Game, M., Carrel, J. E. & Hotrabhavandra, T. (1982). Patch dynamics of plant succession on abandoned surface coal mines: A case history approach. *Journal of Ecology*, **70**, 707–20.

Gilbert, L. E. (1980). Food web organization and the conservation of neotropical diversity. In *Conservation Biology: An Evolutionary – Ecological Perspective*, ed. M. E. Soulé & B. A. Wilcox, pp. 11–33. Massachussetts: Sinauer.

Grubb, P. J. (1977). The maintenance of species richness in plant communities. The importance of the regeneration niche. *Biological Reviews*, **52**, 107–45.

Harper, J. L. (1969). The role of predation in vegetational diversity. *Brookhaven Symposia in Biology*, **22**, 48–62.

Harper, J. L. (1977). *The Population Biology of Plants*. London: Academic Press.

Hollinger, D. Y. (1986). Herbivory and the cycling of nitrogen and phosphorus in isolated California oak trees. *Oecologia*, **70**, 291–7.

Horn, H. S. (1976). Succession. In *Theoretical Ecology*, ed. R. M. May, pp. 187–204. Oxford: Blackwell Scientific Publications.

Hubbell, S. P. (1980). Seed predation and the coexistence of tree species in tropical forests. *Oikos*, **35**, 214–29.

Humphries, R. N. (1980). The development of wildlife interest in limestone quarries. *Reclamation Review*, **3**, 197–207.

Janzen, D. H. (1970). Herbivores and the number of tree species in tropical forests. *American Naturalist*, **104**, 501–28.

Janzen, D. H. (1972). Escape in space by *Sterculia apetala* seeds from the bug *Dysdercus fasciatus* in a Costa Rican deciduous forest. *Ecology*, **53**, 350–61.

Janzen, D. H. (1979). New horizons in the biology of plant defenses. In *Herbivores: Their Interaction with Secondary Plant Metabolites*, ed. G. A. Rosenthal & D. H. Janzen, pp. 331–50. New York: Academic Press.

Jensen, T. S. & Nielsen, O. F. (1986). Rodents as seed dispersers in a heath–oak wood succession. *Oecologia*, **70**, 214–21.

Karr, J. R. (1968). Habitat and avian diversity on strip-mined land in east-central Illinois. *Condor*, **70**, 348–57.

Knapp, R. (1974). Mutual influences between plants, allelopathy, competition and vegetation changes. In *Vegetation Dynamics*, ed. R. Knapp, pp. 119–20. The Hague: Junk.

Lamb, D. (1985). The influence of insects on nutrient cycling in eucalypt forests: a beneficial role? *Australian Journal of Ecology*, **10**, 1–5.

Lambkin, K. (1986). The impact of herbivory on trial plantings in grasslands in the Cordeaux Catchment area. In *Environmental Science 1986*, Chapter 2. Wollongong: University of Wollongong.

Laycock, W. A. (1958). The initial pattern of revegetation of pocket gopher mounds. *Ecology*, **39**, 346–51.

Leigh, J. H. & Holgate, M. D. (1979). The responses to understorey of forests and woodlands of the Southern Tablelands to grazing and burning. *Australian Journal of Ecology*, **4**, 25–45.

Levin, D. A. & Kerster, H. W. (1974). Gene flow in seed plants. *Evolutionary Biology*, **7**, 139–220.

Lubchenco, J. (1983). *Littorina* and *Fucus*: effects of herbivores, substratum heterogeneity and plant escapes during succession. *Ecology*, **64**, 1116–23.

Luxton, M. (1982). The ecology of some soil mites from coal shale tips. *Journal of Applied Ecology*, **19**, 427–42.

McBrien, H., Harmsen, R. & Crowder, A. (1983). A case of insect grazing affecting plant succession. *Ecology*, **64**, 1035–9.

McCormick, J. (1968). Succession. *Via*, **1**, 22–35 & 131–2.

McDonnell, M. J. & Stiles, E. W. (1983). The structural complexity of old field vegetation and the recruitment of bird-dispersed plant species. *Oecologia*, **56**, 109–16.

MacIntosh, R. P. (1980). The relationship between succession and the recovery process in ecosystems. In *The Recovery Process in Damaged Ecosystems*, ed. J. Cairns, pp. 11–62. Ann Arbor: Ann Arbor Science Publications.

Main, A. R. (1981*a*). Ecosystem theory and management. *Journal of the Royal Society of Western Australia*, **4**, 1–4.

Main, A. R. (1981*b*). Plants as animal food. In *The Biology of Australian Plants*, ed. J. S. Pate & A. J. McComb, pp. 342–60. Nedlands: University of Western Australia Press.

Main, A. R. (1987). Management of remnants of native vegetation – a review of the problems and the development of an approach with reference to the Wheatbelt of Western Australia. In *Nature Conservation: The Role of Remnants of Native Vegetation*, ed. D. A. Saunders, G. W. Arnold, A. A. Burbidge & A. J. M. Hopkins, pp. 1–13. Sydney: Surrey Beatty.

Main, B. Y. (1967). *Between Wodgil and Tor.* Brisbane: Jacaranda Press; Perth: Landfall Press.

Maiorana, V. C. (1978). What kinds of plants do herbivores really prefer? *American Naturalist*, **112**, 631–5.

Majer, J. D. (1980). The influence of ants on broadcast and naturally spread seeds in rehabilitated bauxite mined areas. *Reclamation Review*, **3**, 3–9.

Majer, J. D. (1982). Ant-plant interactions in the Darling Botanical District of Western Australia. In *Ant–Plant Interactions in Australia*, ed. R. C. Buckley, pp. 45–61. The Hague: Junk.

Majer, J. D. (1984). The influence of ants on seeding operations in Northern Australian mined areas. *Reclamation and Revegetation Research*, **2**, 299–313.

Marrs, R. H., Granlund, I. H. & Bradshaw, A. D. (1980). Ecosystem development on reclaimed china clay wastes. IV. Recycling of above-ground plant nutrients. *Journal of Applied Ecology*, **17**, 803–13.

Mattson, W. J. & Addy, N. D. (1975). Phytophagous insects as regulators of forest primary production. *Science*, **190**, 515–22.

Miles, J. (1979). *Vegetation Dynamics.* London: Chapman & Hall.

Miller, T. E. (1982). Community diversity and interactions between the size and frequency of disturbance. *American Naturalist*, **120**, 533–6.

Morrow, P. A. (1977). The significance of phytophagous insects in the *Eucalyptus* forests of Australia. In *The Role of Arthropods in Forest Ecosystems*, ed. W. J. Mattsen, pp. 19–29. New York: Springer Verlag.

Morrow, P. A. & La Marche, V. C. (1978). Tree ring evidence for chronic insect suppression of productivity in subalpine *Eucalyptus. Science*, **201**, 1224–6.

Mulcahy, M. J. ed. (1984). *Bauxite Mining in the Jarrah Forest: Impact and Rehabilitation.* Perth, Western Australia: Department of Conservation and Environment. Bulletin 169.

Murdoch, W. W., Evans, F. C. & Peterson, C. H. (1972). Diversity and pattern in plants and insects. *Ecology*, **53**, 819–28.

Murray, D. R. ed. (1986). *Seed Dispersal.* Sydney: Academic Press.

Muston, R. (1986). Fabricate or rehabilitate: an ecological approach to the revegetation of degraded landscapes. In *Xeric Australia: Landscape Design and Management*, pp. 57–67. Adelaide: Australian Institute of Landscape Architects.

Newland, C. (1986). Do herbivores limit the effectiveness of artificial revegetation of sandstone soils? In *Environmental Science 1986*, Chapter 3. Australia: Biology Department, University of Wollongong.

Niering, W. A. & Egler, F. E. (1955). A shrub community of *Viburnum lentago*, stable for 25 years. *Ecology*, **36**, 356–60.

Niering, W. A. & Goodwin, R. H. (1974). Creation of relatively stable shrublands with herbicides: arresting succession on rights of way and pastureland. *Ecology*, **55**, 784–95.

Nimmo, M. & Weatherell, J. (1961). Experiences with leguminous nurses in forestry. In *Report on Forestry Research*, pp. 126–47. London: Her Majesty's Stationery Office.

Noble, I. R. (1981). Predicting successional change. In *Fire Regimes and Ecosystem Properties*, pp. 278–300. USDA Forest Service General Technical Report WO-26.

Odum, E. P. (1953). *Fundamentals of Ecology.* Philadelphia: Saunders.

Otte, D. (1975). Food preference and plant succession. *Oecologia*, **18**, 129–44.

Paine, R. T. (1966). Food web complexity and species diversity. *American Naturalist*, **100**, 65–75.

Pickard, J. (1982). Catastrophic disturbance and vegetation on Little Slope, Lord Howe Island. *Australian Journal of Ecology*, **7**, 161–70.

Pickett, S. T. A. and White, P. S. ed. (1985). *The Ecology of Natural Disturbance and Patch Dynamics*. Orlando: Academic Press.

Pimentel, D. (1961). Species diversity and insect population outbreaks. *Annals of the Entomological Society of America*, **54**, 76–86.

Platt, W. J. (1975). The colonization and formation of equilibrium plant species associations in badger disturbances in a tall grass prairie. *Ecological Monographs*, **45**, 285–305.

Price, P. W. (1975). *Insect Ecology*. New York: John Wiley.

Quinn, R. D. (1986). Mammalian herbivory and resilience in Mediterranean-climate ecosystems. In *Resilience in Mediterranean-Climate Ecosystems*, ed. B. Dell, A. J. M. Hopkins & B. B. Lamont, pp. 113–28. Dordrecht: Junk.

Reeders, A. P. F. (1985). Vertebrate fauna in regenerated mines at Weipa, North Queensland. In *North Australian Mine Rehabilitation Workshop. No.* 8, ed. J. W. Lawrie, pp. 105–18. Weipa: Comalco Aluminium Ltd.

Ringe, J. H., Graves, D. H. & Wittwer, R. F. (1984). Effects of soil amendments and severe cattle grazing on the long-term survival and growth of tree seedlings on eastern Kentucky surface mined land. In *Proceedings of the 1984 Symposium on Surface Mining, Hydrology, Sedimentology, and Reclamation*, ed. D. H. Graves & R. W. De Vore, p. 209. Lexington: University of Kentucky.

Root, R. B. (1973). Organization of a plant-arthropod association in simple and diverse habitats: the fauna of collards (*Brassica oleracea*). *Ecological Monographs*, **43**, 95–124.

Schowalter, T. D. (1981). Insect herbivore relationship to the state of the host plant: biotic regulation of ecosystem nutrient cycling through ecological succession. *Oikos*, **37**, 126–30.

Schowalter, T. D., Webb, J. W. & Crossley, D. A. (1981). Community structure and nutrient content of canopy arthropods in clearcut and uncut forest ecosystems. *Ecology*, **62**, 1010–19.

Schuman, G. E., Booth, D. T., Waggoner, J. W. & Rauzi, F. (1984). The effect of grazing reclaimed mine lands on forage production and composition. *Working Papers: Second International Rangeland Congress*. Adelaide, Australia.

Shea, S. R., Gillen, K. J. & Kitt, R. J. (1978). Variation in sporangial production of *Phytophthora cinnamomi* Rands on Jarrah (*E. marginata* Sm.) forest sites with different understorey compositions. *Australian Forest Research*, **8**, 219–26.

Shea, S. R. & Kitt, R. J. (1976). *The Capacity of Jarrah Forest Native Legumes to Fix Nitrogen*. West Australian Forests Department Research Paper 21.

Shea, S. R., McCormick, J. & Portlock, C. C. (1979). The effect of fires on regeneration of leguminous species in the northern jarrah (*Eucalyptus marginata* Sm.) forest of Western Australia. *Australian Journal of Ecology*, **4**, 195–205.

Shugart, H. H. (1984). Animals and mosaic landscapes. In *A Theory of Forest Dynamics: Theoretical Implications of Forest Succession Models*, ed. H. H. Shugart, pp. 181–95. New York: Springer-Verlag.

Sim, R. (1986). The potential for natural revegetation of grossly disturbed areas. In *Environmental Science 1986*. Chapter 1. Australia: Biology Department, University of Wollongong.

Smart, N. O. E., Hatton, J. C. & Spence, D. H. N. (1985). The effect of long-term exclusion of large herbivores on the vegetation in Murchison Falls National Park, Uganda. *Biological Conservation*, **33**, 229–45.

Smith, C. C. (1940). Succession of animals. *Ecological Monographs*, **10**, 421–84.

Sousa, W. P. (1980). The responses of a community to disturbance: the importance of successional age and species life history. *Oecologia*, **45**, 72–81.

Springett, B. P. (1978). On the ecological role of insects in Australian eucalypt forests. *Australian Journal of Ecology*, **3**, 129–39.

Stanlake, E. A., Eastman, D. S. & Stanlake, M. G. (1978). *Ungulate Use of some Recently Reclaimed*

Strip Mines in Southeastern British Columbia. Wildlife Research and Technical Services Section, Fish and Wildlife Services Report R-1.

Storr, G. M. (1963). Some factors inducing change in the vegetation of Rottnest Island. *West Australian Naturalist*, **9**, 15–22.

Strang, R. M. (1974). Some man-made changes in successional trends on the Rhodesian highveld. *Journal of Applied Ecology*, **11**, 249–63.

Tansley, A. G. & Adamson, R. S. (1925). Studies on the vegetation of the English chalk III. The chalk grasslands of the Hampshire–Sussex border. *Journal of Ecology*, **13**, 177–223.

Temple, S. A. (1977). Plant–animal mutualism: coevolution with dodo leads to near extinction of plant. *Science*, **197**, 885–6.

Waser, N. M. (1983). Competition for pollination and floral character differences among sympatric plant species: a review of evidence. In *Handbook of Experimental Pollination Biology*, ed. C. E. Jones & R. J. Little, pp. 277–93. New York: Van Nostrand, Reinhold.

Watt, A. S. (1960). Population changes in acidophilous grass-heath in Breckland, 1936–57. *Journal of Ecology*, **48**, 605–29.

Watt, A. S. (1962). The effect of excluding rabbits from Grassland A (Xerobrometum) in Breckland, 1936–60. *Journal of Ecology*, **50**, 181–98.

Watt, A. S. (1981). A comparison of grazed and ungrazed Grassland A in East Anglian Breckland. *Journal of Ecology*, **69**, 499–536.

Whelan, R. J. (1978). *The Influence of Insect Grazers on the Establishment of Post-fire Plant Populations*. Ph.D. Thesis, University of Western Australia.

Whelan, R. J. (1986). Seed dispersal in relation to fire. In *Seed Dispersal*, ed. D. R. Murray, pp. 237–71. Sydney: Academic Press.

Whelan, R. J. & Main, A. R. (1979). Insect grazing and post-fire plant succession in south-west Australian woodland. *Australian Journal of Ecology*, **4**, 387–98.

White, P. S. (1979). Pattern, process and natural disturbance in vegetation. *The Botanical Review*, **45**, 230–99.

Williamson, R. L. (1981). Improving species composition and seasonal variety on reclaimed strip mined grasslands through selective grazing intensities and time periods. *Proceedings of the North Dakota Academy of Science*, **35**, 5.

Wilson, D. E. & Janzen, D. H. (1972). Predation on *Scheelea* palm seeds by bruchid beetles: seed density and distance from the parent plant. *Ecology*, **53**, 954–9.

Zeevalking, H. J. & Fresco, L. F. M. (1977). Rabbit grazing and diversity in a dune area. *Vegetatio*, **35**, 193–6.

6

Long-term colonization of fauna in reclaimed land

J. D. MAJER

Introduction

Degraded land is generally rehabilitated with an expected outcome in mind. The objectives of the rehabilitation plan may be prescribed by the relevant regulatory body or they may be drawn up by the original land-user. In either case there is usually an expectation, or hope, that the area will develop and maintain a certain biota after the passage of a period of time.

In some situations the compliance with particular rehabilitation performance standards may be left to the goodwill of the land-user while in others, such as North American coal mines, rehabilitation success is assessed by State or Federal regulatory bodies. There, coal mining companies outlay a performance bond which is generally returned to the company in three instalments as, and when, certain reclamation and rehabilitation standards are met. Phase 1 involves backfilling, regrading, and drainage control while Phase 2 is concerned with the initial establishment of vegetation and the prevention of suspended solids from contributing to streamflow or runoff outside the permit area (U.S. Congress, Office of Technology Assessment, 1986). The remaining bond monies are only released following Phase 3 in which the operator is required to successfully complete all surface coal mining and reclamation activities in accordance with regulatory requirements and with the permit. In cases where the operator has been required to rehabilitate the area for fish or wildlife (Table 1.2) there is clearly an interest in ensuring that in the long-term there continues to be a sustained colonization of wildlife in the mined areas.

To date, few North American regulatory authorities have formulated criteria for the later stages of bond release. Formulation of such criteria for fish or wildlife is hindered by the paucity of long-term studies on fauna recolonization, either in North America or elsewhere. The studies which have been carried out have generally adopted the approach of comparing plots of differing rehabilitation age (see references under 'succession' and 'recolonization' in Chapter 19). This approach can produce ambiguous data since there may have been changes in rehabilitation technology over the timescale represented by these plots (Majer, 1985*a*). The only reliable way to obtain long-term data is to sequentially resample a site over an extended period of time. To illustrate this Chapters 13, 14 and 16 describe some of the long-term (>10 years) arthropod data which have been obtained from German coal mined areas, Dutch polders, and Australian bauxite mines.

The response to this lack of information on fauna colonization has been to assess habitat restoration and use this as an untested indicator of animal return. For instance, permanence and net above-ground annual production are assumed to be measures of the utility of vegetation for livestock and wildlife grazing while plant diversity is considered to be a measure of ecological stability and therefore an indicator of the land's ability to support wildlife (U.S. Congress, Office of Technology Assessment, 1986). Wildlife may also be monitored but scientists generally lack confidence in the utility of the data for reliable assessment of a rehabilitation programme. The problem is also compounded by the fact that the species richness and composition of fauna change throughout the vegetation succession. Early to mid-successional stages of vegetation may benefit different kinds of wildlife than do later successional stages. Also, the plant succession may take decades to reach the climax community, so wildlife may not reach pre-disturbance levels of diversity or density for a considerable period of time.

The purpose of this Chapter is to explore the likely long-term patterns of fauna colonization in rehabilitated land. Inevitably, it must also consider the early stages of fauna colonization. It draws upon the results of published studies and, due to the paucity of these, is augmented by extrapolation from current ecological theory. Since the third phase of bond release in North American coal mines may not be considered until 5 or 10 years after seeding (where annual precipitation is >66 or <66 cm respectively), long-term is here taken to mean longer than the 5–10 year period. This definition will vary according to climate and other circumstances at the site which is under consideration.

The Chapter first considers the characteristics of the newly rehabilitated area and of the species of animals which may colonize it. It then outlines some of the theories of succession which are then applied to published data from disturbed areas to evaluate the applicability of these models to the real-life situation. Next, the long-term outcome of animal succession is examined to see whether the original pre-disturbance community can be achieved. Finally, the influence on the developing animal community of extraneous factors such as fertilizers, climate and rehabilitation technology is reviewed. The influence of animals on the nature of the vegetation succession (Connell & Slatyer, 1977) is not considered in this Chapter since it has already been covered in Chapter 5. It should be realized, however, that any change in the nature of the plant community due to the action of animals will inevitably influence the composition of the present and future animal community.

Rehabilitation and the habitat templet

In order to follow the succession of fauna from the initial to the late stages of rehabilitation it is necessary to characterize the nature of the habitat during this period. The 'habitat templet' proposed by Southwood (1977) provides a useful framework for this. The templet condenses the variety of habitats onto two axes equivalent to their favourableness (particularly for breeding) and predictability (Fig. 6.1). Greenslade & Greenslade (1983) have added a third oblique axis for diversity and biological complexity which increases with predictability and favourableness.

Within each of these zones of the habitat templet there is selection for organisms with particular ecological strategies. For instance, highly fecund pioneer or opportunistic species, more correctly termed r-strategists, are favoured in less predictable environments such as temporary pools, transient decomposition habitats and arable land. On the other hand, more numerically constant species, whose numbers remain near their carrying capacity, are selected for in more predictable, and generally favourable environments such as oligotrophic lakes, rainforest leaf litter and climax forest communities. These are known as K-strategists. Biological diversity is usually high in these habitats as selection is for specialization, and mechanisms to ensure individual and progeny survival. An adversity, or A-selection strategy is selected for in habitats which are predictable but very unfavourable; examples of these are hot springs, montane vegetation and certain deserts. The harshness of these environments leads to low species diversity so there may be little interaction between species. In consequence, selection is for adaptation

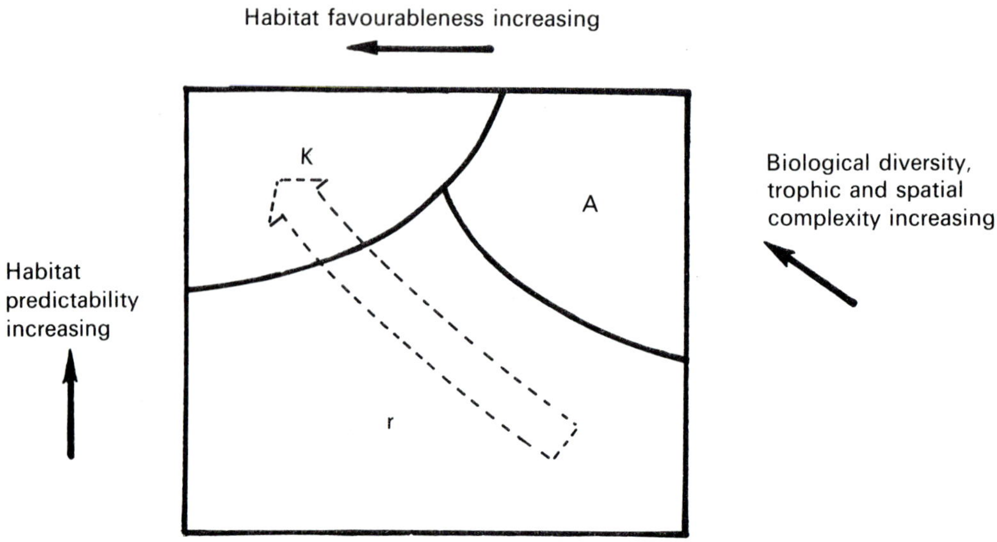

Fig. 6.1. The habitat templet: relationships between habitat predictability, habitat, favourableness, biological unpredictability and selection processes. The zones of K-, r-, and A-selection are areas in which each type of selection tends to predominate (from Greenslade & Greenslade, 1983). The dotted line indicates the change in characteristics of a rehabilitated area as it matures.

to the physical environment, for low energy cost resting stages and mechanisms to ensure synchrony of activity when favourable conditions resume.

It should be stressed that the three types of selection intergrade with each other and that there are no clear boundaries between them. However, to enable the reader to identify the ecological strategy of organisms which might occur in reclaimed land the suites of characteristics of the extremes of the r–K selection spectrum are presented in Table 6.1.

Where does an area of recently disturbed land fall on the habitat templet? Prior to rehabilitation being carried out the degraded land is usually characterized by a low spatial complexity, an absence of biological diversity, a harsh microclimate and, in biological terms, an extremely unpredictable future. These conditions place the degraded area within the r-selecting zone of the habitat templet (Fig. 6.1). However, since the frequency of the disturbance is low (usually once) and its duration short, there is little opportunity for *in situ* selection to take place. The pool of organisms available to colonize the rehabilitated area may therefore be moulded by the type of surrounding environment in

Table 6.1. *The contrasting suites of characteristics of the extremes of the r–K selection spectrum (from Southwood,* 1977)

r-species	K-species
Short generation time	Long generation time
Small size	Large size
High level of dispersal	Low level of dispersal
Much density independent mortality	High survival rate, especially of reproductive stages
High fecundity	Low fecundity with high parental investment *or* Iteroparity, often with synchronous breeding
Panmictic	Territorial
	Clonal
Intraspecific competition 'scramble type'	Intraspecific competition often 'contest type'
Low investment in 'defense' and other interspecific competitive mechanisms	High investment in 'defense' and other interspecific competitive mechanisms
Time efficient	Food and space resource efficient
Populations often 'over-shoot'	Populations seldom 'overshoot'
Population density very variable	Population density relatively constant from generation to generation

which the disturbed land is located. Disturbed land which occurs within an r-selecting environment should have a large pool of suitably adapted species which can colonize the area, while more predictable and favourable surroundings may provide fewer candidate colonizers.

I am unaware of any studies which have been designed to test this hypothesis although work on ant colonization of Western Australian mines indicates that recolonization of iron ore mines in the harsh semi-desert Pilbara region is considerably more rapid than in the bauxite mines situated within the more equable Mediterranean region (Table 6.2). Since the Pilbara climate is often unfavourable and the summer rainfall is unpredictable, the assumed presence of a wide range of r-strategists may explain the rapid colonization of ants. Other explanations based on the apparent ease of restoring a relatively simple desert environment may be advanced so, until further studies are performed, we may only speculate on this hypothesis.

As the rehabilitation matures, the vegetation buffers the microclimate

Table 6.2. *Comparison of the number of ant species in unmined and rehabilitated areas in the mediterranean and harsher semi-desert regions of Western Australia*

	Number of Species Present		
Locality	Unmined area	Rehabilitated area	Percentage return
Mediterranean*	27–33	13–20	39–74
Pilbara**	16	16	100

Notes: * from Majer *et al.*, 1984; ** Dunlop *et al.*, 1985.

and the area moves towards the K-selecting zone of the habitat templet (Fig. 6.1). It then tends to become colonized by K-strategists which occur in the locality. The speed with which this happens may depend on the closeness, in the structural sense, of the rehabilitation to that of the original habitat.

Once again it is difficult to find comparable data with which to test this hypothesis. Comparison of prairie in the western United States, which is relatively easy to reproduce in the structural sense, with forested country in the east provides a clue. Small mammal colonization of rehabilitated coal mines in Campbell County, Wyoming (Prairie) is relatively rapid in comparison to those in coal mines at Danville, Illinois (forest) (Fig. 6.2). This could well be explained by the fact that 5-year-old rehabilitation in prairie may cater for the habitat requirements of many of the original prairie species while that in Illinois forested country, still lacks the structural and floristic complexity of the original vegetation and therefore does not. Van Waggoner (1978) has made a similar observation on small mammals colonizing lignite mined areas in central Texas. Here the proportional return of species to areas which were formerly grassland or fields was greater than that returning to previously wooded country.

The examples given in the above paragraphs are not conclusive; other studies may provide data which contradict what has been said here. There is a need for further investigations, using comparable sampling techniques, on the colonization rates of selected taxa in both r- and K-selecting environments. These should also be performed in habitats of differing structural and floristic complexity. If general trends emerge from such investigations, the nature of the pre-disturbance environment will need to be considered when drawing up short- and long-term rehabilitation performance standards.

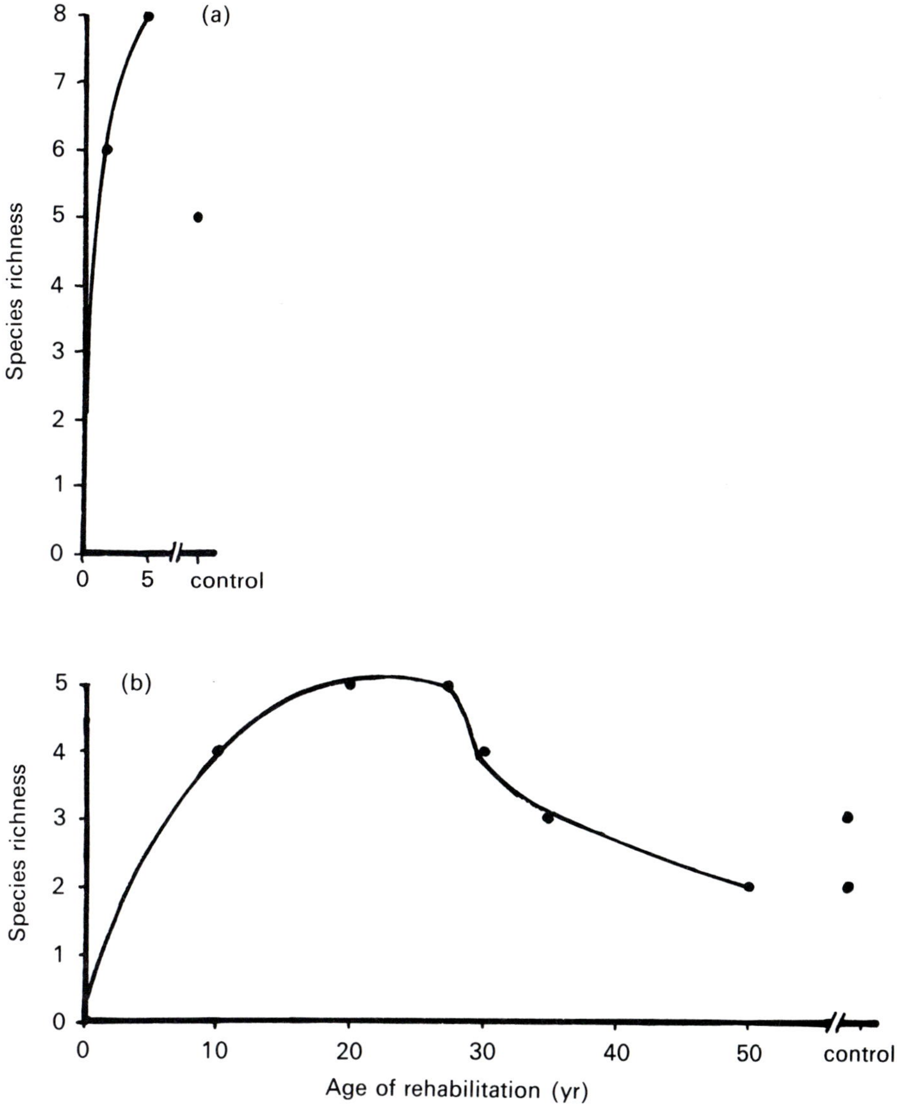

Fig. 6.2. Comparison of the rate of recolonization of small mammal species in rehabilitated coal mines within (a) prairies in Wyoming, and (b) forested country in Illinois (data respectively from Hingten & Clark, 1984; Wetzel, 1958).

Succession of fauna

In this section we will examine the possible long-term sequences of animal colonization in rehabilitated areas. To do this it is of course necessary to consider the earlier events as well. Coverage of this section must inevitably be speculative due to the lack of long-term studies.

The different pathways of succession have already been described in the previous Chapter. To recapitulate, these are:

1. The *facilitation* pathway in which early colonists facilitate the entry of later species by altering the environment so that it is more favourable for later species.
2. The *initial floristic composition* pathway in which most organisms are present in the early stages with various species successively becoming dominant as their life history characteristics are exhibited with the passage of time.
3. The *tolerance* pathway in which there is a change from species which are intolerant of interspecific competition to those which can tolerate competition from earlier species and can grow at lower levels of resources.
4. The *inhibition* pathway in which the earlier species inhibit the establishment of later species and therefore replacement by later species occurs only after death of the earlier species, and
5. The *chronic disturbance* pathway in which one of a variety of final community configurations can result, depending on which species happen to colonize the area first.

These pathways, which are summarized in Fig. 5.1, are not mutually exclusive and elements of more than one type may be recognized within a particular succession. The evidence for these pathways has, until relatively recently, been derived from the study of plant communities (Connell & Slatyer, 1977). Possibly they are not all applicable to heterotroph succession because of, for instance, the mobility of animals. The previous Chapter has indicated some of the ways in which animals can influence the outcome of plant succession. The following subsections will now look at the succession of animals in rehabilitated areas. It will also discuss some of the ways in which plants may influence the nature of the animal succession. First, the animals will be considered at the individual species level to see which of the pathways most accurately describes the colonization of fauna. Second, the overall pattern of species richness curves through time will be looked at. This is relevant to the measurement of the particular reclamation performance standards. Finally the likelihood of the fauna returning to the pre-disturbance composition will be evaluated.

Consideration at the species level

Whichever pathway of succession is followed, animals must first colonize the area which is being rehabilitated. The rate at which species

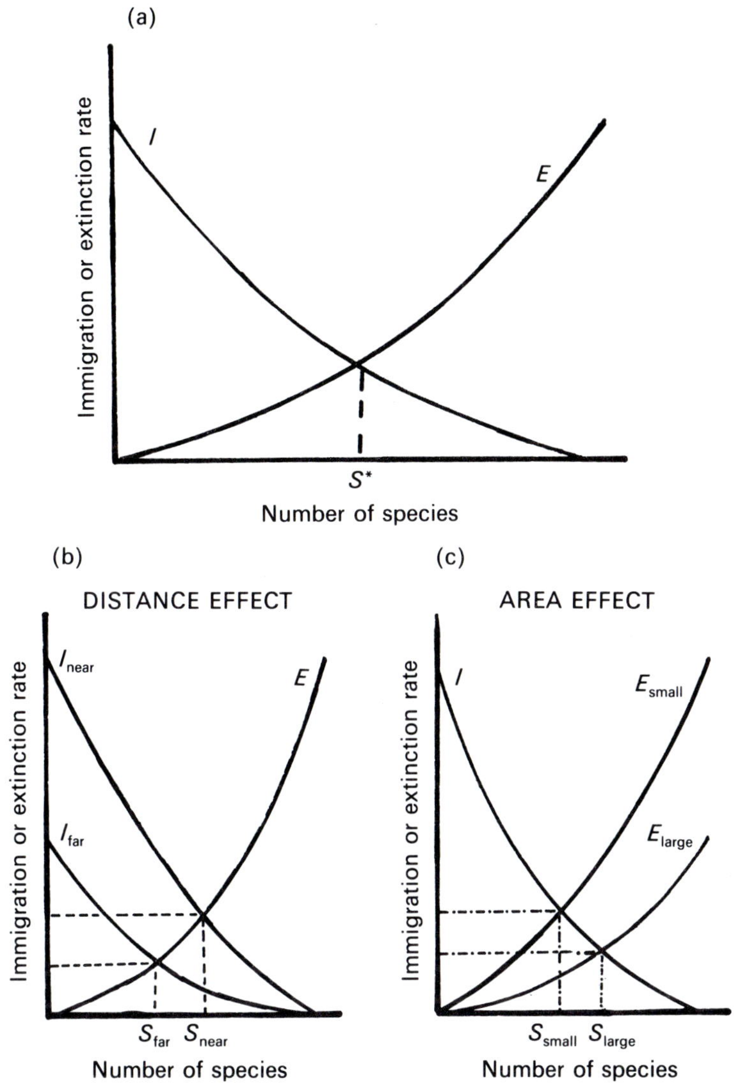

Fig. 6.3. Equilibrium model of island biogeography in which the rate of immigration (*I*) and extinction (*E*) interact to describe the number of species in equilibrium (S^*). (b) and (c) describe how the rate of immigration is influenced by distance from colonizing source and the rate of extinction is influenced by area of the island (adapted from Simberloff, 1974).

colonization takes place depends both on the nature of the rehabilitated area and also the characteristics of the species themselves.

Influence of size and isolation of the area Since the rehabilitated area may differ in its physical and botanical structure from that of the surrounding vegetation, it may function as an ecological 'island'.

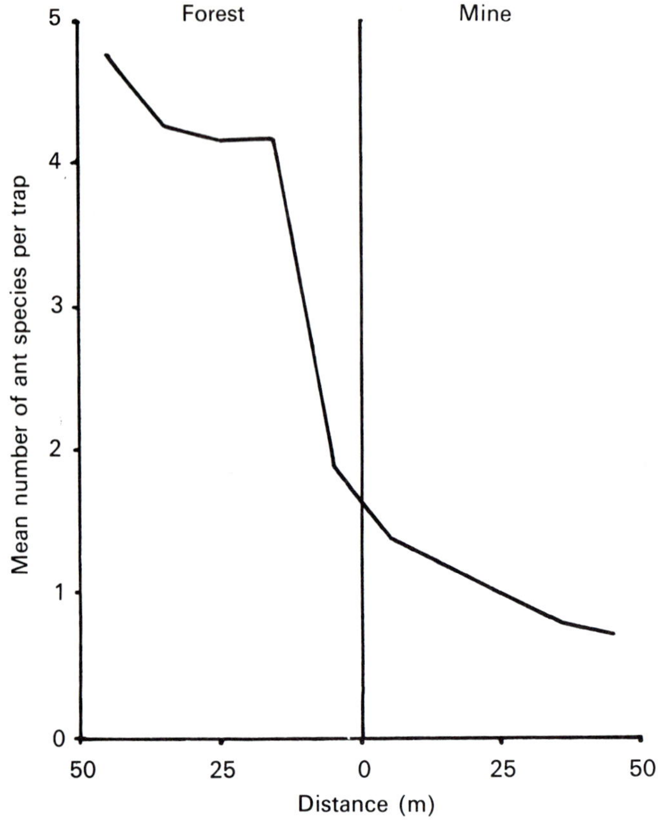

Fig. 6.4. Variation in number of species of ants caught in pitfall traps extending 50 m into a newly topsoiled bauxite mine from the adjacent forest (from Majer. 1980).

Therefore, according to MacArthur & Wilson's (1967) theory of island biogeography, the number of species in the island of rehabilitation is associated with the dynamic equilibrium between immigration and extinction of colonizing species (Fig. 6.3a). The major factors which control the resulting species richness are the degree of isolation and also the area of the island – fewer immigrants are likely to reach isolated areas (Fig. 6.3b), more extinctions are likely in smaller areas (Fig. 6.3c).

There is considerable evidence from the reclamation literature that the degree of isolation influences the number of species occurring in an area. For instance, Majer (1980) found that the number of ant species in recently topsoiled Australian bauxite mines declined with increasing distance from the adjacent forest margin (Fig. 6.4). Patches of unmined vegetation within New Mexico coal mines have also been shown to contribute significantly to bird species richness within rehabilitated areas

(Steele & Grant, 1982). The most striking evidence comes from the study of Dutch polders which have recently been reclaimed from the sea. The colonization by moles (*Talpa europaea*) may clearly be related to the proximity of the original mainland, with animals spreading at the rate of about 2 km yr^{-1} along dykes, roads and banks of ditches (Haeck, 1969). Similar, but more rapid, patterns of colonization have been demonstrated for insects and other arthropods in the polder (see review by Mook, 1971).

I am not aware of any comparative studies of the influence of size of rehabilitated area on the rate of immigration or extinction of colonists. While the rate of immigration may be enhanced by the larger area and longer edge of the rehabilitation, the degree of isolation of its centre is also likely to increase. These two factors are therefore likely to operate in antagonism. There is however, some evidence for high extinction rates operating in small areas of rehabilitation. In a recent 2-year study of an isolated 7 ha area of rehabilitated iron ore waste, Walker, Osborne & Dunlop (1986) found that of 10 reptile species immigrating into the area, only four persisted as stable populations; the other six species became extinct.

Influence of species characteristics The rehabilitated area does not receive a random influx of animals in proportion to its degree of isolation and size. Superimposed upon these variables are the colonization characteristics of the animals themselves. By comparing the species composition of carabid beetles in Dutch polders with that in the surrounding species pool, Ranta & As (1982) found that colonization patterns were non-random. Some species were more successful than others at invading habitat islands and these tend to be characterized by having long wings and a small body size. My own work on ant recolonization of mined areas throughout Western Australia has shown that, regardless of type of disturbance or location, many of the early colonizers are the same species, and are not those which are common in the native vegetation (Majer *et al.*, 1984; Woodroff & Majer, 1981). Vertebrates show similar patterns of non-random colonization. For instance the prairie deer-mouse (*Peromyscus maniculatus*), the prairie vole (*Microtus ochrogaster*), and the house mouse (*Mus musculus*) are all early colonizers of coal mined areas in Wyoming, Indiana and Illinois, even though they are not necessarily the most abundant species in native vegetation (Hingten & Clark, 1984; Sly, 1976; Wetzel, 1958).

To summarize this section so far, the initial colonization of animals in rehabilitated areas is non-random and depends on the characteristics of the individual species. In addition, the rates of immigration and extinction are dependent on the degree of isolation and the size of the area. Putting this in terms of the pathways of succession described earlier (Fig. 5.1), the area becomes colonized by a suite of species consistent with the facilitation, tolerance or inhibition pathways. There does not appear to be much evidence of the initial floristic composition pathway being followed. Different starting points of the chronic disturbance pathway may arise as a result of the variation in composition of colonizing species from the edge to the more isolated centre of the disturbed area.

A number of physical and botanical changes occur as the rehabilitation matures. These include some of the following: a reduction in the percent bare ground, a more buffered microclimate, an increase in plant cover in certain strata, an increase in plant height, an initial increase in plant species richness, an increase in structural complexity of the vegetation, an increased litter layer and an increase in dead wood and logs. As summarized in Table 6.3, there is considerable evidence for coupling between the developing plant and animal communities in reclaimed lands. This survey of arthropod, bird and mammal colonization studies (Table 6.3) indicates numerous positive correlations between animal biomass, abundance, richness and diversity with plant cover, richness and diversity and also structural diversity. The only negative associations are those between soil arthropod diversity and declining plant diversity in maturing quarries in Britain (Usher, 1979) and a reduction in bird species richness in older Virginian surface-mine rehabilitation where canopy closure caused a simplification of the understorey (Chapman, McGinnes & Downing, 1978).

Brenner & Kelly (1981) and Brenner, Kelly & Kelly (1982), among others, have also demonstrated relationships between the density of particular plant and particular bird or mammal species. It therefore seems that the presence of certain plants provides opportunities for the establishment of animal species which benefit from the plant in some way. Although the facilitation pathway was conceived to describe the succession of plant communities, these relationships suggest plant facilitation of animal invasions. Furthermore, animal succession is also linked with changes in the nature of the physical environment and the litter layer. Fox (1982) noted a positive correlation between both ant abundance and small mammal (*Pseudomys*) biomass with a more penetrable substrate while Majer *et al.* (1984) found that ant species

Table 6.3. *Summary of arthropod, bird and mammal studies from rehabilitated lands in which positive statistical associations between animal colonization and plant parameters have been demonstrated*

Dependent variable	Independent variable(s)	Reference
Ant diversity in Australian sand mined land	Floristic composition plus foliage height diversity	Fox & Fox (1982)
Ant diversity or richness in Australian bauxite mined land	Plant species diversity or richness, plus percentage plant cover	Majer *et al.* (1984)
Insect species richness in Alabama surface-mined land	Three-dimensional cover of vegetation	Hawkins & Cross (1982)
Beetle species richness in Wyoming surface-mined land	Plant species richness	Parmenter & MacMahon (1988)
Soil arthropod diversity in disused British quarries	Plant diversity declines with time, while arthropod diversity increases	Usher (1979)
Ground and foliage arthropod abundance (selected orders) on Arizona transmission line corridors	Total plant cover plus percentage cover in selected strata	Johnson *et al.* (1983)
Bird species diversity in Illinois surface-mined land	Foliage height diversity, plus log. percentage plant cover	Karr (1968)
Number of birds in Pennsylvania surface-mined land	Total plant biomass, vegetation height, plus vegetation density	Brenner & Kelly (1981)
Bird species richness in Virginia surface-mined land	Positive relationships with vegetation heterogeneity, plus percent ground cover but negative with canopy density of older areas	Chapman *et al.* (1978)
Small mammal (*Pseudomys*) biomass in Australian sand-mined land	Plant species diversity, plus proportion of heath plant species	Fox (1982)
Selected small mammal numbers in Indiana surface-mined land	No relationship with herbaceous cover but woody cover important for some species	Sly (1976)

Table 6.3. (*cont.*)

Dependent variable	Independent variable(s)	Reference
Selected small mammal numbers in Pennsylvania surface-mined land	Numbers of some species are positively correlated with total plant biomass, vegetation height or foliage density and also with the biomass of individual plant species	Brenner *et al.* 1982

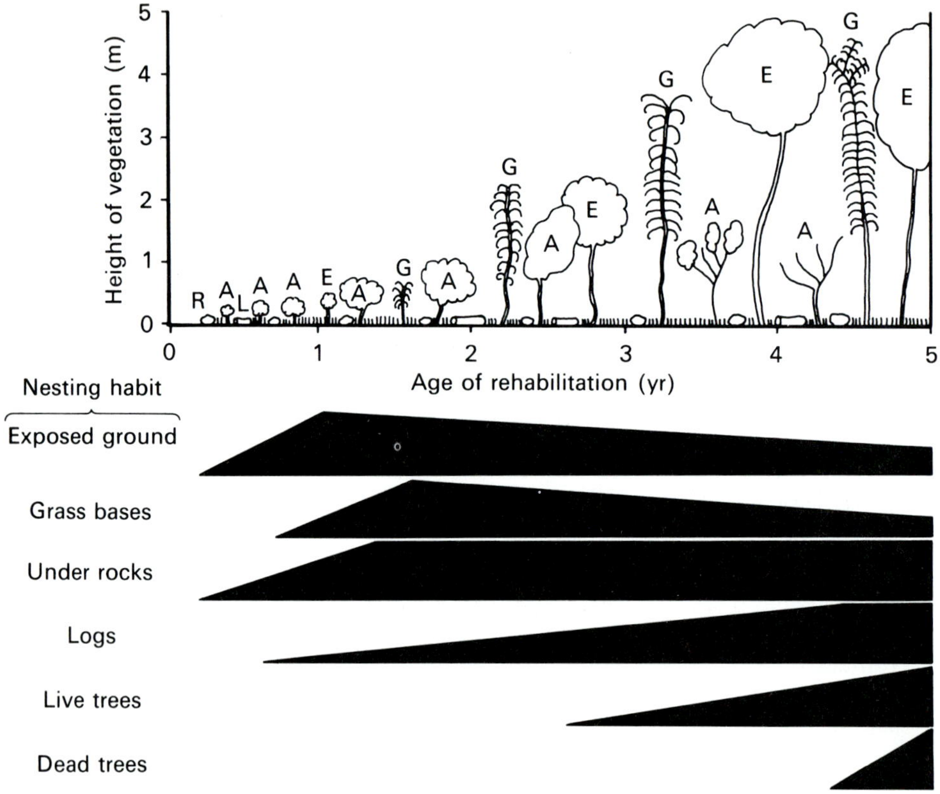

Fig. 6.5. Schematic representation of ant colonization in rehabilitated manganese mines on Groote Eylandt, northern Australia. Key: A, *Acacia* spp; G, *Grevillea* sp.; E, *Eucalyptus* spp.; L, logs, R, rocks (adapted from data in Majer, 1983*a*).

diversity is positively associated with percent litter cover and the presence of larger logs.

The outcome of this coupling between animals with changes in the physical and botanical environment can be seen with the ant fauna in rehabilitated manganese mines in northern Australia (Fig. 6.5). Here, the developing ant community tracks the changes in vegetation, litter, dead wood and bare ground. Rehabilitation personnel should be aware of this coupling since the provision of adequate plant cover is not an adequate measure of the success of rehabilitation. Although plant cover is obviously of importance to the developing fauna, the relationships shown in Table 6.3 indicate that the provision of adequate floristic and structural diversity as well as logs and litter is important if the full range of native animals are to colonize the area.

Brenner *et al.* (1982) have suggested that in addition to the major influence of plant community composition, inter- and intraspecific interactions between animals may also direct the nature of the succession. In other words the tolerance or inhibition pathways may also contribute to the process of animal succession in rehabilitated land.

The importance of interspecific competition in moulding animal community structure has long been recognized (Cody & Diamond, 1975). Until recently, its importance in reclaimed land has rested on supposition. A series of papers by Barry and Marilyn Fox on ants and small mammals in Australian sand mined areas provides the first experimental evidence that interspecific competition influences the fauna replacement process in rehabilitated land. Their surveys of rehabilitation of increasing age revealed that the ant, *Iridomyrmex* sp. C, initially dominated the rehabilitated sites. After 8–9 years of rehabilitation, *Iridomyrmex* sp.C was replaced by *Iridomyrmex* sp. A (Fig. 6.6a). Also, the introduced mouse, *Mus musculus*, was replaced by the native *Pseudomys novaehollandiae* as the rehabilitation aged (Fig. 6.6b).

When the abundance of *Iridomyrmex* sp. A was experimentally reduced, the other ant increased in abundance and spread into the vacated space (Fox, Fox & Archer, 1985). The researchers concluded that species C has its foraging area, and hence abundance, reduced by the presence of species A. Similarly, when *Pseudomys* was experimentally removed from, or added to, 1 ha plots of rehabilitation, *Mus* respectively increased or declined in abundance (Fox & Pople, 1984). Once again this outcome was considered to result from competition between the two species of mouse. Unfortunately the reciprocal experiments were not performed, but it is likely that the early stages of rehabilitation may favour competitive superiority by one species while later on this may change to advantage the other species. The competitive advantages of one species over the other may in turn be associated with changes in the nature of the habitat.

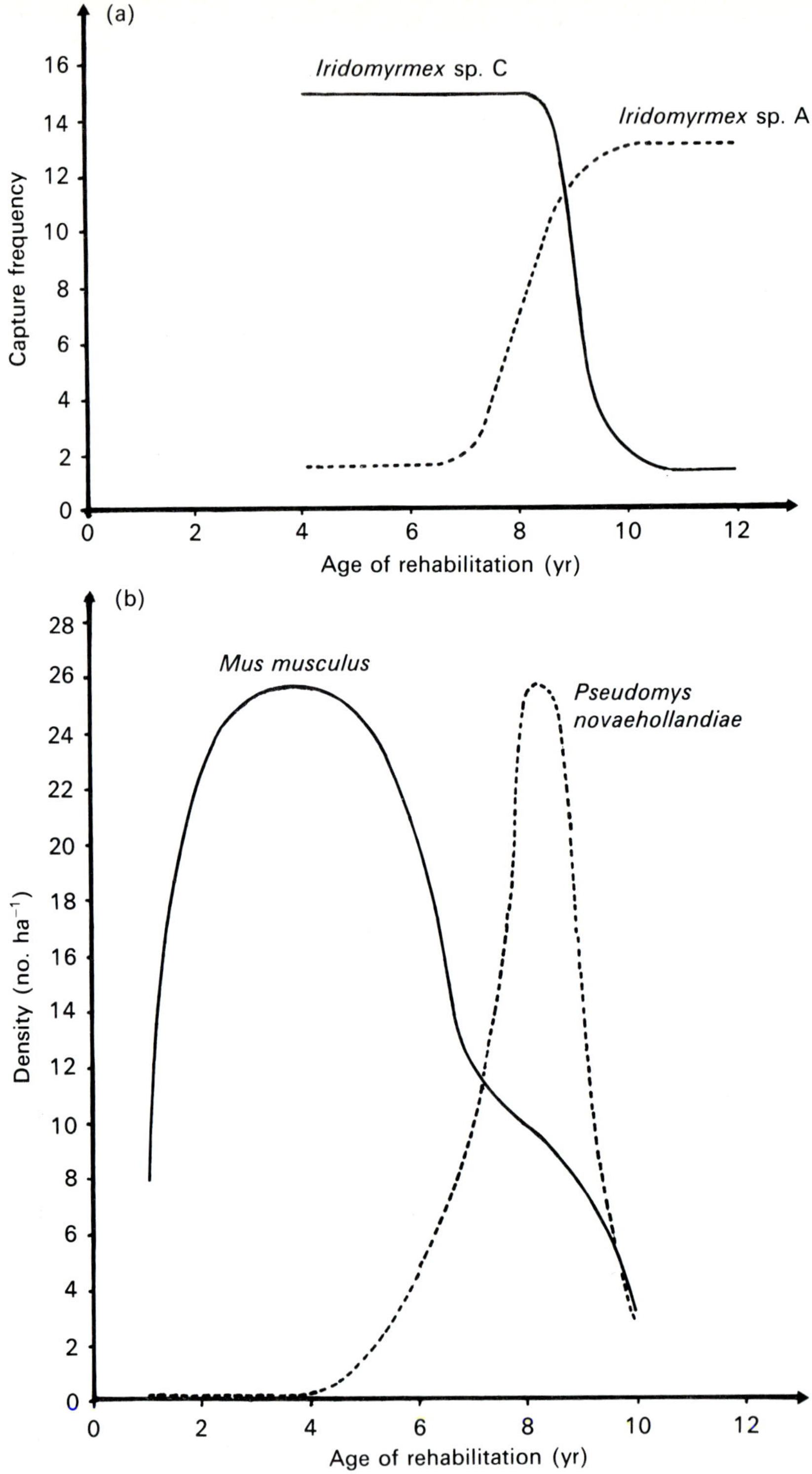

Fig. 6.6. Sequence of replacements of (a) the ant *Iridomyrmex* sp. C by *Iridomyrmex* sp. A and (b) the mouse *Mus musculus* by *Pseudomys novaehollandiae* in rehabilitated mineral sand mines at Myall Lakes, easterr Australia (from Fox & Fox, 1982, 1984).

Fox (1985, 1988) has extended these observations by drawing on data for small mammal communities in Australian heathlands. The eight species which were studied exhibit clear niche separation along habitat, diet and body mass axes. Species from the same genus tend to occupy more similar diet niches than less closely related animals. Therefore, there must be sufficient diversity and abundance of resources to accommodate one member from each genus before a second member from any one dietary group can be accommodated. This assembly order may explain why the accession of species in a particular area may follow a particular pattern, one which is governed by the provision of more appropriate, or more abundant, habitat or diet resources in the rehabilitated area. Following on from this, Fox (1985) suggests that there will be 'forbidden states' in which certain groups of animals cannot occur together in the absence of other genera. This is because 'forbidden state' groupings would result in unused resources and the existence of an 'empty niche', which Fox suggests is an unstable situation.

The foregoing ideas are of considerable importance to the fauna colonization process in rehabilitated areas since they may dictate the type of animal present at a particular time and also the replacement sequence of species. It may be possible to manage the rehabilitation procedure to provide resources for a particular animal species or to accelerate the successional sequence. However, in many cases it may be necessary to wait for the area to pass through a series of species replacements before a particular desired outcome can be achieved.

Consideration at the community level

So far this Chapter has largely considered succession at the species level. I will now look at the pattern of disturbance and recovery at the broader levels of the community and ecosystem. This is of practical importance because success of rehabilitation is often measured in terms of community parameters such as species diversity or proportion of original species present. It is therefore desirable to be able to predict, or at least understand, how these parameters may change over time.

There is a profusion of conflicting terminology relating to this subject so, to avoid confusing the reader, I have adopted that recently put forward by Westman (1986). This has been derived by analogy with the terminology used by physicists when considering the displacement and subsequent recovery of an object such as a coiled spring and is summarized in Fig. 6.7. The vertical axis of Fig. 6.7 could describe the population density of an individual species, the species richness of the

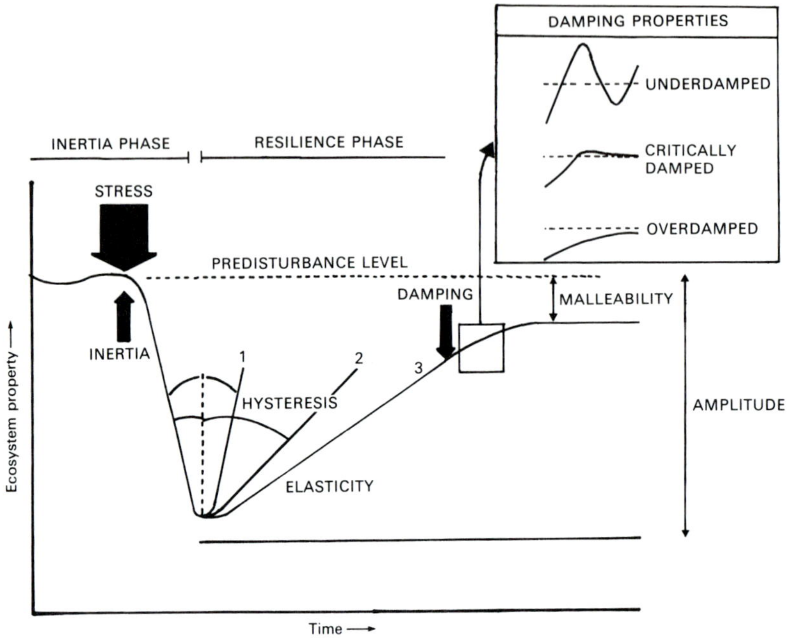

Fig. 6.7. Model illustrating the response of an ecosystem to disturbance showing its resistance to change under stress (inertia) and its five properties of resilience following cessation of the stress (elasticity, amplitude, hysteresis, malleability and damping).

entire community or it could be an index which integrates characteristics of both the physical and biotic environment. Prior to disturbance the ecosystem exists at a pre-disturbance level which, depending on the stability of the ecosystem, fluctuates to a varying extent. The ecosystem, becomes stressed as a disturbance is initiated. The degree of strain inflicted on the ecosystem depends on both the magnitude and period of stress. Thus the strain on a heathland would be larger during open-cut mining than during burning. Also, the strain on a wetland would become greater with the increase of time that it is used as a sanitary landfill site.

The ecosystem often has some ability to resist the external stress applied to it and this is known as the inertia of the system. Westman (1986) cites examples of varying plant community responses to air pollution. However, in terms of magnitude of the stresses discussed in this book, inertia is probably not a significant factor.

The 'disturbance process' and the 'retrogression' which follows is generally followed by a period of recovery which may be characterized

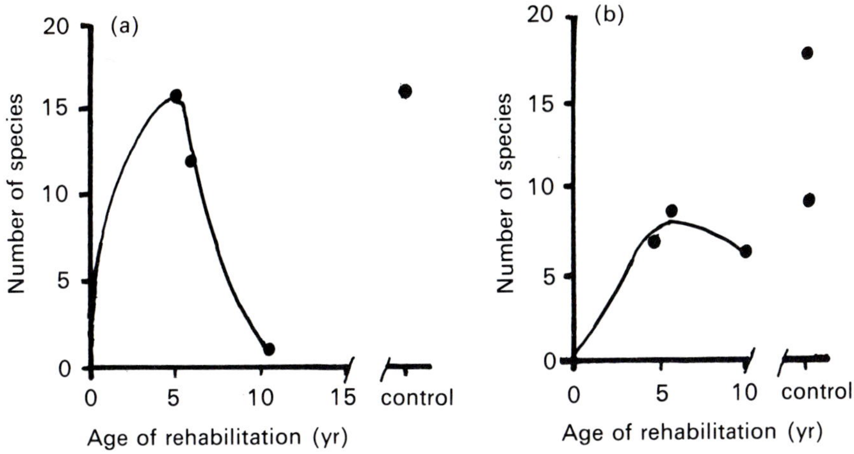

Fig. 6.8. Comparison of the rate of recolonization of (a) birds, and (b) herptiles in rehabilitated bauxite mines in Western Australia (data respectively from Nichols & Watkins, 1984; Nichols & Bamford, 1985).

by a number of parameters. Resilience is the degree, manner, and pace of recovery of the ecosystem to the pre-disturbance level. Westman (1986) has identified four components of resilience: elasticity, amplitude, hysteresis and malleability (Fig. 6.7).

Elasticity is the rate of recovery of an ecosystem property following disturbance. Thus, lines 1, 2 and 3 represent a gradient of decreasing elasticity of ecosystems. I have already given examples where ecosystems may be more elastic than others as a result of their being r-selecting (Table 6.2) or structurally simple (Fig. 6.2a) environments. Climatic pattern may also influence elasticity; this will be discussed later in this Chapter. Certain taxa may also be more elastic than others. For instance, the rate of return of species over the first 5 years in Western Australian bauxite mines is greater for birds than for reptiles and frogs (Fig. 6.8). This may be related to the dispersal abilities of these animals or to the ease with which their habitat requirements may be catered for.

The amplitude component of resilience is the threshold level of strain beyond which return to the original state no longer occurs. This is represented by the difference between the predisturbance level and the solid horizontal line on Fig. 6.7. It is highly probable that major disturbances such as surface-mining may stress some ecosystems beyond the amplitude threshold. However, without long-term recolonization studies, coupled with adequate base-line data, it is not possible to distinguish between an ecosystem which has been stressed beyond the

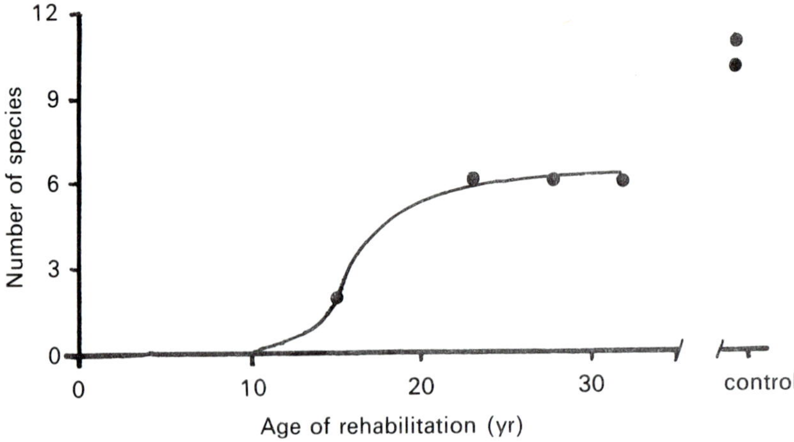

Fig. 6.9. Rate of recolonization of thrips (Thysanoptera) on Illinois surface-mine spoils (data from Stannard, 1967).

amplitude threshold and one which is merely taking a long time to recover. Data on thrip (Thysanoptera) recolonization of Illinois surface-mine spoils indicate that total recovery has not occurred after 32 years (Fig. 6.9). The index of similarity between thrip species in the 32-year-old spoils to those in the forest was 0.32 compared with an inter-forest value of 0.67 (Sorensen's Index of similarity = number of species in common between sites **a** and **b** ÷ number of species at site **a** plus number at site **b**). This, and the levelling off in thrip species richness after 23 years (Fig. 6.9), may indicate that the amplitude threshold of the original ecosystem may have been exceeded.

Hysteresis concerns the degree to which the path of change in an ecosystem property under chronic stress differs from the path of succession upon immediate removal of the stress. Thus, line 1 in Fig. 6.7 is simply the reverse of the path of retrogression and represents a no-hysteresis situation. Lines 2 and 3 represent increasing differences between the retrogression and succession paths and hence lines of increasing hysteresis. The implication of this is that if hysteresis is known to exist in an ecosystem, the techniques used in rehabilitation need not attempt to re-establish the reverse of the sequence of changes observed at the time of disturbance. In view of the abrupt change in ecosystem properties during acute processes such as mining, the hysteresis concept probably has little value in the present context. It may, however, have application to the understanding of rehabilitation after chronic stresses such as pollution and grazing (Westman 1986).

The final characteristic discussed by Westman is the malleability of the ecosystem. This is the difference between the ecosystem's final recovery level and the predisturbance level; the greater the difference, the more malleable is the ecosystem. Undoubtedly when more long-term fauna studies have been performed, some ecosystems will be found to be more malleable than others. However, the nature of the stress also influences the ecosystem's malleability. For instance, in a study of ant recolonization of 30 bauxite mines, one site was selected which had been accidentally cleared although mining never took place. Ten years after rehabilitation the species richness of ants in this plot was disproportionately high in relation to mined plots of similar age (Majer *et al.*, 1984). The preservation of the original soil profile was probably the reason for the reduced malleability of this site.

Continuing with the analogy of a stretched spring, Fox & Fox (1986) described a fifth property of resilience: damping. This is the degree and manner by which a path of restoration is altered by any forces that affect the restoring force. Figure 6.7 shows the path of recovery of overdamped, critically damped and underdamped ecosystems. The first-mentioned ecosystem will take a long time to reach equilibrium but will not oscillate. The underdamped ecosystem will approach equilibrium rapidly but will overshoot and oscillate for some period of time while the critically damped ecosystem will approach equilibrium more slowly, overshoot slightly and then approach equilibrium level. This ecosystem equilibrates most rapidly and is therefore the most resilient of the three examples.

The existence of the damping property has important implications to the ways in which rehabilitation success is assessed. Firstly, a regenerating ecosystem which exhibits high elasticity may give the impression of being highly resilient. However, it may be an underdamped ecosystem which takes longer to equilibrate than a critically damped ecosystem of lower elasticity. Secondly, the existence of oscillations in ecosystem properties creates problems in selecting the interval after rehabilitation when performance standards should be assessed. There are many examples of oscillations in species richness of insects (Fig. 6.10a), birds (Fig. 6.10b) and small mammals (Fig. 6.2b) following rehabilitation so the problem of when to assess performance standards is widespread.

This section has highlighted some of the factors which need to be taken into account when measuring the effectiveness of a rehabilitation programme in encouraging the return of wildlife. The successional patterns which have been described are, to some extent, dependent on intrinsic properties of the ecosystem. However, its resilience may be

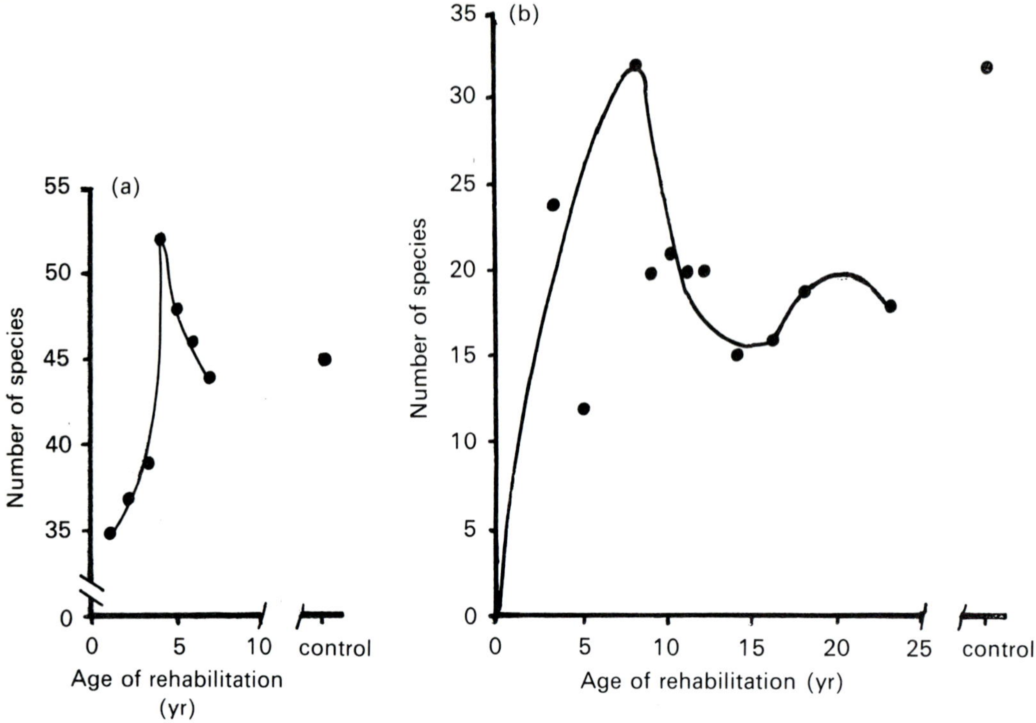

Fig. 6.10. Rate of recolonization of (a) beetles in rehabilitated Wyoming surface-mines, and (b) birds in rehabilitated Virginian strip mines showing oscillations in species richness that can occur during the course of succession (data respectively from Parmenter & MacMahon, in press; Crawford, Hardy & Abler, 1978).

modified by the type of rehabilitation which is carried out. Chapter 8 describes some of the ways in which the rehabilitation programme may be modified to change the resilience of the ecosystem. There are, however, three factors which have a major influence on the ecosystem's resilience – climate, fertilizer application and the post-disturbance land use. The following section deals with the ways in which these factors influence the succession of animals.

Influence of extraneous factors on succession

Climatic factors The climate of an area is governed by such factors as latitude, elevation, prevailing winds, moisture availability and terrain. Agronomists have long recognized the importance of climate to the growth and succession of plant communities. More specifically, the length of the growing season (a measure of the excess of effective rainfall

Table 6.4. *Estimated number of ant species found in 3-year-old rehabilitated plots within mined areas situated throughout Australia (adapted from Majer, 1984)*

Location	Mine type	Climate type	Annual rainfall (mm)	Number of ant species
Eneabba, W.A.	Sand	Mediterranean	550	6
Collie, W.A.	Coal	Mediterranean	973	8
Jarrahdale/Del Park, W.A.	Bauxite	Mediterranean	1155–1287	10
Myall Lakes, N. S. W.	Sand	Warm-temperate	1205–1362	11
Nth Stradbroke Island, Qld.	Sand	Sub-tropical	1645	18
Gove, N. T.; Groote Eylandt, N. T.; Weipa, Qld.	Bauxite or Manganese	Tropical monsoonal	1277–2083	22

over evaporation) has been isolated as the dominant influence, with rapid soil formation and faster, more diverse plant growth occurring in areas with the longest growing seasons.

Although it is well known that rehabilitation of disturbed areas is easier in areas with long growing seasons (National Research Council U.S., 1981), there has been little comment on how climate influences the resilience of the animal community. Mitchell's (1961) work on breeding bird populations in old field succession indicates that the length of growing season may influence the rate of succession of birds. He found only two breeding species and 1.2 breeding pairs per ha in 18-year-old fields on the Anoka Sand Plain, Minnesota compared with Odum's (1950) observation of nine breeding species and 3.2 pairs per ha in 20-year-old fields on the moister and warmer Highlands Plateau of North Carolina.

My own work on ant succession in rehabilitated Australian mines confirms this trend (Majer, 1984). Comparison of species richness of ants in 3-year-old rehabilitation throughout the continent reveals a cline in ant return rate, from lowest in the drier Mediterranean climatic zones to highest in the tropical monsoonal areas (Table 6.4). Variations in return rate are not correlated with species richness of animals in the original ecosystem. In view of the fact that this trend shows up through a backdrop of widely differing rehabilitation techniques, it appears that climate has an overriding influence on the rate of ant return in Australia. Rehabilitation method appears to be of secondary importance. Climate probably directly influences fauna return, by mediating the microclimate in which the animals live, and indirectly, by influencing the plant variables with which the animals are coupled (Table 6.3). These observations apply to ants although there is evidence that other invertebrates may respond to climate in the same way (Majer, 1983*b*).

If this cline is found to apply to animals and plants in general, it may be possible to divide Australia into zones, with each zone having a particular rehabilitation rate, or resilience characteristic. Undoubtedly the same exercise could be applied to either continents or countries. These zones would then provide valuable information on which to base guidelines and timescales for evaluating rehabilitation performance standards through time.

Application of fertilizers Most rehabilitation programmes involve the application of fertilizers and perhaps other compounds to ameliorate the soil at the time of planting or seeding. This greatly increases the primary productivity of the rehabilitated area with increases of up to 600% being reported in American surface-mines fertilized with nitrogen (Richardson *et al.*, 1975). Studies on Australian sand mines indicate that enhanced productivity is not necessarily correlated with enhanced species richness. Clark (1975) found that plant species composition in 3-year-old fertilized plots, when compared with control plots, produced a similarity index of 0.17 while the value for unfertilized plots was 0.27. Since fertilizing may accelerate or retard the succession it may be regarded as a damping force. In view of the close coupling between plants and animals during succession it is highly probable that fertilizing influences the succession of fauna so it may be regarded as an important influence on resilience of the animal community.

Ultimately, the applied fertilizer becomes exhausted. Good rehabilitation should be designed so that naturally occurring decomposer organisms and nitrogen fixers take over the role of fertilizing the area (see Chapter 3). This transition may be associated with a decline in productivity and a change-over to a slower succession, mediated by more natural ecological constraints. There is certainly evidence for this hypothesis based on the transition in species of soil endomycorrhizae during the early stages of sand mine rehabilitation (Jehne & Thompson, 1981).

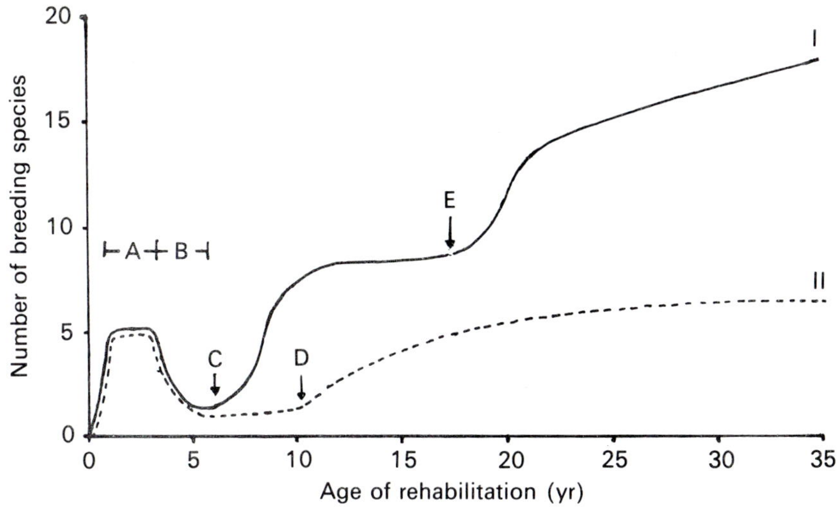

Fig. 6.11. Hypothetical colonization of breeding birds in two surface mines in northern West Virginia. Line I represents the situation where rehabilitation is successful, Line II where the effects of fertilizers and soil ameliorants wear off before plant establishment is complete. Letters A to E are explained in the text (from Whitmore, 1978).

Evidence that this may also influence the fauna has not yet been obtained although Whitmore (1978) has suggested two hypothetical successions of birds in West Virginian surface mines, one based on adequate nutrient availability and the other not (Fig. 6.11). The soil, which has an acidity problem, is initially treated with fertilizers, soil stabilizers and an acid neutralizer. Line I shows what will probably happen during successful rehabilitation. There is an initial peak of grassland nesting species (zone A), which declines in zone B as the grass becomes shaded out by shrub growth. There follows a subsequent increase in the richness of shrub nesting birds (point C) and later, as trees grow up, an increase in tree nesting species (point E). Line II shows the not-uncommon situation in which the effects of fertilizers and soil ameliorants wear off, leading to nutrient starvation and high soil acidity. The succession of birds in zone A is similar to line I but thereafter the poor soil characteristics inhibit plant growth and hence opportunities for nesting birds. At point D a few acid-tolerant shrubs and trees may colonize the area thus enabling a limited number of species of birds to enter and breed. In the long-term, the patterns of colonization of birds in the two areas are markedly different.

Whitmore's model is based upon observations made in numerous

Table 6.5. *Number of invertebrates and species of ants caught per* 140 *pitfall-trap-nights in* Eucalyptus *forest, grazed farmland,* 6*-year-old bauxite mine rehabilitated to native forest and* 6*-year-old sand mine revegetated to grazed pasture (adapted from Majer,* 1985*b*)

	Eucalyptus forest	Forested bauxite mine	Grazed farmland	Pastured sand mine
Total invertebrates (excluding ants)	535	63	238	634
Total ants	753	102	105	92
Total ant species	19	8	4	3

mined areas. Undoubtedly, if more fauna studies were performed in areas subjected to differing fertilizer regimes, fertilizers would be found to have a widespread effect on the pattern and rate of animal succession. This is therefore one other consideration which should be taken into account when assessing the recolonization of fauna in rehabilitated land.

Rehabilitation of native vs agricultural ecosystems The land which is subjected to mining, or other transient use, may have been under native vegetation or it may have supported an agro-ecosystem of considerably lower complexity. Although there are exceptions, the previous land-use tends to dictate the rehabilitation option which is chosen (see Table 1.2). Thus pastures tend to be rehabilitated to pasture and native forests to forest.

The attainability of the pre-disturbance fauna is considerably easier in the former than the latter option. Table 6.5 compares the surface-active invertebrates in a 6-year rehabilitated bauxite mine with that in the original *Eucalyptus* forest. Most taxa are considerably less abundant in the rehabilitation than in the forest and the species richness of ants is only 42 % of the pre-mining level. By comparison, nearby sand mines in old agricultural land produced a fauna which was, for many taxa, more abundant than in the original farm land. Ant species richness had risen to 75 % of the pre-mining level within 6 years of rehabilitation.

These results indicate the relative ease with which rehabilitation may be carried out in agricultural areas when compared with native ecosystems. Clearly this results from the greatly differing complexities of the two ecosystems which are being rehabilitated. It follows then that the time frames for measuring the success of rehabilitation in these two options will differ and this should be recognized by the agencies responsible for monitoring this activity.

The end product of succession It is obviously desirable to know how similar the final community will be in terms of species composition to that of the original forest. The answer to this question is hampered by the paucity of long-term data and also by changes in rehabilitation technology. Older types of rehabilitation are generally less sophisticated than that carried out today and are therefore not representative of current rehabilitation practices.

Table 6.6 shows some similarity index values for fauna in older rehabilitation versus pre-disturbance vegetation and, as a yardstick of variability, between two plots of pre-disturbance vegetation. Species composition differs between the control plots which have been used in all of these studies, indicating the great variation in species composition that occurs between areas. With the exception of the > 56-year-old bird study plots in Illinois, all rehabilitated areas exhibit lower similarity with control plots than do the control plots with each other. Thus, the fauna in all but one of these rehabilitated areas has not converged with that of the original vegetation in the 5–32 year time scale represented by the data. The similarity index of the > 56-year-old plot falls within the same range as the inter-control value although this itself is a low value due to the high variability in fauna composition between control plots.

These data suggest that a considerable period of time is needed for the composition of the fauna of the rehabilitated areas to approach that of the pre-disturbance vegetation. As already mentioned, if the threshold amplitude has been exceeded, the community may never recover to its pre-disturbance composition. Parker *et al.* (1984) describe a succession of soil fauna in New Mexico surface mines in which the successional trend is away from the undisturbed biota. Secondly, if the ecosystem is particularly malleable (Fig. 6.7) the succession will still not converge on the original pre-disturbance level.

Various writers have pointed out the possible existence of multiple stable equilibria in which the developing community may equilibrate at more than one possible combination of species. Demonstration of this possibility in rehabilitated land is not possible due to the absence of long-term studies on paired plots which have been subjected to identical rehabilitation treatments. Neumann (1971) studied the colonization of

Table 6.6. *Comparison of the Sorensen's index of similarity for selected taxa for which long-term data are available*

Taxon	Type of disturbance	Age of oldest rehabilitation (yr)	Index of similarity: Rehab. vs control	Index of similarity: Control vs control	Reference
Slaters (Isopoda)	German coal spoils	28	0.66	0.91	Neumann (1971)
Millipedes (Diplopoda)	German coal spoils	28	0.72	0.90	Neumann (1971)
Beetles (Carabidae)	German coal spoils	28	0.62	0.91	Neumann (1971)
Thrips (Thysanoptera)	Illinois surface-mines	32	0.32	0.67	Stannard (1967)
Herptiles	Australian bauxite mines	10	0.54	0.64	Nichols & Bamford (1985)
Birds	Australian bauxite mines	11	0.57	0.75	Nichols & Watkins (1984)
Birds	Illinois surface-mines	>56	0.59	0.59	Karr (1968)
Small mammals	Wyoming surface-mines	5	0.67	0.75	Hingten & Clark (1984)

Note: Comparisons are made between fauna in oldest rehabilitation and pre-disturbance vegetation and also between two areas of pre-disturbance vegetation.

slaters (Isopoda), millipedes (Diplopoda) and ground beetles (Carabidae) in 28-year-old forested coal mine spoils. If we allow for the fact that the most isolated area had become colonized by fewer species than the area close to the original forest, both areas were populated by a remarkably similar array of species. Indices of similarity for each taxon are respectively 0.66, 0.87 and 0.74. Brenner & Kelly's (1981) work on birds in 15-year-old surface mines suggests a similar trend (index of similarity is 0.76). Thus, based on the limited examples available, it appears that rehabilitated areas of similar age and rehabilitation treatment will tend to become colonized by a similar range of animals.

Summary

This review has shown that in both the long- and short-term the recolonization of fauna is influenced by a wide range of factors. These include the availability of suitable colonizing species in the locality, the order in which species initially colonize the area, the degree of isolation of the rehabilitated area, and also the floristic and structural nature of the surrounding vegetation. Of overriding importance is the prevailing climate, the applied fertilizer regime and whether the area is rehabilitated to native vegetation or to a simple agro-ecosystem.

All of these factors will influence the resilience of the animal community and some also influence the pattern of animal recolonization. Therefore, wide differences exist between the rate and patterns of animal recolonization in different regions. Differences may also exist between different taxa within a particular region.

These complexities point to the need for more baseline surveys of the characteristics of the fauna in areas which are about to be disturbed. This will enable us to understand the factors influencing their ability to recolonize rehabilitated land. Secondly, there is a need for carefully monitored long-term studies of fauna in rehabilitated lands in a variety of vegetation communities and climatic zones. These studies should adopt identical monitoring techniques so that valid comparisons may be made between areas and generalizations constructed on the recolonization characteristics in areas of particular vegetation and climatic type.

The acquisition of such data should enable the agencies responsible for the design, execution and monitoring of rehabilitation to develop sound guidelines for planning and monitoring rehabilitation programmes. In the following Chapter we will look in more detail at the objectives of baseline and rehabilitation monitoring programmes and also the range of techniques which are available to carry these out.

References

Brenner, F. J. & Kelly, J. (1981). Characteristics of bird communities on surface mine lands in Pennsylvania. *Environmental Management*, **5**, 441–9.

Brenner, F. J., Kelly, R. B. & Kelly, J. (1982). Mammalian community characteristics on surface mine lands in Pennsylvania. *Environmental Management*, **6**, 241–9.

Chapman, D. L., McGinnes, B. D. & Downing, R. L. (1978). Breeding bird populations in response to the natural revegetation of abandoned contour mines. In *Surface Mining and Fish/Wildlife Needs in the Eastern United States*, ed. D. E. Samuel, pp. 328–32. Morgantown: West Virginia University.

Clark, S. S. (1975). The effect of sand mining on coastal heath vegetation in New South Wales. *Proceedings of the Ecological Society of Australia*, **9**, 1–16.

Cody, M. L. & Diamond, J. M. ed. (1975). *Ecology and Evolution of Communities*. Cambridge, Mass.: Belknap Press.

Connell, J. H. & Slatyer, R. O. (1977). Mechanisms of succession in natural communities and their role in community stability and organisation. *The American Naturalist*, **111**, 1119–44.

Crawford, H. S., Hardy, D. M. & Abler, W. A. (1978). A survey of bird use of strip-mined areas in southern West Virginia. In *Surface Mining and Fish/Wildlife Needs in Eastern United States*, ed. D. E. Samuel, pp. 241–6. Morgantown: West Virginia University.

Dunlop, J. N., Majer, J. D., Morris, C. J. & Walker, K. J. (1985). A preliminary assessment of minesite rehabilitation in the Pilbara iron ore province using ant communities as ecological indicators. *Mulga Research Centre Journal*, **8**, 25–31.

Fox, B. J. (1982). The influence of disturbance (fire, mining) on ant and small mammal species diversity in Australian heathland. In *Proceedings of a conference on Dynamics and Management of Mediterranean Type Ecosystems*, pp. 213–19. Riverside, California: U.S.D.A. Forest Service.

Fox, B. J. (1985). Mammal species diversity in Australian heath. *Abstracts of Australian Mammal Society Meeting*, University of Melbourne, May 1985, 54.

Fox, B. J. (1988). Small mammal community pattern in Australian heathland: a taxonomic rule for species assembly. In *Patterns in the Structure of Mammalian Communities*, ed. D. W. Morris, Z. Abramsky, B. J. Fox & M. R. Willig. Lubbock: Texas Technical Museum Special Publication Series (in press).

Fox, B. J. & Fox, M. D. (1984). Small-mammal recolonization of open-forest following sand mining. *Australian Journal of Ecology*, **9**, 241–52.

Fox, B. J. & Fox, M. D. (1986). Resilience of animal and plant communities to human disturbance. In *Resilience in Mediterranean Ecosystems*, ed. B. Dell, A. J. M. Hopkins & B. B. Lamont, pp. 39–64. The Hague: Dr W. Junk.

Fox, B. J., Fox, M. D. & Archer, E. (1985). Experimental confirmation of competition between two dominant species of *Iridomyrmex* (Hymenoptera: Formicidae). *Australian Journal of Ecology*, **10**, 105–10.

Fox, B. J. & Pople, A. R. (1984). Experimental confirmation of interspecific competition between native and introduced mice. *Australian Journal of Ecology*, **9**, 323–34.

Fox, M. D. & Fox, B. J. (1982). Evidence for interspecific competition influencing ant species diversity in a regenerating heathland. In *Ant–Plant Interactions in Australia*, ed. R. C. Buckley, pp. 99–110. The Hague: Dr W. Junk.

Greenslade, P. J. M. & Greenslade, P. (1983). Ecology of soil invertebrates. In *Soils, an Australian Viewpoint*, ed. K. Lee, pp. 645–69. Melbourne: CSIRO.

Haeck, J. (1969). Colonization of the mole (*Talpa europaea*) in the Ijsselmeer polders. *Netherlands Journal of Zoology*, **19**, 145–248.

Hawkins, B. A. & Cross, E. A. (1982). Patterns of refaunation of reclaimed strip-mine spoils by nonterricolous arthropods. *Environmental Entomology*, **11**, 762–75.

Hingten, T. M. & Clark, W. R. (1984). Small mammal recolonization of reclaimed coal surface-mined land in Wyoming. *Journal of Wildlife Management*, **48**, 1255–61.

Jehne, W. & Thompson, C. H. (1981). Endomycorrhizae in plant colonization on coastal sand-dunes at Cooloola, Queensland. *Australian Journal of Ecology*, **6**, 221–30.

Johnson, C. D., Beley, J. R., Ditsworth, T. M. & Butt, S. M. (1983). Secondary succession of arthropods and plants in the Arizona Sonoran Desert in response to transmission line construction. *Journal of Environmental Management*, **16**, 125–37.

Karr, J. R. (1968). Habitat and avian diversity on strip-mined land in east-central Illinois. *The Condor*, **70**, 348–57.

MacArthur, R. H. & Wilson, E. O. (1967). *The Theory of Island Biogeography*. Princeton: Princeton University Press.

Majer, J. D. (1980). The influence of ants on broadcast and naturally spread seeds in rehabilitated bauxite mined areas. *Reclamation Review*, **3**, 3–9.

Majer, J. D. (1983*a*). Recolonization by ants in rehabilitated open-cut mines in Northern Australia. *Reclamation and Revegetation Research*, **2**, 279–98.

Majer, J. D. (1983*b*). Ants: bio-indicators of minesite rehabilitation, land-use, and land conservation. *Environmental Management*, **7**, 375–83.

Majer, J. D. (1984). Ant return in rehabilitated mines – an indicator of ecosystem resilience. In *Proceedings of the 4th International Conference on Mediterranean Ecosystems*, ed. B. Dell, pp. 105–8. Perth: University of Western Australia.

Majer, J. D. (1985*a*). Recolonization by ants of rehabilitated mineral sand mines on North Stradbroke Island, Queensland, with particular reference to seed removal. *Australian Journal of Ecology*, **10**, 31–48.

Majer, J. D. (1985*b*). Restoration of pre-mining ecosystems in forested and agricultural areas. *Landline*, **11**, 4.

Majer, J. D., Day, J. E., Kabay, E. D. & Perriman, W. S. (1984). Recolonization by ants in bauxite mines rehabilitated by a number of different methods. *Journal of Applied Ecology*, **21**, 355–75.

Mitchell, M. (1961). Breeding bird populations in relation to grassland succession on the Anoka Sand Plain. *The Flicker*, **33**, 102–8.

Mook, J. H. (1971). Observations on the colonization of the new Ijsselmeer-polders by animals. *Miscellaneous Papers Landbouwhogeschool Wageningen*, **8**, 13–31.

National Research Council U.S. (1981). *Surface Mining: Soil, Coal and Society*. Washington: National Academy Press.

Neumann, U. (1971). Die Sukzession der Bodenfauna (Carabidae: Coleoptera, Diplopoda und Isopoda) in den forstlich rekultivierten Gebieten des Rheinischen Braunkohlenreviers. *Pedobiologica*, **11**, 143–226.

Nichols, O. G. & Bamford, M. J. (1985). Reptile and frog utilization of rehabilitated bauxite minesites and dieback-affected sites in Western Australia's jarrah (*Eucalyptus marginata*) forest. *Biological Conservation*, **34**, 227–49.

Nichols, O. G. & Watkins, D. (1984). Bird utilization of rehabilitated bauxite minesites in Western Australia. *Biological Conservation*, **30**, 109–31.

Odum, E. P. (1950). Bird populations of the Highlands (North Carolina) Plateau in relation to plant succession and avian invasion. *Ecology*, **31**, 587–604.

Parker, L. W., Elkins, N. Z., Aldon, E. F. & Whitford, W. G. (1984). Development of soil biota and some aspects of nitrogen cycling on reclaimed coal mine spoils in the arid southwest. *Journal of Environmental Quality*, **19**, 215–19.

Parmenter, R. R. & MacMahon, J. A. (1988). Arthropod recolonization of reclaimed strip-mines in southwestern Wyoming: 1. The ground-dwelling beetle fauna (Coleoptera). *Environmental Entomology* (in press).

Ranta, E. & As, S. (1982).Non-random colonization of habitat islands by carabid beetles. *Annales Zoologisi Fennici*, **19**, 175–82.

Richardson, B. Z., Farmer, E. E., Brown, R. W. & Packer, P. E. (1975). Rehabilitation research and its application on a surface mined area of eastern Montana. In *Proceedings of the Fort Union Coal Field Symposium*, ed. W. F. Clark, pp. 247–65. Billings: Montana Academy of Science.

Simberloff, D. S. (1974). Equilibrium theory of island biogeography and ecology. *Annual Review of Ecology and Systematics*, **5**, 161–82.

Sly, G. (1976). Small mammal succession on strip-mined land in Vigo County, Indiana. *The American Midland Naturalist*, **95**, 257–67.

Southwood, T. R. E. (1977). Habitat, the templet for ecological strategies? *Journal of Animal Ecology*, **46**, 337–65.

Stannard, L. J. (1967). The redispersal of Thysanoptera (Insecta, thrips) into spoiled strip-mined lands in southern Illinois. *Illinois State Academy of Science Transactions*, **60**, 130–3.

Steele, B. B. & Grant, C. V. (1982). Topographic diversity and islands of natural vegetation: aids in re-establishing bird and mammal communities on reclaimed mines. *Reclamation and Revegetation Research*, **1**, 367–81.

U.S. Congress, Office of Technology Assessment. (1986). *Western Surface Mine Permitting and Reclamation*, OTA-E-279. Washington, D.C: U.S. Government Printing Office.

Usher, M. B. (1979). Natural communities of plants and animals in disused quarries. *Journal of Environmental Management*, **8**, 223–36.

Van Waggoner, K. (1978). The effects of lignite mining on small mammal populations in Texas. In *Surface Mining and Fish/Wildlife Needs in Eastern United States*, ed. D. E. Samuel, pp. 256–66. Morgantown: West Virginia University.

Walker, K. J., Osborne, J. M. & Dunlop, J. N. (1986). Vertebrate colonisation after revegetation of a rehabilitated iron ore waste dump, Newman, Western Australia. In 1986 *Australian Mining Industry Council Environmental Workshop*, pp. 113–50.

Westman, W. E. (1986). Resilience: concepts and measures. In *Resilience in Mediterranean Ecosystems*, ed. B. Dell, A. J. M. Hopkins & B. B. Lamont, pp. 5–19. Dordrecht: Dr W. Junk.

Wetzel, R. M. (1958). Mammalian succession in midwestern floodplains. *Ecology*, **39**, 262–71.

Whitmore, R. C. (1978). Managing reclaimed surface mines in West Virginia to promote non-game birds. In *Surface Mining and Fish/Wildlife Needs in the Eastern United States*, ed. D. E. Samuel, pp. 381–6. Morgantown: West Virginia University.

Woodroff, S. & Majer, J. D. (1981). Colonisation of ants on the exposed banks of the Canning Dam reservoir. *Australian Entomological Magazine*, **8**, 41–6.

Part 3

Practical aspects of fauna in reclaimed land

Introduction

J. D. MAJER

The following six Chapters discuss some of the practical considerations relating to fauna in reclaimed land. The first three apply to the planning and facilitation of rehabilitation while the latter three cover some of the aesthetic or economic after-uses of fauna in reclaimed land.

In Chapter 7, Nick Allen stresses the importance of obtaining baseline predisturbance information on the biota. By doing this, the success of reclamation and rehabilitation programmes can be gauged against these baseline data. He also highlights the need for a standard approach to obtaining such data so that information may be compared between sites or over a period of time for a given site. A standardized baseline fauna survey method is proposed and described in this Chapter.

The various procedures which may be incorporated into the design of the reclamation plan in order to maximize fauna return are discussed by Steve Viert in Chapter 8. This Chapter does not contain an exhaustive list of techniques as many additional recommendations are contained in other Chapters throughout the book. However, Viert particularly stresses the need to incorporate negotiations for wildlife plans early on in the project's design. Such planning should be interdisciplinary in nature and coordinated with the activities of project engineers and specialists from other disciplines.

Chapter 9 discusses the special problem of the inhibitory effects of heavy metals, salt, and extreme alkalinity or acidity on animal recolonization. In this Chapter, Wei-Chun Ma & Herman Eijsackers propose a series of management procedures which should improve the effectiveness of rehabilitation on toxic substrates.

The possibility of introducing livestock grazing as an economically useful land-use following rehabilitation is discussed by William Laycock in Chapter 10. He describes the requirements of the plants which have been introduced as forage on rehabilitated land, the requirements of the livestock in such areas and also the impact of livestock on watershed management and plant species composition.

As wetlands become a rapidly diminishing natural resource throughout the World, mining provides an ideal opportunity for creating artificial wetland habitats. In Chapter 11, Jack Nawrot and Willard Klimstra explore the possibility of creating wetland habitats as a byproduct of mining. They consider legislative obstacles to such developments and recommend a series of planning and management procedures which will establish the biological conditions for wetlands to be created.

Dave Samuel takes this idea a stage further in Chapter 12, by considering the amenity value of reclaimed land and of artificially created wetlands. The possibilities of developing new fishing, hunting, trapping and bird watching areas are outlined and the involvement of community groups in such ventures is suggested. Having considered the suggestions in this series of Chapters, the reader should be aware of the many aesthetically or economically viable land-uses which may be introduced, options which should find greater political and community support than that of merely restoring the environment for its own sake.

7

A methodology for collecting standardized biological data for planning and monitoring reclamation and rehabilitation programmes

N. T. ALLEN

Introduction

The principal aim of most degraded land reclamation programmes conducted by both private organizations and government utilities, is to return the land to a stable state, which can either be used for further ventures or from which the operator can 'walk away' with no further cost involvement. For this latter case to be accomplished effectively, one option – and arguably the most appropriate – is the creation of a self-sustaining ecosystem on the land, whether it be a replica of the original ecosystem or another one particularly suited to the post-reclamation conditions. These options are referred to by Bradshaw (1984) as restoration and replacement, respectively (see Fig. 1.4)

For rehabilitation (biological) options to be exercized in a cost effective and planned manner, it is essential that operators, and their regulatory authorities, have a sound understanding of the landforms and soils, the surface and groundwater hydrology and the biological resources of the sites involved. This knowledge should avoid the implementation of inappropriate programmes which necessitate eventual 're-working' of areas.

There has been a gradual onset of awareness over the last 20 years of the importance of biological processes in successful rehabilitation programmes (see Chapter 1). Concurrently, an increasing body of literature which reflects the importance of fauna interactions in these processes has appeared. However, the material in Chapters 13 to 17, and a review of the bibliography presented in Chapter 19, indicate that the majority of fauna studies to date have covered specific groups of animals and have been conducted 'after the act'. This means that they can not be related back to predisturbance levels (see Westman, 1986, which is discussed in Chapter 6 of this book).

It is the general lack of baseline predisturbance information which forms the subject for this Chapter. Firstly, the importance of standardized baseline fauna surveys is discussed. Then the background to, and methods employed for, comprehensive baseline surveys are given using a recent example from the Australian mining industry. Hopefully this will guide the operator and legislator to the level of detail and sophistication which is practically attainable for gaining baseline fauna data. The success of reclamation and rehabilitation programmes can be gauged against this in the long-term.

Importance of standardized baseline fauna surveys

Government regulatory authorities are increasingly requiring developers of mining, quarrying and other major land disturbance operations to conduct pre-development biological investigations, including fauna surveys, and to propose environmental management plans for the operating and decommissioned projects. The legal developments in relation to these requirements are outlined in Chapter 1. The problems which beset the developers in relation to fauna studies are at least two-fold. Firstly, which of the many variations of fauna survey techniques should be used and which will be acceptable to the regulatory authorities? Secondly, how will the findings be assessed and compared with those of other operators using different techniques, while still enabling adequate definition of environmental management options with any certainty in the long term?

These questions have been dealt with in some detail by the US Congress' Office of Technology Assessment (1986) in their efforts to assess the ability of current mining and reclamation technologies, and of Federal programmes and policies, to meet the statutory mandates for environmental protection during the reclamation of surface coal-mined lands in the western United States. They note that wildlife is one discipline for which standardized data-collection methodologies are lacking and provide detailed summary comparisons of the varying baseline wildlife data requirements and accepted data collection techniques for five western States. The assessment notes that baseline wildlife studies usually collect data on the following factors:

1. species occurrence, including seasonal information,
2. species distribution,
3. relative species abundance or population estimates, including population size indices and species diversity values,
4. reproductive success,
5. food preferences,
6. habitat preferences,
7. delineation of habitats, and
8. habitat quality.

These factors can also be considered when performing monitoring studies, although in the United States these tend to be much less intense, if they are conducted at all. Interestingly, no mention is made of terrestrial invertebrates in the assessment, despite the findings of scientists, conservation agencies and semi-government institutions (see Chapters 1 to 6 of this book). It is also worth noting the Office of Technology Assessment's final statement on wildlife monitoring that the data collected: 'are not used to measure reclamation success directly, but indirectly as a gauge of the use of reclaimed land by wildlife'.

Snyder & Potter (1981) suggest that the implementation of environmental baseline studies is intended to serve two major purposes. They provide a method for obtaining an inventory of the natural resources present on a site and the intensive, and often extensive, data collection effort provides the specific information-base for later development of site-specific restoration plans. They further stress that these purposes should be of major interest to those concerned with the planning, permitting and operation of large-scale surface mining operations.

From their review of a wide range of mining baseline studies and associated restoration plans, these authors also make the point that several important aspects of wildlife and vegetation ecology are often not adequately integrated into the development of site reclamation plans. This is despite the availability of data from extensive baseline and monitoring programmes (Snyder & Potter, 1981). The problems include: (1) poor correlations between the vegetation studies and the habitat utilization rates of important wildlife species; (2) failure to recognize and incorporate the important habitat concepts of plant interspersion, plant succession, diversity and edge into design of the revegetation plans; (3)

a tendency not to use the extensive scientific literature available on the habitat requirements of key wildlife and plant species; and (4) a tendency to generalize or over-simplify the ultimate consequences of mining operations on the biotic resources.

One could continue to review the range of literature relating to baseline biological surveys, but the same general themes come through. Depending on the ultimate objectives for the reclamation proposed, a sound knowledge of the pre-disturbance biological environment will, if interpreted carefully and used in the planning process, allow the implementation of superior rehabilitation programmes which should stand the test of time. However, for these surveys to be useful not only to the particular operator involved, but also to the wider community, some degree of standardization of methods and approaches is essential.

As will be seen in the following Australian example, the regulatory authorities have clearly defined the objectives of the rehabilitation programmes, and the operator has attempted to standardize biological surveys with those of contemporary workers in the area. Close liaison has also been maintained with Government and tertiary education bodies to ensure an acceptable and, importantly, a workable approach.

An Australian example

It is not the intention of this section to provide a comprehensive review of the myriad techniques which have been developed for studying fauna and their interactions with the environment. Rather, it seeks to detail the standardized baseline fauna survey methods which have been developed and successfully put into practice over several years by one Australian company, in response to strict legislative requirements governing environmental management of its varied mining and processing operations. In providing this example, the reader is given an appreciation of the level of baseline fauna surveys which operators of major land disturbance projects can practically attain. The tried and well-tested approach can also be considered as a possible template for other organizations seeking to standardize their survey programmes in the interests of sound and long-term environmental management.

The selected Australian example covers the comprehensive baseline surveys commissioned by Worsley Alumina Pty Ltd, who manage and operate the Worsley Alumina and Boddington Gold Mine projects on behalf of a consortium of joint venturers. Worsley's area of operations covers both private land and sections of the Government-controlled northern jarrah (*Eucalyptus marginata*) forest on the Darling Plateau, between 100 and 160 km south-east of Perth in south-west Western Australia. Annual precipitation ranges from 740 to 1250 mm.

The mining, reclamation and rehabilitation techniques employed at both the bauxite and gold operations are as described for Western Australian bauxite mines in Chapter 16. Environmental impacts of the alumina refining and gold processing and associated waste disposal systems are documented in Reynolds Australia Alumina (1979) and Worsley Alumina (1987), respectively.

Legislative requirements and private commitments

As described in Chapter 1, Western Australia has adopted a system of Agreement Acts for administering major projects. Under the terms of the Alumina Refinery (Worsley) Agreement Act, 1973 (as amended), the Worsley Joint Venturers are required to carry out continuous investigation and research to ascertain the effectiveness of the measures they are taking persuant to their approved Environmental Review and Management Programme (ERMP) (Reynolds Australia Alumina, 1979), for the protection and management of the environment. The Agreement further requires that, for State Forest areas, the Joint Venturers shall, in consultation with the Department of Conservation and Land Management, '... take adequate measures –

1. for the progressive restoration and re-afforestation of the forest destroyed;
2. for the prevention of soil erosion;
3. for the prevention of the formation of deep water pools and other dangers to persons who may use the forest areas.'

With respect to the specific goals for rehabilitation, the broad objectives are to regenerate a stable forest ecosystem, planned to maintain recreation, conservation and other nominated forest values. For conservation values, the long-term aim is to regenerate floral and faunal characteristics compatible with the undisturbed jarrah forests of the area. The provision of written guides to flora, fauna and landscape characteristics is also required (Worsley Alumina & Department of Conservation and Land Management, 1987).

Within this framework, biological investigations for the alumina project were planned as three phases:

1. Phase One – a preliminary biological survey aimed at broad classification of major and significant vegetation types, combined with vertebrate and invertebrate surveys in sites selected for their ecological significance,

2. Phase Two – a more intensive, detailed and quantitative study of both fauna and flora, together with physical and chemical studies of soils of the mining areas, and
3. Phase Three – a monitoring phase in which the information from the earlier phases, particularly from Phase Two, will be employed to gauge both the general biological impacts of project operations, and the success of reclamation and rehabilitation programmes, particularly in mined areas.

The broadly-based Phase One studies (Worsley Alumina & Dames & Moore, 1981) were conducted from 1980–81 and formed the basis for the planning process leading to the Phase Two studies commissioned in 1982 and published in Worsley Alumina (1985). With the later development of the Boddington Gold Mine, the Phase Two methods were also employed for baseline studies conducted in advance of that project (Worsley Alumina, 1989).

In planning the Phase Two fauna studies, Worsley sought the advice of workers from the Western Australian Museum and the Wildlife Section of the Department of Conservation and Land Management – the principal government agencies involved in biological surveys in Western Australia. The detailed programmes were then developed in close consultation with workers at tertiary education institutions and biological consultants, who were also commissioned to carry out the field programme under the supervision of Worsley's staff.

The studies were designed to build systematically on the Phase One data in order to quantify fauna species richness and diversity in the different vegetation community types, and to monitor seasonal trends in animal abundance, behaviour and also breeding strategies and feeding preferences of individual species. The over-riding aim was to ensure that data were collected on a systematic and 'equal-effort' basis, to enable legitimate comparisons between faunal richness and abundance in the various vegetation communities. This work was carried out in conjunction with vegetation research, the total study forming the bases for comparisons during the on-going Phase Three monitoring programmes which have now commenced.

Throughout the studies, regular liaison was maintained with the relevant government agencies, who provided continuing advice and kept the study group abreast of developments in biological survey techniques and methods of data processing.

Methods of baseline data collection

The Worsley studies were conducted during the four seasons of the year and centred on a number of intensive study locations (Fig. 7.1), sited to sample the variations in soils, topography, major vegetation communities (defined in contemporaneous vegetation surveys) and climatic conditions represented throughout the project areas.

At each location, a series of bird observation transects were established, either in the quadrat or line transect configurations, together with a vertebrate fauna trapping grid, an invertebrate sampling transect and a vegetation plot (see Fig. 7.1). These locations formed the basic study units for all systematic fauna investigations. Other areas were sampled by opportunistic collections or observations. Data thus gathered served either to corroborate the findings from systematic surveys or to highlight anomalies and indicate where further work was required. The basic methods employed for both systematic and opportunistic (inventory) fauna data collection were developed from those published by the Biological Surveys Committee of Western Australian (1984) and Majer (1987). They are detailed in the following sections.

Bird Surveys

Systematic effort The use of bird census procedures in Australia has been reviewed in detail by Recher (1984) who notes that most methods employed in the northern hemisphere require modification to account for the ways in which the ecology and behaviour of Australian birds differ from those of their northern counterparts. Pyke & Recher (1984) further detail a range of census procedures and their varying problems, and stress the importance of standardizing methods to ensure comparability of results and to allow other workers to repeat surveys to test the results or monitor changes in the avifauna. The procedures described below have been standardized throughout all the baseline and initial monitoring programmes commissioned by Worsley.

Systematic field data collection for birds centred on a series of 200 m bird observation transects, located around or near the vegetation plots in each intensive study location (Fig. 7.1). Depending on the extent and homogeneity of the habitat being sampled, the transects were layed out in one of the configurations shown in Fig. 7.2. The first example (Fig. 7.2a) was used for narrow strips of vegetation, particularly riparian habitats and small heathland strips surrounded by extensive woodlands. The transects were aligned within, and parallel to the study zone (as illustrated in Fig. 7.1a). The format shown in Fig. 7.2b was used when

(a)

(b)

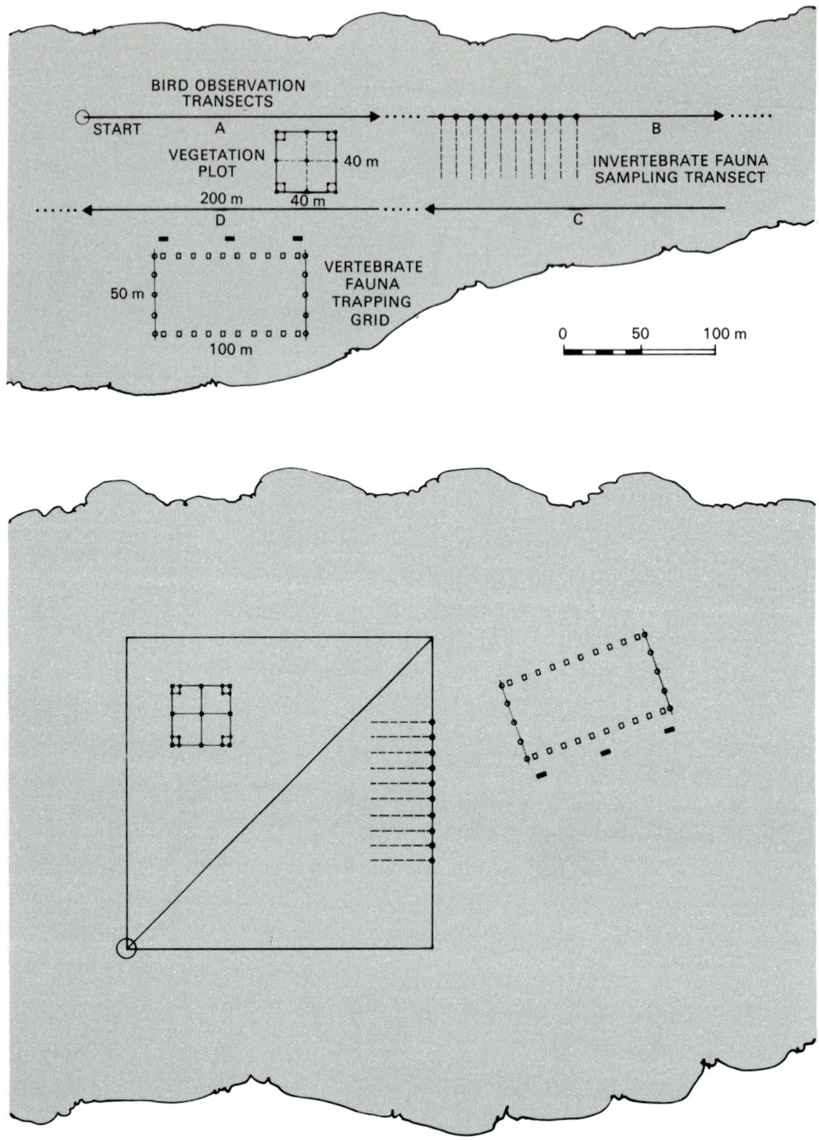

Fig. 7.1. Design of intensive study locations used for systematic flora, vertebrate and invertebrate fauna surveys in (a) narrow habitat strips, and (b) extensive and homogeneous open habitats, in the Worsley Project areas.

the habitat was reasonably extensive and homogeneous, but sub-divided by old forest tracks, exploration lines and other forms of disturbance, thus necessitating the placement of transects in habitat remnants. Finally, the quadrat shown in Fig. 7.2c was used for surveys within extensive and

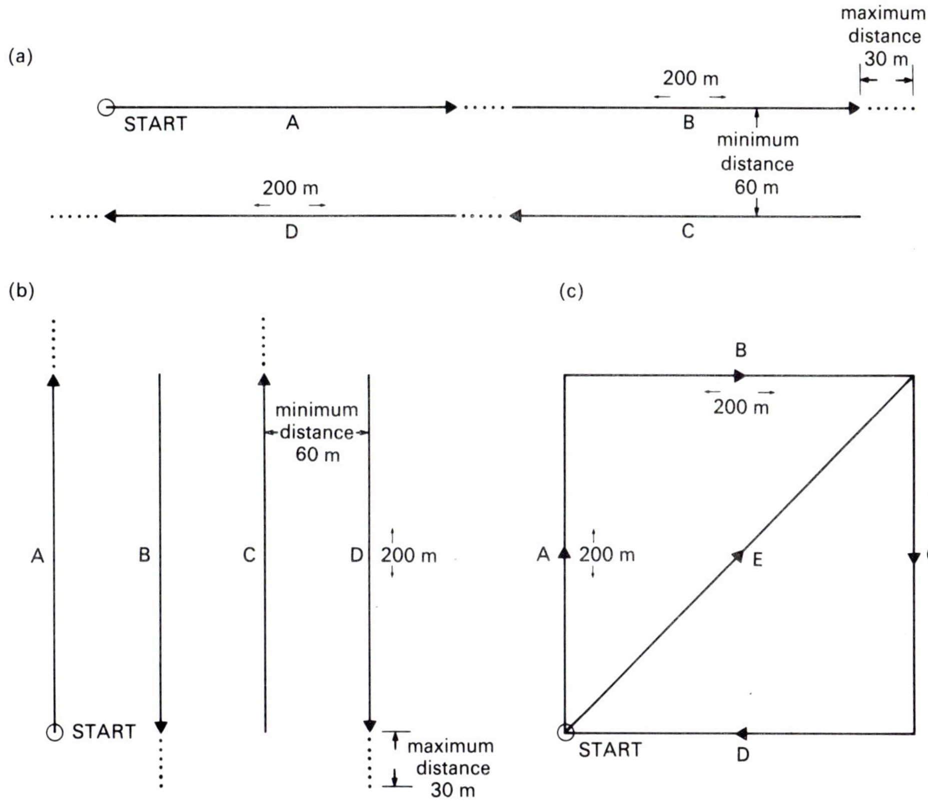

Fig. 7.2. Three formats for bird observation transects used for systematic surveys in (a) narrow habitat strips, (b) extensive and homogeneous habitats, sometimes broken by small patches of other habitats or tracks, etc., and (c) extensive and homogeneous open habitats, in the Worsley Project areas.

homogeneous open habitats (as illustrated in Fig. 7.1 b). All transects were marked with plastic flagging tape every 25 m with various colour combinations used to enable observers to locate sightings accurately.

Birds were recorded up to 50 m either side of the transect lines, depending on the extent and density of vegetation. Bird data were collected between 0700 and 1200 h by systematically traversing each series of transects on foot for an average of 40 min each day. For each intensive study location, all transects were sampled five times during each seasonal survey. Keast (1984) has suggested that, for open forest and woodland habitats in Australia, bird counts should be limited to 2.5–3 h day^{-1}, beginning about an hour after sunrise. During this period bird song is most sustained and the birds are localized as they feed. In the Worsley studies, factors such as distribution and number of study

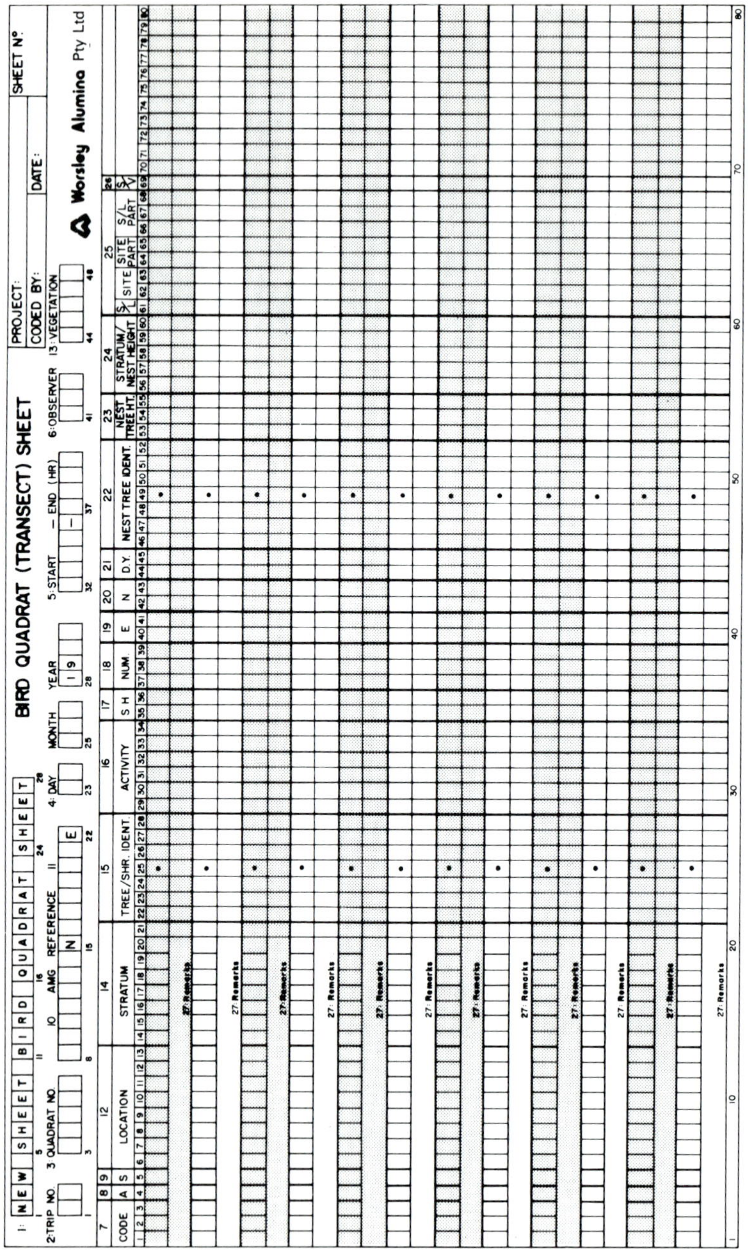
1: NEW SHEET BIRD QUADRAT SHEET
2:TRIP NO.
3 QUADRAT NO.
10 AMG REFERENCE
N
E
4: DAY
MONTH
YEAR
1 9
BIRD QUADRAT (TRANSECT) SHEET
5:START
— END (HR)
6:OBSERVER
13: VEGETATION
PROJECT:
CODED BY:
DATE:
SHEET Nº
Worsley Alumina Pty Ltd
CODE
A
S
LOCATION
STRATUM
TREE/SHR. IDENT.
ACTIVITY
S H
NUM
E
N
D.Y.
NEST TREE IDENT.
NEST TREE HT.
STRATUM/ NEST HEIGHT
S/L
SITE
SITE PART
S/L PART
27: Remarks

(a)

BIRD QUADRAT (TRANSECT) SHEET

Description of Codes

7 : CODE = Australian Bird Atlas Code e.g. Emu - 001

8 : AGE (code)
- A = Adult
- S = Sub-adult
- J = Juvenile
- U = Unknown
- X = Adult + Sub-adult(s)

9 : SEX (code)
- M = Male
- F = Female
- U = Unknown
- X = Male + Female(s)

12: LOCATION (co-ordinates) = Location of bird on quadrat.

e.g. (1) A175 + 025 = 175m along A transect of quadrat, 25m into quadrat (right of transect).
(2) B025 - 010 = 25m along B transect of quadrat, 10m outside quadrat (left of transect.)

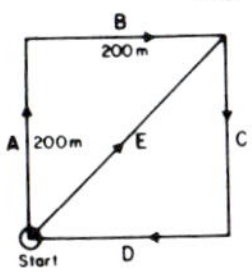

Walk commences at start of Transect A, and continues in a clockwise direction to end of Transect E. Note compass orientation of start corner here:

....

N.B. For discrete TRANSECT data collection, transect length = 200 metres.

13: VEGETATION (code) = Vegetation code from Project maps.

14: STRATUM (code)

Trees
- T1 = Upper canopy
- T2 = Upper branches
- T3 = Lower branches
- T4 = Trunk
- .. = ...
- T9 = Tree (general)

Shrubs
- S1 = Upper canopy
- S2 = Inner branches
- .. = ...
- S9 = Shrub (general)

Herbs
- H1 = Herbaceous layer

Ground
- G1 = On log or stump
- G2 = Ground

N.B. Height of tree or shrub to follow primary stratum code. e.g. T108, T220T4S1

15: TREE/SHR. IDENT. = Tree species in which bird is located. Use first 3 letters of Genus and Species. e.g. ALL.FRA = *Allocasuarina fraseriana*

16: ACTIVITY (code)
- PG = Perching
- CG = Calling
- FG = Feeding
- FB = Feeding on blossom
- DG = Drinking
- FL = Flying
- FT = Flying through
- FO = Flying over
- SG = Swimming
- DP = Displaying
- TC = Territorial conflict
- RT = Copulating
- CN = Constructing nest
- ON = On nest
- FY = Feeding young
- BW = Broken-wing display
- OR = Other

17: SH (code) = Seen and/or heard i.e.
- S = Seen
- H = Heard
- SH = Seen and Heard

18: NUM. - Number of birds observed. If uncertain of actual number, then use the following code:
- 905 = up to 5 birds
- 910 = 5 to 10 birds
- 915 = 11 to 15 birds
- 920 = 16 to 20 birds
- 935 = 21 to 35 birds
- 936 = more than 35 birds
- ... = ...
- ... = ...

BREEDING DATA:

19: E = Number of eggs.

20: N = Number of nestlings [use 99 if not visible].

21: D.Y. = Number of dependant young.

22: NEST TREE/SHR.IDENT. = Tree/shrub species in which nest is located [see 15:].

23: NEST TREE/SHR. HT. = Height of nest tree/shrub in metres.

24: STRATUM/NEST HEIGHT = Stratum code for nest location [see 14:], followed by height of nest above ground in metres.

25: SITE (code)

S/L = Study Location
- A = Mining Area - Saddleback
- B = " " - Quindanning
- C = " " - Marradong
- D = " " - Bunnings
- E = Conveyor One Corridor
- F = " Two Corridor
- G = Refinery Lease Area
- H = Other Minerals Area (part of D)
- J = Mooradung Reserve (No.32448)
- K = ...
- L = ...
- M = ...
- N = ...

Site = Broad Site Description
- 01 = Farm dam
- 02 = Swamp
- 03 = Lake/large dam
- 04 = Gully
- 05 = Lower slope
- 06 = Mid slope
- 07 = Upper slope
- 08 = Ridge
- 09 = Creek bank
- 10 = Mill
- 11 = Rehabilitated mine pit
- 12 = Farm (general)
- 13 = Rehabilitated borrow pit
- 14 = ...
- 99 = Other

Site Part. = Partition within Broad Site Desc.
- 01 = Deep open water
- 02 = Open shallows
- 03 = Sheltered shallows
- 04 = On rocks in shallows
- 05 = Reed beds
- 06 = Open shoreline
- 07 = Open rocky shoreline
- 08 = Regenerating shoreline
- 09 = Revegetated shore
- 10 = Dam wall
- 11 = Associated constructions
- 12 = ...
- 13 = ...
- 99 = Other

S/L Part. = Partition within Study Location
- RCL = Refinery Catchment Lake
- FWL = Freshwater Lake
- PHD = Pipe Head Dam
- RA1 = Residue Area 1
- RA2 = Residue Area 2
- SEP = Solar Evaporation Pond
- DDC = Drainage Diversion Channel
- PSW = Pillow Swamp (O.M.)
- RSW = Round Swamp (O.M.)
- ... = ...
- ... = ...
- ... = ...
- ... = ...

26: S/V = Supplementary vegetation code, to be used when more than one major vegetation association is representated in the Quadrat.
- B = Bullich
- A = Allocasuarina
- F = Flooded gum
- H = Heath
- J = Jarrah
- Y = Yarri
- W = Wandoo
- . = ...

(b)

Fig. 7.3. Examples of (a) the bird quadrat/transect data sheet, and (b) associated coding formats, used for systematic avifaunal surveys in the Worsley Project areas.

locations (up to 11 sites at any one time), travelling time and accessibility necessitated extension of the sampling time by up to 2 h each day. However, sampling times for each site were rotated each day to ensure comparability of data collected over the study period.

On encountering a bird, the following information was recorded: species, location along transect, age, sex, number of individuals, tree/shrub species and stratum used and breeding information. Included in the breeding information were: height of nest above ground, tree/shrub species utilized, tree/shrub height and number of eggs/nestlings/dependent young. These data were coded directly onto the Bird Quadrat/Transect Coding Form illustrated in Fig. 7.3.

Inventory records Opportunistic bird sightings were recorded while travelling or while involved in work relating to other fauna, with

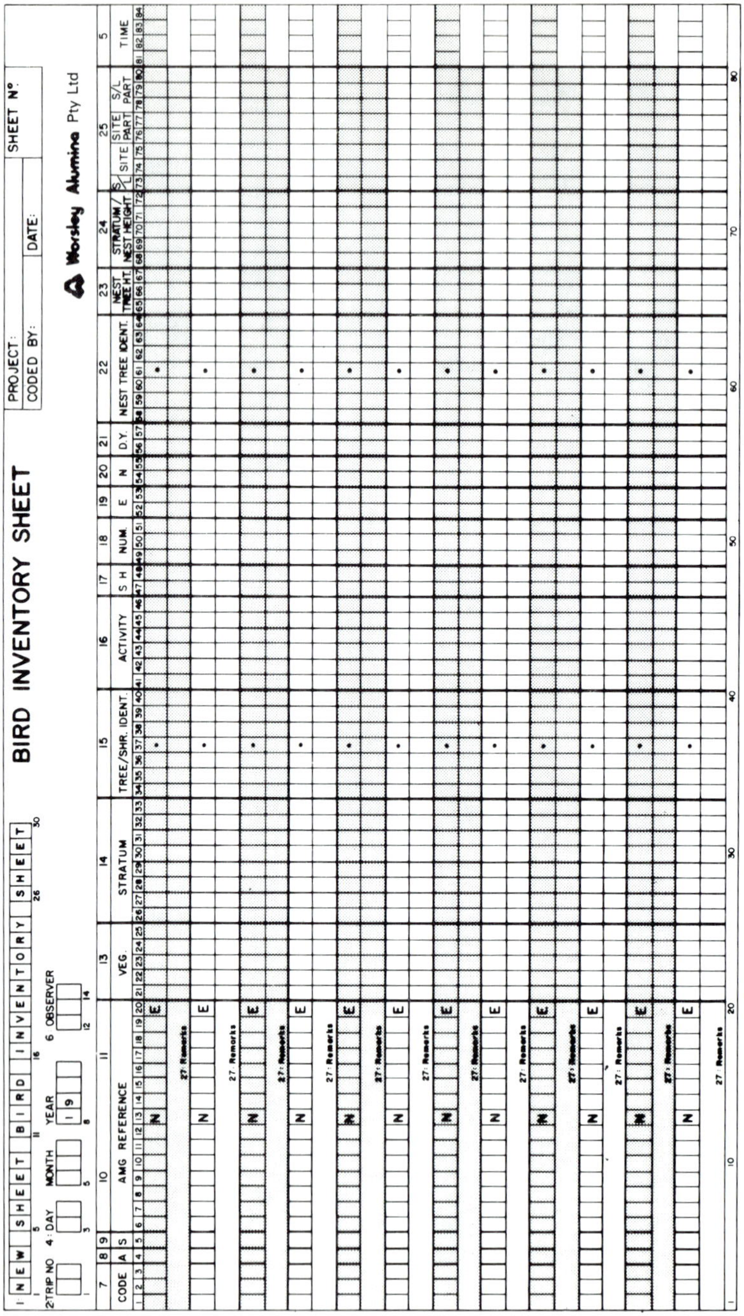
BIRD INVENTORY SHEET
NEW SHEET BIRD INVENTORY SHEET
2 TRIP NO
4: DAY
MONTH
YEAR 19
6 OBSERVER
PROJECT:
CODED BY:
DATE:
SHEET N°
Worsley Alumina Pty Ltd
CODE
A
S
AMG REFERENCE
VEG.
STRATUM
TREE/SHR. IDENT.
ACTIVITY
S H
NUM.
E
N
D.Y.
NEST TREE IDENT.
NEST TREE HT.
STRATUM/ NEST HEIGHT
S/L
SITE
SITE PART
S/L PART
TIME
27: Remarks
N
E

(a)

BIRD INVENTORY SHEET

Description of Codes

5: **TIME** = Time of sighting (24 hour clock).

7: CODE = Australian Bird Atlas Code e.g. Emu = 001.

8: AGE (code)
A = Adult
S = Sub-Adult
J = Juvenile
U = Unknown
X = Adult + Sub-Adult(s)

9: SEX (code)
M = Male
F = Female
U = Unknown
X = Male + Female(s)

10/11: **AMG REFERENCE** - To be taken from maps, or from survey pegs located in the field.

13: **VEGETATION** (code) = Vegetation code from Project maps. If codes are not strictly applicable, use the following:

HEATH = Heathland vegetation
TRE-1 = Trial Rehabilitation Exercise One
REHAB = General rehabilitation
FARM = All sightings on cleared farmland
M/CLD = Cleared areas - Mining area
C/CLD = " " - Conveyor corridor
R/CLD = " " - Refinery lease
OTHER = Accompanied by a description in Field 27:

14: **STRATUM** (code)

Trees
T1 = Upper canopy
T2 = Upper branches
T3 = Lower branches
T4 = Trunk
.. = ...
T9 = Tree(general)

Shrubs
S1 = Upper canopy
S2 = Inner branches
.. = ...
S9 = Shrub(general)

Herbs
H1 = Herbaceous layer

Ground
G1 = On Log or stump
G2 = Ground

N.B. - Height of tree or shrub to follow primary stratum code. e.g. - T108, T220T4S1.

15: TREE/SHR. IDENT. = Tree/shrub species in which bird is located. Use first 3 letters of Genus and Species. e.g. ALL.FRA = *Allocasuarina fraseriana*.

16: ACTIVITY (code)
PG = Perching
CG = Calling
FG = Feeding
FB = Feeding on blossom
DG = Drinking
FL = Flying
FT = Flying through
FO = Flying over
SG = Swimming
DP = Displaying
TC = Territorial conflict
RT = Copulating
CN = Constructing nest
ON = On nest
FY = Feeding young
BW = Broken-wing display
OR = Other

17: SH (code) = Seen and/or heard i.e.
S = Seen
H = Heard
SH = Seen and Heard

18: **NUM.** = Number of birds recorded. If uncertain of actual number, then use the following code:

905 = up to 5 birds
910 = 5 to 10 birds
915 = 11 to 15 birds
920 = 16 to 20 birds
935 = 21 to 35 birds
936 = more than 35 birds
... = ...
... = ...

BREEDING DATA:

19: E = Number of eggs.

20: N = Number of nestlings [use 99 if not visible].

21: D.Y. = Number of dependant young.

22: **NEST TREE/SHR.IDENT.** = Tree/shrub species in which nest is located [see 15:].

23: NEST TREE/SHR.HT. = Height of nest tree/shrub in metres.

24: STRATUM/NEST HEIGHT = Stratum code for nest location [see 14:], followed by height of nest above ground in metres.

25: SITE (code)

S/L = Study Location
A = Mining Area - Saddleback
B = " " - Quindanning
C = " " - Marradong
D = " " - Bunnings
E = Conveyor One Corridor
F = " Two Corridor
G = Refinery Lease Area
H = Other Minerals Area (part of D)
J = Mooradung Reserve (No.32448)
K = ...
L = ...
M = ...
N = ...

Site = Broad Site Description
01 = Farm dam
02 = Swamp
03 = Lake/large dam
04 = Gully
05 = Lower slope
06 = Mid slope
07 = Upper slope
08 = Ridge
09 = Creek bank
10 = Hill
11 = Rehabilitated mine pit
12 = Farm (general)
13 = Rehabilitated borrow pit
14 = ...
99 = Other

Site Part = Partition within Broad Site Desc.
01 = Deep open water
02 = Open shallows
03 = Sheltered shallows
04 = On rocks in shallows
05 = Reed beds
06 = Open shoreline
07 = Open rocky shoreline
08 = Regenerating shoreline
09 = Revegetated shoreline
10 = Dam wall
11 = Associated constructions
12 = ...
13 = ...
99 = Other

S/L part = Partition within Study Location
RCL = Refinery Catchment Lake
FWL = Freshwater Lake
PHD = Pipe Head Dam
RA1 = Residue Area 1
RA2 = Residue Area 2
SEP = Solar Evaporation Pond
DDC = Drainage Diversion Channel
PSW = Pillow Swamp (O.M.)
RSW = Round Swamp (O.M.)
... = ...
... = ...
... = ...
... = ...

(b)

Fig. 7.4. Examples of (a) the bird inventory data sheet, and (b) associated coding formats, used for opportunistic avifaunal surveys in the Worsley Project areas.

nocturnal birds being noted during spot-lighting runs and head-torching activity. The information collected was similar to that recorded for transects, and notes on the vegetation type and a mapping grid reference were ascribed to the points of observation. These data were coded directly onto the Bird Inventory Coding Forms shown in Fig. 7.4.

Mammal, reptile and amphibian surveys

Systematic effort A considered decision was taken to avoid the use of extensive trapping grids involving large numbers of traps set over major portions of particular habitats. Such grids may provide a more detailed assessment of the fauna, but they require much higher personnel commitments and reduce the time available for working on other fauna groups. Also, such sampling can lead to excessive and unacceptable fauna mortality in the rather harsh Australian environment – a factor

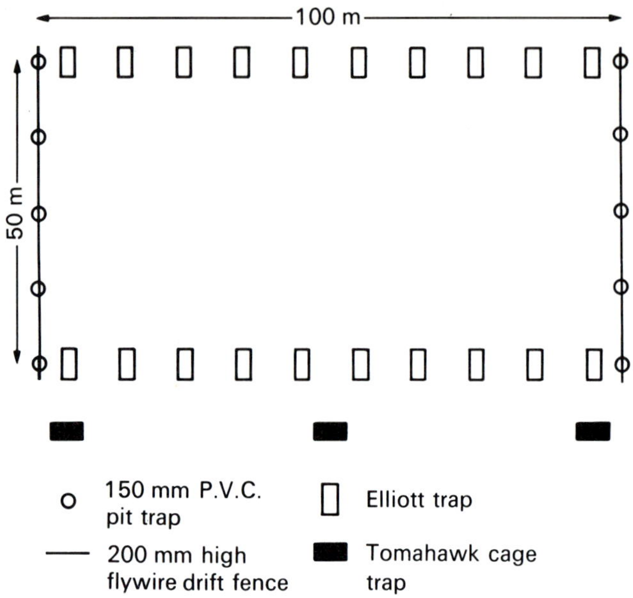

Fig. 7.5. Design of the vertebrate fauna trapping grid used for systematic surveys in the Worsley Project areas.

which could cause serious problems even before a proposed development commences! Importantly, the conditions of the 'License to Take Fauna for Scientific Purposes' issued by the Western Australian Department of Conservation and Land Management for such work generally state that 'only two specimens of any one species are to be sacrificed.' These points are also the principal reasons for the absence of the familiar 'breakback' or 'snap trap' from the trapping grid described here. However, use of fenced pit-lines is becoming more widespread in Australia and the merits of this system for surveying herpetofauna, and vertebrates in general, have been well documented (Cogger, 1985; McKenzie & Youngson, 1983; see How, Humphreys & Dell, 1984, for detailed comparisons of fauna trapability using pit traps, box traps and snap traps in Western Australia).

A vertebrate fauna trapping grid was located in the vicinity of each series of bird observation transects. The 50 × 100 m grid (Fig. 7.5) was designed to minimize variations in trap returns due to differences in habitat size and type, and to provide replication of the traplines within a small area. It consisted of PVC pit-traps, 15 cm in diameter and 50 cm deep, sunk into the ground at 10 m intervals along 50 m lines, bisected at the surface by continuous flywire mesh drift fences. These were buried a few cm into the substrate and rose 20 cm above the ground, thus directing small terrestrial and fossorial vertebrates into the traps. Medium Elliott box traps (32 × 10 × 10 cm; Elliott Scientific Equipment, Victoria, Australia) and Tomahawk cage-traps (66 × 25 × 25 cm; Tomahawk Trap Co., Wisconsin, U.S.A.) were also used. The cage-traps were baited with kitchen scraps, road-killed birds and universal bait; all other traps were baited with universal bait only. Each trapping grid was set for five consecutive nights during each seasonal survey.

During the course of bird transect assessments, all observations of large mammals, such as kangaroos and wallabies, and reptiles and amphibians were recorded. In addition, one systematic person-hour per seasonal survey was spent during the afternoon in each intensive study location actively foraging for reptiles, amphibians and small mammals in the litter and under debris. A further person-hour was devoted at each site to head-torching at night for nocturnally-active fauna such as geckos, frogs, small mammals and nocturnal birds.

Other special studies were conducted where appropriate. For example, pollen grains were sampled from the fur of the nectarivorous honey-possum (*Tarsipes rostratus*) using the method described by Wooler *et al.* (1983). The percentage frequency of particular pollen types identified in this way demonstrated the feeding preferences of the animals throughout the year and this information influences plant species selection for rehabilitation programmes conducted in or adjacent to the Honey-possum's original habitat.

For all animals captured (including birds) or observed, a range of information was coded directly on the Animal Data Sheets shown in Fig. 7.6. Where practicable, captured animals were marked using either ear-tags, wing bands (bats only), or scale/digit removal (Stonehouse, 1978).

Inventory records Two-hour spot-lighting runs covering the study areas were carried out during each seasonal field trip, and opportunistic sightings of mammals, reptiles and amphibians, or recognizable tracks and signs (e.g. diggings, etc.), were recorded while travelling or involved in work on other faunal groups. Additionally, scats of feral cat (*Felis catus*), introduced fox (*Vulpes vulpes*) and indigenous carnivorous fauna were collected to obtain fur samples for analysis and identification by the method of Brunner & Coman (1974). These data were logged onto the Animal Data Sheets used for systematic studies (Fig.7.6).

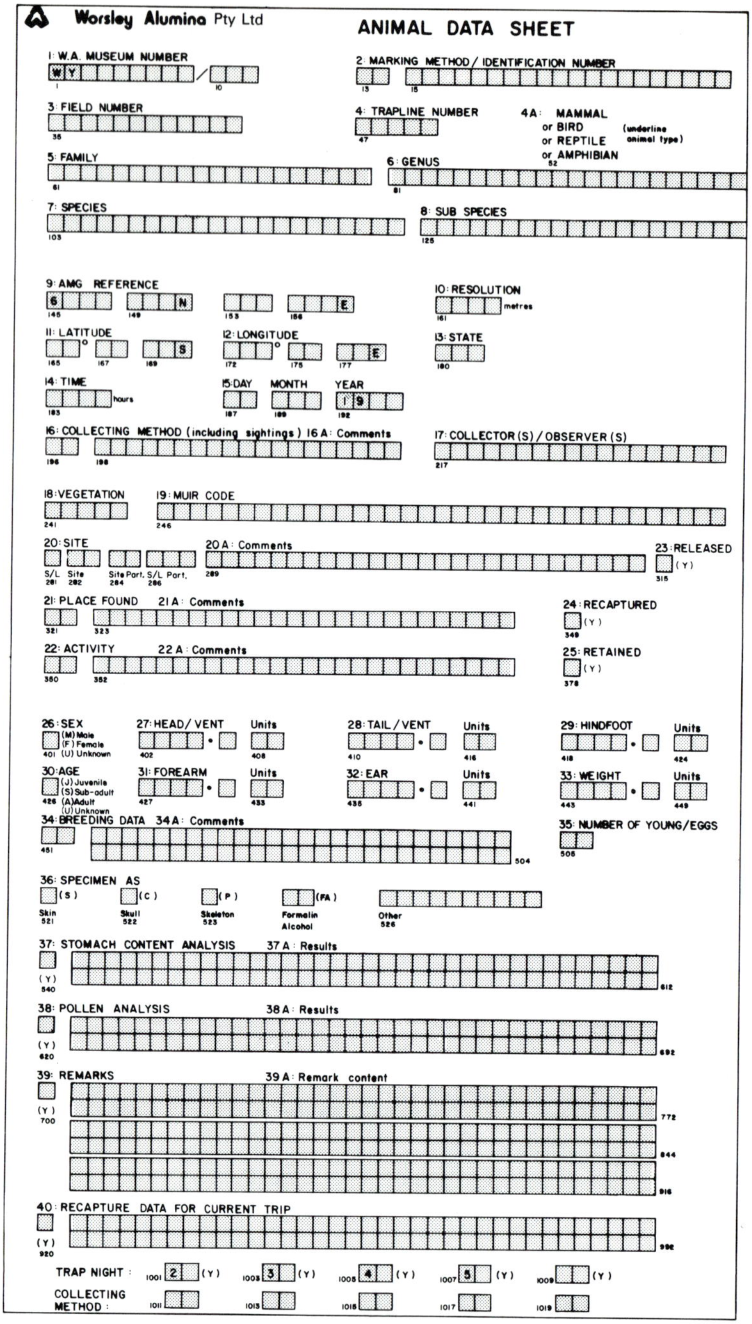

Worsley Alumina Pty Ltd

ANIMAL DATA SHEET

1: W.A. MUSEUM NUMBER W Y / (1) (10)

2: MARKING METHOD / IDENTIFICATION NUMBER (13) (15)

3: FIELD NUMBER (35)

4: TRAPLINE NUMBER (47)

4A: MAMMAL or BIRD or REPTILE or AMPHIBIAN (underline animal type) (52)

5: FAMILY (61)

6: GENUS (81)

7: SPECIES (103)

8: SUB SPECIES (125)

9: AMG REFERENCE 5 (145) N (149) (153) E (156)

10: RESOLUTION metres (161)

11: LATITUDE ° (165) (167) S (169)

12: LONGITUDE ° (172) (175) E (177)

13: STATE (180)

14: TIME hours (183)

15: DAY (187) MONTH (189) YEAR 1 9 (192)

16: COLLECTING METHOD (including sightings) (196) 16 A: Comments (198)

17: COLLECTOR (S) / OBSERVER (S) (217)

18: VEGETATION (241)

19: MUIR CODE (246)

20: SITE S/L (281) Site (282) Site Part. (284) S/L Part. (286)

20 A: Comments (289)

23: RELEASED (Y) (315)

21: PLACE FOUND (321) 21 A: Comments (323)

24: RECAPTURED (Y) (349)

22: ACTIVITY (350) 22 A: Comments (352)

25: RETAINED (Y) (378)

26: SEX (M) Male (F) Female (U) Unknown (401)

27: HEAD/VENT (402) • Units (408)

28: TAIL/VENT (410) • Units (416)

29: HINDFOOT (418) • Units (424)

30: AGE (J) Juvenile (S) Sub-adult (A) Adult (U) Unknown (426)

31: FOREARM (427) • Units (433)

32: EAR (435) • Units (441)

33: WEIGHT (443) • Units (449)

34: BREEDING DATA (451) 34 A: Comments (504)

35: NUMBER OF YOUNG/EGGS (505)

36: SPECIMEN AS

(S) Skin (521) (C) Skull (522) (P) Skeleton (523) (FA) Formalin Alcohol Other (526)

37: STOMACH CONTENT ANALYSIS (Y) (540) 37 A: Results (612)

38: POLLEN ANALYSIS (Y) (620) 38 A: Results (692)

39: REMARKS (Y) (700) 39 A: Remark content (772) (844) (916)

40: RECAPTURE DATA FOR CURRENT TRIP (Y) (920) (992)

TRAP NIGHT: 1001 2 (Y) 1003 3 (Y) 1005 4 (Y) 1007 5 (Y) 1009 (Y)

COLLECTING METHOD: 1011 1013 1015 1017 1019

(a)

ANIMAL DATA SHEET

Description of Codes

2: MARKING METHOD (code)

BB = Bat band	EP = Ear punch	TC = Toe clip	...=..............
DF = Distinctive Feature	ET = Ear tag	VS = Ventral scale removed	...=..............

N.B. If "DF" used, then note feature in place of identification number.

If ear punch, toe clip or ventral scale removal was used, show how this was carried out on the relevant diagram below.

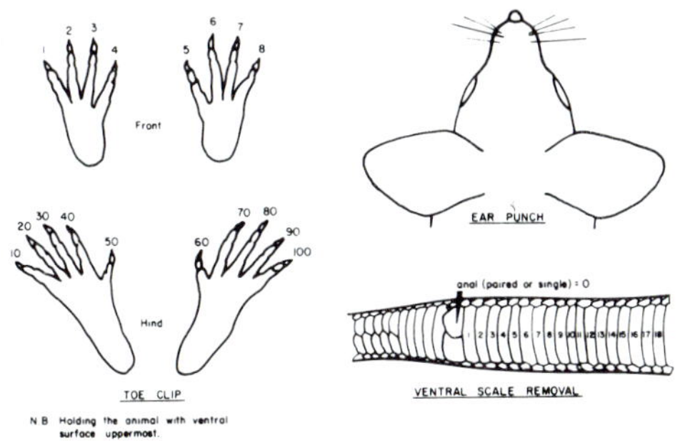

4: TRAPLINE NUMBER

This field must be filled in for ALL animals caught in a trapline or caught/observed in the vicinity of a trapline and its associated bird quadrat, during SYSTEMATIC data collection programmes.

16: COLLECTING METHOD - INCLUDING SIGHTINGS (code)

01 = Elliott trap	06 = Mist net	11 = Driven by fire	16 = SIGHTING ONLY
02 = Pit/fence	07 = Bat trap	12 = Pick-up dead/fresh	17 = Spotlight
03 = Cage trap	08 = Head-torch/hand	13 = Pick-up dead/decomposing	18 = Pick-up scats
04 = Hand catch	09 = Dug from burrow	14 = Predator kill	99 = Other
05 = Stunned	10 = Shot	15 = Car kill	

N.B. If code "16" (SIGHTING ONLY) is used and more than one individual of a species is sighted (or heard in the case of frogs), then include the number, age and sex, where possible, in Field 16A:Comments. e.g. for kangaroo sightings of, say, three adults (A) of unknown sex (U) and one sub-adult (SA) male (M), then record as - 3AU+1SAM (using the same letters as used in Fields 26:Sex and 30:Age).

20: SITE (code)

S/L = Study Location	Site = Broad Site Description	Site Part. = Partition within Broad Site Desc.	S/L Part. = Partition within Study Location
A = Mining Area - Saddleback	01 = Farm dam	01 = Deep open water	RCL = Refinery Catchment Lake
B = " " - Quindanning	02 = Swamp	02 = Open shallows	FWL = Freshwater Lake
C = " " - Marradong	03 = Lake/large dam	03 = Sheltered shallows	PHD = Pipe Head Dam
D = " " - Bunnings	04 = Gully	04 = On rocks in shallows	RA1 = Residue Area 1
E = Conveyor One Corridor	05 = Lower slope	05 = Reed beds	RA2 = Residue Area 2
F = " Two Corridor	06 = Mid slope	06 = Open shoreline	SEP = Solar Evaporation Pond
G = Refinery Lease Area	07 = Upper slope	07 = Open rocky shoreline	DDC = Drainage Diversion Channel
H = Other Minerals Area	08 = Ridge	08 = Regenerating shoreline	PSW = Pillow Swamp (O.M.)
J = Mooradung Reserve (32448)	09 = Creek bank	09 = Revegetated shoreline	RSW = Round Swamp (O.M.)
K = ...	10 = Hill	10 = Dam wall	... = ...
L = ...	11 = Rehabilitated mine pit	11 = Associated constructions	... = ...
M = ...	12 = Farm (general)	12 = ...	... = ...
N = ...	13 = Rehabilitated borrow pit	13 = ...	... = ...
	14 = ...	99 = Other	
	99 = Other		

21: PLACE FOUND (code)

01 = Under bark	05 = Blackboy	09 = Burrow	13 = Log or stump
02 = Hollow log	06 = Tussock	10 = In open	14 = Tree trunk
03 = Hollow limb	07 = Litter	11 = On track	15 = Foliage
04 = Live limb	08 = Rock	12 = On ground	16 = In building
			99 = Other

22: ACTIVITY (code)

01 = Resting	04 = Drinking	07 = Copulating	10 = Inactive
02 = Moving	05 = Basking	08 = Flying	11 = Swimming
03 = Feeding	06 = Aestivating	09 = Active	99 = Other

34: BREEDING DATA (code)

01 = Pregnant/gravid	03 = Pouch young	05 = Young in nest, female not located	07 = Nest preparation
02 = Lactating	04 = Young in nest	06 = Young at heel	08 = Male in breeding condition
			99 = Other

(b)

Fig. 7.6. Examples of (a) the animal data sheet, and (b) associated coding formats used for systematic and opportunistic mammal, reptile, amphibian and bird (only those actually handled) data collected in the Worsley Project areas.

Due to the inherent sampling difficulties with bats, and the possibility of misinterpretation in attempting to accurately compare numbers from different vegetation types, specimens were trapped only at sites where a concentrating effect was produced, e.g. flyways through vegetation and over dams, creeks and rivers. Mist-nets and a modified collapsible bat-

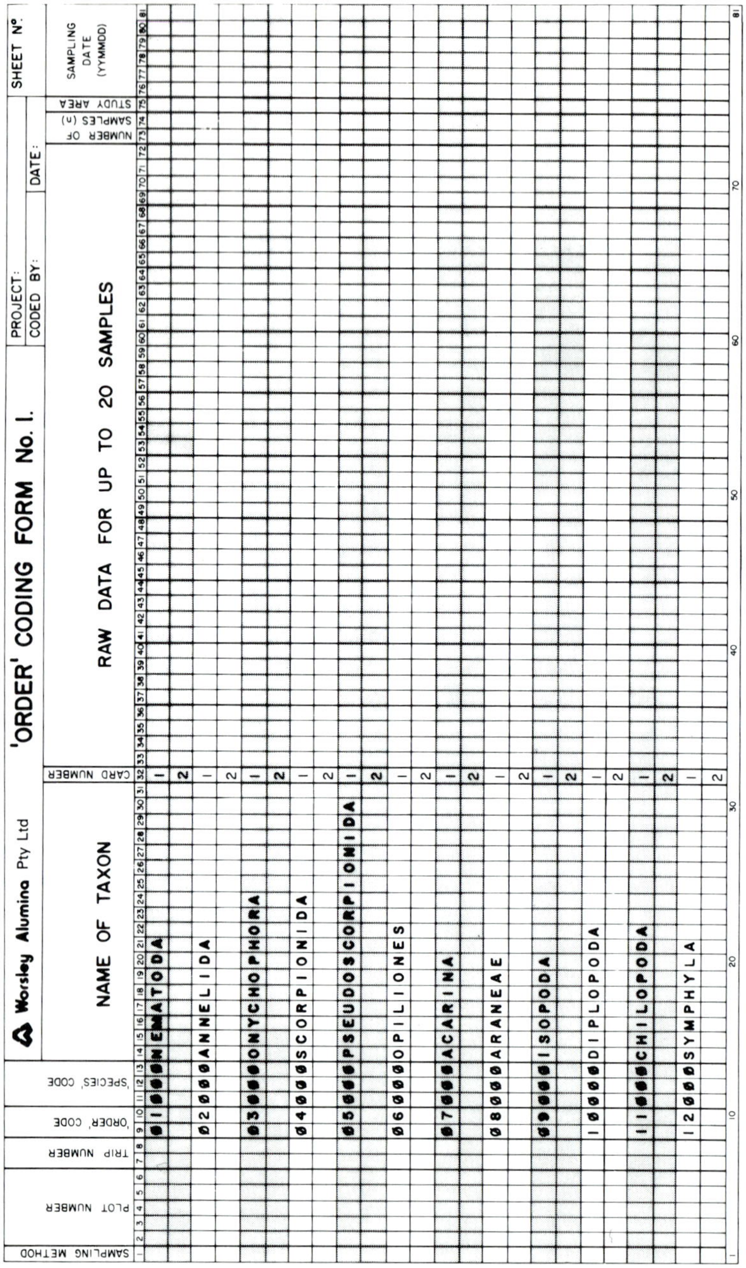
Worsley Alumina Pty Ltd
'ORDER' CODING FORM No. I.
PROJECT:
CODED BY:
DATE:
SHEET Nº.
SAMPLING METHOD
PLOT NUMBER
TRIP NUMBER
'ORDER' CODE
'SPECIES' CODE
NAME OF TAXON
CARD NUMBER
RAW DATA FOR UP TO 20 SAMPLES
NUMBER OF SAMPLES (n)
STUDY AREA
SAMPLING DATE (YYMMDD)
Ø1ØØØNEMATODA
Ø2ØØØANNELIDA
Ø3ØØØONYCHOPHORA
Ø4ØØØSCORPIONIDA
Ø5ØØØPSEUDOSCORPIONIDA
Ø6ØØØOPILIONES
Ø7ØØØACARINA
Ø8ØØØARANEAE
Ø9ØØØISOPODA
1ØØØØDIPLOPODA
11ØØØCHILOPODA
12ØØØSYMPHYLA

(a)

Column	1	Sampling method:	P = pitfall-trap S = sweep B = beat
Columns	2-6	Plot number:	e.g. M01J, R15Y
Columns	7,8	Trip Number:	1-4
Columns	9,10	"Order" code:	listed below

"Order" Codes

01 = NEMATODA	26 = CAELIFERA
02 = ANNELIDA	27 = PHASMATODEA
03 = ONYCHOPHORA	52 = EMBIOPTERA
04 = SCORPIONIDA	28 = PSOCOPTERA
05 = PSEUDOSCORPIONIDA	29 = HOMOPTERA
06 = OPILIONES	30 = HETEROPTERA
07 = ACARINA	31 = THYSANOPTERA
08 = ARANEAE	32 = NEUROPTERA — ADULTS
09 = ISOPODA	33 = NEUROPTERA — LARVAE
53 = AMPHIPODA	34 = STAPHYLINIDAE
10 = DIPLOPODA	35 = CURCULIONOIDEA
11 = CHILOPODA	36 = CARABIDOIDEA
12 = SYMPHYLA	37 = SCARABAEOIDEA
13 = PAUROPODA	38 = TENEBRIONIDAE
14 = COLLEMBOLA	39 = COLEOPTERA — OTHERS
15 = PROTURA	40 = COLEOPTERA — LARVAE
16 = DIPLURA	41 = MECOPTERA
17 = THYSANURA	42 = DIPTERA — ADULTS
18 = ODONATA	43 = DIPTERA — LARVAE
19 = BLATTODEA	44 = LEPIDOPTERA — ADULTS
20 = MANTODEA	45 = LEPIDOPTERA — LARVAE
21 = ISOPTERA	46 = HYMENOPTERA — ANTS
22 = DERMAPTERA	47 = HYMENOPTERA — OTHERS
23 = GRYLLACRIDOIDEA	51 = HYMENOPTERA — LARVAE
24 = TETTIGONIOIDEA	48 = SKINKS
25 = GRYLLOIDEA	49 = FROGS
	50 = MISCELLANEOUS

Column	11,13	"Species" code:	see ant list — Table C4
Columns	14-31	Name of taxon:	names for "order" codes above
Column	32	Card number:	1 or 2
Columns	33-72	Raw data:	twenty samples coded in I4 format on two cards.
Columns	73,74	Number of samples:	up to twenty
Column	75	Study Area:	A-G; parts of Worsley Alumina Project Area.
Columns	76-81	Sampling Date:	(last day of field sampling for each trip — YYMMDD)
Columns	82-86	Vegetation Code:	e.g. 23HDc (see Part A, Section A3.3.3 for codes). Inserted automatically by computer.

(b)

Fig. 7.7. Examples of (a) the 'order' coding form, and (b) associated coding formats, used for systematic invertebrate fauna surveys in the Worsley Project areas.

trap (Tideman & Woodside, 1978) were used. Data were logged onto the Animal Data Sheets (Fig. 7.6).

Invertebrate surveys

An invertebrate sampling transect was located along one of the bird observation transects at each intensive study location (see Fig.7.1), and formed the focal point for the pitfall-trapping, vegetation-sweeping and tree-beating sampling techniques employed for invertebrates.

Along each transect, up to 20 pitfall-traps were located at 10 m intervals. At each trap location a 21 mm internal diameter PVC pitfall-

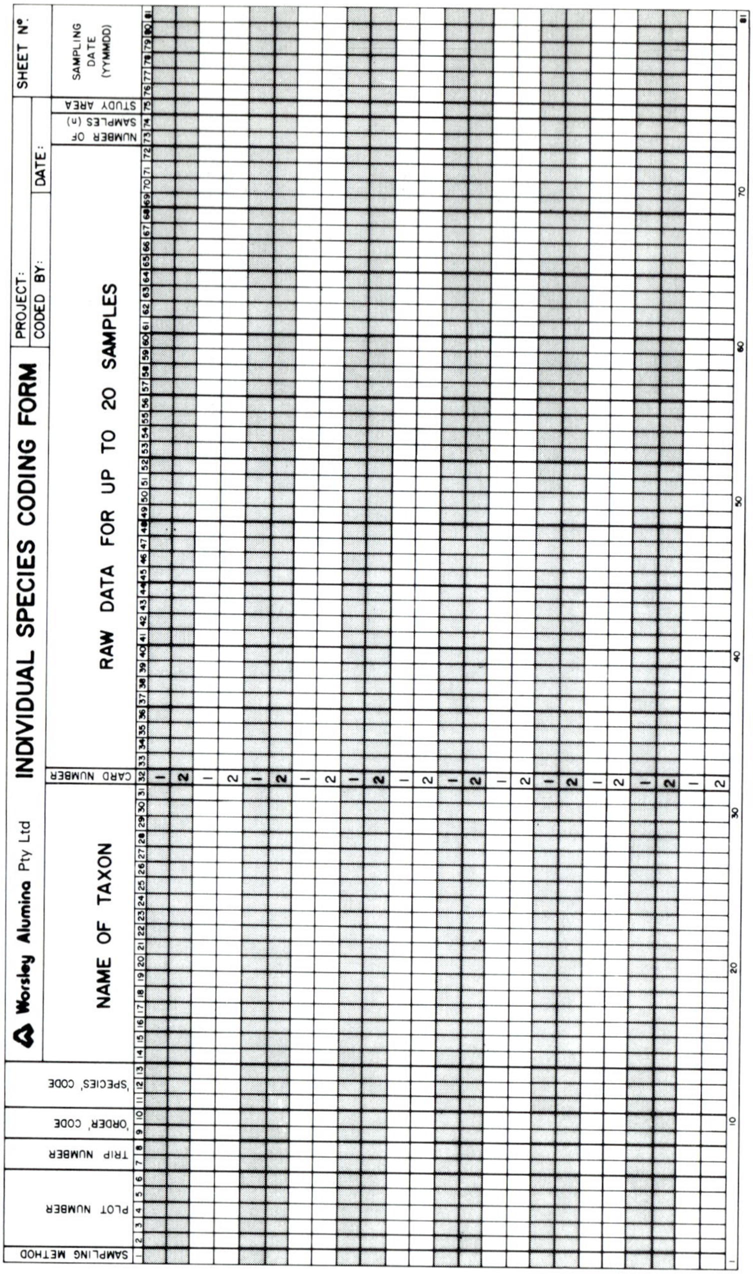

Fig. 7.8. Example of the individual species coding form used for systematic invertebrate fauna surveys in the Worsley Project areas. The coding formats are shown in Fig. 7.7.b.

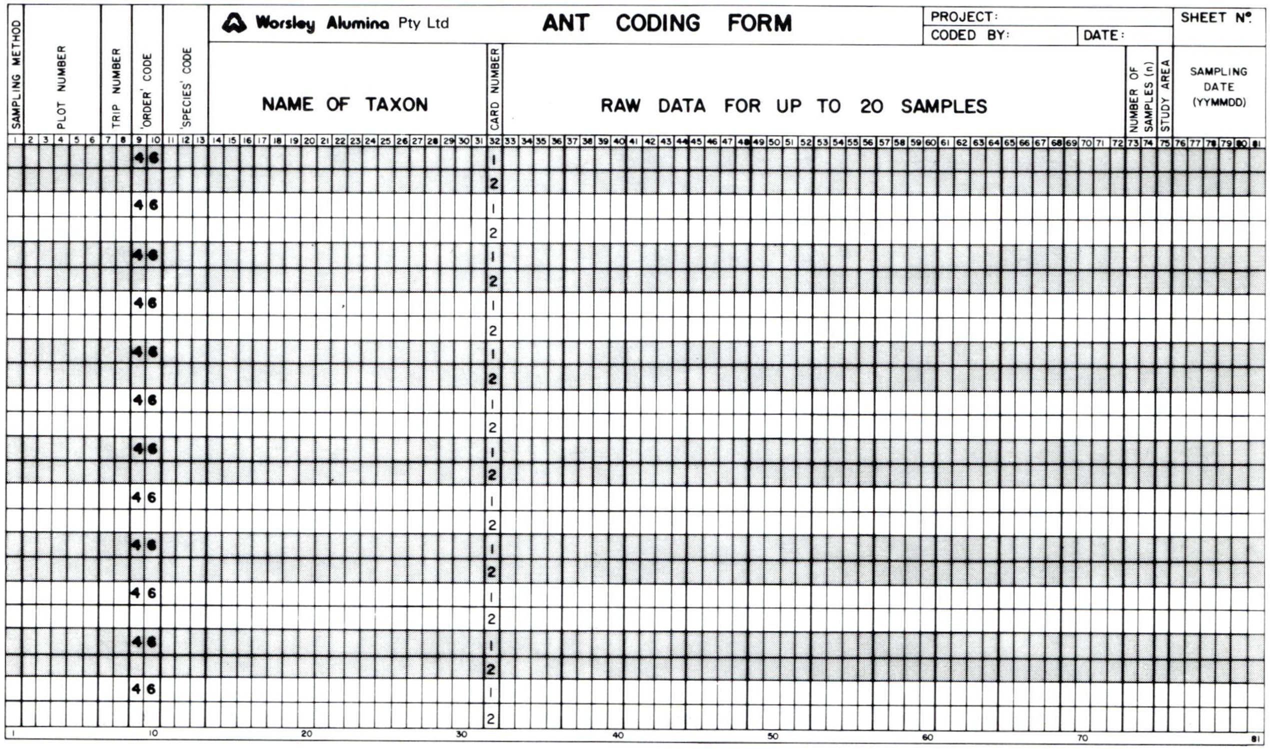

Fig. 7.9. Example of the ant coding form used for systematic invertebrate fauna surveys in the Worsley Project areas of south-western Australia. The coding formats are shown in Fig. 7.7.b.

trap holding-cylinder, described by Majer (1978), was sunk into the ground. At the commencement of each sampling period, pitfall-traps comprising 18 mm internal diameter Pyrex test tubes containing 10 ml 70:30 v/v ethanol:glycerol preservative (the glycerol reducing the rate of evaporation of the alcohol) were placed in each holding cylinder and left open for 5 consecutive days and nights. Greenslade & Greenslade (1971) have demonstrated that the invertebrate preservative used in the traps is non-attractive to ants in Australia, although some other invertebrates may be attracted.

Each pitfall-trap site also formed the starting point for each vegetation-sweep sample, which ran at right angles to the transect and into the core area of habitat being sampled. Herb- and shrub-sweeps were performed using a quadrilateral-shaped sweep-net. Each sweep consisted of a 40 m walk away from the pitfall-trap site and a 40 m return walk, the latter aligned 2 m to the side of the original traverse. Eighty sweeps were performed over the entire 2 × 40 m distance and the contents of the net were placed in containers of 70% ethanol for later sorting.

Up to 20 small trees or large shrubs were flagged for beat sampling in each intensive study location. Generally, the tree nearest to each pitfall-trap was selected, although in all cases a representative mix of tree species was sampled for each location. Beating was performed by jarring each tree with a stout stick, the disloged animals being collected on a 1 m^2 calico beating tray, which was moved four times around the tree canopy during the exercise in order to ensure adequate sampling. Animals were hand picked from the tray and placed in containers of 70% ethanol for sorting.

The three sampling methods provided relative estimates of the abundance and species richness of invertebrates in the ground (pitfall-traps), herb and shrub (sweep samples) and tree (beat samples) strata.

Invertebrate samples were sorted in the laboratory to broad taxonomic levels, with some taxa further sorted to species level when represented by adult forms. Particular emphasis was placed on ant species identification, as a large amount of information is available on this group in relation to rehabilitation of disturbed areas in Australia (Majer, 1984). All results were coded directly onto Invertebrate Data Coding Forms (see Figs. 7.7–7.9 for examples of the three forms used: Order Coding Form, Individual Species Coding Form and Ant Coding Form).

Computer storage and retrieval of data

To enable maximum value to be extracted from the faunal data, Worsley Alumina has developed a comprehensive fauna computing system, a reasonably 'user-friendly' system which is menu-driven and runs on a Fujitsu M180 IIAD machine. The system is described briefly as follows:

1. The programs are written in Fortran 77 with approximately 35,000 lines of code,
2. All files are standard sequential formatted files, although some have very long record lengths – e.g. Animal Master File – 1200 characters/record; Bird Master File – 112 characters/record; Invertebrate Master File – 126 characters/record.
3. The system is relatively free of machine dependencies, except for the programs that require terminal input/output, which make use of the Fujitsu full screen processing capability to display menus and request information – other areas of machine dependency include the job control language and the manner in which the system is menu driven to submit jobs. And
4. the large programs are run in one megabyte of virtual storage, as this is a relatively convenient size, but this can be reduced.

Fauna data are loaded from magnetic tapes, or keyed in directly via a VDU terminal, and subjected to rigorous error checking prior to accession to master files. Bird quadrat/transect and inventory data are combined and stored in one master file, while animal data (mammals, reptiles and amphibians, and birds actually handled) are stored separately in another file. Invertebrate data are stored in a third master file. A flow diagram is presented in Fig. 7.10 to illustrate the computing sequence followed for accession of new data to the Animal Master File.

The interactive system allows rapid sorting and reporting of:

1. Full or partial data sets, catalogues of selected fauna records.
2. Plots of bird locations within quadrats or along transects (for overlays on vegetation maps).
3. Tabulations of sightings and breeding records for birds by season, dates or vegetation types.
4. Hierarchical classifications with totals and percentage occurrences for selected fields (e.g. percentage tree statum utilization by a bird species during a single season, within a particular vegetation type). And
5. quadrat/transect analyses of systematically-collected bird records for use in calculating the number of birds per ha in various vegetation community types.

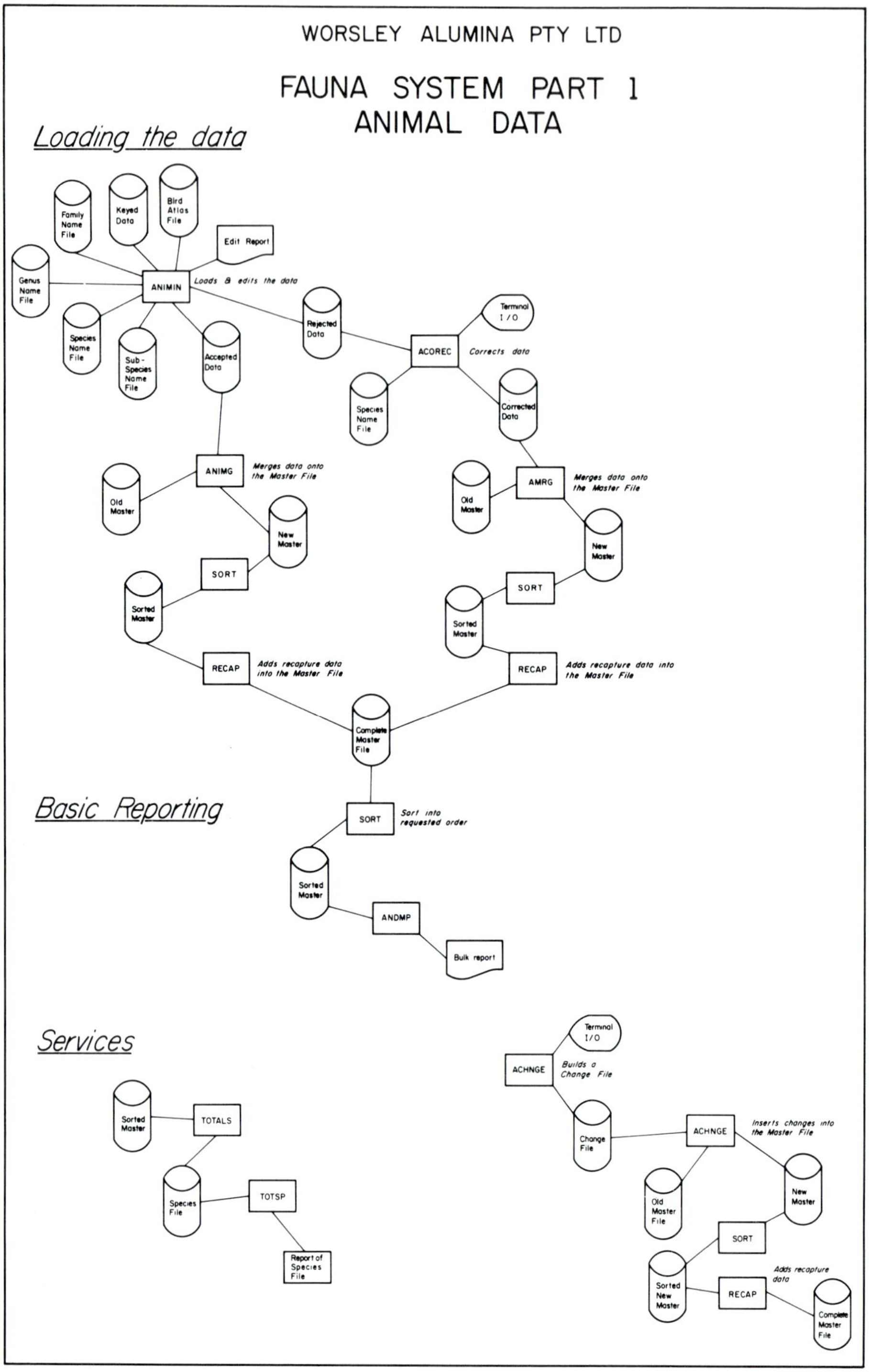

Fig. 7.10. A flow diagram illustrating the computing sequence followed for accession of new fauna data, collected in the Worsley Project areas of south-western Australia, into the Animal Master File of the fauna computing system.

The system also produces tabulations of invertebrate fauna data at the broad taxa or individual species levels, including presence/absence listings for use in ordination analyses and listings of means, standard deviations and ranges for the various sampling methods, sampling periods and replicates used in the studies. In addition, the system provides for bulk changes on fields in the master files (e.g. nomenclature changes) and data editing, both in the interactive mode.

Analysis and presentation of data

The fauna data collected during Worsley Alumina's fauna studies have been analyzed with a variety of statistical techniques including: cluster analysis, division and review of species groupings (e.g. bird 'feeding guilds'), ordination and correlation. These analyses have been employed to evaluate the degree of association between various vegetation communities, based on faunal utilization, and the associations evident between various species of fauna within the Worsley Alumina and Boddington Gold Mine project areas. Knowledge of these associations in the existing vegetation of the area will be essential during the on-going environmental monitoring phase for assessing the success of rehabilitation programmes in disturbed areas and for modifying treatments accordingly. Specific methods for data treatment and various forms of illustrative presentation are given in Worsley Alumina (1985, 1989).

Summary

In providing the detailed example of baseline fauna surveys conducted by one Australian company, this review has not sought to provide the ultimate solution or the best template for those concerned in proposals and projects involving major land disturbance. Rather, the development of a standardized programme has been explained in terms of the specific legal and environmental requirements of one company, and the way that company has approached the issues involved in order to develop a workable and affordable solution. Such a programme will be practically useful in the long-term for monitoring the impacts of operations and the resilience of the restored ecosystems upon decommissioning.

The information provided does, however, emphasize the importance of sound and standardized techniques and the need for close liaison with the controlling regulatory authorities and scientists involved in the biological disiplines. In this respect, if the methods described here serve as a starting point for moves toward industry-wide standardization, the aims of this Chapter will be well served.

Acknowledgements

The programmes described here were developed in close consultation with several scientists and consultants working in Western Australia: K. Youngson and J. Henry of Ninox Wildlife Consulting; J. Majer of Curtin University of Technology; D. Gasmier of Geological Modeling Systems; R. How, J. Dell, L. Smith and R. Johnstone of the Western Australian Museum; L. Mattiske of E. M. Mattiske & Associates, and many other workers from Government agencies and tertiary institutions.

Should the reader be interested in obtaining the Fauna Computing System programs, the software is the property of Worsley Alumina Pty Ltd. Contact should be made to the Environmental Section, Worsley Alumina Pty Ltd, P.O. Box 344, Collie, Western Australia, 6225.

References

Bradshaw, A. D. (1984). Ecological principles and land reclamation practice. *Landscape Planning*, **11**, 35–48.

Biological Surveys Committee of Western Australia. (1984). Biological survey of the Eastern Goldfields of Western Australia. Part 1. Introduction and Methods. *Records of the Western Australian Museum. Supplement Number* 18, 1–25.

Brunner, H. & Coman, B (1974). *The identification of mammalian hair*. Melbourne: Inkata Press.

Cogger, H. G. (1985). *Reptiles and Amphibians of Australia*. Sydney: Reed.

Greenslade, P. & Greenslade, P. J. M. (1971). The use of baits and preservatives in pitfall traps. *Journal of the Australian Entomological Society*, **10**, 253–60.

How, R. A., Humphreys, W. F. & Dell, J. (1984). Vertebrate surveys in semi-arid Western Australia. In *Survey Methods for Nature Conservation*, ed. K. Myers, C. R. Margules & I. Musto, pp. 193–216. Canberra: CSIRO.

Keast, A. (1984). Assessment of community composition and species richness in contrasting habitats. In *Methods of Censusing Birds in Australia*, ed. S. J. J. F. Davies, pp. 19–24. Royal Australasian Ornithologists Union Report 7. Department of Conservation and Environment. Perth, Western Australia: Bulletin No. 153.

Majer, J, D. (1978). An improved pitfall trap for sampling ants and other epigaeic invertebrates. *Journal of the Australian Entomological Society*, **17**, 261–2.

Majer, J. D. (1984). Ant return in rehabilitated mines – an indicator of ecosystem resilience. In *Proceedings of the 4th International Conference on Mediterranean Ecosystems*, ed. B. Dell, pp. 105–8. Perth: University of Western Australia.

Majer, J. D. (1987). The conservation and study of invertebrates in remnants of native vegetation. In *Nature Conservation. The Role of Remnants of Native Vegetation*, ed. D. Saunders, G. Arnold, A. Burbidge & A. Hopkins, pp. 333–5. Sydney: Surrey Beatty.

McKenzie, N. L. & Youngson, W. K. (1983). Mammals. In *The Wildlife of the Great Sandy Desert, Western Australia*, ed. A. A. Burbidge & N. L. McKenzie. Wildlife Research Bulletin of Western Australia Number 12. Perth: Department of Fisheries and Wildlife.

Pyke, G. H. & Recher, H. F. (1984). Censusing Australian birds: a summary of procedures and a scheme for standardization of data presentation and storage. In *Methods of Censusing Birds in Australia*, ed. S. J. J. F. Davies, pp. 55–63. Royal Australasian Ornithologists Union Report 7. Department of Conservation and Environment, Perth: Western Australia. Bulletin No. 153.

Recher, H. F. (1984). Use of bird census procedures in Australia: a review. In *Methods of Censusing Birds in Australia*, ed. S. J. J. F. Davies, pp. 3–14. Royal Australasian Ornithologists Union Report 7. Department of Conservation and Environment, Perth: Western Australia. Bulletin No. 153.

Reynolds Australia Alumina Ltd. (1979). *Worsley Alumina Project Environmental Review and Management Programme*. Perth, Western Australia: Reynolds Australia Alumina Ltd.

Snyder, B. D. & Potter, G. L. (1981). The critical links between baseline studies and development of revegetation and wildlife restoration plans. In *Symposium on Surface Mining Hydrology, Sedimentology and Reclamation*, ed. D. H. Graves, pp. 507–10. University of Kentucky, Lexington, Kentucky. December 7–11, 1981.

Stonehouse, B., ed. (1978). *Animal Marking: Recognition Marking of Animals in Research*. London and Basingstoke: The MacMillian Press.

Tideman, C. R. & Woodside, D. P. (1978). A collapsible bat-trap and a comparison of results obtained with the trap and mist nets. *Australian Wildlife Research*, **5**, 355–62.

U.S. Congress, Office of Technology Assessment. (1986). Baseline and monitoring data. In *Western Surface Mine Permitting and Reclamation*. Washington D.C.: U.S. Government Printing Office.

Westman, W. E. (1986). Resilience: concepts and measures. In *Resilience in Mediterranean Ecosystems*, ed. B. Dell, A. J. M. Hopkins & B. B. Lamont, pp. 5–19. Dordrecht: Dr W Junk.

Wooler, R. D., Russell, E. M., Renfree, M. B. & Towers, P. A. (1983). A comparison of seasonal changes in the pollen loads of nectarivorous marsupials and birds. *Australian Wildlife Research*, **10**, 311–17.

Worsley Alumina Pty Ltd. (1985). *Worsley Alumina Project – Flora and Fauna Studies, Phase Two*. Perth, Western Australia: Worsley Alumina Pty Ltd.

Worsley Alumina Pty Ltd. (1987). *Boddington Gold Mine Environmental Management Programme*. Perth, Western Australia: Prepared by Kinhill Stearns/BHP Engineering.

Worsley Alumina Pty Ltd. (1989). *Boddington Gold Mine Project – Baseline Biological Investigations*. Perth, Western Australia: Worsley Alumina Pty Ltd.

Worsley Alumina Pty Ltd. & Dames & Moore. (1981). *Worsley Alumina Project – Flora and Fauna Studies, Phase One*. Perth, Western Australia: Worsley Alumina Pty Ltd.

Worsley Alumina Pty Ltd & Department of Conservation and Land Management. (1987). *Bauxite Pit Rehabilitation – Saddleback Timber Reserve – Agreed Arrangements between Worsley Alumina Pty Ltd and the Department of Conservation and Land Management*. Perth, Western Australia: Worsley Alumina Pty Ltd & Department of Conservation and Land Management.

8

Design of reclamation to encourage fauna

S. R. VIERT

Introduction

As indicated in the Introduction to this book, the reclamation of disturbed lands has been a practical activity in one form or another for several years. However, where fauna have reinvaded a disturbed area, the return usually has been incidental rather than by design. It has been recognized by many reclamation specialists in the western United States that a unique opportunity to design habitats that will encourage the return of fauna can be provided through rehabilitation.

Although this opportunity exists, a recognized deficiency in current plans to encourage the return of wildlife is the general lack of specific design criteria for wildlife habitat restoration and enhancement (U.S. Congress, Office of Technology Assessment, 1986). As a result, implementation of restoration procedures on behalf of wildlife is difficult. Without specific details and plans regarding how many, what shape, or where to construct features, on-ground construction of habitat features to encourage wildlife has often proven to be less than satisfactory. In cases where specific design criteria have been developed, rehabilitation of wildlife habitats appears to have more potential for success. A further problem is that the lack of proven design criteria hinders negotiations for limited resources which may be sought by other more defensible planning efforts.

Proper design of rehabilitation prescriptions to encourage fauna must necessarily begin at the earliest stages of the planning for rehabilitation. As with scientific studies, careful formulation of objectives, analysis of operating and planning constraints, and definition of goals will increase the chances of a programme's success. If design of rehabilitation methods is not initiated until the time of reclamation, or the environmental, political, and economic conditions in which rehabilitation is to occur are not thoroughly evaluated prior to disturbance, efforts on behalf of wildlife will probably be less than satisfactory. Successful wildlife habitat rehabilitation is highly dependent upon project engineering and other technical aspects of the reclamation planning process. Only an interdisciplinary approach, carried out early in the project planning process, will provide an avenue for equal consideration of habitat requirements and wildlife needs, as well as the more conventionally accepted aspects of rehabilitation.

It is this concept of early, interdisciplinary planning and also the development of detailed design criteria which form the basis of this Chapter. To illustrate this approach, the author has drawn upon his experience with coal mining and associated rehabilitation planning in the western United States.

Importance of early planning and interdisciplinary coordination

Typically, western United States coal mining operations occur in areas where wildlife habitat is a primary or secondary land-use. As a result, restoration of wildlife habitat is usually one of the primary or secondary post-mining land-use goals. This often results in the development of a formal wildlife plan which becomes an inherent component of the overall rehabilitation. Development of a successful wildlife plan starts with early planning and coordination with project engineers and specialists from other disciplines. The key to a successful plan is a thorough understanding of the proposed operation, the budget of the operator, the ecological resources within the project area, and a variety of planning constraints.

Once a mining company has made the internal commitment to restore wildlife habitat during the rehabilitation process, formal interaction between mining engineers and biologists is necessary from the point of conceptual mine planning onward. Early interaction serves to educate each professional group about the concerns and limitations of the other, while creating a forum to facilitate planning for the various situations which may be encountered. In this manner, engineers are made aware of the habitat diversity needs of wildlife (e.g. the form and spatial distribution of water, soil, and topography) which must be incorporated into the rehabilitation plan. In addition, biologists are provided with an overview of proposed operations and are made aware of the budgetary and operational limitations which are controlling the planning procedure.

Following early interaction with project engineers, project biologists must gain a thorough understanding of the ecological resources of the project area. A detailed understanding of the faunal resources which the land is capable of supporting prior to disturbance is critical to the development of a functional wildlife plan. The need for this baseline information, and some of the ways in which it might be collected, are addressed in the previous Chapter.

Concurrently with early planning, biologists must gain an understanding of the constraints acting upon wildlife which are associated with the given project. The extent to which many of the more intensive wildlife restoration measures are implemented is dependent upon: (1) conflicting post-reclamation land-uses; (2) project economics; (3) technological or operational limitations; (4) legal or regulatory requirements (see Chapter 1); and (5) incompatible requirements demanded by other technical disciplines (e.g. hydrology).

In the case of conflicting post-reclamation land-uses, biologists must account for any land management objectives of landowners which are not compatible with the return of wildlife. For example, ranchers typically prefer the return of all disturbed lands to flat pastures, which are specifically suited for producing forage for livestock (see Chapter 10). This conflict may be redressed by careful landscape design which matches the proposed final landforms to the best uses of the land, be they for wildlife habitat, agriculture, or a combination of uses. Such design necessitates detailed interdisciplinary planning and care in the implementation of rehabilitation procedures and post-rehabilitation management. An example of prudent land-use and landscape design can be found at the Trapper Mine of western Colorado. Despite pre-mining land-uses, the steeper, less productive lands are set aside for the return of wildlife habitat, while the flatter, more productive lands are being returned to agriculture.

As far as economic constraints are concerned, the costs and benefits of different wildlife options are difficult to derive due to the inherent problems involved with assigning monetary values to wildlife benefits. When costs and concomitant benefits are not well documented, reclamation managers are hesitant to commit their limited budgets to the pursuits of unknown or unquantifiable benefits. Therefore, biologists must clearly identify benefits of proposed wildlife restoration measures, develop detailed plans and, with the aid of project engineers, determine the associated costs. For example, maintaining a topographic component for wildlife habitat holds significant potential for overall reclamation cost savings (\$1–\$2 saved per m^3 of spoil material left in place) while at the same time offering significant potential benefits to wildlife (topographic diversity). Some further thoughts on the economic benefits of rehabilitation for wildlife are presented at the end of Chapter 12.

Occasionally, technological and/or operational limitations will constrain potential rehabilitation measures. These constraints are often related to the limited ability of current techniques to successfully identify or establish the habitat features that are necessary for survival or welfare of wildlife. For example, many wildlife habitat requirements are associated with physical features of the environment such as topography and surface water. However, identification and reconstruction of the optimum distribution, frequency or variety of topographic features (e.g. escarpments, rock outcrops or water courses) is limited by a lack of data on the ecological requirements of many species and also by the limitations of the available reclamation machinery. Where restoration of habitat features has been attempted it has been based on professional intuition rather than on hard data. Also, the type of equipment has tended to dictate the size and shape of the features which have been constructed. These limitations indicate a practical need for additional research on the contribution of physical habitat features to overall habitat requirements.

A very important constraint upon potential rehabilitation measures in the western United States is the regulatory control of mining and reclamation. Legal requirements designed to protect non-wildlife aspects of the environment often preclude the valuable wildlife potential of the area. For example in the Northern Great Plains, leaving highwalls after cessation of mining is considered by nearly all wildlife personnel to be a viable wildlife habitat enhancement technique. These features have been demonstrated to be technically feasible and ecologically beneficial (Tessmann, 1982*a*). However, development of highwalls as a habitat feature is very difficult to implement due to the regulatory requirement for returning the land to 'approximate original contour'. Another example of regulatory constraints is presented in Chapter 11. As discussed in that Chapter, there are many opportunities for wetland development which are constrained by reclamation requirements for surface water runoff and impoundments. If an innovative and potentially viable rehabilitation method appears feasible, but is in conflict with regulations governing other aspects of reclamation, early planning and coordination with regulatory authorities are essential to demonstrate the need for regulatory flexibility.

The last major constraint to the design and implementation of adequate restoration methods involves conflicts with other environmental disciplines. Rehabilitation which is directed toward restoring optimal wildlife habitat often requires trade-offs with non-wildlife disciplines such as soil science, botany, and hydrology. For example, optimal erosion control might require level terrain, redistribution of soils to a uniform depth, and planting of hardy, introduced grasses. In such a situation, the resulting rehabilitated area will probably provide habitat of limited value for most types of wildlife. However, early interdisciplinary interaction could enable the improved utility of rehabilitated land as wildlife habitat. Negotiations with personnel from other disciplines might highlight the need for specific wildlife habitat features, such as broken terrain, non-uniform redistribution of soil materials, and planting of native shrubs, forbs and grasses. In this manner, trade-offs can then be balanced early in the planning process while options are still available.

Formulation of restoration methods

According to a recent study (U.S. Congress, Office of Technology Assessment, 1986), in comparison to other environmental disciplines, the evaluation and development of rehabilitation methods for wildlife are relatively subjective. The principal means of formulating rehabilitation techniques has been professional judgement based on knowledge of the ecological characteristics of an area and the operational aspects of a proposed disturbance. Despite this subjectivity, rehabilitation methods which have been implemented to date appear to be successful, according to a study recently funded by the US Department of the Interior, Office of Surface Mining (Stoecker, Thompson & Comer, 1984). This study quantitatively evaluated the effectiveness of wildlife-orientated rehabilitation measures which had been used on coal mining operations in the western United States. One of the principal conclusions of this study was that wildlife responses to various habitat features and configurations in the rehabilitation were consistent with professional intuitive judgement.

Because of the lack of analytical means to determine the type of rehabilitation methods which are required, biologists charged with such responsibility typically review specific measures employed at other disturbed sites, along with the degree of success attained by those measures. Such a review must take into account the fact that most techniques employed to encourage wildlife can be classified into four general categories: (1) avoidance measures; (2) operational measures; (3) offsite enhancement; and (4) habitat replacement/enhancement (U.S. Congress, Office of Technology Assessment, 1986). The first three of these categories are measures designed to avoid, lessen, or compensate for direct impacts of a given operation or disturbance. Only the last category, habitat replacement/enhancement, includes measures designed specifically to restore habitat characteristics of disturbed lands. Selected examples of typical rehabilitation techniques and methods are indicated in Table 8.1 and others are described in various Chapters of this book.

When designing an overall habitat rehabilitation plan, it is important to consider individual techniques such as those listed in this Table. However, it is more important to consider them collectively. The combination of basic 'building blocks' of wildlife habitat (water, soil, vegetation, and topography) must be optimized in response to each specific situation or application. Such an optimization is at present necessarily a result of intuitive professional judgement of rehabilitation personnel. These people are equipped with a specific knowledge of the ecological resources of the area, of the technological and operating constraints, and of targeted post-reclamation land uses.

Despite the apparent successes with the 'intuitive approach', many professionals involved with the design of habitat rehabilitation techniques have indicated the need for a more analytical approach. They feel that this would allow for documentation of techniques and greater compatibility of their approaches with personnel from other disciplines. The general lack of design criteria for wildlife habitat restoration and enhancement has been identified as a major deficiency in current wildlife planning. Because of this deficiency, wildlife habitat restoration or enhancement prescriptions found in most formal rehabilitation plans lack specificity. As a result, prescriptions are often difficult to implement on-the-ground during the actual rehabilitation phase. In addition, the lack of proven design criteria makes defence of habitat restoration difficult during early internal negotiations with engineering staff and non-wildlife specialists.

With regard to this difficulty, reclamation professionals indicate that the specificity needed for minimum design standards would answer such questions as: (1) What is the best design or configuration for a given habitat restoration feature?; (2) What size is optimum?; (3) Where is it

Table 8.1. *Examples of typical wildlife rehabilitation techniques and concomitant benefits*

Technique	Principal benefit
Topographic manipulations such as creation of undulating and broken topography, undrained surface depressions, and micro-topographic features.	Provides diversity of physical habitat, visual and thermal cover, and the necessary conditions for development of varied biotic habitat features.
Leave intact or modify remaining highwall.	Provides opportunities for cliff nesting raptor species.
Establish or enhance impoundments.	Provides still-water habitat, and a source of water for wildlife.
Direct-haul topsoil for re-application and spread in non-uniform depths	Native seed and soil microbia aid establishment of floral diversity.
Reconstruct stream channels and streamside habitats.	Predisposes development of riparian systems.
Create large boulder and rock piles randomly distributed about the rehabilitated area.	Provides habitat for small mammals, perch sites for raptors, and substitutes for natural rimrock.
Create large, irregularly shaped brush piles.	Provides temporary substitute for lost micro-habitats and shelters.
Establish shrub/tree plantings along lee sides of rock/brush piles or randomly within rehabilitated area.	Takes advantage of additional snowpack moisture and/or acts as a 'live snowfence'.
Establish shelterbelts within large expanses of rehabilitated grasslands.	Provides wind erosion protection, habitat diversity, and travel routes for wildlife.
Design the shape and spatial distribution of revegetated communities to maximize interspersion and 'edge effect'.	Provides a means to maximize community diversity, wildlife diversity, and wildlife populations.
Select plants to be revegetated based on forage and cover values to wildlife.	Forces consideration of these wildlife needs during seed mix determination.
Improve land management practices to preclude deleterious grazing.	Eliminates habitat degradation due to over-utilization by livestock.
Add special structures to rehabilitated areas as necessary (e.g. implanted dead trees for snags, artificial nesting structures, etc.).	Provides for certain physical habitat needs of wildlife species not provided by other techniques.

most desirable to establish specific features?; and (4) What combination of features would optimize habitat utility for a given species? Wherever details such as these can be provided, justification of the technique is significantly strengthened and the on-the-ground utility of habitat restoration efforts improved.

Because of the lack of information on many aspects of wildlife biology, providing the level of detail indicated above may be difficult in many instances. Additional research is needed to identify and correlate the needs of different wildlife species to discrete physical and biotic habitat features which can be established by rehabilitation efforts. Examples of pioneering efforts in this regard include studies by Fox & Fox (1984) and Majer *et al.* (1984). Research must also address ways in which the design (e.g., size, configuration, etc.), distribution, density, and location of habitat features affect wildlife.

Currently, habitat rehabilitation focuses on the reconstruction of botanical attributes and very seldom stresses replacement of physical characteristics or features of the land. In the western United States coal industry, the principal reasons for this narrow focus are: (1) a limited understanding of which physical features are most important for wildlife; (2) technical and economic limitations on the replacement of many physical features; (3) an absence of regulatory mechanisms for the replacement of physical features necessary for habitat, and (4) little or no flexibility of regulations associated with other reclamation standards. As the state-of-the-art advances, it can be expected that detailed design criteria will substantially improve on-the-ground efforts for the rehabilitation of physical features necessary for wildlife habitat.

Recent innovations in wildlife rehabilitation

When coal mine rehabilitation was first required in the western United States, a variety of measures to encourage wildlife were already available to reclamation specialists. For the most part, these measures were the result of intuition and common sense. The few measures which were aimed specifically at restoration of wildlife habitats were typically applied late in the process with little interdisciplinary planning or detailed design. As increasingly stringent rehabilitation requirements evolved, and environmental awareness on the part of coal operators grew, innovation in wildlife habitat rehabilitation has occurred. In some cases, innovation has meant improvements and updates to existing approaches, while in other cases it has meant completely new approaches. A few examples of these recent innovations in wildlife habitat restoration are provided below to illustrate the concept of early interdisciplinary planning and the development of detailed design criteria to facilitate on-the-ground implementation of the approach.

Fresh topsoiling

It is well known that ecological diversity leads to improved wildlife habitat (Krebs, 1972). However, one of the major difficulties encountered during rehabilitation has been the attainment of adequate floral diversity within a reasonable period of time. Reseeding of soils which have been stockpiled tends to produce a plant community of low diversity because stockpiling for more than a few months essentially sterilizes the seeds in the soil. Fresh topsoiling techniques have been used in recent years to circumvent this problem. The fresh topsoiling technique involves removal of topsoil from directly in front of an advancing pit, with replacement on regraded soil immediately following the pit. In this manner, native soil microbia and plant seed are not lost by sterilization during storage of soil. The result is a healthier growth medium and added floral diversity, which in turn produces higher quality habitat.

The use of this technique must necessarily be introduced and considered by project engineers early in the planning process as it switches the task of retopsoiling from rehabilitation to mine operations. In addition, utilization of the technique introduces a new factor into an already complicated topsoil removal, transport and spreading procedure.

Shrub establishment

Diverse and relatively dense stands of shrubs are of great value to wildlife for a multitude of reasons (Tessmann, 1982*b*). Over the past few years, a variety of new and innovative approaches to the establishment of this aspect of wildlife habitat have occurred. Techniques have ranged from direct transplanting of mature plants with front-end loaders to injection of seed into the soil with hydromulch equipment. In most instances it is too early to conclude that these techniques have been successful. However, interim results have been very encouraging. One example, which appears to be very successful, is the direct transplanting of woody plants in the effort to reestablish 'woody draws' in the Northern Great Plains of the United States (see Fig. 8.1).

Fig. 8.1. Personnel checking tree transplants which have been used in an attempt to revegetate a 'woody draw' in the Northern Great Plains of the United States (photograph by S. R. Viert).

Wetland reclamation

The value of wetlands as terrestrial wildlife habitat is common knowledge. Until recently, rehabilitation for wetlands in the western United States was not encouraged. However, in a few instances, reclamation specialists were able to demonstrate the potential for success of rehabilitation of certain wetlands. In these instances, experimental manipulations of wetlands in coal mining areas have been allowed and the interim results are promising; it is expected that complete restoration will be forthcoming.

Where wetland rehabilitation is being attempted, it has followed the early multidisciplinary planning approach and has involved technical representatives from mining companies, fish and wildlife agencies, mining regulatory agencies, engineers and a variety of experts from other disciplines. These individuals typically meet well in advance of planned wetland disturbance, to determine on a case-by-case basis the rehabilitation potential of each wetland. Wetlands determined to be reclaimable are carefully evaluated by geologists, hydrologists, wildlife ecologists, and botanists, among others, to determine baseline conditions. These data, along with other information, are then used to develop the detailed design criteria which are necessary for the rehabilitation of each wetland.

An excellent example of this process can be found at the Falkirk mine in North Dakota. The Falkirk Mining Company recently (1987) received an 'Excellence in Surface Mining Reclamation' award from the US Department of the Interior for efforts centering on rehabilitation of a 70 ha wetland. At present, the only major topic still under negotiation is that of success criteria; namely at what point does a rehabilitated wetland exhibit successful restoration? The rehabilitation of wetlands is discussed in more detail in Chapter 11.

Architecturally designed post-reclamation land-use

The transient use of land by mining operations offers a unique opportunity not found with more irreversible disturbances such as municipal development, that being the rehabilitation, or possible enhancement, of predisturbance wildlife habitats. This concept has developed into a recent innovation; designing the post-reclamation landscape to reflect the best uses of the reclaimed land. This innovation has seen limited utilization. However, with more stringent environmental regulations governing the conversion of wildlife habitat to other land uses, it is likely that it will be applied more often in future years. As with the previous examples, this technique requires early interdisciplinary planning but also requires involvement of the landowner. The basis of the technique is to optimize the uses of the land over the area of disturbance, and within each optimized unit, maximize the value of this land by the implementation of the most appropriate rehabilitation techniques. The Trapper mine, which was previously mentioned in this Chapter, offers a good example of an area of overgrazed marginal wildlife habitat comprised of both steep and gentle terrain. Following disturbance, the gentle terrain is typically returned to pasture, while the steep terrain is largely set aside for wildlife habitat. Though acreage of pristine land may be lost in the long-term, the quality of habitat is enhanced, compensating for the lost acreage.

Highwall development

In many areas of the western United States, suitable nesting habitat is the limiting factor for raptor populations. In a very few instances (two as of 1985), a major innovation has been attempted on behalf of these animals. This has involved the experimental retention of a remaining highwall. In both instances referred to, significant early planning and interdisciplinary coordination was necessary in order to gain approval and prove the feasibility of this measure. Following

Fig. 8.2. Example of an active golden eagle nest constructed on an experimental highwall in a central Wyoming coal mine. Because of suitable geological characteristics, ledge construction was not necessary in this particular situation (photograph by S. R. Ritter).

approval from the regulatory agencies for a variance from topographic requirements, additional planning, engineering, and study was required to develop the detailed design criteria necessary for implementation. Several factors such as slope stability, erosional stability, and optimal location of nesting structure were investigated to facilitate the development of these design criteria. Fig 8.2 exhibits an active 'highwall' nest constructed on a suitable ledge on one of the experimental highwalls in Wyoming.

Because of the potential economic benefits, this particular innovation has been looked upon favourably by many operators. By 1987, several proposals for highwall development were in various stages of the

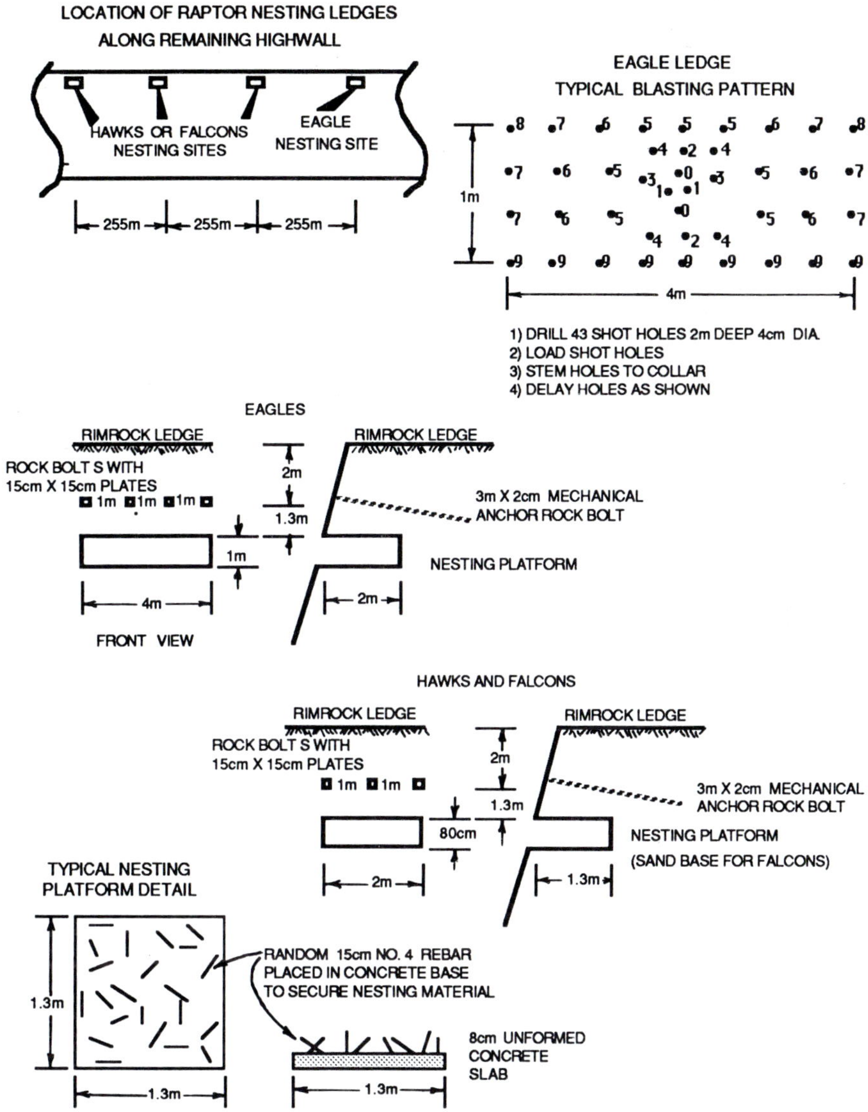

Fig. 8.3. Typical raptor nesting ledge construction detail along remaining highwall.

permitting process. Regardless of approval or denial, early interdisciplinary planning and detailed design criteria are necessary components of each effort.

Figures 8.3 and 8.4 have been provided to give an indication of the level of detail which would be considered adequate for on-the-ground implementation. These conceptual examples show where local raptor

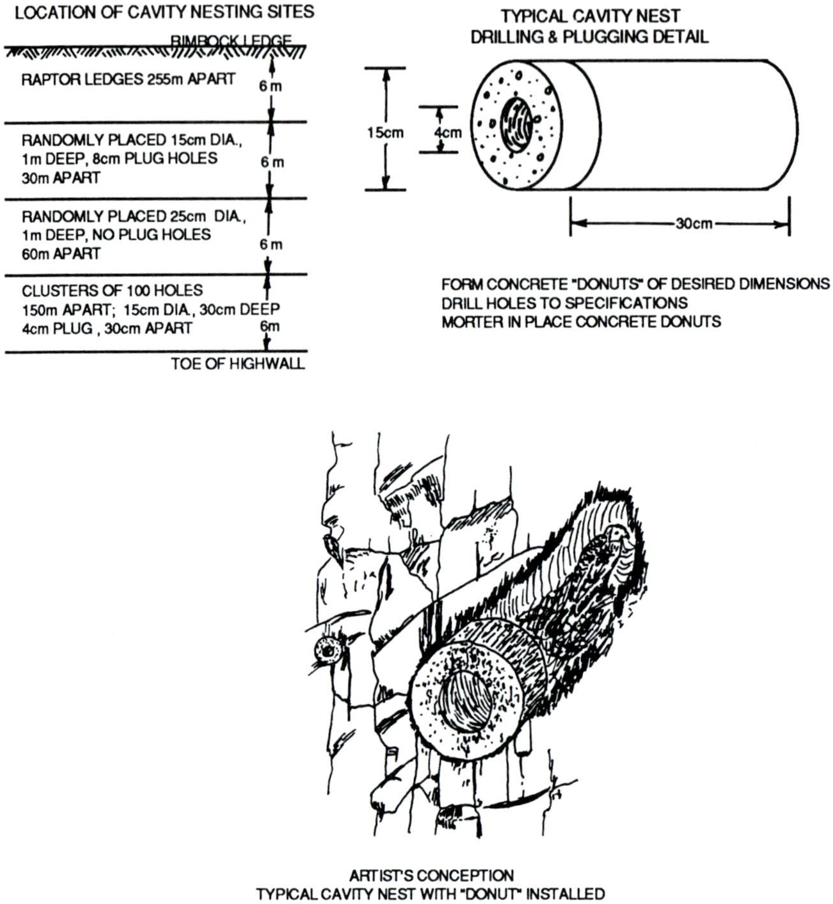

Fig. 8.4. Typical cavity nest construction detail along highwall.

populations would benefit from highwall development. In this conceptual example, prey base and inter- and intra- specific territory size of various species must be taken into account. Once the requirements of a species are determined, details regarding nest-habitat construction and location must be provided. Figure 8.3 indicates where nesting ledges for large raptors should be located along a highwall, how many ledges can be optimally provided, and the dimensions and construction detail which are necessary for implementation. Figure 8.4 indicates where holes could be drilled along a highwall to benefit small-cavity nesting raptors as well as other cavity-nesting avifauna. The latter species could in turn serve as additional prey base for falcons.

Aside from providing raptor nesting habitat, a highwall can provide topographic diversity to otherwise relatively level areas, aid the accumulation of additional surface moisture at the highwall base, and aid in the establishment of greater floral diversity by providing varying slopes and exposures.

These are but a few examples of how early planning and detailed design criteria can facilitate rehabilitation of disturbed lands to satisfactorily encourage fauna recolonization. Without early interdisciplinary planning, there may be a premature exclusion of viable efforts to encourage wildlife, and the inclusion of detailed designs ensures an adequate understanding of required efforts for on-the-ground implementation.

Summary

Although reclamation of disturbed lands has been a practical ongoing activity for several years, rehabilitation to encourage fauna has typically not been a significant effort. This situation has primarily resulted from two failures on the part of the professional biological community: a general lack of involvement of biologists in early interdisciplinary planning, and an absence of specific detailed design criteria for rehabilitation procedures.

To correct the first deficiency, biologists must become more active in land development projects at a much earlier time. This would require a thorough understanding of the proposed project, the budget of the operator, the ecological resources within the project area, and a multitude of planning constraints. With this knowledge, biologists must then become involved in the early interdisciplinary planning efforts of a given project in order to facilitate the consideration of fauna during the rehabilitation process.

Correction of the second deficiency will require biologists to pay attention to developing detailed design criteria for proposed habitat rehabilitation efforts. Detailed designs, as opposed to general concepts, will provide the necessary guidance to engineering and construction personnel to facilitate on-the-ground construction. Development of detailed criteria will necessarily require an analytical approach to the design of habitats and will further require additional research into the habitat needs of targeted wildlife species.

Where these two deficiencies have been corrected, efforts on behalf of wildlife habitat rehabilitation show significant promise. Wetland restoration, highwall development, fresh topsoiling techniques, innovative shrub-land revegetation, and post-reclamation land-use design are but a few examples of how early interdisciplinary planning and detailed design criteria are improving the potential benefits to wildlife and encouraging recolonization of reclaimed lands.

References

Fox, B. J. & Fox, M. D. (1984). Small-mammal recolonization of open-forest following sandmining. *Australian Journal of Ecology*, **9**, 241–52.

Krebs, C. J. (1972). *Ecology – the Experimental Analysis of Distribution and Abundance*. New York: Harper & Row.

Majer, J. D., Day, J. E., Kabay, E. D. & Perriman, W. S. (1984). Recolonization by ants in bauxite mines rehabilitated by a number of different methods. *Journal of Applied Ecology*, **21**, 355–75.

Stoecker, R. E., Thompson, R. W. & Comer, R. D. (1984). *An Evaluation of Wildlife Mitigation Practices at Western Coal Operations*. Boulder: Thorne Ecological Institute.

Tessmann, S. (1982*a*). *The Development of Highwalls as Wildlife Habitats*. Unpublished draft. Wyoming Game and Fish Department. January 1982.

Tessmann, S. (1982*b*). *Measures for Restoring Wildlife Habitat on Reclaimed Mine Lands*. Unpublished draft. Wyoming Game and Fish Department. January 1982.

U.S. Congress, Office of Technology Assessment. (1986). *Western Surface Mine Permitting and Reclamation*, vols. 1 & 2. Washington, D.C.: U.S. Government Printing Office.

9

The influence of substrate toxicity on soil macrofauna return in reclaimed land

WEI-CHUN MA AND H. EIJSACKERS

Introduction

Colonization of derelict waste land by soil fauna is often seriously hampered by the toxicity of the substrate. This, together with other factors including physically adverse conditions and deficiency of nutrients such as nitrogen and phosphorus, may act as a barrier to colonization. The main chemical factors which present problems to rehabilitation of man-made waste land and subsequent fauna return include: (1) the presence of metal ions at toxic levels; (2) adverse osmotic effects due to high salinity; and (3) extreme levels of acidity or alkalinity. The present Chapter deals with the effects of these factors on soil fauna return to non-reclaimed untreated waste sites. Attention has been focused on heavy metals rather than organic pollutants because the former are stable, non-degradable and therefore present a more serious long-term problem than do degradable organic compounds. It is not always possible to pin down substrate toxicity to one single causative factor. In many cases, more than one toxic factor is involved. For the sake of convenience however, examples of substrate toxicity will be presented under the particular heading where it seems most appropriate. Some examples appear under more than one heading.

The emphasis of this Chapter is on soil macrofauna as colonizers of waste sites. Earthworms, together with adults of termites, beetles, ants, and larvae of some flies and beetles, are burrowers whose bioturbation activities may play a key role in soil structuring and pedogenetic processes. As mentioned in Chapter 2, their presence is therefore necessary for a sustained natural recovery of ecosystem function in reclaimed sites. Microarthropods such as mites and springtails are in turn dependent on the pore spaces which are particularly created by these larger soil-structuring invertebrates. Furthermore, lumbricid earthworms together with millipedes, some fly larvae and, to a lesser extent, isopods, amphipods and some beetles, are among the most important soil macroinvertebrates involved in litter fragmentation in temperate zones (see Chapter 3). These litter-fragmenting animals have a specific ecological function in accelerating the chemical and microbiological mineralization processes in the soil environment. Macroinvertebrates such as earthworms are even a valuable management tool in attempts to ameliorate waste land. Once established after introduction, earthworms, in many cases, have been used successfully to improve the fertility of low-productivity range land (Syers & Springett, 1984; Lee, 1985).

Substrate toxicity

Colonizing animals

Colonization depends on the species' capacity to immigrate into a site, rapidly reproduce and to adequately disperse its young. Water and wind-dispersed phoretic microarthropods, and the larvae of flying insects, are the first colonizers of rehabilitated land and they generally arrive prior to the appearance of lumbricid earthworms (Dunger, 1969*a*, *b*; Rusek, 1978; Huhta, 1979). Once established, earthworms take on the dominant functional role in the decomposition of the litter layer. It is emphasized here that the occurrence of flying or walking animals in a given area does not necessarily indicate that they are residents. At most, this can be inferred if the larval stages are also present at the same time. Even the presence of larvae may have resulted from opportunistic behaviour; beetles, for example, may leave the area once they grow into adults. Similarly, the colonization by ants can only be regarded as permanent if ant nests are actually found in the area. Colonization of derelict sites by species of soil fauna with a restricted migratory capability, such as earthworms, may be regarded as a true indication of colonization and of the suitability of such sites to sustain soil fauna populations.

Studies made on earthworms at harsh and polluted environments invariably show that under natural conditions early colonizers mainly consist of opportunistic epigaeic lifeforms with a r-selected life history strategy. These early colonizers are followed by burrowing species that are transitional between r- and K-strategists, and lastly by predominantly K-selected species. Table 9.1 summarizes the attributes distinguishing r-

Table 9.1. *Summary of attributes of r- and K-selected earthworms (from Satchell,* 1980*a*)

	r	K
A. *Directly related to fecundity and length of life cycle*		
1. Number of cocoons produced per worm	Higher	Lower
2. Number of embryos produced per cocoon	Higher	Lower
3. Incubation time of cocoons	Shorter	Longer
4. Maturation time from hatching	Shorter	Longer
5. Duration of reproductive life	Shorter	Longer
6. Time distribution of mortality	Shorter	Longer
7. Form of survivorship curve (Deevey, 1947)	Type III	Type I/II
8. Seasonal stability of population density	Lower	Higher
B. *Related to the feeding behaviour required to sustain different reproductive rates*		
9. Surface or subsurface dwelling	Surface	Subsurface
10. Metabolic rate	Faster	Slower
11. Mobility	Higher	Lower
12. Sensitivity to pH	Lower	Higher
13. Sensitivity to light	Higher	Lower
14. Pigmentation	Pigmented	Unpigmented
15. Avoidance of desiccation	Quiescence	Diapause
16. Size	Smaller	Larger
17. Morphological adaptations to burrowing	Poorly developed	Well developed
18. Form of prostomium	Tanylobic	Epilobic or proepilobic
19. Form of typhlosole	Smaller, unfolded	Larger, folded
C. *Related to metabolic rate*		
20. Nitrogen excretion rate	Faster	Slower
21. Intestinal transit rate	Faster	Slower
22. Oxygen affinity of haemoglobin	Lower	Higher

and K-lifeforms of lumbricid earthworms as related to reproduction, productivity, functional morphology and metabolism. Standen *et al.* (1982), sampling young and old opencast coal mine reclamation sites, observed a dominance of r-selected species of earthworms with a shift to more transitional (r/K) species at the older sites. Dunger (1969*a*, *b*) found in restored coal mining wastes that *Dendrobaena octaedra*, an r-selected surface-living species, was the first colonizing earthworm species and only at a later stage was it followed by burrowing species. At metal-polluted or acidic drier sites, surface-active, r-selected species such as *Lumbricus rubellus*, *Dendrobaena rubida*, *D. octaedra* and *Lumbricus eiseni* and, at wet sites, *Eiseniella tetraedra* are the more important colonizing species (Curry & Cotton, 1983; see also Table 9.4). Studies on colonization of refuse tip sites similarly show that within a period of 3 to 5 years, *L. rubellus* is the most important colonizing species with *E. tetraedra* frequently occurring in zones of percolating water (Brockmann, Koehler & Schriefer, 1980). The immigration rate for *L. rubellus* and *D. octaedra* was estimated at 4 m yr^{-1}. At colliery spoil heaps left to naturally revegetate by woodland plants, populations of *L. rubellus* develop rapidly. The large deep-burrowing and predominantly K-selected *Lumbricus terrestris* appears only after a long time in open grassy sites (Standen, Stead & Dunning, 1982).

It is important to distinguish between natural colonization rate and the colonizing capacity upon experimental introduction. In the latter case, most species appear almost equally successful in establishing themselves given a suitable environment. Predominantly K-selected endogaeic species such as *Aporrectodea caliginosa* are quite successful in many situations, whether introduced in irrigated deserts (El-Duweini & Ghabbour, 1965) or reclaimed polders (van Rhee, 1969). Bengtsson *et al.* (1979) suggested that *A. caliginosa* is a better colonizer upon introduction than *L. rubellus* because of the capability of the former to aestivate during short-term adverse drought conditions and its greater ability to burrow and thus avoid predators. Hoogerkamp, Rogaar & Eijsackers (1983) found that *Aporrectodea longa*, *Aporrectodea rosea* and *A. caliginosa*, which are all more or less K-selected species, as well as the r-selected *L. rubellus*, all had a similar dispersal rate of 9 m yr^{-1}, when inoculated in a favourable environment provided by grassland in silty clay loam polder soil. *L. terrestris*, which is a deep-burrowing anecic K-selected species, had a slower dispersal rate of 4.5 m yr^{-1}, although it was equally successful in establishing itself as a colonizing species. Work by Vimmerstedt (1983) has shown that *L. terrestris* can be successfully introduced in revegetated opencut coal mine sites.

Methods for assessing colonization

Experiments designed to assess the influence of substrate toxicity on soil fauna should ideally be performed on a long-term basis.

Laboratory tests should not be limited to measurements on survivorship; often soil invertebrates may survive during short-term laboratory experiments but this is no warrant for their actual colonization success in the field. Reproduction is a more significant measure of ability to achieve sustainable populations than any other parameter. For instance, heavy metals are generally not acutely toxic to soil invertebrates but may cause a reduced rate of reproduction at sublethal levels. An impaired reproductive rate may result in unsuccessful colonization, even if mortality rate is low. Experiments with earthworms have shown that reproduction is the most sensitive measure of toxicity in comparison with, for example, changes in body weight, food intake rate, or rate of survival (Ma, 1984). Reproductive rate therefore should be included as a test parameter in laboratory experiments designed to study substrate toxicity to soil fauna.

Field experiments may provide valuable information on long-term substrate toxicity to soil fauna. Important insight into the potential of soil fauna to colonize toxic soil substrates can also be derived from retrospective studies on the diversity and abundance of populations present at contaminated sites. The fauna which have developed in the course of time should therefore be studied on sites which have been disturbed or polluted in the past. This may yield interesting information on adaptive strategies and successional stages of soil fauna at such sites.

Examples of different approaches (i.e. laboratory experiments, field experiments and retrospective studies) to assess the significance of substrate toxicity on soil fauna will be discussed in the following sections.

Case studies

Heavy metals

Fecundity in the earthworm *Eisenia fetida* has been reported to be completely inhibited at soil metal concentrations of 50 mg kg^{-1} for Cd, 400 mg kg^{-1} for Ni, 2000 mg kg^{-1} for Cu and $>$ 5000 mg kg^{-1} for Zn and Pb (Malecki, Neuhauser & Loehr, 1982). However, data for other worms do not always agree with these figures, suggesting that soil-dwelling species of earthworms may be more sensitively affected by heavy metals than the compost-inhabiting *E. fetida*. Thus, in experiments with *A. caliginosa*, van Rhee (1975) found a cessation of reproduction by Zn at 1100 mg kg^{-1} and experiments conducted with the earthworm *L. rubellus* show a cessation of reproduction at soil Cu concentrations of

275 mg kg^{-1} or less (Ma, 1984, 1988). The toxic effect of Cu is highly dependent on pH, being much less at a near-neutral pH range than in a more acid soil (Ma, 1988). Similar pH effects on metal toxicity to earthworms have been described by Bengtsson, Gunnarson & Rundgren (1986) in experiments with *D. rubida*. Streit & Jaeggy (1983) demonstrated that soil type is also important. Cu was found to be more acutely toxic to *Octolasium cyaneum* in soil with little organic material than in an organic peat soil. Obviously soil characteristics, metal type, and also inter-species variation in sensitivity all play a role in determining the level of toxicity of heavy metals to soil fauna.

The influence of the soil cationic exchange capacity, pH and organic matter on the availability of heavy metals to earthworms is reflected not only in the toxicity but also in the extent to which the metals accumulate in the earthworm body tissues (Ma, 1982; Ma *et al.*, 1983). This has important consequences to predatory vertebrate animals feeding on earthworms. For instance moles, *Talpa europaea*, in heavy metal polluted areas may show highly elevated metal levels in their body tissues and this depends on whether the soil conditions of a particular site permit a high degree of metal accumulation in their earthworm diet (Ma, 1987).

Relatively little is known about the behavioural responses of soil invertebrates when they encounter chemically polluted substrates. Laboratory experiments suggest that metal-contaminated soil may have a deterrent effect. Isopods such as *Porcellio scaber* have been observed to avoid Pb-contaminated food (van Capelleveen *et al.*, 1986). Earthworms are capable of avoiding Cu-contaminated soil which would prevent active dispersal into Cu-polluted soils (Ma, 1988). Avoidance behaviour in earthworms has also been observed to substrates containing high amounts of freshly deposited pulverized fly ash. Fly ash is a by-product of coal-fired electricity generation stations and is dumped in large quantities at land disposal sites. Pulverized fly ash can be very alkaline and also contains high concentrations of various elements such as Ra-226, Cu, Hg, B, Se, etc. Table 9.2 shows the tendency of *L. rubellus* to avoid burrowing into soil that had been amended above a certain rate with young fly ash taken from 1–2-year-old ash sites. The avoidance was not due to differences in moisture content as the latter was kept at 32.5% in both amended and unamended soil. At low quantities of fly ash (up to 4%), the earthworms tended to show a preference rather than an avoidance of ash-amended soil. This suggests that worm dispersal into ash-amended sites would still occur in the field at low concentrations of ash.

Table 9.2. *Mean (±SE) burrow length made per earthworm* (Lumbricus rubellus) *during* 24 *days when given a choice between a clay loam soil amended with fly ash and unamended soil (from Ma & Eijsackers, unpublished data)*

Percentage ash in soil	Total burrow length (cm)	Mean length in unamended soil (cm)	Mean length in amended soil (cm)	Difference
0	41.3±4.6	25.6±14.6	15.7±12.6	n.s.
2	62.6±38.2	12.8±11.4	49.2±34.5	$P < 0.05$
4	25.3±8.5	8.7±6.5	16.7±4.7	$P < 0.05$
8	44.9±27.6	42.2±26.7	2.7±3.8	$P < 0.05$
16	21.6±19.9	13.8±18.5	7.8±7.5	n.s.
32	39.0±18.4	30.2±10.4	8.8±9.3	$P < 0.05$

Note: n.s. = not significant; P = probability level.

Avoidance behaviour can be considered as just one of the various types of survival strategies exhibited by soil invertebrates to cope with metal-polluted environments. Physiological detoxification mechanisms are important as well. Isopods, for instance, possess a storage-detoxification system for heavy metals in the hepatopancreas (Wieser, 1979; Hopkin & Martin, 1984). Earthworms have chloragogenous tissue which similarly acts as a storage-detoxification organ (Ireland, 1978). Springtails possess an efficient mechanism to excrete assimilated metals (Joosse & Buker, 1979) as do spiders which feed on contaminated isopods (Hopkin & Martin, 1985).

An example of a field study on the effect of heavy metals on populations of lumbricid earthworms is shown in Figure 9.1. The soil of the experimental plots had been amended at various rates with metal-contaminated sewage sludge for a number of years and had subsequently been turned into a ley. Four years later the plots were sampled for colonizing earthworms. The results showed large differences in population size depending on the level of metal concentration present in the soil surface horizon. On the basis of data from this and other experiments, it was concluded that Cu is an especially important toxic factor among the multiple metal contaminants present in sludge-amended sites.

In smelter-influenced mixed-oak stands with 26000 ppm Zn, 2300 ppm Pb, 900 ppm Cd and 340 ppm Cu in the litter layer, a reduction in density of all major taxonomic groups of arthropods has been observed (Strojan, 1978). Ecological implications include a significant

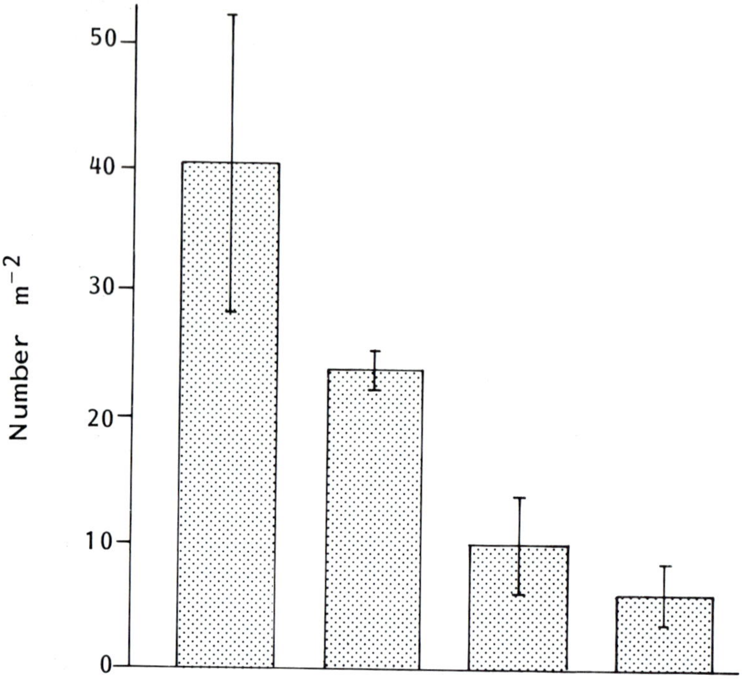

	Soil layer	Soil metal conc. (mg kg^{-1})			
Cd	0 - 3 cm	0.7	15	31	57
	3 - 5	0.5	10	21	31
	5 - 15	0.5	2.6	5.1	5.7
	15 - 25	0.3	0.5	0.7	0.8
Cu	0 - 3 cm	5.5	69	138	265
	3 - 5	4.7	50	99	150
	5 - 15	3.9	13	23	25
	15 - 25	3.0	4.5	5.4	6.0
Pb	0 - 3 cm	23	79	144	254
	3 - 5	22	147	288	438
	5 - 15	19	49	79	89
	15 - 25	16	21	22	24
Ni	0 - 3 cm	3.6	30	57	101
	3 - 5	4.7	24	44	69
	5 - 15	3.9	8.6	15	17
	15 - 25	2.7	4.1	3.5	3.9

Fig. 9.1. Mean population size of the earthworm *Lumbricus rubellus* in grassland which had been amended with metal-contaminated sewage sludge at four rates (from Ma, 1988).

reduction of soil enzymatic activities, leaf litter decomposition rate and a corresponding increase in standing mass of undecayed litter (Strojan, 1978; Tyler, 1984). Similarly, spruce forests containing more than 2500 ppm Cu and 3600 ppm Zn in the litter layer show reduced numbers of species of spiders, harvestmen, ants, beetles and slugs (Bengtsson & Rundgren, 1984). These studies indicate that complete natural recovery of normal ecosystem functioning is not likely to occur at these very high levels of soil metal contamination and that active rehabilitation measures will be necessary to achieve this goal.

Salinity

Salinity is a major factor which limits successful colonization of substrates by soil fauna. Earthworms are especially vulnerable to osmotic stress caused by a high salinity. Electrolytic conductivities exceeding 7 mS cm^{-1} are inhibitory to growth, while still higher values can be lethal (Hartenstein, Neuhauser & Narahara, 1981). Survival and reproduction of springtails are reduced at salinities equivalent to 8 mS cm^{-1} (Hutson, 1978). Whereas water-logging is not an important factor for survival of earthworms such as *L. terrestris* (Rushton, 1986*a*), seawater has been found to exert a direct toxic action on earthworms via the salinity of temporarily flooded sites (Piearce & Piearce, 1979). Most earthworms avoid salinity levels of 1.4% total dissolved salts and 2.9% is considered lethal. Earthworm populations, however, can be successfully introduced into saline environments when salts have been leached out. This has been shown by van Rhee (1963) for *A. longa* when the indigenous populations had been eliminated from a polder by temporary inundation by the sea. Populations of earthworms thus may recover naturally after a few years from periodic flooding by sea water (Piearce & Piearce, 1979). It is clear that attempts to introduce earthworms into waste sites are bound to fail if critical salinity levels are exceeded.

Although excess salinity limits natural or introduced colonization of waste land, the problem may only be temporary. The success of revegetation and the associated survival of soil fauna usually increases rapidly with ageing of the substrate due to leaching of the salts by rains. For instance, fresh pulverized fuel ash has conductivities ranging from 8 to 13 mS cm^{-1}, whereas 2 to 3 years of weathering usually reduces the soluble-salt content of the lagooned ash to a harmless level (Hutnik & Davis, 1973). Hardness and compacted layers in the subsoil, together with levels of soil moisture, organic matter and vegetation cover may then appear as the more serious factors limiting colonization (Parsons &

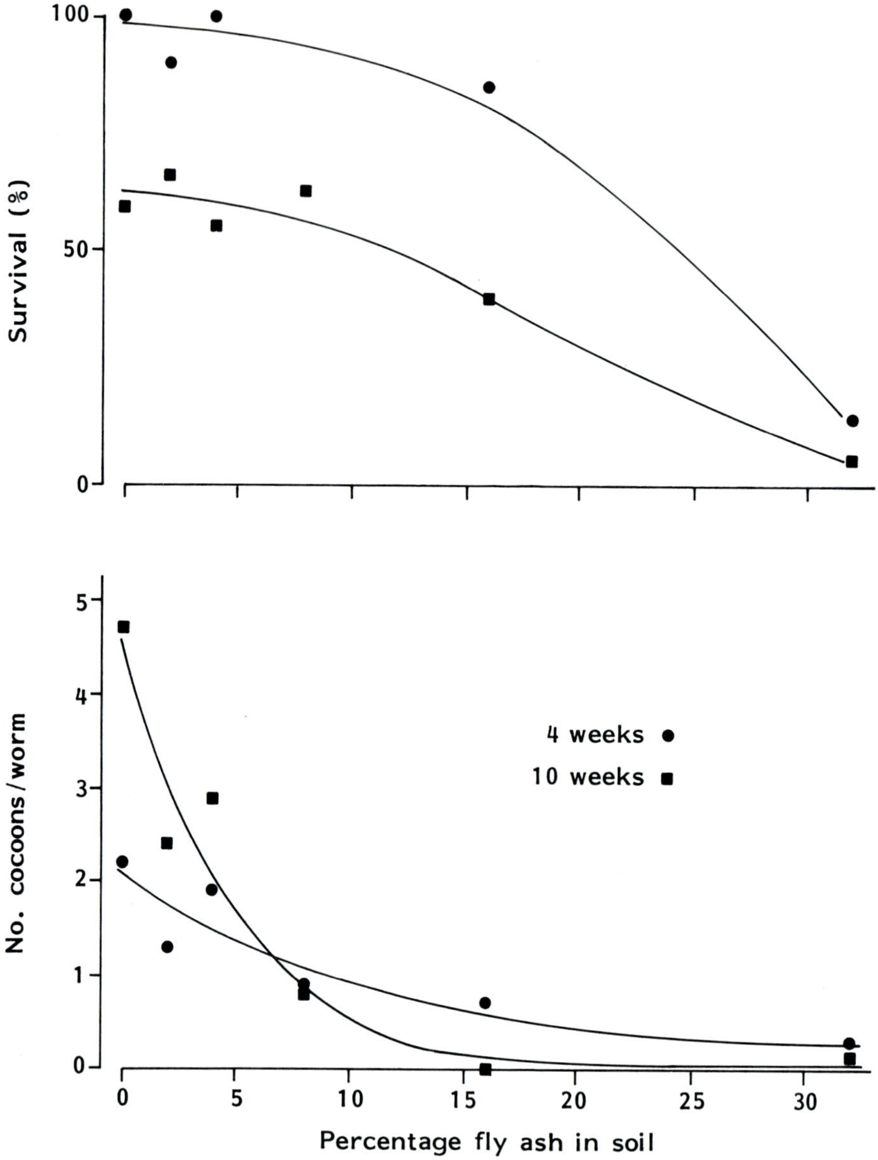

Fig. 9.2. The effect of pulverized fly ash additions to soil substrate on survival and fecundity of the earthworm *Lumbricus rubellus* (from Ma & Eijsackers, unpublished data).

Parkinson, 1986). For example, soil compaction, as measured by bulk density, is a limiting factor to the occurrence of *L. terrestris* on rehabilitated sites (Rushton, 1986*a*; Aritajat, Madge & Gooderham, 1977), although geophagic deep-burrowing species of earthworms such as *A. longa* seem to be less affected (Rushton, 1986*b*). The latter species

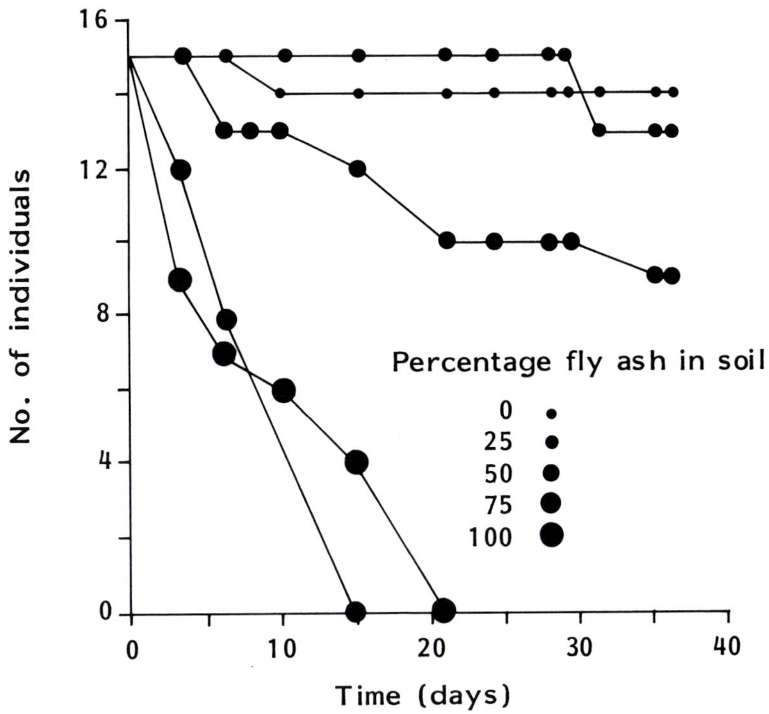

Fig. 9.3. The effect of pulverized fly ash additions to soil substrate on survival of the isopod *Trichoniscus pusillus* (see Table 9.3 for effect on reproduction) (from Ma & Eijsackers, unpublished data).

is in fact rather resistant to soil compaction and is very effective in promoting water infiltration in compacted soil (Piearce, 1984).

Alkalinity

Very alkaline substrates, such as pulverized fly ash, can be strongly toxic to soil fauna (Satchell & Stone, 1977; Eijsackers, Lourijsen & Mentink, 1983). However, as fly ash and other alkaline substrates contain multiple factors which are potentially toxic, the observed effects cannot be attributed unequivocally to alkalinity alone. Figure 9.2 shows the survivability and reproduction of *L. rubellus* in a calcareous loam soil which had been amended with young (1–2-year-old) alkaline (pH 11.6) fly ash. The animals were supplied with uncontaminated leaves of alder as food. Fly ash was injested non-discriminately by the earthworms. The fecundity of the earthworms was especially affected by the presence of fly ash when compared with the survival of the animals. Southwell & Majer (1982) failed to establish

Table 9.3. *The mean number of juveniles per five females of the isopod* Trichoniscus pusillus *exposed to soil substrate amended with fly ash (from Ma & Eijsackers, unpublished data)*

	Percentage fly ash in soil				
Days	0	25	50	75	100
21	15	16	0	0	10
24	51	19	0	0	9
28	40	2	5	0	0
29	53	2	4	0	0
31	52	10	4	0	0
35	47	12	4	0	0
36	49	22	4	0	0

sustained populations of the earthworm *E. fetida* in alkaline residues associated with bauxite refining. It should be noted however, that *E. fetida* is not a soil-dwelling species but one which is specifically confined to man-made organic-rich and nitrogen-rich compost heaps. As such, this species is less suitable for direct inoculation in poor mineral substrates than soil-dwelling species.

The survival rate of the isopod *Trichoniscus pusillus* kept on mixtures of soil and fly ash is shown in Fig. 9.3. The isopods were supplied with clean leaves of alder as food. Survival rate and reproduction were impaired in the presence of fly ash. Also the survival of the resulting juveniles was strongly affected (Table 9.3). The initial numbers of juveniles from the first brood exhibited a mortality ranging from 25 to 100%. On the 25% fly ash mixture, juvenile mortality was followed by the production of a second cohort of juveniles.

The toxicity of a certain factor to soil fauna may be more pronounced if it is not only present in the soil substrate but also in the food. Figure 9.4 shows the results of a laboratory experiment on the effect of young (1–2-year-old) fly ash on the fecundity of the millipede *Glomeris marginata*, an important litter feeding species. The soil substrate had been amended with fly ash, while the food, consisting of air-dried discs of oak leaves, had been dipped into an aqueous ash suspension. Microscopic observations on the presence of fly ash particles in the animal's excrement confirmed that the fly ash was actually ingested. It can be seen that reproduction was totally inhibited, even in the presence of relatively low amounts of fly ash. Mortality was not observed during the

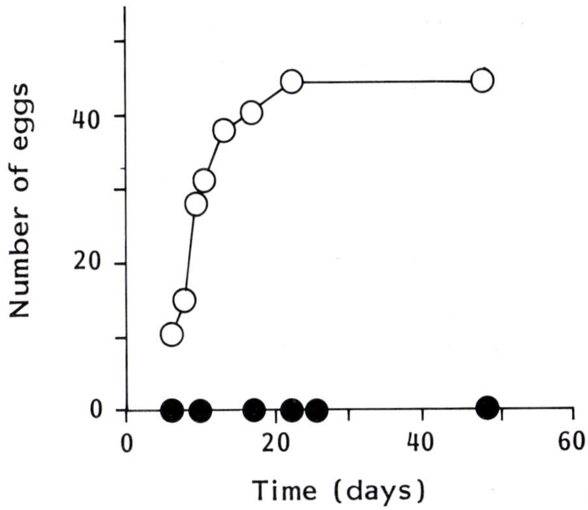

Fig. 9.4. The effect of pulverized fly ash additions to food and soil substrate on fecundity of the millipede *Glomeris marginata*. Black circles: fecundity of millipedes given food containing 5% fly ash and living in soil containing 25% fly ash; open circles: control (no fly ash added to either food or soil) (from Ma & Eijsackers, unpublished data).

experiment, either in the control animals or in those exposed to fly ash.

Leaching may reduce alkalinity. For instance in freshly lagooned pulverized fuel ash, ageing of non-topsoiled ash sites causes a fall in salinity and pH due to leaching. Aged ash sites without topsoil gradually develop an organic mat and, when seeded and adequately fertilized, may develop into a ley. Apparently, the deterrent factor present in fly ash also diminishes with age. Colonizing earthworm populations at aged sites consist of surface-dwelling species such as *L. rubellus*, *Lumbricus castaneus* and *D. rubida*, and at wet places also *Allolobophora chlorotica* (Satchell & Stone, 1977).

Table 9.4 shows the effect of ageing of ash sites on earthworm population size. The natural colonization of ash disposal sites was investigated near a coal-fired power-generating station near the city of Nijmegen, The Netherlands. Young 1–2-year-old sites had a pH of 11.1, contained hard compaction layers at the surface horizons and supported little or no vegetation. Ash sites 4–5 years of age had a pH of 8.2 and a weathered layer of ash and organic matter of about 1 cm depth, whereas sites 15–18 years of age had a pH of 7.9 and a dark organic surface layer of about 5 cm depth. The loam soil of the area adjacent to the ash sites

Table 9.4. *Population size of earthworms measured at disposal sites of pulverized fly ash and at adjacent unamended sites (from Ma & Eijsackers, unpublished data)*

Species	Stage	Age of ash site (yr) 1–2	3–4	4–5	15–18	Unamended site
Lumbricus rubellus	adults	0	1	0	10	0
Lumbricus castaneus	adults	0	0	0	0	68
Pigmented species	immatures	0	0	39	49	20
Allolobophora chlorotica	adults	0	0	0	118	39
Eiseniella tetraedra	adults	0	0	0	10	39
Unpigmented species	immatures	0	0	10	78	167
Total number m^{-2}		0	1	49	265	333

had a pH of 7.6. It can be seen that population size increased with increasing age of the ash sites. Earthworms were lacking in the young ash sites but considerable numbers were found in the older ash sites. *L. rubellus* and *A. chlorotica* occurred in greater numbers on the old ash site than in the surrounding soil.

The occurrence of deep-burrowing species in non-topsoiled sites may be sustained by continuous immigration from the periphery. Topsoiling can greatly improve the chances of survival of deep-burrowing species. Deep-burrowing species, including *L. terrestris*, *A. longa* and *O. cyaneum*, when introduced to ash sites with topsoil, have been observed to penetrate the upper 8 cm of ash; cocoons were found in the ash up to 10 cm depth, and burrows were traced up to 28 cm (Satchell & Stone, 1977). Although these are considerably less than the depths to which these species normally penetrate, burial of surface litter in the ash might be beneficial and promote further development of the ash sites.

Attempts to reduce the pH of alkaline environments may be initially detrimental rather than beneficial. This is firstly because of the mobilization of potentially toxic metals, when present, and secondly because large amounts of mineral acids are required to neutralize very alkaline substrates. This may increase salinity to detrimental levels. Acidification with sulphuric acid of alkaline sediments originating from weathered asbestos-rich serpentinitic rock decreased survival rates of introduced earthworms, probably because of increased soil salinity (Schreier & Timmenga, 1986). In specific cases, for example in alkaline chromate wastes, a lowering of pH may reduce the toxic chromate form to a chromic form which is less toxic to vegetation and probably also to soil fauna (Gemmell, 1976).

Acidity

Acid environments are generally more hostile to fauna recolonization than alkaline environments. Low pH can be caused by the weathering of coal shale, treatment of ores with acid materials, the oxidation of sulphides in tailings ponds, or other reasons. Sulphides, such as iron pyrite (FeS_2), occurring at or near the surface of a waste site may become oxidized, thereby producing an extremely low pH by the formation of sulphuric acid. Excess heavy metals and also Al come into solution in progressing amounts as acidity increases, thereby causing toxic effects on the germination and growth of plants. There is also an increased leaching of basic cations or Ca, Mg, K and Na, while the availability of micronutrients (e.g. Mo) can also be affected. Natural revegetation of unreclaimed acid substrates is therefore difficult and unreclaimed spoil may remain largely bare for at least 40 years (Wali, 1979). In such cases, certain hardy species (e.g. pasture grasses) may be used for rehabilitating strongly acid mine spoils (Foy & Oakes, 1984) or metalliferous wastes (Smith & Bradshaw, 1979). Although leaching may ameliorate toxic conditions in acid substrates, control of acidity will be necessary when rapid establishment and maintenance of vegetation and soil fauna are required.

Earthworms generally do not occur in soil at pH < 3.5, although some acid-tolerant species such as *Dendrobaena* sp. can survive at low densities in rather acid soils. Some species of microarthropods and enchytraeid worms are strongly acid-tolerant and even have a low preferred pH range. Thus, epigaeic and hemiedaphic species of springtails and oribatid mites have been described from bare untreated coal shale heaps with a high concentration of acid-producing pyrite (Luxton, 1982; Moore & Luxton, 1986). In naturally acid environments, such as in coniferous soils, the ecological role of earthworms in litter breakdown is taken over by acidophilic species of enchytraeids, springtails and mites. Liming of such acid environments will reduce these acidophilic species which will be replaced by more basophilic springtails, mites and lumbricid earthworms (Bååth *et al.*, 1980; Hågvar & Amundsen, 1981). Liming has been shown to have a beneficial effect on the establishment of *A. caliginosa* after it had been introduced into an acid coniferous soil (Huhta, 1979).

Dispersal of earthworms into acidic sites is prevented by a strong

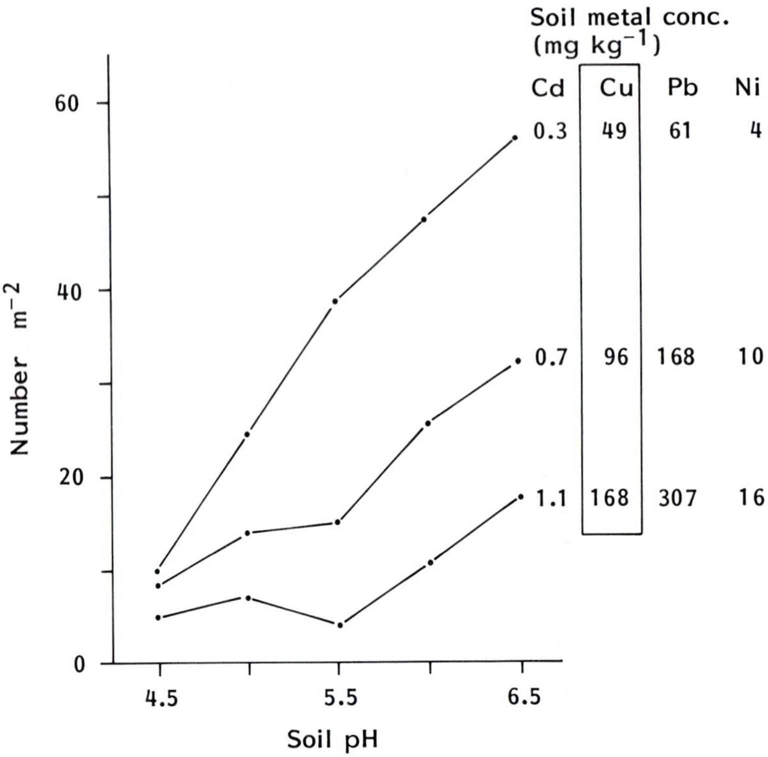

Fig. 9.5. Mean population size of the earthworm *Aporrectodea caliginosa* in an arable soil containing elevated soil metal levels caused by previous amendments with metal-contaminated municipal waste compost. The soil had been adjusted to various pH levels by the application of sulphur or lime (from Ma, 1988).

avoidance behaviour which they display when encountering acid substrates (Satchell, 1955; Laverack, 1961). Acidity has been suggested as a main factor reducing the rate of earthworm colonization of rehabilitated brown coal-lignite opencast workings which had been planted with alder (Dunger, 1969*a*). Furthermore, soil acidity is one of the important factors determining the distribution of earthworms in natural or rehabilitated habitats (Nordström & Rundgren, 1974; Satchell, 1955; Standen, 1979). The colonization rates of springtails, mites and enchytraeid worms is significantly influenced by soil pH, mainly as a result of the species-specific differences in preferred pH range (Hågvar & Abrahamsen, 1980). This influence of pH is mainly due to its effect on reproductive success (Ashraf, 1969; Hutson, 1978; Hågvar & Abrahamson, 1980). It is therefore clear that the ultimate efficiency of ecosystem function in rehabilitated acidic sites will be strongly influenced by the composition and densities of the specifically pH-adapted soil fauna.

In contaminated soil environments, heavy metals may become mobilized by the acidity of the substrate. The influence of soil pH on the toxicity of heavy metals to earthworms has been studied by Ma (1984, 1988) and Bengtsson *et al.* (1986). Figure 9.5 shows the results of a field experiment in grassland on which the soil pH had been adjusted to different levels by applications of either sulphur or lime. Sulphur is transformed microbiologically in soil to sulphuric acid. The soil surface horizon contained elevated levels of various heavy metals as a consequence of previous amendments with heavy metal-enriched composted municipal waste (Ma, 1988). The population size of the acid-sensitive species *A. caliginosa* decreased with increasing soil metal content and decreasing soil pH. Acidification and increased soil levels of heavy metals apparently had an independent effect on the relative size of the earthworm populations present. The field data confirm laboratory evidence indicating that acidification increases, while liming decreases, metal toxicity to earthworms (Ma, 1984; Bengtsson *et al.*, 1986).

Proposed management programme for rehabilitation of toxic land

In principle, germination and growth of plants can occur in the absence of organic matter provided that an adequate moisture and nutrient supply are present. Colonization by soil fauna however, depends on the presence of at least some amounts of organic matter as well as a suitable soil structure, litter layer and vegetation cover. Once a vegetation cover has become established colonization by soil fauna may proceed rapidly. Revegetation must therefore be a first step in any rehabilitation scheme. A rehabilitation management programme should start with the removal of toxic factors which are inhibitory to plants and soil fauna and should be followed by the introduction of organic matter. Cheap sources of organic matter consist of commonly available organic waste materials such as sewage sludge, dredging sludge or municipal wastes. These sources are suitable for reclamation purposes under the condition that they do not contain non-degradable pollutants beyond any critical level. They are a suitable source of nitrogen, which often is a limiting factor to the natural or introduced colonization of derelict sites by soil-dwelling earthworms (Satchell, 1980*b*). A flow scheme for rehabilitation of waste sites, using earthworms as a management tool, is shown in Fig. 9.6.

SITE TREATMENT STEP		SITE DEVELOPMENTAL STAGE
1. None	→	Non-reclaimed toxic site
2. Drainage, leaching, liming, etc.	→	Site with a reduced substrate toxicity
3. (a) Organic amendments (sludge, etc.) (b) Introduction of compostworms (e.g. *Eisenia fetida*)	→	Site amended with stabilized organic matter
4. (a) Seeding (incl. legumes) (b) Fertilization	→	Revegetated site
5. (a) Introduction of surface-active species of soil-dwelling earthworms (e.g. *Lumbricus rubellus, Dendrobaena* sp., etc.) (b) Introduction of endogenic and deep-burrowing species of soil-dwelling earthworms (e.g. *Lumbricus terrestris*)	→	Reclaimed site with stabilized ecosystem functioning

Fig. 9.6. Proposed management scheme for rapid rehabilitation of waste sites with the aid of earthworms.

Addition of organic matter may also help to reduce the toxicity of heavy metals present in the subsoil by reducing their availability through adsorption and chelation. Additions of sewage sludge to provide a nutrient source for the rehabilitation of coal mine spoil have been shown to result in a significant increase in above-ground plant biomass when compared with applications of inorganic fertilizers (Topper & Sabey,

1986). Other organic amendments such as straw mulch or bark may also be effective in developing a soil on substrates such as surface coal mine spoils (Elkins *et al.*, 1984). Also, mulching with leaf compost is an effective way of encouraging the establishment of introduced earthworms at mineral sites (Dunger, 1969*b*). Subsequent seeding with legumes, when provided with adequate lime and phosphorus, can provide a beneficial input of nitrogen into the system (Bradshaw, 1983).

Raw organic sludges and municipal wastes used for topsoiling can be turned into a stable humus with the aid of the compost worm *E. fetida* (Mitchell *et al.*, 1977). *E. fetida* is a typical r-selected species which can easily be bred in mass (Hartenstein, Neuhauser & Kaplan, 1979). However, because this species does not survive in soil under field conditions it will eventually have to be replaced by soil-dwelling species. The latter will have an increased chance of colonization in the topsoil created by *E. fetida* and will perform further activities to improve the structure and organic matter distribution at the site. A management scheme using earthworms has been applied to improve the productivity of deteriorated acid heathland in France (Trehen & Bouché, 1983) and of grassland in New Zealand (Syers & Springett, 1984). First introductions of soil-dwelling species should primarily include surface-active r-selected species, more K-selected endogeic species, and the deep-burrowing K-selected types should follow at some later stage when soil organic matter content has reached a more adequate level.

References

Aritajat, U., Madge, D. S. & Gooderham, P. T. (1977). The effect of compaction of agricultural soils on soil fauna. 1. Field investigations. *Pedobiologia*, **17**, 262–82.

Ashraf, M. (1969). Studies on the biology of Collembola. *Revue d'Ecologie et de Biologie du Sol*, **6**, 337–47.

Bååth, E., Berg, B., Lohm, U., Lundgren, B., Lundkvist, H., Rosswall, T., Söderström, B. & Wiren, A. (1980). Effects of experimental acidification and liming on soil organisms and decomposition in a Scots pine forest. *Pedobiologia*, **20**, 85–100.

Bengtsson, S. A., Nilsson, A., Nordström, S. & Rundgren, S. (1979). Short-term colonization success of lumbricid founder populations. *Oikos*, **33**, 308–15.

Bengtsson, G. & Rundgren, S. (1984). Ground-living invertebrates in metal-polluted forest soils. *Ambio*, **13**, 29–33.

Bengtsson, G., Gunnarson, T. & Rundgren, S. (1986). Effects of metal pollution on the earthworm *Dendrobaena rubida* (Sav.) in acidified soils. *Water, Air and Soil Pollution*, **28**, 361–83.

Bradshaw, A. D. (1983). The reconstruction of ecosystems. *Journal of Applied Ecology*, **20**, 1–17.

Brockmann, W., Koehler, H. & Schriefer, T. (1980). Recultivation of refuse tips: soil ecological studies. In *Soil Biology as Related to Land Use Practices*, ed. D. L. Dindal, pp. 161–8. Washington D.C: Environmental Protection Agency.

Curry, J. P. & Cotton, D. C. F. (1983). Earthworms and land reclamation. In *Earthworm Ecology*, ed. J. E. Stachell, pp. 215–28. London: Chapman & Hall.

Deevey, E. S. (1947). Life tables for natural populations of animals. *Quarterly Review of Biology*, **22**, 283–314.

Dunger, W. (1969*a*). Über den Atneil der Arthropoden an der Umsetzung des Bestandesabfalles in Anfangs-Bodenbildungen. *Pedobiologia*, **9**, 366–71.

Dunger, W. (1969*b*). Fragen der naturlichen und experimentellen Besiedlung kulturfeindliches Böden durch Lumbriciden. *Pedobiologia*, **9**, 146–51.

Eijsackers, H., Lourijsen, N. & Mentink, J. (1983). Effects of fly ash on soil fauna. In *New Trends in Soil Biology*, ed. Ph. Lebrun, H. M. Andre, A. de Medts, C. Gregoire-Wibo & G. Wauthy, pp. 680–1, Louvain-la-Neuve: Dieu-Brichart.

El-Duweini, A. K. & Ghabbour, S. I. (1965). Population density and biomass of earthworms in different types of Egyptian soils. *Journal of Applied Ecology*, **2**, 271–87.

Elkins, N. Z., Parker, L. W., Aldon, E. & Whitford, W. G. (1984). Responses of soil biota to organic amendments in strip mine spoils in northwestern New-Mexico USA. *Journal of Environmental Quality*, **13**, 215–19.

Foy, C. D. & Oakes, A. J. (1984). A winter hardy aluminium tolerant perennial pasture grass for reclamation of acid mine spoils. *Journal of Plant Nutrition*, **7**, 929–52.

Gemmell, R. P. (1976). Revegetation of derelict land polluted by a chromate smelter. 2. Techniques of revegetation of chromate smelter waste. *Environmental Pollution*, **6**, 31–7.

Hågvar, S. & Abrahamsen, G. (1980). Colonisation by Enchytraeidae, Collembola and Acari in sterile soil samples with adjusted pH levels. *Oikos*, **34**, 245–58.

Hågvar, S. & Amundsen, T. (1981). Effects of liming and artificial acid rain on the mite (Acarina) fauna in coniferous forest. *Oikos*, **37**, 7–20.

Hartenstein, R., Neuhauser, E. F. & Kaplan, D. L. (1979). Reproductive potential of the earthworm *Eisenia fetida*. *Oecologia (Berlin)*, **43**, 329–40.

Hartenstein, R., Neuhauser, E. F. & Narahara, A. (1981). Effects of heavy metal and other elemental additives to activated sludge on growth of *Eisenia fetida*. *Journal of Environmental Quality*, **10**, 372–6.

Hoogerkamp, M., Rogaar, H. & Eijsackers, H. J. P. (1983). Effect of earthworms on grassland on recently reclaimed polder sites in the Netherlands. In *Earthworm Ecology*, ed. J. E. Satchell, pp. 85–107. London: Chapman and Hall.

Hopkin, S. P. & Martin, M. H. (1984). Heavy metals in woodlice. *Symposium of the Zoological Society of London*, **53**, 143–66.

Hopkin, S. P. & Martin, M. H. (1985). Assimilation of zinc, cadmium, lead, copper and iron by the spider *Dysdera crocata*, a predator of woodlice. *Bulletin of Environmental Contamination and Toxicology*, **34**, 183–7.

Huhta, V. (1979). Effects of liming and deciduous litter on earthworm (Lumbricidae) populations of a spruce forest, with an inoculation experiment on *Allolobophora caliginosa*. *Pedobiologia*, **19**, 340–5.

Hutnik, R. J. & Davis, G. ed. (1973). *Ecology and Reclamation of Devastated Land*. New York: Gordon and Breach.

Hutson, B. R. (1978). Influence of pH, temperature and salinity on the fecundity of four species of Collembola. *Pedobiologia*, **18**, 163–74.

Ireland, M. P. (1978). Heavy metal binding properties of earthworm chloragosomes. *Acta Biologica Academiae Scientiarum Hungaricae*, **29**, 385–94.

Joosse, E. N. G. & Buker, J. B. (1979). Uptake and excretion of lead by litter dwelling Collembola. *Environmental Pollution*, **18**, 235–40.

Laverack, M. S. (1961). Tactile and chemical perception in earthworms. II. Responses to acid pH solutions. *Comparative Biochemistry and Physiology*, **2**, 22–34.

Lee, K. (1985). *Earthworms: Their Ecology and Relationships with Soils and Land Use*. Sydney: Academic Press.

Luxton, M. (1982). The ecology of some soil mites from coal shale tips. *Journal of Applied Ecology*, **19**, 427–42.

Ma, Wei-Chun. (1982). The influence of soil properties and worm-related factors on the concentration of heavy metals in earthworms. *Pedobiologia*, **24**, 109–19.

Ma, Wei-Chun. (1984). Sublethal toxic effects of copper on growth, reproduction and litter breakdown activity in the earthworm *Lumbricus rubellus* with observations on the influence of temperature and soil pH. *Environmental Pollution (A)*, **33**, 207–19.

Ma, Wei-Chun. (1987). Heavy metal accumulation in the mole *Talpa europaea* and earthworms, as an indicator of metal bioavailability in terrestrial environments. *Bulletin of Environmental Contamination and Toxicology*, **39**, 933–8.

Ma, Wei-Chun. (1988). Toxicity of copper to lumbricid earthworms in sandy agricultural soils amended with Cu-enriched organic waste materials. In *Ecological Implications of Contemporary Agriculture*, ed. A. Quispel & H. Eijsackers. (Proceedings of the Fourth European Ecology Symposium). Ecological Bulletin (Stockholm), Ecological Bulletins (Copenhagen), **39**, 53–6.

Ma, Wei-Chun., Edelman, T. H., Beersum, I. van & Jans, Th. (1983). Uptake of cadmium, lead and copper by earthworms near zinc-smelting complex: influence of soil pH and organic matter. *Bulletin of Environmental Contamination and Toxicology*, **30**, 424–7.

Malecki, M. R., Neuhauser, E. F. & Loehr, R. C. (1982). The effect of metals on the growth and reproduction of *Eisenia fetida* (Oligochaeta, Lumbricidae). *Pedobiologia*, **24**, 129–37.

Mitchell, M. J., Mulligan, R. M., Hartenstein, R. & Neuhauser, E. F. (1977). Conversion of sludges into 'topsoils' by earthworms. *Compost Science*, **18**, 28–32.

Moore, F. R. & Luxton, M. (1986). The collembolan fauna of two coal shale tips in north-west England. *Pedobiologia*, **29**, 359–66.

Nordström, S. & Rundgren, S. (1974). Environmental factors and lumbricid associations in southern Sweden. *Pedobiologia*, **14**, 1–27.

Parsons, W. J. F. & Parkinson, D. (1986). Species composition, distribution and abundance of Collembola colonizing reclaimed mine spoil in Alberta. *Pedobiologia*, **29**, 33–45.

Piearce, T. G. (1984). Earthworm populations in soils disturbed by trampling. *Biological Conservation*, **29**, 241–52.

Piearce, T. G. & Piearce, B. (1979). Responses of Lumbricidae to saline inundation. *Journal of Applied Ecology*, **16**, 461–73.

Rusek, J. (1978). Pedozootische Sukzessionen wahrend der Entwicklung von Okosystemen. *Pedobiologia*, **19**, 133–40.

Rushton, S. P. (1986*a*). The effects of soil compaction on *Lumbricus terrestris* and its possible implications for populations on land reclaimed from open-cast coal mining. *Pedobiologia*, **29**, 85–90.

Rushton, S. P. (1986*b*). Development of earthworm populations on pasture land reclaimed from open-cast coal mining. *Pedobiologia*, **29**, 27–32.

Satchell, J. E. (1955). Some aspects of earthworm ecology. In *Soil Zoology*, ed. D. K. McE. Kevan, pp. 180–99. London: Butterworth.

Satchell, J. E. (1980*a*). r Worms and K worms: a basis for classifying lumbricid earthworm strategies. In *Soil Biology as Related to Land Use Practices*, ed. D. L. Dindal, pp. 848–54. Washington DC: Environmental Protection Agency.

Satchell, J. E. (1980*b*). Potential of the Silpho moor experimental birch plots as a habitat for *Lumbricus terrestris*. *Soil Biology and Biochemistry*, **12**, 317–23.

Satchell, J. E. & Stone, D. A. (1977). Colonisation of pulverized fuel ash sites by earthworms. *Publicaciones del Centro pirenaico de Biologia experimental*, **9**, 59–74.

Schreier, H. & Timmenga, H. J. (1986). Earthworm response to asbestos-rich serpentinitic sediments. *Soil Biology and Biochemistry*, **18**, 85–9.

Smith, R. A. H. & Bradshaw, A. D. (1979). The use of metal-tolerant plant populations for the reclamation of metalliferous wastes. *Journal of Applied Ecology*, **16**, 595–612.

Southwell, L. T. & Majer, J. D. (1982). The survival and growth of the earthworm *Eisenia fetida* (Lumbricidae: Oligochaeta) in alkaline residues associated with the bauxite refining process. *Pedobiologia*, **23**, 42–52.

Standen, V. (1979). Factors affecting the distribution of lumbricids (Oligochaeta) in associations at peat and mineral sites in northern England. *Oecologia (Berlin)*, **42**, 359–74.

Standen, V., Stead, G. B. & Dunning, A. D. (1982). Lumbricid populations in open cast reclamation sites and colliery spoil heaps in County Durham, UK. *Pedobiologia*, **24**, 57–64.

Streit, B. & Jaeggy, A. (1983). Effect of soil type on copper toxicity and copper uptake in *Octolasium cyaneum* (Lumbricidae). In *New Trends in Soil Biology*, ed. Ph. Lebrun, H. M. Andre, A. de Medts, C. Gregoire-Wibo & G. Wauthy, pp. 569–75, Louvain-la-Neuve: Dieu-Brichart.

Strojan, C. L. (1978). The impact of zinc smelter emissions on forest litter arthropods. *Oikos*, **31**, 41–6.

Syers, J. K. & Springett, J. A. (1984). Earthworms and soil fertility. *Plant and Soil*, **76**, 93–104.

Topper, K. F. & Sabey, B. R. (1986). Sewage sludge as a coal mine spoil amendment for revegetation in Colorado. *Journal of Environmental Quality*, **15**, 44–9.

Trehen, P. & Bouché, M. B. (1983). Place des lombriciens dans les processus de restauration des sols de lande. In *New Trends in Soil Biology*, ed. Ph. Lebrun, H. M. Andre, A. de Medts, C. Gregoire-Wibo & G. Wauthy, pp. 471–86. Louvain-la-Neuve: Dieu-Brichart.

Tyler, G. (1984). Pollination on forests: Gusum, Sweden. *Ambio*, **13**, 19–24.

Van Capelleveen, H. E., Straalen, N. M. van, Berg, M. van de & Wachem, E. van. (1986). Avoidance as a mechanism of tolerance for lead in terrestrial arthropods. In *Proceedings of the Third European Congress of Entomology*, ed. H. H. W. Velthuis, pp. 193–207.

van Rhee, J. A. (1963). Earthworm activities and the breakdown of organic matter in agricultural soils. In *Soil Organisms*, ed. J. Doeksen & J. van der Drift, pp. 55–9. Amsterdam: North-Holland Publishing Company.

van Rhee, J. A. (1969). Development of earthworm populations in polder soils. *Pedobiologia*, **9**, 133–40.

van Rhee, J. A. (1975). Copper contamination effects on earthworms by disposal of pig waste in pastures. In *Progress in Soil Zoology*, ed. J. Vanek, pp. 451–7. Prague: Academia Publishing House.

Vimmerstedt, J. P. (1983). Earthworm ecology in reclaimed opencast coal mining sites in Ohio. In *Earthworm Ecology*, ed. J. E. Satchell, pp. 229–40. London: Chapman & Hall.

Wali, M. (1979). *Ecology and Coal Resource Development*, 2. New York: Pergamon Press.

Wieser, W. (1979). Flow of copper through a terrestrial food web. In *Copper in the Environment*, Part 1, Ecological Cycling, ed. J. K. Newall, New York: Wiley.

10

The reclamation of lands for agricultural grazing

W. A. LAYCOCK

Introduction

Most semi-arid and arid areas of the World which have been disturbed by mining or other activities were rangelands (grasslands or shrublands) before disturbance. In such areas, it is generally most feasible to replant adapted native or introduced rangeland species in the rehabilitation process. The post-mining uses of such areas usually include the traditional uses of rangelands such as livestock grazing, wildlife habitat and watershed (Table 1.2). Much of the literature dealing with effects of livestock grazing on rehabilitated lands is from research conducted on rangelands disturbed by mining in the western part of the United States. Strict Federal and State regulations require restoration and maintenance of an adequate cover and production of vegetation on rehabilitated areas and this has necessitated research. One of the reasons why so much research has been conducted in this area is the limited and highly variable rainfall. Because of this, the re-establishment of vegetation on disturbed areas can be difficult and uncertain, forage production and livestock carrying capacity are mostly low and both native and introduced forage species are much more susceptible to damage caused by grazing than are plants in more humid areas.

In more humid areas, disturbed lands can be rehabilitated as cropland, forest land or pasture land (Table 1.2). The proposed post-disturbance land use dictates the type of vegetation to be planted in the rehabilitation process. In humid areas, rehabilitated land that is suitable to be used as cropland can also be seeded to perennial forage species and used as pasture for domestic livestock. Because management of livestock on humid pastures on non-disturbed areas is relatively well understood, there is not much literature specifically dealing with management of livestock grazing or its effects on rehabilitated humid pastures.

Grazing by domestic livestock is a legitimate and proper use of many rehabilitated disturbed lands. Proper choice of forage species to be planted is critical. This Chapter will address: (1) the forage requirements for rehabilitated land used for grazing including vegetation responses to grazing; (2) livestock requirements for grazing including weight gains, forage quality, and water requirements; (3) watershed aspects of grazing; and (4) use of livestock grazing to change species composition, height, or other aspects of vegetation to make a rehabilitated area more suitable for another use.

It is difficult to conduct statistically sound, replicated research on many disturbed areas, especially in arid and semi-arid areas because of large variations in year-to-year weather conditions, soil and other conditions and lack of experimental control of many variables. Because of this, many of the studies summarized in this Chapter, especially from the western United States, have been more in the nature of 'demonstration research' and should be judged as such.

I would like to acknowledge the assistance of the US Department of Agriculture Western Regional Coordinating Committee on Reclamation of Lands Impacted by Mineral Development and Other Drastic Land Disturbances (WRCC-21) in supplying references, unpublished material and other information for this Chapter.

Forage requirements

Land rehabilitated for grazing must produce an appropriate forage base which will: (1) produce enough forage for the planned grazing; (2) be of sufficient quality for the animals and season to be grazed; (3) not contain toxic amounts of heavy metals or trace minerals; and (4) be maintained over time under grazing. To ensure an adequate stand of vegetation, it is usual to protect newly planted forage species from grazing for 1 to 2 years in arid and semi-arid areas. However, at least on mined areas in the western United States, it is often not possible to begin grazing even after 2 years because: (1) small rehabilitated areas next to or intermixed with active mining areas are difficult to manage for grazing; and (2) many State regulations prohibit grazing until the last 2 years of a 10-year bond release waiting period. This means that most rehabilitated mined lands are not grazed for 8–10 years after planting. This is not a desirable situation because, as will be shown later, grazing can be a positive influence in increasing stand production, species diversity, and other beneficial aspects.

Table 10.1. *Stocking rates and forage production in grazed native or rehabilitated areas in North Dakota*

	Steers ha^{-1}	Animal unit months ha^{-1}	Forage production (kg ha^{-1})
Rehabilitated areas			
Heavy grazing	4.2	3.09	1790
Moderate grazing	2.0	1.54	2840
Light grazing	1.4	1.07	3470
Ungrazed	—	—	3250
Grazed native areas	—	0.30–0.70	1340

Vegetation responses to grazing

Northern Great Plains, US and Canada Hofmann, Ries & Lorenz (1981) studied the effects of grazing on coal mined lands near Center, North Dakota. The area was not topsoiled but was seeded to introduce species including smooth brome (*Bromus inermis*), crested wheatgrass (*Agropyron desertorum*), intermediate wheatgrass (*Agropyron intermedium*), alfalfa (*Medicago sativa*) and yellow sweetclover (*Melilotus officinalis*). Spring and early summer grazing (45 days yr^{-1}) at three stocking rates was conducted for 3 years. Forage production in the heavily stocked treatment was significantly less than that in the lightly and moderately stocked and ungrazed areas at the end of the study (Table 10.1).

Grazing was continued each year until one half of the forage on the moderately stocked pasture was utilized. Utilization levels were not reported for the other two stocking treatments. Alfalfa and sweetclover maintained 63–68 % of the total composition during the study, and the different grazing rates had little effect on the composition of the rehabilitated pastures.

Live plant cover was significantly greater on native sites than on all rehabilitated sites. Litter cover was significantly greater on moderately and lightly grazed than on heavily grazed rehabilitated pastures. Total plant cover (live plant plus litter) was 74 % on the heavily grazed rehabilitated pasture. This was significantly less than the total cover of 90–98 % on the moderately and lightly grazed rehabilitated pastures and on the native sites.

On an area 80 km northwest of Bismark, North Dakota, Williamson (1981) studied the effects of grazing in June and early July (42 days) on a mined area resurfaced with a topsoil–subsoil mixture and seeded to a

Table 10.2. *Number of steers and stocking rates in grazed rehabilitated and native areas in southwestern Montana*

		No. of steers ha^{-1}	Total animal days ha^{-1}
Spring grazing	(rehabilitated area)	0.36	104
Summer grazing	(rehabilitated & native areas)	0.12	60
Autumn grazing	(rehabilitated area)	0.32	49

mixture of native grasses. Prior to grazing, slender wheatgrass (*Agropyron trachycaulum*) was the dominant species on all seeded areas. After 1 year of heavy grazing (5 steers ha^{-1}, 85 % utilization), basal ground cover increased and above-ground biomass of vegetation increased from 1590 kg ha^{-1} in 1979 to 1990 kg ha^{-1} in 1980. This was due to a significant increase in sideoats grama (*Bouteloua curtipendula*) and western wheatgrass (*Agropyron smithii*). In pastures grazed moderately (2 steers ha^{-1}, 28 % utilization), lightly (1.2 steers ha^{-1}, 9 % utilization) and with no grazing, herbage production and basal ground cover of plants decreased significantly in 1980. The longer-term effects of the grazing treatments were not reported.

In southeastern Montana, DePuit & Coenenberg (1978, 1980) studied the effects of grazing on the Rosebud coal mine near Colstrip that was not topsoiled but was seeded with native and introduced species. When grazing began, the vegetation was dominated by fairway crested wheatgrass (*Agropyron cristatum*), smooth brome, tall wheatgrass (*Agropyron elongatum*), alfalfa and yellow sweetclover. Three systems of grazing were evaluated over a 3-year period: (1) rotated seasonal grazing (spring, summer and autumn) on rehabilitated mined areas; (2) spring and autumn grazing on rehabilitated areas; and (3) summer grazing on native rangeland. Stocking rates are shown in Table 10.2.

Utilization of perennial grasses and legumes averaged 60 % in the spring and 55 % in autumn. From 1975 through 1978 above-ground standing crop increased from 2490 to 3580 kg ha^{-1} in the spring–autumn grazed pastures and from 2010 to 2910 kg ha^{-1} in the ungrazed areas. The grazed pastures had more crested and tall wheatgrass and alfalfa, and less smooth brome than the ungrazed areas in 1978. Total cover of perennial grasses and legumes also increased substantially from 1975 to 1978, both in spring–autumn grazed pastures and in ungrazed areas.

Litter cover remained about the same in the grazed pastures but increased substantially in the ungrazed areas. Species diversity indices were consistently higher in grazed than in ungrazed areas, indicating an enhancement of stand diversity by spring–autumn grazing. They concluded that grazing, as opposed to non-grazing, is desirable in terms of higher plant community diversity.

In another study near Colstrip, Montana (Young & Rennick, 1983*a*, *b*), coal mined lands were seeded to a mixture of introduced and native species following the return of topsoil. As in the previous study, the resulting stand was dominated by introduced species including fairway crested wheatgrass, pubescent wheatgrass (*Agropyron trichophorum*) and smooth brome. In 1979, summer grazing was initiated to determine if native species could be encouraged by cattle grazing. After 4 years of grazing with utilization levels of approximately 50%, the pasture was still dominated by the three introduced species. Young & Rennick (1983*a*) reported that native species were becoming established and evenness indices were greater in the grazed areas than in exclosures protected from grazing.

At the Decker coal mine in southeastern Montana, yearling cattle were grazed on a topsoiled and seeded area at a grazing intensity averaging 1.48 animal unit months (AUM) ha^{-1} (Kleinman & Layton, 1981; Kleinman, Raphelson & Bendix, 1984). Grazing caused a shift in dominance from pubescent wheatgrass to beardless wheatgrass (*Agropyron inerme*). Number of plant species increased from 33 on non-grazed rehabilitated areas to 64 on comparable grazed rehabilitated areas. Approximately 57 species occurred on comparable native range.

The effect of season of grazing was studied on a disturbed pipeline right-of-way on mixed prairie vegetation in southern Alberta, Canada (Naeth & Bailey, 1986). Late season grazing resulted in a slower return to predisturbed conditions by favouring the establishment of introduced species such as crested wheatgrass and invaders such as flixweed (*Descurainia sophia*). Early season grazing reduced the dominance of introduced and invader species but resulted in more bare ground for a longer period of time. Little club moss (*Selaginella densa*) comprised over 50% of the cover on undisturbed prairie and was eliminated by the disturbance. It had not re-invaded the disturbed area after more than 25 years.

Central Great Plains and Rocky Mountains, United States On a rehabilitated uranium-mined site in central Wyoming, Schuman *et al.*

(1986) studied the effects of light (0.24 steers ha^{-1}) and heavy (0.48 steers ha^{-1}) summerlong grazing on areas seeded to native species including western wheatgrass, thickspike wheatgrass (*Agropyron dasystachyum*), green needlegrass (*Stipa viridula*) and prairie junegrass (*Koeleria cristata*). For the 3 years of grazing, utilization averaged 38% in the lightly grazed pastures and 54% in the heavily grazed pastures. Precipitation was higher during the last 2 years of the study (24–26 cm) than in the first year (17 cm). Herbage production increased from 560 kg ha^{-1} in 1979 to 1420 kg ha^{-1} in 1981 on the ungrazed area, from 580 to 1100 kg ha^{-1} on the lightly grazed area, and from 540 to 1240 kg ha^{-1} on the heavily grazed pastures. Basal plant cover also increased during the study in all areas. Conclusions were that lands successfully rehabilitated, such as those studied, will support livestock grazing without deterioration of the vegetation.

Three different grazing systems were studied for 5 years on rehabilitated coal mined land in northwestern Colorado. Preliminary results for the first 3 years of the study were published by Layden (1984), Laycock & Layden (1986), and Laycock & McGinnies (1985). Results presented from the last 2 years of the study are from unpublished data. The area receives an average of 41 cm of precipitation annually. The vegetation prior to mining was a mountain shrub-sagebrush community with a grass-forb understorey. Topsoil was replaced and the area was seeded in 1976–1978. About 60% of the herbage production on the study area is alfalfa with the rest made up of introduced grasses dominated by intermediate wheatgrass, smooth brome, and fairway and crested wheatgrasses.

Grazing was initiated in 1982. The grazing systems were short duration grazing (SDG) and summerlong grazing at two different rates (heavy and light). The SDG system was divided into eight pastures and cattle were rotated through the pasture three times during each 75–80-day summer grazing season. Stocking rates were quite light in 1982 and were increased to achieve the desired levels of grazing during the last 2 years (1985 and 1986). All treatments were replicated on two mined areas. Data presented in Table 10.3 are from the largest and most productive of the two areas. The 1985 utilization figures shown are the highest achieved during the study. The grazing treatments caused no apparent difference in herbage production over the 5 years of study. Species composition figures have not yet been analyzed.

Table 10.3. *Stocking rates, utilization levels and herbage production on the largest of two replicated areas on rehabilitated coal mined land in northwestern Colorado*

Grazing treatment	Steer days ha^{-1} 1985–86	Utilization (%) 1985	Average herbage production (kg ha^{-1}) 1982–86
Short duration (SDG)	200	61	4890
Summerlong heavy	222	69	5390
Summerlong light	111	44	4790

Southwestern United States No grazing trials on rehabilitated mined land in the southwestern United States were found in the literature. Verma & Thames (1976) discussed potential multiple uses of lands disturbed by coal mining on Black Mesa in northwestern Arizona and included livestock grazing as one of the major uses. Problems cited were limited soil moisture availability for plant growth and limited stock water availability. Jasmer, Donart & Aldon (1983) reported results of a simulated grazing study on rehabilitated lands at the McKinley Mine near Gallup, New Mexico. Plots were mowed at 2 heights (7.5 and 15 cm) at 6 different times during the year from early spring to winter, and evaluated the next year. Conclusions were that thickspike wheatgrass tolerates late spring and summer defoliation better than fairway and crested wheatgrass and that mowing may be useful for converting rank grass plants in old seedings into forms more acceptable to grazing animals.

Western Australia On salt-affected wastelands in Australia, salt-tolerant or halophytic shrubs have been successfully established for grazing by sheep. A total of 32×10^6 ha of non-irrigated land in Australia is salt-affected and 87% of this is naturally saline (Runciman, 1986). Research in Western Australia has determined that salt tolerant or halophytic shrubs that can be used for sheep grazing can be established on these areas. Species used have been bluerush (*Maireana brevifolia*), samphire (*Halosarcia* spp.), wavy leaf saltbush (*Atriplex undulata*), river saltbush (*Atriplex amnicola*) and marsh saltbush (*Atriplex paludosa*) (Clarke, 1982; Malcolm & Pol, 1986). After 6 years of sheep grazing on experimental plots, river saltbush grew well and sheep were able to graze the plants each year. Marsh saltbush had almost disappeared from the

Table 10.4. *Livestock responses in terms of average daily gain (ADG) and gain per hectare on rehabilitated mined areas and native areas in North Dakota*

	ADG (kg)	Total gain (kg ha^{-1})
Rehabilitated areas		
Heavy grazing	0.36	47
Moderate grazing	1.59	121
Light grazing	1.72	86
Native areas		
Heavy grazing	1.36	49
Light grazing	1.59	22

plots, being replaced by dense stands of samphire. Bluebush did well but some plants died from waterlogging. Wavy leaf saltbush persisted with new plants replacing old bushes that died (Malcolm & Pol, 1986). Grazing rates on the experimental plots ranged from 1493 to 1747 grazing days ha^{-1} in a study at Wickepin and 1680 to 1960 grazing days ha^{-1} at Kondut. Grazing periods were 37 to 49 days.

Livestock requirements

Livestock responses to grazing

Two types of performance indices can be calculated using animal weight gain figures: average daily gain (ADG) per animal and total gain of all animals per ha for a given grazing period. Average daily gain is not unit area-specific and thus can be directly compared among grazing systems or rates within a study or between studies. Gain per ha has to be interpreted much more carefully because it is a function of the stocking rate (numbers of animals and the number of days grazed) as well as the average gain of those animals. Thus, a treatment or study that has twice the stocking rate as another treatment or study but with similar ADG, will yield gains per ha twice as high in the highest stocking rate as in the lower rate.

Northern Great Plains, United States On the rehabilitated area near Center, North Dakota, Hofmann, Ries & Lorenz (1981) reported relatively high ADG figures for yearling steers on lightly and moderately grazed rehabilitated areas as well as on native areas (Table 10.4).

The ADG in the heavily grazed rehabilitated pasture was significantly lower than any other treatment; all other ADGs were statistically similar. The ADG figures on native range are quite high; ADGs on native rangeland in drier parts of the western United States often are substantially lower.

The ADG and gain per ha figures illustrate the usual pattern. As stocking rates increased from light to moderate to heavy, ADG decreased. This represents decreased forage availability (and possibly quality) as stocking rate and grazing pressure increased. The total gain per ha increased as the grazing rate increased from light to moderate because of the increased number of animals and only slightly lowered ADG. However, the gain per ha dropped drastically in the heavily grazed (probably overgrazed) treatment because of the low ADG. The total gain per ha in the moderately grazed, rehabilitated treatment was 2.5 times the gain in the heavily grazed native and more than 5 times the gain in the lightly grazed native treatments. This illustrates how productive rehabilitated mined lands can be if they are properly (moderately) grazed.

On the area northwest of Bismark, North Dakota, which was seeded to native species (Williamson, 1981), the light grazing treatment had the highest ADG of yearling steers but the heavy grazing treatment produced more beef per ha than the light treatment. In this case, the heavy treatment did not appear to be too heavy and was not resulting in overgrazing.

In the study near Colstrip in southeastern Montana (DePuit & Coenenberg, 1980), weight responses of yearling steers were used to evaluate forage quality and total productive capability of the rehabilitated mined lands. In both grazing systems tested, all cattle grazing on rehabilitated spoil areas during the spring gained about the same amount (ADG = 1.04–1.08 kg). However, in the summer, ADG of cattle moved to native rangeland was 0.77 kg while ADG for those remaining on spoils was only 0.58 kg. ADG for the combined spring and summer seasons was 0.88 kg for the spoils-range system and 0.73 kg for the spoils-only system. The authors concluded that animal weight response data demonstrated the superiority of the grazing system with complementary minesoil-native vegetation over the exclusive introduced minesoil vegetation. Also, the introduced minesoil vegetation in most years may be best utilized for early spring through early summer grazing, and, in some years, for autumn grazing. The reason for the qualification concerning autumn grazing was that cattle kept on rehabilitated pastures

Table 10.5. *Livestock responses in terms of average daily gain (ADG) and total gain ha^{-1} on rehabilitated mined areas under light and heavy grazing in Central Wyoming*

	ADG (kg)		Gain ha^{-1} (kg)	
Year	Light	Heavy	Light	Heavy
1980	0.60	0.58	12	23
1981	1.12	1.13	32	63
1982	0.99	0.82	27	47
Average	0.90	0.84	24	43

all spring and summer continued to gain in the autumn, while those brought back to the reclaimed pastures from native rangeland lost weight. The abrupt switch in kind and quality of vegetation for the cattle brought from native rangeland to the reclaimed area may have been a factor in the difference.

At the Decker mine in southeastern Montana, a topsoiled and seeded area was grazed to determine if livestock carrying capacity after disturbance by mining and rehabilitation was similar to that before disturbance (Kleinman, Raphelson & Bendix, 1984). Carrying capacity after 4 years of grazing was 3.56 AUM ha^{-1} on the rehabilitated area compared to 0.37 AUM ha^{-1} on the native range. This is nearly a 10-fold increase in carrying capacity on the rehabilitated area.

Central Great Plains and Rocky Mountains, United States On the rehabilitated uranium mined area in central Wyoming (Schuman *et al.*, 1986), ADG of yearling steers varied somewhat from year to year but did not differ between the light (0.24 steers ha^{-1}) and heavy (0.48 steers ha^{-1}) grazing treatments (Table 10.5).

The gain per ha values were not reported and were obtained from the authors. Because ADG was only slightly less, and stocking rate was twice as high in the heavy as in the light treatment, the total gain per ha was almost twice as high in the heavy as in the light treatment. This again illustrates the overwhelming influence that stocking rate has on gain per ha figures. The lack of differences in ADG between treatments indicates that the area was not overstocked under the heavy treatment.

On rehabilitated mined areas in northwestern Colorado (Laycock & McGinnies, 1985), all systems of grazing consistently produced high

Table 10.6. *Livestock responses in terms of average daily gain (ADG) and total gain ha^{-1} on rehabilitated mined areas under three different grazing systems in northwestern Colorado*

Grazing treatment	ADG (kg)	Total gain ha^{-1} (kg)
Short duration (SDG)	1.22	233
Summerlong heavy	1.22	276
Summerlong light	1.31	147

ADG figures for yearling cattle. Over 5 years, mean ADG ranged from 1.18 to 1.32 kg on the larger, more productive mine and 1.13 to 1.27 kg on the smaller mine. Some representative figures for 1985–1986 from the larger mine are shown in Table 10.6.

The high total gains (more than 225 kg ha^{-1}) were achieved only in the last 2 years when the stocking rates were quite high in the SDG and heavy summerlong treatments. The light summerlong treatment had about one-half the total gain per ha as the other two treatments because the stocking rate also was about one-half. Death losses due to bloat were 2–6 animals per year out of approximately 240 animals grazed each year (during 1985 and 1986). Cattle were fed a supplement containing 'poloxalene' to minimize bloat problems.

No native rangeland was included in the Colorado study. However, a 7-year summer grazing study in similar native vegetation (sagebrush-grass) at the same elevation in northeastern Utah (Laycock & Conrad, 1981) yielded ADG of yearlings of 0.68 kg and a total gain of 20 kg ha^{-1}. This is a 44–48 % lower ADG and an 86–93 % lower gain per ha than those on the rehabilitated mined area. This comparison strongly suggests the superiority of the alfalfa and introduced grass vegetation for producing beef.

Midwestern and Eastern United States Few publications were found describing results of grazing on rehabilitated disturbed lands in the more humid midwestern and eastern United States. In southeastern Kansas, both the Pittsburg and Midway and the Sinclair Coal Companies began experimental work in about 1945 to determine the feasibility of converting surface-mined lands to pasture lands (Wells, 1953; Foresman, 1955). Lands were successfully seeded to sweetclover and Korean lespedeza (*Lespedeza stipulacea*). Foresman (1955) reported that the areas were fenced and 'strip grazed', i.e. animals were confined to one part of the pasture for grazing then rotated. Purebred Angus, commercial Herefords, and sheep were grazed. No animal gain figures were given but Wells (1953) repeated information from the Sinclair Coal Co. which stated that 'meat animals produced on stripland pastures and feed are superior to those grazed on average farm forage crops and dress out a higher percentage of quality meat, 62 % and 58 % respectively'.

In Illinois, Hereford, Angus, Shorthorn and crossbred calves (136–181 kg) were raised on ensiled alfalfa ('haylage') grown on rehabilitated coal mined land (Anon., 1968). Average gain of these animals was 0.91 kg day^{-1}. On mined areas with rougher topography, spoils were seeded to a 75 % smooth brome, 25 % alfalfa mixture, and grazed. ADG of calves from these areas was 0.63 kg.

On coal mined lands in Virginia, rehabilitation has been achieved by planting with Kentucky 31 tall fescue (*Festuca arundinacea*), sericea lespedeza (*Lespedeza cuneata*), orchardgrass (*Dactylis glomerata*), timothy (*Phleum pratense*), ladino clover (*Trifolium repens* var. *giganteum*), red clover (*Trifolium pratense*), or alfalfa. All of these species can be used for grazing but the main species used in southwest Virginia have been tall fescue and sericea lespedeza (Baker, Dingus & McCart, 1976). Grazing trials with a beef cow and a calf herd were conducted for 3 years. Grazing was from April 15 through December. Calf weight gains from April 15 through October 31 averaged 179 kg (0.9 kg head^{-1} day^{-1}) for the 3 years. Cows maintained a good body weight and a thrifty condition. Calf crop was approximately 95 %. Weaning weights of calves were higher than the average Virginia feeder calves sold through organized sales. For the vegetation grazed, approximately 1.2 ha of a vigorous stand were required for grazing each cow/calf unit for the 8-month grazing season (approximately 6.7 AUM ha^{-1}).

Western Australia Grazing trials on saltbush and bluebush planted on salt-affected areas in Western Australia indicated that these salt-tolerant shrubs can be grazed without damage to the shrubs or to the sheep. They can provide forage for sheep at a time when need is the greatest, in the autumn and early winter. In trials at two locations, sheep grazed in May initially gained weight, then levelled of and eventually lost weight as food became limiting. Weight loss was about 4 kg per sheep over a 30–44-day period at Kondut and 5 to 9 kg per sheep over a 49-day period at Maya (Malcolm & Pol, 1986).

Forage quality, trace elements and heavy metals

The high ADG and gains ha^{-1} for cattle grazing on most rehabilitated areas indicate that forage quality generally is not a problem. However, the potential does exist for problems on a site-specific basis. In the western United States, most older surface-mines had no topsoil replacement and topsoil depths on newer mines often are shallow enough to allow root penetration into the underlying spoil material. Some spoil material can contain either deficient amounts of essential nutrients or elevated levels of heavy metals or other undesirable elements (see Chapter 9).

Severson & Gough (1983) analyzed both cover-soil (topsoil) and spoil material from 11 coal mines in Colorado, Montana, North Dakota, and Wyoming for 8 different elements (Cd, Co, Cu, Fe, Mn, Ni, Pb and Zn). The large amount of variability, both among mines and among cover-soil and spoil samples collected at a single mine, limited the usefulness of simple average concentrations for characterizing all mines, or all samples collected at a single mine. They suggested that an expected range in concentration would provide a better basis for estimating the concentration of an element in cover-soil or spoil.

Severson & Gough (1984) compared the concentration of these same elements in mine soils and uptake by plants on the same 11 mines. Metal uptake by plants was positively correlated with organic matter content and negatively correlated with soil pH. Correlations between metals in the mine soils and in the plants generally were not statistically significant and showed no consistent pattern for any one metal or plant species. However, according to Gough, Shacklette & Case (1979), Cd and Pb can be absorbed by plants to the extent that plants are toxic to grazing animals. Mammals have no effective mechanism for the elimination of ingested Cd and with time it tends to accumulate in the liver and kidneys. Its relative toxicity to mammals is moderate to high. Pb is not as toxic as Cd but is more toxic than many other elements.

Gough & Severson (1983) determined concentrations of more than 30 elements in crested wheatgrass, intermediate wheatgrass, slender wheatgrass, alfalfa, and fourwing saltbush (*Atriplex canescens*) from the same 11 mine sites in the Northern Great Plains. Concentrations of metals in wheatgrasses from all mines were below the levels considered to be potentially toxic to plants or grazing animals. For alfalfa, concentrations of B at one North Dakota mine might be phytotoxic. Mo levels at one Montana and one North Dakota mine were high enough to be potentially toxic to grazing cattle.

The only study found that determined the effects of intake of heavy metals by livestock on rehabilitated mined lands was reported by Fitzgerald (1979). In Maryland, adult cows were grazed for an entire year on a cross between Sudan grass (*Sorghum sudansis*) and sorghum (*Sorghum vulgare*), corn stubble, and grass–legume pastures, all experimentally sprayed with sewage sludge containing heavy metals. Soils in the pastures where sludge was applied tended to have elevated levels of Zn, Cu, Cd, Cr, Ni, Pb and Hg. The forage plants analyzed also showed elevated levels of these same heavy metals but the analysis was done on plants on which the sludge had been sprayed. Thus, it was not possible to differentiate between the content of the plant tissue and the sludge residue on the surface of the forage. Samples of the diaphragm, heart, liver, kidney, brain, bone and blood were analyzed from experimental animals and animals on untreated pastures. For most tissues and most metals there were only 'moderate differences' between the two groups of animals. However, liver and kidney tissues from animals ingesting sludge-treated forage often contained higher quantities of Cd and Pb. The conclusion was that 'this probably reflects the normal detoxifying or excretory activities of these organs'. This conflicts with information of Gough, Shacklette & Case (1979) who indicated that both Cd and Pb are toxic and that mammals have no effective mechanism to eliminate ingested Cd.

Blood samples from cattle which had eaten sludge-treated forage, contained about the same level of heavy metals as blood from control animals except for Pb. Lead levels were about 4 times higher from animals ingesting the treated forage. No evidence was found of damage to the cows grazing the treated forage or to their calves, and a 95% conception rate was achieved for cows on the sludge-treated pastures. It is not known if longer-term treatment effects would occur if the same cattle were grazed on the sludge-treated pastures over a longer period of several years.

Erdman, Ebens & Case (1978) found high Mo levels and low Cu:Mo ratios in sweetclover on 6 of 8 mined sites in the Northern Great Plains of the United States. Low Cu:Mo ratios in forage can cause molybdenosis, a copper deficiency disease which can cause weight loss, reproductive problems and even death in cattle and other animals. Sweetclover is a known Mo accumulator, but is often planted on mined lands because of its high productivity, N-fixing ability, and palatability. Erdman, Ebens & Case (1978) stated that if the major forage on coal mine spoils is sweetclover or other species with low Cu:Mo ratios, molybdenosis may be expected to occur.

Neuman & Munshower (1984) reported that mixing low Mo overburden with undesirably high Mo material and then covering this with at least 46 cm of topsoil would produce plants with acceptable Cu:Mo ratios. DePuit & Coenenberg (1978) did find low Cu:Mo ratios in yellow sweetclover and alfalfa in southeastern Montana, but no molybdenosis symptoms were noted in cattle and weight gains were satisfactory. The Cu:Mo ratios of most grasses apparently were sufficiently high to offset any problems in the legumes. They recommended against establishment of legume monocultures on mined sites in the Northern Great Plains where molybdenosis might be a problem.

None of the grazing studies in the western United States reported problems with cattle that might have been associated with molybdenosis but liver or other organs were not analyzed to detect any unusual accumulation (or deficiency) of elements. However, Cu deficiencies and molybdenosis intoxication have been found in cattle grazing near uranium mines in Texas (Dollahite *et al.*, 1972) and in Missouri where clay mining operations had contaminated adjacent areas (Ebens *et al.*, 1973).

Most of the studies reviewed used yearling cattle in relatively short-term studies, and problems with mineral imbalances probably would not show up under these conditions. Cow-calf grazing, where the cow and calf would return to the same rehabilitated mined area in subsequent years, may be the only way to determine if any potential mineral imbalance problems exist (Schuman *et al.*, 1980). Examining wild animals that spend a substantial amount of time on disturbed areas may be another approach. King, Leleux & Mulhern (1984) examined livers from white-tailed deer near uranium surface-mines in Texas where molybdenosis had been diagnosed in cattle. Almost two-thirds of 36 deer examined had relatively low levels of Cu in the liver but only one deer had detectable levels of Mo. They concluded that the Cu deficiencies noted in deer in the area were not Mo-induced.

Schuman *et al.* (1980) listed the forage quality indicators and heavy metal contents of a large number of forage species grown on rehabilitated mined lands in the Northern Great Plains of the United States. Other investigators have looked at similar forage quality indicators in other areas. Apart from the potential Mo problem discussed above, few problems or potential threats to livestock grazing have been reported.

Stanley *et al.* (1982) and Schuman (1982) found protein, P and Zn levels were lower in plants from rehabilitated mined areas in Wyoming and Montana than from native range sites. They found no evidence of heavy metal accumulations or toxic levels of other elements. Munshower & Neuman (1978) determined that micro- and macronutrient levels in vegetation growing on rehabilitated mined land in southeastern Montana appeared to be adequate for cattle nutrition, with the possible exceptions of Cu and Zn. Legumes had greater variation in quality between spoils and non-spoils than grasses. On mines of different ages in southeastern Montana, Munshower & Neuman (1980) reported that macronutrient levels generally were reduced in plants growing on non-topsoiled spoils, while plants on topsoiled sites had levels similar to those on native range areas. Micronutrient levels were about the same on non-topsoiled sites as on native range, while levels were elevated in vegetation from topsoiled sites.

Water requirements for livestock

Livestock grazing on rehabilitated lands requires a source of drinking water of reasonably good quality. Proper design of impoundments put in place during the reshaping of the land surface and rehabilitation can supply needed water but water quality can be a problem.

Rumble (1985) measured 27 water quality parameters in coal and bentonite surface-mine impoundments in the Northern High Plains of the United States (Wyoming, South Dakota and North Dakota). Most impoundments were safe for use as a source of livestock drinking water. Eight water quality parameters were different among impoundment types. In some impoundments, concentrations of 4 parameters – sulphate, total dissolved solids (TDS), Pb and Cd – were equal to, or exceeded the recommended safe levels reported in the literature for livestock water. Because many of the anions are present in the dissolved solids, TDS probably is the best indicator of water quality and should be determined in water from surface-mine impoundments prior to use for livestock. Specific conductivity (SC) can be used to estimate TDS by the formula: $\mathrm{TDS} = 1.01\ \mathrm{SC} - 299$.

Table 10.7. *Soil loss, water runoff and protective plant cover on rehabilitated mined areas and native areas in North Dakota*

Treatment	Soil loss ($kg\ ha^{-1}$)	Runoff ($cm\ h^{-1}$)	Total plant and litter cover (%)
Rehabilitated areas			
Ungrazed	18	0.28	99
Lightly grazed	64	0.84	96
Moderately grazed	155	1.37	83
Heavily grazed	1054	2.31	68
Ungrazed native areas	8	0.10	99

Hydrologic responses to grazing

On disturbed and rehabilitated areas to be used for grazing, the planned grazing treatments must maintain the hydrologic integrity of the areas so as not to cause excessive runoff and erosion that will eventually destabilize the site. Only two studies were found that determined the effects of grazing on the hydrologic characteristics of runoff, infiltration, and erosion on rehabilitated lands.

On rehabilitated mined areas and native rangeland near Center, North Dakota, simulated rainfall studies were conducted after 6 years of grazing to determine runoff and soil loss under different treatments (Hofmann, Ries & Gilley, 1983). Simulated rainfall was applied with a sprinkler on plots measuring 4.0×22.1 m in size at an intensity of $4.60\ cm\ h^{-1}$. Some of the results of the runs on pre-wetted plots are shown in Table 10.7.

Ungrazed, lightly grazed and moderately grazed rehabilitated areas and ungrazed native areas had significantly lower soil loss than the heavily grazed rehabilitated pasture (Hofmann *et al.*, 1983; Ries & Hofmann, 1983). The heavily grazed rehabilitated pasture had the most runoff (lowest infiltration), and runoff progressively declined with decreasing grazing intensity.

Ries & Hofmann (1983) concluded that total cover (live vegetation plus litter) was strongly inversely related to soil loss and runoff. However, surface (basal) cover of live vegetation only was a poor measure of site protection. State regulations that use basal cover of plants as the only measure of cover do not provide the best measure of site stability. The authors also concluded that well established rehabilitated grasslands are as erosionally stable as native grasslands so long as the land is not heavily grazed.

On coal-mined land in northwestern Colorado, described by Laycock & McGinnies (1985), the hydrological characteristics of pastures receiving different grazing treatments were measured in the third and fourth years of the study. A rainfall simulator applied water on pre-wetted plots (0.349 m^2 in size) at 10.16 cm h^{-1} for 30 min after runoff began (R. Cogar, unpublished draft of M.Sc. thesis in Watershed Science, Utah State University).

In 1984, a fairly wet year with high forage production, there were no significant differences in infiltration rates or sediment production (soil loss) among grazing treatments. On the largest of the two replications, average sediment production was 230 kg ha^{-1} and average infiltration rate at 30 min was 4.51 cm h^{-1}. On the other replication, sediment production averaged 689 kg ha^{-1} and infiltration rate averaged 3.54 cm h^{-1}.

Forage production was relatively low in 1985 because of low precipitation. Large difference occurred among grazing treatments in levels of forage utilization, infiltration rates and sediment production. On the larger, more productive replication, infiltration was 6.38 cm h^{-1} and sediment was 115 kg ha^{-1} for the SDG system. Both figures were significantly different from the 3.20 cm h^{-1} infiltration rate and 344 kg ha^{-1} of sediment in the heavily grazed summerlong treatment. The lightly grazed summerlong treatment was not measured. On the smaller replication, the same relationships were found with the heavy summerlong treatment producing significantly more sediment and lower infiltration rates than the light summerlong treatment. The SDG system was intermediate for both attributes.

Tentative conclusions were that a moderate rate of summerlong grazing probably would have produced approximately the same runoff and erosion as the SDG system, i.e. intermediate between the heavy and light summerlong treatments, and might be a satisfactory treatment from a hydrological standpoint. It should be pointed out however, that even heavy summerlong grazing produced sediment values (0.38 and 0.92 T ha^{-1}) that were well below what would be considered as a serious erosion loss.

Effect on vegetation desired for other uses

Along the North Sea coast in the Netherlands, large areas of tidal sandflats have been embanked and subsequently turned into nature reserves (Joenje, 1982). These disturbed polders undergo succession starting with salt-tolerant annual plants and eventually form communities dominated by red fescue (*Festuca rubra*) in non-salty, relatively dry areas, with *Agrostis* spp. and *Phragmites* spp. in slightly more moist areas with a higher water table, and *Puccinnelia* spp. in brackish areas with a high water table (see Chapter 14). Experimental grazing was used to create large areas with short grassland vegetation in order to make the area more amenable to grazing by waterfowl and as a breeding area for other birds and waders. Sheep grazing at low stocking rates evenly used all plant communities in early summer. In September, the brackish *Puccinnelia* zone was preferred. Little use was made by sheep of *Phragmites* spp., red fescue, or the annual halophytes. Cattle grazing at higher rates (1 animal per ha) utilized 60 % of the standing vegetation and trampling caused visible damage or loss of some bird nests.

On a rehabilitated polder in the Netherlands used as a meadow bird reserve, a grass mixture was sown which consisted of 20 % perennial ryegrass (*Lolium perenne*), 10 % meadow fescue (*Festuca pratensis*), 10 % timothy (*Phleum pratense*), 10 % roughstalk bluegrass (*Poa trivialis*), 10 % Kentucky bluegrass (*Poa pratensis*), 10 % crested dogtail (*Cynosurus cristatus*), 5 % cocksfoot (orchard grass), 5 % meadow foxtail (*Alopecurus pratensis*), 5 % tall oatgrass (*Arrhenatherum elatius*), 5 % browntop (*Agrostis tenuis*), 5 % red fescue, and 5 % fowl bluegrass (*Poa palustris*) (de Jong, 1977). Cattle were grazed on one-half of the area. Temporary high stocking densities of 4 cows ha^{-1} during the breeding season caused more than 50 % loss of the meadow birds nests. Grazing at 0.6 cows ha^{-1} had little effect on the nesting birds. Under this grazing regime, numbers of breeding birds increased remarkably and were much higher in the grazed areas in the reserve than in surrounding areas. The most favourable conditions for the meadow birds appeared to be a water level within 20 cm of the surface, limited fertilizer use, and a stocking density of not more than 2 cows ha^{-1}.

In Western Australia, Majer (1985) found that pastured ex-sandmines produced an invertebrate fauna that was more than twice as high in numbers of taxa as grazed farmland. After 6 years, ant species richness on the sandmine was 75 % of that in the grazed farmland.

Summary

With proper management, livestock can be grazed successfully on rehabilitated disturbed lands with no apparent detrimental effects on the vegetation or on the hydrological characteristics of the site. Many plants on rehabilitated areas have forage quality similar to native plants. However, some introduced species (e.g. alfalfa) have considerably higher nutritive value than many native species. Some potential elemental imbalance problems have been reported from forages grown on rehabilitated lands, such as low Cu:Mo ratios in some legumes and low Zn content. However, few animal health problems have been reported.

Although most of the data summarized in this Chapter were from studies on rehabilitated mined areas in the western United States, most of the conclusions and recommendations also pertain to other areas of the World.

There are two main schools of thought about rehabilitation of disturbed lands for rangeland uses in the United States. One is to plant native species so that the site returns to something resembling the native landscape as soon as possible. This approach is advocated as the best way to maximize plant species diversity, ensure long-term site stability, and restore wildlife habitat. State and federal regulations for a mined land in the United States generally favour, or impose, this approach on most surface-mined areas.

The second approach is to define the post-disturbance uses of the land very specifically, and then plant the species best adapted to stabilize the site and meet these objectives. Species planted might be native or introduced and the species mixture may be simple or complex. Rehabilitated rangelands seeded to productive introduced species often are superior to similar native rangelands in terms of both plant and animal production. Studies on northern Great Plains and Intermountain rehabilitated mined lands have indicated total animal gains per ha from 2.5 to 13 times higher than those expected from adjacent native rangelands.

In addition, the second approach can also lead to better wildlife habitat than existed in the pre-mining situation. Planting vegetation different from that of surrounding, non-mined lands creates a mosaic of habitats and 'edges' between habitat types, which often benefits wildlife (Thomas, Maser & Rodiek, 1979; Laycock, 1980). This is especially true if individual rehabilitated sites are no larger than 100–200 ha. Using this approach may be difficult or even impossible under many current regulations in the US. However, this approach may often lead to a stand of vegetation better suited to certain post-disturbance land uses than the 'native only' approach.

Recommendations for arid and semi-arid areas

1. Research on grazing management of rehabilitated systems should be conducted for longer periods of time than those of most studies reviewed in this Chapter. Only the North Dakota and Colorado studies conducted were for 5 years or longer.
2. Future studies should include the effects of grazing on all aspects of the system, i.e. soil, hydrology, vegetation and animals, rather than concentrating on one or two of them. The effects of different systems and/or rates of grazing on all of these aspects are also important to include in future research, wherever possible.
3. More emphasis should be placed on analyzing effects of grazing on the combination of rehabilitated and native areas on soils, vegetation and animals in the entire grazing system. The Montana study of DePuit & Coenenberg (1980) was the only one that evaluated such a system – at least in terms of animal response. Proper use of highly productive complementary pastures (such as rehabilitated areas) can increase animal production using less total land area. It can also delay the start of grazing on native rangelands in the spring which usually results in improvement of range condition. Currie (1981) presented details on the planning and benefits of such integrated operations on mined lands.
4. The effects of other classes or kinds of grazing animals need to be evaluated on rehabilitated lands. Almost all studies have used yearling cattle. Cow-calf grazing, where animals are returned to the rehabilitated area for a number of years, may be needed to reveal long-term effects of potential mineral deficiencies or imbalances.
5. Because of their obvious advantages for plant and animal productivity, with no damage to hydrologic characteristics of the site, high-producing introduced plant species should be permitted so as to give a greater flexibility in areas where they are adapted and where they best meet the planned post-mining land uses. Regulations should be changed to permit, or even to encourage, this flexibility. This recommendation agrees with Majer (1985) who concluded that 'success in restoring ecosystems may be achieved more rapidly and more completely, in mines restored to agricultural use than to native-type ecosystems'.
6. Properly controlled grazing should be allowed and encouraged on rehabilitated areas as soon as a stable stand of vegetation is established. State and Federal laws in the western United States

usually prohibit grazing and other management practices for the first 8 years of a bond release period. Managers need these tools to manipulate vegetation toward the desired goals of productivity, cover and species diversity. Both productivity and quality of vegetation often are higher with than without grazing. Species diversity often increases with grazing. Whelan, in Chapter 5, indicated that moderate herbivory is likely to enhance plant species diversity. Lack of grazing for prolonged periods leads to build-up of undecomposed standing litter, which may tie up nutrients, slow down nutrient cycling and lead to improper C:N ratios in the soil. Proper grazing incorporates litter into the soil, which accelerates decomposition and nutrient cycling, and increases protection against runoff and erosion.

References

Anon. (1968). Big beef business flourishes on reclaimed strip land. *Coal Age*, **73**, 71–5.

Baker, C. D., Dingus, G. P. & McCart, G. D. (1976). Producing forages for livestock on surface mined land in southwest Virginia. In *Proceedings Fourth Symposium on Surface Mining and Reclamation, Louisville, Kentucky*, pp. 223–7. Washington, D.C: Environmental Research and Technology Inc.

Clarke, J. A. (1982). The grazing value of saltbushes. *Journal of Agriculture of Western Australia*, **23**, 7–9.

Currie, P. O. (1981). Revegetating mined land for grazing. *Journal of Soil and Water Conservation*, **36**, 213–5.

de Jong, H. (1977). Experiences with the man-made meadow bird reserve "Kievitslander" in Flevoland (The Netherlands). *Biological Conservation*, **12**, 13–31.

DePuit, E. J. & Coenenberg, J. G. (1978). Plant response and forage quality for livestock grazing on coal mine spoils in Montana. In *Ecology and Coal Resource Development*, ed. M. K. Wali, pp. 892–8. New York: Pergamon Press.

DePuit, E. J. & Coenenberg, J. G. (1980). *Plant Response and Forage Quality for Controlled Grazing on Coal Mine Soil Pastures: Project Final Report.* Montana Agricultural Experiment Station. (Report to the U.S. Environmental Protection Agency, Cincinnati, Ohio).

Dollahite, J. W., Rowe, L. D., Cook, L. M., Hightower, D., Bailey, E. M. & Kyzar, J. R. (1972). Copper deficiency and molybedonosis intoxication associated with grazing near a uranium mine. *Southwest Veterinary*, **26**, 47–50.

Ebens, R. J., Erdman, J. A., Feder, G. L., Case, A. A. & Selby, L. A. (1973). *Geochemical Anomalies of a Claypit Area, 111 Callaway County, Missouri, and Related Metabolic Imbalance in Beef Cattle.* U.S. Geological Survey Prof. Paper, 807.

Erdman, J. A., Ebens, R. J. & Case, A. A. (1978). Molybdenosis: a potential problem in ruminants grazing on coal mine spoils. *Journal of Range Management*, **31**, 34–6.

Fitzgerald, P. R. (1979). Recovery and utilization of strip-mined land by application of anaerobically digested sludge and livestock grazing. In: *Utilization of Municipal Sewage Effluent and Sludge on Forest and Disturbed Land*, ed. W. E. Sopper & S. N. Kerr, pp. 497–506. Philadelphia: Pennsylvania State University Press.

Foresman, F. J. (1955). Strip grazing for prize Angus. *Coal Age*, **10**, 74–7.

Gough, L. P. & Severson, R. C. (1983). Rehabilitation materials from surface-coal mines in western U.S.A. II. Biogeochemistry of wheatgrass, alfalfa and fourwing saltbush. *Reclamation and Revegetation Research*, **2**, 103–22.

Gough, L. P., Shacklette, H. T. & Case, A. A. (1979). *Element Concentrations Toxic to Plants, Animals, and Man.* U.S. Geological Survey Bulletin, 1466.

Hofmann, L., Ries, R. E. & Gilley, J. E. (1983). Relationship of runoff and soil loss to ground cover of native and reclaimed grazing land. *Agronomy Journal*, **75**, 599–602.

Hofmann, L., Ries, R. E. & Lorenz, R. J. (1981). Livestock and vegetative performance on reclaimed and non-mined rangeland in North Dakota. *Journal of Soil and Water Conservation*, **36**, 41–4.

Jasmer, G. E., Donart, G. B. & Aldon, R. F. (1983). Responses of re-established vegetation to simulated grazing. In *Symposium on Reclamation of Mined Lands in the Southwest*, ed. E. F. Aldon & W. R. Oaks, pp. 188–91. Soil Conservation Society of America.

Joenje, W. (1982). Nature in New Wadden-polders: conservation by exploitation. In *Polders of the World. Papers of International Symposium, Lelystad*, vol. 3, pp. 68–82, Wagengingen: International Institute for Land Reclamation and Management.

King, K. A., Leleux, J. & Mulhern, B. M. (1984). Molybdenum and calcium levels in white-tailed deer near uranium mines in Texas. *Journal of Wildlife Management*, **48**, 267–70.

Kleinman, L. H. & Layton, E. E. (1981). Livestock operation is born on reclaimed mined lands. *Abstracts of Papers, 34th Annual Management Society of Range Management.* Tulsa, OK. No. 80.

Kleinman, L. H., Raphelson, M. A. & Bendix, D. G. (1984). Livestock grazing on mined land reclamation. *Abstracts of Papers, 37th Annual Meeting, Society of Range Management*, Rapid City, SD. No. 213.

Laycock, W. A. (1980). What is successful reclamation? – A look at the concepts of adaptability, productivity, cover, and diversity of seeded species. In *Proceedings of Northwest Colorado Land Reclamation Seminar II.* Steamboat Springs, CO. p. 1–17.

Laycock, W. A. & Conrad, P. W. (1981). Responses of vegetation and cattle to various systems of grazing on seeded and native mountain rangelands in eastern Utah. *Journal of Range Management*, **34**, 52–8.

Laycock, W. A. & Layden, T. (1986). Cattle production on reclaimed mined lands in northwestern Colorado. In *Rangelands: A Resource under Siege. Proceedings Second International Rangeland Congress*, ed. P. J. Joss, P. W. Lynch & O. B. Williams, pp. 162. Canberra: Australian Academy of Science.

Laycock, W. A. & McGinnies, W. J. (1985). Reclamation and grazing management on a surface coal mine in northwestern Colorado. In *Proceedings Second Annual Meeting of American Society of Surface Mining and Reclamation*, pp. 73–6. Denver.

Layden, T. V. (1984). *Grazing on Reclaimed Mined Land in Northwestern Colorado.* M.Sc. Thesis, Colorado State University.

Majer, J. D. (1985). Restoration of pre-mining ecosystems in forested and agricultural areas. *Landline*, **11**, 4.

Malcolm, C. V. & Pol, J. E. (1986). Grazing and management of saltland shrubs. *Journal of Agriculture of Western Australia*, **2**, 59–63.

Munshower, F. F. & Neuman, D. R. (1978). Trace element concentrations in plants from revegetated strip mined lands and native range in southeastern Montana. In *Ecology and Coal Resource Development*, ed. M. K. Wali, pp. 887–91. New York: Pergamon Press.

Munshower, F. F. & Neuman, D. R. (1980). Elemental concentrations in native plant species growing on minesoils and native range. *Reclamation Review*, **3**, 41–6.

Naeth, M. A. & Bailey, A. W. (1986). The effect of grazing regime on revegetation of rangeland after pipeline construction. In *Abstracts of Papers, 39th Annual Meeting for Society for Range Management*, pp. 66, Kissimmee, Florida.

Neuman, D. R. & Munshower, F. F. (1984). Copper and molybdenum uptake by forages grown on coal mine soils. *Journal of Range Management*, **37**, 517–20.

Ries, R. E. & Hofmann, L. (1983). Re-establishment and use of grasslands on reclaimed soils. In *Can Mined Land Be Made Better Than Before Mining?*, pp. 85–93. Bismark: Bismark, ND.

Rumble, M. A. (1985). Quality of water for livestock in man-made impoundments in the northern high plains. *Journal of Range Management*, **38**, 74–7.

Runciman, H. V. (1986). Forage production from salt affected wasteland in Australia. *Reclamation and Revegetation Research*, **5**, 17–29.

Schuman, G. E. (1982). Forage quality of several grass species grown on reclaimed mined lands. In *Trace Element Mobilization in Western Energy Regions*, ed. L. P. Gough & R. C. Severson, pp. 27. Denver.

Schuman, G. E., Booth, D. T., Waggoner, J. W. & Rauzi, F. (1986). The effect of grazing reclaimed mined lands on forage production and composition. In *Rangelands: A Resource Under Siege. Proceedings Second International Rangeland Congress*, ed. P. J. Joss, P. W. Lynch & O. B. Williams, pp. 163–4. Canberra: Australian Academy of Science.

Schuman, G. E., Munshower, F. F., Ries, R. E. & Robbins, E. G. (1980). *Quality and Heavy Metal Content of Forages Grown on Reclaimed Mined Lands in the Northern Great Plains*. Great Plains Agricultural Council Report No. 98. University of Wyoming Agricultural Experimental Station Research Journal.

Severson, R. C. & Gough, L. P. (1983). Rehabilitation materials from surface-coal mines in western U.S.A. Chemical characteristics of spoil and replaced cover-soil. *Reclamation and Revegetation Research*, **2**, 83–102.

Severson, R. C. & Gough, L. P. (1984). Rehabilitation materials from surface-coal mines in western U.S.A. III. Relations between elements in mine soil and uptake by plants. *Reclamation and Revegetation Research*, **3**, 185–97.

Stanley, M. A., Schuman, G. E., Rauzi, F. & Painter, L. I. (1982). Quality and element content of forages grown on three reclaimed mine sites in Wyoming and Montana. *Reclamation and Revegetation Research*, **1**, 311–26.

Thomas, J. W., Maser, C. & Rodiek, J. E. (1979). *Wildlife Habitats in Managed Rangelands – The Great Basin of Southeastern Oregon: Edges*. U.S. Department of Agriculture, Forest Service, General Technical Report. No. 85.

Verma, T. R. & Thames, J. L. (1976). Multiple uses of surface mined land in the southwest. In *Proceedings of the Fourth Symposium on Surface Mining and Reclamation*, pp. 228–35. Louisville, Kentucky.

Wells, J. R. (1953). The reclamation of stripmined areas in southeastern Kansas. *Kansas Academy Science Transactions*, **56**, 269–92.

Williamson, R. L. (1981). Improving species composition and seasonal variety on reclaimed strip mined grasslands through selective grazing intensities and time periods. *Proceedings of the North Dakota Academy of Science*, **35**, 5.

Young, S. A. & Rennick, R. B. (1983*a*). Using grazing animals to alter vegetation composition of reseeded minesoils. In *Abstracts of Papers, 36th Annual Meeting of the Society of Range Management*, Albuquerque, N.M. pp. 306.

Young, S. A. & Rennick, R. B. (1983*b*). *Use of Grazing Cattle During Post Seeding Management of Revegetated Minesoils*. Final Report, Montana Agricultural Experimental Station, Bozeman.

11

Wetland habitat development on mined lands

J. R. NAWROT AND W. D. KLIMSTRA

Introduction

Wetlands have become a rapidly diminishing natural resource both within the United States and elsewhere. Prior to settlement, more than 87 million ha of wetlands occurred in the US; however, by the mid 1970s less than 40 million ha remained (Tiner, 1984). Not recognized for their inherent hydrologic, biologic, and socio-economic values or functions, wetlands were perceived as a hindrance to agricultural and urban/industrial prosperity. Drainage for agricultural development was identified as being responsible for 87 % of recent wetland losses, with urban development accounting for an additional 8 % loss (Tiner, 1984). Stream and river channelization and dredging by the US Army Corps has also directly impacted the quantity and quality of natural wetlands.

Although mining (all types) and other non-urban developments have accounted for approximately 5 % of recent US wetland losses (Frayer *et al.*, 1983), many of those industries or agencies responsible for these losses have initiated programs to develop additional wetlands and mitigate losses. The US Army Corps of Engineers is now developing wetland habitat in conjunction with its dredging and channelization projects (Newling, 1981). Many examples now exist where wetland habitats are being developed in conjunction with sand and gravel mining (Svedarsky & Crawford, 1982), taconite mining (Dickinson, 1975), coastal urban development (Josselyn & Buchholz, 1982), Polish lignite mining (Janiak, 1979), Australian bauxite mining (Kabay & Nichols, 1981), phosphate mining (Haynes, 1984), and highway construction (Day, 1986). Similarly, wetland habitats have long been a byproduct of mining for coal in the midwest of the US (Klimstra & Nawrot, 1982).

It is only during the last 5 years there has been a nationwide acceptance and promotion of wetlands in the US as a viable and valuable rehabilitation alternative, a trend which has occurred elsewhere in the World (Kabay & Nichols, 1981; Street, 1985).

Wetland habitat – a byproduct of mining

Coal mining in the US has been perceived by the general public as an adverse environmental impact. Mine related waters were generally considered to produce 'acid drainage problems'. Consequently, a disproportionate effort has been directed at emphasizing negative aspects, while little was done to acknowledge the beneficial aspects of mined land wetlands or promote the development of additional wetlands. Environmentalists and legal interests in the United States drafted and lobbied for strict nationwide reclamation legislation (i.e. the Surface Mining Control and Reclamation Act (SMCRA) of 1977 – Public Law 95–87) (see Chapter 1) assuming that federal regulation would benefit forest, agricultural, and wetland resources.

Although wildlife habitat enhancement was a much publicized benefit that could result from the nationwide implementation of SMCRA, habitat diversity decreased in midwestern states. Emphasis on pre-mining land-use and agricultural rehabilitation of prime agricultural lands limited wildlife habitat development. Unfortunately, the many benefits (i.e. fisheries, recreation, municipal water supplies, irrigation, etc.) of pre-law mined land wetlands were ignored in the interest of increasing agricultural productivity. Consequently, regulations for farmland rehabilitation precluded inclusion of wetlands which were once a byproduct of pre-law (i.e. SMCRA P.L. 95–87) reclamation.

Pre-law wetland habitat

Surface mining for coal in Illinois has affected around 97 128 ha of land, resulting in more than 607 ha of water impoundments and shallow water wetlands. Although over 40 % of the area was mined prior to any rehabilitation requirement, less than 5 % of the associated waters were considered adversely affected or undesirable. The Cooperative Wildlife Research Laboratory of Southern Illinois University has long recognized that mined-land wetlands could mitigate the overall wetland decline. Consequently, research was initiated in 1954 to establish baseline data regarding physical, chemical, and biological characteristics of surface-mine-associated waters (Bell, 1956; Arata, 1959; Haynes &

Klimstra, 1975; Coss, 1981; Konik, 1980). Additional investigations, including those performed in other midwest states, have detailed a variety of beneficial aspects of mined land wetlands, including economics of waters as sources for irrigation, fish, wildlife, recreation, industrial and municipal water supplies, and residential development (Riley, 1960; Roseberry & Klimstra, 1964; Deane, 1966; Imhoff, Friz & LaFevers, 1976; Klimstra, Haynes & Filer, 1977; Thompson & Lindahl, 1977; Gibb & Evans, 1978; Grandt, 1978; Christy, Smith & Filer, 1979; Rosso, 1980).

In spite of the fact that most pre-law impoundments were never designed or engineered to meet design criteria of SMCRA P.L. 95–87, the majority of impoundments have provided excellent wetland resources. For instance, many state record fish have been caught from surface mine lakes in the midwest (see Chapter 12). These pre-law lakes and shallow water impoundments have provided outstanding waterfowl habitat in areas where none previously existed (Klimstra, Vohs & Cherry, 1963; Roseberry & Klimstra, 1964; Weber, 1970; Grandt, 1978; Sandusky, 1978; Grandt, 1981; Perkins, 1981; Klimstra & Thornburg, 1982; O'Leary, Klimstra & Nawrot, 1984).

In Illinois, mined land wetlands played a vital role in the reintroduction of the giant Canada goose (*Branta canadensis maxima*), once presumed to be an extinct race (Klimstra & Thornburg, 1982). Beginning in 1967, the Illinois Department of Conservation, in cooperation with Peabody Coal Company, reintroduced this subspecies of goose on pre-law surface-mine areas in west-central Illinois. Initiated with 425 geese, re-establishment has been so successful that those areas are now open to hunting. In 1981, the reintroduction program was expanded to include surface-mine areas in southern Illinois. More than 1500 giant geese have been released in southern Illinois in the past 5 years, with reproduction accounting for more than 200 goslings each year (Butler & Lueth, 1986).

Recent reintroduction programs have successfully established resident flocks of giant Canada geese on surface-mines in both Kentucky and Indiana. At one mine in Indiana more than 300 geese regularly utilize exposed deltas of an active slurry disposal impoundment for loafing during summer moulting.

While construction of islands or artificial nesting structures is now promoted as a management tool, islands in pre-law areas were a consequence of little or no grading of spoilbanks. To enhance nesting success for recently established resident goose flocks, mine operators

Table 11.1. *Wetland-associated avian species utilizing coal surface-mined wetlands in Illinois*[a]

Order	Species	Northwest Illinois	Southern Illinois
Gaviiformes			
	common loon (*Gavia immer*)	R[b]	
	red-throated loon (*G. stellata*)		X
Podicipediformes			
	horned grebe (*Podiceps auritus*)	R	X
	pied-billed grebe (*Podilymbus podiceps*)	C	X
Pelecaniformes			
	American white pelican (*Pelecanus erythrorhynchos*)		X
	double-crested cormorant (*Phalacrocorax auritus*)	X	X
Anseriformes			
	mute swan (*Cygnus olor*)	UB	
	Canada goose (*Branta canadensis*)	CB	X
	greater white-fronted goose (*Anser albifrons*)	U	
	snow goose (*Chen caerulescens*)	U	X
	mallard (*Anas platyrhynchos*)	CB	XB
	American black duck (*A. rubripes*)	U, RB	X
	northern pintail (*A. acuta*)	C	X
	gadwall (*A. strepera*)	C	X
	American wigeon (*A. americana*)	C	X
	northern shoveler (*A. clypeata*)	C	X
	blue-winged teal (*A. discors*)	UB	XB
	green-winged teal (*A. crecca*)	U	X
	wood duck (*Aix sponsa*)	UB	XB
	scaup (*Aythya* spp.)	C	X
	redhead (*A. americana*)	U	X
	canvasback (*A. valisineria*)	U	X
	ring-necked duck (*A. collaris*)	C	X
	common goldeneye (*Bucephala clangula*)	U	X
	bufflehead (*B. albeola*)	C	X
	hooded merganser (*Lophodytes cucullatus*)	RB	X
	common merganser (*Mergus merganser*)	U	X
	ruddy duck (*Oxyura jamaicensis*)	U	X
Ciconiiformes			
	great egret (*Casmerodius albus*)	X	X
	cattle egret (*Bubulcus ibis*)	X	X
	great blue heron (*Ardea herodias*)	X	X
	little blue heron (*Egretta caerulea*)		X
	green-backed heron (*Butorides striatus*)	X	XB

(*continued*)

Table 11.1. (*cont.*)

Order	Species	Northwest Illinois	Southern Illinois
	black-crowned night heron (*Nycticorax nycticorax*)	X	
	American bittern (*Botaurus lentiginosus*)	X	X
	least bittern (*Ixobrychus exilis*)		X
Gruiformes			
	Virginia rail (*Rallas limicola*)	X	X
	king rail (*R. elegans*)	X	
	sora (*Porzana carolina*)	X	X
	common moorhen (*Gallinula chloropus*)	R	X
	American coot (*Fulica americana*)	C,RB	X
Charadriiformes			
	killdeer (*Charadrius vociferus*)	C,RB	XB
	greater yellowlegs (*Tringa melanoleuca*)		X
	lesser yellowlegs (*T. flavipes*)		X
	solitary sandpiper (*T. solitaria*)		X
	upland sandpiper (*Bartramia longicauda*)	X	
	willet (*Cataptrophorus semipalmatus*)		X
	spotted sandpiper (*Actitis macularia*)	X	X
	short-billed dowitcher (*Limnodromus griseus*)	X	X
	common snipe (*Gallinago gallinago*)	X	X
	ruddy turnstone (*Arenaria interpres*)	X	
	pectoral sandpiper (*Calidris melanotos*)	X	X
	least sandpiper (*C. minutilla*)	X	
	dunlin (*C. alpina*)	X	
	herring gull (*Larus argentatus*)		X
	ring-billed gull (*L. delawarensis*)	X	X
	caspian tern (*Sterna caspia*)	X	
	black tern (*Chlidonias niger*)	X	
Coraciiformes			
	belted kingfisher (*Ceryle alcyon*)	X	XB

Notes: [a] From records of the Cooperative Wildlife Research Laboratory of Southern Illinois University 1956–1985.
[b] C = common; U = uncommon; R = rare; X = observed on mined land; B = breeding record.

have included island construction in reclamation plans of final cuts in both Kentucky (Krause, Mahan & Harding, 1985) and Illinois (Klimstra & Nawrot, 1985).

Stricter slope grading requirements of post-law reclamation have greatly reduced or eliminated the formation of islands. However, during the regulatory transition, grading to approximately the original contour contributed to large tracts of rolling pastures interspersed with shallow water wetlands. Although islands were nearly eliminated, the moist soil, diverse aquatic plant communities, and the adjacent graded spoilbank pastures are now important factors in providing winter habitat for migratory waterfowl.

In addition to more than 11000 migratory geese wintering on southern Illinois surface-mines, seasonal migration patterns of ducks and shorebirds have been favourably affected by the abundance and distribution of mined land wetlands. While most waterfowl utilization (excluding the giant Canada goose) of mined land wetland is migratory in nature, the mallard (*Anas platyrhynchos*) and blue-winged teal (*Anas discors*) are the most common breeding species (Sandusky, 1978; Perkins & Lawrence, 1985). Some 59 avian species, representing 8 orders, were recorded as either breeding on, or associated with, mined wetlands in Illinois (Table 11.1). The osprey (*Pandion haliaetus*) has been observed on northern Illinois surface-mine wetlands and the bald eagle (*Haliacetus leucocephalus*) has been observed on both northern and southern mine areas.

Although much emphasis has been placed on surface-mine wetlands as waterfowl habitat, research in the 1950s and '60s examined aspects of habitat development such as water quality (Lewis & Peters, 1954), plant diversity (Brewer & Triner, 1956; Bell, 1956), herpetofauna (Myers & Klimstra, 1963), and aquatic furbearers (Arata, 1959). Wetland habitat enhancement practices were initiated at the Pyatts research area in southern Illinois in the mid 1950s with experimental plantings of moist soil and emergent wetland plants such as smartweed (*Polygonum muhlenbergii*), bulrush (*Scirpus acutus*), and arrowhead (*Sagittaria latifolia* and *S. rigida*) (Arata, 1959). Although the 1950s wildlife rehabilitation program at the Pyatts research area represented some of the earliest wetland plantings on mined lands, it would be decades later before wetland plant establishment would again become a popular reclamation alternative.

During the 1950s through to the late 1970s rehabilitation did not emphasize establishment of wetland vegetation. Moist soil and wetland plant establishment was primarily dependent upon natural colonization in areas with favourable conditions. In fact, during this period the increased stringency of reclamation requirements eventually precluded most wetland habitat establishment except for the retention of deep final cut and haulage road impoundments. With reduced littoral zone development in final cuts, volunteer moist soil and aquatic plant communities are typically limited (Coss, Nawrot & Klimstra, 1985). In contrast, shallow water impoundments associated with graded spoilbank areas yielded larger pond sizes, greater shore lengths, more gradual slopes and wider littoral and moist soil zones with diverse aquatic plant communities (O'Leary *et al.*, 1984).

The high quality emergent and aquatic plant communities of many pre-law shallow impoundments were not those which sceptics would expect to be associated with an industry which tends to produce acid soils and runoff. In fact, characterization of water quality of unplanned pre-law aquatic areas, indicated alkaline conditions with a pH range of 6.2 to 8.3 for most Illinois mine impoundments (Konik, 1980). A biological and chemical investigation of 33 pre-federal law mine impoundments in Illinois and Missouri (Jones, McElligott & Mannz, 1985) further documented the alkalinity of surface-mine waters. Zooplankton and phytoplankton of these impoundments were also similar to impoundments of unmined areas.

Post-law wetland habitat development opportunities

Prior to federal surface-mining regulations (SMCRA P.L. 95–87), wetland/wildlife areas were a common post-mining land-use. However, as advocates of agricultural post-mining land-uses dominated the regulatory process, opportunities for wetland habitat development were severely restricted. Emphasis on agricultural rehabilitation only exacerbates the 161 880 ha annual loss of wetlands converted to cropland (Wildlife Management Institute, 1986). Consequently, every opportunity must be capitalized on to develop wetland habitat in association with any, or all, post-mining land-uses.

While current regulations generally restrict wetland habitat development, permissible opportunities in accordance with SMCRA P.L. 95–87 do exist to establish and enhance wetlands. These opportunities include areas such as final cut and incline lakes, areas between spoil impoundments, sediment ponds, riparian areas associated with stream restoration and water-filled subsidence areas.

Final cut/incline impoundment

Pre-law final cut/incline lakes tended to be diverse in structure as a result of minimal grading standards; however, strict adherence to the requirements of SMRCA P.L. 95–87 does not allow innovativeness or flexibility in wetland habitat development. The diversity and value of

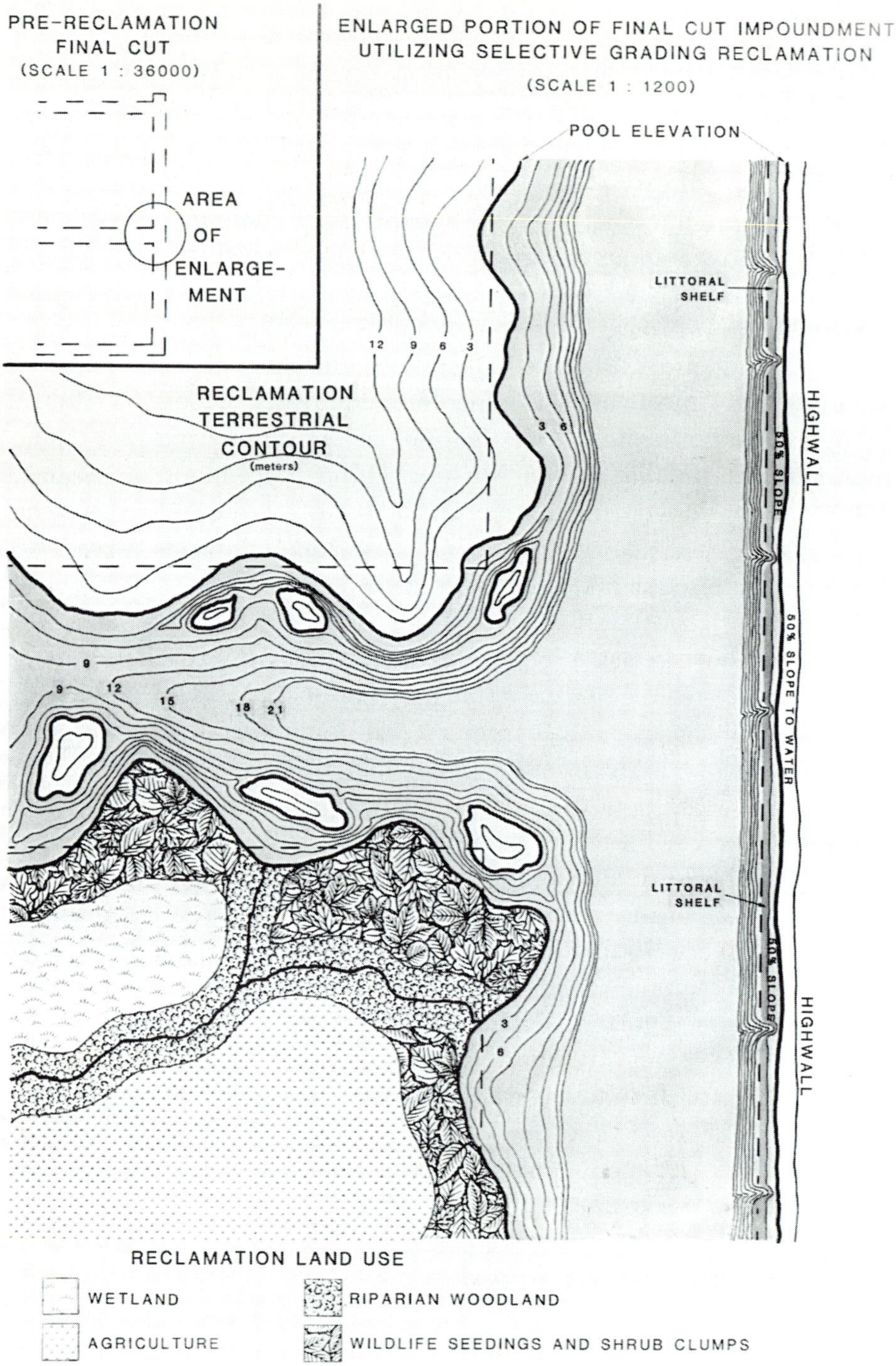

Fig. 11.1. A scheme for enhancement of an incline/final-cut lake for fish and wildlife, emphasizing wetland and terrestrial habitat diversity (from Klimstra & Nawrot, 1985).

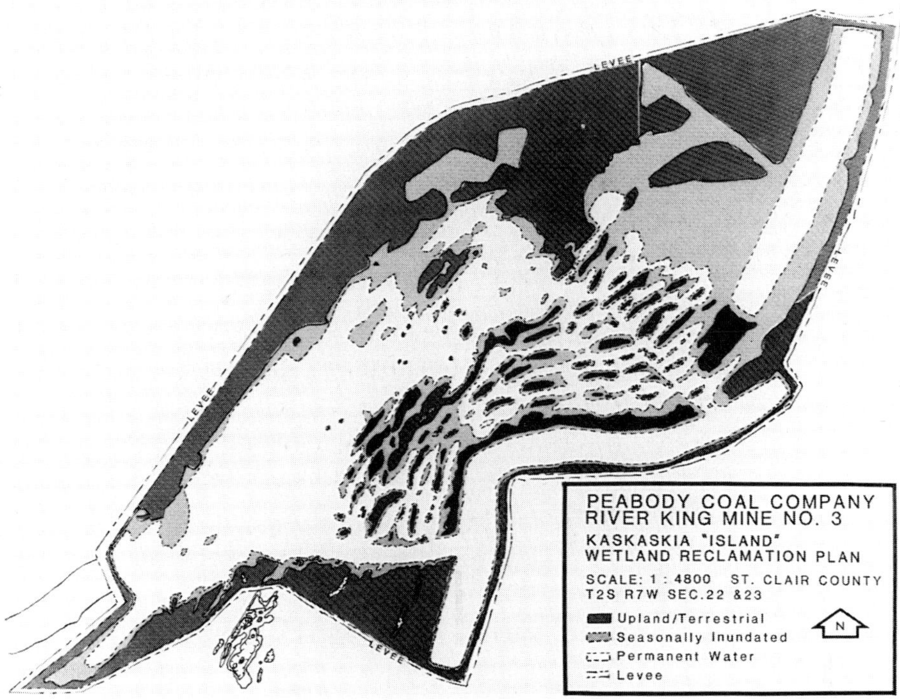

Fig. 11.2. Scheme for the planned rehabilitation of a unique floodplain affected by river channelization, emphasizing aquatic habitat development through water management and selective grading (from Klimstra & Nawrot, 1985).

final cut/incline impoundment wetland habitat can be greatly enhanced through selective grading practices (Fig. 11.1). Construction of benches along the highwall side can increase littoral zone habitat for establishment of submerged and emergent wetland plants. Selective grading along the final cut spoil side and incline can be beneficial for the creation of islands and small peninsulas interspersed with isolated shallow water areas. Planning and implementation of these rehabilitation practices concurrent with active mining provides a cost-effective means for enhancing wetland habit. Equipment associated with coal extraction (i.e. dragline, shovel, bucket/wheel excavator) can efficiently strike-off and grade spoilbanks and construct littoral shelves as a final phase of the active mining operation.

When feasible, significant areas of additional shallow water wetlands associated with between-spoil impoundments can be constructed by planning water level management practices to include flooding of graded spoil areas contiguous with a final cut/incline impoundment (Fig. 11.2).

Water level management facilitates moist soil and wetland plant establishment as well as providing an economical tool for increasing both the area and diversity of waterfowl and shorebird habitat. Sequential drawdowns and reflooding in spring through to autumn can optimize the availability of aquatic wetland foods during autumn and spring bird migrations, or provide a dependable year-long food source for resident species.

Riparian wetlands

When mining through a stream is necessary, the reclamation process can result in a significant contribution to riparian wetland habitat if planned in advance (Robb & Ohlsson, 1982; Gore, 1985). Required reclamation practices for the stream channel include construction of meanders, riffles and pools, and vegetative stabilization of the streambank. Monitoring of recently reclaimed streams suggests fauna and flora of the aquatic community can be rapidly re-established (Bonace, 1983; Thompson, 1985).

Greater opportunity for wetland habitat development exists beyond the confines of the stream channel. Naturally occurring ephemeral and permanent wetland pools associated with the pre-mine floodplain can, and should, be reclaimed to maintain habitat diversity in the rehabilitated areas. During the planning and construction of the stream restoration channel, seasonal overflow wetlands can be included in the reclaimed floodplain (Fig. 11.3). Associated with forested floodplains, these reclaimed wetlands can permanently replace those normally lost due to agricultural drainage in unmined areas.

Sediment ponds

Maintaining water quality in active mining areas requires the construction of sediment ponds. Sediment ponds may be associated with haul roads, coal processing areas, the active pit, and also with topsoil removal and replacement activities. Consequently, there may be many sediment ponds at one mine site. In addition, ponds may be located throughout the mining area. The regulatory usefulness of a sediment pond ends when the immediate watershed is reclaimed and vegetated. However, almost any pond can be a contribution to wetland habitat if it is allowed to remain when mining activities are terminated. Although a sediment pond may eventually become silted in, wetland wildlife can benefit by the successional transition from submerged and emergent wetland plants to moist soil species.

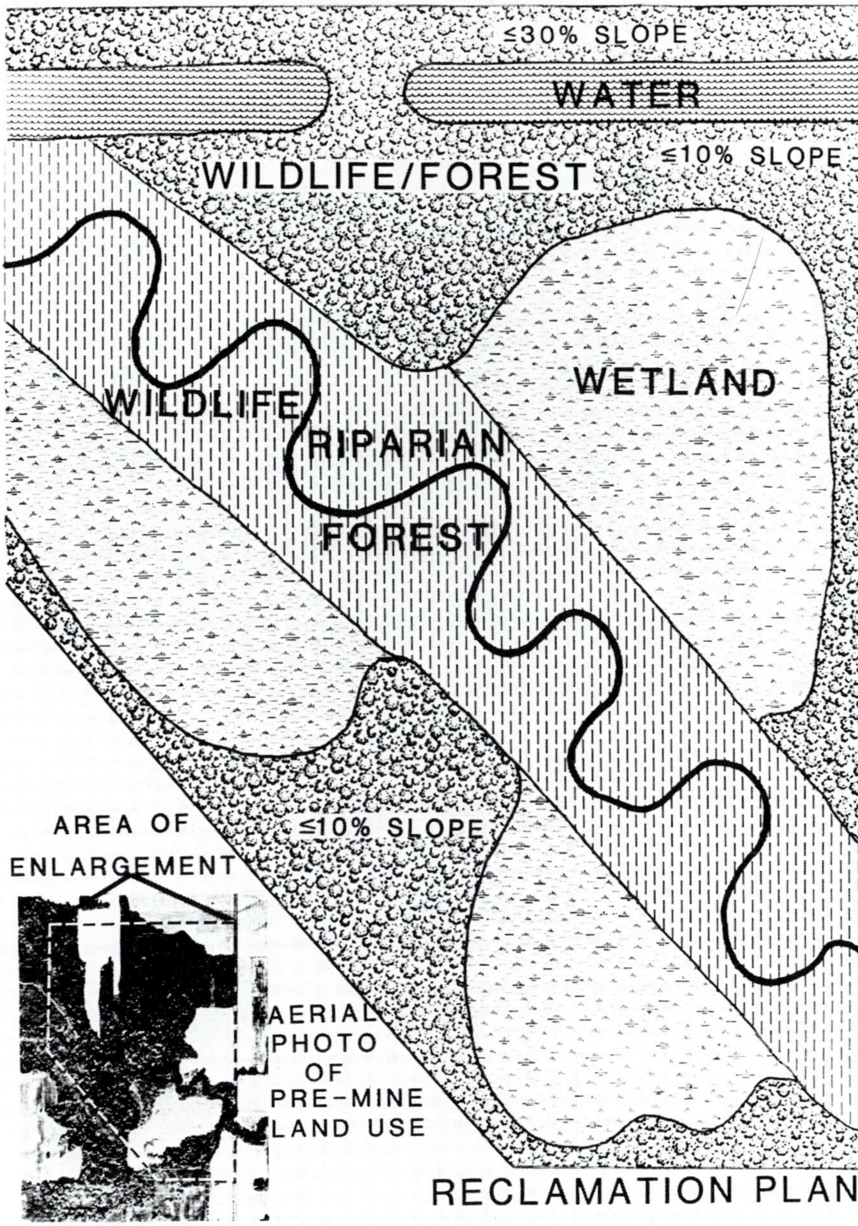

Fig. 11.3. Scheme of planned rehabilitation for stream restoration and wetland enhancement at the Consolidation Coal Burning Star No. 4 Mine in southern Illinois (from Klimstra & Nawrot, 1985).

Fig. 11.4. Slurry pond rehabilitated as a wetland through direct establishment of vegetation without soil cover.

Varying in size, depth and location, sediment ponds present unique opportunities when planned as wetland habitats (Slaski & Fowler, 1980; Fowler, Hill & Fowler, 1985). Recently constructed sediment ponds in southeastern US surface-mine areas were rapidly colonized by aquatic vegetation, aquatic insects, and breeding amphibians (Fowler *et al.*, 1985). Rather than draining and removing these structures, stocking with game and forage fishes has been recommended as a means to capitalize on a valuable existing biological and public resource (Fowler *et al.*, 1985). Although each pond may be less than 1 ha, many ponds can collectively contribute desirable islands of wetland habitat within extensive surface-mine areas.

Slurry ponds

Although the previously discussed types of mine impoundments may not have been fully appreciated as a wetland resource until recently, slurry impoundments have always been, and generally still are, viewed by regulatory agencies as potentially acid-producing coal processing waste disposal areas. Consequently, 1.2 m of soil cover is required for rehabilitation. Ironically, many slurry areas have been identified where wetland habitats have become naturally established on inactive disposal

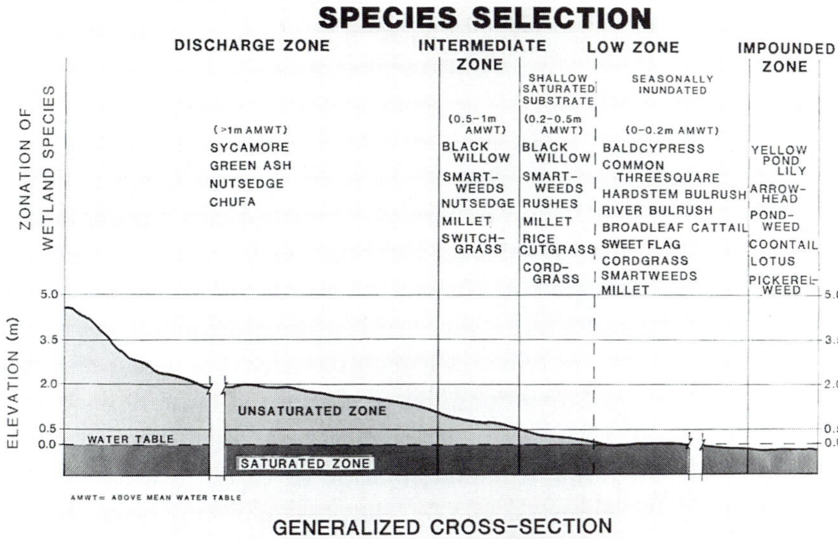

Fig. 11.5. Substrate zonation and adapted wetland species for proposed rehabilitation of a typical single-point discharge slurry impoundment (from Warburton, Klimstra & Nawrot, 1985).

sites (Nawrot, 1981). The Cooperative Wildlife Research Laboratory has demonstrated that, with proper management, slurry disposal areas represent unique opportunities for wetland habitat development (Nawrot & Warburton, 1986) (Fig. 11.4). The success of slurry wetland habitat development depends upon a complete understanding of site-specific physical, chemical and hydrological conditions (Nawrot, 1986) and plant establishment techniques (Warburton, Klimstra & Nawrot, 1985) (Fig. 11.5). Wetland habitat development on slurry impoundments resolves a costly reclamation problem while providing quality shallow water wetlands that are utilized by migratory and resident waterfowl and shorebirds.

Subsidence

Similar to slurry impoundments, subsidence-affected areas are viewed only as a problem requiring reclamation (e.g. impeded drainage of agricultural lands, damage to structures, etc.). However, examples exist in rural areas where past subsidence has contributed to numerous potholes in forested lowlands and extensive shallow wetlands in floodplains. Current mining permit requirements identify the need for a subsidence control plan. While such a plan implies prevention of subsidence, controlled subsidence, as is often the intent of longwall mining, can provide the opportunity to develop diverse shallow water wetland habitat. This is particularly relevant in low-lying rural areas where dwellings are not present and agricultural productivity is normally impaired by frequent flooding.

Summary

Wetland habitat development on mined lands resulting from 'unplanned reclamation' associated with the pre-law era of mining can no longer be depended upon as a means to enhance habitat diversity. The post-law era has brought more stringent reclamation requirements and an obsession to accommodate agricultural rehabilitation interests. Consequently, every opportunity must be capitalized upon to ensure fish and wildlife habitat enhancement is included as a part of every reclamation plan, including the rehabilitation of abandoned mined lands (Nawrot, Woolf & Klimstra, 1982). The loss of natural wetlands has prompted mineral industries throughout the World to include wetlands in reclamation plans. Wetland development of inactive gravel pits for enhancement of breeding bird populations is considered a valuable conservation practice (Catchpole & Tydeman, 1975). Investigations of the value of gravel pits as waterfowl habitat have been pursued since 1972 (Kelcey, 1984). Island development, and establishment of vegetation diversity to enhance waterfowl feeding, resting, and breeding areas were important considerations in British gravel pit reclamation (Street, 1982). Increase of organic matter input in new gravel pit wetlands was identified as a limiting factor affecting the establishment of a macro-invertebrate food source necessary for survival of waterfowl broods (Street, 1982). When properly planned and implemented, wetland habitat reclamation can produce a viable natural resource as a consequence of extracting a valuable energy resource.

References

Arata, A. A. (1959). Ecology of muskrats in strip-mine ponds in southern Illinois. *Journal of Wildlife Management*, **23**, 177–86.

Bell, R. (1956). Aquatic and marginal vegetation of strip-mine waters in southern Illinois. *Transactions Illinois State Academy of Science*, **48**, 85–91.

Bonace, R. T. (1983). *Selected Biological, Physical and Chemical Aspects of Stream Diversions: A Consequence of Surface Mining*. M. A. Thesis, Carbondale: Southern Illinois University.

Brewer, R. E. & Triner, E. D. (1956). Vegetational features of some strip-mined land in Perry County, Illinois. *Transactions Illinois State Academy of Science*, **48**, 73–84.

Butler, R. A. & Lueth, B. K. (1986). Reestablishment of the giant Canada goose on surface mine wetlands. In *Innovative Approaches to Mined Land Reclamation*, ed. C. Carlson & J. H. Swisher, pp. 455–69. Carbondale: Southern Illinois University Press.

Catchpole, C. V. & Tydeman, C. (1975). Gravel pits as new wetland habitats for the conservation of breeding bird communities. *Biological Conservation*, **8**, 47–59.

Christy, P. L., Smith, W. E. & Filer, E. E. (1979). New land uses created by surface mining. *American Mining Congress Journal*, **65**, 40–5.

Coss, R. D. (1981). *Wildlife Habitats Provided by Aquatic Plant Communities of Surface Mine Lakes.* M. A. Thesis, Carbondale: Southern Illinois University.

Coss, R. D., Nawrot, J. R. & Klimstra, W. D. (1985). Wildlife habitats provided by aquatic plant communities of surface mine lakes. In *Surface Mining Hydrology, Sedimentology and Reclamation*, ed. D. H. Graves, pp. 29–39. Lexington: University of Kentucky.

Day, E. A. (1986). *Mitigation for the South Madison Beltline Dane County, Wisconsin.* Wisconsin Department of Transportation.

Deane, J. A. (1966). How strip mining improves midwest water supplies. *Coal Age*, **71**, 66–8.

Dickinson, S. (1975). *Revegetation of Taconite Tailing.* Mineral Waste Stabilization Liaison Committee.

Fowler, D. K., Hill, D. M. & Fowler, L. J. (1985). Colonization of coal surface mine sediment ponds in southern Appalachia by aquatic organisms and breeding amphibians. In *Wetlands and Water Management on Mined Lands*, ed. R. P. Brooks, D. E. Samuel & J. B. Hill, pp. 261–85. University Park: Pennsylvania State University.

Frayer, W. E., Monahan, T. J., Bowden, D. C. & Graybill, F. A. (1983). *Status and Trends of Wetlands and Deepwater Habitats in the Conterminous United States, 1950's to 1970's.* Department of Forest and Wood Sciences. Fort Collins: Colorado State University.

Gibb, J. P. & Evans, R. L. (1978). *Reconnaissance Study of Final Cut Impoundments.* Illinois Institute of Environmental Quality, 78/25.

Gore, J. A. ed. (1985). *The Restoration of Rivers and Streams.* Stoneham: Butterworth.

Grandt, A. F. (1978). Mined-land reclamation in the Interior Coal Province. *Journal of Soil and Water Conservation*, **33**, 62–8.

Grandt, A. F. (1981). Permanent water impoundments. In *Office of Surface Mining Workshop.* St Louis, Missouri.

Haynes, R. J. (1984). Summary of wetlands reestablishment on surface-mined lands in Florida. In *Surface Mining, Hydrology, Sedimentology and Reclamation*, ed. D. H. Graves, pp. 357–62. Lexington: University of Kentucky.

Haynes, R. J. & Klimstra, W. D. (1975). *Illinois Lands Surface Mined for Coal.* Cooperative Wildlife Research Laboratory. Carbondale: Southern Illinois University.

Imhoff, E. A., Friz, T. O. & LaFevers, J. R. (1976). *A Guide to State Programs for the Reclamation of Surface Mined Areas.* U.S. Geological Survey Circular, 731.

Janiak, H. (1979). *Purification of Waters Discharged from Polish Lignite Mines.* U.S. Environmental Protection Agency, Central Research and Design Institute for Open Pit Mining. EPA-600/779-099.

Jones, D. W., McElligott, M. J. & Mannz, R. H. (1985). *Biological, Chemical and Morphological Characterization of 33 Surface Mine Lakes in Illinois and Missouri.* Peabody Coal Company.

Josselyn, M. & Buchholz, J. (1982). Summary of past wetland restoration projects in California. In *Wetland Restoration and Enhancement in California*, ed. M. Josselyn, pp. 1–10. California Sea Grant College Program, Institute of Marine Resources Report No. T-CSGCP-007. La Jolla: University of California.

Kabay, E. D. & Nichols, O. G. (1981). *Formation of Wetlands as a Possible Rehabilitation Option of Open Cut Mining in the Southwest of Western Australia.* Alcoa of Australia Ltd Environmental Research Bulletin, No. 10.

Kelcey, J. G. (1984). The design and development of gravel pits for wildlife in Milton Keynes, England. *Landscape Planning*, **11**, 19–34.

Klimstra, W. D., Haynes, R. J. & Filer, E. E. (1977). Résumé of the utilization of nearly 200,000 acres of surface-mined land in Illinois. In *Proceedings of the 32nd Annual Meeting of the Soil Conservation Society of America*, pp. 135–9.

Klimstra, W. D. & Nawrot, J. R. (1982). Water as a reclamation alternative: an assessment of values. In *Surface Mining, Hydrology, Sedimentology and Reclamation*, ed. D. H. Graves, pp. 37–44. Lexington: University of Kentucky.

Klimstra, W. D. & Nawrot, J. R. (1985). Wetlands as a byproduct of surface mining: Midwest perspective. In *Wetlands and Water Management on Mined Lands*, ed. R. P. Brooks, D. E. Samuel and J. B. Hill, pp. 107–19. University Park: Pennsylvania State University.

Klimstra, W. D. & Thornburg, D. (1982). Mined land wetlands as breeding grounds for the giant Canada goose. In *Surface Mining, Hydrology, Sedimentology and Reclamation*, ed. D. H. Graves, pp. 359–63. Lexington: University of Kentucky.

Klimstra, W. D., Vohs, Jr. P. A. & Cherry, J. D. (1963). Strip-mined lands for recreation. *Illinois Wildlife*, **18**, 6–8.

Konik, J. (1980). *Some Physical, Chemical, and Biological Characteristics of Non-problem Waters Occurring on Lands Surface-mined for Coal.* Illinois Institute of Natural Resources Doc. 80/14.

Krause, J., Mahan, B. & Harding, M. (1985). Waterfowl management and wetland development on Peabody Coal Company surface mines in Illinois, Indiana and Kentucky. Abstract only. In *Wetlands and Water Management on Mined Lands*, ed. R. P. Brooks, D. E. Samuel & J. B. Hill, pp. 327. University Park: Pennsylvania State University.

Lewis, W. M. & Peters, C. (1954). Physico-chemical characteristics of ponds in the Pyatt, Desoto and Elkville strip mined areas of southern Illinois. *Transactions of the American Fisheries Society*, **84**, 117–24.

Myers, C. W. & Klimstra, W. D. (1963). Amphibians and reptiles of an ecologically disturbed (strip-mined) area in southern Illinois. *American Midland Naturalist*, **70**, 126–32.

Nawrot, J. R. (1981). Stabilization of slurry impoundments without soil cover: factors affecting vegetation establishment. In *Surface Mining Hydrology, Sedimentology and Reclamation*, ed. D. H. Graves, pp. 469–76. Lexington: University of Kentucky.

Nawrot, J. R. (1986). Slurry reclamation: soil cover alternatives. In *American Mining Congress Coal Convention*. pp. 6. Pittsburgh, Pennsylvania.

Nawrot, J. R. & Warburton, D. B. (1986). Wetland habitat development techniques for coal mine slurry impoundments. In *National Mined Land Reclamation*, ed. J. H. Swisher & C. Carlson, pp. 586–612. St Louis, Missouri.

Nawrot, J. R., Woolf, A. & Klimstra, W. D. (1982). *A Guide for Enhancement of Fish and Wildlife on Abandoned Mined Lands in the Eastern United States*. U.S. Fish and Wildlife Service, Eastern Energy Land Use Team. FWS/OBS-80/67.

Newling, C. J. (1981). *Monitoring on Dredged Material Research Program Habitat Development Sites.* U.S. Army Corps of Engineers. Dredged Material Research, Volume D-81-1, 1–8.

O'Leary, W. G., Klimstra, W. D. & Nawrot, J. R. (1984). Waterfowl habitats on reclaimed surface-mined lands in southwestern Illinois. In *Surface Mining Hydrology, Sedimentology, and Reclamation*, ed. D. H. Graves, pp. 377–82. Lexington: University of Kentucky.

Perkins, G. A. (1981). *Status of Giant Canada Goose Nesting in West-Central Illinois*. M. A. Thesis. Carbondale: Southern Illinois University.

Perkins, G. A. & Lawrence, J. S. (1985). Bird use of wetlands created by surface mining. *Transactions Illinois Academy of Science*, **78**, 87–95.

Riley, C. V. (1960). The ecology of water areas associated with coal strip-mined lands in Ohio. *Ohio Journal of Science*, **60**, 106–21.

Robb, Jr. A. E. & Ohlsson, K. E. (1982). *Mine-Related Stream Alteration Practices for Protection and Enhancement of Fish and Wildlife*. U.S. Fish and Wildlife Service, Division of Biological Services FWS/OBS82/57.

Roseberry, J. L. & Klimstra, W. D. (1964). Recreational activities on Illinois strip-mined lands. *Journal of Soil and Water Conservation*, **19**, 107–10.

Rosso, W. A. (1980). Reclaim your land with lakes. *Coal Age*, **85**, 76–7.

Sandusky, J. E. (1978). The potential for management of waterfowl nesting habitat on reclaimed mined land. In *Surface Mining and Fish/Wildlife Needs in the Eastern United States*, ed. D. E. Samuel, J. R. Stauffer, C. H. Hocult & W. T. Mason Jr., pp. 325–7. U.S. Fish and Wildlife Service FWS-OBS-78/81.

Slaski, L. J. & Fowler, D. K. (1980). Reclamation of surface mines for wildlife habitat: a case study. In *Surface Mining Hydrology, Sedimentology, and Reclamation*, ed. D. H. Graves, pp. 431–6. Lexington: University of Kentucky.

Street, M. (1982). Management of artificial wetlands for waterfowl – a novel technique to increase the availability of invertebrate food items. *Transactions of the International Congress of Game Biology*, **14**, 159–74.

Street, M. (1985). *The Restoration of Gravel Pits for Wildfowl*. Milton Keynes: A.R.C. Wildlife Centre.

Svedarsky, W. D. & Crawford, R. D. ed. (1982). *Wildlife Values of Gravel Pits*. Agricultural Experiment Station Miscellaneous Publication, 17. St Paul: University of Minnesota.

Thompson, C. S. (1985). Stream relocations under O.S.M. regulations: benefits to fish and invertebrates. In *Wetlands and Water Management on Mined Lands*, ed. R. P. Brooks, D. E. Samuel & J. B. Hill, pp. 305–18. University Park: Pennsylvania State University.

Thompson, D. A. & Lindahll, D. M. (1977). *State Surface Mining Laws: a Survey, a Comparison with the Proposed Federal Legislation and Background Information*. U.S. Government Printing Office Publication 95–25.

Tiner, Jr. R. W. (1984). *Wetlands of the United States: Current Status and Recent Trends*. U.S. Department of the Interior, Fish and Wildlife Service, National Wetlands Inventory.

Warburton, D. B., Klimstra, W. B. & Nawrot, J. R. (1985). Aquatic macrophyte propagation and planting practices for wetland establishment. In *Wetlands and Water Management on Mined Lands*, ed. R. P. Brooks, D. E. Samuel & J. B. Hill, pp. 139–52. University Park: Pennsylvania State University.

Weber, L. S. (1970). The development and use of coal surface-mined lands in Illinois. Illinois State Geological Survey. *Environmental Geology Notes*, **46**, 41–6.

Wildlife Management Institute. (1986). Swampbuster can save taxpayers money. *Outdoor News Bulletin*, **40**, 2–3.

12

The reclamation of lands for outdoor recreation

DAVID E. SAMUEL

Introduction

Surface mining for minerals has devastated thousands of hectares of land throughout the World. However many countries have established new laws which call for reclamation that makes such lands more suitable for many recreational uses (see Chapter 1). There is no better example of this than with the surface mining for coal in the United States. In the past, surface mining caused considerable disturbance to lands, even after rehabilitation had been attempted. Today, reclamation leads to beneficial uses of the land and allows many post-mining recreational applications. These include: hunting, fishing, trapping, hiking, bird watching, nature study, picnicking, riding, and other activities (see Table 1.2).

Much of the following will refer to the surface-mine industry in the United States, especially as it relates to the use of rehabilitated lands by wildlife and fish. The comments, however, apply in general to other regions of the World and to other types of land disturbance. Although some rehabilitation has been carried out to directly benefit fish and wildlife, most has been aimed at promoting good ground cover in order to prevent erosion. In so doing, it has benefited various animal species and the recreational pursuits that are associated with wildlife and fish. I do not have space to discuss all the techniques used to rehabilitate lands for wildlife and fish, and will refer to Chapter 8 and the various handbooks that contain detailed accounts of such techniques. However, one 2-volume set provides the background to many aspects of this problem including: environmental effects; fish and wildlife responses to habitat alteration; and also planning and management activities for fish and wildlife (Leedy, 1981). Other relevant publications include Rafaill &

Vogel (1978), Samuel *et al.* (1978), Fowler & Adkisson (1980), Fowler, Turner & Slaski (1981), Ohlsson *et al.* (1982), and Robb & Ohlsson (1982). Most of these references relate to mining in the eastern United States. There have also been several manuals and symposia aimed at wildlife management on surface-mines in the western United States (see Ecological Consultants, Inc., 1977; Utah State University, 1978; U.S. Department of Interior, 1980; U.S. Department of Agriculture, 1982). Indeed, mining companies in the western United States place considerable emphasis on wildlife (e.g. 86% of western mines which were surveyed had declared wildlife to be a major reclamation goal on at least part of their lease (Scott & Zimmerman, 1984)).

Even though the following will focus on surface mining for coal, other types of mining and subsequent rehabilitation have been accomplished for fish and wildlife. One excellent reference for uses and rehabilitation of gravel pits for fish and wildlife is Svedarsky & Crawford (1982). Wildlife work has also been conducted on limestone quarries (Humphries, 1980), dredged materials (Coastal Zone Resources Division, 1978), bauxite mining (Kabay & Nichols, 1980), tin mines (Holliday, Johnson & Bradshaw, 1979), phosphate mines (Marion & O'Meara, 1983), fluorspar mines (Johnson, 1980), and iron ore mines (Stormer, Chilcote & Shetron, 1978).

Reclamation for fishing

Fishing is a popular pastime in the United States with over 25% of the population involved in this activity (U.S. Department of Interior, & U.S. Department of Commerce, 1982). Reclamation of disturbed land to include fishing can greatly increase its value not only as a public recreation resource, but as a conservation unit.

One of the most common products of surface mining are sediment ponds, which in turn are potential fishing areas. Such ponds may also provide areas which are useful for camping, picnicking or birdwatching. Ponds on mines are created in several ways. Surface mining may create a highwall, with a pond forming at its base. This was especially true in the United States for mining carried out in the 1960s and early 1970s when laws allowed highwalls to remain (see Chapter 11). Today, most fishing opportunities on mined lands occur in sediment ponds. Sediment ponds can provide productive fisheries if water quality is good. Indeed, Gregory, Elser & Lenhart (1984) describe the construction of a northern pike spawning march from a surface-mine and Lewis & Nickum (1964) reared trout in water from the hypolimnion of a surface mine pond.

Some surface mine ponds are characterized by acidity and low fertility, while others exhibit good water quality (Brenner *et al.*, 1985). Even when chemical characteristics seem good, fish production often remains lower than natural non-mined areas (Peltz & Maughan, 1978). In recent years, the entire surface-mine reclamation process has led to much less acidity, due to the separation of acid forming materials during the mining process. This should lead to ponds of much better water quality, and increased potential for fish production.

Several handbooks provide details on construction, maintenance, stocking, and managing ponds for fishing on mined and unmined lands. These include Nelson, Horak & Olson (1978) and Nelson, Ornsborn & Logan (1982).

When large, relatively flat areas are mined, impoundments can be created for wildlife and fish between ridges of spoil (Leedy, 1981). Regardless of where an impoundment is created on the mine, soil should be clay, sediment pH should be at least 5.5, and exposed acid forming rock should be under the water level (Leedy, 1981). Such rock should be covered to prevent oxidation of materials in the coal seam (Riley, 1963; Davis, 1971). Before filling ponds, the bottom soil should be treated with lime to bring its pH up to 7.0. It may also be necessary to add fertilizer periodically to stimulate production of small food organisms. Leedy (1981) suggested that where possible, water level control devices should be installed, allowing drawdowns to remove stunted panfish such as bluegills (*Lepomis* spp.). Stunted panfish take over such ponds leaving poor quality fishing.

Ponds should be a minimum of 0.1 ha in area and be designed to include both shallow and deep water. These variations provide a greater range of niches than a pond of uniform depth and, as a result, encourage a more diverse range of aquatic invertebrates and plants used by fish and other wildlife (Whitehurst & Downing, 1977).

Landowner liability for drownings must be considered before creating or leaving such ponds. In many states in the United States, liability laws have been passed which protect the landowner, giving him or her an incentive for leaving ponds for recreation. For example, West Virginia has just passed such a law which will allow hundreds of sediment ponds to remain in place in order to provide fishing, swimming and waterfowl resting and feeding habitats, as well as habitat for commercially exploited furbearers, other mammals, and non-game birds.

Figure 12.1 (taken from Leedy, 1981, vol. 1, p. 51) depicts management measures which may be taken to enhance fish and recreational use of a

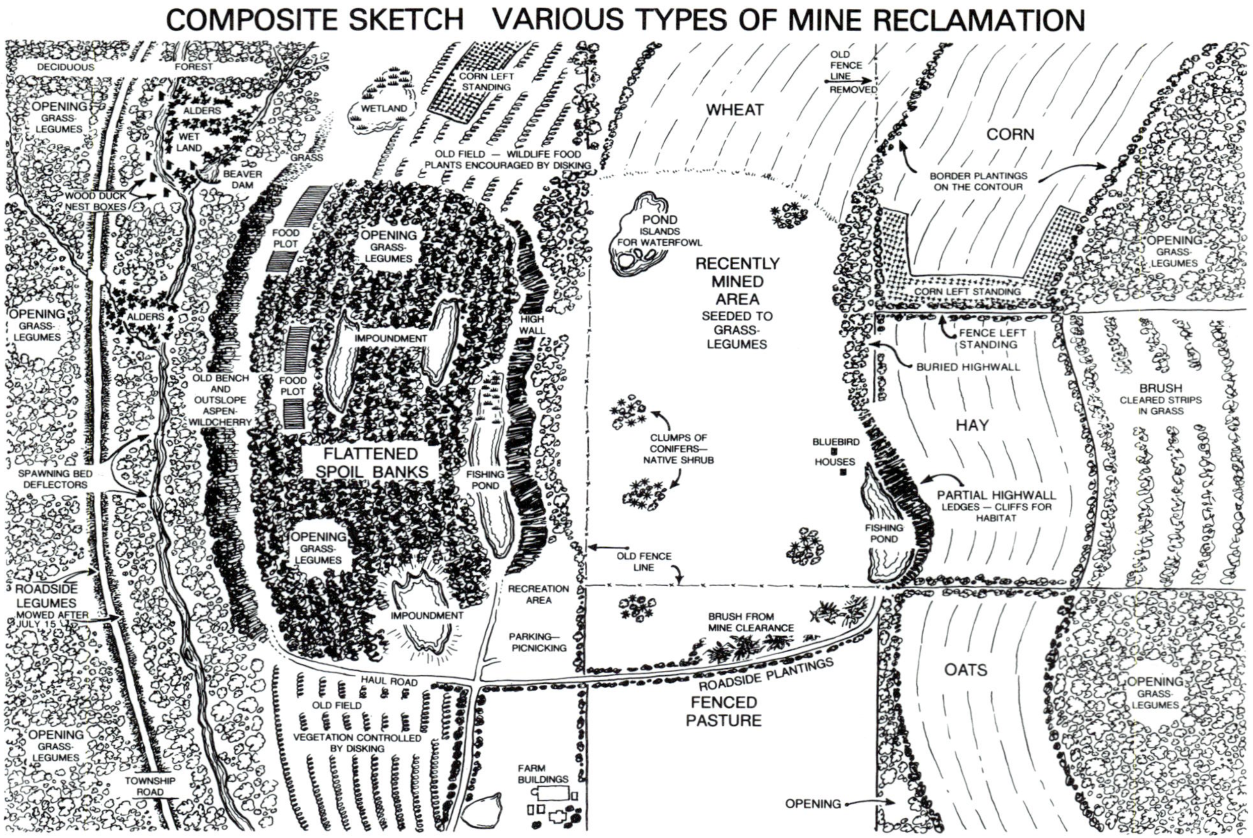

Fig. 12.1. Composite map showing some of the methods for surface mine reclamation which have been suggested for the eastern United States (from Leedy, 1981, vol. 1)

reclaimed area. For example, on an abandoned surface-mine site (left central portion of Fig. 12.1), the area around impoundments left by mining should be flattened to provide easy access to the ponds. Plantings around the ponds make such areas even more attractive to wildlife. The central portion of Fig. 12.1 shows a newly mined site, graded, and seeded to grasses and legumes. Construction of small islands of non-toxic spoil materials within the impoundment will provide nest sites for waterfowl and refuge areas for many terrestrial predators.

Small streams that are altered when land is disturbed can also be converted into valuable recreational fishing areas. For an excellent review of how to protect and enhance such streams before, during and after the reclamation process, the reader is referred to Robb & Ohlsson (1982). Further details on the importance of newly created wetlands, and methods of constructing them, are provided in Chapter 11.

Reclamation for hunting

Hunting provides tremendous economic benefits and rehabilitated lands often provide excellent wildlife habitat for game species. Over 10 % of Americans over 12 years of age hunt (16 million people). The economic benefits from this recreation are far reaching. For example, in Canada the 1981 expenditure by hunting participants was over $1.2 billion (Filion, Jacquemot & Reid, 1985). Each year, in the State of Pennsylvania, over one million deer hunters spend over $10 million on the purchase of hunting licenses. It is not possible to summarize all the wildlife studies that show how game species have benefited from modern rehabilitation, so a sample of studies will be presented.

There has been some question on whether highwalls restrict the movements of large mammals. Knotts & Samuel (1982) found that white-tailed deer (*Odocoileus virginianus*) were abundant around older rehabilitated surface-mines in the eastern United States, and they moved readily over the land bridges left to provide human access across highwalls. Many of these highwalls have been removed during the re-mining process that has occurred during the past 5 years in much of the eastern United States.

Kimmell & Samuel (1978*a*, 1984*a*, *b*) noted that ruffed grouse (*Bonasa umbellus*) chicks had higher insect feeding rates on 20-year-old mines planted to autumn olive (*Eleagnes umbellata*) than on 5-year-old mines rehabilitated with birdsfoot trefoil (*Lotus corniculatus*) and Kentucky 131 fescue (*Festuca arundinacea*). Kimmell & Samuel (1978*b*) and Brenner &

Fig. 12.2. The upper photograph shows a mature aspen tree (*Populus tremuloides*) growing on a surface mine in West Virginia while suckered clones which have resulted from cutting a mature tree are shown in the lower photograph.

Michalski (1984) noted that sites rehabilitated to grasses and legumes were not well utilized, but areas below the spoil area were often utilized by ruffed grouse. Postovit (1981) made recommendations for rehabilitating western surface coal mines for sage grouse (*Centrocercus urophasianus*). Aspen (*Populus tremuloides*) buds provide a major food source for ruffed grouse and Prachar & Samuel (1988) noted that aspen clones could be generated on rehabilitated surface coal mines by cutting mature aspen trees in mid-winter (Fig. 12.2). Such cutting led to as many as 100 new suckers and potential trees during the first growing season, providing habitat for not only grouse, but also many species of wildlife.

Anderson & Samuel (1980) found that surface mines rehabilitated to grasses and legumes, plus tree seedlings and saplings up to 3 m high, provided the best feeding and cover area for wild turkey (*Meleagris gallopavo*) chicks. Bobwhite quail (*Colinus virginianus*) chicks obtained more insects on mines rehabilitated to trefoil (*Lotus corniculatus*) and alfalfa (*Medicago sativa*) that were burnt during spring, as compared to unburned control sites (Brown & Samuel, 1978).

Feierabend (1983) noted that in autumn, migrating mourning doves (*Zenaidura macrura*) utilized coal mines which had been rehabilitated with annual grains, and indicated that land owners and coal industry personnel might consider leasing such lands to hunters during that season. Wheat, rye, oats and buckwheat in that order were the preferred source of food for the doves (refer to the top-center and lower right of Fig. 12.1 for a suggested plan for seeding these grains).

In the eastern United States, surface-mine reclamation benefits deer, wild turkeys, ruffed grouse, bobwhite quail and mourning doves as huntable species. These reclaimed mines are usually in private ownership but hunting there is under the jurisdiction of the state game agency and their laws and seasons.

Reclamation for trapping

Many Americans (1.6% of people over 12 years of age) rely upon trapping as a major source of income (Kellert, 1980). In West Virginia, fur pelts alone brought in more than $2.3 million during the 1978–79 trapping season (Samuel & Bammel, 1980); much of this trapping occurs in areas that have been surface mined. Various furbearers benefit from mine rehabilitation, but of special note are foxes. Red foxes (*Vulpes fulva*) and grey foxes (*Urocyon cineroargentus*) are commonly found on rehabilitated mines (Yearsley & Samuel, 1980), where they utilize insects and small mammals which are abundant there (De Capita & Bookhout, 1975). Muskrats (*Ondatra zibethica*) also take advantage of the wetlands and ponds created by surface-mines.

Reclamation for bird watching

Twenty-five percent of Americans over 12 years of age reported birdwatching as a recreation they have practiced during the previous 2-year period (Kellert, 1985). Even more impressive is the fact that 42% of these spent part, or all, of at least 20 days watching birds during the previous year. Thus, creating non-game habitat via rehabilitation can be very beneficial to this large group of ornithologists.

Raptors

Birds of prey have been the focus of various studies on surface mined lands for several reasons. In the western United States there was concern that nesting sites were being lost, and in the east it was recognized that raptors often utilized rehabilitated surface mined lands. Hawks and owls hold a fascination and provide enjoyment for some people but it is difficult to put an economic value on this facet of their existence. Of more value is their role in the higher trophic levels of the foodchain as controlling agents on some potentially destructive pest species during the more sensitive, unstable phases of rehabilitation. Many hawks and owls feed on reclaimed land and take advantage of temporary prey fluctuations in the initial stages of rehabilitation, but the lack of perches can be a limiting factor in the efficient use of mined sites by raptors. Mining often removes all trees, so perching sites are not available and birds cannot take advantage of the extremely abundant food supply in these areas. Forren & Samuel (1982) erected perches and kestrels (*Falco sparverius*) used them extensively. Wilmers (1982) noted high use of kestrel nest boxes placed around surface-mines rehabilitated to grasses and legumes. Boxes receiving highest utilization were those placed on isolated trees within the rehabilitated mine site, located some 50 m from the woodland edge (Fig. 12.3). Where there are no trees, he suggested erecting 6–10 m posts for placement of nest boxes.

Mindell & Samuel (1984) showed that red-tailed hawks (*Buteo jamaicensis*) perched along the wood borders of mined sites and fed on small mammals living there. In this study small mammal populations were found to be higher on mined sites than on nearby unmined areas. This is probably due to the presence of food and cover on the reclaimed mines. Karr (1968) recorded a similar effect in relation to birds and

Fig. 12.3. A surface mine nest box which is actively used by kestrels (*Falco sparverius*) in West Virginia; active mining is continuing in the foreground.

concluded that mining increased the heterogeneity or patchiness of habitats through alteration in microtopography. Stalmaster *et al.* (1982) suggest management techniques for ferruginous hawks (*Buteo regalis*) nesting on or near coal development areas in Utah and Colorado.

Many studies have shown that insects and small mammals are abundant on mined areas (Brown & Samuel, 1978; Wilmers, 1982; Kimmel & Samuel, 1984*a*). In much of the United States where surface mining is taking place, wooded areas are being mined, and then rehabilitated to grasses and legumes. These 'prairies' provide abundant food resources for raptors, but nest sites and perches are lacking. These can be provided by leaving selected trees along perimeters of mines or by erecting perches (i.e. posts) on the rehabilitated sites. Coal companies in many states in America have already provided such nesting and perching sites on rehabilitated mines in an effort to encourage wildlife use of their mines.

Bluebirds and other passerines

One conspicuous bird which benefits from surface mining is the bluebird (*Sialis sialis*). However, as with raptors, nesting sites are missing on many mined areas. My own unpublished data show that in West

Virginia and Pennsylvania, bluebirds used 51% of 164 boxes recently erected on trees around the periphery of mined areas (see center Fig. 12.1). Other species, such as house wrens (*Troglodytes aedon*), chickadees (*Parus* spp.), and tree swallows (*Iridoprocne bicolor*) also used the boxes. A total of 112% of the boxes were used the first year they were erected, indicating more than one species nested in some boxes. Bluebirds hatched from 73 of the 164 boxes in the first year they were erected. This is a very high first year's utilization rate.

The building and erection of bluebird nest boxes could be the basis for an excellent localized community project. Youth groups, such as the Boy Scouts or Girl Guides, and local citizen groups could work with industry to build and erect bluebird nest boxes. At very minimal costs, such groups could build nest boxes during winter months, and erect them in the spring. Mining companies could provide access to mined lands and the lumber to build the boxes. Citizens groups could assist with the purchase of lumber, and in the erection process. Such a 'nest box project' could become a year-round conservation activity for these volunteer groups. Once boxes are erected they can be checked periodically during the summer for eggs and nestlings. In the autumn the boxes can be cleaned out, and damaged ones replaced in the winter. A cooperative project such as this would be of considerable benefit to wildlife and would also promote interest in the environment. It would also provide good public relations for local industry. The same approach could be used for nest boxes and perches for raptors, but adult help would be needed because the boxes would have to be erected to a height of 15 m.

Many other songbirds benefit from the grassland rehabilitation so common to mined lands in the United States. Many grassland sparrows such as vesper sparrows (*Pooecetes gramineus*), savannah sparrows (*Passerculus sandwichensis*) and grass hopper sparrows (*Ammodramus savannarum*), although rare in some parts of the United States, are common on surface-mines rehabilitated to grasses (Whitmore & Hall, 1978). Many birds that feed on insects or small mammals will do well in such grassland environments. For a good review of bird utilization and opportunities for management on mined lands, see the proceedings of the conference, 'Surface Mining and Fish/Wildlife Needs in the Eastern United States' (Samuel *et al.*, 1978).

Reclamation for wetlands

One common byproduct of disturbing land is wetlands. In fact, it is common for surface mining to lead to the formation of wetlands.

These may be planned and provide excellent wildlife habitat, especially for waterfowl (see Chapter 11). Indeed, Dr Klimstra and his associates have conducted major studies on wetland benefits to wildlife on surface-mines, and on the rehabilitation of aquatic plants in slurry impoundments (Nawrot, 1985). Allaire (1979) noted many species of birds utilizing wet areas on mines and proposed various techniques to increase such use.

Often wetlands form on mines without deliberate planning and these not only benefit wildlife, but perhaps more importantly, they can improve water quality. At present, the main focus on wetlands in mining areas relates to their ability to improve water quality. Today there is a great deal of research conducted on the building of wetlands on disturbed lands for the purpose of water quality improvement; especially removal of Fe and Mn. For a review of such benefits, see the proceedings of a conference, *Wetlands and Water Management on Mined Lands* (Brooks, Samuel & Hill, 1985).

In mining areas where wetlands are uncommon, newly created wetlands not only provide diverse wildlife habitats, but also areas for local nature study. Various species of amphibians and reptiles may utilize such sites, and many non-game birds also will be found there.

Considerable attention has been given to the reclamation of gravel pits to wetlands in England for the past 20 years (for an excellent review of many of these studies see the ARC Wildlife Centre (1983) publication). One example of note is the Great Linford Wildlife Research Project established in 1971 by the Amey Roadstone Corporation (ARC). A 300-ha gravel quarry with no aquatic vegetation was converted to an extensive wetland and waterfowl area since 1972. Modifications include erection of fences and thorn hedges to exclude grazing animals, construction of roosting or 'leafing' sites on the shore of a large man-made lake, planting tall tussocky grasses and other vegetation to provide breeding sites, and the creation of major feeding areas for waterfowl (Street, 1982).

Gravel pits in Great Britain also are important for wintering waterfowl (Owens, 1983). Gravel pit mining has created wetlands which are important for threatened species as well. For example, the sandwich tern (*Sterna sandvicensis*) was lost as a breeding species near Kent, England, until gravel mines were reclaimed (Scott, 1983). Other species such as roseate terns (*Sterna dougalli*), tufted ducks (*Aytha fuligula*), moorhens (*Gallinula chloropus*) and sand martins (*Riparia riparia*) have also benefited. Gravel pit reclamation has played a major role in the conservation of waterfowl and other wetland wildlife species in Great Britain.

Fig. 12.4. Rehabilitated surface mines in the Appalachian Mountains of the United States. Note how the rehabilitated grasslands give diversity to what was formerly a forested area.

Cost of reclamation

Allaire (1980) assessed the costs of game and non-game wildlife management as post-mining land-uses, and found that in most aspects, such reclamation was not only feasible, but economically desirable. However, such assessments are difficult because it is hard to place economic values on hunting, fishing, picnicking, swimming, hiking, birdwatching and the other forms of outdoor recreation.

Several factors appear to be important for fish and wildlife, regardless of the type of disturbance involved. First, is the provision of good ground cover, second is the creation of plant diversity on the site (Fig. 12.4), and third is the provision of good quality water, via wetlands and/or impoundments. Although these factors benefit wildlife they also have other economic benefits in that they often result in reduced soil erosion, enhanced water quality and a generally improved physical environment.

Summary

Since the early 1970s all forms of reclamation (including coal, phosphate, iron ore, gravel, limestone, and other metal mining) have been accomplished in a way that is beneficial to fish, wildlife and the outdoor recreation that accompanies such species. The technological advancements, and the publication of 'how to' handbooks has been rapid and has allowed the rehabilitation of once useless spoiled areas into major wildlife habitats.

In this Chapter I have outlined some success stories on rehabilitated lands, as well as the various ways such lands are utilized by outdoor recreationists. Hunters, fishermen, trappers, hikers, bird watchers and nature enthusiasts all utilize such rehabilitated lands extensively.

Much of the Chapter centered on the surface-mine coal industry in the United States. New reclamation laws passed in the 1970s called for extensive ground cover, reduced erosion problems, and elimination of highwalls. Subsequently, many coal companies have responded to public interest and planted vegetation that gave more plant diversity to benefit wildlife. Water impoundments and the creation of wetlands have also become part of rehabilitation planning. Although improvements still can be made, there is no doubt that the rehabilitation of today is beneficial to wildlife and to the users of such recreation lands.

References

Allaire, P. N. (1979). Coal mining reclamation in Appalachia: low cost recommendations to improve bird/wildlife habitat. In *The Mitigation Symposium: A National Workshop on Mitigating Fish and Wildlife Habitats*, ed. G. Swanson, pp. 245–51. Fort Collins: U.S. Fish and Wildlife Service.

Allaire, P. N. (1980). Comparisons of game and nongame management models: cost analysis of post-mining land use options for wildlife. In *Symposium on Surface Mining Hydrology, Sedimentology, and Reclamation*, ed. D. H. Graves, pp. 315–24. Lexington: University of Kentucky.

Anderson, R. J. & Samuel, D. E. (1980). Evaluation of reclaimed surface mines as wild turkey brood range. In *Fourth National Wild Turkey Symposium*, pp. 186–206. Little Rock: Arkansas.

A.R.C. Wildlife Centre. (1983). *The Proceedings of a Symposium on the Subject of Wildlife on Man-made Wetlands.* Milton Keynes, England: A.R.C. Wildlife Centre.

Brenner, F. J. & Michalski, III. S. (1984). Evaluation of surface coal mines as ruffed grouse habitat. In *Proceedings of the Fourth Annual Better Reclamation with Trees Conference.* Madisonville: Madisonville Community College.

Brenner, F. J., Snyder, W., Schalles, J. F., Miller, J. P. & Miller, C. (1985). Primary productivity of deep-water habitats on reclaimed mined lands. In *Wetlands and Water Management on Mined Lands*, ed. R. P. Brooks, D. E. Samuel & J. B. Hill, pp. 199–210. Pennsylvania: Pennsylvania State University.

Brooks, R. P., Samuel, D. E. & Hill, J. B. (1985). *Wetlands and Water Management on Mined Lands.* Pennsylvania: Pennsylvania State University.

Brown, S. L. & Samuel, D. E. (1978). The effects of controlled burning on potential bobwhite quail brook habitat on surface mines. In *Surface Mining and Fish/Wildlife Needs in the Eastern United States*, ed. D. E. Samuel, J. R. Stauffer, C. N. Hocutt & W. T. Mason, pp. 352–8. Kearneysville: U.S. Fish and Wildlife Service.

Coastal Zone Resources Division. (1978). *Handbook for Terrestrial Wildlife Habitat Development on Dredged Material.* Vicksburg: U.S. Army Engineers, Waterways Experimental Station Environmental Laboratories.

Davis, R. M. (1971). Limnology of a strip-mine pond in western Maryland. *Chesapeake Science*, **12**, 11–20.

De Capita, M. E. & Bookhout, T. A. (1975). Small mammal populations, vegetational cover and hunting use of an Ohio strip-mined area. *Ohio Journal of Science*, **75**, 305–13.

Ecological Consultants, Inc. ed. (1977). *Reclamation for Wildlife Habitat. Proceedings of Reclamation Workshop II, Fort Collins, Colorado, September 1977.* Environmental Research and Technology Inc.

Feierabend, J. S. (1983). *Mourning Doves and Reclaimed Surface Mines: Management and Hunting Potential.* M.Sc. Thesis, West Virginia University.

Filion, F. L., Jacquemot, A. & Reid, R. (1985). *The Importance of Wildlife to Canadians.* Canadian Wildlife Service Environment Canada, No. CW 66-76/1985E.

Forren, J. D. & Samuel, D. E. (1982). Sparrow hawks, perches and reclaimed surface mines. *Green Lands*, **12**, 26–7.

Fowler, D. K. & Adkisson, L. F. (1980). *Survival and Growth of Wildlife Shrubs and Trees on Acid Mine Spoil.* Tennessee Valley Authority Technical Note B37, Norris: Tennessee Valley Authority.

Fowler, D. K., Turner, L. J. & Slaski, L. J. (1981). *Surface Mine Reclamation for Wildlife: A Model Reclamation Plan for Southern Appalachia.* Kearneysville: U.S. Fish and Wildlife Service.

Gregory, R. W., Elser, A. A. & Lenhart, T. (1984). *Utilization of Surface Coal Mine Waste for Construction of a Northern Pike Spawning/Rearing Marsh.* Fort Collins: U.S. Fish and Wildlife Service.

Holliday, R. J., Johnson, M. S. & Bradshaw, A. D. (1979). Wildlife conservation and mining of alluvial tin reserves in southwest England. *Biological Conservation*, **16**, 245–64.

Humphries, R. N. (1980). The development of wildlife interests in limestone quarries. *Reclamation Review*, **3**, 197–207.

Johnson, M. S. (1980). Revegetation and the development of wildlife interest on disused fluorspar tailings dams. *Reclamation Review*, **3**, 209–16.

Kabay, E. D. & Nichols, O. G. (1980). *Use of Rehabilitated Bauxite Mined Areas in the Jarrah Forest by Vertebrate Fauna.* Alcoa of Australia Limited. Environmental Research Bulletin, No. 8.

Karr, J. R. (1968). Habitat and avian diversity on strip-mined land in east-central Illinois. *Condor*, **70**, 348–57.

Kellert, S. R. (1980). Trappers and trapping in American society. In *Worldwide Furbearer Conference Proceedings*, ed. J. A. Chapman & D. Pursley, pp. 1971–2003.

Kellert, S. R. (1985). Bird watching in American society. *Leisure Sciences*, **7**, 343–60.

Kimmel, R. O. & Samuel, D. E. (1978*a*). Feeding behaviour of young ruffed grouse in West Virginia. In *Transactions, 35th Northeast Fish and Wildlife Conference*, White Sulphur Springs.

Kimmel, R. O. & Samuel, D. E. (1978*b*). Ruffed grouse use of a twenty year old surface mine. In *Surface Mining and Fish/Wildlife Needs in the Eastern United States*, ed. D. E. Samuel, J. R. Stauffer, C. N. Hocutt & W. T. Mason, pp. 345–51. Kearneyville: U.S. Fish and Wildlife Service.

Kimmel, R. O. & Samuel, D. E. (1984*a*). Implications of ruffed grouse brood habitat studies in West Virginia. In *Ruffed Grouse Management: State of the Art in the Early 1980's. Proceedings of 45th Midwest Fish and Wildlife Conference, St Louis*, ed. W. L. Robinson, pp. 91–108.

Kimmel, R. O. & Samuel, D. E. (1984*b*). Ruffed grouse brood habitat on reclaimed surface mines in West Virginia. *Minerals and the Environment*, **4**, 150–3.

Knotts, R. W. & Samuel, D. E. (1982). White-tailed deer movement and distribution about surface coal mines in West Virginia, U.S.A. *Minerals and the Environment*, **4**, 17–25.

Leedy, D. L. (1981). *Coal Surface Mining Reclamation and Fish and Wildlife Relationships in the Eastern United States*, vols. I and II. Kearneysville: U.S. Fish and Wildlife Service.

Lewis, W. M. & Nickum, J. (1964). Rearing trout in raceways supplied by waters from the hypolimnion of a strip-mine pond. *Progressive Fish Culturist*, **26**, 27–32.

Marion, W. R. & O'Meara, T. E. (1983). Phosphate reclamation plans and changes in wildlife habitat diversity. In *Reclamation and the Phosphate Industry, Proceedings of Symposium*, ed. D. J. Robertson, Gainesville: Florida Institute of Phosphate Research.

Mindell, D. P. & Samuel, D. E. (1984). Habitat use by red-tailed hawks in surface-mined areas. *Minerals and the Environment*, **4**, 144–9.

Nawrot, J. R. (1985). Wetland development on coal mine slurry impoundments principles, planning, and practices. In *Wetlands and Water Management on Mined Lands*, ed. R. P. Brooks, D. E. Samuel & J. B. Hill, pp. 161–72. Pennsylvania: Pennsylvania State University.

Nelson, R. W., Horak, G. C. & Olson, J. E. (1978). *Western Reservoir and Stream Habitat Improvements Handbook*, Fort Collins: U.S. Fish and Wildlife Service.

Nelson, R. W., Ornsborn, J. F. & Logan, W. T. (1982). *Planning and Management of Mine-Cut Lakes at Surface Coal Mining, Washington, D.C.* Department of the Interior, U.S.D.I. Office of Surface Mining.

Ohlsson, K. E., Robb, Jr. A. E., Guindon, Jr. C. E., Samuel, D. E. & Smith, R. L. (1982). *Best Current Practices for Fish and Wildlife on Surface Mined Lands in the Northern Appalachian Coal Region.* Kearneysville: U.S. Fish and Wildlife Service.

Owens, M. (1983). The importance of man-made wetlands to wildfowl wintering in Britain. In *Wildlife on Man-made Wetlands.* Milton Keynes: A.R.C. Wildfowl Centre.

Peltz, L. R. & Maughan, O. E. (1978). Analysis of fish populations and selected physical and chemical parameters of five strip mine ponds in Wise County, Virginia, with implications for management. In *Surface Mining and Fish/Wildlife Needs in the Eastern United States*, ed. D. E. Samuel, J. R. Stauffer, C. N. Hocutt, W. T. Mason, pp. 171–6. Kearneysville: U.S. Fish and Wildlife Service.

Postovit, B. C. (1981). *Suggestions for Sage Grouse Habitat Reclamation of Surface Mines in Northeastern Wyoming.* M.Sc. Thesis, University of Wyoming.

Prachar, R. E. & Samuel, D. E. (1988). Growing season effectiveness of big game repellent in reducing deer damage. In *Northeast Fish and Wildlife Conference* (in press).

Raifaill, B. L. & Vogel, W. G. (1978). *A Guide for Vegetating Surface-Mined Lands for Wildlife in Eastern Kentucky and West Virginia.* Kearneysville: U.S. Fish and Wildlife Service.

Riley, C. V. (1963). Revegetation and management of critical sites for wildlife. In *Transactions of North American Wildlife Conference*, pp. 269–83.

Robb, Jr. A. E. & Ohlsson, K. E. (1982). *Mine Related Stream Alteration Practices for Protection and Enhancement of Fish and Wildlife.* Kearneysville: U.S. Fish and Wildlife Service.

Samuel, D. E. & Bammel, L. L. (1980). Attitudes and characteristics of independent trappers and National Trappers' Association members in West Virginia. In *Worldwide Furbearer Conference Proceedings*, ed. J. A. Chapman & D. Pursley, pp. 2021–36.

Samuel, D. E., Stauffer, J. R., Hocutt, C. H. & Mason, W. T. (1978). *Surface Mining and Fish/Wildlife Needs in the Eastern United States.* Kearneysville: U.S. Fish and Wildlife Service.

Scott, B. (1983). The management of gravel quarries for wildlife at Dungeness, Kent. In *Wildlife on Man-made Wetlands.* Milton Keynes: A.R.C. Wildfowl Centre.

Scott, M. D. & Zimmerman, G. M. (1984). Wildlife management at surface coal mines in the northwest. *Wildlife Society Bulletin*, **12**, 364–70.

Stalmaster, M., Wright, W., Lockhart, J. M. & Kruegar, R. (1982). Management of nesting ferruginous hawks in relation to coal development in Colorado and Utah. In *Issues and Technology in the Management of Impacted Western Wildlife*, ed. R. D. Comer, J. M. Merino, J. W. Monarch, C. Pustmueller, M. Stalmaster, R. Stoeker, J. Todd & W. Wright, pp. 205–9. Thorne Ecological Institute, Technical paper No. 14.

Stormer, F. A., Chilcote, D. G. & Shetron, S. G. (1978). Wildlife communities of vegetated iron ore tailings basins in northern Michigan. In *Surface Mining and Fish/Wildlife Needs in the Eastern United States*, ed. D. E. Samuel, J. R. Stauffer, C. N. Hocutt, W. T. Mason, pp. 359–68. Kearneysville: U.S. Fish and Wildlife Service.

Street, M. (1982). The Great Linford project: Waterfowl research in a gravel pit wildlife reserve. In *Wildlife Values in Gravel Pits*, ed. W. D. Svedarsky & R. D. Crawford, pp. 170–80. University of Minnesota Miscellaneous Publication No. 17 – 1982.

Svedarsky, W. D. & Crawford, R. D. (1982). *Wildlife Values in Gravel Pits. Symposium Proceedings, Crookston, Minnesota.* Agriculture Experiment Station Publication No. 17-1982. University of Minnesota.

U.S. Department of Agriculture. (1982). *Wildlife User Guide for Mining and Reclamation.* U.S. Department of Agriculture Forest Service. Intermountain Forest and Range Experiment Station General Technical Report. Ogden, Utah.

U.S. Department of Interior. (1980). *A Summary of Fish and Wildlife Needs to Surface Mine Coal in the United States*, Part I. Kearneysville: U.S. Fish and Wildlife Service.

U.S. Department of Interior & U.S. Department of Commerce. (1982). *The 1980 National Survey of Fishing, Hunting, and Wildlife – Associated Recreation.* Washington DC: U.S. Fish and Wildlife Service.

Utah State University, Institute for Land Rehabilitation. (1978). *Rehabilitation of Western Wildlife Habitat: A Review.* Kearneysville: U.S. Fish and Wildlife Service.

Whitehurst, C. A. & Downing, R. L. (1977). *Opportunities for Fish and Wildlife Management on Abandoned Mine Lands in Appalachia.* Kearneysville: U.S. Fish and Wildlife Service.

Whitmore, R. C. & Hall, G. D. (1978). The response of passerine species to a new resource: reclaimed surface mines in West Virginia. *American Birds*, **32**, 6–9.

Wilmers, T. J. (1982). *Kestrel Use of Nest Boxes on Reclaimed Surface Mines in West Virginia and Pennsylvania.* M.Sc. Thesis, West Virginia University.

Yearsley, E. F. & Samuel, D. E. (1980). Use of reclaimed surface mines by foxes in West Virginia. *Journal of Wildlife Management*, **44**, 729–34.

Part 4

Case studies of fauna work in reclaimed land

Introduction

J. D. MAJER

Part 4 consists of a number of case studies, designed to reinforce the content of the preceding Chapters. These studies cover both invertebrates and vertebrates and are taken from three continents and a range of reclamation options. One of the case studies considers the role of animals in the rehabilitation of severely degraded agricultural soils.

In the first Chapter, Wolfram Dunger describes the colonization patterns of soil invertebrates on European coal mine dumps over a 25-year period. Some of the factors which influence colonization and persistence in the rehabilitation are introduced and different stages of rehabilitation, defined on the basis of animal community composition, are identified. Dunger points out that the rate of succession of soil animals may differ from that in the litter layer, leading to the contemporaneous existence of different stages of fauna development in the various strata of the ecosystem. He also convincingly argues that soil zoological studies are a worthwhile component of rehabilitation research because they provide information on the soil microclimate, the dynamics of litter decomposition, and the influence of obstructive factors (e.g. toxic chemicals) on rehabilitation.

In the next Chapter, Jan Meijer describes 16 years of animal invasion of land which was reclaimed from the sea in the Netherlands. He identifies the important factors which govern the rate and type of succession which takes place. Salinity is seen to be an important factor which restricts the rate of development of the terrestrial biota. As well as site factors, Meijer identifies the dispersal abilities of individual species as having an important influence on their occurrence in the reclaimed land.

Returning to coal mines, Dale Fowler describes some vertebrate fauna work in Tennessee coal mined areas. Having identified many of the factors which limit or encourage wildlife return, he proposes a model rehabilitation plan which embodies features which should maximize the colonization of animals in the rehabilitated land. Finally, Fowler comes back to the point expressed earlier by Viert that it is important to convince the mining industry and the regulatory agencies of the values of introducing the necessary changes in rehabilitation technology.

In Chapter 16, Owen Nichols, Boyd Wykes and I describe the extensive studies which have been performed on invertebrates and vertebrates in Western Australian bauxite mines. A wide range of rehabilitation practices which have been introduced in the light of the results of these fauna studies are described.

Finally, in Chapter 17, Lex Parker describes the degradation of agricultural land in semi-arid Western Australia. Deterioration of the soil is in part associated with loss of the soil biota as a result of inappropriate agricultural practices. Parker introduces a model system for rehabilitating such lands which highlights the need for a multidisciplinary approach, involving inputs from hydrologists, microbiologists, botanists and soil zoologists.

13

The return of soil fauna to coal mined areas in the German Democratic Republic

W. DUNGER

Introduction

In the middle and eastern parts of the south of the German Democratic Republic (GDR), deposits of eocene, oligocene or miocene brown coal are heavily utilized for open-cut coal mining. Up to 1982, 107 000 ha of land had been devastated by mining. About two thirds of this area has since been rehabilitated. The recolonization of these areas by the soil fauna has been studied by the author and his collaborators in about 40 dumps (Dunger, 1968, 1979, 1987) and some of the results are summarized in this Chapter.

Study localities

The studies were undertaken at 40 different dumps between 1961 and 1986. The geological and mineralogical properties of the mined material, the open-cut mining techniques, the rehabilitation techniques and the complex relationship of fauna of these areas to that of the natural surroundings are described in the afore-mentioned papers. The investigations concentrated on three regions of open mining.

The Görlitz region (Berzdorf open mining district)

Geology The relatively young (miocene) coal seam is overlaid by fluviatile sands and gravels, materials originated from glacial moraine as well as pleistocene loess. The coal seam and the overlying deposits are folded and displaced by glacial pressure. For that reason easily recultivatable material is often mixed with coal and/or heavy clay, even on the top of the overburden and interburden dumps.

Rehabilitation procedures The efficiency of rehabilitation is correlated with the amount of the sulphide particles. The investigated dumps protrude 20–40 m above their surroundings, and generally are 1–2 km long and about 0.5 km wide. As a result of the narrow shape of the dumps, the conditions for rapid immigration of fauna from the surrounding hilly country are quite good. Prior to rehabilitation, the mined soil is limed at the rate of 4 t ha^{-1}. The pH values are between 5 and 6 in the topsoil and between 4 and 5 at 60 cm depth. Rehabilitation consisted of afforestation with deciduous trees (*Populus* sp. (poplars), *Alnus glutinosa* (alder), *Robinia pseudoacacia* (black locust)) and seeding of *Melilotus officinalis* (yellow sweet clover) and *Lupinus polyphyllus* (perennial lupin). *Pinus sylvestris* (Scots pine) was used for afforestation in only one small (100 × 50 m) site (dump L). The overburden heaping techniques create a regular wavy surface and initial colonization is favoured in the grooves of the dump surface. For further information on rehabilitation of these areas see Dunger (1968).

List of the dumps

N 1961 afforested with deciduous trees; first investigation 1962–1966, then further surveys every 2 to 3 years; second investigation 1985.

T 1959 afforested as for N; period of investigation 1961–1966 only.

E 1955 afforested as for N; period of investigation 1961–1966, controls surveyed 1975–1979.

A 1952 afforested as for N; first investigation 1961–1966, then further surveys every 2 to 3 years; second investigation 1985–1986.

L 1952 afforested with *P. sylvestris*; first investigation 1962–1966; second investigation 1985–1986.

Throughout this Chapter the dumps are designated with indices meaning the years since rehabilitation, for example, N_2 (1962), N_{25} (1985). The natural fauna of the region has been investigated in a nearby Fraxino–Ulmetum (ash–elm association) forest. Throughout the observation period the afforested dumps were not managed with the exception of a thinning some 15 years after rehabilitation. *Populus*, and in some areas, *Alnus* were dominant in dumps N and A by the time of the second period of investigation. On dump L at least half of the *Pinus* trees were dead, and deciduous trees had immigrated into the breaks by the time of the second investigation.

The Leipzig region (Böhlen open mining district)

Geology The brown coal seam is of eocene to middle oligocene age. In the cap deposit, middle oligocene sands with a high content of sulphides dominate, while pleistocene materials are a minor component. The dump soils therefore mainly comprise acidic (pH near or below 2.5), moisture-preventing materials. Only on the dump plot B V (see list below) is cultured topsoil spread over the tertiary material.

Rehabilitation procedures The dumps which were investigated are situated within a large open-cut mining district with minimum dump widths of 2–5 km. The conditions for faunal immigration are thus poorer than in the Berzdorf district. The intensive soil amelioration involves the addition of large amounts of Ca, P and K in the top 60 cm of soil. The pH of the upper layer of ameliorated soils was raised to 4–5 by liming. Beneath the ameliorated horizon or on the surface of unameliorated material, the pH values seldom exceeded 2.5.

List of the dumps

B I Unrehabilitated since mining ceased in 1943. Consists of tertiary material without vegetation except for occasional patches of *Agrostis stolonifera* (creeping bent) and *Calamagrostis epigeios* (bush grass) which cover no more than 2% of the area.

B II Soil ameliorated and rehabilitated to agriculture in 1954 (recultivated with grasses until 1959; 1960/61 with potatoes).

B III Soil ameliorated and afforested with *Populus* sp. in 1955. After 8 years the young trees reached a height of 3–4 m.

B IV Soil ameliorated and initially (1952) agriculturally cultivated with grasses; in 1954 afforested with *Populus*, *Betula* (birch), *Robina* and *Alnus*. After 9 years (1963) the young trees reached a height of 6–7 m.

B V Cultured topsoil (sandy loam), spread 2 m thick over the tertiary material. Rehabilitation for agriculture was commenced in 1950. In 1960, afforestation with *Populus* and *Alnus* was carried out on a margin strip.

Soil zoological investigations were carried out in the Böhlen district between 1961 and 1966 (Brüning, Unger & Dunger, 1965; Dunger, 1968; von Broen & Moritz, 1965; Schiemenz, 1964). Investigations by Zerling (1987) on microarthropods from agriculturally recultivated dumps of the neighbouring Espenhain mining district enable a good comparison to be made.

The Niederlausitz region (Cottbus province)

Geology The brown coal layer is of upper oligocene to miocene age. The overlying cap consists of tertiary gravels and sands with a high content of sulphides as well as glacial and fluviatile sands. Hence, the landscape is characterized by poor-quality *Pinus* forests in this region. The dumps in this area are very large in comparison with dumps in the Berzdorf region.

Rehabilitation procedures In view of the lower quantities of sulphur-containing components here, rehabilitation of the dumps of the Niederlausitz region is generally not so complicated as in the Böhlen district. However, amelioration here is necessary because of the lack of nutrients in the sands and gravels as well as the lack of water in the region.

List of dumps Soil zoological investigations were performed on 22 dumps from 1968–1970. Soils of some of the dumps were ameliorated, others were not. Some were rehabilitated for agriculture, others were afforested 4–40 years prior to the studies reported here.

Colonization by soil animals

Earthworms

The species involved Eleven species of Lumbricidae have been found on the investigated dumps. The most important primary colonizer is *Allolobophora caliginosa*, a mineral soil dweller with perennial development. This species is favoured because of its high tolerance of water deficiency and acidity, rather than its dispersal abilities. On more developed stages of dumps *A. caliginosa* is an effective competitor and continues to maintain a reasonable component of the earthworm biomass for 33 years or longer (Fig. 13.1). Other primary colonizers are species of the genus *Dendrobaena*. They need the accumulation of a litter layer in order to settle. On poor sandy soils, possibly with insufficient nutrients for *A. caliginosa* to occur, *Dendrobaena* spp. are the only species able to colonize. They have a short duration of development (1 year) and a high fertility rate, and they therefore possess a good capacity for colonization. Besides this, their habitat is relatively isolated from the noxious dump soil beneath. The dominant species *Dendrobaena octaedra* is sometimes accompanied by *Dendrobaena subrubicunda* and *Dendrobaena rubida*. Later, immigrating species reduce the abundance of *Dendrobaena* species by competition within the litter layer (*Lumbricus rubellus*) or by burying the

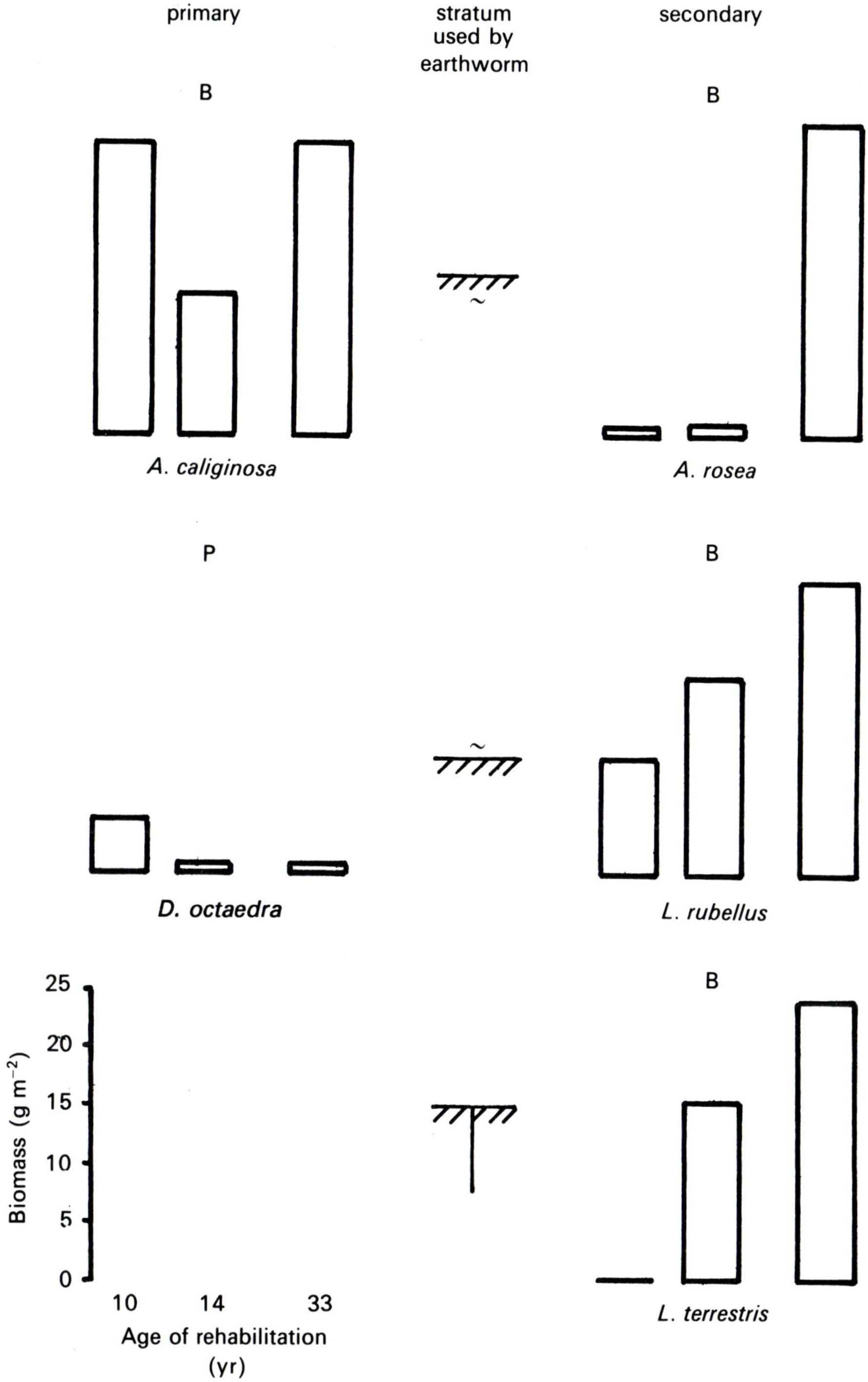

Fig. 13.1. Biomass of primary and secondary colonizing earthworms in 10-, 14- and 33-year-old dumps, rehabilitated with deciduous trees (Berzdorf site A). Key: P = parthenogenetic species with 1 year life span; B = bisexual species with 3 or more years life span (data from Dunger, 1987).

organic material in deeper soil layers (*Lumbricus terrestris*) (Fig. 13.1). Amongst the secondary colonizers, those with the best dispersal power are *Octolasium lacteum* and *Allolobophora rosea*. These species need some 5–10 years to build up a stable population (Fig. 13.1). Other species occurring in the area include *Octolasium cyaneum*, *Allolobophora longa* and *Allolobophora chlorotica*, but only a few specimens were found on dumps. *Lumbricus* species, on the other hand, seemed to have a distinctly smaller power of dispersal. This is true for the litter dwelling *L. rubellus* as well as for the anecic species *L. terrestris*. Once they have immigrated into the area, these species are able to build up stable populations relatively quickly (Fig. 13.1).

Dispersal The dispersal power is determined by the maximum rate of fertility, the possibility of passive transport and by active locomotion. High fertility is evident among colonizing earthworms such as the annual species of *Dendrobaena* as well as the facultative or obligatory parthenogenetic species of *Dendrobaena* and *Octolasium*. The bisexual species which have 3–6 or more years life span are, with regard to fertility, handicapped.

Passive transport may be possible for cocoons or juveniles. Birds, mammals or humans may act as conveyors. Of special importance to this are the techniques used for amelioration and rehabilitation since earthworms may be transported in clods of soil, introduced within plant bundles or by other means. The heaping up of cultured topsoil provides the best conditions for introduction of species. For most species, dispersal by active locomotion is of only limited importance. In practice, the artificial establishment of a 100 m-spaced grid of ameliorated microplots, with added food and introduced worms, is the best method of encouraging species. Quantitative information about the active dispersal of some earthworm species is derived from an investigation on tertiary acid dumps in the Böhlen district between 1961–1966. The dump B IV is situated between dump B V (cultured topsoil heap) which is settled by 5 earthworm species, and a leaf litter compost heap settled by 2 earthworm species (Fig. 13.2). During a 14-year period *A. caliginosa* immigrated 60 m, *O. lacteum* 40 m and *L. terrestris* 22 m into the ameliorated acid dump soil (B IV). The slow spread of *O. lacteum* and *L. terrestris*, and the absence of *A. rosea* and *A. chlorotica*, may be explained by the remaining toxicity of the tertiary soil material. Departing from the compost heap, *D. octaedra* and *L. rubellus* immigrated within the litter stratum. From these data, the linear dispersal power per annum may be

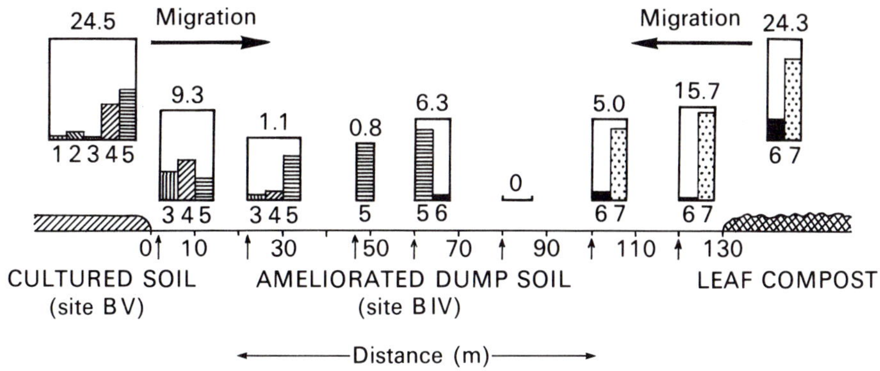

Fig. 13.2. Migration of earthworms over 14 years into tertiary dump soil from cultured top soil (from left to right) and from leaf compost heap (from right to left). The vertical arrows indicate the direction of the migration. Biomass figures are in g m^{-2}. Key to species: 1 = *Allolobophora rosea*; 2 = *Allolobophora chlorotica*; 3 = *Octolasium lacteum*; 4 = *Lumbricus terrestris*; 5 = *Allolobophora caliginosa*; 6 = *Dendrobaena octaedra*; 7 = *Lumbricus rubellus* (data from Dunger, 1969).

calculated. The values are for *D. octaedra* 10.0 m yr^{-1}, *A. caliginosa* and *L. rubellus* 4.3 m yr^{-1}, *O. lacteum* 2.9 m yr^{-1} and *L. terrestris* 1.7 m yr^{-1}. These data do not demonstrate the true capacity for dispersal because of the negative influence of inhibitory factors in the unameliorated soil (Dunger, 1969).

Conditions for persistence in rehabilitation It is probable that afforestation will be accompanied by the accidental introduction of earthworms from the tree-nursery. In the early years the conditions for survival of these specimens will be poor, even in pleistocene dumps, because of the lack of organic matter as food. Only a few specimens of the most tolerant species, *A. caliginosa*, may remain alive because they can tolerate drying and are able to develop dormant stages. If the development of plant cover is quick enough, these specimens can act as primary pioneers and give rise to stable populations. It is not clear whether litter-dwelling populations of *Dendrobaena* species are derived from accidental introductions associated with the rehabilitation process or from immigration.

On dumps rehabilitated for agriculture, the accidental introduction of earthworms is minimal. Here, secondary inoculation may be essential for the successful rehabilitation of the dump. On non toxic dumps, with acidity above pH 3.0–3.5, the food supply and the provision of suitable

Table 13.1. *Earthworm colonization, expressed as g fresh mass* m^{-2}, *on* 30 *to* 40-*year-old afforested dumps in the Niederlausitz region (data from Dunger*, 1979)

Site no.	Woodland type[a]	Humus form[b]	Moisture level[c]	Species of earthworm[d] D o	D r	D s	A c	A r	L r	L t	Total biomass
1	c	r	d	0.03	—	—	—	—	—	—	0.03
2	o	r	d	0.02	0.13	—	—	—	—	—	0.17
3	c	r	d	0.76	0.38	—	—	—	—	—	1.14
4	o	r	d	—	—	—	—	—	1.53	—	1.53
5	c	r	sd	0.08	—	—	—	—	0.80	1.00	1.89
6	od	mo	sd	—	0.02	—	0.74	0.14	1.89	—	2.79
7	do	mo	sm	1.06	—	0.15	—	4.68	—	—	5.89
8	cd	mo	sm	0.70	0.21	—	5.72	—	0.75	—	7.43
9	od	mo	sm	0.30	0.64	—	3.96	—	5.28	—	10.18
10	cd	mo	sm	0.91	0.94	—	15.06	—	—	—	16.91
11	od	mu	m	0.02	—	—	11.23	1.71	2.67	5.42	21.55

Notes: [a] Woodland type: c, coniferous forest; d, deciduous forest (*Alnus*, *Populus*); o, oak mixed forest (*Quercus*, *Fagus*); cd, od, mixed species forest.

[b] Humus form: r, raw humus; mo, moder humus; mu, mull humus.

[c] Moisture level: d, dry; sd, semi-dry; sm, semi-moist; m, moist.

[d] Species: D o, *Dendrobaena octaedra*; D r, *D. rubida*; D s, *D. subrubicunda*; A c, *Allolobophora caliginosa*; A r, *A. rosea*; L r, *Lumbricus rubellus*; L t, *L. terrestris*.

humidity are the main conditions required for the continuous settlement of earthworms. The importance of these factors is most clearly demonstrated on the poor sandy dump soils of the Niederlausitz region (Table 13.1). Among the eleven 30–40-year-old afforested dumps which were investigated, only one (No. 11) exhibited sufficient moisture to allow the build up of a medium abundance of earthworms and therefore the development of a mull humus.

In the Böhlen mining district, the conditions necessary for the continuous establishment of earthworms on ameliorated low pH dump soils were investigated in 1965–1966 (Dunger, 1968, 1969). In this study 7 species (*A. caliginosa*, *A. rosea*, *A. chlorotica*, *O. lacteum*, *L. terrestris*, *L. rubellus*, *Lumbricus castaneus*) were kept in 1 m^2 gauze cages on ameliorated, afforested tertiary dump soils (dump B IV). The survival and/or reproduction of these species were tested under the conditions of additional liming, mulching with *Populus* litter or no treatment. The study showed that *A. caliginosa* is the only species able to colonize such dump soils. Other species may be damaged by the activation of sulphurous acid during the passage of soil particles through the gut and/or by the influence of acids upon the epidermis. Of the remaining species, *A. chlorotica* and *O. lacteum* seem to be the least affected by the adverse pH. Linked with the diminishing of acidity by liming, other lumbricid species are able to colonize these tertiary dump soils (see Fig. 13.2).

Development of earthworm populations The development of earthworm populations was studied on pleistocene dumps (Berzdorf district) over 34 years (Fig. 13.3). *A. caliginosa* is the first species to colonize the moister depressions of the undulating surface of the freshly heaped tips. Possibly these populations build up from specimens which were accidentally introduced during rehabilitation (see T_3, N_5). From the beginning, *A. caliginosa* has burrowed down to 40–50 cm depth. This is possibly because the pH values do not decrease below 4.0 at this depth.

With the increase in vegetation cover, a litter layer becomes evident at about 3–4 years after rehabilitation. This enables the development of populations of *Dendrobaena* species (especially *D. octaedra*). On dumps with thick litter layers (see T_7, E_7), especially under *Pinus* vegetation (see L_{10}), the population of *Dendrobaena* reaches maximum biomass (Fig. 13.3). With the increasing influence of *A. caliginosa* and the beginning of activity of *L. rubellus*, a direct competitor of *Dendrobaena*, the litter layer undergoes increasingly rapid decomposition (see E_{11}). In this period the

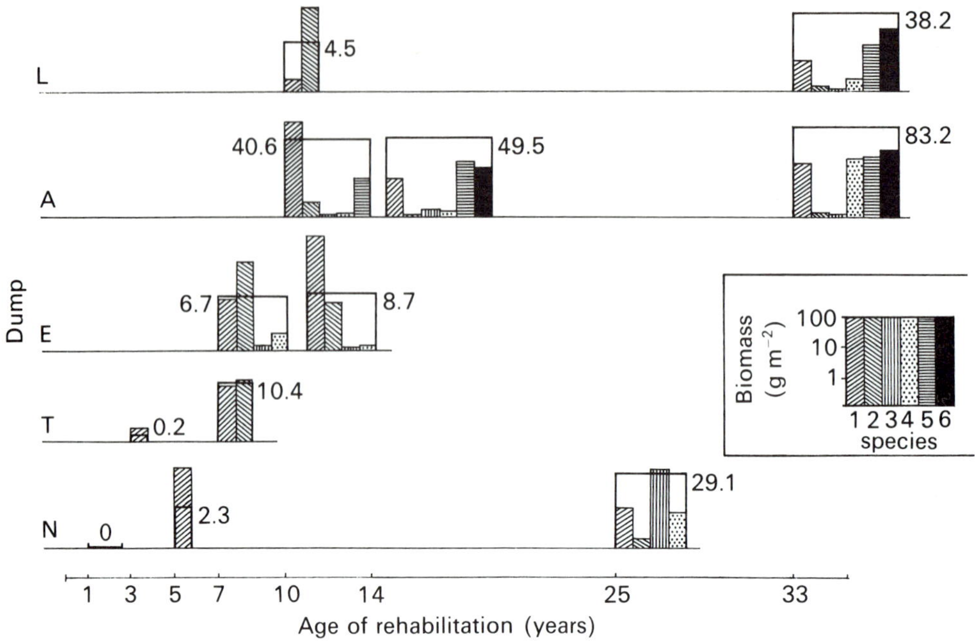

Fig. 13.3. Succession of earthworms expressed as total biomass (see figures) and biomass of single species (see columns 1 to 6), on pleistocene dumps in the Berzdorf mining district rehabilitated with deciduous trees (N–A) or *Pinus* trees (L). Key to species: 1 = *Allolobophora caliginosa*; 2 = *Dendrobaena octaedra*; 3 = *Octolasium lacteum*; 4 = *Allolobophora rosea*; 5 = *Lumbricus rubellus*; 6 = *Lumbricus terrestris* (data from Dunger, 1987).

establishment of other mineral soil species, e.g. *O. lacteum*, *A. rosea* or *A. chlorotica* (the latter being restricted to open soils) may be observed, although there is no change in the predominance of *A. caliginosa*. This situation is unchanged up until the occurrence of *L. terrestris*, an anecic form, which initiates the mixing of organic particles into the deeper soil layers. *L. terrestris* becomes codominant with *A. caliginosa* very quickly (see A_{14}). About 30 years after rehabilitation, on undisturbed deciduous forest dumps, the biomasses of *L. terrestris*, *L. rubellus*, *A. rosea* and *A. caliginosa* are nearly equal (see A_{33-34}). This is not the case in the *Pinus* afforestation (see L_{33-34}). In the cases studied here, the vegetation exhibits a tendency to change towards a mixed woodland (see earlier description under Görlitz region subheading). Under these conditions, the proportion of *A. rosea* seems to be an indicator of the optimization of the decomposition process.

There is an interesting divergence from the described succession. On the dump N, no colonization by the genus *Lumbricus* takes place, even 25 years after the afforestation (see N_{25}, Fig. 13.3). Instead of this genus, *O. lacteum* attained a large population, indicating that *Lumbricus* should be able to survive and reproduce on that dump. It is assumed that the absence of *Lumbricus* is due to a prevention of passive introduction and/ or active immigration. The dump N is a small deciduous wooded area (approximately 100 × 20 m), surrounded by forests of *Pinus* and *Picea* of a minimum width of 120 m. This short distance of pine needle humus is probably enough to act as an effective barrier to immigration by *Lumbricus* species. As a consequence, there is no mull humus development in the dump N, the humus form resting on a moder-like form. Unfortunately, the dump N was destroyed in 1986. Therefore, an experimental introduction of *Lumbricus* species was not possible. Nevertheless, the experience shows that very large dumps, as well as small dump plots surrounded by barriers, may remain uncolonized by some species (*Lumbricus* or others), unless they are artificially introduced.

Enchytraeidae

Enchytraeids were present on all investigated dumps a short time after rehabilitation. Within the initially very thin litter layer (e.g. T_3) small specimens (on average 1 μg fresh mass per individual) are present in great numbers by 3 years (e.g. T_3 has up to 10000 individuals m^{-2}). The increase in colonization by earthworms subsequently decreases the part played by enchytraeids. On dumps with low earthworm populations, the biomass of the Enchytraeidae may reach 2 g m^{-2} under forest or 1 g m^{-2} in dump soils rehabilitated for agriculture. On acid tertiary dump soils, such as those of the Böhlen district, the enchytraeids indicate the type and stage of zootic ripening of the dump soils (e.g. B IV). Investigation on the species composition of this important group of soil animals have not been performed on dumps in the German Democratic Republic.

Gastropoda

Gastropods play virtually no role in the primary colonization of dumps. Among the eventually introduced snail species, *Vallonia pulchella* (on pleistocene dumps), *Vallonia costata* and *Vitrina pellucida* (on acid tertiary dumps), are able to build up stable populations. Very striking are the rapidly increasing numbers of *Helix pomatia* in open, warm plots. In addition, the presence of slugs may indicate more developed stages of dumps. In the Berzdorf mining district, *Deroceras reticulatum*, *Deroceras agreste* (3–4 years after afforestation) and *Arion subfuscus* (after 7 years) are found. In the Böhlen district only *D. reticulatum* occurs.

Microarthropods

General comments In general, a large component of the microarthropod fauna is thought to be introduced by air transport during the pioneer period. The conditions, or features, of the dumps then dictate the survival and breeding of certain species. Under the extreme conditions of unrehabilitated acid tertiary dumps (e.g. B I), microarthropods are the most important early colonizers with Collembola and Trombidiformes (mites) contributing up to 5000 individuals m^{-2}.

On dump soils rehabilitated for agriculture, the microarthropods usually exhibit high population densities which vary between 20000–150000 individuals m^{-2}. Here again, Trombidiformes and Collembola play the most important part whereas oribatid mites are at low densities. To a great extent, the development of microarthropod populations on dumps depends on the management and rehabilitation techniques used. They therefore act as indicators of the type of rehabilitation which has been practiced.

On dumps with deciduous tree afforestation, and with soils which are not subject to toxicity problems, the pioneer population of microarthropods will probably not exceed 20000 individuals m^{-2} (e.g. N_2). By 2 years however, the increasing litter layer may be inhabited by 120000 individuals m^{-2} (e.g. T_3). During the following time period, this pioneer optimum will gradually become reduced to about 25000 individuals m^{-2} (e.g. A_{10}), due in part to immigration of earthworms and other saprophagous macrofauna. At the more developed stages of dump rehabilitation, the population densities of microarthropods increase again up to 84000 (e.g. A_{35}) or even 100000 individuals m^{-2} (e.g. N_{25}) and become more stable in terms of density. In general, low densities of microarthropods in a particular developmental stage suggest that the activity of other saprophagous groups of soil fauna is particularly high. On dumps with restrained earthworm development, the pioneer peak of microarthropod density continues over a longer time. Examples of this are the *Pinus* mixed wood on L_{10} (120000 individuals m^{-2} and L_{33} (275000 individuals m^{-2}) and the 50-year-old *Quercus* (oak) dump forest in the Niederlausitz region (238000 individuals m^{-2}). On rehabilitated, earthworm-free dump soils of the tertiary Böhlen district, the continuing

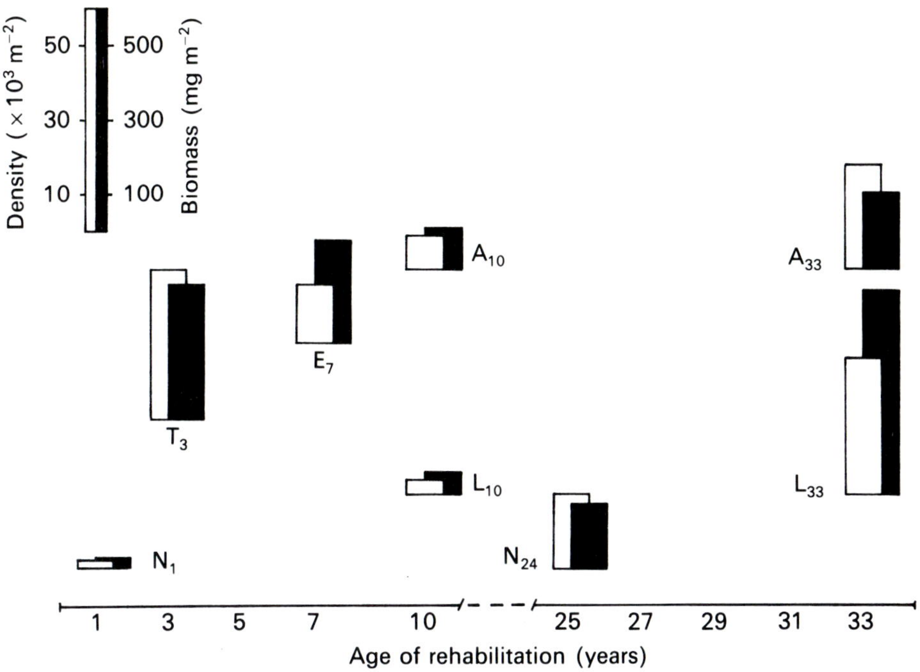

Fig. 13.4. Succession of epigaeic (above-ground) springtails (density and biomass) on rehabilitated pleistocene dumps in the Berzdorf mining district. The bar graphs have been staggered vertically to avoid overlapping (from Dunger, unpublished data).

inhibitory effect of soil acidity only allows microarthropods to build up from 25000 to 50000 individuals m^{-2}.

Collembola The springtails (Collembola) are the best investigated group of microarthropods on dumps in the GDR (Fig. 13.4). In terms of their biomass, they are dominant during the pioneer peak period (e.g. T_3: 360 mg m^{-2} or 50.7% of total microarthropod biomass) and they continue to be dominant up to the tenth year (A_{10}: 106 mg m^{-2} or 57.9% of total microarthropod biomass).

On the other hand, Collembola contribute a low biomass in the first year of colonization (N_2: 30 mg m^{-2} or 27% of total microarthropod biomass). The maximum biomass reached by springtails was in the litter layer of the mixed forest (L_{33}) with a biomass of 542.6 mg m^{-2} (32% of total microarthropod biomass) and an average density of about 36000 individuals m^{-2} (maximum density 51000 individuals m^{-2}).

The composition of the Collembola fauna can best be characterized by examining its community structure (Dunger, 1968; Dunger & Dunger,

1984). Within the hypogaeic (below ground) stratum, development commences with the *Proisotoma minuta*-dominated community. On pleistocene dumps this community type lasts at the most up to the fourth year. On tertiary acid dumps, on the other hand, it may last as long as 9 years or more. In addition to *P. minuta*, characteristic species are *Hypogastrura vernalis* and, in the Böhlen district, *Friesea afurcata. Mesaphorura macrochaeta* are both dominant (Dunger, 1986). At the beginning of the second pioneer stage ('pioneer peak'), the transition to the *Isotomodes productus–Micranurida pygmaea* community takes place. Among others, *Isotoma notabilis* is a characteristic species. A *Willemia intermedia* variant of this community continues to be present on pleistocene dumps from the third up to the sixth year, a *Friesea mirabilis* variant from the sixth up to the tenth year, and the typical community develops at later stages. On *Pinus* afforestation, a poor variant can be found. On tertiary acid dumps the *Willemia intermedia* variant occurs, while on cultured soil dumps (B V) only the *Friesea mirabilis* variant occurs. About 15–20 years after afforestation on pleistocene dumps a gradual transition to the *Mesaphorura tenuisensillata* community takes place (e.g. N_{25}, A_{33}, L_{33}). This community is also characterized by *Mesaphorura hylophila* and may be differentiated by *Isotomiella minor* being absent on N_{25}, recedent on A_{33}, and codominant on L_{33}.

Within the epigaeic (above ground) stratum there is a greater diversity and more rapid succession of springtail communities than in the hypogaeic stratum. On pleistocene dumps (Berzdorf district), the above-ground collembolan communities are dominated by the omnipresent *Lepidocyrtus paradoxus*. Here there is a sequence from the *Entomobrya lanuginosa* community (first to third year), the *Hypogastrura succinea* community (third to fifth year), the *Tomocerus vulgaris* community (on open plots and on *Pinus* afforestation from the seventh up to the thirteenth year) and the typical *Lepidocyrtus paradoxus* community (starting from the seventh year on deciduous woodland afforestation). On tertiary acid dumps in the Böhlen district the corresponding sequence is as follows: *Hypogastrura vernalis* community (unrehabilitated dumps through to the first 6 years of rehabilitation); *Tomocerus vulgaris* community (starting from the eighth year); and the *Deuterosminthurus flavus* community (starting from the eleventh year on culture soil dumps). In general, *Orchesella quinquefasciata* is the dominant species characteristic of tertiary dump soils of the Böhlen district. The above-ground fauna was mainly sampled by pitfall trapping. Therefore exact statements of densities per unit area are not possible. On open plots, such as recently

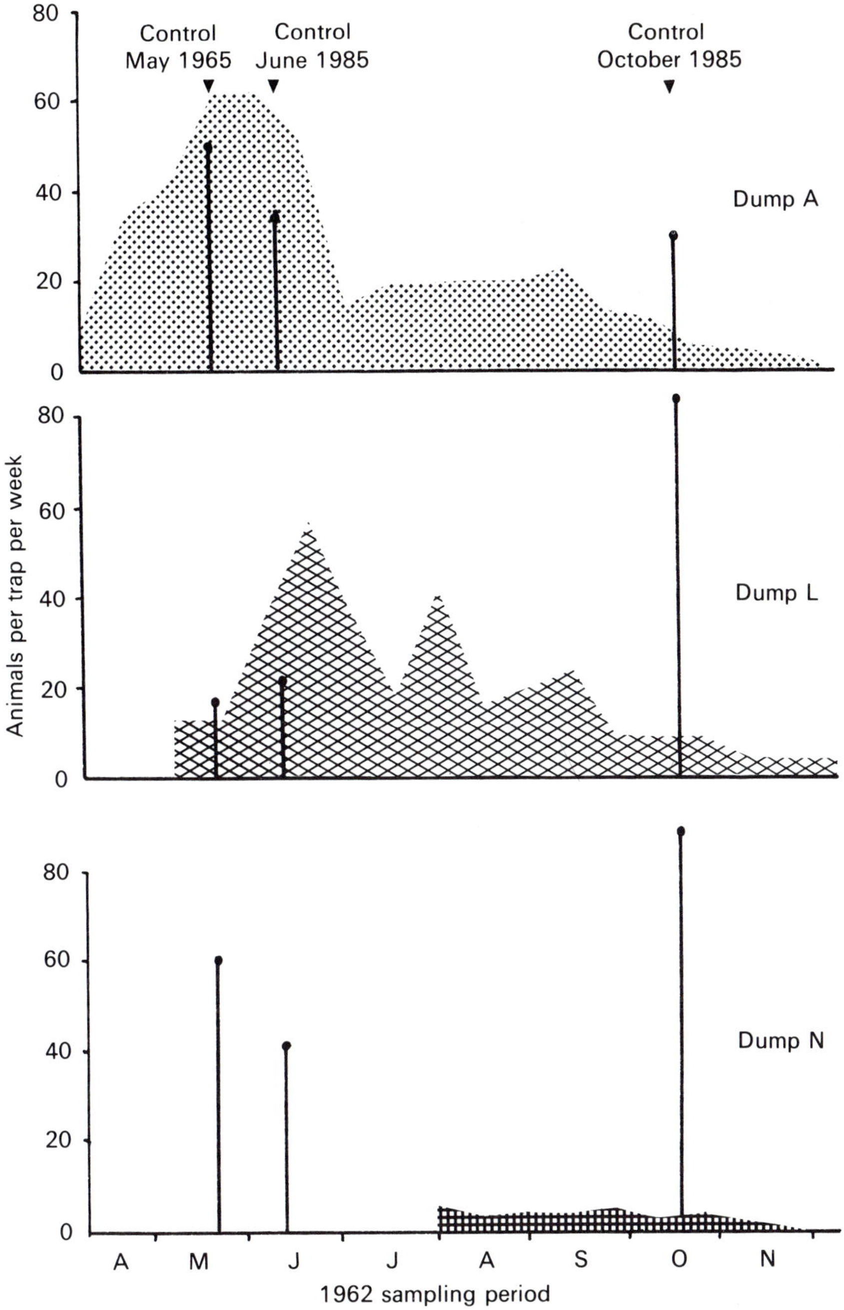

Fig. 13.5. Results of pitfall trapping of epigaeic springtails on rehabilitated pleistocene dumps in the Berzdorf mining district. The trapping period was April until December 1962. The 1965 and 1985 controls of 4-week periods are superimposed on the graph (from Dunger, unpublished data).

rehabilitated dumps and older dumps rehabilitated for agriculture, a high input of animals by air-drifting into the pitfall-traps occurs. Nevertheless, this may be indicative of what is actually happening at these sites. On the dump N_2, there were only 1.2 individuals caught per pitfall-trap per week, while on T_3 there were 157 animals caught per trap per week (pioneer peak period). Extremely high values are reached on the very large agricultural rehabilitation in the Niederlausitz region. Here, up to 2100 animals per trap per week were captured, with *Hypogastrura* species being dominant. Under windless conditions, such extreme numbers are not possible. Figure 13.5 shows that numbers of captured individuals stabilize relatively early and this new level is influenced by microclimatic factors. The comparison of data from 1962 and 1985 (Fig. 13.5) indicates that above-ground springtail activity is nearly equal on dumps A and L over 23 years and surface activity on the dump soil of N_{25} is at a similar level.

Mites To date the mites (Acari) from dumps of the GDR have mostly been studied quantitatively rather than from the systematic point of view. Systematic investigations are currently under way. The oribatids are absent on unameliorated tertiary acid dump soils (B I). On the other hand, they colonize pleistocene dumps as quickly as the springtails and even exceed these in terms of density (Fig. 13.6). Furthermore, their colonization pattern is quite similar to springtails in that they exhibit a pioneer peak on T_3 and a later decrease in abundance as a consequence of the activity of earthworms. Within thick litter, oribatids exceed all other microarthropods in terms of biomass as well as the density (e.g. L_{10}, L_{33}). The same pattern was observed in an investigation of 30–50-year-old forest on dumps in the Niederlausitz region where an abundance of over 200 000 individuals m^{-2} in thick litter was observed.

Some authors are of the opinion that oribatids do not take part in primary colonization. This only seems to be true for sites without accumulated litter. For example, on dump soils rehabilitated for agriculture. Such sites exhibit a very low level of oribatid colonization (e.g. B II; see also Zerling, 1987).

The Parasitiformes mites require more time for colonization and establishment (Fig. 13.7). On unrehabilitated tertiary dumps (e.g. B I), as well as on the first pioneer stage of pleistocene dumps (e.g. N_2), only a few specimens of gamasid mites are present. In the following period, Parasitiformes build up to a pioneer peak and exhibit a decrease in the woodland stage, as do the other microarthropods (e.g. A_{10}).

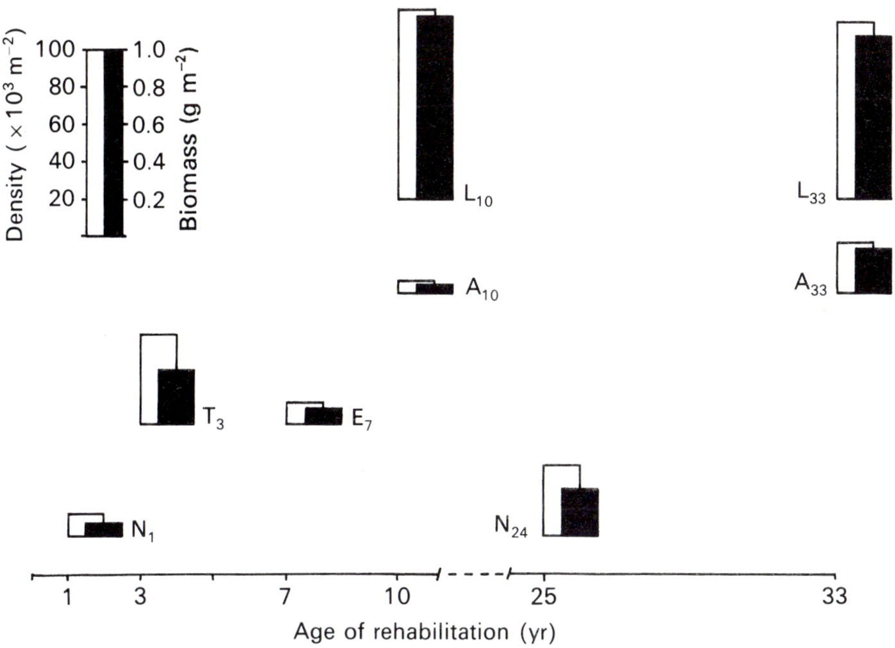

Fig. 13.6. Succession of Oribatei mites (density and biomass) on rehabilitated pleistocene dumps in the Berzdorf mining district. The bar graphs have been staggered vertically to avoid overlapping (from Dunger, unpublished data).

Other mites are represented mostly by Trombidiformes. In general, these exhibit the same tendencies as are shown above. Further systematic studies will no doubt produce interesting findings. For instance, it is now known that the Tydeidae do not colonize the unrehabilitated tertiary acid dumps but are eudominant on ameliorated and afforested dumps.

Other microarthropods The Protura seldom colonize the dumps during the pioneer stage. The first occurrences (e.g. T_3) are of *Eosentomon delicatum* and *Eosentomon transitorium*. Later on (e.g. A_{10}, L_{10}), *Acerentulus exiquus* occurs. A population of Protura, which resembles that of a site with undisturbed vegetation, develops after about 30 years (e.g. A_{33}, L_{33}), with 2000 to 3000 individuals m^{-2} and a biomass of about 20 mg m^{-2}. The colonization of Symphyla is more rapid. They are absent in the pioneer stage, but already attain a stable population after 5 to 7 years (e.g. E_7). *Symphylella vulgaris* is numerous in the thick litter layer of dump L_{10}. Besides this species, only *Symphylellopsis subnuda* was found in the early stages of rehabilitation. The Pauropoda colonize the dumps in

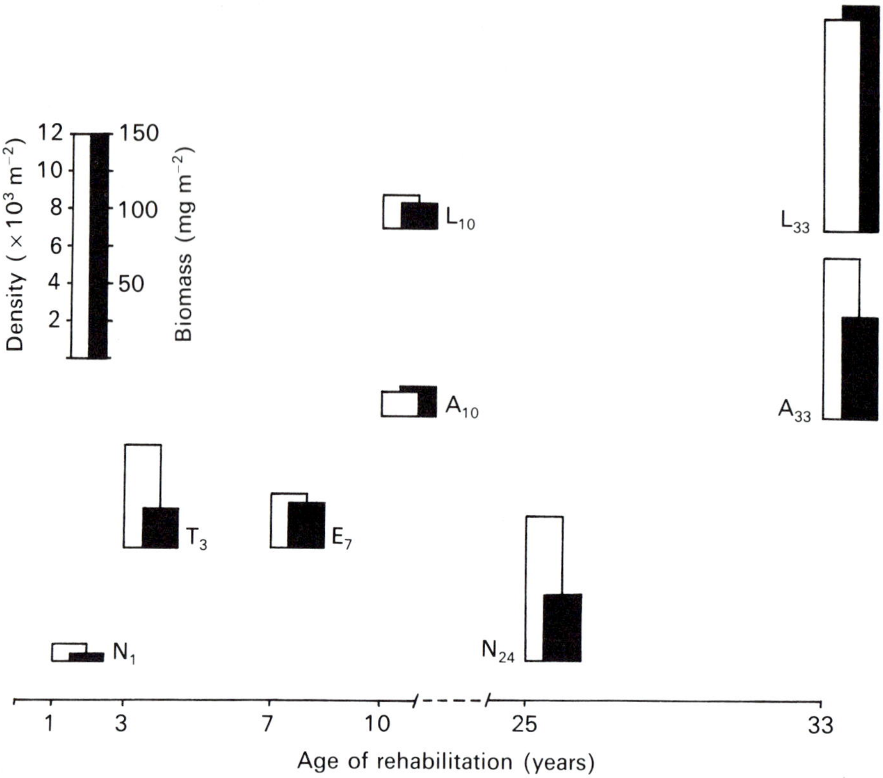

Fig. 13.7. Succession of Parasitiformes mites (density and biomass) on rehabilitated pleistocene dumps in the Berzdorf mining district. The bar graphs have been staggered vertically to avoid overlapping (from Dunger, unpublished data).

low numbers, but in general, with the same pattern of colonization as that of the springtails or oribatids. *Allopauropous gracilis* shows a distinct pioneer peak on T_3 and is also found in tertiary dump soils.

Macroarthropods

The colonization of macroarthropods in the dumps of the GDR (see Dunger, 1968) is partly reported in Broen & Moritz (1965) (spiders), Schiemenz (1964) (Cicadidae), and Dunger & Voigtländer (1989) ('myriapoda'), while other groups are under study. Only a brief overview may be given here. The most important saprophages are the Diptera (fly) larvae. They exhibit high densities in pleistocene dumps at the second pioneer stage (e.g. T_3) and therefore, together with the Collembola, contribute essentially to the pioneer peak. They have a decisive influence on dump soils when earthworm activity is lacking. On the other hand, where earthworms dominate and dictate the dynamics of the humus layer, the density of Diptera larvae is less (< 1 g m^{-2}).

The Diplopoda (millipedes) have little influence on the development of dump soils. Nevertheless, they are already present at the first pioneer stage (e.g. N_2, T_3 occupied by *Craspedosoma rawlinsi*) and in the following years (e.g. E_7, A_{10} by *C. rawlinsi* and *Polydesmus inconstans*). *C. rawlinsi* also colonizes the acid tertiary dump soils before they are ameliorated (e.g. B I). Isopoda (woodlice) are almost entirely absent on dumps during the first 15–20 years.

The succession of zoophagous macroarthropods on dumps has mainly been investigated using pitfall-traps. The Arachnida colonize the dumps immediately after soil is placed on them (see N_2, Fig. 13.8). The high numbers of individuals caught are mainly caused by the smooth surface, which offers little impediment to their movement. Samples on more developed dumps exhibit little differences in catch between different rehabilitation types and ages of plots (e.g. N_{25}, A_{33}, L_{33}). The first colonizers are photophilous and xeroresistant species, which are mainly distributed in open and warm locations (Broen & Moritz, 1965). Further investigations on species succession are to be carried out in the future.

The Carabidae (ground beetles) show similar colonization patterns. Insect larvae and springtail-eating species multiply quickly in the second pioneer stage (e.g. T_3). Later, omniphagous or phytophagous species, such as *Harpalus aeneus*, appear. In general, the pioneers are stenoecious inhabitants of open localities (e.g. *Broscus cephalotes*, *Harpalus pubescens* and *Pterostichus cupreus*) with autumn breeding and larval hibernation during winter or with spring breeding and autumn presence. They are mostly nocturnal animals and prefer xerothermic conditions. Some of these pioneers are able to fly (e.g. *Bembidion* spp.), others are not (e.g. *B. cephalotes*). On pleistocene dumps, littoraea (coastal) species are the first colonizers while on tertiary acid dump soils of the Böhlen district, xerothermophilous continental species colonize first. The population development of carabids up to 14 years after rehabilitation is described in Dunger (1968).

The Staphylinidae (rove beetles) surpass the Carabidae in terms of their density and biomass at all stages of succession. First colonizers are small, flying species from the open fields (e.g. *Tachyporus nitidulus* and *Amischia analis*). These eurytopic, photophilous species are gradually replaced by shrub-dwelling species (e.g. *Ocalea badia*, *Gabrius vernalis*, etc.) in the pre-woodland stage (e.g. E_7, A_{10}), and finally by sylvatic

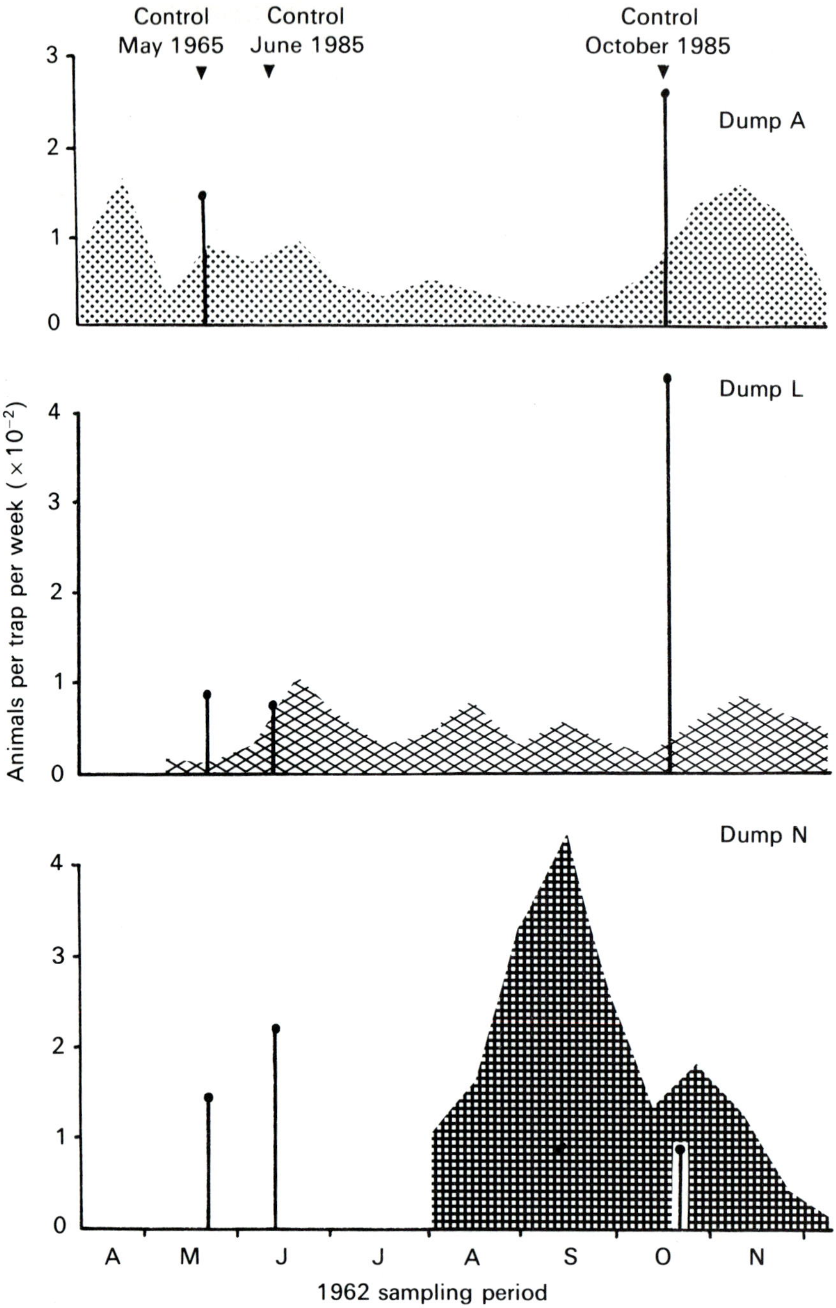

Fig. 13.8. Results of pitfall trapping of Arachnida on rehabilitated pleistocene dumps in the Berzdorf mining district. The trapping period was April until December 1962. The 1965 and 1985 controls of 4-week periods are superimposed on the graph (from Dunger, unpublished data).

species (e.g. *Philonthus decorus*, *Otius punctulatus*, *Ocypus melanarius*, etc.) in the true wooded stage, 20 or more years after rehabilitation. Colonization of staphylinids on dumps of the GDR is described in detail by Vogel & Dunger (1989).

The Chilopoda (centipedes) are represented in the pioneer phase of colonization by the parthenogenetic species, *Lamyctes fulvicornis*. Some time later, *Lithobius forficatus* colonizes the dumps. When the tree canopy closes, *L. fulvicornis* is displaced by *Lithobius microps*. The Geophilomorpha are represented by *Necrophloeophagus flavus* during the first years. On more developed dumps several *Geophilus* species can be found (Dunger & Voigtländer, 1989).

Mammals

Of the vertebrates, only the small mammals have a direct influence on the structural development of dump soils. On 10-year-old afforestated dumps, small mammals have constructed a dense network of subterranean galleries, perforating the pleistocene dump soil (e.g. A_{10}). This activity is very important for breaking up the dump soil surface. Between the seventh and fourteenth year, the species composition of the small mammal community is characterized by *Apodemus sylvaticus*, *Microtus arvalis* and, less often, *Clethrionomys glareolus*. It thus resembles the community composition of unmined woodlands. In addition, *Sorex araneus* always occurs and *Sorex minutus* occasionally occurs.

On afforestated dumps, one or two decades later, populations typical of deciduous woodland (e.g. N_{25}, A_{33}) or mixed *Pinus* woodland (e.g. L_{33}) with different proportions of *C. glareolus* and *Apodemus flavicollis*, are found. *S. araneus* is present on these plots at far smaller densities than on open, pioneer dumps.

Quantitative description of fauna development

A quantitative summary of macro- and microfaunal succession on pleistocene dumps in the Berzdorf district is given in Table 13.2.

The largely above-ground groups such as Arachnida, Coleoptera or 'myriapoda' are not included in the Table because they were mainly sampled by pitfall-traps and therefore can not be quantified with regard to density per unit area. The Enchytraeidae were not studied in 1985/86 and soil microfauna (Nematoda, Testacea etc.) are not considered at all. The figures in Table 13.2 represent average biomass values obtained from investigations throughout 1 year, or at least from different seasons. Since biomass values for different groups (i.e. macrofauna, mesofauna)

Table 13.2. *Quantitative summary of soil faunal succession on pleistocene dumps in the Berzdorf district*

		Dump no./stage							
Fauna group		N1/I	T3/II	E7/III	A10/IV	N25/IV	A33/V	L10/III	L33/V
Lumbricidae	bm[a]	—	0.1	6.7	40.6	29.1	93.7	4.5	38.2
	ME[b]	—	2.7	126.7	487.1	349.2	1124.4	85.5	458.4
Other saprophagous macrofauna	bm	0.1	1.4	1.1	0.4	0.5	0.8	0.6	0.8
	ME	0.2	26.4	20.5	7.8	9.7	15.6	10.6	14.4
Microarthropods	bm	0.1	0.7	0.5	0.2	0.6	0.6	1.1	1.7
	ME (*Ma*)	1.2	7.6	5.1	2.1	6.7	6.0	11.8	17.9
Total saprophagous macrofauna	ME (*Ms*)	0.2	29.1	147.2	484.9	358.9	1140.0	96.1	472.8
Total saprophagous macrofauna and microarthropods	ME	1.4	36.7	152.3	487.0	365.6	1146.0	107.9	490.7
Ms/*Ma* ratio		0.2	3.8	28.9	230.9	53.6	190.0	8.1	26.4

Notes: [a] bm average live mass m^{-2}
[b] ME metabolic equivalents (see text for explanation)
For explanation of dumps (A, E, L, N and T) see section on 'The Görlitz region', for stages of succession (I–V) see section on 'Stages of succession during dump rehabilitation' (from Dunger, unpublished data).

are not truly comparable at the energetic level, the 'metabolic equivalents' (ME) are added. The ME values are basal respiration rates of a specimen of a particular taxon at 10 °C, multiplied by the appropriate biomass (Dunger, 1989). The ME values are thus comparable between groups with different individual biomasses. The dumps mentioned in Table 13.2 are described earlier under the Görlitz region subheading.

Table 13.2 shows that the potential activity of the soil fauna, at least the macrofauna and microarthropods, increases almost continuously throughout the first three stages of dump rehabilitation. During the third stage, differences can be seen between dumps with different rehabilitation conditions (e.g. T, E, A). For instance, dumps surrounded by *Pinus* plantations (e.g. N) or by *Pinus* afforestation itself (e.g. L) can have restrictions to fauna development. Starting from the fourth stage at favourable dumps (e.g. A), the rate of increase becomes lower. On dumps developing into another biological character (L), the activity continues to increase during the fifth stage. Further investigations are now necessary on the quantitative composition of the fauna, the part taken by different groups and the faunal interactions during these various stages of succession.

Influence and interaction of soil fauna in rehabilitated dumps

The development of the biota on freshly heaped dumps involves two correlated processes: the settlement of species (e.g. utilization of ecological niches) and their ecological effects caused by their various activities (e.g. alteration of ecological niches). These activites influence abiotic factors as well as the interrelationship between microflora, macroflora and soil fauna. The result may be advantageous or injurious to the animals involved. As an example, the feeding activity of litter dwelling earthworms (e.g. *Dendrobaena* species) destroys their own food source and environment. Another easily demonstrable interaction is the alteration of litter decomposition by the increasing earthworm population from a fungus-based into a more bacteria-controlled process, which in turn causes a reduction in numbers of fungivorous microarthropods. Apart from the special case of transient flying insects, the initial process of soil recolonization involves the development of microflora and microfauna. This is then followed by the colonization of soil mesofauna feeding on microbes. During these initial phases, the major colonizers are present in the above-ground stratum. The below-ground communities need more time to build up and the rate at which this happens depends on the descending of microbiological activities into the soil stratum. The last step of pioneer colonization involves the settlement of dumps by a stable saprophagous macrofaunal community.

The development of resident faunal communities renders it possible to identify the stages of succession. However, from the above statements, it is clear that succession proceeds at different rates within the above-ground and below-ground stratum. Therefore, a rehabilitated dump may have strata which exhibit two different stages of succession.

During the biotic development of rehabilitated dumps, the succession of animal communities is by no means a uniform process in which there is a steady increase of density and species diversity of all the animal groups involved. As was noted for some micro- and macroarthropods, a 'pioneer peak' may be followed by decreasing densities and also diversities. Superimposed on this, are the individual site factors and rehabilitation procedures on each of the dumps which may act as modifiers of the above-mentioned processes. The most important factor is moisture, e.g. the hydrologic regime and also the influence of vegetation on evapotranspiration. The amount of dead organic matter produced by the developing vegetation is of secondary importance.

Stages of succession during dump rehabilitation

It is now possible to synthesize the aforementioned into five stages of succession.

I. The first pioneer stage is characterized by extremely hot, dry conditions and the absence of organic matter. Present at this stage are air-transported, or flying, predators which may be hunting in the area for flying prey. Many species attempt to colonize at this stage but are hindered by the harsh microclimatic conditions. Species which are r-strategists are the most successful pioneers. Typical communities of microarthropods are characterized by the springtails *Entomobrya lanuginosa* in the above-ground stratum and *Proisotoma minuta* in the below-ground stratum. Under favourable conditions, such as in the Berzdorf district, the first stage may last for only 2 years, but under unfavourable conditions, such as in the Böhlen district (e.g. B I), it may last for over 30 years. In this case, specialized species from dunes or deserts (e.g. the dermapteran *Labidura riparia*) may build up stable populations.

II. The second stage is influenced by both the method of rehabilitation and by the degree of natural colonization by the various organisms. During this stage there is an increasing litter accumulation, mostly arising from the herb layer. The second stage is best represented by the Berzdorf dump (T) (third to fifth year) which has a minimal earthworm population, a high density of Diptera larvae and also of some microarthropod groups (e.g. Collembola). As a consequence, the Ms/Ma-ratio (saprophagous macrofauna/microarthropods) is very low (3.8) (see Table 13.2). The total ME value is about 30-fold the value observed during stage I. The 'pioneer peak' which is exhibited by some microarthropods results in an intensive development of predators (e.g. ground beetles, spiders). The above-ground stratum is settled by a very characteristic *Hypogastrura* community of springtails. The colonization of the below-ground stratum is slower with a gradual change from the *P. minuta* springtail community to the *I. productus–M. pygmaea* community. On pleistocene dumps the second stage may end at the fifth year, while on tertiary acid dumps it may last to the ninth year (e.g. above-ground stratum in B II and B III) or even longer (e.g. below-ground stratum in B IV).

III. The third stage of afforested dumps exhibits a shrub layer without a closed canopy so some otherwise meadow-inhabiting species can live there. The increasing litter layer is now colonized, and diminished, by *Dendrobaena* spp. earthworms and, towards the end of this stage, by *L. rubellus*. The more intensive process of decomposition and humification of soil organic matter manifests itself in the increased Ms/Ma ratio (Table 13.2). This is distinctly higher in leaf litter (e.g. in E_7 it is 28.9) than in needle litter (e.g. in L_{10} it is 8.1). The pioneer peak has now been passed, but the sum of the ME values continues to rise (about 3- to 4-fold above the stage II level). Characteristic of this stage is the invasion of the above-ground stratum by the *Lepidocyrtus cyaneus* springtail community (with some variants) and of the below-ground stratum by the well developed *I. productus–M. pygmaea* springtail community (with certain variants). The duration of the third stage is more variable than the previous one and may last for about 5 years on pleistocene dumps (e.g. E, A) or even 10 or more years on tertiary acid dumps (e.g. B IV).

IV. The fourth woodland-like stage is a precursor to the fully developed woodland stage (stage V). The duration of this stage is very variable. The beginning of the fourth stage is defined by closure of the tree canopy. The humus dynamics are characterized by the shifting of organic matter into the deeper mineral soil layers by the action of anecic earthworms, e.g. *L. terrestris*. In this stage, the preponderance of earthworms over the other saprophagous macrofauna and microarthropods is highest (Ms/Ma ratio 230.9; see Table 13.2).

V. The gradual immigration of further species (especially K-strategists) during the transition to the fifth stage causes a 1.5- to 2-fold increase in the saprophagous activity of the soil fauna and a decrease in the Ms/Ma ratio. This value may be high under favourable conditions (e.g. A_{33}) but low on *Pinus* afforestation, even after secondary colonization of the dump by deciduous trees (e.g. L_{33}). For the characteristic species of stages IV and V, the reader should consult the relevant taxa sections under the 'Colonization by soil animals' heading.

The durations of these stages are extremely variable. Under optimal conditions the beginning of the woodland stage may commence 20 to 25 years after rehabilitation. This stage coincides with the termination of zoological ripening of dump soils. Unfortunately, there are no pedological investigations to give an idea of the duration of the more complex pedological ripening of dump soil profiles.

Practical implications of pedozoology in dump rehabilitation

Pedozoological investigations and stimulation of dump rehabilitation

Soil fauna investigations are very laborious and time consuming. Therefore it is necessary to examine carefully whether pedozoological work is profitable or not in comparison with other lines of investigation. In the case of dump investigations, at least three phenomena are best considered using pedofaunal information: the soil microclimate; the dynamics of humus; and the influence of toxic and other obstructive factors (Dunger, 1968).

From the pedozoological point of view the most important procedures for rehabilitation of coal dumps in the GDR have been found to be: (1) the stabilization of water regime (protection against evaporation and increase of water capacity of the dump soils); (2) nutrient build up (rapid production of organic matter under the herb layer by nitrogen-fixing plants); and (3) the physiological detoxification of dump soils (neutralization of the mineral acids). Useful pedozoological information is derived from the investigations in the GDR and summarized in the following sections.

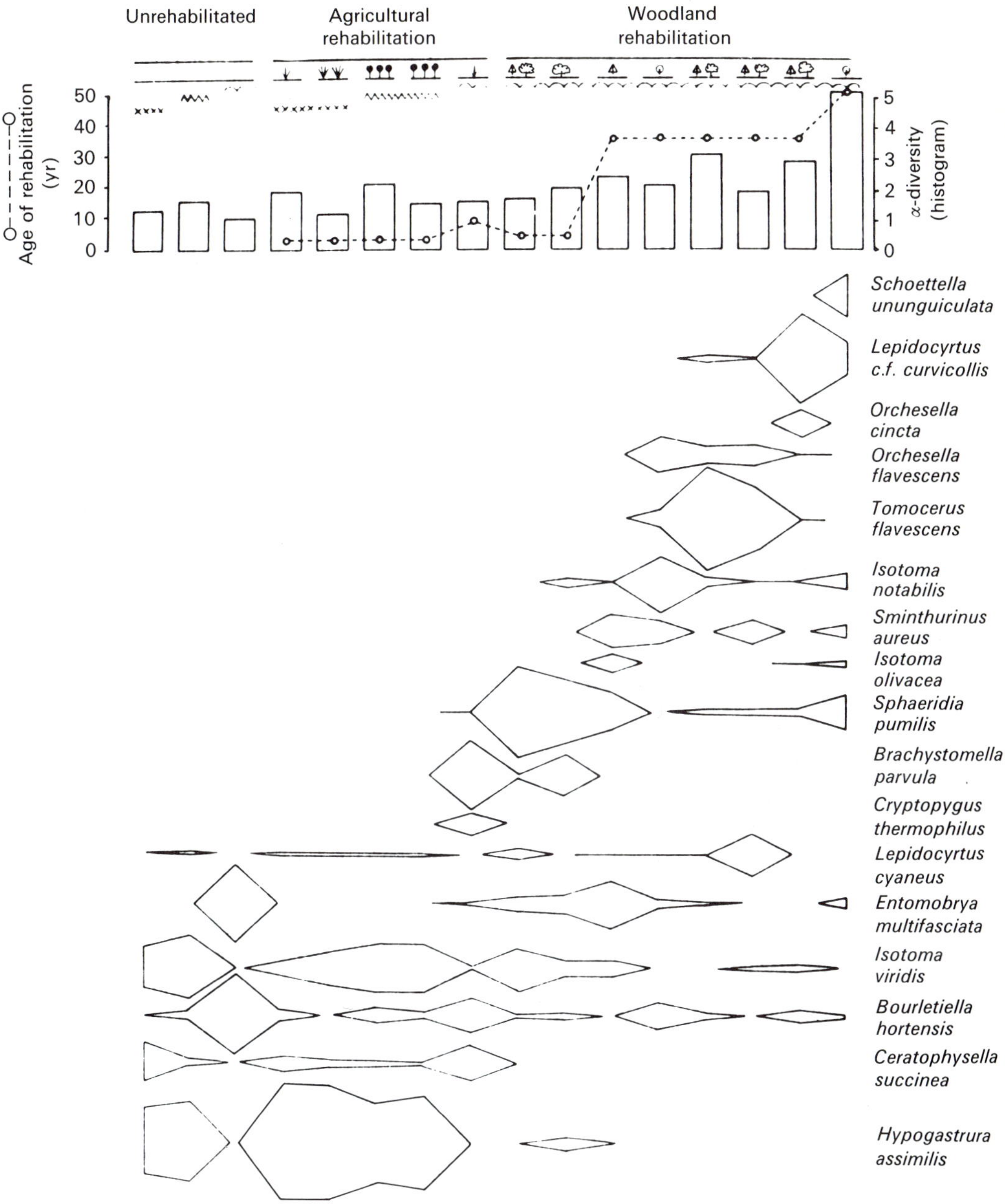

Fig. 13.9. Dominance of epidaeic springtails sampled by pitfall traps on 16 dumps in the Niederlausitz region (from Dunger, 1979).

sand; loamy sands; heavy clay; annuals; perennial fodder-plants; cereals; *Pinus*; deciduous trees; *Fagus, Quercus*

Characterization of dumps by the soil fauna

An overall characterization of dumps may be provided by the quantification of the specific faunal groups which are used in delimitating the colonization stages (see section on 'Stages of succession during dump rehabilitation'). Additional information can be gained from the qualitative analysis of suitable groups. As an example, Fig. 13.9 shows the dominant species of Collembola caught by pitfall-trapping on 16 dumps from the Niederlausitz region. Loamy or sandy unrehabilitated pleistocene dump soils are mainly inhabited by *Hypogastrura assimilis* and *Isotoma viridis* but tertiary coal sandy dump soils are inhabited by *Entomobrya multifasciata* and *Bourletiella hortensis*. An area agriculturally recultivated exhibits no essential alteration over the first 3 years. Perennial fodder-plant cultivation over 8 years promotes the immigration of species that occur quickly in young afforestation (e.g. *Brachystomella parvula*, *Sphaeridia pumilis*). Forty-year-old dumps with a deciduous woodland show another typical community (e.g. *Tomocerus flavescens*, *Orchesella flavescens*). Forests on dumps with a disposition for summer dryness and slow decomposition are characterized by the presence of *Schoettella ununguiculata* and *Orchesella cincta*. One of the drawbacks of this method of characterization is the taxonomical problem. For instance, taxonomy, and therefore the ecological knowledge, of the genus *Lepidocyrtus* is far from clear. Figure 13.9 also shows that, at least of the above-ground Collembola, diversity is not a good measure of the developmental stage of dumps.

Fauna affinity between different dumps

As pointed out in detail by Dunger (1968), fauna communities, such as those of springtails, may be used for site characterization rather than the use of single species. One of the ways to compare the affinity of communities is to use Riedl's index of homogeneity. Its application to dump community comparison (Fig. 13.10) indicates that above-ground microarthropod communities are particularly influenced by vegetation, whereas below-ground communities are influenced to a greater extent by the mineral properties of dump soils. This method thus provides a useful tool for differentiating different areas of rehabilitation. On the other hand, these dominant ecological influences on the soil fauna should be kept in mind when planning pedozoological investigations.

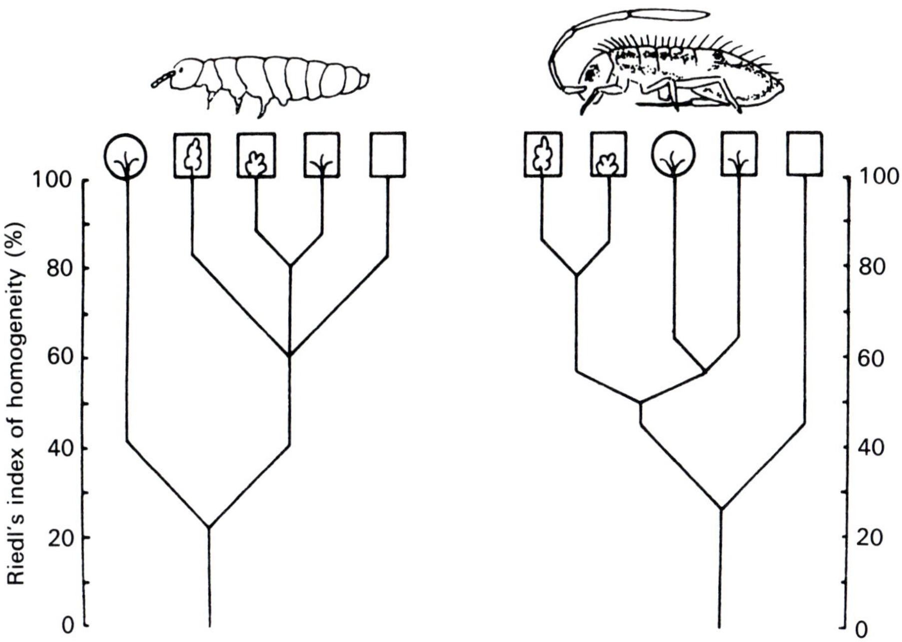

Fig. 13.10. Affinity of hypogaeic (left) and epigaeic (right) springtail communities on a range of rehabilitated tertiary dumps in the Böhlen mining district (from Dunger, 1975). Soil: ○ cultured topsoil; □ untreated tertiary acid dump soil. Rehabilitation method: cereals; deciduous shrubs/trees.

Asynchronism of succession in different strata

For the reasons already given, the rehabilitation stages of dumps cannot be characterized by the synchronous succession of the biota in all strata of dump soils. All investigations performed on dumps in the GDR have reached this conclusion and this asynchrony is summarized in Fig. 13.11 for the dumps N, T, E and A of the Berzdorf district. It is obvious that the above-ground microarthropod community will change even more rapidly than the plant community. Contrary to this, the below-ground microarthropod community needs longer to develop. This phenomenon explains the slowing of the dump rehabilitation processes, which may happen some years after a relatively quick initial development.

Introduction of soil animals

The results of our dump investigations suggest that many soil animal groups have a sufficient potential of dispersal to allow rapid colonization of dumps at the appropriate pedological stage. This is not the case for certain earthworms, snails, Isopoda (woodlice), Geophilo-

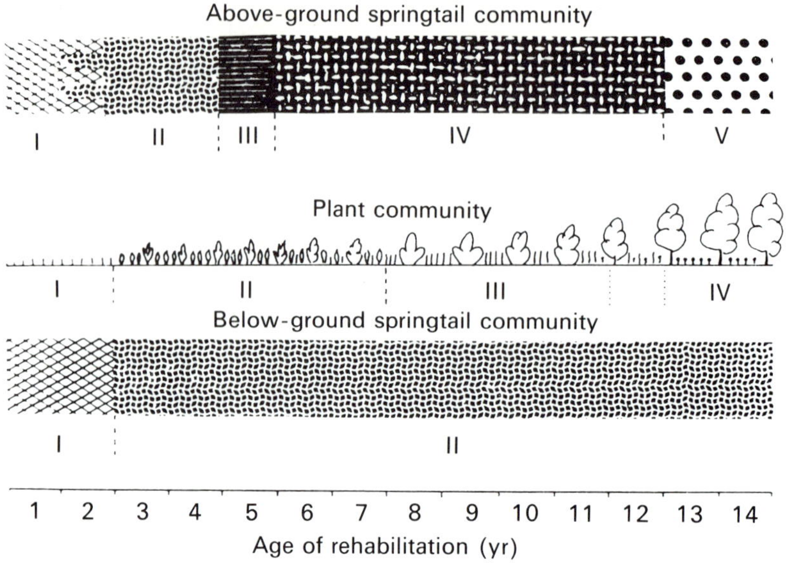

Fig. 13.11. Comparison of the succession plant communities and also epigaeic and hypogaeic communities of springtails on rehabilitated pleistocene dumps in the Berzdorf district (dumps A–N). Stages of vegetation: I = *Melilotus*; II = *Artemisia*; III = *Holcus-Achillea*; IV = *Senecio nemorensis*. Epigaeic communities of springtails: I = *Entomobrya lanuginosa*; II = *Ceratophysella succinea*; III = *Tomocerus vulgaris*; IV = *Lepidocyrtus cyaneus*; V = *Orchesella flavescens*. Hypogaeic communities of springtails: I = *Proisotoma minuta*; II = *Isotomodes productus-Micranurida pygmaea* (from Dunger, 1975).

morpha (centipedes) and some other less important groups. The experience from studies in the GDR and other countries indicates that the introduction of earthworms may often be an essential management consideration in order to accelerate and optimize the rehabilitation process. The international knowledge on the methods of stimulating and/or introducing earthworms has recently been summarized by Brun *et al.* (1988).

Summary

In the GDR, the return of soil fauna to 40 different dumps has been studied between 1961 and 1986. The greatest attention has been paid to the colonization by earthworms and microarthropods, especially Collembola. The populations of animals in some dumps have been studied over 25 years. The dispersal behaviour of important species, and the conditions for stable settlement, are discussed. The results allow some conclusions about the soil faunal action and interaction in rehabilitated dumps as well as about the pedozoological characterization of stages of succession in developing dumps. Finally, experiences with the practical use of pedozoology for the stimulation of dump rehabilitation are mentioned.

References

Broen, B. von & Moritz, M. (1965). Spinnen (Araneae) und Weberknechte (Opiliones) aus Barberfallen von einer tertiaren Rohbodenkippe im Braunkohlenrevier Böhlen. *Abhandlungen und Berichte des Naturkundemuseums Görlitz*, **40** (6), 1–15.

Brun, J. J., Cluzeau, D., Trehen, P. & Bouché, M. B. (1988). Biostimulation: Perspectives et limites de l'amélioration biologique des sols par stimulation ou introduction d'espèces Lombriciennes. *Revue Ecol. Biol. sol. Paris* (in press).

Brüning, E., Unger, H. & Dunger, W. (1965). Untersuchungen zur Frage der biologischen Aktivierung alttertiärer Rohbodenkippen des Braunkohlentagebaues in Abhängigkeit von Bodenmelioration und Rekultivierung. *Zeitschrift für Landeskultur*, **6**, 9–38.

Dunger, W. (1968). Die Entwicklung der Bodenfauna auf rekultivierten Kippen und Halden des Braunkohlentagebaues. Ein Beitrag zur pedozoologischen Standortsdiagnose. *Abhandlungen und Berichte des Naturkundemuseums Görlitz*, **43** (2), 1–256.

Dunger, W. (1969). Fragen der natürlichen und experimentellen Besiedlung kulturfeindlicher Böden durch Lumbriciden. *Pedobiologia*, **9**, 146–51.

Dunger, W. (1975). On the delimitation of soil microarthropod coenoses in time and space. In *Progress in Soil Zoology*, ed. J. Vanek, pp. 43–9. Praha: Academia Publishing House.

Dunger, W. (1979). Bodenzoologische Untersuchungen an rekultivierten Kippböden der Niederlausitz. *Abhandlungen und Berichte des Naturkundemuseums Görlitz*, **52** (11), 1–19.

Dunger, W. (1986). Observations on the ecological behaviour of some species of the *Tullbergia krausbaueri* group. In *Second International Seminar on Apterygota*, ed. R. Dallai, pp. 111–15. Italy: University of Siena.

Dunger, W. (1987). Zur Einwirkung von Kahlschlag, Grundwasserabsenkung und forstlicher Haldenrekultivierung auf die Boden-makrofauna, insbesondere Regenwürmer. *Abhandlungen und Berichte des Naturkundemuseums Görlitz*, **60**, 29–42.

Dunger, W. (1989). Einschätzung produktionsbiologischer Parameter. In *Methoden der Bodenbiologie*, ed. W. Dunger & H. J. Fiedler, pp. 269–81. Jena: Fischer-Verlag.

Dunger, W. & Dunger, I. (1984). Zur Kongruenz von Phytozönosen und Collembolan-Synusien. In *Verhandlungen SIEEC* X, pp. 32–4. Budapest: Z. Kaszab.

Dunger, W. & Voigtländer, K. (1989). Succession of myriapoda in primary colonization of reclaimed land. In *Progress of 7th International Congress of Myriapodology,* ed. A. Minelli, pp. 141–6. Amsterdam.

Schiemenz, H. (1964). Zikaden (Homoptera Auchenorrhyncha) von einer tertiären Rohbodenkippe des Braunkohlentagebaues Böhlen. *Abhandlungen und Berichte des Naturkundemuseums Görlitz*, **39** (16), 1–8.

Vogel, J. & Dunger, W. (1989). Langzeitbeobachtungen zur Sukzession von Staphyliniden und Carabiden auf forstlich rekultivierten Halden des Braunkohlentagebaues. *Abhandlungen und Berichte des Naturkundemuseums Görlitz*, **63** (in press).

Zerling, L. (1987). *Zur Wiederbesiedlung einer landwirtschaftlich genutzten Kippe des Braunkohlentagebaues durch bodenbewohnende Kleinarthropoden unter besonderer Berücksichtigung der Springschwänze* (*Insecta: Collembola*). M.Sc. Thesis, Universität Halle-Wittenberg.

14

Sixteen years of fauna invasion and succession in the Lauwerszeepolder

JAN MEIJER

Introduction

Imagine a 9000-ha tidal estuary that is suddenly cut off from the sea's tidal influence and gradually develops into a 2000-ha freshwater lake surrounded by more or less undisturbed areas. This summarizes the events in the Lauwerszeepolder in the north of the Netherlands, which was closed off from the sea on May 23rd, 1969.

Before its enclosure, the Lauwerszee consisted of a system of gullies, mudflats and sandbanks with saltmarshes along the coast. The mudflats and sandbanks were submerged twice daily by seawater coming in through gullies. This brought in fresh amounts of detritus for the countless bottom-dwelling worms, bivalves and arthropods. An enclosing dam was completed by means of 'open' caissons through which the tide could pass. On May 23rd, 1969, the sliding doors of the caissons were closed at low tide thus completing the enclosure. Since then, sluices have kept the lake at the desired level. The termination of this tidal influence left an enormous area which was completely devoid of animal life. The gullies of the former Lauwerszee formed a lake, the Lauwersmeer, that initially contained salt water with much of its usual fauna. However, rainfall and freshwater runoff from the neighbouring province of Friesland had by September 1969 converted it into a freshwater lake. After enclosure of the polder, the flora and fauna of the original Lauwerszee salt marshes went through a series of changes. However, the newly emerged area developed fauna and flora as a true primary succession, because it started without terrestrial biota.

Although the Dutch have a long-standing tradition of constructing polders, very little data exist on the development of fauna in such areas. Feekes (1936) studied the development of natural vegetation in the Wieringermeerpolder, and made passing reference to the fauna, especially the pests. Haeck (1969) reported on the colonization of the mole in the IJsselmeerpolders and Haeck (1971), Haeck, Hengeveld & Turin (1980) and Mook (1971) undertook investigations on the colonization of the same polders by invertebrates, mostly carabid beetles. All these studies were short-term and dealt with polders constructed in what was formerly a freshwater lake. Moreover, these polders were immediately put to agricultural use.

There was clearly a need for a long-term study of the colonization and development of polder fauna. The Lauwerszeepolder was especially suited for such work since large parts of the polder were to become a nature reserve. The government bodies responsible for the reclamation and development of the polder planned to monitor the changes in the bird fauna but no broader zoological investigations were planned. Around that time I obtained a grant to conduct a 4-year study of the colonization of arthropods in this polder. After this period the study continued on a smaller scale in my spare time. The aim of this study essentially was to monitor the colonization of the polder by certain groups of arthropods. Data on the abiotic environment, as well as the vegetation of this polder are scarce and reflect the interests of other workers. Joenje (1978) presented detailed information on the soil and the development of the vegetation during the first 6 years of reclamation, while Ente & De Glopper (1985) described various aspects (soil, vegetation, birdlife, land-use, etc.) over the first 15 years. A discussion of the invertebrate fauna in 4 study areas during the first 6 years after reclamation can be found in Meijer (1980).

Large areas of the Lauwerszeepolder were designed to be nature reserves (2800 ha), or military training grounds (2500 ha). The remainder, apart from the 2000-ha lake, was designated for agricultural (especially along the fertile borders) or recreational use (1800 ha) (Ente & De Glopper, 1985). Except for localized sowing of grasses to stabilize wind-blown sand, no attempts were made to vegetate the nature reserves. The prime function of these nature reserves was the protection of bird life, so access by the general public was forbidden. The Lauwerszeepolder as a whole has now become a very important feeding area for geese and other waterfowl (Van Eerden, 1985).

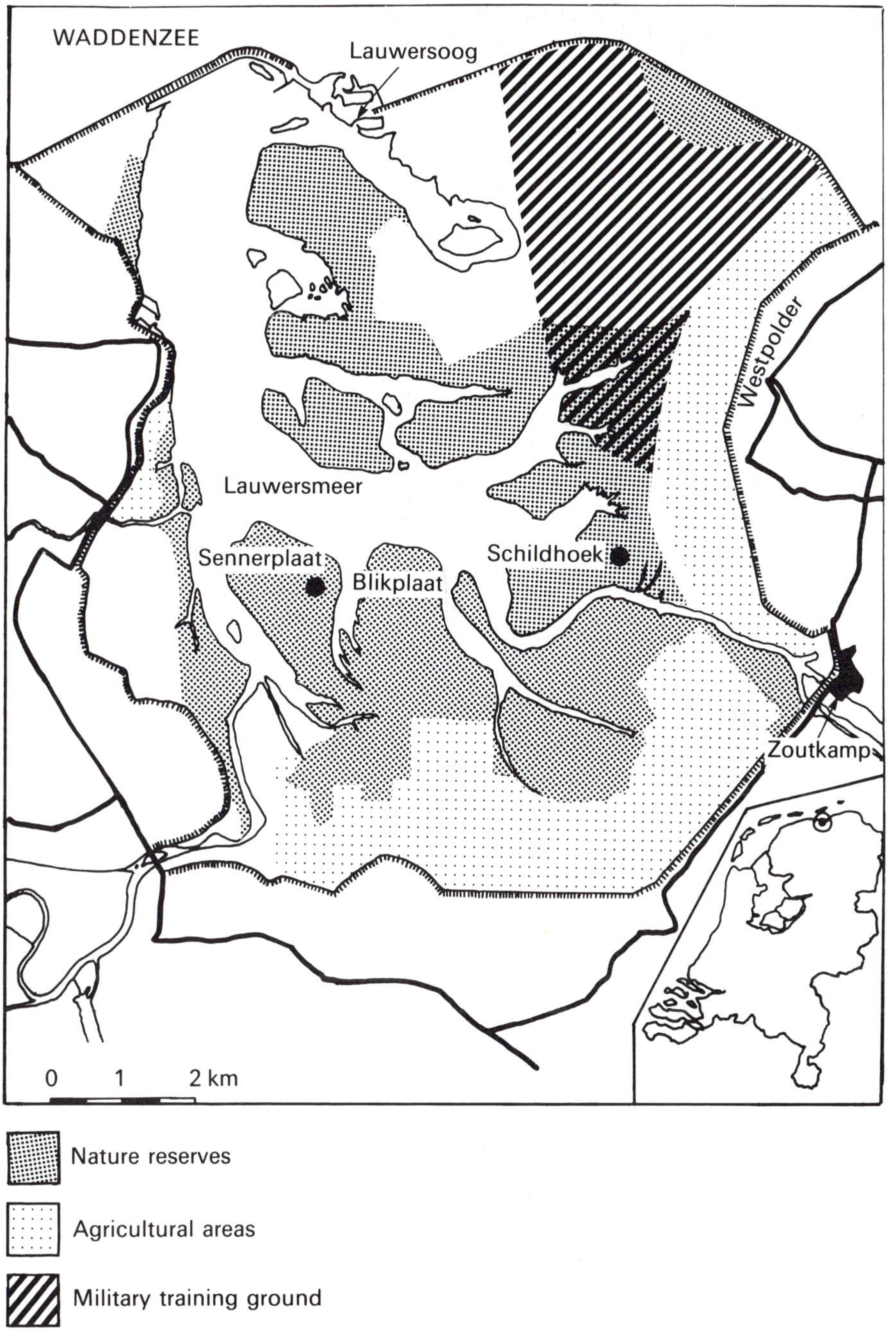

Fig. 14.1. Map of the Lauwerszeepolder showing the location of the study areas (●). Inset shows position of this polder (○) within the Netherlands (adapted from Ente & De Glopper, 1985).

The invertebrate study

The study sites

Initially four study areas were selected on the basis of soil type. These were: 'Silt A, B and C' and 'Fine Sand' in the Schildhoek area, 'Coarse Sand' in the Blikplaat area, and 'Silty Sand A, B and C' in the Sennerplaat, all situated in the nature reserve in the centre of the polder (Fig. 14.1). These were areas in which no revegetation was carried out (the neglect option, see Chapter 1). The invertebrate sampling was discontinued in 2 of these areas after 4 years. In the remaining areas, the Schildhoek A and B and the Sennerplaat A and B, 2 groups of 2 pitfall-traps have been continuously operated from 1969 until 1987 and I hope to be able to continue this study for a longer time span. In 1982 the growth of reeds became so dense that large areas became unsuitable as grazing ground for geese. Therefore cattle were introduced into the Schildhoek area to restore a grassy vegetation. Clearly this influenced the succession of vegetation and fauna. An 80 m circular patch was fenced off around both groups of traps in the Schildhoek area to prevent cattle damaging the traps.

Soil and vegetation of the study sites

The Sennerplaat study area was located on the waterfront, 80 cm above the mean lake level, and had a rather sandy soil with a clay layer at a depth of *ca.* 1 m. It was initially better drained and less often inundated than the Schildhoek study area, which was located away from the lake in flat terrain on heavier soil, some 40 cm above the lake level. Initially the barren Schildhoek study area was frequently submerged by water blown across the area from the lake, even when the lake level was relatively low. As the vegetation developed, rainwater runoff was impeded, as was the blowing in of water from the lake. Therefore, after some years, both areas tended to be waterlogged throughout winter as well as after excessive rainfall in summer. During periods of drought, the water table sank appreciably and the terrain became dry (Fig. 14.2). Hoekstra (1979) showed that the soil of areas like the Sennerplaat has a low storage capacity for water; a net rainfall (precipitation minus evaporation) of 57 mm causes the water table to rise by 1 m. This also explains why the study areas are virtually always waterlogged in winter.

The slow desalination of the undisturbed polder soil (Fig. 14.3) was probably influenced by the nature of the underlying soil strata. This permitted very little horizontal transport of water, as demonstrated by

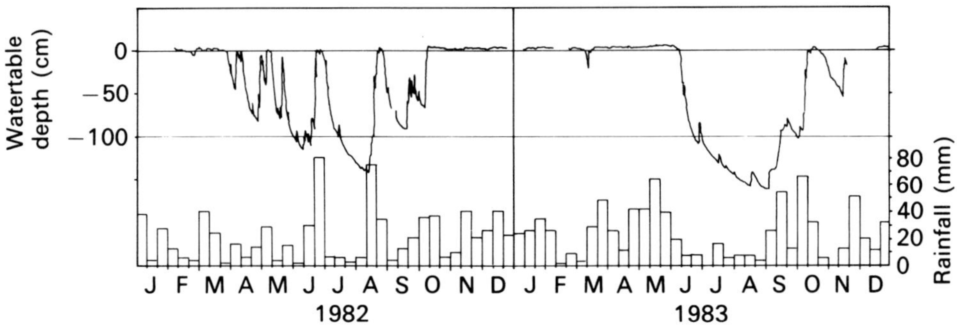

Fig. 14.2. Relationship between rainfall and water table in the Sennerplaat A during 1982 and 1983 (from Slager, 1985).

the position of the water table. Movement of water in the soil was essentially vertical (Slager, 1985). Consequently, desalination in areas like the Schildhoek was very slow. In some places the soil may stay permanently saline with, in dry periods, increasing salinity occurring in the upper layer due to capillary movement of saline water (De Glopper, 1985). The slightly coarser soil of the Sennerplaat permitted more percolation of water, and hence more rapid desalination. Initially this percolation of saline water was manifested by the extreme salinity along the border of the Sennerplaat (data in Meijer, 1980). This was probably caused by a surface flow of percolating saline water from the centre of the area. In later years the soil of the Sennerplaat appeared to become clogged and even the driest parts of the study area became waterlogged during winter (Fig. 14.2).

The seed bank in the soil of the study areas was initially very small. A loose growth of luxuriant halophytes developed in 1969 near the former salt marshes. This produced considerable amounts of seed which were dispersed by water into the centre of the polder. In 1970 a number of very well developed halophytes occurred in the study area and these produced sufficient seeds to enable them to become widespread by 1971. However, at this stage, total plant cover was at the most only 10 % (Fig. 14.4). The ongoing desalination of the soil, together with the arrival of other species, produced different plant communities in the Sennerplaat to those in the Schildhoek area (Table 14.1).

The vegetation of the Schildhoek may be characterized as follows: initially there was a cover of halophitic herbs that was gradually invaded by grasses such as *Puccinellia maritima*, a salt marsh species, and later by *Agrostis stolonifera* (Fig. 14.5). In the Schildhoek B area the latter species was accompanied by other grasses, while *Salicornia europaea* had

Table 14.1. *Vegetation of two sites in the Lauwersmeer where the pitfall traps were operated from* 1969 *to* 1984

	Schildhoek A															
	69	70	71	72	73	74	75	76	77	78	79	80	81	82	83	84
1. *Salicornia europaea*	+	1	2m	2m	2m	2m	2a	2a	2a	2a	2a	1	+			
2. *Atriplex hastata*			+	+	1	+	+	+	1	+	+	1	+			
3. *Suaeda maritima*				+	1	1	1	2a	2a	1	1		+			
4. *Spergularia marina*				1	1	2a	2a	2a	2a	+	1	+				
5. *Aster tripolium*						+	+	+	1	+	1	1	1	+	+	+
6. *Puccinellia maritima*								2a	2a	2m	2m	3	2m	2a	1	1
7. *P. capillaris/distans*							+	+	1	1	1	2a	2a	1		
8. *Spergularia media*							+	+		+	1	1	+			
9. *Agrostis stolonifera*								+	1	2a	2a	2m	3	3	4	5
10. *Juncus bufonius*														+		+
11. *Odontites verna serotina*													+			+
12. *Poa annua/trivialis*																
13. *Epilobium parviflorum*																
14. *Sagina nodosa*																
15. *Sagina maritima*																
16. *Taraxacum* sp.																
17. *Phragmitis australis*																
18. *Calamagrostis epigejos*																
19. *Mentha aquatica*																
20. *Leontodon autumnalis*																
21. *Epilobium adnatum*																
22. *Carex otrubae*																
23. *Angelica sylvestris*																
24. *Lycopus europaeus*																
25. *Holcus lanatus*																
26. *Heleocharis palustris*																
27. *Cardamine pratensis*																
Mosses											+	+	+	+	+	+
Total cover (%)	< 1	< 1	6	10	18	15	15	20	20	30	35	45	60	80	80	90
Height of vegetation (*d*)	3	4	3	3	3	2	2	2	3	2	2	2	3	3	3	4

Note: These data were gathered by Dr W. Joenje. (Explanation of symbols: + = some specimens, cover < 1 %; 1 = a number of specimens, cover < 5 %; 2a = many specimens, cover < 5 %; 2m = cover 5–25 %; 3 = cover 25–50 %; 4 = cover 50–75 %; 5 = cover 75–100 %)

disappeared 2 years earlier. Initially the vegetation on the Sennerplaat was the same as on the Schildhoek, but the turnover of species with time was much quicker and resulted in a taller vegetation which mostly consisted of reeds (*Phragmitis australis*) with an undergrowth of smaller grasses of the same species composition as on the Schildhoek. These reed beds (Fig. 14.6) have a different association of herb species in the Sennerplaat A and B areas. It is anticipated that in another 10 or 15 years the Sennerplaat will be covered by an open woodland of *Salix* and other species with an undergrowth of grasses and herbs. The undisturbed areas on the Schildhoek will also be covered with reeds, which, in the longer term, may be invaded by *Salix* species. The grazing of cattle in certain

Sennerplaat B															
69	70	71	72	73	74	75	76	77	78	79	80	81	82	83	84
	+	2m	2m	2m	1		+								
		1	1	2a	1										
		1	2m	2m	1	+									
			+	1	2a	1	+								
				+	1	2a	2m	2a	2a	1	1	+			
			+	2a	2m	2m	2m	1	+						
						+	2m	2m	3	2m	2m	2m	2m	2m	2m
							+	+	1	1	+				
									+		+	+	+		+
				2m	3	3	2m	2m	2m	2a	1	2m			
										+	+	+			
											+	+			
					1	1		+			+				
										+	+	+			
								+	2a	2a	2m	3	3	3	3
									+	1	2a	2m	2m	2m	3
											+	2a	2m	2a	2m
											+	+	+		
									+	+	+	+			
												+	+	+	+
									+	1	2a	2m	2m	2m	2m
											+	1	1	1	2a
												+	+	1	1
				1	3	2m	2a	2m	2m	2m	2m	2m	2m	2m	2m
0	0	10	15	30	55	45	40	50	65	60	80	90	90	95	95
		4	3	3	2	1	2	2	5	5	8	13	15	15	16

areas of the Schildhoek since 1982 had no visible effects on the vegetation around the traps due to the fencing out of the cattle. Outside the fences the vegetation was generally reduced to a short turf of *Agrostis* and the growth of reeds was inhibited (Fig. 14.7). Figure 14.8 schematically illustrates the development of the vegetation in the Lauwerzeepolder at various levels above the lake; the approximate location of the sites is indicated.

The vegetation data, together with data on the soil and weather, were submitted to a canonical correspondence analysis (CANOCO; Ter Braak, 1985) by Mr. H. Siepel of the Rijks Instituut voor Natuurbeheer. This produced a first axis (49.1% of total variance) which was correlated with the age of the polder (correlation coefficient = 0.62) and a second axis (29.2% of total variance) which was correlated with the soil salinity

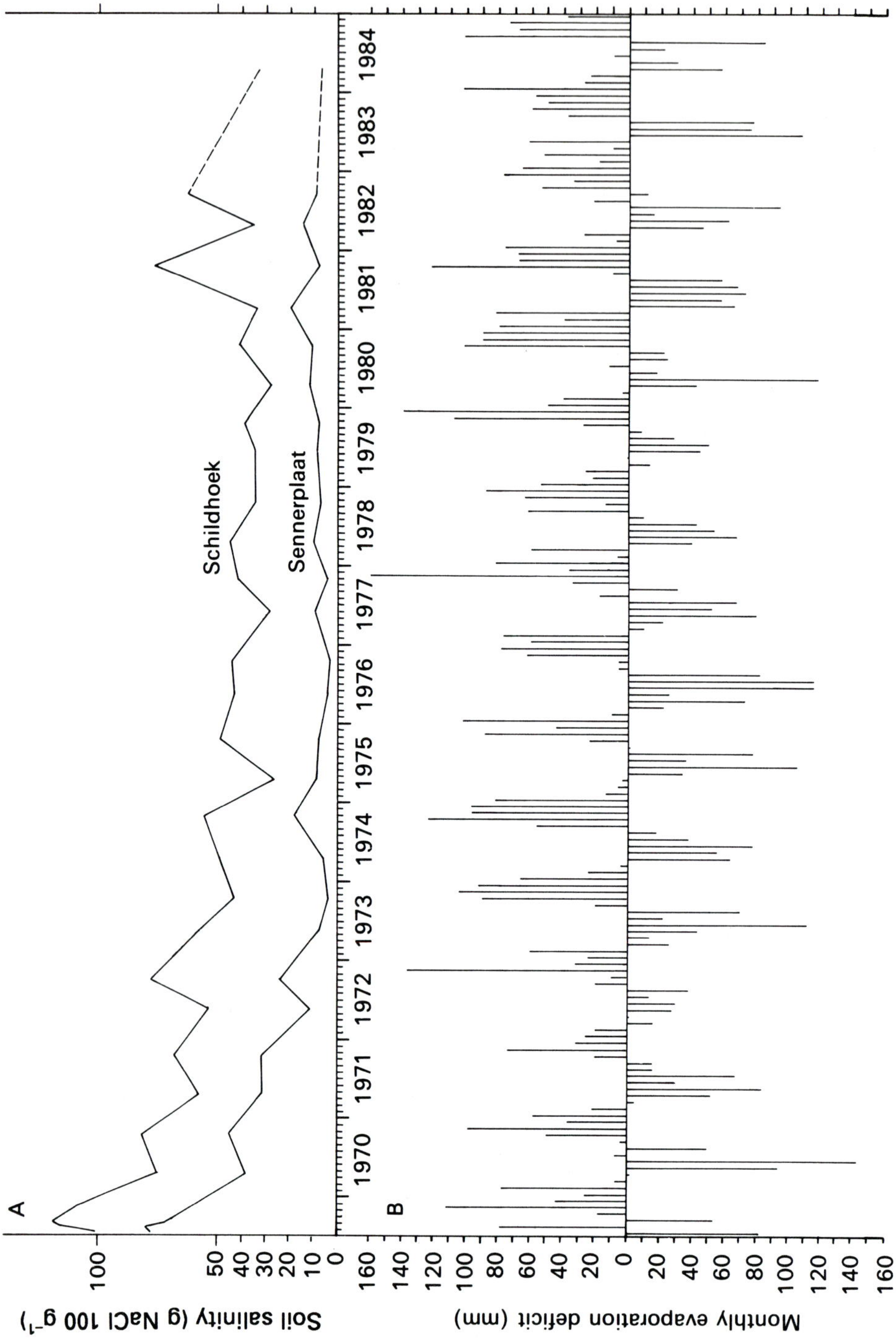

Fig. 14.3. Soil salinity (g NaCl 100 g^{-1} dry soil) in relation to the monthly evaporation deficit (precipitation minus evaporation) (data gathered by the Rijksdienst voor de IJsselmeer Polders, Lelystad)

Fig. 14.4. The Schildhoek B in 1971 where pitfall trapping was performed. The vegetation is mainly *Salicornia europaea*. In the right foreground is a specimen of *Atriplex hastata*, to the left is a specimen of *Suaeda maritima*.

Fig. 14.5. The Sennerplaat A in 1972 where pitfall trapping was performed. The vegetation is *Salicornia europaea* and *Atriplex hastata* which is being invaded by grasses.

Fig. 14.6. The Sennerplaat B in 1984. The reeds have an undergrowth of *Mentha aquatica* and *Lycopus europaeus*.

Fig. 14.7. The Schildhoek B in 1984 showing pitfall trapping area fenced off against cattle.

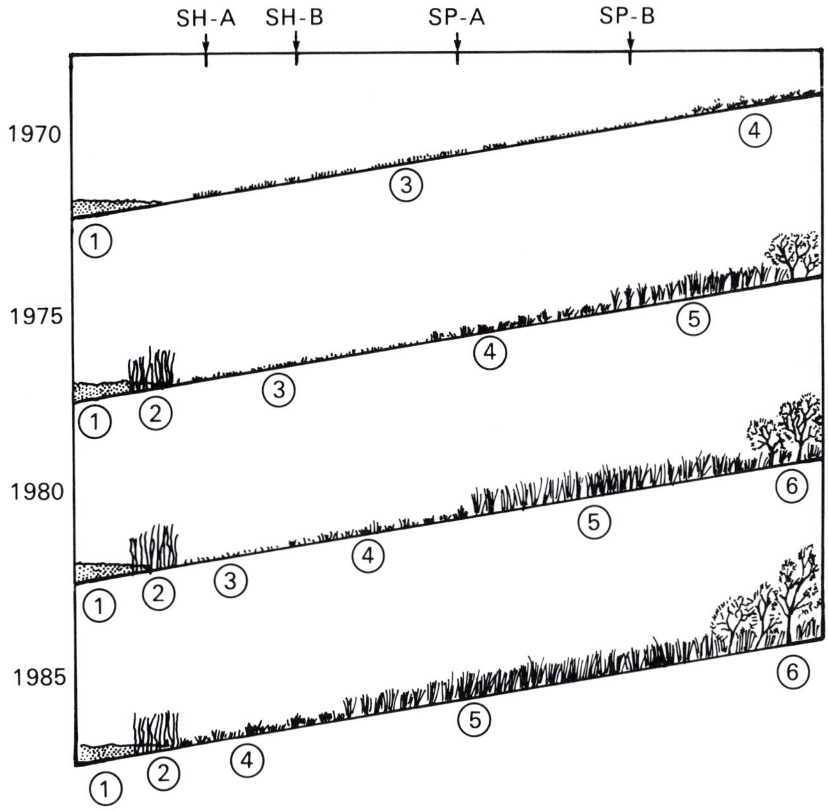

Fig. 14.8. Schematic representation of the development of vegetation in the Lauwerszeepolder. Key: 1, water; 2, reeds; 3, *Salicornia*/*Puccinellia*; 4, *Agrostis*; 5, reeds; 6, *Salix* shrubbery. The approximate positions of the Schildhoek (SH) and Sennerplaat (SP) A and B study sites are also indicated (from Prop, 1985).

in spring (correlation coefficient $= -0.57$). It is likely that these variables are correlated, but CANOCO is designed to discern between the variance in the data caused by each variable. The first axis therefore represents the influence of age and the second axis the influence of the different desalination in the 4 sites.

Borrius (1986) studied the impact of grazing on the breeding of meadow birds, and also looked at the soil and soil-surface fauna during 1983. All the vegetation associations recorded in my Sennerplaat and Schildhoek sites until 1984 were found in Borrius' 1983 study areas. He found that, depending on soil salinity and elevation, the development of the vegetation succession was either retarded or speeded up. Close to the lake *Salicornia* and reed beds were present, while on the higher ground *Salix* shrubbery was the dominant vegetation.

Invertebrate sampling methods

The surface-active invertebrates were sampled by two pitfall-traps at each site. The pitfall-traps consisted of a sunken metal container within which a 15-cm-wide plastic funnel was suspended. Animals were trapped in a plastic vial, containing 4% formalin, screwed onto the lid attached to the nozzle of the funnel. Catches were gathered each week from April through October by changing the plastic vial. The 2 traps at each site were placed 6 m apart. The catches were sorted and identified at a later stage; all material is kept in the Rijksmuseum van Natuurlijke Historie at Leiden.

To save space, data for only some of the catches and for only 2 sites, the Schildhoek A and the Sennerplaat B, the least and the most developed sites respectively, are presented and discussed.

A number of statistical measures were used to describe certain aspects of the invertebrate community. These are number of species, William's index of diversity (α), an index of dominance (d) (Southwood, 1978), and indices measuring the change of the population from year to year, namely mean log R and annual variability (AV) (Wolda, 1978). R is the net reproductive rate of one species from one year to the next assuming no dispersal. AV is the variance of mean log R and describes the range of increases and decreases that occurred in the population. I have followed Wolda in accepting all data pairs in which the catch of at least one of the pair is 2 or more. The catch of the other member of the pair may be as low as zero, and in that case is arbitrarily set at 0.5.

The trapping results, together with data on the soil, weather, and vegetation, were submitted to a canonical correspondence analysis. Unfortunately, the data on the vegetation were unsuitable for comparison with the carabid and spider data due to colinearity of the matrix of vegetation data, therefore only the total cover of the vegetation was used in the analysis. Apart from soil salinity, only one set of soil data (for 1969) was available, therefore only the percentage of clay was used as a soil variable in the analysis. Data on soil salinity were gathered in spring and autumn by the Rijksdienst voor de IJsselmeerpolders; these data were also used in the analysis.

Invasion and succession of the arthropod fauna

I have concentrated on 2 major groups of arthropod; the carabid beetles and the spiders. Both groups are numerous and readily identifiable and, moreover, they are mostly non-specialized carnivores which might be expected to appear early on in the succession. Thanks to the help of

Mr Carl Schulz, leafhoppers (Homoptera, Auchenorrhyncha) were identified and these will be mentioned also. Late colonizers such as opilionids (harvestmen), isopods (slaters) and certain other groups were also noted. The results of Borrius' (1986) study of soil and surface invertebrates will also be discussed in the context of my investigation.

Carabids (*Coleoptera*, *Carabidae*)

A total of 124347 specimens of 78 species were caught in the duration of the study. The catches of the 22 most common species, as well as some statistics on the total catch, are shown in Table 14.2; these are the species used in the CANOCO analysis. The species not mentioned in the Table were caught in very low numbers and represented 779 specimens, or 0.6% of the total catch.

Although the total number of species caught was high, the highest number of species caught in any one year was only 27 (Sennerplaat B, 1976). It is apparent that even after 16 years, the carabid fauna is still very unstable in terms of species composition. In most years, catches are strongly dominated by one species; even in 1983 the index of dominance in the Sennerplaat areas was about 0.8. Of the 32 annual catches mentioned in Table 14.2, only 7 have an index of diversity greater than 4. Annual variance is also high, although mean log R was usually close to zero. This indicates a rapid change in the composition of the fauna with a decrease in numbers of one species balanced by an increase in another. During the first 6 years of the study, the fauna exhibited an increasing stability of species composition (Meijer, 1980). The present data, from a different set of traps to that described in Meijer (1980), also indicate this trend of decreasing annual variance during the early period, but in later years, instability of species composition increased again.

The 1969 catch was very small indeed and comprised only 22 specimens, probably all immigrants. In 1970, numbers had strongly increased in the Schildhoek sites and in 1971 in the Sennerplaat sites also. The lag in the latter case was caused by the environmental differences – the vegetation in the Sennerplaat developed rather more slowly. The steep increase in numbers of *Bembidion minimum* and *Bembidion varium* (Table 14.2) was caused by the ability of these species to produce more than their normal one generation per year (Meijer, 1980).

Figure 14.9 shows the yearly distribution of the catches of some important species. A gradual replacement of species is apparent. From this Figure and Table 14.2, one might get the impression that in some years conditions were changing so quickly that the carabid fauna was

Table 14.2. *Annual catches from* 1969 *through* 1984 *in the Schildhoek* A *and the Sennerplaat* B *of those carabid species that were used for canonical correspondence analysis*

	Schildhoek A												
	69	70	71	72	73	74	75	76	77	78	79	80	81
1. *Bembidion varium*		1462	1579	358	109	102	323	3	1	3	1		1
2. *Trechus quadristriatus*	1	8		2	2			3	14	10	1		
3. *Bembidion aeneum*		10	48	123	5	16	16	26	8	8	16	38	74
4. *B. normannum*		6	10	25	12	16	15			11	3	1	
5. *Dichirotrichus pubescens*	1	44	1577	1655	2107	368	55	641	3695	5616	2631	687	26
6. *Pogonus chalceus*	1	11	10	226	567	406	150	283	644	847	167	65	3
7. *Bembidion minimum*		356	205	234	96	103	594	242	31	90	194	166	102
8. *Dyschirius salinus*		2	40	72	510	1876	4258	2208	1038	1471	279	52	8
9. *Loricera pilicornis*				5	2	1	3	1		2	135	18	2
10. *Agonum marginatum*	1	1		2	3	1		2	2	1	4	8	42
11. *Dyschirius luedersi*													222
12. *Amara convexiuscula*			1		1			3	1				
13. *Bembidion properans*						2		4				1	1
14. *Amara communis*						1					1		
15. *Acupalpus dorsalis*			1			2	3	23	1	1	1	1	1
16. *Harpalus pubescens*			8	1			1	2	5	5			1
17. *Bembidion bipunctatum*								2		1	4	3	5
18. *Pterostichus strenuus*				14	3	7	2	1	1	1	2	6	20
19. *P. vernalis*					1	2	1						2
20. *Clivina fossor*		2	42	18	6	13	3	5	2	7	7	2	2
21. *Dyschirius globosus*													1
22. *Bembidion assimile*												2	7
Total catch	5	1907	3530	2753	3493	2921	5432	3456	5450	8083	3448	1052	526
Total number of species	5	15	14	18	23	19	17	21	17	19	17	16	20
Dominance index (*d*)	0.20	0.77	0.45	0.60	0.60	0.64	0.78	0.64	0.68	0.70	0.76	0.65	0.42
Index of diversity (α)		2.21	1.85	2.58	3.30	2.72	2.17	2.98	2.17	2.33	2.33	2.68	4.12
Mean log R		0.17	0.05	0.02	−0.00	−0.01	−0.00	0.00	0.01	0.01	−0.01	−0.01	0.01
Annual variability		3.01	0.95	0.68	0.54	0.48	0.32	0.69	0.46	0.31	0.59	0.28	1.15
Number of data pairs for AV		9	1	16	15	17	16	20	18	17	16	12	14

unable to adapt, especially when the development was temporarily reversed between 1975 and 1977. During this period the early colonizer, *Dichirotrichus pubescens*, strongly increased in numbers in the Schildhoek, and to a lesser extent in the Sennerplaat, while no other species decreased in abundance at that time.

A strip of uncultivated land in the 100-year-old Westpolder (Fig. 14.1), adjacent to the Lauwersmeer, has a number of carabid species (Meijer, 1976) that could be expected to be able to survive in the Sennerplaat environment. However, they are still not caught after 16 years, or are present in very small numbers (e.g. *Bembidion guttula* and *Bembidion ustulatum*) in the Lauwerszeepolder sites. Probably some of the 56 species which are still rare in the Lauwerszeepolder may become common; for instance, in 1978 no one could have foretold the eventual success of *Bembidion assimile* or *Dyschirius globosus* in the Sennerplaat (Fig. 14.9).

			Sennerplaat B															
82	83	84	69	70	71	72	73	74	75	76	77	78	79	80	81	82	83	84
				23	864	78	16	1	1	1							1	
1	2		1	3	7	2	2		2		1	4			1		1	
131	52	257		4	8	8	4	5	27	1	6	16	6	15	9	6	2	4
		1			6				1								1	
1				1	845	573	2873	1091	9	118	587	2	3					
	1	1					1		2	1	1	3		1				
171	155	158		8	20	55	37	74	76	51	2	13	6	2	2	2	3	1
12	2	3				2	138	83	73	76	56	53	29	5	1	2	1	
	1	2				27	10	3	2			20	8	5		2		
3		1	1		4	21	5		1		1	2				1		1
223	39	1														5	31	2
					1	129	15		2	4	1							
3						4	29	86	24	44	19	4	3	13	2		1	
								1	67	7	9	14	8	10	1	2		1
	4	8					1	17	196	214	67	18	9	2	2	9	5	22
					2		4		1	9	14	9	3					
							428	650	166	83	18	30	172	39	2			
62	6	19	3		1	41	25	16	23	40	29	95	99	223	77	125	53	39
17	9	16						3	2	3		3	50	95	64	48	36	16
1		3				2				10	2	18	28	80	54	130	80	72
1	19	156										10	23	69	415	1229	3264	1622
23	27	118										1	4	12	147	758	679	1591
656	318	758	8	41	1781	944	3605	2039	695	710	858	353	493	594	788	2336	4170	3399
16	13	19	5	7	16	14	22	19	25	27	26	29	22	24	18	22	22	21
0.34	0.49	0.34	0.38	0.56	0.49	0.61	0.80	0.54	0.28	0.30	0.68	0.27	0.35	0.38	0.53	0.53	0.78	0.48
2.95	2.72	3.54		2.43	2.42	2.33	3.12	2.90	5.07	5.56	5.06	7.49	4.72	5.02	3.28	3.36	3.05	2.98
−0.01	−0.02	0.02		0.08	0.09	0.01	0.02	−0.01	0.01	0.01	−0.01	0.00	−0.00	−0.00	−0.02	0.01	−0.01	0.00
0.57	0.56	0.62		1.16	1.53	0.91	1.17	0.60	0.72	0.64	0.43	0.70	0.41	0.52	0.51	0.32	0.22	0.35
17	14	13		6	1	16	17	20	18	25	23	26	24	23	17	18	18	16

Macropterous individuals of di- or polymorphic winged species seem to have the best chance of reaching new habitats. Moreover, the view is commonly held that the incidence of macroptery in a carabid population declines with the age of the population (cf. Thiele, 1977, pp. 289–97). Table 14.3 lists the occurrence of the macropterous morph of two species in the Sennerplaat B and Schildhoek A areas between 1978 and 1984. The only species in which the micropterous form was caught before the macropterous form is *D. globosus*, probably transported into the polder on floating tufts of shore plants. In the above-mentioned Westpolder, only micropterous specimens of this species were found in a catch of 526 specimens. Den Boer (in Thiele, 1977) also reported the absence of macropterous specimens in his Dutch material of this species. However, the apparently rare macropterous form suddenly appeared in the Lauwerszeepolder, even later than the micropterous forms, indicating the dispersal power of the macropterous form. Haeck (1971) also

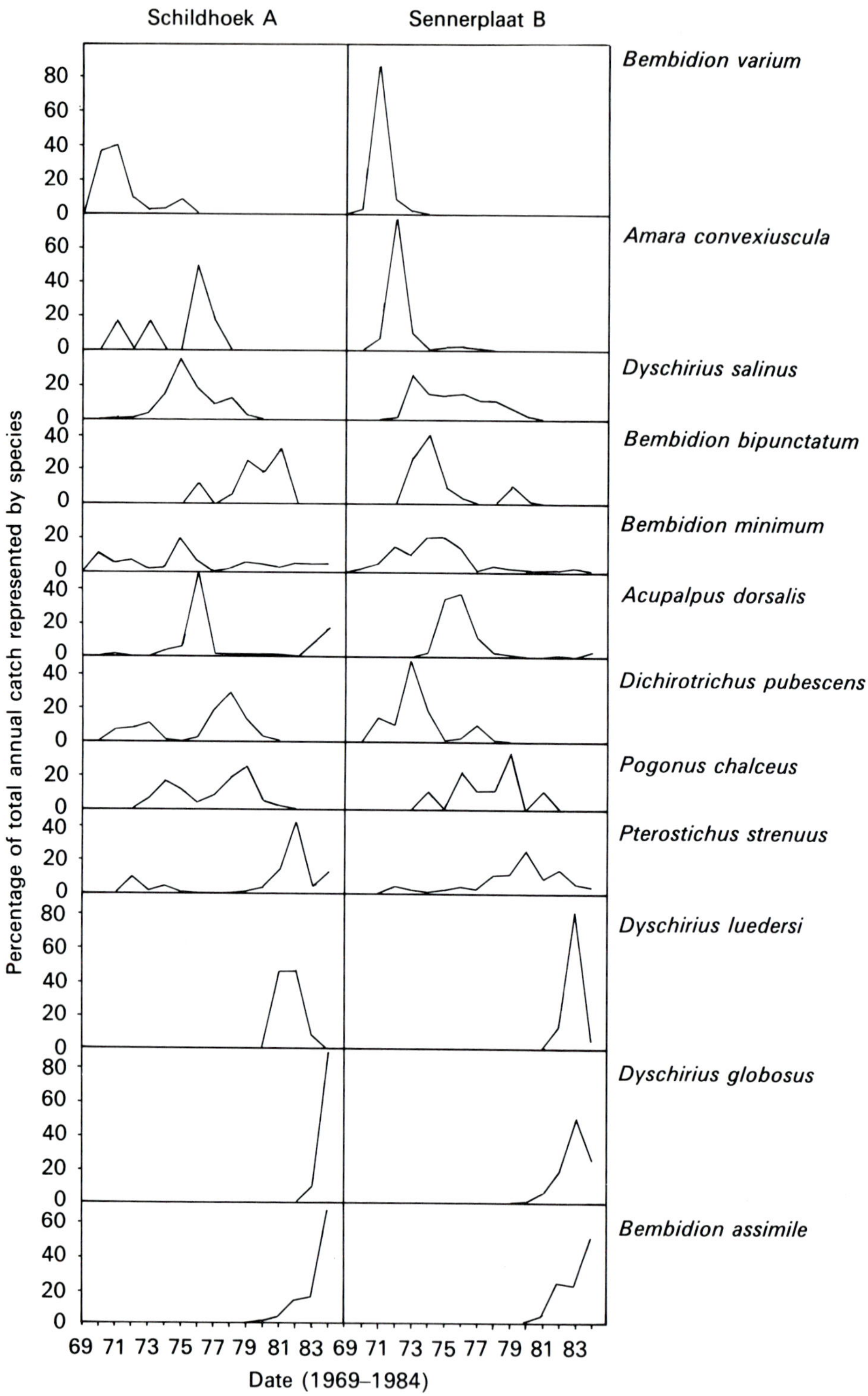

Fig. 14.9. Distribution of the catches of some carabid beetle species over the study period in the Schildhoek A and the Sennerplaat B study sites.

Table 14.3. *Percentage of macroptery in two species of carabid beetles in the Lauwerszeepolder*

Species	Year	Schildhoek		Sennerplaat	
		A	B	A	B
Bembidion assimile					
	1978	—	—	—	100
	1979	—	100	—	100
	1980	100	—	75	41.7
	1981	100	89.2	89.5	56.5
	1982	87.0	85.9	79.0	34.8
	1983	81.5	78.9	63.3	33.0
	1984	97.5	94.7	51.8	24.6
Dyschirius globosus					
	1978	—	—	0	0
	1979	—	0	—	0
	1980	—	0	0	1.4
	1981	0	0	0.8	1.0
	1982	0	7.4	0.9	0.2
	1983	10.5	15.0	0.4	0.3
	1984	10.9	7.5	0.8	0.4

The total catches in the Schildhoek A and the Sennerplaat B are given in Table 14.2 (— means that the species was not caught).

reported macropterous specimens of this species in the IJsselmeer polders.

Even over short distances, the percentage of macropterous specimens may be different: there were about twice as many macropterous *B. assimile* in the Sennerplaat A than in the Sennerplaat B, although the sites are less than 100 m apart. It seems that such differences are due to differential reproductive success rather than dispersal. In general, the percentage of the macropterous morph of this species declined in the study area. All specimens of *Clivina fossor* caught in the Schildhoek A and B, contrary to those caught in the Sennerplaat, were macropterous. Either they were all immigrants, unable to survive the wet winter in those sites, or the adverse environment allowed successful breeding of the macropterous morph only.

The results of the canonical analysis are plotted in the way indicated by Ter Braak (1986) in Fig. 14.10. Table 14.4 shows the first axis, which accounts for 49.0% of the total variance, to be strongly correlated with the number of years since the reclamation, and the second axis, which accounts for a further 21.5% of the total variance, to be correlated with

Table 14.4 *Correlation coefficients of environmental variables with the first two axes of canonical correspondence analysis for the carabid beetle data shown in Table* 14.2 *and Fig.* 14.10 *and the spider data shown in Table* 14.5 *and Fig.* 14.13

	Carabids		Spiders	
Environmental factor	axis 1	axis 2	axis 1	axis 2
A. Age of polder (+)	0.76	0.10	0.85	−0.20
B. Excess of rain 1 (*)	0.27	−0.00	0.14	−0.33
C. Excess of rain 2 (*)	0.68	0.29	0.49	−0.32
D. Excess of rain 3 (*)	−0.05	0.30	−0.04	−0.19
E. Excess of rain 4 (*)	0.11	0.16	0.12	0.06
F. Total cover (○)	−0.24	0.03	−0.18	0.05
G. Salinity (autumn) (●)	−0.40	−0.06	−0.16	0.26
H. Salinity (spring) (●)	−0.41	0.75	−0.14	0.62
I. % of clay (×)	−0.40	0.74	−0.14	0.62

Notes: + Years since reclamation
* Excess of rainfall over evaporation (cf. Fig. 14.3) in: 1 = Jan.–Mar.; 2 = April–June; 3 = July–Sept.; 4 = Oct.–Dec.
○ Total cover of vegetation (Table 14.1)
● g NaCl 100 g^{-1} dry soil × Particles < 2μm (clay).

salinity in spring and the percentage of clay of the soil. Thus the most important factor seems to be the age of the polder, which itself is an expression of the maturity of the polder ecosystem.

For clarity, only the data on the 2 extreme sites, the Schildhoek A and the Sennerplaat B are plotted in Fig. 14.10. In this Figure the arrangement of the catches relative to the axes indicates to what extent the catches have changed from year to year. It is apparent that the development of the carabid fauna was not gradual. For instance, in the Schildhoek the catches from 1971 through 1979 are grouped closely together, while in the Sennerplaat B the development of the carabid community slowed down between 1974 and 1977. In 1984, the development of the fauna seems to have stopped.

The relative lengths of the arrows in Fig. 14.10 represent the importance of environmental factors in determining the occurrence of carabid species over the years and the sites. The arrangement of the species relative to the arrows indicates to which degree each factor played a role in determining their occurrence. Indeed the arrangement of the species, or rather the centres of the distributions of the species, relative to the arrow 'age of the polder' (A) reflects the order in which

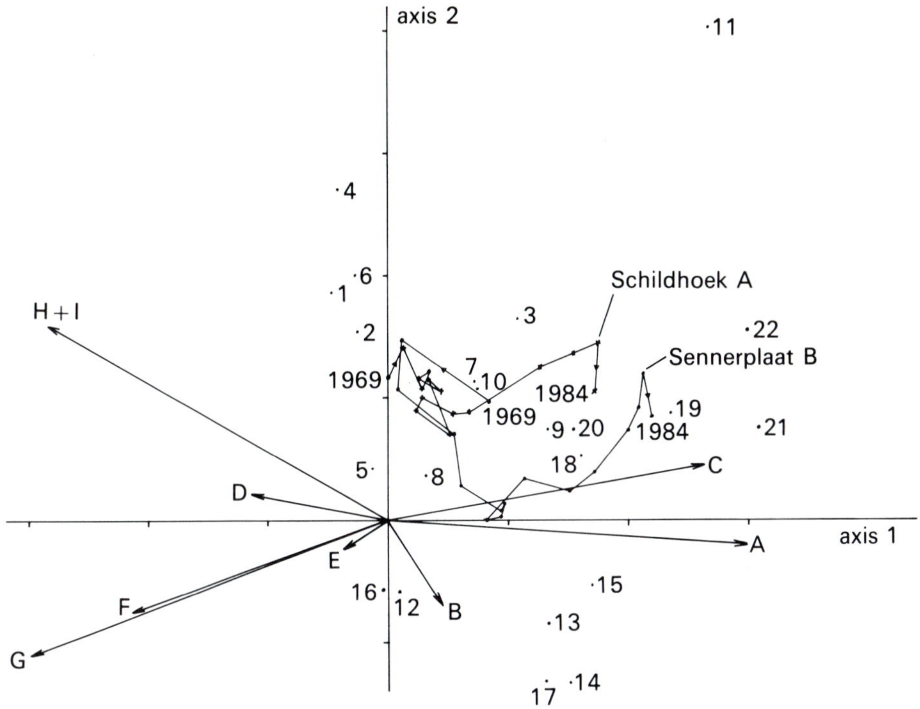

Fig. 14.10. Results of the CANOCO analysis of the carabid beetle data. The species are numbered (1–22), as in Table 14.2; the environmental factors are labelled (A–I), as in Table 14.4.

the species appeared in the polder. The arrangement with respect to the second axis reflects the distribution of the species over the sites/years, as ranked by salinity. Considering the variance (70.5%) explained by the first 2 axes of the ordination, and the correlation of environmental factors with those axes (Table 14.4), it appears that age of the polder, excess of rain in spring, percentage of clay and soil salinity are the major factors determining the development of the carabid fauna.

Elsewhere in the Schildhoek, Borrius (1986) found that even in 1983, there were carabid species which had disappeared from the Schildhoek A and B sites in the early 1970s, while other species which I had recorded had not yet arrived. His pitfall-trap catches from different vegetation types illustrate this for 3 common species of *Dyschirius*. *Dyschirius salinus* was mostly caught in low lying *Salicornia*, *Dyschirius luedersi* in *Agrostis* turf and reedbeds, and *D. globosus* in situations comparable to the 1983/84 vegetation of the Sennerplaat B. The same holds for various *Bembidion* species. Although vegetation type obviously influences the carabid

Table 14.5. *Year catches from* 1969 *through* 1984 *in the Schildhoek A and the Sennerplaat B of those spider species that were used for canonical correspondance analysis*

	Schildhoek A												
	69	70	71	72	73	74	75	76	77	78	79	80	81
1. *Bathyphantes gracilis*	40	32	57	33	20	15	12	2	2	13	10	9	10
2. *Erigone atra*	50	32	40	17	13	8	31	26	12	8	15	22	18
3. *E. dentipalpis*	18	12	18	2	6	6	8	4	14	4			1
4. *Hypomma bituberculatum*		91	4	9	24	12	10	44	51	24	3		2
5. *Clubiona stagnatilis*		28	17	18	9	7	5	5					
6. *Erigone longipalpis*	40	351	1699	582	1068	548	616	264	146	204	355	314	358
7. *Lepthyphantes tenuis*	29	73	187	126	164	79	126	33	74	88	3	2	13
8. *Oedothorax apicatus*	3	1	117	82	36	20	30	66	26	12		1	
9. *Porrhomma microphthalmum*	2	10	132	293	79	181	84	237	97	64	4	3	4
10. *Enoplognatha schaufussi*		9	6	1	3	4	7	37	15	6	2		
11. *Argenna patula*						5	46	74	94	32	4	14	2
12. *Erigone arctica*	1	4	251	152	233	143	147	95	104	387	144	5	
13. *Centromerita bicolor*			1	3		3	1	2		1	13	6	
14. *Pachygnatha degeeri*		1	1			1	2		2		2	9	2
15. *Silometopus ambiguus*							1	3	11	47	58	9	1
16. *Walckenaeria kochi*		2	1	16	2		1		3		5	47	24
17. *Leptorhoptrum robustum*			1	7	1	2	1	4	10	6	70	61	15
18. *Robertus arundineti*			38	5	17	55	15	10	3	45	6	32	11
19. *Gongylidiellum vivum*		1									1	1	8
20. *Allomengea scopigera*												1	
21. *Pardosa purbeckensis*		1	180	50	51	58	100	279	78	30	90	711	442
22. *Oedothorax retusus*	1	2	11	35	10	50	44	96	64	79	90	72	75
23. *Trochosa ruricola*				1			3	8	5	8	7	4	1
24. *Oedothorax fuscus*	9	17	28	9	4	5	16	56	136	185	35	7	1
25. *Pachygnatha clercki*			9	30	7	56	46	22	7	19	64	172	64
26. *Bathyphantes approximatus*										1	11	10	35
27. *Pardosa amentata*			1									2	8
28. *Pirata piraticus*					2	4	15	3	7	12	18	152	544
Total catch	200	689	2815	1478	1757	1256	1371	1374	964	1275	1012	1673	1644
Total number of species	11	23	26	24	23	22	27	26	25	22	25	27	26
Dominance index (d)	0.25	0.51	0.60	0.39	0.61	0.43	0.45	0.20	0.15	0.30	0.35	0.43	0.33
Index of diversity (α)	2.50	4.58	3.96	4.07	3.74	3.79	4.77	4.55	4.68	3.78	4.64	4.57	4.38
Mean log R		0.03	0.02	−0.01	−0.01	0.00	0.00	0.00	−0.00	0.00	−0.00	0.00	−0.01
Annual variability		0.99	1.01	0.33	0.25	0.26	0.19	0.23	0.24	0.25	0.66	0.42	0.38
Number of data pairs for AV		17	23	24	22	23	24	25	25	22	24	25	25

communities, grazing has less influence. In 1982 cattle were introduced in the Schildhoek but Borrius found only slight differences between the carabid populations of grazed and ungrazed parts of the Schildhoek in 1983.

Spiders (Araneae)

A total of 77403 spiders of 79 species were caught. The 28 commonest species (Table 14.5) represent 99.2% of the total catch; these species were used in the CANOCO analysis. The nature of the spider fauna is comparable to the carabid fauna: low numbers of species, a low diversity, sometimes strongly dominated by one species (e.g. $d = 0.91$ in

			Sennerplaat B															
82	83	84	69	70	71	72	73	74	75	76	77	78	79	80	81	82	83	84
3	5	3	12	19	31	17	17	10	2	3	7	12	4	3	5	17	10	17
2	36	40	29	46	23	9	14	10	5	4	17	18	8	17	3		2	15
		1	8	19	21	2	3			1	7	1	2	2				
8	1	35		1		1	11	26	24	55	19	10	16	4	11	8	3	5
	3			2		1	1	2	4	10	3	2	1			1		
321	41	104	16	22	604	639	397	246	469	64	22	11	3	2				
1	19	5	5	24	190	38	44	21	22	15	27	21		3	1	6	5	8
					151	95	21	2			6	3						
1		1	1	10	37	19	11	4	5	23	17	1	2					
		1				12	2	4		13	27	6						
6		15				2	5	9	12	55	47	42	15	5		1		
			2	9	167	209	385	97	55	6	39	13	1	1				
2	3	10				1			4	3	3	6	1	7		5		3
7	6	40	1			1	2	3	13	31	103	94	36	56	2	2		1
1									2	34	5	2			1			
18	6	19		1	4		1	5	2	1			9	19	7	2		
	1	7			1	4	3	2		1	11	40	37	8	6	10	2	8
11		11			1		5	12	1	3	2	4	1	7		3	1	
7	1	2													1	11	9	14
1		2													1	3		1
859	48	546		5	35	138	92	251	415	590	177	230	245	301	43	31	1	14
129	56	251			3	31	32	65	223	179	32	48	108	81	8	1	2	
26	12	47							3	4	9	18	42	64	29	32	15	73
	11	8	3	32	57	9	93	61	97	382	107	88	10	1		1		
83	27	250			2	165	108	125	138	112	93	147	196	219	57	52	13	39
2		5						1			2	7	20	64	68	33	17	20
9		6					1			3	6	13	127	225	188	266	9	66
407	461	288							1	6	1	248	419	924	868	525	540	276
1925	742	1705	79	198	1334	1404	1249	960	1502	1605	812	1092	1310	2034	1317	1034	651	581
28	21	30	10	14	19	22	22	24	23	28	33	31	27	31	23	30	21	22
0.45	0.62	0.32	0.37	0.23	0.45	0.46	0.32	0.26	0.31	0.37	0.22	0.23	0.32	0.45	0.66	0.51	0.83	0.48
4.64	4.02	5.17	3.03	3.43	3.14	3.70	3.79	4.46	3.85	4.82	6.91	5.94	4.81	5.19	3.96	5.78	4.15	4.52
0.00	−0.01	0.02		0.05	0.03	0.00	0.00	−0.00	0.00	0.01	0.00	−0.00	−0.01	0.00	−0.02	0.00	−0.01	0.02
0.50	0.68	0.64		0.41	0.89	0.69	0.22	0.20	0.29	0.38	0.31	0.27	0.46	0.27	0.49	0.32	0.49	0.44
27	28	25		12	17	19	18	20	24	25	29	28	28	27	26	25	25	20

1983 in the Sennerplaat A), and strongly varying from year to year (annual variability as high as 0.83 in 1983 in the Schildhoek B).

In 1969 more than 600 spiders were caught, mostly in the Schildhoek sites. As with the carabids, smaller numbers were caught at the Sennerplaat sites because of the more adverse environment. Evidently the bulk of the spiders were immigrants. Young spiders, and adults of some small species, have a special technique of aerial dispersal known as ballooning. This involves drifting by wind while suspended by a thread of silk. Sometimes this aeronautic behaviour is simultaneously displayed by great numbers of spiders (Fig. 14.11) and new habitats can be rapidly colonized. In 1970, spider populations rapidly built up, again due to the

Fig. 14.11. Mass aeronautic behaviour of immature *Pirata piraticus* in spring 1986. The top of the fencepost is covered with silk.

capacity of species such as *Erigone longipalpis* to produce a continuous succession of generations. Table 14.5 and Fig. 14.12 show some of the species replacements in the Schildhoek A and Sennerplaat B sites. Some species (e.g. *Pirata piraticus*) were rather restricted in time, while others (e.g. *Pardosa purbeckensis*) were present from the outset and were still present in appreciable numbers in 1984.

The results of the canonical correspondence analysis are presented in Table 14.4 and Fig. 14.13. The sites/years are plotted in Fig. 14.13 on a scale double that used for the species and environmental factors. As in carabids, the first axis, which here represents 60.0% of the total variance, is strongly correlated with the age of the polder (Table 14.4) and to a lesser extent with the excess of rain over evaporation during spring. The second axis accounts for 11.2% of the total variance and, as for the carabids, is correlated with the percentage of soil clay and spring salinity.

The arrangement of the sites/years with respect to the first 2 principal components in Fig. 14.13 shows that the spider fauna developed gradually, especially in the Sennerplaat B, but that development seems to have stopped after 1981–82. For the carabids, development was retarded in 1974–77 in the Sennerplaat and in 1971–78 in the Schildhoek (Fig.

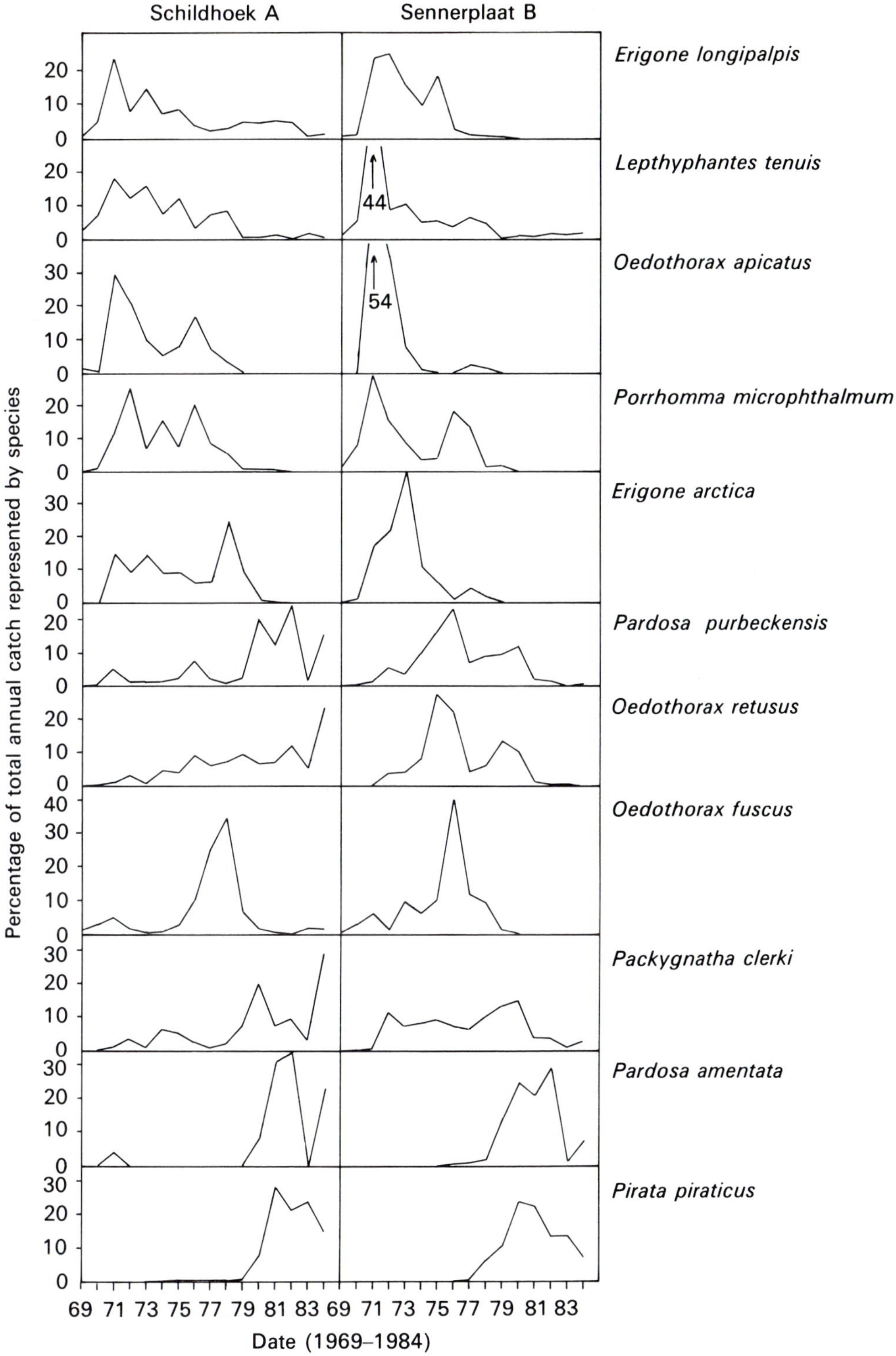

Fig. 14.12. Distribution of the catches of some spider species over the study period in the Schildhoek A and the Sennerplaat B study sites.

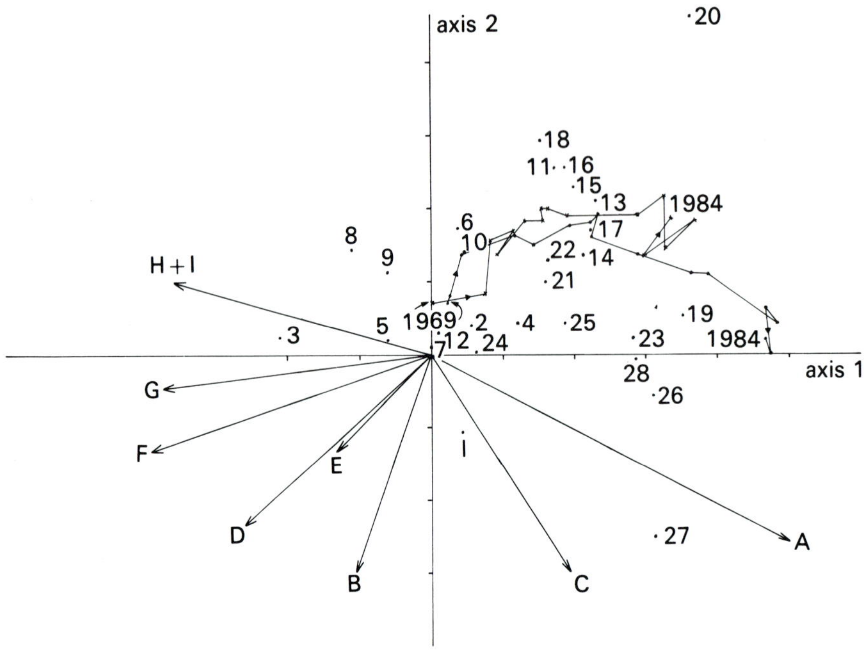

Fig. 14.13. Results of the CANOCO analysis of the spider data. The species are numbered (1–28), as in Table 14.5; the environmental factors are labelled (A–I) as in Table 14.4.

14.10). However, this is less evident with spiders. Evidently spiders and carabids respond to the environment in different ways. Nevertheless, the same factors seem to govern certain aspects of the development of both the carabid and the spider fauna.

Borrius (1986) only mentioned lycosid spiders in his study of the effects of cattle grazing in the Schildhoek. Their distribution in relation to the vegetation types he sampled in 1983 was the same as I found in the corresponding vegetation types in the 4 sites during the course of the study period. Borrius found that *P. piraticus* was the only spider caught in greater numbers in ungrazed than grazed areas.

Other groups

Schulz & Meijer (1978) reported on the leafhopper catches from my polder sites up to 1972; 13 species were recorded by that time. Table 14.6 lists the trapping data up to 1980 in the same way as for carabids and spiders. In all, 6426 specimens of 17, mostly rare, species were trapped. Most leafhopper species are restricted to certain plants, hence the low number of species in the rather uniform polder vegetation.

Table 14.6. *Annual catches from* 1969 *through* 1980 *of leafhoppers (Homoptera, Auchenorrhyncha) in two sites in the Lauwerszeepolder*

	Schildhoek A												Sennerplaat B											
	69	70	71	72	73	74	75	76	77	78	79	80	69	70	71	72	73	74	75	76	77	78	79	80
1. *Macrostelis sexnotatus*	2												5	2										
2. *Javesella pellucida*	1	3															2		1					2
3. *Psammotettix confinis*											1							7	5	7	7	3		
4. *Aphrodes makarovi*								1	2											2	6	18	4	2
5. *Streptanus aemulans*								1		3								1						1
6. *Deltocephalus pulicaris*																				5				
7. *Aphrodes serratulae*																				2	6	1	65	14
8. *Streptanus sordidus*											3	16								1			1	3
9. *Psammotettix putoni*	1	4					79	512	238	22	13	112					2	1	6	18	13	2	1	2
10. *Macrostelis sordidipennis*							29	252	228	505	477	1471						2	4	65	1			
11. *Philaenus spumarius*																						1	2	4
12. *Javesella dubia*												2											1	4
13. *Aphrodes flavostriatus*																							16	21
14. *Neophilaenus lineatus*																								1
Number of species	3	2	0	0	0	0	2	4	3	3	4	4	1	1	0	0	2	4	5	7	5	5	7	10
Total catch	4	7	0	0	0	0	108	766	468	530	494	1601	5	2	0	0	4	11	18	104	45	11	89	52

The bulk of the catch is made up of *Psammotettix putoni* and *Macrostelis sordidipennis*. The last species was present in enormous numbers. Pitfall-traps, as used in this study, are not entirely appropriate for catching leafhoppers. However, occasional net-trapping did not yield species other than those found in the pitfall traps. D-VAC suction sampling was carried out on June 23, 1981 in very poor (146.5 g dry matter m^{-2}) *P. maritima* vegetation close to the Schildhoek A. It revealed densities of 1844 *M. sordidipennis* m^{-2}. During the week of sampling the 2 pitfall-traps in the Schildhoek A caught more than 3000 leafhoppers. Although still unsorted, the 1981 and 1982 Schildhoek A pitfall-trap data indicate a further increase and subsequent decline in leafhopper abundance (10913 and 371 respectively). This seems to be associated with a decline in *P. maritima*.

Harvestmen (Arachnida, Opiliones) were recorded for the first time in July 1974. Although harvestmen do not seem to have great dispersal powers, compared to other groups they were able to reach the polder rapidly. Probably they were transported into the polder on floating tufts of shore vegetation. The species recorded are epigaeic species, whose young live in the litter layer; dispersal by wind is therefore unlikely.

Table 14.7 lists the catches of harvestmen. It indicates 352 specimens in 3 species, although another species, *Lacinius ephippiatus*, was present in the 1986 catches in the Sennerplaat B. Again, the Sennerplaat B has the richest fauna. *Nemastoma lugubre*, a litter-dwelling species very common in the Westpolder (maximum year's catch in 5 pitfall traps was 5857 specimens in 1977), has not been trapped in the Lauwerzeepolder during the study period. This species lives in vegetation along ditches and could easily be transported into the polder on floating tufts of vegetation.

Two species of Orthoptera were present in 1984; the tettigoniid *Conocephalus dorsalis* in the reedbeds of the Sennerplaat and the acridid *Chorthippus albomarginatus* in both the Sennerplaat and the Schildhoek. These species are the only 2 orthopterans expected in the study areas.

The first woodlouse (Isopoda), a single specimen of *Trachelipus rathkei*, was caught in 1984 in the Sennerplaat B. In 1986 two more specimens were found there as well as 9 specimens of *Trichoniscus pusillus*. No centipedes or millipedes have yet been found, but the catches in 1985 and 1986 have only superficially been inspected. However, Chilopoda and Diplopoda (Meijer, 1984), and probably Isopoda, are very common in the Westpolder (Meijer, 1984).

Gastropoda were present soon after completion of the polder but their presence was only casually noted. In the Schildhoek a small species of

Table 14.7. *Catches of adult harvestmen (Arachnida, Opiliones) in the Lauwersmeer*

	Schildhoek A								Sennerplaat A							
Species	77	78	79	80	81	82	83	84	77	78	79	80	81	82	83	84
1. *Oligolophus tridens*	—	—	2	1	1	—	1	—	—	1	10	9	14	12	1	16
2. *Mitopus morio*	—	—	—	—	—	—	—	—	—	1	2	—	—	—	—	1
3. *Paroligolophus agrestis*	—	—	—	—	—	—	—	—	—	—	—	—	—	—	—	—

	Schildhoek B								Sennerplaat B							
Species	77	78	79	80	81	82	83	84	77	78	79	80	81	82	83	84
1. *Oligolophus tridens*	—	1	1	3	20	3	2	23	1	5	2	44	32	46	25	58
2. *Mitopus morio*	—	—	—	—	1	—	—	—	—	—	—	2	1	3	1	1
3. *Paroligolophus agrestis*	—	—	—	—	—	—	—	—	—	—	—	—	—	—	3	2

slug, *Deroceras reticulatum*, and a snail, *Succinea* sp., are both extremely common but are far less so in the Sennerplaat. In the Sennerplaat B 2 more species of snail are regularly found in the pitfall traps: *Cochlicopa lubrica* and *Vallonia costata*.

Borrius (1986) also sampled the soil fauna. He found earthworms, (*Eiseniella tetraedra*) in reedbeds. The distribution of earthworms over the vegetation types which he sampled was negatively correlated with soil salinity. On the other hand, the occurrence of enchytraeid worms was positively correlated with soil salinity, the highest densities occurring in *Puccinellia* vegetation. The biomass of these worms was very low compared to normal Dutch meadows (5.8 g m^{-2} of earthworms and 0.22 g m^{-2} of enchytraeids). By 1971, I (Meijer, 1972) found 2 species of earthworms (*Allolobophora caliginosa* and *Lumbricus castaneus*) in the Lauwersmeer close to a deserted colony of black-headed gulls on high and well drained former salt marshes. At both the Schildhoek and the Sennerplaat, earthworms and enchytraeid worms are regularly found in the traps.

Discussion

After 16 years of undisturbed development the polder fauna still seems to be in a relatively immature state. Some groups of arthropods that play an important role in degrading dead vegetable matter have not yet, or have only just, arrived and major groups such as carabids and spiders are still poorly represented. By comparison, the Westpolder carabid fauna comprises at least 63 species and its spider fauna 62 species (Meijer, 1976), while in 1984 in the Lauwerszeepolder the total number for all sites was 33 and 45 respectively. A number of the polder species most certainly will disappear again as development proceeds, but on the other hand quite a number of the Westpolder species have yet to reach the Lauwerszeepolder sites (Meijer, 1984).

This slow development of the fauna is partly caused by the slow desalination of the soil, e.g. in some places in the Schildhoek the pioneer vegetation which occurred in 1971 could still be found in 1984. The analysis of the data showed that the age of the polder, perhaps to be interpreted as the maturity of the ecosystem, is the overriding factor in the development of the vegetation as well as the carabid and spider fauna. As the information on the abiotic and biotic environment of the carabids and spiders is very incomplete, it is impossible to say exactly which factors determine the maturity of the polder ecosystem. Duffey (1975), when discussing the spider fauna of British fens of different ages, referred to an as yet unexplainable 'historical factor'. There, certain marshland spiders were absent in meadows which had reverted to marshland during the last 30–40 years 'in spite of what appears to be a very suitable habitat'. The absence of the harvestman, *N. lugubre*, diplopods and chilopods in the Sennerplaat in 1984 is very much the same case, although the Sennerplaat obviously is a much younger habitat than Duffey's fen sites. In October 1987 a specimen of *N. lugubre* was trapped in the Sennerplaat B, indicating that the habitat was indeed suitable. (By the autumn of 1988 this species had become established.)

Some species simply may have been unable to reach the Sennerplaat as it is quite isolated, being surrounded by low lying, saline and frequently submerged areas. Flightless arthropods may be unable either to cross that barrier, even between flooded periods, or to withstand the salinity, which increases in dry, non-flooded periods. But some of the harvestmen, isopods and gastropods found in the Sennerplaat also have poor powers of dispersal. The fact that some species, e.g. the micropterous *D. globosus*, isopods, gastropods and opilionids, did reach the polder (probably via the water) suggests that the habitat is still unsuitable for many others.

Maybe the development of the fauna could be accelerated by artificially introducing various species of soil- and litter-dwelling arthropods which probably have poor dispersal powers. For instance, this might be done by introducing topsoil and vegetation in the Sennerplaat area from areas like the site in the Westpolder. An experiment like that could answer the question of whether the absence of certain species or groups of arthropods is caused by the unsuitability of the habitat or by the lack of immigrants with poor powers of dispersal. However, such procedures would destroy the possibility of monitoring the natural development of the fauna and certainly would be incompatible with the goals set by the Rijksdienst voor de IJsselmeerpolders.

Summary

When the Lauwerszeepolder was enclosed in 1969 a long-term study was started to monitor the development of part of the arthropod fauna. This was carried out by means of pitfall trapping in 2 areas in the centre of the polder that were designed to be nature reserves. On the slowly desalinating soil, a vegetation of halophitic herbs developed that changed into reed and grass land. The slow development of the carabid and spider fauna is described and observations are presented on leafhoppers and harvestmen as well as occasional observations on other groups.

After 16 years of development the number of species (a total of 33 carabids and 45 spiders in 1984 at all sites) is still low compared to that in a 100-year-old polder, the Westpolder, next to the Lauwerszeepolder (63 and 62 species respectively). Over the course of years, an increasing total of species was found, while once common species had disappeared and had been replaced by others following desalination. In 1984 both faunas still have a low diversity, are strongly dominated by one species and are exhibiting a high annual variability. Canonical correspondence analysis (CANOCO) showed time and differences in desalination to be major factors in governing the development of the fauna as well as the vegetation.

The absence of some groups and/or species may either be due to a low dispersal power of the animals or to the unsuitability of the habitat. Some Westpolder carabid and spider species that are likely to be able to live in the Lauwerszeepolder by 1984 still are not present; on the other hand, the unwinged form of one carabid species was present before the winged form. Some species of snails and earthworms had soon arrived, isopods arrived only after 15 years while centipedes and millipedes had still not arrived. It is therefore difficult to decide whether the suitability or the accessibility of the habitat is the overriding factor. The study is still continuing.

I would like to dedicate this Chapter to Professor Dr L. Vlijm, my original supervisor, in honour of his retirement in September 1987.

References

References marked * contain the nomenclature on which the species identifications were based. They are not cited in the text.

Borrius, K. (1986). *Direkte en indirekte invloeden van extensieve begrazing met pinken op de broedende steltloperweidevogels van de Schildhoek/Pampusplaat in 1982 en 1983*. University of Groningen, unpublished report.

De Glopper, R. J. (1985). De ontzilting. In *The Lauwerszee Fifteen Years Closed Off. Results of Research and Experience with Development and Management*, ed. P. J. Ente & R. J. De Glopper, pp. 81–99. Lelystad: Rijksdienst voor de IJsselmeerpolders.

Duffey, E. (1975). Habitat selection by spiders in man-made environments. In *Proceedings of the 6th International Arachnological Congress, Amsterdam*, pp. 53–67.

*Duijm, M. & Kruseman, G. (1983). *De krekels en sprinkhanen in de Benelux*. Amsterdam: Koninklijke Nederlandse Natuurhistorische Vereniging.

Ente, P. J. & De Glopper, R. J. ed. (1985). *The Lauwerszee Fifteen Years Closed Off. Results of Research and Experience with Development and Management*. Lelystad: Rijksdienst voor de IJsselmeerpolders.

Feekes, W. (1936). *De ontwikkeling van de natuurlijke vegetatie in de Wieringermeerpolder, de eerste grote droogmakerij van de Zuiderzee*. Ph.D. Thesis, University of Wageningen.

*Gittenberger, E., Backhuys, W. & Ripken, Th. E. J. (1970). *De Landslakken van Nederland*. Amsterdam: Koninklijke Nederlandse Natuurhistorische Vereniging.

*Gravestein, W. H. (1976). Naamlijst van de in Nederland voorkomende Cicaden. *Entomologische Berichten, Amsterdam*, **36**, 51–7.

Haeck, J. (1969). Colonization of the mole (*Talpa europaea L.*) in the IJsselmeerpolders. *Netherlands Journal of Zoology*, **19**, 145–248.

Haeck, J. (1971). The immigration and settlement of carabids in the new IJsselmeerpolders. *Miscellaneous Papers Landbouwhogeschool Wageningen*, **8**, 33–52.

Haeck, J., Hengeveld, R. & Turin, H. (1980). Colonization of road verges in three Dutch polders by plants and ground beetles (Coleoptera: Carabidae). *Entomologia Generalis*, **6**, 201–15.

*Heukels, H. & Van Ooststroom, S. J. (1977). *Flora van Nederland*, 19-de druk. Groningen: Wolters-Noordhoff.

Hoekstra, H. (1979). *Het verloop van de grondwaterstanden op de Zuidelijke Ballastplaat (Lauwerszeegebied) in 1978.* Werkdocument 1979 – 193 Abw. Lelystad: Rijksdienst voor de IJsselmeerpolders.

Joenje, W. (1978). *Plant Colonization and Succession on Embanked Sandflats.* Ph.D. Thesis, University of Groningen.

*Locket, G. H., Millidge, A. F. & Merrett, P. (1974). *British Spiders*, Vol. III. London: The Ray Society.

Meijer, J. (1972). An isolated earthworm population in the recently reclaimed Lauwerszeepolder. *Pedobiologia*, **12**, 409–11.

Meijer, J. (1976). Carabids (Coleoptera: Cicindelidae and Carabidae) and spiders (Araneida) in the Lauwerszee area. *Entomologische Berichten, Amsterdam*, **36**, 128.

Meijer, J. (1980). *Colonization of the Lauwerszeepolder by some Elements of the Arthropod Fauna.* Ph.D. Thesis, Free University, Amsterdam.

Meijer, J. (1984). Different phenological strategies in two nemastomid harvestmen (Arachnida, Opilionida, Nemastomatidae). *Bulletin of the British Arachnological Society*, **6**, 211–16.

Mook, J. H. (1971). Observations on the colonization of the new IJsselmeerpolders by animals. *Miscellaneous Papers Landbouwhogeschool Wageningen*, **8**, 13–31.

Prop, D. (1985). *De broedvogels van de Lauwerszee in 1983.* Rijksdienst voor de RIJP Rapport 1985 – 16 abw. Lelystad: Rijksdienst voor de IJsselmeerpolders.

*Sankey, J. H. P. & Savory, T. H. (1974). *British Harvestmen* (*Arachnida*, *Opiliones*). London and New York: Academic Press.

Schulz, C. A. & Meijer, J. (1978). Migration of leafhoppers (Homoptera: Auchenorrhyncha) into a new polder. *Holarctic Ecology*, **1**, 73–8.

Slager, H. (1985). *Grondwaterregimes in natuurterreinen in het Lauwerszeegebied.* Rijksdienst voor de IJsselmeerpolders, unpublished report.

Southwood, T. R. E. (1978). *Ecological Methods.* Second edn. London: Chapman & Hall.

*Sutton, S. L. (1972). *Woodlice.* London: Ginn & Company Limited.

Ter Braak, C. J. F. (1985). *CANOCO – A FORTRAN Program for Canonical Correspondence Analysis and Detrended Correspondence Analysis.* Wageningen, The Netherlands: IWIS-TNO.

Ter Braak, C. J. F. (1986). Canonical correspondence analysis: a new eigenvector technique for multivariate direct gradient analysis. *Ecology*, **67**, 1167–79.

Thiele, H. U. (1977). *Carabid Beetles in their Environment.* Berlin, Heidelberg, New York: Springer-Verlag.

*Turin, H., Haeck, J. & Hengeveld, R. (1977). *Atlas of the Carabid Beetles of the Netherlands.* Amsterdam: North-Holland Publishing Company.

Van Eerden, M. R. (1985). Aspects of development of bird populations. In *The Lauwerszee Fifteen Years Closed Off. Results of Research and Experience with Development and Management*, ed. P. J. Ente & R. J. De Glopper, pp. 84–100. Lelystad: Rijksdienst voor de IJsselmeerpolders.

Wolda, H. (1978). Fluctuations in abundance of tropical insects. *The American Naturalist*, **112**, 1017–45.

15

The return of vertebrate fauna to surface coal mined areas in Tennessee

DALE K. FOWLER

Introduction

Scope of project

In 1979, the Tennessee Valley Authority (TVA) and the US Fish and Wildlife Service (USFWS) undertook a large-scale coal surface-mine demonstration project designed to benefit the fish and wildlife resource within the southern Appalachian coalfield of North America. Rehabilitation of surface-mined lands within this region is often difficult due to toxic spoil conditions, low soil fertility and excessively steep terrain. In addition, sandstones and shales overlying the coal strata are of Pennsylvanian age and frequently acidic. Portions of the mine spoil along some coal seams may have a pH of 2.5 or lower (Fowler & Adkisson, 1980).

The study areas, located on Massengale and Brushy Mountains in Campbell County, Tennessee, are within the mixed Mesophytic Forest region (Braun, 1950), and are representative of conditions in much of Ecoregion Province 2210, Eastern Deciduous Forest (Bailey, 1978). This region is characterized by low mountains, where less than 20% of the land is gently sloping and the local relief is 305–941 m (Hammond, 1964).

The main focus of the project was to devize and implement a practical, cost effective, post-mining land use option featuring fish and wildlife (Fowler & Turner, 1981) on an active surface mining operation located on Brushy Mountain (Fig. 15.1). Another project objective was to compare the degree of biological recovery resulting from the wildlife-

Fig. 15.1. The study mine on Brushy Mountain was two growing seasons old when the terrestrial monitoring data presented in this Chapter was collected. Note the predominately grassland nature of the habitat and the general lack of a well developed shrub component.

oriented rehabilitation approach to that of conventional surface-mine rehabilitation techniques.

Terrestrial monitoring data were subsequently collected on the demonstration mine and an adjacent mining operation rehabilitated conventionally. Both of these sites were located on Brushy Mountain. In addition, an older minesite on nearby Massengale Mountain (Fig. 15.2), which had been rehabilitated conventionally, was monitored during the 1980 season. The older Massengale Mountain surface-mine, operated experimentally from 1972 to 1974 under TVA contract, was the first back-to-contour surface mining effort in the United States (Allen, 1973). This rehabilitation approach later became a Nation-wide requirement for coal surface mine operators with the enactment of the Surface Mining Control and Reclamation Act (SMCRA) of 1977 (Public Law 95–87). The Brushy Mountain minesite was one of the first to be reclaimed under provisions of this new legislation.

Although not the original intent of the research, faunal comparisons between these different aged mines and adjacent undisturbed woodland

Fig. 15.2. In 1972, the first back-to-contour surface mine rehabilitation effort in the United States was initiated on Massengale Mountain in Campbell County, Tennessee.

provide insight into the colonization of such disturbed sites by terrestrial fauna. The information collected during the 1980 season on the conventionally rehabilitated minesites on Brushy Mountain and Massengale Mountain will be used as the basis of this Chapter.

The author is indebted to the many TVA and USFWS personnel who assisted with this project. Special thanks are due to TVA biologists, Linda J. Fowler and Donald W. Allsbrooks, who had major project responsibilities, and to Lawrence J. Slaski, project coordinator for the USFWS. All photographs used in the Chapter were contributed by TVA.

Methods

Mine rehabilitation

Revegetation of the Massengale and Brushy Mountain study sites occurred during 1973 and 1978, respectively. Thus, the vegetation established on these mines was 7 and 2 years of age at the time of this 1980 study. The portions of these mines discussed in this Chapter were rehabilitated with plants that are commonly used throughout the coal fields of southern Appalachia. Both mine sites were graded back to the original contour, previously stored topsoil was redistributed, and a

Table 15.1. *Revegetation materials used to establish ground cover on the study coal surface-mines*

Materials	Application rates ($kg\ ha^{-1}$) Brushy Mt.	Massengale Mt.
Grasses		
Kentucky 31 fescue (*Festuca elatior*)	7.3	22.9
annual ryegrass (*Lolium multiflorum*)	1.7	
perennial ryegrass (*Lolium perenne*)	2.8	
balbo rye (*Secale* sp.)	5.1	11.7
weeping lovegrass (*Eragrostis curvula*)		0.9
Legumes		
sericea lespedeza (*Lespedeza cuneata*)	3.7	5.3
yellow sweet clover (*Meliotus officinalis*)	0.4	0.4
Mulch		
Hydro-mulch	98	138
Straw mulch	75 bales	
Plant nutrients[a]		
Nitrogen	21	23
Phosphorus	13	19
Potassium	11	

Note: [a] Ammonium nitrate (33% N) and triple superphosphate (45% P) fertilizers were used during reclamation operations on the Massengale Mountain mine. Triple superphosphate (46% P), urea (56% N), and muriate potash (60% K) fertilizers were used on the mine at Brushy Mountain.

conventional rehabilitation mixture of fertilizer, sericea lespedeza (*Lespedeza cuneata*), Kentucky-31 fescue (*Festuca elatior*) and other perennial and annual grasses and legumes were hydroseeded over the disturbed area (Table 15.1). Black locust trees (*Robinia pseudoacacia*), interspersed with Virginia pine (*Pinus virginiana*), were planted over the Massengale Mountain mine while black locust tree seedlings were established on the Brushy Mountain site. These trees were grid planted 2.1 m apart at rates of approximately 2224 trees ha^{-1} on both mine sites.

Two undisturbed woodland sites, adjacent to the mined lands, were also selected for study. Both woodland sites were located far enough from the adjacent minesites to prevent any effect on monitoring operations from mining disturbances. The Massengale Mountain woodland site was characterized by steeply sloping terrain dissected by rugged ravines on a predominantly south-facing slope with one wet weather stream present.

The Brushy Mountain woodland site was located on a predominantly east-facing slope with gently rolling to moderately steep terrain.

Habitat analysis

Vegetation was sampled on both study mines and the adjacent woodland sites during August 1980 to determine the species composition and density of ground, shrub, and overstorey strata within ten 0.04-ha circular plots using a modification of the technique developed by James & Shugart (1970). On both minesites, circular plots were equally divided between upper and lower slopes. Percent total ground cover was measured along either side of each sampling station using a comb-like apparatus having 10 teeth (75 cm long) spaced 8 cm apart. This tool was placed perpendicular to the ground, and the number of points touching vegetation was recorded and expressed as a percentage (Mueller-Dombois & Ellenberg, 1974). Multiple-vegetation hits on the same tooth were recorded as one observation.

Percent canopy cover was determined in each woodland plot by visually estimating the percentage of the canopy that was closed at each sampling station (10 sampling stations per plot). In each circular plot, these 10 estimates of canopy cover were averaged to yield a mean canopy cover.

In addition, all vegetation was clipped to ground level within 0.5-m^2 sampling frames systematically located at 50 points within each study mine and the Brushy Mountain woodland site. Biomass samples were not collected on the Massengale Mountain woodland site due to budget constraints. These biomass samples were then separated into grass and herbaceous components, oven dried at 100 °C, and weighed to the nearest gram.

Breeding bird census

Breeding birds were censused for 10 days between May 28 and June 19, 1980, on the Brushy Mountain and Massengale Mountain minesites using the spot mapping technique (Williams, 1936). Territories were designated as 'mine' or 'edge', based on whether sightings of individual territorial males clustered on the mine proper or the wood–mine interface. In addition, two woodland sites (one in the forest above the Brushy Mountain mine, and one in the forest below the Massengale Mountain mine) were censused for breeding birds between

May 28 and June 10, 1980. Each woodland site was 800 m in length and 100 m in width (8 ha) and was subdivided into clearly marked 50-m square sections. Positions of breeding birds observed within the study sites were then plotted on data sheets. Censusing was conducted between 0630 and 1030 hours.

Small mammal trapping

Small mammals were snap-trapped for 10 consecutive days in spring 1980 (June 17–26) on Brushy Mountain and Massengale Mountain study sites. Traplines on all sites were located in woodland, mine, and forest–mine interface habitats. Two traplines were established within each habitat type, making a total of 6 lines per site. Fifty trap stations, spaced at 10-m intervals, were established on each line. Two Victor mouse-traps® were placed at each station, and one Museum Special trap® was placed at alternate stations, for a total of 125 traps per line. Thus, each individual habitat type represented 2500 trap-nights of effort. Traps were placed either in a triangular configuration (3-trap set) or facing each other (2-trap set), with approximately 0.25-m spacing, and were baited daily with a peanut butter and rolled oat mixture. All traplines were checked once each morning during the 10-day trapping period. Captured animals were identified, weighed, measured, sexed, and aged.

Data analysis

Relative dominance and relative frequency were calculated for ground cover vegetation on each site. These two values were summed for each species (Mueller-Dombois & Ellenberg, 1974). Species diversity (Shannon & Weaver, 1963) and evenness indices (Pielou, 1975), and similarity coefficients (Tramer, 1974) were calculated to compare the shrub, overstorey, avian and mammalian data collected on the study sites. Natural logarithms ($\log_e$) were employed in the species diversity and evenness indices calculations.

General wildlife observations

During routine fieldwork on the Brushy Mountain and Massengale Mountain study areas, general sightings of wildlife were recorded, as well as tracks, scats, or other signs of wildlife. Raptors, game birds, waterfowl, furbearers, and game mammals were among those species of special interest.

Results and discussion

Habitat analysis

Ground cover Percent ground cover (Table 15.2) was high on both Massengale and Brushy Mountain minesites (92 and 80%, respectively). However, herbaceous biomass (75% of total) was markedly higher than grass (25% of total) on the Massengale Mountain mine (Table 15.2), but somewhat lower than the grass component (51% of total) on the younger Brushy Mountain site. The estimated biomass production compares very well with similar data collected on other disturbed lands. For example, Fowler, Turner & Adkisson (1981) reported annual oven-dried biomass weights of 2949 kg ha^{-1} for transmission line rights of way in eastern Tennessee, while Duvall & Hillmon (1965) calculated biomass weights as high as 3362 kg ha^{-1} for forest clear cuts in the South.

A species similarity coefficient of 51% indicated that ground cover species composition on the two minesites was relatively dissimilar. A total of 25 different species of ground cover plants was identified on the Brushy Mountain minesite (Table 15.2). Of these, 20 were volunteer species and 5 were rehabilitation species. Kentucky 31 fescue had the highest percent dominance and frequency of occurrence, while sericea lespedeza was second in both categories. None of the volunteer species had a high frequency of occurrence. Red maple (*Acer rubrum*), ash (*Fraxinus* sp.), and yellow poplar (*Liriodendron tulipifera*) were the only tree seedlings in the volunteer species category.

Eighteen species of ground cover plants were recorded on the Massengale Mountain mine site (Table 15.2) with sericea lespedeza having the highest percent frequency of occurrence and dominance, and Kentucky-31 fescue having the second highest values. Although 14 of the 18 species were volunteer species, none had high frequency or dominance values.

Results of the ground cover analysis for the mine sites suggest that fescue grass dominates on surface-mines rehabilitated with these seed mixtures for the first two growing seasons. The perennial sericea lespedeza then emerges as the dominant plant species and biomass component of the ground cover.

A total of 59 and 64 species of ground cover plants were respectively identified on the Brushy Mountain and Massengale Mountain woodland sites (Table 15.2). Of these, the most frequently occurring plants on Brushy Mountain were aster (*Aster* sp.), thoroughwort (*Eupatorium* sp.), bedstraw (*Galium* sp.), and Virginia creeper (*Parthenocissus quinque-folia*).

Table 15.2. *Summary of habitat analysis conducted on the surface-mine and woodland study sites*

		Brushy Mountain		Massengale Mountain	
Stratum[a]	Component	Minesite	Woodland	Minesite	Woodland
Ground cover	Herbaceous biomass (kg ha^{-1})	1192	180	2704	n.d.[b]
	Grass biomass	1222		914	n.d.
	Total biomass	2414	180	3618	n.d.
	Percent cover	80	48	92	39
	No. volunteer species	20		14	
	No. planted species	5		4	
	Total species	25	59	18	64
Shrub	Density (no. ha^{-1})	1158	9716	642	6355
	No. volunteer species	6		2	
	No. planted species	1		4	
	Total species	7	20	6	24
	Species diversity	0.43	2.27	0.76	2.18
	Species evenness	0.22	0.76	0.42	0.69
Overstorey	Density (no. ha^{-1})		563		605
	Total basal area (m^2 ha^{-1})		21		24
	Mean tree height (m)		24		23
	Total canopy cover (%)		85		85
	Total species		16		17
	Species diversity		2.50		2.25
	Species evenness		0.90		0.79

Notes: [a] Ground cover is defined as herbaceous vegetation, grasses, and woody seedlings on the woodland floor. Shrubs are woody plants (excluding vines) less than 8 cm in diameter at breast height (d.b.h.) and 5 m in height. Overstorey is defined as woody plants greater than 8 cm in d.b.h. and 5 m in height.
[b] n.d. = not determined.

The most frequently occurring plants on Massengale Mountain were red maple, aster, thoroughwort, and beggar's ticks (*Desmodium* sp.). Total percent ground cover was 48 % and 39 %, respectively, for the Brushy and Massengale Mountain sites. A species similarity coefficient of 70 % indicates that the ground cover vegetations on the two woodland sites are very similar in species composition.

Shrub stratum Seven tree species were recorded on the Brushy Mountain mine site with all occurring within the shrub stratum (Table 15.2). Of these, 6 species were volunteer while the other species (black locust) was planted. Total density was 1158 individuals ha^{-1}. The average height of planted black locust trees was 1.6 m. Average heights of volunteer tree species ranged from 0.07 to 1.30 m reflecting the young age of the mine vegetation. On the Massengale Mountain mine site, 5 tree and one shrub species were recorded with yellow-poplar and ash being the only volunteer tree species. Total density was 642 trees ha^{-1} (Table 15.2). The average height of black locust trees was 1.6 m and the average height of volunteer tree species ranged from 0.3 to 1.7 m. Thus, all woody vegetation sampled on the Massengale Mountain minesite was also within the shrub stratum (Table 15.2).

The Brushy Mountain woodland site had approximately 9716 shrub stems ha^{-1} (Table 15.2), comprised principally of black gum (*Nyssa sylvatica*), sassafras (*Sassafras albidum*), green ash (*Fraxinus pennsylvanica*), and sugar maple (*Acer saccharum*). By contrast, the Massengale Mountain woodland site had approximately 6355 shrubs ha^{-1} (Table 15.2), consisting primarily of flowering dogwood (*Cornus florida*), red maple, and black gum.

Overstorey stratum Total estimated woodland overstorey density was 563 trees ha^{-1} on Brushy Mountain, with sugar maple, white oak (*Quercus alba*), northern red oak (*Quercus rubra*), mockernut hickory (*Carya tomentosa*), and shagbark hickory (*Carya ovata*) being the most frequent trees. However, the most dominant trees (based on total basal area) were chestnut oak (*Quercus prinus*) and white oak. Total basal area of the Brushy Mountain woodland was estimated as 21 m^2 ha^{-1} (Table 15.2). Height of forest canopy on the Brushy Mountain woodland site ranged from 21 to 28 m and averaged 24 m while total estimated canopy cover was 85 %. Sixteen overstorey tree species were recorded on Brushy Mountain with the species diversity and evenness indices, respectively, being 2.50 and 0.90 (Table 15.2). The latter index indicates a relatively even distribution of tree species densities and is typical of the mixed mesophytic forest type (Braun, 1950).

Total woodland tree density on Massengale Mountain was estimated at 605 ha^{-1} with 17 tree species being recorded (Table 15.2). Most frequently occurring trees were red maple, chestnut oak, northern red oak, and black oak (*Quercus velutina*). Trees with highest densities were red maple, chestnut oak, northern red oak, and white oak. Based on basal area, chestnut oak, red maple, and northern red oak were the most dominant trees. Total basal area for the Massengale Mountain site was estimated at 24 m^2 ha^{-1} (Table 15.2). The height of the forest canopy on the Massengale Mountain woodland sites ranged from 21 to 30 m with an average of 23 m while total canopy cover was estimated at 85%. Species diversity and evenness indices of 2.25 and 0.79 for the Massengale Mountain site are comparable to those calculated for Brushy Mountain.

Discussion The woodland sites on Brushy Mountain and Massengale Mountain, with very similar vegetative profiles, are typical of the pre-mining vegetation. The species composition and condition of these woodland habitats are also generally typical of the second growth forest found over most of the southern Appalachian region. The associated habitat is generally considered to be of acceptable quality for most species of resident wildlife in the area.

In general, both the Brushy Mountain and Massengale Mountain minesites contained very few species of volunteer vegetation, percent ground cover and biomass production was high, the shrub stratum was sparse and there was no overstorey component. By contrast, the native ground flora of the adjacent woodland sites had much higher species diversity indices, much lower percent ground cover and biomass estimates, and a well developed shrub and overstorey stratum. For example, shrubs on the woodland sites occurred at densities of 7 and 12 times greater, on average, than the respective shrub densities observed on the Brushy and Massengale Mountain minesites.

Although the number and densities of volunteer plant species on the Brushy and Massengale minesites was quite low compared to the woodland sites, the data of Farmer, Cunningham, & Barnhill (1981) indicate the presence of a diverse seed bank in the adjacent forest topsoil. These investigators spread fresh forest topsoil collected near our study mines over surface minesoils. They found that first-year plant communities were comprised of some 134 different taxa.

As the forest topsoil on the Brushy and Massengale mines had been

Fig. 15.3. The dense mat of sericea lespedeza and fescue grass was probably a major factor in retarding the growth of established tree plantations and inhibiting subsequent invasion of natural vegetation onto the study mines from the adjacent forest. This photograph was taken on the Massengale Mountain minesite. Conventional revegetation procedures, which involved establishment of Kentucky-31 fescue, sericea lespedeza, and black locust seedlings, were used on the majority of the minesite. Note the shrubland appearance of the mine habitat.

stored and then respread over the mine sites, one must assume that most of the forest seedbank was damaged or buried during grading operations. In addition, the Kentucky-31 fescue grass planted on the minesites is known to have allelopathic properties (Walters & Gilmore, 1976; Larson & Swartz, 1980). These properties may inhibit growth and survival of planted trees and shrubs and also volunteer plant species invading the mine sites from the adjacent woodland (Fig. 15.3). The invading plants that did become established encountered increasingly intense competition from the aggressive grasses and legumes seeded over the mined area. Thus, a very simple plant community on the rehabilitated mines forms the habitat available for colonization by terrestrial vertebrate fauna.

Table 15.3. *Breeding bird species composition, diversity and evenness indices, and territorial densities recorded in June* 1980 *on the Brushy and Massengale Mountain study sites*

	Brushy Mountain			Massengale Mountain		
Species	Mine	Edge	Forest	Mine	Edge	Forest
ruffed grouse (*Bonasa umbellus*)						1.0
yellow-billed cuckoo (*Coccyzus americanus*)			1.0			
downy woodpecker (*Picoides pubescens*)			5.0			
northern flicker (*Colaptes auratus*)					1.0	
pileated woodpecker (*Dryocopus pileatus*)						2.5
eastern wood peewee (*Contopus virens*)		5.0	1.0		1.0	
northern rough-winged swallow (*Stelgidopterx ruficollis*)				1.5		
Carolina chickadee (*Parus carolinensis*)			1.0			
tufted titmouse (*P. bicolor*)			1.0			1.0
white-breasted nuthatch (*Sitta carolinensis*)			5.0			5.0
eastern bluebird (*Sialia sialis*)		2.5				
wood thrush (*Hylocichla mustelina*)			5.0			5.0
white-eyed vireo (*Vireo griseus*)					1.0	
solitary vireo (*V. solitarius*)						5.0
yellow-throated vireo (*V. flavifrons*)			5.0			5.0
red-eyed vireo (*V. olivaceus*)			25.0			15.0
golden-winged warbler (*Vermivora chrysoptera*)		5.0		4.0	2.5	
chestnut-sided warbler (*Dendroica pubescens*)		2.5			1.5	
cerulean warbler (*D. cerulea*)			15.0			15.0
common yellowthroat (*Geothlypis trichas*)	5.0	5.0		3.5		

black-and-white warber (*Mnitilta varia*)			5.0		1.5	5.0
American redstart (*Setophaga ruticilla*)			15.0			10.0
worm-eating warbler (*Helmitheros vermivorus*)			5.0			15.0
ovenbird (*Seiurus aurocapillis*)			35.0			27.5
Kentucky warbler (*Opornus formosus*)			5.0			5.0
hooded warbler (*Wilsonia citrinia*)		2.5	15.0		1.5	10.0
yellow-breasted chat (*Icteria virens*)				1.5	1.5	
scarlet tanager (*Piranga olivacea*)			5.0			5.0
indigo bunting (*Passerina cyancea*)	2.5	32.5		6.0	6.0	
rufous-sided towhee (*Pipilo erythophthalmus*)			5.0	3.5	1.5	
field sparrow (*Spizella pusilla*)	7.5	2.5		11.5		
song sparrow (*Melospiza melodia*)	10.0	15.0		5.0		
brown-headed cowbird (*Molothrus ater*)	2.5	5.0	1.0	1.0	1.0	
American Goldfinch (*Carduelis tristis*)	30.0			1.5		
Number of species with territories	6	11	19	10	11	16
Total territories per 40 ha	57.5	77.5	155.0	39.0	18.5	132.0
Species diversity indices	1.40	1.84	2.46	2.05	1.99	2.47
Species evenness indices	0.78	0.80	0.83	0.89	0.83	0.89

Breeding bird census

Mine habitat In 1980, a total of 11 breeding birds species was recorded on the Brushy Mountain minesite while 16 species established territories on the older Massengale Mountain minesite (Table 15.3). The highest total density of breeding birds was recorded on the Brushy Mountain minesite, largely because of the high numbers of species with edge territories (Table 15.3).

However, when the total number of within-site territories was examined, the younger Brushy Mountain minesite had more edge territories than minesite interior territories while the reverse situation was observed on the Massengale Mountain minesite (Table 15.3). The avian species diversity indices computed for the two minesites also followed this pattern while species evenness indices ranged from 0.78 on the Brushy Mountain mine interior to 0.84 on the Massengale Mountain mine interior (Table 15.3). In addition, both edge and mine interior territory densities were higher on Brushy than Massengale Mountain.

Within-site species similarity coefficients indicated relatively weak relationships of mine interior and mine edge breeding bird species to forest avifauna. On Brushy Mountain, comparisons of breeding bird species on mine interior to mine edge, mine interior to forest, and mine edge to forest habitats resulted in similarity coefficients of 59, 8, and 20 %, respectively. Similarity coefficients on Massengale Mountain (in the same order) were 48, 0, and 15 %. Thus, the relationship of mine interior and mine edge to forest habitats was very weak on the older minesite. On both minesites, the relationship of mine interior to mine edge habitat was moderately strong. By contrast, between-site species similarity coefficients indicated strong relationships between the same habitat types on the different minesites. Calculations for mine interior and mine edge resulted in strong between-minesite similarity coefficients of 75 and 64 %, respectively. However, some of the typically shrubland bird species with territories on the older Massengale Mountain mine habitat (such as golden-winged warblers, and yellow-breasted chats) were observed only as visitors to the Brushy Mountain site. This difference in species composition is attributed to the more developed shrubland type habitat on the older Massengale site (Table 15.2).

Woodland habitat Breeding bird populations on the two woodland sites were generally similar (Table 15.3), with 19 and 16 breeding bird species recorded on the Brushy Mountain and Massengale Mountain woodland sites, respectively. Total number of territorial males per 40 ha was 8% higher on the Brushy Mountain woodland site. However, species diversity indices for the two sites were essentially identical, there was little difference in the species evenness indices (Table 15.3), and the species similarity coefficient (74%) indicated a strong relationship of avifauna on the two sites. Since the composition of the shrub and overstorey strata on both of these sites differs very little (Table 15.2), the similarity in breeding bird species composition was not surprising.

Discussion The breeding bird fauna associated with the Brushy Mountain minesite was typical of the early grassland stage of the vegetation. However, such shrubland species as golden-winged warblers and common yellowthroats were already nesting in the dense, 1.5-m-tall stands of black locust trees. By contrast, the breeding avifauna on the older Massengale Mountain sites included bird species requiring more mature, shrubland habitat. Such birds as the white-eyed vireo and the yellow-breasted chat, typically associated with thick undergrowth and dense shrubby thickets, were thus present on Massengale, but not Brushy Mountain.

In general, avian species are highly mobile and select breeding territories on the basis of the structural components of the vegetation (Wilson, 1974). Thus, the species best suited to the successional stage of a given habitat establish breeding territories on these sites as they become available. As the habitat changes from grassland to shrubland and eventually to forest, the composition and densities of respective species shift with the changing conditions of the habitat. This observation is strengthened by the weak species similarity coefficient comparisons of forest to mine interior and forest to mine edge habitats. On relatively narrow contour surface-mines, it is unlikely that alterations in the species and mixes of reclamation vegetation can be used to change the breeding bird species composition and sequence of occurrence on surface mined lands. However, breeding bird densities and the rate at which the habitat changes to conditions favoring a particular group of avian species can probably be manipulated through judicious selection and placement of plant materials. Certainly structural components, such as increased 'edge', can be planned and established (Fowler & Turner, 1981).

In our study area, the overall breeding bird species composition increased due to the creation of mine edge and mine interior habitats in what before was predominately a forest habitat. As the coal surface mines in the mountainous terrain of southern Appalachia are relatively narrow bands of disturbance following the coal seams around the contour, the amount of woodland breeding bird habitat lost during mining operations may be relatively insignificant to the regional populations of breeding woodland birds.

Small mammal colonization

Mine habitat A total of 48 small mammals (representing 4 species) was captured on the Brushy Mountain mine interior habitat. These captures are summarized by study site in Table 15.4. The southern bog lemming, house mouse, white-footed mouse, and short-tailed shrew, comprised 54, 25, 17, and 4 % of the total catch respectively. Trapping operations yielded 107 animals on the mine interior habitat of Massengale Mountain (Table 15.4). Trap success was 4.3 animals per 100 trap-nights with white-footed mice comprising 78 % of the total catch on the Massengale Mountain mine interior. Small populations of cotton rats, southern bog lemmings, and house mice, associated with small predominately grassy areas, were also present on the Massengale Mountain minesite.

As only a few short-tailed shrews were captured on either minesite, it appears that mine interior habitat conditions were very marginal for these species. Insectivores characteristically occupy habitats with high soil moisture and litter cover, and relatively constant temperature (Getz, 1961; Pruitt, 1959). The general absence of these conditions on newly reclaimed surface-mines no doubt limits colonization of mine interiors by shrews.

Major differences in densities of small mammal species on Brushy Mountain and Massengale Mountain mine interior habitats are evident in Table 15.4. Higher populations of southern bog lemmings and house mice were found on the younger Brushy Mountain site, while the older Massengale Mountain minesite supported higher populations of white-footed mice and hispid cotton rats (Table 15.4). White-footed mice are typically associated with older, more developed vegetation (Verts, 1957; Sly, 1976; Mumford & Bramble, 1969). For example, LoBue & Darnell (1959) found that white-footed mouse populations were significantly correlated with vegetation height. Thus the low initial density of white-footed mice within the young mine interior habitat and the much higher populations on the older mine interior habitat are not surprising. These differences in species densities are to be expected because of the age differential between the two minesites and the associated ground cover and shrub stratum differences (Table 15.2). The higher populations of

Table 15.4. *Numbers of small mammals captured, trap success indices, and species diversity and evenness indices for mine interior, edge, and woodland habitats on Brushy and Massengale Mountain*

Species	Brushy Mountain			Massengale Mountain		
	Mine	Edge	Forest	Mine	Edge	Forest
	No. (%)	No. (%)	No. (%)	No. (%)	No. (%)	No. (%)
short-tailed shrew (*Blarina brevicauda*)	2(4)	11(14)	18(29)	1(*)	37(23)	33(41)
pine vole (*Microtus pinetorum*)	0(0)	4(5)	0(0)	0(0)	0(0)	0(0)
mouse mouse (*Mus musculus*)	12(25)	0(0)	0(0)	3(3)	0(0)	0(0)
woodland jumping mouse (*Napeozapus insignis*)	0(0)	4(5)	4(6)	1(*)	2(1)	7(9)
eastern woodrat (*Neotoma floridana*)	0(0)	0(0)	0(0)	0(0)	2(1)	0(0)
white-footed mouse (*Peromyscus leucopus*)	8(17)	56(73)	28(45)	83(78)	106(67)	31(38)
hispid cotton rat (*Sigmodon hispidus*)	0(0)	0(0)	0(0)	12(11)	0(0)	0(0)
smoky shrew (*Sorex fumeus*)	0(0)	2(3)	11(18)	0(0)	9(6)	10(12)
southern bog lemming (*Synaptomys cooperi*)	26(54)	0(0)	1(2)	6(6)	2(1)	0(0)
eastern chipmunk (*Tamias striatus*)[a]	0(0)	0(0)	0(0)	1(*)	1(*)	0(0)
Total individuals	48	77	62	107	159	81
Number of species	4	5	5	7	7	4
Trap success/100 trap-nights	1.9	3.1	2.5	4.3	6.4	3.2
Species diversity	1.11	0.91	1.27	0.83	0.97	1.20
Species evenness	0.80	0.57	0.79	0.43	0.50	0.87

Notes: * Less than 1 % of catch.
[a] Chipmunks were probably common in edge and forest habitats on both Brushy and Massengale Mountain. The snap traps used were too small to effectively capture the larger chipmunks.

southern bog lemmings on Brushy Mountain were attributed to the presence of dense, low-growing ground cover, comprised principally of grasses and legumes, surrounding localized wet seeps on the minesite. This situation differs on Massengale Mountain due to the tall dense stands of sericea lespedeza. However, a few grassy areas still allow remnant populations of southern bog lemmings to persist in marginal habitat. The high house mouse populations on Brushy Mountain are expected to decline as the vegetation develops to more mature shrubland, as evidenced by the low numbers of house mice captured on the older Massengale site.

A between-site species similarity coefficient of 73% indicated a strong relationship between the species composition of the two mine interior sites. Although there were more small mammal species captured on the Massengale Mountain mine interior, the species diversity index was lower, principally because of the greater unevenness of the species distributions on the older site (Table 15.4).

Edge habitat Seventy-seven animals (5 species) were captured on the Brushy Mountain edge habitat (Table 15.4). Overall, white-footed mice (73%) and short-tailed shrews (14%), were the most common species encountered. The woodland jumping mouse, smoky shrew, and the pine vole were represented by a few individuals. On Massengale Mountain, a total of 159 animals (7 species) was captured on the edge habitat (Table 15.4). White-footed mice (106 animals) and short-tailed shrews (37 animals) represented the predominant species captured (90% of the total catch). When between-site similarity coefficients were computed, the younger Brushy and older Massengale 'edge' exhibited a high degree of similarity (67%), with white-footed mice and short-tailed shrews representing the dominant species occupying both edge habitats. Major differences in species densities between the areas included an increase in the number of white-footed mice and a dramatic increase in short-tailed shrews captured on the Massengale site (Table 15.4). Both species diversity and evenness indices were relatively similar for the two mine edge habitats.

Woodland habitat On Brushy Mountain, 62 animals were captured representing 5 species (Table 15.4). The 3 most common species were white-footed mice (28 animals), short-tailed shrews (18 animals), and smoky shrews (11 animals). The woodland site on Massengale Mountain yielded 81 animals representing 4 species (Table

15.4). White-footed mice (31 animals) and short-tailed shrews (33 animals) accounted for 79 % of the animals captured.

A between-site species similarity coefficient of 89 % indicated a strong relationship between the small mammal species present on both areas. Small mammal species diversity indices for Brushy and Massengale woodland sites were nearly identical and species evenness indices were relatively similar (Table 15.4).

Discussion The strong relationship of small mammal species composition on mine interior and mine edge habitats to adjacent woodland habitats indicated that the majority of species colonizing the rehabilitated surface-mines have their origin in the forest. However, typically grassland species such as the southern bog lemming and hispid cotton rat were also present on the mine habitats. In addition, high populations of house mice were observed on the younger Brushy Mountain site. One would not expect to find any of these species in the adjacent mountainous woodland and it is interesting to speculate on their origin. It is most likely that these small mammal species initially reached the mine habitat as stowaways in the trucks used to transport the seed. In any case, it is clear that the surface-mines are now serving as colonization corridors for small mammal species which normally would rarely be found in the immediate area.

Although addition of the grassland small mammal component to the local mammalian fauna increased the total species base of the area, the reader should bear in mind that the trapping techniques used in the study were suitable for detecting only a portion of the mammalian species present in the woodland habitats.

The study suggests that small mammal populations begin to colonize surface-mine habitats shortly after initial rehabilitation vegetation is established, and that population numbers rapidly approach and/or exceed those in the adjacent undisturbed forest. The rapidity of this colonization is important because small mammals comprise the major food base of economically important furbearers and federally protected hawks and owls in the region. Thus, prompt revegetation should have a positive impact on these predator populations within the vicinity of coal surface mines in southern Appalachia. Additional research is needed to determine how vegetation mixtures and planting patterns influence the density, species composition, and predator susceptibility of small mammal populations on reclaimed surface-mines in the region.

Table 15.5. *Wildlife observed on Brushy Mountain and Massengale Mountain sites*

Species	Brushy Mountain		Massengale Mountain	
	General vicinity	Minesite	General vicinity	Minesite
Birds				
pied-billed grebe (*Podilymbus podiceps*)		X		
mallard (*Anas platyrhynchos*)		X		
blue-winged teal (*A. discors*)	X	X		
hooded merganser (*Lophodytes cucullatus*)		X		
turkey vulture (*Cathartes aura*)	X	X	X	
sharp-shinned hawk (*Accipiter striatus*)	X			
red-tailed hawk (*Buteo jamaicensis*)	X	X	X	X
broad-winged hawk (*B. platypterus*)	X			
American kestrel (*Falco sparverius*)		X		
ruffed grouse (*Bonasa umbellus*)	X	X	X	
bobwhite quail (*Colinus virginianus*)	T			
turkey (*Meleagris gallopavo*)	X	X		
Mammals				
raccoon (*Procyon lotor*)	T	T		T
spotted skunk (*Spilogale putorius*)				X
red fox (*Vulpes vulpes*)		X		
bobcat (*Lynx rufus*)	T	T		
woodchuck (*Marmota monax*)		X	X	T
eastern cottontail (*Sylvilagus floridanus*)			X	X
white-tailed deer (*Odocoileus virginianus*)		X	X	X

Notes: X denotes actual sighting, and T denotes tracks or other evidence of presence.

Fig. 15.4. Wild turkeys were observed on the Brushy Mountain minesite during spring, summer, and winter of 1980–81. During the warmer months, young turkey poults gleaned insects from the surface-mine ground cover plants. During the cold weather, adult birds were attracted to seed heads of the rehabilitation grasses, particularly during times of snow cover, when other food sources were concealed.

General wildlife observations

From August 1979 to January 1981, 4 species of waterfowl were observed on the Brushy Mountain sediment ponds (Table 15.5). The absence of waterfowl on Massengale Mountain is attributed to the lack of sediment ponds on the study portion of the minesite. Five species of raptors were observed in the general vicinity of Brushy Mountain. Although most of the raptors were observed soaring over the study sites, the American kestrel and the red-tailed hawk were seen perching on dead snags on the Brushy Mountain mine on several occasions. The red-tailed hawk was the only raptor observed on the Massengale Mountain study site.

Three species of game birds were recorded in the general vicinity of the Brushy Mountain minesites (Table 15.5). In May 1980 a wild turkey nest containing 12 eggs was found in the woodland area above the Brushy Mountain minesite. Subsequently during summer 1980, young turkey poults were frequently seen on the Brushy Mountain mine site. A group of 13 adult birds (2 males, 11 females) were seen on the demonstration site in February 1981 (Fig. 15.4).

A ruffed grouse, flushed from beneath one of the fruiting autumn olive shrubs during December, was the only game bird recorded on, or in the general vicinity of, the Massengale Mountain minesites. Examination of fecal droppings indicated that the bird had been consuming the autumn olive berries lying in the ground litter beneath the shrubs.

Five species of game mammals were recorded from the Brushy Mountain mine sites or in their general vicinity (Table 15.5). White-tailed deer occasionally visited the Brushy Mountain sites but were not particularly common in the area (judging from the low numbers of tracks). Fecal droppings of the eastern cottontail were frequently observed beneath the dense stands of sericea lespedeza on Massengale Mountain. The red fox (from sightings) and bobcat (from tracks) were known to be present on the Brushy Mountain mine site. A spotted skunk (which was disturbing small mammal trapping operations), was trapped in the Massengale Mountain woodland area. Although the range of this species extends throughout Tennessee, it is known not to be common (Kennedy & Harvey, 1979).

Implications of findings to mine rehabilitation

The low numbers and densities of volunteer plant species observed on the study minesites has significant implications to rehabilitation operations. It is particularly noteworthy that heavy-seeded tree species, such as oaks, which one would expect to germinate from respread topsoil, were not observed on the study mines. Haynes (1983) examined the vegetation on a 43-year-old coal surface mine in Illinois and also noted a limited and slow appearance of heavy-seeded tree species on his study site. Leftwich (1974), who studied plant succession on coal surface-mines in Illinois, noted that most volunteer tree species observed bore light seeds and that oaks were still a minor component of the vegetation on mine sites excavated during the 1920s. It thus appears that heavier-seeded plant species must be established during rehabilitation operations if they are to be a component of the vegetation within a reasonable timeframe. For some post-mining land-use, such as pasture land, this situation would not matter. However, for other land-uses, such as forestry production or wildlife habitat establishment, heavy-seeded tree species are highly desirable.

It is likely that most of the volunteer plant species invaded the mine sites from the adjacent forest in the 2-year interval before the dense grass/herbaceous ground cover developed. This hypothesis is strengthened by our documentation of more volunteer plant species at higher densities on the younger Brushy Mountain mine site than on the older Massengale mine site. If most of the colonization occurred within the first 2 years after rehabilitation, then subsequent mortality of the colonizing plants with the passage of time would account for the lower numbers on the older minesite.

If this hypothesis is correct, the vegetation on these reclaimed mine sites will remain relatively static until the condition is broken by a physical perturbation, such as wildfire, or by eventual shading out of the ground cover by the surviving trees on the mine sites and along the forest periphery. In view of the sparse tree cover on the mine sites, the latter case will be a long slow process as will the invasion of heavy-seeded tree species from the adjacent forest. Haynes (1983) made a similar observation in his study of natural vegetation development on a 43-year-old coal surface mine in Illinois: 'Once established, the dense ground and understorey cover may have prohibited or precluded the establishment of many new trees; thereby, suppressing the rate of succession on the mine site.'

Results of the study also suggest that most of the natural seedbank present in the characteristically shallow Appalachian topsoils is damaged and/or buried during removal, storage, and regrading operations on minesites. Seedbank plants which do manage to propagate, apparently compete poorly with the aggressive grasses and legumes sown abundantly over the mined area during reclamation operations.

Results of this study thus indicate that natural plant invasion has been severely inhibited on our study sites. This condition may well be the rule, rather than the exception, on conventionally rehabilitated surface mines in the southern Appalachian coalfields.

The preceding scenario has serious implications to rehabilitation of coal surface-mines for post-mining land uses featuring wildlife:

1. The rehabilitation planner cannot assume that beneficial wildlife food and cover plants will 'volunteer' onto the minesite from the adjacent vegetation, particularly in relatively level terrain where mine interiors can be great distances from adjacent undisturbed woodland. Consequently, these types of plants must be established on the minesite during rehabilitation operations.
2. The allelopathic properties of some grasses and legumes may make it unfeasible to use them in combination with certain tree

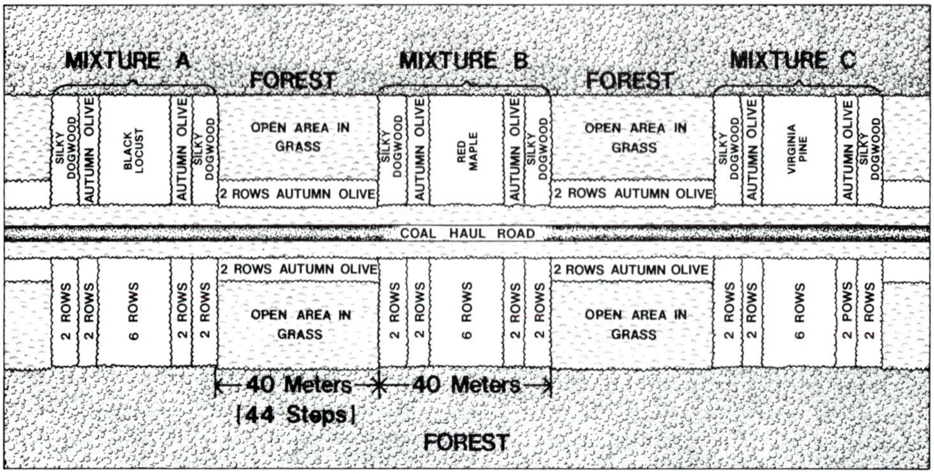

Fig. 15.5. The planting arrangement of wildlife food and cover plants along a hypothetical contour surface mine. The planting design provides habitat diversity on the mine and distributes food and cover uniformly over the site. Tree and shrub rows are spaced 4 steps apart (3 m), while plants within rows are 3 steps apart (2 m).

and shrub species in rehabilitation plans. Allelopathic interactions of various mixes of plant species used for mine rehabilitation are not well understood and clearly warrant additional research.

3. Rehabilitation practices need to go beyond simple erosion control and focus on the long-term end product that is desired on the rehabilitated minesite. For example, rehabilitation of minesites for wildlife should include food production and habitat structure development provisions for specific wildlife species or groups of species. In short, mine rehabilitation needs to be viewed as a long-term ecosystem repair process rather than a 'quick fix' of short-term erosion problems.

A note of caution is in order. The common denominator in any rehabilitation operation is the personnel who must establish the plant materials. For widespread application, the rehabilitation approach cannot be of a complexity that requires the close supervision of a resource specialist. Therein lies the challenge. Complex ecological repairs must be performed on drastically disturbed minesites using relatively simple approaches that can be readily applied by average rehabilitation personnel. In addition, the techniques used must be comparable in cost to the conventional rehabilitation requirements required by regulatory agencies.

The rehabilitation plan constraints outlined in the preceding paragraph quickly lead one to the concept of thoroughly tested and accepted generic model rehabilitation plans for various post-mining land uses in a given region. Such a plan was developed as one of the products of this project. This model rehabilitation plan features establishment of wildlife food and cover plants in a simple repeating pattern along a regraded back-to-contour minesite (Fig. 15.5). The approach was easily applied by rehabilitation personnel from a cooperating coal mining company and is offered here as one way in which wildlife habitat can be restored to a coal surface-mine using the generic model concept. For complete details the reader is referred to Fowler & Turner (1981).

Finally, it is the author's experience that the biological problems associated with rehabilitation of mined lands are by far the easiest part of the issue to address. The most difficult challenge lies in convincing the mining industry and regulatory agencies to accept and apply the technology after it has been developed.

References

Allen, Jr. N. (1973). Experimental multiple seam mining and reclamation on steep mountain slopes. In *Research and Applied Technology Symposium on Mine-land Reclamation*, pp. 98–104. Pittsburgh: Bituminous Coal Research, Inc.

Bailey, R. G. (1978). *Description of the Ecoregions of the United States.* Forest Service, U.S. Department of Agriculture.

Braun, E. L. (1950). *Deciduous Forests of Eastern North America.* New York: Hafer Press.

Duvall, V. L. & Hillmon, J. B. (1965). New grazing research programs for southern forest ranges. *Journal of Range Management*, **18**, 132–6.

Farmer, Jr. R. E., Cunningham, M. & Barnhill, M. A. (1981). First-year development of plant communities originating from forest topsoils placed on southern Appalachian minesoils. *Journal of Applied Ecology*, **19**, 283–94.

Fowler, D. K. & Adkisson, L. F. (1980). *Survival and Growth of Wildlife Shrubs and Trees on Acid Mine Spoil.* Technical Note B37. Division of Land and Forest Resources, Tennessee Valley Authority.

Fowler, D. K. & Turner, L. J. (1981). *Surface Mine Reclamation for Wildlife: A Contour Surface Mine Reclamation Plan Featuring Fish and Wildlife.* Fish and Wildlife Service, U.S. Department of the Interior. Biological Services Program No. FWS/OBS 81/09.

Fowler, D. K., Turner, L. T. & Adkisson, L. F. (1981). Utilization of volunteer vegetation in ground cover operations on new transmission line rights of way. In *Environmental Concerns in Rights of Way Management*, pp. 2701–22. Palo Alto: Electric Power Research Institute.

Getz, L. L. (1961). Factors influencing the local distribution of shrews. *American Midland Naturalist*, **65**, 67–88.

Hammond, E. H. (1964). Classes of Land-Surface Form in the 48 States, U.S.A. *Annals Association of American Geographers*, 54.

Haynes, R. J. (1983). Natural vegetation development on a 43-year-old surfaced mine site in Perry County, Illinois. In *Proceedings 1983 Symposium on Surface Mining Hydrology, Sedimentology, and Reclamation,* ed. D. H. Graves & R. W. DeVore, pp. 457–66.

James, F. C. & Shugart, Jr. H. H. (1970). A quantitative method of habitat description. *Audubon Field Notes*, **24**, 727–36.

Kennedy, M. L. & Harvey, M. J. (1979). *Tennessee Mammals: Capsular Description of Twenty-four Selected Species*. Nashville: Tennessee Wildlife Resources Agency.

Larson, M. M. & Swartz, E. L. (1980). Allelopathic inhibition of black locust, red clover, and black alder by six common herbaceous species. *Forest Science*, **26**, 511–20.

Leftwich, V. P. (1974). *Plant Succession on Missouri Strip-mined Land*. Ph.D. Dissertation, Columbia: University of Missouri.

LoBue, J. & Darnell, R. M. (1959). Effect of habitat disturbance on a small mammal population. *Journal of Mammalogy*, **40**, 425–37.

Mueller-Dombois, D. & Ellenberg, H. E. (1974). *Aims and Methods of Vegetation Ecology*. New York: John Wiley and Sons, Inc.

Mumford, R. E. & Bramble, W. C. (1969). Small mammals on surface-mined land in southwestern Indiana. In *Ecology and Reclamation of Devastated Land*, ed. R. J. Hutnik & G. Davis, pp. 377–88. London: Scientific Publications Ltd.

Pielou, E. C. (1975). *Ecological Diversity*. New York: Wiley.

Pruitt, Jr. W. O. (1959). *Microclimates and Local Distribution of Small Mammals on the George Reserve, Michigan*. Miscellaneous Publication 109, Museum of Zoology. Ann Arbor: University of Michigan.

Public Law 95–87 (1977). *Surface Mining Control and Reclamation Act of 1977*. 30 U.S.C. 1201 Note. Washington, D.C.

Shannon, C. E. & Weaver, W. (1963). *The Mathematical Theory of Communication*. Urbana: University of Illinois Press.

Sly, G. R. (1976). Small mammal succession on strip-mined land in Vigo County, Indiana. *American Midland Naturalist*, **95**, 257–67.

Tramer, E. J. (1974). On latitudinal gradients in avian diversity. *Condor*, **76**, 123–30.

Verts, B. J. (1957). The population and distribution of two species of *Peromyscus* on some Illinois strip-mined land. *Journal of Mammalogy*, **38**, 53–9.

Walters, D. T. & Gilmore, A. R. (1976). Allelopathic effects of fescue on the growth of sweetgum. *Journal of Chemical Ecology*, **2**, 469–79.

Williams, A. B. (1936). The composition of a beech–maple climax community. *Ecological Monographs*, **15**, 173–305.

Wilson, M. F. (1974). Avian community organization and habitat structure. *Ecology*, **42**, 594–8.

16

The return of vertebrate and invertebrate fauna to bauxite mined areas in south-western Australia

O. NICHOLS, B. J. WYKES AND J. D. MAJER

Introduction

The jarrah *Eucalyptus marginata* forest of Western Australia occupies approximately 1.75 million ha of the south-western corner of the State. Most of the State's population of 1.5 million people live in the south-west and there are pressures on the forest to support a number of land-uses, such as water catchment, timber production, recreation, honey production and conservation. All of these land-uses depend on conservation of the forest ecosystem, including the soil, flora and fauna.

The forest also overlays significant deposits of bauxite. These deposits occur in pockets (average area 10–20 ha) covering some 7–8 % of the forest area, although only 3–4 % is ever likely to be mined because of various environmental and economic constraints.

The mining process involves clearing the forest, removing topsoil (50 mm) and overburden (0–1 m), blasting a cemented caprock layer and mining to an average depth of 2–5 m. This leaves a compacted clay pit floor, ready for rehabilitation. In this Chapter we discuss the evolution of Government environmental requirements, together with fauna research and rehabilitation techniques since the commencement of mining.

Early environmental requirements and research

Alcoa of Australia Ltd. commenced mining bauxite in the forest in 1963. The Alumina Refinery Agreement Act (1961) only required the company to leave the site tidy, and to remove, stockpile and replace overburden material (up to 1 m depth) when mining was completed. The Forests Department was responsible for revegetation. A variety of trees, including several *Eucalyptus* and *Pinus* spp. were planted, but no understorey seeding or soil ripping was carried out. There was only minimal fertilizing and re-shaping of pits.

The results were less than satisfactory, with poor understorey development, pines blowing over as a result of roots not being able to penetrate compacted subsoil, and some species showing signs of nutrient deficiencies. No research on fauna was conducted, either before or after mining.

With the gradual expansion of mining operations, a parliamentary act was passed in 1969, *viz*, the Alumina Refinery (Pinjarra) Agreement Act (1969). This required Alcoa 'to take adequate measures at the Company's expense for the progressive restoration and re-afforestation of the forest destroyed, the prevention of soil erosion and formation of deep water pools, and safety measures'. Regulations were not very specific and contained no requirements for research or detailed rehabilitation requirements. Nevertheless, the Company commenced its environmental programme in 1973 by employing a forester, by introducing deep ripping of the soil (to aid tree root penetration), and by initiating tree nutrition and growth studies. Research on understorey development from seeding and fresh topsoil return commenced in 1975, and in this period (1975–76), the first fauna studies were carried out. The impetus for introducing understorey seeding and fresh topsoil return was to increase the conservation value of rehabilitated minesites, and early fauna surveys were designed to assess whether these improvements in rehabilitation technology were likely to provide suitable fauna habitat.

Development of current legislative requirements

By 1977, the alumina industry had developed to the point where Alcoa wished to develop a third refinery and a fourth minesite. Preparation of the Wagerup Environmental Review and Management Programme (ERMP) – a type of Environmental Impact Statement, described in Chapter 1 – commenced in 1977. To deal with the extra work required, the Company employed a full-time zoologist and a number of consultants. Extensive fauna studies were undertaken in unmined forest (to predict likely effects of mining) and in mined and rehabilitated sites (to assess recolonization). Pre-mining studies included surveys of all vertebrate groups, ants, termites, earthworms, general invertebrate collections and associated habitat studies.

The ERMP process and subsequent Wagerup Agreement Act (1978) introduced a number of legally binding requirements which had implications for ongoing fauna research and protection of fauna habitat. The Company had to submit 5- and 25-year mine plans to the Government for approval. They were also required to carry out continuous investigations and research to ascertain the effectiveness of the ameliorative measures being taken in the mining and rehabilitation programme. Annual reports had to be submitted to Government outlining all research being conducted, and summarizing research findings. The Company also agreed to rehabilitate any areas of forest adjacent to mining activities which had been affected by dieback disease (*Phytophthora cinnamomi* – a fungus which kills jarrah trees and many understorey species), and not to mine in eastern areas of the lease until it could be shown that mining can be carried out in those areas without affecting the salinity of water supplies.

The current requirements for rehabilitation and research result from the cumulative requirements of the various Agreements. Many of these requirements were already being practiced when legislation was introduced in 1978. Although there are no specific guidelines for fauna research or monitoring, both have progressed considerably and the remainder of this Chapter describes the evolution of fauna research in conjunction with the development of rehabilitation techniques.

Fauna research

General invertebrate fauna studies

Some of the earliest work involved the long-term monitoring of surface-active invertebrates in mined areas for 3 years after rehabilitation which had been carried out in 1976 (Majer, 1981). One area was not revegetated, one was planted with a single species of tree (marri, *Eucalyptus calophylla*), and the other was seeded with a diverse mixture of native plants. Ants were identified to species level, while other invertebrates were recorded to either order or family.

The seeded plot was rapidly colonized by a rich and abundant invertebrate fauna while the area planted to monoculture and the unvegetated area were colonized by relatively few invertebrates (Table 16.1). These differences were tentatively attributed to the formation of a

Table 16.1. *Summary of long- and short-term studies of invertebrates in bauxite mined areas*

	Long-term study of three areas		
	Unplanted	Planted	Seeded
Invertebrate abundance	1	2	3
Ant abundance	1	2	3
Ant richness	1	2	3
% Similarity to forest ant fauna	1	2	3

	Short-term study of three areas	
Factors with which ant richness and/or diversity is positively associated	Species richness (S)	Shannon's Diversity Index (H^1)
Plant species richness	24%	—
Plant species diversity	—	28%
Time since rehabilitation	13%	27%
% plant cover (especially 50–125 cm strata)	13%	38%
% litter cover	—	13%
Presence of large logs	+ + +[a]	+ + +

Notes: The ranks for the long-term study are from 1 (lowest value) to 3 (highest) and the percentage values given under the short-term study indicate the amount of variance in the ant parameter accounted for by the corresponding variables within the rehabilitated area. All variables are significantly correlated ($P < 0.05$) with one or both of the ant parameters (data from Majer, 1981; Majer *et al.*, 1984).

[a] The contribution of large logs to the return of ant fauna was assessed by a separate analysis and found to be important. The variance which logs contribute to the ant parameters is not quantified here.

species-rich understorey in the seeded plot and to the relatively simple ecosystem which resulted from the planting option.

In order to obtain more information on the effect of rehabilitation on invertebrate colonization, a 1-year short-term study was undertaken during 1979 on 30 bauxite mines differing in age and method of rehabilitation. The physical and botanical conditions in each plot were measured and invertebrates associated with the soil, litter and vegetation were sampled. Statistical analysis indicated that the return of a rich ant fauna was enhanced by the provision of a dense litter cover, large logs and also a dense and diverse plant cover (Table 16.1) (Majer *et al.*, 1984).

Concurrent studies on the rate of litter decomposition indicated that leaves were decomposing at rates approaching that in the original forest and that this was in part associated with the restoration of a diverse and dense soil and litter fauna (Majer, unpublished data). These studies in rehabilitated mines indicate how rehabilitation procedures may be directed so as to favour a return of invertebrates important to the welfare of the ecosystem.

A comparative study of ant recolonization of rehabilitated mines throughout Australia showed that increases in richness per mine plot were generally linear over the first 3–5 years (Majer, 1984). By comparison, the number of ant species in 3-year-old mines throughout Australia showed that there was a gradient in return rates from the driest mediterranean areas to a high value in tropical monsoonal areas (Table 6.4). This gradient probably reflects climatic factors rather than differences in ant species richness throughout Australia. Because the ant fauna provides a good bio-indicator of the abundance and richness of plants and other invertebrate animals (Majer, 1983), this gradient of ant return rates may reflect a gradient of ecosystem resilience (see Chapter 6 for a more detailed discussion). The positions of the Western Australian bauxite mines along this gradient suggest that it may take longer to achieve a certain level of ecosystem development than in moister parts of Australia.

General vertebrate fauna studies

Studies on the return of vertebrate species following rehabilitation were carried out concurrently with this early invertebrate research. The objectives of the vertebrate surveys were:

1. To determine what species were using rehabilitated minesites in order to provide a measure of early success of the rehabilitation programme.
2. To determine what locally occurring species were absent from minesites, and suggest reasons for their absence. And
3. to identify any rare species, either in rehabilitated sites or unmined forest, and determine whether further studies are necessary to assess the likely effects of mining on their conservation status.

These studies (Kabay & Nichols, 1980) demonstrated that over 80 % of birds, 75 % of reptiles and all mammal species which use upland jarrah forest were using rehabilitated sites. In this context, 'using' referred to animals feeding, resting, sheltering or breeding in rehabilitated sites.

Species which were not present included bird species which require tall tree canopies or bark of mature trees, and reptiles which need abundant logs for shelter. Such species should return as appropriate trees mature. Bark type is now one of the characteristics used in selecting tree species. Two rare animal species, the carpet snake, *Python spilotes*, and the red-eared firetail, *Emblema oculata*, were recorded in forest near minesites and studies commenced on them. These are described later.

These early studies did not conclusively demonstrate that rehabilitation fulfilled all the requirements of the returning species. They only showed that these species were making some use of rehabilitated areas and that these areas were not forming migratory barriers to species from adjacent unmined areas. Also, the results represent surveys over a large number of areas, and not detailed studies of particular sites. Large differences in the suitability of different rehabilitation types for particular species were known to exist.

Therefore, further studies were considered to be necessary in order to determine:

1. The extent of vertebrate use of minesites, i.e. population numbers, breeding records, similarity of communities to those recorded in unmined forest, etc.
2. Detailed information on the rare species in the locality.
3. What characteristics of rehabilitation were best suited to fauna needs. And
4. what potential improvements could be introduced to better cater for fauna.

Detailed studies commenced in 1980–81 on birds and reptiles in order to provide data and answers to these questions.

Bird studies in rehabilitated minesites

The first detailed studies of bird communities in rehabilitated areas were carried out in summer 1980–81 (Nichols & Watkins, 1984). These surveys compared bird numbers, densities, species diversity and community similarity indices in a range of rehabilitated areas and also in both good and poor (i.e. degraded by dieback) quality forest.

The studies demonstrated that rehabilitation of mined areas with *Pinus* or *Eucalyptus* species alone, with no provision for understorey, resulted in low bird species richness, diversity and density. By contrast, 4 or 5 year old areas seeded with a wide range of understorey species, or given fresh topsoil containing seeds, can support similar bird species numbers, densities and species diversities to unmined areas of forest (Table 16.2).

Table 16.2. *Bird community parameters for seven plots censused in summer*

	Healthy jarrah forest		Dieback-affected forest		Rehabilitated minesite		
	I[a]	II	III	IV	V	VI	VII
Total density ha^{-1}	13.2	12.6	6.0	13.5	6.9	0.17	12.0
Species richness	16	16	9	14	16	1	12
Diversity H^1	1.09	1.08	0.88	0.99	1.08	—	0.96
Standard error of H^1	0.01	0.01	0.02	0.01	0.01	—	0.02

Notes: Diversity is calculated using the Shannon-Weiner Index $H^1 = -\Sigma p_i \log_{10} p_i$ where p_i is the proportion of the *i*th species.

[a] Description of plots:

I – healthy jarrah forest.

II – healthy jarrah forest.

III – severely dieback affected forest.

IV – dieback affected forest with some marri (*Eucalyptus calophylla*) regrowth.

V – planted 1976 with *Eucalyptus resinifera*, marri, and seeded with thick *Acacia* understorey.

VI – planted 1970 with *Eucalyptus microcorys*. No understorey added. Atypical of current rehabilitation methods.

VII – planted 1975 with marri and received fresh topsoil.

Comparisons between minesite bird faunas showed that different vegetation communities supported characteristic bird populations. For example, sites possessing thick *Acacia* stands supported bird species which are characteristic of dense stream-zone vegetation. By contrast, sites given fresh topsoil produced a good understorey and bird populations which were more similar to those from unmined upland forest. This study suggested that a variety of rehabilitation techniques was desirable in order to support a wider range of species than would any single technique.

Similarity indices provide additional information to that provided by density and diversity values. Indices may be used to compare communities between sites on the basis of species composition (e.g. Sorensen's Index) or they may take into account relative abundances of species (Canberra Metric Index). Both indices were applied to the bird data and the results showed that current rehabilitated sites were more similar to healthy, unmined forest than was poor quality, dieback-affected forest (Table 16.3). This indicated that the composition of the avian community had developed to the point where it was within the range of species found in unmined forest.

Whilst these results were encouraging, two further studies were needed: all year round bird studies, to detect seasonal variation in bird populations, and longer term studies to determine the stability of bird populations in rehabilitated areas.

Detailed avian research

Alcoa has encouraged ecological research in the jarrah forest by sponsoring projects by students and staff at Perth's tertiary education institutions. In this way Wykes (1983) undertook a detailed study of the forest avifauna, between 1980 and 1983, as a Research Fellow at the Western Australian Institute of Technology (now Curtin University).

The research was conducted in all seasons over 2 years in relatively undisturbed, mature forest. The principal site, of 8 km^2 and within a larger forest tract, incorporated a range of microhabitats from ridge to valley. A strip-transect technique was used to estimate seasonal bird densities and these data were also used to test for correlations of bird distributions with floristic and structural vegetation characteristics. Sticky-board and sweep-net arthropod sampling along the transects, and visual estimates of flower abundance were used to monitor relative seasonal food abundance. Foraging observations were systematically collected on all bird species. Statistical comparisons were used as a basis for interpretation wherever possible.

Table 16.3. *Similarity indices for comparisons between plots described in Table* 16.2

Plot description		Sorensen's Index						Canberra Metric Index					
		I	II	III	IV	V	VII	I	II	III	IV	V	VII
I	Healthy forest	—	—	—	—	—	—	—	—	—	—	—	—
II	Healthy forest	0.75	—	—	—	—	—	0.42	—	—	—	—	—
III	Severe dieback forest	0.40	0.40	—	—	—	—	0.23	0.27	—	—	—	—
IV	Dieback forest rehabilitated	0.47	0.53	0.43	—	—	—	0.32	0.38	0.34	—	—	—
V	1976, rehab. tall acacias	0.56	0.44	0.08	0.33	—	—	0.30	0.28	0.16	0.16	—	—
VII	1975, rehab. fresh topsoil	0.57	0.57	0.28	0.46	0.50	—	0.34	0.30	0.15	0.27	0.36	—

Note: Plot VI was omitted from analyses because only one bird was recorded during surveys.

The same techniques were used over 1 year to investigate the avifauna of 6-year-old vegetation in a rehabilitated mine-pit and surrounding forest. However, undertaking avian studies in small patches of rehabilitated vegetation presents sampling difficulties. The effect of edges and small sample sizes on technique and interpretation must be taken into account.

The average density of birds in the forest was found to be 52 per 10 ha, with a total of 44 species and an average of 27 species per bimonthly census. This was depauperate compared to avian communities of structurally similar forests in south-eastern Australia (Collins, Wykes & Nichols, 1986). The research suggested that this may have resulted from three possible contributing factors. Firstly, there is a poor geographic potential for migratory avoidance of limiting periods of food supply (low winter arthropod abundance and only periodic abundance of nectar). Secondly, there have been prehistoric pressures of periodic isolation and contraction of the small south-west forest region, with concurrent climatic variation. Finally, specific ecological constraints such as poor potential for litter and bark foraging exist. Consequently, most of the insectivores, nectarivores and granivorous parrots of the forest are either habitat generalists throughout southern Australia or are widespread in the south-west and utilize most woodland and forest types.

None of the bird species was confined to the jarrah forest and distributions within the forest revealed only a few habitat correlations. These were of species which favoured the denser vegetation of streams and lower slopes rather than the laterite capped ridges where mining occurs. Thus the outlook for successful conservation and management of the avifauna in mined regions is promising. The adaptable, opportunistic characteristics of the bird species were verified by their readiness to colonize the rehabilitated plot of young eucalypts with a dense understorey. Avian species richness and overall density values were indistinguishable between the forest and rehabilitation sites, and seasonal patterns were similar (Fig. 16.1). As in previous studies, avifauna of the rehabilitation was found to be closest in composition to that of jarrah stream vegetation, reflecting the structural affinities of these habitats.

This detailed research project did not generate any recommendations for major changes in rehabilitation practice. However, it provided basic information on the jarrah forest avifauna that will be invaluable in interpreting developments as the rehabilitation matures. It also provided reassuring confirmation that the avifauna appears well capable of adapting to changes of the type generated by mining. Some of the reasons for this may be peculiar to the jarrah forest bird community.

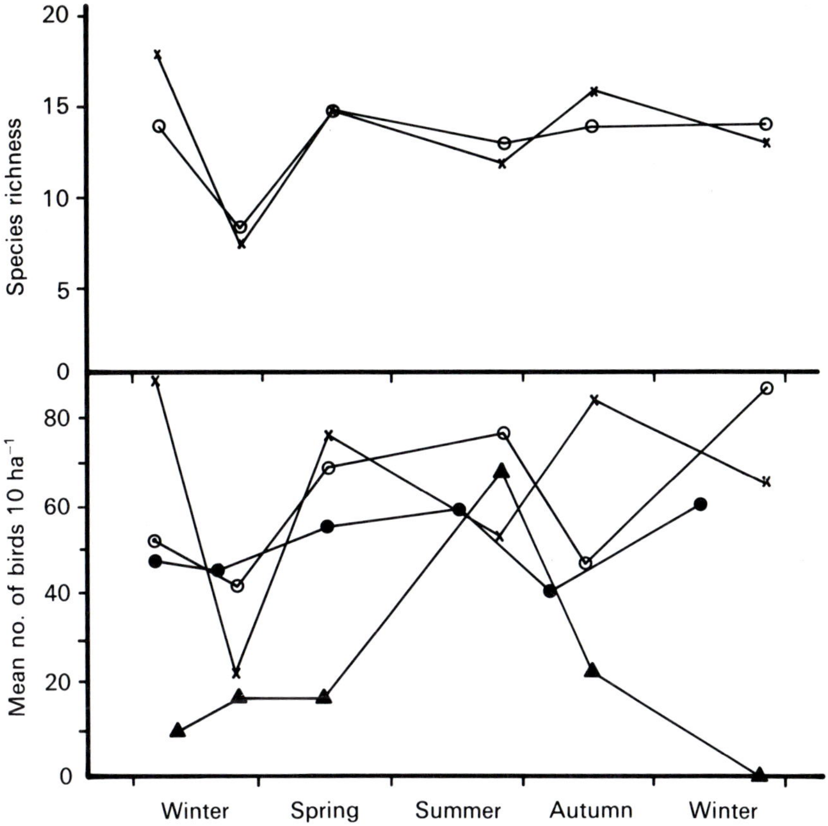

Fig. 16.1. Comparison of avian communities in jarrah forest and rehabilitated areas. Key o = 6-year-old rehabilitated minepit; x = jarrah (*Eucalyptus marginata*) forest adjacent to rehabilitation; ▲ = jarrah forest degraded by dieback disease (*Phytophthora cinnamomi*); ● = principal jarrah forest sampling site. Mean number of birds 10 ha^{-1} $visit^{-1}$: rehabilitation, 62.7; adjacent forest, 65.5; degraded forest, 17.5; principal jarrah forest, 52.4

Further research needs were identified in the study. In particular, two parrot species (the red-tailed black cockatoo *Calyptorhynchus magnificus* and the red-capped parrot *Purpureicephalus spurius*) were recommended for attention. Both have populations restricted to the south-west and their primary food resource of eucalypt seeds is very specific. Furthermore, they may be traditional in their use of particular tree hollows for nesting. In view of the apparently vulnerable behaviour of these birds, more needs to be known of the factors which regulate their populations.

Reptile studies in rehabilitated minesites

A detailed study of reptile utilization of rehabilitated areas was conducted concurrently with the bird study of summer 1980/81. The same sites were surveyed for reptiles, using pitfall-traps, hand collection and visual sightings. Results are described in Nichols & Bamford (1985).

Rehabilitated sites as young as 4–6 years old were found to support similar numbers of reptile species to unmined sites, although densities were comparable to values recorded in only the poorer quality forest. The results were interpreted in terms of the known ecologies of jarrah forest reptiles. Some species, such as the skink, *Lerista distinguenda*, require heavy leaf litter and were only present in minesites in low numbers, while others, such as the skink, *Menetia greyii*, prefer shaded open ground and were common in the 4–5-year-old site. Other skinks, such as *Egernia napoleonis* and *Ctenotus labillardieri*, were uncommon in the rehabilitated ares due to the scarcity of large logs with suitable cracks for shelter. By contrast, the skink, *Leiolopisma trilineata*, was much more common in rehabilitated areas than in the forest sites. The species normally occurs near streams where conditions are cool, moist and heavily shaded. In this respect, the vegetation of rehabilitated areas where *L. trilineata* was common was more like that of stream zones than upland forest.

As sites mature, most of the vegetation parameters mentioned so far approach values which are characteristic of the mature forest. Leaf litter accumulates while the amount of open ground generally declines. Observations in older rehabilitated areas have shown that as the dense, short-lived acacia's senesce, sites with dense vegetation cover gradually become more open and more like typical upland forest. This in turn allows smaller understorey species to proliferate. Finally, in the longer term, larger logs and stumps will eventually build up.

Sufficiently large numbes of three skink species were obtained to allow examination of differences in size and body weight between rehabilitation and forest populations. It was thought that any differences in either body weight or length would reflect differences in the well-being or age distribution of species in the two habitats. This might therefore provide information on the suitability of rehabilitated areas for the species.

The results (Nichols & Bamford, 1985) indicated little difference in either size or body weight distribution of *Tiliqua rugosa* or *M. greyii* in rehabilitation and forest. No small individuals of *Morethia obscura* were recorded in rehabilitated areas. This may indicate that the species is not yet breeding in these areas but needs to be confirmed by further studies which have recently commenced. This technique proved very useful in performing detailed comparisons of populations of individual species in the two sites, and could easily be extended to other species.

Specific invertebrate studies

Predatory invertebrate studies in rehabilitated areas Detailed research was performed on predatory invertebrates concurrently with the above-mentioned vertebrate studies. The rationale for selecting this group for detailed investigation was that, because they constitute the higher levels of food chains, they ought to reflect the taxonomic composition and abundance of taxa from the lower trophic levels.

Predatory invertebrates were surveyed in four rehabilitated areas and two unmined forest control sites. The results are described in Nichols & Burrows (1985). As with other studies, most quantitative indices such as species numbers, densities and diversities indicated that populations in rehabilitated sites compared favourably with the range of populations found in both poor and good quality forest (Table 16.4). Species similarity dendrograms indicated that the predatory fauna (either derived with or without dominant ant taxon) of the best rehabilitated site was at least as similar to that of the two forest sites as they were to each other (Fig. 16.2). By considering the ecology of individual invertebrate taxa, a number of potential improvements to the rehabilitation programme were suggested. These included introduction or retention of topsoil and logs, encouragement of a leaf litter and ground cover, and the elimination of haul roads around pits wherever possible. Introduction of these practices into the rehabilitation programme should encourage the return of these animals and in turn accelerate the re-establishment of the ecosystem processes with which they are associated.

At the same time as this study, detailed studies on particular invertebrate taxa were carried out. Since it was not possible to examine all invertebrate species, several factors were considered in determining which taxa would yield the most information on the success of rehabilitation. It was anticipated that the resulting studies would provide useful information which could be used for making improvements to rehabilitation.

The factors which were considered were:

1. The ecological role of the animal, e.g. decomposer, predator, pollinator, etc.,
2. The current state of taxonomic knowledge about it. And

Table 16.4. *Summary of quantitative data obtained from the predatory invertebrate survey of rehabilitated sites*

	Rehabilitated sites[a]				Dieback site	Healthy forest
	1	2	3	4	5	6
Species numbers in quantitative survey	29	37	29	22	38	43
Total individuals	338	551	736	314	246	467
Diversity (H^1)[b]	0.79	1.10	0.82	0.87	1.23	1.19
Numbers of herb/shrub species	7	9	8	6	10	7
Abundance of herb/shrub species[c]	31	47	44	42	79	11
Species in pitfall traps	20	21	20	13	17	26
Number of non-ant species	20	27	19	15	29	33
Abundance of non-ant species	59	95	97	64	115	99

Notes: [a] Rehabilitated site descriptions:

Site 1: 6 years old, received fresh topsoil but no seeding. Tree and understorey development good.

Site 2: 7 years old, received fresh topsoil, trees planted and seeded with dense understorey.

Site 3: 6 years old, seeded with dense understorey and received bitumenized straw mulch.

Site 4: 13 years old, trees planted but no fresh topsoil or seeds applied. Sparse understorey.

[b] Diversity (H^1), was calculated using the Shannon-Wiener Index (see Table 16.2). No correction has been made for the fact that Site 5 only possessed one transect. Therefore, for this site, abundance data would be disproportionately low although richness and diversity data are probably not greatly affected.

[c] Species abundance is the total number of individuals. Herb/shrub refers to invertebrates found on herbs and shrubs.

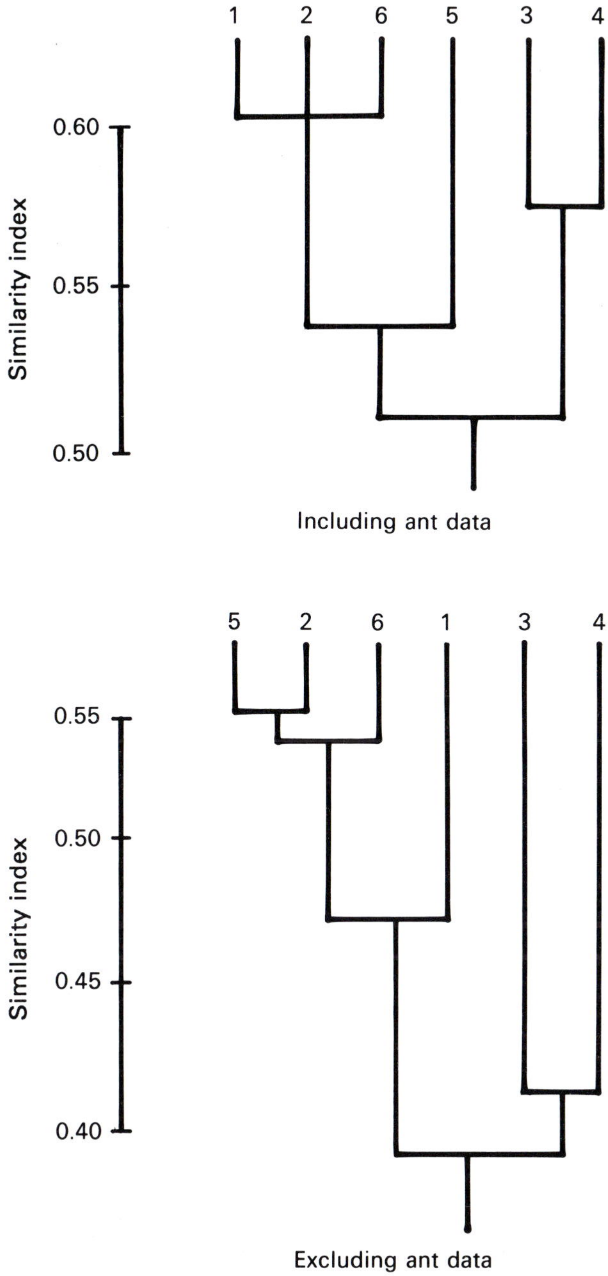

Fig. 16.2. Dendrograms constructed using Sorensen's similarity index showing the interrelationships between predatory invertebrate populations of plots described in Table 16.4.

3. the availability of a researcher with appropriate expertise to study it.

Using these guidelines, three groups were selected for study, namely springtails, termites and arachnids.

Springtails Most springtails (Collembola) feed on micro-organisms such as fungi, algae, bacteria and lichens. A few feed on dead or live plant material or pollen. There is evidence that they play a role in the maintenance of soil fertility by helping the breakdown of organic matter as a result of their grazing on fungi and possibly also vegetable matter.

In view of their importance in nutrient cycling and as indicators of soil conditions, Greenslade & Majer (1980) examined springtails from the three rehabilitated plots and the forest control plot used in the long-term invertebrate study (described earlier in this Chapter under 'General invertebrate fauna studies'). Over the first 3 years of rehabilitation the numbers of individuals were highest on the seeded plot and lowest on the planted and unvegetated plots. Species richness was lowest on the unplanted plot while equal numbers of species had colonized the planted and seeded plots. However, even in the planted and seeded plots, there were still only one third of the number of species on the forest control plot.

It was concluded that the rehabilitated plots lacked winter-active species. This was probably because the poor soil structure and low organic matter content of the recently rehabilitated plots limited the availability of refuges in which these animals could escape harsh summer conditions. This seems to retard the development of a balanced springtail fauna and needs to be considered when improving future rehabilitation techniques.

Termites Termites (Isoptera) were selected for investigation because of their role in breaking down woody material and as primary decomposers in the nutrient cycling process. Surveys of termites were carried out in a range of rehabilitated areas (Nichols & Bunn, 1980). These revealed that most species which are present in the jarrah forest were also recorded in rehabilitated minesites. However, successional trends were apparent. In young rehabilitated areas (0–4 years), only species capable of feeding on sound wood were present. With increasing age of rehabilitation, species which feed on decaying wood began to appear and finally, litter- and grass-eaters colonized. Most species were present after 12 years.

The study provided two reasons for optimism. Firstly, it indicated that none of the tree species used in the rehabilitation were attacked to any degree by termites. Secondly, the encouraging return of species suggests that the wood and litter decomposition processes will re-establish over time. Several possible improvements to rehabilitation were suggested and these conform with findings from other studies. These included the provision of more logs and ground litter in order to increase termite abundance and diversity. This in turn should bring secondary benefits such as improved nutrient cycling, and the provision of food and shelter for species which feed on termites.

Arachnids Mawson (1986) recently completed a study of spiders (Araneae), harvestmen (Opiliones) and scorpions (Scorpiones) in rehabilitated minesites. As a result of their behaviour, density and population dynamics, spiders and allied groups are important stabilizing agents in terrestrial arthropod communities (Coyle, 1981). Altogether, they may serve as useful indicators of whether the sites are developing towards ecological stability.

The arachnid study involved detailed investigation of the density and diversity of animals and related this to physical (presence of rocks, logs, etc.) and vegetation parameters.

Results showed that it takes a minimum of 8 years for populations of arachnids in rehabilitated minesites to reach a level of complexity comparable with that of the surrounding unmined jarrah forest. Overall, the species richness and also the numbers of ground-dwelling species were within the range recorded in two forest control sites, although densities were lower. Comparison of communities using Sorensons Similarity Index (Table 16.5) showed that highest similarity values were recorded when comparing rehabilitated sites with each other (e.g. plot 1 with 2) or with the forest plots (e.g. plot 2 with 4, or plot 3 with 5), while the similarity between the two forest sites was lower. These results indicate that the rehabilitated sites are providing a habitat which can already support an arachnid community comparable to that found in unmined forest.

In the case of ground-dwelling arachnids, both numbers of species and absolute variance values were positively correlated with both percentage ground cover and leaf litter depth. There was a positive correlation between both species numbers and absolute numbers of individuals of

Table 16.5. *Comparison of arachnid species similarity between sites using Sorensen's Similarity Index*

	Site no.				
Site no.	1	2	3	4	5
1	—	0.53	0.34	0.46	0.31
2	—	—	0.46	0.50	0.36
3	—	—	—	0.39	0.48
4	—	—	—	—	0.41

Note: Sites 1, 2 and 3 are respectively 1-, 2- and 8-year-old rehabilitation sites. Sites 4 and 5 represent unmined forest controls; site 4 has been heavily logged while site 5 is more typical of undisturbed jarrah forest.

web weavers, ambushers, stalkers and pouncers, with the number of plants greater than 1.5 m in height.

On the basis of these correlations, and other observations, a number of suggestions for improving the rehabilitation were made. It was suggested that the provision of more rocks and logs would provide refuges for animals and would promote the return of greater numbers and diversities of ground dwelling species. Provision of a substitute leaf litter such as cleared understorey vegetation, early on in the rehabilitation, would increase microclimate stability and further promote general arthropod community development.

Mawson (1986) also considered the consequences of a possible fire in the rehabilitation, and suggested that burning the 8-year-old site would open up the tree canopy and provide better conditions for web-building spiders. Fire is a recurrent, natural ecological factor in the jarrah forest. Finally, the study recommended increasing the diversity of plant species to provide a more continuous availability of flowering plants. This would increase the diversity of those arthropod species which are directly or indirectly dependent on flowers.

Specific issues

Whilst many of the earlier studies were designed to produce a general picture of fauna recolonization, they were also designed to provide answers to other important questions – for example, what specific areas require more detailed examination, and which species require in-depth studies? Three examples of specific investigations are outlined below.

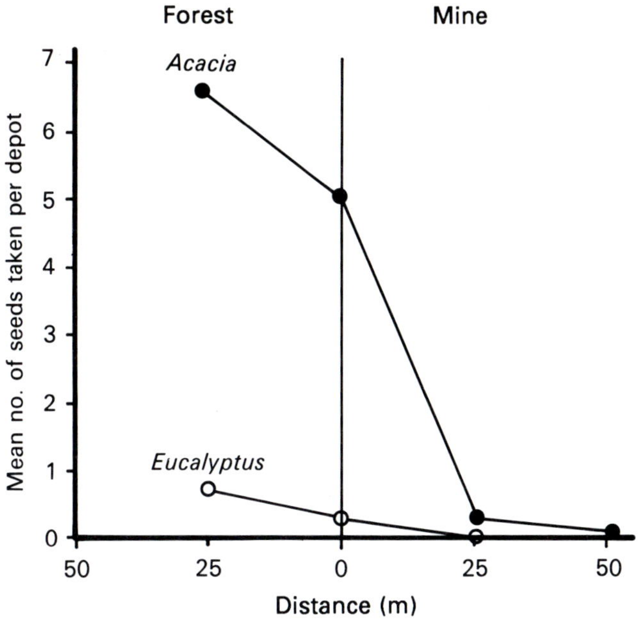

Fig. 16.3. Mean numbers of *Acacia extensa* and jarrah (*Eucalyptus marginata*) seeds taken by ants after 30 h from depots situated along transects from jarrah forest into a bauxite mine. Ten seeds of each species were placed at the depots (from Majer, 1980).

Ant seed removal A number of jarrah forest ants are important seed-associated species. Three basic types of ant–seed relationships are found in the jarrah forest: seed predation, in which ants take the seed and consume the kernel; elaiosome collection, in which ants take the seed, consume the oily elaiosome appendage and leave the kernel intact; and nest decoration, in which ants incorporate seeds into the structure of their nests. This relationship between ants and seeds in the jarrah forest is of particular relevance to bauxite mine reseeding operations.

One technique which is used to rehabilitate bauxite mines is the broadcasting of seeds of native species over the prepared topsoil. Because ants can remove 66% of seeds from the forest floor in a 24-h period (Majer, 1980), they could obviously impede the minesite seeding operation. Transects of pitfall-traps from the forest into bauxite mines, together with corresponding seed sources, were therefore monitored over 24-h periods for seed removal. It was found that ants had colonized the edges of the mines prior to seeding and that seed removal was a slight problem up to 25 m into the mined area (cf. Figs. 6.4 and 16.3). Elaiosome possessing *Acacia extensa* seeds were removed by ants 7.5

times faster than jarrah seeds, which do not possess an elaiosome. Seed removal in newly seeded bauxite mines therefore does not seem to be a serious problem. However, there is likely to be a slight alteration by ants of the ratio of germinating seeds around the edges of bauxite mines.

Bird breeding A survey was undertaken to determine which bird species bred in rehabilitated sites (Curry & Nichols, 1986). Although the earlier studies appeared to demonstrate favourable bird recolonization, this is not complete until species are breeding in mine sites and rearing young. Although the rehabilitated sites are still comparatively young, this study was designed to see whether species which could be expected to breed at this stage were doing so, and whether other species may be encouraged to commence breeding.

Twelve bird species were found nesting in rehabilitated minesites and 21 species used these sites for foraging or hunting, whilst breeding elsewhere. As expected, most breeding species were those which nested in shrubs or lower canopy foliage, tree forks, etc. It was concluded that birds breeding in rehabilitated sites represent virtually all of those locally occurring species that could be expected to find suitable nest sites in early stages of regrowth. Tree growth, suitable for canopy nesting species, will only occur with time, although nest boxes are being evaluated in an attempt to encourage birds that nest in tree hollows to breed.

Carpet snake study The carpet snake, *P. spilotes*, is a reptile on Western Australia's official rare fauna list. The species has suffered considerable decline in the cleared, wheat-growing farmland east of the jarrah forest. In our early fauna studies, carpet snakes were recorded in forest within the mining envelope, and also in planned mining areas. This prompted a detailed study of its distribution, abundance and habitat requirements.

Results (Nichols, unpublished data) confirmed the decline in geographic distribution of this species (Smith, 1981). However, this species was still found to be relatively common in the jarrah forest. Although the majority of sightings were made close to streams (< 200 m), carpet snakes were also recorded in upland forest, where mining generally takes place. This prompted more detailed studies of habitat requirements, using radiotracking as an investigative tool.

Whilst being tracked, snakes used logs, rocks, holes and small shrubs for shelter. All of these are present in rehabilitated areas although logs are less common and usually smaller than in the forest. During the course of the study, two snakes were recorded in rehabilitated areas.

The study concluded that, although mining was being carried out in carpet snake habitat, the snakes were relatively common in the forest, and that they use rehabilitated areas. How and for what purpose they use these areas is not yet clear. With mining in the long-term occupying a maximum of 3–4 % of the forest area, it appears unlikely that it will pose a serious threat to the species.

Red-eared firetail study Like the carpet snake, the red-eared firetail, *E. oculata*, was placed on the rare fauna list because it was believed that its range had declined significantly and it had become uncommon. As with the carpet snake, red-eared firetails were recorded within possible future mining envelopes, and studies commenced to determine what possible threat mining posed to the species.

Calls of the species were recorded and, when played back, attracted males if they were present. This method enabled extensive surveys of forest areas to be effectively carried out. This information was combined with data from the Royal Australasian Ornithologists Union *Atlas of Australian Birds (1981)* to provide a detailed picture of distribution and abundance, relative to that of other species (Nichols, Watkins & Kabay, 1981).

The study showed that there had only been a minimal decline in the species' distribution and that it was one of the most common bird species in areas of prime habitat. In the northern jarrah forest (the only area where any possible future mining could occur) birds were only recorded near stream zone vegetation, and this habitat is not mined. Some pairs were recorded in stream zone vegetation adjacent to rehabilitated minepits, suggesting that the mining and rehabilitation operations do not cause permanent disturbance to resident birds.

This study led to the conclusion that mining operations would not significantly affect the distribution or status of this species.

Changes to rehabilitation resulting from fauna studies

As a result of these detailed investigations, a large number of suggested improvements to rehabilitation have been made in order to benefit fauna re-establishment. Since the fauna research programme was conducted with the aim not just of evaluation but also of improving rehabilitation, it is appropriate to review these suggestions and show how this research has influenced rehabilitation practices. Table 16.6 summarizes the recommendations arising from fauna studies and notes what action to change rehabilitation practices has followed that study. It should be noted that changes to rehabilitation practice range from those

Table 16.6. *Changes to rehabilitation resulting from fauna studies*

Suggested change to rehabilitation	Source of recommendation	Action
1. Deep rip soil to facilitate tree root penetration.	Suggested from early observations.	Ripping techniques and equipment have progressed considerably since ripping was commenced in 1971.
2. Control feral animals – mainly cats	General vertebrate survey.	Cat trapping is occasionally carried out when needed. Feral pig control programmes are implemented by the W. A. Water Authority.
3. Increase return of woody material.	Nichols & Bunn (1980)	Raking is no longer used to remove root material, sticks etc.
4. Return larger logs for fauna shelter.	Nichols & Bunn (1980), Nichols & Burrows (1985), Nichols & Bamford (1985), Mawson (1986)	See 3. Also, logs on the side of pits are pushed back after rehabilitation in some areas.
5. Increased litter cover.	Nichols & Bunn (1980), Nichols & Bamford (1985), Mawson (1986), Majer *et al.* (1984), Greenslade & Majer (1980)	Woodchips and chopped vegetation waste are being investigated on an experimental basis. Accumulation of litter in an area seeded using current methods is more rapid than in older style rehabilitation (Nichols & Ward, in prep.).

6. Seed dense understorey.	Kabay & Nichols (1980), early observations.	Seed is spread on all areas at a rate of 1–2 kg ha^{-1}.
7. Increase diversity of understorey species.	Mawson (1986), Majer *et al.* (1984)	Approximately 70 species are now used in the seed mix. Fresh topsoil is used wherever practicable.
8. Bark of trees to be more like jarrah bark.	Wykes (1983), Mawson (1986)	Bark characteristics is one factor considered in tree selection.
9. Increase cover of understorey species.	Majer *et al.* (1984), Nichols & Burrows (1985)	Covered by 6 and 7.
10. Manage understorey by thinning, make areas more representative of upland forest.	Nichols & Watkins (1984), Collins, Wykes & Nichols (1986)	Understorey slashing has been tried experimentally but is not an operational technique. Species used in rehabilitation no longer form such a tall, dense canopy.
11. Introduce fire to change structure of sites to become more like unmined, upland forest.	Mawson (1986)	A burning plan has been developed and several experimental burns carried out.
12. Reduce number of perimeter roads around pits.	Nichols & Burrows (1985)	Perimeter roads are rehabilitated unless they are needed for a specific management purpose.
13. Introduce nest boxes for hollow nesting birds and mammals.	Curry & Nichols (1986), Wykes (1983), Nichols & Watkins (1984)	Nest boxes have been installed on an experimental basis, and are monitored.

which were implemented exclusively for the benefit of fauna (e.g. experimental nest boxes and feral animal control) to those introduced primarily for other purposes but which benefit some fauna species (e.g. scrub management, deep ripping of soil).

The fauna studies, together with other areas of research, have resulted in the adoption of the following rehabilitation sequence, which is described in detail in Nichols *et al.* (1985). Rehabilitation commences with reshaping of the pit walls and construction of erosion control grade banks and sumps. Stockpiled overburden material is then respread, followed by topsoil. Wherever possible, this is obtained fresh from a pit which is about to be mined. Deep ripping along contours is performed over the whole pit to a depth of 1.5 m, using a winged tyne. Trees are planted at a density of 625 ha^{-1}, and native seed (70 species) is sown at a rate of 1–2 kg ha^{-1}. Fertilizer is spread in two forms: each tree receives 200 g of N–P slow release fertilizer in pellet form, and 450 kg ha^{-1} superphosphate (with trace elements) is broadcast by helicopter. Monitoring of rehabilitation success is carried out at age 9 months and any unsatisfactory sites are replanted or re-seeded as required.

Conclusions

The fauna research programme conducted by Alcoa, Government and academic personnel has evolved over a period of 10 years from basic presence/absence studies to detailed radiotracking projects, breeding studies, and investigations on complex species/habitat interactions. At all stages, the depth of these studies has been beyond legal requirements because the importance of conservation as one of several compatible land uses was recognized by the Company.

The research programme has demonstrated several techniques and generalizations which would be useful for any mining company wishing to commence fauna research. Initially, presence/absence studies are necessary to determine what species are not being catered for by rehabilitation. The absence of some species will be obvious and predictable, for example, birds which forage in canopies of tall trees. Others, such as centipede species, may be limited by the availability of log or rock shelters, and their absence cannot be predicted.

If the aim of rehabilitation is simply to avoid local extinction of species, presence/absence studies may be all that is required. However, if rehabilitation is designed to produce a functioning forest ecosystem, more detailed quantitative and individual species studies are needed. These often uncover other questions which need answering. For example, in Alcoa's minesites, some rehabilitation techniques produced vegetation and bird communities more similar to stream-side vegetation than to upland forest. This therefore raises the question of how successional processes and management (through fire, etc.) will affect bird populations. These studies have now commenced.

Despite the fact that the research conducted has been comprehensive, there are always more questions to answer. Hopefully, the research will gradually become a process of fine-tuning the rehabilitation process. In the case of Alcoa, longer term fauna research requirements will be determined by the success of rehabilitation, economic management practices in rehabilitation, fire management, water supply needs and other factors.

It is possible to summarize by stating that the fauna research programme of mining companies should be determined by two factors; the clearly defined aims of rehabilitation, and the findings of fauna studies as they progress. Studies on fauna return following bauxite mining have shown that successful fauna return can be achieved, provided aims are clearly defined and rehabilitation methods changed when results demonstrate the need for more suitable techniques.

References

Collins, B. G., Wykes, B. J. & Nichols, O. (1986). Recolonization of restored bauxite minelands in southwestern Australian forests. In *Birds of Eucalypt Forests and Woodlands: Ecology, Conservation, Management,* ed. A. Keast, H. F. Recher, H. Ford & D. Saunders, pp. 341–54. Sydney: Surrey Beatty & Son Ltd.

Coyle, F. A. (1981). Effects of clearcutting on the spider community of a southern Appalachian forest. *Journal of Arachnology*, **9**, 258–98.

Curry, P. J. & Nichols, O. G. (1986). Early regrowth in rehabilitated bauxite minesites as breeding habitat for birds in the jarrah forest of south-western Australia. *Australian Forestry*, **49**, 112–14.

Greenslade, P. & Majer, J. D. (1980). Collembola of rehabilitated mine sites in Western Australia. In *Soil Biology as Related to Land Use Practices*, ed. D. L. Dindal, pp. 397–408. Washington: Environmental Protection Agency.

Kabay, E. D. & Nichols, O. G. (1980). *Use of Rehabilitated Bauxite Mined Areas in the Jarrah Forest by Vertebrate Fauna.* Alcoa of Australia Ltd Environmental Research Bulletin No. 8.

Majer, J. D. (1980). The influence of ants on broadcast and naturally spread seeds in rehabilitated bauxite mined areas. *Reclamation Review*, **3**, 3–9.

Majer, J. D. (1981). *The Role of Invertebrates in Bauxite Mine Rehabilitation.* Forests Department of Western Australia Bulletin No. 93.

Majer, J. D. (1983). Ants: Bio-indicators of minesite rehabilitation, land-use and land conservation. *Environmental Management*, **7**, 375–83.

Majer, J. D. (1984). Ant return in rehabilitated mines – an indicator of ecosystem resilience. In *Proceedings of the 4th International Conference on Mediterranean Ecosystems held at Perth, Western Australia,* ed. B. Dell, pp. 105–8. Perth: University of Western Australia.

Majer, J. D., Day, J. E., Kabay, E. D. & Perriman, W. S. (1984). Recolonization by ants in bauxite mines rehabilitated by a number of different methods. *Journal of Applied Ecology*, **21**, 355–75.

Mawson, P. R. (1986). *A Comparative Study of Arachnid Community Structure in Rehabilitated Bauxite Mines*. Western Australian Institute of Technology Biology Bulletin No. 46.

Nichols, O. G. & Bamford, M. J. (1985). Reptile and frog utilisation of rehabilitated bauxite minesites and dieback affected sites in Western Australia's jarrah (*Eucalyptus marginata*) forest. *Biological Conservation*, **34**, 227–49.

Nichols, O. G. & Bunn, S. (1980). *Termite Utilisation of Rehabilitated Bauxite Mine Sites*. Alcoa of Australia Ltd Environmental Research Bulletin No. 9.

Nichols, O. G. & Burrows, R. (1985). Recolonisation of revegetated bauxite minesites by predatory invertebrates. *Forest Ecology and Management*, **10**, 49–64.

Nichols, O. G. & Watkins, D. (1984). Bird utilisation of rehabilitated bauxite minesites in Western Australia. *Biological Conservation*, **30**, 109–31.

Nichols, O. G., Watkins, D. & Kabay, E. D. (1981). The distribution of the red-eared firetail (*Emblema oculata*) in relation to bauxite mining in the northern jarrah forest. *Emu*, **82**, 169–72.

Nichols, O. G., Carbon, B. A., Colquhoun, I. J., Croton, J. T. & Murray, N. J. (1985). Rehabilitation after bauxite mining in south-western Australia. *Landscape Planning*, **12**, 75–92.

Royal Australasian Ornithologist's Union. (1981). *Atlas of Australian Birds*. Melbourne: R.A.O.U.

Smith, L. A. (1981). A revision of the python genus *Aspidites* and *Python* (Serpentes, Boidae) in Western Australia. *Records of the Western Australian Museum*, **81**, 211–12.

Wykes, B. J. (1983). *The Jarrah Forest Avifauna and its Re-establishment after Bauxite Mining*. Western Australian Institute of Technology Biology Bulletin No. 11.

17

Soil biota and plants in the rehabilitation of degraded agricultural soils

C. A. PARKER

Introduction

For ten years (1936–46) I farmed in the dry wheatbelt of Western Australia (WA), where the average rainfall is around 325 mm, most falling in the 5 months from mid-May to mid-October. The summer is hot and dry, with occasional thunderstorms. Land was still being cleared for agriculture; low scrub was rolled or chained, while forests were felled by axe. By late-summer it was legally permitted to burn the dried material. After the fire, unburned logs and sticks were heaped and burned to ashes. It was then common practice to lightly plow the land ready for seeding. Since there were no weeds in the virgin forest, wheat was often 'sown dry' to germinate with the first rains. If the soils were moderately fertile, the land would be resown to wheat 2–3 years in succession, whereupon it would be left to grow 'volunteer pasture', i.e. grow weeds and grasses for 2–3 years. Sheep would be grazed on these fields, so earth-dams were sunk in natural waterways to supply them with water. The dams seldom filled during the first few years. Neighbouring farmers would commiserate 'don't worry, the dam will fill when the land consolidates'.

The point of this story is that the soils in their virgin state were soft and friable and permeable to water. After a few years they 'consolidated', which meant that voids and drainage channels were greatly reduced so that permeability was also reduced. Within a very few years it was impossible to plow these soils in the dry condition.

Soil degration continued more slowly so heavier plows and tractors were needed to till the soils. Now, many surface soils which were cleared between the 1920s and 1960s are hard setting and their subsoils are compacted. Waterlogging, flooding and soil erosion resulted and the Department of Agriculture of WA promoted contour farming with clover-pasture leys in rotation with wheat. This was successful in the wetter and more dissected districts, but was difficult to apply in the flatter and drier areas where clover grew poorly and wheat farming was more profitable. It is the less sandy of these soils that I briefly describe below.

Thirty years after leaving the farm for the University, Professor M. S. Chowdhury showed me the beneficial effect of a litter layer on a degraded soil near Morogoro in Tanzania. Subterranean termites appeared to be largely responsible for breaking down the litter and improving the structure of the surface soil, but faunal information was not available. My next stop was at IITA (International Institute of Tropical Agriculture) in Nigeria where soil physicist, Dr R. Lal had established a well designed and elaborate experiment on a hillside subject to massive sheet erosion. He was seeking a control for erosion using different soil treatments in an area receiving about 2500 mm annual rainfall. After 2 years, all runoff and erosion had stopped in the treatment area where litter was laid on the surface. In stark contrast, severe erosion had washed the topsoil away from areas where the same amount of litter had been plowed into the soil. The soils where litter was laid were completely covered with the faecal pellets of earthworms and, when these were removed, the surface was honeycombed with their burrows. This result was so dramatic that beginning in 1977, zoologist Dr Ian Abbott and I began studies on the soil fauna in virgin forests and adjacent farmland in WA (Abbott, Parker & Sills, 1979). We then looked for earthworms and found them to be widespread in farmland soils in WA (Abbott and Parker, 1980). There is a moral in this story that will not have escaped professional soil zoologists.

In the following discussion I will outline methods for the rehabilitation of 'broad-acre' farmland soils in a harsh, semi-arid environment.

The Soil Biota

The Soil Microbiota

This group contains the bacteria, including filamentous actinomycetes, the fungi and the Protozoa. Certain nitrogen-transforming specialists among the bacteria are well known. For instance, there are the free-living nitrogen fixers, both aerobic and anaerobic, which are likely to be important in soils where the C/N ratio is high and where the oxygen supply is limited.

Oxygen deficit is caused by high soil water content (not necessarily saturation) when, in the presence of readily decomposable soil organic matter, the rate of biological oxygen demand exceeds the rate of oxygen diffusion in soil. Areas of anaerobiosis in soil-water then favour nitrogen-fixing bacteria because oxygen and nitrogen compete for substrate electrons (Parker & Scutt, 1958). Symbiotic nitrogen-fixing bacteria are present in soils where legumes, *Allocasuarina* spp., or other root nodulating plants are present. In WA the native legumes use slow-growing bacteria (Lange, 1961) which will not nodulate; for example clover, peas, faba beans or *Medicago* species. For this reason, inoculation of the seed with appropriate rhizobia is widely practiced and may be essential in some soils (Chatel, Greenwood & Parker, 1968). The lucerne bacteria are not as acid-tolerant as clover bacteria, but more acid-tolerant strains have recently been found in acidic soils in the mediterranean region (Howieson & Ewing, 1986). Their introduction may permit several million ha of land in the dry WA wheatbelt to grow vigorous, soil-enriching annual *Medicago* species or lucerne (*Medicago sativa*).

Ammonia-oxidizing bacteria (nitrifiers) appear to be widely present, even in acidic soils, so that nitrate is formed. This can be lost by denitrification under conditions of oxygen deficit or by leaching into the groundwater (Alexander, 1977).

Organic matter decomposition can be carried out by bacteria and fungi alone (Skinner, 1986). Some bacteria and many fungi can hydrolize cellulose, but only certain basidiomycete fungi can attack lignin. As Abbott has written in Chapter 2, the breakdown of organic matter releases mineral nutrients for plant growth, and ultimately the amorphous dark organic residue known as humus which is essential for good soil structure is formed. Humus can be decomposed by micro-organisms but only slowly.

The soil fungi have certain advantages over bacteria. For example, they can grow across voids between soil aggregates or pores caused by the larger soil fauna. Thus, the fungi are more mobile than bacteria. Certain species are also able to grow at moisture levels well below the wilting point of plants, so that during rainy periods they can commence the attack on plant materials in the litter layer, or on standing straw in wheatfields. The ability of certain basidiomycetes to decompose lignin is of supreme importance in the evolution of plants and land animals, because without them lignin could not be decomposed, and essential nutrients would soon be locked up in semi-decomposed plant debris. Without these fungi, there could be no woody trees or shrubs and our cereals could not stand up to be harvested. On the debit side, certain basidiomycetes cause root diseases in crop plants, trees and shrubs, while others are destructive as wood rotting fungi.

While bacteria and fungi can decompose most, if not all, organic matter by themselves, the process is accelerated by predation (see Chapter 3). The amoebae have long been known as bacteriovores but, within the last 10 years, many have been shown to be active mycophages (Old, 1986). Mesofauna, e.g. mites nematodes and springtails, may also be mycophagous and, together with amoebae, play an important role in regulating microbial populations and in speeding the decomposition of organic matter in the soil (Chapter 3).

The soil macrofauna

The macrofauna, defined as greater than 2 mm in length (Chapter 2), embrace enchytraeid worms, beetle larvae, millipedes, ants, termites, spiders, earthworms and others. All these, with the exception of enchytraeids, make burrows greater than 1 mm in diameter, which are important for water, air and root penetration in soil. Their faecal pellets can behave as stable aggregates which, if deposited on the soil surface, are resistant to rain-drop impact thus protecting the surface of the soil beneath. They mix different layers of soil, characteristically bringing fine subsurface soil to the surface. In the case of earthworms (Satchel, 1967), they frequently defaecate in their burrows, thus transferring surface soil enriched with organic matter into the subsoil. Animals burrowing from the surface downwards, are most important for the percolation of water, for aeration, and for the penetration of the soil by plant roots. An experiment conducted in our laboratory to examine the effect of soil compaction on earthworm burrowing and upon the penetration of compacted soil by wheat roots, showed that roots grew very much faster down the burrows (Campbell-Clause, 1986).

Energy-rich organic matter drives the soil biotic system. For instance, 1 t of average leaf litter contains roughly the same thermal energy as 400 litres of diesel fuel, with which one can plow 20 ha of land to 20 cm depth. Thus, despite the vagaries of climate and other limitations, it is not surprising that the soil fauna can achieve important effects in soil. However, for the larger soil fauna to flourish, such material must be located on the soil surface as litter. This is because these animals feed on and, in doing so, comminute the surface litter. While working above-ground, they are subject to desiccation and predation, which they can avoid by digging and retreating into vertical burrows. Lal and his associates on Nigeria have clearly shown the benefits of surface litter and

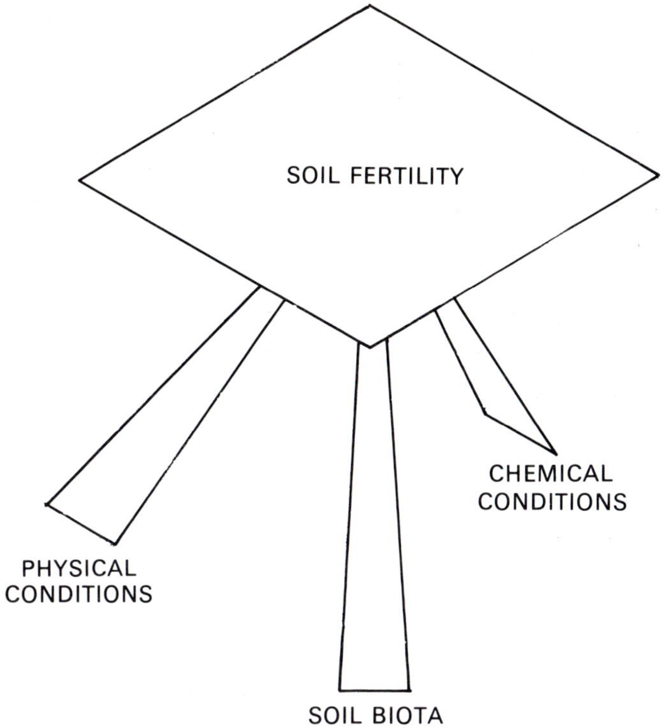

Fig. 17.1. The tripod of soil conditions which maintain soil fertility.

its associated soil fauna. His success in completely controlling water erosion in a tropical soil with high rainfall can only be described as extraordinary (Lal, 1979, 1987).

Altogether, the soil biota, covering both micro-organisms and fauna, perform essential physical and chemical functions in the soil. They are an important, though much neglected, contributor to soil fertility. I hesitate to define the term 'fertile soil' – generally it is described as the ability to produce a satisfactory yield of economic plants. Soils may be naturally fertile and, as such, they will support good plant growth, at least initially, with little or no added fertilizer. Such soils are well endowed with essential elements and also with exchangeable Ca, they have expanding clays, a well developed structure, are well drained and can be cultivated for long periods without becoming seriously degraded.

Soil fertility can also be human-made, for instance by the application of farmyard manures, or other organic wastes. The application of essential major and/or minor elements can also increase soil organic matter by promoting more vigorous plant growth. Thus, such procedures may transform a chemically poor and unproductive soil into a productive one. In Australia, millions of hectares of virgin sandy soils have been improved in this way, but in such cases leguminous pastures were established before the cropping phase to increase the quantity and quality of soil organic matter. Because earthworms have been found to be widely distributed in the WA wheatbelt it seems likely that soil fauna may have played an unrecognized part in the process of soil improvement. This was especially so on farms where cereals were grown in rotation with clover pastures (Abbott & Parker, 1980).

Thinking on soil fertility problems has for too long been dominated by chemical and physical 'solutions'. The role and fate of the soil biota, particularly that of the macrofauna, has been abysmally neglected in Australian agriculture (Abbott *et al.*, 1979). The soil biota must be recognized as an essential leg of the biological physical and chemical tripod which sustains a fertile soil (Fig. 17.1).

Causes of land degradation in agriculture

When virgin forest or shrubland is cleared for cropping, major disruptions are inflicted upon the original ecosystem. They are listed below for agricultural land-use.

1. The soil is immediately exposed to the disruptive forces of wind, water and sun. Water and wind erosion can cause serious loss of soil at this early stage.
2. A number of direct consequences of cultivation then occur. These include:
   (a) structural breakdown of the surface soil causing it to be difficult to cultivate, and reducing the vigour of sown crops;
   (b) disruption of the continuity of surface channels leading to the subsurface soil. This immediately reduces the rates of water percolation;
   (c) soil is compacted down to 40–45 cm depth with maximum compaction around 20–25 cm (Soane, Dickson & Campbell, 1982). This is made worse the heavier the machinery, the wetter the soil, and the more frequently the machines are employed. Compaction further reduces rates of percolation of water and of penetration by plant roots, while increasing the energy required to plow the soil;
   (d) elimination of the larger soil animals, thus destroying a major factor in the maintenance of the permeability and friability of surface soils in the virgin state (Abbott *et al.*, 1979),
3. The hardy perennial deep-rooted trees and shrubs are replaced by shallow-rooted annual crop plants such as cereals, oil seeds, or

lupins which are considerably deeper-rooted than most other annuals. Pasture plants such as annual clovers and medics, grasses and weeds are even more shallow-rooted. Two major changes arise from these substitutions:

(a) the nutrient cycling system of the original vegetation is broken (Nye & Greenland, 1960), so that plant nutrients are leached below the root zone of pastures and crops. The loss of cations such as K, Ca and Mg to the deep groundwater may also be a major cause of the acidification of surface soils (Dimbleby, 1952);

(b) the former balance between rainfall, evapotranspiration, soil water and stream-flow is drastically upset because less water is transpired by crop and pasture plants (with rare exceptions, e.g. lucerne) than by the original flora. Secondly, the rate of water percolation into the soil is reduced, so that more water evaporates from the surface, and/or runs off the surface, thus causing waterlogging, soil erosion and flooding. Thirdly, water which infiltrates below the root zone of the shallow-rooted crop and pasture plants raises the soil water table, thus increasing stream flow and eutrophication. Where salt is stored in the deep subsoil and was held in place by the original vegetation, salination of the valley soils and streams takes place (Nulsen & Baxter, 1982; Malcolm & Pol, 1984).

The costs to farmers of land degradation are found in lower yields per millimetre of rainfall, in higher energy and fertilizer inputs, the need for liming, in flood and erosion control and, where salt is stored in the deep subsoil, in the salination of the valleys and streams (Mulcahy, 1983). The cost to urban areas may also be high, especially where sources of fresh water are limited.

Rehabilitation of agricultural land

The importance of climate and soil type

Climate It is almost axiomatic that a benign climate (reliable rainfall, long growing season, favourable temperatures for growth) makes for easier land rehabilitation, whereas the opposite is true of harsh climates (hot, near-arid areas with irregular rain such as are found in many of the pastoral areas in Australia). Here, in harsh climates, sheep or cattle are grazed on native shrubs and grasses which are vulnerable to overgrazing and are slow and difficult to re-establish (Williams, Suijendorp & Wilcox, 1980). Production of plant biomass is poor because of low and irregular rainfall. As a result, soil faunal activity is limited by available energy sources as well as by spasmodic droughts and floods. The overall result is serious land degradation caused by wind and water erosion, the latter because the rains, although irregular, are often delivered in heavy rainstorms. The difficulty of land rehabilitation in areas between these two climatic extremes will vary with the degree of aridity and the reliability of rainfall.

Soils Again, for convenience, one can describe the characteristics of two extremes. 'Tractable' soils are widespread in the Old World, where certain soils have been farmed for 2,000 years or more (e.g. in Greece and Italy). A tractable soil is young, rich in Ca and other essential plant nutrients, it has a deep soil-organic layer, expanding clays, good structural strength and therefore allows good drainage and root penetration. Such a soil supports high yields of plant material and, where farming systems include pasture leys, soil faunal activity is also high. Where the tractable soil has suffered subsoil compaction due to heavy wheeled machines, it will respond to deep-ripping although, because of the high cost of ripping, 'gantry farming' is becoming popular (Tillet *et al.*, 1986). More promising for broadacre farming are lightweight rubber-tracked vehicles being developed for pulling farm implements (Culshaw, 1987).

'Intractable' soils, of which we have large areas in the WA wheatbelt between the 300 mm and 350 mm isohyets, are ancient and are characterized by high Na, non-expanding clays (mainly kaolin), low soil organic matter (often below 1% in the surface 10 cm), and a shallow layer of organically enriched soil (Churchward, 1986). After cultivation and cropping, the surface soils rapidly lose structure and become hard-setting, the subsoils become compacted by heavy wheeled machinery and rates of water infiltration are reduced. In the virgin state, these surface soils are friable, well-drained and sustain an active soil macrofauna. Ants and termites appear to be the main burrowers in virgin soils in the low rainfall fringe of the south-eastern wheatbelt (Abbott *et al.*, 1979), but indigenous earthworms have recently been found in similar virgin soils (McCredie, Woollett & Parker, unpublished data).

Some of these ancient soils contain huge amounts of salt, mostly sodium chloride of cyclic origin, stored in their deep kaolinitic subsoil. This gives rise to serious salination of valley soils and streams after the land is cleared. Mulcahy (1983) provides a succinct and expert review of salination in Western Australia.

An approach to soil rehabilitation

Overall, the main problems to be addressed in rehabilitating such soils as are described above are: (1) hard setting surface soil; (2) compacted subsoil; (3) the disrupted soil–plant–water balance; and (4) the disrupted process of nutrient recycling. Because of the vast areas of degraded soils, such studies should be made in the context of landscape ecology (Naveh & Liebermann, 1984), and should embrace the original flora and fauna, different soil properties, drainage lines, water sheds and water catchment areas. This is especially so in view of the large size of wheat farms, the average being around 1000 ha in the wetter areas and about 2000 ha in the drier zones. Such farms will have their own ridges and valleys, with drainage lines entering from farms up-slope, while on the lower side the surface runoff affects farms further down. Many streams which were once sources of fresh water, have become too salty for rural or urban use.

Agroforestry

In zones with higher rainfall (possibly 500 mm and upward), there are more options for the farmer than where rainfall is limiting. Agroforestry is one option of increasing interest and, assuming growth periods of 20–40 years, substantial improvement of physically degraded soils could be expected through increases in soil organic matter and soil faunal activity (the natural fauna may need to be reintroduced). It seems ecologically sensible to use mixtures of indigenous trees, and perhaps understorey shrubs, with different rooting behaviour and different leaf characteristics to maintain or improve soil pH and to encourage a richer and more vigorous soil fauna. Again, knowledge as to how the stability of the original ecosystem was maintained should be sought.

A model system for the rehabilitation of degraded agricultural soils in the wheatbelt of Western Australia

I use a WA example because it is familiar to me and because it embraces most of the principles and problems that might be encountered under difficult climatic and soil conditions elsewhere. A valley with a distinct drainage line is selected in an average annual rainfall of about 350 mm, around 70% of which falls between May and October. Some 50–100 ha of land is chosen in this valley to include part of the ridge and of the valley bottom, and to provide space for the various rotations. The ridge is between 10 and 20 m higher than the bottom of the valley, and about 500 m from it (Fig. 17.2). The pH of the surface soil should

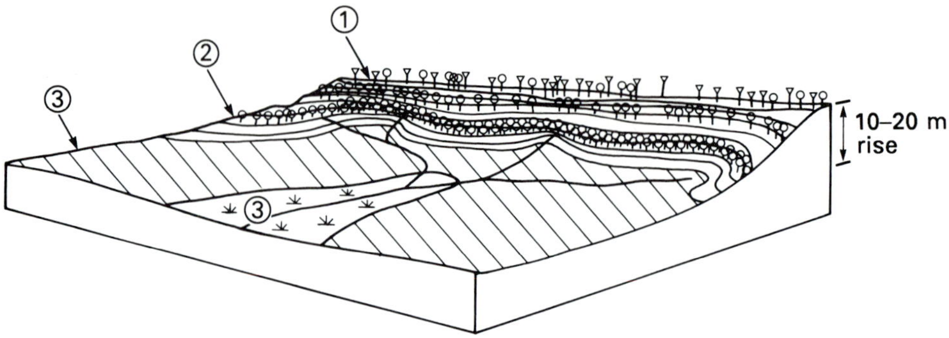

Area ①–ridge. This area is revegetated with native flora.

Area ②–slopes. This area is contour plowed to reduce runoff and is planted with deep-rooted legumes and shrubs plus annual medics.

Area ③–lower slope and valley floor. Sown with crops, medics or lucerne.

Area ③–saline flats. Planted with salt-tolerant browse shrubs.

Fig. 17.2. Map of a hypothetical degraded area in the Western Australian wheatbelt illustrating areas where the model system of rehabilitation could applied. Y = shrubs, ♀ = trees.

be around 6.0 ± 0.4 (measured in 0.002 M $CaCl_2$; Dolling & Ritchie, 1985). The area had been cleared for about 60 years. Minimum tillage, preferably zero tillage, should be practised where possible because this method improves hard-setting surface soil (Hamblin, 1984) and favours the introduction of soil animals. The land is divided into three major areas.

Area 1 – *the ridge*

The ridge could be rather rocky (the rocks may be lateritic, granitic or mixed in origin), but most of the surface is covered with a greyish-brown silty, loamy sand. It was wooded before settlement and was made bare either by clearing or by frequent wild-fires and overgrazing by sheep. This area may be separated by a contour bank from the area below.

Area 1 should be planted to trees and browse shrubs, preferably legumes such as tagasaste (*Chamaecytisus palmensis*) (Snook, 1986) or *Allocasuarina* spp. to provide soil nitrogen and nitrogen-rich litter for soil fauna. When established, controlled grazing by sheep might be permissible. In time, the trees are cut for firewood, buildings, for fences, or furniture. A portion of this area could be set aside to commence the re-establishment of the original plant and faunal community.

The functions of the plants are to help stabilize the water balance, to reduce water erosion and to provide the landholder with a financial return from land which is often of little value. Where possible an area of native trees and shrubs should be planted to increase the habitat areas of native birds, small animals and soil fauna. Good fencing is essential.

Area 2 – the upper valley slope

Area 2 embraces land which is easily erodible by water, has hard-setting surface soil and may have deeply compacted subsoil. This would be planted to deep-rooted leguminous browse shrubs such as tagasaste and annual clover or medic pasture species to provide ground cover. It could be grazed heavily but with care, as perennial plants cannot survive continuous grazing. It may perhaps be cropped again when the shrubs begin to decline because of age. Lucerne could be used as an alternative deep-rooted perennial plant in this area. When the deep soil water is reduced, the soil organic matter is increased, the surface structure is improved and deep drainage channnels have been formed by the roots. A cropping phase could be considered using chemical weed control and the least possible tillage.

Area 3 – the lower slope and valley floor

Soil of Area 3 is usually of better quality but is nevertheless also hard-setting and deeply compacted. This croppable soil may comprise 60–80 % of the total area of the model land system.

The remainder of the valley could be divided into three or four fields, with lucerne being grown for 3–4 years in rotation with cereals and/or annual medics/clover in the other two or three fields. The function of the lucerne with respect to soil would be to drive its strong and thick tap-roots through the compacted soil and through natural hard-pans down to the saturated soil level. At the end of the 3–4 year period that water level should be substantially lowered (Nulsen & Baxter, 1982). The fleshy taproots of lucerne are easily decomposed and should form well-defined deep drainage channels for water and root penetration. Lucerne would provide high quality protein-rich fodder, which can be fed with roughage to livestock, and nitrogen-rich organic matter for the soil fauna. The functions of the annual medics and/or clovers would be for winter stockfeed, when most lucerne varieties tend to be dormant, for soil organic matter enrichment, and to encourage soil faunal activity. They would also produce large quantities of seed which would help to establish dense populations of seed-eating and hole-drilling ants, of which several native species are known to invade farmlands that are seldom tilled (Twigg, Majer & Stynes, 1982).

In wetter areas, perennial grasses should be considered in conjunction with legumes, as grass roots are more fibrous and therefore ultimately more interactive with the soil. Salty soils at the bottom of the valleys can be fenced and planted with salt-tolerant perennial grasses and shrubs (Malcolm & Pol, 1984). It should be stressed that skillful grazing management is essential to maintain perennial plants in satisfactory production.

The selection of suitable plants and soil fauna

Plants

Plants must, of course, be established before soil animals can be introduced to degraded soils (Chapter 6) and, where possible, some of these should be nitrogen fixing for soil enrichment, enhanced soil faunal activity and for grazing livestock. The model system outlined here could be helpful in providing a detailed understanding of the area's characteristics which would then enable effective search and selection of plants adapted to the areas of concern.

When British farmers settled in Australia two centuries ago, they found no domesticated crop or pasture plants, so they farmed with European plants on the richer and better-watered soils. In time they began to seek crop plants, and much later, pasture plants, which were better adapted to Australian conditions. This process has continued, especially with pasture plants, but become very much more 'target oriented'. For example, since World War II many useful annual grass, clover and medic species have been collected and introduced from the Mediterranean and other regions of similar climate and soil. In the case of the medics, it was found that their rhizobial symbionts are less tolerant of soil acidity than their hosts. This motivated agricultural scientists to travel to Mediterranean areas where islands of granitic soil are known to be acidic. There they discovered acid-tolerant medic bacteria which will enable some 2 million ha of WA land to grow productive leguminous pastures (Howieson & Ewing, 1986).

Trees also need to be selected to provide breaks against wind erosion, to lower the groundwater levels, as habitats for birds and insects, and to recycle cations. The slow acidification of certain Australian soils after clearing (Porter, 1984), may well be due to the breakdown of the recycling system of trees. Certain deep-rooted native trees can recycle cations such as Ca, K or Mg and can decrease the acidity of the soils on which they grow (Dimbleby, 1952; Nye & Greenland, 1960). This would explain how relatively stable soil pH values are maintained in natural forests. Different species of European plants vary in their ability to recycle cations from depth (Dimbleby, 1952), a capability which should be explored in the search for suitable deep-rooted plants.

Soil fauna

Soil fauna, both indigenous and introduced, should be studied to determine the most appropriate species for a given climate and soil type.

Many earthworms were introduced to WA accidentally, probably in potted flower plants from Britain or in balled-earth fruit trees from South Africa (Milewski, 1986; Lee, 1985). At least one lumbricid species, *Aporrectodea caliginosa*, as well as *Microscolex dubius*, are widely scattered on farms in the dry wheatbelt where cropping is not intense (Abbott & Parker, 1980). Indigenous earthworms are widely present in virgin lands in the 300 mm rainfall fringe of the WA wheatbelt. There may be 'ecotypes' among these which are adapted to different soils, and also able to adapt to soils in agricultural use. Such earthworms may be of value in rehabilitating compacted and poorly structured soils. For example, the indigenous earthworm *Megascolex imparicystis*, which burrows to 5 m deep in sandy soils (Milewski & Parker, unpublished data), has adapted to agricultural land (Abbott, Ross & Parker, 1985).

In addition to earthworms, there are many species of seed-eating ants in the Western Australian wheatbelt, some being opportunist (Twigg *et al.*, 1982) and able to colonize agricultural soils where there is a supply of grass or leguminous seeds. Like earthworms, they would be unlikely to tolerate intensive cereal farming where the soil is regularly tilled, but might well become established with zero-tillage agriculture, even where there are short rotations such as 1 year cereal and 1 year pasture. Subterranean termites are widely present in wheatbelt soils, but their effects on the fertility of these soils is unknown (Wells, 1984).

Cultivation practices may negate colonization by soil fauna, so zero tillage should be considered with this in mind. Where soils are physically degraded to the point where cultivation is required to produce strong seedlings (Jarvis, Hamblin & Delroy, 1986), a narrow seedbed could be made leaving about 80% of the surface soil untilled. This would enable soil fauna to survive, grow and recolonize the cultivated strips.

Summary and conclusions

The interactions of soil fauna and plants have been seriously neglected in the study of degraded broadacre farmlands.

The problems of soil degradation will be more easily solved where the soils are inherently fertile and the climate is benign. They will be more difficult to solve where the soils are infertile and the climate is harsh.

In all cases, the water balance is the first casualty when the land is cleared of deep-rooted perennial plants, tilled and planted with shallow-rooted annuals. The surface runoff water is lost to production and causes problems downslope such as waterlogging, soil erosion, flooding, leaching of nutrients, and salinity in cases where salt is stored in the deep subsoil.

Soil health is the second casualty. Deprived of the protection of a forest canopy and surface litter, and of the restorative activities of the soil fauna, its former structure is destroyed by tillage. Further insult is added by the use of heavy, wheeled machines on wet soils, which cause deep compaction.

Disruption of the deep nutrient cycling system of the forest is the third and least recognized of the harmful effects of clearing land for agriculture. Its consequences can be seen in the eutrophication of streams and lakes, and probably acidification of surface soils.

The original ecosystem maintained a complex and functional balance. Remnants of virgin ecosystems which have not been cleared for agriculture should be sought for study in order to learn from them about the interactions of the different members of the plant and animal community, and also their combined effects upon the maintenance of good soil conditions and water balance. For this reason, if no other, we should jealously protect areas of native woodland, especially where most of the land has been cleared. Degraded woodland, invaded by weeds, perhaps grazed by livestock, should also be preserved and studied. It is here and along road verges that opportunist species of soil fauna may be established – fauna which could be useful in agricultural soils. In many parts of the World there remain tracts of pristine virgin forest or shrubland which maintain stable ecosystems suitable for study. In the Old World, secondary forest ecosystems will have to suffice. The ultimate success of the 'biological plow' (soil fauna and plant roots) depends on the abandonment of the mechanical plow and use of the lightest possible machinery. Soil zoologists working with degraded agricultural soils should think in terms of landscape ecology, and seek the collaboration of botanists, foresters, soil scientists, agronomists, hydrologists and ecologists. To cure the ills caused by agricultural land-use calls for complex thinking and experimentation, and will require the intellectual resources that only collaborative research can bestow.

Acknowledgements

My thanks are due to R. A. Nulsen and R. J. Jarvis for their valuable field studies and for discussions, and to T. A. McCredie for his help.

References

Alexander, M. (1977). *Introduction to Soil Microbiology*. New York: Wiley.

Abbott, I., Parker, C. A. & Sills, I. D. (1979). Changes in the abundance of large soil animals and the physical properties of soils following cultivation. *Australian Journal of Soil Research*, **17**, 343–53.

Abbott, I. & Parker, C. A. (1980). The occurrence of earthworms in the Wheatbelt of Western Australia in relation to land use and rainfall. *Australian Journal of Soil Research*, **18**, 343–53

Abbott, I., Ross, J. S. & Parker, C. A. (1985). Ecology of the large indigenous earthworm *Megascolex imparicystis* in relation to agriculture near Lancelin, Western Australia. *Journal of the Royal Society of Western Australia*, **68**, 13–15.

Campbell-Clause, J. M. (1986). *The Contribution of Earthworms* (Megascolex imparicystis *and* Aporrectodea caliginosa) *in Mitigating the Adverse Effects on Hydraulic Properties and Plant Root Growth caused by Compaction in a Sandy Soil and a Sandy Clay Loam*. Research Report to the School of Agriculture, The University of Western Australia.

Chatel, D. L., Greenwood, R. M. & Parker, C. A. (1968). Saprophytic competence as an important character in the selection of *Rhizobium* for inoculation. In *Proceedings of the Ninth International Congress of Soil Science*, Adelaide, Vol. 11, pp. 65–73.

Churchward, H. M. (1986). The Mediterranean Southwest. In *Australian Soils: The Human Impact*, ed. J. S. Russell & R. F. Isbell, pp. 65–89. Brisbane: University of Queensland.

Culshaw, D. (1987). *Tractive Performance of a Small Rubber Tracked Vehicle*. AFRC Institute of Engineering Research, Silsoe, UK, Div. Note DN1389.

Dimbleby, G. W. (1952). Soil regeneration in the North East Yorkshire Moors. *Journal of Ecology*, **40**, 331–41.

Dolling, P. J. & Ritchie, G. S. P. (1985). Estimates of soil solution, ionic strength and the determination of pH in West Australian soils. *Australian Journal of Soil Research*, **23**, 309–14.

Hamblin, A. D. (1984). The effect of tillage on soil surface properties and the water balance of a xeralfic alfisol. *Soil and Tillage Research*, **4**, 543–59.

Howieson, J. G. & Ewing, M. A. (1986). Acid tolerance in the R. meliloti–Medicago symbiosis. *Australian Journal of Agricultural Research*, **37**, 55–64.

Jarvis, R. J., Hamblin, A. P. & Delroy, N. D. (1986). Cereal Cropping with Alternative Tillage Systems in Western Australia. *Technical Bulletin No.* 71, *Western Australia.*

Lal, R. (1979). The role of physical properties in maintaining the productivity of soils in the tropics. In *Soil Physical Properties and Crop Production in the Tropics*, ed. R. Lal & D. J. Greenland, pp. 7–44. Berlin: John Wiley. N.Y.

Lal. R. (1987). Earthworms. In *Tropical Ecology and Physical Edaphology*. ed. R. Lal, pp. 285–336. New York: John Wiley.

Lange, R. T. (1961). Nodule bacteria associated with the indigenous Leguminosae of South-Western Australia. *Journal of General Microbiology*, **26**, 351–59.

Lee, K. E. (1985). *Earthworms: Their Ecology and Relationships with Soils and Land Use*. Sydney: Academic Press.

Malcolm, C. V. & Pol, J. E. (1984). Grazing and management of saltland shrubs. *Journal of Agriculture of Western Australia*, **27** (4th series), 59–63

Milewski, A. (1986). Earthworms in the Low Rainfall Farming Areas of Southern South Africa. *Unpublished report*.

Mulcahy, M. J. (1983). Learning to live with salinity. *Journal of the Australian Institute of Agricultural Science*, **49**, 11–16

Naveh, Z. & Liebermann, A. S. (1984). *Landscape Ecology: Theory and Applications*. New York: Springer-Verlag.

Nulsen, R. A. & Baxter, I. N. (1982). The potential of agronomic manipulation for controlling salinity in Western Australia. *Journal of the Australian Institute of Agricultural Science*. **48**, 222–6.

Nye, P. H. & Greenland, D. J. (1960). *The Soil under Shifting Cultivation*. Farnham: Commonwealth Agricultural Bureau.

Old, K. M. (1986). Mycophagous soil amoebae: their biology and significance in the ecology of soil-borne plant pathogens. *Progress in Protistology*, **1**, 301–32.

Parker, C. A. & Scutt, P. B. (1958). Competitive inhibition of nitrogen fixation by oxygen. *Biochemica Biophysica Acta*, **29**, 662.

Porter, W. M. (1984). Causes of soil acidity and ten other articles by different authors about causes and effects of acidity. *Journal of Agriculture of Western Australia*, **25**, No. 4 (4th series), 119–47.

Satchell, J. E. (1967). Lumbricidae. In *Soil Biology*, ed. A. Burgess & F. Raw, pp. 257–322. London: Academic Press.

Skinner, F. A. (1986). The potential for microbial manipulation of soil structure. *Journal of Applied Bacteria*, Symposium Supplement, pp. 595–695.

Snook, L. C. (1986). *Tagasaste, Tree Lucerne: High Production Fodder Crop*. Shepparton: Victoria Publishers.

Soane, B. D., Dickson, J. W. & Campbell, D. J. (1982). Compaction by agricultural vehicles: a review. III. Incidence and control of compaction in crop production. *Soil and Tillage Research*, **2**, 3–36.

Tillet, N. D., Holt, J. B., Chestney, A. A. W. & Reed, J. N. (1986). The Design Reconstruction of an Experimental Field Gantry for Leaf Vegetable Production. *AFRC Institute of Engineering Research*, Silsoe, UK. Div. Note DN1373.

Twigg, L., Majer, J. D. & Stynes, B. A. (1982). The influence of seed harvesting ants in annual ryegrass and their possible effects on the epidemicology of ryegrass toxidity. In *Proceedings of the Third Australasian Conference on Grassland Invertebrate Ecology*, ed. K. E. Lee, pp. 235–43. Adelaide: South Australian Government Printer.

Wells, J. (1984). *Subterranean Termites in Wheatbelt Soils*. School of Agriculture, University of Western Australia. Unpublished report.

Williams, D. B., Suijendorp, H. & Wilcox, D. G. (1980). The Gascoyne Basin. In *Desertification: Associated Case Studies*, ed. M. R. Biswas & A. K. Biswas, pp. 3–106. Oxford: Pergamon Press.

Part 5

Synthesis of book

18

Colonization of reclaimed land by animals: an ecologist's overview

HARRY F. RECHER

Introduction

Animals are key components of all terrestrial and aquatic ecosystems and their presence is necessary for the successful rehabilitation of disturbed or derelict land. By facilitating the recycling of nutrients and by moderating harsh physical conditions, animals encourage the establishment of complex plant communities. In planning the rehabilitation process it is therefore important to consider both the kinds of animals required for establishing a functional ecosystem – one that can perpetuate itself – and the kinds of plant and animal communities desired as the final product of the rehabilitation process.

The colonization and growth of plant and animal populations are interactive processes. Each new component to the system interacts with established populations creating conditions favouring the colonization of additional species and leading to the extinction of others. As documented throughout the preceding Chapters, there is a precise pattern to these events. They occur in ways explained by well known ecological principles and can therefore be used either to facilitate the establishment of desired communities or to hold plant and animal communities at a particular stage of development.

Role of fauna

The activities of animals are important in the establishment and maintenance of complex plant communities. Where the objective of rehabilitation is the return of the original community or a close approximation of that community, it is necessary to restore the associated fauna. Even where the objective is the establishment of highly modified ecosystems, such as grazing lands, animals are important in creating and maintaining high levels of productivity.

Animals are important regulators of nutrient cycles. They do this in several ways. Firstly, insect grazers stimulate plant growth and the redirection of nutrients. Their frass is an important input of nutrients and energy to soil and litter communities (Springett, 1980). Secondly, soil organisms fragment and disperse plant litter. This facilitates the process of decomposition by micro-organisms and assists in the recycling of nutrients and energy. Thirdly, the burrowing activities of animals improve soil structure. Litter which is incorporated in the soil acts as a nutrient sink, binding and slowly releasing nutrients and energy into the ecosystem. Incorporation of litter within the soil profile (e.g. in termite mounds) may be particularly important in arid or exposed environments, such as are associated with post-mining activities, where there is a high risk of loss from erosion and fire. Conversely, nutrients which lie below the roots of plants may be brought to the surface. In addition animal burrows and nests facilitate the percolation of water into the soil.

No one group of animals dominates these activities. In temperate areas earthworms are generally considered the most important of the soil fauna. In drier and/or warmer environments ants and termites play significant roles in increasing soil friability and incorporating organic matter within the soil profile, while earthworms are absent or insignificant. Vertebrates may also move large amounts of soil when constructing burrows or dens and by turning over litter searching for food.

There are other ways that animals facilitate the rehabilitation process. Animals, for example, are important in the dispersal of seeds and spores. Dispersal of spores may be particularly important in the establishment of mycorrhizal associations necessary for the establishment of many plant species. Many animals act as pollinators and, by grazing, may increase the number of plant species able to co-exist on a site. Grazing can increase plant species diversity by creating space for new plants to establish and by preventing any one plant species from dominating the available resources.

Conversely, too-heavy grazing during the early stages of rehabilitation can retard plant growth or decrease plant species diversity by selective feeding. It is easiest to counter such effects by ensuring the rapid establishment of a diverse array of animal species. Predation and competitive interactions between species can inhibit the dominance of any one species or group of species, thereby minimizing the risk of pest outbreaks. In this sense diversity contributes to community stability by reducing the amplitude of changes in plant growth rates, vegetative cover and species abundances.

Pattern of fauna establishment

Although the species present and rate of recovery vary with location, extent and kind of disturbance, type of vegetation, physical conditions and proximity of potential colonizers, the pattern of fauna establishment is the same on all rehabilitation sites. As discussed by Majer in Chapter 1, Bradshaw (1984) has illustrated this pattern as one of increasing number of species and trophic complexity with time. For fauna, species richness and trophic complexity are linked to increases in the complexity of the vegetation and moderating environmental conditions. The occurrence of any particular species and its abundance may also depend on the presence or absence of other species of animals.

Colonization

Depending on the type of disturbance all surface and soil organisms and associated vegetation may be destroyed. In addition soil structure is modified and organic matter depleted. The resulting physical environment is often characterized by extremes of temperature and moisture. Soil nutrients may be removed or be initially unavailable and the pH of the soil may be changed. Toxic substances in the form of heavy metals or chemicals may be present. The severity of these conditions restricts the numbers of animals able to re-colonize. Rehabilitation attempts to correct these deficiencies by adding nutrients in the form of fertilizer, sowing or planting vegetation, returning organic matter in the form of soil and/or brushed vegetation, introducing fauna and by regulating further disturbance (e.g. grazing).

Successful colonization depends on organisms reaching the site and then being able to tolerate or withstand the harsh conditions encountered. Colonization is not a random process. It depends on the characteristics of the species available for colonization, the type and extent of disturbance, and distance to a source of potential colonizers.

For all groups of organisms, the first colonizers are species which are inherently mobile or easily dispersed, tolerant of a wide range of conditions and have a high intrinsic rate of increase (r-selected species).

As conditions moderate, less tolerant species can colonize. These are often less easily dispersed and are less fecund (K-selected species). Table 6.1 describes the attributes of r- and K-selected species.

Colonization of rehabilitated lands may be more rapid in regions normally subject to harsh conditions (e.g. arid habitats) than where the climate is more equitable. Majer has suggested that this may be due to the prevalence of r-selected species in the former and to K-strategists in the latter. This hypothesis merits testing as it has important implications for the management of rehabilitation sites. Where K-strategists predominate, for example in forest environments, there may be greater justification in moderating site conditions by the addition of organic matter or cover and in the introduction of suitable pioneer species, than where potential early colonizing species are abundant. This may also be an argument for introducing species able to tolerate harsh conditions where there is reason to believe such animals may be scarce in the original environment retained near or around the rehabilitation site.

The rate and pattern of colonization of rehabilitation by fauna may also be affected by the size of the area disturbed and distance to potential sources of colonists. To some extent these variables can be modelled on the theory of island biogeography, which predicts that small islands or those far from the mainland will have fewer species than large islands or those near the mainland. The model is based on equilibrium values between rates of colonization, which are high for large and near islands, and rates of extinction, which are high for small and far islands. Extinction rates are also considered to be dependent on population size. Areas of rehabilitation can be considered as 'islands', with small or isolated patches having the same attributes as small or remote islands. There is some evidence from forestry studies (e.g. Recher *et al.*, 1987; Kavanagh *et al.*, 1985) and mine rehabilitation that species richness on disturbed sites is affected by patch size.

Very small patches of rehabilitation may be unable to sustain animals which require large territories or home ranges. These animals may also be affected by the shape of the patch. Fauna restricted to linear patches or strips of vegetation with a high edge to area relationship may be exposed to increased predation and competition from species inhabiting adjoining areas (e.g. Recher *et al.*, 1987). Soil fauna and animals with limited dispersal powers may not be able to colonize the central portions of very large areas of rehabilitation.

To the extent that isolation affects colonization of rehabilitation, it can be corrected by the introduction of appropriate propagules. Patch size and associated edge problems are less easily corrected with mining disturbance, but can be adjusted in forestry areas where both logged and unlogged areas can be increased or decreased in size and the shape varied as required (e.g. Harris, 1985). This may also be possible for sites used for spoil or waste disposal or where wetlands are drained. Dispersal within aquatic habitats retained within, or created as part of, the rehabilitation process is probably not a problem in the establishment of a diverse ecosystem.

Succession

Colonization of rehabilitation is a continuing process. As increasingly complex communities are established, and the structural diversity of the vegetation increases, opportunities are created for different species to invade the site. Among vertebrates, for example, colonization may not occur until there are suitable prey organisms or places in which they may den and nest.

Plant species diversity has been taken as a measure of 'ecological stability' and indicates the capacity of the land to support fauna (U.S. Congress, Office of Technology Assessment, 1986). The relationship between stability and diversity in ecological communities is far from established and there are many examples of diverse plant communities which support few animals (e.g. coastal heaths in Australia). In the context of rehabilitation of disturbed lands, however, an increased number of plant species is probably associated with more moderate conditions and greater stability. There is also a positive correlation between plant species richness and structural diversity of the vegetation. Plant species richness can therefore be used to monitor the success of rehabilitation as habitat for fauna and the probability of colonization by desired species.

As colonization proceeds, species which have already colonized the site may decline in abundance or become extinct. The apparent replacement of one species or group of species by another on disturbed sites through time is known as 'succession'. Successional events are common on disturbed sites and can be observed as readily in a post-fire community as in post-mining regeneration.

The processes by which succession takes place are less clear. In part, changes in species abundances and community composition reflect the varying lengths of time it takes different species to colonize or to increase in numbers. Changing environmental conditions, such as more moderate temperatures or improved moisture relations, may favour one species over another. Alternatively, new colonizers may displace other species through competition for resources, by predation or by modifying site conditions. A combination of these processes is probably the best explanation of successional events.

Southwood's (1977) 'habitat templet' model, which is described in Chapter 6, uses habitat co-ordinates ('favourableness' and 'predictability') to predict the pattern of succession. The addition of a third axis by Greenslade & Greenslade (1983), representing diversity and biological complexity, increases the value of the model as a tool to conceptualize the successional changes associated with rehabilitation processes (see Fig. 6.1). For example, habitats which are predictable and/or favourable will tend to have high biological diversity. Less-favourable or unpredictable habitats will support fewer species. It is also possible to describe the attributes of species associated with differing conditions. As mentioned above, harsh environments will be dominated by a few species able to tolerate such conditions. These will tend to be r-selected species and most of their energy and resources will be directed at coping with the physical environment. K-strategists, which will dominate in more favourable and predictable habitats, will expend a greater part of their resources and energy on interactions (e.g. competition, predation) with other species.

Concluding principles

Regardless of the ultimate objective, the goal of rehabilitation is the creation of a stable and often self-perpetuating ecosystem. Achieving this requires the establishment of an appropriate plant community and the colonization of this community by animals. Both require the development of a functional soil ecosystem. The procedures followed in attaining these objectives are relatively simple, but depend upon the ultimate goal of rehabilitation. Simple agricultural systems are easier to develop than re-creating an original, complex ecosystem. In human terms, the latter is probably not possible, but a reasonable approximation can be made given sufficient time.

As has tended to occur in the past, colonization by animals can be left largely to chance. Animals are relatively mobile and disturbed sites may be quickly colonized, regardless of human intervention. If vegetation or organic matter is present, colonizing animals rapidly establish populations and create conditions favouring colonization by other organisms. Although this process may be satisfactory for most purposes, it can be facilitated by thoughtful management. A benefit of facilitating early colonization by a diverse range of animals is the greater stability which results, and the increased rate of development of a complex vegetation requiring less maintenance. The following actions are critical:

1. *SURVEY* of the plants and animals inhabiting the area to be disturbed is necessary if the original fauna is to be restored. A knowledge of the species available for recolonizing retained habitats is also necessary in deciding what animals, if any, should be introduced to the rehabilitation and the kinds of resources they may require.
2. *PROVISION* of organic matter by return of soil and litter and/or brushing is important. These provide an early source of nutrients and energy for soil fauna and moderate environmental extremes. Brush, for example, shades the soil surface.
3. *INOCULATION* of fauna will assist rapid colonization by a diverse array of organisms. Many animals will be introduced with soil or brush, but consideration should be given to inoculating sites with active soil communities from undisturbed environments similar to that removed by the original disturbance.
4. *DIVERSIFICATION* of the habitat provided by the provision of shelter and den or nest sites will increase the rate at which many animals can return to a site. This is particularly true for vertebrates which must otherwise rely on the development of a mature plant community. Retention of logs and the return of these to sites being rehabilitated is a simple procedure capable of rapidly diversifying the habitat for fauna.
5. *RETENTION* of patches or strips of undisturbed vegetation within areas affected by mining or other activities, which disturb the soil surface and destroy the original plant cover, provides a refuge for fauna. These are then available to repopulate the rehabilitated site as vegetation is restored. This is a basic principle for managing wildlife in forestry areas (e.g. Recher, Rohan-Jones & Smith, 1980; Harris, 1985) and could be applied with good effect on other disturbed lands.

The successful rehabilitation of disturbed or degraded lands relies as much on the establishment of animal communities as it does on the growth of plants. It is necessary first to ensure the colonization of the site by soil animals and the development of a functional soil ecosystem. This is necessary for nutrient recycling, moderation of site conditions favouring the establishment of a diverse plant community, and providing an important food resource for vertebrates. Secondly it is important to create a structurally diverse environment which provides appropriate foraging areas as well as den and nest sites for a wide range of animals. Structural diversity contributes to the moderation of site conditions, further enhancing the development of complex plant and animal communities. Regardless of the ultimate goals of rehabilitation, the establishment of balanced plant and animal communities is necessary to optimize the options available for future uses of the land.

References

Bradshaw, A. D. (1980). Ecological principles and land reclamation practice. *Landscape Planning*, **11**, 35–48.

Greenslade, P. J. M. & Greenslade, P. (1983). Ecology of soil invertebrates. In *Soils, an Australian Viewpoint*, ed. K. Lee, pp. 645–69. Melbourne: CSIRO.

Harris, L. D. (1985). *The Fragmented Forest. Island Biogeography Theory and the Preservation of Biotic Diversity*. Chicago: University of Chicago Press.

Kavanagh, R. P., Shields, J. M., Recher, H. F. & Rohan-Jones, W. (1985). Bird populations of a logged and unlogged forest mosaic at Eden, New South Wales. In *Birds of Eucalypt Forests and Woodlands: Ecology, Conservation and Management*, ed. A. Keast, H. F. Recher, H. Ford & D. Saunders, pp. 273–81. Sydney: Surrey Beatty.

Recher, H. F., Rohan-Jones, W. & Smith, P. (1980). *Effects of the Eden Woodchip Industry on Terrestrial Vertebrates with Recommendations for Management*. Forestry Commission of NSW Research Note No. 42.

Recher, H. F., Shields, J., Kavanagh, R. & Webb, G. (1987). Retaining remnant mature forest for nature conservation at Eden, New South Wales: a review of theory and practice. In *Nature Conservation. The Role of Remnants of Native Vegetation*, ed. D. A. Saunders, G. W. Arnold, A. A. Burbidge & A. J. M. Hopkins, pp. 177–94. Sydney: Surrey Beatty.

Southwood, T. R. E. (1977). Habitat, the templet for ecological strategies? *Journal of Animal Ecology*, **46**, 337–65.

Springett, B. P. (1980). On the ecological role of insects in Australian eucalypt forests. *Australian Journal of Ecology*, **3**, 129–40.

U.S. Congress, Office of Technology Assessment. (1986). *Western Surface Mine Permitting and Reclamation*. OTA-E-279. Washington D.C.: U.S. Government Printing Office.

Part 6

Bibliography

19

Bibliography of fauna studies in reclaimed lands

J. BUTCHER, J. D. MAJER AND P. UNSWORTH

Introduction and use of bibliography

This bibliography has been compiled to provide a general data base for persons interested in fauna aspects of land reclamation. It spans the period when publications first emerged on the topic through to July 1987.

Included here are primary and secondary journal papers, reviews, conference papers, government handbooks and popular articles. In view of the fact that much of the relevant information is included in unpublished theses or mimeographed documents with a limited circulation, we have decided to include these as well.

The coverage of this bibliography is largely confined to areas where land disturbance has resulted in a primary succession (e.g. mining). However, where publications are thought to have special relevance to this subject (e.g. old-field succession) they have been included.

Each publication is indexed by the continent which the work relates to, the type of land disturbance which is involved (e.g. mine), the type of investigation reported on (e.g. soil structure) and finally the taxonomic group which was studied.

While every effort has been taken to ensure comprehensive and accurate compilation of this bibliography, notification of errors and omissions to the authors will be gratefully received.

Key to references by subject

Europe

12, 28, 29, 34, 43, 44, 52, 54, 55, 59, 74, 75, 76, 79, 82, 83, 86, 87, 88, 89, 91, 112, 113, 114, 124, 127, 128, 148, 173, 175, 196, 198, 200, 205, 223, 227, 228, 234, 235, 236, 237, 238, 239, 240, 241,

242, 244, 245, 246, 247, 248, 249, 250, 256, 257, 258, 263, 266, 275, 277, 279, 299, 307, 308, 323, 324, 325, 326, 332, 336, 338, 343, 344, 345, 347, 348, 357, 371, 372, 378, 379, 380, 402, 412, 413, 414, 428, 430, 432, 444, 449, 450, 470, 471, 472, 486, 489, 490, 492, 494, 502, 526, 530, 531, 541, 555, 600, 602, 603, 604, 605, 606, 607, 608, 609, 614, 615, 621, 622, 631, 648, 649, 650, 665, 666, 694, 695, 696, 707, 712, 714, 721, 723, 725, 726, 744, 747, 752, 766, 767, 774, 775, 778, 779, 781, 830, 840, 854, 864, 867, 868, 885, 887, 888, 896, 897, 901, 921, 922, 923, 948, 963, 965, 971, 992.

United Kingdom

19, 20, 31, 34, 47, 49, 56, 69, 85, 89, 90, 91, 156, 157, 160, 188, 192, 203, 204, 205, 206, 207, 226, 253, 276, 314, 315, 316, 317, 326, 354, 355, 356, 384, 385, 401, 411, 412, 415, 416, 417, 418, 419, 420, 421, 422, 425, 436, 437, 455, 456, 460, 518, 538, 554, 581, 585, 623, 672, 691, 716, 743, 753, 754, 768, 791, 822, 832, 841, 842, 843, 844, 845, 846, 847, 848, 849, 850, 861, 873, 889, 890, 891, 900, 901, 904, 920, 945, 949.

North America

2, 3, 4, 5, 6, 7, 8, 9, 13, 14, 15, 16, 17, 18, 21, 22, 23, 24, 25, 26, 27, 32, 33, 35, 36, 37, 38, 39, 40, 41, 42, 45, 46, 50, 51, 53, 57, 60, 61, 62, 63, 65, 66, 67, 68, 70, 72, 73, 77, 78, 80, 81, 92, 93, 94, 95, 96, 97, 98, 99, 100, 101, 102, 103, 104, 105, 106, 107, 108, 109, 110, 111, 115, 116, 118, 119, 120, 122, 123, 125, 130, 131, 132, 133, 134, 135, 136, 139, 140, 141, 142, 143, 144, 145, 146, 147, 150, 151, 152, 153, 154, 155, 158, 159, 162, 163, 164, 165, 166, 167, 168, 169, 170, 171, 172, 174, 176, 179, 180, 181, 182, 183, 184, 185, 186, 187, 189, 190, 191, 193, 194, 195, 197, 199, 201, 208, 209, 212, 213, 214, 215, 216, 217, 218, 219, 220, 221, 222, 223, 224, 225, 229, 230, 231, 232, 233, 251, 252, 253, 254, 255, 260, 261, 262, 264, 265, 267, 268, 269, 270, 271, 278, 280, 281, 282, 283, 284, 285, 286, 287, 288, 289, 290, 291, 292, 293, 294, 295, 299, 304, 305, 306, 309, 310, 311, 312, 313, 320, 321, 322, 327, 328, 329, 330, 331, 334, 335, 339, 340, 341, 349, 350, 352, 353, 358, 359, 360, 361, 362, 363, 364, 365, 366, 367, 368, 369, 370, 373, 374, 375, 376, 381, 382, 383, 386, 387, 388, 389, 390, 391, 392, 394, 395, 396, 397, 398, 399, 400, 403, 405, 406, 408, 409, 410, 414, 423, 424, 426, 427, 429, 431, 433, 434, 435, 438, 439, 440, 441, 442, 443, 445, 451, 452, 453, 454, 459, 461, 462, 463, 464, 465, 466, 467, 468, 469, 473, 474, 475, 476, 477, 478, 479, 480, 481, 482, 483, 484, 485, 487, 488, 491, 493, 495, 496, 497, 498, 499, 500, 501, 503, 504, 506, 507, 508, 509, 510, 511, 512, 513, 514, 515, 516, 517, 519, 520, 522, 523, 524, 525, 527, 528, 529, 532, 533, 534, 536, 540, 542, 543, 544, 545, 546, 547, 550, 551, 553, 556, 582, 583, 584, 586, 587, 588, 589, 590, 591, 592, 593, 594, 595, 596, 597, 598, 599, 601, 610, 611, 612, 613, 616, 617, 618, 619, 620, 624, 625, 626, 627, 628, 629, 630, 632, 633, 634, 635, 636, 637, 638, 639, 640, 641, 642, 643, 644, 645, 646, 647, 663, 664, 667, 668, 669, 670, 671, 674, 675, 676, 677, 678, 679, 680, 681, 682, 683, 684, 685, 686, 687, 688, 689, 692, 693, 697, 698, 699, 700, 701, 702, 703, 704, 705, 706, 708, 709, 710, 711, 713, 715, 720, 722, 724, 727, 728, 729, 731, 732, 733, 734, 735, 736, 737, 738, 739, 740, 741, 742, 745, 748, 749, 750, 751, 755, 757, 758, 759, 760, 761, 762, 763, 764, 765, 767, 769, 770, 772, 773, 776, 777, 780, 782, 783, 784, 785, 786, 787, 788, 789, 793, 794, 795, 797, 799, 800, 802, 803, 804, 805, 806, 807, 808, 809, 810, 812, 813, 814, 815, 817, 818, 819, 820, 821, 823, 824, 825, 826, 827, 828, 829, 831, 833, 834, 835, 836, 837, 838, 839, 851, 852, 853, 855, 856, 857, 858, 859, 860, 861, 862, 863, 865, 866, 872, 874, 875, 876, 877, 878, 879, 880, 881, 882, 883, 884, 886, 892, 893, 894, 895, 898, 899, 902, 903, 905, 906, 907, 908, 909, 910, 911, 912, 913, 914, 915, 916, 917, 918, 919, 924, 925, 926, 927, 928, 929, 930, 931, 932, 933, 934, 935, 936, 937, 938, 939, 943, 944, 946, 947, 950, 952, 953, 954, 955, 956, 957, 958, 959, 960, 961, 962, 964, 966, 967, 968, 969, 973, 974, 975, 976, 978, 979, 980, 981, 982, 983, 984, 985, 986, 987, 988, 989, 990, 991, 993.

South America

Africa

Asia

Australasia

Islands

General

Mine or quarry

737, 739, 741, 742, 743, 745, 749, 750, 751, 753, 754, 758, 759, 760, 761, 762, 763, 764, 767, 771, 773, 774, 776, 777, 782, 783, 784, 785, 786, 787, 789, 790, 791, 792, 793, 794, 795, 798, 802, 804, 805, 808, 810, 812, 814, 815, 817, 819, 820, 821, 822, 823, 824, 825, 826, 827, 828, 829, 830, 834, 835, 836, 837, 839, 841, 843, 844, 845, 846, 847, 848, 849, 850, 851, 852, 853, 855, 857, 858, 860, 861, 862, 863, 865, 866, 869, 871, 872, 874, 875, 876, 877, 878, 879, 880, 881, 882, 889, 890, 891, 892, 893, 895, 898, 899, 900, 901, 903, 905, 907, 908, 909, 910, 911, 912, 915, 916, 917, 918, 920, 924, 925, 926, 927, 929, 930, 932, 933, 934, 936, 937, 939, 940, 941, 942, 943, 944, 950, 954, 955, 956, 957, 958, 959, 960, 961, 962, 964, 965, 966, 967, 968, 969, 970, 973, 974, 975, 976, 977, 978, 979, 980, 981, 982, 983, 984, 985, 986, 987, 988, 991, 992, 993.

Polluted area

19, 20, 39, 40, 61, 78, 80, 89, 91, 95, 119, 170, 193, 198, 211, 229, 230, 248, 326, 328, 329, 414, 425, 428, 429, 468, 501, 524, 543, 552, 584, 600, 662, 671, 689, 722, 729, 731, 770, 801, 811, 818, 859, 911, 947, 965.

Drained land

12, 28, 29, 89, 148, 149, 157, 159, 205, 256, 257, 266, 277, 309, 343, 344, 357, 378, 379, 380, 402, 409, 444, 494, 503, 522, 530, 531, 537, 539, 591, 602, 603, 604, 605, 621, 631, 666, 673, 694, 695, 714, 722, 723, 725, 726, 747, 767, 809, 832, 864, 885, 921, 948, 972.

Road-verge

2, 89, 205, 345, 433, 612, 613, 856, 965.

Urban or industrial site

34, 113, 173, 198, 205, 223, 259, 342, 347, 371, 416, 417, 418, 419, 420, 490, 535, 778, 803, 840, 888, 901.

Agricultural area

43, 58, 121, 205, 209, 232, 258, 299, 307, 308, 335, 358, 377, 489, 541, 544, 549, 586, 587, 588, 618, 620, 632, 636, 681, 696, 707, 732, 741, 767, 807, 833, 854, 857, 868, 897, 935.

Off-road vehicles

60, 136, 140, 225, 311, 312, 313, 434, 894.

Spoil or slag heap

19, 21, 22, 33, 35, 52, 59, 63, 74, 75, 76, 85, 87, 92, 93, 102, 108, 114, 124, 147, 150, 151, 152, 169, 179, 189, 192, 193, 201, 205, 208, 215, 219, 223, 227, 228, 233, 234, 235, 236, 237, 238, 239, 240, 241, 242, 243, 244, 245, 246, 247, 255, 260, 267, 269, 282, 285, 301, 317, 323, 324, 335, 338, 349, 361, 367, 400, 408, 409, 413, 414, 415, 416, 421, 422, 436, 448, 449, 450, 451, 467, 469, 470, 471, 472, 481, 491, 496, 497, 502, 506, 524, 525, 527, 536, 538, 543, 585, 609, 622, 623, 637, 640, 641, 645, 648, 649, 650, 674, 678, 683, 684, 692, 697, 708, 721, 729, 734, 737, 740, 748, 756, 766, 768, 772, 774, 779, 784, 788, 806, 808, 812, 816, 822, 827, 828, 838, 871, 879, 887, 896, 898, 902, 904, 928, 931, 938, 945, 946, 955, 971, 988, 989, 990, 992.

Artificially produced or disturbed land

31, 34, 49, 79, 119, 200, 279, 318, 323, 334, 341, 357, 369, 407, 409, 430, 455, 492, 500, 606, 607, 608, 712, 738, 765, 781, 799, 800, 862, 873, 883.

Naturally cleared land

3, 48, 318, 319, 405, 431, 435, 485, 526, 690, 691, 797, 813, 972.

Cleared for experimental purposes

175, 300, 348, 417, 418, 548, 724, 849.

Soil sterilization

56.

Surface drilling

137, 138, 856.

General

4, 9, 14, 15, 27, 29, 35, 42, 44, 50, 54, 57, 62, 63, 68, 73, 81, 86, 87, 88, 89, 91, 93, 95, 100, 111, 117, 119, 138, 142, 143, 145, 146, 160, 161, 166, 167, 168, 170, 172, 177, 181, 190, 199, 201, 210, 220, 229, 230, 251, 262, 268, 300, 301, 302, 311, 312, 313, 326, 336, 341, 342, 350, 351, 363, 364, 368, 374, 376, 383, 388, 389, 390, 391, 392, 395, 414, 423, 424, 425, 427, 429, 451, 453, 462, 477, 478, 479, 481, 497, 505, 514, 515, 516, 519, 535, 542, 558, 561, 565, 567, 569, 572, 575, 577, 578, 579, 596, 625, 627, 628, 629, 630, 634, 642, 643, 644, 651, 652, 654, 659, 663, 671, 688, 689, 693, 695, 704, 706, 709, 717, 734, 739, 757, 759, 763, 765, 790, 793, 812, 815, 817, 827, 831, 835, 861, 862, 863, 864, 870, 871, 875, 876, 881, 893, 898, 903, 906, 912, 914, 915, 916, 918, 940, 944, 962, 964, 973, 974, 989, 990, 991.

Soil structure

1, 37, 43, 53, 85, 99, 101, 115, 124, 145, 147, 151, 169, 174, 188, 196, 203, 207, 213, 215, 216, 218, 233, 248, 253, 260, 266, 269, 280, 309, 317, 318, 335, 348, 359, 367, 396, 397, 402, 412, 415, 416, 443, 476, 492, 501, 525, 527, 528, 536, 556, 559, 560, 639, 640, 641, 645, 648, 666, 669, 687, 692, 698, 725, 726, 737, 739, 741, 744, 747, 748, 754, 767, 768, 780, 782, 783, 784, 787, 788, 789, 801, 816, 822, 824, 829, 833, 858, 861, 870, 872, 902, 904, 910, 929, 930, 950, 954, 967, 988.

Decomposition and nutrient turnover

84, 85, 89, 174, 215, 224, 239, 240, 241, 243, 266, 416, 420, 501, 506, 528, 581, 656, 674, 721, 744, 768, 777, 782, 789, 844, 870, 931, 954, 955.

Impact on plant diversity

188, 206, 411, 544, 571, 636, 690, 718, 826, 966, 985.

Impact on plant growth

35, 70, 125, 171, 188, 206, 215, 218, 370, 386, 402, 498, 590, 620, 690, 697, 708, 728, 740, 750, 751, 767, 826, 930, 954, 966.

Succession

34, 52, 58, 59, 60, 74, 88, 104, 108, 109, 126, 140, 162, 178, 185, 197, 200, 205, 217, 223, 225, 227, 228, 237, 246, 279, 286, 293, 296, 298, 299, 301, 303, 307, 308, 322, 336, 346, 352, 354, 358, 361, 369, 377, 400, 414, 416, 422, 433, 434, 448, 449, 450, 457, 490, 522, 553, 555, 559, 560, 561, 563, 565, 575, 585, 588, 608, 615, 623, 635, 648, 649, 650, 653, 674, 675, 678, 707, 727, 752, 771, 790, 799, 800, 805, 814, 846, 854, 867, 868, 879, 920, 922, 952.

Description of recolonization

2, 11, 28, 34, 35, 52, 56, 88, 104, 106, 108, 110, 113, 119, 126, 127, 128, 153, 154, 158, 171, 175, 181, 189, 193, 203, 205, 211, 223, 226, 227, 228, 259, 265, 281, 284, 288, 294, 297, 303, 305, 314, 323, 327, 333, 336, 337, 346, 349, 358, 369, 381, 384, 387, 393, 394, 400, 414, 435, 436, 439, 441, 442, 446, 448, 449, 450, 458, 460, 466, 486, 488, 513, 549, 553, 555, 557, 562, 566, 568, 570, 574, 575, 576, 580, 582, 583, 585, 586, 597, 608, 613, 616, 617, 623, 632, 635, 655, 657, 658, 665, 672, 676, 678, 683, 684, 691, 699, 703, 718, 719, 753, 754, 756, 772, 777, 791, 799, 800, 808, 820, 823, 825, 828, 830, 837, 838, 857, 861, 862, 865, 867, 873, 874, 878, 890, 894, 895, 901, 907, 926, 938, 939, 943, 954, 977, 979, 980, 984.

Long-term effects of fauna

159, 239, 295, 564.

Pest problems

41, 71, 272, 273, 274, 306, 353, 404, 532, 550, 560, 573, 722, 773, 775, 798, 905, 908.

Grazing

12, 13, 22, 44, 46, 88, 98, 159, 163, 188, 194, 212, 214, 255, 263, 267, 282, 332, 362, 396, 397, 398, 399, 426, 430, 444, 467, 473, 474, 496, 500, 504, 508, 509, 510, 511, 512, 541, 543, 611, 631, 638, 701, 702, 733, 735, 741, 744, 778, 785, 786, 832, 902, 943, 950, 966, 985, 986.

Forestry

10, 20, 34, 102, 116, 123, 129, 141, 145, 197, 208, 223, 280, 292, 295, 298, 316, 414, 446, 486, 491, 500, 551, 562, 564, 586, 588, 633, 650, 735, 739, 746, 761, 792, 804, 867, 869, 927, 934, 941, 942, 950, 970, 987.

Agriculture

22, 34, 62, 209, 216, 219, 232, 255, 280, 282, 314, 316, 362, 372, 489, 501, 575, 581, 639, 753, 754, 778, 782, 783, 897, 950, 966.

Amenity

8, 18, 21, 22, 24, 26, 31, 32, 34, 38, 61, 78, 89, 92, 96, 119, 120, 133, 134, 144, 146, 154, 155, 173, 176, 179, 195, 204, 209, 254, 261, 276, 278, 290, 314, 315, 316, 325, 332, 334, 341, 357, 362, 371, 410, 355, 464, 467, 475, 483, 493, 503, 507, 523, 530, 531, 557, 566, 667, 670, 673, 680, 699, 715, 745, 761, 762, 860, 861, 882, 885, 890, 899, 924, 925, 926, 936, 944, 949, 950, 982.

Conservation

3, 16, 23, 28, 30, 31, 34, 36, 40, 45, 55, 67, 69, 77, 97, 119, 136, 137, 138, 139, 144, 146, 148, 156, 157, 171, 173, 178, 180, 181, 204, 221, 222, 252, 263, 271, 275, 276, 291, 304, 314, 316, 321, 325,

329, 339, 340, 347, 356, 371, 382, 384, 401, 405, 406, 431, 437, 438, 445, 455, 456, 459, 461, 482, 484, 485, 488, 507, 517, 522, 529, 534, 540, 554, 591, 600, 601, 612, 614, 615, 626, 660, 662, 664, 672, 676, 679, 686, 710, 711, 720, 722, 732, 737, 755, 758, 791, 795, 797, 803, 809, 813, 821, 827, 836, 839, 846, 851, 852, 853, 855, 859, 861, 873, 880, 886, 890, 901, 909, 913, 923, 924, 937, 949, 963, 965, 968, 969, 981.

Wetlands

21, 31, 34, 40, 51, 55, 64, 65, 72, 80, 83, 94, 107, 110, 118, 119, 134, 139, 144, 145, 146, 148, 149, 150, 151, 155, 156, 165, 182, 183, 186, 200, 231, 252, 254, 256, 264, 270, 276, 279, 287, 288, 289, 293, 310, 314, 315, 316, 320, 325, 327, 330, 334, 354, 356, 360, 365, 366, 384, 385, 406, 409, 410, 439, 440, 445, 447, 455, 460, 468, 480, 493, 495, 507, 518, 533, 591, 595, 614, 615, 619, 631, 637, 646, 667, 668, 669, 670, 672, 677, 680, 685, 686, 696, 703, 716, 722, 731, 736, 738, 739, 742, 745, 749, 760, 791, 806, 809, 834, 841, 843, 845, 846, 847, 848, 849, 850, 861, 866, 873, 877, 883, 884, 886, 889, 890, 891, 899, 900, 901, 932, 933, 946, 949, 963, 978, 988.

Migration

48, 121, 256, 319, 327, 343, 345, 487, 521, 538, 546, 603, 606, 607, 621, 781, 958.

Trophic structure

17, 19, 33, 48, 79, 82, 105, 122, 285, 369, 373, 386, 408, 463, 465, 513, 518, 598, 617, 700, 743, 825, 842, 844, 847, 850, 854, 856, 891, 935, 946, 983.

Colonization

5, 6, 7, 12, 24, 25, 31, 34, 47, 48, 49, 66, 74, 75, 76, 94, 103, 107, 112, 114, 118, 119, 121, 130, 148, 149, 150, 152, 164, 178, 184, 187, 192, 202, 231, 234, 235, 236, 238, 240, 242, 244, 245, 246, 247, 257, 258, 264, 286, 288, 298, 299, 318, 324, 328, 338, 343, 344, 348, 355, 361, 370, 377, 382, 384, 402, 413, 416, 417, 418, 421, 422, 430, 432, 440, 444, 452, 457, 469, 470, 471, 472, 490, 494, 499, 502, 507, 526, 537, 538, 539, 547, 548, 563, 575, 599, 602, 604, 605, 609, 618, 621, 622, 661, 672, 675, 694, 703, 712, 714, 716, 723, 724, 726, 731, 748, 756, 758, 761, 764, 766, 767, 771, 779, 791, 810, 819, 845, 849, 861, 862, 868, 887, 888, 890, 896, 909, 921, 922, 945, 948, 951, 956, 957, 959, 960, 961, 969, 971, 972, 975, 976, 981.

Fisheries

13, 34, 89, 97, 99, 131, 132, 135, 191, 211, 231, 280, 314, 316, 331, 375, 403, 414, 440, 454, 523, 552, 589, 592, 593, 610, 624, 662, 682, 713, 769, 770, 776, 794, 802, 818, 829, 841, 861, 890, 892, 911, 917, 919, 947, 953, 993.

Platyhelminthes

33, 100, 150, 200, 526, 832, 907.

Aschelminthes

56, 89, 211, 226, 260, 269, 280, 342, 535, 537, 597, 598, 674, 770, 777, 826, 832, 894, 954, 955.

Annelida

1, 19, 34, 37, 43, 47, 56, 85, 86, 89, 112, 113, 119, 121, 124, 125, 150, 151, 152, 170, 177, 192, 196, 200, 205, 211, 223, 226, 234, 237, 239, 242, 243, 247, 248, 253, 258, 293, 318, 319, 342, 343, 348, 349, 400, 402, 412, 414, 415, 416, 419, 420, 422, 425, 439, 451, 463, 465, 489, 490, 501, 535, 537, 539, 567, 569, 577, 581, 592, 597, 603, 633, 691, 692, 725, 726, 730, 731, 744, 747, 753, 754, 766, 767, 768, 777, 779, 792, 796, 816, 822, 832, 843, 844, 847, 850, 854, 867, 868, 890, 891, 894, 904, 928, 929, 930, 931, 932, 947, 954, 965.

Arthropoda

3, 10, 11, 18, 19, 27, 28, 35, 40, 41, 48, 51, 54, 56, 60, 63, 66, 71, 74, 75, 79, 80, 81, 82, 89, 91, 113, 114, 119, 121, 122, 124, 125, 126, 127, 129, 137, 140, 145, 149, 150, 151, 152, 161, 168, 170, 171, 172, 174, 175, 177, 186, 187, 189, 191, 192, 193, 198, 200, 202, 203, 205, 207, 210, 211, 223, 224, 225, 226, 334, 235, 236, 237, 238, 239, 240, 241, 243, 244, 245, 246, 247, 249, 250, 253, 259, 260, 272, 273, 277, 283, 286, 288, 293, 296, 299, 301, 303, 306, 307, 324, 326, 327, 328, 330, 333, 334, 338, 341, 342, 343, 344, 345, 347, 348, 353, 358, 360, 361, 369, 372, 378, 379, 380, 393, 400, 401, 404, 405, 406, 407, 408, 410, 413, 414, 415, 416, 417, 418, 419, 420, 421, 422, 429, 431, 433, 434, 435, 436, 439, 451, 456, 457, 462, 463, 465, 466, 470, 471, 472, 485, 490, 498, 502, 505, 515, 518, 521, 531, 532, 535, 537, 539, 541, 544, 548, 553, 556, 557, 558, 559, 560, 561, 562, 563, 564, 565, 566, 567, 568, 569, 570, 571, 572, 573, 574, 576, 577, 578, 579, 580, 586, 588, 592, 595, 597, 598, 599, 600, 602, 604, 605, 606, 607, 608, 609, 620, 621, 623, 633, 636, 648, 649, 650, 652, 653, 656, 657, 665, 671, 674, 675, 678, 689, 697, 707, 712, 721, 730, 731, 732, 735, 739, 750, 756, 770, 771, 774, 775, 777, 780, 781, 790, 792, 796, 797, 798, 799, 800, 801, 807, 813, 814, 825, 828, 832, 833, 834, 841, 842, 843, 844, 845, 847, 849, 850, 854, 861, 867, 868, 869, 871, 874, 878, 883, 887, 888, 889, 891, 894, 896, 904, 905, 908, 920, 922, 923, 932, 933, 936, 940, 942, 945, 947, 948, 952, 954, 955, 963, 965, 968, 971, 972, 977, 988, 990, 992.

Mollusca

35, 49, 56, 100, 119, 137, 150, 151, 152, 168, 192, 193, 200, 211, 239, 342, 400, 414, 415, 416, 420, 439, 463, 465, 492, 518, 531, 535, 552, 592, 595, 597, 598, 621, 633, 662, 731, 744, 841, 843, 844, 904, 990.

Arachnida

19, 28, 48, 56, 79, 113, 114, 124, 137, 151, 152, 161, 175, 187, 189, 192, 202, 203, 205, 207, 211, 223, 234, 239, 260, 269, 277, 301, 308, 327, 342, 348, 358, 361, 369, 372, 377, 378, 380, 400, 414, 415, 416, 419, 420, 422, 433, 434, 451, 463, 485, 490, 526, 535, 537, 538, 539, 555, 557, 561, 563, 565, 566, 575, 578, 580, 586, 588, 597, 600, 602, 607, 608, 633, 653, 657, 674, 707, 744, 770, 771, 777, 790, 792, 799, 800, 801, 832, 841, 854, 874, 884, 894, 904, 907, 936, 941, 954, 955, 965, 972, 977.

Insecta

3, 10, 11, 18, 19, 27, 28, 34, 35, 41, 48, 49, 54, 56, 63, 66, 71, 74, 75, 76, 79, 80, 82, 89, 113, 119, 121, 122, 123, 124, 125, 126, 129, 137, 149, 150, 151, 152, 161, 168, 170, 172, 175, 177, 187, 189, 191, 192, 193, 198, 200, 202, 203, 205, 207, 210, 211, 223, 226, 234, 237, 239, 241, 243, 247, 250, 259, 260, 263, 269, 272, 273, 274, 277, 283, 286, 288, 293, 296, 299, 301, 303, 306, 307, 324, 327, 328, 330, 331, 333, 334, 341, 342, 343, 344, 345, 346, 347, 348, 353, 361, 369, 372, 377, 379, 380, 384, 393, 400, 401, 404, 405, 413, 414, 415, 416, 417, 418, 419, 420, 421, 422, 428, 431, 433, 434, 435, 436, 439, 451, 456, 457, 462, 463, 465, 466, 471, 472, 485, 490, 498, 515, 518, 521, 526, 532, 535, 537, 538, 539, 541, 544, 548, 553, 555, 556, 557, 558, 559, 560, 561, 562, 563, 564, 565, 566,

568, 569, 570, 571, 572, 573, 574, 576, 577, 578, 579, 580, 586, 588, 592, 595, 597, 598, 599, 600, 602, 605, 606, 608, 620, 621, 623, 633, 635, 648, 649, 650, 652, 656, 657, 658, 665, 674, 675, 678, 689, 697, 707, 712, 714, 731, 732, 735, 750, 752, 756, 761, 770, 771, 774, 775, 777, 780, 781, 790, 792, 796, 797, 799, 800, 801, 813, 825, 828, 829, 832, 833, 834, 841, 842, 843, 844, 845, 846, 847, 849, 850, 854, 861, 867, 868, 869, 874, 878, 883, 884, 889, 890, 891, 894, 901, 904, 905, 907, 908, 922, 923, 932, 933, 936, 940, 941, 942, 945, 947, 948, 954, 955, 963, 964, 965, 968, 971, 972, 977, 988, 990, 992.

Diplopoda

56, 89, 124, 239, 342, 369, 400, 414, 415, 416, 485, 535, 563, 565, 577, 649, 650, 730, 744, 799, 800, 854, 894, 904, 907.

Chilopoda

56, 124, 177, 187, 223, 226, 239, 369, 380, 400, 414, 415, 416, 420, 485, 563, 565, 575, 633, 657, 744, 799, 854, 867, 894, 904, 907, 941, 972.

Pauropoda

342, 372, 416, 419, 420, 535, 563, 622.

Symphyla

137, 342, 416, 419, 420, 531, 535, 894.

Crustacea

49, 56, 89, 150, 151, 152, 168, 193, 211, 301, 331, 334, 342, 369, 414, 415, 416, 419, 429, 518, 535, 565, 575, 597, 631, 649, 650, 657, 662, 689, 731, 744, 799, 818, 834, 841, 843, 844, 850, 854, 871, 884, 891, 904, 907, 947, 988.

Collembola

56, 63, 76, 79, 89, 113, 124, 161, 175, 177, 187, 189, 202, 203, 205, 207, 210, 211, 223, 226, 234, 239, 241, 243, 247, 250, 260, 269, 301, 324, 333, 342, 348, 372, 377, 380, 400, 413, 414, 415, 416, 417, 418, 419, 420, 421, 422, 433, 451, 485, 490, 535, 537, 538, 539, 556, 557, 561, 563, 565, 566, 569, 570, 572, 575, 576, 577, 580, 588, 623, 652, 658, 674, 678, 707, 744, 752, 771, 777, 780, 796, 799, 801, 832, 841, 854, 878, 894, 904, 907, 941, 954, 955, 965, 972, 992.

Protura

348, 372, 490, 563, 771, 894.

Diplura

415, 416, 420, 771, 894, 904, 907, 965.

Thysanura

56, 202, 342, 400, 414, 416, 419, 420, 535, 563, 799, 907.

Ephemeroptera

49, 119, 150, 151, 152, 168, 193, 200, 211, 286, 288, 293, 327, 328, 330, 331, 334, 414, 439, 518, 597, 598, 599, 731, 770, 828, 829, 841, 883, 907, 932, 933, 947, 977, 988.

Odonata

35, 48, 49, 54, 119, 150, 151, 152, 168, 200, 211, 286, 288, 293, 325, 327, 330, 331, 334, 401, 414, 436, 439, 548, 595, 598, 731, 832, 834, 841, 847, 883, 907, 963, 977, 988, 990.

Blattodea

202, 301, 561, 563, 565, 575, 657, 771, 972, 977.

Isoptera

48, 121, 129, 137, 177, 202, 210, 269, 272, 273, 274, 301, 343, 393, 400, 404, 414, 505, 561, 563, 565, 569, 572, 577, 578, 579, 652, 656, 658, 756, 790, 796, 799, 894, 941, 954, 955, 977.

Mantodea

202, 657, 977.

Zoraptera

942.

Dermaptera

48, 202, 234, 301, 415, 416, 561, 565, 575, 657, 771, 904, 972, 977.

Plecoptera

151, 152, 168, 193, 211, 327, 331, 414, 597, 599, 770, 828, 829, 933, 947, 977, 988.

Orthoptera

18, 56, 125, 145, 149, 170, 202, 210, 283, 301, 361, 400, 414, 415, 416, 463, 465, 561, 563, 565, 569, 570, 572, 575, 578, 586, 697, 732, 739, 771, 777, 792, 799, 800, 904, 907, 936, 968, 972, 977.

Phasmatodea

977.

Embidoptera

799, 800.

Psocoptera

56, 260, 361, 372, 400, 414, 415, 416, 537, 539, 563, 588, 674, 771, 799, 800, 904, 907, 954, 955.

Phthiraptera

744.

Hemiptera

28, 48, 49, 56, 79, 119, 124, 151, 152, 161, 168, 187, 192, 198, 200, 202, 205, 288, 293, 301, 306, 334, 361, 372, 380, 400, 414, 415, 416, 419, 420, 433, 439, 463, 465, 466, 485, 518, 521, 537, 539, 548, 561, 563, 565, 575, 586, 588, 595, 602, 636, 707, 731, 771, 774, 777, 781, 790, 799, 800, 829, 833, 841, 844, 854, 878, 883, 894, 904, 907, 936, 947, 965, 977.

Thysanoptera

56, 187, 400, 414, 415, 416, 433, 463, 465, 485, 563, 575, 588, 707, 771, 799, 800, 825, 841, 878, 904, 907, 965, 977.

Megaloptera

151, 152, 211, 293, 414, 598, 828, 829, 834, 844, 850, 947.

Neuroptera

124, 400, 414, 415, 416, 588, 771, 799, 904, 907, 977.

Coleoptera

18, 19, 44, 48, 49, 54, 56, 66, 79, 119, 124, 127, 150, 151, 152, 161, 168, 177, 187, 192, 193, 200, 202, 205, 211, 223, 226, 234, 239, 277, 286, 288, 293, 301, 307, 327, 334, 342, 344, 345, 353, 361, 372, 379, 380, 400, 414, 415, 416, 419, 420, 439, 463, 465, 471, 485, 490, 518, 526, 532, 535, 537, 539, 544, 548, 553, 555, 557, 561, 563, 565, 566, 569, 573, 575, 577, 586, 588, 595, 597, 599, 600, 602, 604, 606, 608, 633, 648, 649, 650, 675, 707, 712, 714, 731, 732, 744, 770, 771, 775, 777, 780, 790, 797, 799, 800, 829, 832, 834, 841, 844, 847, 850, 854, 867, 868, 883, 894, 904, 907, 908, 932, 933, 936, 945, 947, 965, 971, 972, 977.

Mecoptera

771, 907, 977.

Siphonaptera

274.

Diptera

18, 19, 34, 35, 48, 56, 79, 80, 89, 91, 119, 124, 150, 151, 152, 161, 168, 170, 187, 192, 193, 200, 202, 211, 223, 226, 234, 237, 239, 243, 277, 286, 288, 293, 301, 326, 327, 330, 331, 334, 342, 361, 372, 380, 384, 400, 414, 415, 416, 419, 420, 422, 428, 439, 463, 465, 466, 485, 490, 515, 518, 526, 535, 537, 539, 548, 561, 563, 565, 575, 586, 588, 592, 595, 597, 598, 599, 621, 770, 771, 777, 798, 799, 828, 829, 834, 841, 842, 843, 844, 845, 846, 847, 849, 850, 854, 861, 867, 883, 889, 890, 891, 894, 904, 907, 932, 933, 947, 965, 972, 977, 990.

Trichoptera

119, 151, 152, 168, 193, 200, 211, 293, 327, 328, 330, 331, 414, 415, 416, 439, 463, 465, 518, 588, 599, 731, 813, 828, 829, 841, 844, 847, 850, 904, 932, 933, 947, 988.

Lepidoptera

34, 44, 54, 56, 71, 150, 161, 168, 187, 192, 202, 277, 301, 325, 332, 369, 400, 401, 404, 405, 414, 415, 416, 419, 420, 436, 456, 463, 465, 526, 537, 539, 561, 563, 565, 586, 588, 605, 621, 735, 744, 771, 799, 800, 901, 904, 907, 948, 977.

Hymenoptera

3, 10, 11, 18, 48, 56, 74, 75, 79, 89, 124, 126, 161, 170, 177, 187, 192, 198, 200, 202, 210, 223, 296, 299, 301, 303, 346, 347, 361, 369, 377, 380, 393, 400, 414, 415, 416, 431, 433, 435, 451, 457, 463, 465, 485, 498, 539, 541, 557, 558, 559, 560, 561, 562, 563, 564, 565, 566, 568, 569, 570, 571, 572, 573, 574, 575, 577, 578, 579, 580, 586, 588, 620, 621, 635, 652, 657, 658, 665, 707, 732, 744, 756, 771, 777, 780, 790, 792, 796, 799, 800, 841, 867, 868, 869, 878, 894, 904, 907, 922, 923, 940, 941, 972, 977.

Pisces

13, 21, 22, 25, 26, 34, 40, 50, 54, 61, 78, 80, 81, 89, 95, 96, 99, 100, 110, 116, 119, 131, 132, 133, 134, 135, 142, 145, 146, 150, 155, 157, 168, 182, 191, 193, 195, 208, 220, 229, 230, 231, 254, 255, 261, 270, 288, 291, 292, 293, 311, 312, 313, 314, 320, 321, 330, 331, 334, 356, 366, 409, 410, 414, 429, 439, 440, 441, 468, 475, 477, 494, 515, 516, 517, 523, 524, 526, 531, 533, 592, 594, 597, 598, 610, 619, 625, 631, 633, 662, 667, 680, 689, 706, 710, 716, 729, 734, 739, 742, 745, 749, 769, 804, 817, 828, 829, 834, 841, 843, 846, 861, 877, 883, 884, 892, 899, 906, 911, 915, 944, 982, 988.

Amphibia

10, 16, 25, 34, 54, 79, 94, 116, 118, 119, 138, 145, 182, 192, 220, 255, 270, 286, 288, 292, 293, 311, 312, 313, 320, 321, 337, 400, 414, 441, 442, 446, 456, 494, 533, 592, 594, 625, 633, 637, 652, 655, 681, 683, 684, 698, 717, 718, 719, 739, 742, 748, 749, 769, 790, 857, 861, 864, 874, 875, 876, 899, 901, 906, 915, 941, 944, 964, 970, 991.

Reptilia

10, 11, 16, 94, 116, 118, 119, 136, 137, 138, 145, 182, 220, 270, 301, 311, 312, 313, 320, 321, 332, 337, 374, 400, 414, 435, 441, 442, 446, 456, 533, 561, 578, 579, 592, 594, 625, 633, 637, 652, 653, 654, 655, 656, 658, 681, 683, 717, 718, 719, 732, 739, 742, 748, 749, 769, 825, 857, 861, 869, 874, 875, 878, 899, 906, 910, 915, 918, 941, 944, 970, 991.

Aves

4, 5, 6, 7, 9, 10, 11, 12, 14, 15, 16, 17, 18, 23, 24, 25, 27, 28, 31, 32, 34, 35, 36, 38, 42, 44, 48, 50, 52, 54, 57, 59, 64, 66, 72, 78, 83, 89, 92, 94, 99, 100, 103, 105, 108, 109, 116, 118, 119, 122, 123, 130, 137, 138, 139, 141, 142, 145, 146, 148, 154, 156, 157, 159, 162, 164, 170, 171, 172, 176, 177, 178, 179, 181, 182, 184, 185, 186, 190, 192, 195, 197, 204, 208, 209, 220, 221, 227, 228, 232, 251, 254, 256, 257, 261, 262, 263, 264, 265, 270, 278, 279, 283, 284, 286, 290, 291, 292, 294, 299, 301, 311, 312, 313, 314, 315, 320, 321, 323, 325, 332, 334, 337, 341, 354, 355, 356, 357, 360, 371, 374, 381, 382, 384, 385, 388, 389, 390, 391, 392, 393, 400, 401, 409, 414, 424, 432, 437, 441, 444, 446, 447, 448, 449, 450, 452, 453, 455, 456, 460, 462, 463, 464, 465, 466, 475, 480, 482, 486, 491, 494, 495, 498, 499, 500, 503, 507, 513, 516, 517, 526, 528, 531, 533, 542, 547, 549, 561, 569, 577, 578, 579, 582, 589, 592, 594, 613, 614, 615, 616, 617, 618, 619, 621, 625, 631, 633, 637, 640, 652, 653, 654, 658, 660, 661, 663, 665, 667, 668, 669, 670, 672, 673, 681, 683, 684, 686, 692, 694, 696, 699, 704, 705, 706, 709, 710, 711, 715, 717, 718, 719, 723, 724, 732, 734, 735, 739, 742, 748, 749, 758, 759, 761, 762, 764, 769, 772, 790, 791, 802, 806, 808, 810, 817, 819, 820, 821, 825, 830, 836, 838, 842, 843, 845, 846, 848, 855, 857, 858, 860, 861, 862, 864, 866, 869, 873, 874, 875, 876, 877,

878, 879, 881, 882, 890, 895, 897, 899, 900, 901, 903, 906, 910, 914, 915, 916, 918, 921, 925, 935, 936, 939, 941, 944, 946, 948, 949, 951, 956, 957, 958, 959, 960, 961, 964, 968, 969, 970, 973, 974, 975, 976, 977, 979, 980, 988, 990, 991.

Mammalia

2, 10, 11, 12, 13, 14, 15, 16, 19, 20, 22, 25, 27, 28, 30, 32, 33, 34, 35, 39, 42, 46, 50, 57, 66, 78, 82, 89, 92, 93, 94, 98, 99, 100, 104, 106, 108, 116, 118, 119, 128, 136, 137, 138, 141, 142, 145, 153, 154, 157, 158, 159, 163, 169, 170, 171, 172, 177, 181, 182, 188, 190, 191, 192, 194, 195, 201, 204, 206, 208, 212, 214, 216, 219, 220, 221, 226, 251, 254, 255, 262, 263, 267, 268, 270, 271, 280, 281, 282, 283, 286, 290, 291, 292, 294, 295, 296, 297, 298, 299, 300, 301, 305, 311, 312, 313, 320, 321, 325, 332, 337, 343, 351, 352, 356, 360, 370, 374, 384, 386, 387, 388, 389, 390, 391, 392, 393, 394, 396, 397, 398, 399, 400, 401, 408, 414, 424, 430, 435, 436, 437, 441, 443, 444, 446, 456, 457, 458, 467, 469, 473, 474, 475, 477, 480, 487, 488, 491, 494, 498, 500, 504, 508, 509, 510, 511, 512, 516, 517, 528, 533, 543, 545, 546, 547, 550, 561, 569, 577, 578, 579, 582, 592, 594, 601, 605, 611, 612, 613, 616, 617, 620, 621, 625, 631, 632, 633, 637, 638, 652, 653, 654, 658, 662, 663, 668, 669, 681, 683, 684, 689, 690, 692, 697, 700, 701, 702, 704, 705, 706, 708, 709, 710, 711, 715, 717, 718, 719, 724, 728, 732, 733, 734, 735, 739, 742, 744, 748, 749, 750, 751, 756, 758, 759, 769, 773, 775, 778, 790, 802, 803, 805, 817, 823, 825, 830, 836, 837, 838, 855, 857, 859, 860, 861, 862, 865, 866, 871, 874, 875, 876, 877, 878, 881, 882, 893, 899, 903, 906, 908, 910, 915, 916, 918, 921, 925, 926, 927, 938, 941, 943, 944, 946, 951, 952, 962, 964, 966, 968, 970, 982, 983, 984, 985, 986, 988, 989, 990, 991.

List of references

1. Abbott, I. (1985). Influence of some environment factors on indigenous earthworms in the northern jarrah forest of Western Australia. *Australian Journal of Soil Research*, **23**, 271–90.
2. Adams, L. W. (1984). Small mammal use of an interstate highway median strip. *Journal of Applied Ecology*, **21**, 175–8.
3. Akre, R. D., Hansen, L. D., Reed, H. C. and Corpus, L. D. (1981). Effects of ash from Mt. St. Helens on ants and yellowjackets. *Melanderia*, **37**, 1–19.
4. Allaire, P. N. (1978*a*). Effects on avian populations adjacent to an active strip-mine site. In *Surface Mining and Fish/Wildlife Needs in the Eastern United States*, ed. D. E. Samuel, J. R. Stauffer, C. H. Hocutt and W. T. Mason, pp. 232–40. Morgantown: West Virginia University.
5. Allaire, P. N. (1978*b*). Reclaimed surface mines: new potential for some North American birds. *American Birds*, **32**, 3–5.
6. Allaire, P. N. (1979). Coal mining reclamation in Appalachia: low cost recommendations to improve bird/wildlife habitat. In *The Mitigation Symposium: A National Workshop on Mitigating Losses of Fish and Wildlife Habitats*, ed. G. A. Swanson, pp. 245–51. Rocky Mountain Forest and Range Experiment Station (Forest Service).
7. Allaire, P. N. (1980*a*). *Bird Species on Mined Lands: Assessment and Utilization in Eastern Kentucky*. Kentucky: Institute for Mining and Minerals Research.
8. Allaire, P. N. (1980*b*). Comparisons of game and nongame management models: cost analysis of post-mining land use options for wildlife. In *Symposium on Surface Mining Hydrology, Sedimentology and Reclamation*, ed. D. H. Graves, pp. 315–24. University of Kentucky.
9. Allan, P. F. and Steiner, W. W. (1959). *Autumn Olive for Wildlife and other Conservation Uses.* US Department of Agriculture Leaflet No. 458.
10. Allen, N. T. (1984). Collection, storage and analysis of biological data from fauna in the

Worsley Alumina project area, Western Australia. In *Australian Mining Industry Council Environmental Workshop Papers*, pp. 291–313.

11. Allen, N. T. (1987). Welcoming back the native wildlife. In *Mining and the Return of the Living Environment*, pp. 15–19. Canberra: Australian Mining Industry Council.
12. Altenburg, W., Beemster, N., Dijk, K. van, Esselink, P., Prop, D. and Visser, H. (1985). *Development of Breeding Bird Population of the Lauwerszee in the Period 1978–1983*. Rijksdienst voor de Ijsselmeerpolders, RIJP Rapport 1985 – 15 ABW.
13. Alverson, K. (1973). Cattle and catfish on surface-mined land. *Soil Conservation*, **39**, 4–6.
14. Ambrose, R. E., Hinkle, C. R. and Wenzel, C. R. (1983*a*). *Practices for Protecting and Enhancing Fish and Wildlife on Coal Surface-Mined Land in Central and Southern Appalachia*. US Fish and Wildlife Service Report.
15. Ambrose, R. E., Hinkle, C. R. and Wenzel, C. R. (1983*b*). *Practices for Protecting and Enhancing Fish and Wildlife on Coal Surface-Mined Land in the Southcentral U.S.* US Fish and Wildlife Service Report.
16. Anderson, B. W. and Ohmart, R. D. (1979). Riparian revegetation: an approach to mitigating for a disappearing habitat in the southwest. In *The Mitigation Symposium: A National Workshop on Mitigating Losses of Fish and Wildlife Habitats*, ed. G. A. Swanson, pp. 481–7. Rocky Mountain Forest Range Experiment Station (Forest Service).
17. Anderson, R. J. (1980). *Evaluation of Reclaimed Surface Mines as Wild Turkey Brood Range*. M.Sc. Thesis, West Virginia University (unpublished manuscript).
18. Anderson, R. J. and Samuel, D. E. (1980). Evaluation of reclaimed surface mines as wild turkey brood range. In *4th National Wild Turkey Symposium, March 1980, Littlerock, Arkansas*, pp. 186–206.
19. Andrews, S. M., Cooke, J. A. and Johnson, M. S. (1982). Fluoride in small mammals and their potential food sources in contaminated grasslands. *Fluoride*, **15**, 56–63.
20. Andrews, S. M., Johnson, M. S. and Cooke, J. A. (1985). *Cadmium in Small Mammals from Grassland Established on Metalliferous Mine Waste* (unpublished manuscript).
21. Anon. (1951). More strip pits for fishin'! *Outdoor Indiana*, **18**, 18–19.
22. Anon. (1968*a*). Big beef business flourishes on reclaimed strip land. *Coal Age*, **73**, 71–5.
23. Anon. (1968*b*). Giant geese find new home on mined land. *Mined-Land Conservation*, **4**, 4.
24. Anon. (1969*a*). Canadian geese transplanted in Illinois. *Coal Mining and Processing*, **6**, 33.
25. Anon. (1969*b*). Reclaimed land is for the birds and fish and beaver and deer and possums and frogs. *Mined-Land Conservation*, **5**, 1–4.
26. Anon. (1971). Limnology of a strip mine pond in western Maryland. *Chesapeake Science*, **12**, 111–20.
27. Anon. (1978). *Handbook for Terrestrial Wildlife Habitat Development on Dredged Material*. US Army Engineer Waterways Experimental Station Technical Report.
28. Anon. (1979). *De Lauwersmeer*. Herlingen: Werkgroep Lauwersmeer.
29. Anon. (1982). *Study of Possible Establishment of the Oostvaardersplassen Area*. Ministerie van Verkeer en Waterstaat, Rijksdienst Voor de Ijsselmeerpolders, Flevobericht no. 185.
30. Anon. (1980). Game reserve grows out of disused quarry. Blue Circle Ltd gives back to the land. *Pit and Quarry*, **72**, 138–40, 144.
31. Appleton, T. P. (1983). The management of Rutland Water reservoir for wildlife. In *The Proceedings of a Symposium on the Subject of Wildlife on Man-Made Wetlands*, pp. 121–33. Milton Keynes, England: ARC Wildlife Centre.
32. Arata, A. A. (1957). Trapping the striplands. *Illinois Wildlife*, **12**, 5–6.
33. Arata, A. A. (1959). Ecology of muskrats in strip-mine ponds in southern Illinois. *Journal of Wildlife Management*, **23**, 177–86.
34. ARC Wildlife Centre. (1983). *The Proceedings of a Symposium on the Subject of Wildlife on Man-Made Wetlands*. Milton Keynes, England: ARC Wildlife Centre.

35. Argonne National Laboratory. (1979). *Land Reclamation Program Annual Report 1978.* Argonne National Laboratory Report ANL/LRP-5.

36. Armbruster, J. S., ed. (1983). *Impacts of Coal Surface Mining on 25 Migratory Bird Species of High Federal Interest.* US Fish and Wildlife Service Report.

37. Armstrong, M. J. and Bragg, N. C. (1984). Soil physical parameters and earthworm populations associated with opencast coal workings and land restoration. *Agriculture, Ecosystems and Environment*, **11**, 131–43.

38. Arnold, C. W. Jr and Samuel, D. E. (1975). *Occurrence of Ruffed Grouse Near Strip Mines in West Virginia.* Grouse Workshop, September 1973, Longlake, New York.

39. Arthur, W. J., Markham, O. D., Groves, C. R., Keller, B. L. and Halford, D. K. (1986). Radiation dose to small mammals inhabiting a solid radioactive waste disposal area. *Journal of Applied Ecology*, **23**, 13–26.

40. Arway, J. A., Kepler, S. R. and Spotts, D. E. (1985). Stream substrate analysis – a pollution assessment technique. In *Wetlands and Water Management on Mined Lands. Proceedings of a Conference October 23–24, 1985*, ed. R. P. Brooks, D. E. Samuel and J. B. Hill, pp. 377–80. Pennsylvania State University.

41. Ashby, W. C. (1964). Vegetation development on a strip-mined area in southern Illinois. *Transactions of the Illinois State Academy of Science*, **57**, 78–83.

42. Ashby, W. C., Kolar, C., Guerke, M. L., Pursell, C. F. and Ashby, J. (1978). *Our Reclamation Future with Trees.* Coal Extraction and Utilization Research Center Report. Carbondale: Southern Illinois University.

43. Atlavinyte, O. (1976). The effect of land reclamation and field management on the change in specific composition and densities of Lumbricidae. *Polish Ecological Studies*, **2**, 147–52.

44. Baines, J. C. (1986). Design considerations at establishment. In *Ecology and Design in Landscape*, ed. A. D. Bradshaw, D. A. Goode and E. H. P. Thorpe, pp. 73–81. Oxford: Blackwell.

45. Baker, B. (1977). Some mined land reclamation problems of the Uravan mineral belt. In *Reclamation for Wildlife Habitat. Proceedings of Reclamation Workshop II, Fort Collins, Colorado, September 1977*, ed. Ecology Consultants Inc., pp. 64–8. Environmental Research and Technology Inc.

46. Baker, C. D., Dingus, G. P. and McCart, G. D. (1976). Producing forages for livestock on surface mined land in southwest Virginia. In *Proceedings of the Fourth Symposium on Surface Mining and Reclamation, Louisville, Kentucky*, pp. 223–7. Washington, DC: Environmental Research and Technology Inc.

47. Baker, G. H. (1985). The distribution, abundance and species associations of earthworms (Lumbricidae) in a reclaimed peat soil in Ireland. *Holarctic Ecology* (in press).

48. Ball, E. and Glucksman, J. (1975). Biological colonization of Motmot, a recently-created tropical island. *Proceedings of the Royal Society London B*, **190**, 421–42.

49. Barnes, R. S. K. and Jones, J. M. (1972). Observations on the fauna and flora of reclaimed land at Calshot, Hampshire. *Hampshire Field Club and Archaeological Society Proceedings*, **29**, 81–91.

50. Barr Engineering Company. (1980). *Enhancement of Fish and Wildlife Resources in the Reclamation of Hard Rock Mined Areas in the Upper Midwest: A Guidelines Manual.* US Fish and Wildlife Service Report.

51. Barr, W. C. (1979). TVA strip mine aquatic assessment model: benthic invertebrate module. In *Proceedings Symposium on Surface Mining Hydrology, Sedimentology and Reclamation*, ed. S. D. Carpenter and R. W. de Vore, pp. 243–9. Lexington: University of Kentucky.

52. Barta, Z., Bejcek, V. and Tyrner, P. (1971). Initial avian synusy in the plant community *Atriplicetum Nitensis* Knapp 1945, found on brown coal waste dump in 1970. *Biologia*, **26**, 835–40.

53. Bauer, A. M. (1982). Manipulating mining operations to create wildlife habitats: a pre-mining planning process. In *The Wildlife Values of Gravel Pits Symposium, Crookston, Minnesota, June 1982*, ed. W. D. Svedarsky and R. D. Crawford, pp. 41–3. University of Minnesota Miscellaneous Publication 17–1982.
54. Bauer, H. J. (1973). Ten years' studies of biocenological succession in the excavated mines of the Cologne lignite district. In *Ecology and Reclamation of Devastated Land*, vol. 1, ed. R. J. Hutnik and G. Davis, pp. 271–83. New York: Gordon and Breach.
55. Bauer, H. J. (1976). *Artificial Wetlands: Damage to the Landscape or Valuable Natural Assets?* Council of Europe Information Centre for Nature Conservation Report 46.221.
56. Baweja, K. D. (1939). Studies of the soil fauna, with special reference to the recolonization of sterilized soil. *Journal of Animal Ecology*, **8**, 120–61.
57. Beck, R. E. (1978). Surface coal mining law in the western United States: how does it provide for wildlife? *North Dakota Law Review*, **54**, 337–74.
58. Beckworth, S. L. (1954). Ecological succession on abandoned farmlands and its relationship to wildlife. *Ecological Monographs*, **24**, 349–76.
59. Bejcek, V. and Tyrner, P. (1980). Primary succession and species diversity of avian communities on spoil banks after surface mining of lignite in the Most Basin (north-western Bohemia). *Folia Zoologica*, **29**, 67–77.
60. Beley, J. R., Ditsworth, T. M., Butt, S. M. and Johnson, C. D. (1982). Arthropods, plants and transmission lines in Arizona: community dynamics during secondary succession in a pinyon-juniper chaparral habitat. *Southwestern Naturalist*, **27**, 325–32.
61. Bell, A. V. and Nancarrow, D. R. (1974). Salmon and mining in northeastern New Brunswick. *Canadian Metal and Mining Bulletin*, **67**, 44–53.
62. Bengson, S. A. (1986). Holistic reclamation of copper tailings in the southwest. In *New Horizons for Mined Land Reclamation. Proceedings American Society for Surface Mining and Reclamation. 1986 National Meeting, Jackson, Mississippi*, pp. 41–4.
63. Benner, B. L. and Kannowski, P. B. (1984). Collembola of southwestern North Dakota: species composition and habitat distribution. *Prairie Naturalist*, **16**, 79–90.
64. Bennison, S. (1984). Utilization of open-cut pits and residue ponds following mining at Greenbushes. In *Australian Mining Industry Council Environmental Workshop Papers*, pp. 134–6.
65. Bhowmik, N. G. (1985). Hydrolic considerations in the alteration of streams. In *Fish and Wildlife Relationships to Mining: Symposium Proceedings*, ed. L. B. Starnes, pp. 74–102. American Fisheries Society (Water Quality Section).
66. Birkenholz, D. (1958). Notes on a wintering flock of long-eared owls. *Transactions of the Illinois State Academy of Science*, **51**, 83–6.
67. Birkenholz, D. (1958). *Reclamation of Spoil-Bank Area for Wildlife Purposes.* M.Sc. Thesis, Southern Illinois University.
68. Bjugstad, A. J. (1979). Management plan and monitoring. In *User Guide to Vegetation.* USDA Forest Service General Technical Report INT-64, pp. 67–75.
69. Bjugstad, A. J., Kissel, K. and Whitman, W. (1982). Germination of shrubs for wildlife habitat restoration on mined lands in the Northern Great Plains. In *Issues and Technology in the Management of Impacted Western Wildlife*, ed. R. D. Comer, J. M. Merino, J. W. Monarch, C. Pustmueller, M. Stalmaster, R. Stoeker, J. Todd and W. Wright, pp. 238–48. Thorne Ecological Institute Technical Publication No. 14.
70. Bjugstad, A. J. and Whitman, W. C. (1982). Perennial forbs for wildlife habitat restoration on mined lands in the Northern Great Plains. In *Proceedings of the Annual Conference of the Western Association of Fish and Wildlife Agencies, July 18–22, Las Vegas, Nevada.*
71. Black, R. F. (1979). Mineral sands rehabilitation at Eneabba, W.A. In *Mining Rehabilitation*, ed. I. Hore-Lacy, pp. 69–74. Canberra: Australian Mining Industry Council.

72. Blomberg, G. E. (1982). Duck use of gravel pits near Fort Collins, Colorado. In *The Wildlife Values of Gravel Pits Symposium, Crookston, Minnesota, June 1982*, ed. W. D. Svedarsky and R. D. Crawford, pp. 162–9. University of Minnesota Miscellaneous Publication 17–1982.
73. Boccardy, J. A. and Spaulding, V. M. (1968). *Effects of Surface Mining on Fish and Wildlife in Appalachia*. US Department of the Interior, Bureau of Sport Fish and Wildlife Resource Publication no. 65.
74. Bode, E. (1973). *Beiträge zu den Erscheinungen einer Sukzession der Terricolen Zoozonose auf Rekultivierungsflachen*. Dissertation Braunschweig.
75. Bode, E. (1975*a*). Ein Beitrag zue Ameisenbesiedlung (Formicidae [Hymenoptera, Hexapoda]) forstlicher Rekultivierungsflachen des Braunkohlentagebaues. *Wald Hygiene*, **11**, 13–20.
76. Bode, E. (1975*b*). Pedozoologische Sukzessionsuntersuchungen auf Rekultivierungsflächen des Braunkohlentagebaues. *Pedobiologia*, **15**, 284–9.
77. Bookhout, T. (1977). Lessons from mid-Appalachia strip mine reclamation efforts. In *Reclamation for Wildlife Habitat. Proceedings of Reclamation Workshop II, Fort Collins, Colorado, September 1977*, ed. Ecology Consultants Inc., pp. 154–67. Environmental Research and Technology Inc.
78. Bookhout, T. A., Stone, C. P., Bittner, J. D., Tubb, R. A., Taub, S. H. and Dies, R. E. (1968). *Potential of a Strip-Mined Area for Fish and Wildlife Reclamation*. The Ohio State University Research Foundation Project 2296.
79. Boomsma, J. J. and Arntzen, J. W. (1985). Abundance, growth and feeding of natterjack toads (*Bufo calamita*) in a 4-year-old artificial habitat. *Journal of Applied Ecology*, **22**, 395–405.
80. Bosserman, R. W. and Hill, P. L. (1985). Community ecology of three wetland ecosystems impacted by acid mine drainage. In *Wetlands and Water Management on Mined Lands. Proceedings of a Conference October 23–24, 1985*, ed. R. P. Brooks, D. E. Samuel and J. B. Hill, pp. 287–303. Pennsylvania State University.
81. Bosserman, R. W., Mitsch, W. J., Hill, P. L., Smith, F. L. and Taylor, J. R. (1981). Modelling and management of wetlands in the western coalfields of Kentucky. In *Progress in Wetlands Utilization and Management, Orlando, Florida, June 1981*.
82. Bostrom, U. and Hansson, L. (1981). Small rodent communities on mires: implications for population performance in other habitats. *Oikos*, **37**, 216–24.
83. Bournaud, M., Ledant, J. P., Broyer, J. and Richoux, M. (1982). The pond area as related to the avifauna during the nesting period. *Bulletin D'Écologie*, **13**, 125–44.
84. Bowen, G. D. (1984). The roles of microorganisms in ecological processes. In *Australian Mining Industry Council Environmental Workshop Papers*, pp. 47–58.
85. Bradshaw, A. D. (1982). The landscape reborn, *New Scientist*, **95**, 901–4.
86. Bradshaw, A. D. (1983). The reconstruction of ecosystems: presidential address to the British Ecological Society, December 1982. *Journal of Applied Ecology*, **20**, 1–17.
87. Bradshaw, A. D. (1984). Ecological principles and land reclamation practice. *Landscape Planning*, **11**, 35–48.
88. Bradshaw, A. D. (1986). Ecological principles in landscape. In *Ecology and Design in Landscape*, ed. A. D. Bradshaw, D. A. Goode and E. H. P. Thorpe, pp. 15–36. Oxford: Blackwell.
89. Bradshaw, A. D. and Chadwick, M. J. (1980). *The Restoration of Land: The Ecology and Reclamation of Derelict and Degraded Land*. Oxford: Blackwell.
90. Bradshaw, A. D., Goode, D. A. and Thorpe, E. H. P., ed. (1986). *Ecology and Design in Landscape*. Oxford: Blackwell.
91. Bradshaw, A. D., McNeilly, T. S. and Gregory, R. P. G. (1965). Industrialization, evolution and the development of heavy metal tolerance in plants. In *Ecology and the Industrial*

Society, ed. G. T. Goodman, R. W. Edwards and J. M. Lambert, pp. 327–43. Oxford: Blackwell.

92. Bramble, W. C. and Ashley, R. H. (1955). Natural revegetation of spoil banks in central Pennsylvania. *Ecology*, **36**, 417–23.
93. Bramble, W. C. and Sharp, W. M. (1949). Rodents as a factor in direct seeding on spoil bank in central Pennsylvania. *Journal of Forestry*, **47**, 477–8.
94. Branch, W. L. (1985). Design and construction of replacement wetlands on lands mined for sand and gravel. In *Wetlands and Water Management on Mined Lands. Proceedings of a Conference October 23–24, 1985*, ed. R. P. Brooks, D. E. Samuel and J. B. Hill, pp. 173–9. Pennsylvania State University.
95. Branson, B. A. and Batch, D. L. (1971). Effects of strip mining on small-stream fishes in east-central Kentucky. *Proceedings of the Biological Society of Washington*, **84**, 507–18.
96. Brenner, F. J. (1974). *Ecology and Productivity of Strip-Mine Areas in Mercer County, Pennsylvania*. Institute for Research on Land and Water Resources, Research Project Technical Completion Report 1 July 1971–31 May 1974. University Park: Pennsylvania State University.
97. Brenner, F. J. (1978*a*). *Evaluation of Factors Promoting the Preservation of Aquatic Ecosystems in Reclaimed Strip Mine Areas*. Institute for Research on Land and Water Resources Report. University Park: Pennsylvania State University.
98. Brenner, F. J. (1978*b*). Food and cover evaluation of strip mine plants as related to wildlife management. In *Surface Mining and Fish/Wildlife Needs in the Eastern United States*, ed. D. E. Samuel, J. R. Stauffer, C. H. Hocutt and W. T. Mason, pp. 294–305. Morgantown: West Virginia University.
99. Brenner, F. J. (1981). Surface coal mine reclamation – a need for re-evaluation. In *Energy, Environment and the Economy*, ed. S. J. Majumdar, pp. 81–92. Pennsylvania Academy of Science Publications.
100. Brenner, F. J. (1983). Environmental aspects of coal production in Pennsylvania and a guide to reclamation. In *Pennsylvania Coal: Resources, Technology and Utilization*, ed. S. K. Majumdar and E. W. Miller, pp. 405–24. Pennsylvania Academy of Science Publications.
101. Brenner, F. J. (1984*a*). Management strategies and ecological approaches to facilitate reclamation on mined lands. In *Strategies for Environmentally Sound Development in the Mining and Energy Industries, Crete, October 21–28*.
102. Brenner, F. J. (1984*b*). Restoration of natural ecosystems on surface coal mine lands in the northeastern United States. In *The Biosphere: Problems and Solutions*, ed. T. N. Veriroglu, pp. 211–25. Amsterdam: Elsevier Science.
103. Brenner, F. J. and Kelly, J. (1981). Characteristics of bird communities on surface mine lands in Pennsylvania. *Environmental Management*, **5**, 441–9.
104. Brenner, F. J., Kelly, R. B. and Kelly, J. (1982). Mammalian community characteristics on surface mine lands in Pennsylvania. *Environmental Management*, **6**, 241–9.
105. Brenner, F. J. and Michalski, III, S. (1984). Evaluation of surface coal mines as ruffed grouse habitat. In *Proceedings of the Fourth Annual Better Reclamation with Trees Conference, Owensboro, Kentucky June 7–8 1984*, ed. P. E. Pope. Kentucky: Madisonville Community College.
106. Brenner, F. J. and Musaus, M. J. (1977). Selectivity of browse species by white-tailed deer on strip-mine lands in Mercer County, Pennsylvania. *Proceedings of the Pennsylvania Academy of Science*, **51**, 105–8.
107. Brenner, F. J., Snyder, W., Schalles, J. F., Miller, J. P. and Miller, C. (1985). Primary productivity of deep-water habitats on reclaimed mined lands. In *Wetlands and Water Management on Mined Lands. Proceedings of a Conference October 23–24, 1985*, ed. R. P. Brooks, D. E. Samuel and J. B. Hill, pp. 199–209. Pennsylvania State University.

108. Brenner, F. J., Werner, M. and Pike, J. (1984). Ecosystem development and natural succession in surface coal mine reclamation. *Minerals and the Environment*, **6**, 10–22.

109. Brewer, R. (1958). Breeding-bird populations of strip-mined land in Perry County, Illinois. *Ecology*, **39**, 543–5.

110. Brice, J. R. and Boody, O. C. (1983). Fish populations in reclaimed and natural lakes in central Florida's phosphate region: a preliminary report. In *Reclamation and the Phosphate Industry. Proceedings of the Symposium, Clearwater Beach, Florida 26–28 January 1983*, ed. D. J. Robertson. Florida Institute of Phosphate Research.

111. Bridges, J. M. and Thompson, C. S. (1978). Including wildlife management in surface mine reclamation. In *The Annual Meeting of the Illinois State Academy of Science, April 21–22.*

112. Brock, A. M. (1965). *Earthworm Studies on Opencast Coal Sites in the East Midlands.* Agricultural Research Council Internal Report no. 312/65.

113. Brockmann, W., Koehler, H. and Schriefer, T. (1980). Recultivation of refuse tips: soil ecological studies. In *Soil Biology as Related to Land Use Practices*, ed. D. L. Dindal, pp. 161–8. Washington: Environmental Protection Agency.

114. Broen, B. von and Moritz, M. (1965). Spinnen (Araneae) und Weberknechte (Opiliones) aus Barberfallen von einer tertiaren Rohbodenkippe im braunkohlenrevier Bohlen. *Abhandlungen und Berichte des Naturkundemuseums Gorlitz*, **40**, 1–15.

115. Brohard, E. (1961). Strip-mine land can be salvaged. *Soil Conservation*, **27**, 66–7.

116. Bromley, P. T. and Cushwa, C. T. (1985). *Wildlife and Fish Habitat on Reclaimed Surface-Mined Lands.* Powell River Project Series, Publication no. 460–109.

117. Brooks, D. R. (undated). *Our Natural Heritage: Are some Elements Renewable?* Queensland Chamber of Mines Seminar, 4th September, Brisbane (unpublished manuscript).

118. Brooks, R. P., Hill, J. B., Brenner, F. J. and Capets, S. (1985). Wildlife use of wetlands on coal surface mines in western Pennsylvania. In *Wetlands and Water Management on Mined Lands. Proceedings of a Conference October 23–24, 1985*, ed. R. P. Brooks, D. E. Samuel and J. B. Hill, pp. 337–52. Pennsylvania State University.

119. Brooks, R. P., Samuel, D. E. and Hill, J. B., ed. (1985). *Wetlands and Water Management on Mined Lands. Proceedings of a Conference October 23–24, 1985.* Pennsylvania State University.

120. Brookshire, D. S., Eubanks, L. S., Rowe, R. D. and Livingston, D. B. (1977). The impact of coal development on wildlife: an economic analysis. In *Energy and the Environment: Fifth National Conference, Cincinnati, Ohio*, pp. 480–6.

121. Brown, M. C. (1964). The occupation of newly irrigated lands by earthworms. *Australian Journal of Science*, **26**, 290–2.

122. Brown, S. L. (1981). *Effects of Controlled Burning on Potential Bobwhite Quail Brood Habitat on Surface Mines in West Virginia.* M.Sc. Thesis, West Virginia University (unpublished manuscript).

123. Brown, S. L. and Samuel, D. E. (1978). The effects of controlled burning on potential bobwhite quail brood habitat on surface mines. In *Surface Mining and Fish/Wildlife Needs in the Eastern United States*, ed. D. E. Samuel, J. R. Stauffer, C. H. Hocutt and W. T. Mason, pp. 352–8. Morgantown: West Virginia University.

124. Bruning, E., Unger, H. and Dunger, W. (1965). Untersuchungen zue Frage der biologischen Aktivierung alttertiarer Rohbodenkippen des Braukohlentagebaues in Abhangigkeit von Bodenmelioration und Rekultivierung. *Zeitschrift für Landeskultur*, **6**, 9–38.

125. Buckalew, D. W., Riley, R. K., Yoder, W. A. and Vail, W. J. (1982). Invertebrates as vectors of endo mycorrhizal fungi and *Rhizobium* on surface mine soils. *Proceedings of the West Virginia Academy of Science*, **54**, 1.

126. Buckley, R. C., ed. (1982). *Ant–Plant Interactions in Australia.* The Hague: Junk.

127. Bulavintsev, V. I. (1979*a*). Formation of ground beetle complexes Coleoptera Carabidae in areas disrupted by open cut mining. *Vestnik Zoologii*, **5**, 65–71.

128. Bulavintsev, V. I. (1979*b*). Formation of the population of small vertebrates on the territories disturbed by the open mineral fields. *Zoologicheskiĭ Zhurnal*, **58**, 386–92.

129. Bunn, S. E. (1983). Termite (Isoptera) fauna of jarrah forest in the Wagerup-Willowdale region, Western Australia: relevance to the rehabilitation of bauxite minesites. *Forest Ecology and Management*, **6**, 169–77.

130. Burley, J. B. and Hopkins, R. B. (1984). Potential for enhancing nongame bird habitat values on abandoned mine lands of western North Dakota. In *Proceedings 1984 Symposium on Surface Mining, Hydrology, Sedimentology, and Reclamation, Lexington, December 2–7, 1984*, ed. H. Graves. Lexington: University of Kentucky.

131. Burner, C. C. (1954). Report on progress of fish management studies on strip mine lakes. *Kansas Fish and Game*, **11**, 5–8.

132. Burner, C. C. (1958). Let's fish the strip lakes. *Kansas Fish and Game*, **16**, 3–6.

133. Burner, C. C. (1973). Fishery management in strip mine lakes. In *Research and Applied Technology Symposium on Mined Land. Proceedings*, pp. 304–18. Monroeville, Pennsylvania: Bitumen Coal Research Incorporated.

134. Burner, C. C. and Leist, C. (1953). A limnological study of the College Farm strip-mine lake. *Transactions of the Kansas Academy of Science*, **56**, 78–85.

135. Burner, C. C. and Schoonover, R. (1953). *Progress Report on Fisheries Management Projects in Linn County, Kansas*. Kansas Fish and Game Commission.

136. Bury, R. B. and Luckenbach, R. A. (1983). Vehicular recreation in arid land dunes: biotic responses and management alternatives. In *Environmental Effects of Off-Road Vehicles. Impacts and Management in Arid Regions*, ed. R. H. Webb and H. G. Wilshire, pp. 207–21. New York: Springer-Verlag.

137. Butler, W. H. (1975). Additions to the fauna of Barrow Island, W.A. *The Western Australian Naturalist*, **13**, 78–80.

138. Butler, W. H. (1983). The Barrow Island experience. In *The 53rd Anzaas Congress, Perth 1983, on Behalf of Wapet*, ed. B. G. Collins. WAIT School of Biology Bulletin no. 7.

139. Butler, R. A. and Lueth, B. K. (1986). Re-establishment of the giant Canada goose on surface mine wetlands. In *National Mined Land Reclamation*, ed. J. H. Swisher and C. Carlson. pp. 455–69. St. Louis, Missouri.

140. Butt, S. M., Beley, J. R., Ditsworth, T. M., Johnson, C. D. and Balda, R. P. (1980). Arthropods, plants and transmission lines in Arizona: community dynamics during secondary succession in a desert grassland. *Journal of Environmental Management*, **11**, 267–84.

141. Byrnes, W. R. and Miller, J. H. (1973). Natural revegetation and cast overburden properties of surface-mined coal lands in southern Indiana. In *Ecology and Reclamation of Devastated Land*, vol. 1, ed. R. J. Hutnik and G. Davis, pp. 285–306. New York: Gordon and Breach.

142. Cairns, Jr. J. (1978). Summary, recommendations and the future of reclaiming surface mines for fish and wildlife. In *Surface Mining and Fish/Wildlife Needs in the Eastern United States*, ed. D. E. Samuel, J. R. Stauffer, C. H. Hocutt and W. T. Mason, pp. 15–22. Morgantown: West Virginia University.

143. Cairns, Jr. J. (1979). Ecological considerations in reclaiming surface mined lands. *Minerals and the Environment*, **1**, 83–9.

144. Cairns, Jr. J. (1985). Keynote address: facing some awkward questions concerning rehabilitation management practices on mined lands. In *Wetlands and Water Management on Mined Lands. Proceedings of a Conference October 23–24, 1985*, ed. R. P. Brooks, D. E. Samuel and J. B. Hill, pp. 9–17. Pennsylvania State University.

145. Cairns, Jr. J., Dickson, K. L. and Herricks, E. E., ed. (1977). *Recovery and Restoration of Damaged Ecosystems*. Charlottesville: University Press of Virginia.

146. Cairns, Jr. J. and Pratt, J. R. (1985). Aquatic environments on mined lands: progress and

future needs. In *Wetlands and Water Management on Mined Lands. Proceedings of a Conference October 23–24, 1985*, ed. R. P. Brooks, D. E. Samuel and J. B. Hill, pp. 1–8. Pennsylvania State University.

147. Cammen, L. M., Seneca, E. D. and Copeland, B. J. (1974). *Animal Colonization of Salt Marshes Artificially Established on Dredge Spoil.* NTIS Report COM-75-10247.

148. Candian, P. and Caniglia, G. (1981). The reclaimed land areas of the Middle Lagoon south of Venice. VIII. Floristic catalog of polder A. *Societa Veneziana di Scienze Naturali Lavori*, **6**, 3–12.

149. Canestrelli, P. (1979). The reclaimed lands of the Middle Lagoon, south of Venice. IV. The orthopteroid fauna of the artificial island D-E. *Societa Veneziana di Scienze Naturali Lavori*, **4**, 92–114.

150. Canton, S. P. (1982). Comparative limnology and biota of mine spoils ponds in Colorado. *The Southwestern Naturalist*, **27**, 33–42.

151. Canton, S. P. and Ward, J. V. (1979). *Environmental Effects of Western Coal Surface Mining. Part 1. The Limnology and Biota of Mine Spoils Ponds in Northwest Colorado.* Ecological Protection Agency Ecological Research Series 600/3-79-124.

152. Canton, S. P. and Ward, J. V. (1981). Benthos and zooplankton of coal strip mine ponds in the mountains of northwestern Colorado, U.S.A. *Hydrobiologia*, **85**, 23–31.

153. Capita, M. E. de. (1975). *Evaluation of Strip-Mine Reclamation for Terrestrial Wildlife Restoration.* M.Sc. Thesis, Ohio State University (unpublished manuscript).

154. Capita, M. E. de and Bookhout, T. A. (1975). Small mammal populations, vegetational cover, and hunting use of an Ohio strip-mined area. *Ohio Journal of Science*, **75**, 305–13.

155. Cash, S. L. (1968). *Limnology and Fisheries Productivity of Acid and Alkaline Strip-Mine Lakes.* M.Sc. Thesis, Kansas State College of Pittsburg.

156. Catchpole, C. K. and Tydeman, C. F. (1975). Gravel pits as new wetland habitats for the conservation of breeding bird communities. *Biological Conservation*, **8**, 47–59.

157. Caufield, C. (1981). Britain's heritage of wildlife drains away. *New Scientist*, **91**, 583–6.

158. Chabwela, H. N. and Gilbert, F. F. (1982). The effects of aggregate extraction on deer habitat preferences in southern Ontario. In *The Wildlife Values of Gravel Pits Symposium, Crookston, Minnesota, June 1982*, ed. W. D. Svedarsky and R. D. Crawford, pp. 148–54. University of Minnesota Miscellaneous Publication 17-1982.

159. Chalk, D. E. (1979). Predicting impacts of a proposed irrigation water conservation project on wildlife habitat. In *The Mitigation Symposium: A National Workshop on Mitigating Losses of Fish and Wildlife Habitats*, ed. G. A. Swanson, pp. 305–9. Rocky Mountain Forest Range Experiment Station (Forest Service).

160. Challenger, G. P. (1980). Reclamation for wildlife: setting the scene. *Reclamation Review*, **3**, 193–6.

161. Chandler, C. S. (1982). *The Possible Use of Invertebrates as an Additional Source of Information on Land Use and of the Success of Revegetation Attempts.* B. Sc. Honours Thesis, University of Queensland (unpublished manuscript).

162. Chapman, D. L., McGinnes, B. S. and Downing, R. L. (1978). Breeding bird populations in response to the natural revegetation of abandoned contour mines. In *Surface Mining and Fish/Wildlife Needs in the Eastern United States*, ed. D. E. Samuel, J. R. Stauffer, C. H. Hocutt and W. T. Mason, pp. 328–32. Morgantown: West Virginia University.

163. Charton, P. and Carr, Jr., E. A. (1977). A stock answer to reclamation of surface-mined lands. *Appalacia*, **10**, 10–13.

164. Childress, D. A. and Eng, R. L. (1979). Dust abatement project with wildlife enhancement on Canyon Ferry reservoir, Montana. In *The Mitigation Symposium: A National Workshop on Mitigating Losses of Fish and Wildlife Habitats*, ed. G. A. Swanson, pp. 282–8. Rocky Mountain Forest Range Experiment Station (Forest Service).

165. Chimney, M. J. (1976). *The Niche Relationships Among the Planktonic Rotifers in a Small Strip Mine Lake.* PhD Dissertation, Southern Illinois University.

166. Clark, J. R. (1979). Mitigation and grassroots conservation of wetlands – urban issues. In *The Mitigation Symposium: A National Workshop on Mitigating Losses of Fish and Wildlife Habitats*, ed. G. A. Swanson, pp. 141–51. Rocky Mountain Forest and Range Experiment Station (Forest Service).

167. Clark, W. F., ed. (1975*a*). *Proceedings of the Fort Union Coal Field Symposium*, vol. 1, Administration Section, Montana Academy of Sciences.

168. Clark, W. F., ed. (1975*b*). *Proceedings of the Fort Union Coal Field Symposium*, vol. 2, Aquatic Ecosystems Section, Montana Academy of Sciences.

169. Clark, W. F., ed. (1975*c*). *Proceedings of the Fort Union Coal Field Symposium*, vol. 3, Reclamation Section, Montana Academy of Sciences.

170. Clark, W. F., ed. (1975*d*). *Proceedings of the Fort Union Coal Field Symposium*, vol. 5, Terrestrial Ecosystems Section, Montana Academy of Sciences.

171. Cleaves, D. T., Miller, R. M. and Taylor, J. D. (1979). Jim Bridger mine project. In *Land Reclamation Program Annual Report 1978.* Argonne National Laboratory Report ANL-LRP-5, pp. 18–30.

172. Coastal Zone Resources Division. (1978). *Handbook for Terrestrial Wildlife Habitat Development on Dredged Material.* U.S. Army Engineer Waterways Experiment Station Environmental Laboratory, Vicksburg, Missouri. Technical Report.

173. Cole, L. (1986). Urban opportunities for a more natural approach. In *Ecology and Design in Landscape*, ed. A. D. Bradshaw, D. A. Goode and E. H. P. Thorpe, pp. 417–31. Oxford: Blackwell.

174. Coleman, D. C. (1985). Role of soil fauna in pedogenesis and nutrient cycling in natural and human-managed ecosystems. In *Soil Ecology and Management*, ed. J. H. Cooley, pp. 19–80. Athens, Georgia: International Association for Ecology (Institute of Ecology, University of Georgia).

175. Coleman, D. C. and MacFadyen, A. (1966). The recolonization of gamma-irradiated soil by small arthropods: a preliminary study. *Oikos*, **17**, 62–70.

176. Colenso, B. E., Boyce, M. S. and Tate, J. Jr. (1980). Developing criteria for reclamation of sage grouse habitat on a surface coal mine in northeastern Wyoming. In *Symposium on Surface Mining Hydrology, Sedimentology and Reclamation*, ed. D. H. Graves, pp. 27–31. University of Kentucky.

177. Collins, B. G., ed. (1983). *Mined Land Rehabilitation – Is It Sustainable? The 53rd Anzaas Congress Symposium, Perth, 1983.* WAIT School of Biology Bulletin no. 7.

178. Collins, B. G., Wykes, B. J. and Nichols, O. G. (1986). Re-colonization of restored bauxite minelands by birds in southwestern Australian forests. In *Birds of Eucalypt Forests and Woodlands: Ecology, Conservation, Management*, ed. A. Keast, H. F. Recher, H. Ford and D. Saunders, pp. 341–54. Sydney: Surrey Beatty and Sons.

179. Collins, F. W. (1957). Game management practices on strip-mined land. *South Eastern Association of Game and Fish Commission. Proceedings. Annual Conference*, **10**, 213–21.

180. Collins, F. W. (1958). *Reclaimed Lands and Wildlife Potential.* Department of Conservation, Kentucky, Division of Strip Mine Reclamation Report.

181. Comer, R. D., Merino, J. M., Monarch, J. W., Pustmueller, C., Stalmaster, M., Stoeker, R., Todd, J. and Wright, W., ed. (1982). *Issues and Technology in the Management of Impacted Western Wildlife.* Thorne Ecological Institute Technical Paper no. 14.

182. Coss, R. D. (1981). *Wildlife Habitats Provided by Aquatic Plant Communities of Surface Mine Lakes.* MA Thesis, Southern Illinois University, Cabondale (unpublished manuscript).

183. Crawford, B. T. (1942). *Ecological Succession in a Series of Stripmine Lakes in Central Missouri.* M.A. Degree, University of Missouri (unpublished manuscript).

184. Crawford, H. S., Abler, W. A. and Hardy, D. M. (1982). Nesting of birds on strip-mined

highwalls in southern West Virginia. In *Transactions of the 35th Northeast Fish and Wildlife Conference, White Sulphur Springs, West Virginia*, pp. 87–91.

185. Crawford, H. S., Hardy, D. M. and Abler, W. A. (1978). A survey of bird use of strip-mined areas in southern West Virginia. In *Surface Mining and Fish/Wildlife Needs in the Eastern United States*, ed. D. E. Samuel, J. R. Stauffer, C. H. Hocutt and W. T. Mason, pp. 241–6. Morgantown: West Virginia University.

186. Crawford, R. D. and Rossiter, J. A. (1982). General design considerations in creating artificial wetlands for wildlife. In *The Wildlife Values of Gravel Pits Symposium, Crookston, Minnesota, June 1982*, ed. W. D. Svedarsky and R. D. Crawford, pp. 44–7. University of Minnesota Miscellaneous Publication 17-1982.

187. Crawford, R. L. (1979). Autumn population of spiders and other arthropods in an urban landfill. *Northwest Science*, **53**, 51–3.

188. Crompton, G. and Sheail, J. (1975). The historical ecology of Lakenheath Warren in Suffolk, England: a case study. *Biological Conservation*, **8**, 299–313.

189. Cross, E. A. and Wilman, J. M. (1982). A preliminary study of the refaunation of alkaline shale coal surface mine spoil by soil arthropods. In *1982 Symposium on Surface Mining Hydrology, Sedimentology and Reclamation*, ed. D. H. Graves, pp. 353–8. University of Kentucky.

190. Crowl, J. M. (1957). Recovering striplands in Kentucky. *Coal Age*, **62**, 77–9.

191. Crunkilton, R. L. (1982). An overview of gravel mining in Missouri and fish and wildlife implications. In *The Wildlife Values of Gravel Pits Symposium, Crookston, Minnesota, June 1982*, ed. W. D. Svedarsky and R. D. Crawford, pp. 80–8. University of Minnesota Miscellaneous Publication 17-1982.

192. Cullimore, C. and Johnson, M. S. (1978). *A Survey of the Flora and Fauna on the Reclaimed Tailings Dam of Laporte Industries Limited, Derbyshire*. Department of Botany, University of Liverpool (unpublished manuscript).

193. Cumming, K. B. and Hill, D. M. (1971). *Stream Faunal Recovery after Manganese Strip Mine Reclamation*. (Virginia Polytechnic Institute and State University. Virginia Cooperative Fishery Unit). Environmental Protection Agency Project no. WP-01530.

194. Currie, P. O. (1981). Revegetating mined land for grazing. *Journal of Soil and Water Conservation*, **36**, 213–15.

195. Curry, J. A. and Fowler, D. K. (1978). Coal strip mining in Appalachia – fish and wildlife considerations: an overview. In *Surface Mining and Fish/Wildlife Needs in the Eastern United States*, ed. D. E. Samuel, J. R. Stauffer, C. H. Hocutt and W. T. Mason, pp. 23–8. Washington: US Fish and Wildlife Service.

196. Curry, J. P. and Cotton, D. C. F. (1983). Earthworms and land reclamation. In *Earthworm Ecology: From Darwin to Vermiculture*, ed. J. E. Satchell, pp. 215–28. London: Chapman and Hall.

197. Curtis, R. L., Fowler, D. K., Nicholson, C. H. and Adkisson, L. F. (1978). Breeding bird populations on three contour surface mines reclaimed under differing intensities and types of treatment. In *Surface Mining and Fish/Wildlife Needs in the Eastern United States*, ed. D. E. Samuel, J. R. Stauffer, C. H. Hocutt and W. T. Mason, pp. 369–75. Washington: US Fish and Wildlife Service.

198. Czechowski, W. (1980). The ants *Lasius Niger* (L.) (Hymenoptera, Formicidae) as indicators of the degree of environment pollution in a city. *Przeglad Zoologiczny*, **24**, 113–21.

199. Dames, O. and Moore, R. (1983). *A Survey of Wetland Reclamation Projects in the Florida Phosphate Industry*. Final Report Prepared for Florida Institute of Phosphate Research, Publication no. 03-019-011.

200. Danell, K. and Sjoberg, K. (1982). Successional patterns of plants, invertebrates and ducks in a man-made lake. *Journal of Applied Ecology*, **19**, 395–409.

201. Davidson, W. H. (1970). *Deer Prefer Pine Seedlings Growing Near Black Locust.* Northeast Forest Experimental Station Research Note NE-111.

202. Davieson, G. and Majer, J. D. (1983). Recolonization by invertebrates in rehabilitated mineral sand mines at Capel, south-west Western Australia. *Mulga Research Centre Annual Report 1982*, pp. 89–95. Bentley: WAIT School of Biology.

203. Davis, B. N. K. (1963). A study of micro-arthropod communities in mineral soils near Corby, Northants. *Journal of Animal Ecology*, **32**, 49–71.

204. Davis, B. N. K. (1979). Chalk and limestone quarries as wildlife habitats. *Minerals and the Environment*, **1**, 48–56.

205. Davis, B. N. K. (1986). Colonization of newly created habitats by plants and animals. *Journal of Environmental Management*, **22**, 361–71.

206. Davis, B. N. K., Lakhani, K. H., Brown, M. C. and Park, D. G. (1985). Early seral communities in a limestone quarry: an experimental study of treatment effects on cover and richness of vegetation. *Journal of Applied Ecology*, **22**, 473–90.

207. Davis, B. N. K. and Murphy, P. W. (1961). An analysis of the Acarina and Collembola fauna of land reclaimed from opencast iron-stone mining. *University of Nottingham, School of Agriculture Report*, pp. 85–9.

208. Davis, G. and Davidson, W. H. (1968). Coal-mine spoil banks offer good potential for timber and wildlife production. *Pennsylvania Forests*, **58**, 20–1.

209. Davison, V. E. (1954). *Lespedezas for Quail and Good Land Use.* US Department of Agriculture Leaflet no. 373.

210. Dell, B., ed. (1984). *Proceedings of the 4th International Conference On Mediterranean Ecosystems Held at Perth, Western Australia.* Perth: Botany Department, University of Western Australia.

211. Department of National Development. (1978). *Biological Release and Recycling of Toxic Metals from Lake and River Sediments.* Australian Water Resources Council Technical Paper no. 33. Canberra: AGPS.

212. Depuit, E. J. (1983). Potential feasibility and utility of complementary grazing systems on reclaimed lands in the Northern Great Plains. *Bulletin of the Ecological Society of America*, **64**, 88.

213. Depuit, E. J. (1984). Potential topsoiling strategies for enhancement of vegetation diversity on mined lands. *Minerals and the Environment*, **6**, 115–20.

214. Depuit, E. J. and Coenenberg, J. G. (1978). Plant response and forage quality for livestock grazing on coal mine spoils in Montana. In *Ecology and Coal Resource Development*, vol. 2, ed. M. K. Wali, pp. 892–8. New York: Pergamon Press.

215. Depuit, E. J. and Coenenberg, J. G. (1979). *Responses of Revegetated Coal Strip Mine Spoils to Variable Fertilization Rates, Longevity of Fertilization Program and Season of Seeding.* Montana Agricultural Experiment Station Research Report 150. Montana State University.

216. Depuit, E. J. and Coenenberg, J. G. (1980). *Plant Response and Forage Quality for Controlled Grazing on Coal Mine Soil Pastures: Project Final Report.* Montana Agricultural Experiment Station Report to US Environmental Protection Agency, Cincinnati, Ohio.

217. Depuit, E. J., Coenenberg, J. G. and Skilbred, C. L. (1980). *Establishment of Diverse Native Plant Communities on Coal Surface-Mined Lands in Montana as Influenced by Seeding Method, Mixture and Rate.* Montana Agricultural Experiment Station Research Report no. 163. Montana State University.

218. Depuit, E. J., Skilbred, C. L. and Coenenberg, J. G. (1982). Effects of two years of irrigation on revegetation of coal surface-mined land in southeastern Montana. *Journal of Range Management*, **35**, 67–73.

219. Depuit, E. J., Willmuth, W. H. and Coenenberg, J. G. (1977). *Plant Response and Forage Quality for Controlled Grazing on Coal Mine Spoils Pastures: 1976 Progress Report.* Montana

Agricultural Experimental Station, Reclamation Research Unit Report no. 115. Montana State University.

220. Dick, J. H. and Thirgood, J. V. (1976). Development of land reclamation in British-Columbia. In *Practices and Problems of Land Reclamation in Western North America*, ed. M. K. Wali, pp. 65–78. Grand Forks: University of North Dakota Press.

221. Dickson, K. L. and Vance, D. (1981). *Revegetating Surface Mined Lands for Wildlife in Texas and Oklahoma*. US Fish and Wildlife Service Report.

222. Dierks, E. (1961). Strip-mine becomes a wildlife haven. *Soil Conservation*, **27**, 66.

223. Dindal, D. L., ed. (1980). *Soil Biology as Related to Land Use Practices: Proceedings of The VII International Colloquium of Soil Zoology*. Washington: Environmental Protection Agency.

224. Dindall, D. L. and Wray, C. C. (1977). Community structure and role of macroinvertebrate decomposers in the rehabilitation of a limestone quarry. In *Limestone Quarries: Response to Land Use Pressures*, ed. E. J. Perry and N. A. Richards, pp. 72–99. Syracuse, New York: Sunny College of Environmental Science and Forestry.

225. Ditsworth, T. M., Butt, S. M., Beley, J. R., Johnson, C. D. and Balda, R. P. (1982). Arthropods, plants and transmission lines in Arizona: community dynamics during secondary succession in a pinyon-juniper woodland. *Southwestern Naturalist*, **27**, 167–81.

226. Dixon, J. M. and Hambler, D. J. (1984). An experimental approach to the reclamation of a limestone quarry floor: the first three years. *Environmental Conservation*, **11**, 19–28.

227. Dorsch, H. and Dorsch, I. (1968). Avifaunistische Untersuchungen im Braunkohlentagebau Kulkwitz I. *Mitteilung der Interessengemeinschaft Avifauna DDR der biologischen Gesellschaft in der DDR*, **1**, 51–79.

228. Dorsch, H. and Dorsch, I. (1969). Avifaunistische Untersuchungen im Braunkohlentagebau Kulkwitz II Mitteilung. *Der Falke*, **9**, 296–302.

229. Doudoroff, P. and Katz, M. (1950). Critical review of literature on the toxicity of industrial wastes and their components to fish. I Alkalies, acids, and inorganic gases. *Sewage and Industrial Wastes*, **22**, 1432–58.

230. Doudoroff, P. and Katz, M. (1953). Critical review of literature on the toxicity of industrial wastes and their components to fish. II The metals, as salts. *Sewage and Industrial Wastes*, **25**, 802–39.

231. Doxtater, G. D. (1985). Potential of future mine-cut lakes. In *Fish and Wildlife Relationships to Mining: Symposium Proceedings*, ed. L. B. Starnes, pp. 1–6. American Fisheries Society (Water Quality Section).

232. Duebbert, H. F., Jacobson, E. T., Higgins, K. F. and Podell, E. P. (1981). *Establishment of Seeded Grasslands for Wildlife Habitat in the Prairie Pothole Region*. US Department of the Interior, Fish and Wildlife Service, Special Scientific Report, Wildlife no. 234.

233. Dugan, P. R. (1975). Bacterial ecology of strip mine areas and its relationship to the production of acidic mine drainage. *Ohio Journal of Science*, **75**, 266–79.

234. Dunger, W. (1964*a*). Bericht über die Tagung der Arbeitsgruppe Bodenzoologie der biologischen Gesellschaft in der DDR. *Pedobiologia*, **4**, 123–7.

235. Dunger, W. (1964*b*). Erste Ergebnisse bodenzoologischer Untersuchungen auf einer Braunkohlenkippe bei Bohlen. *Pedobiologia*, **4**, 124–5.

236. Dunger, W. (1965). Studien über die faunistische Neubesiedlung von Rohbodenkippen des Braunkohlentagebaues. *Publikace Slezskeho Muzea V Opave*, **14**, 33–43.

237. Dunger, W. (1967*a*). Die Entwicklung der Makro- und Megafauna in rekultivierten Haldenboden. In *Progress in Soil Biology*, ed. O. Graff and J. E. Satchell, pp. 340–52. Amsterdam: North-Holland Publishing.

238. Dunger, W. (1967*b*). Studien zue tierischen Besiedelung aufgeforsteter Kippenboden. *Wissenschaftliche Zeitschrift der Technischen Universitaet (Dresden)*, **16**, 591–2.

239. Dunger, W. (1968*a*). Die Entwicklung der Bodenfauna auf rekultivierten Kippen und

Halden des Braunkohlentagebaues. Ein Beitrag zur pedozoologischen Standortsdiagnose. *Abhandlungen und Berichte des Naturkundemuseums Gorlitz*, **43**, 1–256.

240. Dunger, W. (1968*b*). Ein Vergleich der Zellulolytischen und der bodenzoologischen Aktivitat einiger Boden. *Tagungsberichte. Deutsche Akademie der Landwirtschaftswissenschaften zu Berlin*, **98**, 183–7.

241. Dunger, W. (1968*c*). Produktionsbiologische Untersuchungen an der Collembolen-Fauna gestörter Böden. *Pedobiologia*, **8**, 16–22.

242. Dunger, W. (1969*a*). Fragen der Naturlichen und experimentellen Besiedlung kulturfeindlicher Böden durch Lumbriciden. *Pedobiologia*, **9**, 146–51.

243. Dunger, W. (1969*b*). Über den Anteil der Arthropoden an der Umsetzung des Bestandesabfalles in Anfangs-bodenbildungen. *Pedobiologia*, **9**, 366–71.

244. Dunger, W. (1974*a*). Ekogeneza i zasiedlanie przykopalnianych hald wegla brunatnego przez glebowe organismy zwierzece (Okogenese und bodenzootische Besiedlung auf Halden des Braunkohlenbergbaues; Polnisch). *Sprawozdania Wroclawskiego Towarzystwa Naukowego*, **27**, 80–4.

245. Dunger, W. (1974*b*). Anthropogene Okogenesen als pedozoologische untersuchungs Objekte. *Transactions. 10th International Congress of Soil Science, Moscow* **3**, 87–92.

246. Dunger, W. (1975). Sukzession von Pflanzen- und Tiegeme inschaften auf bergbaulich genutzten Flächen. *Mitteilungen der Sektion Schulbiologie der Biologischen Gesellschaft der DDR*, **16**, 25–32.

247. Dunger, W. (1979). Bodenzoologische Untersuchungen an rekultivierten Kippboden der Nierderlausitz. *Abhandlungen und Berichte des Naturkundemuseums Gorlitz*, **52**, 1–19.

248. Dunger, W. (1982). Die Tiere des Bodens als Leitformen für anthropogene Umweltveränderungen. *Decheniana-Beihefte (Bonn)*, **26**, 151–7.

249. Dunger, W. (1987). Zur Einwirkung von Kahlschlag, Grundwasserabsenkung und fortlicher Haldenrekultivierung auf die Boden-Makrofauna, insbesondere Regenwürmer. *Abhandlungen und Berichte des Naturkundemuseums Gorlitz*, **60**, 29–42.

250. Dunger, W. and Dunger, J. (1984). Zur Kongruenz von Phytozonosen und Collembolan-Synusien. In *Verhandlungen Sieed*, ed. Z. Kaszab, pp. 32–4. Budapest.

251. Eagles, P. F. J. (1982). Gravel pit reclamation in Ontario: the Crieff case study. In *The Wildlife Values of Gravel Pits Symposium, Crookston, Minnesota, June 1982*, ed. W. D. Svedarsky and R. D. Crawford, pp. 141–7. University of Minnesota Miscillaneous Publication 17–1982.

252. Ecology Consultants Inc., ed. (1977). *Reclamation for Wildlife Habitat. Proceedings of Reclamation Workshop II, Fort Collins, Colorado, September 1977.* Environmental Research and Technology Inc.

253. Edwards, C. A. (1979). Soil biology and land use. *Nature*, **281**, 339–40.

254. Edwards, H. K. (1982). The chain of lakes project along interstate 80 in Nebraska. In *The Wildlife Values of Gravel Pits Symposium, Crookston, Minnesota, June 1982*, ed. W. D. Svedarsky and R. D. Crawford, pp. 211–14. University of Minnesota Miscillaneous Publication 17-1982.

255. Edwards. J. H. (1947). Unleveled spoil bring good return as farming project. *Coal Age*, **52**, 92–7.

256. Eerden, M. R. van (1984). Waterfowl movements in relation to food stocks. In *Coastal Waders and Wildfowl in Winter*, ed. P. R. Evans, pp. 84–100. (Overdruk no. 335).

257. Eerden, M. R. van (1985). Aspects of development of bird populations. In *The Lauwerszee Fifteen Years Closed-Off. Results of Research and Experience with Development and Management*, ed. P. J. Ente and R. J. de Glopper, pp. 145–72. Rijksdienst voor de Ijsselmeerpolders. Flevobericht no. 247.

258. Eijsackers, H. J. P. (1983). Development of earthworm populations in abandoned arable

fields under grazing management. In *Earthworm Ecology: From Darwin to Vermiculture*, ed. J. E. Satchell, pp. 241–6. London: Chapman and Hall.

259. El-Kifl, A. H., Wahab, A. E. A. and Metwally, A. M. (1974). Some insects in a newly reclaimed area in Nasr city. *Bulletin de la Société Entomologique D'Égypte*, **58**, 227–42.

260. Elkins, N. Z., Parker, L. W., Aldon, E. and Whitford, W. G. (1984). Responses of soil biota to organic amendments in stripmine spoils in northwestern New Mexico. *Journal of Environmental Quality*, **13**, 215–19.

261. Ellers, R. G. (1971). Biologist brings strip mine back to life. *The Plain Dealer*, November 10, 1971. (Ohio Reclamation Association).

262. Elliott, C. L. and McKendrick, J. D. (1982). Strip mine reclamation and wildlife in Alaska. *Agroborealis*, **14**, 4–6.

263. Emanuelsson, U. and Bergendorff, C. (1986). History as a guideline to nature conservation and urban park management in Scania, southern Sweden. In *Ecology and Design in Landscape*, ed. A. D. Bradshaw, D. A. Goode and E. H. P. Thorpe, pp. 237–55. Oxford: Blackwell.

264. Emerick, N. R. (1985). Nesting islands for giant Canada geese on west-central Illinois strip-mine lands. In *Wetlands and Water Management on Mined Lands. Proceedings of a Conference October 23–24, 1985*, ed. R. P. Brooks, D. E. Samuel and J. B. Hill, pp. 381. Pennsylvania State University.

265. Eng, R. L., Pitcher, E. J., Scott, S. J. and Greene, R. J. (1979). Minimizing the effect of surface coal mining on a sage grouse population by a directed shift of breeding activities. In *The Mitigation Symposium: A National Workshop on Mitigating Losses of Fish and Wildlife Habitats*, ed. G. A. Swanson, pp. 464–8. Rocky Mountain Forest Range Experiment Station (Forest Service) General Technical Report.

266. Ente, P. J. (1985). Soil reclamation with special emphasis on polders. In *Soil Ecology and Management*, ed. J. H. Cooley, pp. 81–8. Athens, Georgia: International Association for Ecology. Institute of Ecology, University of Georgia.

267. Erdman, J. A., Ebens, R. J. and Case, A. A. (1978). Molybdenosis: a potential problem in ruminants grazing on coal mine spoils. *Journal of Range Management*, **31**, 34–6.

268. Etter, H. M. (1973). Mined-land reclamation studies on Bighorn sheep range in Alberta, Canada. *Biological Conservation*, **5**, 191–5.

269. Ettershank, G., Elkins, N. Z., Santos, P. F., Whitford, W. G. and Aldon, E. F. (1978). *The Use of Termites and Other Soil Fauna to Develop Soils on Strip Mine Spoils.* US Department of Agriculture, Forest Service Research Note.

270. Ettinger, W. and Yuill, C. (1982). Sand and gravel pit reclamation in Louisiana: creation of wetlands habitats and its integration into adjacent undisturbed bayou. In *The Wildlife Values of Gravel Pits Symposium, Crookston, Minnesota, June 1982*, ed. W. D. Svedarsky and R. D. Crawford. University of Minnesota Miscillaneous Publication 17-1982.

271. Falkenberg, K. (1984). Reclaiming western mines for wildlife. *Coal Mining and Processing*, **21**, 59–62.

272. Farnell, G. (1979*a*). *Notes Compiled on the Study of Termites Undertaken by Leigh Miller*. CSIRO Division of Forest Research, Darwin (unpublished manuscript).

273. Farnell, G. (1979*b*). Rehabilitation of mined land on Groote Eylandt, N.T. In *Mining Rehabilitation*, ed. I. Hore-Lacy, pp. 33–40. Canberra: Australian Mining Industry Council.

274. Farnell, G. (1981). The Groote Eylandt Mining Company rehabilitation programme. In *North Australian Mine Rehabilitation Workshop, Gove N.T.*, ed. D. A. Hinz, pp. 57–68. NABALCO.

275. Farnham, R. S. (1979). Peatland reclamation. In *IGT Peat as an Energy Resource Symposium, Arlington, July 1979*, pp. 205–7.

276. Fawcett, F. (1983). Local authority planning for wildlife in gravel quarries. In *The Proceedings of a Symposium on the Subject of Wildlife on Man-Made Wetlands*, pp. 176–85. Milton Keynes, England: A.R.C. Wildlife Centre.

277. Feekes, W. (1936). *De Ontwikkeling van de Natuurlijke Vegetatie in de Wieringermeer Polder, de Eerste Groote Droogmakerij van de Zuiderzee*. Ph.D. Thesis, University of Wageningen.

278. Feierabend, J. S. (1983). *Mourning Doves and Reclaimed Surface Mines: Management and Hunting Potential*. MS Thesis, West Virginia University (unpublished manuscript).

279. Feriancova-Masarova, Z. and Ferianc, O. (1979). Further knowledge derived from the succession of bird settlements on the recently built water tank Liptovska Mara in 1978. *Biologia*, **34**, 655–61.

280. Fitzgerald, P. R. (1979). Recovery and utilization of strip-mined land by application of anaerobically digested sludge and livestock grazing. In *Utilization of Municipal Sewage Effluent and Sludge on Forest and Disturbed Land. Proceedings of Symposium, Philadelphia, Pennsylvania, March 21–23, 1977*, ed. W. E. Sopper and S. N. Kerr, pp. 497–506. Pennsylvania State University Press.

281. Flick, W. C. (1977). *Small Mammal Use of Abandoned Surface Mines in Southwestern Virginia*. M.Sc. Thesis, Virginia Polytechnic Institute and State University (unpublished manuscript).

282. Foresman, F. J. (1955). Strip grazing for prize Angus. *Coal Age*, **60**, 74–7.

283. Forren, J. D. (1981). *Artificial Perch Use by Raptors on Reclaimed Surface Mines in West Virginia*. M.Sc. Thesis, West Virginia University (unpublished manuscript).

284. Forren, J. D. and Samuel, D. (1982). Sparrow hawks, perches, and reclaimed surface mines. *Green Lands*, **12**, 26–7.

285. Fowler, D. K. and Adkisson, L. F. (1980). *Survival and Growth of Wildlife Shrubs and Trees on Acid Mine Spoil*. NTIS Report NP-25246.

286. Fowler, D. K. and Fowler, L. J. (1981). *Rapid Restoration of Biological Productivity to Coal Surface Mines: Annual Biological Monitoring Report*. Prepared by the Tennessee Valley Authority for the US Fish and Wildlife Service.

287. Fowler, D. K. and Hammer, D. A. (1976). Techniques for establishing vegetation on reservoir inundation zones. *Journal of Soil and Water Conservation*, **31**, 116–18.

288. Fowler, D. K., Hill, D. M. and Fowler, L. J. (1985). Colonization of coal surface mine sediment ponds in southern Appalachia by aquatic organisms and breeding amphibians. In *Wetlands and Water Management on Mined Lands. Proceedings of a Conference October 23–24, 1985*, ed. R. P. Brooks, D. E. Samuel and J. B. Hill, pp. 261–85. Pennsylvania State University.

289. Fowler, D. K. and Maddox, J. B. (1974). Habitat improvement along reservoir inundation zones by barge hydroseeding. *Journal of Soil and Water Conservation*, **29**, 263–5.

290. Fowler, D. K. and Peery, III, C. H. (1973). Three years development of a public use wildlife area on a mountain coal surface mine in southwest Virginia. In *Resource and Applied Technology Symposium on Mined-Land Reclamation Proceedings*, pp. 319–33. Monroeville, Pennsylvania: Bituminous Coal Resources Inc.

291. Fowler, D. K. and Turner, L. J. (1981). *A Contour Surface Mine Reclamation Plan Featuring Fish and Wildlife*. US Fish and Wildlife Service Report.

292. Fowler, D. K., Turner, L. J. and Slaski, L. J. (1981). *Surface Mine Reclamation for Wildlife: A Model Reclamation Plan for Southern Appalachia*. Fish and Wildlife Service, US Department of the Interior. Biological Services Program.

293. Fowler, L. J. and Fowler, D. K. (1982). *Rapid Restoration of Biological Productivity to Coal Surface Mines: Second Annual Biological Monitoring Report: Aquatics Section*. Prepared by the Tennessee Valley Authority for the US Fish and Wildlife Service.

294. Fowler, L. J., Fowler, D. K. and Allsbrooks, D. W. (1981). *Rapid Restoration of Biological Productivity to Coal Surface Mines: Second Annual Biological Monitoring Report*. Prepared by the Tennessee Valley Authority for the US Fish and Wildlife Service.

295. Fowler, L. J., Fowler, D. K. and Thomas, J. E. (1982). Dispersal of autumn olive seeds by foxes on coal surface mines in east Tennessee. *Journal of the Tennessee Academy of Science*, **57**, 83–5.
296. Fox, B. J. (1982). The influence of disturbance (fire, mining) on ant and small mammal species diversity in Australian heathland. In *Proceedings of a Conference on Dynamics and Management of Mediterranean Type Ecosystems*. Riverside, California: USDA Forest Service.
297. Fox, B. J. and Fox, M. D. (1978). Recolonization of coastal heath by *Pseudomys novaehollandiae* (Muridae) following sand mining. *Australian Journal of Ecology*, **3**, 447–65.
298. Fox, B. J. and Fox, M. D. (1984). Small-mammal recolonization of open-forest following sand mining. *Australian Journal of Ecology*, **9**, 241–52.
299. Fox, B. J. and Fox, M. D. (1986). Resilience of animal and plant communities to human disturbance. In *Resilience of Mediterranean-Type Ecosystems*, ed. B. Dell, A. J. M. Hopkins and B. B. Lamont, pp. 39–64. Dordrecht: Junk.
300. Fox, B. J. and Pople, A. R. (1984). Experimental confirmation of interspecific competition between native and introduced mice. *Australian Journal of Ecology*, **9**, 323–34.
301. Fox, J. E. D., ed. (1978). *Rehabilitation of Mined Lands in Western Australia*. Bentley, W.A.: WAIT.
302. Fox, J. E. D. (1984). Rehabilitation of mined lands. *Forestry Abstracts*, **45**, 565–600.
303. Fox, M. D. and Fox, B. J. (1982). Evidence for interspecific competition influencing ant species diversity in a regenerating heathland. In *Ant–Plant Interactions in Australia*, ed. R. C. Buckley, pp. 99–110. The Hague: Junk.
304. Frischknecht, N. C. (1977). Use of shrubs for mined land reclamation and wildlife habitat. In *Reclamation for Wildlife Habitat. Proceedings of Reclamation Workshop II, Fort Collins, Colorado, September 1977*, ed. Ecology Consultants Inc., pp. 113–29. Environmental Research and Technology Inc.
305. Frohlich, R. K. and Marion, W. R. (1984). Mammal populations on phosphate-mined and natural areas of north Florida. *Journal of Wildlife Management*, **48**, 1398–402.
306. Funk, D. T. (1973). Growth and development of alder plantings on Ohio strip-mine banks (discussion at end of paper). In *Ecology and Reclamation of Devastated Lands*, vol. 2, ed. R. J. Hutnik and G. Davis, pp. 483–91. New York: Gordon and Breach.
307. Gack, C. and Kobel-Lamparski, A. (1983). Die Pillenkäfer alten und neuen Rebgelandes im Kaiserstuhl (südbaden). *Veroeffentlichungen Naturschutz Landschaftspflege*, 57/58, 325–40.
308. Gack, C. and Kobel-Lamparski, A. (1985). Wiederbesiedlung und Sukzession auf neuen Rebboschungen im Kaiserstuhl am Beispiel epigaischer Spinnen. *Verhandlungen der Gesellschaft Okologie* (in press).
309. Gallagher, J. L. (1980). Salt marsh soil development. In *Rehabilitation and Creation of Selected Coastal Habitats. Proceedings of a Workshop, Sapelo Island, Georgia, May 1976*, ed. J. C. Lewis and E. W. Bunce, pp. 28–34. Washington, DC: US Fish and Wildlife Service, Biological Services Program.
310. Gallagher, T. J. (1982). Eugene's urban wildlife area: Delta Ponds. In *The Wildlife Values of Gravel Pits Symposium, Crookston, Minnesota, June 1982*, ed. W. D. Svedarsky and R. D. Crawford, pp. 122–6. University of Minnesota Miscellaneous Publication 17–1982.
311. Galvin, M., Hoover, K. D. and Avery, M. L. (1979*a*). *Management of Transmission Line Rights-Of-Way for Fish and Wildlife*. US Fish and Wildlife Service, Biological Services Program Report, vol. 1.
312. Galvin, M., Hoover, K. D. and Avery, M. L. (1979*b*). *Management of Transmission Line Rights-Of-Way for Fish and Wildlife*. US Fish and Wildlife Service, Biological Services Program Report, vol. 2.
313. Galvin, M., Hoover, K. D. and Avery, M. L. (1979*c*). *Management of Transmission Line*

Rights-Of-Way for Fish and Wildlife. US Fish and Wildlife Service, Biological Services Program Report, vol. 3.

314. Garfitt, W. (1983). Conversion, development and management of a small worked-out gravel pit for sport, fish production and wildlife. In *The Proceedings of a Symposium on the Subject of Wildlife on Man-Made Wetlands*, pp. 151–63. Milton Keynes, England: ARC Wildlife Centre.
315. Garr, J. J. (1981). Extraction, reclamation and after use at gravel quarries near Canterbury, Kent. *Minerals and the Environment*, **2**, 124–43.
316. Gawn, P. E. (1983). Landscape planning for wildlife in conjunction with recreational activity and other land uses in gravel quarries. In *The Proceedings of a Symposium on the Subject of Wildlife on Man-Made Wetlands*, pp. 164–75. Milton Keynes, England: ARC Wildlife Centre.
317. Gemmell, R. P. and Connell, R. K. (1984). Conservation and creation of wildlife habitats on industrial land in Greater Manchester. *Landscape Planning*, **11**, 175–86.
318. Ghabbour, S. I. and Shakir, S. H. (1982). Population density and biomass of earthworms in agro-ecosystems of the Mariut coastal desert region in Egypt. *Pedobiologia*, **23**, 189–98.
319. Ghilarov, M. S. and Perel, T. S. (1984). Transfer of earthworms (Lumbricidae, Oligochaeta) for soil amelioration in the USSR. *Pedobiologia*, **27**, 107–13.
320. Gilbert, T., King, T. and Barnett, B. (1981). *An Assessment of Wetland Habitat Establishment at a Central Florida Phosphate Mine Site*. US Fish and Wildlife Service Report.
321. Giles, R. H., Jr. (1981). *Endangered Species: An Aid to Land-Use Planning in the Coal Producing Countries of Southwestern Virginia*. Department of Fisheries and Wildlife Sciences Occasional Paper.
322. Giles, R. H., Jr. (1982). Mitigation and management of damaged ecosystems or damaged habitat: overview. In *Research on Fish and Wildlife Habitat*, ed. W. T. Mason Jr, pp. 165–9. US Environmental Protection Agency Report EPA-600/8-82-022.
323. Giller, F. (1965). Avifaunistische Untersuchungen im links-rheinischen Braunkohlengebiet. *Bonner Zoologische Beitraege*, 1/2, 36–49.
324. Gladis, T., Dunger, W., Joost, W., Klausnitzer, B. and Richter, K. (1987). Faunistisch-ökologische Untersuchungen auf dem neuen Müllberg Leipzig-Möckern, S. Beitrag: Collembola. *Entomologische Nachrichten* (Dresden).
325. Goode, D. A. and Smart, P. J. (1986). Designing for wildlife. In *Ecology and Design in Landscape*, ed. A. D. Bradshaw, D. A. Goode and E. H. P. Thorpe, pp. 219–35. Oxford: Blackwell.
326. Goodman, G. T., Edwards, R. W. and Lambert, J. M., ed. (1965). *Ecology and the Industrial Society*. Oxford: Blackwell.
327. Gore, J. A. (1979). Patterns of initial benthic recolonization of a reclaimed coal strip-mined river channel. *Canadian Journal of Zoology*, **57**, 2429–39.
328. Gore, J. A. (1982). Benthic invertebrate colonization: source distance effects on community composition. *Hydrobiologia*, **94**, 183–93.
329. Gore, J. A., ed. (1985). *The Restoration of Rivers and Streams*. Boston: Butterworth.
330. Gore, J. A. and Johnson, L. S. (1979). Biotic recovery of a reclaimed river channel after coal strip mining. In *The Mitigation Symposium: A National Workshop on Mitigating Losses of Fish and Wildlife Habitats*, ed. G. A. Swanson, pp. 239–44. Rocky Mountain Forest and Range Experiment Station (Forest Service) General Technical Report.
331. Gore, J. A. and Swartley, W. A. (1985). Distribution of mayfly nymphs in relation to water quality of streams draining coal surface-mined areas on the Cumberland Plateau. In *Fish and Wildlife Relationships to Mining: Symposium Proceedings*, ed. L. B. Starnes, pp. 59–73. American Fisheries Society (Water Quality Section).

332. Green, B. H. (1986). Controlling ecosystems for amenity. In *Ecology and Design in Landscape*, ed. A. D. Bradshaw, D. A. Goode and E. H. P. Thorpe, pp. 195–209. Oxford: Blackwell.

333. Greenslade, P. and Majer, J. D. (1980). Collembola of rehabilitated mine sites in Western Australia. In *Soil Biology as Related to Land Use Practices*, ed. D. L. Dindal, pp. 397–408. Washington: Environmental Protection Agency.

334. Gregory, R. W., Elser, A. A. and Lenhart, T. (1984). *Utilization of Surface Coal Mine Waste for Construction of a Northern Pike Spawning/Rearing Marsh.* US Fish and Wildlife Service Report.

335. Grigg, N. S. (1976). *Precipitation Management for Reclamation of Overgrazed Areas in Arid and Semi-Arid Regions. Completion Report.* Colorado State University, Department of Civil Engineering, Fort Collins, Colorado. OWRT Project no. A-026-COLO.

336. Grubb, P. J. (1986). The ecology of establishment. In *Ecology and Design in Landscape*, ed. A. D. Bradshaw, D. A. Goode and E. H. P. Thorpe, pp. 83–97. Oxford: Blackwell.

337. Gunness, A. (1984). Vertebrate fauna survey in regenerate mines at Weipa. *Landline*, **10**, 1–2.

338. Guttmann, R. (1979). *Untersuchungen zur Entwicklung der Bodenfauna rekultivierter Schutthalden eines Stahlwerkes.* Dissertation Braunschweig.

339. Gwynn, T. (1965). *Reclaiming Strip-Mined Land in North Dakota by Establishing Game Management Areas.* Bismark, North Dakota: Knife River Coal Company.

340. Gwynn, T. (1966). *Reclaiming Strip-Mined Land by Establishing Game Management Areas.* Bismark, North Dakota: Knife River Coal Company.

341. Hackett, B. (1971). *Landscape Planning: An Introduction to Theory and Practice.* Newcastle upon Tyne: Oriel Press.

342. Hackett, B., ed. (1977). *Landscape Reclamation Practice.* Guildford, England: IPC Science and Technology Press.

343. Haeck, J. (1969). Colonization of the mole (*Talpa Europaea* L.) in the Ijsselmeer polders. *Netherlands Journal of Zoology*, **19**, 145–248.

344. Haeck, J. (1971). The immigration and settlement of carabids in the new Ijsselmeer-polders. *Miscellaneous Papers Landbouwhogeschool Wageningen*, **8**, 33–52.

345. Haeck, J., Hengeveld, R. and Turin, H. (1980). Colonization of road verges in three Dutch polders by plants and ground beetles (Coleoptera: Carabidae). *Entomologia Generalis*, **6**, 201–15.

346. Haering, R. and Fox, B. J. (1987). Short-term coexistence and longterm competitive displacement of two dominant species of *Iridomyrmex*: the successional response of ants to regenerating habitats. *Journal of Animal Ecology*, **56**, 495–508.

347. Haeseler, V. V. (1972). Man-made habitats (deforested area, gravel-pit, city gardens and parks) as refuges for insects, exemplified by the Hymenoptera Aculeata. *Zooloogische Jahrbüecher*, **99**, 133–212.

348. Hagvar, S. and Abrahamsen, G. (1980). Colonisation by Enchytraediae, Collembola and Acari in sterile soil samples with adjusted pH levels. *Oikos*, **34**, 245–58.

349. Hamilton, W. E. and Vimmerstedt, J. P. (1980). Earthworms on forested spoil banks. In *Soil Biology as Related to Land Use Practices*, ed. D. L. Dindal, pp. 409–17. Washington: Environmental Protection Agency.

350. Hampton, E. L., Pennington, W. L., Lackey, J. L., North, J. C. and McCoy, V. W. (1979). Stream biological surveys – self-defence for coal mine operators. In *Proceedings. Symposium on Surface Mining Hydrology, Sedimentology and Reclamation*, ed. S. D. Carpenter and R. W. de Vore, pp. 251–3. Lexington: University of Kentucky.

351. Hannan, J. C. (1984). Mine rehabilitation. In *A Handbook for the Coal Mining Industry*, pp. 87–8. New South Wales Coal Association.

352. Hansen, L. P. and Warnock, J. E. (1978). Response of two species of *Peromyscus* to vegetational succession on land strip-mined for coal. *The American Midland Naturalist*, **100**, 416–23.

353. Harman, D. M., Tyne, M. A. van and Thompson, W. A. (1985). Comparison of locust borer *Megacyllene robiniae* Forster (Coleoptera: Cerambycidae) attacks on coal strip-mined lands and lands not mined. *Annals of the Entomological Society of America*, **78**, 50–3.

354. Harrison, J. (1972). Creating a wetland habitat. *Bird Study*, **17**, 111–22.

355. Harrison, J. (1974). *The Sevenoaks Gravel Pit Reserve.* Chester: W.A.G.B.I.

356. Harrison, J. G. (1976). *Wetlands for Waterfowl – An Example of the Management of a Man-Made Wetland.* Council of Europe Information Centre for Nature Conservation Report.

357. Hartmut, J. von (1972). Establishing a water biotope. Reclamation with biological-ecological objectives along the Elbe canal. *Landschaft und Stadt*, **4**, 168–76.

358. Haskins, M. F. and Shaddy, J. H. (1986). The ecological effect of burning, mowing, and plowing on ground-inhabiting spiders (Arancae) in an old-field ecosystem. *Journal of Arachnology*, **14**, 1–13.

359. Haufler, J. B., Downing, R. L. and McGinnes, B. S. (1978). Factors influencing the revegetation success of orphan mines in southwest Virginia. In *Surface Mining and Fish/Wildlife Needs in the Eastern United States*, ed. D. E. Samuel, J. R. Stauffer, C. H. Hocutt and W. T. Mason, pp. 287–93. Morgantown: West Virginia University.

360. Hawkes, C. L. (1978). Aquatic habitat of coal and bentonite clay strip mine ponds in the Northern Great Plains. In *Ecology and Coal Resource Development*, vol. 2, ed. M. K. Wali, pp. 609–14. New York: Pergamon Press.

361. Hawkins, B. A. and Cross, E. A. (1982). Patterns of refaunation of reclaimed strip mine spoils by nonterricolous arthropods. *Environmental Entomology*, **11**, 762–75.

362. Hayes, L. A. (1981). *Changes in Habitat Diversity Associated with Recent Surface Mine Reclamation Practices in Illinois.* MA Thesis, Southern Illinois University.

363. Hayes, L. A., Nawrot, J. R. and Klimstra, W. D. (1984). Habitat diversity change associated with reclamation in Illinois. In *The National Symposium on Surface Mining Hydrology, Sedimentology, and Reclamation, Lexington, December 1984*, ed. H. Graves, pp. 363–8. Lexington: University of Kentucky.

364. Haynes, R. J. (1978). The surface mining control and reclamation act of 1977 and potential impacts on fish and wildlife. *Proceedings. Annual Conference of the Southeastern Association of Fish and Wildlife Agencies*, **32**, 790–8.

365. Haynes, R. J. (1984). Summary of wetlands reestablishment on surface-mined lands in Florida. In *Proceedings 1984 Symposium on Surface Mining, Hydrology, Sedimentology, and Reclamation, Lexington, December 2–7, 1984*, ed. H. Graves, pp. 357–62. Lexington: University of Kentucky.

366. Haynes, R. J. (1986). Surface mining and wetland reclamation strategies. In *New Horizons for Mined Land Reclamation.* Proceedings. American Society for Surface Mining and Reclamation, 1986 National Meeting, Jackson, Mississippi, pp. 209–13.

367. Haynes, R. J. and Klimstra, W. D. (1975). *Illinois Land Surface Mined for Coal.* Southern Illinois University, Cooperative Wildlife Research Laboratory Report.

368. Haynes, R. J. and Klopatek, J. M. (1979). Reclamation of abandoned mine lands and fish and wildlife mitigation needs. In *The Mitigation Symposium: A National Workshop on Mitigating Losses of Fish and Wildlife Habitats*, ed. G. A. Swanson, pp. 256–63. Rocky Mountain Forest Range Experiment Station (Forest Service) General Technical Report.

369. Heatwole, H. and Levins, R. (1972). Trophic structure stability and faunal change during recolonization. *Ecology*, **53**, 531–4.

370. Hemler, D. and Samuel, D. E. (1983). Aspen propagation on abandoned surface mines and use by white-tailed deer. In *Proceedings of the Third Annual Better Reclamation with Trees Conference*, pp. 48–54.

371. Henke, H. and Sukopp, H. (1986). A natural approach to cities. In *Ecology and Design in Landscape*, ed. A. D. Bradshaw, D. A. Goode and E. H. P. Thorpe, pp. 307–24. Oxford: Blackwell.

372. Hermosilla, W. (1976). Beobachtungen an der Bodenfauna von rekultivierten Boden im Braunkohlentagebaugebiet der ville. *Decheniana (Bonn)*, **129**, 73–5.

373. Herricks, E. E. (1980). *A Review of Technologies Available for Surface Mine Reclamation for Fish and Wildlife*. US Fish and Wildlife Service, Eastern Energy and Land Use Report.

374. Herricks, E. E., Krzysik, A. J., Szafoni, R. E. and Tazik, D. J. (1982). *Best Current Practices for Fish and Wildlife on Surface-Mined Lands in the Eastern Interior Coal Region*. US Fish and Wildlife Service Report.

375. Herricks, E. E. and Sale, M. J. (1978). Preplanning mining operations for mitigation of impacts on fish and wildlife resources. In *Surface Mining and Fish/Wildlife Needs in the Eastern United States*, ed. D. E. Samuel, J. R. Stauffrer, C. H. Hocutt and W. T. Mason, pp. 69–75. Morgantown: West Virginia University.

376. Herricks, E. E., Tazik, D. J., Karr, J. R. and Leary, M. E. (1980). *Planning for Fish and Wildlife on Illinois Coal Surface Mines*. US Fish and Wildlife Service Report.

377. Herrera de Fournier, M. E. and Fournier, O. L. A. (1977). Production, decomposition and invertebrates of forest litter in several stages of succession in Ciudad Colon, Costa Rica. *Revista de Biologia Tropical*, **25**, 275–88.

378. Heydemann, B. (1960). Die biozonotische Entwicklung vom Vorland zum Koog, I. Teili Spinners (Arancae). *Abhandlungen der Akademie der Wissenschaften und der Literatur in Mainz, Mathematisch – Naturwissenschaft – Lichen Klasse*, **11**, 1–169.

379. Heydemann, B. (1962). Die biozonotische Entwicklung vom Vorland zum Koog, II. Teili Käfer (Coleoptera). *Abhandlungen der Akademie der Wissenschaften und der Literatur in Mainz, Mathematisch – Naturwissenschaft – Lichen Klasse*, **11**, 170–370.

380. Heydemann, B. (1967). *Die biologische Grenze Land-Meer im Bereich der Salzwiesen*. Wiesbaden: Franz Steiner.

381. Hickey, J. J. and Mikol, S. A. (1979). *Estimating Breeding Bird Densites on Coal Lands in Montana and Wyoming*. US Fish and Wildlife Service Report.

382. Hicks, R. R. and Samuel, D. E. (1985). *Surface Mine Reclamation: A Wildlife Habitat Opportunity*. Coraopolis, PA: The Ruffled Grouse Society.

383. Higgins, K. F. (1982). Principal flora and wildlife use of a 35–40 year old gravel pit in North Dakota. In *The Wildlife Values of Gravel Pits Symposium, Crookston, Minnesota, June 1982*, ed. W. D. Svedarsky and R. D. Crawford, pp. 159–61. University of Minnesota Miscellaneous Publication 17-1982.

384. Hill, D. A. (1983). Habitat requirements of breeding waterfowl on gravel quarries: management implications. In *The Proceedings of a Symposium on the Subject of Wildlife on Man-Made Wetlands*, pp. 43–72. Milton Keynes, England: ARC Wildlife Centre.

385. Hill, D. A. (1984). Population regulation in the mallard (*Anas platyrhynchos*). *Journal of Animal Ecology*, **53**, 191–202.

386. Hingtgen, T. M. and Clark, W. R. (1984*a*). Impact of small mammals on the vegetation of reclaimed land in the Northern Great Plains. *Journal of Range Management*, **37**, 438–41.

387. Hingtgen, T. M. and Clark, W. R. (1984*b*). Small mammal recolonization of reclaimed coal surface-mined land in Wyoming. *Journal of Wildlife Management*, **48**, 1255–61.

388. Hinkle, C. R., Ambrose, R. E. and Wenzel, C. R. (1981*a*). *A Handbook for Meeting Fish and Wildlife Information Needs to Surface Mine Coal – OSM Region 1*. US Fish and Wildlife Service Report.

389. Hinkle, C. R., Ambrose, R. E. and Wenzel, C. R. (1981 *b*). *A Handbook for Meeting Fish and Wildlife Information Needs to Surface Mine Coal – OSM Region 2*. US Fish and Wildlife Service Report.

390. Hinkle, C. R., Ambrose, R. E. and Wenzel, C. R. (1981 *c*). *A Handbook for Meeting Fish and Wildlife Information Needs to Surface Mine Coal – OSM Region 3*. US Fish and Wildlife Service Report.

391. Hinkle, C. R., Ambrose, R. E. and Wenzel, C. R. (1981 *d*). *A Handbook for Meeting Fish and Wildlife Information Needs to Surface Mine Coal – OSM Region 4*. US Fish and Wildlife Service Report.

392. Hinkle, C. R., Ambrose, R. E. and Wenzel, C. R. (1981 *e*). *A Handbook for Meeting Fish and Wildlife Information Needs to Surface Mine Coal – OSM Region 5*. US Fish and Wildlife Service Report.

393. Hinz, D. A. (1980). Return to nature after mining. *Landline*, **3**, 4–5.

394. Hoag, A., Dahmer, T. and Schrad, M. (1982). Rodent colonization of revegetated surface-mined lands in Montana and Wyoming. In *Issues and Technology in the Management of Impacted Western Wildlife*, ed. R. D. Comer, J. M. Merino, J. W. Monarch, C. Pustmueller, M. Stalmaster, R. Stoeker, J. Todd and W. Wright, pp. 201–4. Thorne Ecological Institute Technical Publication no. 14.

395. Hoey, A. F. van (undated). *Wildlife Management with Herbaceous Cover*. Indiana Department of Natural Resources, Division of Fish and Wildlife. Management Series no. 4.

396. Hofmann, L., Ries, R. E. and Gilley, J. E. (1983). Relationship of runoff and soil loss to ground cover of native and reclaimed grazing land. *Agronomy Journal*, **75**, 599–602.

397. Hofmann, L., Ries, R. E. and Lorenz, R. J. (1981). Livestock and vegetative performance on reclaimed and nonmined rangeland in North Dakota. *Journal of Soil and Water Conservation*, **36**, 41–4.

398. Hofmann, L., Ries, R. E., Power, J. F. and Lorenz, R. J. (1977). Effects of grazing intensity on vegetation and animal performance on reclaimed strip-mined land. In *Fifth Symposium on Surface Mining and Reclamation, Louisville, Kentucky*, pp. 306–10.

399. Hofmann, L., Ries, R. E., Power, J. F. and Lorenz, R. J. (1978). Grazing reclaimed strip-mined sites. *North Dakota Farm Research*, **36**, 3–5.

400. Holland, F. R. (1973). Wildlife benefits from strip-mine reclamation. In *Ecology and Reclamation of Devastated Land*, vol. 1, ed. R. J. Hutnik and G. Davis, pp. 377–88. New York: Gordon and Breach.

401. Holliday, R. J., Johnson, M. S. and Bradshaw, A. D. (1979). Wildlife conservation and mining of alluvial tin reserves in southwest England. *Biological Conservation*, **16**, 245–64.

402. Hoogerkamp, M., Rogaar, H. and Eijsackers, H. J. P. (1983). Effect of earthworms on grassland on recently reclaimed polder soils in The Netherlands. In *Earthworm Ecology: From Darwin to Vermiculture*, ed. J. E. Satchell, pp. 85–105. London: Chapman and Hall.

403. Horack, G. C. and Olson, J. E. (1980). *Proceedings of the Uranium Mining and Milling Workshop*. (Western Energy and Land Use Team). US Fish and Wildlife Service Report.

404. Hore-Lacy, I., ed. (1979). *Mining Rehabilitation*. Canberra: Australian Mining Industry Council.

405. Howell, J. F. (1981). Codling moth: the effects of volcanic ash from the eruption of Mount St. Helens on egg, larval, and adult survival. *Melanderia*, **37**, 50–5.

406. Hudy, M. and Gregory, R. W. (1984). *Some Limnological Characteristics of Eight Limestone Excavations in South Florida*. US Fish and Wildlife Service Report.

407. Huhta, V., Ikonen, E. and Vilkamaa, P. (1977). Animal succession in artificial soil made of

sewage sludge and crushed bark. In *Soil Organisms as Components of Ecosystems*, ed. U. Lohm and T. Persson, Ecological Bulletins no. 25, pp. 573–7.

408. Hull, C. L. (1981). *Biota of Bentonite Mine Spoils*. M.Sc. Thesis, Colorado State University.
409. Hull, R. J. and Shann, J. (1981). Using dredging spoils to build tidal marshes. *Maritimes*, **24**, 7–9.
410. Humphrey, S. A. (1970). *A Qualitative Survey of the Macroinvertebrates from Three Acid-Polluted Strip-Mine Lakes*. Southern Illinois University Research Report.
411. Humphries, R. N. (1980). The development of wildlife interest in limestone quarries. *Reclamation Review*, **3**, 197–207.
412. Hunter, F. and Currie, J. A. (1956). Structural changes during bulk soil storage. *Journal of Soil Science*, **7**, 75–80.
413. Huther, W. (1971). Collembolen von einen Hamburger Mullplatz. *Entomologische Mitteilungen (Hamburg)*, **72**, 157–65.
414. Hutnik, R. J. and Davis, G., ed. (1973). *Ecology and Reclamation of Devastated Land*, vol. 1, New York: Gordon and Breach.
415. Hutson, B. R. (1972). The invertebrate fauna of a reclaimed pit heap. In *Landscape Reclamation*, vol. 2, pp. 64–79. Guildford, England: IPC Science and Technology Press.
416. Hutson, B. R. (1974). *The Invertebrate Fauna Influencing Soil Development on Industrial Reclamation Sites*. PhD Thesis, University of Newcastle upon Tyne (unpublished manuscript).
417. Hutson, B. R. (1978*a*). Effects of variations of the plaster–charcoal culture method on a collembolan, *Folsomia candida*. *Pedobiologia*, **18**, 138–44.
418. Hutson, B. R. (1978*b*). Influence of pH, temperature and salinity on the fecundity and longevity of four species of Collembola. *Pedobiologia*, **18**, 163–79.
419. Hutson, B. R. (1980*a*). Colonization of industrial reclamation sites by Acari, Collembola and other invertebrates. *Journal of Applied Ecology*, **17**, 255–75.
420. Hutson, B. R. (1980*b*). The influence on soil development of the invertebrate fauna colonizing industrial reclamation sites. *Journal of Applied Ecology*, **17**, 277–86.
421. Hutson, B. R. (1981). Age distribution and the annual reproductive cycle of some Collembola colonizing reclaimed land in Northumberland, England. *Pedobiologia*, **21**, 410–16.
422. Hutson, B. R. and Luff, M. L. (1978). Invertebrate colonization and succession on industrial reclamation sites. *Scientific Proceedings of The Royal Dublin Society, Series A*, **6**, 165–74.
423. Illinois Department of Conservation. (1984). *Habitat Restoration Notes for Mined Land*. Illinois Department of Conservation Report.
424. Institute for Land Rehabilitation. (1978). *Rehabilitation of Western Wildlife Habitat: A Review*. US Fish and Wildlife Service Report.
425. Ireland, M. P. (1979). Metal accumulation by the earthworms *Lumbricus rubellus*, *Dendrobaena veneta* and *Eiseniella tetraedra* living in heavy metal polluted sites. *Environmental Pollution*, **19**, 201–6.
426. Iverson, L. R. and Wali, M. K. (1982). Buried, viable seeds and their relation to revegetation after surface mining. *Journal of Range Management*, **35**, 648–52.
427. Jahn, L. R. (1979). Summary of the symposium, and recommendations for improving mitigation in the future. In *The Mitigation Symposium: A National Workshop on Mitigating Losses of Fish and Wildlife Habitats*, ed. G. A. Swanson, pp. 6–18. Rocky Mountain Forest and Range Experiment Station (Forest Service) General Technical Report.
428. Jankovic, M. (1976). Chironomid populations in the Pek river Yugoslavia after mine pollution. *Arhiv Bioloskih Nauka*, **28**, 153–60.

429. Jewell, M. E. (1922). The fauna of an acid stream. *Ecology*, **3**, 22–6.
430. Joenje, W. (1982). Nature in new Wadden-polders: conservation by exploitation. In *Polders of the World. Papers International Symposium, Lelystad*, vol. 3, pp. 68–82. Wageningen: International Institute for Land Reclamation and Improvement.
431. Johansen, C. A., Eves, J. D., Mayer, D. F., Bach, J. C., Nedrow, M. E. and Kious, C. W. (1981). Effects of ash from Mt. St. Helens on bees. *Melanderia*, **37**, 20–9.
432. Johansen, O. (1984). Sand martin *Riparia riparia* in Hordaland County. *Fauna (Oslo)*, **37**, 53–5.
433. Johnson, C. D., Beley, J. R., Ditsworth, T. M. and Butt, S. M. (1983). Secondary succession of arthropods and plants in the Arizona sonoran desert in response to transmission line construction. *Journal of Environmental Management*, **16**, 125–37.
434. Johnson, C. D., Ditsworth, T. M., Beley, J. R., Butt, S. M. and Balda, R. P. (1981). Arthropods, plants and transmission lines in Arizona: secondary succession in a sonoran desert habitat. *Journal of Environmental Management*, **13**, 151–63.
435. Johnson, D. W. (1986). Desert buttes: natural experiments for testing theories of island biogeography. *National Geographic Research*, **2**, 152–66.
436. Johnson, M. S. (1980). Revegetation and the development of wildlife interest on disused fluorspar tailings dams. *Reclamation Review*, **3**, 209–16.
437. Johnson, M. S., Putwain, P. D. and Holliday, R. J. (1978). Wildlife conservation value of derelict metalliferous mine workings in Wales. *Biological Conservation*, **14**, 131–47.
438. Jolly, W. C. (1979). Congress and the mitigation of fish and wildlife habitat losses. In *The Mitigation Symposium: A National Workshop on Mitigating Losses of Fish and Wildlife Habitats*, ed. W. D. Svedarsky and R. D. Crawford, pp. 520–6. Rocky Mountain Forest Range Experiment Station (Forest Service) General Technical Report.
439. Jones, D., McElligott, M. J. and Mannz, R. H. (1985). Summary of biological, chemical, and morphological characterizations of 33 surface-mine lakes in Illinois and Missouri. In *Wetlands and Water Management on Mined Lands. Proceedings of a Conference October 23–24, 1985*, ed. R. P. Brooks, D. E. Samuel and J. B. Hill, pp. 211–38. Pennsylvania State University.
440. Jones, D. W., Mannz, R. H., McElligott, M. J. and Imboden, B. (1985). A rotenone survey to determine the standing crop of fishes in a 212 acre surface-mine lake in St. Clair, Illinois. In *Fish and Wildlife Relationships to Mining: Symposium Proceedings*, ed. L. B. Starnes, pp. 7–13. American Fisheries Society (Water Quality Section).
441. Jones, G. S. (1967). *Vertebrate Ecology of a Strip-Mined Area in Southern Indiana*. M.Sc. Thesis, Purdue University (unpublished manuscript).
442. Jones, G. S. (1971). Notes on the reptiles and amphibians observed on a strip-mined study area in Pike County, Indiana. *Indiana Audubon Quarterly*, **49**, 48–50.
443. Jones, G. S. and Zeigler, F. W. (1975). Salamanders (*Plethodon cinereus* and *P. glutinosus*) on a strip mine in west-central Indiana. *Indiana Audubon Quarterly*, **53**, 8–10.
444. Jong, H. de (1977). Experiences with the man-made meadow bird reserve Kievitslanden in Flevoland (The Netherlands). *Biological Conservation*, **12**, 13–31.
445. Joseph, T. (1977). The importance of a lake's littoral zone and its relationship to mine pond reclamation. In *Reclamation for Wildlife Habitat. Proceedings of Reclamation Workshop II, Fort Collins, Colorado, September 1977*, ed. Ecology Consultants Inc., pp. 50–63. Environmental Research and Technology Inc.
446. Kabay, E. D. and Nichols, O. G. (1980). *Use of Rehabilitated Bauxite Mined Areas in the Jarrah Forest by Vertebrate Fauna*. Environmental Research Bulletin no. 8. Perth: Alcoa of Australia Limited.
447. Kabay, E. D. and Nichols, O. G. (1981). *Formation of Wetlands as a Possible Rehabilitation Option for Open Cut Mining in the South West of Western Australia*. Environmental Research Bulletin no. 10. Perth: Alcoa of Australia Limited.

448. Kalbe, L. (1957). Zur Vogelwelt stillgelegter Braunkohlengebiete in der Leipziger Tieflandsbucht. *Beitraege zur Vogelkunde*, 1/3, 177–82.

449. Kalbe, L. (1959). Zur Verbreitung und Okologie der Wirbeltiere in stillgelegten Braunkohlegebieten im süden Leipzigs. *Wissenschaftlische Zeitschrift. Karl-Marx-Universitaet (Leipzig), Mathematisch-Naturwissenschaftliche Reihe.*

450. Kalbe, L. (1961). Die Vogelwelt stillgelegter Braunkohlengruben in oberlausitzer Niederung. *Der Falke*, **3**, 84–7.

451. Kannowski, P. B. and Benner, B. L. (1978). *Invertebrates of Southwestern North Dakota: Soil Fauna.* Institute for Ecological Studies, University of North Dakota. Regional Environmental Assessment Program Report no. 79-3.

452. Karr, J. R. (1968). Habitat and avian diversity on strip-mined land in east-central Illinois. *The Condor*, **70**, 348–57.

453. Karr, J. R. (1980). *Strip-Mine Reclamation and Bird Habitats.* USDA Forest Service, Intermountain Forest and Range Experiment Station General Technical Report, INT-86, pp. 88–97.

454. Kays, C. E. (undated). *Wildlife – Recreation: Strip Mined Lands as Fish and Wildlife Habitat.* Frankfort, Kentucky: Department of Fish and Wildlife Resources.

455. Kelcey, J. G. (1983). The management of Milton Keynes city balancing lakes for wildlife. In *The Proceedings of a Symposium on the Subject of Wildlife on Man-Made Wetlands*, pp. 142–50. Milton Keynes, England: ARC Wildlife Centre.

456. Kelcey, J. G. (1984). The design and development of gravel pits for wildlife in Milton Keynes, England. *Landscape Planning*, **11**, 19–34.

457. Kentish, K. M. (1983). *Mine Rehabilitation: A Study of Revegetation and Fauna Return at Anglesea, Victoria (1980–1982).* MSc Thesis, Deakin University.

458. Kentish, K. M. and Bourne, A. R. (1982). Return of fauna to heathland regenerating after mining. *Bulletin of the Australian Mammal Society*, **7**, 29.

459. Kerr, R. (1977). Ideas about reclaiming western mined lands for wildlife. In *Reclamation for Wildlife Habitat. Proceedings of Reclamation Workshop II, Fort Collins, Colorado, September 1977*, Ecology Consultants Inc., pp. 69–73. Environmental Research and Technology Inc.

460. Keywood, K. P. and Melluish, W. D. (1953). A report on the bird population of four gravel pits in the London area. *London Bird Report*, **17**, 43–72.

461. Kickson, K. L. and Vance, D. (1981). *Revegetating Surface Mined Lands for Wildlife in Texas and Oklahoma.* Denton: Department of Biological Sciences, North Texas State University.

462. Kimmel, R. O. (1982). *Ruffed Grouse Brood Habitat on Reclaimed Surface Mines in West Virginia.* PhD Dissertation, West Virginia University (unpublished manuscript).

463. Kimmel, R. O. and Samuel, D. E. (1978*a*). Feeding behaviour of young ruffed grouse in West Virginia. In *Transactions 35th Northeast Fish and Wildlife Conference, White Sulphur Springs, West Virginia*, pp. 43–51.

464. Kimmel, R. O. and Samuel, D. E. (1978*b*). Ruffed grouse use of a twenty year old surface mine. In *Surface Mining and Fish/Wildlife Needs in the Eastern United States*, ed. D. E. Samuel, J. R. Stauffer, C. H. Hocutt and W. T. Mason, pp. 345–51. Morgantown: West Virginia University.

465. Kimmel, R. O. and Samuel, D. E. (1984*a*). Implications of ruffed grouse brood habitat studies in West Virginia. In *Ruffed Grouse Management: State of the Art in the Early 1980's. Proceedings. 45th Midwest Fish and Wildlife Conference*, ed. W. L. Robinson, pp. 91–108.

466. Kimmel, R. O. and Samuel, D. E. (1984*b*). Ruffed grouse brood habitat on reclaimed surface mines in West Virginia. *Minerals and the Environment*, **4**, 150–3.

467. King, K. A., Leleux, J. and Mulhern, B. M. (1984). Molybdenum and copper levels in white-tailed deer near uranium mines in Texas. *Journal of Wildlife Management*, **48**, 267–70.

468. Kinney, E. C. (1964). *Extent of Acid Mine Pollution in the United States Affecting Fish and Wildlife*. US Fish and Wildlife Service Circular 191.

469. Kirkland Jr, G. L. (1976). Small mammals of a mine waste situation in the Central Adirondacks, New York: a case of opportunism by *Peromyscus maniculatus*. *The American Midland Naturalist*, **95**, 103–10.

470. Kirmse, W. (1955). *Okologische Untersuchungen an der Espenhainer Halde*. Diplomabeit Zoologisches Institute der Universität Leipzig.

471. Klausnitzer, B. (1983). Faunistische-okologische Untersuchungen auf dem neuen Müllberg Leipzig-Möckern, 2 Beitrag: Carabidae. *Hercynia*, **20**, 392–402.

472. Klausnitzer, B., Joost, B., Wolff, W. and Wolff, H. (1980). Faunistisch-okologische Untersuchungen auf dem neuen Müllberg Leipzig-Mockern, 1 Beitrag: Gesamtmaterial. *Wissenschaftliche Zeitschrift der Karl-Marx Universität. Mathematischnaturwissenschaftliche*, **29**, 646–52.

473. Kleinman, L. H. and Layton, D. E. (1981). Livestock operation is born on reclaimed mined lands. Abstracts of papers. *34th Annual Meeting, Society of Range Management*. Tulsa, Oklahoma. no. 80.

474. Kleinman, L. H., Raphelson, M. A. and Bendix, D. S. (1984). Livestock grazing on mined land reclamation. Abstracts of papers. *37th Annual Meeting, Society of Range Management*. Rapid City, South Dakota. no. 213.

475. Klimstra, W. D. (1959). The potential of wildlife management on strip-mined areas. *Illinois Wildlife*, **14**, 5–9.

476. Klimstra, W. D. (1962). Wildlife management and recreation on strip-mined areas. *Outdoor Illinois*, **1**, 14–19.

477. Klimstra, W. D. (1982). Wildlife reclamation of mined lands. In *Research on Fish and Wild Habitat*, ed. W. T. Mason and S. Ikers, pp. 183–94. Environmental Protection Report no. 600/8-82-022.

478. Klimstra, W. D. and Jewell, S. R. (1973). Strip mining resources in conflict impacts and reclamation efforts in Illinois. In *North American Wildlife and Natural Resources Conference (38th) March 1973*, ed. J. B. Trefethen, pp. 121–31. Washington, D.C.: Wildlife Management Institute.

479. Klimstra, W. D. and Nawrot, J. R. (1984). Reclamation for wildlife: state of art. In *Proceedings of the Fourth Annual Better Reclamation with Trees Conference, Owensboro, Kentucky June 7–8 1984*, ed. P. E. Pope. Kentucky: Madisonville Community College.

480. Klimstra, W. D. and Nawrot, J. R. (1985). Wetlands as a byproduct of surface mining: midwest perspective. In *Wetlands and Water Management on Mined Lands. Proceedings of a Conference October 23–24, 1985*, ed. R. P. Brooks, D. E. Samuel and J. B. Hill, pp. 107–20. Pennsylvania State University.

481. Klimstra, W. D. and Nawrot, J. R. (1988). Mineral extraction: reclamation. *CLRA Reclamation Newsletter* (in press).

482. Klimstra, W. D. and Thornburg, D. (1982). Mined land wetlands as breeding grounds for the giant Canada goose. In *The National Symposium on Surface Mining Hydrology, Sedimentology, and Reclamation, Lexington, Kentucky, December 1982*, ed. D. H. Graves, pp. 359–63. Lexington: University of Kentucky.

483. Klimstra, W. D., Vohs, Jr, P. A. and Cherry, J. D. (1963). Strip-mined lands for recreation. *Illinois Wildlife*, **18**, 6–8.

484. Kling, C. L. (1977). Reclamation planning for wildlife prior to mining. In *Reclamation for Wildlife Habitat. Proceedings of Reclamation Workshop II, Fort Collins, Colorado, September 1977*, ed. Ecology Consultants Inc., pp. 167–74. Environmental Research and Technology Inc.

485. Klostermeyer, E. C., Corpus, L. D. and Campbell, C. L. (1981). Population changes in arthropods in wheat following volcanic ash fall-out. *Melanderia*, **37**, 45–9.

486. Kniess, H. and Wegner, H. (1980). The bird communities of brown coal surface mining areas in northern Bavaria recultivated by forestry. *Anzeiger der Ornithologischen Gesellschaft in Bayern*, **19**, 37–46.

487. Knotts, R. W. (1975). *White-Tailed Deer Movement and Distribution about Surface Mines in Preston County, West Virginia*. M.Sc. Thesis, West Virginia University (unpublished manuscript).

488. Knotts, R. W. and Samuel, D. E. (1982). White-tailed deer movement and distribution about surface coal mines in West Virginia, USA. *Minerals and the Environment*, **4**, 17–25.

489. Kobel-Lamparski, A. and Lamparski, F. (1983). Die Wiederbesiedlung flurbereinigten Rebgelandes im Kaiserstuhl durch Lumbriciden. *Mitteilungen der Deutschen Bodenkundl Gesellschaft*, **38**, 337–42.

490. Koehler, H. (1984). *Methodische, Ökologische und experimentelle Untersuchungen zur Sukzession der Mesofauna der Abdeckschicht einer Brauschuttdeponie unter besonderer Berücksichtigung der Gamasina (Acari, Parasitiformes)*. Erlangung des Doktorgrades Dissertation, Universität Bremen.

491. Kolar, C. and Ashby, W. C. (1978). Potential for woodland habitat from surface-mine tree plantings. In *The Wildlife Management Institutes 43rd North American Wildlife and Natural Resources Conference, Phoenix, March 1978*, pp. 323–30.

492. Kooistra, M. J. (1981). The interpretation and classification of features produced by pelecypods (Mollusca) in marine inter-tidal deposits in The Netherlands. *Geoderma*, **26**, 83–94.

493. Koopmann, R. W. (1982). Pits, ponds and people: reclamation and public use. In *The Wildlife Values of Gravel Pits Symposium, Crookston, Minnesota, June 1982*, ed. W. D. Svedarsky and R. D. Crawford, pp. 127–31. University of Minnesota Miscillaneous Publication 17-1982.

494. Koridon, A. H., Polman, G. K. R., Poorter, E. P. R., Ven, G. A. and Zijlstra, M. (1980). *Development and Research of a New Nature Reserve in Flevoland: The Oostvaardersplassen*. Rijksdienst voor de Ijsselmeerpolders. Flevobericht no. 169.

495. Krause, J., Mahan, B. and Harding, M. (1985). Waterfowl management and wetland development on Peabody Coal Company surface mines in Illinois, Indiana, and Kentucky. In *Wetlands and Water Management on Mined Lands. Proceedings of a Conference October 23–24, 1985*, ed. R. P. Brooks, D. E. Samuel and J. B. Hill, pp. 327. Pennsylvania State University.

496. Krause, R. R. (1964). Spoil bank goes from waste to fodder. *Coal Mining and Processing*, **1**, 29–31.

497. Krause, R. R. (1973). Strip mining: conflict or competition? In *North American Wildlife and Natural Resources Conference (38th) March 1973*, ed. J. B. Trefethen, pp. 142–7. Washington, D.C.: Wildlife Management Institute.

498. Krementz, D. G. (1980). *Granivore Communities in a Northern Desert Shrub–Steppe Versus an Adjacent Reclaimed Mine Spoils*. M.Sc. Thesis, University of Wyoming (unpublished manuscript).

499. Krementz, D. G. and Sauer, J. R. (1982). Avian communities on partially reclaimed mine spoils in south central Wyoming. *Journal of Wildlife Management*, **46**, 761–5.

500. Kruse, W. H., Balda, R. P., Simono, M. J., Macrander, A. M. and Johnson, C. D. (1979). Community development in two adjacent pinyon–juniper eradication areas twenty-five years after treatment. *Journal of Environmental Management*, **8**, 237–47.

501. Laird, J. M. and Korger, M. (1981). Earthworms. *Critical Reviews in Environmental Control*, **11**, 189–218.

502. Lambrecht, P. (1970). *Ein Beitrag zur Kenntnis der Makrofauna auf Rekultivierten Kippen und Halden des Braunkohlkenbergbaues der Niederlausitz unter Besonderer Berucksichtigung der Lumbriciden*. Diplomarbeit (Gorlitz) Ph Potsdam.

503. Landin, M. C. (1978). National perspective of colonial waterbirds nesting on dredged material islands. In *The Wildlife Management Institutes 43rd North American Wildlife and Natural Resources Conference, Phoenix, March 1978*, pp. 89–100.

504. Lang, R. (1982). Biomass and forage production from reclaimed stripmined land and adjoining native range in central Wyoming. *Journal of Range Management*, **35**, 754–5.

505. Langkamp, P. J., Dalling, M. J., Calder, D. M. and Farrell, T. P. (1979). Studies on the rehabilitation of mined areas on Groote Eylandt, Northern Territory 1. Vegetation description. *Reclamation Review*, **2**, 147–55.

506. Lawrence, J. M. (1984). Soil fauna recolonization of high elevation mine spoils in southeastern British Columbia. In *Reclamation of Land Disturbed by Mining, Proceedings of the Eighth Annual British Columbia Mine Reclamation Symposium*. Mining Association of British Columbia.

507. Lawrence, J. S., Klimstra, W. D., Cleary, W. G. and Perkins, G. A. (1985). Contribution of surface-mined wetlands to selected avifauna in Illinois. In *Wetlands and Water Management on Mined Lands, Proceedings of a Conference October 23–24, 1985*, ed. R. P. Brooks, D. E. Samuel and J. B. Hill, pp. 317–25. Pennsylvania State University.

508. Laycock, W. A. and Layden, T. (1986). Cattle production on reclaimed mined lands in northwestern Colorado. In *Rangelands: A Resource Under Siege. Proceedings Second International Rangeland Congress*, ed. P. J. Joss, P. W. Lynch and O. B. Williams, pp. 162. Canberra: Australian Academy of Science.

509. Laycock, W. A. and McGinnies, W. J. (1985). Reclamation and grazing management on a surface coal mine in northwestern Colorado. In *Proceedings Second Annual Meeting American Society of Surface Mining and Reclamation, Denver, Colorado*, pp. 73–6.

510. Laycock, W. A. and McGinnies, W. J. (undated). *Grazing on Reclaimed Mined Land in Northwestern Colorado*. Report to The Bureau of Land Management.

511. Layden, T. V. (1984). *Grazing on Reclaimed Mined Land in Northwestern Colorado*. MSc Thesis, Colorado State University, Fort Collins, Colorado (unpublished manuscript).

512. Layden, T. V. and Laycock, W. A. (1985). Grazing on reclaimed mined lands in northwestern Colorado. Abstracts. *38th Annual Meeting, Society of Range Management*. Saltlake City, Utah.

513. Leclerc, M. G. (1982). *Bird Communities and Vegetation Structure of Reclaimed Surface Mines in West Virginia*. M.Sc. Thesis, West Virginia University (unpublished manuscript).

514. Leedy, D. L. (1981). *Coal Surface Mining Reclamation and Fish and Wildlife Relationships in the Eastern United States*, vol. 1. US Fish and Wildlife Service Report.

515. Leedy, D. L. (1982). Research needs and priorities of gravel pits. In *The Wildlife Values of Gravel Pits Symposium, Crookston, Minnesota, June 1982*, ed. W. D. Svedarsky and R. D. Crawford, pp. 221–6. University of Minnesota Miscillaneous Publication 17-1982.

516. Leedy, D. L. and Franklin, T. M. (1981). *Coal Surface Mining Reclamation and Fish and Wildlife Relationships in the Eastern United States*, vol. 2. US Fish and Wildlife Service Report.

517. Leedy, D. L., Mason, W. T. and Cushwa, C. T. (1981). *Coal Surface Mining Reclamation and Fish and Wildlife Relationships in the Eastern United States*, vol. 1. US Fish and Wildlife Service, Department of the Interior (Office of Biology Services), Eastern Energy and Land Use Team Report.

518. Lees, P. R. and Street, M. (1973). The feeding ecology of young mallard (*Anas platyrhynchos platyrhynchos*) and tufted duck (*Aythya fuligula*) in wet gravel quarries. In: *Proceedings XII Congress International Union of Game Biologists, Stockholm*, pp. 249–53.

519. Leitch, J. A. (1982). Economics of gravel pit reclamation for wildlife. In *The Wildlife Values of Gravel Pits Symposium, Crookston, Minnesota, June 1982*, ed. W. D. Svedarsky and R. D. Crawford, pp. 24–30. University of Minnesota Miscillaneous Publication 17-1982.

520. Lesser, M. A. (1975). Symposium on biological implications of strip mining. *Ohio Journal of Science*, **75**, 265–319.

521. Leston, D. (1957). Spread potential and the colonisation of islands. *Systematic Zoology*, **6**, 41–6.

522. Lewis, J. C. and Bunce, E. W., ed. (1980). *Rehabilitation and Creation of Selected Coastal Habitats. Proceedings of a Workshop, Sapelo Island, Georgia, May 1976*. Washington, DC: US Fish and Wildlife Service, Biological Services Program.

523. Lewis, W. M. and Nickum, J. (1964). Rearing trout in raceways supplied by waters from the hypolimnion of a strip-mine pond. *Progressive Fish Culturist*, **26**, 27–32.

524. Lewis, W. M. and Peters, C. (1956). Coal mine slag drainage. *Industrial Wastes*, **1**, 145–7.

525. Lind, O. T. and Campbell, R. S. (1970). Community metabolism in acid and alkaline strip-mine lakes. *Transactions of the American Fish Society*, **99**, 577–82.

526. Lindroth, C. H. (1971). Biological investigations on the new volcanic island Surtsey, Iceland. *Miscellaneous Papers Landbouwhogeschool Wageningen*, **8**, 65–70.

527. Lindsay, S. F. and Bookhout, T. A. (1978). Lead and mercury levels in vegetation from strip-mined areas in eastern Ohio. In *Bulletin of Environmental Contamination and Toxicology*, pp. 360–4. New York: Springer-Verlag.

528. Lindsay, S. F., Bookhout, T. A. and White, G. C. (1978). Nutritive levels in plants from stripmined areas in eastern Ohio. *The Ohio Journal of Science*, **78**, 70–9.

529. Lobe, N. K. (1979). Energy resource development and wildlife preservation. In *The Mitigation Symposium: A National Workshop on Mitigating Losses of Fish and Wildlife Habitats*, ed. G. A. Swanson, pp. 424–8. Rocky Mountain Forest Range Experiment Station (Forest Service) General Technical Report.

530. Loenen, M. and Pinkers, M. (1979). *Development of Grevelingenbekken as Nature and Recreation Reserve*. Rijksdienst voor de Ijsselmeerpolders. Flevobericht no. 156.

531. Loenen, M. and Pinkers, M. (1980). *The Development of the Grevelingen-Area for Nature and Recreation*. Rijksdienst voor de Ijsselmeerpolders. Flevobericht no. 155.

532. Loerch, C. R. and Cameron, E. A. (1984). Within-tree distributions and seasonality of immature stages of the bronze birch borer, *Agrilus anxius* (Coleoptera: Buprestidae). *Canadian Entomologist*, **116**, 147–52.

533. Lomax, J. L. (1982). Wildlife use of mineral extraction industry sites in the coastal plains of New Jersey. In *The Wildlife Values of Gravel Pits Symposium, Crookston, Minnesota, June 1982*, ed. W. D. Svedarsky and R. D. Crawford, pp. 115–21. University of Minnesota Miscillaneous Publication 17-1982.

534. Ludwig, A. and Martin, S. (1977). The importance of non-game wildlife in mined land reclamation. In *Reclamation for Wildlife Habitat. Proceedings of Reclamation Workshop II, Fort Collins, Colorado, September 1977*, ed. Ecology Consultants Inc., pp. 40–5. Environmental Research and Technology Inc.

535. Luff, M. L. and Hutson, B. R. (1977). Soil fauna populations. In *Landscape Reclamation Practice*, ed. B. Hackett, pp. 125–47. Guildford, England: IPC Science and Technology Press.

536. Lundgren, D. G. (1975). Microbiological problems in strip mine areas: relationship to the metabolism of *Thiobacillus ferroloxidans*. *Ohio Journal of Science*, **75**, 280–7.

537. Luxton, M. (1982*a*). Studies on the invertebrate fauna of New Zealand peat soils IV. Pasture soils on rukuhia peat. *Pedobiologia*, **24**, 297–308.

538. Luxton, M. (1982*b*). The ecology of some soil mites from coal shale tips. *Journal of Applied Ecology*, **19**, 427–42.

539. Luxton, M. (1983). Studies on the invertebrate fauna of New Zealand peat soils V. Pasture soils on Kaipaki peat. *Pedobiologia*, **25**, 135–48.

540. Lynott, B. (1977). Critical wildlife habitats of North Dakota. In *Reclamation for Wildlife*

Habitat. Proceedings of Reclamation Workshop II, Fort Collins, Colorado, September 1977, ed. Ecology Consultants Inc., pp. 28–39. Environmental Research and Technology Inc.

541. Mabelis, A. A. and Mabelis-Jonkers, J. C. F. (1978). Verspreiding van mieren in kalkrijke gebieden van Zuid-Limburg (Hym., Formicidae). *Entomologische Berichten*, **38**, 165–8.

542. McArthur, L. B. and Whitmore, R. C. (1979). *Passerine Community Composition and Diversity in Man-Altered Environments*. West Virginia University, West Virginia Forestry Notes, no. 7.

543. McBride, F. D., Chavensaksongkaram, C. and Urie, D. H. (1977). *Sludge-Treated Coal Mine Spoils Increase Heavy Metals in Cover Crops*. USDA Forest Service, North Central Experimental Station Research Note no. NC-221.

544. McBrien, H., Harmsen, R. and Crowder, A. (1983). A case of insect grazing affecting plant succession. *Ecology*, **64**, 1035–9.

545. McCann, S. A. (1975). The movement and occurrence of small mammals on a reclamation plot. In *Proceedings of the Fort Union Coal Field Symposium*, vol. 3, Reclamation Section, ed. W. F. Clark, pp. 318–24. Montana Academy of Sciences.

546. McCann, S. A. (1976). Home ranges of the meadow vole and deer mouse (on a reclamation test pit in eastern Montana). *Proceedings of the Montana Academy of Science*, **36**, 11–17.

547. McConnell, D. L. and Samuel, D. E. (1985). Small mammal and avian populations utilizing cattail marshes on reclaimed surface mines in West Virginia. In *Wetlands and Water Management on Mined Lands. Proceedings of a Conference October 23–24, 1985*, ed. R. P. Brooks, D. E. Samuel and J. B. Hill, pp. 329–36. Pennsylvania State University.

548. McDonald, G. and Buchanan, G. A. (1981). The mosquito and predatory insect fauna inhabiting fresh-water ponds, with particular reference to *Culex annulirostris* Skuse (Diptera: Culicidae). *Australian Journal of Ecology*, **6**, 21–7.

549. McDonnell, M. J. and Stiles, E. W. (1983). The structural complexity of old field vegetation and the recruitment of bird-dispersed plant species. *Oecologia*, **56**, 109–16.

550. McGowan, K. J. (1980). *Populations of Small Mammals and their Utilization of Arthropods on Ohio Strip-Mined Lands*. M.Sc. Thesis, Ohio State University.

551. MacInko, G. (1984). Trees and wildlife: a commentary on British practice with some notes from America. In *Proceedings of the Fourth Annual Better Reclamation with Trees Conference, Owensboro, Kentucky June 7–8 1984*, ed. P. E. Pope. Kentucky: Madisonville Community College.

552. McKay, T. R., McNally, P. and Kvasnicka, J. (1984). Overview of Recent Studies on Radium in Mussels in the Alligator River Region. Department of Mines and Energy, Mines Division, Northern Territory (unpublished manuscript).

553. MacMahon, J. A. (1988). Arthropod recolonization of reclaimed strip-mines in south-western Wyoming: 1. The ground-dwelling beetle fauna (Coleoptera). *Environmental Entomology* (in press).

554. McRae, S. G. (1986). Opportunities for creative reclamation following sand and gravel extraction. In *New Horizons for Mine Land Reclamation*. Proceedings. American Society for Surface Mining and Reclamation. 1986 National Meeting, Jackson, Mississippi, pp. 51–3.

555. Mader, H. (1984). The succession of carabid and spider communities in a charcoal mining area and the influence of different afforestation practices. In *The XVII International Congress of Entomology, Hamburg*. (abstracts).

556. Mahoney, C. L. (1976). Soil insects as indicators of use patterns in recreation areas. *Journal of Forestry*, **74**, 35–7.

557. Majer, J. D. (1977). The possible use of ant faunas as indicators of land use and of the success of revegetation attempts. In *Proceedings, Western Australian Entomology Workshop, Perth*, ed. J. D. Majer, pp. 8.

558. Majer, J. D. (1978*a*). Preliminary survey of the epigaeic invertebrate fauna with particular reference to ants, in areas of different land use at Dwellingup, Western Australia. *Forest Ecology and Management*, **1**, 321–34.

559. Majer, J. D. (1978*b*). *Studies on Invertebrates in Relation to Bauxite Mining Activities in the Darling Range – A Review of the First Eighteen Months Research*. Environmental Research Bulletin no. 3. Perth: Alcoa of Australia Limited.

560. Majer, J. D. (1978*c*). The ecology of ants in relation to bauxite mine rehabilitation. *Bulletin of the Ecological Society of Australia*, **8**, 9.

561. Majer, J. D. (1978*d*). The importance of invertebrates in successful land reclamation with particular reference to bauxite mine rehabilitation. In *Rehabilitation of Mined Lands in Western Australia*, ed. J. E. D. Fox, pp. 47–61. Bentley, W.A.: WAIT.

562. Majer, J. D. (1980*a*). *A Preliminary Ecological Survey of the Wagerup Ant Fauna*. Environmental Research Bulletin no. 7. Perth: Alcoa of Australia Limited.

563. Majer, J. D. (1980*b*). Soil and litter invertebrates of the jarrah forest and woodland regions. In *Managing Nitrogen Economies of Natural and Man Made Forest Ecosystems*, ed. R. A. Rummery and F. J. Hingston, pp. 47–59. Perth: CSIRO.

564. Majer, J. D. (1980*c*). The influence of ants on broadcast and naturally spread seeds in rehabilitated bauxite mined areas. *Reclamation Review*, **3**, 3–9.

565. Majer, J. D. (1981). *The Role of Invertebrates in Bauxite Mine Rehabilitation*. Forests Department of Western Australia Bulletin no. 93.

566. Majer, J. D. (1983*a*). Ants: bio-indicators of minesite rehabilitation, land-use, and land conservation. *Environmental Management*, **7**, 375–83.

567. Majer, J. D. (1983*b*). Assessing and promoting invertebrate return in revegetated minesites. In *Australian Mining Industrial Council Environmental Workshop, Hunter Valley*, pp. 18–36.

568. Majer, J. D. (1983*c*). Recolonisation by ants in rehabilitated open-cut mines in northern Australia. *Reclamation and Revegetation Research*, **2**, 279–98.

569. Majer, J. D. (1983*d*). *The Role of Fauna in Mined Land Rehabilitation*. WAIT School of Biology Bulletin no. 7, pp. 40–55.

570. Majer, J. D. (1983*e*). Zones of potential land rehabilitation rate for the Australian subcontinent. *Landline*, **8**, 5–6.

571. Majer, J. D. (1983*f*). The influence of ants on seeding operations in northern Australian mined areas. *Reclamation and Revegetation Research*, **2**, 299–313.

572. Majer, J. D. (1984*a*). Ant return in rehabilitated mines – an indicator of ecosystem resilience. In *Proceedings of the 4th International Conference on Mediterranean Ecosystems held at Perth, Western Australia*, ed. B. Dell, pp. 105–8. Perth: Botany Department, University of Western Australia.

573. Majer, J. D. (1984*b*). Ant return in rehabilitated mineral sand mines on North Stradbroke island. In *Focus on Stradbroke: new information on North Stradbroke island and surrounding areas*, ed. R. J. Coleman, J. Covacevich and P. Davies, pp. 325–32. Brisbane: Boolarong Press.

574. Majer, J. D. (1985*a*). Recolonization by ants of rehabilitated mineral sand mines on North Stradbroke Island, Queensland, with particular reference to seed removal. *Australian Journal of Ecology*, **10**, 31–48.

575. Majer, J. D. (1985*b*). Restoration of pre-mining ecosystems in forested and agricultural areas. *Landline*, **11**, 4.

576. Majer, J. D., Day, J. E., Kabay, E. D. and Perriman, W. S. (1984). Recolonization by ants in bauxite mines rehabilitated by a number of different methods. *Journal of Applied Ecology*, **21**, 355–75.

577. Majer, J. D. and Dunlop, J. N. (1984). *The Importance of Animals and Fauna Studies in Relation*

to Minesite Rehabilitation. 1984 Northern Australia Mine Rehabilitation Workshop (unpublished manuscript).

578. Majer, J. D. and Kabay, E. D. (1979). A case for considering invertebrates in rehabilitated mines. In *Australian Mining Industry Council Environmental Workshop, Bunbury.*
579. Majer, J. D. and Kabay, E. D. (1980). Invertebrate studies in rehabilitated areas. *Landline,* **3**, 7.
580. Majer, J. D., Sartori, M., Stone, R. and Perriman, W. S. (1982). Recolonisation by ants and other invertebrates in rehabilitated mineral sand mines near Eneabba, Western Australia. *Reclamation and Revegetation Research*, **1**, 63–81.
581. Maltby, E. (1975). Numbers of soil microorganisms as ecological indicators of changes resulting from moorland reclamation on Exmoor, U.K. *Journal of Biogeography*, **2**, 117–36.
582. Marion, W. R., Maehr, D. S. and Frolich, R. K. (1981). Phosphate mine reclamation and habitats for wildlife. In *Surface Mining Hydrology and Reclamation Symposium, Lexington, Kentucky, December 1981*, pp. 501–6.
583. Marion, W. R. and O'Meara, T. E. (1983). Phosphate reclamation plans and changes in wildlife habitat diversity. In *Reclamation and the Phosphate Industry. Proceedings of the Symposium, Clearwater Beach, Florida 26–28 January, 1983*, ed. D. J. Robertson, Florida Institute of Phosphate Research.
584. Markham, O. D., ed. (1983). *Idaho National Engineering Laboratory Radioecology and Ecology Programs: 1983 Progress Report.* US Department of Energy (Radiological and Environmental Sciences Laboratory) Report no. DOE/ID-12098.
585. Marrs, R. H., Granlund, I. H. and Bradshaw, A. D. (1980). Ecosystem development on reclaimed china clay wastes. IV. Recycling of above-ground plant nutrients. *Journal of Applied Ecology*, **17**, 803–13.
586. Martin, J. L. (1964). The insect ecology of red pine plantations in central Ontario III. Soil-surface fauna as indicators of stand change. *Proceedings of the Entomological Society of Ontario*, **95**, 87–102.
587. Martin, J. L. (1964). The insect ecology of old-field red pine plantations in central Ontario I. Description of the study area. *Proceedings of the Entomological Society of Ontario*, **95**, 70–87.
588. Martin, J. L. (1966). The insect ecology of red pine plantations in central Ontario IV. The crown fauna. *The Canadian Entomologist*, **98**, 10–27.
589. Martin, S. B. and Platts, W. S. (1981). *Influence of Forest and Rangeland Management on Anadromous Fish Habitat in Western North America: Effects of Mining.* US Forestry Service General Technical Report.
590. Marx, D. H. (1975). Mycorrhizae and establishment of trees on strip-mined land. *Ohio Journal of Science*, **75**, 288–97.
591. Mason, H. L. (1980). Deterioration of marsh in San Francisco Bay. In *Rehabilitation and Creation of Selected Coastal Habitats. Proceedings of a Workshop, Sapelo Island, Georgia, May 1976*, ed. J. C. Lewis and E. W. Bunce, pp. 53–4. US Fish and Wildlife Service, Biological Services Program.
592. Mason, W. T. Jr., ed. (1978). *Methods for Assessment and Prediction of Mineral Mining Impacts on Aquatic Communities: A Review and Analysis.* US Fish and Wildlife Service, Biological Services Program.
593. Mason, W. T. Jr. (1980). *Guidance Documents for Protection and Enhancement of Fish and Wildlife on Midwestern Surface Mined Lands – Illinois.* US Fish and Wildlife Service, Eastern Energy and Land Use Team.
594. Mason, W. T. and Ikers, S. (1982). *Research on Fish and Wildlife Habitat.* Washington: US Environmental Protection Agency.

595. Master, W. A. (1979). *Characterization of Three Acid Strip Mine Lakes in Grundy County, Illinois*, Argonne National Laboratory Report ANL/LRP-TM-15.

596. Mathisen, J. E. (1982). Wildlife values and management of gravel pits in forest ecosystems. In *The Wildlife Values of Gravel Pits Symposium, Crookston, Minnesota, June 1982*, ed. W. D. Svedarsky and R. D. Crawford, pp. 155–8. University of Minnesota Miscellaneous Publication 17–1982.

597. Matter, W. J. and Ney, J. J. (1981). The impact of surface mine reclamation on headwater streams in southwest Virginia. *Hydrobiologia*, **78**, 63–71.

598. Maupin, J. K., Wells, J. R. and Leist, C. (1954). A preliminary survey of food habits of the fish and physio-chemical conditions of the water of three strip-mine lakes. *Transactions of the Kansas Academy of Science*, **57**, 164–71.

599. Maurer, M. A. and Brusven, M. A. (1983). Insect abundance and colonization rate in *Fontinalis neo-mexicana* (Bryophyta) in an Idaho batholith stream, U.S.A. *Hydrobiologia*, **98**, 9–15.

600. Maurer, R. (1974). Die Vielfalt der Käfer- und Spinnenfauna des Wiesenbodens im Einflussbereich von Verkehrsimmissionen. *Oecologia (Berlin)*, **14**, 327–51.

601. Medcraft, J. R. and Clark, W. R. (1986). Big game habitat use and diets on a surface mine in northeastern Wyoming. *Journal of Wildlife Management*, **50**, 135–42.

602. Meijer, J. (1971). Immigration of arthropods into the new Lauwerszee polder. *Miscellaneous Papers Landbouwhogeschool Wageningen*, **8**, 53–64.

603. Meijer, J. (1972). An isolated earthworm population in the recently reclaimed Lauwerszeepolder. *Pedobiologia*, **12**, 409–11.

604. Meijer, J. (1974*a*). A comparative study of the immigration of carabids (Coleoptera, Carabidae) into a new polder. *Oecologia (Berlin)*, **16**, 185–208.

605. Meijer, J. (1974*b*). De afloop van een rupsenplaag. *Entomologische Berichten (Amsterdam)*, **34**, 121–2.

606. Meijer, J. (1975). Carabid (Coleoptera, Carabidae) migration studied with Laboulbeniales (Ascomycetes) as biological tags. *Oecologia (Berlin)*, **19**, 99–103.

607. Meijer, J. (1977). The immigration of spiders (Araneida) into a new polder. *Ecological Entomology*, **2**, 81–90.

608. Meijer, J. (1980). The development of some elements of the arthropod fauna of a new polder. *Oecologia (Berlin)*, **45**, 220–35.

609. Messner, B. (1963). Über das Vorkommen von Labidura Riparia (pall.) auf den Abraumhalden der Braunkohlentagebaue um Trobitz und Lauchhammer. *Entomologische Berichten*, **1**, 24–8.

610. Mettee, M. F. and O'Neil, P. E. (1985). Changes in the water quality and fish community structure as related to surface mining for coal in the Tyro Creek watershed, Alabama. In *Fish and Wildlife Relationships to Mining: Symposium Proceedings*, ed. L. B. Starnes, pp. 51–8. American Fisheries Society (Water Quality Section).

611. Meyn, R. L. and Young, R. P. (1976). *Reseeding Mine Spoil to Sustain Itself and Support Livestock*. Montana Agricultural Experiment Station, Montana State University, Bozeman, Montana. Capsule Information Series no. 11.

612. Michael, E. D. (1976). *Effects of Highways on Wildlife*. West Virginia Department of Highways Research Project no. 42. West Virginia University Agricultural Experimental Station, Division of Forestry Report no. 72-1B.

613. Michael, E. D. (1980). *Use of Different Roadside Cover Plantings by Wildlife*. NTIS Report PB81-108458.

614. Milne, B. (1969). Birds and gravel pits. *Countryside*, 1969, pp. 164–8.

615. Milne, B. (1974). Ecological succession and bird life at a newly excavated gravel pit. *Gravel Study*, **21**, 263–78.

616. Mindell, D. P. (1978). *Habitat Use by Red-Tailed Hawks in Surface Mined Areas.* M.Sc. Thesis, West Virginia University (unpublished manuscript).

617. Mindell, D. P. and Samuel, D. E. (1984). Habitat use by red-tailed hawks in surface-mined areas. *Minerals and the Environment*, **4**, 144–9.

618. Mitchell, M. J. (1961). Breeding bird populations in relation to grassland succession on the Anoka sand plain. *Flicker*, **33**, 102–8.

619. Mitsch, W. J., Cardamone, M. A., Taylor, J. R. and Hill, Jr., P. L. (1985). Wetlands and water quality management in the eastern interior coal basin. In *Wetlands and Water Management on Mined Lands. Proceedings of a Conference October 23–24, 1985*, ed. R. P. Brooks, D. E. Samuel and J. B. Hill, pp. 121–37. Pennsylvania State University.

620. Mittelbach, G. G. and Gross, K. L. (1984). Experimental studies of seed predation in old-fields. *Oecologia (Berlin)*, **65**, 7–13.

621. Mook, J. H. (1971). Observations on the colonization of the new Ijssselmeer-polders by animals. *Miscellaneous Papers Landbouwhogeschool Wageningen*, **8**, 13–31.

622. Moore, F. R. (1982). Ecology of the Pauropoda from a coal shale heap in Lancashire, England. *Pedobiologia*, **24**, 309–17.

623. Moore, F. R. and Luxton, M. (1986). The collembolan fauna of two coal shale tips in north-west England. *Pedobiologia*, **29**, 359–66.

624. Moore, R. and Mills, T. (1977). *An Environmental Guide to Western Surface Mining. Part One: Federal Leasable and Locatable Mineral Regulations.* US Fish and Wildlife Service Report.

625. Moore, R. and Mills, T. (1977). *An Environmental Guide to Western Surface Mining. Part Two: Impacts, Mitigation and Monitoring.* US Fish and Wildlife Service Report.

626. Morgan, R. L. (1973). Environmental impact of surface mining: the biologist's viewpoint. In *Some Environmental Aspects of Strip Mining in North Dakota*, ed. M. K. Wali, pp. 61–71. North Dakota Geological Survey, Educational Series 5.

627. Morin, M. D. (1984). A computerized approach to fish and wildlife planning on mine lands. In *Proceedings of the Fourth Annual Better Reclamation with Trees Conference, Owensboro, Kentucky June 7–8 1984*, ed. P. E. Pope. Kentucky: Madisonville Community College.

628. Morris, R. A. (1982). Regulatory and land use aspects of sand and gravel mining as they affect reclamation for wildlife habitat and open space: a national perspective. In *The Wildlife Values of Gravel Pits Symposium, Crookston, Minnesota, June 1982*, ed. W. D. Svedarsky and R. D. Crawford, pp. 16–23. University of Minnesota Miscellaneous Publication 17-1982.

629. Morris, R. G. and Nichols, B. E. (1979). Wildlife habitat evaluation and mitigation. In *The Mitigation Symposium: A National Workshop on Mitigating Losses of Fish and Wildlife Habitats*, ed. G. A. Swanson, pp. 369–77. Rocky Mountain Forest Range Experiment Station (Forest Service) General Technical Report.

630. Morrison, D. G. (1982). Principles of revegetating mined lands. In *The Wildlife Values of Gravel Pits Symposium, Crookston, Minnesota, June 1982*, ed. W. D. Svedarsky and R. D. Crawford, pp. 51–8. University of Minnesota Miscellaneous Publication 17-1982.

631. Moss, B. (1986). Restoration of lakes and lowland rivers. In *Ecology and Design in Landscape*, ed. A. D. Bradshaw, D. A. Goode and E. H. P. Thorpe, pp. 399–415. Oxford: Blackwell.

632. Moulton, M. P., Choate, J. R. and Bissell, S. J. (1981). Small mammals on revegetated agricultural land in eastern Colorado. *The Prairie Naturalist*, **13**, 99–104.

633. Mumford, R. E. and Bramble, W. C. (1973). Small mammals on surface-mined land in southwestern Indiana. In *Ecology and Reclamation of Devastated Land*, vol. 1, ed. R. J. Hutnik and G. Davis, pp. 369–76. New York: Gordon and Breach.

634. Muncy, J. A. (1984). Reclamation of 'orphan' or abandoned coal surface mines in southern Appalachia for productive forest and wildlife use. In *The Better Reclamation with Trees*

Conference, June 1984, Owensboro, Kentucky, ed. P. E. Pope. Kentucky: Madisonville Community College.

635. Munsee, J. R. and Schrock, J. R. (1983). A comparison of ant faunae on unreclaimed stripmines in Indiana. *Proceedings of the Indiana Academy of Science*, pp. 257–61.

636. Murdoch, W. W., Evans, F. C. and Peterson, C. H. (1972). Diversity and patterns in plants and insects. *Ecology*, **53**, 819–29.

637. Myers, C. W. and Klimstra, W. D. (1963). Amphibians and reptiles of an ecologically disturbed (strip-mined) area in southern Illinois. *American Midland Naturalist*, **70**, 126–32.

638. Naeth, M. A. and Bailey, A. W. (1986). The effect of grazing regime on revegetation of rangeland after pipeline construction. Abstracts of papers. *39th Annual Meeting Society for Range Management*. Kissimee, Florida. no. 66.

639. National Research Council U.S. (1981). *Surface Mining: Soil, Coal, and Society*. Committee on soil as a resource in relation to surface mining for coal. Washington: National Academy Press.

640. Nawrot, J. R. (1985). Wetland development on coal mine slurry impoundments: principles, planning, and practices. In *Wetlands and Water Management on Mined Lands, Proceedings of a Conference October 23–24, 1985*, ed. R. P. Brooks, D. E. Samuel and J. B. Hill, pp. 161–72. Pennsylvania State University.

641. Nawrot, J. R., Fuson, M. L. and Downing, D. M. (1981). Reedgrass and slurry pond reclamation. *Mining Congress Journal*, **23**, 23–8.

642. Nawrot, J. R. and Klimstra, W. D. (1977). Illinois mined land inventories: their implementation and utilization. In *The Fifth Symposium of Surface Mining and Reclamation, NCA/BCR Conference and Expo IV, Louisville, Kentucky*, pp. 54–60.

643. Nawrot, J. R. and Klimstra, W. D. (1983). Fish and wildlife: permit requirements and assessment practices. In *Surface Mining Environment Monitoring and Reclamation Handbook*, ed. L. V. A. Sendlein, H. Yazicigil and C. L. Carlson, pp. 509–33. New York: Elsevier.

644. Nawrot, J. R., Woolf, A. and Klimstra, W. D. (1982). *A Guide for Enhancement of Fish and Wildlife on Abandoned Mined Lands in the Eastern United States*. US Fish and Wildlife Service, Eastern Energy Land Use Team Report.

645. Nawrot, J. R. and Yaich, S. C. (1982*a*). Habitat enhancement of coal mine slurry impoundments through wetland development. In *Proceedings of the Ninth Annual Conference of Wetland Restoration and Creation, Tampa, Florida, May 1982*, pp. 228–42.

646. Nawrot, J. R. and Yaich, S. C. (1982*b*). Wetland development potential of coal mining tailings basins. In *Wetlands*, vol. 2, Third Annual Meeting of Society of Wetland Scientists, Wrightsville, North Carolina, May 1982, pp. 179–90.

647. Neuman, D. R. and Munshower, F. F. (1984). Copper and molybdenum uptake by forages grown on coalmine soils. *Journal of Range Management*, **37**, 517–20.

648. Neumann, U. (1971*a*). Die Ausbreitungsfähigkeit von Carabiden in den forstlichen Rekultivierungen des rheinischen Braunkohlenreviers. *Miscellaneous Papers Landbouwhogeschool Wageningen*, **8**, 89–104.

649. Neumann, U. (1971*b*). Die Sukzession der Bodenfauna (Carabidae [Coleoptera], Diplopoda und Isopoda) in den forstlich rekultivierten Gebieten des rheinischen Braunkohlenreviers. *Pedobiologia*, **11**, 193–226.

650. Neumann, U. (1973). Succession of soil fauna in afforested spoil banks of the brown-coal mining district of Cologne. In *Ecology and Reclamation of Devastated Land*, vol. 1, ed. R. J. Hutnik and G. Davis, pp. 335–48. New York: Gordon and Breach.

651. Nicholls, A. O., ed. (1978). *Report on a Workshop on Succession and Disturbance in Australia held at Shoalwater Bay, Queensland, 26 November to 2 December 1977*. CSIRO Division of Land Use Research (Canberra) Technical Memorandum 78/26.

652. Nichols, O. G. (1981). Fauna research in relation to bauxite mining and rehabilitation. *Landline*, **5**, 4–5.

653. Nichols, O. G. (1983). Succession and long-term stability in fauna populations in revegetated minesites. In *Australian Mining Industry Council Environmental Workshop, Hunter Valley*, pp. 100–12.

654. Nichols, O. G. (1984). *Designing and Conducting Fauna Studies for Mine Rehabilitation Programmes*. 1984 Northern Australia Mine Rehabilitation Workshop (unpublished manuscript).

655. Nichols, O. G. and Bamford, M. J. (1985). Reptile and frog utilisation of rehabilitated bauxite minesites and dieback-affected sites in Western Australia's jarrah (*Eucalyptus marginata*) forest. *Biological Conservation*, **34**, 227–49.

656. Nichols, O. G. and Bunn, S. (1980). *Termite Utilisation of Rehabilitated Bauxite Mined Areas*. Environmental Research Bulletin no. 9. Perth: Alcoa of Australia Limited.

657. Nichols, O. G. and Burrows, R. (1985). Recolonisation of revegetated bauxite minesites by predatory invertebrates. *Forest Ecology and Management*, **10**, 49–64.

658. Nichols, O. G., Carbon, B. A., Colquhoun, I. J., Croton, J. T. and Murray, N. J. (1985). Rehabilitation after bauxite mining in south-western Australia. *Landscape Planning*, **12**, 75–92.

659. Nichols, O. G. and Michaelsen, D. V. (1985). An inventory of fauna research and surveys in relation to mining and revegetation in Australia. *Landline*, **11**, 1–13.

660. Nichols, O. G. and Watkins, D. (1984). Bird utilization of rehabilitated bauxite minesites in Western Australia. *Biological Conservation*, **30**, 109–31.

661. Nichols, O. G., Watkins, D. and Kabay, E. D. (1982). The distribution of the red-eared firetail *Emblema oculata* in relation to bauxite mining in the northern jarrah forest. *Emu*, **82**, 169–72.

662. Nijhoff, P. (1979). Lake Baikal endangered by pollution. *Environmental Conservation*, **6**, 111–15.

663. Ohlsson, K. E., Robb, Jr., A. E. and Guindon, Jr., C. E. (1982). *Best Current Practices for Fish and Wildlife on Surface-Mined Lands in the Northern Appalachian Coal Region*. US Fish and Wildlife Service Report.

664. Ohlsson, K. E., Robb, Jr., A. E., Guiddon, Jr., C. E., Baker, Jr., M. and Slaski, L. J. (1981). Management practices for fish and wildlife on surface mined lands in northern Appalachia. In *The Surface Mining Hydrology and Reclamation Symposium, Lexington, Kentucky, December 1981*, pp. 517–20.

665. Oinonen, E. A. (1956). On the ants of the rocks and their contribution to the afforestation of rocks in southern Finland. *Acta Entomologica Fennica*, **12**, 179–213.

666. Okruszko, H. (1985). Decession in the natural evolution of low peatlands. In *Soil Ecology and Management*, ed. J. H. Cooley, pp. 89–132. Athens, Georgia: International Association for Ecology, Institute of Ecology, University of Georgia.

667. Oldenburg, R. C., Montgomery, R. A., Harder, T. M. and Burger, G. V. (1982). Gravel pits as fish and wildlife habitat at the Max McGraw Wildlife Foundation. In *The Wildlife Values of Gravel Pits Symposium, Crookston, Minnesota, June 1982*, ed. W. D. Svedarsky and R. D. Crawford, pp. 215–20. University of Minnesota Miscillaneous Publication 17-1982.

668. O'Leary, W., Klimstra, W. D. and Nawrot, J. R. (1984). Waterfowl habitats on reclaimed surface mined lands in southwestern Illinois. In *The National Symposium on Surface Mining Hydrology, Sedimentology, and Reclamation, Lexington, December 2–7 1984*, ed. H. Graves, pp. 377–82. Lexington: University of Kentucky.

669. Olson, R. A. (1981). *Wetland Vegetation, Environmental Factors, and their Interaction in Stripmine Ponds, Stockdams, and Natural Wetlands*. Rocky Mountain Forest and Range Experiment Station (Forest Service) General Technical Report.

670. Olson, R. A. and Barker, W. T. (1979). Strip-mine impoundments for the birds. *Rangelands*, **1**, 248–9.

671. Osborne, L. L., Davies, R. W. and Linton, K. J. (1979). Effects of limestone strip mining on benthic macroinvertebrate communities. *Water Research*, **13**, 1285–90.

672. Owen, M. (1983). The importance of man-made wetlands to wildfowl wintering in Britain. In *The Proceedings of a Symposium on the Subject of Wildlife on Man-Made Wetlands*, pp. 73–97. Milton Keynes, England: ARC Wildlife Centre.

673. Pandit, A. K. and Fotedar, D. N. (1982). Restoring damaged wetlands for wildlife. *Journal of Environmental Management*, **14**, 359–68.

674. Parker, L. W., Elkins, N. Z., Aldon, E. F. and Whitford, W. G. (1984). Development of soil biota and some aspects of nitrogen cycling on reclaimed coal mine spoils in the arid southwest. *Journal of Environmental Quality*, **19**, 215–19.

675. Parmenter, R. R. and MacMahon, J. A. (1986). Early-successional patterns of arthropod recolonization on reclaimed strip-mines in southwestern Wyoming: the ground-dwelling beetle fauna (Coleoptera). *Environmental Entomology*, **16**, 168–75.

676. Parmenter, R. R., MacMahon, J. A., Waaland, M. E., Stuebe, M. M., Landers, P. and Crisafulli, C. M. (1985). Reclamation of surface coal mines in western Wyoming for wildlife habitat: a preliminary analysis. *Reclamation and Revegetation Research*, **4**, 93–115.

677. Parsons, D. (1964). Comparative limnology of strip-mine lakes. *Verhandlungen der Internationalen Vereinigung fuer Theoretische und Angewandte Limnologie*, **15**, 293–8.

678. Parsons, W. F. J. and Parkinson, D. (1986). Species composition, distribution and abundance of Collembola colonizing reclaimed mine spoil in Alberta. *Pedobiologia*, **29**, 33–45.

679. Patchett, J. M. (1982). Conservation, recreation, mining, and rehabilitation in Story County, Iowa. In *The Wildlife Values of Gravel Pits Symposium, Crookston, Minnesota, June 1982*, ed. W. D. Svedarsky and R. D. Crawford, pp. 89–94. University of Minnesota Miscillaneous Publication 17-1982.

680. Peltz, L. R. and Maughan, O. E. (1978). Analysis of fish populations and selected physical and chemical parameters of five strip mine ponds in Wise County, Virginia with implications for management. In *Surface Mining and Fish/Wildlife Needs in the Eastern United States*, ed. D. E. Samuel, J. R. Stauffer, C. H. Hocutt and W. T. Mason, pp. 171–6. Morgantown: West Virginia University.

681. Pentecost, E. D. (1979). Wildlife investigations. In *Staunton I Reclamation Demonstration Project Progress Report II*. Argonne National Laboratory Report ANL/LRP-4, pp. 81–93.

682. Pentecost, E. D. (1980). Habitat restoration and wildlife utilization of coal surface-mine lands. In *AUA-ANL Biology Symposium April 21–22, 1980*. Argonne National Laboratory.

683. Pentecost, E. D. and Stupka, R. C. (1978). Wildlife investigations at a coal refuse reclamation site in southern Illinois. In *Surface Mining and Fish/Wildlife Needs in the Eastern United States*, ed. D. E. Samuel, J. R. Stauffer, C. H. Hocutt and W. T. Mason, pp. 107–18. Morgantown: West Virginia University.

684. Pentecost, E. D. and Stupka, R. C. (1978). Wildlife Investigations at a Coal Refuse Reclamation Site in Southern Illinois. NTIS Report CONF-781240-1.

685. Perkins, M. P., Klimstra, W. D. and Nawrot, J. R. (1988). *Potential hazards and benefits of highwalls* (in press).

686. Perkins, G. A. and Lawrence, J. S. (1985). Bird use of wetlands created by surface mining. *Transactions Illinois Academy of Science*, **78**, 87–95.

687. Perry, E. F. (1978). Minesoil properties that influence fish and wildlife habitats. In *Surface Mining and Fish/Wildlife Needs in the Eastern United States*, ed. D. E. Samuel, J. R. Stauffer, C. H. Hocutt and W. T. Mason, pp. 223–31. Morgantown: West Virginia University.

688. Peterson, E. B. and Etter, H. M. (1970). *A Background for Disturbed Land Reclamation and Research in the Rocky Mountain Region of Alberta.* Canadian Forestry Service, Department of Fisheries and Forestry. Forest Research Laboratory, Edmonton, Alberta, Information Report A-X-34.

689. Phillips, G. R. and Russo, R. C. (1978). *Metal Bioaccumulation in Fishes and Aquatic Invertebrates: A Literature Review.* US Environmental Protection Agency Report EPA-600/3-78-103.

690. Pickard, J. (1982). Catastrophic disturbance and vegetation on Little Slope, Lord Howe Island. *Australian Journal of Ecology*, **7**, 161–70

691. Piearce, T. G. (1982). Recovery of earthworm populations following salt-water flooding. *Pedobiologia*, **24**, 91–100.

692. Pietz, R. I., Peterson, J. R., Prater, J. E. and Zenz, D. R. (1984). Metal concentrations in earthworms from sewage sludge – amended soils at a strip mine reclamation site. *Journal of Environmental Quality*, **13**, 651–4.

693. Plass, W. T. (1975). Reclamation of surface mined land. *Ohio Journal of Science*, **75**, 298–304.

694. Polman, G. K. R. (1980). *The Meadowbird Areas of the 'Kievitslanden' in Flevoland Management Plan.* Rijksdienst voor de Ijsselmeerpolders. Flevobericht no. 160.

695. Polman, G. K. R. (1981). Water management in the Oostvaarderplassen. Rijksdienst voor de Ijsselmeerpolders. *Overdruk* no. 205, pp. 165–76.

696. Polman, G. K. R. (1982). The Oostvaardersplassen, the developing of marshy ecosystems especially for waterfowl. Rijksdienst voor de Ijsselmeerpolders. *Overdruk* no. 287, pp. 614–25.

697. Ponder, Jr., F. (1980). *Rabbits and Grasshoppers: Vectors of Endomycorrhizal Fungi on New Coal Mine Spoil.* USDA Forest Service Research Note no. NC-250.

698. Porter, K. J. and Hakanson, D. E. (1976). Toxicity of mine drainage to embryonic and larval boreal toads (Bufonidae: *Bufo boreas*). *Copeia*, **2**, 327–31.

699. Postovit, B. C. (1981). *Suggestions for Sage Grouse Habitat Reclamation on Surface Mines in Northeastern Wyoming.* MSc Thesis University of Wyoming.

700. Prachar, R. E. (1984). *Establishment, Survival, and Growth of Aspen on an Abandoned Surface Mine.* M.Sc. Thesis, West Virginia University (unpublished manuscript).

701. Prachar, R. E. and Samuel, D. E. (1988*a*). Deer aspen relationship on an abandoned surface mine. *Journal of Applied Forestry* (in press).

702. Prachar, R. E. and Samuel, D. E. (1988*b*). Growing season effectiveness of big game repellent in reducing deer damage. *Transactions Northeast Fish and Wildlife Conference* (in press).

703. Pratt, J. R., Cairns, Jr., J. and Stewart, P. M. (1985). Development of microbial communities in mined lakes. In *Wetlands and Water Management on Mined Lands. Proceedings of a Conference October 23–24, 1985*, ed. R. P. Brooks, D. E. Samuel and J. B. Hill, pp. 239–59. Pennsylvania State University.

704. Proctor, B. R., Thompson, R. W., Bunin, J. E., Fucik, K. W., Tamm, G. R. and Wolf, E. G. (1983*a*). *Practices for Protecting and Enhancing Fish and Wildlife on Coal Surface-Mined Land in the Green River–Hams Fork Region.* US Fish and Wildlife Service Report.

705. Proctor, B. R., Thompson, R. W., Bunin, J. E., Fucik, K. W., Tamm, G. R. and Wolf, E. G. (1983*b*). *Practices for Protecting and Enhancing Fish and Wildlife on Coal Surface-Mined Land in the Powder River–Fort Union Region.* US Fish and Wildlife Service Report.

706. Proctor, B. R., Thompson, R. W., Bunin, J. E., Fucik, K. W., Tamm, G. R. and Wolf, E. G. (1983*c*). *Practices for Protecting and Enhancing Fish and Wildlife on Coal Mined Land in the Uinta-Southwestern Utah Region.* US Fish and Wildlife Service Report.

707. Purvis, G. and Curry, J. P. (1980). Successional changes in the arthropod fauna of a new

ley pasture established on previously cultivated arable land. *Journal of Applied Ecology*, **17**, 309–21.

708. Radvanyi, A. (1980). Control of small mammal damage in the Alberta Oil sands reclamation and afforestation program. *Forest Science*, **26**, 687–702.

709. Rafaill, B. L. and Vogel, W. G. (1978). *A Guide for Vegetating Surface-Mined Lands for Wildlife in Eastern Kentucky and West Virginia*. US Fish and Wildlife Service Report.

710. Ramani, R. V. and Clar, M. L. (1978). *User's Manual for Premining Planning of Eastern Surface Coal Mining, vol. 1: Executive Summary*. US Environmental Protection Agency Report no. EPA-600/7-78-180.

711. Randall, D. (1982). Behemoths at Black Butte. *Defenders* **57**, 18–27.

712. Ranta, E. and As, S. (1982). Nonrandom colonization of habitat islands by carabid beetles. *Annales Zoologici Fennici*, **19**, 175–82.

713. Rappoport, A. (1979). The mitigation problem: a background for the mitigation symposium. In *The Mitigation Symposium: A National Workshop on Mitigating Losses of Fish and Wildlife Habitats*, ed. G. A. Swanson, pp. 1–5. Rocky Mountain Forest and Range Experiment Station (Forest Service) General Technical Report.

714. Ratti, E. (1979). The reclaimed lands of the Middle Lagoon, south of Venice. V. The Coleoptera of the artificial island D-E. *Societa Veneziana di Scienze Naturali Lavori*, **4**, 115–69.

715. Rawson, J. W. (1971). Surface mining and wildlife. In *Proceedings of the Revegetation and Economic Use of Surface-mined Land and Mine Refuse Symposium*, ed. D. M. Bondurant, pp. 37–9. West Virginia University, College of Agriculture and Forestry.

716. Reed, E. B. (1975). Limnological characteristics of strip mine ponds in northwestern Colorado, U.S.A. *Verhandlungen der Internationalen Vereinigung fuer Theoretische und Angewandte Limnologie*, **19**, 856–65.

717. Reeders, A. P. (1983). Assessing and promoting vertebrate re-colonisation of regenerated minesites at Weipa. In *Australian Mining Industrial Council Environmental Workshop, Hunter Valley*. pp. 37–49.

718. Reeders, A. P. (1985). Vertebrate fauna in regenerated mines at Weipa, north Queensland. In *North Australian Mine Rehabilitation Workshop no. 9, Weipa, June 1985*, ed. J. W. Lawrie, pp. 105–18. Comalco Aluminium Limited.

719. Reeders, A. P. F. and Morton, A. G. (1983). *Vertebrate Fauna in the Regenerated Mines at Weipa*. Report to Comalco Limited, Perth.

720. Reeves, P. and Deland, M. (1977). Implementation of the surface mining control and reclamation act of 1977. In *Reclamation for Wildlife Habitat. Proceedings of Reclamation Workshop II, Fort Collins, Colorado, September 1977*, ed. Ecology Consultants Inc., pp. 86–112. Environmental Research and Technology Inc.

721. Rehwald, R. (1970). *Versuche zur Anwendung des Netzbeuteltestes mit Blattmaterial zur Charakterisierung der bobenbiologischen Aktivität von Standorten mit Roh, Moder und Müllhumus*. Diplomarbeit (Gorlitz) PH Potsdam.

722. Reimold, R. J. (1980). Marsh creation: impact of pesticides on the fauna, use of infrared photography, ditching and diking. In *Rehabilitation and Creation of Selected Coastal Habitats. Proceedings of a Workshop, Sapelo Island, Georgia, May 1976*, ed. J. C. Lewis and E. W. Bunce, pp. 132–5. Washington DC: US Fish and Wildlife Service, Biological Services Program.

723. Renno, O. (1974). Bird fauna of the West-Estonian fens and its variation during the reclamation of fens. In *Estonian Wetlands and their Life*, ed. E. Kumari, pp. 204–13. Academy of Sciences of the Estonian SSR. International Biological Programme. Estonian Contribution no. 7.

724. Reynolds, T. D. and Trost, C. E. (1979). The effect of crested wheatgrass plantings on

wildlife on the Idaho National Engineering Laboratory site. In *The Mitigation Symposium: A National Workshop on Mitigating Losses of Fish and Wildlife Habitats*, ed. G. A. Swanson, pp. 665–6. Rocky Mountain Forest Range Experiment Station (Forest Service) General Technical Report.

725. Rhee, J. A. van (1969*a*). Inoculation of earthworms in a newly drained polder. *Pedobiologia*, **9**, 128–32.

726. Rhee, J. A. van (1969*b*). Development of earthworm populations in polder soils. *Pedobiologia*, **9**, 133–40.

727. Rhodes, M. J., Cloud, T. J. Jr. and Haag, D. (1983). Habitat evaluation procedures for planning surface mine reclamation in Texas. *The Wildlife Society Bulletin*, **11**, 222–32.

728. Richards, T. W., McComb, W. C. and Vogel, W. G. (1984). Seed predation reduces stocking of northern red oak direct-seeded on revegetated mine sites. In *Proceedings of the Fourth Annual Better Reclamation with Trees Conference, Owensboro, Kentucky June 7–8 1984*, ed. P. E. Pope. Kentucky: Madisonville Community College.

729. Richardson, B. Z. (1980). Reclamation of acid-producing spoils on a western surface mine. In *Proceedings High-Altitude Revegetation Workshop no. 4, February 26–27, 1980*, ed. C. L. Jackson and M. A. Schuster, pp. 107–12. Colorado State University, Water Resources Institute.

730. Richardson, J. A. (1977). High-performance plant species in reclamation. In *Landscape Reclamation Practice*, ed. B. Hackett, pp. 148–72. Guildford, England: IPC Science and Technology Press.

731. Rickard, W. H., Fitzner, R. E. and Cushing, C. E. (1981). Biological colonization of an industrial pond: status after two decades. *Environmental Conservation*, **8**, 241–7.

732. Rickard, W. H. and Rogers, L. E. (1983). Industrial land-use and the conservation of native biota in the shrub-steppe region of western North America. *Environmental Conservation*, **10**, 205–11.

733. Ries, R. E. and Depuit, E. J. (1984). Perennial grasses for mined land. *Journal of Soil and Water Conservation*, **39**, 26–9.

734. Riley, C. V. (1954). The utilization of reclaimed coal striplands for the production of wildlife. *North American Wildlife Conference. Transactions*, **19**, 324–37.

735. Riley, C. V. (1957). Reclamation of coal strip-mined lands with reference to wildlife plantings. *Journal of Wildlife Management*, **21**, 402–13.

736. Riley, C. V. (1960). The ecology of water areas associated with coal strip mined lands in Ohio. *Ohio Journal of Science*, **60**, 106–21.

737. Riley, C. V. (1963). Revegetation and management of critical sites for wildlife. In *28th North American Wildlife Conference. Transactions, 1963*, pp. 269–83.

738. Riley, C. V. (1965). Limnology of acid mine water impoundments. In *Acid Mine Drain Research Symposium 1. Proceedings*, pp. 175–87. Monroville, Pennsylvania: Bitumen Coal Research Inc.

739. Riley, C. V. (1977). Ecosystem development on coal surface-mined lands, 1918–75. In *Recovery and Restoration of Damaged Ecosystems*, ed. J. Cairns Jr., K. L. Dickson and E. E. Herricks, pp. 303–46. Charlottesville: University Press of Virginia.

740. Riley, R. K. and Brown, M. L. (1978). The natural invasion of wildlife food and cover species onto bituminous coal spoil. In *Surface Mining and Fish/Wildlife Needs in the Eastern United States*, ed. D. E. Samuel, J. R. Stauffer, C. H. Hocutt and W. T. Mason, pp. 376–80. Morgantown: West Virginia University.

741. Ringe, J. M., Graves, D. H. and Wittwer, R. F. (1984). Effects of soil amendments and severe cattle grazing on the long-term survival and growth of tree seedlings on eastern Kentucky surface mined land. In *Proceedings 1984 Symposium on Surface Mining, Hydrology, Sedimentology, and Reclamation, Lexington, December 2–7, 1984*, ed. H. Graves. Lexington: University of Kentucky.

742. Robb, A. E. Jr. and Ohlsson, K. E. (1982). *Mine Related Stream Alteration Practices for Protection and Enhancement of Fish and Wildlife*. US Fish and Wildlife Service Report.

743. Roberts, R. D. and Johnson, M. S. (1978). Dispersal of heavy metals from abandoned mine workings and their transference through terrestrial food chains. *Environmental Pollution*, **16**, 293–310.

744. Roberts, R. D. and Roberts, J. M. (1986). The selection and management of soils in landscape schemes. In *Ecology and Landscape*, ed. A. D. Bradshaw, D. A. Goode and E. H. P. Thorpe, pp. 99–126. Oxford: Blackwell.

745. Rockett, L. C. (1976). *The Limnology of Existing Strip Mine Ponds in Northeastern Wyoming as Related to Mined Land Reclamation*. Wyoming Game and Fish Department. Project no. 3075-05-7401.

746. Rodiek, J. (1982). Wildlife habitat management and landscape architecture. *Landscape Journal*, **1**, 3–8.

747. Rogaar, H. and Boswinkel, J. A. (1978). Some soil morphological effects of earthworm activity; field data and X-ray radiography. *Netherlands Journal of Agricultural Science*, **26**, 145–60.

748. Roseberry, J. L. and Klimstra, W. D. (1963). *Re-Establishment of Fauna and Flora on Spoilbanks Resulting from Wheel and Shovel Excavators: A Comparative Analysis*. Southern Illinois University Job Completion Report P-R Project W-64-R-5 (unpublished manuscript).

749. Rosso, W. A. (1980). Reclaim your land with lakes. *Coal Age*, **85**, 76–7.

750. Rothwell, F. M. and Holt, C. (1978). *Vesicular–Arbuscular Mycorrhizae Established with Glomus fasciculatus Spores Isolated from the Feces of Cricetine Mice*. Forest Service, Northeastern Forest Experiment Station Research Note NE-259.

751. Rothwell, F. M. and Victor, B. J. (1984). A new species of Endogonaceae: *Glomus botryoides*. *Mycotaxon*, **20**, 163–7.

752. Rusek, J. (1978). Succession of soil animals during ecosystem development. *Pedobiologia*, **18**, 426–33.

753. Rushton, S. P. (1986*a*). Development of earthworm populations on pasture land reclaimed from open-cast coal mining. *Pediobiologia*, **29**, 27–32.

754. Rushton, S. P. (1986*b*). The effects of soil compaction on *Lumbricus Terrestris* and its possible implications for populations on land reclaimed from open-cast coal mining. *Pedobiologia*, **29**, 85–90.

755. Russell, K. R. (1982). Wildmis – a system for estimating the cost to remedy habitat losses. In *Research on Fish and Wildlife Habitat*, ed. W. T. Mason, pp. 170–82. US Environmental Protection Agency Report EPA-600/8-82-022.

756. Ryan, P. (1986). Rehabilitation at Rum Jungle, Northern Territory. *Landline*, **12**, 1–3.

757. Rykiel, Jr., E. J. (1979). Ecological disturbances. In *The Mitigation Symposium: A National Workshop on Mitigating Losses of Fish and Wildlife Habitats*, ed. G. A. Swanson, pp. 624–6. Rocky Mountain Forest Range Experiment Station (Forest Service) General Technical Report.

758. Samuel, D. E. (1979*a*). Utilization of surface mines by wildlife in northern West Virginia. *West Virginia Academy of Science*. pp. 182–6.

759. Samuel, D. E. (1979*b*). Surface mine reclamation for wildlife: the new federal law and you. In *Proceedings of the 1979 Northeastern Fish and Wildlife Conference, Providence, Rhode Island*, pp. 145–50.

760. Samuel, D. E. and McConnell, D. (1988). An inventory of wetlands on surface mines in West Virginia. In *The Freshwater, Wetlands and Wildlife Conference, Charleston, SC*. (in press).

761. Samuel, D. E., Stauffer, J. R., Hocutt, C. H. and Mason, W. T., ed. (1978). *Surface Mining and Fish/Wildlife Needs in the Eastern United States*. Morgantown: West Virginia University.

762. Samuel, D. E. and Whitmore, R. C. (1985). Reclamation and management of surface mined areas for game and non-game birds in West Virginia. In *Proceedings International Hill Lands Symposium, October 1976, Morgantown, West Virginia*, pp. 578–82.

763. Sanders, W. D., Burley, J. B. and Churchward, C. A. (1982). A study for vegetation and wildlife habitat on Lower Grey Cloud Island. In *The Wildlife Values of Gravel Pits Symposium, Crookston, Minnesota, June 1982*, ed. W. D. Svedarsky and R. D. Crawford, pp. 102–8. University of Minnesota Miscellaneous Publication 17-1982.

764. Sandusky, J. E. (1978). The potential for management of waterfowl nesting habitat on reclaimed mined land. In *Surface Mining and Fish/Wildlife Needs in the Eastern United States*, ed. D. E. Samuel, J. R. Stauffer, C. H. Hocutt and W. T. Mason, pp. 325–7. Morgantown: West Virginia University.

765. Sapa, A. J. (1979). Restoration of wildlife habitat to offset project losses, Garrison Diversion Unit, North Dakota. In *The Mitigation Symposium: A National Workshop on Mitigating Losses of Fish and Wildlife Habitats*, ed. G. A. Swanson, pp. 318-21. Rocky Mountain Forest Range Experiment Station (Forest Service) General Technical Report.

766. Satchell, J. E., ed. (1983). *Earthworm Ecology: From Darwin to Vermiculture*. London: Chapman and Hall.

767. Satchell, J. E. and Stone, D. A. (1972). *Earthworm Activity in Pulverised Fuel Ash Sites Restored to Agriculture*, Final Report. Grange-over-Sands, Lancashire, England: The Nature Conservancy, Merlewood Research Station.

768. Satchell, J. E. and Stone, D. A. (1977). Colonization of pulverized fuel ash sites by earthworms. *P. Cent. Pir. Biol. Exp.*, **9**, 59–74.

769. Sather, J. H. and Smith, R. D. (1984). *An Overview of Major Wetland Functions and Values*. US Fish and Wildlife Service Report.

770. Savage, N. L. and Rabe, F. W. (1973). The effects of mine and domestic wastes on macroinvertebrate community structure in the Coeur d'Alene river. *Northwest Science*, **47**, 159–68.

771. Schagen, J. van (1986). *Recolonisation by Ants and other Invertebrates in Rehabilitated Coal Mine Sites near Collie, Western Australia*. WAIT School of Biology Bulletin no. 13.

772. Schaid, T. A., Uresk, D. W., Tucker, W. L. and Linder, R. L. (1983). Effects of surface mining on the vesper sparrow in the Northern Great Plains. *Journal of Range Management*, **36**, 500–3.

773. Schavilje, J. P. (1941). Reclaiming Illinois strip mined coal lands with trees. *Journal of Forestry*, **39**, 714–19.

774. Schiemenz, H. (1964). Zikaden (Homoptera Auchenorrhyncha) von einer tertiaren Rohbodenkippe des Braunkohlenta gebaues Bohlen. *Abhandlungen und Berichte des Naturkundemuseums Gorlitz*, **39**, 1–8.

775. Schlatzer, G. (1973). Some experiences with various species in Danish reclamation work. In *Ecology and Reclamation of Devastated Lands*, vol. 1, ed. R. J. Hutnik and G. Davis, pp. 33–64. New York: Gordon and Beach.

776. Schnell, G. D., Johnson, F. L. and Risser, P. G. (1981). *Model Fish and Wildlife Plan for a Surface Coal Mine in Oklahoma*. Prepared for Oklahoma Department of Mines. University of Oklahoma.

777. Schoedf, W. H. (1987). *Impact of Sewage Sludge, Fertilizer and Revegetation on Biotic Microcommunities in Reclaimed Pennsylvannia Strip-Mine Sites*. M.Sc. Thesis, State University of New York, College of Environmental Science and Forestry (unpublished manuscript).

778. Schreiber, K. F. (1980). Development of a derelict land programme in Baden-Wurttemberg under the influence of different land-use regimes. In *Verhandlungen Gesellschaft für Okologie, band VIII, Freising-Weihenstephan 1979*, ed. W. Haber, pp. 185–203. Göttingen: Gesellschaft für Okologie.

779. Schriefer, T. (1981). Regenwürmer (Lumbricidae) auf unterschiedlich abgedeckten Mülldeponien. Bestandsaufnahme und Besiedlungsmechanismen. *Pedobiologia*, **22**, 153–66.

780. Schrock, J. R. (1983). *The Succession of Insects on Unreclaimed Coal Strip Mine Spoil Banks in Indiana*. PhD Dissertation, University of Kansas.

781. Schulz, C. A. and Meijer, J. (1978). Migration of leafhoppers (Homoptera: Auchenorrhyncha) into a new polder. *Holartic Ecology*, **1**, 73–8.

782. Schuman, G. E. (1981). Management of arid and semi-arid mined lands. In *International Conference on Aridic Soils – Properties, Genesis and Management*. Abstracts. Jerusalem: International Soil Science Society.

783. Schuman, G. E. (1982). Forage quality of several grass species grown on reclaimed mined lands. In *Proceedings of a Symposium Stressing the Application of Data to Problems in Arid and Semi-Arid Environments*, ed. L. P. Gough and R. C. Severson. US Department of the Interior Geological Survey.

784. Schuman, G. E., Berg, W. A. and Power, J. F. (1976). Management of mine wastes in the western United States. In *Land Application of Waste Materials*, pp. 180–94. Ankeny, Iowa: Soil Conservation Society of America.

785. Schuman, G. E., Booth, D. T., Waggoner, J. W. and Rauzi, F. (1984). The effect of grazing reclaimed mined lands on forage production and composition. In *Rangelands: A Resource Under Siege*. Proclamation. Second International Rangeland Congress, ed. P. J. Joss, P. W. Lynch and O. B. Williams, pp. 163–4. Canberra: Australian Academy of Science.

786. Schuman, G. E., Munshower, F. F., Ries, R. E. and Robbins, E. G. (1980). *Quality and Heavy Metal Content of Forages Grown on Reclaimed Mined Lands in the Northern Great Plains*. University of Wyoming Agricultural Experimental Station Research Journal, Great Plains Agricultural Council Report no. 98.

787. Schuman, G. E., Rauzi, F. and Taylor, Jr., E. M. (1980). *The Effect of Topsoil Depth on Forage Production, Water Infiltration, and Water Storage of Reclaimed Mined Land*. Agronomy abstracts. 1980 Annual Meeting American Society of Agronomy.

788. Schuman, G. E. and Sedbrook, T. A. (1984). Sawmill wood residue for reclaiming bentonite spoils. *Forest Products Journal*, **34**, 65–8.

789. Schuman, G. E., Taylor, E. M., Rauzi, F. and Pinchak, B. A. (1985). Revegetation of mined land: influence of topsoil depth and mulching method. *Journal of Soil and Water Conservation*, **40**, 249–52.

790. Schur, B. (1985). *Jarrah Forest or Bauxite Dollars? A Critique of Bauxite Mine Rehabilitation in the Jarrah Forests of Southwestern Australia*. Perth, Western Australia: Campaign to Save Native Forests.

791. Scott, B. (1983). The management of gravel quarries for wildlife at Dungeness, Kent. In *The Proceedings of a Symposium on the Subject of Wildlife on Man-Made Wetlands*, pp. 134–41. Milton Keynes: ARC Wildlife Centre.

792. Scott, J. (1974). *The Regeneration of Bauxite Mining Sites*. B.Sc. Honours Dissertation. Department of Zoology, University of Western Australia (unpublished manuscript).

793. Scott, M. D. (1979). Computerized evaluation of the wildlife habitat option as a post-mining land use. In *The Mitigation Symposium: A National Workshop on Mitigating Losses of Fish and Wildlife Habitats*, ed. G. A. Swanson, pp. 622–3. Rocky Mountain Forest Range Experiment Station (Forest Service) General Technical Report.

794. Scott, M. D. and Zimmerman, G. M. (1984). Wildlife management at surface coal mines in the northwest. *Wildlife Society Bulletin*, **12**, 364–70.

795. Scott, S. (1977). Practical applications of wildlife habitat restoration programs. In *Reclamation for Wildlife Habitat. Proceedings of Reclamation Workshop II, Fort Collins, Colorado, September 1977*, ed. Ecology Consultants Inc., pp. 137–47. Environmental Research and Technology Inc.

796. Seymour, J. (1981). The dunes of Cooloola. *Ecos*, **30**, 3–11.

797. Shanks, Jr., C. H. and Chase, D. L. (1981). Effect of volcanic ash on adult *Otiorhynchus* (Coleoptera: Curculionidae). *Melanderia*, **37**, 63–6.

798. Shaw, W. J., Pekin, J. F. and Philliskirk, A. (1982). *Rehabilitation Objectives and Work Plan for the 1982/1983 Budget Year of the Environmental Section of the Groote Eylandt Mining Company* (unpublished manuscript).

799. Simberloff, D. S. and Wilson, E. O. (1969). Experimental zoogeography of islands: the colonization of empty islands. *Ecology*, **50**, 278–96.

800. Simberloff, D. S. and Wilson, E. O. (1970). Experimental zoogeography of islands: a two-year record of colonization. *Ecology*, **51**, 934–7.

801. Singh, U. R., Pandey, P. N. and Tripathi, B. D. (1979). Ecophysiology of a polluted pedo-ecosystem with special reference to its mesofaunal composition. *Indian Journal of Ecology*, **6**, 53–60.

802. Skelly and Loy, Engineers Consultants. (1978). *Surface Coal Mining and Reclamation for Fish/Wildlife in Appalachia*. US Fish and Wildlife Service Report.

803. Slack, H. A. (1979). Wildlife mitigation and energy development: balancing our heritage and our future. In *The Mitigation Symposium: A National Workshop on Mitigating Losses of Fish and Wildlife Habitats*, ed. G. A. Swanson, pp. 49–53. Rocky Mountain Forest and Range Experiment Station (Forest Service) General Technical Report.

804. Slaski, L. J. and Fowler, D. K. (1980). Reclamation of surface mines for wildlife habitat: a case study. In *Symposium on Surface Mining Hydrology, Sedimentology and Reclamation*, ed. D. H. Graves, pp. 431–6. Lexington: University of Kentucky.

805. Sly, G. (1976). Small mammal succession on strip-mined land in Vigo County, Indiana. *The American Midland Naturalist*, **95**, 257–67.

806. Smith, B. M. and Hill, E. P. (1979). Waterfowl use of strip mine ponds in Alabama. *Annual Conference of the South Eastern Association of Fish and Wildlife Agencies, Proceedings*, **33**, 1–10.

807. Smith, C. C. (1940). Succession on abandoned eroded farmland. *Ecological Monographs*, **10**, 421–84.

808. Smith, H. G. (1964). Spoilbanks and birdlife. *Soil Conservation*, **30**, 77.

809. Smith, H. K. (1980). Coastal habitat development in the dredged material research program. In *Rehabilitation and Creation of Selected Coastal Habitats: Proceedings of a Workshop, Sapelo Island, Georgia, 1976*, ed. J. C. Lewis and E. W. Bunce, pp. 117–25. US Fish and Wildlife Service Report.

810. Smith, J. R. (1984). Breeding bird response to forest development on surface-mined land in southern Illinois. In *Proceedings of the Fourth Annual Better Reclamation with Trees Conference, Owensboro, Kentucky June 7–8 1984*, ed. P. E. Pope. Kentucky: Madisonville Community College.

811. Smith, M. A., ed. (1985). *Contaminated Land: Reclamation and Treatment*. New York: Plenum Press.

812. Smith, R. L. (1973). Strip mining impacts and reclamation efforts in Appalachia. In *North American Wildlife and Natural Resources Conference (38th) March 1973*, ed. J. B. Trefethen, pp. 132–42. Washington, DC: Wildlife Management Institute.

813. Smith, S. D. (1981). Preliminary report effects of Mt. St. Helens ashfall on lotic Trichoptera. *Melanderia*, **37**, 56–62.

814. Smith Davidson, V. G. (1928). Animal communities of a deciduous forest succession. *Ecology*, **9**, 479–500.

815. Snyder, B. D. and Potter, G. L. (1981). The critical links between baseline studies and development of revegetation and wildlife restoration plans. In *Symposium on Surface Mining Hydrology, Sedimentology and Reclamation*, ed. D. H. Graves, pp. 507–10. University of Kentucky.

816. Southwell, L. T. and Majer, J. D. (1982). The survival and growth of the earthworm *Eisenia foetida* (Lumbricidae: Oligochaeta) in alkaline residues associated with the bauxite refining process. *Pedobiologia*, **23**, 42–52.

817. Spaulding, W. M., Jr. and Ogden, R. D. (1968). *Effects of Surface Mining on the Fish and Wildlife Resources of the United States*. US Department of the Interior, Bureau of Sport Fish and Wildlife Resource Publication no. 68.

818. Spencer, D. F., Yeung, H. Y. and Greene, R. W. (1983). Alterations in the zooplankton community of a fly ash treated lake. *Hydrobiologia*, **107**, 123–30.

819. Sponsler, M. (1976). *A Regional Comparison of Avian Populations on Unmined and Reclaimed Lands in Illinois*. M.Sc. Thesis, Southern Illinois University.

820. Sponsler, M., Klimstra, W. D. and Nawrot, J. R. (1984). Comparison of avian populations on unmined and reclaimed lands in Illinois. In *The National Symposium on Surface Mining Hydrology, Sedimentology, and Reclamation, Lexington, December 1984*, ed. H. Graves. Lexington: University of Kentucky.

821. Stalmaster, M., Wright, W., Lockhart, J. M. and Kruegar, R. (1982). Management of nesting ferruginous hawks in relation to coal development in Colorado and Utah. In *Issues and Technology in the Management of Impacted Western Wildlife*, ed. R. D. Comer, J. M. Merino, J. W. Monarch, C. Pustmueller, M. Stalmaster, R. Stoeker, J. Todd and W. Wright, pp. 205–9. Thorne Ecological Institute Technical Publication no. 14.

822. Standen, V., Stead, G. B. and Dunning, A. (1982). Lumbricid populations in open cast reclamation sites and colliery spoil heaps in County Durham, U.K. *Pedobiologia*, **24**, 57–64.

823. Stanlake, E. A., Eastman, D. S. and Stanlake, M. G. (1978). *Ungulate use of some Recently Reclaimed Strip Mines in Southeastern British Columbia*. Wildlife Research and Technical Services Section, Fish and Wildlife Services Report.

824. Stanley, M. A., Schuman, G. E., Rauzi, F. and Painter, L. I. (1982). Quality and element content of forages grown on three reclaimed mine sites in Wyoming and Montana. *Reclamation and Revegetation Research*, **1**, 311–26.

825. Stannard, L. J. (1967). The redispersal of Thysanoptera (Insecta, thrips) into spoiled strip-mined lands in southern Illinois. *Illinois State Academy of Science. Transactions.* **60**, 130–3.

826. Stanton, N. L. and Krementz, D. (1982). Nematode densities on reclaimed sites on a cold desert shrub-steppe. *Reclamation and Revegetation Research*, **1**, 233–41.

827. Starnes, L. B., ed. (1985). *Fish and Wildlife Relationships to Mining: Symposium Proceedings*. American Fisheries Society (Water Quality Section).

828. Starnes, L. B., Maddox, J. B. and Zarger, T. G. (1978). Effects of remedial reclamation treatments on terrestrial and aquatic ecosystems – a progress report. In *Surface Mining and Fish/Wildlife Needs in the Eastern United States*, ed. D. E. Samuel, J. R. Stauffer, C. H. Hocutt and W. T. Mason, pp. 276–86. Morgantown: West Virginia University.

829. Starnes, L. B. and Scanlon, D. H. (1985). Results of remedial reclamation treatments on terrestrial and aquatic ecosystems in Ollis Creek watershed, Tennessee. In *Fish and Wildlife Relationships to Mining: Symposium Proceedings*. ed. L. B. Starnes, pp. 14–23. American Fisheries Society (Water Quality Section).

830. Steele, B. B. and Grant, C. V. (1982). Topographic diversity and islands of natural vegetation: aids in re-establishing bird and mammal communities on reclaimed mines. *Reclamation and Revegetation Research*, **1**, 367–81.

831. Stevens, R. (1983). Selecting species mixtures for optimum wildlife habitat development on disturbed lands. In *The Third Annual Reclamation Symposium, Salt Lake City, Utah, January 10–11*.

832. Stewart, A. J. A. and Lance, A. N. (1983). Moor-draining: a review of impacts on land use. *Journal of Environmental Management*, **17**, 81–99.

833. Stinner, B. R., Crossley, Jr., D. A., Odum, E. P. and Todd, R. L. (1984). Nutrient budgets and internal cycling of N, P, K, Ca, and Mg in conventional tillage, no-tillage, and old-field ecosystems on the Georgia piedmont. *Ecology*, **65**, 354–69.

834. Stockinger, N. F. and Hays, H. A. (1960). Plankton, benthos and fish in three strip-mined lakes with varying pH values, *Transactions of the Kansas Academy of Science*, **63**, 1–11.

835. Stoecker, R. E. (1982). Improving wildlife habitat as mined land reclamation. In *Shaping Land for Tomorrow: A Land Reclamation Forum Conference Proceedings*, ed. A. M. Bauer, pp. 99–102. East Lansing, Michigan: Michigan State University.

836. Stoecker, R. E., Thompson, R. W. and Comer, R. D. (1984). *An Evaluation of Wildlife Mitigation Practices at Western Coal Operations.* Final Draft Report USDI-OSM. Boulder, Colorado: Thorne Ecological Institute.

837. Stoner, P. J. (1980). *An Evaluation of Mammal Use on Reclaimed Surface Mined Lands and Unmined Lands in Two Regions in Illinois.* M.Sc. Thesis, Southern Illinois University, Carbondale (unpublished manuscript).

838. Stormer, F. A., Chilcote, D. G. and Shetron, S. G. (1978). Wildlife communities of vegetated iron ore tailings basins in northern Michigan. In *Surface Mining and Fish/Wildlife Needs in the Eastern United States*, ed. D. E. Samuel, J. R. Stauffer, C. H. Hocutt and W. T. Mason, pp. 359–68. Morgantown: West Virginia University.

839. Stranathan, S. and Bruggeman, L. (1977). Developing plants compatible to wildlife and revegetation. In *Reclamation for Wildlife Habitat. Proceedings of Reclamation Workshop II, Fort Collins, Colorado, September 1977*, ed. Ecology Consultants Inc., pp. 148–53. Environmental Research and Technology Inc.

840. Street, H. E. and Goodman, G. T. (1967). Revegetation techniques in the Lower Swansea Valley. In *The Lower Swansea Valley Report*, ed. K. J. Hilton, pp. 71–110. London: Longmans.

841. Street, M. (1977). The food of mallard ducklings in a wet gravel quarry, and its relation to duckling survival. *Wildfowl*, **28**, 113–25.

842. Street, M. (1978). The role of insects in the diet of mallard ducklings – an experimental approach. *Wildfowl*, **29**, 93–100.

843. Street, M. (1979). Research on the improvement of gravel pits for waterfowl by adding straw. *Game Conservancy Annual Review for 1978*, pp. 56–61.

844. Street, M. (1982*a*). Management of artificial wetlands for waterfowl – a novel technique to increase the availability of invertebrate food items. *Transactions. International Congress of Game Biologists*, **14**, 159–74.

845. Street, M. (1982*b*). The Great Linford Project: waterfowl research in a gravel pit wildlife reserve. In *The Wildlife Values of Gravel Pits Symposium, Crookston, Minnesota, June 1982*, ed. W. D. Svedarsky and R. D. Crawford, pp. 170–80. University of Minnesota Miscillaneous Publication 17-1982.

846. Street, M. (1983*a*). The Great Linford Wildfowl Research Project – a case history. In *The Proceedings of a Symposium on the Subject of Wildlife on Man-Made Wetlands*, pp. 21–42. Milton Keynes, England: ARC Wildlife Centre.

847. Street, M. (1983*b*). The use of waste straw to promote the production of invertebrate foods for waterfowl in manmade wetlands. In *Managing Wetlands and their Birds*, pp. 98–103. Gloucestershire, UK: I.W.R.B. Slimbridge.

848. Street, M. (1985). *The Restoration of Gravel Pits for Wildfowl.* Milton Keynes, England: ARC Wildlife Centre.

849. Street, M. and Titmus, G. (1979). The colonisation of experimental ponds by Chironomidae (Diptera). *Aquatic Insects*, **1**, 233–44.

850. Street, M. and Titmus, G. (1982). A field experiment on the value of allochthonous straw as food and substratum for lake macro-invertebrates. *Freshwater Biology*, **12**, 403–10.

851. Streeter, R. (1977). Wildlife and western mining: where do our priorities lie? In *Reclamation for Wildlife Habitat. Proceedings of Reclamation Workshop II, Fort Collins, Colorado, September 1977*, ed. Ecology Consultants Inc., pp. 1–8. Environmental Research and Technology Inc.

852. Streeter, R. G., Moore, R. T., Skinner, J. J., Martin, S. G., Terrel, T. L., Klimstra, W. D., Tate, J. J. and Nolde, M. J. (1979). Energy mining impacts and wildlife management: which way to turn. In *44th North American Wildlife and Natural Resources Conference. Transactions*, pp. 26–65. Washington, DC: Wildlife Management Institute.

853. Strickland, D. (1977). An overview of mining and its impact on wildlife in Wyoming. In *Reclamation for Wildlife Habitat. Proceedings of Reclamation Workshop II, Fort Collins, Colorado, September 1977*, ed. Ecology Consultants Inc., pp. 9–27. Environmental Research and Technology Inc.

854. Strüve-Kusenberg, R. (1981). Succession and trophic structure of the soil fauna of fallow land. *Pedobiologia*, **21**, 132–41.

855. Stuart, R. W. (1974). Surface mining and wildlife. *North Dakota Outdoors*, **37**, 2–7.

856. Stubbs, C. W. B. and Markham, B. J. (1979). Wildlife mitigation measures for oil and gas activity in Alberta. In *The Mitigation Symposium: A National Workshop on Mitigating Losses of Fish and Wildlife Habitats*, ed. G. A. Swanson, pp. 264–9. Rocky Mountain Forest Range Experiment Station (Forest Service) General Technical Report.

857. Suchecki, J. L. and Evans, A. K. (1979). Wildlife populations associated with natural, agricultural, and abandoned surface mined lands in southern Indiana: a site specific study. In *Ecology and Coal Resource Development. Proceedings of the International Congress for Energy and the Ecosystem*, ed. M. K. Wali, pp. 576–83. New York: Pergamon Press.

858. Sullivan, K. B. (1969). From open pit mines to grass and pines. *Soil Conservation*, **35**, 29–30.

859. Sullivan, T. P. and Sullivan, D. S. (1982). Responses of small-mammal populations to a forest herbicide application in a 20-year-old conifer plantation. *Journal of Applied Ecology*, **19**, 95–106.

860. Svedarsky, W. D. (1982). The Red River Valley Natural History Area: wildlife management and environmental education in an abandoned gravel pit in northwest Minnesota. In *The Wildlife Values of Gravel Pits Symposium, Crookston, Minnesota, June 1982*, ed. W. D. Svedarsky and R. D. Crawford, pp. 132–40. University of Minnesota Miscellaneous Publication 17-1982.

861. Svedarsky, W. D. and Crawford, R. D., ed. (1982). *Wildlife Values of Gravel Pits. Symposium Proceedings, Crookston, Minnesota, June 1982.* Agricultural Experiment Station, University of Minnesota Miscellaneous Publication 17-1982.

862. Swanson, G. A., ed. (1979). *The Mitigation Symposium: A National Workshop on Mitigating Losses of Fish and Wildlife Habitats, July 1979, Fort Collins, Colorado.* USDA, Forest Service, General Technical Report.

863. Swanson, G. A. (1982). Summary of wildlife values of gravel pits symposium. In *The Wildlife Values of Gravel Pits Symposium, Crookston, Minnesota, June 1982*, ed. W. D. Svedarsky and R. D. Crawford, pp. 1–5. University of Minnesota Miscellaneous Publication 17-1982.

864. Swart, D. (1983). *Management Plan. Handbook.* Rijksdienst voor de Ijsselmeerpolders. Flevobericht no. 188.

865. Sweeney, B. A. (1982). Small mammal community structure in reclaimed and unreclaimed gravel mining areas. *Ohio Journal of Science*, **82**, 95.

866. Szafoni, R. E. (1982). Wildlife considerations in the development of riparian communities. In *The Wildlife Values of Gravel Pits Symposium, Crookston, Minnesota, June 1982*, ed. W. D. Svedarsky and R. D. Crawford, pp. 59–66. University of Minnesota Miscellaneous Publication 17-1982.

867. Szujecki, A., Mazur, S., Szyszko, J., Perlinski, S. and Tracz, H. (1980). Studies on requirements and possibilities of zooamelioration of afforested arable land. In *Soil Biology as Related to Land Use Practices*, ed. D. L. Dindal, pp. 444–9. Washington: Environmental Protection Agency.

868. Szujecki, A., Szyszko, J., Mazur, S. and Perlinski, S. (1977). Changes in the structure of macrofauna communities on afforested arable land. In *Soil Organisms as Components of Ecosystems*, ed. U. Lohm and T. Persson, pp. 580–4. Ecological Bulletins no. 25.

869. Tacey, W. H. (1982). *Strategies for Understorey Return to Bauxite Mined Sites in the Darling Range*. Alcoa of Australia (unpublished report).

870. Tate, III, R. L. and Klein, D. A., ed. (1985). *Soil Reclamation Processes: Microbiological Analyses and Applications*. New York: Marcel Dekker.

871. Taylor, E. G. (1978). Rehabilitation after mineral sands mining at Western Titanium Limited, Capel, W. A. In *Rehabilitation of Mined Lands in Western Australia*, ed. J. E. D. Fox, pp. 93–8. Bentley, W.A.: WAIT.

872. Taylor, Jr., E. M., Schuman, G. E., Howard, G. S. and Rauzi, F. R. (1979). *The Effectiveness and Cost of Standing Stubble Compared to Crimped Residue for Establishing Grass Cover on Mined Lands*. Society for Range Management Annual Meeting Abstracts, Feb. 12–15, 1979.

873. Taylor, K. (1983). Linear wetlands as habitat for birds and population changes following engineering modifications to river systems. In *The Proceedings of a Symposium on the Subject of Wildlife on Man-Made Wetlands*, pp. 101–8. Milton Keynes, England: ARC Wildlife Centre.

874. Tennessee Valley Authority. (1969). *Reclamation Brings Wildlife Back to Paradise Strip Mine*. Tennessee Valley Authority, Division of Forestry, Fisheries, and Wildlife Development.

875. Tennessee Valley Authority. (1970). *Sequatchie Strip Mine Reclamation Benefits Wildlife*. Tennessee Valley Authority, Division of Forestry, Fisheries, and Wildlife Development.

876. Tennessee Valley Authority Office of Natural Resources. (1980*a*). Selected plantings coax wildlife to move in after miners move out. *Impact*, **3**, 2–4.

877. Tennessee Valley Authority Office of Natural Resources. (1980*b*). Strip mine sediment ponds – sow's ears or silk purses? *Impact*, **3**, 5–7.

878. Teraguchi, S., Teraguchi, M. and Upchurch, R. (1977). Structure and development of insect communities in an Ohio old-field. *Environmental Entomology*, **6**, 247–57.

879. Terrel, T. L. and French, T. (1976). Wintering bird populations on coal strip-mines in north central Alabama. *The Journal of the Alabama Academy of Science*, **46**, 1–13.

880. Terrel, T. L. and Shinn, R. (1977). Problems of reclaiming for wildlife on private lands. In *Reclamation for Wildlife Habitat. Proceedings of Reclamation Workshop II, Fort Collins, Colorado, September 1977*, ed. Ecology Consultants Inc., pp. 74–85. Environmental Research and Technology Inc.

881. Tessmann, S. (1982). Habitat reclamation procedures for surface mines in Wyoming. In *Issues and Technology in the Management of Impacted Western Wildlife*, ed. R. D. Comer, J. M. Merino, J. W. Monarch, C. Pustmueller, M. Stalmaster, R. Stoeker, J. Todd and W. Wright, pp. 185–94. Thorne Ecological Institute Technical Publication no. 14.

882. Thames, J. L. and Verma, T. R. (1975). Coal mine reclamation on the Black Mesa and the Four Corners areas of northeastern Arizona. In *Practices and Problems of Land Reclamation in Western North America*, ed. M. K. Wali, pp. 48–64. Grand Forks: University of North Dakota Press.

883. Thompson, C. S. (1985*a*). Stream relocations under OSM regulations: benefits to fish and invertebrates. In *Wetlands and Management on Mined Lands. Proceedings of a Conference October 23–24, 1985*, ed. R. P. Brooks, D. E. Samuel and J. B. Hill, pp. 305–16. Pennsylvania State University.

884. Thompson, C. S. (1985*b*). Stream relocation on surface mining land. In *Fish and Wildlife*

Relationships to Mining: Symposium Proceedings, ed. L. B. Starnes, pp. 39–50. American Fisheries Society (Water Quality Section).

885. Tigelaar, L. (1985). Recreation development in the Lauwerszee. In *The Lauwerszee Fifteen Years Closed-Off. Results of Research and Experience with Development and Management*, ed. P. J. Ente and R. J. de Glopper, pp. 185–94. Rijksdienst voor de Ijsselmeerpolders. Flevobericht no. 247.
886. Tiner, Jr., R. W. (1985). An overview of freshwater wetlands. In *Wetlands and Water Management on Mined Lands. Proceedings of a Conference October 23–24, 1985*, ed. R. P. Brooks, D. E. Samuel and J. B. Hill, pp. 57–63. Pennsylvania State University.
887. Tischler, W. (1952). Biozonötische Untersuchungen an Ruderalstellen. *Zoologische Jahrbücher (Systematik)*, **81**, 122–74.
888. Tischler, W. (1973). Ecology of arthropod fauna in man-made habitats: the problem of synanthropy. *Zoologischer Anzeiger (Leipzig)*, **191**, 157–61.
889. Titmus, G. (1979). The emergence of midges (Diptera: Chironomidae) from a wet gravel-pit. *Freshwater Biology*, **9**, 165–79.
890. Titmus, G. (1983). The biology of fish populations in gravel quarries. In *The Proceedings of a Symposium on the Subject of Wildlife on Man-Made Wetlands*, pp. 109–20. Milton Keynes, England: ARC Wildlife Centre.
891. Titmus, G. and Badcock, R. (1981). Distribution and feeding of larval Chironomidae in a gravel pit lake. *Freshwater Biology*, **11**, 263–71.
892. Tolbert, V. R., Goldsmith, A. M. and Beauchamp, J. J. (1985). Comparison of growth responses of creek chubs from an undisturbed and a surface-mining-disturbed stream. In *Fish and Wildlife Relationships to Mining: Symposium Proceedings*, ed. L. B. Starnes, pp. 24–38. American Fisheries Society (Water Quality Section).
893. Tolin, W. A. (1975). *Toxic Heavy Metals in Wildlife Inhabiting Strip-Mined Areas in Ohio*. M.Sc. Thesis, Ohio State University (unpublished manuscript).
894. Tomlin, A. D. (1977). Pipeline construction – impact on soil micro and mesofauna (Arthropoda and Annelida) in Ontario. *Proceedings of the Entomological Society of Ontario*, **108**, 13–17.
895. Tompkins, D. B. (1981). *Natural Revegetation of Abandoned Surface Mines in Northern West Virginia*. M.Sc. Thesis, West Virginia University (unpublished manuscript).
896. Topp, W. (1971). Zur ökologie der müllhalden. *Annales Zoologici Fennici*, **8**, 194–222.
897. Tormala, T. (1980). The bird community of reserved fields in central Finland. *Ornis Fennica*, **37**, 161–6.
898. Trefethen, J. B., ed. (1973). *North American Wildlife and Natural Resources Conference (38th) March 1973*. Washington, DC: Wildlife Management Institute.
899. Turner, L. J. and Fowler, D. K. (1981). *Utilization of Surface Mine Ponds in East Tennessee by Breeding Amphibians*. Fish and Wildlife Service, US Department of the Interior, Biological Services Program.
900. Tydeman, C. (1982). Gravel pits as wildlife habitat in Great Britain. In *The Wildlife Values of Gravel Pits Symposium, Crookston, Minnesota, June 1982*, ed. W. D. Svedarsky and R. D. Crawford, pp. 6–12. University of Minnesota Miscellaneous Publication 17-1982.
901. Tydeman, C. (1983). The general value of man-made wetlands for the wildlife in Europe. In *The Proceedings of a Symposium on the Subject of Wildlife on Man-Made Wetlands*, pp. 5–22. Milton Keynes, England: ARC Wildlife Centre.
902. Tyner, E. H. and Smith, R. M. (1945). The reclamation of the strip-mined coal lands of West Virginia with forage species. *Soil Science Society of America. Proceedings*, **10**, 429–36.
903. Tyus, H. M. and Lockhart, J. M. (1979). Mitigation and research needs for wildlife on western surface mined lands. In *The Mitigation Symposium: A National Workshop on Mitigating Losses of Fish and Wildlife Habitats*, ed. G. A. Swanson, pp. 252–5. Rocky

Mountain Forest Range Experiment Station (Forest Service) General Technical Report.

904. University of Newcastle upon Tyne. (1972). *Landscape Reclamation*, vol. 2. Guildford, England: IPC Science and Technology Press.

905. Ueckert, D. W. (1979). Impact of a white grub (*Phyllophaga crinita*) on a shorgrass community and evaluation of selected rehabilitation practices. *Journal of Range Management*, **32**, 445–8.

906. Urbanek, R. P. (1976). *Vertebrate and Floral Diversity on Strip-Mined Land in Williamson and Saline Counties, Illinois.* M.A. Thesis, Southern Illinois University, Carbondale.

907. Urbanek, R. P. (1982). *Arthropod Community Structure on Strip-Mined Lands in Ohio*, PhD Dissertation, Ohio State University.

908. Urbanek, R. P. and Klimstra, W. D. (1986). Vertebrates and vegetation on a surface-mined area in southern Illinois. *Transactions of the Illinois State Academy of Science*, **79**, 175–87.

909. US Congress, Office of Technology Assessment. (1986). *Western Surface Mine Permitting and Reclamation.* OTA Report no. OTA-E-279. Washington: US Printing Office, June, 1986.

910. USDA Forest Service. (1979*a*). Biological problems/treatments. In *User Guide to Soils.* USDA Forest Service General Technical Report INT-68, pp. 56–70.

911. USDA Forest Service. (1979*b*). Stream fisheries. In *User Guide to Hydrology*. USDA Forest Service General Technical Report INT-74, pp. 22–3.

912. USDA Forest Service. (1982). *Wildlife: User Guide for Mining and Reclamation.* USDA Forest Service, General Technical Report INT-126.

913. US Department of the Interior. (1979). *Permanent Regulatory Program Implementing Section 501(B) of the Surface Mining Control and Reclamation Act of 1977.* Final Environmental Statement OSM-EIS-1.

914. US Fish and Wildlife Service (1978*a*). *Impacts of Transmission Lines on Birds in Flight.* Proceedings of a Conference, January 31–February, Oak Ridge Associated Universities, Tennessee. US Department of the Interior.

915. US Fish and Wildlife Service. (1978*b*). *Mined Land Reclamation for Fish and Wildlife in the Eastern United States.* US Fish and Wildlife Service Report.

916. US Fish and Wildlife Service. (1978*c*). *Rehabilitation of Western Wildlife Habitat: A Review.* US Fish and Wildlife Service Report.

917. US Fish and Wildlife Service. (1980). *A Summary of Fish and Wildlife Information Needs to Surface Mine Coal in the United States. Part 1. Fish and Wildlife Information Needs in the Federal Surface Mining Permanent Regulations.* US Fish and Wildlife Report.

918. US Fish and Wildlife Service. (1982). *Best Current Practices for Fish and Wildlife on Surface-Mined Lands in the Eastern Interior Coal Region.* US Fish and Wildlife Service Report.

919. US Fish and Wildlife Service. (1984). *Proceedings of the National Wetland Values Assessment Workshop.* US Department of the Interior. US Fish and Wildlife Service Report.

920. Usher, M. B. (1979). Natural communities of plants and animals in disused quarries. *Journal of Environmental Management*, **8**, 223–36.

921. Veen, J. (1980). Breeding behavior and breeding success of a colony of little gulls *Larus minutus* in The Netherlands. *Limosa*, **53**, 73–83.

922. Vepsäläinen, K. and Pisarski, B. (1982). Assembly of island ant communities. *Annales Zoologici Fennici*, **19**, 327–35.

923. Vepsäläinen, K. and Wuorenrinne, H. (1978). Ecological effects of urbanization on the mound-building *Formica* L. species. *Memorabilia Zoologica*, **29**, 191–202.

924. Verts, B. J. (1956*a*). *An Evaluation of Wildlife and Recreational Values of a Strip-Mined Area.* MS Thesis, Southern Illinois University.

925. Verts, B. J. (1956*b*). A place to hunt. *Illinois Wildlife*, **12**, 4.

926. Verts, B. J. (1957). The population and distribution of two species of *Peromyscus* on some Illinois strip-mined land. *Journal of Mammalogy*, **38**, 53–9.

927. Verts, B. J. (1959). Notes on the ecology of mammals of a strip-mined area in southern Illinois. *Transactions of the Illinois State Academy of Science*, **52**, 134–9.

928. Vimmerstedt, J. P. (1969). Earthworms speed leaf decay on spoilbanks. *Ohio Report on Research and Development*, **54**, 3–5.

929. Vimmerstedt, J. P. (1970). Stripmine reclamation. *Ohio Report on Research and Development*, **55**, 60–1.

930. Vimmerstedt, J. P. (1983). Earthworm ecology in reclaimed opencast coal mining sites in Ohio. In *Earthworm Ecology: From Darwin to Vermiculture*, ed. J. E. Satchell, pp. 229–40. London: Chapman and Hall.

931. Vimmerstedt, J. P. and Finney, J. H. (1973). Impact of earthworm introduction on litter burial and nutrient distribution in Ohio strip-mine spoil banks. *Soil Science Society of America. Proceedings*, **37**, 388–91.

932. Vinikour, W. S. (1980). Biological consequences of stream routing through a final-cut strip mine pit: benthic macroinvertebrates. *Hydrobiologia*, **75**, 33–43.

933. Vinikour, W. S. (1981). Aquatic insect drift through a final-cut strip mine pit, with emphasis on drift distances. *Hydrobiologia*, **77**, 225–32.

934. Vogel, W. G. (1979). Are trees neglected plants for reclaiming surface mines? *Proceedings of the West Virginia Academy of Science*, **51**, 127–38.

935. Vohs, Jr., P. A. (1957). A combination salad for wildlife. *Illinois Wildlife*, **13**, 3–5.

936. Vohs, P. A. and Birkenholz, D. E. (1962). Response of bobwhite quail to management on some Illinois strip-mined lands. *Transactions of the Illinois State Academy of Science*, **55**, 13–19.

937. Voight, J. A. and Nagy, J. G. (1979). Wildlife mitigations at bureau of reclamation projects: a case study approach. In *The Mitigation Symposium: A National Workshop on Mitigating Losses of Fish and Wildlife Habitats*, ed. G. A. Swanson, pp. 414–18. Rocky Mountain Forest Range Experiment Station (Forest Service) General Technical Report.

938. Voight, J. R. and Glenn-Lewin, D. C. (1979). Strip mining, *Peromyscus* and other small mammals in southern Iowa. *Proceedings. Iowa Academy of Science*, **86**, 133–6.

939. Wackenhut, P. B. (1980). *The Breeding Biology of the Horned Lark on Coal Surface Mines in Northern West Virginia*. MS Thesis, West Virginia University (unpublished manuscript).

940. WA Department of Conservation and Environment. (1978). *Bauxite Mining in the Darling Range, Western Australia*. Perth: Department of Conservation and Environment Bulletin no. 44.

941. WA Department of Conservation and Environment. (1984). *Bauxite Mining in the Jarrah Forest: Impact and Rehabilitation*. Perth: Department of Conservation and Environment Bulletin no. 169.

942. WA Education Department. (1984). Ants are invertebrates. *Nature Walkabout*, **20**, 10–13.

943. Waggoner, K. van (1978). The effects of lignite mining and reclamation on small mammal populations in Texas. In *Surface Mining and Fish/Wildlife Needs in the Eastern United States*, ed. D. E. Samuel, J. R. Stauffer, C. H. Hocutt and W. T. Mason, pp. 256–66. Morgantown: West Virginia University.

944. Wali, M. K., ed. (1975). *Practices and Problems of Land Reclamation in Western North America*. Grand Forks: University of North Dakota Press.

945. Walsh, G. B. (1910). Coleoptera of the Grangetown slag-heaps. *Naturalist*, **35**, 339–40.

946. Warburton, D. B., Klimstra, W. B. and Nawrot, J. R. (1985). Aquatic macrophyte propagation and planting practices for wetland establishment. In *Wetlands and Water*

Management on Mined Lands. Proceedings of a Conference October 23–24, 1985, ed. R. P. Brooks, D. E. Samuel and J. B. Hill, pp. 139–52. Pennsylvania State University.

947. Warner, R. W. (1973). Acid coal mine drainage effects on aquatic life. In *Ecology and Reclamation of Devastated Land*, vol. 1, ed. R. J. Hutnik and G. Davis, pp. 227–37. New York: Gordon and Breach.
948. Wathern, P. (1977). Some aspects of habitat building in the Dutch polderland. *Landscape Design*, **118**, 20–1.
949. Watmough, B. (1983). The effects of recreational activities on wintering wildfowl. In *The Proceedings of a Symposium on the Subject of Wildlife on Man-Made Wetlands*, pp. 186–209. Milton Keynes, England: ARC Wildlife Centre.
950. Wells, J. R. (1953). The reclamation of stripmined areas in southeastern Kansas. *Kansas Academy of Science. Transactions*, **56**, 269–92.
951. Wesley, D. E., Perkins, C. J. and Sullivan, A. D. (1976). Preliminary observations of cottonwood plantations as wildlife habitat. In *Proceedings: Symposium on Eastern Cottonwood and Related Species*, pp. 460–76.
952. Wetzel, R. M. (1958). Mammalian succession on midwestern floodplains. *Ecology*, **39**, 262–71.
953. Whitehurst, C. A. and Downing, R. L. (1977). *Opportunities for Fish and Wildlife Management on Abandoned Mine Lands in Appalachia.* US Fish and Wildlife Service Report.
954. Whitford, W. G. and Elkins, N. Z. (1987). The importance of soil ecology in strip mine reclamation. In *Diverse Scientific Perspectives on Reclamation in the San Juan Basin.* New Mexico: University of New Mexico Press (in press).
955. Whitford, W. G., Elkins, N. Z. and Parker, L. W. (1981). *The Role of Arthropods in Developing Soils on Mine Spoils*, New Mexico Energy R and D Program Report EMD 2-68-3212. New Mexico State University.
956. Whitmore, R. C. (1978). Managing reclaimed surface mines in West Virginia to promote nongame birds. In *Surface Mining and Fish/Wildlife Needs in the Eastern United States*, ed. D. E. Samuel, J. R. Stauffer, C. H. Hocutt and W. T. Mason, pp. 381–6. Morgantown: West Virginia University.
957. Whitmore, R. C. (1979*a*). Short-term change in vegetation structure and its effect on grasshopper sparrows in West Virginia. *The Auk*, **96**, 621–5.
958. Whitmore, R. C. (1979*b*). Temporal variation in the selected habitats of a guild of grassland sparrows. *Wilson Bulletin*, **91**, 592–8.
959. Whitmore, R. C. (1980). Reclaimed surface mines as avian habitat islands in the eastern forest. *American Birds*, **34**, 13–14.
960. Whitmore, R. C. (1981). Structural characteristics of grasshopper sparrow habitat. *Journal of Wildlife Management*, **45**, 811–14.
961. Whitmore, R. C. and Hall, G. A. (1978). The response of passerine species to a new resource: reclaimed surface mines in West Virginia. *American Birds*, **32**, 6–9.
962. Whitt, D. M. (1968). Surface-mined land. How much? What next? *Soil Conservation*, **33**, 123–5.
963. Wildermuth, H. and Krebs, A. (1983). Man-made wetlands as habitats for dragonflies. *Vierteljahrsschrift der Naturforschenden Gesellschaft in Zuerich*, **128**, 21–42.
964. Wilkey, M. and Zeller, S. (1979). Land reclamation at abandoned deep coal mine. *Journal of the Environmental Engineering Division, Proceedings of the American Society of Civil Engineers*, **105**, 843–53.
965. Williamson, P. and Evans, P. R. (1973). A preliminary study of the effects of high levels of inorganic lead on soil fauna. *Pedobiologia*, **13**, 16–21.
966. Williamson, R. L. (1981*a*). Improving species composition and seasonal variety on reclaimed strip mined grasslands through selective grazing intensities and time periods. *Proceedings of the North Dakota Academy of Science*, **35**, 5.

967. Williamson, R. L. (1981*b*). Reclaiming seasonally balanced native grasslands after strip mining. *Proceedings of the North Dakota Academy of Science*, **35**, 3.

968. Wilmers, T. J. (1982). *Kestrel Use of Nest Boxes on Reclaimed Surface Mines in West Virginia and Pennsylvania*. M.Sc. Thesis, West Virginia University (unpublished manuscript).

969. Wilmers, T. and Samuel, D. E. (1986). Kestrel use of nest boxes on surface mines in West Virginia and Pennsylvania. *Raptor Research (Foundation)* (in review).

970. Winter, J. W. and Atherton, R. G. (1985). Fauna survey of the natural habitats in the Weipa region. In *North Australian Mine Rehabilitation Workshop no. 9, Weipa, June 1985*, ed. J. W. Lawrie, pp. 95–104. Comalco Aluminium Limited.

971. Wolfe, E. (1969). *Okologisch-Faunistische Untersuchungen über den Einfluss Klimatolo. Klimatologischer Faktoren auf die Carabidenfauna von Schachtund Stollenhalden Sowie der Angrenzenden Feldflur*. Dissertation Zoologisches Institut der Martin-Luther-Universität Halle/Saale.

972. Woodroff, S. and Majer, J. D. (1981). Colonisation of ants on the exposed banks of the Canning Dam reservoir. *Australian Entomological Magazine*, **8**, 41–6.

973. Wray II, T. (1979). *Breeding Biology and Reproductive Success of Three Sparrow Species on Reclaimed Surface Mines in West Virginia*. M.Sc. Thesis, West Virginia University (unpublished manuscript).

974. Wray II, T. (1982). *Breeding Bird Community Structure and Function on a Reclaimed Surface Mine*. PhD Dissertation, West Virginia University (unpublished manuscript).

975. Wray II, T., Strait, K. A. and Whitmore, R. C. (1982). Reproductive success of grassland sparrows on a reclaimed surface mine in West Virginia. *The Auk*, **99**, 157–64.

976. Wray II, T., Wackenhut, P. B. and Whitmore, R. C. (1978). The reproductive biology of passerine birds breeding on reclaimed surface mines in northern West Virginia. In *Surface Mining and Fish/Wildlife Needs in the Eastern United States*, ed. D. E. Samuel, J. R. Stauffer, C. H. Hocutt and W. T. Mason, pp. 333–4. Morgantown: West Virginia University.

977. Wykes, B. J. (1985). *The Jarrah Avifauna and its Re-establishment After Bauxite Mining*. Bentley: WAIT School of Biology Bulletin no. 11.

978. Wyngaard, G. A. (1985). Ethical and ecological concerns in land reclamation policy: an analysis of the surface mining control and reclamation act of 1977. In *Wetlands and Water Management on Mined Lands. Proceedings of a Conference October 23–24, 1985*, ed. R. P. Brooks, D. E. Samuel and J. B. Hill, pp. 75–81. Pennsylvania State University.

979. Yahner, R. H., Garton, A. W. and Howell, J. C. (1975). Breeding avifauna associated with two strip mine areas. *Journal of the Tennessee Academy of Science*, **50**, 95–8.

980. Yahner, R. H. and Howell, J. C. (1975). Habitat use and species composition of breeding avifauna in a deciduous forest altered by strip mining. *Journal of Tennessee Academy of Science*, **50**, 142–7.

981. Yeager, L. E. (1940). Wildlife management on coal stripped land. *North American Wildlife Conference. Transactions*, **5**, 348–53.

982. Yeager, L. E. (1942). Coal-stripped land as a mammal habitat, with special reference to fur animals. *The American Midland Naturalist*, **27**, 613–35.

983. Yearsley, E. F. (1976). *Use of Reclaimed Surface Mines by Foxes in Preston County, West Virginia*. MS Thesis, West Virginia University (unpublished manuscript).

984. Yearsley, E. F. and Samuel, D. E. (1980). Use of reclaimed surface mines by foxes in West Virginia. *Journal of Wildlife Management*, **44**, 729–34.

985. Young, S. A. and Rennick, R. B. (1983*a*). *Using Grazing Animals to Alter Vegetation Composition of Reseeded Minesoils*. Abstracts of papers. 36th Annual Meeting, Society of Range Management. Alburquerque, New Mexico. no. 306.

986. Young, S. A. and Rennick, R. B. (1983*b*). *Use of Grazing Cattle During Post Seedling Management of Revegetated Minesoils*. Montana Agricultural Experimental Station (Bozeman, Montana) Final Report.

987. Zarger, T. G. (1980). Forestation of surface mines for wildlife. In *Trees for Reclamation*

Symposium Proceedings. USDA Forest Service General Technical Report NE-61, pp. 71–4.

988. Zarger, T. G., Maddox, J. B., Starnes, L. B. and Seawell, W. M. (1979). *Ecological Recovery After Reclamation of Toxic Spoils Left by Coal Surface Mining*. US Environmental Protection Authority Report no. EPA-600/7-79-209.

989. Zellmer, S. D. (1979). A coal refuse reclamation project. *Proceedings of the West Virginia Academy of Science*, **51**, 115–27.

990. Zellmer, S. D., Bernard, J. R., Dvorak, A. J., Master, W. A., Miller, R. M., Olsen, R. D., Pentecost, E. D., Schubert, J. P., Sobek, A. A., Vinikour, W. S. and Wilkey, M. L. (1979). Macoupin County refuse reclamation project. In *Land Reclamation Program Annual Report 1978*. Argonne National Laboratory Report ANL/LRP-5, pp. 37–48.

991. Zellmer, S. D. and Wilkey, M. L. (1979). A reclamation demonstration project at an abandoned deep mine. *Minerals and the Environment*, **1**, 57–63.

992. Zerling, L. (1987). *Zur Wiederbesiedlung einer Landwirtschaftlich Genutzten Kippe des Braunkohlentagebaues durch Bodenbewohnende Kleinarthropoden unter Besonderer Berücksichtigung der Springschwanze (Insecta: Collembola)*. M.Sc. Thesis, Universität Halle-Wittenberg (unpublished manuscript).

993. Zurbuch, P. E. (1978). Eastern fish and wildlife habitat unsuitable for surface mining. In *Surface Mining and Fish/Wildlife Needs in the Eastern United States*, ed. D. E. Samuel, J. R. Stauffer, C. H. Hocutt and W. T. Mason, pp. 39–41. Morgantown: West Virginia University.

Appendix 1

This Appendix presents a taxonomic list of protozoans and animals which may occur in terrestrial ecosystems and freshwater impoundments or streams. The major feeding habits and other aspects of their ecology which may be of interest are given. Each taxon is only defined at the taxonomic level necessary to place it in a relatively homogeneous ecological unit.

Phylum	Class	Order
MASTIGOPHORA	PHYTOMASTIGOPHOREA	—
	ZOOMASTIGOPHOREA	—
SARCODINA	RHIZOPODEA	—
	ACTINOPODEA	—
SPOROZOA	—	—
CNIDOSPORA	—	—
CILIOPHORA	—	—
PORIFERA	Demospongiae	Haplosclerida
CNIDARIA	Hydrozoa	Limnomedusae (sub-order)
		Anthomedusae (sub-order)
PLATYHELMINTHES	Turbellaria	Neorhabdocoela
		Seriata
		Temnocephalida
	Trematoda	Monogenea
		Digenea
		Aspidobothrea
	Cestoda	Eucestoda (sub-class)
	Cestoda	Cestodaria (sub-class)
	Mesozoa	—
RHYNCHOCOELA	—	—
GASTROTRICHA	—	—
ROTIFERA	—	—
NEMATODA	Adenophorea	—
	Secernentea	—
NEMATOMORPHA	Gordioidea	—
ACANTHOCEPHALA	—	—
MOLLUSCA	Gastropoda	—
	Bivalvia	—

Common name(s)	Roles in ecosystem	Terrestrial or aquatic
Flagellates	Most have chromoplasts and are autotrophic. Many are classed as plants.	A
Flagellates	Some parasites, others free-living or commensals in guts of animals.	A
Amoebas	Some parasites, others free-living detritus feeders.	A, T
Heliozoans	Parasites of algae and higher plants.	A
Sporozoans	Parasites of vertebrates and invertebrates.	A, T
Sporozoans	Parasites of vertebrates and invertebrates.	A, T
Ciliates	Many are suspension feeders. Some are herbivores.	A
Freshwater sponges	Filter feeders, mostly on particles smaller than bacteria. Some have symbiotic zoochlorellae.	A
Freshwater jellyfish	Carnivores	A
Freshwater hydras	Carnivores, especially on small crustaceans.	A
—	Free-living and parasitic species (e.g. on snails).	A
Triclads, Planarians	Free-living, mostly carnivores.	A, T
—	Commensals, especially on crustaceans.	A
Flukes	Parasites, especially of fish and amphibians. Have a single host.	A
Flukes	Parasites of vertebrates. Have dual hosts.	A, T
—	Parasites, especially of fish and reptiles.	A, T
Tapeworms	Parasites of vertebrates. Have intermediate hosts.	T, A
—	Parasites of fish.	A
—	Parasites of worms, echinoderms and molluscs.	A
Nemerteans	Carnivores, feed on worms, molluscs and crustaceans.	A, T
—	Part of the decomposer community on the bottom of still water.	A
Wheel animals	Very small, mainly filter feeding in freshwater.	A
Roundworms	Mostly free-living.	A, T
Roundworms	Mostly parasites of invertebrates and vertebrates.	A, T
Hairworms	Adults free-living, juveniles parasitic in insects and crustaceans.	A, T
Spiny-headed worms	Parasites, adults in vertebrates, larvae in invertebrates.	A, T
Slugs, Snails, Freshwater snails	Freshwater and terrestrial herbivores and detritivores. Some are vectors of parasites.	A, T
Clams, Mussels	Filter feeders on live and dead material. Some have larval stage which parasitizes fish.	A

Appendix 1 (*cont.*)

Phylum	Class	Order
ANNELIDA	OLIGOCHAETA	—
	HIRUDINEA	—
ONYCHOPHORA	—	—
ARTHROPODA (sub-phylum Chelicerata)	ARACHNIDA	Scorpiones
		Pseudoscorpiones
		Solifugae
		Palpigradi
		Uropygi
		Amblypygi
		Araneae
		Ricinulei
		Opiliones
		Acarina
ARTHROPODA (sub-phylum Crustacea)	BRANCHIPODA	—
	MAXILLOPODA	Branchiura (sub-class) Copepoda (sub-class)

Common name(s)	Roles in ecosystem	Terrestrial or aquatic
Earthworms	Free-living in soil and water, mostly scavengers or detritivores.	T, A
Leeches	Mostly ectoparasitic blood-suckers of vertebrates, some predators.	T, A
Peripatus	Found in humid areas, on ground, in rotting logs and near streams. Predators.	T
Scorpions	Mostly tropical or sub-tropical. Predators, feeding on insects, millipedes, spiders, etc.	T
Pseudoscorpions	Live in crevices, under bark and in plant litter. Feed on collembolans, beetles, booklice, bugs, etc.	T
Windscorpions	Desert dwellers, usually crepuscular or nocturnal, a few are diurnal. Predators, mostly on termites, sometimes beetles and earthworms.	T
Microwhip scorpions	Live in humid areas, in the soil under stones, avoid light. Predators.	T
Whipscorpions, Schizomids	In tropics and warm humid areas. Live in litter, tree bark or under stones. Food unknown but feed on insects in captivity.	T
Tailless whipscorpions, Whipspiders	In tropics and subtropics. Live under stones, logs, bark and litter. Feed on insects and arachnids.	T
Spiders	Predators. Found in all terrestrial habitats. A few are aquatic.	T, A
Ricinuleids	Secretive animals, found in forest litter, under logs and in caves. Predators, feed on small termites, fly larvae and spiders.	T
Harvestmen, Daddy long legs	Predators. Found in a variety of habitats including caves.	T
Ticks, Mites	Some are parasitic or predators, others are important decomposer organisms in the soil and litter layers. Occupy all terrestrial habitats, some are aquatic.	T, A
Water fleas, Shell-backed shrimps, Shell-less fairy shrimps	Have drought-resistant eggs and live in ephemeral waters. Filter feeders, are important food for vertebrates.	A
Fish lice	Parasitic, almost entirely on fish.	A
Copepoda, Oar-footed crustacea	Free-swimming filter feeders. Some are commensals or parasites of other invertebrates or vertebrates. Important food for vertebrates.	A

Appendix 1 (*cont.*)

Phylum	Class	Order
	OSTRACODA	Podocopida
	MALOCOSTRACA	Anaspidacea
		Bathynellacea Thermosbaenacea Spelaeogriphacea Amphipoda
		Isopoda
		Decapoda
ARTHROPODA (subphylum Uniramia)	CHILOPODA	—
	DIPLOPODA	—
	SYMPHYLA	—
	PAUROPODA	—
	COLLEMBOLA	—
	DIPLURA	—
	PROTURA	—
	INSECTA	Archaeognatha
		Thysanura
		Ephemeroptera
		Odonata

Common name(s)	Roles in ecosystem	Terrestrial or aquatic
—	Benthic or living on plants. Omnivores, some are commensals with other Crustacea.	A
—	Benthic. Can filter feed or eat other small invertebrates.	A
—	Filter feeders on algae. Some live in ground water.	A
—	Benthic detritus feeders.	A
—	Live in caves.	A
Sandhoppers, Sand or Beach fleas	Aquatic forms are benthic, a few are parasites or commensals. Terrestrial forms are scavengers and are involved in litter decomposition.	A, T
Woodlice, Pill bugs, Sow bugs	Scavengers, some are parasitic. Terrestrial forms are involved in litter decomposition.	A, T
Crayfish, Crabs, Prawns	Found in wide range of aquatic and terrestrial habitats. Omnivores, carnivores or herbivores. Some feed on particles suspended in water.	A, T
Centipedes	Mostly in tropics and subtropics. Predators.	T
Millipedes	Live in rotten logs, leaf litter, under stones and bark. Feed on rotten vegetation and wood.	T
—	Live in moist habitats such as soil, litter and under wood. Feed on plant rootlets.	T
—	Live in moist habitats such as soil, litter and under wood. Feed on fungal hyphae.	T
Springtails, Collembolans	Found in soil, leaf litter, under bark and on plants. Feed on fungi, decaying or live vegetation and some are scavengers.	T
—	Found in soil, leaf litter, decaying wood and under stones. Feed on decaying vegetation, some are carnivores.	T
—	Found in soil, leaf litter, under bark and stones. Mainly feed on vegetable material.	T
—	Found in soil, leaf litter, under bark and stones. Feed on algae, lichens, vegetable debris or are scavengers.	T
Silverfish	Found in soil, leaf litter, under stones and in ant or termite nests. Omnivores or hervibores.	T
Mayflies	Aquatic nymphs, adults found close to water. Immatures are herbivores or scavengers, adults do not feed.	A, T
Dragonflies, Damselflies	Aquatic larvae, adults often found close to water. Adults and immatures are carnivores.	A, T

Appendix 1 (*cont.*)

Phylum	Class	Order
		Blattodea
		Isoptera
		Mantodea
		Zoraptera
		Grylloblattodea
		Dermaptera
		Plecoptera
		Orthoptera
		Phasmatodea
		Embioptera
		Psocoptera
		Phthiraptera
		Hemiptera
		Thysanoptera
		Neuroptera
		Coleoptera
		Strepsiptera

Common name(s)	Roles in ecosystem	Terrestrial or aquatic
Cockroaches	Found in logs, under bark, in caves and in leaf litter. Some are domestic pests. Omnivores.	T
Termites, White ants	Social insects. Nests in ground or timber. Feed on sound or decaying wood, grass, fungi and other vegetable matter. Some are serious timber pests.	T
Praying mantids	Found on vegetation and tree trunks. Predators.	T
Zorapterons	Found in dead wood, leaf litter, under bark or in termite nests. Mainly fungivores.	T
Grylloblattids	Found beneath stones. Feed on moss and may scavenge insect material.	T
Earwigs	Found in soil, leaf litter, under bark and on foliage and flowers. Feed on a wide range of living and dead plant or animal matter.	T
Stoneflies	Aquatic larvae, adults found on stones and vegetation close to water.	A, T
Locusts, Grasshoppers, Crickets	Found in and on vegetation, in or on ground, in logs etc. Herbivores, scavengers or omnivores. Some herbivorous species are serious plant pests.	T
Stick insects	Found on vegetation. Herbivores, some are serious defoliators.	T
Web-spinners	Found in silken tunnels under rocks, bark and in leaf litter. Feed on living and dead vegetable material.	T
Psocids, Booklice	Found under bark, stones, in leaf litter, on vegetation and in stored products. Feed on algae, fungi, lichens and fragments of plant and insect tissue.	T
Lice	Ectoparasites on vertebrates, mainly birds.	T
Bugs	Mainly on vegetation but representatives in most habitats. Some are aquatic. Plant sap feeders. Some are predators or parasites. Many of the herbivorous species are serious pests.	T, A
Thrips	Found on dried leaves, fresh vegetation and on flowers. Feed on fungi, flowers and some prey on small arthropods. Some damage flowers.	T
Lacewings, Alder flies, Antlions	Usually on vegetation but found in other habitats. Some have aquatic larvae. Predators.	T, A
Beetles	Found in all terrestrial habitats, some aquatic. Full range of feeding habits. Some are serious pests.	T, A
Stylops	Larvae parasitic on other insects, adults free-living.	T

Appendix 1 (*cont.*)

Phylum	Class	Order
		Mecoptera
		Siphonaptera
		Diptera
		Trichoptera
		Lepidoptera
		Hymenoptera
TARDIGRADA	—	—
PENTASTOMIDA	—	—
BRYOZOA	—	—
ENTOPROCTA	—	—
CHORDATA	Elasmobranchiomorphi	—
	Osteichthyes	—
	Amphibia	Anura
		Urodela
		Gymnophiona
	Reptilia	Squamata
		Crocodilia
		Testudines
	Aves	Struthioniformes
		Rheiformes
		Casuariformes
		Apterygiformes
		Gaviiformes
		Podicipediformes

Common name(s)	Roles in ecosystem	Terrestrial or aquatic
Scorpionflies	Found amongst vegetation. Predators, some also feed on nectar.	T
Fleas	Ectoparasites on warm-blooded vertebrates, larvae in vegetable and animal debris.	T
Two-winged flies	Found in all terrestrial habitats, some have aquatic larvae. Full range of feeding habits. Some are serious pests.	T, A
Caddis flies	Aquatic larvae, adults found near water. Larvae are herbivores or omnivores, adults may take nectar.	A, T
Moths, Butterflies	Found in most terrestrial habitats. Larvae mostly herbivores, some are predators, adults may take nectar. Many species are pests.	T
Wasps, Bees, Ants	Many are social insects, found in most habitats, many wasp larvae are parasitic in insects. Larvae feed in a variety of ways, adults often take nectar. Ants feed on prey, saps and seeds.	T
Water bears	Herbivores, carnivores or detritus feeders. Can withstand extreme desiccation in an anabiotic state.	T, A
Tongue worms	Parasites in respiratory passages of vertebrates.	T
Moss animals	Filter feeders on phytoplankton and bacteria.	A
—	One freshwater genus, filter feeders on organic particles and protozoans.	A
Sharks	A few freshwater species, carnivores.	A
Bony fish	Wide range of herbivores and carnivores.	A
Frogs, Toads	Predominantly insectivores, larvae feed on algae.	T, A
Salamanders, Newts	Mostly insectivores, larval stages omnivores.	T, A
—	Worm-like burrowing animals.	T
Lizards, Snakes	Carnivores, insectivores, a few are herbivores.	T
Crocodiles, Alligators	Carnivores.	A
Turtles, Tortoises	Some carnivores, others herbivores.	T, A
Ostriches	Herbivores, insectivores.	T
Rheas	Herbivores, insectivores.	T
Cassowaries, Emus	Frugivores, herbivores, insectivores.	T
Kiwis	Feed on soil and litter invertebrates.	T
Loons	Piscivores.	T, A
Grebes	Feed on fish and aquatic invertebrates.	T, A

Appendix 1 (*cont.*)

Phylum	Class	Order
		Pelicaniformes
		Ciconiiformes
		Anseriformes
		Falconiformes
		Galliformes
		Gruiformes
		Charadriiformes
		Columbiformes
		Psittaciformes
		Cuculiformes
		Strigiformes
		Caprimulgiformes
		Apodiformes
		Trogoniformes
		Coraciiformes
		Piciformes
		Passeriformes

Common name(s)	Roles in ecosystem	Terrestrial or aquatic
Pelicans, Gannets, Cormorants, etc.	Piscivores.	T, A
Herons, Storks, Flamingos, etc.	Herons take mainly fish, but also other vertebrates and invertebrates. Flamingos are plankton feeders, others take a range of aquatic and terrestrial prey.	T, A
Ducks, Geese, Swans, etc.	Species differ in feeding habits from purely grazing terrestrial herbivores to diving ducks which feed on fish and aquatic invertebrates.	T, A
Vultures, Hawks, Falcons, etc.	Scavengers or predators on other vertebrates and large insects.	T
Quails, Turkeys, Wildfowl, etc.	Herbivores, granivores with insects taken while young.	T
Cranes, Rails, Bustards, etc.	Primarily invertebrate-feeders, but probably also take some vegetable matter.	T, A
Gulls, Terns, Auks, etc.	Scavengers, piscivores and feeders on aquatic invertebrates.	T, A
Pigeons, Doves	Granivores.	T
Parrots	Feed on fruit, seeds, nectar and pollen as well as insects.	T
Cuckoos, Roadrunners	Insectivores, but roadrunners also take small vertebrates.	T
Owls	Mainly feed on small to medium-sized vertebrates, but also large nocturnal insects; fruit also eaten by some species.	T
Frogmouths, Nightjars, etc.	Primarily feed on nocturnal insects.	T
Swifts, Hummingbirds	Insectivores and nectar-feeders.	T
Trogons	Frugivores.	T
Kingfishers, Bee-eaters, Hornbills	Insectivores, some small vertebrates may be taken.	T
Toucans, Woodpeckers, etc.	Insectivores, some also feed on sap and fruit.	T
All other birds including Finches, Thrushes, Swallows, Corvants, Wrens, etc.	Species range from being purely insectivores to frugivores. Many species change diet with season, feeding on insects during nesting period then switching to fruit and/or seeds during non-feeding phase of life cycle.	T

Appendix 1 (*cont.*)

Phylum	Class	Order
	MAMMALIA	Monotremata
		Marsupalia
		Insectivora
		Chiroptera
		Primates
		Carnivora
		Hyracoidea
		Proboscidea
		Perissodactyla
		Artiodactyla
		Edentata
		Pholidota
		Tubulidentata
		Rodentia
		Lagomorpha

Taxonomy based on Borror & Delong (1970) and Waterhouse (1970) for insects, Bowman & Abele (1982) for crustaceans, Barnes (1980) for other invertebrates, Welty (1979) for birds and Romer & Parsons (1977) for other vertebrates.

References

Barnes, R. D. (1980). *Invertebrate Zoology*. Philadelphia: W. B. Saunders.

Borror, D. J. & Delong, D. M. (1970). *An Introduction to the Study of Insects*. New York: Holt, Rinehart and Winston.

Common name(s)	Roles in ecosystem	Terrestrial or aquatic
Platypus, Echidna	Platypus largely feeds on freshwater invertebrates, the echidna feeds on ants and other insects.	T
Pouched mammals	Herbivores, carnivores, insectivores, omnivores, nectarivores.	T
Moles, Shrews, Hedgehogs	Insectivores.	T
Bats	Nectarivores, frugivores, insectivores, piscivores, carnivores and sanguivores.	T
Lemurs, Monkeys, Apes	Omnivores, herbivores or frugivores.	T
True carnivores, Cats, Dogs, Bears	Carnivores; bears are omnivores.	T
Conies	Herbivores.	T
Elephants	Herbivores.	T
Odd-toed ungulates, Horses, Rhinoceroses	Herbivores.	T
Even-toed ungulates, Deer, Bison, Sheep, Goats	Herbivores.	T
Sloths, Armadillos, Anteaters	Sloths are hervibores; armadillos are insectivores and omnivores; anteaters feed on ants and termites.	T
Pangolin	Insectivores, feed on ants and termites.	T
Aardvark	Insectivores, feed on ants and termites.	T
Squirrels, Rats, Mice	Omnivores, granivores, herbivores, insectivores.	T
Rabbits	Herbivores.	T

Bowman, T. E. & Abele, L. G. (1982). Classification of the recent Crustacea. In *The Biology of Crustacea*, vol. 1, ed D. E. Bliss, pp. 1–27. New York: Academic Press.

Romer, A. S. & Parsons, T. S. (1977). *The Vertebrate Body*. Philadelphia: W. B. Saunders.

Waterhouse, D. F., ed. (1970). *The Insects of Australia*. Melbourne: Melbourne University Press.

Welty, J. C. (1979). *The Life of Birds*. Philadelphia: Saunders College Publishing.

Index

Page numbers in bold type denote illustration or figure.

.V.

ORTHOGONAL DESIGNS

01-29

PURE AND APPLIED MATHEMATICS

A Program of Monographs, Textbooks, and Lecture Notes

Contributions to *Lecture Notes in Pure and Applied Mathematics* are reproduced by direct photography of the author's typewritten manuscript. Potential authors are advised to submit preliminary manuscripts for review purposes. After acceptance, the author is responsible for preparing the final manuscript in camera-ready form, suitable for direct reproduction. Marcel Dekker, Inc. will furnish instructions to authors and special typing paper. Sample pages are reviewed and returned with our suggestions to assure quality control and the most attractive rendering of your manuscript. The publisher will also be happy to supervise and assist in all stages of the preparation of your camera-ready manuscript.

LECTURE NOTES
IN PURE AND APPLIED MATHEMATICS

1. *N. Jacobson,* Exceptional Lie Algebras
2. *L.-Å. Lindahl and F. Poulsen,* Thin Sets in Harmonic Analysis
3. *I. Satake,* Classification Theory of Semi-Simple Algebraic Groups
4. *F. Hirzebruch, W. D. Newmann, and S. S. Koh,* Differentiable Manifolds and Quadratic Forms
5. *I. Chavel,* Riemannian Symmetric Spaces of Rank One
6. *R. B. Burckel,* Characterization of C(X) Among Its Subalgebras
7. *B. R. McDonald, A. R. Magid, and K. C. Smith,* Ring Theory: Proceedings of the Oklahoma Conference
8. *Y.-T. Siu,* Techniques of Extension of Analytic Objects
9. *S. R. Caradus, W. E. Pfaffenberger, and B. Yood,* Calkin Algebras and Algebras of Operators on Banach Spaces
10. *E. O. Roxin, P.-T. Liu, and R. L. Sternberg,* Differential Games and Control Theory
11. *M. Orzech and C. Small,* The Brauer Group of Commutative Rings
12. *S. Thomeier,* Topology and Its Applications
13. *J. M. López and K. A. Ross,* Sidon Sets
14. *W. W. Comfort and S. Negrepontis,* Continuous Pseudometrics
15. *K. McKennon and J. M. Robertson,* Locally Convex Spaces
16. *M. Carmeli and S. Malin,* Representations of the Rotation and Lorentz Groups: An Introduction
17. *G. B. Seligman,* Rational Methods in Lie Algebras
18. *D. G. de Figueiredo,* Functional Analysis: Proceedings of the Brazilian Mathematical Society Symposium
19. *L. Cesari, R. Kannan, and J. D. Schuur,* Nonlinear Functional Analysis and Differential Equations: Proceedings of the Michigan State University Conference
20. *J. J. Schäffer,* Geometry of Spheres in Normed Spaces
21. *K. Yano and M. Kon,* Anti-Invariant Submanifolds
22. *W. V. Vasconcelos,* The Rings of Dimension Two
23. *R. E. Chandler,* Hausdorff Compactifications
24. *S. P. Franklin and B. V. S. Thomas,* Topology: Proceedings of the Memphis State University Conference
25. *S. K. Jain,* Ring Theory: Proceedings of the Ohio University Conference
26. *B. R. McDonald and R. A. Morris,* Ring Theory II: Proceedings of the Second Oklahoma Conference
27. *R. B. Mura and A. Rhemtulla,* Orderable Groups
28. *J. R. Graef,* Stability of Dynamical Systems: Theory and Applications
29. *H.-C. Wang,* Homogeneous Banach Algebras
30. *E. O. Roxin, P.-T. Liu, and R. L. Sternberg,* Differential Games and Control Theory II
31. *R. D. Porter,* Introduction to Fibre Bundles
32. *M. Altman,* Contractors and Contractor Directions Theory and Applications
33. *J. S. Golan,* Decomposition and Dimension in Module Categories
34. *G. Fairweather,* Finite Element Galerkin Methods for Differential Equations
35. *J. D. Sally,* Numbers of Generators of Ideals in Local Rings
36. *S. S. Miller,* Complex Analysis: Proceedings of the S.U.N.Y. Brockport Conference
37. *R. Gordon,* Representation Theory of Algebras: Proceedings of the Philadelphia Conference
38. *M. Goto and F. D. Grosshans,* Semisimple Lie Algebras
39. *A. I. Arruda, N. C. A. da Costa, and R. Chuaqui,* Mathematical Logic: Proceedings of the First Brazilian Conference
40. *F. Van Oystaeyen,* Ring Theory: Proceedings of the 1977 Antwerp Conference

41. *F. Van Oystaeyen and A. Verschoren,* Reflectors and Localization: Application to Sheaf Theory
42. *M. Satyanarayana,* Positively Ordered Semigroups
43. *D. L. Russell,* Mathematics of Finite-Dimensional Control Systems
44. *P.-T. Liu and E. Roxin,* Differential Games and Control Theory III: Proceedings of the Third Kingston Conference, Part A
45. *A. V. Geramita and J. Seberry,* Orthogonal Design: Quadratic Forms and Hadamard Matrices

Other Volumes in Preparation

ORTHOGONAL DESIGNS

Quadratic Forms and Hadamard Matrices

ANTHONY V. GERAMITA
Department of Mathematics and Statistics
Queen's University
Kingston, Ontario, Canada

JENNIFER SEBERRY
Department of Applied Mathematics
The University of Sydney
New South Wales, Australia

MARCEL DEKKER, INC. New York and Basel

Library of Congress Cataloging in Publication Data

Geramita, A V
Orthogonal designs.

(Lecture notes in pure and applied mathematics ; v. 45)
Bibliography: p.
Includes index.
1. Hadamard matrices. 2. Forms, Quadratic.
I. Seberry, Jennifer, [Date] joint author.
II. Title.
QA166.4.G47 511'.6 79-12112
ISBN 0-8247-6774-8

MARCEL DEKKER, INC.
270 Madison Avenue, New York, New York 10016

Current printing (last digit):
10 9 8 7 6 5 4 3 2 1

PRINTED IN THE UNITED STATES OF AMERICA

PREFACE

Problems concerned with the structure and existence of various kinds of matrices with entries from $\{0, 1, -1\}$, for example, Hadamard matrices and their generalizations to weighing matrices, have long been of interest to workers in combinatorics and also to applied statisticians, coding theorists and other applied mathematicians. This volume is devoted to the study of a new approach to these questions, which has illiminated their connections and exposed new ones between these questions and several other areas of mathematics. This new approach is through the investigation of Orthogonal Designs: roughly speaking, special matrices with indeterminate entries.

Until now this subject has been discussed in our research papers and in those of our colleagues and students. The discovery of the intimate relationship between orthogonal designs and rational quadratic forms, which has not yet appeared in print, and thè finding of numerous new constructions and interesting objects that appear fundamental to the study of Hadamard matrices (and their generalizations) were what finally prompted us to look afresh at the work that had already appeared. We decided to recast our work and that of our collaborators and students in the light of these new discoveries. This work is the result of that effort.

As will be clear in the text, Orthogonal Designs is a heavy "borrower" of mathematics. The reader will find us using results from, for example, algebraic number theory, quadratic forms, difference sets, representation theory, coding theory, finite geometry and cyclotomy. We have not expected the reader to be conversant with all of these areas; indeed, we have tried to present the material in such a way that even the novice to these areas will understand why and how we intend to use the results stated, even if the proofs of some of these results remain a mystery. We have, in those cases where detailed explanation would take too long, given references where the interested reader can fill out his or her background. As a result, we believe the book can be read profitably both by experts and by people new to this area of combinatorics.

To dispel any notion that this book closes the area for further research, we have included a list of well over 100 problems, all unsolved, and suggested directions in which we feel further research is possible. These problems vary in depth, and we have indicated those we feel are major.

Some comments on how this volume is organized: The organization is, in part, dictated by the Janus-like features of the study. In the first three chapters we describe the nature of the problem at hand and make some remarks on the ingredients of a solution. After some preliminaries we inquire rather deeply into the algebraic side of the problem. In broad terms, the algebra allows us to describe the first set of nontrivial necessary conditions on the constituents of a solution. In Chapter 4 we turn about and, with the necessary conditions as a guide, attempt to satisfy these conditions. It is here that many different methods of construction are used and discussed. Chapter 5 concentrates on one of the construction methods and analyzes it in detail, both algebraically and combinatorially. Here again the interplay between classical algebra and combinatorics is shown to have striking consequences. In Chapter 6 a very strong nonexistence theorem for orthogonal designs is proved, and in Chapter 7 we give some applications to the problems which began the entire investigation; i.e., we give the strongest results so far obtained on the existence of Hadamard matrices. Chapter 8 is largely a summarizing of results from other chapters and a filling-in, in special cases, of results not obtained yet by the more general methods. Finally, in the Appendices, we tabulate the numerous calculations we have made in specific orders.

No book of this size could be the child of only two parents, and we feel a deep sense of gratitude to the many people who have helped us. We are greatly indebted to Dan Shapiro (Ohio State) for patiently explaining his work on similarities to us. His work has had an enormous impact on this study. We would next like to thank our students, Peter Eades and Peter Robinson (Canberra), and Warren Wolfe (Kingston). Their continuing interest and good ideas have broadened this subject far beyond what we had ever envisaged. Their contributions will be evident to any reader. We also thank: Joan Geramita, I. Blake, S. R. Cavior, M. Griffin, L. G. Kovács, T. Y. Lam, R. Magoon, R. C. Mullin, N. J. Pullman, L. G. Roberts, and C. Small. They all helped at various stages and, in some instances, saved us from writing utter nonsense. In the proofreading we owe special thanks to Joan Geramita and Chuck Weibel. They gave more generously of their time than we could ever indicate here. The beautiful typing, from a difficult manuscript, was done by Evelyn Laurent with the appendices done by Irene Gaskill. We sincerely thank them also. We'd also like to thank the Institute for Advanced Study (Princeton) for its generosity to the first author during the academic year 1977-78, which allowed us the time to put this book in final shape.

Finally, a word on our experience in writing this book. We approached the problems herein from quite different backgrounds. It became evident that Geramita was more interested in the connections with algebra and how the algebra explains nonexistence, while Seberry was more interested in actually making the objects that could be made. The inevitable tensions that arise from these differing viewpoints will, we hope, make this book more interesting to read.

Anthony V. Geramita
Jennifer Seberry

CONTENTS

ORTHOGONAL DESIGNS

CHAPTER 1

ORTHOGONAL DESIGNS--THE PROBLEM AND SOME REMARKS ON ITS SOLUTION

An orthogonal design of order n and type $(s_1, \ldots, s_\ell)$, s_i positive integers, is an $n \times n$ matrix X, with entries from $\{0, \pm x_1, \ldots, \pm x_\ell\}$ (the x_i commuting indeterminates) satisfying

$$XX^t = (\sum_{i=1}^{\ell} s_i x_i^2) I_n .$$

Alternatively, each row of X has s_i entries of the type $\pm x_i$ and the rows are orthogonal under the Euclidean inner product.

We may view X as a matrix with entries in the field of fractions of the integral domain $Z[x_1, \ldots, x_\ell]$, (Z the rational integers), and then if we let $f = (\sum_{i=1}^{\ell} s_i x_i^2)$, X is an invertible matrix with inverse $\frac{1}{f} X^t$. Thus, $X^t X = f I_n$, and so our alternative description of the rows of X applies equally well to the columns of X.

The task to which this text is addressed can be simply described as follows: Find necessary and sufficient conditions on the set of integers $\{n; s_1, \ldots, s_\ell\}$ such that there exists an orthogonal design in order n of type $(s_1, \ldots, s_\ell)$.

The generality of the question is such that it includes many other problems which have been extensively studied and provides an umbrella under which these problems may be considered simultaneously. Also, in this generality the connections between these classical combinatorial problems and some of the great mathematics of the past century have been illuminated. More particularly, the general approach has shown the close connection the combinatorial problems studied have with the classification theorems of quadratic forms over $\mathbb{Q}$, the rational numbers. These classification theorems, largely the work of Minkowski, are among the few complete mathematical triumphs of this century. The fact that a partial solution to our general question is embedded in this beautiful theory gives us hope that there will now continue a deeper investigation of these combinatorial problems and the still unresolved problem of the classification of quadratic forms over Z.

§1.1. GENERALITIES ON HURWITZ-RADON FAMILIES TO PROVE THERE EXIST ORTHOGONAL DESIGNS OF TYPE (1, 1, ..., 1)

If X is an orthogonal design of order n and type $(s_1, \ldots, s_\ell)$ on the indeterminates $x_1, \ldots, x_\ell$, we may write

$$X = A_1 x_1 + \ldots + A_\ell x_\ell \qquad (\dagger)$$

where the A_i are $\{0,1,-1\}$ matrices of size $n \times n$. If we let $A*B$ denote the Hadamard product of the matrices A and B (the (i, j) entry of this product is the product of the (i, j) entry of A with the (i, j) entry of B), then the fact that the entries in X were linear monomials in the x_i (or zero) gives us

(i) $A_i * A_j = 0$ if $i \neq j$.

If we write out the fact that $XX^t = (\sum_{i=1}^{\ell} s_i x_i^2) I_n$, using ($\dagger$), and compare coefficients, we find that

(ii) $A_i A_i^t = s_i I_n$, $1 \leq i \leq \ell$,

(iii) $A_i A_j^t + A_j A_i^t = 0$, $1 \leq i \neq j \leq \ell$.

We can obviously reverse this procedure, and we state that precisely.

1.1. PROPOSITION. A necessary and sufficient condition that there exists an orthogonal design of order n and type $(s_1, \ldots, s_\ell)$ is that there exist matrices $\{A_1, \ldots, A_\ell\}$ satisfying:

(0) the A_i are $\{0,1,-1\}$ matrices, $1 \leq i \leq \ell$;

(i) $A_i * A_j = 0$ for $1 \leq i \neq j \leq \ell$;

(ii) $A_i A_i^t = s_i I_n$, $1 \leq i \leq \ell$;

(iii) $A_i A_j^t + A_j A_i^t = 0$, $1 \leq i \neq j \leq \ell$.

The first focus of our attack on the general problem concerns the maximum number of distinct variables that can appear in an orthogonal design of order n.

The last proposition shows that if the orthogonal design involves ℓ-variables, we get a collection of ℓ matrices, $\{A_1, \ldots, A_\ell\}$ satisfying (0)-(iii). Form the real matrices $B_i = \frac{1}{\sqrt{s_i}} A_i$. Then the $\{B_i\}$ satisfy

(a) $B_i B_i^t = I_n$, $1 \leq i \leq \ell$,

(b) $B_i B_j^t + B_j B_i^t = 0$, $1 \leq i \neq j \leq \ell$.

If we normalize the collection $\{B_1, B_2, \ldots, B_\ell\}$ by multiplying each member, on the right, by B_1^t and let $C_i = B_i B_1^t$, then $C_1 = I$ and the remaining $C_2, \ldots, C_\ell$ are orthogonal, anti-commuting, skew-symmetric matrices, as is easily checked.

The question of how many real orthogonal, anti-commuting skew-symmetric matrices there can be in order n is a question that was completely settled in the early part of the century by J. Radon [1]. Radon's work (which extended earlier work by A. Hurwitz [1]) was centered around a proposition concerning the composition of quadratic forms. The question Radon dealt with was: Given n, find the maximal m so that

$$(x_1^2 + \ldots + x_n^2)(y_1^2 + \ldots + y_m^2) = f_1^2 + \ldots + f_n^2$$

where the f_i are real bilinear functions of the x_i and y_j. We shall not go into the connections between this problem and its relation to orthogonal matrices here, but shall just be content to quote the relevant facts. For a discussion of this Radon-Hurwitz problem, we suggest (Herstein [1], Curtis [1] or Lam [1]).

For ease in exposition we make the following definition.

1.2. DEFINITION. A family $\{A_1, \ldots, A_s\}$ of real orthogonal matrices of order n satisfying:

(1) $A_i = -A_i^t$, $1 \leq i \leq s$,

(2) $A_i A_j = -A_j A_i$, $1 \leq i \neq j \leq s$,

will be called a Hurwitz-Radon family (H-R family).

Now, if n is a positive integer, write $n = 2^a b$, b odd, and then set $a = 4c+d$, $0 \leq d < 4$. If we denote by $\rho(n)$ the number $8c+2^d$, the main theorem of Radon [1] states:

1.3. THEOREM. (1) Any H-R family of order n has fewer than $\rho(n)$ members.

(2) There is an H-R family of order n having exactly $\rho(n)-1$ members.

We thus immediately have:

1.4. COROLLARY. The maximum number of variables in an orthogonal design of order n is $\leq \rho(n)$.

Unfortunately, Radon's theorem is not suitable, directly, for use with Proposition 1.1 since we need $\{0, 1, -1\}$ matrices.

If, however, we look for integer H-R families, this will automatically give us $\{0, 1, -1\}$ matrices, since an integer orthogonal matrix can only have entries from $\{0, 1, -1\}$. The fact that (2) of Theorem 1.3 could be improved in this direction was noted independently by Geramita-Pullman in [2] and Gabel in [1].

We need to make a few remarks about the function $\rho(n)$. First observe that when $n = 2^a b$, b odd, then $\rho(n) = \rho(2^a)$. Let $n_1 = 2^{4s+3}$, $n_2 = 2^{4(s+1)}$, $n_3 = 2^{4(s+1)+1}$, $n_4 = 2^{4(s+1)+2}$ and $n_5 = 2^{4(s+1)+3}$; then

$$\rho(n_2) = \rho(n_1)+1,$$
$$\rho(n_3) = \rho(n_1)+2,$$
$$\rho(n_4) = \rho(n_1)+4,$$
$$\rho(n_5) = \rho(n_1)+8.$$

We shall assume that the reader is familiar with the elementary properties of tensor products for matrices (see Marcus and Minc [1]).

Let

$$A = \begin{bmatrix} 0 & 1 \\ -1 & 0 \end{bmatrix}, \quad P = \begin{bmatrix} 0 & 1 \\ 1 & 0 \end{bmatrix}, \quad Q = \begin{bmatrix} 1 & 0 \\ 0 & -1 \end{bmatrix}.$$

1.5. PROPOSITION. (a) $\{A\}$ is an H-R family of $\rho(2)-1$ integer matrices of order 2.

(b) $\{A\otimes I_2, P\otimes A, Q\otimes A\}$ is an H-R family of $\rho(4)-1$ integer matrices of order 4.

(c) $\{I_2\otimes A\otimes I_2, I_2\otimes P\otimes A, Q\otimes Q\otimes A, P\otimes Q\otimes A, A\otimes P\otimes Q, A\otimes P\otimes P, A\otimes Q\otimes I_2\}$ is an H-R family of $\rho(8)-1$ integer matrices of order 8.

Proof. A tedious check using the symmetries of A, P, Q and I_2 and the behavior of these symmetries with respect to $\otimes$.

1.6. THEOREM. There is an H-R family of integer matrices of order n having $\rho(n)-1$ members.

Proof. The proposition above handles the cases $n = 2t, 4t, 8t$, t odd (just tensor with I_t). The reader may easily verify that if $\{M_1, \ldots, M_s\}$ is an H-R family of integer matrices of order n, then

(1) $\{A\otimes I_n\}\cup\{Q\otimes M_i \mid i = 1, \ldots, s\}$ is an H-R family of s+1 integer matrices of order 2n.

If, in addition, $\{L_1, \ldots, L_m\}$ is an H-R family of integer matrices of order k, then

(2) $\{P\otimes I_k\otimes M_i \mid 1\le i\le s\}\cup\{Q\otimes L_j\otimes I_n \mid 1\le j\le m\}\cup\{A\otimes I_{nk}\}$ is an H-R family of s+m+1 integer matrices of order 2nk.

Let $n_1, \ldots, n_5$ be as before Proposition 1.5, and we may proceed by induction: Starting with the fact that (c) in Proposition 1.5 gives us the case $n_1 = 2^3$. (Note that the nature of $\rho(n)$ allows us to only consider $n = 2^\ell$.)

Now (1) gives us the transition from n_1 to n_2; if we now use (2), letting $k = n_1$, $n = 2$ (and hence by (a) of Proposition 1.5, $s = 1$), we get the transition from n_1 to n_3. We then use (2) (this time with (b) and (c) of the Proposition) to get the two remaining transitions n_1 to n_4, n_1 to n_5. That completes the proof.

To apply this theorem to orthogonal designs, we need one observation. If A, B are two members of an integral H-R family, then A and B are each $\{0, 1, -1\}$ matrices and $A*B = 0$. The orthogonality of A (and

B) and the fact that the entries are integers show immediately that A and B are $\{0, 1, -1\}$ matrices with precisely one nonzero entry in each row and column. From the anticommutativity of A and B it follows that $(A+B)(A+B)^t = 2I$, and so $A+B$ has exactly two nonzero entries in each row and column, and hence $A*B = 0$. With the aid of Proposition 1.1, we sum this up by stating:

1.7. THEOREM. Given any natural number n, then

(a) any orthogonal design in order n can involve at most $\rho(n)$ variables;

(b) there is an orthogonal design in order n involving $\rho(n)$ variables.

<u>Proof.</u> Only (b) requires a small comment. If $A_1, \ldots, A_{\rho(n)-1}$ is an integral H-R family of order n, then $\{I_n, A_1, \ldots, A_{\rho(n)-1}\}$ is a family of matrices satisfying the conditions of Proposition 1.1 and gives an orthogonal design of type $\underbrace{(1, 1, \ldots, 1)}_{\rho(n)\text{-tuple}}$ in order n.

Thus, we have found the restrictions that must be placed on the number of variables in an orthogonal design, and we may rephrase our original question: Given n and $\ell \leq \rho(n)$, find necessary and sufficient conditions on $\{s_1, \ldots, s_\ell\}$ such that there exists an orthogonal design of order n and type $(s_1, \ldots, s_\ell)$.

REMARK. The reader will probably have noticed that the H-R integer families in orders 2, 4 and 8 are well known. In case $n = 2$, or $n = 4$, they come from the usual representations of the complex numbers and quaternions in 2×2 and 4×4 matrices respectively. For $n = 8$ the matrices are derived from the usual method of describing the multiplication table for the Cayley numbers (see, for example, Schafer [1]).

CHAPTER 2

SOME ALGEBRAIC AND COMBINATORIAL NONEXISTENCE RESULTS

In this section we intend to explain some easily obtained nonexistence theorems for orthogonal designs. Many of these results will be generalized in later sections, but we feel that these simpler special cases will give the reader an idea as to how the subject developed and what sorts of propositions might be expected.

To help in the development, and because of independent interest, we make a new definition.

2.1. DEFINITION. A weighing matrix of weight k and order n is an $n \times n$ $\{0, 1, -1\}$ matrix A such that $AA^t = kI_n$. (Note: $A^tA = AA^t = kI_n$.)

Such matrices have already appeared naturally as the "coefficient" matrices of an orthogonal design. (See Raghavarao [1] or [2] for why these are called weighing matrices and why there is interest in them by statisticians.) We shall refer to such a matrix as a $W(n, k)$.

If $n = k$, then a $W(n, n)$ is what is usually referred to in the literature as an Hadamard matrix.

It is an easy exercise to show:

2.2. PROPOSITION. In order that a $W(n, n)$ exist, $n = 1, 2$ or $4 \mid n$.

The proof uses, in an essential way, the fact that the entries in an Hadamard matrix are $\{\pm 1\}$, and the statement would be false without that, since $(3I_9)(3I_9)^t = 9I_9$, for example.

Now, when n is odd, $\rho(n) = 1$, and an orthogonal design on one variable is nothing more than a weighing matrix.

We shall next attempt to find some necessary conditions on the type of an orthogonal design in order n. We shall only consider a few special cases here: namely, n odd and $n = 2b$, b odd. We shall come back to the general problem later.

§2.1. ODD ORDER

We have already seen that $\rho(n) = 1$, and we need only consider orthogonal designs on one variable, that is, weighing matrices.

Proposition 2.2 already tells us something: If n is odd and a $W(n,k)$ exists, then $k \neq n$.

2.3. PROPOSITION. If X is a $W(n,k)$, n odd, then $k = a^2$ for some $a \in Z$.

<u>Proof.</u> More generally, if X is a matrix of odd order n with rational entries and $XX^t = qI_n$, then $q = r^2$ with $r \in \mathbb{Q}$; for $\det(X)^2 = q^n$ and since q^n is a square and n is odd, q is already a square. The proposition follows from the observation that if an integer is the square of a rational number, it is the square of an integer.

This proposition by itself does not begin to tell the whole story in odd order, as the following example shows:

2.4. EXAMPLE. There is no $W(5,4)$.

The property of being a weighing matrix is unaffected by row (or column) permutations. Multiplication of a row (or column) by -1 also does not affect the property of being a weighing matrix. Thus, there is no loss in generality if we assume a $W(5,4)$ has first row $[1\ 1\ 1\ 1\ 0]$. The inner product of rows 1 and 2 of our matrix is zero, and so there are an even number of nonzero entries under the 1's of the first row. There must, then, be a zero in the second row, last column. This then allows only three nonzero entries in the last column; a contradiction.

This example can be generalized.

2.5. PROPOSITION. If n is odd, then a necessary condition that a $W(n,k)$ exists is that $(n-k)^2-(n-k) \geq n-1$.

Proof. (We are indebted to P. Eades for this proof, which is much more illuminating than the proof we gave in Geramita-Geramita-Wallis [1].)

We start with a preliminary observation: If M is an $n \times n$ $\{0,1\}$ matrix with exactly k nonzero entries in each row and column, and if we number the rows of M by $r_1, \ldots, r_n$, then, for any $1 \leq j \leq n$,

$$\sum_{\substack{i=1 \\ i \neq j}}^{n} r_i \cdot r_j = k^2 - k. \qquad (+)$$

To see this let J be the $n \times n$ matrix of ones. Then $MJ = kJ = M^tJ$, and hence $MM^tJ = k^2J$, and so $\sum_{i=1}^{n} r_i \cdot r_j = k^2$. Since $r_j \cdot r_j = k$, (+) is then clear.

To see the relevance of this observation, suppose n is odd and W is a $W(n,k)$, and set $X = W*W$. Then X is a $\{0,1\}$-matrix with exactly k nonzero entries in every row and column. An additional fact about X is that the inner product of any two of its rows is even. Let $Y = J-X$, and note that Y has exactly $n-k$ nonzero entries in every row and column and, since n is odd, the inner product of any pair of rows of Y is odd and so ≥ 1. Applying our preliminary observation to Y, we obtain

$$(n-k)^2-(n-k) \geq n-1,$$

as was to be shown.

REMARK. We conclude that $W(5,4)$, $W(11,9)$, $W(19,16)$, $W(29,25)$, $W(41,36)$, and so on, do not exist. This remark does not exclude

$W(7,4)$, $W(13,9)$, $W(31,25), \ldots, W(111,100)$,

among others.

This proposition is far from the last word on nonexistence, as the following example shows:

2.6. EXAMPLE. There is no $W(9,4)$. (This proof is due to J. Verner.) (Note that here we do have $(n-k)^2-(n-k) = 5^2-5 = 20 > 8 = n-1$.)

As always, we may assume the first row is [1 1 1 1 0 0 0 0 0].

CLAIM. There is a row with no nonzero entry in the first four columns.

Assuming the validity of the claim, let us see how we then arrive at a contradiction.

We may, in view of the claim, assume the first two rows of the matrix are

$$\begin{bmatrix} 1\ 1\ 1\ 1 & \vdots & 0\ 0\ 0\ 0 & \vdots & 0 \\ 0\ 0\ 0\ 0 & \vdots & 1\ 1\ 1\ 1 & \vdots & 0 \end{bmatrix}.$$

Now, any row must have an even number of nonzero entries in the first four columns (to be orthogonal to the first row) and an even number in the second four columns (for the same reason). This could only happen if every row has a zero in the last column. That is the desired contradiction.

It remains only to prove the claim. Assume the contrary; that is, we suppose every other row has some nonzero entries in the first four columns. The number of such entries must be even and hence is either two or four. Let k_1 rows have two nonzero entries in the first four columns and k_2 rows have four nonzero entries in the first four columns. Then

$$k_1+k_2 = 8.$$

Since every column must have four nonzero entries, we must have $16 = 2k_1+4k_2$ nonzero entries in the first four columns.

Solving these equations, we find that $k_1 = 10$. This is a contradiction and so establishes the claim.

REMARK. No one has yet been able to fit this example into a more general nonexistence theorem.

The "boundary" values of Proposition 2.5 are of special interest, that is, when $n = (n-k)^2-(n-k)+1$. An inspection of the proof shows that if A is a $W(n,k)$ for such an n and k and if we set $B = J-A*A$ (where J is the $n \times n$ matrix of 1's), then $BB^t = (n-k-1)I_n + J$; that is, B is the incidence matrix of a projective plane of order $n-k-1$ (see Hall [1]).

Thus, for example, the existence of $W(111, 100)$ would imply the existence of a projective plane of order 10. There is a deep theorem of Bruck-Ryser-Chowla on the nonexistence of projective planes (see Hall [1] or Ryser [1]), and this may be used to assert nonexistence for certain weighing matrices. We shall not go into that here but will sum up this discussion by stating:

2.7. PROPOSITION. A $W(m^2+m+1, m^2)$ exists only if a projective plane of order m exists.

It is worth mentioning now, and we shall prove in §4.7, that for those m where it is currently known that a projective plane of order m exists, then it is also known that a $W(m^2+m+1, m^2)$ also exists. It is hard to believe that the additional structure in a weighing matrix will not make it more difficult to exist than a projective plane, yet there is no evidence to the contrary. Clearly the existence problem for $W(m^2+m+1, m^2)$ merits considerable attention.

There is little more we can add to the nonexistence problem for weighing matrices of odd order, and so we shall leave that subject now. This last tantalizing connection between weighing matrices of odd order and projective planes has made us wonder what other combinatorial structures may be related to the nonboundary values of Proposition 2.5. This area is wide open for further study.

§2.2. $n = 2b$, b ODD

We have seen that $\rho(n) = 2$ in this case, and so our investigation of orthogonal designs in these orders must center about one-variable designs (that is, weighing matrices) and two-variable designs.

For our first theorem in these orders, we shall need a classical theorem about quadratic forms. The theorem can be stated for matrices without any reference to quadratic forms, however, and since we intend to come back to quadratic forms later, we shall for now just state the theorem in its unmotivated form.

2.8. DEFINITION. Let R be any commutative ring with identity, and let A, B be two $n \times n$ symmetric matrices with entries in R. We say that A and B are <u>congruent</u> if there is an invertible matrix P, with entries in R, such that $PAP^t = B$.

NOTATION. If A is an $n \times n$ matrix over the ring R and B is an $m \times m$ matrix over R, then $A \oplus B$ is the $(n+m) \times (n+m)$ matrix $\begin{bmatrix} A & 0 \\ 0 & B \end{bmatrix}$ where 0 stands for the appropriate-sized matrix of zeros.

2.9. THEOREM. (Witt Cancellation Theorem). Let F be a field of characteristic $\neq 2$, and let A and B be symmetric $n \times n$ matrices over F. Let X be any symmetric matrix over F. If $A \oplus X$ is congruent to $B \oplus X$, then A is congruent to B.

<u>Proof.</u> See Lam [1].

We now apply this to orthogonal designs.

2.10. THEOREM. (Raghavarao-van Lint-Seidel). Let $n \equiv 2 \pmod 4$, and let A be a rational matrix of order n with $AA^t = kI_n$, $k \in \mathbb{Q}$. Then $k = q_1^2+q_2^2$, $q_1, q_2 \in \mathbb{Q}$.

<u>Proof.</u> (This theorem was first proved by Raghavarao in [1] and another proof later given by van Lint-Seidel in [1]. The proof we give here is different from both and is based on a suggestion of H. Ryser.)

It is a well-known theorem of Lagrange that every rational number is the sum of four squares of rational numbers, so let $k = k_1^2+k_2^2+k_3^2+k_4^2$. Form the matrix

$$M = \begin{bmatrix} k_1 & k_2 & k_3 & k_4 \\ -k_2 & k_1 & -k_4 & k_3 \\ -k_3 & k_4 & k_1 & -k_2 \\ -k_4 & -k_3 & k_2 & k_1 \end{bmatrix}.$$

It is easy to check that

$$MM^t = kI_4. \qquad (+)$$

The hypothesis of the theorem asserts that

$$AA^t = kI_n. \qquad (*)$$

Thus, from (†) we obtain that I_4 is congruent to kI_4 over $\mathbb{Q}$ and from (*) that I_n is congruent to kI_n over $\mathbb{Q}$. By Witt's Cancellation Theorem, applied $\frac{n-2}{4}$ times, and the fact that $n \equiv 2 \pmod 4$, we obtain I_2 is congruent to kI_2 over $\mathbb{Q}$; that is, there is a 2×2 rational matrix B such that $BB^t = kI_2$. From this it is obvious that k is a sum of two squares in $\mathbb{Q}$.

2.11. COROLLARY. Let $n \equiv 2 \pmod 4$. If X is

(a) a $W(n,k)$ or

(b) an orthogonal design of type (s_1, s_2) in order n,

then each of k, s_1, s_2, s_1+s_2 is a sum of two squares in Z.

<u>Proof.</u> (a) From the theorem we obtain that k is a sum of two squares of rational numbers. It is a famous theorem of Fermat (see, for example, Samuel [1]) that if n is an integer and $n = c^2 d$, d square-free ($c, d \in Z$), then n is a sum of two squares of integers if and only if whenever p is a prime and $p | d$, then $p = 2$ or $p \equiv 1 \pmod 4$. It follows easily from this that if an integer is the sum of two squares of rational numbers, then it is the sum of two squares of integers.

(b) If X is an orthogonal design of type (s_1, s_2) in order n, then $X = A_1x_1 + A_2x_2$. Then A_i is a $W(n, s_i)$, and setting $x_1 = x_2 = 1$ we find $A_1 + A_2$ is a $W(n, s_1+s_2)$. The result now follows from (a).

2.12. PROPOSITION. Let $n \equiv 2 \pmod 4$, and let A be a rational matrix of order n satisfying:

(i) $A = -A^t$;

(ii) $AA^t = kI_n$.

Then $k = r^2$, $r \in \mathbb{Q}$.

<u>Proof.</u> Since $AA^t = kI_n$, we have $\det A = (k^n)^{1/2}$. Since $n \equiv 2 \pmod 4$, $n/2 = s$ is odd. Now, since A is skew-symmetric, $\det A = q^2$ where q = Pfaffian of A (see Artin [1]). Thus, $q^2 = k^s$, and since s is odd, $k = r^2$ for some $r \in \mathbb{Q}$.

2.13. COROLLARY. If $n \equiv 2 \pmod 4$ and X is an orthogonal design in order n of type (s_1, s_2), then $s_1 s_2$ is a square in $\mathbb{Z}$.

Proof. Let $X = A_1 x_1 + A_2 x_2$; then $A_i A_i^t = s_i I_n$ and $A_1 A_2^t + A_2 A_1^t = 0$. Let $B_1 = \frac{1}{s_1} A_1^t A_1$, $B_2 = \frac{1}{s_1} A_1^t A_2$. Then $B_1 = I_n$ and $B_1 B_2^t + B_2 B_1^t = 0$; that is, $B_2 = -B_2^t$. Also, $B_2 B_2^t = \frac{s_2}{s_1} I_n$. Thus, by the proposition, $\frac{s_2}{s_1}$ is a square in $\mathbb{Q}$; but then so is $s_1^2 \left(\frac{s_2}{s_1}\right) = s_1 s_2$. However, if an integer is the square of a rational number, it is the square of an integer.

So far, all the conditions that we have found necessary for the existence of an orthogonal design in order $n \equiv 2 \pmod 4$ have not depended on the fact that the matrices we want should have entries $\{0, 1, -1\}$. In fact, we have proved half of the following:

2.14. THEOREM. Let $n \equiv 2 \pmod 4$. A necessary and sufficient condition that there exist two rational matrices A and B of order n such that $AA^t = q_1 I_n$, $BB^t = q_2 I_n$, and $AB^t + BA^t = 0$ is that q_1, q_2 each be a sum of two squares in $\mathbb{Q}$ and that $q_1 q_2$ be a square in $\mathbb{Q}$.

Proof. We have already seen the necessity of these conditions, and so it remains only to show that they are sufficient.

Write $q_1 = r_1^2 u$, $q_2 = r_2^2 v$ where $r_1, r_2 \in \mathbb{Q}$, $u, v \in \mathbb{Z}$, and u and v are square-free. Since $q_1 q_2$ is a square, we obtain $u = v$. Since q_1 and q_2 are each a sum of two squares, we find that $u = v = s^2 + t^2$. Now $q_1 = (r_1 s)^2 + (r_1 t)^2$ and $q_2 = (r_2 s)^2 + (r_2 t)^2$.

Let

$$A_1 = \begin{bmatrix} r_1 s & r_1 t \\ -r_1 t & r_1 s \end{bmatrix}, \quad B_1 = \begin{bmatrix} -r_2 t & r_2 s \\ -r_2 s & -r_2 t \end{bmatrix};$$

then if $n = 2m$, m odd, one easily checks that $A = A_1 \otimes I_m$, $B = B_1 \otimes I_m$ are the required matrices.

We can get one additional fact from knowing that in an orthogonal design the coefficient matrices are special.

2.15. PROPOSITION. If $n \equiv 2 \pmod 4$ $(n > 2)$ and X is an orthogonal design in order n of type (s_1, s_2), then $s_1+s_2 < n$.

Proof. If $s_1+s_2 = n$, then by setting the variables in the design equal to one, we would obtain an Hadamard matrix, that is, a $W(n,n)$. This contradicts Proposition 2.2.

This last proposition says, for example, that there is no orthogonal design of type (1, 9) in order 10, although the existence of such an orthogonal design would not contradict Corollary 2.11 or Corollary 2.13.

§2.3. STATEMENT OF THE ALGEBRAIC PROBLEM FOR ORTHOGONAL DESIGNS

It has already become apparent that two separate kinds of theorem are being proved. If we glance back at the statement of Proposition 1.1, we see that the existence of an orthogonal design in order n and of type $(s_1, \ldots, s_\ell)$ depended on finding a collection of ℓ matrices, $A_1, \ldots, A_\ell$, satisfying two rather different types of conditions, which we shall label combinatorial and algebraic; namely,

$$\left\{\begin{matrix}\text{combinatorial}\\ \text{conditions}\end{matrix}\right\} \leftrightarrow \left\{\begin{array}{ll}(0) & \text{the } A_i \text{ are } \{0,1,-1\} \text{ matrices, } 1 \le i \le \ell, \\ (i) & A_i * A_j = 0, \quad 1 \le i \ne j \le \ell;\end{array}\right.$$

$$\left\{\begin{matrix}\text{algebraic}\\ \text{conditions}\end{matrix}\right\} \leftrightarrow \left\{\begin{array}{ll}(ii) & A_i A_i^t = s_i I_n, \quad 1 \le i \le \ell, \\ (iii) & A_i A_j^t + A_j A_i^t = 0, \quad 1 \le i \ne j \le \ell.\end{array}\right.$$

When we looked at an orthogonal design in odd order (that is, a weighing matrix), the algebraic conditions yielded the simple statement that the weight had to be a square integer, while the combinatorial condition, in this case only (0), turned out to be the more mystifying and to have the deeper significance (re: the connection with finite projective planes).

In orders $n \equiv 2 \pmod 4$ the algebraic conditions are somewhat more substantial, and the only general combinatorial fact we know to date arose from the simple result that (apart from order 2) an Hadamard matrix can only exist in orders divisible by 4. In Geramita-Verner [1], by a rather tedious and uninstructive argument, it is shown that there is no orthogonal design in order 18 of type (1, 16). This fact is not covered by any general theorem and appears to us to be but the tip of an iceberg, indicating what promises to be a rich source of possible combinatorial relations. The entire question of what combinatorial facts prohibit the existence of orthogonal designs in these orders $\equiv 2 \pmod 4$ is virtually untouched.

Theorem 2.14, however, opens up the possibility that the algebraic part of the problem may be tractable. This turns out to be the case and will occupy much of our efforts.

We state the algebraic problem after a definition.

2.16. DEFINITION. A rational family of order n and type $[s_1, \ldots, s_\ell]$, where the s_i are positive rational numbers, is a collection of ℓ rational matrices of order n, $A_1, \ldots, A_\ell$, satisfying:

(a) $A_iA_i^t = s_iI_n$, $1 \leq i \leq \ell$;

(b) $A_iA_j^t + A_jA_i^t = 0$, $1 \leq i \neq j \leq \ell$.

ALGEBRAIC PROBLEM. Find necessary and sufficient conditions on n and $\{s_1, \ldots, s_\ell\}$ in order that there exists a rational family of order n and type $[s_1, \ldots, s_\ell]$.

Clearly, an orthogonal design gives rise to a rational family, but the converse is obviously not true. Nonetheless, if one wants to know if there is an orthogonal design in order n of type $(s_1, \ldots, s_\ell)$ and one has proved there can be no rational family in order n of type $[s_1, \ldots, s_\ell]$, then one knows there can be no orthogonal design of that order and type.

We may, in fact, rephrase many of the results so far proved for orthogonal designs in terms of rational families. For example, we have proved:

2.17. PROPOSITION. A rational family in order n cannot consist of more than $\rho(n)$ members; furthermore, there are rational families in order n consisting of ℓ members for every $\ell \leq \rho(n)$.

Proof. Examine the discussion after Proposition 1.1 and the proof of Theorem 1.6.

We have also shown

2.18. PROPOSITION. A necessary and sufficient condition that there exists a rational family of

(a) type $[s]$ in order n, if n is odd, is that s be a square in $\mathbb{Q}$;

(b) type $[s]$ in order n, if $n \equiv 2 \pmod 4$, is that s be a sum of two squares in $\mathbb{Q}$;

(c) type $[s_1, s_2]$ in order n, if $n \equiv 2 \pmod 4$, is that s_1, s_2 each be a sum of two squares in $\mathbb{Q}$ and $s_1 s_2$ be a square in $\mathbb{Q}$.

We shall pursue this algebraic problem in greater depth in the next chapter. For now we shall concentrate on trying to find other combinatorial facts about orthogonal designs.

§2.4. THE GERAMITA-VERNER THEOREM AND SOME CONSEQUENCES

The major combinatorial result so far found is motivated by the following theorem.

2.19. THEOREM. (Delsarte-Goethals-Seidel). Let A be a $W(n, n-1)$ with the rows reordered so that the diagonal consists of zeros.

(a) If $n \equiv 2 \pmod 4$, then multiplication by -1 of rows (or columns) of A as necessary yields a matrix $\overline{A}$ with $\overline{A} = \overline{A}^t$.

(b) If $n \equiv 0 \pmod 4$, then multiplication by -1 of rows (or columns) of A as necessary yields a matrix $\overline{A}$ with $\overline{A} = -\overline{A}^t$.

Proof. See Delsarte-Goethals-Seidel [1].

This result may be generalized to orthogonal designs as follows.

2.20. THEOREM. (A. Geramita-J. Verner). Let X be an orthogonal design of order n and type $(s_1, \dots, s_\ell)$ with $\sum_{i=1}^{\ell} s_i = n-1$.

(a) If $n \equiv 2 \pmod 4$, there is an orthogonal design $\overline{X}$ of order n

and type $(s_1, \ldots, s_\ell)$ where $\overline{X}$ has 0-diagonal and $\overline{X} = \overline{X}^t$.

(b) If $n \equiv 0 \pmod 4$, there is an orthogonal design $\overline{X}$ of order n and type $(s_1, \ldots, s_\ell)$ where $\overline{X}$ has 0-diagonal and $\overline{X} = -\overline{X}^t$.

Proof. If necessary, reorder the rows (or columns) so that the orthogonal design X has 0-diagonal. In this form if x_1 (say) occurs in position (i, j), $i < j$, then position (j, i) contains $\pm x_1$. For suppose not, and assume, without loss of generality, that position (j, i) contains $\pm x_2$. Consider the various incidences between the i-th and j-th rows. Count all occurrences of $\begin{pmatrix} \pm x_1 \\ \pm x_1 \end{pmatrix}$, and assume there are t_1 of these; similarly, assume there are a total of t_2 occurrences of $\begin{pmatrix} \pm x_1 \\ \pm x_2 \end{pmatrix}$ and $\begin{pmatrix} \pm x_2 \\ \pm x_1 \end{pmatrix}$ and a total of t_3 occurrences of $\begin{pmatrix} \pm x_1 \\ \pm x_k \end{pmatrix}$ and $\begin{pmatrix} \pm x_k \\ \pm x_1 \end{pmatrix}$, $k \neq 1, 2$. Since rows i and j are orthogonal, it follows that each of t_1, t_2 and t_3 must be even. Observe also that these incidences account for all but one of the x_1's in rows i and j, namely, that occurring as $\begin{pmatrix} x_1 \\ 0 \end{pmatrix}$. Thus, $2t_1+t_2+t_3 = 2s_1-1$, and this is a contradiction.

Now suppose $n \equiv 2 \pmod 4$, and multiply rows and columns of the orthogonal design by -1, as necessary, so that each variable in the first row and column appears with coefficient $= +1$. Call the resulting matrix $\overline{X}$, and replace every variable in it by $+1$ to obtain the $W(n, n-1)$,

$$\begin{bmatrix} 0 & 1 & \cdots & 1 \\ 1 & \ddots & & * \\ \vdots & & \ddots & \\ 1 & * & & 0 \end{bmatrix}.$$

By Theorem 2.19(a), multiplication of appropriate rows and columns of this matrix by -1 will make it symmetric. However, as the first row and column are already symmetric, it follows that the entire matrix is symmetric. Hence, $\overline{X} = \overline{X}^t$, as was to be shown.

For $n \equiv 0 \pmod 4$, multiply the rows and columns of X so that each variable in the first row appears with coefficient $= +1$ and each variable in the first column appears with coefficient $= -1$. Call the resulting matrix $\overline{X}$, and set each variable $= +1$. The argument above, with

Theorem 2.19(b), implies $\overline{X} = -\overline{X}^t$.

2.21. COROLLARY. Let $n \equiv 0 \pmod 4$. There is an orthogonal design of order n and type $(s_1, \ldots, s_\ell)$ with $\sum_{i=1}^{\ell} s_i = n-1$ if and only if there is an orthogonal design of order n and type $(1, s_1, s_2, \ldots, s_\ell)$ with $1+\sum_{i=1}^{\ell} s_i = n$.

Proof. The sufficiency is evident.

To establish the necessity, one observes that in view of Theorem 2.20 (b) if there is an orthogonal design of the type described, then there is one $\overline{X}$ where $\overline{X} = -\overline{X}^t$ on the variables $x_1, x_2, \ldots, x_\ell$. It is then easily verified that $Y = yI + \overline{X}$ is an orthogonal design of type $(1, s_1, s_2, \ldots, s_\ell)$.

2.22. COROLLARY. If $n \neq 1, 2, 4, 8$, then there is a $\rho(n)$-tuple $(s_1, \ldots, s_{\rho(n)})$ with $s_i > 0$ and $\sum_{i=1}^{\rho(n)} s_i \leq n$ which is not the type of an orthogonal design of order n.

Proof. If n is odd, $n > 1$, there is no $W(n,2)$ since 2 is not a square.

If $n \equiv 2 \pmod 4$, $n > 2$, there is no orthogonal design of type $(1, 2)$ in order n, by Corollary 2.13.

So let $n \equiv 0 \pmod 4$, $n \neq 4, 8$. In this case $n - \rho(n) > 0$. So we may consider the $\rho(n)$-tuple $(1, 1, \ldots, n-\rho(n))$. The sum of the entries in this tuple is $n-1$, and so by Corollary 2.21 there is an orthogonal design on $\rho(n)+1$ variables. This contradicts Theorem 1.7 (a).

The full strength of Corollary 2.21 will best be realized after the algebraic question of orthogonal designs is dealt with in greater depth. We come back to this theorem again at the end of the next chapter.

CHAPTER 3

THE ALGEBRAIC THEORY OF ORTHOGONAL DESIGNS

As we saw in the last chapter, it is possible to obtain some nontrivial necessary conditions for the existence of orthogonal designs just by considering the equations that the coefficient matrices of an orthogonal design must satisfy. The _ad-hoc_ procedures we gave in the last chapter can, with difficulty, be pursued further. However, these procedures quickly become inadequate. To properly describe the solution to the "algebraic problem of orthogonal designs", it is necessary to discuss the theory of quadratic and bilinear forms. Only then can the "algebraic problem" be put in proper perspective. It is possible to highlight this theory fairly quickly, and we do that in this chapter. References for all omitted proofs are included.

Following the general discussion of quadratic and bilinear spaces, we restrict ourselves to the field of rational numbers and develop the theory of similarities of a bilinear space in detail. It is in the study of similarities that the connections with orthogonal designs is revealed.

The major source for this work is the Berkeley thesis of Dan Shapiro (Shapiro [2] and the papers Shapiro [3], [4], [5]) and his subsequent unpublished manuscript "Rational Spaces of Similarities". Shapiro first pointed out the connections between the problems we were studying and his work on similarities. He subsequently solved the algebraic problem of orthogonal designs. Special cases of the solution had first been given in Geramita-Geramita-Wallis [1], Geramita-Wallis [3], and Wolfe [1], [3].

§3.1. GENERALITIES ON QUADRATIC AND BILINEAR FORMS

We recall here some elementary notions about quadratic and bilinear forms. In addition to reminding the reader of these ideas, this introduction will serve to establish the notation we shall use. Most proofs are omitted, and we refer the reader to any of the following excellent sources for further information: (Lam [1], Scharlau [1], O' Meara [1], Serre [1], Bourbaki [1]).

3.1. DEFINITION. Let k denote a field and V a vector space over k. A function $B : V \times V \to k$ is called a <u>bilinear form</u> if, for every $v \in V$, the maps $B_v : V \to k$ and ${}_vB : V \to k$, defined by $B_v(x) = B(x, v)$ and ${}_vB(x) = B(v, x)$, are linear maps. If, in addition, $B_v = {}_vB$ for all $v \in V$, then B is called a <u>symmetric bilinear form</u>.

We shall be exclusively concerned with symmetric bilinear forms and so shall usually drop the adjective "symmetric" when referring to them.

3.2. DEFINITION. A function $q : V \to k$ is called a <u>quadratic form</u> on V if

(i) $q(\alpha v) = \alpha^2 q(v)$, for all $\alpha \in k$, $v \in V$, and

(ii) the map from $V \times V \to k$ given by
$(v_1, v_2) \to q(v_1+v_2)-q(v_1)-q(v_2)$
is a (symmetric) bilinear form.

The pair (V, B) will be called a <u>bilinear space</u>, and the pair (V, q) will be called a <u>quadratic space</u>. If k is a field of characteristic not equal to 2, these two notions are intimately connected.

To see this, let (V, q) be a quadratic space, and define $B_q : V \times V \to k$ by $B_q(x, y) = (1/2)(q(x+y)-q(x)-q(y))$. Then B_q is a bilinear form on V by definition of (V, q). Thus, to the quadratic space (V, q) we can make correspond the bilinear space (V, B_q).

On the other hand, let (V, B) be a bilinear space, and define $q_B : V \to k$ by $q_B(x) = B(x, x)$. It is easy to see that q_B is a quadratic form on V since $q_B(\alpha x) = B(\alpha x, \alpha x) = \alpha^2 B(x, x) = \alpha^2 q_B(x)$ and the map from $V \times V \to k$ given by $(x, y) \to q_B(x+y)-q_B(x)-q_B(y)$ is a bilinear form.

It is not hard to show that these two processes are inverse to each other; that is, given (V, B), form (V, q_B), and then construct (V, B_{q_B}). Then (V, B) and (V, B_{q_B}) are the same bilinear space; that is, $B = B_{q_B}$. To see this, observe that $B_{q_B}(x, y) = (1/2)(q_B(x+y) - q_B(x) - q_B(y))$, where $q_B(v) = B(v, v)$. Thus,

$$B_{q_B}(x, y) = (1/2)[B(x+y, x+y) - B(x, x) - B(y, y)] = B(x, y).$$

On the other hand, starting with (V, q), form (V, B_q) and then (V, q_{B_q}). Then (V, q) and (V, q_{B_q}) are the same quadratic space; that is, $q = q_{B_q}$. To see this, note that

$$q_{B_q}(x) = B_q(x, x) = (1/2)(q(x+x) - q(x) - q(x)) = q(x).$$

We shall now assume characteristic not 2 and thus, in view of the correspondence outlined above, freely interchange the notions of quadratic and bilinear space.

§3.2. THE MATRIX FORMULATION

Let V be a finite-dimensional vector space over k (say of dimension n), and let $\{e_i\}$, $1 \leq i \leq n$, be a basis for V. If $v \in V$ and

$$v = \sum_{i=1}^{n} \alpha_i e_i,$$

then we shall denote v by the column vector

$$\begin{bmatrix} \alpha_1 \\ \vdots \\ \alpha_n \end{bmatrix} = \mathrm{col}(\alpha_1, \ldots, \alpha_n).$$

3.3. DEFINITION. If B is a bilinear form on V, then we form the $n \times n$ matrix $A = (a_{ij})$ as follows: $a_{ij} = B(e_i, e_j)$. It is easy to check that if $x = \mathrm{col}(\alpha_1, \ldots, \alpha_n)$, $y = \mathrm{col}(\beta_1, \ldots, \beta_n)$, then $B(x, y) = x^t A y$ ("t" denotes matrix transpose). We call A <u>the matrix of the form B with respect to the basis $\{e_i\}$</u>. (Note that A is a symmetric matrix.)

If (V, q) is a quadratic space, then the matrix of q with respect to the basis $\{e_i\}$ is the matrix of B_q (as defined earlier) with respect to the basis $\{e_i\}$.

On the other hand, if V is an n-dimensional vector space over k, with a fixed basis $\{e_i\}$, $1 \leq i \leq n$, and A is any $n \times n$ symmetric matrix, then we can define a bilinear form on V by means of A. Namely, if $x = \mathrm{col}(\alpha_1, \ldots, \alpha_n)$, $y = \mathrm{col}(\beta_1, \ldots, \beta_n)$, $x, y \in V$, define a map from $V \times V \to k$ by $(x, y) \to x^t Ay$. This is clearly a bilinear form on V, and its matrix with respect to the basis $\{e_i\}$ is clearly A.

Thus, for a fixed basis of V the distinct bilinear forms on V are in one-to-one correspondence with the $n \times n$ symmetric matrices.

If (V, B) is a bilinear space and $\{e_i\}$, $1 \leq i \leq n$, and $\{e_i'\}$, $1 \leq i \leq n$, are two sets of bases for V, it is natural to ask how the matrices of B with respect to each of these bases are related. The formulation turns out to be relatively simple.

Let P be the (invertible) $n \times n$ matrix whose i-th column expresses the coordinates of e_i with respect to the bases $\{e_i'\}$. Then, if $v \in V$ and $v = \mathrm{col}(\alpha_1, \ldots, \alpha_n)$ with respect to the basis $\{e_i\}$, then

$$P \begin{bmatrix} \alpha_1 \\ \vdots \\ \alpha_n \end{bmatrix} = \begin{bmatrix} \alpha_1' \\ \vdots \\ \alpha_n' \end{bmatrix}$$

gives the coordinates of v with respect to the basis $\{e_i'\}$.

If we let A and A' denote the matrices of B with respect to the bases $\{e_i\}$ and $\{e_i'\}$, respectively, then it is easy to check that

$$A = P^t A' P. \qquad (*)$$

Thus A and A' are congruent (see Definition 2.8).

On the other hand, if (V, B) is a bilinear space whose matrix, with respect to some basis, is X and if P is any invertible matrix, then $Y = P^t XP$ is also the matrix of (V, B), but with respect to a different basis.

The notion of congruence is easily seen to be an equivalence relation on the set of symmetric $n \times n$ matrices.

3.4. DEFINITION. The bilinear space (V, B) is called <u>nondegenerate</u>

if the map from $V \to V^*$ (dual space) given by $x \to B(x,-)$ is an isomorphism of vector spaces.

In matrix terms, (V, B) is nondegenerate if the matrix of B with respect to any basis is invertible, as is easily shown. (In view of (*), if the matrix of B with respect to one basis is invertible, so is the matrix of B with respect to any basis.)

§3.3. MAPPINGS BETWEEN BILINEAR SPACES

Let (V_1, B_1) and (V_2, B_2) be two bilinear spaces.

3.5. DEFINITION. A linear transformation $f : V_1 \to V_2$ is called a <u>similarity</u> with <u>similarity factor</u> $\sigma(f) \in k$ if for any pair of vectors $u, w \in V_1$ we have

$$B_2(f(u), f(w)) = \sigma(f)B_1(u, w).$$

In matrix terms, if we choose bases for V_1 and V_2, respectively, and, with respect to these chosen bases, let A be the matrix for B_1, A' the matrix for B_2, and X the matrix for f, then

$$\sigma(f)A = X^t A' X.$$

If $\sigma(f) = 1$, then f is usually called a <u>linear morphism</u>, and if in addition $V_1 = V_2$, f is called an <u>isometry</u>.

If (V, B) is nondegenerate, then the isometries of (V, B) form a group under composition which is called the <u>orthogonal group</u> of the bilinear space. Also, the similarities of a nondegenerate bilinear space, with nonzero similarity factors, form a group. We shall investigate this in more detail later.

Finally, if (V, B_1) and (V, B_2) are two nondegenerate bilinear spaces, then (V, B_1) is <u>isometric</u> to (V, B_2) if and only if there is an isometry from (V, B_1) to (V, B_2); equivalently, if and only if the matrices of B_1 and B_2, with respect to any basis of V, are congruent.

Thus the classification of nondegenerate bilinear spaces of dimension n over a field k (up to isometry) is equivalent to classifying the congruence classes of invertible $n \times n$ symmetric matrices over k.

We shall spend some time in understanding this classification over a few fields. In general, such a classification involves a deep understanding of the arithmetic structure of k.

§3.4. NEW SPACES FROM OLD

Let (V, B) be a bilinear space, and let $W \subset V$ be a subspace. Then, since $B : W \times W \to k$ is a bilinear map, we may consider (W, B) as a bilinear space in this way. The inclusion map from W to V is then a linear morphism. In the same way, if (V, q) is a quadratic space, then $q|_W : W \to k$ is a quadratic form on W and $(W, q|_W)$ is a quadratic space.

Let (V_1, B_1) and (V_2, B_2) be two bilinear spaces.

3.6. DEFINITION. Let $V = V_1 \oplus V_2$; we make V into a bilinear space as follows: Let $x, y \in V$, $x = x_1 \oplus x_2$, $y = y_1 \oplus y_2$, $x_i, y_i \in V_i$, and define $B(x, y) = B_1(x_1, y_1) + B_2(x_2, y_2)$. It is easy to verify that (V, B) is a bilinear space, and we call (V, B) the <u>orthogonal sum</u> of (V_1, B_1) and (V_2, B_2) and write $(V, B) = (V_1, B_1) \perp (V_2, B_2)$.

If (V, B) is any bilinear space, then we say that $x, y \in V$ are <u>orthogonal</u> if $B(x, y) = 0$ and extend this definition to two subspaces V_1, V_2 of V if every vector in V_1 is orthogonal to every vector in V_2. In the orthogonal sum the subspaces that we can naturally identify with V_1 and V_2 are orthogonal.

Now suppose that (V, B) is a bilinear space and that (U, B) and (W, B) are subspaces which are orthogonal and $U+W = V$. It is easy to see that:

3.7. PROPOSITION. (V, B) and $(U, B) \perp (W, B)$ are isometric bilinear spaces.

We may carry through this entire discussion for quadratic spaces by defining $(V_1, q_1) \perp (V_2, q_2) = (V, q)$ by setting $q(v_1+v_2) = q_1(v_1) + q_2(v_2)$ where $v_i \in V_i$. It is routine to check that (V, q) is a quadratic space and that the correspondence between bilinear and quadratic spaces respects orthogonal sums.

In matrix terms: If we choose bases for V_1 and V_2 and let A_i be the matrix for B_i with respect to this basis of V_i $(i = 1, 2)$, then, with respect to the basis for $V_1 \oplus V_2$ obtained by taking the "union" of the two given bases, the matrix of B (the orthogonal sum) is

$$\begin{bmatrix} A_1 & 0 \\ 0 & A_2 \end{bmatrix}.$$

(Recall that this is the matrix we have denoted $A_1 \oplus A_2$.)

There is one further construction we need to consider. As before, let (V_1, B_1) and (V_2, B_2) be two bilinear spaces, and now set $V = V_1 \otimes V_2$. We make V into a bilinear space by defining $B : V \times V \to k$ to be the unique bilinear map satisfying: $B(u_1 \otimes u_2, w_1 \otimes w_2) = B_1(u_1, w_1)B_2(u_2, w_2)$, where $u_i, w_i \in V_i$, and call this space the <u>product</u> of (V_1, B_1) and (V_2, B_2).

In matrix terms: Let $\{e_i\}$, $1 \le i \le n$, be a basis for V_1 and $\{f_j\}$, $1 \le j \le m$, a basis for V_2. Suppose $A = (a_{ij})$ is the matrix for B_1 with respect to the $\{e_i\}$ and $C = (c_{rs})$ is the matrix for B_2 with respect to the $\{f_j\}$. Then, if we consider the ordered basis for $V_1 \otimes V_2$ to be

$$\{e_1 \otimes f_1, \ldots, e_1 \otimes f_m, \ldots, e_n \otimes f_1, \ldots, e_n \otimes f_m\},$$

the matrix for B with respect to this basis is

$$\begin{bmatrix} a_{11}C & a_{12}C & & a_{1n}C \\ \vdots & \vdots & & \vdots \\ a_{n1}C & & \cdots & a_{nn}C \end{bmatrix}.$$

In similar fashion we can define the <u>tensor product of quadratic</u> spaces in such a way that the correspondence between bilinear and quadratic forms is respected (see Lam [1] for a complete discussion).

§3.5. CLASSIFICATION THEOREMS FOR BILINEAR SPACES

We have already seen that the classification (up to isometry) of non-degenerate quadratic or bilinear spaces of dimension n over a field k is equivalent to classifying equivalence classes (under congruence) of invertible symmetric $n \times n$ matrices over k.

Let k^* denote the nonzero elements of k, and $(k^*)^2$ the subgroup of squares of k^*. Then one invariant of a congruence class lies in the quotient group, $k^*/(k^*)^2$. More specifically:

3.8. DEFINITION. If A is an invertible $n \times n$ matrix over k and $\det A = d$, then $\bar{d} \in k^*/(k^*)^2$ is called the discriminant of A and denoted $\operatorname{disc} A$.

A major step toward the classification of bilinear spaces is the following:

3.9. THEOREM. If A is an invertible symmetric $n \times n$ matrix, then A is congruent to a diagonal matrix, $\operatorname{diag}(a_1, \ldots, a_n)$.

Proof. See any of the references on quadratic forms mentioned at the beginning of §3.1.

Thus every congruence class contains at least one diagonal matrix. So the problem of classifying quadratic spaces reduces to the question of when two invertible diagonal matrices are congruent.

NOTATION. In view of the previous theorem, we shall adopt the following standard notation. Let (V, q) be a quadratic space, and suppose that with respect to some basis for V the associated matrix is $\operatorname{diag}(a_1, \ldots, a_n)$. We shall refer to the quadratic space (or the isometry class of the quadratic space) as $\langle a_1, \ldots, a_n \rangle$ and drop the space V when no confusion (hopefully) can occur. If $\operatorname{diag}(a_1, \ldots, a_n)$ is congruent to $\operatorname{diag}(b_1, \ldots, b_n)$, then $\langle a_1, \ldots, a_n \rangle = \langle b_1, \ldots, b_n \rangle$ (and conversely).

Clearly $\langle a_1, \ldots, a_n \rangle = \langle a_{\sigma(1)}, \ldots, a_{\sigma(n)} \rangle$ for any permutation σ, since, on the matrix level, this just amounts to re-ordering the basis, which is an isometry. Also $\langle a_1 s_1^2, \ldots, a_n s_n^2 \rangle = \langle a_1, \ldots, a_n \rangle$ for any $s_i \in k^*$.

This last observation has as immediate corollaries:

3.10. THEOREM. If k is any algebraically closed field, then any two invertible symmetric $n \times n$ matrices are congruent.

Proof. In an algebraically closed field everything is a square, so $\langle a_1, \ldots, a_n\rangle = \langle 1.b_1^2, \ldots, 1.b_n^2\rangle = \langle 1, \ldots, 1\rangle$. Thus every nondegenerate bilinear space is congruent to a fixed bilinear space of the same dimension.

3.11. THEOREM. If $k = \mathbb{R}$ (the real numbers), then

$$\langle a_1, \ldots, a_n\rangle = \langle \underbrace{1, \ldots, 1}_{\ell}, \underbrace{-1, \ldots, -1}_{s}\rangle$$

for some $0 \leq \ell,\ s \leq n$.

Proof. In $\mathbb{R}$ everything ($\neq 0$) is either a square or the negative of a square.

Now if $k = \mathbb{R}$ and if an invertible symmetric $n \times n$ matrix A is congruent to $\langle \underbrace{1, \ldots, 1}_{\ell}, \underbrace{-1, \ldots, -1}_{s}\rangle$, we call the integer $\ell - s$ the signature of A and write $\operatorname{sgn}(A)$.

3.12. THEOREM. (Sylvester's Law of Inertia). The signature of a real symmetric invertible matrix A is well defined and is an invariant of the congruence class containing A. Furthermore, if A and B are real symmetric invertible $n \times n$ matrices, A is congruent to B if and only if $\operatorname{sgn}(A) = \operatorname{sgn}(B)$.

§3.6. CLASSIFICATION OF QUADRATIC FORMS OVER $\mathbb{Q}$

Let F be the field with p elements (p a prime). We shall describe a procedure for deciding when an element of F is a square. We may as well assume $p \neq 2$ since in any finite field of characteristic two, everything is a square.

Now $F^* = F\setminus\{0\}$ is a cyclic group of order $p-1$ (which is even) and hence has a unique subgroup of index 2 which consists of the squares of F^*. Let the two-element group $F^*/(F^*)^2$ be denoted by the group $\{1, -1\}$ under multiplication.

The map $\chi : F^* \to \{1, -1\}$, which is the natural map onto a quotient, satisfies

$$\chi(a) = \begin{cases} 1 & \text{if } a \text{ is a square in } F^*, \\ -1 & \text{if } a \text{ is not a square in } F^*. \end{cases}$$

The map χ is usually denoted $\left(\frac{-}{p}\right)$; that is, $\left(\frac{-}{p}\right)(a) = \chi(a) = (a/p)$ and is called the <u>Legendre character</u>.

We would like a way of calculating $\chi(a)$. The decisive procedure for such a calculation is given by Gauss's celebrated <u>Law of Quadratic Reciprocity</u>.

3.13. THEOREM. (Gauss). Let p and q be distinct primes, both different from 2. Then

(i) $(p/q) = (q/p).(-1)^{((p-1)/2).((q-1)/2)}$ (reciprocity),

(ii) $(2/p) = (-1)^{(p^2-1)/8} = \begin{cases} 1 & \text{if } p \equiv \pm 1 \pmod 8, \\ -1 & \text{if } p \equiv \pm 5 \pmod 8, \end{cases}$

(iii) $(-1/p) = \begin{cases} 1 & \text{if } p \equiv 1 \pmod 4, \\ -1 & \text{if } p \equiv 3 \pmod 4. \end{cases}$

We illustrate how to use this theorem by an example.

EXAMPLE. Is $n = 2^3.3.7$ a square in Z_{67}? Let $p = 67$, $n = 2^3.3.7$. Find (n/p).

Since the Legendre character is a group homomorphism, we have

$$(n/p) = (2^3/p)\ (3/p)(7/p).$$

Now $(2^3/p) = (2^2/p)(2/p)$, 2^2 is clearly a square, mod p, so $(2^3/p) = (2/p)$. Since $67 \equiv -5 \pmod 8$, $(2/67) = -1$. Now

$$(3/67) = (67/3)(-1)^{33} = -(67/3) = -(1/3) = -1.$$

Now

$$(7/67) = (67/7)(-1)^{99} = -(67/7) = -(4/7) = -(2^2/7) = -1.$$

Therefore $(n/67) = -1$, and so n is not a square in Z_{67}.

Let p be a prime integer, and let $\mathbb{Q}_p$ denote the field of p-adic numbers. If $a, b \in \mathbb{Q}_p$, $a, b \neq 0$, then the <u>p-adic Hilbert symbol</u> $(a,b)_p$ is defined by:

$$(a,b)_p = \begin{cases} 1 & \text{if there are p-adic numbers, } x, y, \text{ with } ax^2+by^2 = 1, \\ -1 & \text{otherwise.} \end{cases}$$

The Hilbert symbols can be shown to have the following properties:

(i) $(a,b)_p = (b,a)_p$, $(a,c^2)_p = 1$;

(ii) $(a,-a)_p = 1$ and $(a,1-a)_p = 1$;

(iii) $(aa',b)_p = (a,b)_p(a',b)_p$ (bilinearity);

(iv) $(a,b)_p = (a,-ab)_p = (a,(1-a)b)_p$.

The importance of the Hilbert symbols is illustrated by the following theorem.

3.14. THEOREM. Let $\langle a_1,\ldots,a_n\rangle$ be a quadratic form over the field $\mathbb{Q}_p$, and let $s_p(\langle a_1,\ldots,a_n\rangle) = \prod_{i<j}(a_i,a_j)_p$; then

(1) $s_p(\langle a_1,\ldots,a_n\rangle)$ is an invariant of the quadratic form and is called the Hasse-invariant at p;

(2) the nondegenerate quadratic forms $\langle a_1,\ldots,a_n\rangle$ and $\langle b_1,\ldots,b_n\rangle$ over $\mathbb{Q}_p$ are congruent (over $\mathbb{Q}_p$) if and only if they have the same discriminant and the same Hasse-invariant.

We are now ready to state the crowning achievement of the theory. If $\langle a_1,\ldots,a_n\rangle$ is a quadratic form over $\mathbb{Q}$, then it may be viewed as a quadratic form over $\mathbb{Q}_p$ for any prime p and over $\mathbb{R}$. the real numbers. If two quadratic forms over $\mathbb{Q}$ are congruent over $\mathbb{Q}$, they are obviously also congruent over $\mathbb{Q}_p$ for every p and over $\mathbb{R}$. The amazing thing is that the converse is true.

3.15. THEOREM. (Hasse-Minkowski). If $\langle a_1,\ldots,a_n\rangle$ and $\langle b_1,\ldots,b_n\rangle$ are nondegenerate quadratic forms over $\mathbb{Q}$, then they are congruent over $\mathbb{Q}$ if and only if they are congruent over $\mathbb{Q}_p$ (for every p) and over $\mathbb{R}$.

We have already seen how easy it is to decide congruence over $\mathbb{R}$; we just need to calculate the signature. All that is left to make this classification theorem work is an algorithm for calculating the Hasse-invariants, that is, to calculate Hilbert symbols.

So suppose $<a_1, \ldots, a_n>$ is a quadratic form over $\mathbb{Q}$; we can multiply each a_i by squares in $\mathbb{Q}$ and not change the congruence class of our form. Thus there is no loss in assuming the a_i are square-free integers.

Property (iii) of the Hilbert symbols implies that all we really need is a way to calculate $(r, s)_p$, where r, s are ± 1 or primes, and p is a prime. The prime $p = 2$ will require special comment.

3.16. THEOREM. Let $p \neq 2$ be a prime. Then

(i) $(r, s)_p = 1$ if r and s are relatively prime to p;

(ii) $(r, p)_p = (r/p)$, the Legendre symbol, if r and p are relatively prime;

(iii) $(p, p)_p = (-1/p)$.

REMARKS. (iii) is easily derived from (ii) and the formal properties of the Hilbert symbols; the major part of this theorem is (i) and (ii).

Note also that (i) has as a corollary the comforting fact that in checking if $<a_1, \ldots, a_n>$ and $<b_1, \ldots, b_n>$ are congruent over $\mathbb{Q}$, we need only check the Hasse-invariants at the finite collection of primes that divide the a_i's and the b_j's.

We are left only with the calculation of $(r, s)_2$. This is just a bit more complicated to explain (see Serre [1], for example). But, for our purposes, this calculation will not be necessary. To see why, we first define $(r, s)_\infty$ by:

$$(r, s)_\infty = \begin{cases} 1 & \text{if there are real numbers } x \text{ and } y \text{ such that } rx^2 + sy^2 = 1, \\ -1 & \text{otherwise.} \end{cases}$$

Note that $(r, s)_\infty = 1$ unless r and s are both negative.

With this definition we have:

3.17. THEOREM. (Product Formula)

$$\prod_{\substack{p \text{ prime} \\ p = \infty}} (r, s)_p = 1$$

for any r and s.

3.18. COROLLARY

$$\prod_{\substack{p \text{ prime} \\ p=\infty}} s_p(<a_1, \ldots, a_n>) = 1.$$

It follows from this that if two quadratic forms over $\mathbb{Q}$ have the same discriminant and the same Hasse invariant at every prime p (including $p = \infty$) except perhaps one, then they have the same Hasse invariant at that prime also. Thus we may, with impunity, ignore the prime $p = 2$.

We would like to show, by considering some examples, how easy it is to use this criterion for congruence over $\mathbb{Q}$.

EXAMPLE. (a) $<1,1,5,5> = <1,1,1,1>$ over $\mathbb{Q}$.

Proof. They have the same discriminant since $1.1.5.5 = 25 = 1.1.1.1 \pmod{(\mathbb{Q}^*)^2}$. Also both have the same signature and the same Hasse invariant at ∞. The only other prime at which we need check is $p = 5$. Now $s_5(<1,1,1,1>) = 1$, clearly, and

$$s_5(1,1,5,5) = \underbrace{(1,1)_5}_{=1}\underbrace{(1,5)_5(1,5)_5(1,5)_5(1,5)_5}_{=1=[(1,5)_5]^4}(5,5)_5$$

$$= (5,5)_5 = (-1/5) = 1 \quad \text{since} \quad 5 \equiv 1 \pmod 4).$$

Therefore, the two quadratic forms are congruent over $\mathbb{Q}$.

(b) $<1,3,6,8> = <1,1,1,1>$ over $\mathbb{Q}$.

The only primes we need consider are $2, 3, \infty$. The two forms have the same discriminant and signature and the same Hasse-invariant at $p = \infty$. By our remarks we can ignore $p = 2$ and just consider $p = 3$.

Now $<1,3,6,8> = <1,3,6,2>$ by eliminating squares, and

$$s_3(<1,3,6,2>) = (1,3)_3(1,6)_3(1,2)_3(3,6)_3(3,2)_3(6,2)_3.$$

Since 1 is a square,

$$s_3(<1,3,6,2>) = (3,6)_3(3,2)_3(6,2)_3.$$

Using bilinearity, we have

$$s_3(<1,3,6,2>) = (3,2)_3(3,3)_3(3,2)_3(3,2)_3(2,2)_3.$$

Since the Hilbert symbol only takes on the value ± 1,

$$s_3(\langle 1,3,6,2\rangle) = (3,2)_3(3,3)_3(2,2)_3.$$

Now $(2,2)_3 = 1$ since 2 and 3 are relatively prime. Also $(3,3)_3 = (-1/3) = -1$ since $3 \equiv 3 \pmod 4$, and $(3,2)_3 = (2/3) = -1$ since $3 \equiv -5 \pmod 8$. Therefore, $s_3(\langle 1,3,6,2\rangle) = 1$.

Thus by the Corollary to the Product Formula, we have $s_p(\langle 1,3,6,2\rangle) = 1$ for every prime p (including $p = 2$), and so we are done.

(c) $\langle 1,2,7,14\rangle \not\cong \langle 1,1,1,1\rangle$.

Now $s_p(\langle 1,1,1,1\rangle) = 1$ for every p (including $p = \infty$), $\mathrm{disc}(\langle 1,1,1,1\rangle) = \bar{1}$ and $\mathrm{sgn}(\langle 1,1,1,1\rangle) = 4$. Now $s_\infty(\langle 1,2,7,14\rangle) = 1$, $\mathrm{disc}(\langle 1,2,7,14\rangle) = \bar{1}$ and $\mathrm{sgn}(\langle 1,2,7,14\rangle) = 4$. The only primes we need check are $p = 2, 7$ (and we ignore 2).

Now $s_7(\langle 1,2,7,14\rangle) = (2,7)_7(2,2)_7(7,7)_7$ after some easy reductions. But

$(2,7)_7 = (2/7) = 1$ since $7 \equiv -1 \pmod 8$,

$(2,2)_7 = 1$ since 2 and 7 are relatively prime,

and

$(7,7)_7 = (-1/7) = -1$ since $7 \equiv 3 \pmod 4$.

Therefore,

$$s_7(\langle 1,2,7,14\rangle) = -1 \neq s_7(\langle 1,1,1,1\rangle).$$

Thus the two forms are <u>not</u> congruent over $\mathbb{Q}$.

It is hoped that these few examples illustrate the power and ultimate simplicity of the classification theory for quadratic forms over $\mathbb{Q}$.

§3.7. THE SIMILARITIES OF A BILINEAR SPACE

From now on our discussion will always assume that our field k is $\mathbb{Q}$, the rational numbers. This hypothesis is not necessary for most of what we say, but it does avoid some difficulties, and it is the only field with which we are, in this text, eventually concerned. The interested reader should consult D. Shapiro's papers [3, 4] for the general discussion.

Recall that if (V_1, q_1) and (V_2, q_2) are two nondegenerate quadratic spaces with associated bilinear spaces (V_1, B_1) and (V_2, B_2), then the linear transformation $f : V_1 \to V_2$ is a similarity with similarity factor $\sigma(f) \in \mathbb{Q}$ if for any $x, y \in V_1$ we have $B_2(f(x), f(y)) = \sigma(f). B_1(x, y)$.

The set of all similarities of a fixed nondegenerate bilinear (quadratic) space (V, B) will be abusively denoted Sim(V) when there is no danger of losing sight of the fact that B gives the underlying bilinear structure on V. Clearly, Sim(V) is closed under composition and scalar multiplication.

Define $\mathrm{Sim}^*(V) = \{f \in \mathrm{Sim}(V) \mid \sigma(f) \neq 0\}$. Then it is easy to see that $\mathrm{Sim}^*(V)$ is a group under composition, and the map $\sigma : \mathrm{Sim}^*(V) \to \mathbb{Q}^*$ by

$$\sigma : f \to \sigma(f)$$

is a homomorphism of groups, and $\ker \sigma = O(V)$, the orthogonal group of (V, B).

3.19. DEFINITION. The bilinear spaces (V_1, B_1) and (V_2, B_2) are similar if there is a similarity $f : V_1 \to V_2$ with $\sigma(f) \neq 0$.

It is not, in general, true that the sum of two similarities of the bilinear space (V, B) is again a similarity.

3.20. EXAMPLE. Let $V = \mathbb{Q}^2$, and, with respect to the usual basis of $\mathbb{Q}^2$, assume B has matrix I_2. Let $g : V \to V$ be the identity map; then g is a similarity, and $\sigma(g) = 1$.

Let $f : V \to V$ have matrix $\begin{bmatrix} 1 & 0 \\ 0 & -1 \end{bmatrix}$ with respect to the fixed basis. Then f is a similarity; for if $x = \mathrm{col}(\alpha_1, \alpha_2)$, $y = \mathrm{col}(\beta_1, \beta_2)$, then $f(x) = \mathrm{col}(\alpha_1, -\alpha_2)$ and $f(y) = \mathrm{col}(\beta_1, -\beta_2)$ and

$$B(x, y) = \alpha_1\beta_1 + \alpha_2\beta_2 = B(f(x), f(y)) = \alpha_1\beta_1 + (-\alpha_2)(-\beta_2).$$

Now f+g has matrix $\begin{bmatrix} 2 & 0 \\ 0 & 0 \end{bmatrix}$, and we leave it to the reader to check that this is not a similarity.

The question as to when a subset of Sim(V) is closed under addition is of paramount importance in our investigation. We consider that question in some detail in the next section.

§3.8. LINEAR SUBSPACES OF Sim(V)

Let (V, B) be a fixed nondegenerate bilinear space, and let End(V) be the $\mathbb{Q}$-algebra of all endomorphisms of V, that is, all $\mathbb{Q}$-linear maps from V to V.

3.21. DEFINITION. The adjoint map $\sim$: End(V) $\to$ End(V) is defined by: If $f \in$ End(V), then $\tilde{f}$ is the unique map such that $B(f(x), y) = B(x, \tilde{f}(y))$, for all $x, y \in V$.

In matrix terms, choose a basis for V, and let X be the matrix for B with respect to this basis, A the matrix for f; then $\sim : A \to X^{-1}A^tX = \tilde{A}$. To see this let $x = \text{col}(\alpha_1, \ldots, \alpha_n)$, $y = \text{col}(\beta_1, \ldots, \beta_n)$; then

$$\begin{aligned} B(f(x), y) &= (Ax)^tXy \\ &= x^tA^tXy \\ &= x^t(XX^{-1})A^tXy \\ &= x^tX(X^{-1}A^tXy) \\ &= B(x, \tilde{f}(y)). \end{aligned}$$

It is easy to check that $\sim$ is a $\mathbb{Q}$-algebra antihomomorphism; that is,

3.22. PROPOSITION. In matrix terms,

(i) $\widetilde{(A+C)} = \tilde{A}+\tilde{C}$;

(ii) $\widetilde{(\alpha A)} = \alpha\tilde{A}$ and $\tilde{I} = I$;

(iii) $\widetilde{(AC)} = \tilde{C}\tilde{A}$;

(iv) $\tilde{\tilde{A}} = A$.

The adjoint map gives an easy way to identify similarities.

3.23. PROPOSITION. Let (V, B) be a nondegenerate bilinear space and $f \in$ End(V). f is a similarity with similarity factor $\sigma(f) = \sigma$ if and only if $\tilde{f}f = \sigma.1_V$ (where 1_V denotes the identity map of V).

Proof. f is a similarity with similarity factor σ

$\iff B(f(x), f(y)) = \sigma.B(x, y)$ for all $x, y \in V$,

$\iff B(x, \tilde{f}f(y)) = \sigma.B(x, y)$ for all $x, y \in V$.

Since B is nondegenerate, this last holds if and only if $\tilde{f}f = \sigma.1_V$.

NOTE. If $\sigma \neq 0$, then f is invertible and $\tilde{f} = \left(\frac{1}{\sigma}f\right)^{-1}$, and so $f\tilde{f} = \sigma 1_V$ also.

From this proposition we can see how the adjoint map also allows a method for deciding when the sum of two elements of Sim(V) is again in Sim(V).

3.24. PROPOSITION. Let (V, B) be a nondegenerate bilinear space, and suppose $f, g \in \mathrm{Sim}(V)$. Then the following are equivalent:

(1) $f+g \in \mathrm{Sim}(V)$;

(2) $\alpha f+\beta g \in \mathrm{Sim}(V)$ for all $\alpha, \beta \in \mathbb{Q}$;

(3) $\tilde{f}g+\tilde{g}f = c1_V$ for some $c \in \mathbb{Q}$.

Proof. (1) <=> (3). $f+g \in \mathrm{Sim}(V)$ if and only if $\widetilde{(f+g)}(f+g) = d1_V$, where $d = \sigma(f+g)$ by Proposition 3.23.

Using Proposition 3.22, this in turn is equivalent to $\tilde{f}f+\tilde{g}f+\tilde{f}g+\tilde{g}g = d1_V$ if and only if $\tilde{f}g+\tilde{g}f = (d-\sigma(f)-\sigma(g))1_V$.

(2) => (1). Obvious.

(1) and (3) => (2).

$$\begin{aligned}\widetilde{(\alpha f+\beta g)}(\alpha f+\beta g) &= \alpha^2\tilde{f}f+\alpha\beta\tilde{f}g+\beta\alpha\tilde{g}f+\beta^2\tilde{g}g \\ &= \alpha^2\sigma(f)1_V+\alpha\beta c1_V+\beta^2\sigma(g)1_V \\ &= (\alpha^2\sigma(f)+\alpha\beta c+\beta^2\sigma(g))1_V,\end{aligned}$$

and so, by Proposition 3.23, $\alpha f+\beta g \in \mathrm{Sim}(V)$.

NOTE. If, in addition, $\sigma(f+g) \neq 0$, we could then add $g\tilde{f}+f\tilde{g} = c1_V$ to the list of equivalences above.

3.25. DEFINITION. If $f, g \in \mathrm{Sim}(V)$ satisfy any of the equivalent conditions of the preceding proposition, then we say that f and g are compatible similarities.

REMARK. If $f_1, \ldots, f_r \in \mathrm{Sim}(V)$ are mutually compatible similarities, then the entire $\mathbb{Q}$-linear span of the f_i lie in $\mathrm{Sim}(V)$, and we obtain a linear subspace of $\mathrm{Sim}(V)$.

3.26. EXAMPLE. Let V be an n-dimensional vector space over $\mathbb{Q}$ with basis $\{e_i\}$, $1 \leq i \leq n$, and suppose that with respect to this basis the matrix for the bilinear form B on V is I_n. Then $\mathrm{End}(V)$ is the set of $n \times n$ matrices over $\mathbb{Q}$, which we denote $M_n(\mathbb{Q})$, and the adjoint map $\sim : M_n(Q) \to M_n(Q)$ is nothing more than $\sim : A \to A^t$; that is, $\tilde{A} = A^t$.

Applying the last two propositions, we find that A is a similarity if and only if $A^tA = dI_n$ for some $d \in \mathbb{Q}$, and if A and B are similarities, $A+B$ is a similarity if and only if $A^tB + B^tA = cI$ for some $c \in \mathbb{Q}$.

In this case, the only similarity with similarity factor equal to 0 is the matrix of all zeros. So, if A and B are similarities and A+B is a similarity and $A+B \neq 0$, then $BA^t+AB^t = cI$ also.

If we look back at the definition of a rational family (Definition 2.16), we see that a rational family gives rise to a collection of mutually compatible similarities of the bilinear space of this example.

Linear subspaces of Sim(V) have a naturally defined quadratic structure which relates to the quadratic structure on V. More precisely:

3.27. PROPOSITION. Let (V, B) be a nondegenerate bilinear space. The similarity factor map $\sigma : \mathrm{Sim}(V) \to \mathbb{Q}$ is a quadratic form on any linear subspace of Sim(V).

Proof. Let $S \subseteq \mathrm{Sim}(V)$ be a linear subspace. If $f \in S$ and $\alpha \in \mathbb{Q}$, then $\sigma(\alpha f) = \alpha^2\sigma(f)$ is clear, and so it remains only to show that the map $B : S \times S \to \mathbb{Q}$, given by $B(f, g) = \sigma(f+g) - \sigma(f) - \sigma(g)$, is bilinear. It obviously suffices to show that

$$B(\alpha_1 f_1+\alpha_2 f_2, g) = \alpha_1 B(f_1, g) + \alpha_2 B(f_2, g). \tag{*}$$

We need to compute $\sigma(\alpha_1 f_1+\alpha_2 f_2+g)$, $\sigma(\alpha_1 f_1+\alpha_2 f_2)$, $\sigma(f_1+g)$ and $\sigma(f_2+g)$. We may use Propositions 3.23 and 3.24. Set $\tilde{f}_1 g+\tilde{g}f_1 = c_1 1_V$, $\tilde{f}_2 g+\tilde{g}f_2 = c_2 1_V$ and $\tilde{f}_1 f_2+\tilde{f}_2 f_1 = c_3 1_V$; then a simple computation shows:

$$\sigma(\alpha_1 f_1+\alpha_2 f_2+g) = \alpha_1^2\sigma(f_1) + \alpha_2^2\sigma(f_2) + \sigma(g) + \alpha_2 c_2 + \alpha_1 c_1 + \alpha_1\alpha_2 c_3,$$

$$\sigma(\alpha_1 f_1+\alpha_2 f_2) = \alpha_1^2\sigma(f_1) + \alpha_2^2\sigma(f_2) + \alpha_1\alpha_2 c_3,$$

$$\sigma(f_1+g) = \sigma(f_1) + \sigma(g) + c_1,$$

$$\sigma(f_2+g) = \sigma(f_2) + \sigma(g) + c_2.$$

The verification of (*) is now complete after a routine further calculation.

Thus (S, σ) is a quadratic space, and the associated bilinear space is (S, B_σ), where $B_\sigma : S \times S \to \mathbb{Q}$ is given by $B_\sigma(f, g) = (1/2)(\sigma(f+g)-\sigma(f)-\sigma(g))$.

Note that $\tilde{f}g+\tilde{g}f = 2B_\sigma(f, g)1_V$ for any f, g in a linear subspace of Sim(V).

From now on, whenever we speak of a subspace of Sim(V), we shall assume it is a quadratic (or bilinear) space with the quadratic structure we have just described.

3.28. EXAMPLE. This is a continuation of Example 3.26. Let $A_1, \dots, A_\ell$ be a rational family of type $[s_1, \dots, s_\ell]$ in order n. As we have already noted, $A_1, \dots, A_\ell$ span a linear subspace of Sim(V). We first observe that the subspace spanned by these matrices has dimension equal to ℓ. For suppose $\alpha_1 A_1 + \dots + \alpha_\ell A_\ell = 0$, where $\alpha_i \in \mathbb{Q}$.

CLAIM. The α_i are all equal to 0.

Proof. It will be enough to show $\alpha_1 = 0$. The claim then follows by induction on ℓ.

Since

$$\alpha_1 A_1 + \alpha_2 A_2 + \dots + \alpha_\ell A_\ell = 0, \tag{1}$$

we may multiply (1) by A_1^t, on the left, to obtain

$$\alpha_1 A_1^t A_1 + \alpha_2 A_1^t A_2 + \dots + \alpha_\ell A_1^t A_\ell = 0. \tag{2}$$

By definition of a rational family, this is equal to

$$\alpha_1 s_1 I_n - \alpha_2 A_2^t A_1 - \dots - \alpha_\ell A_\ell^t A_1 = 0. \tag{3}$$

Now take the transpose of (1), and multiply the result on the right by A_1 to get

$$\alpha_1 s_1 I_n + \alpha_2 A_2^t A_1 + \dots + \alpha_\ell A_\ell^t A_1 = 0. \tag{4}$$

Adding (3) and (4) gives $2\alpha_1 s_1 I_n = 0$, and so $2\alpha_1 s_1 = 0$. But since $s_1 \neq 0$, we get $\alpha_1 = 0$, as claimed.

Now we have seen that this subspace of Sim(V) has a quadratic structure. The A_i are a basis for this subspace, and since $A_i A_j^t + A_j A_i^t = 0$ $(i \neq j)$ we have that the A_i are mutually orthogonal $(\tilde{f}g + \tilde{g}f = 2B_\sigma(f,g)1_V$, so $B_\sigma(A_i, A_j) = 0)$. Furthermore, the A_i have similarity factors s_i, and thus we see that the rational family gives rise to the quadratic space (S, q), where $q = \langle s_1, \dots, s_\ell \rangle$, and S is the subspace of Sim(V) spanned (independently) by $A_1, \dots, A_\ell$. Note further that since the $s_i > 0$, this is a nondegenerate quadratic space and the only 0-similarity is the zero matrix.

Since 0-similarities are troublesome in the general theory, we shall from now on consider only linear subspaces S of $\mathrm{Sim}(V)$ which are non-degenerate quadratic spaces and whose only 0-similarity is the zero matrix. As the example above shows, these are, in fact, the linear subspaces of $\mathrm{Sim}(V)$ that we are most interested in. (For the more general discussion, see Shapiro [3].)

So now let (V, B) be a nondegenerate bilinear space and (S, B_σ) a nondegenerate subspace of $\mathrm{Sim}(V)$ as specified above. How are these two spaces related? The next proposition is crucial to our understanding (S, B_σ).

3.29. PROPOSITION. Let (V, B) be a nondegenerate bilinear space of dimension n, and (S, B_σ) a nondegenerate subspace of $\mathrm{Sim}(V)$. Then (S, B_σ) is similar to a subspace of (V, B).

Proof. Pick $x \in V$ with $B(x, x) \neq 0$, and define $\phi : S \to V$ by $\phi(f) = f(x)$. Clearly ϕ is a linear transformation from S to V, and since S is a nondegenerate subspace of $\mathrm{Sim}(V)$ in which the 0-map is the only 0-similarity, we have that the only noninvertible similarity in S is the 0-map. Thus ϕ is one-to-one. If we let $W = \{f(x) \mid f \in S\} \subseteq V$, then W is a subspace of V, and we want to show that ϕ is a similarity between (S, B_σ) and (W, B).

Choose a basis for V, and let A denote the matrix of B with respect to this basis. Let $f, g \in S$, and suppose they have matrices X and Y, respectively, with respect to the basis chosen for V. We have seen that $\tilde{f}g+\tilde{g}f = 2B_\sigma(f, g)1_V$, so in matrix terms we have:

$$A^{-1}X^tAY + A^{-1}Y^tAX = 2B_\sigma(f, g)I_n;$$

that is,

$$X^tAY = 2B_\sigma(f, g)A - Y^tAX.$$

Now let $x \in V$ be as above, and write $x = \mathrm{col}(\alpha_1, \ldots, \alpha_n)$ (with respect to the chosen basis); then

$$x^t(X^tAY)x = 2B_\sigma(f, g)x^tAx - x^t(Y^tAX)x;$$

that is,

$$B(f(x), g(x)) = 2B_\sigma(f, g)q_B(x) - B(g(x), f(x)).$$

Since B is symmetric, we have $B(f(x), g(x)) = q_B(x)B_\sigma(f, g)$. This proves the proposition.

REMARK. We have gotten a bit more out of the proof than is stated in the proposition. Not only do we have a similarity between the two spaces considered, we even have a nice form for the similarity factor. In particular, if there were an $x \in V$ such that $q_B(x) = 1$, we would have an isometry between (S, B_σ) and (W, B). This fact is important enough to isolate, but first a definition.

3.30. DEFINITION. Let (V, B) be a bilinear space and (V, q_B) the associated quadratic space. Let D equal the range of q_B, $q_B : V \to k$. We say that q_B <u>represents</u> $\alpha \in k$ if $\alpha \in D$.

REMARK. This definition is a bit unorthodox since 0 always belongs to the range of q_B. It is more usual to say that q_B represents 0 if there is a $v \neq 0$ with $q_B(v) = 0$. This should cause no confusion, however, since if we say "q_B represents 0", we shall mean in the nontrivial sense.

3.31. COROLLARY. Let (V, B) be a nondegenerate bilinear space and (S, B_σ) a nondegenerate subspace of $\mathrm{Sim}(V)$. If q_B represents 1, then (S, B_σ) is isometric to a subspace of (V, B).

3.32. EXAMPLE. This is a continuation of Examples 3.26 and 3.28. In those examples, $q_B = \langle 1, 1, \ldots, 1\rangle$, and q_B, therefore, always represents 1. Thus the corollary above applies to the examples. We state that fact separately.

3.33. PROPOSITION. If there is a rational family of type $[s_1, \ldots, s_\ell]$ in order n, then there is a subspace $(S, \langle s_1, \ldots, s_\ell\rangle)$ of $\mathrm{Sim}(V)$, and $(S, \langle s_1, \ldots, s_\ell\rangle)$ is isometric to a subspace of $(V, \langle 1, 1, \ldots, 1\rangle)$.

§3.9. RELATIONS BETWEEN RATIONAL FAMILIES IN THE SAME ORDER

In Proposition 2.18 we completely characterized rational families in odd orders and in orders $n \equiv 2 \pmod 4$. Thus our interest will be mainly in orders n such that $4 | n$.

In this case we can make a simplification of our problem which will be useful later.

3.34. PROPOSITION. Let $n \equiv 0 \pmod 4$. There is a rational family of type $[s_1, \ldots, s_\ell]$ in order n if and only if there is a rational family of type

$$\left[1, \frac{s_2}{s_1}, \ldots, \frac{s_\ell}{s_1}\right]$$

in order n.

Proof. Let $A_1, \ldots, A_\ell$ be the rational family of type $[s_1, \ldots, s_\ell]$ in order n. It is easy to see that

$$\frac{1}{s_1}A_1^tA_1, \frac{1}{s_1}A_1^tA_2, \ldots, \frac{1}{s_1}A_1^tA_\ell$$

is a rational family in order n of type

$$\left[1, \frac{s_2}{s_1}, \ldots, \frac{s_\ell}{s_1}\right].$$

Conversely, let $B_1, \ldots, B_\ell$ be the rational family in order n of type

$$\left[1, \frac{s_2}{s_1}, \ldots, \frac{s_\ell}{s_1}\right].$$

It is easy to see that if X is a rational matrix of order n and $XX^t = s_1I_n$, then $XB_1, XB_2, \ldots, XB_\ell$ is a rational family of type $[s_1, s_2, \ldots, s_\ell]$ in order n. So it suffices to construct such a matrix X.

By a theorem of Lagrange, every positive rational number is the sum of four squares of rational numbers; so let $s_1 = q_1^2+q_2^2+q_3^2+q_4^2$. Consider the 4×4 matrix

$$M = \begin{bmatrix} q_1 & q_2 & q_3 & q_4 \\ -q_2 & q_1 & -q_4 & q_3 \\ -q_3 & q_4 & q_1 & -q_2 \\ -q_4 & -q_3 & q_2 & -q_1 \end{bmatrix};$$

then $MM^t = s_1 I_4$. By hypothesis, $4|n$, and so $n = 4u$. If we let $X = M \otimes I_u$, then X is a matrix of the desired type. This completes the proof of the proposition.

With this proposition proved we see that the study of rational families in orders divisible by 4 is equivalent to the study of rational families of type $[1, \ldots]$.

But given a rational family in order n of type $[1, s_2, \ldots, s_\ell]$, there is certainly another rational family (maybe the one we already have) of the same order and type for which the first matrix is the identity matrix.

Thus we may always assume we are dealing with a rational family in which the first matrix is the identity.

§3.10. CLIFFORD ALGEBRAS

Let (V, q) be a quadratic space over the field k, and let A be a k-algebra which contains V as a linear subspace. A is said to be <u>compatible</u> with q if whenever $x \in V$, $x^2 = q(x).1$.

3.35. DEFINITION. An algebra $C \supset (V, q)$ which is compatible with q is called a <u>Clifford Algebra for</u> (V, q) if it has the following universal property: If $A \supset V$ is compatible with q, then there is a unique k-algebra homomorphism $\phi : C \to A$ such that $\phi(x) = x$ for all $x \in V$.

The usual nonsense for universal objects shows that if the Clifford Algebra exists, it is unique up to canonical isomorphism. Thus existence is the only problem.

Clifford Algebras always exist, and their algebra structure has been extensively studied. It would take us very far afield to go into these matters here. We shall just be content to list a few properties and to refer the reader to an excellent account of this subject in Lam [1].

Let (V, q) be a quadratic space over k where $\dim V < \infty$, and let $C(V)$ be its Clifford Algebra.

3.36. THEOREM. (1) $C(V)$ is a finite-dimensional algebra over k and is generated by V.

(2) If $\dim_k V = n$, then $\dim_k C(V) = 2^n$.

(3) If $\dim_k V$ is odd, then $C(V)$ is a semisimple algebra (that is, a direct sum of matrix rings over division rings containing k).

(4) If $\dim_k V$ is even, then $C(V)$ is a simple algebra with center k (that is, a matrix ring over a division ring whose center is k).

(5) All irreducible (left) modules (that is, no proper submodules) for a Clifford Algebra have the same dimension (over k) which is a power of 2.

3.37. EXAMPLES. (a) Let V be a 1-dimensional vector space, and let $q = \langle a \rangle$. Then

$$C(V) \simeq \frac{k[x]}{(x^2-a)}.$$

(Note that if "a" is a square in k, then $C(V) \simeq k \times k$, and if "a" is not a square, x^2-a is irreducible in $k[x]$, and so $C(V)$ is a field which is a 2-dimensional vector space over k.)

(b) Let $a, b \in k$, and consider the 4-dimensional algebra, denoted $\left(\frac{a,b}{k}\right)$, with basis $1, i, j, ij$ where $i^2 = a$, $j^2 = b$, $ij = -ji$.

This algebra is called a <u>quaternion</u> <u>algebra</u> ($a = b = -1$, $k = \mathbb{R}$ (reals) gives the usual quaternions). If V is 2-dimensional and $q = \langle a,b \rangle$, then $C(V) \simeq \left(\frac{a,b}{k}\right)$.

Now, by (4) of the theorem above, $\left(\frac{a,b}{k}\right)$ is either a division ring with center k or 2×2 matrices over k. We shall need to be able to distinguish these two general cases, and we state the relevant fact here.

3.38. PROPOSITION. $\left(\frac{a,b}{k}\right)$ is <u>not</u> a division ring if and only if there are $x, y \in k$ such that $ax^2+by^2 = 1$.

Notice that if $k = \mathbb{Q}_p$, $\left(\frac{a,b}{k}\right)$ is <u>not</u> a division ring if and only if $s_p(a,b) = 1$.

For a thorough discussion of quaternion algebras, including a proof of the fact that they are Clifford Algebras, we refer the reader again to Lam [1]. A proof of Wedderburn's theorem on semisimple algebras (which we shall use implicitly) may be found, for example, in Scharlau [1].

§3.11. SIMILARITY REPRESENTATIONS

3.39. DEFINITION. Let (V, B) be a nondegenerate bilinear space over $\mathbb{Q}$ of dimension n, and let (V, q_B) be the associated quadratic space. We say that (V, q_B) (or (V, B)) is <u>positive definite</u> if $q_B(x) > 0$ for every $x \neq 0$, $x \in V$.

Clearly, if $q_B = \langle a_1, \ldots, a_n \rangle$, then (V, q_B) is positive definite if and only if $a_i > 0$, $i = 1, \ldots, n$; alternatively, the signature of $q_B = n$.

Note also that in our investigation of rational families, the only spaces we have seen are positive definite. Thus from now on, unless we specifically state otherwise, all quadratic spaces we consider will be positive definite over $\mathbb{Q}$.

3.40. DEFINITION. Let (V, B) be an arbitrary bilinear space, and let $W \subset V$ be a subspace. The <u>orthogonal complement</u> of W, denoted $W^{\perp}$, is $\{x \in V \mid B(x, w) = 0 \text{ for all } w \in W\}$.

Warning. In general, the orthogonal complement of a subspace is no "complement" in the usual sense; it is quite possible, even in a nondegenerate bilinear space, for $W = W^{\perp}$.

For positive definite spaces, though, this cannot occur.

3.41. PROPOSITION. Let (V, q) be a positive definite space and $W \subset V$ a subspace. Then

(i) $W \cap W^{\perp} = 0$;

(ii) $W + W^{\perp} = V$;

(iii) $(W, q|_W) \perp (W^{\perp}, q|_{W^{\perp}}) \simeq (V, q)$, where all spaces are again positive definite.

Now if (V, B) is a positive definite space and $f \in \mathrm{Sim}(V, B)$ is not a 0-similarity, then $\sigma(f) > 0$. Thus any nondegenerate subspace of $\mathrm{Sim}(V, B)$ is automatically positive definite. So let S be a nondegenerate linear subspace of $\mathrm{Sim}(V, B)$, where (V, B) is positive definite, and let us suppose $1_V \in S$. (By Section 3.9, this is all we need consider for rational families.) We denote, as usual by σ, the quadratic form on S, and let $S_1 \subset S$ be the orthogonal complement of 1_V in S.

3.42. PROPOSITION. The inclusion $S_1 \to \mathrm{End}(V)$ is compatible with the quadratic form $-\sigma$ on S_1.

Proof. Now since $(S_1, \sigma|_{S_1})$ is a quadratic space, so is $(S_1, -\sigma)$. Let $f \in S_1$; then f is orthogonal to 1_V, so $B_\sigma(f, 1_V) = 0$; that is, $f+\tilde{f} = 0$; that is, $f = -\tilde{f}$. Now $f\tilde{f} = \sigma(f).1_V$ in general, and so $f^2 = -\sigma(f).1_V$, proving the proposition.

In view of this proposition and the universal property of Clifford Algebras, there is a unique $\mathbb{Q}$-algebra homomorphism $\pi : C(S_1, -\sigma) \to \mathrm{End}(V)$. However, this is just another way of saying that V is a (left)-module for the algebra $C(S_1, -\sigma)$; that is, the algebra $C(S_1, -\sigma)$ is represented on the vector space V.

3.43. PROPOSITION. Let (V, B) and S_1 be as above, and let $W \subseteq V$ be a subspace which is also a submodule of the $C(S_1, -\sigma)$-module V. Then $W^\perp$ is also a submodule of V.

Proof. Let $\pi : C(S_1, -\sigma) \to \mathrm{End}(V)$ be as above. To say that W is a submodule of V means that for every $z \in C(S_1, -\sigma)$, $\pi(z) : W \to W$. In view of the fact that S_1 generates $C(S_1, -\sigma)$, this is equivalent to saying that $\pi(x) : W \to W$ for all $x \in S_1$. Thus to show that $W^\perp$ is also a submodule, it will be enough to show that

$$\pi(x) : W^\perp \to W^\perp \quad \text{for all } 0 \neq x \in S_1;$$

that is,

$$B(\pi(x)(v), w) = 0 \quad \text{for every } w \in W,\ v \in W^\perp.$$

Now $x \in S_1$ implies $\widetilde{\pi(x)} = -\pi(x)$; thus $\widetilde{\pi(x)} : W \to W$, and so $\widetilde{\pi(x)}(w) = w' \in W$. We thus have $B(v, \widetilde{\pi(x)}(w)) = B(v, w') = 0$.

The importance of this proposition to our discussion is now evident: We started with (V, B) a positive definite space and (S, σ) a positive definite subspace of $\mathrm{Sim}(V)$. We then showed that V could be considered a module for the Clifford Algebra $C(S_1, -\sigma)$. Now if W is a submodule of V, then the proposition showed that $W^\perp$ is also a submodule for V. Thus, by Proposition 3.41, we can write $(V, B) = (W, B) \perp (W^\perp, B)$, and each summand is a module for $C(S_1, -\sigma)$, and both summands are again positive definite spaces. We may continue this process, as long as any of the summands has a $C(S_1, -\sigma)$-submodule. We conclude, then, that $(V, B) = (V_1, B) \perp \ldots \perp (V_r, B)$, where the (V_i, B) are each $C(S_1, -\sigma)$-modules which have no nontrivial $C(S_1, -\sigma)$-submodules; that is, the (V_i, B) are <u>irreducible</u> $C(S_1, -\sigma)$-modules.

It is fairly easy to understand what we have done here: for, let $1_V, f_1, \ldots, f_k$ be a basis for S; these are a collection of endomorphisms of V. The subspaces V_i we have found are nothing more than subspaces of V that are invariant under all the f_i (any subspace is invariant under 1_V). We have used the fact that V is a positive definite space to show that the complement of an invariant subspace is again invariant. In matrix terms this amounts to being able to choose a basis for V in such a way that the f_i become block diagonal matrices. The fact that we can recognize these invariant subspaces of V as irreducible modules for a certain Clifford Algebra will be important because then we shall be able to discuss the sizes of these blocks, the irreducible modules for Clifford Algebras being so completely understood.

With this discussion and the notation above, the following proposition is evident.

3.44. PROPOSITION. (S, B_σ) is isometric to a linear subspace of $\mathrm{Sim}(V_i, B)$.

NOTATION. (1) If (V_1, B_1) is isometric to a subspace of (V_2, B_2), we write $(V_1, B_1) < (V_2, B_2)$. The analogous notation will be used for quadratic spaces.

(2) Let $(V_1, B_1) < (V_2, B_2)$. We say that (V_1, B_1) <u>divides</u> (V_2, B_2) if $(V_2, B_2) = (V_1, B_1) \otimes (U, B)$ for some (U, B). In the quadratic case, when there can be no misunderstanding about the spaces being considered, we abusively write $q_1 | q_2$.

§3.12. SOME FACTS ABOUT POSITIVE DEFINITE FORMS OVER $\mathbb{Q}$

We write $n\langle a\rangle$ for the quadratic form $\underbrace{\langle a, a, \ldots, a\rangle}_{n\text{-times}}$.

3.45. PROPOSITION. Let (V, q) be a positive definite quadratic space over $\mathbb{Q}$.

(1) If $\dim V \geq 4$, q represents every element of $\mathbb{Q}^+$ ($\mathbb{Q}^+$ equals positive rationals).

(2) If $q \simeq \langle 1, a, b, ab\rangle$, $a, b \in \mathbb{Q}^+$, then, for every $x \in \mathbb{Q}^+$, $xq = \langle x, xa, xb, xab\rangle \simeq \langle 1, a, b, ab\rangle$.

(3) If $4 \mid n$ and $q \simeq n\langle 1\rangle$, then $xq \simeq q$, for all $x \in \mathbb{Q}^+$.

(4) For any $a, b, x, y \in \mathbb{Q}^+$, $\langle 1, a, b, ab\rangle \otimes \langle x, y\rangle \simeq 8\langle 1\rangle$.

<u>Proof.</u> (1) This generalizes the theorem of Lagrange that $\langle 1, 1, 1, 1\rangle$ represents every element of $\mathbb{Q}^+$ and can be found in Serre [1].

(2), (3), and (4) all follow from the Hasse-Minkowski Theorem 3.15.

3.46. COROLLARY. Let $(S, \sigma) < \mathrm{Sim}(V, q)$, where q is positive definite and $\dim S = \dim V = 4$. Then $(S, \sigma) \simeq (V, q)$.

<u>Proof.</u> By 3.45 (1) above, q represents 1. Thus, by Corollary 3.31, (S, σ) is isometric to a subspace of (V, q). But, since $\dim S = \dim V$, that subspace must be all of (V, q).

The next proposition is a special case of something true over general fields (see Shapiro [3]). However, over $\mathbb{Q}$, an <u>ad hoc</u> proof can be given.

3.47. PROPOSITION. Let $n = 2^m t$, t odd, $a, b \in \mathbb{Q}^+$. Then

(i) $n\langle 1\rangle \simeq n\langle a\rangle \iff 2^m\langle 1\rangle \simeq 2^m\langle a\rangle$,

(ii) $\langle 1, a\rangle \mid n\langle 1\rangle \iff \langle 1, a\rangle \mid 2^m\langle 1\rangle$,

(iii) $\langle 1, a, b, ab\rangle \mid n\langle 1\rangle \iff \langle 1, a, b, ab\rangle \mid 2^m\langle 1\rangle$.

<u>Proof.</u> (i) If $m \geq 2$, then all isometries are true by (3) of Proposition 3.45. So we need only consider $m = 0, 1$.

$m = 0$: $t\langle 1\rangle \simeq t\langle a\rangle \iff \langle 1\rangle \simeq \langle a\rangle$.

Proof. If: Obvious.

Only If: The isometry implies both have the same discriminant; that is, a^t is a square in $\mathbb{Q}$. But t odd implies $a = q^2$ for some $q \in \mathbb{Q}$.

$m = 1 : 2t\langle 1\rangle \simeq 2t\langle a\rangle \iff 2\langle 1\rangle \simeq 2\langle a\rangle$.

Proof. If: Obvious.

Only If: In this case we must have $s_p(2t\langle a\rangle) = 1$ for every prime p. But $s_p(2t\langle a\rangle) = ((a,a)_p)^x$, where

$$x = \sum_{j=1}^{2t-1} j = \frac{(2t-1)(2t)}{2} = t(2t-1),$$

which is odd since t is odd. Thus $((a,a)_p)^x = (a,a)_p = 1$ for every prime p. This is enough to prove that $\langle 1,1\rangle \simeq \langle a,a\rangle$, since the signatures and discriminants are obviously the same.

(ii) We first note that $\langle 1,a\rangle \mid n\langle 1\rangle$ always if $m \geq 3$, since $8\langle 1\rangle \simeq \langle 1,a\rangle \otimes \langle 1,b\rangle \otimes \langle x,y\rangle$ for any $a,b,x,y \in \mathbb{Q}^+$ by (4) of Proposition 3.45. Also $\langle 1,a\rangle \mid n\langle 1\rangle$ implies that $m \geq 1$. So we need only consider $m = 1$, $m = 2$.

$m = 1 : \langle 1,a\rangle \mid 2t\langle 1\rangle \iff \langle 1,a\rangle \mid 2\langle 1\rangle$.

Proof. If: Obvious.

Only If: The discriminant of $\langle 1,a\rangle \otimes \alpha$ is $a^t = 1$. Since t is odd, "a" is a square, and so $\langle 1,a\rangle \simeq \langle 1,1\rangle$.

$m = 2 : \langle 1,a\rangle \mid 4t\langle 1\rangle \iff \langle 1,a\rangle \mid 4\langle 1\rangle$.

Proof. If: Obvious.

Only If: $\langle 1,a\rangle \mid 4t\langle 1\rangle$ implies $\langle 1,a\rangle \otimes \langle \alpha_1,\ldots,\alpha_{2t}\rangle = 4t\langle 1\rangle$. If $t > 1$, we have, by repeated application of Proposition 3.45 (i), that

$$\langle \alpha_1,\ldots,\alpha_{2t}\rangle \simeq \langle 1,1,\ldots,1,u,v,w\rangle = \alpha.$$

Comparing the Hasse-Invariants, we find $(a,uvwa)_p(-1,uvw)_p = 1$ for every prime p. Let $d = uvw$. Then $d = \operatorname{disc}\alpha$, and we have that $(a,da)_p(-1,d)_p = 1$ for every prime p. But this is enough to guarantee that $\langle 1,a\rangle \otimes \langle 1,d\rangle \simeq 4\langle 1\rangle$.

(iii) Notice that by (4) of Proposition 3.45 we have $\langle 1,a,b,ab\rangle \mid n\langle 1\rangle$

always, if $m \geq 3$. Also $\langle 1,a,b,ab\rangle \mid n\langle 1\rangle$ implies $m \geq 2$. So we need only consider the case $m = 2$; that is, $\langle 1,a,b,ab\rangle \mid 4t\langle 1\rangle \iff \langle 1,a,b,ab\rangle \mid 4\langle 1\rangle$.

Proof. If: Obvious.

Only If: Suppose $\langle 1,a,b,ab\rangle \otimes \langle \alpha_1,\ldots,\alpha_t\rangle \simeq 4t\langle 1\rangle$, where t is odd, $t > 1$. By (4) of Proposition 3.45 and the Witt Cancellation Theorem 2.9, we obtain $\langle 1,a,b,ab\rangle \otimes \langle \alpha\rangle \simeq 4\langle 1\rangle$, which completes the proof.

3.48. PROPOSITION. Let $(S,\sigma) < \mathrm{Sim}(V,q)$ where $\dim V = n$, q is positive definite and $\dim S \geq 5$. Then $8 \mid n$.

Proof. We already know that if $q = n\langle 1\rangle$, then $8 \mid n$ by the Radon-Hurwitz Theorem.

Since $\dim S \geq 5$, we know, by Proposition 3.45(1), that σ represents 1; that is, there is a similarity in S with similarity factor equal to 1, which we call f. It is an easy exercise to show that $(\tilde{f} \circ S,\sigma)$ is another subspace of $\mathrm{Sim}(V,q)$ which is isometric to (S,σ) and that $\tilde{f} \circ S$ contains 1_V. So with no loss of generality we may assume $1_V \in S$.

Let $(T,\tau) < (S,\sigma)$, where $\dim T = 5$ and $1_V \in T$, and let τ_1 be the form on the orthogonal complement of 1_V in T. Then, as we have seen, $C(V,-\tau_1)$ is represented on V. By Theorem 3.36 we know that $C(V,-\tau_1)$ is a simple algebra with center equal to $\mathbb{Q}$ and of dimension 16 over $\mathbb{Q}$. By Wedderburn's theorem $C = C(V,-\tau_1)$ is either

(a) a 16-dimensional division ring with center equal to $\mathbb{Q}$,

(b) 2×2 matrices over K, where K is a division ring, center $(K) = \mathbb{Q}$ and $[K : \mathbb{Q}] = 4$,

(c) 4×4 matrices over $\mathbb{Q}$.

Now, in each case we know precisely the dimensions of the irreducible modules; in cases (a) and (b) we know that every irreducible module has dimension divisible by 8, so if C is either (a) or (b), we shall have that $8 \mid n$.

It remains only to show that (c) cannot occur. In case (c) every irreducible module has dimension 4 over $\mathbb{Q}$. So if $(W,\tilde{q})$ is an irreducible submodule of (V,q), we have, by Proposition 3.44, that (T,τ) is isometric to a linear subspace of $\mathrm{Sim}(W,\tilde{q})$. But, since $\dim W = 4$ and

$\tilde{q}$ is positive definite, we have, by Proposition 3.45(1), that q represents 1 and so, by Corollary 3.31, that (T,τ) is isometric to a subspace of $(W,\tilde{q})$. But $\dim T = 5 > \dim W = 4$, which is a contradiction. Thus we have that $8 | n$.

§3.13. REDUCTION OF THE ALGEBRAIC PROBLEM OF ORTHOGONAL DESIGNS TO ORDERS A POWER OF 2

As we have already seen, the existence of a rational family of type $[1, s_2, \ldots, s_\ell]$ in order n is equivalent to the existence of a linear subspace (S,σ) of $\mathrm{Sim}(\mathbb{Q}^n, n\langle 1\rangle)$ where $1_{\mathbb{Q}^n} \in S$ and $\sigma = \langle 1, s_2, \ldots, s_\ell\rangle$.

3.49. PROPOSITION. Let (S,σ) be a positive definite quadratic space. Then $(S,\sigma) < \mathrm{Sim}(\mathbb{Q}^4, 4\langle 1\rangle) \iff (S,\sigma) < (\mathbb{Q}^4, 4\langle 1\rangle)$.

<u>Proof.</u> <u>Only If</u>: Since $4\langle 1\rangle$ represents 1, obviously we are done by Corollary 3.31.

<u>If</u>: Since $(S,\sigma) < (\mathbb{Q}^4, 4\langle 1\rangle)$, we know $\dim S \le 4$. So we have to consider separately the cases $\dim S = 1, 2, 3, 4$. They are all proved in the same way, so we shall just exhibit the proof for $s = 2$. Let $\sigma = \langle a, b\rangle$. Then we can find u, v such that $\langle a, b, u, v\rangle \simeq \langle 1, 1, 1, 1\rangle$. In matrix terms, there is a 4×4 matrix P with $PP^t = \mathrm{diag}(a, b, u, v)$. Now, by Proposition 1.5, there is an integer Radon-Hurwitz family of order 4. Call the matrices in that family A_1, A_2, A_3, A_4. If $P = (p_{ij})$, let

$$X_1 = \sum_{j=1}^{4} p_{1j} A_j, \quad X_2 = \sum_{j=1}^{4} p_{2j} A_j.$$

It is easy to check that X_1, X_2 form a rational family of type $[a, b]$ in order 4 and thus span the appropriate linear subspace of $\mathrm{Sim}(\mathbb{Q}^4, 4\langle 1\rangle)$.

3.50. PROPOSITION. Let (S,σ) be a positive definite quadratic space.

(i) $(S,\sigma) < \mathrm{Sim}(\mathbb{Q}^8, 8\langle 1\rangle) \iff (S,\sigma) < (\mathbb{Q}^8, 8\langle 1\rangle)$.

(ii) $(S,\sigma) < \mathrm{Sim}(\mathbb{Q}^2, 2\langle 1\rangle) \iff (S,\sigma) < (\mathbb{Q}^2, 2\langle 1\rangle)$.

<u>Proof.</u> Exactly like Proposition 3.49, using the existence of a Radon-Hurwitz family of eight integer matrices of order 8 (or for (ii) the same fact for order 2).

3.51. PROPOSITION. Let $a, b > 0$, and let q be positive definite.

(i) $(S, <1, a>) < Sim(V, q) \Rightarrow <1, a> | q$.

(ii) $(S, <1, a, b>) < Sim(V, q) \Leftrightarrow (S', <1, a, b, ab>) < Sim(V, q)$.

(iii) $(S, <1, a, b>) < Sim(V, q) \Rightarrow <1, a, b, ab> | q$.

Proof. (i) Since $(S, <1, a>) < Sim(V, q)$, we have a representation of $C(S_1, <-a>)$ on V. Now, we noted that $C(S_1, <-a>) \simeq \frac{\mathbb{Q}[x]}{(x^2+a)}$. Since x^2+a is irreducible in $\mathbb{Q}[x]$, $\frac{\mathbb{Q}[x]}{(x^2+a)} = K$, a field, and $\dim_{\mathbb{Q}} K = 2$. Now, an irreducible K-module is a 1-dimensional K vector space. So $(V, q) \simeq (V_1, q_1) \perp \dots \perp (V_t, q_t)$, where the V_i are 1-dimensional over K; that is, 2-dimensional over $\mathbb{Q}$. By Propositions 3.44 and 3.29 we conclude that $(S, <1, a>)$ is similar to a subspace of (V_i, q_i). Since both S and V_i are 2-dimensional, we must have $q_i = \alpha_i <1, a>$. Hence $q = <1, a> \otimes <\alpha_1, \dots, \alpha_t>$, as was to be shown.

(ii) If: Obvious.

Only If: There is no loss in assuming that $1_V \in S$, $f_1, f_2 \in S$, $\sigma(f_1) = a$, $\sigma(f_2) = b$ and $B_\sigma(f_1, f_2) = 0 = B_\sigma(1_V, f_1) = B_\sigma(1_V, f_2)$. It is an easy matter, then, to check that 1_V, f_1, f_2, $f_1 f_2$ span a space, S', and that they are orthogonal (under B_σ) and that $\sigma(f_1 f_2) = ab$.

(iii) Now $(S, <1, a, b>) < Sim(V, q)$ implies that V is a module for $C = C(S_1, <-a, -b>)$. By Proposition 3.38 we have that C is a division ring which is 4-dimensional over $\mathbb{Q}$. Now an irreducible module for a division ring is always 1-dimensional, hence 4-dimensional over $\mathbb{Q}$. Thus

$$(V, q) \simeq (V_1, q_1) \perp \dots \perp (V_t, q_t),$$

where each V_i is 4-dimensional over $\mathbb{Q}$. Now, by Proposition 3.44 again, we have that $(S, <1, a, b>) < Sim(V_i, q_i)$, $1 \le i \le t$. Now, by (ii) of this proposition, we obtain $(S', <1, a, b, ab>) < Sim(V_i, q_i)$. Now use Proposition 3.45(1) to note that q_i represents 1, and so, by Corollary 3.31, we have that $(S', <1, a, b, ab>) < (V_i, q_i)$. But since $\dim_{\mathbb{Q}} S' = \dim_{\mathbb{Q}} V_i$, we have $q_i \simeq <1, a, b, ab>$. Thus $q \simeq \underbrace{<1, \dots, 1>}_{t\text{-times}} \otimes <1, a, b, ab>$.

REMARK. We have obtained a bit more out of the proof of (iii). Notice that if t is even, we may invoke Proposition 3.45 to assert that $q \simeq 4t<1>$. With this remark in mind we may improve Proposition 3.48.

3.52. PROPOSITION. Let $(S,\sigma) < \mathrm{Sim}(V,q)$ where $\dim V = n$, q is positive definite and $\dim S \geq 5$. Then $8|n$ and $q \simeq n<1>$.

Proof. The interesting fact here is that we have forced q to be $n<1>$. No other positive definite form has a 5-dimensional space of similarities.

We have already proved $8|n$, and, in the proof of Proposition 3.48, we have seen that $(U,<1,a,b>) < \mathrm{Sim}(V,q)$. The remark before this Proposition finishes the proof, since if $8|n$, t must be even.

3.52A. THEOREM. (Shapiro [1]). Let $n = 2^m t$ where t is odd, and let (S,σ) be a positive definite quadratic space where σ represents 1. Then

$$(S,\sigma) < \mathrm{Sim}(\mathbb{Q}^n, n<1>) <=> (S,\sigma) < \mathrm{Sim}(\mathbb{Q}^{2^m}, 2^m<1>).$$

Proof. First note that in Proposition 2.18 we proved this for $m = 0,1$. So we may assume $m \geq 2$. The fact that we are interested in rational families and the discussion in §3.9 shows that our assumption that σ represents 1 is no drawback. Note also that "If" is obvious, since, in matrix terms, we can just tensor with I_t.

$m = 2$: We need only consider $\dim S = 2, 3$ and 4.

Case 1. $\sigma = <1,a>$. By Proposition 3.51 (i), we have that $<1,a>|n<1>$ and hence, by Proposition 3.47, $<1,a>|4<1>$; that is, $(S,<1,a>) < (\mathbb{Q}^4, 4<1>)$. Now use Proposition 3.49 to assert $(S,<1,a>) < \mathrm{Sim}(\mathbb{Q}^4, 4<1>)$.

Case 2. $\sigma = <1,a,b>$. The route is pretty much the same as in Case 1. By Proposition 3.51 (iii), we have $<1,a,b,ab>|n<1>$ and hence, by Proposition 3.47, $<1,a,b,ab>|4<1>$. Hence $(S,<1,a,b>) < (\mathbb{Q}^4, 4<1>)$. Now use Proposition 3.49 to get $(S,<1,a,b>) < \mathrm{Sim}(\mathbb{Q}^4, 4<1>)$.

Case 3. $\sigma = <1,a,b,c>$. We postpone this case for a moment.

$m \geq 3$, $\dim\sigma = 2$. Let $\sigma = <1,a>$. Now $<1,a>|8<1>$ for any "a" by Proposition 3.45 (4). Thus $(S,\sigma) < (\mathbb{Q}^8, 8<1>)$. Now use Proposition 3.50

to get $(S,\sigma) < \mathrm{Sim}(\mathbb{Q}^8, 8<1>)$. But clearly, then $(S,\sigma) < \mathrm{Sim}(\mathbb{Q}^{2^m}, 2^m<1>)$ for any $m \geq 3$.

$m \geq 3$, $\dim\sigma = 3$. Let $\sigma = <1, a, b>$. Again using Proposition 3.45(4), we have $<1, a, b, ab> | 8<1>$ and so $(S, <1, a, b>) < (\mathbb{Q}^8, 8<1>)$. By Proposition 3.50 we get $(S, <1, a, b>) < \mathrm{Sim}(\mathbb{Q}^8, 8<1>)$ and thus isometric to a linear subspace of $\mathrm{Sim}(\mathbb{Q}^{2^m}, 2^m<1>)$ for any $m \geq 3$. We are left with considering the case where $\dim\sigma \geq 4$.

$\dim\sigma \geq 5$. Let $(S,\sigma) < \mathrm{Sim}(\mathbb{Q}^{2^m t}, 2^m t<1>) = \mathrm{Sim}(V, q)$. The usual Clifford Algebra techniques give $(V, q) \simeq (V_1, q_1) \perp \ldots \perp (V_\ell, q_\ell)$. By Proposition 3.52 we know $8 | \dim V_i$ and $q_i = n_i<1>$, $n_i = \dim V_i$. Since the dimension (over $\mathbb{Q}$) of an irreducible module for a Clifford Algebra is always a power of 2 (Theorem 3.36), we know that $n_i = 2^s$, $s \geq 3$. Since $2^s \ell = \sum_{i=1}^{\ell} n_i = 2^m t$, we have $s \leq m$. We now invoke Propositions 3.44 and 3.52 to finish.

We are left only with the case $\dim\sigma = 4$.

We proceed as in the case for $\dim\sigma \geq 5$, and, using the notation there, we get $(S,\sigma) < \mathrm{Sim}(V_i, q_i)$ where $\dim V_i = 2^s$ and $s \leq m$.

Case 1. $s \geq 3$. Let $\sigma = <1, a, b, c>$. We may proceed exactly as in the case $m \geq 3$, $\dim\sigma = 3$, to conclude that $q_i = 2^s<1>$, in which case we are done.

Case 2. $s = 2$, $\sigma = <1, a, b, c>$. Now we get $(S', <1, a, b>) < \mathrm{Sim}(V_i, q_i)$, but by Proposition 3.51 (ii), that implies $(T, <1, a, b, ab>) < \mathrm{Sim}(V_i, q_i)$. By Proposition 3.45(1), q_i represents 1, so $(T, <1, a, b, ab>) < (V_i, q_i)$. Since $\dim V_i = \dim T = 4$, we have $q_i \simeq <1, a, b, ab>$. But, for exactly the same reasons, we have $q_i \simeq \sigma$; that is, $\sigma \simeq <1, a, b, ab>$.

Thus we are reduced to considering the case $\dim\sigma = 4$ and $\sigma = <1, a, b, ab>$. But now, by Proposition 3.51, we have

$$(S', <1, a, b, ab>) < \mathrm{Sim}(\mathbb{Q}^n, n<1>) \iff (S, <1, a, b>) < \mathrm{Sim}(\mathbb{Q}^n, n<1>),$$

reducing us to the case where $\dim\sigma = 3$, which we have already handled.

3.53. COROLLARY. There is a rational family in order $n = 2^m t$, t odd, of type $[s_1, \ldots, s_\ell]$ if and only if there is a rational family of the same type in order 2^m.

§3.14. SOLUTION OF THE ALGEBRAIC PROBLEM OF ORTHOGONAL DESIGNS IN ORDERS 4t, 8t (t ODD)

In this section we propose to classify the rational families that exist in orders 4t and 8t (t odd). By Theorem 3.52A, this amounts to a classification of the rational families that can exist in orders 4 and 8.

ORDER 4. Since $\rho(4) = 4$, we have, by the results of Section 3.9, only to consider the conditions which allow there to exist rational families of types $[1, s_1]$, $[1, s_1, s_2]$ and $[1, s_1, s_2, s_3]$ in order 4.

Case 1. $[1, s_1]$.

3.54. PROPOSITION. (Geramita-Wallis [3]). A necessary and sufficient condition that there be a rational family of type $[1, s_1]$ in order 4 is that s_1 be a sum of three squares in $\mathbb{Q}$.

Proof. To show the necessity, observe that by the results of Section 3.9 we may assume the rational family is $\{I_4, A\}$, where $A = -A^t$ and $AA^t = s_1 I$. The fact that A is skew-symmetric forces its diagonal entries to all be zero, and hence, evidently, s_1 is a sum of three rational squares.

For the sufficiency, let $s_1 = q_1^2 + q_2^2 + q_3^2$ where $q_i \in \mathbb{Q}$. If

$$A = \begin{bmatrix} 0 & q_1 & q_2 & q_3 \\ -q_1 & 0 & -q_3 & q_2 \\ -q_2 & q_3 & 0 & -q_1 \\ -q_3 & -q_2 & q_1 & 0 \end{bmatrix},$$

then it is easily checked that $\{I_4, A\}$ is a rational family of type $[1, s_1]$.

3.55. COROLLARY. A necessary and sufficient condition that there be a rational family of type $[s_1, s_2]$ in order 4 is that s_1/s_2 be a sum of three squares in $\mathbb{Q}$.

Proof. See Section 3.9.

REMARK. Note that $m/n = mn/n^2 = (1/n)^2(mn)$, and so m/n is a sum of three rational squares if and only if mn is a sum of three rational squares. Now it is well known (see, for example, Serre [1]) that a positive integer is the sum of three rational squares if and only if it is the sum of three integer squares if and only if (Gauss) it is not of the form $4^a(8b+7)$.

Thus, for example, there is no rational family of type $[k, 7k]$ or $[3s, 5s]$ in order 4.

Case 2. $[1, s_1, s_2]$.

We will not have to treat this case separately, as the following proposition shows.

3.56. PROPOSITION. There is a rational family of type $[1, s_1, s_2]$ in order 4 if and only if there is a rational family of type $[1, s_1, s_2, s_1s_2]$ in order 4.

Proof. If: Obvious.

Only If: By Section 3.9 there is no loss in assuming the rational family is $\{I_4, A, B\}$. It is a routine verification that $\{I_4, A, B, AB\}$ is a rational family of type $[1, s_1, s_2, s_1s_2]$.

Case 3. $[1, s_1, s_2, s_3]$.

This situation is completely settled by

3.57. PROPOSITION. There is a rational family of type $[1, s_1, s_2, s_3]$ in order 4 if and only if $\langle 1, s_1, s_2, s_3\rangle \simeq \langle 1,1,1,1\rangle$ over $\mathbb{Q}$.

Proof. By Proposition 3.49 we see that there is a rational family of type $[1, s_1, s_2, s_3]$ in order 4 if and only if $\langle 1, s_1, s_2, s_3\rangle < 4\langle 1\rangle$. (Note the abusive notation.) Since both forms have dimension four, this is true if and only if $\langle 1, s_1, s_2, s_3\rangle \simeq 4\langle 1\rangle$ over $\mathbb{Q}$.

3.58. COROLLARY. There is a rational family of type $[1, s_1, s_2]$ in order 4 if and only if $\langle 1, s_1, s_2, s_1s_2\rangle \simeq 4\langle 1\rangle$.

REMARK. It is now clear how the Hasse-Minkowski classification of quadratic forms comes into play. The algorithmic nature of the criteria for congruence over $\mathbb{Q}$ makes it extremely amenable to machine analysis, in specific orders. (The print-outs in the appendices verify this.)

We now list a few corollaries which had been obtained by quite different methods in Geramita-Wallis [3].

3.59. COROLLARY. Let $a, b \in \mathbb{Q}^+$. There is a rational family of type $[a, a, a, b]$ in order 4 if and only if b/a is a square in $\mathbb{Q}$.

Proof. By Section 3.9 such a rational family exists if and only if a rational family of type $[1, 1, 1, b/a]$ exists in order 4. By Proposition 3.57 this occurs if and only if $\langle 1,1,1,b/a\rangle \simeq 4\langle 1\rangle$ over $\mathbb{Q}$. By Witt Cancellation, this occurs if and only if $\langle b/a\rangle \simeq \langle 1\rangle$, that is, if and only if b/a is a square in $\mathbb{Q}$.

3.60. COROLLARY. Let $a, b \in \mathbb{Q}^+$. There is a rational family of type $[a, a, b]$ in order 4 if and only if b/a is a sum of two squares in $\mathbb{Q}$.

Proof. As in the previous corollary, such a rational family exists if and only if there is a rational family of type $[1, 1, b/a]$ in order 4. By Corollary 3.58, this exists if and only if $\langle 1,1,b/a,b/a\rangle \simeq 4\langle 1\rangle$. By the Witt Cancellation Theorem this is true if and only if $\langle b/a, b/a\rangle \simeq \langle 1,1\rangle$.

Now two isometric forms represent the same elements of $\mathbb{Q}^*$, and so b/a is represented by $\langle 1,1\rangle$; hence b/a is a sum of two squares.

On the other hand, if b/a is a sum of two squares in $\mathbb{Q}$, then b/a is represented by $\langle 1,1\rangle$, and so $\langle 1,1\rangle \simeq \langle b/a, x\rangle$ for some $x \in \mathbb{Q}^+$. Thus if we compare discriminants, we see that $(b/a)x = u^2$, $u \in \mathbb{Q}$; that is, $x = (b/a)v^2$, $v \in \mathbb{Q}$, and so $\langle b/a, x\rangle \simeq \langle b/a, b/a\rangle$, as was to be shown.

REMARK. There is another way to see that $\langle b/a, b/a\rangle \simeq \langle 1, 1\rangle$ if and only if b/a is a sum of two squares in $\mathbb{Q}$.

First note that $\langle b/a, b/a\rangle \simeq \langle (b/a)a^2, (b/a)a^2\rangle \simeq \langle ab, ab\rangle$.

Since the signatures and discriminants of the forms $\langle ab, ab\rangle$ and $\langle 1, 1\rangle$ are the same, the two forms are congruent if and only if $s_p(\langle ab, ab\rangle) = 1$ for all primes p.

Now $(ab, ab)_p = (-1, ab)_p$. Write

$$ab = 2^{\alpha} p_1^{\alpha_1} \cdots p_s^{\alpha_s} q_1^{\beta_1} \cdots q_t^{\beta_t},$$

p_i, q_j primes, and where the $p_i \equiv 1 \pmod 4$ and the $q_j \equiv 3 \pmod 4$. Now, since -1 is a square mod p only for $p = 2$ and $p \equiv 1 \pmod 4$, we see that the α_i may be even or odd. Since -1 is not a square mod q if $q \equiv 3 \pmod 4$, we shall have $(-1, ab)_{q_j} = -1$ for $q_j | ab$, $q_j \equiv 3 \pmod 4$ unless β_j is even. But this, then, is precisely the condition that ensures that ab is a sum of two squares (Fermat).

ORDER 8. We know that $\rho(8) = 8$, so we must consider rational families having at most 8 members. The key result we need here is Proposition 3.50, which asserts that there is a rational family of type $[1, s_1, \ldots, s_\ell]$ $(1 \le \ell \le 7)$ in order 8 if and only if $<1, s_1, \ldots, s_\ell> = \sigma < 8<1>$.

Case 1. Much of the work is covered by the following

3.61. PROPOSITION. For any $s_1, s_2, s_3, s_4 \in \mathbb{Q}^+$ there is a rational family of type $[1, s_1, s_2, s_3, s_4]$ in order 8.

Proof. We need to show that

$$<1, s_1, s_2, s_3, s_4> = \sigma < 8<1>.$$

By Proposition 3.45 (4),

$$<1, s_1, s_2, s_1s_2> \otimes <1, s_3> \simeq 8<1>;$$

that is,

$$\tau = <1, s_1, s_2, s_1s_2, s_3, s_1s_3, s_2s_3, s_1s_2s_3> \simeq 8<1>.$$

But

$$\tau = <1, s_1, s_2, s_3> \perp <s_1s_2, s_1s_3, s_2s_3, s_1s_2s_3> = <1, s_1, s_2, s_3> \perp q.$$

Now q is positive definite of dimension 4, and so, by Proposition 3.45 (1) q represents s_4. Thus $q \simeq <s_4, u, v, w>$ for some $u, v, w \in \mathbb{Q}^+$. Thus $8<1> \simeq <1, s_1, s_2, s_3, s_4> \perp <u, v, w>$.

Case 2. $[1, s_1, s_2, s_3, s_4, s_5, s_6]$.

We need not consider this case separately from the case of 8-members, as the following proposition shows.

3.62. PROPOSITION. Let $s_i \in \mathbb{Q}^+$, $1 \le i \le 6$. There is a rational family of type $[1, s_1, \ldots, s_6]$ in order 8 if and only if there is a rational family of type

$$[1, s_1, \ldots, s_6, \prod_{i=1}^{6} s_i]$$

in order 8.

Proof. If: Obvious.

Only If: As always, we may assume the rational family is $\{I_8, A_1, \ldots, A_6\}$. It is routine to check that

$$\{I_8, A_1, \ldots, A_6, \prod_{i=1}^{6} A_i\}$$

is the required rational family.

Case 3. $[1, s_1, s_2, s_3, s_4, s_5, s_6, s_7]$.

3.63. PROPOSITION. Let $s_i \in \mathbb{Q}^+$, $1 \leq i \leq 7$. There is a rational family of type $[1, s_1, \ldots, s_7]$ in order 8 if and only if $\langle 1, s_1, \ldots, s_7 \rangle \simeq 8\langle 1 \rangle$ over $\mathbb{Q}$.

Proof. Exactly as in Proposition 3.57.

3.64. COROLLARY. Let $s_i \in \mathbb{Q}^+$, $1 \leq i \leq 6$. There is a rational family of type $[1, s_1, \ldots, s_6]$ in order 8 if and only if

$$\langle 1, s_1, \ldots, s_6, \prod_{i=1}^{6} s_i \rangle \simeq 8\langle 1 \rangle$$

over $\mathbb{Q}$.

Proof. Exactly as in Corollary 3.58.

We can further deduce, as in Corollaries 3.59 and 3.60,

3.65. COROLLARY. There is a rational family of type

(i) $[a, a, a, a, a, a, a, b]$ in order 8, $(a, b \in \mathbb{Q}^+)$ if and only if $b/a \in (\mathbb{Q}^*)^2$;

(ii) $[a, a, a, a, a, a, b]$ in order 8, $(a, b \in \mathbb{Q}^+)$ if and only if b/a is a sum of two squares in $\mathbb{Q}$.

Case 4. $[1, s_1, s_2, s_3, s_4, s_5]$.

As we have noted before, this amounts to deciding when

$<1, s_1, \ldots, s_5> = \sigma < 8<1>$.

3.66. PROPOSITION. If $\sigma = <1, s_1, \ldots, s_5>$, $s_i \in \mathbb{Q}^+$, and d equals the discriminant of σ, $\sigma < 8<1>$ if and only if there is a $u \in \mathbb{Q}^+$ such that $s_p(\sigma) = (-1,d)_p(-1,u)_p(u,d)_p$ for every prime p.

Proof. $\sigma < 8<1>$ if and only if there are $u, v \in \mathbb{Q}^+$ with $\sigma \perp <u,v> \simeq 8<1>$. Comparing discriminants, we have $duv = 1$ in $\mathbb{Q}^*/(\mathbb{Q}^*)^2$; that is, $v = du$ in $\mathbb{Q}^*/(\mathbb{Q}^*)^2$. Thus $\sigma \perp <u, du> \simeq 8<1>$.

By the Hasse-Minkowski Theorem, this can happen if and only if $s_p(8<1>) = 1 = s_p(\sigma \perp <u, du>)$ for all primes p. An easy calculation shows that $s_p(\sigma \perp <u, du>) = s_p(\sigma)(d,d)_p(u,u)_p(u,d)_p$. This product is 1 at every prime p if and only if $s_p(\sigma) = (-1,d)_p(-1,u)_p(u,d)_p$.

REMARK. In practice, this proposition seems difficult to use. One may be lucky in finding the appropriate u, but it seems difficult to assert when no u exists. The next two propositions give some easy criteria when no u exists.

3.67. PROPOSITION. There is a rational family of type $[a,a,a,a,a,b]$ ($a, b \in \mathbb{Q}^+$) in order 8 if and only if b/a is a sum of three squares in $\mathbb{Q}$.

Proof. Such a family exists if and only if one of type $[1,1,1,1,1,b/a]$ exists in order 8. By Proposition 3.66 this happens if and only if there is a $u \in \mathbb{Q}^+$ such that $<1,1,1,1,1,b/a,u,(b/a)u> \simeq 8<1>$. Applying Witt's Cancellation Theorem, this is equivalent to saying

$<b/a, u, (b/a)u> \simeq <1,1,1>$.

The congruence implies that b/a is represented by $<1,1,1>$, so b/a is a sum of three squares in $\mathbb{Q}$. This proves "Only If".

To show "If", we have that b/a is represented by $<1,1,1>$; hence $<1,1,1> \simeq <b/a, x, y>$ for some $x, y \in \mathbb{Q}^+$. Comparing discriminants, we have that $y = (b/a)x$ in $\mathbb{Q}^*/(\mathbb{Q}^*)^2$, which finishes the proof. More generally, we have:

3.68. PROPOSITION. (Wolfe [1]). There is a rational family of type $[a_1, a_2, a_3, a_4, a_5, a_6]$ in order 8 where $a_1a_2a_3a_4 = q^2$ for some $q \in \mathbb{Q}$ and $s_p(\langle a_1, a_2, a_3, a_4 \rangle) = 1$ for all primes p if and only if a_5a_6 is a sum of three squares in $\mathbb{Q}$.

Proof. The hypothesis gives that $\langle a_1, a_2, a_3, a_4 \rangle \simeq 4\langle 1 \rangle$. Thus $\sigma = \langle a_1, \ldots, a_6 \rangle < 8\langle 1 \rangle$ if and only if $4\langle 1 \rangle \perp \langle a_5, a_6 \rangle < 8\langle 1 \rangle$. By Witt's Cancellation Theorem this happens if and only if $\langle a_5, a_6 \rangle < 4\langle 1 \rangle$ if and only if $\langle 1, a_5a_6 \rangle < 4\langle 1 \rangle$ if and only if $\langle a_5a_6 \rangle < 3\langle 1 \rangle$ (Witt again!) if and only if a_5a_6 is a sum of three squares in $\mathbb{Q}$.

REMARK. This last applies notably to rational families of type $[a, a, a, a, b, c]$ in order 8.

There are other special cases one can handle easily for families of six members in order 8. We mention, and leave the proof to the reader, one more.

3.69. PROPOSITION. Let $a, b, c \in \mathbb{Q}^+$. There is a rational family of type $[a, a, a, b, b, c]$ in order 8 where ab is a sum of two squares in $\mathbb{Q}$ if and only if c is a sum of three squares in $\mathbb{Q}$.

§3.15. SOLUTION OF THE ALGEBRAIC PROBLEM OF ORTHOGONAL DESIGNS IN ORDERS 16t (t ODD)

Up to this point we have asked the patient reader to accept many unproved and deep statements from the classical theory of quadratic forms over $\mathbb{Q}$. This, we have felt, is not unreasonable, since many excellent sources for the material exist, and a full exposition here would have taken us too far afield.

In this section, however, the discussion of full proofs is even more difficult. Vast new amounts of background must be introduced to get each new proposition. Thus, we shall, albeit reluctantly, usually not give proofs of these deeper statements but just refer the reader to the imaginative arguments of Shapiro.

In spite of this disclaimer, there is still much we can say for these orders with only a small number of additional facts.

3.70. DEFINITION. (1) Let (V,q) be a quadratic space over the field k. If $x \neq 0$, $x \in V$ and $q(x) = 0$, then x is called an isotropic vector.

(2) If there is an isotropic vector in (V,q), then (V,q) is called isotropic (not isotropic equals anisotropic).

(3) We say that (V,q) is universal if q represents every element $a \in k$.

3.71. PROPOSITION. Let (V,q) be a 2-dimensional nondegenerate quadratic space. The following are equivalent:

(i) (V,q) is isotropic;

(ii) there is a basis for V such that the matrix of B_q with respect to this basis is $\begin{bmatrix} 0 & 1 \\ 1 & 0 \end{bmatrix}$;

(iii) there is a basis for V such that the matrix of B_q with respect to this basis is $\begin{bmatrix} -1 & 0 \\ 0 & 1 \end{bmatrix}$;

(iv) the discriminant of q is equal to -1.

Proof. The proof is elementary; see Scharlau [1].

This unique (up to isometry) 2-dimensional isotropic space is usually called a hyperbolic plane and denoted H.

3.72. PROPOSITION. A hyperbolic plane is universal.

Proof. Again, elementary; see Scharlau [1].

3.73. COROLLARY. If (V,q) is a nondegenerate isotropic quadratic space, then (V,q) is universal.

Proof. (V,q) has a hyperbolic plane as an orthogonal summand, and so the corollary is clear from Proposition 3.72.

3.74. DEFINITION. Let (V,q) be a nondegenerate quadratic space over $\mathbb{Q}$. We say that (V,q) is indefinite if there are $x, y \in V$ with $q(x) > 0$ and $q(y) < 0$.

One of the nicer consequences of the Hasse-Minkowski theory is the following.

3.75. THEOREM. (Meyer's Theorem). Any indefinite quadratic space (nondegenerate) over $\mathbb{Q}$ of dimension greater than or equal to 5 is isotropic.

With these facts in hand, we can now consider rational families in order 16t (t odd). By Theorem 3.52A, we must, equivalently, consider linear subspaces $(S,\sigma) < \mathrm{Sim}(\mathbb{Q}^{16}, 16\langle 1\rangle)$. We shall continue to abusively refer to this as considering $\sigma < \mathrm{Sim}(16\langle 1\rangle)$.

We proved, Proposition 2.17, that $\rho(16) = 9$, and so a rational family in order 16 cannot involve more than 9 members.

Case 1. 9-member rational families.

3.76. LEMMA. Let $V = \mathbb{Q}^2$, and let $q = \langle 1, a\rangle$, $a \in \mathbb{Q}^+$. Then $\begin{bmatrix} 0 & -a \\ 1 & 0 \end{bmatrix}$ and $\begin{bmatrix} 1 & 0 \\ 0 & -1 \end{bmatrix} = A \in \mathrm{Sim}(V,q)$. Furthermore, $\tilde{A} = A$.

Proof. A simple application of the definitions.

3.77. PROPOSITION. Let $a \in \mathbb{Q}^+$, and let $\sigma = 8\langle 1\rangle \perp \langle a\rangle$. Then $\sigma < \mathrm{Sim}(16\langle 1\rangle)$.

Proof. Let $A_1, \ldots, A_7$ be the H-R family of order 8, constructed in Proposition 1.5. Recall that $A_i \in \mathrm{Sim}(\mathbb{Q}^8, 8\langle 1\rangle)$ and that $\tilde{A}_i = -A_i$ (in this example "$\sim$" was "t"). It is then an easy matter to check that

$$\{I_2 \otimes I_8 = I_{16},\ \begin{bmatrix} 0 & -a \\ 1 & 0 \end{bmatrix} \otimes I_8 = f_1,\ \begin{bmatrix} 1 & 0 \\ 0 & -1 \end{bmatrix} \otimes A_i = f_{i+1},\ i = 1,\ldots,7\}$$

is in $\mathrm{Sim}(\mathbb{Q}^2 \otimes \mathbb{Q}^8 = V, \langle 1, a\rangle \otimes 8\langle 1\rangle = q)$ and that $f_i\tilde{f}_j + f_j\tilde{f}_i = 0$, $i \neq j$, and $\tilde{f}_i = -f_i$, $i = 1, \ldots, 8$. Also $f_1\tilde{f}_1 = a1_V$ and $f_j\tilde{f}_j = 1_V$ for $j = 2, \ldots, 8$. Thus we have $\sigma = 8\langle 1\rangle \perp \langle a\rangle < \mathrm{Sim}(q)$. But, by Proposition 3.45 (4), $q \simeq 16\langle 1\rangle$, and so we are done.

REMARK. This is a generalization of (1) in the proof of Theorem 1.6 and was noted by Shapiro [1].

3.78. THEOREM. (Shapiro). Let $\sigma < \mathrm{Sim}(16\langle 1\rangle)$, $\dim\sigma = 9$. Then there is an $a \in \mathbb{Q}^+$ such that $\sigma \simeq 8\langle 1\rangle \perp \langle a\rangle$.

Proof. We shall postpone our remarks about this proof until the end of this section.

3.79. COROLLARY. There is a rational family of type $[s_1, \ldots, s_9]$ in order 16 if and only if $s_p(\langle a_1, \ldots, a_9\rangle) = 1$ for every prime p.

EXAMPLE. There is a rational family of type $[a, a, a, a, a, a, a, b, b]$ in order 16 if and only if ab is a sum of two squares in $\mathbb{Q}$.

Case 2. 7-member rational families.

3.80. PROPOSITION. Any 7-member rational family exists in order 16.

Proof. By Proposition 3.77 it would be enough to show that if σ is positive definite and $\dim\sigma = 7$, then there is an $a \in \mathbb{Q}^+$ with $\sigma < 8\langle 1\rangle \perp \langle a\rangle$.

By repeated use of Proposition 3.45 (1), we may assume $\sigma = 4\langle 1\rangle \perp \langle u, v, w\rangle$ for some $u, v, w \in \mathbb{Q}^+$. By Witt's Cancellation Theorem it will be enough to find $a \in \mathbb{Q}^+$ with $\langle u, v, w\rangle < 4\langle 1\rangle \perp \langle a\rangle$.

Let $\tau = \langle u, v, w\rangle$; then $u\tau \simeq \langle 1, y, z\rangle$, and suppose $u\tau < 4\langle 1\rangle \perp \langle d\rangle$ for some $d \in \mathbb{Q}^+$. Then $\tau < 4\langle u\rangle \perp \langle ud\rangle$. But $4\langle u\rangle \simeq 4\langle 1\rangle$, and letting $a = ud$ would finish the proof.

Thus there is no loss in assuming $\tau = \langle 1, y, z\rangle$. Consider $\langle 1, 1, 1, -y, -z\rangle$. By Meyer's Theorem this is isotropic, and so $\langle 1, 1, 1, -y, -z\rangle \simeq \langle 1, -1\rangle \perp \langle -e, s, t\rangle$ for some $e, s, t \in \mathbb{Q}^+$. Add $\langle y, z, e\rangle$ to both sides, and observe that $\langle \ell, -\ell\rangle \simeq \langle 1, -1\rangle$ (unique hyperbolic plane!) for any $\ell \in \mathbb{Q}$. Hence

$$\begin{array}{ccc} \langle 1,1,1,-y,-z\rangle \perp \langle y,z,e\rangle & \simeq & \langle 1,-1\rangle \perp \langle -e,s,t\rangle \perp \langle y,z,e\rangle \\ \wr\| & & \wr\| \\ \langle 1,1,1\rangle \perp H \perp H \perp \langle e\rangle & \simeq & H \perp H \perp \langle s,t,y,z\rangle. \end{array}$$

Using Witt's Cancellation Theorem again, to cancel the hyperbolic planes, gives $\langle 1, 1, 1, e\rangle \simeq \langle s, t, y, z\rangle$. Adding $\langle 1\rangle$ to both sides gives the desired result.

Case 3. 8-member rational families.

This is the only case we have yet to consider. There is one easy part of this discussion and one difficult part. We prove the easy part.

3.81. PROPOSITION. Let σ be a positive definite 8-dimensional form over $\mathbb{Q}$, and suppose $\operatorname{disc}\sigma = 1$. Then $\sigma < \operatorname{Sim}(16\langle 1\rangle)$.

<u>Proof.</u> Let $\sigma = 5\langle 1\rangle \perp \langle u, v, w\rangle$, without loss of generality. Since $\operatorname{disc}\sigma = 1$, $w = uv$ in $\mathbb{Q}^*/(\mathbb{Q}^*)^2$, and so $\sigma = 5\langle 1\rangle \perp \langle u, v, uv\rangle$.

Let $\tau = 5\langle 1\rangle \perp \langle u, v\rangle$; by Proposition 3.80, $\tau < \operatorname{Sim}(16\langle 1\rangle)$; that is, there is a rational family of type $[1,1,1,1,1,u,v]$ in order 16. By Section 3.9 we may assume this rational family is $\{I_{16}, A_1, \ldots, A_6\}$. It is an easy matter to check that $\{I_{16}, A_1, \ldots, A_6, \prod_{i=1}^{6} A_i\}$ is a rational family of type $[1,1,1,1,1,u,v,uv]$, and hence $\sigma < \operatorname{Sim}(16\langle 1\rangle)$.

The case where the discriminant is not equal to 1 is deeper.

3.82. THEOREM. Let (S,σ) be a nondegenerate linear subspace of $\operatorname{Sim}(\mathbb{Q}^{16}, 16\langle 1\rangle)$ where $\dim\sigma = 8$ and $\operatorname{disc}\sigma \neq 1$. Then $(S,\sigma) \subset (S',\sigma')$ where $\dim\sigma' = 9$ and $(S',\sigma') < \operatorname{Sim}(\mathbb{Q}^{16}, 16\langle 1\rangle)$.

<u>Proof.</u> See Shapiro [3].

REMARK. Thus 8-dimensional subspaces of $\operatorname{Sim}(16\langle 1\rangle)$ are <u>not</u> maximal if their discriminant is not equal to 1.

3.83. COROLLARY. Let σ be a positive definite quadratic form where $\dim\sigma = 8$ and $\operatorname{disc}\sigma \neq 1$. Then $\sigma < \operatorname{Sim}(16\langle 1\rangle)$ if and only if there is an $a, b \in \mathbb{Q}^+$ with $\sigma \perp \langle a\rangle \simeq 8\langle 1\rangle \perp \langle b\rangle$.

<u>Proof.</u> This is immediate from Theorems 3.82 and 3.78.

REMARK. If $\sigma \perp \langle a\rangle \simeq 8\langle 1\rangle \perp \langle b\rangle$, and if $d = \operatorname{disc}\sigma \neq 1$, then $da = b$ in $\mathbb{Q}^*/(\mathbb{Q}^*)^2$; that is, $\sigma \perp \langle db\rangle \simeq 8\langle 1\rangle \perp \langle b\rangle$. By Hasse-Minkowski $s_p(\sigma \perp \langle db\rangle) = 1$ for all primes p. An easy calculation shows that

$$s_p(\sigma \perp \langle db\rangle) = s_p(\sigma)(d,-b)_p = 1.$$

Thus $s_p(\sigma) = (d,-b)_p$ for all primes p. Thus, we may restate the last corollary.

3.84. COROLLARY. Let σ be a positive definite quadratic form where $\dim\sigma = 8$ and $\operatorname{disc}\sigma = d \neq 1$. Then $\sigma < \operatorname{Sim}(16\langle 1\rangle)$ if and only if there is a $b \in \mathbb{Q}^+$ with $s_p(\sigma) = (d, -b)_p$ for any prime p.

We conclude this section with a short discussion of what is involved in the proof of Theorem 3.78. Let $\dim \sigma = 9$, $(S, \sigma) < \mathrm{Sim}(V, 16\langle 1\rangle)$, and write $\sigma = \langle 1\rangle \perp \sigma_1$, and let C = Clifford Algebra of $-\sigma_1$. Since $\dim(-\sigma_1) = 8$, C is a simple algebra. Let $\phi : C \to \mathrm{End}(V)$ be the usual similarity representation. Then C simple implies ϕ is one-to-one, and a dimension count ($\dim C = 2^8$, $\dim_{\mathbb{Q}} \mathrm{End}(V) = 16 \times 16 = 2^8$) shows ϕ is an isomorphism. If we call the <u>Witt Invariant</u> of $-\sigma_1$ the class of C in the Brauer group of $\mathbb{Q}$ and denote it $c(-\sigma_1)$, $C \cong \mathrm{End}_{\mathbb{Q}} V$ implies that this is the trivial element of the Brauer group; that is, $c(-\sigma_1) = 1$. (See Lam [1], p. 120.) Now the Hilbert symbol $(a, b)_p$ may also be interpreted as the class of the quaternion algebra $\left(\frac{a, b}{\mathbb{Q}_p}\right)$ in the Brauer group of $\mathbb{Q}_p$ in the following way: Since there is only one nontrivial quaternion algebra over $\mathbb{Q}_p$, $\left(\frac{a, b}{\mathbb{Q}_p}\right) = 1$ if the quaternion algebra is 2×2 matrices over $\mathbb{Q}_p$ and -1 if it is the other quaternion algebra over $\mathbb{Q}_p$. The p-adic Hasse-invariant s_p of the quadratic form is then just the appropriate product of quaternion algebras in the Brauer group of $\mathbb{Q}_p$. We can define a global Hasse-invariant of the quadratic form $\sigma = \langle a_1, \ldots, a_n\rangle$, $a_i \in \mathbb{Q}$ as $s(\sigma)$, where $s(\sigma)$ is in the Brauer group of $\mathbb{Q}$ by $s(\sigma) = \prod_{i<j} \left[\left(\frac{a_i, a_j}{\mathbb{Q}}\right)\right]$ where $\left[\left(\frac{a_i, a_j}{\mathbb{Q}}\right)\right]$ denotes the class of the quaternion algebra in the Brauer group. This global invariant (it is invariant of σ) is closely related to the Witt Invariant, and, in fact, in (Lam, [1]) the exact connections are given. In the case we're considering, $\dim(-\sigma_1) = 8$, the calculation gives

$$c(-\sigma_1) = 1 = s(-\sigma_1) \cdot \left(\frac{-1, \delta}{\mathbb{Q}}\right)$$

where δ = disc of $-\sigma_1$. In our usual terminology $1 = s_p(-\sigma_1)(-1, \delta)_p$ for all primes p.

Now since $\delta = \mathrm{disc}(-\sigma_1) = \mathrm{disc}(\sigma_1)$ (since $\dim(-\sigma_1) = 8$) and since $\sigma = \langle 1\rangle \perp \sigma_1$, we have $\delta = \mathrm{disc}\,\sigma$.

Furthermore, $s_p(-\sigma_1) = (-1, \delta)_p s_p(\sigma_1)$, as a simple calculation shows. Now $s_p(\sigma_1) = s_p(\sigma)$. Thus we have

$$1 = s_p(-\sigma_1)(-1,\delta)_p = (-1,\delta)_p s_p(\sigma_1)\cdot(-1,\delta)_p$$
$$= s_p(\sigma_1) = s_p(\sigma)$$

for all primes p. Hence, if $\tau = 8<1> \perp <\delta>$, we see that $s_p(\sigma) = s_p(\tau)$ for all primes p; they have the same discriminant and same signature, and by the Hasse-Minkowski Theorem they are isometric.

§3.16. SOLUTION OF THE ALGEBRAIC PROBLEM OF ORTHOGONAL DESIGNS IN ORDERS 32t (t ODD)

As usual, we are interested in finding conditions on the positive definite form σ in order that $\sigma < \mathrm{Sim}(32<1>)$. Since $\rho(32) = 10$, we know $\dim\sigma \leq 10$.

3.85. PROPOSITION. For any $a, b \in \mathbb{Q}^+$, $\sigma = 8<1> \perp <a,b> < \mathrm{Sim}(32<1>)$.

Proof. We have already seen that for any $b \in \mathbb{Q}^+$, $8<1> \perp <b> < \mathrm{Sim}(16<1>)$. By the same proof as in Proposition 3.77 we conclude that

$$8<1> \perp <b> \perp <a> < \mathrm{Sim}(<1,a> \otimes 16<1> \simeq 32<1>).$$

With just this proposition we may prove

3.86. PROPOSITION. If σ is positive definite, $\dim\sigma = 9$, then $\sigma < \mathrm{Sim}(32<1>)$.

Proof. We may write $\sigma = 6<1> \perp \tau$ where $\dim\tau = 3$ and τ is positive definite. Consider $\tau \perp <-1,-1> = \mu$. Since μ is indefinite and $\dim\mu = 5$, we know, by Meyer's Theorem, that μ represents 0 nontrivially. But this is just another way of saying that τ and $<1,1>$ represent a common value c. In that case $\tau = <a,b> \perp <c> = <a,b,c>$ and $<1,1> \simeq <c,?>$, but by checking the invariants we see $<1,1> \simeq <c,c>$. Thus

$$\tau < <a,b,c> \perp <c> = <a,b,c,c> \simeq <a,b,1,1>$$

for some $a, b \in \mathbb{Q}^+$. Thus $6<1> \perp \tau < 8<1> \perp <a,b>$ for some $a, b \in \mathbb{Q}^+$.

Thus we are left only with considering the case where $\dim\sigma = 10$.

3.87. THEOREM. (Shapiro [1]). If $\dim\sigma = 10$ and σ is positive definite, $\sigma < \mathrm{Sim}(32<1>)$ if and only if $\sigma \simeq 8<1> \perp <a,b>$.

Proof. We have shown "If". We again omit the proof. It is similar, in spirit, to the (omitted) proof of Theorem 3.78.

§3.17. THE PERIODICITY THEOREM AND THE GENERAL SOLUTION OF THE ALGEBRAIC PROBLEM OF ORTHOGONAL DESIGNS

In this section we omit all proofs and refer the reader to Shapiro [3].

3.88. PROPOSITION. Let σ be a positive definite form and $\sigma = \langle 1\rangle \perp \sigma_1$. Let C be the Clifford Algebra of $-\sigma_1$, and let V be any C-module. Then there is a positive definite form q on V where $\sigma < \mathrm{Sim}(V, q)$.

REMARK. If $\dim\sigma \geq 5$, then $\sigma = \langle 1\rangle \perp \sigma_1$ automatically, and by Proposition 3.52 this gives that $q \simeq n\langle 1\rangle$ where $8 \mid n$ and $n = \dim V$.

3.89. THEOREM. (Periodicity). Let σ be positive definite and suppose $\sigma \simeq 8\langle 1\rangle \perp \tau$, where τ represents 1. Then $\sigma < \mathrm{Sim}(2^{m+4}\langle 1\rangle)$ if and only if $\tau < \mathrm{Sim}(q)$ for some 2^m-dimensional positive definite quadratic form q.

Proof. We shall only remark that the proof rests on Proposition 3.88 and the classical periodicity theorems for Clifford Algebras.

3.89. COROLLARY. Suppose σ is positive definite and $\sigma = 8\langle 1\rangle \perp \tau$ where $\dim\tau \geq 5$; then

$$\sigma < \mathrm{Sim}(2^{m+4}\langle 1\rangle) \iff \tau < \mathrm{Sim}(2^{m}\langle 1\rangle).$$

Proof. This corollary just comes from the theorem and the remark preceding it.

We are still not in a position to use the full force of this periodicity theorem. The fact that we still have restrictions on 10-dimensional σ in order 2^5 gives trouble in order 2^6.

The case of 2^6. Now $\rho(2^6) = 12$. Suppose $\dim\sigma = 12$. Write $\sigma = 8\langle 1\rangle \perp \tau$, $\dim\tau = 4$. By periodicity, $\tau < \mathrm{Sim}(q)$, where q is 4-dimensional. By Corollary 3.46 we have $\tau < \mathrm{Sim}(q)$ if and only if $\tau \simeq q$, and an easy application of Proposition 3.51(ii) shows $\tau \simeq q \simeq \langle 1, a, b, ab\rangle$ for some $a, b \in \mathbb{Q}^+$, and so $\operatorname{disc}\tau = 1$. In summary, $\tau < \operatorname{Sim}(q)$ if and only if $\operatorname{disc}\tau = 1$. For later reference:

3.90. PROPOSITION. Let σ be a positive definite form of dimension 12. $\sigma < \operatorname{Sim}(2^6\langle 1\rangle)$ if and only if $\operatorname{disc}\sigma = 1$.

If $\dim\sigma = 11$, then, since $11 \equiv 3 \pmod 4$, we may use the proofs of Corollary 3.58 or (equivalently) Corollary 3.64 to see that if d equals the discriminant of σ, then $\sigma < \operatorname{Sim}(2^6\langle 1\rangle)$ if and only if $\sigma \perp \langle d\rangle < \operatorname{Sim}(2^6\langle 1\rangle)$. Since the discriminant of $\sigma \perp \langle d\rangle = 1$, we may apply Proposition 3.90 to obtain:

3.91. PROPOSITION. Let σ be a positive definite form of dimension 11. Then $\sigma < \operatorname{Sim}(2^6\langle 1\rangle)$.

The Case of 2^7. Now $\rho(2^7) = 16$. The periodicity handles $\dim\sigma = 16, 15, 14, 13$. So we need only consider $\dim\sigma = 12$. If we can show that every σ of dimension 12 is $< \operatorname{Sim}(2^7\langle 1\rangle)$, then we need never consider dimension 12 (or less) again. Thus we may always use the periodicity.

But for $\dim\sigma = 12$, write $\sigma = 8\langle 1\rangle \perp \tau$, the periodicity implies $\sigma < \operatorname{Sim}(2^7\langle 1\rangle)$ if and only if $\tau < \operatorname{Sim}(q)$ where $\dim q = 8$. But τ represents 1, so $\langle 1, a, b\rangle < \tau$ for some $a, b \in \mathbb{Q}^+$, and hence, by Proposition 3.51(iii), $\langle 1, a, b, ab\rangle \mid q$. Applying Proposition 3.45(4), we obtain $q \simeq 8\langle 1\rangle$. We may now invoke Proposition 3.61 to see that since $\dim\tau = 4$, $\tau < \operatorname{Sim}(8\langle 1\rangle)$ is always possible.

We summarize all the major results from Sections 3.15 on.

3.92. THEOREM. Let σ be a positive definite form over $\mathbb{Q}$, d equal the discriminant of σ, and $m \geq 3$.

(A) If $\dim\sigma < 2m$, then $\sigma < \operatorname{Sim}(2^m\langle 1\rangle)$.

(B) If $2m \leq \dim\sigma \leq \rho(2^m)$, then:

(i) if $m \equiv 0 \pmod 4$, $\rho(2^m) = 2m+1$ and

(a) $\dim\sigma = 2m+1$; $\sigma < \operatorname{Sim}(2^m\langle 1\rangle)$ if and only if $s_p(\sigma) = 1$ for all primes p;

(b) $\dim\sigma = 2m$; if $d = 1$, $\sigma < \mathrm{Sim}(2^m\langle 1\rangle)$ always; if $d \neq 1$, $\sigma < \mathrm{Sim}(2^m\langle 1\rangle)$ if and only if there is a $b \in \mathbb{Q}^+$ with $s_p(\sigma) = (d,-b)_p$ for all primes p;

(ii) if $m \equiv 1 \pmod 4$, $\rho(2^m) = 2m$; $\dim\sigma = 2m$; $\sigma < \mathrm{Sim}(2^m\langle 1\rangle)$ if and only if there are $a, b \in \mathbb{Q}^+$ with $\sigma \simeq 2(m-1)\langle 1\rangle \perp \langle a,b\rangle$ if and only if $s_p(\sigma) = (-d,a)_p$ for some $a \in \mathbb{Q}^+$, and all primes p;

(iii) if $m \equiv 2 \pmod 4$, $\rho(2^m) = 2m$; $\dim\sigma = 2m$; $\sigma < \mathrm{Sim}(2^m\langle 1\rangle)$ if and only if $d = 1$;

(iv) if $m \equiv 3 \pmod 4$, $\rho(2^m) = 2m+2$;

(a) $\dim\sigma = 2m+2$, $\sigma < \mathrm{Sim}(2^m\langle 1\rangle)$ if and only if $d = 1$ and $s_p(\sigma) = 1$ for all primes p;

(b) $\dim\sigma = 2m+1$, $\sigma < \mathrm{Sim}(2^m\langle 1\rangle)$ if and only if $s_p(\sigma) = (-1,d)_p$ for all primes p;

(c) $\dim\sigma = 2m$, $\sigma < \mathrm{Sim}(2^m\langle 1\rangle)$ if and only if there is an $a \in \mathbb{Q}^+$ with $s_p(\sigma) = (d,-a)_p(a,-1)_p$ for every prime p.

§3.18. COMBINING THE ALGEBRAIC SOLUTION WITH COMBINATORIAL FACTS

One of the strongest combinatorial facts we discovered was Theorem 2.20. It, combined just with the algebraic restriction on the number of variables in an orthogonal design of order n, was enough to eliminate many $\rho(n)$-tuples as the types of orthogonal designs in order n. We now combine it with the new algebraic information to eliminate more tuples (even those of length less than $\rho(n)$).

We illustrate, by some examples, how the two types of information may be combined.

3.93. EXAMPLE. We want to know if it is possible for an orthogonal design of type $(1,1,17)$ to exist in order 20. Since $\rho(20) = 4$, we get no contradiction just from the number of variables involved. Since $17 = 1^2+4^2$, we know there is a rational family of type $[1,1,17]$ in order 20. Now use Corollary 2.21 to note that if an orthogonal design of type $(1,1,17)$ exists in order 20, then there is an orthogonal design of type $(1,1,1,17)$

in order 20. Hence there is a rational family of type $[1, 1, 1, 17]$ in order 20 and thus, by Theorem 3.52A, in order 4. Thus $\langle 1, 1, 1, 17\rangle \simeq \langle 1, 1, 1, 1\rangle$, which is a contradiction since the discriminants are not equal. So there is no orthogonal design of type $(1, 1, 17)$ in order 20.

Exactly the same procedure may be used to show that there is no orthogonal design of type $(7, 12)$ in order 20.

Similarly there are no orthogonal designs of type $(1, 1, 1, 1, 1, 66)$, $(1, 1, 1, 1, 1, 1, 65)$ or $(1, 1, 1, 1, 1, 1, 1, 64)$ in order 72.

These examples indicate how we may modify Theorem 3.92. The examples indicate how the proof should be constructed and so shall be omitted.

3.94. THEOREM. Let $n \equiv 0 \pmod 4$, $n = 2^m b$, b odd. Let $\Delta = (a_1, \ldots, a_\ell)$ where $\ell < \rho(n)$ and $\sum_{i=1}^{\ell} a_i = n-1$. The following give necessary conditions for Δ to be the type of an orthogonal design in order n. (Let $\sigma = \langle a_1, \ldots, a_\ell\rangle$; $d = \operatorname{disc}\sigma$.)

(A) $m = 2$. $\ell = 2$; $(-1, a_1a_2)_p(a_1, a_2)_p = 1$ for all primes p.
$\ell = 3$; $a_1a_2a_3$ is a square, and $1 = (a_1, a_2)_p(-1, a_1a_2)_p$ for all primes p.

(B) $m \geq 3$.

(i) $m \equiv 0 \pmod 4$; $\ell = 2m$; $s_p(\sigma) = 1$ for all primes p.
$\ell = 2m-1$; if $d \neq 1$, then there is a $b \in \mathbb{Q}^+$ with $s_p(\sigma) = (d, -b)_p$ for all primes p.

(ii) $m \equiv 1 \pmod 4$; $\ell = 2m-1$; there is a $b \in \mathbb{Q}^+$ with $s_p(\sigma) = (-d, b)_p$ for all primes p.

(iii) $m \equiv 2 \pmod 4$; $\ell = 2m-1$; $d = 1$.

(iv) $m \equiv 3 \pmod 4$; $\ell = 2m+1$; $d = 1$ and $s_p(\sigma) = 1$ for all primes p.
$\ell = 2m$; $s_p(\sigma) = (-1, d)_p$ for all primes p.
$\ell = 2m-1$; there is an $a \in \mathbb{Q}^+$ with $s_p(\sigma) = (d, -a)_p(a, -1)_p$ for all primes p.

3.95. EXAMPLE

(A) As we have seen, (A) eliminates $(7, 12)$ and $(1, 1, 17)$ in order 20.

(B) (i) Consider whether there is an orthogonal design of type $(1,1,1,1,3,3,36)$ in order $48 = 3.16$. This is not eliminated by Theorem 3.92(i)(b) since if $\sigma = 5\langle 1\rangle \perp \langle 3,3,36\rangle$, $\operatorname{disc}\sigma = 1$. But $s_3(\sigma) = (3,3)_3 = (-1,3)_3 = -1$. So it is eliminated by Theorem 3.94(B)(i) above.

(ii) We have not found anything yet that this eliminates.

(iii) This one is easy to use; for example, there is no orthogonal design of type $(2,2,2,2,2,2,2,2,2,2,43)$ in order 64.

(iv) Parts of this are easy to use, especially $\ell = 2m+1$, $\ell = 2m$; the other part is not so easy.

We have not explored yet which designs are eliminated (in general) by this theorem. Later, when we consider the possibilities of orthogonal designs in various orders, we shall make use of these theorems.

In this chapter, then, we have dealt at some length with finding necessary conditions for the existence of orthogonal designs in various orders. The conditions, of course, then amount to eliminating various tuples as the types of orthogonal designs. In general, what we have found is (algebraically) if the length of tuple is "small" compared to $\rho(n)$, the tuple exists as the type of a rational family. Hence, in general, we have no algebraic conditions on small tuples. On the other hand, the combinatorial results tell us to exercise care when the orthogonal design we are after is almost full (that is, few zeros per row).

In the next chapter we take a more positive (depending on your point of view) approach. We give many constructions and examples of orthogonal designs and give some indication of the scope of possibilities that exist.

CHAPTER 4
GENERAL CONSTRUCTIONS FOR ORTHOGONAL DESIGNS

In previous chapters we have studied some necessary conditions for the existence of orthogonal designs. We now turn to the task of actually constructing such designs. The ideas and methods we use are quite varied, and many have been used in the construction of Hadamard matrices.

There is one unifying theme in the constructions presented in this chapter. They revolve, in the main, around finding circulant matrices with prescribed properties or discovering the obstructions to finding such matrices. There are several methods of obtaining the appropriate collections of circulants. The ways they may be used often depend on how we obtained them. In general, the more control we attempt to exert on the internal structure of the circulants, the more interesting the ways we can use them.

§4.1. SOME ORTHOGONAL DESIGNS EXIST

Proposition 1.5 actually gives a construction for orthogonal designs. We review that proposition and add a remark about uniqueness in the following.

4.1. THEOREM. There are orthogonal designs of type (1) in order 1, (1,1) in order 2, (1,1,1,1) in order 4 and (1,1,1,1,1,1,1,1) in order 8. These are equivalent under the equivalence operations

(a) interchange rows or columns;

(b) multiply rows or columns by -1;

(c) replace any variable by its negative throughout the design;

to one of the arrays of appropriate order in Display 4.1.

$$[x], \begin{bmatrix} x & y \\ y & -x \end{bmatrix}, \begin{bmatrix} a & b & c & d \\ -b & a & d & -c \\ -c & -d & a & b \\ -d & c & -b & a \end{bmatrix} \text{ or } \begin{bmatrix} a & b & c & d \\ -b & a & -d & c \\ -c & d & a & -b \\ -d & -c & b & a \end{bmatrix},$$

$$\left[\begin{array}{cccc|cccc} a & b & c & d & e & f & g & h \\ -b & a & d & -c & f & -e & -h & g \\ -c & -d & a & b & g & h & -e & -f \\ -d & c & -b & a & h & -g & f & -e \\ \hline -e & -f & -g & -h & a & b & c & d \\ -f & e & -h & g & -b & a & -d & c \\ -g & h & e & -f & -c & d & a & -b \\ -h & -g & f & e & -d & -c & b & a \end{array}\right].$$

Display 4.1

Proof. By tedious systematic elimination.

Here we leave completely the question of equivalence of orthogonal designs, except to say that Lakein and Wallis [1] have briefly considered inequivalence of Baumert-Hall arrays of small order (see §4.4 for definition), and Hain [1] conjectured and Eades [4] established that there are exactly two equivalence classes of circulant weighing matrices of order 13, but otherwise the question is completely open.

We believe equivalence of orthogonal designs is an area worthy of study. We refer any interested reader to the work of M. Hall, W. D. Wallis and others described in J. Cooper [4], J. Wallis [25, pp. 408-425], Basil Gordon [1] on equivalence of Hadamard matrices.

In Chapter 1 we gave a construction for H-R families (see Theorem 1.6). It is possible to generalize that result to orthogonal designs.

4.2. THEOREM. If there exists an orthogonal design of type $(u_1, u_2, \ldots, u_s)$ in order n, then there exists an orthogonal design of type

(i) $(u_1, u_2, \ldots, u_{s-1}, u_s, u_s)$ in order $2n$ with $s+1$ variables,

(ii) $(u_1, u_2, \ldots, u_{s-1}, u_s, u_s, u_s)$ in order $4n$ with $s+2$ variables,

(iii) $(u_1, u_2, \ldots, u_{s-1}, u_s, u_s, u_s, u_s, u_s)$ in order $8n$ with $s+4$ variables,

(iv) $(u_1, u_2, \ldots, u_{s-1}, u_s, u_s, u_s, u_s, u_s, u_s, u_s, u_s, u_s)$ in order $16n$ with $s+8$ variables.

Proof. In each case we replace each of the first $s-1$ variables by $x_i I_m$, where $m = 2, 4, 8, 16$, respectively. In cases (i), (ii), (iii) and (iv) the last variable is replaced by

$$\begin{bmatrix} x & y \\ y & -x \end{bmatrix}, \quad \begin{bmatrix} x & y & z & 0 \\ y & -x & 0 & -z \\ z & 0 & -x & y \\ 0 & -z & y & x \end{bmatrix}, \quad X \text{ and } W,$$

respectively, where X and W are given in Display 4.2.

$$W = \begin{bmatrix} X & Y \\ Y^t & Z \end{bmatrix} = \left[\begin{array}{cccc|cccc|cccc|cccc}
x & y & z & 0 & a & 0 & 0 & -b & c & 0 & 0 & d & 0 & -e & f & 0 \\
y & -x & 0 & -z & 0 & -a & b & 0 & 0 & -c & -d & 0 & e & 0 & 0 & -f \\
z & 0 & -x & y & 0 & -b & -a & 0 & 0 & d & -c & 0 & -f & 0 & 0 & -e \\
0 & -z & y & x & b & 0 & 0 & a & -d & 0 & 0 & c & 0 & f & e & 0 \\
\hline
a & 0 & 0 & b & -x & y & z & 0 & 0 & -e & f & 0 & -c & 0 & 0 & -d \\
0 & -a & -b & 0 & y & x & 0 & -z & e & 0 & 0 & -f & 0 & c & d & 0 \\
0 & b & -a & 0 & z & 0 & x & y & -f & 0 & 0 & -e & 0 & -d & c & 0 \\
-b & 0 & 0 & a & 0 & -z & y & -x & 0 & f & e & 0 & d & 0 & 0 & -c \\
\hline
c & 0 & 0 & -d & 0 & e & -f & 0 & -x & y & z & 0 & a & 0 & 0 & -b \\
0 & -c & d & 0 & -e & 0 & 0 & f & y & x & 0 & -z & 0 & -a & b & 0 \\
0 & -d & -c & 0 & f & 0 & 0 & e & z & 0 & x & y & 0 & -b & -a & 0 \\
d & 0 & 0 & c & 0 & -f & -e & 0 & 0 & -z & y & -x & b & 0 & 0 & a \\
\hline
0 & e & -f & 0 & -c & 0 & 0 & d & a & 0 & 0 & b & x & y & z & 0 \\
-e & 0 & 0 & f & 0 & c & -d & 0 & 0 & -a & -b & 0 & y & -x & 0 & -z \\
f & 0 & 0 & e & 0 & d & c & 0 & 0 & b & -a & 0 & z & 0 & -x & y \\
0 & -f & -e & 0 & -d & 0 & 0 & -c & -b & 0 & 0 & a & 0 & -z & y & x
\end{array}\right]$$

Display 4.2

4.3. COROLLARY. There exists an orthogonal design of type $(1, 1, 1, \ldots, 1)$ with $\rho(n)$ variables in order $n = 2^a . b$ (b odd).

Proof. This follows immediately from Theorem 1.6.

We now note that orthogonal designs of the same order but different types can be easily made by setting variables equal to zero or to one another. For easy reference, this is stated in the following lemma:

4.4. LEMMA. If A is an orthogonal design of order n and type $(u_1, \ldots, u_s)$ on the variables $x_1, \ldots, x_s$, then there is an orthogonal design of order n and types $(u_1, \ldots, u_i+u_j, \ldots, u_s)$ and $(u_1, \ldots, u_{j-1}, u_{j+1}, \ldots, u_s)$ on the $s-1$ variables $x_1, \ldots, \hat{x}_j, \ldots, x_s$.

Proof. Set the variable $\hat{x}_j = x_i = x_j$ in the first case and $\hat{x}_j = 0$ in the second.

4.5. COROLLARY. For any integer n, an orthogonal design of order n and type $(u_1, \ldots, u_s)$ exists if $\sum_{i=1}^{s} u_i \le \rho(n)$, and so if $n = 2, 4, 8$, orthogonal designs of order n and any type exist.

Proof. The proof follows by using the designs of type $(1, 1, \ldots, 1)$ in Corollary 4.3.

4.6. EXAMPLE

$$\begin{bmatrix} x & y & z & w \\ -y & x & w & -z \\ -z & -w & x & y \\ -w & z & -y & x \end{bmatrix}$$

is an orthogonal design of type $(1,1,1,1)$ in order 4. We can make designs of type $(1,1,2)$ by (for example) setting $z = w = v$ and of type $(1,1,1)$ by (for example) setting $y = 0$.

$$\begin{bmatrix} x & y & v & v \\ -y & x & v & -v \\ -v & -v & x & y \\ -v & v & -y & x \end{bmatrix}$$

is of type $(1,1,2)$, and

$$\begin{bmatrix} x & 0 & z & w \\ 0 & x & w & -z \\ -z & -w & x & 0 \\ -w & z & 0 & x \end{bmatrix}$$

is of type $(1,1,1)$.

Another method of finding orthogonal designs, already foreshadowed by the proof of Theorem 4.2, is to replace variables by suitable matrices of variables. Similar methods were first used extensively by J. Wallis (see Wallis [25]) in constructing Hadamard matrices. The results now quoted are due to Joan Murphy Geramita, P. J. Robinson and ourselves.

The next lemma is given for easy reference. The remaining lemmas of this section are of far-reaching consequences and great power in constructing orthogonal designs.

4.7. LEMMA. If A is an orthogonal design of order n and type $(u_1, \ldots, u_s)$ on $x_1, \ldots, x_s$, then there exists an orthogonal design of order mn and of type $(u_1, \ldots, u_s)$ on $x_1, \ldots, x_s$ for any integer $m \geq 1$.

Proof. Replace each variable x_i of A by $x_i I_m$.

The next result is most useful, and part of it first appeared in Geramita-Geramita-Wallis [1]. It was the start of what is now amicable orthogonal designs (see Chapter 5).

4.8. LEMMA. If there is an orthogonal design of type (a, b) in order n, there is an orthogonal design of type (a, a, b, b) in order 2n, of type (a, a, 2a, b, b, 2b) in order 4n, and (a, a, 2a, 2a, 2a, 8b) and (a, 2a, 2a, 3a, 2b, 6b) in order 8n.

Proof. To obtain the required designs in order 2n, 4n and 8n, respectively, the two variables of the design of type (a, b) in order n should be replaced by the matrices of commuting variables (we use $\bar{x}_i$ for $-x_i$ and $\bar{y}_j$ for $-y_j$) given in Display 4.3.

$$X_1 = \begin{bmatrix} x_1 & x_2 \\ \bar{x}_2 & x_1 \end{bmatrix}, \quad \begin{bmatrix} y_1 & y_2 \\ y_2 & \bar{y}_1 \end{bmatrix} = Y_1$$

$$X_2 = \begin{bmatrix} x_1 & x_2 & x_3 & x_3 \\ \bar{x}_2 & x_1 & x_3 & \bar{x}_3 \\ \bar{x}_3 & \bar{x}_3 & x_1 & x_2 \\ \bar{x}_3 & x_3 & \bar{x}_2 & x_1 \end{bmatrix}, \quad \begin{bmatrix} y_1 & y_2 & y_3 & y_3 \\ y_2 & \bar{y}_1 & y_3 & \bar{y}_3 \\ y_3 & y_3 & \bar{y}_2 & \bar{y}_1 \\ y_3 & \bar{y}_3 & \bar{y}_1 & y_2 \end{bmatrix} = Y_2$$

$$X_3 = \begin{bmatrix} x_0 & x_1 & x_2 & x_3 & x_2 & x_4 & x_3 & x_4 \\ \bar{x}_1 & x_0 & x_3 & \bar{x}_2 & x_4 & \bar{x}_2 & x_4 & \bar{x}_3 \\ \bar{x}_2 & \bar{x}_3 & x_0 & x_1 & x_3 & \bar{x}_4 & \bar{x}_2 & x_4 \\ \bar{x}_3 & x_2 & \bar{x}_1 & x_0 & \bar{x}_4 & \bar{x}_3 & x_4 & x_2 \\ \bar{x}_2 & \bar{x}_4 & \bar{x}_3 & x_4 & x_0 & x_1 & x_2 & \bar{x}_3 \\ \bar{x}_4 & x_2 & x_4 & x_3 & \bar{x}_1 & x_0 & \bar{x}_3 & \bar{x}_2 \\ \bar{x}_3 & \bar{x}_4 & x_2 & \bar{x}_4 & \bar{x}_2 & x_3 & x_0 & x_1 \\ \bar{x}_4 & x_3 & \bar{x}_4 & \bar{x}_2 & x_3 & x_2 & \bar{x}_1 & x_0 \end{bmatrix}, \quad y_1 \begin{bmatrix} 1 & 1 & - & 1 & - & - & 1 & - \\ 1 & - & 1 & 1 & - & 1 & - & - \\ - & 1 & - & - & - & 1 & - & - \\ 1 & 1 & - & 1 & 1 & 1 & - & 1 \\ - & - & - & 1 & 1 & 1 & 1 & - \\ - & 1 & 1 & 1 & 1 & - & - & - \\ 1 & - & - & - & 1 & - & - & - \\ - & - & - & 1 & - & - & - & 1 \end{bmatrix} = Y_3$$

and

$$X_4 = \begin{bmatrix} x_0 & x_2 & x_3 & x_3 & x_3 & x_2 & \bar{x}_1 & \bar{x}_1 \\ \bar{x}_2 & x_0 & x_3 & \bar{x}_3 & x_2 & \bar{x}_3 & \bar{x}_1 & x_1 \\ \bar{x}_3 & \bar{x}_3 & x_0 & x_2 & \bar{x}_1 & \bar{x}_1 & \bar{x}_2 & \bar{x}_3 \\ \bar{x}_3 & x_3 & \bar{x}_2 & x_0 & \bar{x}_1 & x_1 & \bar{x}_3 & x_2 \\ \bar{x}_3 & \bar{x}_2 & x_1 & x_1 & x_0 & x_2 & x_3 & x_3 \\ \bar{x}_2 & x_3 & x_1 & \bar{x}_1 & \bar{x}_2 & x_0 & x_3 & \bar{x}_3 \\ x_1 & x_1 & x_2 & x_3 & \bar{x}_3 & \bar{x}_3 & x_0 & x_2 \\ x_1 & \bar{x}_1 & x_3 & \bar{x}_2 & \bar{x}_3 & x_3 & \bar{x}_2 & x_0 \end{bmatrix}, \quad \begin{bmatrix} y_2 & y_1 & \bar{y}_2 & \bar{y}_2 & \bar{y}_2 & y_1 & \bar{y}_2 & \bar{y}_2 \\ y_1 & \bar{y}_2 & y_2 & \bar{y}_2 & \bar{y}_1 & \bar{y}_2 & \bar{y}_2 & y_2 \\ \bar{y}_2 & y_2 & \bar{y}_1 & y_2 & \bar{y}_2 & \bar{y}_2 & \bar{y}_2 & y_1 \\ \bar{y}_2 & \bar{y}_2 & y_2 & y_1 & \bar{y}_2 & y_2 & \bar{y}_1 & \bar{y}_2 \\ \bar{y}_2 & \bar{y}_1 & \bar{y}_2 & \bar{y}_2 & \bar{y}_2 & y_1 & y_2 & y_2 \\ y_1 & \bar{y}_2 & \bar{y}_2 & y_2 & y_1 & y_2 & \bar{y}_2 & y_2 \\ \bar{y}_2 & \bar{y}_2 & \bar{y}_2 & \bar{y}_1 & y_2 & \bar{y}_2 & \bar{y}_1 & \bar{y}_2 \\ \bar{y}_2 & y_2 & y_1 & \bar{y}_2 & y_2 & y_2 & \bar{y}_2 & y_1 \end{bmatrix} = Y_4,$$

Display 4.3

respectively. This is possible because $X_i Y_i^t = Y_i X_i^t$, $i = 1, 2, 3, 4$.

4.9. COROLLARY. If there are orthogonal designs of type $(1, k)$, $1 \le k \le j$, in order n, then there are orthogonal designs of type $(1, m)$ in order $2n$ for $1 \le m \le 2j+1$. In particular, if there are orthogonal designs of type $(1, k)$, $1 \le k \le n-1$, in order n, then there are orthogonal designs of type $(1, m)$, $1 \le m \le 2^t n-1$, in order $2^t n$, t a positive integer.

4.10. EXAMPLE. Since there is an orthogonal design of type (1,1) in order 2, there exist, using Corollary 4.9, orthogonal designs (1, k), $1 \le k \le 2^t-1$, in every order 2^t, t a positive integer.

The following lemma is crucial to the powerful results on Hadamard matrices we will obtain later.

4.11. LEMMA. If there exists an orthogonal design in order n and of type $(s_1, s_2, \ldots, s_u)$, then there exist orthogonal designs of type

(i) $(e_1 s_1, e_2 s_2, \ldots, e_u s_u)$ where $e_i = 1$ or 2,

(ii) $(s_1, s_1, fs_2, \ldots, fs_u)$ where f = 1 or 2,

in order 2n.

<u>Proof.</u> (i) Replace each variable by

$$\begin{bmatrix} x_i & 0 \\ 0 & x_i \end{bmatrix} \text{ if } e_i = 1 \text{ and by } \begin{bmatrix} x_i & x_i \\ x_i & -x_i \end{bmatrix} \text{ if } e_i = 2.$$

(ii) Replace the variable x_1 by

$$\begin{bmatrix} x_0 & x_1 \\ -x_1 & x_0 \end{bmatrix}$$

and the variable x_i, $i \ne 1$, by

$$\begin{bmatrix} 0 & x_i \\ x_i & 0 \end{bmatrix} \text{ or } \begin{bmatrix} x_i & x_i \\ x_i & -x_i \end{bmatrix}$$

according as f is 1 or 2.

§4.2. SOME MATRIX RESULTS AND CONSTRUCTIONS FOR HADAMARD MATRICES, W(h, h), AND WEIGHING MATRICES, W(h, h-1)

One of the most useful constructive methods for orthogonal designs has been that using two or more circulant matrices. In this section we give some results about circulant matrices starting with the more general concept of type 1; then we develop some existence results.

First we give some definitions and elementary results. We use the following notation:

4.12. NOTATION. A (1, -1) matrix is a matrix whose only entries are +1 or -1. We use similar notation for a (0, 1, -1) matrix, (a, b, c) matrix, etc. We use J_n for the $n \times n$ matrix with every entry +1. (We shall sometimes drop the subscript if the order is obvious.)

4.13. DEFINITION

a) Let G be an additive abelian group of order t, and order the elements of G as $z_1, \ldots, z_t$. Let ψ and ϕ be two functions from G into a commutative ring. We define two matrices $M = (m_{ij})$ and $N = (n_{ij})$, of order t, as follows:

$$m_{ij} = \psi(z_j - z_i) \text{ and } n_{ij} = \phi(z_j + z_i).$$

M and N are called <u>type 1</u> and <u>type 2</u> matrices, respectively.

REMARK. The words "type i" used to describe these matrices leave out information: the way the elements of G are ordered and which functions ψ and ϕ are being used. One should say, e.g., in describing M, "type 1 with respect to the following ordering of G and the function ϕ"; however, this cumbersome phrase will be omitted since the ordering for G is usually understood and fixed, while the functions ψ and ϕ are usually explicit.

b) Let G be as above with its elements ordered as above. Let X be a subset of G, and suppose $0 \notin X$. If ψ and ϕ are defined by:

$$\psi(x) = \begin{cases} a, & x = 0 \\ b, & x \in X \\ c, & x \notin X \cup \{0\} \end{cases}, \quad \phi(x) = \begin{cases} d, & x = 0 \\ e, & x \in X \\ f, & x \notin X \cup \{0\}, \end{cases}$$

then M will be called the <u>type 1 (a, b, c) incidence matrix generated by X</u>, and N the <u>type 2 (d, e, f) incidence matrix generated by X</u>.

REMARK. If we drop the restriction that $0 \notin X$ and let

$$\psi(x) = \phi(x) = \begin{cases} 1 & \text{if } x \in X \\ -1 & \text{if } x \notin X, \end{cases}$$

we obtain the type i (i = 1, 2) <u>(1, -1) incidence matrix generated by X</u>, and if we let

$$\psi(x) = \phi(x) = \begin{cases} 1 & \text{if } x \in X \\ 0 & \text{if } x \notin X, \end{cases}$$

then we obtain the type i (i = 1,2) <u>(1,0) incidence matrix generated by X</u>.

Notice that these latter two "incidence" matrices are really special cases of Definition 4.13 b) where we let $a = b$ or $a = c$ depending on whether $0 \in X$ or $0 \notin X$.

4.14. EXAMPLE. Consider the field $\dfrac{Z_3[x]}{(x^2-x-1)} = GF(3^2)$. We order the elements $g_1 = 0$, $g_2 = 1$, $g_3 = 2$, $g_4 = x$, $g_5 = x+1$, $g_6 = x+2$, $g_7 = 2x$, $g_8 = 2x+1$, $g_9 = 2x+2$. Define the set

$$\begin{aligned} X &= \{y : y = z^2 \text{ for some } z \in GF(3^2),\ z \neq 0\} \\ &= \{x+1, 2, 2x+2, 1\}. \end{aligned}$$

Then the type 1 and type 2 (0,1,-1) incidence matrices generated by X are given by A and B, respectively:

$$A = \begin{bmatrix} 0 & 1 & 1 & -1 & 1 & -1 & -1 & -1 & 1 \\ 1 & 0 & 1 & -1 & -1 & 1 & 1 & -1 & -1 \\ 1 & 1 & 0 & 1 & -1 & -1 & -1 & 1 & -1 \\ -1 & -1 & 1 & 0 & 1 & 1 & -1 & 1 & -1 \\ 1 & -1 & -1 & 1 & 0 & 1 & -1 & -1 & 1 \\ -1 & 1 & -1 & 1 & 1 & 0 & 1 & -1 & -1 \\ -1 & 1 & -1 & -1 & -1 & 1 & 0 & 1 & 1 \\ -1 & -1 & 1 & 1 & -1 & -1 & 1 & 0 & 1 \\ 1 & -1 & -1 & -1 & 1 & -1 & 1 & 1 & 0 \end{bmatrix},$$

$$B = \begin{bmatrix} 0 & 1 & 1 & -1 & 1 & -1 & -1 & -1 & 1 \\ 1 & 1 & 0 & 1 & -1 & -1 & -1 & 1 & -1 \\ 1 & 0 & 1 & -1 & -1 & 1 & 1 & -1 & -1 \\ -1 & 1 & -1 & -1 & -1 & 1 & 0 & 1 & 1 \\ 1 & -1 & -1 & -1 & 1 & -1 & 1 & 1 & 0 \\ -1 & -1 & 1 & 1 & -1 & -1 & 1 & 0 & 1 \\ -1 & -1 & 1 & 0 & 1 & 1 & -1 & 1 & -1 \\ -1 & 1 & -1 & 1 & 1 & 0 & 1 & -1 & -1 \\ 1 & -1 & -1 & 1 & 0 & 1 & -1 & -1 & 1 \end{bmatrix},$$

If the additive abelian group in Definition 4.13 is the cyclic group Z_t of integers modulo t with the usual ordering $0,1,2,\ldots,t-1$, then the type 1 and type 2 matrices are very familiar.

4.15. DEFINITION

a) A <u>circulant matrix</u> $A = (a_{ij})$ of order n is one for which

$a_{ij} = a_{1,j-i+1}$ where $j-i+1$ is reduced modulo n to $0,1,2,\ldots,n-1$. For example:

$$\begin{bmatrix} 1 & 2 & 3 & 4 \\ 4 & 1 & 2 & 3 \\ 3 & 4 & 1 & 2 \\ 2 & 3 & 4 & 1 \end{bmatrix}.$$

b) A set $D = \{x_1, x_2, \ldots, x_k\} \subseteq \{0,1,2,\ldots,n-1\}$ will be said to <u>generate a circulant</u> (1, -1) matrix if the first row of the circulant matrix is defined by

$$a_{1x} = \begin{cases} +1, & x \in D, \\ -1, & x \notin D. \end{cases}$$

c) A matrix $A = (a_{ij})$ of order n will be called <u>back circulant</u> if $a_{ij} = a_{1,i+j-1}$ where $i+j-1$ is reduced modulo n. For example:

$$\begin{bmatrix} 1 & 2 & 3 & 4 \\ 2 & 3 & 4 & 1 \\ 3 & 4 & 1 & 2 \\ 4 & 1 & 2 & 3 \end{bmatrix}.$$

REMARKS

i) Any type 1 matrix defined on Z_t (with its usual ordering) is circulant since:

$$m_{ij} = \psi(j-i) = \psi(j-i+1-1) = m_{1,j-i+1}.$$

ii) Any type 2 matrix defined on Z_t (with its usual ordering) is back circulant since:

$$n_{ij} = \phi(i+j) = \phi(i+j-1+1) = n_{1,i+j-1}.$$

Clearly, any circulant matrix is a type 1 matrix, and any back circulant matrix is a type 2 matrix. In any case:

<u>A type 1 matrix is analogous to a circulant matrix;</u>

<u>a type 2 matrix is analogous to a back circulant matrix.</u>

Thus, all propositions proved about type 1 and type 2 matrices have corollaries about circulant and back circulant matrices.

4.16. LEMMA. Suppose G is an additive abelian group of order v with elements ordered $z_1, z_2, \ldots, z_v$. Let ϕ, ψ, and μ be functions from G to some commutative ring R.

Define $A = (a_{ij}), B = (b_{ij})$ and $C = (c_{ij})$ by $a_{ij} = \phi(z_j - z_i)$, $b_{ij} = \psi(z_j - z_i)$ and $c_{ij} = \mu(z_j + z_i)$. Then (i) $C^t = C$, (ii) $AB = BA$, (iii) $AC^t = CA^t$.

Proof

(i) $c_{ij} = \mu(z_j + z_i) = \mu(z_i + z_j) = c_{ji}$.

(ii) $(AB)_{ij} = \sum_{g \in G} \phi(g - z_i)\psi(z_j - g)$.

Putting $h = z_i - z_j - g$, it is clear that as g ranges through G, so does h, and the above expression becomes

$$\sum_{h \in G} \phi(z_j - h)\psi(h - z_i)$$
$$= \sum_{n \in G} \psi(h - z_i)\phi(z_j - h)$$

(since R is commutative); this is $(BA)_{ij}$.

$$\text{(iii)}\quad (AC^t)_{ij} = \sum_{g \in G} \phi(g - z_i)\mu(z_j + g)$$
$$= \sum_{h \in G} \phi(h - z_j)\mu(z_i + h) \qquad (h = z_j + g - z_i)$$
$$= (CA^t)_{ij}.$$

4.17. COROLLARY. If X and Y are type 1 matrices and Z is a type 2 matrix, all defined on the same abelian group with a fixed ordering, then (i) $Z^t = Z$, (ii) $XY = YX$, (iii) $XZ^t = ZX^t$.

4.18. LEMMA. i) If X is a type i, $i = 1, 2$, matrix, then so is X^t.

ii) If X and Y are type i matrices, $i = 1, 2$, both defined on the same abelian group with a fixed ordering, then so is $X+Y$ and $X-Y$.

Proof. ia) If $X = (x_{ij})$ is type 2 defined using a function ϕ, then $x_{ij} = \phi(z_i + z_j) = \phi(z_j + z_i) = x_{ji}$. So X^t is also defined as type 2 using ϕ.

ib) If $X = (x_{ij})$ is of type 1 defined using ψ, then $x_{ij} = \psi(z_j - z_i)$. Now define a type 1 matrix $M = (m_{ij})$ using μ, where $\mu(x) = \psi(-x)$. Then $m_{ij} = \mu(z_j - z_i) = \psi(z_i - z_j) = x_{ji}$. Thus $M = X^t$.

iia) If X and Y are type 2 defined using ϕ_1 and ϕ_2, then the type 2 matrices defined using $\phi_1+\phi_2$ and $\phi_1-\phi_2$, respectively, give $X+Y$ and $X-Y$, respectively.

iib) Similarly, if X and Y are type 1 defined using ψ_1 and ψ_2, define type 1 matrices using $\mu_1+\mu_2$ and $\mu_1-\mu_2$, respectively, to obtain $X+Y$ and $X-Y$, respectively, where $\mu_i(x) = \psi_i(-x)$.

Combining this lemma with Corollary 4.17 gives:

4.19. COROLLARY. i) If X and Y are type 1 matrices, defined on the same abelian group with a fixed ordering, then

$$XY = YX \qquad XY^t = Y^tX$$

$$X^tY = YX^t \qquad X^tY^t = Y^tX^t.$$

ii) If P is a type 1 matrix and Q is a type 2 matrix, both defined on the same abelian group with a fixed ordering, then

$$PQ^t = QP^t \qquad P^tQ^t = QP$$

$$PQ = Q^tP^t \qquad P^tQ = Q^tP.$$

We now summarize the most used results for circulants.

4.20. COROLLARY. i) Two circulant matrices of the same order commute.

ii) A back circulant matrix is symmetric.

iii) The product of a back circulant matrix with a circulant matrix of the same order is symmetric. In particular, if B is back circulant and A is circulant,

$$AB^t = BA^t.$$

REMARK. From now on, whenever we refer to a collection of type 1 and type 2 matrices all defined on the same abelian group G, we shall assume that the ordering of the group elements has been fixed.

4.21. LEMMA. i) Let X and Y be type 2 (d, e, f) incidence matrices generated by subsets A and B of an additive abelian group G. Suppose, further, that

$$a \in A \Rightarrow -a \in A \quad \text{and} \quad b \in B \Rightarrow -b \in B.$$

Then,

$$XY = YX \quad \text{and} \quad XY^t = YX^t.$$

ii) The same result holds if X and Y are type 1.

Proof. i) Since X and Y are symmetric, we only have to prove that $XY^t = YX^t$.

Suppose $X = (x_{ij})$ and $Y = (y_{ij})$ are defined by

$$x_{ij} = \phi(z_i+z_j),$$
$$y_{ij} = \psi(z_i+z_j),$$

where $z_1, z_2, \ldots$ are the elements of G.

Then

$$\begin{aligned}
(XY^t)_{ij} &= \sum_k \phi(z_i+z_k)\psi(z_k+z_j) \\
&= \sum_k \phi(-z_i-z_k)\psi(z_k+z_j) && \text{since } a \in A \Rightarrow -a \in A \\
&= \sum_\ell \phi(z_j+z_\ell)\psi(-z_\ell-z_i-z_j+z_j) && z_\ell = -z_k-z_i-z_j \\
&= \sum_\ell \phi(z_j+z_\ell)\psi(z_\ell+z_i) && \text{since } b \in B \Rightarrow -b \in B \\
&= (YX^t)_{ij}.
\end{aligned}$$

ii) The additional hypotheses on A and B force X and Y to be symmetric. The proof, then, is similar to i), and we leave it to the reader as an easy exercise.

4.22. LEMMA. Let $R = (r_{ij})$ be the permutation matrix of order n, defined on an additive abelian group $G = \{g_i\}$ of order n by

$$r_{k,j} = \begin{cases} 1 & \text{if } g_k+g_j = 0 \\ 0 & \text{otherwise}. \end{cases}$$

1) If M is a type 1 matrix defined on G, then MR is a type 2 matrix defined on G.

2) If N is a type 2 matrix defined on G, then NR is a type 1 matrix defined on G.

3) If X is a subset of G where $0 \notin X$ and M is the type 1 (a,b,c) incidence matrix generated by X, then MR is the type 2 (a,b,c) incidence matrix generated by $-X$.

4) If X is as in 3) and N is the type 2 (a,b,c) incidence matrix generated by X, then NR is the type 1 (a,b,c) incidence matrix generated by $-X$.

<u>Proof.</u> 1) Let $M = (m_{ij})$ be defined by $m_{ij} = \psi(g_j - g_i)$, and let $\mu(x) = \psi(-x)$. We claim that MR is the type 2 matrix defined by μ, for

$$(MR)_{ij} = \sum_k m_{ik} r_{kj} = m_{i\ell}, \text{ where } g_\ell + g_j = 0,$$

$$= \psi(g_\ell - g_i) = \psi(-g_j - g_i) = \mu(g_j + g_i).$$

2) follows from a similar argument.

3) and 4) are clear from 1) and 2) and the relationship between ψ and μ.

4.23. COROLLARY. Let G be an additive abelian group and X a subset of G where $0 \notin X$. Let M be the type 1 (a,b,c) incidence matrix generated by X, and N the type 2 (a,b,c) incidence matrix generated by $-X$.

Then

$$MM^t = NN^t.$$

<u>Proof.</u> Lemma 4.22 gives that $M = NR$, where R is the permutation matrix appropriate to G. The corollary follows since $RR^t = I$.

SUPPLEMENTARY DIFFERENCE SETS, THEIR INCIDENCE MATRICES AND THEIR USES AS SUITABLE MATRICES

4.24. DEFINITION. Let $S_1, S_2, \ldots, S_n$ be subsets of V, an additive abelian group of order v. Let $|S_i| = k_i$ and $S_i = \{s_{i1}, s_{i2}, \ldots, s_{ik_i}\}$. If the equation

$$g = s_{ij} - s_{im}$$

has exactly λ solutions for each nonzero element g of V, then $S_1, S_2, \ldots, S_n$ will be called n-$\{v; k_1, k_2, \ldots, k_n; \lambda\}$ <u>supplementary</u>

difference sets or sds. If $k_1 = k_2 = \ldots = k_n = k$, we write n-$\{v;k;\lambda\}$ sds.

4.25. LEMMA. Suppose $A_1, \ldots, A_n$ are the type 1 (0,1) incidence matrices generated by $S_1, \ldots, S_n$, where $S_1, \ldots, S_n$ are n-$\{v;k_1,k_2,\ldots,k_n;\lambda\}$ supplementary difference sets. Then

$$\sum_{i=1}^{n} A_i A_i^t = \left(\sum_{i=1}^{n} k_i - \lambda\right) I + \lambda J.$$

Proof. This follows from the definition by a simple counting argument. (See Wallis [25, p.290] for a fuller proof.)

4.26. EXAMPLE. $S_1 = \{0,2,3\}$ and $S_2 = \{0,1,4\}$ are 2-$\{5;3;3\}$ sds in Z_5. Their type 1 (1,0) incidence matrices are the circulants

$$A_1 = \begin{bmatrix} 1 & 0 & 1 & 1 & 0 \\ 0 & 1 & 0 & 1 & 1 \\ 1 & 0 & 1 & 0 & 1 \\ 1 & 1 & 0 & 1 & 0 \\ 0 & 1 & 1 & 0 & 1 \end{bmatrix} \quad \text{and} \quad A_2 = \begin{bmatrix} 1 & 1 & 0 & 0 & 1 \\ 1 & 1 & 1 & 0 & 0 \\ 0 & 1 & 1 & 1 & 0 \\ 0 & 0 & 1 & 1 & 1 \\ 1 & 0 & 0 & 1 & 1 \end{bmatrix}$$

which satisfy

$$A_1 A_1^t + A_2 A_2^t = 3I + 3J.$$

We observe that for these subsets of Z_5, $x \in S_i \Rightarrow -x \in S_i$. So, if R is the back diagonal matrix of order 5 (see Lemma 4.22), we see $A_1 R = B_1$ and $A_2 R = B_2$ are the type 2 (1,0) incidence matrices generated by S_1 and S_2, respectively.

$$B_1 = A_1 R = \begin{bmatrix} 1 & 0 & 1 & 1 & 0 \\ 0 & 1 & 1 & 0 & 1 \\ 1 & 1 & 0 & 1 & 0 \\ 1 & 0 & 1 & 0 & 1 \\ 0 & 1 & 0 & 1 & 1 \end{bmatrix} \quad \text{and} \quad B_2 = A_2 R = \begin{bmatrix} 1 & 1 & 0 & 0 & 1 \\ 1 & 0 & 0 & 1 & 1 \\ 0 & 0 & 1 & 1 & 1 \\ 0 & 1 & 1 & 1 & 0 \\ 1 & 1 & 1 & 0 & 0 \end{bmatrix}.$$

Using Lemma 4.21 (or directly), we obtain

$$B_1 B_2 = B_2 B_1 \quad \text{and} \quad B_1 B_2^t = B_2 B_1^t.$$

Also

$$A_1 A_2 = A_2 A_1 \quad \text{and} \quad A_1 A_2^t = (B_1 R)(B_2 R)^t = B_2 B_1^t.$$

We also observe that

$$A_1B_2^t = B_2A_1^t \text{ and } A_iA_i^t = B_iB_i^t.$$

4.27. LEMMA. Let $A_1,\ldots,A_n$ be the type 1 (1,0) incidence matrices generated by $S_1,\ldots,S_n$ where $S_1,\ldots,S_n$ are n-$\{v;k_1,\ldots,k_n;\lambda\}$ sds. Let $B_i = A_i - J$. Then

$$\sum_{i=1}^{n} B_iB_i^t = 4\left(\sum_{j=1}^{n} k_j-\lambda\right)I + \left[nv-4\left(\sum_{j=1}^{n} k_j-\lambda\right)\right]J.$$

We are constantly searching for (0, 1, -1) matrices to substitute for the variables in an orthogonal design. We shall be precise about what is needed.

4.28. DEFINITION. A set of m (0, 1, -1) matrices $A_1, A_2, \ldots, A_m$ of order n will be called <u>suitable matrices</u> for the orthogonal design of type $(s_1, s_2, \ldots, s_m)$ if

1) $A_iA_j^t = A_jA_i^t$, $1 \le i, j \le m$;

2) $\sum_{i=1}^{m} s_iA_iA_i^t = kI_n$.

4.29. THEOREM. Let S_1,S_2,S_3,S_4 be 4-$\{t;k_1,k_2,k_3,k_4;\sum_{j=1}^{4} k_j-t\}$ sds for which $x \in S_i \Rightarrow -x \in S_i$, and let A_1, A_2, A_3, A_4 be the type 1 (1, -1) incidence matrices of these sets. Then A_1, A_2, A_3, A_4 are suitable matrices for an orthogonal design of type (s,s,s,s) giving a weighing matrix of weight $4st$.

<u>Proof.</u> Since $x \in S_i \Rightarrow -x \in S_i$ we see that A_i, $i = 1,2,3,4$, are symmetric and commuting. Using Lemma 4.27, we have

$$\sum_{i=1}^{4} A_iA_i^t = 4\left(\sum_{j=1}^{4} k_j - \sum_{j=1}^{4} k_j+t\right)I + \left(4t-4\left(\sum_{j=1}^{4} k_j - \sum_{j=1}^{4} k_j+t\right)\right)J = 4tI.$$

In particular,

$$\sum_{i=1}^{4} sA_iA_i^t = 4stI.$$

If the variables of the orthogonal design are replaced by the A_i, $i = 1,2,3,4$, we have a weighing matrix of weight $4st$.

REMARK. If the orthogonal design of the theorem is of order $4s$, then the weighing matrix obtained will be of order $4st$ and weight $4st$, in other words, an Hadamard matrix of order $4st$. In this case the symmetric matrices of Theorem 4.29 are a special kind of what will be called Williamson matrices (see Definition 4.69).

SOME W(h,h) AND W(h,h-1)

4.30. DEFINITION. A matrix $A = I+S$ will be called <u>skew-type</u> if $S^t = -S$.

We recall the following:

4.31. DEFINITION. A $(0,1,-1)$ matrix $W = W(p,k)$ of order p satisfying

$$WW^t = kI_p$$

is called a <u>weighing matrix of order</u> p and <u>weight</u> k or simply a <u>weighing matrix</u>. A $W(p,p)$ is called an <u>Hadamard matrix</u>. A $W = W(p,k)$ for which $W^t = -W$ is called a <u>skew-weighing matrix</u>, and an Hadamard matrix $H = I+S$ for which $S^t = -S$ is called a <u>skew-Hadamard matrix</u>. A $W = W(p,p-1)$ satisfying $W^t = W$, $p \equiv 2 \pmod 4$ is called a <u>symmetric conference matrix</u>.

Weighing matrices have long been studied in order to find optimal solutions to the problem of weighing objects whose weights are small relative to the weights of the moving parts of the balance being used. It was shown by Raghavarao [1], [2] that if the variance of the errors in the weights obtained by individual weighings is σ^2 (it is assumed the balance is not biased and the errors are mutually independent and normal), then using a $W(p,k)$ to design an experiment to weigh p objects will give a variance of $\frac{\sigma^2}{k}$. Indeed, for an Hadamard matrix the variance is $\frac{\sigma^2}{p}$, which is optimal for $p \equiv 0 \pmod 4$, and for a symmetric conference matrix the variance is $\frac{\sigma^2}{p-1}$, which is optimal for $p \equiv 2 \pmod 4$.

Sloane and Harwitt [1] survey the application of weighing matrices to improve the performance of optical instruments such as spectrometers.

Spectrometers measure the intensity of a dispersed spectrum at a finite number (n, say) of wavelengths. According to Ibbett, et al [1], either one detector scans the screen, making the n measurements sequentially, or else the n measurements are made simultaneously by a detector with spatial resolution. The first method has the disadvantage of not being able to compensate for variations in the intensity of the signal, while the second approach suffers the disadvantage of a lower signal-to-noise ratio (Ibbett, et al [1]).

A modification can be made to the second system which improves the signal-to-noise ratio. This is achieved by using a weighing matrix as square mask, where 1 is clear, 0 is opaque and -1 is a mirror (180° phase shift). Again the variance of the estimates of the wavelengths made using a mask of weight is $1/n$ of the estimates when measured separately.

Sloane and Harwitt [11] also indicate that weighing designs are applicable to other problems of measurements (such as lengths, voltages, resistances, concentrations of chemicals, etc.) in which the measure of several objects is the sum (or a linear combination) of the individual measurements.

The following properties of Hadamard matrices and weighing matrices are easily proved.

4.32. LEMMA. Let $U = U(p_1, k_1)$ and $V = V(p_2, k_2)$ be weighing matrices. Then $W = U \times V$ is a weighing matrix of order p_1p_2 and weight k_1k_2.

4.33. COROLLARY. Since $\begin{bmatrix} 1 & 1 \\ 1 & - \end{bmatrix}$ is a $W(2,2)$, there are Hadamard matrices of order 2^t, t a positive integer.

4.34. LEMMA. Let p be a prime power. Then there is a $W = W(p+1, p)$ for which $W^t = (-1)^{\frac{1}{2}(p-1)}W$. If $p \equiv 3 \pmod 4$, then $W+I_p$ is a $W(p+1, p+1)$.

Proof. Let $a_0, a_1, \ldots, a_{p-1}$ be the elements of $GF(p)$ numbered so that

$$a_0 = 0, \quad a_{p-i} = -a_i, \quad i = 1, \ldots, p-1.$$

Define $Q = (x_{ij})$ by

$$x_{ij} = \chi(a_j - a_i) = \begin{cases} 0 & \text{if } i = j, \\ 1 & \text{if } a_j - a_i = y^2 \text{ for some } y \in GF(p), \\ -1 & \text{otherwise.} \end{cases}$$

Now Q is a type 1 matrix with the properties that

$$QQ^t = pI - J,$$
$$QJ = JQ = 0,$$
$$Q^t = (-1)^{\frac{1}{2}(p-1)}Q.$$

This follows since exactly half of $a_1, \ldots, a_{p-1}$ are squares, -1 is a square for $p \equiv 1 \pmod 4$ but not for $p \equiv 3 \pmod 4$, and

$$\sum_y \chi(y)\chi(y+c) = \sum_y \chi(y^2)\chi(1+cy^{-1}) = \sum_{z \neq 1} \chi(x) = -1.$$

Let e be the $1 \times p$ vector of all ones. Then

$$W = \begin{bmatrix} 0 & e \\ (-1)^{\frac{1}{2}(p-1)}e^t & Q \end{bmatrix}$$

is the required matrix. If $p \equiv 3 \pmod 4$, $W + I_p$ is a $W(p+1, p+1)$.

4.35. COROLLARY. There are Hadamard matrices of order $p+1$ where $p \equiv 3 \pmod 4$ is a prime power, and of order $2(p+1)$ where $p \equiv 1 \pmod 4$ is a prime power.

Proof. For $p \equiv 3 \pmod 4$ use $W+I$; for $p \equiv 1 \pmod 4$ use

$$\begin{bmatrix} W+I & W-I \\ W-I & -W-I \end{bmatrix}.$$

4.36. COROLLARY. There are Hadamard matrices of order $2^t \Pi(p_i^{r_i}+1)$ where $p_i^{r_i}$ are prime powers and t, an integer, is > 0 if $p_j^{r_j} \equiv 1 \pmod 4$, for some j, and ≥ 0 otherwise.

Proof. Use Lemma 4.32 and Corollary 4.35.

It is conjectured that:

4.37. HADAMARD CONJECTURE. There exists an Hadamard matrix of order 4t for every positive integer t.

4.38. CONJECTURE. (Wallis). There exists a weighing matrix $W(4t, k)$, $k = 0, 1, \ldots, 4t$, for every positive integer t.

This conjecture, of course, includes the Hadamard Conjecture.

4.39. DEFINITION. We say that the weighing matrix $W = W(2n, k)$ is <u>constructed from two circulant matrices</u> M, N of order n if

$$W = \begin{bmatrix} M & N \\ N^t & -M^t \end{bmatrix}.$$

4.40. EXAMPLE

$$M = \begin{bmatrix} 0 & 1 & 1 \\ 1 & 0 & 1 \\ 1 & 1 & 0 \end{bmatrix} \quad \text{and} \quad N = \begin{bmatrix} - & 1 & 1 \\ 1 & - & 1 \\ 1 & 1 & - \end{bmatrix}$$

of order 3 satisfy

$$MM^t + NN^t = 5I.$$

Then

$$W = \begin{bmatrix} M & N \\ N^t & -M^t \end{bmatrix} = \left[\begin{array}{ccc|ccc} 0 & 1 & 1 & - & 1 & 1 \\ 1 & 0 & 1 & 1 & - & 1 \\ 1 & 1 & 0 & 1 & 1 & - \\ \hline - & 1 & 1 & 0 & - & - \\ 1 & - & 1 & - & 0 & - \\ 1 & 1 & - & - & - & 0 \end{array}\right]$$

is a $W(6, 5)$ constructed from two circulant matrices.

4.41. THEOREM. (Goethals and Seidel). Let $q \equiv 1 \pmod 4$ be a prime power; then there is a $W(q+1, q)$ of the form

$$S = \begin{bmatrix} A & B \\ B & -A \end{bmatrix}$$

with zero diagonal and square circulant submatrices A and B.

<u>Proof.</u> Let z be any primitive element of $GF(q^2)$, the quadratic

extension of GF(q). We choose any basis of V the vector space of dimension 2 over $GF(q^2)$. With respect to this basis, v is defined by the matrix

$$(v) = \tfrac{1}{2}\begin{bmatrix} z^{q-1}+z^{1-q} & (z^{q-1}-z^{1-q})z^{\frac{1}{2}(q+1)} \\ (z^{q-1}-z^{1-q})z^{-\frac{1}{2}(q+1)} & z^{q-1}+z^{1-q} \end{bmatrix},$$

which actually has its elements in GF(q). Then det(v) = 1, and the eigenvalues of v are z^{q-1} and z^{1-q}, both elements of $GF(q^2)$ whose $\frac{1}{2}(q+1)$-th power, and no smaller, belongs to GF(q). Hence v acts on the projective line PG(1,q) as a permutation with period $\frac{1}{2}(q+1)$ without fixed points. This divides the points of PG(1,q) into two sets of transitivity, each containing $\frac{1}{2}(q+1)$ points. In addition, w defined by the matrix

$$(w) = \begin{bmatrix} 0 & z^{q+1} \\ 1 & 0 \end{bmatrix}$$

has $\chi\det(w) = -\chi(-1)$. The eigenvalues of w are $\pm z^{\frac{1}{2}(q+1)}$ elements of $GF(q^2)$ whose square is in GF(q). Hence w acts on PG(1,q) as a permutation of period 2, which maps any point of one set of transitivity, defined above by v, into the other set. Indeed, for $i = 1,\dots,\frac{1}{2}(q+1)$, the mapping $v^i w$ has no eigenvalue in GF(q). Note vw = wv.

Finally, we represent the q+1 points of PG(1,q), $x_0, x_1, \dots, x_q$, by the following q+1 vectors in V:

$$x, v(x), v^2(x), \dots, v^{\frac{1}{2}(q-1)}(x), w(x), vw(x), \dots, v^{\frac{1}{2}(q-1)}w(x).$$

We define

$$S = \chi\det(x_i, x_j).$$

Observing that any linear mapping $u : V \to V$ satisfies

$$\det(u(x), u(y)) = \det u \cdot \det(x, y),$$

for all $x, y \in V$, we see that

$$\det(v^i w(x), v^j w(x)) = \det(w)\cdot\det(v^i(x), v^j(x)) = \det(w)\cdot\det(x, v^{j-i}(x)),$$

$$\det(v^i(x), v^j w(x)) = -\det(v^i w(x), v^j(x)) = \det(v^j(x), v^i w(x)),$$

$$\det(v^i(x), v^j(x)) = -\det(v^{\frac{1}{2}(q+1)+i}, v^j(x)),$$

and so S has the required form.

4.42. EXAMPLE. Let $q = 5$ and z be a root of $z^2+z+2 = 0$ (a primitive polynomial over $GF(5^2)$). Then

$$z^4 = 3z+2,\ z^{-4} = z^{20} = 2z+4,\ z^3 = 4z+2,\ z^{-3} = z^{21} = 2z+1,\ z^6 = 2.$$

Hence

$$(v) = \frac{1}{2}\begin{bmatrix} z^4+z^{-4} & (z^4-z^{-4})z^3 \\ (z^4-z^{-4})z^{-3} & z^4+z^{-4} \end{bmatrix} = \begin{bmatrix} 3 & 4 \\ 2 & 3 \end{bmatrix}$$

and

$$(w) = \begin{bmatrix} 0 & z^6 \\ 1 & 0 \end{bmatrix} = \begin{bmatrix} 0 & 2 \\ 1 & 0 \end{bmatrix}.$$

We now choose some vector x, say, $x = \begin{bmatrix} 1 \\ 0 \end{bmatrix}$. Then

$$x_0 = x = \begin{bmatrix} 1 \\ 0 \end{bmatrix},\ x_1 = v(x) = \begin{bmatrix} 3 \\ 2 \end{bmatrix},\ x_3 = v^2(x) = \begin{bmatrix} 2 \\ 2 \end{bmatrix},$$

$$x_4 = w(x) = \begin{bmatrix} 0 \\ 1 \end{bmatrix},\ x_5 = vw(x) = \begin{bmatrix} 4 \\ 3 \end{bmatrix},\ x_6 = v^2w(x) = \begin{bmatrix} 4 \\ 2 \end{bmatrix}.$$

Since $\chi(1) = \chi(4) = 1$ and $\chi(2) = \chi(3) = -1$,

$$\det(x_i, x_j) = \left[\begin{array}{ccc|ccc} 0 & 2 & 2 & 1 & 3 & 2 \\ 3 & 0 & 2 & 3 & 1 & 3 \\ 3 & 3 & 0 & 2 & 3 & 1 \\ \hline 4 & 2 & 3 & 0 & 1 & 1 \\ 2 & 4 & 2 & 4 & 0 & 1 \\ 3 & 2 & 4 & 4 & 4 & 0 \end{array}\right] \text{ and } \chi\det(x_i, x_j) = \left[\begin{array}{ccc|ccc} 0 & - & - & 1 & - & - \\ - & 0 & - & - & 1 & - \\ - & - & 0 & - & - & 1 \\ \hline 1 & - & - & 0 & 1 & 1 \\ - & 1 & - & 1 & 0 & 1 \\ - & - & 1 & 1 & 1 & 0 \end{array}\right].$$

4.43. COROLLARY. Let $p \equiv 1 \pmod 4$ be a prime power. Then there exist four circulant symmetric matrices

$$X_1 = I+A,\ X_2 = I-A,\ X_3 = X_4 = B$$

of order $\frac{1}{2}(p+1)$ which satisfy

$$\sum_{i=1}^{4} X_i X_i^t = 2(p+1)I_{\frac{1}{2}(p+1)}.$$

Proof. Construct A and B as in the theorem.

4.44. COROLLARY. (J. Wallis). Let $p \equiv 1 \pmod 4$ be a prime power; then there exist four (1, -1) matrices X_1, X_2, X_3, X_4 of order $\frac{1}{2}p(p+1)$ which satisfy

$$\sum_{i=1}^{4} X_i X_i^t = 2p(p+1)I_{\frac{1}{2}p(p+1)}, \quad X_i X_j^t = X_j X_i^t.$$

Equivalently, there are Williamson matrices of order $\frac{1}{2}p(p+1)$. (See Definition 4.69.)

Proof. Construct A and B as in the theorem, and Q of order p as in the proof of Lemma 4.34. Then

$$X_1 = (I \times J) + (A \times (I+Q)),$$
$$X_2 = B \times (I+Q),$$
$$X_3 = (I \times J) + (A \times (I-Q)),$$
$$X_4 = B \times (I-Q)$$

are the required matrices.

We will not prove the next results here, but they are included for the sake of completeness, as all skew-Hadamard matrices are orthogonal designs of type (1,n-1) in order n (where no reference is given, Wallis [25] should be consulted; amicable Hadamard matrices are defined in Example 5.5).

4.45. SUMMARY. There are skew-Hadamard matrices of the following orders:

SI	$2^t \Pi k_i$	t, r_i all positive integers $k_i = p_i^{r_i}+1 \equiv 0 \pmod 4$, p_i a prime.
SII	$(p-1)^u+1$	p the order of a skew-Hadamard matrix, $u > 0$ an odd integer.
SIII	$2(q+1)$	$q \equiv 5 \pmod 8$ a prime power.
SIV	$2^s(q+1)$	$q = p^t$ is a prime power such that $p \equiv 5 \pmod 8$, $t \equiv 2 \pmod 4$, $s \geq 1$ an integer.
SV	$4m$	$m \in \{$odd integers between 3 and 25 inclusive$\}$.

SVI	$m'(m'-1)(m-1)$	m and m' the orders of amicable Hadamard matrices, where $(m-1)m'/m$ is the order of a skew-Hadamard matrix.
SVII	$4(q+1)$	$q = 8f+1$ (f odd) is a prime power.
SVIII	$(\|t\|+1)(q+1)$	$q = s^2+4t^2 \equiv 5 \pmod 8$ is a prime power, and $\|t\|+1$ is the order of a skew-Hadamard matrix (Wallis [19]).
SIX	$4(1+q+q^2)$	where q is a prime power and $\begin{cases} 1+q+q^2 \text{ is a prime} \equiv 3, 5 \text{ or } 7 \pmod 8; \text{ or} \\ 3+2q+2q^2 \text{ is a prime power (Spence [2]).} \end{cases}$
SX	hm	h the order of a skew-Hadamard matrix, m the order of amicable Hadamard matrices.

§4.3. THE GOETHALS-SEIDEL ARRAY AND OTHER CONSTRUCTIONS USING CIRCULANT MATRICES--CONSTRAINTS ON CONSTRUCTIONS USING CIRCULANT MATRICES

In studying skew-Hadamard matrices (orthogonal designs of type $(1, n-1)$ in order n), Szekeres realized that none were known for quite small orders, including 36. To find this matrix Goethals and Seidel gave an array (described in this section) which uses circulant matrices. This and its generalization by Wallis and Whiteman have proved invaluable in the construction of Hadamard matrices, and we will see that they play a major role in constructing orthogonal designs. Here we have another example of a method devised to give a single case having far-reaching uses.

We now consider the use of circulant matrices in constructing orthogonal designs. All the constructions using circulants require that we find circulants $A_1, \ldots, A_s$ of order n satisfying

$$\sum_{i=1}^{s} A_i A_i^t = fI, \text{ where } f = \sum_{j=1}^{r} s_j x_j^2.$$

One question we shall explore in this section is the restrictions that must be placed on $(s_1, \ldots, s_r)$ in order that such circulant matrices exist.

This problem is analogous to the problems we discussed in Chapter 3 when we discovered algebraic limitations on orthogonal designs.

Conditions imposed on $(s_1, \ldots, s_r)$ in order to construct orthogonal designs from circulants is closer to the combinatorial spirit of the subject. Although there is no reason to believe that all the orthogonal designs we look for in orders $4n$ or $8n$, n odd, can be expected to come from circulants, we will find they usually do. Thus circulant matrices are important constructive tools, and we should decide what limitations there are on their use. We also note that circulant matrices are amenable to algebraic assault because of their relationship to roots of unity. This aspect to circulants will become more apparent when we discuss Griffin's work on Golay sequences in Section 4.6.

We first give constructions using circulants and then consider restrictions on their use.

4.46. PROPOSITION. Suppose there exist two circulant matrices B of order n satisfying

$$AA^t + BB^t = fI_n .$$

Further suppose that R is the back diagonal matrix; then

$$H = \begin{bmatrix} A & B \\ -B^t & A^t \end{bmatrix} \quad \text{or} \quad G = \begin{bmatrix} A & BR \\ -BR & A \end{bmatrix} \quad \text{is a } W(2n, f),$$

when A, B are $(0,1,-1)$ matrices, and an orthogonal design of order $2n$ and type $(u_1, u_2, \ldots, u_s)$ on $x_1, \ldots, x_s$ when $f = \sum_{i=1}^{s} u_i x_i^2$.

Further, H and G are skew or skew-type if A is skew or skew-type.

Proof. A straightforward verification.

REMARK. We note here that these properties remain true if A and B are type 1 matrices and R is the appropriately chosen matrix (see Lemmas 4.16 to 4.22).

4.47. DEFINITION. We say that an orthogonal design is constructed from two circulant matrices M, N of order n if

$$W = \begin{bmatrix} M & N \\ -N^t & M^t \end{bmatrix} \quad \text{or} \quad W = \begin{bmatrix} M & NR \\ -NR & M \end{bmatrix}.$$

4.48. EXAMPLE

$$A = \begin{bmatrix} x_1 & x_2 & -x_2 \\ -x_2 & x_1 & x_2 \\ x_2 & -x_2 & x_1 \end{bmatrix} \quad \text{and} \quad B = \begin{bmatrix} 0 & x_2 & x_2 \\ x_2 & 0 & x_2 \\ x_2 & x_2 & 0 \end{bmatrix}$$

of order 3 satisfy

$$AA^t + BB^t = (x_1^2 + 4x_2^2)I.$$

Thus

$$H = \begin{bmatrix} A & B \\ -B^t & A^t \end{bmatrix} = \left[\begin{array}{ccc|ccc} x_1 & x_2 & -x_2 & 0 & x_2 & x_2 \\ -x_2 & x_1 & x_2 & x_2 & 0 & x_2 \\ x_2 & -x_2 & x_1 & x_2 & x_2 & 0 \\ \hline 0 & -x_2 & -x_2 & x_1 & -x_2 & x_2 \\ -x_2 & 0 & -x_2 & x_2 & x_1 & -x_2 \\ -x_2 & -x_2 & 0 & -x_2 & x_2 & x_1 \end{array}\right]$$

and

$$G = \begin{bmatrix} A & BR \\ -BR & A \end{bmatrix} = \left[\begin{array}{ccc|ccc} x_1 & x_2 & -x_2 & 0 & x_2 & x_2 \\ -x_2 & x_1 & x_2 & x_2 & x_2 & 0 \\ x_2 & -x_2 & x_1 & x_2 & 0 & x_2 \\ \hline 0 & -x_2 & -x_2 & x_1 & x_2 & -x_2 \\ -x_2 & -x_2 & 0 & -x_2 & x_1 & x_2 \\ -x_2 & 0 & -x_2 & x_2 & -x_2 & x_1 \end{array}\right]$$

are orthogonal designs of order 6 and type (1,4) on x_1, x_2. H and G are constructed from two circulants.

4.49. THEOREM. (Goethals-Seidel). Suppose there exist four circulant matrices A, B, C, D of order n satisfying

$$AA^t + BB^t + CC^t + DD^t = fI_n.$$

Let R be the back diagonal matrix. Then

$$GS = \begin{bmatrix} A & BR & CR & DR \\ -BR & A & D^tR & -C^tR \\ -CR & -D^tR & A & B^tR \\ -DR & C^tR & -B^tR & A \end{bmatrix}$$

is a $W(4n, f)$ when A, B, C, D are $(0,1,-1)$ matrices, and an orthogonal design of order $4n$ and type $(u_1, u_2, \ldots, u_s)$ on $x_1, x_2, \ldots, x_s$ when A, B, C, D have entries from $\{0, \pm x_1, \ldots, \pm x_s\}$ and

$$f = \sum_{j=1}^{s} u_j x_j^2 .$$

Further, GS is skew or skew-type if A is skew or skew-type.

This theorem was modified by Wallis and Whiteman to allow the circulant matrices to be generalized to types 1 and 2.

4.50. LEMMA. (Wallis-Whiteman). Let A, B, D be type 1 matrices and C a type 2 matrix defined on the same abelian group of order n. Then if

$$AA^t + BB^t + CC^t + DD^t = fI_n ,$$

$$H = \begin{bmatrix} A & B & C & D \\ -B^t & A^t & -D & C \\ -C & D^t & A & -B^t \\ -D^t & -C & B & A^t \end{bmatrix}$$

is a $W(4n, f)$ when A, B, C, D are $(0,1,-1)$ matrices, and an orthogonal design of order $4n$ and type $(u_1, u_2, \ldots, u_s)$ on $x_1, x_2, \ldots, x_s$ when A, B, C, D have entries from $\{0, \pm x_1, \ldots, \pm x_s\}$ and

$$f = \sum_{j=1}^{s} u_j x_j^2 .$$

Further, H is skew or skew-type if A is skew or skew-type.

4.51. EXAMPLE. We construct an orthogonal design of order 12 and type $(3, 3, 3, 3)$ in order 12 by using the circulant matrices with first rows, respectively, abc, -bad, -c-da, -dc-b,

in the Goethals-Seidel array (Theorem 4.49; see also Example 4.76).

To illustrate the use of Lemma 4.50 recall that in Example 4.14 we gave a type 1 matrix A and a type 2 matrix B defined on the additive group G of $GF(3^2)$ ($G = Z_3 \times Z_3$). Let R be the matrix defined on G by Lemma 4.22. Notice that by Lemma 4.22 $AR = B$ and $BR = A$. Also, R is a type 2 matrix.

Let $W = \begin{bmatrix} 0 & I_3 & 0 \\ 0 & 0 & I_3 \\ I_3 & 0 & 0 \end{bmatrix}$; then W is the type 1 (1, 0) incidence matrix generated by $\{2x\}$ (we are using the notation of Example 4.14).

Let $L = \begin{bmatrix} J_3 & 0 & 0 \\ 0 & J_3 & 0 \\ 0 & 0 & J_3 \end{bmatrix}$; then L is the type 1 (1, 0) incidence matrix generated by the subgroup $\{0,1,2\}$ of G. Since the identity matrix is always a type 1 matrix for a group G (for which the identity element of the group is the first element of G), we obtain that $U = L-I+W$ and $V = I+W^t$ are type 1 matrices on G. (See Lemma 4.18.)

Set $X_1 = aI+bA$, $X_2 = b(U+V)$, $X_4 = b(U-V)$ and $X_3 = bB-aR$. Then X_1, X_2 and X_4 are type 1 matrices and X_3 is a type 2 matrix. By inspection, all entries are from $\{0, \pm a, \pm b\}$. Also

$$\sum_{i=1}^{4} X_i X_i^t = (2a^2+26b^2)I_9.$$

Thus these matrices may be used in place of A, B, C, D in Lemma 4.50 to give an orthogonal design of type (2,26) in order 36.

The most general theorem we can give on using circulant matrices in the construction of orthogonal designs is

4.52. THEOREM. Suppose there is an orthogonal design of type $(u_1, u_2, \ldots, u_s)$ on the variables $x_1, x_2, \ldots, x_s$ in order m. Let $X_1, X_2, \ldots, X_s$, where $s \le \rho(n)$, be circulant (type 1) matrices of order n with entries from $\{0, \pm y_1, \ldots, \pm y_r\}$ which satisfy

$$u_1X_1X_1^t + u_2X_2X_2^t + \ldots + u_sX_sX_s^t = fI_n.$$

Further suppose

1) all X_i are symmetric, or
2) at most one is not symmetric, or
3) $X_1, \ldots, X_{j-1}$ are symmetric and $X_j, \ldots, X_s$ are skew-symmetric.

Then if $f = v_1y_1^2 + v_2y_2^2 + \ldots + v_ry_r^2$, there is an orthogonal design of type $(v_1, v_2, \ldots, v_r)$ on the variables $y_1, y_2, \ldots, y_r$ in order mn.

Proof. The main difficulty arises because the variables of the orthogonal design are commutative. When we replace commuting variables by matrices Y_i, $i = 1, \ldots, s$, we have to ensure that the matrices pairwise satisfy

$$Y_iY_j^t = Y_jY_i^t. \tag{*}$$

We established in Section 2 that if the Y_i are circulant and symmetric, (*) is satisfied. Also if Y_iR is back circulant (type 2) and Y is circulant (type 1), then (*) is satisfied. We also note that if Y_i and Y_j are skew-symmetric, the back circulant matrices Y_iR and Y_jR satisfy (*) since

$$(Y_iR)(Y_jR)^t = Y_iRR^tY_j^t = -Y_iY_j = Y_jY_i^t = (Y_jR)(Y_iR)^t.$$

Thus the result can be obtained in the first case by replacing each variable x_i, $i \neq j$, in the orthogonal design of order m by the circulant symmetric matrix X_i; in the second case the variable x_j is replaced by the back circulant matrix X_jR. The third result is obtained by replacing x_i, $i \neq j, j+1, \ldots, s$, by X_i, and $x_j, \ldots, x_s$ by $X_jR, \ldots, X_sR$.

4.53. EXAMPLE. There is an orthogonal design of type $(1,1,1,1,3,3,3,3)$ in order 16 (we will see this in §5.8, Example 5.101(c)). Consider the circulant matrices X_i, with first rows

$$y_1y_2y_2{-y_2}{-y_2}\,,\quad y_2{-y_2}y_2y_2{-y_2}\,,\quad {-y_2}y_2y_2y_2y_2\,,\quad {-y_2}y_2y_2y_2y_2\,,$$

$$y_3y_4{-y_4}{-y_4}y_4\,,\quad {-y_4}y_3{-y_3}{-y_3}y_3\,,\quad {-y_3}y_3y_3y_3y_3\,,\quad {-y_4}y_4y_4y_4y_4\,,$$

and call them respectively $X_1, \ldots, X_8$. Then

$$X_1X_1^t + X_2X_2^t + X_3X_3^t + X_4X_4^t + 3X_5X_5^t + 3X_6X_6^t + 3X_7X_7^t + 3X_8X_8^t$$
$$= (y_1^2 + 19y_2^2 + 30y_3^2 + 30y_4^2)I_5.$$

We use 2) of Theorem 4.52 to assert the existence of an orthogonal design of type (1, 19, 30, 30) in order 80; the matrix X_1R is used to replace the first variable, and the circulant symmetric matrices $X_2, \ldots, X_8$ are used to replace the other variables.

We have noted that in Theorem 4.49 the only requirement was to have circulant matrices, but in Theorem 4.52 the internal structure of the circulant matrices was restricted severely. If the matrices are circulant and symmetric, we will loosely call this Williamson criteria, and if merely circulant, we will call this Goethals-Seidel criteria. Thus in Theorem 4.49 we only had Goethals-Seidel criteria operating, but in Theorem 4.52 we were almost entirely limited to Williamson criteria.

Fortunately, in trying to find designs of order $n \equiv 0 \pmod 8$ constructed using eight circulant matrices, we will not be as restricted as in Theorem 4.52. The difficulty of finding matrices to replace the variables has led to the following lemma using part Williamson and part Goethals-Seidel criteria.

4.54. LEMMA. Suppose $X_1, X_2, \ldots, X_8$ are eight circulant (type 1) matrices of order n satisfying

1) X_i, $1 \le i \le 8$, have entries from $\{0, \pm x_1, \ldots, \pm x_s\}$

and 2) $\sum_{i=1}^{8} X_iX_i^t = fI$.

Further suppose

i) $X_1, X_2, \ldots, X_8$ are all symmetric or all skew, or

ii) $X_1 = X_2 = \ldots = X_i$ and $X_{i+1}, \ldots, X_8$ are all symmetric or all skew, $1 \le i \le 8$, or

iii) $X_2 = X_3 = X_4$ and X_5, X_6, X_7, X_8 are all symmetric (skew), or

iv) $X_1X_2^t = X_2X_1^t$, $X_3 = X_4$ and X_5, X_6, X_7, X_8 are all symmetric, or

(v) $X_1, \ldots, X_i$ are all skew and $X_{i+1}, \ldots, X_8$ all symmetric, or

(vi) X_2, X_3, X_4 are all skew and X_5, X_6, X_7, X_8 all symmetric, or

(vii) $X_i X_{i+4}^t = X_{i+4} X_i^t$, $i = 1, 2, 3, 4$.

Then, with

$$f = \sum_{i=1}^{s} u_i x_i^2 I,$$

there exists an orthogonal design of order $8n$ and type $(u_1, u_2, \ldots, u_s)$.

Proof. As in the proof of Theorem 4.52, the main difficulty is ensuring the matrices $Y_1, \ldots, Y_8$ used to replace the commuting variables of the basic design pairwise satisfy

$$Y_i Y_j^t = Y_j Y_i^t.$$

The results of the lemma may be obtained, recalling the results of §4.2, by using the following constructions:

i) Use the circulant matrices to replace the variables in design 1.

ii) Use a back circulant matrix $X_1R = X_iR$ to replace the first i variables in design 1.

iii) We use design 2 which needs A, B, E, F, G, H all circulant, B repeated three times, and E, F, G, H all symmetric.

iv) We use design 4 for which X, A, B, C, D, E, F must all be circulant, B repeated twice, C, D, E, F symmetric, and $XA^t = AX^t$.

v) We use $X_1R, \ldots$ the back circulant matrices $X_1R, \ldots, X_iR$ to replace the first i variables of design 1 and $X_{i+1}, \ldots, X_8$ to replace the last $8-i$ variables.

vi) Use design 5 with $B = X_2$, $C = X_3$, $D = X_4$, $E = X_5$, $F = X_6$, $G = X_7$, $H = X_8$, there is no symmetry restriction on $A = X_1$.

vii) Use design 6.

A	B	C	D	E	F	G	H
-B	A	D	-C	F	-E	-H	G
-C	-D	A	B	G	H	-E	-F
-D	C	-B	A	H	-G	F	-E
-E	-F	-G	-H	A	B	C	D
-F	E	-H	G	-B	A	-D	C
-G	H	E	-F	-C	D	A	-B
-H	-G	F	E	-D	-C	B	A

Design 1

AR	B	B	B	E	F	G	H
-B	AR	B	-B	F	-E	-H	G
-B	-B	AR	B	G	H	-E	-F
-B	B	-B	AR	H	-G	F	-E
-E	-F	-G	-H	AR	$-B^t$	$-B^t$	$-B^t$
-F	E	-H	G	B^t	AR	B^t	$-B^t$
-G	H	E	-F	B^t	$-B^t$	AR	B^t
-H	-G	F	E	B^t	B^t	$-B^t$	AR

Design 2

AR	B	B	B	E	F	G	H
-B	AR	B	-B	-F	-E	-H	G
-B	-B	AR	B	G	H	-E	-F
-B	B	-B	AR	H	-G	F	-E
-E	-F	-G	-H	AR	B^t	B^t	B^t
-F	E	-H	G	$-B^t$	AR	$-B^t$	B^t
-G	H	E	-F	$-B^t$	B^t	AR	$-B^t$
-H	-G	F	E	$-B^t$	$-B^t$	B^t	AR

Design 3

XR	AR	B	B	C	D	E	F
-AR	XR	B	-B	D	-C	-F	E
-B	-B	XR	AR	E	F	-C	-D
-B	B	-AR	XR	F	-E	D	-C
-C	-D	-E	-F	XR	AR	B^t	B^t
-D	C	-F	E	-AR	XR	$-B^t$	B^t
-E	F	C	-D	$-B^t$	B^t	XR	-AR
-F	-E	D	C	$-B^t$	$-B^t$	AR	XR

Design 4

$$\left[\begin{array}{cccc|cccc}
AR & B & C & D & E & F & G & H \\
-B & AR & D & -C & F & -E & -H & G \\
-C & -D & AR & B & G & H & -E & -F \\
-D & C & -B & AR & H & -G & F & -E \\
\hline
-E & -F & -G & -H & AR & B^t & C^t & D^t \\
-F & E & -H & G & -B^t & AR & -D^t & C^t \\
-G & H & E & -F & -C^t & D^t & AR & -B^t \\
-H & -G & F & E & -D^t & -C^t & B^t & AR
\end{array}\right]$$

Design 5

$$\left[\begin{array}{cccc|cccc}
A & BR & CR & DR & E & FR & GR & HR \\
-BR & A & D^tR & -C^tR & FR & -E & -H^tR & G^tR \\
-CR & -D^tR & A & B^tR & GR & H^tR & -E & -F^tR \\
-DR & C^tR & -B^tR & A & HR & -G^tR & F^tR & -E \\
\hline
-E & -FR & -GR & -HR & A & BR & CR & DR \\
-FR & E & -H^tR & G^tR & -BR & A & -D^tR & C^tR \\
-GR & H^tR & E & -F^tR & -CR & D^tR & A & -B^tR \\
-HR & -G^tR & F^tR & E & -DR & -C^tR & B^tR & A
\end{array}\right]$$

Design 6

4.55. EXAMPLE. The following orthogonal designs in order 24 are constructed by using this lemma. The reader may refer to the Table of the Appendix of Orthogonal Designs in order 24 to find the first rows of the circulant matrices which should be used as indicated:

for (1,1,1,1,6,6) use part (i);

for (1,1,1,1,2,10) use part (ii);

for (1,1,2,2,5,8) use part (iii);

for (1,1,1,3,4,9) use part (iv);

for (1,2,2,8,11) use part (v);

for (1,1,4,4,5) use part (vi);

for (1,2,5,5,8) use part (vii);

for (1,2,2,4,13) use part (viii).

CONSTRAINTS ON CONSTRUCTION USING CIRCULANT MATRICES

Of course we would like to use these constructions to form orthogonal designs, but first we must consider some combinatorial limitations on these methods (algebraic limitations on the types of orthogonal designs were discussed earlier).

4.56. LEMMA. Let A_i, $i = 1,2,3,\ldots,m$, be circulant matrices of order n where

$$\sum_{i=1}^{m} A_i A_i^t = \left(\sum_{j=1}^{r} s_j x_j^2\right) I_n.$$

Suppose $A_i = \sum_{j=1}^{r} x_j A_{ij}$ and that $A_{ij} J = y_{ij} J$. Then

$$s_j = \sum_{j=1}^{m} y_{ij}^2.$$

Proof. By definition

$$\sum_{i=1}^{m} (x_1 A_{i1} + x_2 A_{i2} + \ldots)(x_1 A_{i1}^t + x_2 A_{i2}^t + \ldots) = (s_1 x_1^2 + s_2 x_2^2 + \ldots) I.$$

So

$$\sum_{i=1}^{m} x_1^2 (A_{i1} A_{i1}^t) + \sum_{i=1}^{m} x_2^2 (A_{i2} A_{i2}^t) + \ldots = (s_1 x_1^2 + s_2 x_2^2 + \ldots) I,$$

and setting $x_j = 1$, $x_i = 0$ for $i \neq j$ we have

$$\sum_{i=1}^{m} A_{ij} A_{ij}^t = s_j.$$

Post-multiplying by J gives

$$\sum_{i=1}^{m} y_{ij}^2 J = s_j J,$$

and equating coefficients gives the results.

REMARK. If $m = 4$ and we have four circulants A_1, A_2, A_3, A_4 such that

$$\sum_{i=1}^{4} A_i A_i^t = (x_1^2 + s x_2^2) I_n,$$

n odd, then s must be the sum of three squares. For we may assume $A_1 = x_1 I + x_2 A_{12}$, $A_2 = x_2 A_{22}$, $A_3 = x_2 A_{32}$ and $A_4 = x_2 A_{42}$. Then by the lemma we have

$$y_{12}^2 + y_{22}^2 + y_{32}^2 + y_{42}^2 = s.$$

But $A_{12} = -A_{12}^t$, and the order of A_{12} is n(odd). So $y_{12} = 0$, and consequently s is the sum of three squares. This should be compared with Proposition 3.54.

A METHOD FOR CONSTRUCTING ORTHOGONAL DESIGNS BY USING CIRCULANT MATRICES

The method outlined in this section has been used successfully to compute four variable orthogonal designs of order 20 and two variable orthogonal designs of order 28. Some success has been achieved with weighing matrices of orders 18, 22, 26, 30, 44 and 52. The results of this computation are included in the tables of numerical results in the Appendices. We believe that the method can be extended to construct three and four variable orthogonal designs of order 28 and two variable orthogonal designs of order 36, but so far this has not been done.

The method is presented as it applies to the Goethals-Seidel construction (Theorem 4.49), but there are no difficulties in extending the results for more general circulant constructions, such as those mentioned in §4.8.

Specifically, for positive integers $s_1, s_2, \ldots, s_u$ and odd v, the method searches for four circulant matrices X_1, X_2, X_3, X_4 of order v with entries from $\{0, \pm x_1, \pm x_2, \ldots, \pm x_u\}$ such that

$$\sum_{i=1}^{4} X_i X_i^t = \left(\sum_{i=1}^{u} s_i x_i^2 \right) I. \tag{3.1}$$

The existence of an orthogonal design of type $(s_1, s_2, \ldots, s_u)$ and order 4v follows from the Goethals-Seidel construction (Theorem 4.49).

REMARK. The restriction that v is odd is not necessary for most of the results which follow. However, the restriction is made because we are principally interested here in constructing orthogonal designs of order not divisible by 8. Orthogonal designs of order divisible by a large power of 2 can often be constructed using other methods (see Chapter 7).

Equation (3.1) has v^2 components, but since $X_i X_i^t$ is circulant and symmetric, at most $\frac{1}{2}(v+1)$ of these components are independent. The next two definitions are made to isolate the independent components.

4.57. DEFINITION. If A_1, A_2, A_3, A_4 are $v \times v$ circulant matrices with entries from $\{0, \pm x_1, \pm x_2, \ldots, \pm x_u\}$ and the first row of A_j has m_{ij} entries of the kind $\pm x_i$, then the $u \times 4$ matrix $M = (m_{ij})$ is called the <u>entry matrix</u> of (A_1, A_2, A_3, A_4).

4.58. DEFINITION. Suppose that A is a $v \times v$ circulant matrix with rows $r_1, r_2, \ldots, r_v$, and denote $\frac{1}{2}(v-1)$ by w. Then the <u>IPV (Inner Product Vector) of A</u> is $[r_1 r_2^t, r_1 r_3^t, \ldots, r_1 r_w^t]$. Note that if $(d_1, d_2, \ldots, d_v)$ is the first row of AA^t, then the IPV of A is $(d_2, d_3, \ldots, d_w)$.

It is clear that $(X_1, X_2, X_3, X_4) = (A_1, A_2, A_3, A_4)$ is a solution of (3.1) if and only if

$$\sum_{j=1}^{4} m_{ij} = s_i \quad \text{for} \quad 1 \le i \le u, \tag{3.2}$$

and

$$\sum_{j=1}^{4} b_j = 0, \text{ where } b_j \text{ is the IPV of } A_j. \tag{3.3}$$

In other words, to find a solution of (3.1) we need four circulant matrices with entries from $\{0, \pm x_1, \pm x_2, \ldots, \pm x_u\}$ whose entry matrix has i-th row adding to s_i for $1 \le i \le u$ and whose IPV's add to zero.

4.59. DEFINITION. The <u>content</u> of a circulant matrix A with entries from $\{0, \pm x_1, \pm x_2, \ldots, \pm x_u\}$ is the set of pairs $(\varepsilon x_i, m)$ where εx_i $(\varepsilon = \pm 1)$ occurs a nonzero number m times in the first row of A. Our next task is to show how the contents of solutions of (3.1) may be determined from the knowledge of the parameters $v, s_1, s_2, \ldots, s_u$.

4.60. DEFINITION. Suppose that the rowsum of A_j is $\sum_{i=1}^{u} p_{ij} x_j$ for $1 \le j \le 4$. Then the $u \times 4$ integral matrix $P = (p_{ij})$ is called the <u>sum matrix</u> of (A_1, A_2, A_3, A_4). The <u>fill matrix</u> of (A_1, A_2, A_3, A_4) is M-abs(P). The content of A_i is determined by the i-th columns of the sum and fill matrices.

The following theorem may be used to find the sum matrix of a solution of (3.1).

4.61. THEOREM. (Eades Sum Matrix Theorem). The sum matrix P of a solution of (3.1) satisfies

$$PP^t = \mathrm{diag}(s_1, s_2, \ldots, s_u). \tag{3.5}$$

Proof. Suppose that A is a $v \times v$ circulant matrix with rowsum a, and denote by b the sum of the squares of the first row of A, and by c the sum of the entries of the IPV of A. Then

$$(JA)(A^tJ^t) = a^2JJ^t = a^2vJ.$$

But also

$$\begin{aligned}(JA)(A^tJ^t) &= J(AA^t)J^t\\ &= (b+2c)JJ^t\\ &= v(b+2c)J.\end{aligned}$$

Hence $a^2 = b+2c$. Thus if (p_{ij}) and (m_{ij}) are the sum and entry matrices of a solution of (3.1), then since the sum of the sums of the entries of the IPV's is zero, it follows that

$$\sum_{j=1}^{4}\left[\left(\sum_{i=1}^{u} p_{ij}x_i\right)^2 - \left(\sum_{i=1}^{u} m_{ij}x_j\right)^2\right] = 0.$$

Expanding this equation and equating coefficients of x_ix_j gives (3.5).

REMARKS. (a) Note that the Sum Matrix Theorem (Theorem 4.61) implies that a necessary condition for the existence of an orthogonal design of type $(s_1, s_2, \ldots, s_u)$ and order 4v constructed by using the Goethals-Seidel array is the existence of a $u \times 4$ integral matrix P satisfying (3.5). In fact this theorem says that the only time we can hope to construct an orthogonal design using the Goethals-Seidel array in order $n \equiv 0 \pmod 4$ of type (s_1, s_2, s_3, s_4) is when there is a 4×4 integer matrix P such that $PP^t = \mathrm{diag}(s_1, s_2, s_3, s_4)$. This is analogous to Theorem 3.57 of Chapter 3, which says that in orders $n \equiv 4 \pmod 8$ a rational family of type $[s_1, s_2, s_3, s_4]$ exists in order n if and only if there is a 4×4 rational matrix Q with $QQ^t = \mathrm{diag}(s_1, s_2, s_3, s_4)$. This also shows that, for four variable designs, the Goethals-Seidel approach will be less useful in orders divisible by a large power of 2.

(b) Suppose that P and Q are the sum and fill matrices of a solution $(X_1, X_2, X_3, X_4) = (A_1, A_2, A_3, A_4)$ of (3.1). If B and C are permutation matrices of orders u and 4, respectively, then BPC and BQC are the sum and fill matrices of another solution of (3.1) formed by permuting the indices of the A_i and the x_j. Hence BPC and BQC are regarded as essentially the same as P and Q. Similarly, if P' is formed from P by multiplying some rows and columns by -1, then P' is regarded as essentially the same as P.

We state the first step of the method.

<u>Step</u> 1. Use the Sum Matrix Theorem to find a sum matrix of a solution of (3.1).

If the algebraic necessary conditions (Theorem 3.57) for the existence of an orthogonal design of order of type $(s_1, s_2, \ldots, s_u)$ and order 4v hold, then the existence of a solution to (3.5) is guaranteed by a recent result of Pall (see Eades and Pall [1]).

In most cases, if the s_i are small (for instance, $s_1 + \ldots + s_u \leq 28$), then the solution of (3.5) is essentially unique and can be found easily by hand.

It is clear that if Q is the fill matrix of a solution of (3.1), then

the entries of Q are even nonnegative integers, (3.7)

and if $M = (m_{ij}) = \text{abs}(P)+Q$, then M satisfies (3.2) and

the sum of a column of M is at most v. (3.8)

There may be a large number of matrices which satisfy (3.2), (3.7) and (3.8) (see Example 4.64), but the next two lemmata may be used to reduce the number of possibilities.

4.62. LEMMA. (Eades). Suppose that A is a circulant matrix of odd order v, with entries from $\{0, 1, -1\}$ and with k nonzero entries in each row.

(a) If $k \geq v-1$, then each entry of the IPV of A is odd.

(b) If each entry of the IPV of A is even, then $v \geq k+\sqrt{k}+1$.

Proof. Part (a) can be proved by an elementary parity check. For part (b), a standard counting argument may be employed as follows. Suppose that the ij-th entry of A is a_{ij}, and denote by B_i the set

$$\{j : 1 \le j \le v \text{ and } a_{ij} = 0\},$$

for $1 \le i \le v$. Each B_i contains $v-k$ elements. Also, since each column of A contains k nonzero entries, each integer in $\{1,2,\ldots,v\}$ occurs in $v-k$ of the B_i. It follows that each element of B_1 occurs in $v-k-1$ of the B_i for $i \ge 2$; hence

$$\sum_{i=2}^{v} |B_1 \cap B_i| = (v-k)(v-k-1).$$

But since the inner product of each pair of distinct rows of A is even and v is odd, $|B_1 \cap B_i|$ is odd for $2 \le i \le v$. In particular, $|B_1 \cap B_i| \ge 1$. Hence

$$\sum_{i=2}^{v} |B_1 \cap B_i| \ge v-1,$$

and so

$$(v-k)^2 - (v-k) \ge v-1.$$

Completing the square gives

$$(v-k-1)^2 \ge k.$$

By part (a), $v > k \ge 0$, and so $v \ge k + \sqrt{k+1}$.

4.63. LEMMA. (Eades). Suppose that the entry matrix of a solution $(X_1, X_2, X_3, X_4) = (A_1, A_2, A_3, A_4)$ of (3.1) is

$$\begin{bmatrix} V & \\ & W \end{bmatrix}$$

where V is $\ell \times r$ and W is $(u-\ell) \times (4-r)$. Then

$$\sum_{j=1}^{r} A_j A_j^t = \left(\sum_{i=1}^{\ell} s_i x_i^2 \right) I$$

and

$$\sum_{j=r+1}^{4} A_j A_j^t = \left(\sum_{i=\ell+1}^{u} s_i x_i^2 \right) I.$$

The proof of this lemma is straightforward and thus omitted.

Before the use of these lemmas is illustrated with an example, the second step of the method is stated explicitly.

Step 2. Using (3.2), (3.7), (3.8) and Lemmas 4.62 and 4.63, find all possible fill matrices which could accompany the sum matrix found in Step 1.

If v and the s_i are small, then there are usually very few possible fill matrices, and they can be found easily without a computer.

4.64. EXAMPLE. The existence of an orthogonal design of type (1, 5, 5, 9) and order 20 is listed in Geramita and Wallis [3] as being undetermined. To construct such an orthogonal design, we require four 5×5 circulant matrices B_1, B_2, B_3, B_4, with entries from $\{0, \pm x_1, \pm x_2, \pm x_3, \pm x_4\}$, such that

$$\sum_{i=1}^{4} B_i B_i^t = (x_1^2 + 5x_2^2 + 5x_3^2 + 9x_4^2) I. \tag{3.9}$$

$1 = 1^2$, $5 = 1^2 + 2^2$, $9 = 3^2 = 2^2 + 2^2 + 1^2$ are essentially the only ways of writing 1, 5, 9 as sums of at most four squares, and so it is not difficult to show that (essentially) the only 4×4 integral matrix P which satisfies $PP^t = \text{diag}(1, 5, 5, 9)$ is

$$P = \begin{bmatrix} 1 & & & \\ & 1 & 2 & \\ & -2 & 1 & \\ & & & 3 \end{bmatrix}. \tag{3.10}$$

(See Remark (b) after Theorem 4.61.)

Now there are eight 4×4 integral matrices which, on the basis of (3.2), (3.7), and (3.8) could be fill matrices.

$$
\text{(a)}\begin{bmatrix} & & & \\ 2 & & & \\ 2 & & & \\ & 2 & 2 & 2 \end{bmatrix},\
\text{(b)}\begin{bmatrix} & & & \\ 2 & & & \\ & 2 & & \\ 2 & & 2 & 2 \end{bmatrix},\
\text{(c)}\begin{bmatrix} & & & \\ 2 & & & \\ & & 2 & \\ 2 & 2 & & 2 \end{bmatrix},\
\text{(d)}\begin{bmatrix} & & & \\ 2 & & & \\ & & & 2 \\ 2 & 2 & 2 & \end{bmatrix},
$$

$$
\text{(e)}\begin{bmatrix} & & & \\ & 2 & & \\ & & 2 & \\ 4 & & & 2 \end{bmatrix},\
\text{(f)}\begin{bmatrix} & & & \\ & & 2 & \\ & 2 & & \\ 4 & & & 2 \end{bmatrix},\
\text{(g)}\begin{bmatrix} & & & \\ & 2 & & \\ & & & 2 \\ 4 & & 2 & \end{bmatrix},\
\text{(h)}\begin{bmatrix} & & & \\ & & 2 & \\ & & & 2 \\ 4 & 2 & & \end{bmatrix}. \qquad (3.11)
$$

However, four of these matrices can be discounted as possible fill matrices by using Lemmas 4.62 and 4.63.

Suppose that (B_1, B_2, B_3, B_4) has sum matrix P above (3.10) and fill matrix (3.11)(b). Then the entry matrix is

$$
\begin{bmatrix} 1 & & & \\ 2 & 1 & 2 & \\ & 4 & 1 & \\ 2 & & 2 & 5 \end{bmatrix},
$$

which satisfies (3.2) and (3.8). But the (3,2)-th entry of this entry matrix indicates by Lemma 4.62 that every entry of the IPV of B_2 has a term in x_3^2 with odd coefficient. But x_3 occurs at most once in each row of each of the other circulant matrices, and it follows that the IPV's of the other circulant matrices have no terms in x_3^2. Hence it is impossible for the IPV's of the B_i to add to zero; so (3.11)(b) is not the fill matrix of the B_i.

Suppose that (3.11)(f) is the fill matrix of (B_1, B_2, B_3, B_4); this gives entry matrix

$$
\begin{bmatrix} 1 & & & \\ & 1 & 4 & \\ & 4 & 1 & \\ 4 & & & 5 \end{bmatrix}.
$$

If this is the entry matrix of (B_1, B_2, B_3, B_4), then

$$
\begin{bmatrix} 1 & & & \\ 4 & 5 & & \\ & & 1 & 4 \\ & & 4 & 1 \end{bmatrix}
$$

is the entry matrix of another solution (C_1, C_2, C_3, C_4) of (3.10) (see Remark (b) after Theorem 4.61). It follows by Lemma 4.63 that

$$C_1C_1^t + C_2C_2^t = (x_1^2+9x_2^2)I_5,$$

and thus, using the two-circulant construction, there is an orthogonal design of type (1, 9) and order 10. This is impossible, as it implies the existence of an Hadamard matrix of order 10, and so (3.11)(f) is not the fill matrix of (B_1, B_2, B_3, B_4).

Similarly it can be shown that (3.11)(h) and (3.11)(e) are not possible.

Each of the possible fill matrices (3.11)(a), (c), (d), (g) could specify the contents of a solution of (3.9). For each of these possibilities, we need to search through the circulant matrices whose contents are thus specified until we find a combination whose IPV's add to zero. For instance, for (3.11)(a) we need to find four 5×5 permutation matrices M_1, M_2, M_3, M_4 such that

$$(x_1, x_2, -x_2, x_3, -x_3)M_1,$$

$$(x_2, -x_3, -x_3, x_4, -x_4)M_2,$$

$$(x_2, x_2, x_3, x_4, -x_4)M_3,$$

$$(x_4, x_4, x_4, x_4, -x_4)M_4$$

are the first rows of circulant matrices whose IPV's add to zero. If this search fails, then we consider circulant matrices with contents specified by (3.11)(c), and so on. Note that there are a large number (about 2×10^8) of 4-tuples (M_1, M_2, M_3, M_4) of 5×5 permutation matrices; however, only a small proportion of these need be considered, as we shall presently see.

Once the sum and fill matrices have been chosen, the final steps of the method may be executed.

<u>Step</u> 3. For each $i \in \{1, 2, 3, 4\}$ write down a circulant matrix A_i with contents specified by the i-th columns of the sum and fill matrices.

Step 3 can be executed easily either by hand or by computer. Of course, the circulant matrices A_i can be represented by their first rows.

4.64A. DEFINITION. Two circulant matrices with the same content are isometric if they have the same IPV.

Step 4. For each $i \in \{1,2,3,4\}$, write a list L_i of nonisometric circulant matrices with the same contents as A_i. Attach to each circulant matrix its IPV.

The problem of executing the fourth step is considered next. Given two circulant matrices with the same content, how do we determine whether they are isometric (without the time-consuming calculation of IPV's)? How large are the lists L_i? Useful necessary and sufficient conditions for isometry are, in general, unknown, but one obvious sufficient condition can be described as follows. Denote by S_v the group of $v \times v$ permutation matrices, and suppose that $T \in S_v$ represents the v-cycle $(12\ldots v)$. Let R denote the $v \times v$ back diagonal matrix (see Section 4.2). The subgroup of S_v generated by T and R is denoted by $\langle T, R\rangle$. If A and B are $v \times v$ circulant matrices with first rows a and aK for some $K \in \langle T, R\rangle$, then it can be seen immediately that A and B are isometric. It follows that the number of nonisometric circulant matrices with the same content is at most the index of $\langle T, R\rangle$ in S_v, that is, $(v-1)!/2$. Thus the lists L_i in Step 4 contain at most $(v-1)!/2$ entries. A complete set of distinct coset representatives of $\langle T, R\rangle$ in S_v is easily seen to be $E = \{M \in S_v : M$ represents a permutation θ on $\{1,2,\ldots,v\}$ which satisfies $v\theta = v$ and $1\theta \leq \frac{1}{2}(v-1)\}$. Thus to compute the list L_i in Step 4, we first write out the elements of $S = \{B : B$ is a circulant matrix with first row a_iM for some $M \in E\}$, where a_i denotes the first row of the circulant matrix A_i chosen at Step 3. This can be done easily either automatically or by hand.

Of course S may contain isometric elements. But it can be shown (as follows) that if $a_i = (x_1, x_2, \ldots, x_v)$, then no two distinct elements of S are isometric.

4.65. LEMMA. If $a_i = (x_1, x_2, \ldots, x_v)$ and B_1 and B_2 are elements of S with first rows a_iM_1 and a_iM_2 where M_1 and M_2 are $v \times v$ permutation matrices, then B_1 and B_2 are isometric if and only if they are equal.

Proof. The first entries of the IPV's of B_1 and B_2 are equal; that is,

$$a_i M_1 T^{-1} M_1^{-1} a_i^t = a M_2 T^{-1} M_2^{-1} a_i^t.$$

Symmetrizing gives

$$a_i M_1 (T+T^{-1}) M_1^{-1} a_i^t = a_i M_2 (T+T^{-1}) M_2^{-1} a_i^t.$$

Since $a_i = (x_1, x_2, \ldots, x_v)$, we obtain

$$T+T^{-1} = MTM^{-1} + MT^{-1}M^{-1}$$

where M denotes $M_1^{-1}M_2$. A simple combinatorial argument using the fact that v is odd shows that $T+T^{-1}$ can be written uniquely as a sum of two permutation matrices. Hence either $T = MTM^{-1}$ or $T^{-1} = MTM^{-1}$. In either case, since the subgroup of S_v generated by T is self-centralizing (see Wielandt [1]), we can deduce that $M \in \langle T, R \rangle$. Thus M_1 and M_2 are in the same coset of $\langle T, R \rangle$, but both are elements of S, so $M_1 = M_2$.

The converse is immediate.

This lemma implies that sometimes the list L_i achieves its maximum size $(v-1)!/2$. However this is rare. For instance, if the content of A_i is $\{(\varepsilon x_i, n_{\varepsilon i}) : 1 \le i \le u,\ \varepsilon = \pm 1\}$, then the subgroup

$$L = \{M \in S_v : a_i M = a_i\}$$

of S_v has order

$$m = \left(\prod_{i=-u}^{u} n_i!\right)\left(v - \sum_{i=-u}^{u} n_i\right)!.$$

Hence there are at most $v!/m$ entries of the list L_i, and often $v!/m < (v-1)!/2$. However, the coset representatives of L in S_v are more difficult to deal with by computer than the representatives of $\langle T, R \rangle$. Hence L is used only in hand calculations. When a computer is used, the sort-merge package program may be used to eliminate isometric elements of the set S.

The final step of the method is to search the lists L_i for an answer.

Step 5. Search for one circulant matrix C_i with IPV c_i from each list L_i $(1 \leq i \leq 4)$ such that $c_1+c_2+c_3+c_4 = 0$.

In the implementations for orthogonal designs of orders 20 and 28, there was no difficulty in using a naive algorithm for the search at Step 5 because the lists L_i were relatively small. However, to extend the method to higher orders, it seems that a sophisticated search algorithm would need to be employed.

Two notes on the execution of Steps 4 and 5 are presented next.

Firstly, suppose that C_1, C_2, C_3, C_4 are circulant matrices whose sum and fill matrices satisfy (3.2), (3.5), (3.7) and (3.8). Then the sum of the sums of the entries of the IPV's of the C_i is zero (see proof of Theorem 4.61). That is, if $(c_{i1}, c_{i2}, \ldots, c_{iw})$ is the IPV of C_i $(1 \leq i \leq 4)$, then

$$\sum_{i=1}^{4} \sum_{j=1}^{w} c_{ij} = 0.$$

Hence if

$$\sum_{i=1}^{4} c_{ij} = 0 \quad \text{for} \quad 1 \leq j \leq w-1,$$

then

$$\sum_{i=1}^{4} c_{ij} = 0 \quad \text{for} \quad 1 \leq j \leq w.$$

Hence only $\frac{1}{2}(v-3)$ of the $\frac{1}{2}(v-1)$ components of the IPV's need to add to zero for (3.1) to hold. This saves time and space in computer implementation and provides a simple error-checking device for hand calculations.

Secondly, we note that the IPV's of nonisometric circulant matrices may be dependent in the following way. Suppose that $N \in S_v$ normalizes the subgroup $<T>$ of S_v generated by T. Note that there is an integer d prime to v such that $NT^iN^{-1} = T^{id}$ for $0 \leq i < v$. Now if the circulant matrix A has first row a, then the i-th entry of the IPV of A is $aT^{-i}a^t$. Hence the IPV of the circulant matrix B with first row aN has i-th entry $aNT^{-i}N^{-1}a^t$, that is, $aT^{-id}a^t$. Hence the IPV of B is a permutation of the IPV of A, described as follows. Suppose that the IPV of

A is $(h_1, h_2, \ldots, h_w)$ and $(id)^*$ denotes the image of id in $\{0, 1, \ldots, v-1\}$ modulo v. Then the IPV of B is $(h_{1\theta}, h_{2\theta}, \ldots, h_{w\theta})$ where θ is the permutation on $\{1, 2, \ldots, w\}$ defined by

$$\theta : i \mapsto \begin{cases} (id)^* & \text{if } 1 \le (id)^* \le w, \\ v-(id)^* & \text{otherwise.} \end{cases} \qquad (3.12)$$

Note that $\theta = 1$ if and only if $N \in \langle T, R \rangle$. Hence the index of the normalizer of $\langle T \rangle$ in S_v is $v\phi(v)$, where ϕ is the Euler function. If v is prime, then the set E' of $v \times v$ permutation matrices which represent a permutation on $\{1, 2, \ldots, v\}$ which fixes v and $v-1$ is a complete set of distinct coset representatives of the normalizer of $\langle T \rangle$ in S_v.

For automatic computation this means that one of the lists, say L_1, may consist of elements $S' = \{B : B$ is a circulant matrix with first row a_1M for some $M \in E'\}$. This produces a considerably shorter list, and the search (Step 5) may be proportionally shorter in time.

The use of the normalizer of $\langle T \rangle$ in hand calculations is illustrated in the completion of Example 4.63 below. First, however, we show how the facts above may be used to construct a certain four-variable orthogonal design of order 28.

4.66. EXAMPLE. An orthogonal design of type $(1, 1, 1, 25)$ and order 28 can be constructed as follows. We want four 7×7 circulant matrices V_1, V_2, V_3, V_4 with entries from $\{0, \pm x_1, \pm x_2, \pm x_3, \pm x_4\}$ such that

$$\sum_{i=1}^{4} V_i V_i^t = (x_1^2 + x_2^2 + x_3^2 + 25x_4^2)I. \qquad (3.13)$$

The conditions (3.2), (3.5), (3.7), (3.8) imply that the sum and fill matrices of (V_1, V_2, V_3, V_4) must be $\mathrm{diag}(1, 1, 1, 5)$ and

$$\begin{bmatrix} & & & \\ & & & \\ & & & \\ 6 & 6 & 6 & 2 \end{bmatrix},$$

respectively. Hence V_4 must be $(J-2I)x_4$ up to isometry (see (3.3)); thus V_4 has IPV $(3x_4^2, 3x_4^2, 3x_4^2)$. Choose a skew-symmetric 7×7

matrix C_1 with entries from $\{0,1,-1\}$ and precisely one zero in each row; denote its IPV by (d_1, d_2, d_3). Now the normalizer of $\langle T\rangle$ in S_7 acts cyclically on (d_1, d_2, d_3) by (3.12), and further, it preserves skew-symmetry. Hence there are skew-symmetric circulant matrices C_2 and C_3 with IPV's (d_2, d_3, d_1) and (d_3, d_1, d_2), respectively. For $1 \le i \le 3$, denote $x_i I + x_4 C_i$ by V_i. It is clear that the IPV's of the V_i, $1 \le i \le 4$, add to (f, f, f), where $f = (d_1+d_2+d_3+3)x_4^2$. But since the sum and fill matrices of (V_1, V_2, V_3, V_4) satisfy (3.2), (3.5), (3.7), (3.8), it follows that $f+f+f = 0$; that is, $f = 0$. Hence the IPV's of the V_i add to zero, and thus the V_i satisfy (3.13).

Example 4.63 completed: The index of the normalizer of $\langle T\rangle$ in S_5 is 6, and so there are at most six circulants of order 5 with the same contents whose IPV's differ by more than just a permutation. A complete set of distinct coset representatives of this subgroup is

$$F = \{1, (12), (23), (34), (45), (51)\}.$$

Suppose that a solution (B_1, B_2, B_3, B_4) of (3.9) has sum matrix P (3.10) and fill matrix (3.11)(a). Using the set F, a list L_i of circulants with contents thus specified and essentially different IPV's can be made for each $i \in \{1, 2, 3, 4\}$. A short search reveals that if B_1, B_2, B_3, B_4 have first rows

$$(x_1, x_2, -x_3, x_3, -x_2),$$

$$(x_2, x_4, -x_3, -x_3, -x_4),$$

$$(x_3, x_2, x_4, -x_4, x_2),$$

$$(-x_4, x_4, x_4, x_4, x_4),$$

respectively, then the B_i satisfy (3.9).

Using similar methods it is possible to show that it is impossible to construct a (1, 3, 6, 8), (2, 2, 5, 5), or (3, 7, 8) in order 20 by using four circulants. It can also be shown that, while a (4, 9) exists in order 14, it is impossible to construct it from two circulants.

For ease of reference we summarize these results as:

4.67. LEMMA. (Eades). It is not possible to find four circulant matrices A_1, A_2, A_3, A_4 of order 5 with entries the commuting variables x_1, x_2, x_3, x_4, and 0 which satisfy

$$\sum_{i=1}^{4} A_i A_i^t = (\sum_{j=1}^{4} s_j x_j^2) I_5,$$

where (s_1, s_2, s_3, s_4) is (1,4,4,9), (1,3,6,8), (2,2,5,5) or (3,7,8). Equivalently, it is not possible to use four circulant matrices in the Goethals-Seidel array to construct orthogonal designs of these types in order 20.

4.68. LEMMA. (Eades). It is not possible to construct the orthogonal design of type (4, 9) in order 14 using two circulant matrices.

REMARK. It would be interesting to know if orthogonal designs of types (1,4,4,9), (1,3,6,8), (2,2,5,5) or (3,7,8) are impossible to construct by <u>any</u> method in order 20. If that were so, it would make the construction method by circulants assume even greater importance. We shall not even hazard a guess here, although experience should indicate that some of these designs will be impossible to construct by any method.

§4.4. BAUMERT-HALL ARRAYS

In 1933 Paley wrote a most important paper on the construction of Hadamard matrices which he called 'orthogonal matrices'. At the same time J. A. Todd realized that these matrices gave symmetric balanced incomplete block designs--of great interest in the design and analysis of experiments for agriculture and medicine.

Thus Paley opened the way for R. C. Bose's fundamental and path-finding use of Galois fields in the construction of balanced incomplete block designs--a most valuable contribution to applied statistics.

Yet it was not until Williamson's 1944 and 1947 papers that more Hadamard matrices were found. Williamson used what we would now call orthogonal designs of types (1,n-1) and (2,n-2) in order n.

Paley listed the orders less than 200 for which Hadamard matrices were not known, viz., 92, 116, 148, 156, 172, 184, and 188. Williamson suggested using what we will call the Williamson Array

$$\begin{bmatrix} A & B & C & D \\ -B & A & D & -C \\ -C & -D & A & B \\ -D & C & -B & A \end{bmatrix}$$

to find Hadamard matrices and in fact obtained the matrices of orders 148 and 172 by finding suitable matrices (using the theory we now call cyclotomy; see Storer [1] and §4.9) to replace the variables of the array. Thus we define

4.69. DEFINITION. Four (1,-1) matrices X_1, X_2, X_3, X_4 of order n which satisfy

$$\sum_{i=1}^{4} X_i X_i^t = 4nI, \quad X_i X_j^t = X_j X_i^t$$

will be called Williamson matrices. Eight (1,-1) matrices $X_1, \ldots, X_8$ of order n which satisfy

$$\sum_{i=1}^{8} X_i X_i^t = 8nI, \quad X_i X_j^t = X_j X_i^t$$

will be called eight Williamson matrices.

Baumert, Golomb and Hall [1] found Williamson matrices of order 23 giving the Hadamard matrices of orders 92 and 184. We can appreciate their excitement when on the night of September 27, 1961, after an hour of computer calculation, the output arrived. In fact, there turned out to be one and only one example of Williamson matrices of order 23.

Later, Baumert [2,3] was to find Williamson matrices giving the Hadamard matrix of order 116. We shall give the Hadamard matrix of order 188 in Section 4.6 (Lemma 4.114).

The remainder of this section is devoted to the exciting results that have come from Baumert and Hall's search for the Hadamard matrix of order 156. But first a definition.

4.70. DEFINITION. An orthogonal design of type (t,t,t,t) and order 4t will be called a Baumert-Hall array of order t.

Now Baumert and Hall realized that since Williamson matrices of order 13 were known, if a Baumert-Hall array of order 3 could be found, then the Hadamard matrix of order 156 would be found. In fact, they realized:

4.71. THEOREM. (Baumert-Hall). If there exists a Baumert-Hall array of order t and Williamson matrices of order n, then there exists an Hadamard matrix of order $4nt$; equivalently, if there exists an orthogonal design of type (t, t, t, t) in order $4t$ and Williamson matrices of order n, then there exists an Hadamard matrix of order $4nt$.

Proof. Replace the variables of the Baumert-Hall array by the Williamson matrices.

In 1965 Baumert and Hall [1] published the first Baumert-Hall array of order 3 (Display 4.4):

$$\left[\begin{array}{cccccccccccc}
A & A & A & B & -B & C & -C & -D & B & C & -D & -D \\
A & -A & B & -A & -B & -D & D & -C & -B & -D & -C & -C \\
A & -B & -A & A & -D & D & -B & B & -C & -D & C & -C \\
B & A & -A & -A & D & D & D & C & C & -B & -B & -C \\
B & -D & D & D & A & A & A & C & -C & B & -C & B \\
B & C & -D & D & A & -A & C & -A & -D & C & B & -B \\
D & -C & B & -B & A & -C & -A & A & B & C & D & -D \\
-C & -D & -C & -D & C & A & -A & -A & -D & B & -B & -B \\
D & -C & -B & -B & -B & C & C & -D & A & A & A & D \\
-D & -B & C & C & C & B & B & -D & A & -A & D & -A \\
C & -B & -C & C & D & -B & -D & -B & A & -D & -A & A \\
-C & -D & -D & C & -C & -B & B & B & D & A & -A & -A
\end{array}\right]$$

Display 4.4

Many attempts were made to generalize this array, but none were successful until in 1971 L. R. Welch found a Baumert-Hall array of order 5 (Display 4.5):

-D	B	-C	-C	-B	C	A	-D	-D	-A	-B	-A	C	-C	-A	A	-B	-D	D	-B
-B	-D	B	-C	-C	-A	C	A	-D	-D	-A	-B	-A	C	-C	-B	A	-B	-D	D
-C	-B	-D	B	-C	-D	-A	C	A	-D	-C	-A	-B	-A	C	D	-B	A	-B	-D
-C	-C	-B	-D	B	-D	-D	-A	C	A	C	-C	-A	-B	-A	-D	D	-B	A	-B
B	-C	-C	-B	-D	A	-D	-D	-A	C	-A	C	-C	-A	-B	-B	-D	D	-B	A
-C	A	D	D	-A	-D	-B	-C	-C	B	-A	B	-D	D	B	-B	-A	-C	C	-A
-A	-C	A	D	D	B	-D	-B	-C	-C	B	-A	B	-D	D	-A	-B	-A	-C	C
D	-A	-C	A	D	-C	B	-D	-B	-C	D	B	-A	B	-D	C	-A	-B	-A	-C
D	D	-A	-C	A	-C	-C	B	-D	-B	-D	D	B	-A	B	-C	C	-A	-B	-A
A	D	D	-A	-C	-B	-C	-C	B	-D	B	-D	D	B	-A	-A	-C	C	-A	-B
B	-A	-C	C	-A	A	B	-D	D	B	-D	-B	C	C	B	-C	A	-D	-D	-A
-A	B	-A	-C	C	B	A	B	-D	D	B	-D	-B	C	C	-A	-C	A	-D	-D
C	-A	B	-A	-C	D	B	A	B	-D	C	B	-D	-B	C	-D	-A	-C	A	-D
-C	C	-A	B	-A	-D	D	B	A	B	C	C	B	-D	-B	-D	-D	-A	-C	A
-A	-C	C	-A	B	B	-D	D	B	A	-B	C	C	B	-D	A	-D	-D	-A	-C
-A	-B	-D	D	-B	B	-A	C	-C	-A	C	A	D	D	-A	-D	B	C	C	-B
-B	-A	-B	-D	D	-A	B	-A	C	-C	-A	C	A	D	D	-B	-D	B	C	C
D	-B	-A	-B	-D	-C	-A	B	-A	C	D	-A	C	A	D	C	-B	-D	B	C
-D	D	-B	-A	-B	C	-C	-A	B	-A	D	D	-A	C	A	C	C	-B	-D	B
-B	-D	D	-B	-A	-A	C	-C	-A	B	A	D	D	-A	C	B	C	C	-B	-D

Display 4.5

For future reference we define:

4.72. DEFINITION. A Baumert-Hall-Welch array of order t is a Baumert-Hall array of order t constructed from sixteen circulant matrices.

The circulant structure of Welch's array gave the clue to generalizing Baumert-Hall arrays. First we consider:

4.73. DEFINITION. Four circulant (type 1) (0,1,-1) matrices X_i, $i = 1,2,3,4$, of order n which are nonzero for each of the n^2 entries for exactly one i, i.e., $X_i * X_j = 0$ for $i \neq j$, and which satisfy

$$\sum_{i=1}^{4} X_i X_i^t = nI$$

will be called T-matrices of order n.

A type 1 matrix has constant row (and column) sum; so:

4.74. LEMMA. Let X_i, $i = 1,\ldots,4$, be T-matrices with row sum (and column sum) x_i, respectively. Then

$$\sum_{i=1}^{4} x_i^2 = n.$$

Proof. $X_i J = x_i J$; so considering $\sum_{i=1}^{4} X_i X_i^t J = nJ$ gives the result.

The following result, in a slightly different form, was also discovered by R. J. Turyn.

4.75. THEOREM. (Cooper-Wallis). Suppose there exist T-matrices X_i, $i = 1,\ldots,4$, of order n. Let a, b, c, d be commuting variables. Then

$$A = aX_1+bX_2+cX_3+dX_4$$

$$B = -bX_1+aX_2+dX_3-cX_4$$

$$C = -cX_1-dX_2+aX_3+bX_4$$

$$D = -dX_1+cX_2-bX_3+aX_4$$

can be used in the Goethals-Seidel (or Wallis-Whiteman) array to obtain a Baumert-Hall array of order n; equivalently, if there exist T-matrices of order n, there exists an orthogonal design of type (n,n,n,n) in order $4n$.

Proof. By straightforward verification.

4.76. EXAMPLE. Let

$$X_1 = \begin{bmatrix} 1 & 0 & 0 \\ 0 & 1 & 0 \\ 0 & 0 & 0 \end{bmatrix}, \quad X_2 = \begin{bmatrix} 0 & 1 & 0 \\ 0 & 0 & 1 \\ 1 & 0 & 0 \end{bmatrix}, \quad X_3 = \begin{bmatrix} 0 & 0 & 1 \\ 1 & 0 & 0 \\ 0 & 1 & 0 \end{bmatrix}, \quad X_4 = 0.$$

Then X_1, X_2, X_3, X_4 are T-matrices of order 3, and the Baumert-Hall array of order 3 is in Display 4.6.

a	b	c	-b	a	d	-c	-d	a	-d	c	-b
c	a	b	a	d	-b	-d	a	-c	c	-b	-d
b	c	a	d	-b	a	a	-c	-d	-b	-d	c
b	-a	-d	a	b	c	-d	-b	c	c	-a	d
-a	-d	b	c	a	b	-b	c	-d	-a	d	c
-d	b	-a	b	c	a	c	-d	-b	d	c	-a
c	d	-a	d	b	-c	a	b	c	-b	d	a
d	-a	c	b	-c	d	c	a	b	d	a	-b
-a	c	d	-c	d	b	b	c	a	a	-b	d
d	-c	b	-c	a	-d	b	-d	-a	a	b	c
-c	b	d	a	-d	-c	-d	-a	b	c	a	b
b	d	-c	-d	-c	a	-a	b	-d	b	c	a

Display 4.6

We will not give the proofs here which can be found in Wallis [25, p. 360] and Hunt and Wallis [1] but will just quote the results given there. More results on Baumert-Hall arrays (see Proposition 4.114) are given in §4.6 after some new concepts have been introduced. In §4.9 we show how cyclotomy may be used in constructing these arrays, including the previously unpublished array of Hunt of order 61.

4.77. LEMMA. There exist Baumert-Hall arrays of order t, $t \in X$, $X = \{x : x$ is an odd integer, $0 \leq x \leq 25, 31, 37, 41, 61\}$.

4.78. COROLLARY. There exist Hadamard matrices of order $4tq$ where $t \in X$, X given in the previous corollary, and q is the order of Williamson matrices. In particular, there exist Hadamard matrices of order $4tq$, $q = \frac{1}{2}(p+1)$ or $\frac{1}{2}p(p+1)$ where $p \equiv 1 (\mathrm{mod}\ 4)$ is a prime power.

Proof. The required matrices are given in Corollaries 4.43 and 4.44.

A recent list of orders for which Williamson matrices are known is given in Wallis [23].

A most important theorem which shows how Baumert-Hall-Welch arrays can be used is now given. To date, the only such array known is that of order 5.

4.79. THEOREM. (Turyn). Suppose there is a Baumert-Hall-Welch array BHW of order s constructed of sixteen circulant (or type 1) $s \times s$

blocks. Further suppose there are T-matrices of order t. Then there is a Baumert-Hall array of order st.

Proof. Since BHW is constructed of sixteen circulant (or type 1) blocks, we may write $BHW = (N_{ij})$, $i, j = 1, 2, 3, 4$, where each N_{ij} is circulant (or type 1).

Since $(BHW)(BHW)^t = s(a^2+b^2+c^2+d^2)I_{4s}$ where a, b, c, d are the commuting variables, we have

$$N_{i1}N_{j1}^t + N_{i2}N_{j2}^t + N_{i3}N_{j3}^t + N_{i4}N_{j4}^t = \begin{cases} s(a^2+b^2+c^2+d^2)I_s, & i=j,\ i=1,2,3,4 \\ 0 & , i \neq j. \end{cases}$$

Suppose the T-matrices are T_1, T_2, T_3, T_4. Then form the matrices

$$A = T_1 \times N_{11} + T_2 \times N_{21} + T_3 \times N_{31} + T_4 \times N_{41}$$

$$B = T_1 \times N_{12} + T_2 \times N_{22} + T_3 \times N_{32} + T_4 \times N_{42}$$

$$C = T_1 \times N_{13} + T_2 \times N_{23} + T_3 \times N_{33} + T_4 \times N_{43}$$

$$D = T_1 \times N_{14} + T_2 \times N_{24} + T_3 \times N_{34} + T_4 \times N_{44}.$$

Now

$$AA^t + BB^t + CC^t + DD^t = st(a^2+b^2+c^2+d^2)I_{st},$$

and since A, B, C, D are type 1, they can be used in the Wallis-Whiteman generalization of the Goethals-Seidel array to obtain the desired result. (See also Lemma 4.22.)

Since the Baumert-Hall array of order 5 given by Welch is constructed of sixteen circulant blocks, we have:

4.80. COROLLARY. Suppose there are T-matrices of order t. Then there is a Baumert-Hall array of order $5t$; equivalently, there is an orthogonal design of type $(5t, 5t, 5t, 5t)$ in order $20t$.

As we have seen, Baumert and Hall's array of order 3, discovered to obtain the Hadamard matrix of order 156, has led to one of the most powerful constructions for Hadamard matrices.

In fact, to prove the Hadamard conjecture it would be sufficient to prove:

4.81. CONJECTURE. There exists a Baumert-Hall array of order t for every positive integer t, or equivalently, there exists an orthogonal design of type (t, t, t, t) in order $4t$ for every positive integer t.

§4.5. PLOTKIN ARRAYS

Following the exciting results on Baumert-Hall arrays, which if they all exist, would answer the Hadamard conjecture in the affirmative, it became clear that similar designs in order $8n$ would give results of great import. Alas, as we shall now see, such designs of order $8n$, n odd, are very hard to find.

These classes of orthogonal designs are of great interest and worthy of further study.

4.82. DEFINITION. An orthogonal design of type (t, t, t, t, t, t, t, t) in order $8t$ will be called a <u>Plotkin array</u>.

REMARK. Matrices with elements $\{1, -1\}$ which can be used in Plotkin arrays to give Hadamard matrices (eight Williamson matrices) have been found by J. Wallis [23], and of course Williamson matrices (each used twice) will also suffice. Still the problem of finding suitable matrices to replace the variables in designs to give Hadamard matrices or weighing matrices is largely untouched.

We first see that if an Hadamard matrix exists, then Plotkin arrays exist in four times the order. In view of our results (given later), this means that Plotkin arrays always exist in orders $2^t q$, q odd, and $t > [2 \log_2 (q-3)]+2$.

4.83. THEOREM. (Plotkin). Suppose there exists an Hadamard matrix of order $2t$. Then there exists an orthogonal design of type (t, t, t, t, t, t, t, t) in order $8t$.

<u>Proof.</u> Let H be an Hadamard matrix of order $2t$. Let

$$S = \frac{1}{2}\begin{pmatrix} I & -I \\ I & I \end{pmatrix} H, \quad T = \frac{1}{2}\begin{pmatrix} I & I \\ -I & I \end{pmatrix} H,$$

$$U = \frac{1}{2}\begin{pmatrix} I & -I \\ -I & -I \end{pmatrix} H, \quad V = \frac{1}{2}\begin{pmatrix} I & I \\ I & -I \end{pmatrix} H.$$

Then define

$$H_{2t}(a,b) = (S \times a) + (T \times b),$$

$$H_{4t}(a,b,c,d) = \begin{bmatrix} H_{2t}(a,b) & H_{2t}(c,d) \\ H_{2t}(-c,d) & H_{2t}(a,-b) \end{bmatrix},$$

and

$$B_{4t}(a,b,c,d) = \begin{bmatrix} S \times a + T \times b & U \times c + V \times d \\ U \times (-c) + V \times (-d) & S \times a + T \times b \end{bmatrix}.$$

Then

$$H_{8t}(x_1,x_2,x_3,x_4,x_5,x_6,x_7,x_8) = \begin{bmatrix} H_{4t}(x_1,x_2,x_3,x_4) & B_{4t}(x_5,x_6,x_7,x_8) \\ B_{4t}(x_5,x_6,x_7,x_8) & -H_{4t}(-x_1,x_2,x_3,x_4) \end{bmatrix}$$

is the required Plotkin array.

Only two Plotkin arrays are known of order $8t$, t odd! The 8×8 matrix of Theorem 4.1, which is unique under the equivalence operations, (i) multiply any row or column by -1, (ii) interchange any pair of rows or columns, (iii) replace any variable by its negative throughout, is a design of type $(1,1,1,1,1,1,1,1)$. Plotkin found that the following matrix is equivalent under (i), (ii) and (iii) to the Baumert-Hall array of the previous section.

$$A(x,y,z,w) = \left[\begin{array}{cccc|cccc|cccc} y & x & x & x & -z & z & w & y & -w & w & z & -y \\ -x & y & x & -x & w & -w & z & -y & -z & z & -w & -y \\ -x & -x & y & x & w & -y & -y & w & z & z & w & -z \\ -x & x & -x & y & -w & -w & -z & w & -z & -y & -y & -z \\ -y & -y & -z & -w & z & x & x & x & -w & -w & z & -y \\ -w & -w & -z & y & -x & z & x & -x & y & y & -z & -w \\ w & -w & w & -y & -x & -x & z & x & y & -z & -y & -z \\ -w & -z & w & -z & -x & x & -x & z & -y & y & -y & w \\ -y & y & -z & -w & -z & -z & w & y & w & x & x & x \\ z & -z & -y & -w & -y & -y & -w & -z & -x & w & x & -x \\ -z & -z & y & z & -y & -w & y & -w & -x & -x & w & x \\ z & -w & -w & z & y & -y & y & z & -x & x & -x & w \end{array}\right] \qquad (5.1)$$

Also, the next matrix is a Baumert-Hall array of order 12, but is not equivalent to (5.1).

$$B(x,y,z,w) = \left[\begin{array}{rrrrrrrrrrrr}
y & x & x & x & -w & w & z & y & -z & z & w & -y \\
-x & y & x & -x & -z & z & -w & -y & w & -w & z & -y \\
-x & -x & y & x & -y & -w & y & -w & -z & -z & w & z \\
-x & x & -x & y & w & w & -z & -w & -y & z & y & z \\
 & & & & & & & & & & & \\
-w & -w & -z & -y & z & x & x & x & -y & -y & z & -w \\
y & y & -z & -w & -x & z & x & -x & -w & -w & -z & y \\
-w & w & -w & -y & -x & -x & z & x & z & y & y & z \\
z & -w & -w & z & -x & x & -x & z & y & -y & y & w \\
 & & & & & & & & & & & \\
z & -z & y & -w & y & y & w & -z & w & x & x & x \\
y & -y & -z & -w & -z & -z & -w & -y & -x & w & x & -x \\
z & z & y & -z & w & -y & -y & w & -x & -x & w & x \\
-w & -z & w & -z & -y & y & -y & z & -x & x & -x & w
\end{array}\right]. \qquad (5.2)$$

Then we have

4.84. LEMMA. There is a Plotkin array of order 24, i.e., an orthogonal design of type (3, 3, 3, 3, 3, 3, 3, 3) in order 24.

Proof.

$$\begin{bmatrix} A(x_1,x_2,x_3,x_4) & B(x_5,x_6,x_7,x_8) \\ B(-x_5,x_6,x_7,x_8) & -A(-x_1,x_2,x_3,x_4) \end{bmatrix}$$

is the required design.

These results lead to:

4.85. CONJECTURE. (Plotkin). There exist Plotkin arrays in every order 8n, n a positive integer.

§4.6. GOLAY SEQUENCES AND OTHER SEQUENCES WITH ZERO AUTO-CORRELATION FUNCTION--CONSTRUCTIONS USING COMPLEMENTARY SEQUENCES TO FORM ORTHOGONAL DESIGNS AND WEIGHING MATRICES

Sequences with zero or small auto-correlation function have long been of interest to engineers working on signal processing--for example, in radar and sonar.

R. J. Turyn [2] was the first to apply an idea common in the fields of radar pulse compression and very recent work in surface wave encodings in constructing Baumert-Hall arrays.

The early work of Golay was concerned with two (1, -1) sequences, but Welti approached the subject from the point of view of two orthonormal vectors, each corresponding to one of two orthogonal waveforms. Later work, including Turyn's [6], used four or more sequences.

Other authors who have worked in this area are Jaurequi [1], Kruskal [1], Squire, Whitehouse and Aleup [1], Taki, et al., [1] and Tseng [1].

Since we are concerned with orthogonal designs, we shall consider sequences of commuting variables.

Let $X = \left\{\{a_{11}, \ldots, a_{1n}\}, \{a_{21}, \ldots\} \ldots \{a_{2n}, \ldots\} \ldots \{a_{m1}, \ldots, a_{mn}\}\right\}$ be m sequences of commuting variables of length n.

4.86. DEFINITION. (1) The <u>nonperiodic auto-correlation function of the family of sequences</u> X (denoted N_X) is a function defined by

$$N_X(j) = \sum_{i=1}^{n-j} (a_{1,i}a_{1,i+j} + a_{2,i}a_{2,i+j} + \ldots + a_{m,i}a_{m,i+j}).$$

Note that if the following collection of m matrices of order n is formed,

$$\begin{bmatrix} a_{11}a_{12}\ldots a_{1n} \\ a_{11} \quad a_{1,n-1} \\ \ddots \\ \bigcirc \quad a_{11} \end{bmatrix}, \begin{bmatrix} a_{21}a_{22}\ldots a_{2n} \\ a_{21} \quad a_{2,n-1} \\ \ddots \\ \bigcirc \quad a_{21} \end{bmatrix}, \ldots, \begin{bmatrix} a_{m1}a_{m2}\ldots a_{mn} \\ \ddots \quad a_{m,n-1} \\ \bigcirc \quad a_{m1} \end{bmatrix}$$

then $N_X(j)$ is simply the sum of the inner products of rows 1 and $j+1$ of these matrices.

(2) The <u>periodic auto-correlation function of the family of sequences</u> X (denoted P_X) is a function defined by

$$P_X(j) = \sum_{i=1}^{n} (a_{1,i}a_{1,i+j} + a_{2,i}a_{2,i+j} + \ldots + a_{m,i}a_{m,i+j}),$$

where we assume the second subscript is actually chosen from the complete set of residues mod n.

We can interpret the function P_X in the following way: Form the

m circulant matrices which have first rows, respectively, $[a_{11}a_{12}\dots a_{1n}], [a_{21}a_{22}\dots a_{2n}], \dots, [a_{m1}a_{m2}\dots a_{mn}]$; then $P_X(j)$ is the sum of the inner products of rows 1 and $j+1$ of these matrices.

We say the <u>weight</u> of X is the number of nonzero entries in X.

If X is as above with $N_X(j) = 0$, $j = 1, 2, \dots, n-1$, then we will call X <u>m-complementary sequences</u> of length n.

If $X = \{A_1, A_2, \dots, A_m\}$ are m-complementary sequences of length n and weight $2k$ such that $Y = \{(A_1+A_2)/2, (A_1-A_2)/2, \dots, (A_{2i-1}+A_{2i})/2, (A_{2i-1}-A_{2i})/2, \dots\}$ are also m-complementary sequences (of weight k), then X will be said to be <u>m-complementary disjointable sequences</u> of length n. X will be said to be <u>m-complementary disjoint sequences</u> of length n if all $\binom{m}{2}$ pairs of sequences are disjoint, i.e., $A_i * A_j = 0$ for all i, j, where $*$ is the Hadamard product.

NOTATION. We use - for -1, and $\bar{x}$ for $-x$, and A^* to mean the order of the entries in the sequence A are reversed.

One more piece of notation is in order. If g_r denotes a sequence of integers of length r, then by xg_r we mean the sequence of integers of length r obtained from g_r by multiplying each member of g_r by x.

4.87. PROPOSITION. Let X be a family of sequences as above; then

$$P_X(j) = N_X(j) + N_X(n-j), \quad j = 1, \dots, n-1.$$

4.88. COROLLARY. If $N_X(j) = 0$ for all $j = 1, \dots, n-1$, then $P_X(j) = 0$ for all $j = 1, \dots, n-1$.

NOTE. $P_X(j)$ may equal 0 for all $j = 1, \dots, n-1$, even though the $N_X(j)$ are not.

4.89. DEFINITION. If $X = \left\{\{a_1, \dots, a_n\}, \{b_1, \dots, b_n\}\right\}$ are two sequences where $a_i, b_j \in \{1, -1\}$ and $N_X(j) = 0$ for $j = 1, \dots, n-1$, then the sequences in X are called <u>Golay complementary sequences of length</u> n. E.g., n = 2   11 and 1-

n = 10   1--1-1---1 and 1------11-

n = 26   111--111-1-----1-11--1----
---11---1-11-1-1-11--1---- .

We note that if X is as above and A is the circulant matrix with first row $\{a_1,\ldots,a_n\}$ and B the circulant matrix with first row $\{b_1,\ldots,b_n\}$, then

$$AA^t+BB^t = \Sigma(a_i^2+b_i^2)I_n.$$

Consequently, such matrices may be used to obtain Hadamard matrices constructed from two circulants.

We would like to use Golay sequences to construct other orthogonal designs, but first we consider some of their properties.

4.90 LEMMA. Let $X = \left\{\{a_1,\ldots,a_n\},\{b_1,\ldots,b_n\}\right\}$ be Golay complementary sequences of length n. Suppose k_1 of the a_i are positive and k_2 of the b_i are positive; then

$$n = (k_1+k_2-n)^2 + (k_1-k_2)^2,$$

and n is even.

Proof. Since $P_X(j) = 0$ for all j, we may consider the two sequences as $2-\{n;k_1,k_2;\lambda\}$ supplementary difference sets with $\lambda = k_1+k_2-\frac{1}{2}n$. But the parameters (counting difference two ways) satisfy $\lambda(n-1) = k_1(k_1-1) + k_2(k_2-1)$. On substituting λ in this equation we obtain the result of the enunciation.

The smaller values of n, k_1, k_2 of the lemma are considered in Table 6.1. We note that the lemma says there is "no solution" when the length n is not the sum of two squares. "OK" in the table indicates an order for which Golay sequences exist (see later in this section). Malcolm Griffin has shown (see Theorem 4.94) no Golay sequences can exist for lengths $n = 2.9^t$. The value n = 18 had previously been excluded by a complete search but is now theoretically excluded by Griffin's theorem and independently by a result of Kruskal and C. H. Yang. Recent work of Andres and Stanton [1] has led to a greatly improved computer algorithm which has shown there are no Golay sequences of length $n = 34, 36, 50$ or 58.

n	k_1	k_2	(2)	n	k_1	k_2	(2)
2	1	0	OK	32	12	16	OK
4	1	1	OK	34	16	13	do not exist
6			no solution	36	15	15	do not exist
8	4	2	OK	38			no solution
10	4	3	OK	40	18	16	OK
12			no solution	42			no solution
14			no solution	44			no solution
16	6	6	OK	46			no solution
18	9	6	do not exist	48			no solution
20	9	7	OK	50	25	20	do not exist
22			no solution	52	31	27	OK
24			no solution	54			no solution
26	11	10	OK	56			no solution
28			no solution	58	34	31	do not exist
30			no solution				

Table 6.1

4.91. LEMMA. For Golay sequences $X = \left\{\{x_i\}, \{y_i\}\right\}$ of length n

$$x_{n-i+1} = e_i x_i \iff y_{n-i+1} = -e_i y_i,$$

where $e_i = \pm 1$. That is,

$$x_{n-i+1} x_i = -y_{n-i+1} y_i.$$

4.92. EXAMPLE. The sequences of length 10 are

1--1-1---1

1------11- .

Clearly, $e_1 = 1$, $e_2 = 1$, $e_3 = 1$, $e_4 = -1$, and $e_5 = -1$.

Proof. We use the fact that if x, y, z are ± 1, $(x+y)z \equiv x+y \pmod 4$ and $x+y \equiv xy+1 \pmod 4$.

Let $i = 1$. Clearly the result holds. We proceed by induction. Suppose the result is true for every $i \le k-1$. Now consider $N(j) = N(n-k) = 0$, and we have

$$\begin{aligned} 0 &= x_1 x_{n+1-k} + x_2 x_{n+2-k} + \dots + x_k x_n + y_1 y_{n+1-k} + y_2 y_{n+2-k} + \dots + y_k y_n \\ &= x_1 e_k x_k + x_2 e_{k-1} x_{k-1} + \dots + x_k e_1 x_1 + y_1 y_{n+1-k} - y_2 e_{k-1} y_{k-1} - \dots - y_k e_1 y_1 \\ &\equiv e_1 + e_2 + \dots + e_k + y_1 y_{n+1-k} - e_{k-1} - \dots - e_2 - y_k e_1 y_1 \pmod 4 \\ &\equiv e_1 + e_k + y_1 y_{n+1-k} - y_k e_1 y_1 \pmod 4 \\ &\equiv e_k + y_k y_{n+1-k} \pmod 4 \\ &\equiv 0 \pmod 4. \end{aligned}$$

So

$$y_{n+1-k} = -e_k y_k.$$

4.93. LEMMA. (Griffin). Let $a_0, a_1, \dots, a_m$ and $b_0, b_1, \dots, b_m$ be Golay sequences of length $2n = m+1$. We recall that $a_{2n-1-i} = e_i a_i \iff b_{2n-1-i} = -e_i b_i$ and write $S = \{0 \le i < n : e_i = 1\}$ and $D = \{0 \le i < n : e_i = -1\}$. Let η be a $2n$-th root of unity and ζ be a $4n$-th root of unity. Then

$$\left|\sum_{i=0}^{m} a_i \eta^i\right|^2 + \left|\sum_{i=0}^{m} b_i \eta^i\right|^2 = 4n, \tag{1}$$

and

$$\begin{aligned} &\left|\sum_{i\in S} a_i \operatorname{Re}(\zeta^{2i+1})\right|^2 + \left|\sum_{i\in D} a_i \operatorname{Im}(\zeta^{2i+1})\right|^2 + \\ &+ \left|\sum_{i\in S} b_i \operatorname{Im}(\zeta^{2i+1})\right|^2 + \left|\sum_{i\in D} b_i \operatorname{Re}(\zeta^{2i+1})\right|^2 = n. \end{aligned} \tag{2}$$

REMARK. Condition (2) is not equivalent to being a Golay sequence, as can be seen from the sequences $X = \left\{\{1,1,1,1,-1,-1,1,1\}, \{1,-1,-1,1,1,-1,1,-1\}\right\}$ which satisfy (2) but for which $N_X \neq 0$.

<u>Proof.</u> We recall that for Golay sequences X, $N_X = 0 \implies P_X = 0$. Thus

if $T = (t_{ij})$ is defined by $t_{12} = t_{i,i+1} = 1$ and all other elements zero, we have, writing

$$A = \sum_{i=0}^{m} a_i T^i \quad \text{and} \quad B = \sum_{i=0}^{m} b_i T^i,$$

$$AA^t + BB^t = 4nI.$$

Since $TT^t = T^{2n} = I$ and T has characteristic polynomial $\lambda^{2n}-1$, there exists a unitary matrix U such that $USU^* = \Delta = \mathrm{diag}(1, \xi, \xi^2, \ldots, \xi^m)$, where ξ is a primitive $2n$-th root of unity. Consequently

$$\left(\sum_{i=0}^{m} a_i \Delta^i\right)\left(\sum_{i=0}^{m} a_i \Delta^i\right)^* + \left(\sum_{i=0}^{m} b_i \Delta^i\right)\left(\sum_{i=0}^{m} b_i \Delta^i\right)^* = 4nI.$$

Let $\eta = \xi^{j-1}$; then the j-th term on the main diagonal gives

$$\begin{aligned} 4n &= \left(\sum_{i=0}^{m} a_i \eta^i\right)\left(\sum_{i=0}^{m} a_i \eta^{-i}\right) + \left(\sum_{i=0}^{m} b_i \eta^i\right)\left(\sum_{i=0}^{m} b_i \eta^{-i}\right) \\ &= \left|\sum_{i=0}^{m} a_i \eta^i\right|^2 + \left|\sum_{i=0}^{m} b_i \eta^i\right|^2. \end{aligned} \tag{1}$$

We now substitute $a_{2n-i-1} = e_i a_i$ and $b_{2n-i-1} = e_i b_i$, $0 \le i < n$, in (1) with $\eta = \zeta^2$. Thus

$$\begin{aligned} 4n &= \left|\sum_{i=0}^{n-1} (a_i \eta^i + e_i a_i \eta^{-(i+1)})\right|^2 + \left|\sum_{i=0}^{n-1} (b_i \eta^i - e_i b_i \eta^{-(i+1)})\right|^2 \\ &= \left|\sum_{i\in S} a_i(\zeta^{2i} + \zeta^{-(2i+2)}) + \sum_{i\in D} a_i(\zeta^{2i} - \zeta^{-(2i+2)})\right|^2 \\ &\quad + \left|\sum_{i\in S} b_i(\zeta^{2i} - \zeta^{-(2i+2)}) + \sum_{i\in D} b_i(\zeta^{2i} + \zeta^{-(2i+2)})\right|^2 \\ &= \left|(\sum_{i\in S} a_i(\zeta^{2i+1} + \zeta^{-(2i+1)}) + \sum_{i\in D} a_i(\zeta^{2i+1} - \zeta^{-(2i+1)})\right|^2 \\ &\quad + \left|(\sum_{i\in S} b_i(\zeta^{2i+1} - \zeta^{-(2i+1)}) + \sum_{i\in D} b_i(\zeta^{2i+1} + \zeta^{-(2i+1)})\right| \end{aligned}$$

$$= 4\left|\sum_{i\in S} a_i \mathrm{Re}(\zeta^{2i+1})\right|^2 + 4\left|\sum_{i\in D} a_i \mathrm{Im}(\zeta^{2i+1})\right|^2$$

$$+ 4\left|\sum_{i\in S} b_i \mathrm{Im}(\zeta^{2i+1})\right|^2 + 4\left|\sum_{i\in D} b_i \mathrm{Re}(\zeta^{2i+1})\right|^2$$

giving (2).

We now use condition (2) to prove:

4.94. THEOREM. (Griffin). There are no Golay sequences of length 2.9^t.

<u>Proof.</u> We show that equation (2) cannot be satisfied for all 12-th roots when n is a power of 3.

Setting $\xi = -1$ in (2), we get

$$(\sum_S a_i)^2 + (\sum_D b_i)^2 = n. \tag{3}$$

If n is a product of primes, each $\equiv 3 \pmod 4$, then by elementary number theory, n must be a square, $n = q^2$, and solutions to this equation must be unique (up to order and sign), yielding

$$\sum_S a_i + \sum_D b_i = \alpha q, \quad \alpha = \pm 1. \tag{4}$$

If $3|q$, we use $\xi = \frac{1}{2} + \frac{1}{2}\sqrt{-3}$ ($\xi^3 = -1$, $\xi^5 = \xi^{-1}$) in (2) to give a Diophantine equation of the form

$$3(c^2+d^2) + e^2+f^2 = 4q^2, \tag{5}$$

and since the summation in forming c involves only $i \in D$ with $i \not\equiv 1 \pmod 3$ and the summation for d involves $i \in S$ with $i \not\equiv 1 \pmod 3$, c and d involve $2/3$ of the q^2 elements, and consequently $c \equiv d \pmod 2$. Taking residues mod 8 shows that c, d, e, f are all even, giving

$$3(c_0^2+d_0^2) + e_0^2 + f_0^2 = q^2.$$

When q is a power of 3, this equation has a unique solution (up to order and sign), so $c = d = 0$ and $e+f = 2\beta q$, $\beta = \pm 1$. From now on assume that q is a power of 3. Let $S(0) = \{i \in S \mid i \equiv 0, 5 \pmod 6\}$, $S(1) = \{i \in S \mid i \equiv 1, 4 \pmod 6\}$ and $S(2) = \{i \in S \mid i \equiv 2, 3 \pmod 6\}$ with a similar notation for D. Then $e+f = 2\beta q$ gives

$$\sum_S a_i - 3\sum_{S(1)} a_i + \sum_D b_i - 3\sum_{D(1)} b_i = 2\beta q. \qquad (6)$$

The same argument with $\xi = \frac{1}{2}(\sqrt{3}+\sqrt{-1})$ a primitive 12-th root again yields an equation of type (5), and $c = d = 0$ gives $c+d = 0$; so

$$\sum_{S(0)} a_i - \sum_{S(2)} a_i + \sum_{D(0)} b_i - \sum_{D(2)} b_i = 0. \qquad (7)$$

Adding (6), twice (4), and thrice (7), we get

$$6\Big(\sum_{S(0)} a_i + \sum_{D(0)} b_i\Big) = 2q(\alpha+\beta) \equiv 0 \pmod 4.$$

But this is a contradiction since $S(0) \cup D(0)$ has an odd number of elements, $q^2/3$.

OTHER SEQUENCES WITH ZERO AUTO-CORRELATION FUNCTION

We now discuss other sequences with zero auto-correlation function.

4.95. LEMMA. Suppose $X = \{X_1, X_2, \ldots, X_m\}$ is a set of $(0,1,-1)$ sequences of length n for which $N_X = 0$ or $P_X = 0$. Further suppose the weight of X_i is x_i and the sum of the elements of X_i is a_i. Then

$$\sum_{i=1}^{m} a_i^2 = \sum_{i=1}^{m} x_i.$$

Proof. Form circulant matrices Y_i for each X_i. Then

$$Y_i J = a_i J \quad \text{and} \quad \sum_{i=1}^{m} Y_i Y_i^t = \sum_{i=1}^{m} x_i.$$

Now considering

$$\sum_{i=1}^{m} Y_i Y_i^t J = \sum_{i=1}^{m} a_i^2 J = \sum_{i=1}^{m} x_i J,$$

we have the result.

Now a few simple observations are in order, and for convenience we put them together as a lemma.

4.96. LEMMA. Let $X = \{A_1, A_2, \ldots, A_m\}$ be m-complementary sequences of length n. Then

(i) $Y = \{A_1^*, A_2^*, \ldots, A_i^*, A_{i+1}, \ldots, A_m\}$ are m-complementary sequences of length n;

(ii) $W = \{A_1, A_2, \ldots, A_i, -A_{i+1}, \ldots, -A_m\}$ are m-complementary sequences of length n;

(iii) $Z = \{\{A_1, A_2\}, \{A_1, -A_2\}, \ldots, \{A_{2i-1}, A_{2i}\}, \{A_{2i-1}, -A_{2i}\}, \ldots\}$ are m- (or m+1 if m was odd when we let A_{m+1} be n zeros) complementary sequences of length 2n;

(iv) $U = \{\{A_1/A_2\}, \{A_1/-A_2\}, \ldots, \{A_{2i-1}/A_{2i}\}, \{A_{2i-1}/-A_{2i}\}, \ldots\}$, where A_j/A_k means $a_{j1}a_{k1}a_{j2}a_{k2}\ldots a_{jn}a_{kn}$, are m- (or m+1 if m was odd when we let A_{m+1} be n zeros) complementary sequences of length 2n.

By a lengthy but straightforward calculation, it can be shown that:

4.97. THEOREM. Suppose $X = \{A_1, \ldots, A_{2n}\}$ are 2m-complementary sequences of length n and weight ℓ and $Y = \{B_1, B_2\}$ are 2-complementary disjointable sequences of length t and weight 2k. Then there are 2m-complementary sequences of length nt and weight $k\ell$.

The same result is true if X are 2m-complementary disjointable sequences of length n and weight 2ℓ and Y are 2-complementary sequences of weight k.

<u>Proof.</u> Using an idea of R. J. Turyn, we consider

$$A_{2i-1} \times (B_1+B_2)/2 + A_{2i} \times (B_1^*-B_2^*)/2 \text{ and}$$

$$A_{2i-1} \times (B_1-B_2)/2 - A_{2i} \times (B_1^*+B_2^*)/2,$$

for $i = 1, \ldots, m$, which are the required sequences in the first case, and

$$(A_{2i-1}+A_{2i})/2 \times B_1 + (A_{2i-1}-A_{2i})/2 \times B_2^* \text{ and}$$

$$(A_{2i-1}+A_{2i})/2 \times B_2 - (A_{2i-1}-A_{2i})/2 \times B_1^*,$$

for $i = 1, \ldots, m$, which are the required sequences for the second case.

The proof now follows by an exceptionally tedious but straightforward verification.

4.98. COROLLARY. Since there are Golay sequences of lengths 2, 10, and 26, there are Golay sequences of length $2^a 10^b 26^c$ for a, b, c non-negative integers.

4.99. COROLLARY. There are 2-complementary sequences of lengths $2^a 6^b 10^c 14^d 26^e$ of weights $2^a 5^b 10^c 13^d 26^e 2$, where a, b, c, d, e are non-negative integers.

Proof. Use the sequences of Tables 5 and 6 of Appendix H.

We now propose to use R. J. Turyn's idea of using m-complementary sequences to construct orthogonal designs.

CONSTRUCTIONS USING COMPLEMENTARY SEQUENCES TO FORM BAUMERT-HALL ARRAYS

4.100. LEMMA. Consider four (1, -1) sequences $A = \{X, U, Y, W\}$, where

$$X = \{x_1 = 1, x_2, x_3, \ldots, x_m, h_m x_m, \ldots, h_3 x_3, h_2 x_2, h_1 x_1 = -1\},$$

$$U = \{u_1 = 1, u_2, u_3, \ldots, u_m, f_m u_m, \ldots, f_3 u_3, f_2 u_2, f_1 u_1 = 1\},$$

$$Y = \{y_1, y_2, \ldots, y_{m-1}, y_m, g_{m-1} y_{m-1}, \ldots, g_3 y_3, g_2 y_2, g_1 y_1\},$$

$$V = \{v_1, v_2, \ldots, v_{m-1}, v_m, e_{m-1} v_{m-1}, \ldots, e_3 v_3, e_2 v_2, e_1 v_1\}.$$

Then $N_A = 0$ implies that $h_i = f_i$ and $g_j = e_j$. Here

$$8m-2 = \left(\sum_{i=1}^{m} x_i + x_i h_i\right)^2 + \left(\sum_{i=1}^{m} u_i + u_i f_i\right)^2 + \left(y_m + \sum_{i=1}^{m-1} y_i + y_i g_i\right)^2 +$$

$$+ \left(v_m + \sum_{i=1}^{m-1} v_i + v_i e_i\right)^2 .$$

Proof. We note that if a, b, x, y, z are all ± 1, $a+b \equiv ab+1 \pmod 4$, and so $x+xyz \equiv y+z \pmod 4$. Clearly, $N_A(2m-1) = 0$ gives $-h_1 = f_1 = 1$, and

$$\begin{aligned} N_A(2m-2) &= x_1x_2h_2 + x_2h_1x_1 + f_2u_1u_2 + u_2f_1u_1 + g_1y_1^2 + e_1v_1^2 \\ &\equiv h_1+h_2+f_1+f_2+g_1+e_1 \pmod 4 \\ &\equiv h_2f_2 + g_1e_1 + 2 \pmod 4 \\ &\equiv 0 \pmod 4). \end{aligned}$$

This gives $h_2f_2 = g_1e_1$. We proceed by induction to show that $h_if_i = g_{i-1}e_{i-1}$ for all $i \leq m$.

Assume $h_if_i = g_{i-1}e_{i-1}$, i.e., $h_i+f_i+g_{i-1}+e_{i-1} \equiv 0 \pmod 4$, for all $i < k \leq m$. Now consider

$$\begin{aligned} N_A(2m-k) &= (x_1x_kh_k+x_2x_{k-1}h_{k-1} + \dots + x_{k-1}x_2h_2+x_kx_1h_1) \\ &\quad + (u_1u_kf_k+u_2u_{k-1}f_{k-1} + \dots + u_{k-1}u_2f_2+u_ku_1f_1) \\ &\quad + (y_1y_{k-1}g_{k-1}+y_2y_{k-2}g_{k-2} + \dots + y_{k-2}y_2g_2+y_{k-1}y_1g_1) \\ &\quad + (v_1v_{k-1}e_{k-1}+v_2v_{k-2}e_{k-2} + \dots + v_{k-2}v_2e_2+v_{k-1}v_1e_1) \\ &\equiv h_1 + \dots + h_k+f_1 + \dots + f_k+g_1 + \dots + g_{k-1}+e_1 + \dots + \\ &\qquad e_{k-1} \pmod 4 \\ &\equiv h_kf_k + g_{k-1}e_{k-1} + 2 \pmod 4 \\ &\equiv 0 \pmod 4). \end{aligned}$$

This gives the result for all $i \leq m$.

Suppose $k = m+j < m$. Then

$$\begin{aligned} N_A(2m-k) &= (x_1x_{m-j+1}+\dots+x_jx_m) + (x_{j+1}h_mx_m+\dots+x_mh_{j+1}x_{j+1}) \\ &\quad + (h_mx_mh_jx_j+\dots+h_{m-j+1}x_{m-j+1}h_1x_1) \\ &\quad + (u_1u_{m-j+1}+\dots+u_ju_m) + (u_{j+1}f_mu_m+\dots+u_mf_{j+1}u_{j+1}) \\ &\quad + (f_mu_mf_ju_j+\dots+f_{m-j+1}u_{m-j+1}f_1u_1) \\ &\quad + (y_1y_{m-j+1}+\dots+y_jy_m) + (y_{j+1}g_{m-1}y_{m-1}+\dots+ \\ &\qquad y_{m-1}g_{j+1}y_{j+1}) \\ &\quad + (y_mg_jy_j+\dots+g_{m-j+1}y_{m-j+1}g_1y_1) \\ &\quad + (v_1v_{m-j+1}+\dots+v_jm_j) + (v_{j+1}e_{m-1}v_{m-1}+\dots+ \\ &\qquad y_{m-1}e_{j+1}v_{j+1}) \end{aligned}$$

$$+ (v_m e_j v_j + \ldots + e_{m-j+1} v_{m-j+1} e_1 v_1)$$

$$\equiv (h_1 + \ldots + h_m + h_{m-j+1} + \ldots + h_m) + (f_1 + \ldots + f_m + f_{m-j+1} + \ldots + f_m)$$

$$+ (g_1 + \ldots + g_{m-1} + g_{m-j+1} + \ldots + g_{m-1} + 1)$$

$$+ (e_1 + \ldots + e_{m-1} + e_{m-j+1} + \ldots + e_{m-1} + 1) \pmod 4$$

$$\equiv h_1 f_1 + h_{m-j+1} f_{m-j+1} \pmod 4$$

$$\equiv h_{m-j+1} f_{m-j+1} - 1 \pmod 4.$$

Hence $h_{m-j+1} f_{m-j+1} = 1$. So in general $h_i f_i = 1$ and $e_i g_i = 1$.

The last result follows by Lemma 4.95.

4.101. COROLLARY. Consider four (1, -1) sequences $A = \{X, U, Y, V\}$, where

$$X = \{x_1 = 1, x_2, x_3, \ldots, x_m, -x_m, \ldots, -x_3, -x_2, -x_1 = -1\},$$

$$U = \{u_1 = 1, u_2, u_3, \ldots, u_m, f_m u_m, \ldots, f_3 u_3, f_2 u_2, f_1 u_1 = 1\},$$

$$Y = \{y_1, y_2, \ldots, y_{m-1}, y_m, y_{m-1}, \ldots, y_3, y_2, y_1\},$$

$$V = \{v_1, v_2, \ldots, v_{m-1}, v_m, e_{m-1} v_{m-1}, \ldots, e_3 v_3, e_2 v_2, e_1 v_1\}.$$

Then $N_A = 0$ implies that all e_i are +1 and all f_i are -1. Here $8m-6$ is the sum of two squares.

Similarly we can prove

4.102. COROLLARY. Consider four (1, -1) sequences $A = \{X, U, Y, V\}$ where

$$X = \{x_1 = 1, x_2, x_3, \ldots, x_m, x_{m+1}, x_m, \ldots, x_3, x_2, x_1 = 1\},$$

$$U = \{u_1 = 1, u_2, u_3, \ldots, u_m, u_{m+1}, f_m u_m, \ldots, f_3 u_3, f_2 u_2, -1\},$$

$$Y = \{y_1, y_2, \ldots, y_m, -y_m, \ldots, -y_2, -y_1\},$$

$$V = \{v_1, v_2, \ldots, v_m, e_m v_m, \ldots, e_2 v_2, e_1 v_1\},$$

which have $N_A = 0$. Have $e_i = -1$ for all i and $f_i = +1$ for all i. Here $8m+2$ is the sum of two squares.

4.103. DEFINITION. Sequences such as those described in the last lemma will be called Turyn sequences of length ℓ (the four sequences are of weights ℓ, ℓ, $\ell-1$ and $\ell-1$). A sequence $X = \{x_1, \ldots, x_n\}$ will be called skew if n is even and $x_i = x_{n-i+1}$, and symmetric if n is odd and $x_i = x_{n-i+1}$.

4.104. LEMMA. There exist Turyn sequences of lengths 2, 4, 6, 8, 3, 5, 7, 13 and 15.

Proof. Consider

$\ell = 2$: X = {{1-}, {11}, {1}, {1}}
$\ell = 4$: X = {{11--}, {11-1}, {111}, {1-1}}
$\ell = 6$: X = {{111---}, {11-1-1}, {11-11}, {11-11}}
$\ell = 8$: X = {{11-1-1--}, {1111---1}, {111-111}, {1--1--1}}
$\ell = 3$: X = {{111}, {11-}, {1-}, {1-}}
$\ell = 5$: X = {{11-11}, {1111-}, {11--}, {11--}}
$\ell = 7$: X = {{111-111}, {11---1-}, {11-1--}, {11-1--}}
$\ell = 13$: X = {{1111-1-1-1111}, {111--1-1--11-}, {111-11--1---}, {111--1-11---}}

or

X = {{111-11-11-111}, {111--1-1--11-}, {11---1-111--}, {1111-1-1----}}

$\ell = 15$: X = {{11-111-1-111-11}, {111-11---11-11-}, {1111--1-11----}, {1----1-1-1111-}}.

REMARK. These sequences were constructed using the Research School of Physical Sciences, Australian National University, DEC-10 System.

A complete computer search in the case of $\ell = 9$, 10, 14 and 16 gave no solution for any decomposition into squares. In fact, we found

ℓ	$4\ell-6 = x^2+y^2$	Result
6	$18 = 3^2+3^2$	yes
8	$26 = 1^2+5^2$	yes
10	$34 = 3^2+5^2$	none
12	$42 \neq x^2+y^2$	no
14	$50 = 1^2+7^2$	none
	$= 5^2+5^2$	none
16	$58 = 3^2+7^2$	none
18	$66 \neq x^2+y^2$	no
20	$74 = 5^2+7^2$	ran out of time

Table 6.2

In order to satisfy the conditions of Theorem 4.75, we are led to study sequences of a more restricted type.

4.104A. DEFINITION. Four complementary disjoint (0, 1, -1) sequences of length t and total weight t will be called T-sequences.

4.105. EXAMPLE. Consider

$$T = \{\{10000\}, \{01100\}, \{0001-\}, \{00000\}\}.$$

The sequences are disjoint, as the i-th entry is nonzero in one and only one of the four sequences. The total weight is 5, and $N_T = 0$.

Another example is obtained by using the Golay sequences

$$X = 1--1-1---1 \text{ and } Y = 1------11-.$$

Let $\underline{0}$ be the vector of 10 zeros; then

$$T = \{\{1, \underline{0}\}, \{0, \tfrac{1}{2}(X+Y)\}, \{0, \tfrac{1}{2}(X-Y)\}, \{0, \underline{0}\}\}$$

are T sequences of length 11.

4.106. THEOREM. (Turyn). Suppose A = {X, U, Y, V} are Turyn sequences of length ℓ, where X is skew and Y is symmetric for ℓ

even and X is symmetric and Y is skew for ℓ odd. Then there are T-sequences of length $2\ell-1$ and $4\ell-1$.

Proof. We use the notation A/B as before to denote the interleaving of two sequences $A = \{a_1, \ldots, a_m\}$ and $B = \{b_1, \ldots, b_{m-1}\}$.

$$A/B = \{a_1, b_1, a_2, b_2, \ldots, b_{m-1}, a_m\}.$$

Let 0_t be a sequence of zeros of length t. Then

$$T_1 = \{\{\tfrac{1}{2}(X+Y), 0_{\ell-1}\}, \{\tfrac{1}{2}(X-Y), 0_{\ell-1}\}, \{0_\ell, \tfrac{1}{2}(Y+V)\}, \{0_\ell, \tfrac{1}{2}(Y-V)\}\}$$

and

$$T_2 = \{\{1, 0_{4\ell-2}\}, \{0, X/Y, 0_{2\ell-1}\}, \{0, 0_{2\ell-1}, U/0_{\ell-1}\}, \{0, 0_{2\ell-1}, 0_\ell/V\}\}$$

are the T-sequences of lengths $2\ell-1$ and $4\ell-1$, respectively.

4.107. COROLLARY. There are T-sequences constructed from Turyn sequences of lengths 3, 5, 7, 9, 11, 13, 15, 19, 23, 25, 27, 29, 31, 51, 59.

4.108. THEOREM. If X and Y are Golay sequences of length r, then writing 0_r for the vector of r zeros, $T = \{\{1, 0_r\}, \{0, \tfrac{1}{2}(X+Y)\}, \{0, \tfrac{1}{2}(X-Y)\}, \{0_{r+1}\}\}$ are T-sequences of length r+1.

4.109. COROLLARY. (Turyn). There exist T-sequences of lengths $1+2^a 10^b 26^c$, where a, b, c are nonnegative integers.

4.110. COROLLARY. There exist T-sequences of lengths 3, 5, 7, ..., 33, 41, 51, 53, 59, 65, 81, 101.

A desire to fill the gaps in the list in Corollary 4.110 leads to the following idea.

4.111. LEMMA. Suppose $X = \{A, B, C, D\}$ are 4-complementary sequences of length ℓ, ℓ, $\ell-1$, $\ell-1$, respectively, and weight k. Then

$$Y = \{\{A, C\}, \{A, -C\}, \{B, D\}, \{B, -D\}\}$$

are 4-complementary sequences of length $2\ell-1$ and weight 2k. Further, if $\tfrac{1}{2}(A+B)$ and $\tfrac{1}{2}(C+D)$ are also (0,1,-1) sequences, then, with 0_t the sequence of t zeros,

$$Z = \{\{\tfrac{1}{2}(A+B), 0_{\ell -1}\}, \{\tfrac{1}{2}(A-B), 0_{\ell -1}\}, \{0_{\ell}, \tfrac{1}{2}(C+D)\}, \{0_{\ell}, \tfrac{1}{2}(C-D)\}\}$$

are 4-complementary sequences of length $2\ell-1$ and weight k. If A, B, C, D are $(1, -1)$ sequences, then Z consists of T-sequences of length $2\ell-1$.

In fact, Turyn has found sequences satisfying these conditions.

4.112. COROLLARY. The following four $(1, -1)$ sequences are of lengths 24, 24, 23, 23:

1 -1 -1 -1 1 -1 1 -1 -1 -1 -1 1 1 1 1 1 1 -1 -1 1 -1 -1 -1 1

1 -1 -1 1 -1 -1 1 -1 1 1 1 -1 -1 -1 -1 -1 1 -1 -1 -1 1 -1 -1 -1

1 1 1 -1 -1 -1 1 1 -1 -1 1 -1 -1 -1 -1 1 -1 -1 -1 -1 1 -1 1

1 1 -1 -1 1 -1 1 1 -1 1 -1 1 1 1 -1 1 -1 -1 1 -1 -1 -1 1 .

Hence there are T-sequences of length 47.

We may summarize these results in one theorem.

4.113. THEOREM. If there exist T-sequences of length t, then

i) there exist T-matrices of order t;

ii) there exists a Baumert-Hall array of order t;

iii) there exists an orthogonal design of type (t, t, t, t) in order $4t$.

<u>Proof.</u> i) follows by using the T-sequences as first rows of circulant matrices. The rest follows from Theorem 4.75.

Hence we have:

4.114. PROPOSITION. Combining the results of Section 4.4 and this section, we have Baumert-Hall arrays of order

i) $A = \{1+2^a 10^b 26^c$, a, b, c nonnegative integers$\}$,

ii) $B = \{1, 3, \ldots, 33, 37, 41, 47, 51, 53, 59, 61, 65, 81, 101\}$,

iii) $5b$ where $b \in A \cup B$.

We note there is a Baumert-Hall array of order 47, and hence, as promised before, there is an Hadamard matrix of order 188.

CONSTRUCTIONS USING COMPLEMENTARY SEQUENCES

We now give some results which are useful in constructing new sequences from old. In particular, we want to use complementary sequences to construct orthogonal designs.

REMARK. Since our interest is in orthogonal designs, we shall not be restricted to sequences with entries only ± 1, but shall allow 0's and variables too. One very simple remark is in order. If we have a collection of sequences X (each having length n) such that $N_X(j) = 0$, $j = 1, \ldots, n-1$, then we may augment each sequence at the beginning with k zeros and at the end with ℓ zeros so that the resulting collection (say $\overline{X}$) of sequences having length $k+n+\ell$ still has $N_{\overline{X}}(j) = 0$, $j = 1, \ldots, k+n+\ell-1$.

4.115. LEMMA. Suppose there exist 2-complementary sequences X_i, Y_i of length n_i which give an orthogonal design of type (u_{1i}, u_{2i}) constructed from two circulants, $i = 1, 2$.

Then there exist 4-complementary sequences of length n which can be used in the Goethals-Seidel array to give orthogonal designs of type $(u_{11}, u_{12}, u_{21}, u_{22})$ in order $4n$ where $n \geq \max(n_1, n_2)$.

If in addition X_2, Y_2 are disjointable, then there exists an orthogonal design of type $(u_{11}, u_{21}, \frac{1}{2}u_{12}, \frac{1}{2}u_{22})$ in order $4n$ where $n \geq \max(n_1, n_2)$.

Proof. Very straightforward.

4.116. COROLLARY. Let r be any number of the form $2^a 10^b 26^c 5^d 13^e$, and let n be any integer $\geq 2^a 10^b 26^c 6^d 14^e$, a, b, c, d, e nonnegative integers. Then there exist orthogonal designs of order $4n$ and types (i) $(1, 1, r, r)$ and (ii) $(1, 4, r, r)$; and order $2n$ and type (r, r).

Proof. Sequences of the weights r are by Corollary 4.98 and Theorem 4.97 disjointable, and the $(1,4)$ sequences are $\{ab\bar{a}, a0a\}$.

4.117. COROLLARY. There exist orthogonal designs of type (i) $(1,1,5,5)$ in order $4m$, $m \geq 6$; (ii) $(1,1,13,13)$ in order $4m$, $m \geq 14$; (iii) $(1,1,10,10)$ in order $4m$, $m \geq 10$; (iv) $(1,1,20,20)$ in order $4m$, $m \geq 20$.

Another construction based on the existence of 2-complementary sequences which is extremely useful is:

4.118. LEMMA. Let $X = \{U, V\}$ be 2-complementary sequences of length n giving a design constructed of two circulants of type (a, b) such that $N_X(j) = 0$. Then, with U^* and V^* their reverse sequences and w, x, y, z variables,

i) $Y = \{x, y, zU, zV\}$,

ii) $Y = \{yx\text{-}y, y0y, zU, zV\}$,

iii) $Y = \{\{zU, 0, zU^*\}, \{zU, x, -zU^*\}, \{zV, 0, zV^*\}, \{zV, y, -zV^*\}\}$

have $N_Y(j) = 0$. Furthermore, they may be used in the Goethals-Seidel array to give orthogonal designs of type (i) $(1,1,a,b)$, $(1,4,a,b)$ in every order $4m$, $m \geq n$, and (ii) $(1,1,4a,4b)$ in every order $4m$, $m \geq 2n+1$.

We see that in some cases this lemma is a considerable improvement on the previous lemma. For example:

4.119. COROLLARY. There exist orthogonal designs of types (i) $(1,1,20)$ in every order $4m$, $m \geq 7$; (ii) $(1,1,40)$ in $4m$, $m \geq 13$; (iii) $(1,1,16,16)$ in every order $4m$, $m \geq 9$; (iv) $(1,1,8,8)$ in every order $4m$, $m \geq 5$; (v) $(1,1,52,52)$ in every order $4m$, $m \geq 29$.

Proof. Use $\{101, 11-\}$, $\{10111-, 101--1\}$, $\{aab\bar{b}, a\bar{a}bb\}$, $\{ab, \bar{a}b\}$, $\{aaab\bar{a}aa\bar{b}\,\bar{a}a\bar{a}b0b, bbb\bar{a}\,\bar{b}bbab\bar{b}b\bar{b}\,\bar{a}0\bar{a}\}$.

We give one other method for constructing orthogonal designs using complementary sequences.

4.120. LEMMA. Let $X = \{A, B, Z, Z\}$ be 4-complementary sequences of length n and weight k. Then writing Z^* for the reverse of Z and with x, y, z variables,

$$Y = \{\{yA, 0, yB\}, \{yA, 0, -yB\}, \{yZ, 0, yZ^*\}, \{yZ, x, -yZ^*\}\}$$

are 4-complementary sequences of length $2n+1$ which may be used to give an orthogonal design of type $(1, 2k)$ in every order $4m$, $m \geq 2n+1$.

Proof. Use the four sequences of Y to generate matrices which can be used in the Goethals-Seidel array.

4.121. COROLLARY. There exist orthogonal designs of types: (i) (1,20) in order 4m, $m \geq 7$; (ii) (1,24) in order 4m, $m \geq 7$; (iii) (1, 32) in order 4m, $m \geq 13$; (iv) (1,44) in order 4m, $m \geq 13$; (v) (1,52) in order 4m, $m \geq 15$; (vi) (1, 36) in order 4m, $m \geq 11$; (vii) (1,2,66) in order 4m, $m \geq 19$.

Proof. From Table 6 of Appendix H there exist 4-complementary sequences {A,B,Z,A} for lengths and weights: 3 and 10, 3 and 12, 5 and 18, 5 and 20, 6 and 16, 6 and 22, 7 and 26, length 9 and type (1, 33).

4.122. SUMMARY. Two sequences $x_1, \ldots, x_n$ and $y_1, \ldots, y_n$ are called Golay complementary sequences of length n if all their entries are ±1 and

$$\sum_{i=1}^{n-j} (x_i x_{i+j} + y_i y_{i+j}) = 0 \quad \text{for every } j \neq 0,\ j = 1, \ldots, n-1,$$

i.e., $N_X = 0$.

These sequences have the following properties:

1. $\sum_{i=1}^{n} (x_i x_{i+j} + y_i y_{i+j}) = 0$ for every $j \neq 0$, $j = 1, \ldots, n-1$ (where the subscripts are reduced modulo n),

   i.e., $P_X = 0$.
2. n is even and the sum of two squares.
3. $x_{n-i+1} = e_i x_i \iff y_{n-i+1} = -e_i y_i$ where $e_i = \pm 1$.
4. $$\left[\sum_{i \in S} x_i \operatorname{Re}(\zeta^{2i+1})\right]^2 + \left[\sum_{i \in D} x_i \operatorname{Im}(\zeta^{2i+1})\right]^2 + \left[\sum_{i \in S} y_i \operatorname{Im}(\zeta^{2i+1})\right]^2 + \left[\sum_{i \in D} y_i \operatorname{Re}(\zeta^{2i+1})\right]^2 = \tfrac{1}{2}n,$$

   where $S = \{i : 0 \leq i < n,\ e_i = 1\}$, $D = \{i : 0 \leq i < n,\ e_i = -1\}$ and ζ is a 2n-th root of unity.
5. Exist for orders $2^a 10^b 26^c$, a, b, c nonnegative integers.
6. Do not exist for orders $2 \cdot 9^c$ (c a positive integer), 34, 36, 50 or 58.

§4.7. EXISTENCE OF WEIGHING MATRICES

At Queen's University in 1972 we became interested in weighing matrices. We observed that in order to establish existence in all orders for a given weight we needed to consider weighing matrices in odd orders.

We noticed that there was a circulant $W(7,4)$. It has first row

-110100.

Then D. Gregory found a noncirculant $W(13,9)$. After observing that the zeros of this matrix give the incidence matrix of a finite projective plane, we found a circulant $W(13,9)$ with first row

0010111--01-1.

At the Fifth Southeastern Conference on Combinatorics, Graph Theory and Computing in Boca Raton, Florida, in 1972, Rick Wilson and R. C. Mullin said they thought $W(q^2+q+1,q^2)$ might exist when q was a prime power.

Mullin was writing a book on Coding Theory with Ian Blake, who quickly saw the possibilities of using weighing matrices and especially circulant matrices W to form generator matrices $[I, W]$ of codes over $GF(3)$ which would generalize the Pless symmetry codes.

Wallis and Whiteman (Theorem 4.124) finally showed that circulant $W(q^2+q+1,q^2)$ existed when q was a prime power.

In writing this section it seemed that the proof of Wallis and Whiteman was too circuitous and a prettier, more direct proof was desirable. Our colleague, L. G. Kovács, has now given us three proofs; the second is illustrated by Example 4.127 and we feel is intrinsically very beautiful. The first proof is illustrated by Example 4.128. The proof we have given here is the shortest, but hides to some extent the delightful intimacy between circulant weighing matrices and cyclic projective planes. Recent work of David Glynn gives more insight. (See Glynn [1].)

It was Kovács' work that allowed Hain and Eades to establish that there are only two equivalence classes of circulant $W(13, 9)$.

If A is a $W(n,k)$, then $(\det A)^2 = k^n$. Thus if n is odd and a $W(n, k)$ exists, then k must be a perfect square.

In Proposition 2.5 it is shown that

$(n-k)^2-(n-k)+2 > n$

must also hold. It is noted there that the "boundary" values of this condition are of special interest, that is, if

$$(n-k)^2-(n-k)+1 = n,$$

for in this case the zeros of A occur such that the incidence between any pair of rows is exactly one. So if we let $B = J_n - A*A$, B satisfies

$$BB^t = (n-k-1)I_n + J_n, \quad BJ_n = (n-k)J_n;$$

that is, B is the incidence matrix of the projective plane of order $n-k-1$.

Thus the nonexistence of the projective plane of order $n-k-1$ implies the nonexistence of the $W(n, k)$ when $n = (n-k)^2-(n-k)+1$. So rewrite the Bruck-Ryser-Chowla Theorem from Hall [1, p.107-112] to allow us to consider the nonexistence of projective planes.

4.122A. THEOREM. (Bruck-Ryser). If there exists a projective plane of order s, then the Diophantine equation

$$x^2 = sy^2 + (-1)^{(s^2+s)/2} z^2$$

has a solution in the integers not all zero. That is, the Hilbert symbol $((-1)^{(s^2+s)/2}, s)_p = +1$ for all primes p, including $p = \infty$.

4.122B. EXAMPLE. Consider $s = n-k-1 = 6$, $s^2+s+1 = n = 43$, $s^2 = k = 36$. The Bruck-Ryser Theorem says that there is a projective plane only if

$$(-1^{21}, 6)_p = (-1, 6)_p = +1 \text{ at all primes } p.$$

But at $p = 3$,

$$(-1, 6)_3 = (-1, 2)_3(-1, 3)_3 = \left(\frac{2}{3}\right) = -1.$$

So there is no projective plane of order 6 and no $W(43, 36)$.

Similarly, if $s = 2t = n-k-1$, $s^2+s+1 = n$, $s^2 = k$, where $t \equiv 3 \pmod 4$ is a prime, there is no projective plane of order $2t$ and no $W(4t^2+2t+1, 4t^2)$.

Before we prove our result on circulant weighing matrices, we prove the following more general result.

4.123. LEMMA. (Blake). Let q be the power of an odd prime and k any integer $k \geq 3$. Then there exists a

$$W((q^k-1)/(q-1), q^{k-1}).$$

Proof. Let G be a $k \times (q^k-1)$ matrix whose columns contain all the distinct nonzero k-tuples over the finite field $GF(q)$. In coding terms, the row space of G, denoted by C, is equivalent to a maximum length cyclic code. It is known that the weight of every nonzero codeword in C is $(q-1)q^{k-1}$. If G^1 is the $k \times (q^k-1)$ matrix whose rows are any set of k linearly independent codewords of C, then every nonzero k-tuple over $GF(q)$ appears as a column of G^1.

Let H be a $k \times n$ submatrix of G, $n = (q^k-1)/(q-1)$, with the property that any two of its columns are linearly independent. We assume that H is normalized in the sense that the first nonzero element in each column is unity. Let A be an $n \times n$ matrix whose rows are chosen from the nonzero vectors of the row space of H and have the property that any two distinct rows are linearly independent. Assume for convenience that the first k rows of A are rows of H. It follows readily from observations on G that every row of A has weight q^{k-1}. It is not difficult to show that if H^1 is the $(0,1)$ matrix obtained from H by replacing each nonzero element by unity, then the rows of H^1 are the incidence vectors of the complements of the hyperplanes of the projective geometry $PG(k-1,q)$.

Let x_1 and x_2 be two distinct rows of A. Since they are independent, they can be extended to a basis x_i, $i = 1,\dots,k$, each vector of which is a row of A. Let B be the $k \times n$ matrix whose i-th row is x_i, $i = 1,\dots,k$. Assume B has been normalized by multiplying each column so that the first nonzero element in each column is unity. Let B^1 be the $k \times q^{k-1}$ submatrix of B consisting of those columns with unity in the first row. Every (k-1)-tuple over $GF(q)$, including the all-zeros (k-1)-tuple, appears in the columns of B in rows 2 through k. Each element of $GF(q)$ appears q^{k-2} times in the second row of B. In the matrix A, replace $\alpha \in GF(q)$ by $\chi(\alpha)$, where χ is the usual quadratic character, and call the resulting matrix $S(q^{k-1})$. We now show that over the real numbers

$$S(q^{k-1})S(q^{k-1})^t = q^{k-1}I_n$$

and thus that $S(q^{k-1})$ is the required $W((q^k-1)/(q-1), q^{k-1})$.

Since every row of A is of weight q^{k-1} and each nonzero element of $GF(q)$ is either a square or a nonsquare, the inner product over the reals of any row of $S(q^{k-1})$ with itself is q^{k-1}. Let $x_1 = (\alpha_1, \ldots, \alpha_n)$, $x_2 = (\beta_1, \ldots, \beta_n)$ be two distinct rows of A. If $y_1 = (\chi(\alpha_1), \ldots, \chi(\alpha_n))$ and $y_2 = (\chi(\beta_1), \ldots, \chi(\beta_n))$ are the corresponding rows of $S(q^{k-1})$, then the inner product of y_1 and y_2 over the reals is the number of nonzero coordinate positions for which $\chi(\alpha_i) = \chi(\beta_i)$ less the number of nonzero coordinate positions for which $\chi(\alpha_i) \neq \chi(\beta_i)$. Since χ is multiplicative, i.e., $\chi(\alpha)\chi(\beta) = \chi(\alpha\beta)$, multiplication of a coordinate position by a nonzero element of $GF(q)$ does not change the agreement or disagreement between coordinate positions of y_1 and y_2. As before, assume that x_1 and x_2 are the first two rows of the matrix B, which is assumed in normalized form. In the nonzero positions of x_1, each element of $GF(q)$ appears in x_2, q^{k-2} times. Thus the inner product of the corresponding vectors y_1 and y_2 is zero, which completes the lemma.

We now show how to construct circulant weighing matrices. These will be used extensively in later theorems.

4.124. THEOREM. (Wallis-Whiteman, proof by L. G. Kovács). Let q be a prime power. Then there is a circulant $W(q^2+q+1, q^2)$.

Proof. Let D be a cyclic planar difference set with parameters $(q^2+q+1, q+1, 1)$. (See Baumert [1] for definition.) These always exist for q a prime power, and the incidence matrix of D is the incidence matrix of the projective plane of order q.

Without loss of generality, we assume $0 \in D$. We note that d and $-d$ cannot both be in D because $d-0 = 0-(-d)$, contradicting the uniqueness of differences in D.

Let

$$\psi(x) = \sum_{d \in D} x^d$$

be the Hall polynomial of D. (See Baumert [1, p. 8].) Then

$$\psi^2(x) = \sum_{d \in D} x^{2d} + 2\sum_{\substack{e, f \in D \\ e \neq f}} x^{e+f}.$$

We wish to show the coefficients of x^i in $\psi^2(x)$ are 0, 1, 2, i.e., that $2d \neq 2e$ unless $d = e$, $e+f \neq e'+f'$ unless $e = e'$ and $f = f'$, and $2d \neq e+f$ unless $d = e = f$.

Clearly, $2d \neq 2e$ for $d \neq e$. If $e+f = e'+f'$, then $e-e' = f-f'$, and by the uniqueness of differences in D either $e = f$ and $e' = f'$ or $e = -f'$ and $e' = -f$. In the first case $2e = 2e'$, $e = e'$, $f = f'$, and in the second case $e+f = -(e+f)$, i.e., $e+f = 0$ and e and $-e \in D$, which is not possible. If $2d = e+f$, then $d-e = f-d$, and by the uniqueness of differences in D, either $d = f$, $e = d$ or $d = -d$, $e = -f$. In the first case there is nothing to prove, and in the second case e and $-e \in D$, which is not possible.

Hence if B is the cyclic incidence matrix of D, then B^2 has elements 0, 1, 2, and B^2-J has elements 0, 1, -1.

Now

$$\begin{aligned}(B^2-J)(B^2-J)^t &= BBB^tB^t - JB^tB^t - BBJ + J^2, \\ &= (qI+J)^2 - 2(q+1)^2J + (q^2+q+1)J, \\ &= q^2 I.\end{aligned}$$

So B^2-J is the required $W(q^2+q+1, q^2)$.

4.125. EXAMPLE. $\{0,1,3,9\}$ is a difference set mod13, whose circulant incidence matrix B has first row

1101000001000.

B^2 is a circulant matrix with first row

1212211002202,

and B^2-J is the required circulant matrix with first row

0101100--11-1.

4.126. EXAMPLE. David Glynn of the University of Adelaide has further generalized this construction by observing that if A and B are the circulant incidence matrices of two projective planes of the same order and $C = AB-J$ is a $(0,1,-1)$ matrix, then C is a circulant weighing matrix.

In the above example, B, of order 13, has Hall polynomial

$$\psi(x) = x^0+x^1+x^3+x^9,$$

and

$$\psi(x^2) = x^0+x^2+x^5+x^6.$$

We can form the two inequivalent weighing matrices of order 13 by forming

$$B^2-J \quad\text{and}\quad AB^t-J,$$

where A and B^t have Hall polynomials $\psi(x^2)$ and $\psi(x^{-1})$, respectively. Hence we obtain circulant weighing matrices with Hall polynomials

$$\alpha(x) = x^1+x^3+x^4-x^7-x^8+x^9+x^{10}-x^{11}+x^{12}$$

and

$$\beta(x) = x^2+x^4+x^5+x^6-x^7-x^8+x^{10}-x^{11}+x^{12}.$$

The first rows of the weighing matrices for $\alpha(x)$ and $\beta(x^2)$ are

0101100--11-1

and

0-0-10011-111,

which are clearly inequivalent.

4.127. EXAMPLE. (Kovács' second method). (We refer the reader to Hughes and Piper [1] or Dembowski [1] for any unexplained terms in this and the next example.) $L_i = \{0+i, 1+i, 3+i, 9+i\}$ are the lines of a projective geometry. $L^2 = \{0,2,5,6\}$ is an oval with the property that any two of its translates $\{0+i, 2+i, 5+i, 6+i\}$ have precisely one point in common. We form a circulant matrix W with first row (a_{1j}) by choosing

$$a_{1j} = |L_j \cap L^2| - 1.$$

Hence the first row of W is

010111--01-10.

4.128. EXAMPLE. (Kovács' first method--for q odd). (0,1,3,9) is a difference set mod13, so $L_i = \{0+i, 1+i, 3+i, 9+i\}$ are the lines of the projective geometry of order 3. Now $L_0^2 = \{0,2,5,6\}$ is an oval and $L_j^2 = \{0+j, 2+j, 5+j, 6+j\}$ are also ovals, any two of which have precisely one common tangent. The tangents of L_0^2 are L_0, L_1, L_3 and L_9, so 1, 3, 4, 9, 10, 12 are exterior points, 0, 2, 5, 6 are on the oval, while 7, 8, 11 are interior to the oval.

We form our circulant weighing matrix by choosing the first row to have -1, 0, 1 in the (0, i) position (i = 0, 1, ..., 12) according as i is interior on or exterior to the oval, i.e.,

0 1 2 3 4 5 6 7 8 9 10 11 12,
0 1 0 1 1 0 0 - - 1 1 - 1. (*)

The translates of the oval $L_0^{-1} = \{0,4,10,12\}$ also satisfy the unique common tangent condition; from this oval, we get the first row

0 1 - 1 0 - - 1 1 1 0 1 0. (**)

Map $i \to -2i$; then (**) becomes

0 - 0 - 1 0 0 1 1 - 1 1 1. (***)

Now (*) and (***) are inequivalent.

REMARK. Choosing r = 2 and -1 in the example gives inequivalent W(13, 9). It is interesting to consider what values of r will give different solutions.

§4.8. MORE SPECIFIC CONSTRUCTIONS USING CIRCULANT MATRICES

The constructions of this section will be used extensively later to discuss existence of orthogonal designs.

In any of the following constructions, similar results may be obtained by replacing the words circulant and back circulant by type 1 and type 2, respectively (see Section 4.2).

4.129. CONSTRUCTION. Suppose there is a W(n, k) constructed from two circulant matrices M, N of order n/2 with the property that $M*N = 0$ (* denotes Hadamard product). Then $A = x_1M + x_2N$, $B = x_1N - x_2M$ may be used in $\begin{bmatrix} A & BR \\ -BR & A \end{bmatrix}$ to obtain an orthogonal design of order n and type (k, k) on x_1, x_2.

Proof. A straightforward verification. One need only observe that since M, N are circulant, MR, NR are back circulant, and if X is circulant and Y is back circulant, then $XY^t = YX^t$.

4.130. EXAMPLE. Write T for the circulant matrix of order n whose first row is nonzero only in the second column, the entry there being 1. Now

$$M = T+T^2 \quad \text{and} \quad N = T^3-T^4$$

may be used to give a $W(2n, 4)$ constructed from two circulants, $(M*N = 0)$. Then, using the construction $A = x_1M + x_2N$, $B = x_1N - x_2M$ gives an orthogonal design of type $(4,4)$ in order $2n$ constructed from circulants.

4.131. CONSTRUCTION. Suppose there exist $W(n, k_i)$, $i = 1,2$, constructed from circulant matrices M_i, N_i, $i = 1,2$, of order $n/2$ where $M_1*M_2 = N_1*N_2 = 0$ and $M_1M_2^t + M_2M_1^t = N_1N_2^t + N_2N_1^t = 0$; then

$$A = x_1M_1 + x_2M_2, \quad B = x_1N_1 + x_2N_2$$

may be used as two circulants to give an orthogonal design of order n and type (k_1, k_2) on the variables x_1, x_2.

4.132. EXAMPLE. With T as in the previous example and $n = 2k+1$, let

$$M_1 = T^{k-1} - T^{k+2}, \quad N_1 = T^{k-1} + T^{k+2},$$

$$M_2 = T^k + T^{k+1}, \quad N_2 = T^k - T^{k+1},$$

which satisfy the conditions of the construction. Then

$$A = x_1M_1 + x_2M_2, \quad B = x_1N_1 + x_2N_2$$

give an orthogonal design of type $(4,4)$ in order n.

4.133. CONSTRUCTION. Suppose there exist orthogonal designs X_1, X_2 of order $2n$ and type $(u_{i1}, u_{i2}, \ldots, u_{im_i})$ on the variables $x_{i1}, x_{i2}, \ldots, x_{im}$, $i = 1,2$, each of which is constructed using two circulants.

Then there exists an orthogonal design of order $4n$ and type $(u_{11}, u_{12}, \ldots, u_{1m_1}, u_{21}, u_{22}, \ldots, u_{2m_2})$ on the variables $x_{11}, x_{12}, \ldots, x_{1m_1}, x_{21}, x_{22}, \ldots, x_{2m_2}$.

Proof. Let A_i, B_i be the matrices used to form the orthogonal design X_i. Then use A_1, B_1, A_2, B_2 in the Goethals-Seidel array to get the result.

4.134. COROLLARY. Suppose there exist $W(n, k_i)$, $i = 1, 2$, constructed from circulant matrices M_i, N_i, $i = 1, 2$, of order $n/2$. Then there exists an orthogonal design of order $2n$ and type (k_1, k_2) and a $W(2n, k_1+k_2)$.

Proof. Set $A = x_1 M_1$, $B = x_1 N_1$, $C = x_2 M_2$, $D = x_2 N_2$ in the Goethals-Seidel array.

4.135. EXAMPLE. The circulant matrices $A(a)$, $B(a)$, with first rows

$$aaa{-}a0_{n-4}, \quad aa{-}aa0_{n-4}, \quad \text{respectively,}$$

give a $W(2n, 8)$ constructed from circulants for every $n \geq 4$, and the circulant matrices $C(c, d)$, $D(d)$ with first rows

$$dc{-}d0_{m-3}, \quad d0d0_{m-3}, \quad \text{respectively,}$$

give an orthogonal design of type $(1, 4)$ in every order $2m$, $m \geq 3$, where 0_t is a sequence of t zeros.

Hence

$$\{A(a), B(a), A(b), B(b)\}$$
$$\{C(c, d), D(d), C(a, b), D(b)\}$$
$$\{A(a), B(a), C(c, d), D(d)\}$$

can be used as four circulant matrices in the Goethals-Seidel array to give orthogonal designs in order $4s$, $s \geq 4$, of types $(8, 8)$, $(1, 1, 4, 4)$ and $(1, 4, 8)$, respectively.

4.136. COROLLARY. Suppose there exist orthogonal designs of order $2n$ and type (k, j) constructed from two circulants; then there exists a $(1, 1, k, j)$ and a $(1, 4, k, j)$ in order $4n$.

Proof. Let 0_s be a sequence of s zeros. Then the circulant matrices whose first rows follow give (1,1) and (1,4) designs in order 2n:

$$a0_{n-1} \quad \text{and} \quad b0_{n-1},$$

$$ba{-}b0_{n-3} \quad \text{and} \quad b0b0_{n-3}.$$

The existence of (1,1,k,j) and (1,4,k,j) now follows from Construction 4.133.

The following lemma is along the same lines.

4.137. LEMMA. Suppose A and B are circulant matrices from which a W(2n,k) may be constructed. Then if the number of zeros in the first row of

i) A is > 0, there exist orthogonal designs of types (2,2k) and (k+1,k+1) in order 4n and a W(4n,2k+2);

ii) A is > 0 and B is > 0, there exist orthogonal designs of types (2,2,2k) and (k+2,k+2) in order 4n and a W(4n,2k+4);

iii) A is $\geq$ 2 and B is $\geq$ 2 in the (1,i+1) and (1,1+i+j) positions (possibly after permuting), then there exist orthogonal designs of types (8,2k) and (k+4,k+4) and a W(4n,2k+8);

iv) A is > 2 and B is > 2 in the (1,1+i), (1,1+i+j), (1,1+i+2j) positions (possibly after permuting), then there exist orthogonal designs of types (10,2k) and (k+5,k+5) and a W(4n,2k+10).

Proof. Circulate the first row of A and B (if possible) to form matrices A' and B' with zero diagonal. Now, using the following matrices in the Goethals-Seidel array, we get the result.

i) $aA'+bI, aA'-bI, aB', aB'$ and $aA'+bI, bA'-aI, aB', bB'$;

ii) $aA'+bI, aA'-bI, aB'+cI, aB'-cI$ and $aA'+bI, bA'-aI, aB'+bI, bB'-aI$;

iii) $aA+bT^i+bT^{i+j}, aA-bT^i-bT^{i+j}, aB+bT^i-bT^{i+j}, aB-bT^i+bT^{i+j}$ and $aA+bT^i+bT^{i+j}, bA-aT^i-aT^{i+j}, aB+bT^i-bT^{i+j}, bB-aT^i+aT^{i+j}$;

iv) $aA+bT^i+bT^{i+2j}, aA-bT^i-bT^{i+2j}, aB+bT^i+bT^{i+j}-bT^{i+2j}$, $aB-bT^i-bT^{i+j}+bT^{i+2j}$ and $aA+bT^i+bT^{i+2j}, bA-aT^i-aT^{i+2j}$, $aB+bT^i+bT^{i+j}-bT^{i+2j}, bB-aT^i-aT^{i+j}+aT^{i+2j}$

where T^{i+k} is the circulant matrix with 1's in the positions $(j, j+i+k)$, $1 \le j \le n$, and zeros elsewhere.

The next lemma uses a method similar to that used with disjoint sequences to form new orthogonal designs.

4.138. LEMMA. Suppose A, B, C, D are circulant matrices from which a $W(4n, k)$ may be constructed. Further suppose $A+C$, $A-C$, $B+D$, $B-D$ are also $(0,1,-1)$ matrices. Then $aA+bC$, $bA-aC$, $aB+bD$, $bB-aD$ may be used in the Goethals-Seidel array to form a $W(4n, 2k)$ and an orthogonal design of type (k,k) in order $4n$.

The next few results, which are constructively most useful, use the ideas of previous lemmas of this section with the circulant $W(q^2+q+1, q^2)$ established in the last section.

4.139. THEOREM. Let q be a prime power. Then there exist orthogonal designs of types (i) $(1, 3, 6, 2q^2)$, (ii) $(2, 2, 4, 4q^2)$, (iii) $(2, 2, q^2+1, q^2+1)$, (iv) $(3, 4, 6, 2q^2)$, (v) $(3, 3, 3, 3q^2)$, (vi) $(4, 4, 4, 4q^2)$, (vii) $(4, 4, 2(q^2+1), 2(q^2+1))$, (viii) $(q^2+1, q^2+1, q^2+1, q^2+1)$, (ix) $(q^2+2, q^2+2, q^2+2, q^2+2)$, (x) $(q^2+3, q^2+3, q^2+3, q^2+3)$, (xi) $(1, 1, q^2+1, q^2+1)$, (xii) $(1, 4, q^2+1, q^2+1)$, (xiii) $(1, 4, q^2, q^2)$, (xiv) (q^2, q^2, q^2+1, q^2+1), (xv) $(1, q^2, q^2+1, q^2+1)$, (xvi) $(1, 6, 3q^2)$, (xvii) $(1, 2, q^2, 2q^2)$, (xviii) $(2, 2, 2(q^2+1), 2(q^2+1))$, (xix) $(1, 1, 1, q^2)$, (xx) $(1, 1, q^2, q^2)$, (xxi) $(1, q^2, q^2, q^2)$, (xxii) (q^2, q^2, q^2, q^2), (xxiii) $(1, 1, q^2+4, q^2+4)$, in order $4(q^2+q+1)$.

Proof. Let $W = W(q^2+q+1, q^2)$ be the circulant weighing matrix found in Section 4.7. Let T be the shift matrix of order q^2+q+1. Since $P = J - W*W$ ($*$ the Hadamard product) is the circulant incidence matrix of the projective plane of order q, $P = \sum_i T^{d_i}$ where the set of the d_i, D, is a $(q^2+q+1, q+1, 1)$ planar difference set. Now there exist $x, y = x+1$ in D. Let $X = T^x$, $Y = T^y$ and $Z = T^z$ for some $z \in D$, $z \ne x$, $z \ne y$. Let $m = (q^2+q)/2$.

Then if we use

(i) $aX+bY+cW$, $aX+bY-cW$, $bI-aT+bT^2$, $bI+dT-bT^2$

(ii) $aX+bY+dW$, $aX-bY+dW$, $aX+cY-dW$, $aX-cY-dW$

(iii) $aX+bY$, $aX-bY$, $cX+dW$, $dX-cW$

(iv) $aX+bY+cW$, $aX+bW-cW$, $dI+bT-bT^3+dT^4$, $-dI+bT-aT^2+bT^3+dT^4$

(v) $aX+bY+cW$, $aX+dZ-cW$, $bY-dZ-cW$, $aX-bY-dZ$

(vi) $aX+bY+cZ+dW$, $aX+bY-cZ-dW$, $aX-bY+cZ-dW$, $aX-bY-cZ+dW$

(vii) $cW+dZ+aX+bY$, $cW+dZ-aX-bY$, $dW-cZ+aX-bY$, $dW-cZ-aX+bY$

(viii) $aX+bW$, $aX-bW$, $cX+dW$, $cX-dW$

(ix) $aX+bY+dW$, $-bX+aY-cW$, $-cX-dY+dW$, $-dX+cY+aW$

(x) $aX+bY+cZ+dW$, $-bX+aY+dZ-cW$, $-cX-dY+aZ+dW$, $-dX+cY-dZ+aW$

(xi) aX, bX, $cX+dW$, $cX-dW$

(xii) $aI+bT^m-bT^{m+1}$, bT^m+bT^{m+1}, $dX+cW$, $cX-dW$

(xiii) $aI+bT^m-bT^{m+1}$, bT^m+bT^{m+1}, cW, dW

(xiv) aW, bW, $cX+dW$, $dX-cW$

(xv) aX, bW, $cX+dW$, $dX-cW$

(xvi) $aX+aY-bW$, $aX+bW$, $aY+bW$, $cI+aT^m-aT^{m+1}$

(xvii) $aX+bW$, $aX-bW$, cW, dI

(xviii) $aW+bX+cY$, $aW+bX-cY$, $dW-aX+dY$, $bW-aX-dY$

(xix) aI, bI, cI, dW

(xx) aI, bI, cW, dW

(xxi) aI, bW, cW, dW

(xxii) aW, bW, cW, dW

(xxiii) $aX+aY+bW$, $bX+bY-aW$, $cI+aT^m-aT^{m+1}$, $dI+bT^m-bT^{m+1}$,

respectively, in the Goethals-Seidel array, we obtain the designs of the enunciation.

4.140. LEMMA. Let n be odd. If there exists a circulant $W(n,k)$ [and necessarily $(n-k)^2-(n-k)+2 > n$], then there exist

i) $W(4n,\ell)$ for $\ell \in \{k+i,\ i = 0,1,\ldots,6\}$;

ii) $W(4n,\ell)$ for $\ell \in \{2k+i,\ i = 0,1,\ldots,10,13\}$;

iii) $W(4n,\ell)$ for $\ell \in \{3k+i,\ i = 0,1,2,3,6,7,9\}$;

iv) $W(4n,\ell)$ for $\ell \in \{4k+i,\ i = 2,4,6,8,12\}$

(among others), all constructed using four circulant matrices in the Goethals-Seidel array. A circulant $W(q^2+q+1, q^2)$ exists for q a prime power. (See Section 4.7.)

4.141. LEMMA. Suppose q is a prime power and q^2+q+1 is a prime. Then there exist orthogonal designs of types $(2q^2, 2(q^2+2q+2))$ and $(2(q^2+q+1), q(q^2+q+1))$ in order $4(q^2+q+1)$ and consequently an Hadamard matrix of order $4(q^2+q+1)$. Further, if $q^2+q+1 \equiv 3 \pmod 4$, there exist orthogonal designs of order $4(q^2+q+1)$ and types $(1,1,2q^2,2(q+1)^2)$, $(1,1,(q+1)^2,(q+1^2))$, $(1,1,2q^2+2q+1,2q^2+2q+1)$, $(1,1,q^2+2q+2,q^2+2q+2)$, $(1,q^2,q^2,(q+1)^2)$, $(1,q^2+1,q^2+1,(q+1)^2)$.

Proof. Form the matrix W which is the circulant $W(q^2+q+1, q^2)$. Let $A = J-W*W$, where $*$ is the Hadamard product. Then A is the incidence matrix of the $(q^2+q+1, q+1, 1)$ difference set and satisfies

$$AA^t = qI+J, \quad A*W = 0.$$

Let $X = Q+I$ and $Y = Q-I$ where Q is the matrix of order q^2+q+1 constructed in the proof of Lemma 4.34. Then

$$XX^t+YY^t = 2(q^2+q+2)I - 2J.$$

Then

$$aW+bA, \quad aW-bA, \quad bX, \quad bY,$$

$$aW+bA, \quad bW-aA, \quad bX, \quad aY,$$

may be used in the Goethals-Seidel array to form the required orthogonal designs. Setting $a = b = 1$ gives the Hadamard matrix.

For $q^2+q+1 \equiv 3 \pmod 4$, $Q^t = -Q$. Hence, with T as before and x chosen so that $T^x*A = 0$, the following matrices give the required designs:

$$aI+bQ, \quad bA+dW, \quad cI+bQ, \quad bA-dW,$$

$$aI+bQ, \quad bA, \quad cI+dQ, \quad dA,$$

$$aI+bQ, \quad bA+dW, \quad cI+dQ, \quad dA-bW,$$

$$aI+bQ, \quad bA+dT^x, \quad cI+dQ, \quad dA-bT^x,$$

$$aI+bQ, \quad bA, \quad cW, \quad dW,$$

$$aI+bQ, \quad bA, \quad cW+dT^x, \quad dW-cT^x.$$

4.142. LEMMA. Suppose q is a prime power. Then there exist orthogonal designs $(1,q^2)$, (q^2,q^2), $(2,2q^2)$, (q^2+1,q^2+1) in order $2(q^2+q+1)$. Further, if $q^2+q+1 \equiv 3 \pmod 4$ is a prime, then there exists an orthogonal design of type $(1,(q+1)^2)$ in order $2(q^2+q+1)$.

Proof. With A, W, Q as above, the results can be constructed using the following two circulants respectively: aI, bW; aW, bW; $aI+bW, aI-bW$; $aI+bW, bI-aW$; $aI+bQ, bA$.

RESULTS OF A MORE NUMERIC CHARACTER

In combinatorial construction we are often motivated by the desire to "fill in gaps" left unresolved by other methods or to find "prettier" constructions.

We would hope to see 'prettier' results to cover the cases of the remainder of this section, but we are tied by a kind of natural law: "This is the way things work." Nevertheless, the building tools we have do combine to give strong results. For example, the next lemma surely gives only a glimpse of what must be true.

4.143. LEMMA. There exist orthogonal designs constructed from two circulants for orders $2n$ with

i) type $(1,1)$ and $n \geq 1$,
ii) type $(2,2)$ and $n \geq 2$,
iii) type $(1,4)$ and $n \geq 3$,
iv) type $(4,4)$ and $n \geq 4$,
v) type $(2,8)$ and $n \geq 6$,
vi) type $(5,5)$ and $n \geq 6$,
vii) type $(8,8)$ and $n \geq 8$,
viii) type $(4,16)$ and $n \geq 10$,
ix) type $(10,10)$ and $n \geq 10$,
x) type (13) and $n \geq 11$,
xi) type $(13,13)$ and $n \geq 14$.

Proof. Use the first rows given in Table 3 of Appendix H to give designs constructed from two circulants.

The next lemma is of a similar type:

4.144. LEMMA. There exist $W(2n,k)$ constructed from two circulants for

i) $n \geq 3$, $k \in \{0,1,2,4,5\}$;

ii) $n \geq 5$, $k \in \{0,1,2,4,5,8\}$;

iii) $n \geq 7$, $k \in \{0,1,2,4,5,8,10\}$;

iv) $n \geq 9$, $k \in \{0,1,2,4,5,8,10,16\}$;

v) $n \geq 11$, $k \in \{0,1,2,4,5,8,10,13,16,20\}$;

vi) $n \geq 15$, $k \in \{0,1,2,4,5,8,10,13,16,20,26\}$.

Proof. Let 0_s be the zero sequence of length s. Then we may generate the required matrices by using the circulant matrices with the indicated first rows.

$k = 0$: $0_n, 0_n$

$k = 1$: $10_{n-1}, 0_n$

$k = 2$: $10_{n-1}, 10_{n-1}$

$k = 4$: $110_{n-2}, 1\text{-}0_{n-2}$

$k = 5$: $1010_{n-3}, 11\text{-}0_{n-3}$

$k = 8$: $111\text{-}0_{n-4}, 11\text{-}10_{n-4}$

$k = 10$: $101\text{--}10_{n-6}, 10111\text{-}0_{n-6}$

$k = 13$: $1110\text{-}110\text{-}1\text{-}0_{n-11}, 101000\text{-}00010_{n-11}$

$k = 16$: $111\text{-}11\text{-}10_{n-8}, 111\text{---}1\text{-}0_{n-8}$

$k = 20$: $1\text{--}1\text{-}1\text{---}10_{n-10}, 1\text{------}11\text{-}0_{n-10}$

$k = 26$: $111\text{--}111\text{-}1\text{--}0\text{-}0_{n-14}, \text{-}0\text{-}1\text{-}11\text{--}1\text{----}0_{n-14}$.

The results of the next corollary will be used later in this section.

4.145. COROLLARY. There exist orthogonal designs of types (1,1,13,13), (1,4,9,9), (1,2,4,18), (1,8,8,9), (1,9,9,9), (2,4,4,18), (2,8,9,9), (4,4,9,9) and (5,5,9,9) in order 52.

Proof. First we note there exists $B = W(13,9)$, which is circulant. So from Lemma 4.142 there exist (1,9), (9,9) and (2,18) in 26 constructed using two circulant matrices. This fact, together with the results of Lemma 4.133 when used in Construction 4.49, give all the designs except

(1,1,13,13). This last design comes from Theorem 4.139.

The next three results indicate that our knowledge of what exists and why is very incomplete. More results of the kinds in Lemmata 4.146, 4.147, 4.148 can undoubtedly be found and used as in the remainder of this section. An existence theory for sequences as those used may be very deep.

4.146. LEMMA. Sequences of commuting variables with zero nonperiodic auto-correlation function exist which give the following orthogonal designs in order 4n, $n \geq 7$:

(1,1,1,1)	(1,2,2,4)	(2,2,5,5)	(3,3,6,6)
(1,1,1,4)	(1,2,3,6)	(2,2,8,8)	(4,4,4,4)
(1,1,1,9)	(1,2,4,8)	(2,2,9,9)	(4,4,4,16)
(1,1,2,2)	(1,2,6,12)	(2,2,10,10)	(4,4,5,5)
(1,1,2,8)	(1,4,4,4)	(2,3,4,6)	(4,4,8,8)
(1,1,4,4)	(1,4,5,5)	(2,4,4,8)	(4,4,10,10)
(1,1,4,16)	(2,2,2,2)	(2,4,6,12)	(5,5,5,5)
(1,1,5,5)	(2,2,2,8)	(2,5,5,8)	(6,6,6,6)
(1,1,8,8)	(2,2,4,4)	(2,8,8,8)	(7,7,7,7)
(1,1,9,9)	(2,2,4,16)	(3,3,3,3).	

Proof. Table 1 of Appendix H has the sequences which should be used as first rows of circulant matrices in the Goethals-Seidel array to obtain the result.

4.147. LEMMA. The following orthogonal designs in order 4n can be constructed by using sequences of commuting variables with zero non-periodic auto-correlation function to form circulant matrices which are then used in the Goethals-Seidel array:

i) (1,2,3,6), (1,1,4,4), (1,1,2,8), (7) for $n \geq 3$;
ii) (1,1,8,8), (1,1,9,9), (1,4,4,4) for $n \geq 5$;
iii) (1,1,4,16), (1,2,6,12) for $n \geq 6$;
iv) (1,24), (23), (26), (1,27) for $n \geq 7$;
v) (29), (30) for $n \geq 8$;
vi) (1,1,32), (1,35), (31) for $n \geq 9$;

vii) (1, 36), (38) for $n \geq 10$;

viii) (1, 43), (1, 2, 2, 36), (39), (42) for $n \geq 11$;

ix) (1, 44), (48) for $n \geq 12$;

x) (47), (50), (1, 51) for $n \geq 13$;

xi) (1, 1, 17, 17) for $n \geq 16$;

xii) (1, 2, 66) for $n \geq 18$;

xiii) (10, 10, 13, 13) for $n \geq 14$.

Proof. Use the sequences from Tables 1 and 2 of Appendix H for (i)-(x). For xiii) use the appropriate sequences of Table 3 of Appendix H.

4.148. LEMMA. The following orthogonal designs exist in orders 4n as stated:

i) (1, 14) for $n(\text{odd}) \geq 5$;

ii) (1, 22), (1, 1, 13, 13), (1, 1, 1, 16), (1, 1, 26) for $n(\text{odd}) \geq 7$;

iii) (1, 25), (1, 30) for $n(\text{odd}) \geq 9$;

iv) (1, 4, 13), (1, 1, 40) for $n(\text{odd}) \geq 11$;

v) (1, 1, 50), (1, 1, 25) for $n(\text{odd}) \geq 13$;

vi) (1, 4, 26), (1, 1, 29), (1, 1, 58), (1, 52) for $n(\text{odd}) \geq 15$.

All may be constructed using four circulant matrices in the Goethals-Seidel array.

Proof. See the table of sequences with zero periodic auto-correlation function in Appendix H to get the required first rows to use in the Goethals-Seidel array.

These results are used in the next theorem. For results quoted in orders 16, 28, 32, 48 and 56 refer to the appropriate appendix.

4.149. THEOREM. There exist orthogonal designs of order 4n and type (1, k) when

i) for $n \geq t$, $t = 3, 5, 7$, $k \in \{x : x \leq 4t-1,\ x = a^2+b^2+c^2\}$;

ii) for $n \geq 9$, $k \in \{x : x \leq 35,\ x = a^2+b^2+c^2,\ x \neq 34\}$;

iii) for $n \geq 11$, $k \in \{x : x \leq 43,\ x = a^2+b^2+c^2,\ x \neq 34, 37, 42\}$;

iv) for $n \geq 13$, $k \in \{x : x \leq 51,\ x = a^2+b^2+c^2,\ x \neq 34, 37, 42, 45, 46, 48, 49\}$;

v) for $n \geq 15$, $k \in \{x : x \leq 59,\ x = a^2+b^2+c^2,\ x \neq 34, 37, 42, 45, 46, 48, 49, 53, 54, 56, 57\}$;

vi) for $n \geq 16$, $k \in \{x : x \leq 63,\ x = a^2+b^2+c^2,\ x \neq 37, 42, 45, 46, 48, 49, 53, 54, 56, 57, 60, 61, 62, 63\}$.

In all cases, when n is odd, four circulant matrices exist which give the required design.

Proof. Use the tables of Appendix H as follows: We use Table 1 to obtain the result for $n \geq 3$ and $k = 12, 16, 17, 18, 19$ for $n \geq 5$. Table 4 gives (1,14) for n(odd) ≥ 5 and (1,13) for n(odd) ≥ 7. We noted above that (1,13) exists in 20 from Appendix B and both (1,13) and (1,14) exist in 16 and 24, which gives the result.

For $n \geq 7$, Table 2 gives the result immediately except for (1,22), (1,25) and (1,26). Table 4 gives (1,25) for n(odd) ≥ 9 and (1,22) and (1,26) for n(odd) ≥ 7. Their existence in 32, 40, 48, 56 follows because (1,1,11,11), (1,1,12,12) and (1,1,13,13) exist in these orders. The existence of (1,25) in order 28 gives the result for (1,25).

Table 2 gives the result immediately for (1, 32), (1, 33) and (1, 35) in orders $n \geq 9$. Table 4 gives (1, 30) for n(odd) ≥ 9 and (1,29) for n(odd) ≥ 15. Both (1, 30) and (1,29) exist in orders 32, 40, 48, and 56. We exhibit first rows of four circulant matrices which give (1,29) in orders 36, 44 and 52:

36 : $ab b\bar{b} b\bar{b} b\bar{b}\bar{b}$, $0bbb\bar{b}b\bar{b}\bar{b}\bar{b}$, $b\bar{b}0bbbb0b$, $b\bar{b}0b0b0\bar{b}b$;

44 : $ab\bar{b}bbb\bar{b}\bar{b}\bar{b}b\bar{b}$, $\bar{b}b0bbb000b0$, $bbb0\bar{b}bb0\bar{b}b\bar{b}$, $b0b000\bar{b}000b$;

52 : $b0b0_{10}$, $ba\bar{b}0_{10}$, $0b\bar{b}\bar{b}\bar{b}b\bar{b}\bar{b}b\bar{b}\bar{b}\bar{b}b$, $b\bar{b}bb\bar{b}\bar{b}\bar{b}\bar{b}\bar{b}\bar{b}bb\bar{b}$.

This gives the result.

For $n \geq 11$, we consider $k \in \{36, 38, 40, 41, 43\}$. Tables 2, 4 and the existence of (1, 20) in order 4n, $n \geq 6$, which gives (1,1,20,20), and hence (1, 41) in orders 8n, $n \geq 6$, gives the result.

For $n \geq 13$ we consider $k \in \{44, 50, 51\}$. Table 2 gives the result for $k = 44, 51$. Table 4 gives (1, 50) for n(odd) ≥ 13, so the existence of

(1,50) in 56, 64, 72,... (which follows since (1,25) exists in $4m$, $m \geq 7$) gives the result.

Table 4 gives $(1,k)$ for $k \in \{52, 58, 59\}$ in orders $4n$, $n(\text{odd}) \geq 15$. (1,26) and (1,29) exist in orders 32 and $4m$, $m \geq 9$. Hence we have (1,52), (1,58) and (1,59) in orders 64, 72, 80,... which gives the result.

Table 2 gives (1,34) in all $4m$, $m \geq 16$.

We can now repeatedly use the doubling construction ($(1,k)$ in order $n \Rightarrow (1,1,k,k)$ in order $2n$) to say something about orders divisible by 8.

4.150. COROLLARY. There exist orthogonal designs of order $8n$ and type $(1,k)$ when

i) for $n \geq 3$, $k \in \{x : 1 \leq x \leq 23\}$;
ii) for $n \geq 5$, $k \in \{x : 1 \leq x \leq 39,\ x \neq 30, 31\}$;
iii) for $n \geq 7$, $k \in \{x : 1 \leq x \leq 55,\ x \neq 30, 31, 46, 47\}$;
iv) for $n \geq 9$, $k \in \{x : 1 \leq x \leq 71,\ x \neq 30, 31, 46, 47, 56, 57, 61, 62, 68, 69\}$.

The next theorem indicates that we may be able to prove theorems of the type, "If $(s_1, \ldots, s_r)$ satisfies all the existence criteria for an orthogonal design, then $(s_1, \ldots, s_r)$ is the type of an orthogonal design in some large enough order tn and every order un, $u \geq t$." We will give, in a later chapter, the results that Eades has found in this direction.

4.151. THEOREM. Suppose (s_1, s_2, s_3, s_4) satisfies Wolfe's necessary conditions for the existence of orthogonal designs in order $n \equiv 4 \pmod 8$ given by Proposition 3.57:

i) If $s_1+s_2+s_3+s_4 \leq 12$, there is an orthogonal design of type (s_1, s_2, s_3, s_4) and order $4t$ for all $t \geq 3$.
ii) If $s_1+s_2+s_3+s_4 \leq 16$, there is an orthogonal design of type (s_1, s_2, s_3, s_4) and order $4t$ for all $t \geq 4$, with the possible exception of (2,2,5,5) which exists in order $4t$, $t \geq 4$, $t \neq 5$.
iii) If $16 < s_1+s_2+s_3+s_4 \leq 28$, the Table 4.8.1 gives the smallest N such that (s_1, s_2, s_3, s_4) is the type of an orthogonal design which exists for all $4t \geq N$.

Table 4.8.1

N is the order such that the indicated design exists in every order $4t \geq N$.

$12 < s_1+s_2+s_3+s_4 \leq 16$	N	$16 < s_1+s_2+s_3+s_4 \leq 20$	N	$20 < s_1+s_2+s_3+s_4 \leq 24$	N	$24 < s_1+s_2+s_3+s_4 \leq 28$	N
(1,1,4,9)	16	(1,1,1,16)	24	(1,1,2,18)	48	(1,1,1,25)	56
(1,2,2,9)	16	(1,1,8,8)	20	(1,1,4,16)	24	(1,1,5,20)	144
(1,2,4,8)	16	(1,1,9,9)	20	(1,1,10,10)	40	(1,1,8,18)	56
(1,4,4,4)	16	(1,2,8,9)	40	(1,2,2,16)	48	(1,1,9,16)	312
(1,4,5,5)	16	(1,3,6,8)	48	(1,2,6,12)	24	(1,1,13,13)	48
(2,2,2,8)	16	(1,4,4,9)	48	(1,4,8,8)	32	(1,2,4,18)	80
(2,2,5,5)	24	(1,5,5,9)	40	(1,4,9,9)	72	(1,3,6,18)	468
(2,3,4,6)	16	(2,2,4,9)	40	(2,2,2,18)	48	(1,4,4,16)	40
(4,4,4,4)	16	(2,2,8,8)	20	(2,2,4,16)	24	(1,4,10,10)	40
		(2,3,6,9)	40	(2,2,9,9)	24	(1,8,8,9)	80
		(2,4,4,8)	20	(2,2,10,10)	24	(1,9,9,9)	80
		(2,5,5,8)	20	(2,4,6,12)	24	(2,4,4,18)	80
		(3,3,6,6)	20	(2,4,8,9)	160	(2,8,8,8)	28
		(4,4,5,5)	20	(3,3,3,12)	48	(2,8,9,9)	80
		(5,5,5,5)	20	(3,4,6,8)	56	(3,6,8,9)	952
				(4,4,4,9)	112	(4,4,4,16)	28
				(4,4,8,8)	24	(4,4,9,9)	48
				(4,5,5,9)	168	(4,4,10,10)	28
				(6,6,6,6)	24	(5,5,8,8)	32
						(5,5,9,9)	80
						(7,7,7,7)	28

Proof. Appendices A and D give i) and ii) immediately except for (1,1,4,9) and (1,2,2,9). Both these exist in orders 16, 20, 24 and 28 (Appendices B, C, D, E), and so we have the results of i) and ii).

Now (1,1,1,16) does not exist in order 20, and (1,3,6,8) and (1,4,4,9) are not known in order 20. All (s_1, s_2, s_3, s_4) with $16 < s_1+s_2+s_3+s_4 \leq 20$ exist in order 20. The existence of (1,8), (4,5) and (5,5) in order 12, (1,1,2,4,9) in order 24 and the results of Appendix E give all these (s_1, s_2, s_3, s_4) in 24 except (1,3,6,8) and (2,3,6,9). The existence of (1,2,3,6) in 4n, $n \geq 3$, gives (1,2,3,6,6) and hence (1,3,6,8) in 8n, $n \geq 3$. These results plus those in Table 1 of Appendix H give the result immediately except for (1,1,1,16), (1,2,8,9), (1,3,6,8), (1,4,4,9), (1,5,5,9), (2,2,4,9) and (2,3,6,9); and of these all except the last two exist in 28 from Appendix C. A (1,1,1,16) and a (1,4,4,9) exist in 40, as a (1,1,16) and a (1,4,9) exist in 20 from Appendix B. (2,3,6,9) exists in 24 and 32 giving the result. These, together with the existence of (1,1,1,16) in 4n, n(odd) $\geq$ 7, and the results in 32, give the results of the table.

For $20 < s_1+s_2+s_3+s_4 \leq 24$, Table 2 of Appendix H plus the (1,2,8) in 12 giving a (1,1,4,16) in 24, gives the result immediately for those with N = 24.

From Appendix C (1,1,2,18), (1,1,10,10), (1,2,2,16), (2,2,2,18) and (3,3,3,12) exist in 28. From Appendix E, the results in order 32 and the existence of appropriate results in order 20, we have these designs existing in 24, 32 and 40. Hence N = 48. Table 3 of Appendix H shows (1,1,10,10) exists in 4n, $n \geq 10$, so N = 40 for this 4-tuple.

Table 3 of Appendix H shows (1,4,8,8) exists in 4n, $n \geq 8$, so N = 32 for this 4-tuple.

Now (1,4,9,9) exists in, from Appendix E, 24, 32, 40 (since (1,4,9) exists in 20 from Appendix B), and 52 since there exists a circulant W(13,9) and a (1,4) made of two circulant matrices (from Table 3 of Appendix H) in 26. This gives N = 72.

Appendix E gives a (2,4,8,9) in 24; it exists in 32, and the (1,2,8,9) in 20 gives it in 40 and 7×20 (this latter by replacing the first variable by a circulant W(7,4) and the other variables by the identity matrix of order 7). Hence N = 160.

Similarly, a (4,4,4,9) exists in 7×12, and a (4,5,5,9) exists in 7×20. Clearly, (1,4,4,4,4,4,4,4) and (1,1,1,1,4,4,4,4) exist in 56 by replacing some of the variables of the (1,1,1,1,1,1,1,1) design in 8 by the circulant W(7,4). Hence (4,4,4,9) and (4,5,5,9) exist in 56. (4,4,9) and (4,5,9) exist in 20 and 24 (Appendices B and E), so (4,4,4,9) and (4,5,5,9) exist in 40 and 48. They both exist in 32. Hence N = 112 for (4,4,4,9), and N = 168 for (4,5,5,9).

(2,3,4,6) exists in 4n, $n \geq 4$, so (2,3,4,6,6) and (3,4,6,8) exist in 8n, $n \geq 4$. Appendix C gives (3,4,6,8) in 28 and so N = 56.

Table 2 of Appendix H gives the result immediately for (2,8,8,8), (4,4,4,16), (4,4,10,10) and (7,7,7,7). Table 3 gives the result for (5,5,8,8), (1,4,4,16) and (1,4,10,10).

From above, (1,2,4,18), (1,8,8,9), (1,9,9,9), (2,4,4,18), (2,8,9,9), (4,4,9,9) and (5,5,9,9) exist in 32 and 52. From results in 20, 24 and 28 (Appendices B, E and C), they all exist in 40, 48 and 56. Hence N = 80.

From Table 3 of Appendix H (1,1,13,13) exists in 4n, $n \geq 14$; it is constructed above for order 52. (1,13) exists in 24, so (1,1,13,13) exists in 48, and we have N = 48.

(1,1,1,25) exists in 28 (Appendix C), 32, 40 and 48 (since (1,1,1,1,12) exists in 24). Hence N = 56.

(1,1,5,20), (1,1,8,18) and (1,1,9,16) exist in 32, 40 (from above and the existence of (1,4,9) in 20 which gives (1,1,8,18) in 40) and 48 from results in 24 from Appendix E. Now (1,1,8,18) exists in 28 (see Appendix E), so N = 56 for this 4-tuple. The (1,1,5,5) in 12 gives a (1,1,5,20) in 7×12, so N = 144. Now (1,1,1,9) exists in 12, so (1,1,9,16) exists in 21×12, and N = 312.

(1,3,6,18) exists in 16n, $n \geq 2$, because (1,2,3,6) in 4n, $n \geq 3$, gives (1,2,3,6,12) in 16n, $n \geq 3$, and (1,3,6,18) also exists in 32. The (1,2,3,6) in 12 means a (1,3,6,18) exists in 12×13. Hence N = 468.

(3,6,8,9) exists in 32, and since (2,3,6,9) exists in 20, it also exists in 20×7. Hence N = 952.

§4.9. USING CYCLOTOMY TO CONSTRUCT ORTHOGONAL DESIGNS

We remarked in §4.4 that Williamson [1] used (implicitly) the theory of cyclotomy to construct Hadamard matrices of order 148 and 172.

Subsequently, many authors have used this theory to construct Hadamard and skew-Hadamard matrices. Some of these are mentioned in §5.7.

R. M. Wilson [1] and others have used the theory to make major advances in studying balanced incomplete block designs.

In this section we will give the constructions for Baumert-Hall arrays promised in §4.4 and indicate some other constructions for orthogonal designs. We remark that the possibilities for using cyclotomy in the theory of orthogonal designs has by no means been exhausted here.

4.152. DEFINITION. Let x be a primitive root of $F = GF(q)$, where $q = p^{\alpha} = ef+1$ is a prime power. Write $G = \langle x \rangle$. The <u>cyclotomic classes</u> C_i in F are:

$$C_i = \{x^{es+i} : s = 0,1,\dots,f-1\},\ i = 0,1,\dots,e-1.$$

We note that the C_i are pairwise disjoint and their union is G.

For fixed i and j, the <u>cyclotomic number</u> (i,j) is defined to be the number of solutions of the equation

$$z_i+1 = z_j \quad (z_i \in C_i,\ z_j \in C_j),$$

where $1 = x^0$ is the multiplicative unit of F. That is, (i,j) is the number of ordered pairs s,t such that

$$x^{se+i}+1 = x^{et+j} \quad (0 \le s,t \le f-1).$$

Note that the number of times

$$x^{es+i}-x^{et+k} \in C_j$$

is the cyclotomic number $(k-j, i-j)$.

NOTATION. Let $A = \{a_1,a_2,\dots,a_k\}$ be a k-set; then we will use ΔA for the collection of differences between distinct elements of A, i.e.,

$$\Delta A = [a_i-a_j : i \ne j,\ 1 \le i,j \le k].$$

Now

$$\Delta C_i = (0,0)C_i + (1,0)C_{i+1} + (2,0)C_{i+2} + \dots$$

and

$$\begin{aligned}\Delta(C_i - C_j) = {} & (0,0)C_j + (1,0)C_{j+1} + \dots \\ & \dots + (0,0)C_i + (1,0)C_{i+1} + \dots \\ & \dots + (0,i-j)C_j + (1,i-j)C_{j+1} + \dots \\ & \dots + (0,j-i)C_i + (1,j-i)C_{i+1} + \dots \ .\end{aligned}$$

NOTATION. We use $C_a \& C_b$ to denote the adjunction of two sets with repetitions remaining. If $A = \{a,b,c,d\}$ and $B = \{b,c,e\}$, then $A \& B = [a,b,b,c,c,d,e]$. $C_a \sim C_b$ is used to denote adjunction, but with the elements of the second set becoming signed. So $A \sim B = [a,b,-b,c,-c,d,-e]$.

The transpose of a cyclotomic class C_i^T will be defined as $-C_i$, where

$$\begin{aligned} -C_i &= -\{x^{ej+i} : 0 \le j \le f-1\} \\ &= \{-x^{ej+i} : 0 \le j \le f-1\} \\ &= \{x^{em+i+k} : 0 \le j \le f-1\} \end{aligned}$$

with $k = e/2$ for f odd and $k = 0$ for f even.

We define $[C_i]$ the incidence matrix of the cyclotomic class C_i by

$$c_{jk} = \begin{cases} 1, & \text{if } z_j - z_k \in C_i \\ 0, & \text{otherwise.} \end{cases}$$

As $G = C_0 \cup C_1 \cup \dots \cup C_{e-1} = GF(p^\alpha)/\{0\}$, its incidence matrix is $J-I$ (i.e., $\Sigma_{s=0}^{e-1}[C_s] = J-I$), and the incidence matrix of $GF(p^\alpha)$ is J. Therefore, the incidence matrix of $\{0\}$ will be I.

Now if $[C_i]$ is the incidence matrix of a cyclotomic class, then $[C_i^T] = [C_i]^T$, and

i) $[C_i]^T = [C_i]$ if f is even;

ii) $[C_i]^T = [C_{i+e/2}]$ if f is odd.

The term $[C_i][C_j]$ will be taken to mean the ordinary matrix product of the incidence matrices of the cosets C_i and C_j.

$$[C_i][C_j] = \begin{cases} \Sigma_{s=0}^{e-1} a_s [C_s], & \text{if } C_j \neq C_i^T, \text{ and} \\ \Sigma_{s=0}^{e-1} a_s [C_s] + fI, & \text{if } C_j = C_i^T \end{cases}$$

where the a_s are integers giving the coefficients of the matrices.

The incidence matrices of $C_a \& C_b$ and $C_a \sim C_b$ will be given by

$$[C_a \& C_b] = [C_a] + [C_b] \quad \text{and} \quad [C_a \sim C_b] = [C_a] - [C_b].$$

In order to illustrate the method and use of cyclotomy, we will now prove the result quoted in Theorem 5.85, viz.,

4.153. LEMMA. Let $q \equiv 5 \pmod 8$ be a prime power and $q = s^2 + 4t^2$ be its proper representation with $s \equiv 1 \pmod 4$. Then with C_i the cyclotomic class defined above,

$$C_0 \& C_1 \text{ and } |t| \text{ copies of } C_0 \& C_2$$

are $(|t|+1)$ - $\{q; (q-1)/2; (|t|+1)(q-3)/4\}$ supplementary difference sets with

$$x \in C_0 \& C_1 \Rightarrow -x \notin C_0 \& C_1$$

and

$$y \in C_0 \& C_1 \Rightarrow -y \in C_0 \& C_2.$$

<u>Proof.</u> We see from Storer that the cyclotomic numbers for $q = ef+1 = 4f+1$ (f odd) are given by

	0	1	2	3
0	A	B	C	D
1	E	E	D	B
2	A	E	A	E
3	E	D	B	E

$$2A + 2E = f - 1,$$
$$B + D + 2E = f,$$
$$A + B + C + D = f,$$

where for f odd we have

$$16A = q - 7 + 2s, \qquad 16D = q + 1 + 2s + 8t,$$
$$16B = q + 1 + 2s - 8t, \qquad 16E = q - 3 - 2s.$$
$$16C = q + 1 - 6s,$$

Now $-1 \in C_2$, so

$$x \in C_0 \& C_1 \Rightarrow -x \in C_2 \& C_3 \Rightarrow -x \notin C_0 \& C_1,$$

and

$$y \in C_0 \& C_2 \Rightarrow -y \in C_0 \& C_2.$$

Clearly, $|C_0 \& C_1| = |C_0 \& C_2| = (q-1)/2$. So it remains to show that

$$\Delta(C_0 \& C_1) \ \& \ |t| \Delta(C_0 \& C_2) = [(|t|+1)(q-3)/4]G.$$

Now

$$\begin{aligned}\Delta(C_0 \& C_1) &= (A+E)G \ \& \ BC_0 \ \& \ EC_1 \ \& \ EC_2 \ \& \ DC_3 \ \& \ EC_0 \ \& \ DC_1 \ \& \\ &\qquad BC_2 \ \& \ EC_3 \\ &= [(q-5)/8]G \ \& \ (q-1-4t)/8(C_0 \& C_2) \ \& \ (q-1+4t)/8(C_1 \& C_2);\end{aligned}$$

$$\begin{aligned}\Delta(C_0 \& C_2) &= 2AC_0 \ \& \ 2EC_1 \ \& \ 2AC_2 \ \& \ 2EC_3 \ \& \ (C+A)(C_0 \& C_2) \ \& \\ &\qquad (D+B)(C_1 \& C_3) \\ &= (q-5)/4(C_0 \& C_2) \ \& \ (q-1)/4(C_1 \& C_3).\end{aligned}$$

Since the sign of t is ambiguously determined, we choose a generator of F which gives $|t| = -t$. Then

$$\begin{aligned}\Delta(C_0 \& C_1) \ \& \ |t| \Delta(C_0 \& C_2) &= (q-5+q-1-4t-2qt+10t)/8(C_0 \& C_2) \ \& \\ &\qquad (q-5+q-1+4t-2qt+2t)/8(C_1 \& C_3) \\ &= [(q-3-qt+3t)/4]G \\ &= [(q-3)(1+|t|)/4]G\end{aligned}$$

as required.

We can adapt the cyclotomic arrays of Storer [1] in order to consider the matrices

$$P = \Sigma_{i=1}^{s} a_i[C_i]$$

where the a_i are commuting variables (see Hunt and Wallis [1]). For the following results the arrays are often hard to use, and an individual calculation of the arrays was made. The result for 61 is previously unpublished and due to David C. Hunt.

4.154. LEMMA. There exist T-matrices of order $t \in \{13, 19, 25, 31, 41, 61\}$.

Proof. Use the matrices X_1, X_2, X_3, X_4 given below:

t		X_1, X_2, X_3, X_4
13 = 4.3+1	$3^2+2^2+0^2+0^2$	$[C_0], [C_1 \cup \{0\}], [C_2 \cup C_3], [\phi]$
19 = 6.3+1	$3^2+3^2+1^2+0^2$	$[C_0], [C_2], [\{0\}\ \&\ C_3 \cup C_4], [C_1 \cup C_5]$
25 = 8.3+1	$5^2+0^2+0^2+0^2$	$[C_0\ \&\ C_5 \cup \{0\}], [C_1 \cup C_7], [C_2 \cup C_3],$ $[C_4 \cup C_6]$
31 = 10.3+1	$3^2+3^2+3^2+2^2$	$[C_0\ \&\ C_3 \cup C_2], [C_4\ \&\ C_5 \cup C_9],$ $[C_7\ \&\ C_8 \cup C_6], [C_1 \cup \{0\}]$
37 = 12.3+1	$6^2+1^2+0^2+0^2$	$[C_0\ \&\ C_1 \cup C_2 \cup C_3\ \&\ C_4\ \&\ C_5], [\{0\}],$ $[C_6 \cup C_7\ \&\ C_8 \cup C_9\ \&\ C_{10} \cup C_{11}], [\phi]$
41 = 8.5+1	$5^2+4^2+0^2+0^2$	$[C_0 \cup C_2 \cup C_3], [C_4\ \&\ C_6 \cup C_1 \cup \{0\}],$ $[C_5 \cup C_7], [\phi]$
61 = 20.3+1	$6^2+5^2+0^2+0^2$	$[\{0\}\ \&\ C_1\ \&\ C_{11}\ \&\ C_5], [C_7\ \&\ C_{10} \cup C_0 \cup C_9],$ $[C_2\ \&\ C_4\ \&\ C_6 \cup C_3 \cup C_8], [\phi]$

In the remainder of this section we shall consider various matrices obtained by taking linear combinations of the incidence matrices of cyclotomic classes. Part of this work appeared in the Ph.D. thesis of Joan Cooper [2].

We always consider p to be a prime power.

Case 1. $p = 2f+1$, f odd. Consider

$$P = aI+b[C_0]+c[C_1] \tag{9.1}$$

where a, b, c are commuting variables. Now

$$PP^T = a^2I+a(b[C_0]+c[C_1]+b[C_0]^T+c[C_1]^T)+b^2[C_0][C_0]^T+c^2[C_1][C_1]^T+ bc([C_0][C_1]^T+[C_0]^T[C_1]).$$

Using the adapted cyclotomic array of Hunt and Wallis [1] for $e = 2$ given in Table 9.1 we can calculate the coefficients of the incidence matrices in PP^T.

	C_0, C_1	$\{0\}$
00	$A = (f-1)/2$	f
11	$A = (f-1)/2$	f
01	$A+B = f$	0

Table 9.1

where ii is short for $[C_i][C_i]^T$ and ij is short for $([C_i][C_j]^T + [C_i]^T[C_j])$.

The coefficient of $[C_0]$ and $[C_1]$ is $(b^2+c^2+bc)A+bcB+ab+ac$. Hence, in this case we have

$$PP^T = (a^2-ab-ac+(b^2+c^2)/2+(b-c)^2/2f)I + \\ + (ab+ac-(b^2+c^2)/2+(b+c)^2/2f)J. \qquad (9.2)$$

Summarizing:

4.155. THEOREM. Suppose $p = 2f+1$ (f odd) is a prime power and G is the associated cyclic group $GF(p)/\{0\}$ of order $p-1$ with cyclotomic classes C_0 and C_1 of order f. Then

$$P = aI+b[C_0]+c[C_1],$$

a, b, c commuting variables, is a square matrix satisfying (9.2).

4.156. COROLLARY. Suppose $p = 2f+1$ (f odd) is a prime power. Then there exists a nontrivial orthogonal integer matrix of order p.

Proof. Set $a = -(f-1)c/2$, $b = 0$, and c any integer in Theorem 4.155.

We now use P to obtain orthogonal designs. Let X, Y, Z and W be derived from P given in (9.1) by setting

i) $a = c = 0$, iii) $a = b$, $c = 0$,

ii) $a = -b$, $c = 0$, iv) $c = -b$,

respectively; then

$$XX^T = [b^2(f+1)/2]I + [b^2(f-1)/2]J,$$
$$YY^T = [b^2(5+f)/2]I + [b^2(f-3)/2]J,$$
$$ZZ^T = [b^2(f+1)/2]I + [b^2(f+1)/2]J,$$
$$WW^T = (a^2+b^2+2b^2f)I - b^2J.$$

Then we have:

4.157. THEOREM. Let $I+S$ be a skew-Hadamard matrix of order

i) $(f+1)/2$,

ii) $(f-1)/2$,

iii) $(f+3)/2$,

respectively, where $p = 2f+1$ (f odd) is a prime power. Then with X, Y, Z, W as above,

i) $I \times XR + S \times W$,

ii) $I \times YR + S \times W$,

iii) $I \times ZR + S \times W$,

respectively (R is the back circulant unit), are orthogonal designs of order

i) $(f+1)(2f+1)/2$ and type $((f-1)/2, f^2)$,

ii) $(f-1)(2f+1)/2$ and type $((f-3)/2, (f-1)^2)$,

iii) $(f+3)(2f+1)/2$ and type $((f+1)/2, (f+1)^2)$,

respectively.

Proof. Straightforward verification.

EXAMPLE. With $f = 5$ we see that an orthogonal design of type $(1, 16)$ exists in order 22.

Case 2. $p = 2f+1$, $e = 2$, f even. Again we use

$$P = aI+b[C_0]+c[C_1]$$

and obtain PP^T as before. However, as f is even, $C_i^T = C_i$, $i = 1, 2$, and

$$PP^T = a^2I+2ab[C_0]+2ac[C_1]+b^2[C_0][C_0]+c^2[C_1][C_1]+2bc[C_0][C_1].$$

From Storer [1] the cyclotomic matrix for $e = 2$, f even, is

	0	1
0	B	A
1	A	A

$B = (f-2)/2,$

$A = f/2.$

Using Hunt and Wallis [1] we see that the expression $[C_i][C_j]$ is easily determined (see Table 9.2) and, since for f even $C_i^T = C_i$, the cyclotomic arrays can be used immediately to evaluate PP^T.

	C_0	C_1	$\{0\}$
00	B	A	f
11	A	B	f
01	A	A	0

(writing ii for $[C_i]^2$ and ij for $[C_i][C_j]$).

Table 9.2

Hence

$$PP^T = (a^2+(b^2+c^2)f)I+(2ab+b^2(f-2)/2+c^2f/2+bcf)[C_0]$$
$$+ (2ac+c^2(f-2)/2+b^2f/2+bcf)[C_1].$$

The coefficients of $[C_0]$ and $[C_1]$ are equal when

i) $b = c$, or

ii) $2a = b+c$.

The case $b = c$ is trivial, and the other case gives

$$PP^T = ((b-c)^2/4+(b-c)^2f/2)I+((b+c)^2f/2+bc)J.$$

Thus we have:

4.159. THEOREM. Let $p = 2f+1$ (f even) be a prime power and G the associated cyclic group of $GF(p)$ of order $p-1$ with cyclotomic classes C_0 and C_1 of order f. Then

$$P = ((b+c)/2)I+b[C_0]+c[C_1],$$

where b and c are commuting variables, is a matrix satisfying

$$PP^T = ((b-c)^2/4+(b-c)^2f/2)I+((b+c)^2f/2+bc)J.$$

To obtain orthogonal designs we will consider

$$P = aI+b[C_0]+c[C_1]$$
$$Q = aI+c[C_0]+b[C_1]. \qquad (9.3)$$

Now

$$PP^T+QQ^T = ((a-b-c)^2+a^2-2bc+(b-c)^2f)I+(2a(b+c)-(b^2+c^2)+(b+c)^2f)J.$$

Setting $c = -b$ we get M, N satisfying

$$MM^T+NN^T = 2(a^2+b^2+2b^2f)I-2b^2J.$$

Choosing $X = dI+b(J-I)$ and $Y = -dI+b(J-I)$ we have

$$XX^T+YY^T = 2(d^2+b^2)I+2b^2(2f-1)J,$$

and hence, since N, M, X, Y are all symmetric, we have:

4.160. THEOREM. Let $p = 2f+1$ (f even) be a prime power. Suppose there exists a skew-Hadamard matrix $I+S$ of order $2f$. Then

$$I \times \begin{bmatrix} X & Y \\ -Y & X \end{bmatrix} + S \times \begin{bmatrix} M & N \\ N & -M \end{bmatrix}$$

is an orthogonal design of order $4f(2f+1)$ and type $(2, 2(f-1), 8f^2)$.

Proof. By straightforward verification.

Case 3. Now we consider $p = 4f+1$ (f even or odd) a prime power. As before, G is the associated cyclic group of $GF(p)$ of order $p-1$ with cyclotomic classes C_0, C_1, C_2 and C_3 of order f. Then

$$P = [(b+c)/2]I+b[C_0]+c[C_1]+b[C_2]+c[C_3],$$

where b and c are commuting variables, is a matrix satisfying

$$PP^T = ((b-c)^2/4+(b-c)^2f)I+((b+c)^2f+bc)J.$$

To obtain orthogonal designs we will consider two cases,

a) $P = [(b+c)/2]I+b[C_0]+c[C_1]+b[C_2]+c[C_3]$

$Q = [(b+c)/2]I+c[C_0]+b[C_1]+c[C_2]+b[C_3],$

b) $P = aI+b[C_0]+c[C_1]+b[C_2]+c[C_3]$

$Q = aI+c[C_0]+b[C_1]+c[C_2]+b[C_3].$

In case a)

$$PP^T+QQ^T = ((b-c)^2/2+2(b-c)^2f)I+2((b+c)^2f+bc)J.$$

Setting $b = -c$, we get M, N satisfying

$$MM^T+NN^T = 2(b^2+4b^2f)I-2b^2J.$$

Choosing

$$X = dI+b(J-I) \text{ and } Y = -dI+b(J-I),$$

we have

$$XX^T+YY^T = 2(d^2+b^2)I+2b^2(4f-1)J,$$

and hence, since M, N, X, Y are all symmetric, we have:

4.161. THEOREM. Let $p = 4f+1$ (f even or odd) be a prime power. Suppose there exists a skew-Hadamard matrix $I+S$ of order $4f$. Then

$$I \times \begin{bmatrix} X & Y \\ -Y & X \end{bmatrix} + S \times \begin{bmatrix} M & N \\ N & -M \end{bmatrix}$$

is an orthogonal design of order $8f(4f+1)$ and type $(2, 32f^2)$.

Proof. By straightforward verification.

In case b) we set $b = -c$, and we get M, N satisfying

$$MM^T+NN^T = 2(a^2+b^2+4fb^2)I-2b^2J.$$

Choosing

$$X = dI+b(J-I) \text{ and } Y = -dI+b(J-I),$$

we have

$$XX^T+YY^T = 2(d^2+b^2)I+2b^2(4f-1)J,$$

and hence since M, N, X, Y are all symmetric, we have:

4.162. THEOREM. Let $p = 4f+1$ (f even or odd) be a prime power. Suppose there exists a skew-Hadamard matrix $I+S$ of order $4f$. Then

$$I \times \begin{bmatrix} X & Y \\ -Y & X \end{bmatrix} + S \times \begin{bmatrix} M & N \\ N & -M \end{bmatrix}$$

is an orthogonal design of order $8f(4f+1)$ and type $(2, 2(4f-1), 32f^2)$.

Clearly, we have by no means exploited all the possibilities for constructing orthogonal designs by these means. In particular, these methods could easily be used to obtain similar results for other e.

§4.10. GENERALIZED GOETHALS-SEIDEL ARRAYS

Denote by U_v the multiplicative group of generalized permutation matrices of order v; that is, the elements of U_v are $v \times v$ matrices with entries from $\{0, 1, -1\}$ such that each row and column contains precisely one nonzero entry. If T denotes the permutation matrix which represents $(1\ 2 \dots v)$, then the circulant matrices of order v over a commutative ring K with identity are the elements of the group ring $K\langle T\rangle$.

4.163. DEFINITION. If H is an abelian subgroup of U_v and there is an element R of U_v such that $R^2 = I$ and $R^{-1}AR = A^{-t}$ for all $A \in H$, then we shall call KH a <u>GC-ring</u> (generalized circulant ring).

The elements of a GC-ring may be used in the Goethals-Seidel array in the same way as circulant matrices. That is, if A_1, A_2, A_3, A_4 are elements of a GC-ring such that

$$\sum_{i=1}^{4} A_i A_i^t = mI, \tag{10.1}$$

then the rows of

$$\begin{bmatrix} A_1 & A_2R & A_3R & A_4R \\ -A_2R & A_1 & A_4^tR & -A_3^tR \\ -A_3R & -A_4^tR & A_1 & A_2^tR \\ -A_4R & A_3^tR & -A_2^tR & A_1 \end{bmatrix}$$

are mutually orthogonal.

Wallis and Whiteman [1] showed essentially that if H is an abelian group of permutation matrices, then KH is a GC-ring. The elements of KH are called type 1 matrices on H (see §4.2).

Delsarte, Goethals and Seidel [1] introduced another GC-ring. If D denotes the $v \times v$ matrix $\mathrm{diag}(1, 1, \dots, 1, -1)$, then DT generates a cyclic subgroup L of U_v of order 2v. The group ring KL is a GC-ring.

REMARKS

a) Mullin and Stanton [1] use the term group matrix rather than type 1 matrix.

b) The definition of type 1 matrix by Wallis and Whiteman in fact only includes the case where H represents a transitive permutation group. However, the extension to the intransitive case is not difficult (see Wielandt [1], p. 1-10).

c) Suppose that b is odd and N denotes the $b \times b$ matrix $\mathrm{diag}(1,-1,1,-1,\ldots,-1,1)$. Then a $b \times b$ matrix A is circulant if and only if $N^{-1}AN$ is negacyclic. Hence an equation of the form (10.1) has a solution consisting of negacyclic matrices of order b if and only if it has a solution consisting of circulant matrices of order b.

The Goethals-Seidel array itself may be generalized as follows.

4.164. DEFINITION. Let G denote the group

$$<r, x_1, x_1^t, x_2, x_2^t, \ldots \mid x_i x_j = x_j x_i, x_i x_j^t = x_j^t x_i \text{ for } i,j \in \{1,2,\ldots\}, \quad r^2 = 1, rx_i r = x_i^t>.$$

Denote by S the subset

$$\{0, \pm x_1, \pm x_1^t, \pm rx_1, \pm rx_1^t, \pm x_2, \pm x_2^t, \pm rx_1, \pm rx_2^t, \ldots\}$$

of the integral group ring $\mathbb{Z}G$. The notion of transpose may be abstracted by defining an operation $(\ )^t$ on $\mathbb{Z}G$ by $(x_i)^t = x_i^t$, $(x_i^t)^t = x_i$, $r^t = r$, and extending to $\mathbb{Z}G$ in the obvious fashion. If $A = (a_{ij})$ is an $n \times n$ matrix with entries from $\mathbb{Z}G$, then A^* denotes the $n \times n$ matrix with ij-th entry a_{ji}^t. If A has entries from S and

$$AA^* = \left(\sum_{i=1}^{u} s_i x_i x_i^t\right)I,$$

then A is called a GGS array (generalized Goethals-Seidel array) of type $(s_1, s_2, \ldots, s_u)$ and order n.

For example, the Goethals-Seidel array itself, written as

$$\begin{bmatrix} x_1 & rx_2^t & rx_3^t & rx_4^t \\ -rx_2^t & x_1 & rx_4 & -rx_3 \\ -rx_3^t & -rx_4 & x_1 & rx_2 \\ -rx_4^t & rx_3 & -rx_2 & x_1 \end{bmatrix},$$

is a GGS array of type $(1,1,1,1)$ and order 4.

The essential use of GGS arrays is immediate. Suppose that there is a GGS array A of type $(s_1, s_2, \ldots, s_u)$ and order n, and $X_1, X_2, \ldots, X_u$ are $v \times v$ matrices from some GC-ring such that the entries of the X_i are from $\{0, \pm y_1, \pm y_2, \ldots, \pm y_\ell\}$ and

$$\sum_{i=1}^{u} s_i X_i X_i^t = \left(\sum_{j=1}^{\ell} m_j y_j^2 \right) I.$$

Then replacing the entries of A with the appropriate matrices yields an orthogonal design of type $(m_1, m_2, \ldots, m_\ell)$ and order nv. Examples of orthogonal designs constructed in this way are given later in this section.

More importantly, GGS arrays may be used to produce more GGS arrays.

4.165. THEOREM. (Eades). Suppose that there is a GGS array of type $(s_1, s_2, \ldots, s_u)$ and order n, and the $v \times v$ matrices $A_1, A_2, \ldots, A_u$ are from some GC-ring and have entries from $\{0, \pm x_1, \pm x_2, \ldots, \pm x_u\}$. If

$$\sum_{i=1}^{u} s_i A_i A_i^* = \left(\sum_{j=1}^{\ell} m_j x_j x_j^t \right) I,$$

then there is a GGS array of type $(m_1, m_2, \ldots, m_\ell)$ and order nv.

Proof. Suppose that A is a GGS array of type $(s_1, s_2, \ldots, s_u)$ and order v, and the following replacements are made:

$0 \mapsto$ zero matrix of order v;

$\pm x_i \mapsto \pm A_i$;

$\pm x_i^t \mapsto \pm A_i^t$;

$\pm rx_i \mapsto \pm rRA_i$;

$\pm rx_i^t \mapsto \pm rRA_i^*$.

Then the resulting matrix B has entries from S and

$$BB^* = \left(\sum_{i=1}^{u} s_i A_i A_i^*\right) \times I_n$$

$$= \left(\sum_{j=1}^{\ell} m_j x_j x_j^t\right) I_{nv}.$$

To illustrate this theorem, a GGS array of type (2,2) and order 6 is constructed. The 2-circulant construction (see Example 4.48) gives a GGS array of type (1,1) and order 2:

$$\begin{bmatrix} x_1 & rx_2^t \\ -rx_2^t & x_1 \end{bmatrix}.$$

The circulant matrices

$$A_1 = \begin{bmatrix} x_1 & x_2 & \\ & x_1 & x_2 \\ x_2 & & x_1 \end{bmatrix} \quad \text{and} \quad A_2 = \begin{bmatrix} x_1 & -x_2 & \\ & x_1 & -x_2 \\ -x_2 & & x_1 \end{bmatrix}$$

satisfy $A_1A_1^* + A_2A_2^* = 2(x_1x_1^t + x_2x_2^t)I$. Following the replacements in the proof of Theorem 4.165, a GGS array of type (2,2) and order 6 is obtained:

$$\begin{bmatrix} x_1 & x_2 & & & -rx_2^t & rx_1^t \\ & x_1 & x_2 & -rx_2^t & rx_1^t & \\ x_2 & & x_1 & rx_1^t & & -rx_2^t \\ & rx_2^t & -rx_1^t & x_1 & x_2 & \\ rx_2^t & -rx_1^t & & & x_1 & x_2 \\ -rx_1^t & & rx_2^t & x_2 & & x_1 \end{bmatrix}.$$

Note that the theorem could be applied a times to obtain a GGS array of type $(2^a, 2^a)$ and order $3^a.2$.

The existence of a GGS array clearly implies the existence of an orthogonal design of the same type and order, but the converse is false (see Remark 4.172). In many cases, however, the converse is true. An important fact is that every orthogonal design on 2-variables can be made into a GGS array by replacing the second variable x_2 by rx_2^t. The following proposition gives some infinite families of GGS arrays with 4-variables.

4.166. PROPOSITION. (Eades). Suppose that a is a positive integer and ℓ is a product of at least a positive integers; that is, $\ell = \ell_1 \ell_2 \ldots \ell_j$ where $j \geq a$.

a) If $\ell_i \geq 2$ for $1 \leq i \leq j$, then there is a GGS array of type $(2^a, 2^a, 2^a, 2^a)$ and order 4ℓ.

b) If $\ell_i \geq 4$ for $1 \leq i \leq j$, then there are GGS arrays of type $(3^a, 3^a, 3^a, 3^a)$ and $(4^a, 4^a, 4^a, 4^a)$ and order 4ℓ.

Proof. For $\ell_1 \geq 2$ consider the sequences $a_1 = (x_1, x_2, 0_{\ell_1 - 2})$, $a_2 = (x_1, -x_2, 0_{\ell_1 - 2})$, $a_3 = (x_3, -x_4, 0_{\ell_1 - 2})$, $a_4 = (x_3, x_4, 0_{\ell_1 - 2})$, where $0_{\ell_1 - 2}$ denotes a sequence of $\ell_1 - 2$ zeros. These sequences are complementary, and, further, if A_i is the circulant matrix with first row a_i, then

$$\sum_{i=1}^{4} A_i A_i^* = 2\left(\sum_{i=1}^{4} x_i x_i^t\right) I.$$

Using Theorem 4.165 and the Goethals-Seidel array, a GGS array of type $(2, 2, 2, 2)$ and order $4\ell_1$ may be obtained. Repeating this procedure a times gives a). For b) the following complementary sequences may be used in a similar fashion:

$$(3, 3, 3, 3) : (0, -x_2, -x_3, -x_4), (x_1, 0, -x_3, x_4), (x_1, x_2, 0, -x_4), (x_1, -x_2, x_3, 0),$$

$$(4, 4, 4, 4) : (x_1, -x_2, -x_3, -x_4), (x_1, x_2, -x_3, x_4), (x_1, x_2, x_3, -x_4), (x_1, -x_2, x_3, x_4).$$

A numerical investigation of GGS arrays of order 12 has been made, and the results are listed in Eades [4]. These GGS arrays have been used to construct orthogonal designs of orders 36 and 60.

GGS arrays with 2-variables have been used successfully for constructing orthogonal designs of highly composite orders congruent to 2 modulo 4. Examples are given later.

It seems that GGS arrays are the most powerful method for constructing orthogonal designs from circulants in orders not divisible by 8.

SOME INFINITE FAMILIES OF ORTHOGONAL DESIGNS

The Goethals-Seidel array and its generalizations have been used to construct many infinite families of orthogonal designs. The theorems below illustrate some of the techniques involved.

4.167. THEOREM. (Eades). If there is a GGS array of type $(s_1, s_2, \ldots, s_u)$ and order n, then there is an orthogonal design of type $(s_1, s_1, s_2, s_2, \ldots, s_u, s_u)$ and order $2n$.

Proof. The negacyclic matrix

$$X_i = \begin{bmatrix} x_i & y_i \\ -y_i & x_i \end{bmatrix}$$

is an orthogonal design of type $(1, 1)$. Hence

$$\sum_{i=1}^{u} s_i X_i X_i^t = \left(\sum_{i=1}^{u} s_i (x_i^2 + y_i^2) \right) I.$$

The combination of Theorem 4.167 and Proposition 4.166 gives a large collection of orthogonal designs. For example, for each $a > 0$ there is an orthogonal design of type $(4^a, 4^a, 4^a, 4^a, 4^a, 4^a, 4^a, 4^a)$ and order 8.5^a.

4.168. THEOREM. (Eades). Suppose that q is a prime power of the form $3m+1$. Then there is a skew symmetric weighing matrix of weight q^2 and order $4(q^2+q+1)/3$.

Proof. For each v denote the $v \times v$ permutation matrix which represents $(1\ 2 \ldots v)$ by T_v. Since $q = 3m+1$, 3 divides q^2+q+1, but 9 does not divide q^2+q+1. Hence the group S_1 of permutation matrices generated by T_{q^2+q+1} is isomorphic to the group generated by $T_{3m^2+3m+1} \times T_3$. Both these groups **represent** transitive permutation groups, and it follows that they are conjugate. Hence for each circulant matrix W of order q^2+q+1 there is a permutation matrix N of order q^2+q+1 such that

$N^{-1}WN$ has the form

$$\begin{bmatrix} X_1 & X_2 & X_3 \\ X_3 & X_1 & X_2 \\ X_2 & X_3 & X_1 \end{bmatrix},$$

where the X_i are circulant matrices of order $3m^2+3m+1$. If W is a circulant weighing matrix of weight q^2 (see Section 4.7), then the X_i have entries from $\{0, 1, -1\}$ and

$$X_1X_1^t + X_2X_2^t + X_3X_3^t = q^2I.$$

Thus x_1I, x_2X_1, x_2X_2, x_2X_3 may be used in the Goethals-Seidel array to obtain an orthogonal design of type $(1, q^2)$ and order $4(3m^2+3m+1)$.

4.169. THEOREM. (Eades). Suppose that there is a skew symmetric weighing matrix of weight k and order n, and $4k-1$ is prime. Then there is a skew symmetric weighing matrix of weight $4k^2$ and order $(4k-1)n$.

Proof. Since $4k-1$ is prime, there is a cyclic difference set D with parameters $(v, k, \lambda) = (4k-1, 2k, k)$. Further, D may be chosen so that $0 \in D$, and for $x \neq 0$, $x \in D$ if and only if $-x \notin D$. (Baumert [1] has details on difference sets.) Denote the incidence matrix of this difference set by B. The circulant matrices $A_1 = x_2B$ and $A_2 = x_1I + x_2(2B-J-I)$ satisfy

$$A_1A_1^t + kA_2A_2^t = (x_1^2+4k^2x_2^2)I_{4k-1}.$$

The theorem follows since a skew symmetric weighing matrix of weight k may be used as a GGS array of type $(1, k)$.

REMARK. Some examples given by this theorem are listed below:

$(1,1)$	in order n	$\Rightarrow$ $(1,4)$	in order	$3n$
$(1,2)$	in order n	$\Rightarrow$ $(1,16)$	in order	$7n$
$(1,3)$	in order n	$\Rightarrow$ $(1,36)$	in order	$11n$
$(1,5)$	in order n	$\Rightarrow$ $(1,100)$	in order	$19n$

$(1,6)$ in order $n \Rightarrow (1,144)$ in order $23n$

$(1,8)$ in order $n \Rightarrow (1,256)$ in order $31n$

$(1,11)$ in order $n \Rightarrow (1,484)$ in order $43n$, etc.

4.170. PROPOSITION. (Eades). Suppose that q is a prime power and q^2+q+1 is a prime of the form $4m-1$. Then there are orthogonal designs of types $(1,1,2q^2,2(q+1)^2)$ and $(1,1,q^2+(q+1)^2,q^2+(q+1)^2)$ and order $4(q^2+q+1)$.

Proof. (This is a special case of Lemma 4.141. We have included it because the proof is so different from 4.141.) Suppose that W is a circulant weighing matrix of weight q^2 and order q^2+q+1. Denote J-abs(W) by B. It can be shown that B is the incidence matrix of a cyclic projective plane of order q; hence $BB^t = qI+J$. If $q^2+q+1 = 4m-1$ and is prime, then there is a cyclic difference set E with parameters $(4m-1,2m-1,m-1)$ such that $0 \notin E$, and for $x \neq 0$, $x \in E$ if and only if $-x \notin E$. (In fact, E is the complement of the difference set in the proof of the previous theorem.) If F is the incidence matrix of E, then $A = 2F-J$ is skew-symmetric and $AA^t = (q^2+q+1)I-J$. The circulant matrices

$$X_1 = x_1 I + x_3 A,$$

$$X_2 = x_2 I + x_4 A,$$

$$X_3 = x_3 W + x_4 B,$$

$$X_4 = x_4 W - x_3 B,$$

satisfy

$$\sum_{i=1}^{4} X_i X_i^t = (x_1^2+x_2^2+(2q^2+2q+1)(x_3^2+x_4^2))I.$$

The circulant matrices

$$Y_1 = x_1 I + x_4 A,$$

$$Y_2 = x_2 I + x_4 A,$$

$$Y_3 = x_3 W + x_4 B,$$

$$Y_4 = x_3 W - x_4 B,$$

satisfy

$$\sum_{i=1}^{4} Y_i Y_i^t = (x_1^2+x_2^2+2q^2x_3^2+2(q+1)^2x_4^2)I.$$

REMARK. It is not known whether Proposition 4.170 generates an infinite family of orthogonal designs. The first few primes in

$$\{q^2+q+1 : q \text{ is a prime power and } q^2+q+1 \equiv 3 (\mathrm{mod}\ 4)\}$$

are 7, 31, 307, 1723, 8011, 9507.

4.171. THEOREM. (Eades). Suppose that q is a prime power and $v \geq 6$. Then there are orthogonal designs of types $(5,5q^2)$, $(5q^2,5q^2)$, $(10,10q^2)$, $(5(q^2+1),5(q^2+1))$ and order $2v(q^2+q+1)$.

Proof. The existence of an orthogonal design of type $(5,5)$ and order $2v$ is established in §4.8. Hence there is a GGS array of type $(5,5)$ and order $2v$. The circulant weighing matrices of weight q^2 and order q^2+q+1 may be used to complete the proof.

LIMITATIONS

4.172. REMARK. There are two ways in which the use of GGS arrays for constructing orthogonal designs is limited.

First, little is known about the existence of GGS arrays. A numerical investigation of GGS arrays of order 12 shows that existence of a GGS array is harder to establish than existence of the corresponding orthogonal design. Further, it can be deduced from Theorem 4.167 that the number of variables of a GGS array of order n is at most $[\frac{1}{2}\rho(2n)]$. If 8 divides n, then $[\frac{1}{2}\rho(2n)] < \rho(n)$, and so there are many orthogonal designs for which a corresponding GGS array does not exist. Note also that if 16 divides n, then $[\frac{1}{2}\rho(2n)] > 4$, but no GGS array with more than four variables is known.

Second, it can be proved that not all orthogonal designs can be constructed using GGS arrays. There is an orthogonal design of type $(4,9)$ and order 14 (see Chapter 8). However, using the methods of

Section 4.3, it can be shown that there is no orthogonal design of type (4, 9) and order 14 constructed by using two 7×7 circulant matrices in the two-circulant construction.

§4.11. BALANCED WEIGHING MATRICES

A most important concept in the design and analysis of experiments is that of a (v, k, λ) configuration. This is equivalent to a (0, 1) matrix A (the incidence matrix of the configuration) of order v satisfying

$$AA^t = (k-\lambda)I+\lambda J, \quad AJ = JA = kJ, \tag{11.1}$$

where

$$\lambda(v-1) = k(k-1). \tag{11.2}$$

It is natural, then, to ask when such a matrix can be signed in order to produce a weighing matrix $M = W(v, k)$. The work of this section is due to Mullin [1], [2] and Mullin and Stanton [1], [2].

4.173. DEFINITION. A <u>balanced weighing matrix</u> M is a square (0, 1, -1) matrix such that squaring all its entries gives the incidence matrix of a (v, k, λ) configuration. That is,

$$MM^t = kI_v$$

and $A = M * M$ satisfies (11.1) with $\lambda(v-1) = k(k-1)$. We write M is a BW(v, k).

REMARK. Although we will not study it here, balanced weighing matrices have proved most useful in providing previously unknown balanced incomplete block designs (see Mullin and Stanton [1], [2]).

NECESSARY CONDITIONS FOR THE EXISTENCE OF BALANCED WEIGHING MATRICES

Since a BW(v, k) implies the existence of a (v, k, λ) configuration, the following conditions are known to be necessary:

i) if v is even, then $(k-\lambda)$ must be a perfect square;

ii) if v is odd, then the equation

$$x^2 = (k-\lambda)y^2 + (-1)^{[(v-1)/2]}\lambda z^2 \qquad (11.3)$$

must have a solution in integers other than $x = y = z = 0$. (See Ryser [1, p.111].)

It is also trivial that for a $BW(v,k)$ to exist,

iii) $\lambda = k(k-1)/(v-1)$ must be even.

Further we saw in §10 that for a $W(v,k)$ to exist,

iv) if v is odd, then k must be a perfect square, and

v) if v is odd, then $(v-k)^2-(v-k)+2 > v$;

and in Chapter 2,

vi) if $v \equiv 2 \pmod 4$, then k must be the sum of two squares.

In the event that $v \equiv 1 \pmod 4$, we note that iv) is stronger than ii) since if $k = \alpha^2$, then $x = \alpha$, $y = z = 1$, is a solution of (11.3), while for the parameters $v = 27$, $k = 13$, $\lambda = 6$ (12.3) has a solution, but k is not a perfect square. (Just note that $<7,6> = <1,42>$, and so (11.3) has a rational, hence integral, solution.) For $v \equiv 3 \pmod 4$, iv) implies that v) has a solution if and only if $k-\lambda$ is the sum of two squares.

CONSTRUCTION METHOD FOR BALANCED WEIGHING DESIGNS

The direct sum of two matrices $W(n_1,k)$ and $W(n_2,k)$ is a $W(n_1+n_2,k)$, and the Kronecker product of matrices $W(n_1,k_1)$ and $W(n_2,k_2)$ is a $W(n_1n_2,k_1k_2)$, but this is not true for balanced weighing designs since in general the property of balance is lost under these operations. This fact alone makes the construction of balanced designs difficult. This is further emphasized by the fact that the conditions i), ii), and the condition that $(v-1) \mid k(k-1)$ need not hold in general for an unbalanced design. Here we discuss the generation of balanced weighing designs from group difference sets.

Let G be a finite Abelian group of order v. If G admits a difference set $D = \{d_1,d_2,\ldots,d_k\}$, then choose $M(\chi)$ (or M) to be a type 1 incidence matrix of D obtained from the map χ.

Strictly speaking, $M(\chi)$ is determined only up to a permutation of rows and columns, but this is in no way relevant to the present discussion. Type 1 matrices have an interesting property, which we now discuss.

4.174. DEFINITION. Let r_g denote the g-th row of a type 1 incidence matrix M defined on an Abelian group G. We say M has the invariant scalar product property (ISP property) if for all $g, h, \theta \in G$,

$$r_g \cdot r_h = r_{g+\theta} \cdot r_{h+\theta},$$

where $\cdot$ denotes the usual scalar product of vectors.

4.175. LEMMA. Any type 1 matrix defined by χ on G has the ISP property.

Proof. Note that

$$\begin{aligned} r_g \cdot r_h &= \sum_{k \in G} \chi(k-g)\chi(k-h) \\ &= \sum_{k \in G} \chi((k-\theta)-g)\chi((k-\theta)-h) \\ &= \sum_{k \in G} \chi(k-(g+\theta))\chi(k-(h+\theta)) \\ &= r_{g+\theta} \cdot r_{h+\theta} \end{aligned}$$

as required.

A similar result holds for column scalar products.

4.176. LEMMA. A type 1 (0,1,-1) incidence matrix is a $W(v,k)$ matrix if and only if the following equation holds for all $g \in G$:

$$\sum_{\theta \in G} \chi(\theta)\chi(\theta+g) = k\delta_{0,g}, \tag{11.4}$$

where $\delta_{0,g}$ is the Kronecker delta.

Proof. This is clear because of the ISP property

$$r_0 \cdot r_g = \sum_{\theta \in G} \chi(\theta)\chi(\theta-g) = \sum_{\theta \in G} \chi(\theta+g)\chi(\theta).$$

The equality of these two summations is of practical importance since it saves calculation in verifying (11.3). In particular, if v is odd, one need only check $(v-1)/2$ equations since the nonzero elements of G can be partitioned into inverse pairs.

4.177. LEMMA. Let D be a difference set in G. Let $M = M(\chi)$ be a type 1 $(0,1,-1)$ incidence matrix. Then M^*M is the incidence matrix of a (v,k,λ) configuration if $\chi(g) = 0$ if and only if $g \in G-D$.

Proof. This is evident.

4.178. DEFINITION. We refer to a function χ satisfying the condition of Lemma 4.177 as a D-function. If the image of χ is $\{0,1,-1\}$, we call χ a restricted function. Putting these results together, we obtain:

4.178A. THEOREM. (Mullin). There is a matrix $BW(v,k)$ if there is a D-function χ on an Abelian group of order v such that

$$\sum_{\theta \in G} \chi(\theta)\chi(\theta+g) = k\delta_{0,g}.$$

This theorem can be used as a basis for a computer algorithm.

For notational convenience, given a restricted function χ on an Abelian group G, we denote $\sum_{\theta \in G} \chi(\theta)\chi(\theta+g)$ by $F(\chi,g)$. We demonstrate a limitation of the construction of Theorem 4.178A in the next theorem. (This can also be obtained from Lemma 4.184.)

4.179. THEOREM. (Mullin). If there is a D-function χ in an Abelian group G of order v such that $F(\chi,g) = k\delta_{0,g}$ for all $g \in G$ and v is even, then $\lambda = k(k-1)/(v-1)$ satisfies $\lambda \equiv 0 \pmod 4$.

Proof. Since v is even, there exists an element $\bar{g} \neq 0$ in G such that $\bar{g} = -\bar{g}$. Let $(a_1,b_1)(a_2,b_2),\ldots,(a_t,b_t)$ be the pairs of elements of D whose difference is $\bar{g}$. Here $t = \lambda/2$, since if $a_i - b_i = \bar{g}$, then $b_i - a_i = \bar{g}$. Now consider

$$F(\chi,\bar{g}) = \sum_{\theta \in G} \chi(\theta)\chi(\theta+\bar{g}).$$

The only nonzero terms in this expression arise when both θ and $\theta+\bar{g}$ belong to D, since χ is a D-function. Thus

$$F(\chi,\bar{g}) = \sum_{i=1}^{t} [\chi(a_i)\chi(b_i)+\chi(b_i)\chi(a_i)]$$

$$= 2\sum_{i=1}^{t} \chi(a_i)\chi(b_i) = 0.$$

Since each of the t terms in the latter sum is either 1 or -1, this expression must have $t/2$ terms of each value, and t must be even. This shows that $\lambda \equiv 0 \bmod 4$ as required.

There is a (4, 3,2) configuration C which is derivable from a difference set in the group of integers mod 4; however, there is no D-function for any difference set which will produce a BW(4, 3). It is possible to sign the matrix of C to produce an orthogonal matrix nonetheless. More generally, there is a cyclic $((3^{2n}-1)/2, 3^{2n-1}, 2.\ 3^{2n-2})$ configuration (since this is the complementary configuration of the set of hyperplanes in PG(2n-1, 3)), but there is no way of signing these matrices cyclically to make them orthogonal in view of Lemma 4.179. The results of Mullin show that all of these can be signed to produce orthogonal matrices. Not all incidence matrices of (v, k, λ) configurations with v even can be signed to produce orthogonal matrices. It can be shown that the matrix of the self-dual (16, 6, 2) configuration cannot be signed (Schellenberg [1]).

We introduce new concepts which provide a labour-saving device in the calculation associated with Theorem 4.178A in some applications.

4.180. DEFINITION. Let R be a finite ring with unit. A restricted function χ on the additive group of R with the property that $\chi(1) = 1$ is called a <u>normal function.</u> Let U(R) denote the group of units of R. Let $N(R,\chi) = N(\chi)$ be defined by $N(\chi) = \{g : g \in U(R) | \chi(g,\theta) = \chi(g)\chi(\theta), \forall\theta \in R\}$.

Because of the importance of $N(\chi)$ in the next theorem, we demonstrate a structural property of this set.

4.181. PROPOSITION. $N(\chi)$ is a subgroup of U(R).

<u>Proof.</u> Let g and h be members of $M(\chi)$. Then for every $\theta \in R$,

$\chi(gh\theta) = \chi(g)\chi(h\theta) = \chi(g)\chi(h)\chi(\theta)$.

Since R is finite, the result follows.

It is clear that χ is a linear representation of $N(\chi)$ under these circumstances.

4.182. THEOREM. (Mullin). Let R be a finite ring with unit and χ a normal function on R. Let $M(\chi)$ be defined as above.

If $g \in N(\chi)$, then $F(\chi, g) = F(\chi, 1)$.

Proof

$$F(\chi, g) = \sum_{\theta \in R} \chi(\theta)\chi(\theta+g).$$

Let $\tau = g^{-1}\theta$ or equivalently $\theta = g\tau$. Then, since this mapping is 1-1, we have

$$\begin{aligned} F(\chi, g) &= \sum_{\tau \in R} \chi(g\tau)\chi(g\tau+g) \\ &= \sum_{\tau \in R} \chi(g\tau)\chi(g(\tau+1)) \\ &= \sum_{\tau \in R} (\chi(g))^2 \chi(\tau)\chi(\tau+1). \end{aligned}$$

Since $\chi(g)\chi(g^{-1}) = \chi(1) = 1$, $\chi(g) \neq 0$ and $(\chi(g))^2 = 1$. This yields

$$F(\chi, g) = \sum_{\tau \in R} \chi(\tau)\chi(\tau+1) = F(\chi, 1).$$

As an application of this result, let us consider $G = GF(7)$. Let $\chi(0) = -1$, $\chi(1) = \chi(2) = \chi(4) = 1$ and $\chi(3) = \chi(5) = \chi(6) = 0$. Since the field marks 1, 2 and 4 are the quadratic residues and since $7 \equiv 3 \pmod 4$, $N(\chi) = \{1, 2, 4\}$. Now $F(\chi, 2) = F(\chi, 4) = F(\chi, 1) = \chi(0)\chi(1) + \chi(1)\chi(2) = 0$, and since $G = \{0\} \cup N(\chi) \cup -N(\chi)$, we have

$$F(\chi, g) = 4\delta_{0,g}, \qquad g \in G.$$

Thus M is a $W(7,4)$ matrix. But $\{0, 1, 2, 4\}$ is a difference set, and therefore M is also a $BW(7,4)$ matrix. Thus the vector

$$(-1\ 1\ 1\ 0\ 1\ 0\ 0)$$

when developed cyclically mod 7, generates a $BW(7,4)$.

REGULAR BALANCED WEIGHING MATRICES

4.183. DEFINITION. If a $BW(v,k)$ matrix is such that the number of -1's per row is constant, we say that it is regular.

In a BW matrix we denote the number of -1's per row by $a(-1)$ and the number of 1's per row by $a(1)$. Since if M is regular, then $-M$ is also regular, we may assume that we are dealing with matrices for which $a(1) \geq k/2$. Clearly, every group-generated $BW(v,k)$ is regular, as is its transpose. Using this fact, Mullin [1] proved, using a somewhat different method, a generalization of a result of Schellenberg [1] which applies these to matrices $BW(v,k)$.

4.184. LEMMA. (Mullin). If a $W(v,k)$ matrix is a regular type 1 matrix, then $a(1) = (k \pm \sqrt{k})/2$ and $a(-1) = (k \pm \sqrt{k})/2$.

Proof. The proof is a slight generalization of a result in Ryser [1, p. 134]. Let $e = a(1)-a(-1)$, and J denote the $v \times v$ matrix all of whose entries are 1. Clearly, we have

$$HJ = eJ = H^T J,$$

and hence

$$HH^T J = e^2 J = kJ.$$

Thus

$$e^2 = k,$$
$$a(1)+a(-1) = k,$$
$$a(1)-a(-1) = \pm\sqrt{k},$$

and the result follows.

4.185. COROLLARY. If a $W(v,k)$ matrix is a regular type 1 matrix, then k is a perfect square.

4.186. COROLLARY. If a $BW(v,k)$ matrix is a type 1 matrix, then $a(-1) \geq \lambda/4$ with equality if and only if $v = k = 4$.

Proof. Let us first note that in any BW(v,k) matrix, if $v = k$, then $k = \lambda$.

Now in any BW(v,k) matrix, we observe that $4(v-k-1)+\lambda \geq 0$, with equality only for $v = k = \lambda = 4$. This is immediate from the fact that in any (v, k, λ) configuration, as defined earlier, we have $v \geq k$ with equality only for $v = k = \lambda$.

The above inequality implies that the inequality

$$4(\lambda v-\lambda+k)-4k\lambda+\lambda^2 \geq 4k$$

is also valid, with equality only for $v = k = \lambda = 4$. But by the definition of λ, we have

$$k^2 = \lambda v-\lambda+k,$$

and therefore

$$(2k-\lambda)^2 \geq 4k,$$

with equality as above.

Now let us assume that $a(-1) < \lambda/4$. Since

$$k+\lambda k/2 > k/2 \geq \lambda/2,$$

the corollary is true unless

$$a(-1) = (k-\sqrt{k})/2.$$

Let us assume that $(k-\sqrt{k})/2 < \lambda/4$. Then

$$(2k-\lambda)^2 \leq 4k,$$

which is impossible unless equality holds in which case $v = k = \lambda$ as required.

The design generated by

$$(-1, 1, 1, 1) \bmod 4$$

satisfies the corollary with equality.

APPLICATION OF THE FROBENIUS GROUP DETERMINANT THEOREM TO BALANCED WEIGHING MATRICES

For the theory of group characters, the reader is referred to Speiser [1]. For Abelian groups, the Frobenius group determinant theorem (Speiser [1, p. 178]), in the notation employed here, becomes the following:

4.187. THEOREM. (Frobenius Group Determinant Theorem). Let M be a type 1 matrix over an Abelian group G of order v. Then

$$\det M(\chi) = \prod_{j=1}^{v} \sum_{g \in G} \alpha^{(j)}(g)\chi(g),$$

where $\alpha^{(j)}$ denotes the j-th irreducible character of G.

For the cyclic group of order v (written as the residues modulo v), this becomes

$$\det M(\chi) = \prod_{j=0}^{v-1} \sum_{k=0}^{v-1} \omega^{jk}\chi(k),$$

where ω is a primitive v-th root of unity.

Any group G of order v admits the main character

$\alpha^{(1)}(g) = 1, \quad g \in G.$

Every group determinant can be factored into forms in the indeterminates $\chi(g)$, which are irreducible over the integers, since it is clear that the expansion of the group determinant is a form with integer coefficients.

To illustrate the use of this theorem, we tackle the problem of finding a cyclic BW(10, 9) matrix M. Using the integers mod 10, we can, without loss of generality, assume that $\chi(g) = 0$ if and only if $g = 0$. Without any further theory, except for Lemma 4.184, there are $\binom{9}{3} = 84$ functions χ to consider. By the Frobenius group determinant theorem,

$$G(\chi) = \sum_{j=1}^{9} (-1)^j \chi(j),$$

corresponding to the character α defined by $\alpha(j) = (-1)^j$, is a divisor of $\det M = 3^{10}$.

Now let c denote the number of even residues j such that $\chi(j) = -1$. Then $G(\chi)$ can be determined in terms of c as follows:

c	$G(\chi)$
0	6
1	1
2	-3
3	-7 .

There are 40 functions with $c = 1$ and 30 with $c = 2$; therefore, the number of functions to be investigated has been reduced. Moreover, we have some structural information. As we shall see, the structural information is extremely important. In the following, if $\chi(g) = x$, we say that x appears in position g.

We note now that the inner product of absolute values of any pair of distinct rows is 8, since $\lambda = 8$ in the associated symmetric design. Thus the number of terms with value -1 in $r_0 \cdot r_j$ must be 4 for $j = 1, 2, \ldots, 9$. In particular, this means that in r_0 and r_j the number of times 0 opposes -1 must be even, that is, 0 or 2. Hence if translation (of row 0) by j units moves 0 to a position containing -1, then there must be a -1 in position $-j$ which is translated to column zero. Thus -1's occur in pairs of inverse positions. Now let us consider the case of $c = 2$. There is exactly one -1 on an odd residue. But since the parity of inverse pairs is equal, this -1 must be in position 5; that is, $\chi(5) = 1$. Now it is easily verified (considering row 1) that the three -1's cannot be consecutive in any event, and thus the remaining -1's occur in inverse pairs of positions $\chi(4) = \chi(6) = 1$. Also since $c = 2$, $\chi(1) = \chi(3) = \chi(7) = \chi(9) = 1$, and $\chi(2) = \chi(8) = -1$. We have determined the only possible function χ with $c = 2$. However, for this function $r_0 \cdot r_1 = -4$, and the matrix is not orthogonal.

Let us now consider the case $c = 1$. Clearly, no solution exists in this case since there is only one self-inverse element 5, which is odd. Hence there is no cyclic $BW(10, 9)$ matrix.

BALANCED WEIGHING MATRICES WITH $v \leq 25$

4.188. COMMENT. We conclude with a list of all triples v, k, λ with $k > \lambda > 0$ which satisfy $\lambda(v-1) = k(k-1)$ and $\lambda \equiv 0 \bmod 2$. We also list a function $E(v, k) = E$ where $E(v, k) = 1$ if a matrix $B(v, k)$ exists, and $E(v, k) = 0$ otherwise. In this regard it is useful to note that the matrices $BW(4n, 4n-1)$ are coexistent with skew Hadamard matrices of order $4n$ and that matrices $BW(4n+2, 4n+1)$ are coexistent with symmetric Hadamard matrices. The list of values for which such designs are known to exist are listed in Wallis [25].

In view of our earlier remarks about the usefulness of BW(v, k)'s, it would be of interest to establish the existence of more of these matrices. This will now be discussed.

	v	k	λ	E	Reason or Reference
1)	4	3	2	1	Mullin.
2)	6	5	4	1	*, Complement PG(1, 5).
3)	7	4	2	1	Circulant with first row [-110100].
4)	8	7	6	1	*, Complement PG(1, 7).
5)	10	9	8	1	*, Complement PG(1, 9).
6)	11	5	2	0	Condition (iv).
7)	12	11	10	1	*, Complement PG(1, 11).
8)	13	9	4	1	Condition (v).
9)	14	13	12	1	*, Complement PG(1, 13).
10)	15	8	4	0	Condition (iv).
11)	16	6	2	0	Schellenberg [1].
12)	16	10	6		
13)	16	15	14	1	*
14)	18	17	16	1	*, Complement PG(1, 17).
15)	19	9	4		
16)	20	19	18	1	*, Complement PG(1, 19).
17)	21	16	12	1	*, Complement PG(2, 4).
18)	22	7	2	0	Condition (i).
19)	22	15	10	0	Condition (i).
20)	22	21	20	0	Condition (vi).
21)	23	12	6	0	Condition (iv).
22)	24	23	22	1	*, Complement PG(1, 23).
23)	25	16	10		

* See Comment 4.188.

THERE ARE NO CIRCULANT BALANCED WEIGHING MATRICES BW(v, v-1) BASED ON (v, v-1, v-2) CONFIGURATIONS

Without loss of generality we assume that in such matrices the element 0 occurs down the main diagonal.

4.189. LEMMA. In any circulant orthogonal matrix based on a (v, k, λ) configuration, the parameter k is a perfect square.

Proof. Suppose that the first row of the orthogonal matrix contains a entries of 1 and b entries of -1. By the circulant property, every row and column has sum $a-b$. If the matrix is denoted by N, we have

$$NJ = N^T J = (a-b)J.$$

But

$$NN^T = kI.$$

Hence

$$NN^T J = kIJ = kJ.$$

But

$$NN^T J = N(a-b)J = (a-b)^2 J.$$

Thus $k = (a-b)^2$.

In the following we assume without loss of generality that $a > b$; otherwise we multiply the entire matrix by -1. For convenience we set $a-b = t$.

4.190. LEMMA. If a denotes the number of 1's in the first row of an orthogonal circulant matrix, and b the number of -1's, then

$$a = (t^2+t)/2 \quad \text{and} \quad b = (t^2-t)/2.$$

Proof. This is immediate since

$$a+b = k = t^2$$
$$a-b = t.$$

Thus in looking for circulant orthogonal matrices based on trivial designs, we need only consider (t^2+1, t^2, t^2-1) configurations where t is odd.

4.191. DEFINITION. An orthogonal circulant based on a (t^2+1, t^2, t^2-1) configuration will henceforth be called a trivial circulant.

Let $x_{\alpha\beta}$ $(\alpha, \beta = 0, 1, 2)$ represent the number of times that α in row i is in the same column as β in row j (we use 2 to represent the entry -1). We require

4.192. LEMMA. Either $x_{01} = x_{10} = 1$, $x_{02} = x_{20} = 0$, or vice versa.

Proof. We actually determine all $x_{\alpha\beta}$. It is clear that

i) $\Sigma x_{\alpha\beta} = t^2+1$,

ii) $\Sigma x_{0j} = \Sigma x_{i0} = 1$,

iii) $\Sigma x_{1j} = \Sigma x_{i1} = (t^2+t)/2$,

iv) $\Sigma x_{2j} = \Sigma x_{i2} = (t^2-t)/2$.

Finally, orthogonality gives

v) $x_{11}+x_{22} = x_{12}+x_{21}$.

But $x_{11}+x_{22}+x_{12}+x_{21} = \lambda$, and thus each expression in v) equals $\frac{1}{2}(t^2-1)$. From iii) and iv), addition gives

$$x_{10}+x_{20} = 1 = x_{01}+x_{02};\ x_{00} = 0.$$

Also

$$x_{10}-x_{01} = x_{02}-x_{20} = x_{21}-x_{12}$$

$$= \text{an even number.}$$

This proves that $x_{10} = x_{01}$, $x_{02} = x_{20}$, as required. It is useful to record the table of values following.

	Case A	Case B
x_{00}	0	0
$x_{01} = x_{10}$	1	0
$x_{02} = x_{20}$	0	1
$x_{12} = x_{21}$	$\frac{t^2-1}{4}$	$\frac{t^2-1}{4}$
x_{11}	$\frac{(t+3)(t-1)}{4}$	$\frac{(t-1)^2}{4}$
x_{22}	$\frac{(t+1)^2}{4}$	$\frac{(t-3)(t+1)}{4}$.

In the light of Lemma 4.192, we note that we can write $v = 4m+2$, $k = 4m+1 = t^2$, $\lambda = 4m$, and the typical circulant has the following form (illustration for $m = 6$).

Row 1: $0 a_1 a_2 \ldots a_{11} a_{12} \theta a_{12} a_{11} \ldots a_3 a_2 a_1$.

(The symmetry of the sequence is guaranteed by the fact that $x_{0i} = x_{i0}$ for $i = 1, 2$.)

Row j is obtained by a cyclic shift through $j-1$ places to the right.

We now prove:

4.193. LEMMA. $\theta = 1$ if $t \equiv 1 \pmod 4$; $\theta = -1$ if $t \equiv 3 \pmod 4$. Also

$$\sum_{j=1}^{m} a_{2j} = 0, \quad \sum_{i=0}^{m-1} a_{2i+1} = \frac{t-\theta}{2}.$$

<u>Proof.</u> Take the scalar products of row 1 with rows $2, 4, 6, \ldots, 2m+2$; add, and re-arrange. We have

$$2(a_1+a_3+\ldots+a_{2m-1})(a_2+a_4+\ldots+a_{2m}) + \theta(a_2+a_4+\ldots+a_{2m}) = 0.$$

Thus $(2a_1+\ldots+2a_{2m-1}+\theta)(a_2+a_4+\ldots+a_{2m}) = 0$; since only the second integer is even, we get $\Sigma a_{2j} = 0$. Also, since $2\Sigma a_{2n} + 2\Sigma a_{2j} + \theta = t$, we find that $\Sigma a_{2i} = \frac{t-\theta}{2}$.

Finally, note that the sum Σa_{2i} is an even integer (m terms); thus $t-\theta$ is divisible by 4, and $\theta = 1$ for $t \equiv 1 \pmod 4$, $\theta = -1$ for $t \equiv 3 \pmod 4$. This completes the proof.

Actually, if one takes scalar products of row 1 with rows $3, 5, \ldots, 2m+1$, and adds, one gets the identity

$$(\Sigma a_{2i})(\Sigma a_{2i}+\theta) + (\Sigma a_{2j})^2 = m,$$

which also produces the desired results.

EXAMPLE. At this stage, it is most instructive to look at the example for $m = 6$. The scalar products for rows 2, 4, 6, 8, 10, 12 are written down as follows.

1) $\sum_{i=1}^{12} a_i a_{i+1} = 0 (a_{13} = \theta)$.

In the sequence $a_1, a_2, a_3, \ldots, a_{13}$ there must be exactly six sign changes to produce a zero sum in 1). Hence a_1 has the same sign as a_{13}; that is, $a_1 = \theta = 1$.

2) $a_1 a_2 + \sum_{1}^{10} a_i a_{i+3} + a_{11} a_{12} = 0$.

Write the sequence

$$a_3, a_6, a_9, a_{12}, a_{11}, a_8, a_5, a_2, a_1, a_4, a_7, a_{10}, a_{13}.$$

By the same argument, $a_3 = \theta$.

3) $a_1 a_4 + a_2 a_3 + \sum_{1}^{8} a_i a_{i+5} + a_9 a_{12} + a_{10} a_{11} = 0$.

Consider the sequence

$$a_5, a_{10}, a_{11}, a_6, a_1, a_4, a_9, a_{12}, a_7, a_2, a_3, a_8, \theta,$$

and we get $a_5 = \theta$.

4) $a_1 a_6 + a_2 a_5 + a_3 a_5 + \sum_{1}^{6} a_i a_{i+7} + a_7 a_{12} + a_9 a_{10} = 0$.

The relevant sequence is

$$a_7, a_{12}, a_5, a_2, a_9, a_{10}, a_3, a_4, a_{11}, a_8, a_1, a_6, \theta,$$

and the result is $a_7 = \theta$.

5) $a_1 a_8 + a_2 a_7 + a_3 a_6 + a_4 a_5 + \sum_{1}^{4} a_i a_{i+9} + a_5 a_{12} + a_6 a_{11}$

$$+ a_7 a_{10} + a_8 a_9 = 0.$$

Here, the sequence

$$a_9, a_8, a_1, a_{10}, a_7, a_2, a_{11}, a_6, a_3, a_{12}, a_5, a_4, \theta$$

proves $a_9 = \theta$.

Similarly, $a_{11} = \theta$, and $\sum a_{2i+1} = 6\theta = 6$ (a contradiction of Lemma 4.193).

The method just outlined is completely general and gives:

4.194. LEMMA. In an orthogonal circulant of the type we have been considering, with one zero per row, we have

$$a_1 = a_3 = \dots = a_{2m-1} = \theta.$$

Thus $\Sigma a_{2i+1} = m\theta = \frac{1}{2}(t-\theta)$.

It is easy to deduce, from Lemma 4.194, that

$$t = \theta(2m+1).$$

But $t^2 = 4m+1$, and so $4m+1 = (2m+1)^2$.

We conclude that $m = 0$ and state:

4.195. THEOREM. An orthogonal circulant with one zero per row only exists for $m = 0$; in this case, it is the identity matrix of order 2 or, equivalently, the transposition matrix of order 2.

§4.12. NEGACYCLIC MATRICES

A very recent paper of G. Berman [2] has brought attention to another type of weighing matrix. The results of this section, where not specifically referenced, have arisen from ideas of L. G. Kovács and Peter Eades.

4.196. DEFINITION. Let P be the square matrix of order v, whose elements p_{ij} are defined as follows:

$$p_{i,\,i+1} = 1, \qquad i = 0, 1, \dots, v-2,$$

$$p_{v-1,\,0} = -1,$$

$$p_{ij} = 0, \qquad \text{otherwise.}$$

Any matrix of the form $\Sigma a_i P^i$, with a_i commuting coefficients, will be called <u>negacyclic</u>.

4.197. LEMMA. P of order v satisfies

$$P^v = -I, \quad P^t = -P^{v-1}, \quad PP^t = I.$$

If $A = \Sigma a_i P^i$, $B = \Sigma b_j P^j$ and R is the back diagonal matrix, then

$$AB = BA \quad \text{and} \quad A(BR)^t = BRA^t.$$

We now note some other properties of negacyclic matrices:

4.198. LEMMA. If $A = \Sigma a_i P^i$ is a negacyclic matrix of odd order v, then $-XAX$, where $X = \text{diag}(1, -1, 1, -1, \ldots, 1)$, is a circulant matrix.

4.199. LEMMA. Suppose $v \equiv 0 (\text{mod} 2)$. The existence of a negacyclic $C = W(v, v-1)$ is equivalent to the existence of a $W(v, v-1)$ of the form

$$\begin{bmatrix} A & B \\ B^t & -A^t \end{bmatrix}$$

where A and B are negacyclic and $A^t = (-1)^{\frac{1}{2}v+1} A$.

Proof. First we suppose there is a negacyclic matrix $N = W(2n, 2n-1)$ of order $2n$ which is used to form two negacyclic matrices A and B of order n which satisfy

$$AA^T + BB^T = (an-1)I_n. \tag{12.1}$$

Let the first row of the negacyclic matrix N be

$$0 x_1 y_1 x_2 y_2 \ldots y_{n-1} x_n.$$

We choose A and B to be negacyclic matrices with first rows

$$0 y_1 y_2 \ldots y_{n-1} \quad \text{and} \quad x_1 x_2 \ldots x_n,$$

respectively. If the order n is odd and the first rows of A and B are

$$0 a_1 \ldots a_t (\varepsilon_t a_t) \ldots (\varepsilon_1 a_1) \quad \text{and} \quad 1 b_1 b_2 \ldots b_t (\delta_t b_t) \ldots (\delta_1 b_1),$$

with $\varepsilon_i = \pm 1$, $\delta_j = \pm 1$, then taking the dot product of the first and $(i+1)$st rows, $i \le t$ (reducing using $xy \equiv x+y-1 (\text{mod } 4)$), we obtain

$$2t - 2i + \varepsilon_i (\text{mod } 4) \quad \text{and} \quad 2t - 2i + 1 (\text{mod } 4),$$

respectively. Hence using (12.1),

$$\varepsilon_i + 1 \equiv 0 (\text{mod } 4),$$

we have $\varepsilon_i = -1$.

If the order n is even and the first rows of A and B are

$$0a_1 \ldots a_{t-1}a_t(\varepsilon_{t-1}a_{t-1}) \ldots (\varepsilon_1 a_1)$$

and

$$1b_1b_2 \ldots b_t(\delta_{t-1}b_{t-1}) \ldots (\delta_1 b_1),$$

with $\varepsilon_i = \pm 1$, $\delta_j = \pm 1$, then taking the dot product of the first and (i+1)st rows, $i \le t-1$ (reducing modulo 4), we obtain

$$2t-2t-1+\varepsilon_i \pmod 4 \qquad \text{and} \qquad 2t-2i+2b_i-2 \pmod 4,$$

respectively. Hence, using (12.1),

$$\varepsilon_i + 2b_i - 3 \equiv 0 \pmod 4,$$

and since $b_i \ne 0$, we have $\varepsilon_i = 1$.

This means the first row of the original negacyclic matrix of order $2n$ can be written as

$$0x_1a_1x_2a_2 \ldots x_ta_t1\bar{a}_t(\delta_t x_t)\bar{a}_{t-1} \ldots \bar{a}_2(\delta_2x_2)\bar{a}_1(\delta_1x_1) \quad \text{for } n \text{ odd}$$

and

$$0x_1a_1x_2a_2 \ldots a_{t-1}x_ta_t(\delta_t x_t)a_{t-1} \ldots a_2(\delta_2x_2)a_1(\delta_1x_1) \quad \text{for } n \text{ even}$$

with $\delta_j = \pm 1$ and $\bar{a}_i = -a_i$.

The inner product of the first and (2i-1)th rows, $i \le t$ and $t-1$ respectively, is

$$-\delta_i + 1 \equiv 0 \pmod 4 \qquad \text{and} \qquad \delta_i + 1 \equiv 0 \pmod 4.$$

So we have the first rows of A and B

$$0a_1 \ldots a_t\bar{a}_t \ldots \bar{a}_1 \quad \text{and} \quad b_1b_2 \ldots b_t1b_t \ldots b_2b_1 \quad \text{for } n \text{ odd} \tag{12.2}$$

and

$$0a_1 \ldots a_{t-1}a_ta_{t-1} \ldots a_1 \quad \text{and} \quad b_1b_2 \ldots b_t\bar{b}_t \ldots \bar{b}_2\bar{b}_1 \quad \text{for } n \text{ even}, \tag{12.3}$$

as required.

It is straightforward to check that negacyclic matrices A and B, which satisfy $AA^t + BB^t = (2n+1)I_n$ and are of the form (12.2) and (12.3), give a negacyclic matrix $W(2n, 2n-1)$ when formed into first rows

$$0b_1a_1b_2 \ldots b_ta_t1\bar{a}_tb_t \ldots \bar{a}_1b_1, \quad \text{for } n \text{ odd},$$

or

$$0b_1a_1b_2\dots b_ta_t\bar{b}_t\dots a_1\bar{b}_1, \text{ for } n \text{ even.}$$

4.200. EXAMPLE. The negacyclic matrices with first rows

01-111 and 011-111-

of orders 6 and 8 are equivalent to the existence of

$$\begin{bmatrix} 0 & - & 1 \\ - & 0 & - \\ 1 & - & 0 \end{bmatrix}, \begin{bmatrix} 1 & 1 & 1 \\ - & 1 & 1 \\ - & - & 1 \end{bmatrix} \text{ and } \begin{bmatrix} 0 & 1 & 1 & 1 \\ - & 0 & 1 & 1 \\ - & - & 0 & 1 \\ - & - & - & 0 \end{bmatrix}, \begin{bmatrix} 1 & - & 1 & - \\ 1 & 1 & - & 1 \\ - & 1 & 1 & - \\ 1 & - & 1 & 1 \end{bmatrix}.$$

The following are first rows of negacyclic matrices $W(v, v-1)$ of orders 4 and 10, respectively:

0 1 1 -,

0 1 1 - 1 - - - - 1.

4.201. COMMENT. Peter Eades and Delsarte, Goethals and Seidel [1] have determined that the only negacyclic $W(v, v-1)$ of order $v < 1000$ have $v = p^r+1$ where p^r is an odd prime power. On the positive side we know (we omit the proof):

4.202. THEOREM. (Delsarte-Goethals-Seidel [1]). There is a negacyclic $W(p^r+1, p^r)$ whenever p^r is an odd prime power.

Very recent work of G. Berman [2] has led us to believe that many results of a similar type to those in this chapter can be obtained for negacyclic as well as circulant matrices, for he has shown:

4.203. THEOREM. (Berman [2]). There are negacyclic weighing matrices

i) $W\left(\frac{(p^{(t+1)r}-1)}{2(p^r-1)}, p^{(t-1)r}\right)$ for t(odd integer) ≥ 0, r(integer) > 0, p odd prime;

ii) $W\left(\frac{p^{tr}-1}{2}, p^{(t-1)r}\right)$ for t(integer) ≥ 0, r(integer) > 0, p odd prime;

iii) $W\left(\frac{(p^{tr}-1)}{2(p^{d}-1)}, p^{(t-1)r}\right)$ for $d \mid r$, t(odd integer) ≥ 0, r(integer) > 0,

$p \equiv 3 \pmod 4$ an odd prime.

EXAMPLE. The following is the first row of a negacyclic $W(20, 9)$:

1100-1--0-000-000010.

CHAPTER 5

AMICABLE ORTHOGONAL DESIGNS

§5.1. INTRODUCTION

In our first paper, Geramita-Geramita-Wallis [1], we quoted the following remarkable pairs of orthogonal designs:

$$\text{I.}\quad X = \begin{bmatrix} x_1 & x_2 \\ x_2 & -x_1 \end{bmatrix}; \quad Y = \begin{bmatrix} y_1 & y_2 \\ -y_2 & y_1 \end{bmatrix}$$

$$\text{II.}\quad X = \begin{bmatrix} x_1 & x_2 & x_3 & x_3 \\ -x_2 & x_1 & x_3 & -x_3 \\ x_3 & x_3 & -x_1 & -x_2 \\ x_3 & -x_3 & x_2 & -x_1 \end{bmatrix}; \quad Y = \begin{bmatrix} y_1 & y_2 & y_3 & y_3 \\ y_2 & -y_1 & y_3 & -y_3 \\ -y_3 & -y_3 & y_2 & y_1 \\ -y_3 & y_3 & y_1 & -y_2 \end{bmatrix}.$$

They have the property that $XY^t = YX^t$.

The existence of just these two pairs of orthogonal designs had the following useful consequences.

5.1. PROPOSITION. If there is an orthogonal design of order n and type (a,b), then

i) there is an orthogonal design in order $2n$ of type (a,a,b,b);

ii) there is an orthogonal design in order $4n$ of type $(a,a,2a,b,b,2b)$.

Proof. i) Let the orthogonal design in order n be on the variables z_1, z_2. Using X and Y as in I above and setting $z_1 = X$ and $z_2 = Y$ gives the result. ii) is proved analogously using the pair X and Y from II above.

We obtain the following replication result.

5.2. COROLLARY. If all orthogonal designs $(1,k)$, $1 \le k \le n-1$, exist in order n, then all orthogonal designs $(1,\ell)$, $1 \le \ell \le 2n-1$, exist in order $2n$.

5.3. EXAMPLE. There is an orthogonal design of type $(1,1)$ in order 2, and so if $n = 2^s$, there are orthogonal designs of type $(1,k)$, $1 \le k \le 2^s-1$, in order 2^s. In particular, there are weighing matrices $W(n,k)$, for all $k \le n$, when $n = 2^s$.

This very simple application of the existence of pairs of orthogonal designs like I and II and other replications that come from them convinced us that such pairs would be extremely important in constructing orthogonal designs themselves. Consequently, we suggested in Geramita-Geramita-Wallis [1] that these examples be studied in greater detail than we had done.

This suggestion was taken up by Warren Wolfe in his Queen's University dissertation [1] and was a major breakthrough not only to our study of such pairs of orthogonal designs but to the study of orthogonal designs themselves. A major portion of this chapter will be devoted to an exposition of this aspect of Wolfe's contributions.

Subsequent to Wolfe's work on this problem, he and D. Shapiro had an opportunity to meet and discuss Wolfe's work. Shapiro was greatly interested in Wolfe's handling of the problem and saw that what had begun as two interesting examples could in fact be the basis for the study of an interesting aspect of the theory of quadratic forms. Using a different approach, Shapiro has brilliantly analyzed this problem about quadratic forms in (Shapiro [5]). His work has had deep implications back into the combinatorial problems that originally motivated the entire discussion. We shall briefly discuss this happy state of affairs in this chapter.

§5.2. DEFINITIONS AND ELEMENTARY OBSERVATIONS

5.4. DEFINITION. Let X and Y be orthogonal designs of the same order n where X is of type $(u_1,\ldots,u_s)$ on the variables $\{x_1,\ldots,x_s\}$ and Y is of type $(v_1,\ldots,v_t)$ on the variables $\{y_1,\ldots,y_t\}$. We say that

X and Y are amicable orthogonal designs if $XY^t = YX^t$.

REMARK. We observe that writing $Z = XY^t$, we must have $ZZ^t = (u_1x_1^2+\ldots+u_sx_s^2)(v_1y_1^2+\ldots+v_ty_t^2)I_n$. So amicable orthogonal designs are related to symmetric Z which have inner product factorization into quadratic forms.

NOTATION. With X, Y as above, we'll say that there are amicable orthogonal designs of type $((u_1,\ldots,u_s);(v_1,\ldots,v_t))$ in order n.

With this notation, examples I and II of §5.1 are amicable orthogonal designs of type $((1,1);(1,1))$ in order 2 and of type $((1,1,2);(1,1,2))$ in order 4, respectively.

5.5. EXAMPLE. Before christening these pairs, and after finding a few, we realized that such things had already been anticipated in the study of Hadamard matrices. Namely, in J. Wallis [8] the study of pairs of Hadamard matrices $W = I+S$ and M, where W, M have order = n, $S = -S^t$, $M = M^t$ and $WM^t = MW^t$, was extensively pursued and many examples discovered. These amicable Hadamard matrices are an important special case of amicable orthogonal designs. They have type $((1,n-1);(n))$ in order n.

Now, let X and Y be amicable orthogonal designs of type $((u_1,\ldots,u_s);(v_1,\ldots,v_t))$ in order n. Write $X = \sum_{i=1}^{s} A_ix_i$, $Y = \sum_{j=1}^{t} B_jy_j$. The fact that the x's and y's are assumed to commute and that $\mathbb{Z}$ is an infinite integral domain easily yields:

0) The A_i and B_j are all $\{0,1,-1\}$ matrices, and
$A_i * A_\ell = 0,\ 1 \le i \ne \ell \le s;$
$B_j * B_k = 0,\ 1 \le j \ne k \le t.$

i) $A_iA_i^t = u_iI_n,\ 1 \le i \le s;$
$B_jB_j^t = v_jI_n,\ 1 \le j \le t.$

ii) $A_iA_\ell^t+A_\ell A_i^t = 0,\ 1 \le i \ne \ell \le s;$
$B_jB_k^t+B_kB_j^t = 0,\ 1 \le j \ne k \le t.$

iii) $A_i B_j^t = B_j A_i^t$, $1 \leq i \leq s$, $1 \leq j \leq t$.

Conditions 0), i) and ii) are not new. They are just the assertions that X and Y exist (i.e., orthogonal designs of that order and type exist). Condition iii) is the new one. We see, again, the dichotomy between algebraic and combinatorial properties: i), ii) and iii) can, and will, be treated separately from 0). This was, in fact, Wolfe's approach to the problem, i.e., to just consider (at first) conditions i), ii) and iii). (Oddly enough, it was first in the study of amicable orthogonal designs that the algebraic and combinatorial distinctions became clear. The treatment we are giving in this text, especially in Chapter 3, is a classic story of hindsight! It is to Wolfe's credit that he forced this distinction on us.)

It is clear that conditions 0) - iii) are necessary and sufficient for the existence of amicable orthogonal designs of type $((u_1, \ldots, u_s);(v_1, \ldots, v_t))$. For reference, we state this precisely.

5.6. PROPOSITION. A necessary and sufficient condition that there exist amicable orthogonal designs X and Y of type $((u_1, \ldots, u_s);(v_1, \ldots, v_t))$ in order n is that there exist a family of matrices $\{A_1, \ldots, A_s; B_1, \ldots, B_t\}$ of order n satisfying 0) - iii) above.

Following the route in Chapter 3, we make the following definition.

5.7. DEFINITION. An <u>amicable family of type</u> $[[u_1, \ldots, u_s];[v_1, \ldots, v_t]]$ in order n is a collection of rational matrices of order n, $\{A_1, \ldots, A_s; B_1, \ldots, B_t\}$ satisfying i), ii), iii) above, where the u_i and v_j are positive rational numbers.

5.8. PROPOSITION. Let $\{A_1, \ldots, A_s; B_1, \ldots, B_t\}$ be an amicable family of type $[[u_1, \ldots, u_s];[v_1, \ldots, v_t]]$ in order n, and let P and Q be rational matrices of order n satisfying $PP^t = aI_n$, $QQ^t = bI_n$. Then $\{PA_1Q, \ldots, PA_sQ; PB_1Q, \ldots, PB_tQ\}$ is an amicable family of order n and type $[[u_1ab, \ldots, u_sab];[v_1ab, \ldots, v_tab]]$.

<u>Proof.</u> Direct verification.

We shall now investigate the existence problem for amicable families and amicable orthogonal designs in some special cases. Observe that if $\{A_1,\dots,A_s;B_1,\dots,B_t\}$ is an amicable family in order n, then $\{A_1,\dots,A_s\}$ and $\{B_1,\dots,B_t\}$ are each rational families in order n, and so $s,t \le \rho(n)$.

n ODD. In this case $\rho(n) = 1$, and so an amicable family of type $[[u];[v]]$ is $\{A;B\}$, A, B rational matrices, where $AA^t = uI_n$, $BB^t = vI_n$ and $AB^t = BA^t$. We know from Proposition 2.3 that u and v are each squares in $\mathbb{Q}$. With that observation we can prove the following, somewhat disappointing, proposition.

5.9. PROPOSITION. Let n be odd. A necessary and sufficient condition that there be an amicable family of type $[[u];[v]]$ in order n is that there exist rational families of type $[u]$ and $[v]$ in order n.

Proof. The necessity is obvious since an amicable family is made up of two rational families (plus more). To show the sufficiency, let $u = a^2$, $v = b^2$, $a,b \in \mathbb{Q}^+$; then $\{aI_n;bI_n\}$ is an amicable family of the desired order and type.

Thus, there are no new algebraic restrictions which prohibit amicable orthogonal designs from existing in order n, n odd.

5.10. PROBLEM. Let n be odd, and suppose there are $W(n,k_1)$ and $W(n,k_2)$. Are there amicable orthogonal designs of type $((k_1);(k_2))$ in order n?

It is hard to believe that the answer to this question will be yes. Yet, we know of no case where it is not true. In fact, if A is a $W(n,k)$, then $X = Ax$ and $Y = Ay$ form amicable orthogonal designs of type $((k);(k))$ in order n.

In the special case that $k_1 = 1$ in Problem 5.10, that problem becomes:

5.11. PROBLEM. Let n be odd, and suppose there is a $W(n,k)$. Is there a symmetric $W(n,k)$?

We can make a small contribution to the question of existence for symmetric $W(n, k)$ by considering the following.

5.12. LEMMA. Let $\{M_1, \ldots, M_s; N_1, \ldots, N_r\}$ be a family of weighing matrices of order n and length (s, r) which satisfy

a) $M_i^t = -M_i$, $i = 1, \ldots, s$; $N_j = N_j^t$, $j = 1, \ldots, r$;

b) $M_i M_i^t = iI_n$, $N_j N_j^t = jI_n$, $1 \le i \le s$, $1 \le j \le r$;

c) $M_i N_j^t = M_j N_i^t$, $1 \le i \le s$, $1 \le j \le r$.

Then

i) if there is a family of order n and length $(s, 1)$, then there is a family of order $2n$ and length $(2s+1, 1)$;

ii) there is a family of order 2^s and length $(2^s-1, 1)$;

iii) there exist skew-symmetric $W(2^s, i)$ for all $i = 0, 1, \ldots, 2^s-1$ where s is a positive integer;

iv) if there is a family of order n and length $(s, 1)$, there is a family of order $2n$ and length $(0, 2s+2)$;

v) there is a family of order 2^s and length $(0, 2^s)$;

vi) there exist symmetric $W(2^s, i)$ for all $i = 0, 1, \ldots, 2^s$.

<u>Proof.</u> i) Let $\{M_1, \ldots, M_s; N_1\}$ be the family of length $(s, 1)$. Then with

$$A = \begin{bmatrix} 0 & 1 \\ -1 & 0 \end{bmatrix}, \quad H = \begin{bmatrix} 1 & 1 \\ 1 & -1 \end{bmatrix}, \quad M = \begin{bmatrix} 0 & 1 \\ 1 & 0 \end{bmatrix}$$

$$\{I_2 \otimes M_1, \ldots, I_2 \otimes M_s, A \otimes N_1, A \otimes N_1 + I_2 \otimes M_1, \ldots, A \otimes N_1 + I_2 \otimes M_s; H \otimes M_s\}$$

is the family of length $(2n+1, 1)$.

ii) Clearly $\{A;H\}$ is a family in order 2 and with length $(1;1)$, and so we have the result by repeated application of i).

iii) The skew-symmetric matrices of the family of ii) give the result.

iv) Consider

$$\{A \otimes M_1, \ldots, A \otimes M_s, M \otimes N_1, I \otimes N_1 + A \otimes M_1, \ldots, I \otimes N_1 + A \otimes M_s, H \otimes N_1\}.$$

v) The symmetric matrices of the family of iv) give the result.

The lemma shows that we only need symmetric weighing matrices of odd order to ensure the existence of symmetric weighing matrices for most large orders and most weights. Clearly, by taking the back circulant matrix one obtains from Theorem 4.124 (see Lemma 4.22 also), we obtain a symmetric $W(q^2+q+1, q^2)$ whenever q is a prime power. We also note that the $W(p+1, p)$ constructed for prime powers p in Lemma 4.34 of Chapter 4 is symmetric for $p \equiv 1 \pmod 4$. Further observe that using circulant matrices M, N of order $2n+1$ with first rows

$$0_n\ 1\ 1\ 0_{n-1} \quad \text{and} \quad 0_n\ 1\ {-1}\ 0_{n-1}$$

in $\begin{bmatrix} M & N \\ -N^t & M^t \end{bmatrix}$, where 0_k is the zero vector of length k, gives a symmetric $W(4n+2, 4)$ for every integer $n \geq 1$.

We may summarize all this in the following.

5.13. PROPOSITION

i) There is a symmetric $W(n, 4)$ for all $n \geq 10$.

ii) There is a symmetric $W(n, 9)$ for all $n \geq 68$.

iii) Let k be a square integer; then there exists an integer $N(k)$ such that there is a symmetric $W(n, k)$ for every $n \geq N(k)$.

Since fewer symmetric weighing matrices than weighing matrices are known and since they always give amicable orthogonal designs, this area merits further study, as does the entire question of amicable orthogonal designs in odd orders.

$n = 2b$, b ODD. In this case $\rho(n) = 2$, and so the largest possible size for an amicable family is $\{A_1, A_2; B_1, B_2\}$.

5.14. PROPOSITION. Let $n = 2b$, b odd. A necessary and sufficient condition that there be an amicable family in order n of type $[[u_1, v_1]; [u_2, v_2]]$ is that there be rational families of type $[u_1, v_1]$ and $[u_2, v_2]$ in order n.

Proof. The necessity is obvious. To prove sufficiency we remark that a glance at the proof of Theorem 2.14 of Chapter 2 should convince the reader that it is enough to consider the case where $u_1 = v_1 = u$ and $u_2 = v_2 = v$, with u and v each a sum of two squares in $\mathbb{Q}$, and $n = 2$. In this case, suppose $u = \alpha^2+\beta^2$, $v = \gamma^2+\Delta^2$; then

$$A_1 = \begin{bmatrix} \alpha & \beta \\ -\beta & \alpha \end{bmatrix}, \quad A_2 = \begin{bmatrix} -\beta & \alpha \\ -\alpha & -\beta \end{bmatrix}, \quad B_1 = \begin{bmatrix} \gamma & \Delta \\ \Delta & -\gamma \end{bmatrix}, \quad B_2 = \begin{bmatrix} \Delta & -\gamma \\ -\gamma & -\Delta \end{bmatrix}$$

give the required amicable family.

Thus, the only algebraic restrictions on the existence of amicable families in orders $n = 2b$, b odd, come from the restrictions on rational families that we already knew.

That there are new combinatorial restrictions on amicable orthogonal designs in orders $n = 2b$, b odd, is evident from the following example.

5.15. EXAMPLE. (W. Wolfe). There is no amicable orthogonal design of type $((1);(2,2))$ in order 6. (Wolfe's result is more general; see Wolfe [1].)

We know that there are orthogonal designs of type (1) and (2,2) in order 6 (see Geramita-Geramita-Wallis [1]), so this example will show the inadequacy of Proposition 5.14.

There is no loss in assuming the first orthogonal design in the (presumed) pair of type $((1);(2,2))$ is zI_6 and the second is $X = Ax+By$. Then A and B are symmetric weighing matrices of weight 2 and order 6.

5.16. LEMMA. If a symmetric $W(n,2)$ exists with zero diagonal, then $n \equiv 0 \pmod 4$.

Proof. It is easy to check that a symmetric $W(n,2)$ with zero diagonal is (up to simultaneous row and column permutation and simultaneous multiplication of row i and column i by "-1") the direct sum of blocks

$$\begin{bmatrix} 0 & 1 & 1 & 0 \\ 1 & 0 & 0 & 1 \\ 1 & 0 & 0 & -1 \\ 0 & 1 & -1 & 0 \end{bmatrix}.$$

Now, it is possible to show that there is no symmetric orthogonal design of type (2,2) in order 6.

By the lemma, each variable must appear on the diagonal. Thus, we may assume that X looks like

$$\begin{bmatrix} x & x & y & y & 0 & 0 \\ x & -x & a & b & & \\ y & & & & & \\ y & & & & & \\ 0 & & & & & \\ 0 & & & & & \end{bmatrix}.$$

CASE 1. $a \neq 0$. Then we may assume $a = y$, $b = -y$, and so

$$X = \begin{bmatrix} x & x & y & y & 0 & 0 \\ x & -x & y & -y & 0 & 0 \\ y & y & u & v & & \\ y & -y & w & s & & \\ 0 & 0 & & & & \\ 0 & 0 & & & & \end{bmatrix}.$$

Checking columns 1 and 3, we see that u, w are $\neq 0$, and hence the rest of column 3 consists of zeros, and that gives too many zeros in the last two rows.

CASE 2. $a = 0$. Then $b = 0$, also, and we have

$$X = \begin{bmatrix} x & x & y & y & 0 & 0 \\ x & -x & 0 & 0 & y & y \\ y & 0 & & & & \\ y & 0 & & & & \\ 0 & y & & & & \\ 0 & y & & & & \end{bmatrix}.$$

Now there must be exactly two y's in the third column, but putting the second y in either the third or fourth rows gives the wrong inner product with columns one and three, while putting the second y in either the fifth or sixth rows gives trouble with the inner product of columns two and three.

Thus, such an X cannot exist.

REMARK. Wolfe's result is that there is no symmetric orthogonal design of type (2,2) in any order $n \equiv 2 \pmod 4$. We have not been able to place this example in any general framework but will see later that the symmetry conditions come into play in many places.

§5.3. THE NUMBER OF VARIABLES IN AN AMICABLE ORTHOGONAL DESIGN

Given a positive integer n, we may ask about the maximum number of variables that may appear in an amicable orthogonal design of order n. Clearly, the number of variables in both designs cannot exceed $2\rho(n)$. Indeed, when n is odd or if $n = 2b$, b odd, then the examples of §5.2 show that $2\rho(n)$ is possible. A bit of reflection shows that there are two separate questions that can, and should, be considered.

5.17. PROBLEM. Given an orthogonal design X involving s variables, let $A(X) = \{$orthogonal designs $Y | X$ and Y are amicable$\}$. ($A(X) = \phi$ is entirely possible!) Define $m(X) = \max_{Y \in A(X)} \{$number of variables in $Y\}$. What is $m(X)$?

5.18. PROBLEM. Fix an integer $s \leq \rho(n)$, and let ${}_n0_s = \{$orthogonal designs of order n involving s variables$\}$.

Define $m(s) = \max_{X \in {}_n0_s} \{s+m(X)\}$; then what is $m(s)$?

Problem 5.17 seems extremely difficult, in general. In fact, deciding whether or not $A(X) = \phi$ seems, in general, intractable. We shall have little to say about this problem directly.

Problem 5.18, on the other hand, can be given a complete solution. The presentation of the solution to Problem 5.18 that we give follows, for the most part, that of W. Wolfe in [1]. A different, and more general, solution is given by D. Shapiro in [5].

We begin by imitating some of the notions of Chapter 1. Suppose X and Y form an amicable orthogonal design in order n of type $((u_0, u_1, \ldots, u_s); (v_1, \ldots, v_t))$. [Note the numbering of the variables in the first design.] The coefficient matrices of the two orthogonal designs will be denoted

$$\{A_i,\ 0 \leq i \leq s;\ B_j,\ 1 \leq j \leq t\}.$$

Now, consider the <u>real</u> matrices $\alpha_i = \frac{1}{\sqrt{u_0 u_i}} A_i A_0^t$, $0 \le i \le s$, $\beta_j = \frac{1}{\sqrt{u_0 v_j}} B_j A_0^t$, $1 \le j \le t$. It is an easy matter to check that $\alpha_0 = I_n$ and $\{\alpha_i, 1 \le i \le s; \beta_j, 1 \le j \le t\}$ form a collection of real matrices of order n satisfying:

0) $\alpha_i = -\alpha_i^t$, $1 \le i \le s$; $\beta_j = \beta_j^t$, $1 \le j \le t$;

i) $\alpha_i^2 = -I_n$, $1 \le i \le s$; $\beta_j^2 = I_n$, $1 \le j \le t$;

ii) $\alpha_i\alpha_j + \alpha_j\alpha_i = 0$, $1 \le i \ne j \le s$; $\beta_k\beta_\ell + \beta_\ell\beta_k = 0$, $1 \le k \ne \ell \le t$;

iii) $\alpha_i\beta_j + \beta_j\alpha_i = 0$ for all $1 \le i \le s$, $1 \le j \le t$.

By analogy with Definition 1.2, we make the following definition.

5.19. DEFINITION. A family of real matrices $\{\alpha_i, 1 \le i \le s; \beta_j, 1 \le j \le t\}$ of order n satisfying 0)-iii) above will be called a <u>Hurwitz-Radon family of type</u> (s, t) <u>in order</u> n (or simply an H-R(s, t) family in order n).

So, an H-R(s, 0) family is what we discussed in Chapter 1. If we can show, for a given s and t, that there cannot be an H-R(s, t) family in order n, then there cannot be amicable orthogonal designs in order n formed by X and Y, where X involves s+1 variables and Y involves t variables.

If we neglect condition 0), we see that i)-iii) amount to saying that we have a representation of the Clifford algebra of the real quadratic form $s\langle -1\rangle \perp t\langle 1\rangle$ on the vector space $\mathbb{R}^n$; i.e., $\mathbb{R}^n$ is a module for this Clifford algebra. Now, the irreducible modules for a Clifford algebra always have dimension a power of 2, and if $n = 2^a b$, b odd, the maximum dimension of an irreducible module is $\le 2^a$; i.e., we get a representation of the Clifford algebra of $s\langle -1\rangle \perp t\langle 1\rangle$ on $\mathbb{R}^{2^a}$.

Let $C^{s,t}$ denote the Clifford algebra of $s\langle -1\rangle \perp t\langle 1\rangle$. If $n = 2^a$, we want to know whether $C^{s,t}$ can be nontrivially represented on $\mathbb{R}^{2^a}$. If it cannot, then we know that there is no H-R(s, t) family in any order $2^a b$, b odd. This will, at least, provide upper bounds for Problem 5.18 (and, consequently Problem 5.17). We shall then have to deal with the sharpness of these bounds.

Fortunately, the representation theory of the algebras $C^{s,t}$ is very well known. The earliest complete discussion we have found is in the 1950 paper of Kawada and Iwahori [1]. A more modern treatment can be found in [Lam [1], pp. 126-139].

If we let d be the degree of an irreducible real matrix representation of $C^{s,t}$ of minimal degree > 1, then the following are restatements of the Kawada-Iwahori theorems.

5.20. THEOREM. If $s+t = 2k$, then $C^{s,t}$ is a central simple $\mathbb{R}$-algebra, and d (as defined above) is given by:

a) if $t-k \equiv 0,1 \pmod 4$, then $d = 2^k$;

b) if $t-k \equiv 2,3 \pmod 4$, then $d = 2^{k+1}$.

5.21. THEOREM. If $s+t = 2k+1$, then $C^{s,t}$ is a semisimple algebra over $\mathbb{R}$, and d is given by:

a) if $t-k \equiv 0,2$ or $3 \pmod 4$, then $d = 2^{k+1}$;

b) if $t-k \equiv 1 \pmod 4$, then $d = 2^k$.

5.22. DEFINITION. Let $n = 2^a$, t be integers, $t > 0$. Define

$$\rho_t(n) = \max\{s \mid C^{s-1,t} \text{ has an irreducible real matrix representation of degree} \le n\}.$$

The relevance of ρ_t to our discussion is obvious, for if $m = 2^a b$, b odd, we could never find an amicable orthogonal design formed by X and Y in order m, where X involves s variables and Y involves t variables if $s > \rho_t(2^a)$.

The exact values of $\rho_t(n)$ are given by Kawada and Iwahori.

5.23. THEOREM. Let $n = 2^{4a+b}$, $0 \le b < 4$. Then $\rho_t(n)-1 = 8a-t+\delta$, where the values of δ are given in the table below:

t(mod 4) \ b	0	1	2	3
0	0	1	3	7
1	1	2	3	5
2	-1	3	4	5
3	-1	1	5	6 .

Some easy (but tedious) consequences of this theorem are:

5.24. COROLLARY

i) $\rho_0(n) = \rho(n)$,

ii) $\rho_t(2n) = \rho_{t-1}(n)+1$,

iii) $\rho_t(n) = \rho_{t+8}(2^4 n)$.

In the next example we shall illustrate how to use these theorems in some low orders.

5.25. EXAMPLE. Let $n = 4x$, x odd; then $\rho(n) = 4$, and so an orthogonal design in order n cannot involve more than 4 variables. Let X and Y be amicable orthogonal designs in order n, and suppose Y involves t variables ($t \leq 4$). Then, the discussion before Theorem 5.23 asserts that X cannot involve more than $\rho_t(4)$ variables.

We enumerate the possibilities in a table:

Table 5.1: $n = 4x$, x odd

t = number of variables in Y	The number of variables in X is $\leq$
4	$\rho_4(4) = 0$
3	$\rho_3(4) = 3$
2	$\rho_2(4) = 3$
1	$\rho_1(4) = 3$
0	$\rho_0(4) = 4$

We see already that there is a marked change from n odd and $n = 2x$, x odd, which we investigated in the previous section. There is no orthogonal design which is "amicable" with a 4-variable design in such an order. Furthermore, the maximum total number of variables we can ever hope to involve between X and Y is $6 < 2\rho(4) = 8$.

For future reference, we include a similar table for $n = 8x$ and $n = 16x$, x odd.

Table 5.2: n = 8x, x odd

t = number of variables in Y	The number of variables in X is ≤
8	$\rho_8(8) = 0$
7	$\rho_7(8) = 0$
6	$\rho_6(8) = 0$
5	$\rho_5(8) = 1$
4	$\rho_4(8) = 4$
3	$\rho_3(8) = 4$
2	$\rho_2(8) = 4$
1	$\rho_1(8) = 5$
0	$\rho_0(8) = 8$

Table 5.3: n = 16x, x odd

t = number of variables in Y	The number of variables in X is ≤
9	$\rho_9(16) = 1$
8	$\rho_8(16) = 1$
7	$\rho_7(16) = 1$
6	$\rho_6(16) = 2$
5	$\rho_5(16) = 5$
4	$\rho_4(16) = 5$
3	$\rho_3(16) = 5$
2	$\rho_2(16) = 6$
1	$\rho_1(16) = 9$
0	$\rho_0(16) = 9$

We would like to show, at least for these three special cases, that these bounds are achieved by an H-R(s, t) family (condition 0) is now considered); moreover, we would like to have the α_i (respectively the β_j) be disjoint $(0,1,-1)$ matrices. As in the discussion which preceded Theorem 1.7 of Chapter 1, that will follow immediately if we know that all matrices have integer entries.

Now, we saw in Proposition 1.5 that there is an H-R(3,0) family of integer matrices in order 4 (and by tensoring with I_x, in order $4x$ for any odd x), and so, in Table 5.1, we have taken care of the possibility that $t = 4$ = number of variables in Y, 0 = number of variables in X. To show that all other possibilities in Table 5.1 are actually achieved, it suffices just to construct an H-R(2,3) integer family in order 4. The rest of the table comes from omitting some members of this family.

Let $A = \begin{bmatrix} 0 & 1 \\ -1 & 0 \end{bmatrix}$, $P = \begin{bmatrix} 0 & 1 \\ 1 & 0 \end{bmatrix}$, $Q = \begin{bmatrix} 1 & 0 \\ 0 & -1 \end{bmatrix}$.

5.26. PROPOSITION. (Wolfe's Slide Lemma). If $\{M_i, 1 \leq i \leq s; N_j, 1 \leq j \leq t\}$ is an H-R(s,t) family in order n (of integer matrices), then

$$\{P \otimes M_i,\ 1 \leq i \leq s,\ A \otimes I_n;\ P \otimes N_j,\ 1 \leq j \leq t,\ Q \otimes I_n\}$$

is an H-R(s+1,t+1) family in order $2n$ (of integer matrices).

Proof. The verifications are routine.

It is an easy matter to check that $\{A;Q,P\}$ is an integer H-R(1,2) family in order 2, and so, by the Slide Lemma, we obtain an H-R(2,3) family of integer matrices in order 4.

In considering orders $8x$, x odd, we first note that Display 5.1

$$\left[\begin{array}{cccc|cccc}
0 & a & b & c & d & e & f & g \\
-a & 0 & c & -b & e & -d & -g & f \\
-b & -c & 0 & a & f & g & -d & -e \\
-c & b & -a & 0 & g & -f & e & -d \\
\hline
-d & -e & -f & -g & 0 & a & b & c \\
-e & d & -g & f & -a & 0 & -c & b \\
-f & g & d & -e & -b & c & 0 & -a \\
-g & -f & e & d & -c & -b & a & 0
\end{array}\right]$$

Display 5.1

gives an H-R(7, 0) family of integer matrices in order 8 (and hence in order 8x, x odd). Now there is a symmetric (1,1,1,1,1) design in order 8 (see the proof of Theorem 4.2 of Chapter 4) and hence an H-R(0,5) family of integer matrices in order 8. Finally, we use the Slide Lemma with the H-R(2, 3) and H-R(3, 0) families in order 4 to obtain an H-R(3,4) integer family in order 8. These give all possibilities in Table 5.2.

We complete the special cases by observing that there is an H-R(0, 9) integer family in order 16 (given by the symmetric (1,1,1,1,1,1,1,1,1) design in order 16 which appears in the proof of Theorem 4.2 of Chapter 4). We use the Slide Lemma with the integer matrices of the H-R(0,5), H-R(3,4), H-R(4,1) and H-R(7,0) families in order 8 to obtain the integer H-R(1,6), H-R(4,5), H-R(5,2) and H-R(8,1) families in order 16. These give all the possibilities for Table 5.3.

Wolfe has given the general result in [1], and so we shall not pursue the proofs here. We shall be content to just state Wolfe's theorem and the three lemmata that were used to obtain it (in addition to the Slide Lemma and the explicit computations in orders 4, 8 and 16). The lemmata are of independent interest and provide some construction methods. They shall be generalized later.

5.27. THEOREM. For any integer n and any $t \leq \rho(n)$, there is an $H\text{-}R(\rho_t(n)-1, t)$ family of integer matrices of order n.

5.28. COROLLARY. For any integer n there are amicable orthogonal designs X and Y involving s and t variables, respectively, X of type $(\underbrace{1, \ldots, 1}_{s\text{-tuple}})$, Y of type $(\underbrace{1, \ldots, 1}_{t\text{-tuple}})$ for any $s \leq \rho_t(n)$.

5.29. LEMMA. Conversely, we note that the existence of amicable orthogonal designs X and Y of order n and types $(u_0, u_1, \ldots, u_s)$ and $(v_1, v_2, \ldots, v_t)$ on variables x_i and y_j, respectively, implies the existence of an H-R(s,t) family.

We summarize all the relevant facts in the proof of Theorem 5.27.

5.30. PROPOSITION. (Wolfe). Let A, P, Q be as above.

1) If $\{M_i, 1 \le i \le s\}$ is an H-R(s, 0) (integer) family in order n, then $\{A \otimes M_i, 1 \le i \le s, P \otimes I_n, Q \otimes I_n\}$ is an H-R(0, s+2) (integer) family in order 2n.

2) If $\{N_j, 1 \le j \le t\}$ is an H-R(0, t) (integer) family in order n, then

$$\{A \otimes Q \otimes I_2 \otimes N_j, 1 \le j \le t, I_2 \otimes A \otimes I_{2n}, A \otimes P \otimes Q \otimes I_n,$$
$$Q \otimes Q \otimes A \otimes I_n, P \otimes Q \otimes A \otimes I_n, I_2 \otimes P \otimes A \otimes I_n, A \otimes P \otimes P \otimes I_n\}$$

is an H-R(t+6, 0) (integer) family in order 8n.

3) (Jump) If $\{M_i, 1 \le i \le s; N_j, 1 \le j \le t\}$ is an H-R(s, t) (integer) family in order n, then

$$\{A \otimes P \otimes A \otimes Q \otimes M_i, 1 \le i \le s; A \otimes P \otimes A \otimes Q \otimes N_j, 1 \le j \le t,$$
$$Q \otimes Q \otimes Q \otimes Q \otimes I_n, P \otimes I_4 \otimes Q \otimes I_n, Q \otimes P \otimes I_2 \otimes Q \otimes I_n,$$
$$Q \otimes Q \otimes P \otimes Q \otimes I_n, I_8 \otimes P \otimes I_n, P \otimes Q \otimes A \otimes A \otimes I_n,$$
$$I_2 \otimes P \otimes A \otimes A \otimes I_n\}$$

is an H-R(s, t+8) (integer) family in order $2^4 \cdot n$.

Wolfe proves one more fact which will be of interest to us.

5.31. THEOREM. If $n = 2^a n_0$, n_0 odd. If $t \le \rho(n)$, then $t + \rho_t(n) \le 2a+2$. Furthermore, there is a $t_0 \le \rho(n)$ with $t_0 + \rho_{t_0}(n) = 2a+2$.

5.32. COROLLARY. If X and Y are amicable orthogonal designs in order $n = 2^a n_0$, n_0 odd, then the total number of variables in X and Y is $\le 2a+2$, and that bound is achieved.

§5.4. THE ALGEBRAIC THEORY OF AMICABLE ORTHOGONAL DESIGNS

We continue to pursue the analogy we have already seen between the study of orthogonal designs and the study of amicable orthogonal designs. The algebraic theory, naturally, will rest on the study of amicable families.

5.33 PROPOSITION. (Wolfe). Let $\{A_1,\dots,A_s;B_1,\dots,B_t\}$ be an amicable family of type $[[u_1,\dots,u_s];[v_1,\dots,v_t]]$ in order n. Then $\{A_1A_2^tB_i,\ 1\le i\le t,\ A_3,\dots,A_s\}$ is a rational family of type $[u_1u_2v_1, u_1u_2v_2,\dots,u_1u_2v_t,u_3,\dots,u_s]$ in order n.

Proof. A routine check.

5.34. COROLLARY. Let $n \equiv 4 \pmod 8$. In order that there be amicable families of type $[[a_1,a_2,a_3];[b_1,b_2,b_3]]$ in order n, one must have

a) rational families of type $[a_1,a_2,a_3]$ and $[b_1,b_2,b_3]$ in order n

and

b) $a_1a_2a_3b_1b_2b_3$ a square in $\mathbb{Z}$.

Proof. Use Proposition 5.33 and the results of Chapter 3.

In the same vein:

5.35. COROLLARY. Let $n \equiv 4 \pmod 8$. Necessary conditions that there be amicable families in order n of type

1) $[[a_1,a_2,a_3];[b_1,b_2]]$ are that
   a) there be rational families of type $[a_1,a_2,a_3]$ and $[b_1,b_2]$ in order n, and
   b) at every prime p, $(-1,a_1a_2a_3b_1b_2)_p = (b_1,b_2)_p(a_1a_2a_3,b_1b_2)_p$;
2) $[[a_1,a_2,a_3];[b_1]]$ are that
   a) there be rational families of type $[a_1,a_2,a_3]$ and $[b_1]$ in order n, and
   b) $a_1a_2a_3b_1$ be a sum of fewer than four squares in $\mathbb{Z}$.

REMARK. Shapiro has given a different formulation of these two corollaries and generalized this to other n. We shall give his results later in this section.

5.36. EXAMPLE

i) There are no amicable orthogonal designs of type $((1,1,1);(1,1,2))$ in any order $4n_0$, n_0 odd, although designs of both types exist in every order 4n.

ii) There are no amicable orthogonal designs of type $((1,1,1);(1,3))$ in any order $4n_0$ by Corollary 5.35. This also eliminates amicable orthogonal designs of type $((1,11);(4,4,4))$ in order 12 and, in fact, in any order $4n_0$, n_0 odd, $n_0 \geq 3$.

iii) There is no amicable family of type $[[1,1,1];[7]]$ in any order $4n_0$, n_0 odd, $n_0 > 1$. So, in particular, there are no amicable orthogonal designs of type $((4,4,4);(7))$ in order 12.

Daniel Shapiro has recently solved the entire algebraic problem of amicable orthogonal designs. This is in Shapiro [5]. He has continued his study of the space of similarities and extended his earlier work to handle amicable families. These techniques are too complicated for us to attempt to analyze here. Suffice it to say that his work requires a deep understanding of the modern theory of quadratic forms and applies to a much more general setting than we have considered here. A source of great pleasure to us has been the realization that the combinatorial problem we were considering inspired an interesting new investigation in quadratic forms. Not only has this problem borrowed heavily from the theory of quadratic forms for answers, but it has suggested new problems in that theory.

We shall quote here a few of Shapiro's theorems (in a form suitable for our discussion).

The first theorem, though not surprising, is surprisingly complicated to prove and is the analogue of Theorem 3.52A.

5.37. THEOREM. Let $n = 2^m n_0$, n_0 odd. There is an amicable family of type $[[a_1,\dots,a_s];[b_1,\dots,b_t]]$ in order $n \iff$ there is an amicable family of the same type in order 2^m.

The next theorem also has its counterpart in Theorem 3.92, part A.

5.38. THEOREM. Let $n = 2^m$ $(m \geq 3)$, and choose $t \leq \rho(n)$ and $s \leq \rho_t(n)$. If $s+t \leq 2m-1$, then for any positive rationals $a_1,\dots,a_s$, $b_1,\dots,b_t$ there is an amicable family of type $[[a_1,\dots,a_s];[b_1,\dots,b_t]]$.

Thus, for s, t as above ($m \geq 3$), the only time there are algebraic restrictions are when $2m \leq s+t \leq 2m+2$. In the next proposition we shall abstract a portion of Shapiro's theorem dealing with the interval $[2m, 2m+2]$ for the special cases of $n = 2^2$ and $n = 2^3$.

5.39. THEOREM. Let $n \equiv 4 \pmod 8$. Necessary and sufficient conditions that there be an amicable family in order n of type

1) $[[a_1, a_2, a_3];[b_1, b_2, b_3]]$ are that
   a) $<a_1, a_2, a_3>$ and $<b_1, b_2, b_3>$ be isometric to subforms of $4<1>$, and
   b) $<a_1, a_2, a_3> \simeq <b_1, b_2, b_3>$;
2) $[[a_1, a_2, a_3];[b_1, b_2]]$ is that $<b_1, b_2> < <a_1, a_2, a_3> < 4<1>$;
3) $[[a_1, a_2, a_3];[b_1]]$ is that $<b_1> < <a_1, a_2, a_3> < 4<1>$.

If $n \equiv 8 \pmod{16}$, necessary and sufficient conditions that there be an amicable family in order n of type

4) $[[a_1, a_2, a_3, a_4];[b_1, b_2, b_3, b_4]]$ is that $<a_1, a_2, a_3, a_4> \simeq <b_1, b_2, b_3, b_4>$;
5) $[[a_1, a_2, a_3, a_4];[b_1, b_2, b_3]]$ is that $<b_1, b_2, b_3> < <a_1, a_2, a_3, a_4>$;
6) $[[a_1, a_2, a_3, a_4, a_5];[b_1]]$ is that $b_1 \prod_{i=1}^{5} a_i$ is a square in $\mathbb{Q}$;
7) $[[a_1, a_2, a_3, a_4];[b_1, b_2]]$ is that $<b_1, b_2> < <a_1, a_2, a_3, a_4>$.

5.40. EXAMPLE. We have already seen possibilities that are eliminated by this theorem in orders $\equiv 4 \pmod 8$.

To see some possibilities eliminated in orders $n \equiv 8 \pmod{16}$, observe:

1. There is no amicable family of type $[[1,1,1,2];[1,1,1,1]]$ in order $8n_0$, n_0 odd. Thus, since $<1,1> \simeq <2,2>$, we have no amicable orthogonal designs of type $((2,2,2,2);(1,2,2,2))$ in order 8.

2. There is no amicable family of type $[[1,1,3];[1,1,1,1]]$ in any order $n \equiv 8 \pmod{16}$ since $<1,1,3> \not< <1,1,1,1>$. This is clear, for if $<1,1,3> < <1,1,1,1>$, we would have, by Witt Cancellation, that $<3> < <1,1>$, but 3 is not the sum of two squares in $\mathbb{Q}$.

In particular, there are no amicable orthogonal designs of type $((2,2,2,2);(2,2,3))$ in order 8.

3. There is no amicable family of type $[[1,1,1,1,2];[1]]$ in any order $n \equiv 8 \pmod{16}$, so, in particular, there are no amicable orthogonal designs of type $((1,1,2,2,2);(1))$ in order 8.

4. There is no amicable family of type $[[1,1,1,1];[1,k]]$ when k is not the sum of three squares in $\mathbb{Q}$ for any order $n \equiv 8 \pmod{16}$. So, in particular, there are no amicable orthogonal designs of type $((2,2,2,2);(1,7))$ in order 8.

§5.5. THE COMBINATORIAL THEORY OF AMICABLE ORTHOGONAL DESIGNS

We have already seen that there is more at stake in deciding if there are amicable orthogonal designs of a certain order and type than just the existence of amicable families of the same order and type: first, because there may not be orthogonal designs of a certain type, much less amicable ones, and second, because even when orthogonal designs of the appropriate types exist, none may be amicable.

Unfortunately, in this area we have only a few theorems and a few examples. There has not yet been discovered anything analogous to Corollary 2.21 for amicable orthogonal designs.

Warren Wolfe has given two useful combinatorial theorems which exclude possibilities not eliminated by the algebraic theory. The second of these has been extended by Peter J. Robinson.

5.41. THEOREM. (Wolfe). Suppose X and Y are amicable orthogonal designs in order $n \equiv 0 \pmod 4$, where X is of type $(1,1,1,a_1,\ldots,a_s)$ and Y is of type $(b_1,\ldots,b_t)$. Then, there exists an orthogonal design in order n of type $(1,b_1,\ldots,b_t)$.

<u>Proof.</u> Let $X = A_1x_1+A_2x_2+A_3x_3+\sum_{j=1}^{s} B_jx_{j+3}$. By applying row and column operations to X and Y simultaneously, we can assume

$$A_1 = \bigoplus_{n/4} \begin{bmatrix} 0 & 1 & & \\ -1 & 0 & & \\ & & 0 & 1 \\ & & -1 & 0 \end{bmatrix}, \quad A_2 = \bigoplus_{n/4} \begin{bmatrix} & & 1 & 0 \\ & & 0 & -1 \\ -1 & 0 & & \\ 0 & 1 & & \end{bmatrix},$$

and

$$A_3 = \bigoplus_{n/4} \begin{bmatrix} & & 0 & 1 \\ & & 1 & 0 \\ 0 & -1 & & \\ -1 & 0 & & \end{bmatrix}.$$ The patient reader will then discover that the relations $A_iY^t = YA_i^t$, $i = 1,2,3$, force Y to be skew-symmetric. Then zI_n+Y is the required orthogonal design.

REMARK. This theorem is, in some sense, only "$\frac{1}{2}$" combinatorial. An examination of the proof shows that all we really need to get Y skew-symmetric is that A_1, A_2, A_3 have integer entries (and hence $\{0,1,-1\}$ entries). This indicates that there may be some merit in defining and studying integer families and integer amicable families (by analogy with rational families and amicable families).

5.42. EXAMPLE. We may use Theorem 5.41 to observe that there are no amicable orthogonal designs of type $((1,1,1);(1,1,16))$ in order 20 since there is no orthogonal design of type $(1,1,1,16)$ in order 20. This type is not eliminated by any other theorems. This also eliminates anything of type $((1,1,1,\ldots);(b_1,\ldots,b_t))$ in order n where $\sum_{i=1}^{t} b_i = n$, e.g., $((1,1,1,5);(8))$ in order 8.

Before we proceed to the next theorem, we give a lemma:

5.43. LEMMA. Suppose X and Y are amicable orthogonal designs in order n, where X is of type $(1,u_1,\ldots,u_s)$ on the variables $x_0,\ldots,x_s$ and Y is of type $(v_1,\ldots,v_t)$ on the variables $y_1,\ldots,y_t$; then there exist monomial matrices P and Q (i.e., with elements 0, ± 1 and only one nonzero element per row and column) so that

$$PXQ = x_0I + \Sigma x_iM_i, \quad PYQ = \Sigma y_iN_i,$$

where

0) $M_i * M_j = 0$ for $i \neq j$; $N_\ell * N_k = 0$ for $\ell \neq k$;

i) $M_i^t = -M_i$, $\forall i$; $N_j^t = N_j$, $\forall j$;

ii) $M_iM_i^t = u_iI_n$, $\forall i$; $N_jN_j^t = v_jI_n$, $\forall j$;

iii) $M_iN_j^t = N_jM_i^t$;

iv) $M_iM_j^t + M_jM_i^t = 0$, $i \neq j$; $N_\ell N_k^t + N_kN_\ell^t = 0$, $k \neq \ell$.

Proof. Choose P and Q so that the variable x_0 occurs on the diagonal of PXQ, and the rest follows immediately.

5.44. LEMMA. (Robinson). If A and $C = xB+yD$ are amicable orthogonal designs of order $n \equiv 0 \pmod 4$ and types $((1, n-1); (1, m))$, $m \in \{0,1,\ldots,n-1\}$, then we may assume

$$B = \bigoplus_{n/2-1} Y \oplus X,$$

where $X = \begin{bmatrix} - & 0 \\ 0 & 1 \end{bmatrix}$ and $Y = \begin{bmatrix} 0 & 1 \\ 1 & 0 \end{bmatrix}$.

Proof. From Lemma 5.42 we may assume B is symmetric and $A = Ix_1+Wx_2$, where W is a skew weighing matrix of weight n-1. It is obvious that we can find a monomial matrix P such that

$$PBP^t = \bigoplus_{1 \leq i \leq n/2} X_i,$$

where $X_i = X$ or Y or $\pm I_2$.

If, however, $X_i = \pm I_2$, for any i, then positions (2i, 2i-1) and (2i-1, 2i) of AB cannot be equal. Therefore, none of the X_i's are $\pm I_2$. This also means that, at most, one of the X_i's is $\pm X$, for if two X_i's are $\pm X$, we can find another monomial matrix which produces $\pm I_2$ somewhere on the diagonal of B.

We now assume $X_i = Y$, $1 \leq i \leq n/2$, and

$$A = \begin{bmatrix} \begin{smallmatrix} 0 & 1 \\ - & 0 \end{smallmatrix} & A_1 & A_2 & \cdots \\ \bar{A}_1^t & & & \\ \bar{A}_2^t & & Z & \\ \vdots & & & \end{bmatrix},$$

where A_i are (2×2) matrices with entries ± 1.

Since A and B are amicable, we have that W and B are amicable, and therefore WB is symmetric. Let

$$A_1 = \begin{bmatrix} x_1 & x_2 \\ x_3 & x_4 \end{bmatrix}, \quad x_i = \pm 1,$$

$$A_1 Y = \begin{bmatrix} x_2 & x_1 \\ x_4 & x_3 \end{bmatrix} \quad \text{and} \quad \overline{A}_1^t Y = \begin{bmatrix} \overline{x}_3 & \overline{x}_1 \\ \overline{x}_4 & \overline{x}_2 \end{bmatrix}.$$

Because AB is symmetric, we must have $x_2 = -x_3$ and $x_1 = -x_4$; i.e.,

$$A_1 = \pm \begin{bmatrix} 1 & 1 \\ - & - \end{bmatrix} \quad \text{or} \quad \pm \begin{bmatrix} 1 & - \\ 1 & - \end{bmatrix}.$$

This reasoning is also true for the other A_i's.

Now $\sum_{i=1}^{n/2-1} A_i A_i^t = (n-2)I_2$, by the orthogonality of A, and

$$A_i A_i^t = \begin{bmatrix} 2 & -2 \\ -2 & 2 \end{bmatrix} \quad \text{or} \quad \begin{bmatrix} 2 & 2 \\ 2 & 2 \end{bmatrix}.$$

But since there is an odd number of A_i's,

$$\Sigma A_i A_i^t \neq (n-2)I_2.$$

Therefore, at least one of the X_i's is X.

We now assume $X_1 = \pm X_2 = X$. Then we see that the product AB is not symmetric. Therefore,

$$X_1 = X \text{ and } X_i = Y, \; 2 \leq i \leq n/2.$$

5.45. THEOREM. (Wolfe-Robinson). Let X and Y be amicable orthogonal designs of type $((1, a, n-a-1); (u_1, \ldots, u_s))$ in order n; let $n \equiv 0 \pmod 4$, $n \neq 4$, $a = 1$ or $n \equiv 0 \pmod 8$, $a = 2, 3, 4, 5$. Then $u_i \neq 1$ for any $1 \leq i \leq s$.

Proof. The case for $a = 1$. Write $X = A_1x_1 + A_2x_2 + A_3x_3$, $Y = \sum_{j=1}^{s} B_j y_j$. With no loss of generality we may assume

$$A_1 = \bigoplus_{n/2} \begin{bmatrix} 1 & 0 \\ 0 & -1 \end{bmatrix}, \quad A_2 = \bigoplus_{n/2} \begin{bmatrix} 0 & 1 \\ 1 & 0 \end{bmatrix}.$$

The conditions $A_1A_3^t + A_3A_1^t = 0$ and $A_iB_j^t = B_jA_i^t$, $i = 1, 2$, $1 \leq j \leq s$, will show that A_3 and the B_j are all block matrices with every 2×2 block of the form $\begin{bmatrix} a & b \\ -b & a \end{bmatrix}$. We may think of such a block as the complex number a+bi, where $a, b \in \{0, 1, -1\}$. Let $\overline{A}_3$, $\overline{B}_j$ be the $n/2 \times n/2$ complex matrices so obtained.

If we let "*" denote conjugate transpose, then

i) $\overline{A}_3 = \overline{A}_3^t$, $\overline{B}_j = \overline{B}_j^t$, $1 \le j \le s$;

ii) $\overline{A}_3\overline{A}_3^* = (n-2)I_{n/2}$, $\overline{B}_j\overline{B}_j^* = u_jI_{n/2}$, $1 \le j \le s$;

iii) $\overline{A}_3\overline{B}_j^* = \overline{B}_j\overline{A}_3^*$, $1 \le j \le s$.

The fact that A_3 had weight n-2 implies that the off-diagonal elements of $\overline{A}_3$ are neither pure real nor pure imaginary.

Now, let us assume that one of the u's = 1; with no loss we assume $u_1 = 1$. Then, the entries in $\overline{B}_1$ are from $\{0, \pm 1, \pm i\}$. We claim that $\overline{B}_1$ is a diagonal matrix, for suppose there is a nonzero entry in the (i, j)-position of $\overline{B}_1$ for $i \ne j$ (there is no loss in assuming it's in the (1,2) position); then if we let z be the (1,2) entry in $\overline{A}_3$, we obtain from (i) and (iii)

$$\begin{bmatrix} 0 & z \\ -z & 0 \end{bmatrix}\begin{bmatrix} 0 & \overline{x} \\ \overline{x} & 0 \end{bmatrix} = \begin{bmatrix} 0 & x \\ x & 0 \end{bmatrix}\begin{bmatrix} 0 & -\overline{z} \\ \overline{z} & 0 \end{bmatrix}$$ ("-" denotes complex conjugation);

i.e., $z\overline{x} = x\overline{z}$.

Since x is pure real or pure imaginary, this gives that z is pure real or pure imaginary. This, however, contradicts the fact that all non-diagonal entries in $\overline{A}_3$ were <u>not</u> pure real or pure imaginary.

Thus, we are forced to assume that $\overline{B}_1$ is a diagonal matrix. We may as well assume the first diagonal entry in $\overline{B}_1$ is +1 (if not, multiply $\overline{A}_3$ and $\overline{B}_1$ by one of $\pm iI_{n/2}$ or $-I_{n/2}$; the resulting matrices still satisfy i), ii), iii)). The next diagonal entry is ± 1 or $\pm i$. Considering just the top 2×2 block of $\overline{A}_3$ and $\overline{B}_1$, we have (by iii))

$$\begin{bmatrix} 0 & z \\ -z & 0 \end{bmatrix}\begin{bmatrix} 1 & 0 \\ 0 & \overline{x} \end{bmatrix} = \begin{bmatrix} 1 & 0 \\ 0 & x \end{bmatrix}\begin{bmatrix} 0 & -\overline{z} \\ \overline{z} & 0 \end{bmatrix},$$

and so $\overline{z} = \overline{x}z$. If x were real, we would have $\overline{z} = \pm z$, which implies z is pure real or pure imaginary (a contradiction). Thus $x = \pm i$, and so the (1,2) entry of $\overline{A}_3$ is $\overline{z}$, where $\overline{z} = \pm iz$. Now by assumption $n \ne 4$, so there is a third diagonal entry in $\overline{B}_1$. Call it y. If we let z_{ij} denote the (i, j)-th entry of $\overline{A}_3$, we find (by iii)) that

a) $z_{13}\bar{y} = -\bar{z}_{13}$ and

b) $z_{23}\bar{y} = \pm i z_{23}$.

Hence, if y is real, a) implies that z_{13} is pure real or pure imaginary, while if y is pure imaginary, then b) implies that z_{23} is pure real or pure imaginary; in either case, a contradiction. This completes the proof for $a = 1$.

CASES $a = 2, 3$ or 4. Assume that A and B are amicable orthogonal designs of type $(1, a, n-a-1)$ and (1), respectively.

We may assume B is of the form given in Lemma 5.43, and this implies that the first two rows of A have the following form:

$$x_1 a_2 a_3 a_4 a_5 a_6 \ldots a_{n/2-3} a_{n/2-2} a_{n/2-1} a_{n/2},$$

$$\bar{a}_2 x_1 \bar{a}_4 \bar{a}_3 \bar{a}_6 \bar{a}_5 \ldots \bar{a}_{n/2-2} \bar{a}_{n/2-3} \bar{a}_{n/2-1} a_{n/2},$$

for some a_i's. The inner product of these two rows gives

$$-2(a_3 a_4 + \ldots + a_{n/2-3} a_{n/2-2}) - a_{n/2-1}^2 + a_{n/2}^2 = 0. \qquad \text{(I)}$$

Let x_2 be the variable which appears a times per row and column, and x_3 the remaining variable.

From (I) we have

i) $a_{n/2-1} = \pm a_{n/2}$,

ii) $a_i = \pm x_2$ for an even number of i's, $z \le i \le n/2-1$,

iii) $a_{2i-1} = a_{2i} = \pm x_2$ for an even number of i's, $2 \le i \le n/4-1$.

We note that similar properties exist for the other rows of A, except the last two.

Now we consider the matrix A_1 obtained from A by putting $x_1 = x_3 = 0$ and $x_2 = 1$ and show that no such matrix can exist if $a = 2, 3$ or 4.

CASE 1. $a = 2$ or 3. We can find a monomial P such that $PBP^t = B$, and if $n \ge 16$, PA_1P^t is given in Display 5.2.

$$
\left[\begin{array}{cc|cccc|c|cccc|cc}
0 & a_1 & & 0 & & & 0 & & 0 & & & 1 & 1 \\
\bar{a}_1 & 0 & & & & & & & & & & - & 1 \\
\hline
 & & 0 & a_2 & & & & 1 & 0 & 1 & 0 & & \\
0 & & \bar{a}_2 & 0 & & & 0 & 0 & - & 0 & - & 0 & \\
 & & & & 0 & a_2 & & 1 & 0 & - & 0 & & \\
 & & & & \bar{a}_2 & 0 & & 0 & - & 0 & 1 & & \\
\hline
0 & & & 0 & & & A_2 & & 0 & & & 0 & \\
\hline
 & & - & 0 & - & 0 & & 0 & a_2 & & & & \\
0 & & 0 & 1 & 0 & 1 & 0 & \bar{a}_2 & 0 & & & 0 & \\
 & & - & 0 & 1 & 0 & & & & 0 & a_2 & & \\
 & & 0 & 1 & 0 & - & & & & \bar{a}_2 & 0 & & \\
\hline
- & + & & & & & & & & & & 0 & a_1 \\
- & - & & 0 & & & 0 & & 0 & & & \bar{a}_1 & 0
\end{array}\right]
$$

Display 5.2

where $a_i = \begin{cases} \pm 1 & \text{if } a = 3, \\ 0 & \text{if } a = 2. \end{cases}$

By deleting rows and columns 3, 4, 5, 6, n-2, n-3, n-4 and n-5, we obtain a matrix similar in structure to A_1 but in order n-8. Hence the existence of A_1 in order n implies the existence of a similar matrix in order 8. It is easy to see, however, that no such design exists in order 8.

Therefore, there is no A_1 in order n. This means that there are no amicable orthogonal designs of types ((1); (1, a, n-a-1)), a = 2 or 3, in order $n \equiv 0 \pmod 8$.

CASE 2. a = 4. We can find a monomial Q such that $QBQ^t = B$ and, if $n \geq 24$,

$$
QA_1Q^t = \left[\begin{array}{cc|c|ccc}
 & 0 & 0 & & Z_1 & L \\
 & & & Z_1 & Z_2 & \\
\hline
 & 0 & A_2 & & 0 & \\
\hline
 & \bar{Z}_1^t & & & & \\
\bar{Z}_1^t & \bar{Z}_2^t & 0 & & 0 & \\
\bar{L}^t & & & & &
\end{array}\right],
$$

where

$$L = \begin{bmatrix} 1 & 1 \\ - & 1 \\ 1 & 1 \\ - & 1 \end{bmatrix}, \quad Z_1 = \begin{bmatrix} 1 & 0 & 1 & 0 \\ 0 & - & 0 & - \\ - & 0 & - & 0 \\ 0 & 1 & 0 & 1 \end{bmatrix}, \quad Z_2 = \begin{bmatrix} 1 & 0 & - & 0 \\ 0 & - & 0 & 1 \\ 1 & 0 & - & 0 \\ 0 & - & 0 & 1 \end{bmatrix}.$$

By deleting rows and columns 1, 2, 3, 4, n, n-1, n-2 and n-3 of QA_1Q^t and replacing the first two columns (rows) of the $Z_2(\overline{Z}_2^t)$ in the corners by $L(\overline{L}^t)$, we obtain a matrix with similar structure to A_1 but in order n-8. Hence, the existence of A_1 in order n implies the existence of a similar matrix in order 16. It is easy to see that no such designs can exist in order 8 or 16, and so there can be no amicable designs of types ((1); (1, 4, n-5)) in order $n \equiv 0 \pmod 8$.

CASE 3. a = 5. We proceed in a similar fashion to that of Case 2 but with diagonal 2×2 blocks of the form $\begin{bmatrix} 0 & a_1 \\ \bar{a}_1 & 0 \end{bmatrix}$ inserted. We note that this makes no difference to the proof.

5.46. COROLLARY. There are no amicable orthogonal designs of order $n \equiv 0 \pmod 4$ and types ((1); (1, a, b, c)), a+b+c = n-1, a, b, c $\neq$ 0, and abc odd.

<u>Proof.</u> By considering (I) in the proof of Theorem 5.43, we can see that each variable appears an even number of times off the diagonal 2×2 block, and therefore only one of a, b and c is odd.

REMARK. This theorem uses, at every stage, the full force of the fact that the matrices we're considering have {0, 1, -1} entries and very strongly the fact that X had no zeros. Wolfe was led to prove the case a = 1 of this theorem after a frustrating attempt to construct pairs of amicable orthogonal designs in orders $n = 2^a$ of type $((1, 1, 2, \ldots, 2^{a-1}); (1, 1, 2, \ldots, 2^{a-1}))$. If they had existed, they would have involved the greatest number of variables (2a+2) and would have no zeros. In orders n = 2 and n = 4 they exist, but those are the only times.

This theorem eliminates many types not previously eliminated; e.g., there are no amicable designs of type ((1, 1, 6); (1, 7)) in order 8.

The only other general combinatorial theorems we know which shed some light on the existence problem for amicable orthogonal designs come out of work of P. J. Robinson which we shall now discuss briefly.

5.47. THEOREM. (Robinson). There are no amicable orthogonal designs of order $n \equiv 0 \pmod 4$, $n \geq 8$, and types $((1,1,k);(n))$, k odd.

<u>Proof.</u> Assume that such an amicable pair, A and B, exists.

By multiplying by suitable monomial matrices if necessary, we may assume

$$A = x_1 I + x_2 X + x_3 Y,$$

with

$$X = \oplus_{n/2} \begin{bmatrix} 0 & 1 \\ - & 0 \end{bmatrix}.$$

By considering $XY^t = -YX^t$, $Y^t = -Y$, $XB^t = BX^t$, and $B^t = B$, we see that Y and B are made up of 2×2 blocks of the form

$$\begin{bmatrix} a & b \\ b & \bar{a} \end{bmatrix}, \quad a, b \in \{0, 1, -1\}.$$

Now $(AB)^t = B^t A^t = BA^t = AB^t = AB$, and therefore A and B are amicable only if AB is symmetric and hence only if the top left hand 2×2 block of AB, and therefore YB, is symmetric.

The top left hand 2×2 block of YB, however, is made up of the sum of 2×2 blocks of the form

$$\begin{bmatrix} x & y \\ y & \bar{x} \end{bmatrix} \begin{bmatrix} a & b \\ b & \bar{a} \end{bmatrix} = \begin{bmatrix} xa+yb & xb-ya \\ -xb+ya & xa+yb \end{bmatrix},$$

with $a, b = \pm 1$ and $x, y \in \{0, 1, -1\}$.

This sum cannot produce a symmetric block unless the sum of the $(ya-xb)$'s is zero, which is clearly impossible with an odd number of nonzero x and y's.

Therefore, there can be no designs A and B of the required types.

An indication of the gap between the algebraic theory of amicable families and the actual existence of amicable orthogonal designs has been highlighted by some recent work of P. J. Robinson and of Robinson in collaboration with J. S. Wallis.

Robinson [3] has shown that the following are not the types of amicable orthogonal designs in order 8, although these types are not eliminated by the theorems of Wolfe and Robinson or by the algebraic theory:

$((1);(2,2,2,2))$,

$((1,2,2,3);(4,4))$,

$((1,7);(5))$.

Robinson and Seberry [1] have also shown that there is no amicable orthogonal design of type $((1,15);(1))$ in order 16. This is a most surprising result, since the number of variables is so much less than the number allowed by Theorem 5.27.

§5.6. CONSTRUCTION OF AMICABLE ORTHOGONAL DESIGNS

In the first section of this chapter we gave two examples of amicable orthogonal designs and gave a small indication (Proposition 5.1) of the strength of this notion. In this section we turn from the limitations on existence to the actual construction of amicable orthogonal designs. As will soon be apparent, the construction problem is an exceedingly difficult one and, as with so many aspects of this area, deserves further study.

We shall pay particular attention to amicable orthogonal designs which are full (i.e., neither matrix has any zeros) because of their special relevance to the Hadamard Matrix Problem.

One class of important full amicable orthogonal designs is the amicable Hadamard matrices (see Example 5.5). There is a substantial literature on these objects, and in view of their importance to the Hadamard Matrix Problem, we shall deal with their construction in the next section.

5.48. LEMMA. (Wolfe). Suppose there are amicable orthogonal designs $X = x_1X_1 + x_2X_2 + \dots + x_rX_r$ and $Y = y_1Y_1 + y_2Y_2 + \dots + y_sY_s$ in order n of types $((u_1, u_2, \dots, u_r);(v_1, v_2, \dots, v_s))$. Further suppose there are amicable orthogonal designs $W = w_1W_1 + w_2W_2 + \dots + w_tW_t$ and Z in order m of types $((p_1, p_2, \dots, p_t);(z))$. Then there exist amicable orthogonal designs in order mn of types

$((zu_1, zu_2, \dots, zu_{i-1}, p_1u_i, p_2u_i, \dots, p_tu_i, zu_{i+1}, \dots, zu_r);(zv_1, zv_2, \dots, zv_s))$.

Proof. The required matrices in the variables $a_1, a_2, \dots, a_{i-1}$, $c_1, \dots, c_t$, $a_{i+1}, \dots, a_r$ and $b_1, \dots, b_s$ are

$$a_1X_1 \times Z + \dots + a_{i-1}X_{i-1} \times Z + \sum_{j=1}^{t} c_jX_i \times W_j + a_{i+1}X_{i+1} \times Z + \dots + a_rX_r \times Z,$$

and

$$\sum_{j=1}^{s} b_jY_j \times Z.$$

5.49. COROLLARY. Suppose there exist amicable orthogonal designs of types $((u_1, u_2, \dots, u_t);(v_1, v_2, \dots, v_s))$ in order n. Further suppose there exist amicable Hadamard matrices of order m. Then there exist amicable orthogonal designs of types $((u_1, (m-1)u_1, mu_2, \dots, mu_t);(mv_1, mv_2, \dots, mv_s))$ in order mn.

5.50. COROLLARY. There exist amicable orthogonal designs of types $((1, 1, 2, 4, \dots, 2^t);(2^t, 2^t))$ and $((1, 2^{t+1}-1);(2^t, 2^t))$ in order 2^{t+1}.

Proof. In §5.1 we gave amicable orthogonal designs of type $((1, 1);(1, 1))$ in order 2. Setting the variables in the second design equal to each other, we get amicable Hadamard matrices of order 2. This corollary then follows by repeated application of Corollary 5.49.

As one special case of Corollary 5.50, we obtain the well-known fact that amicable Hadamard matrices exist in orders which are a power of 2. Consequently, we have:

5.51. COROLLARY. Suppose there exist amicable orthogonal designs of types $((u_1, u_2, \dots, u_r);(v_1, v_2, \dots, v_s))$ in order n. Then there exist amicable orthogonal designs of types $((u_1, u_1, 2u_1, \dots, 2^{t-1}u_1, 2^tu_2, \dots, 2^tu_r);$

$(2^t v_1, 2^t v_2, \ldots, 2^t v_s))$ in order $2^t n$.

We now turn to a different construction for amicable orthogonal designs. A similar construction was first used to construct amicable Hadamard matrices.

5.52. THEOREM. Let $p \equiv 3 \pmod 4$ be a prime power; then there exist amicable orthogonal designs of order $p+1$ and types $((1,p);(1,p))$.

Proof. Let Q be the matrix constructed in Lemma 4.34. Recall that Q is a type 1 matrix with the properties:

$$QQ^t = pI - J,$$

$$QJ = JQ = 0,$$

$$Q^t = (-1)^{\frac{1}{2}(p-1)} Q.$$

Now let $U = cI + dQ$, where c, d are commuting variables. Define $R = (r_{ij})$ by

$$r_{ij} = \begin{cases} 1 & a_i + a_j = 0, \\ 0 & \text{otherwise.} \end{cases}$$

Then UR is a (symmetric) type 2 matrix.

Let a, b be commuting variables. Then for $p \equiv 3 \pmod 4$

$$A = \begin{bmatrix} a & b \ldots b \\ -b & \\ \vdots & aI+bQ \\ -b & \end{bmatrix} \quad \text{and} \quad B = \begin{bmatrix} -c & d \;\ldots\; d \\ d & \\ \vdots & (cI+dQ)R \\ d & \end{bmatrix},$$

are the required amicable designs of types $((1,p);(1,p))$ in order $p+1$.

$$A = \begin{bmatrix} a & b & b & b & b & b & b & b \\ -b & a & b & b & -b & b & -b & -b \\ -b & -b & a & b & b & -b & b & -b \\ -b & -b & -b & a & b & b & -b & b \\ -b & b & -b & -b & a & b & b & -b \\ -b & -b & b & -b & -b & a & b & b \\ -b & b & -b & b & -b & -b & a & b \\ -b & b & b & -b & b & -b & -b & a \end{bmatrix} \quad \text{and}$$

$$B = \begin{bmatrix} -c & d & d & d & d & d & d & d \\ d & -d & -d & d & -d & d & d & c \\ d & -d & d & -d & d & d & c & -d \\ d & d & -d & d & d & c & -d & -d \\ d & -d & d & d & c & -d & -d & d \\ d & d & d & c & -d & -d & d & -d \\ d & d & c & -d & -d & d & -d & d \\ d & c & -d & -d & d & -d & d & d \end{bmatrix}$$

Display 5.3

5.53. EXAMPLE. Display 5.3 shows amicable orthogonal designs of types $((1,7);(1,7))$ in order 8.

5.54. COROLLARY. Let $p \equiv 3 \pmod 4$ be a prime power. Then there exist amicable orthogonal designs of order $2^s(p+1)$ and type $((1,1,2,\ldots,2^{s-1},2^s p);(2^s,2^s p))$.

Proof. Construct the amicable orthogonal designs $x_1 I + x_2 S$ and $y_1 R + y_2 P$ of types $((1,p);(1,p))$ in order $(p+1)$. Now repeatedly replace the variable x_1 by $\begin{bmatrix} x_1 & x_{s+2} \\ -x_{s+2} & x_1 \end{bmatrix}$ and any other variables x_i or y_j by $\begin{bmatrix} x_i & x_i \\ x_i & -x_i \end{bmatrix}$ or $\begin{bmatrix} y_j & y_j \\ y_j & -y_j \end{bmatrix}$, respectively.

In a similar vein we have:

5.55. LEMMA. Let $n+1 \equiv 2 \pmod 4$ be the order of a symmetric conference matrix N (i.e., $N^t = N$, $NN^t = nI_{n+1}$, and N has entries $0, 1, -1$). Then there exist amicable orthogonal designs of order $2(n+1)$ and types (i) $((2,2n);(2,2n))$, (ii) $((1,n);(1,n))$.

Proof. Let a, b, c, d be commuting variables. Then for (i) the required designs are

$$\begin{bmatrix} aI+bN & aI-bN \\ aI-bN & -aI-bN \end{bmatrix} \text{ and } \begin{bmatrix} cI+dN & cI-dN \\ -cI+dN & cI+dN \end{bmatrix},$$

while for (ii) they are

$$\begin{bmatrix} aI & bN \\ bN & -aI \end{bmatrix} \text{ and } \begin{bmatrix} cI & dN \\ -dN & cI \end{bmatrix}.$$

The fact that N can be chosen to have zero diagonal is guaranteed by Theorem 2.20(a).

5.56. COROLLARY. Let $q \equiv 1 \pmod 4$ be a prime power. Then there exist amicable orthogonal designs of order $2(q+1)$ and types (i) $((2,2q);(2,2q))$, (ii) $((1,q);(1,q))$.

Proof. Choose N of Lemma 5.55 as $N = \begin{bmatrix} 0 & 1 \dots 1 \\ 1 & \\ \vdots & Q \\ 1 & \end{bmatrix}$ where Q is formed as in the proof of Theorem 5.52.

5.57. COROLLARY. Let $q \equiv 1 \pmod 4$ be a prime power. Then there exist amicable orthogonal designs of order $2^{s+1}(q+1)$ and types

i) $((2,2,4,\dots,2^s,2^{s+1}q);(2^{s+1},2^{s+1}q))$,

ii) $((1,1,2,\dots,2^{s-1},2^s q);(2^s,2^s q))$.

Part i) of the next theorem is a corollary to Lemma 5.48, while part ii) is not. The theorem is motivated by Theorem 5.52.

5.58. THEOREM. Suppose there are amicable orthogonal designs of order n and of type $((1,s);(r,p))$. Suppose further that there are amicable orthogonal designs of order m and type $((1,a);(b))$. Then there are amicable orthogonal designs of order mn and types

i) $((1,a,sb);(rb,pb))$,

ii) $((1,ra,pa);(b,sa))$.

Proof. From Lemma 5.43 we may choose the orthogonal design of type $(1,s)$ to be x_1I+x_2S, where $S^t = -S$, and the orthogonal design of type (r,p) to be x_3R+x_4P, where $R^t = R$ and $P^t = P$. Further, the design of type $(1,a)$ can be chosen as v_1I+v_2A, where $A^t = -A$, and the design of type (b) as v_3B, where $B^t = B$. Then

i) $y_1 I \times I + y_2 I \times A + y_3 S \times B$ and $y_4 R \times B + y_5 P \times B$,

ii) $y_1 I \times I + y_2 R \times A + y_3 P \times A$ and $y_4 I \times B + y_5 S \times A$,

may be used to give the required amicable orthogonal designs.

5.59. COROLLARY. Let $q \equiv 3 \pmod 4$ be a prime power. Suppose there are amicable orthogonal designs of order m and type $((1,a);(b))$. Then there are amicable orthogonal designs of order $m(q+1)$ and types

i) $((1,a,bq);(b,bq))$,

ii) $((1,a,qa);(b,qa))$.

Consider

$$\left.\begin{array}{ll}
X_1 = \begin{bmatrix} x_1 & x_2 & x_3 & x_3 \\ -x_2 & x_1 & x_3 & -x_3 \\ x_3 & x_3 & -x_1 & -x_2 \\ x_3 & -x_3 & x_2 & -x_1 \end{bmatrix}, & Y_1 = \begin{bmatrix} y_1 & y_2 & y_3 & y_3 \\ y_2 & -y_1 & y_3 & -y_3 \\ -y_3 & -y_3 & y_2 & y_1 \\ -y_3 & y_3 & y_1 & -y_2 \end{bmatrix} \\
= x_1X_1 + x_2X_2 + x_3X_3 & = y_1Y_1 + y_2Y_2 + y_3Y_3 \\
X_2 = \begin{bmatrix} x_1 & x_2 & x_3 & x_3 \\ -x_2 & x_1 & x_3 & -x_3 \\ -x_3 & -x_3 & x_1 & -x_2 \\ -x_3 & x_3 & -x_2 & x_1 \end{bmatrix}, & Y_2 = \begin{bmatrix} y_1 & y_2 & y_3 & y_3 \\ y_2 & -y_1 & y_3 & -y_3 \\ y_3 & y_3 & -y_2 & -y_1 \\ y_3 & -y_3 & -y_1 & y_2 \end{bmatrix} \\
= x_1U_1 + x_2U_2 + x_3U_3 & = y_1V_1 + y_2V_2 + y_3V_3
\end{array}\right\} \quad (*)$$

Then we have:

5.60. LEMMA. Suppose there is a set of pairwise amicable weighing matrices (or orthogonal designs) $\{M_1, \ldots, M_s, N_1, \ldots, N_t\}$ of order m and weights $(m_1, \ldots, m_s, n_1, \ldots, n_s)$ where

$$M_i^t = -M_i, \ \forall_i, \quad \text{and} \quad N_j^t = N_j, \ \forall_j.$$

Then there are amicable orthogonal designs of order $4m$ and types

i) $(1, m_a, m_b, 2m_c)$ and $(n_d, m_e, 2m_f)$,

ii) $(1, m_a, n_b, 2n_c)$ and $(n_d, n_e, 2n_f)$,

iii) $(1, m_a, n_b, 2m_c)$ and $(n_d, n_e, 2m_f)$,

iv) $(1, m_a, m_b, 2n_c)$ and $(n_d, m_e, 2n_f)$,

where $m_i \in \{m_1, \ldots, m_s\}$, $i \in \{1, \ldots, s\}$ and $n_j \in \{n_1, \ldots, n_t\}$, $j \in \{1, \ldots, t\}$.

Proof. Use the matrices defined by (*) above. The required designs are:

i) $u_1 I \times I + u_2 V_1 \times M_a + u_3 V_2 \times M_b + u_4 V_3 \times M_c$ and

$v_1 U_1 \times N_d + v_2 U_2 \times M_e + v_3 U_3 \times M_f$,

ii) $u_1 I \times I + u_2 U_1 \times M_a + u_3 U_2 \times N_b + u_4 U_3 \times N_c$ and

$v_1 V_1 \times N_d + v_2 V_2 \times N_e + v_3 V_3 \times N_f$,

iii) $u_1 I \times I + u_2 U_1 \times M_a + u_3 U_2 \times N_b + u_4 U_3 \times M_c$ and

$v_1 V_1 \times N_d + v_2 V_2 \times N_e + v_3 V_3 \times M_f$,

iv) $u_1 I \times I + u_2 V_1 \times M_a + u_3 V_2 \times M_b + u_4 V_3 \times N_c$ and

$v_1 U_1 \times N_d + v_2 U_2 \times M_e + v_3 U_3 \times N_f$.

We now look specifically at what happens for orders which are a power of 2.

SPECIFIC ORDERS 2^n

5.61. LEMMA. There are amicable orthogonal designs X, Y of order 2 for the types indicated in the table (by symmetry we need only consider the upper triangular block):

X \ Y	(1)	(2)	(1,1)
(1)	*	*	*
(2)		*	*
(1,1)			*

Table 5.4

Proof. All these can be obtained by equating and killing variables in

$$X = \begin{bmatrix} x_1 & x_2 \\ -x_2 & x_1 \end{bmatrix} \quad \text{and} \quad Y = \begin{bmatrix} y_1 & y_2 \\ y_2 & -y_1 \end{bmatrix}.$$

5.62. LEMMA. There are amicable orthogonal designs X, Y of order 4 for the types indicated in Table 5.5 (those entries left blank correspond to designs which do not exist by Theorem 5.41).

X \ Y	(1)	(2)	(3)	(4)	(1,1)	(1,2)	(1,3)	(2,2)	(1,1,1)	(1,1,2)
(1)	*	*	*	*	*	*	*	*	*	*
(2)		*	*	*	*	*	*	*	*	*
(3)			*	*	*	*	*	*	*	*
(4)				*	*	*	*	*		*
(1,1)					*	*	*	*	*	*
(1,2)						*	*	*	*	*
(1,3)							*	*		*
(2,2)								*		*
(1,1,1)									*	
(1,1,2)										*

Table 5.5

Proof. Let a, b, c, x, y, z be commuting variables. Then

$$\begin{bmatrix} a & b & c & 0 \\ -b & a & 0 & -c \\ -c & 0 & a & b \\ 0 & c & -b & a \end{bmatrix} \text{ and } \begin{bmatrix} x & y & z & 0 \\ y & -x & 0 & -z \\ z & 0 & -x & y \\ 0 & -z & y & x \end{bmatrix},$$

or

$$\begin{bmatrix} 0 & a & b & c \\ -a & 0 & c & -b \\ -b & -c & 0 & a \\ -c & b & -a & 0 \end{bmatrix} \text{ and } \begin{bmatrix} 0 & x & y & z \\ -x & 0 & -z & y \\ -y & z & 0 & -x \\ -z & -y & x & 0 \end{bmatrix},$$

are amicable orthogonal designs of types ((1,1,1);(1,1,1)); amicable orthogonal designs of types ((1,1,2);(1,1,2)) are given in Chapter 5, §1; now, equating and killing variables in these designs gives the result.

5.63. THEOREM. There exist amicable orthogonal designs of types $((1,1,2,4,\ldots,2^{s-1});(2^{s-2},2^{s-2},2^{s-1}))$ in order 2^s, $s \geq 1$.

Proof. Use Lemma 5.48 with the ((1,1);(2)) and ((1,1,2);(1,1,2)) amicable orthogonal designs in orders 2 and 4, respectively.

REMARK. We noticed with quite some surprise that we can only find "full" (i.e., no zeros) amicable orthogonal designs in order 8 for types ((1,1,2,4);(2,2,4)), ((1,1,2,2,2);(8)), ((1,2,2,3);(2,6)) and ((1,7);(1,7)). Peter Robinson has established that there are no other "full" amicable designs of types $((1,a,b,c,d);(e,f,g))$ in order 8. (There are other amicable orthogonal designs, e.g., $((1,1,2);(1,1,2))$ and $((1,1,1,1,1);(1))$.) These results indicate that the algebraic theory comes nowhere near explaining the existence or nonexistence of amicable orthogonal designs.

The precise situation for amicable orthogonal designs in order 8 of type $((1,u_1,\ldots,u_t);(v_1,\ldots,v_s))$ with $\sum_{i=1}^{t} u_i = 7$ and $\sum_{i=1}^{s} v_i = 8$ is given by:

5.64. THEOREM. (Robinson-Wolfe). The following amicable orthogonal designs (and those which can be derived from them) exist in order 8:

$$((1,1,2,2,2);(8)), \qquad ((1,2,2,3);(2,6)),$$
$$((1,1,2,4);(2,2,4)), \qquad ((1,7);(1,7)).$$

All others, of the type $((1,a,b,c,d);(e,f,g))$, where $a+b+c+d = 7$ and $e+f+g = 8$, do not exist.

Proof. Theorem 5.63 gives the $((1,1,2,4);(2,2,4))$; Theorem 5.52, the $((1,7);(1,7))$. The amicable orthogonal designs of types $((1,1,2,2,2);(8))$ and $((1,2,2,3);(2,6))$ are given in (Robinson [3]).

Theorems 5.45, 5.41 and 5.38 eliminate many possibilities.

We shall show below that there are no amicable orthogonal designs of type $((1,7);(5))$ in order 8; this eliminates several remaining possibilities.

Example 5.40 (4) eliminates the $((1,7);(2,2,2,2))$. Theorem 5.39 (5) eliminates $((1,1,3,3);(2,2,4))$.

The remaining possibilities are not excluded by the algebraic theory, but P. Robinson has shown (combinatorially) that the $((1,1,3,3);(8))$ and the $((1,2,2,3);(4,4))$ are impossible, and that takes care of all the rest.

To finish the proof of Theorem 5.64, we need to show nonexistence of $((1,7);(5))$, $((1,1,3,3);(8))$ and $((1,2,2,3);(4,4))$ in order 8. We shall only show here the argument for the $((1,7);(5))$. The other two require longer but similar arguments.

Note that by Lemma 5.43 it is enough to prove:

5.65. THEOREM. There are no amicable orthogonal designs A, B of type ((7);(5)) in order 8 where $A^t = -A$ and $B^t = B$. (Hence there are no amicable orthogonal designs of type ((1, 7); (5)) in order 8.)

In order to prove this theorem, we need the following two lemmas.

5.66. LEMMA. If a and b are ±1, then

1) $a+b \equiv ab+1 \pmod 4$,

2) $-a-b \equiv a+b \pmod 4$.

5.67. LEMMA. Let

$$\mathcal{B} = \begin{bmatrix} 1 & 1 & 1 & 1 \\ 1 & - & 1 & - \\ 1 & 1 & - & - \\ 1 & - & - & 1 \end{bmatrix} \quad \text{and} \quad \mathcal{B}' = \begin{bmatrix} 1 & 1 & 1 & 1 \\ 1 & 1 & - & - \\ 1 & - & 1 & - \\ 1 & - & - & 1 \end{bmatrix}.$$

If B is a symmetric (0,1,-1) matrix in order 8 such that $BB^t = 5I_8$, then we can find a monomial matrix R such that

$$RBR^t = \begin{bmatrix} P & \mathcal{B} \\ \mathcal{B} & Q \end{bmatrix} \quad \text{or} \quad \begin{bmatrix} \mathcal{B}_1 & I \\ I & -\mathcal{B}_1 \end{bmatrix},$$

where P and Q are monomials and $\mathcal{B}_1 = \mathcal{B}$ or $\mathcal{B}'$.

Proof. Let the i-th row of B be $b_{i1}, b_{i2}, \ldots, b_{i8}$, and define

$$0_B(i,j) = \sum_{k=1}^{8} |b_{ik}|\,|b_{jk}|.$$

Since $BB^t = 5I_8$, we have $0_B(i,j) = 2$ or 4 for all i, j, $i \neq j$.

If $0_B(i,j) = 4$ for at most one j, then any column containing a zero from row i has at least six ±1's. Therefore, for all i there exist j_1 and j_2 such that

$$0_B(i,j_1) = 0_B(i,j_2) = 4. \qquad (*)$$

We also note that $0_B(j_1,j_2) = 4$ for the j_1 and j_2 given in (*).

Now we define an equivalence relation $\sim$ on the rows of B as follows:

$$\text{row } i \sim \text{row } j \text{ if and only if } 0_B(i,j) = 4,$$

and consider the equivalence classes of $\sim$.

Since each equivalence class contains at least three rows, it can be seen that there are at most two equivalence classes, each with four rows.

If all the rows were in the same equivalence class, then $0_B(i, j) = 4$ for all i and j, which is clearly impossible.

We now consider a permutation matrix R_1 such that the first four rows of R_1B are in the same equivalence class. Now let $R_1BR_1^t = B_1$.

Clearly, the first four rows of B_1 are in the same equivalence class, and B_1 is symmetric, and hence it can be shown that

$$B_1 = \begin{bmatrix} A_1 & A_2 \\ A_2^t & A_3 \end{bmatrix},$$

where either A_1 and A_3 are symmetric Hadamard matrices in order 4 and A_2 is a monomial matrix, or A_1 and A_3 are symmetric monomial matrices and A_2 is an Hadamard matrix.

If A_2 is an Hadamard matrix, then there exist monomial matrices D and D' such that $DA_2D'^t = \mathcal{B}$.

Now let

$$R_2 = \begin{bmatrix} D & 0 \\ 0 & D' \end{bmatrix},$$

and therefore

$$R_2B_1R_2^t = \begin{bmatrix} P & \mathcal{B} \\ \mathcal{B} & Q \end{bmatrix}.$$

Hence

$$RBR^t = \begin{bmatrix} P & \mathcal{B} \\ \mathcal{B} & Q \end{bmatrix},$$

where $R = R_2R_1$. Now, if A_1 and A_3 are symmetric Hadamard matrices, then there exist monomial matrices C and D such that

$$CA_1C^t = B_1 \quad \text{and} \quad CA_2D^t = I.$$

Now, let

$$R_2' = \begin{bmatrix} C & 0 \\ 0 & D \end{bmatrix};$$

then

$$R_2'B_1R_2'^t = \begin{bmatrix} C & 0 \\ 0 & D \end{bmatrix} \begin{bmatrix} A_1 & A_2 \\ A_2^t & A_3 \end{bmatrix} \begin{bmatrix} C^t & 0 \\ 0 & D^t \end{bmatrix}$$

$$= \begin{bmatrix} \mathcal{B}_1 & I \\ I & DA_3D^t \end{bmatrix} .$$

But $BB^t = 0$, so $DA_3D^t = -\mathcal{B}_1$. Therefore, upon putting $R = R_2'R_1$, we have the required result.

<u>Proof of Theorem.</u> Assume there exist amicable orthogonal designs A and B of type ((7);(5)) with $A^t = -A$, $B^t = B$.

We may assume B is one of the forms given in Lemma 5.67. First we assume

$$B = \begin{bmatrix} P & \mathcal{B} \\ \mathcal{B} & Q \end{bmatrix} .$$

We can further assume

$$P = \begin{bmatrix} 1 & 0 & 0 & 0 \\ 0 & 1 & 0 & 0 \\ 0 & 0 & 0 & a \\ 0 & 0 & a & 0 \end{bmatrix} \text{ or } \begin{bmatrix} 1 & 0 & 0 & 0 \\ 0 & 1 & 0 & 0 \\ 0 & 0 & a & 0 \\ 0 & 0 & 0 & a \end{bmatrix} \text{ or } \begin{bmatrix} 0 & 1 & 0 & 0 \\ 1 & 0 & 0 & 0 \\ 0 & 0 & 0 & a \\ 0 & 0 & a & 0 \end{bmatrix} ,$$

where $a = \pm 1$, since, given any other P, either no Q can be found or we can apply permutations to B which leave $\mathcal{B}$ fixed but transform P into one of the above forms.

Assume

$$P = \begin{bmatrix} 1 & 0 & 0 & 0 \\ 0 & 1 & 0 & 0 \\ 0 & 0 & 0 & a \\ 0 & 0 & a & 0 \end{bmatrix} ,$$

and let the first five rows of A be

$$\begin{matrix} 0 & a_1 & a_2 & a_3 & a_4 & a_5 & a_6 & a_7 \\ \bar{a}_1 & 0 & b_2 & b_3 & b_4 & b_5 & b_6 & b_7 \\ \bar{a}_2 & \bar{b}_2 & 0 & c_3 & c_4 & c_5 & c_6 & c_7 \\ \bar{a}_3 & \bar{b}_3 & \bar{c}_3 & 0 & d_4 & d_5 & d_6 & d_7 \\ \bar{a}_4 & \bar{b}_4 & \bar{c}_4 & \bar{d}_4 & 0 & e_5 & e_6 & e_7 \end{matrix} .$$

For (A, B) to be amicable orthogonal designs, AB must be symmetric.

Consider positions (1,4) and (4,1) in AB.

$$aa_2+a_4-a_5-a_6+a_7 = -a_3+d_4+d_5+d_6+d_7.$$

Hence, by Lemma 5.66,

$$aa_2a_3a_4a_5a_6a_7d_4d_5d_6d_7 = -1. \qquad (1)$$

But $-a_1b_3-a_2c_3+a_4d_4+a_5d_5+a_6d_6+a_7d_7 = 0$ (by the orthogonality of A); that is,

$$a_1a_2a_4a_5a_6a_7b_3c_3d_4d_5d_6d_7 = -1. \qquad (2)$$

On multiplying (1) and (2) we get

$$aa_1a_3b_3c_3 = 1. \qquad (3)$$

Now we consider positions (2,4) and (4,2) of AB. By reasoning as above, we obtain $aa_1a_3b_3c_3 = -1$, which contradicts (3).

By using reasoning similar to that of the above case, it can be shown that none of the possible B's can be used to produce amicable orthogonal designs of type ((7);(5)) in order 8 with $A^t = -A$ and $B^t = B$.

A complete determination of exactly which (nonfull) amicable orthogonal designs may exist in order 8 has not yet been made. In fact, full amicable orthogonal designs, when neither design is of type (1,...), have hardly been looked at in order 8.

§5.7. AMICABLE HADAMARD MATRICES

Skew-Hadamard matrices have played a significant part in the search for Hadamard matrices (see Williamson [2], Spence [2], and Wallis [1,2,3]). Szekeres [2] realized that skew-Hadamard matrices were of interest in themselves and, in fact, equivalent to doubly regular tournaments.

Once attention was focused on skew-Hadamard matrices, Wallis [4] realized that in order to form skew-Hadamard matrices of order mn from ones of orders m and n, a notion of "amicability" could be decisive. This was the origin of the idea of amicable Hadamard matrices.

Although this section is embedded in the chapter on amicable orthogonal designs, it will be clear that the ideas here, which historically precede the rest of the ideas of this chapter, have exerted a strong influence on the development of "amicability" for orthogonal designs.

5.68. DEFINITION. The $s \times s$ matrices W and M are called amicable Hadamard matrices if

i) $W = I+S$, where $S^t = -S$ and $M = M^t$ are both Hadamard matrices, i.e., weighing matrices of weight s,

and

ii) $WM^t = MW^t$.

These pairs of matrices exist for many orders, some of which are discussed below. We conjecture:

5.69. CONJECTURE. There exist amicable Hadamard matrices in all orders $n \equiv 0 \pmod 4$.

Amicable Hadamard matrices and amicable orthogonal designs of types $((1,n-1);(n))$ in order n are equivalent notions (see Lemma 5.43). Hence they may be used to construct other orthogonal designs as follows (see Lemma 5.48):

5.70. LEMMA. Let $y_1A_1 + y_2A_2 + \ldots + y_pA_p$ be an orthogonal design of order n and type $(a_1, a_2, \ldots, a_p)$ on the p variables $y_1, y_2, \ldots, y_p$. Let $W = I_s+S$ and M be amicable Hadamard matrices of order s. Then

$$x_0A_1 \times I_s + x_1A_1 \times S + x_2A_2 \times M + \ldots + x_pA_p \times M$$

is an orthogonal design of order sn and type $(a_1, a_1(s-1), sa_2, \ldots, sa_p)$ on the variables $x_0, x_1, \ldots, x_p$.

5.71. COROLLARY. Let W and M (as above) be amicable Hadamard matrices of order s, and let n be any integer. Then there is an orthogonal design of order ns of type $(1, s-1, s, s, \ldots, s)$ on the variables $x_0, x_1, x_2, \ldots, x_{\rho(n)}$.

Proof. We use the $(1,1,1,\ldots,1)$ design on $\rho(n)$ variables that exist in every order n. The first variable is replaced by $x_0 I + x_1 W_1$; the i-th variable $i > 1$ is replaced by $x_i M$.

5.72. EXAMPLE. The most interesting consequence of this corollary is to note that there is an orthogonal design of type $(1,1,2,\ldots,2)$ on $\rho(n)+1$ variables in order $2n$. This follows because there is an orthogonal design of type $(1,1,1,\ldots,1)$ on $\rho(n)$ variables in order n and amicable Hadamard matrices of order 2.

In particular, there is an orthogonal design of type $(1,1,2,2,2,2,2,2,2)$ in order 16.

5.73. COROLLARY. If there is a Baumert-Hall array of order n and amicable Hadamard matrices of order s, then there is an orthogonal design of order $4ns$ and of type $(n, n(s-1), ns, ns, ns)$.

5.74. EXAMPLE. We saw in Section 4 of Chapter 4 that there is a Baumert-Hall array of order 3. Hence there exists an orthogonal design of order 24 and type $(3,3,6,6,6)$ by using amicable Hadamard matrices of order 2.

5.75. COROLLARY. Let W, M be amicable Hadamard matrices of order s. If there is a Plotkin array of order $8n$, then there is an orthogonal design of order $8ns$ and type $(n, n(s-1), ns, \ldots, ns)$ on the variables $x_1, x_2, x_3, \ldots, x_9$.

5.76. EXAMPLE. Plotkin found an array of order 24 and type $(3,3,3,3,3,3,3,3)$. Hence, using the amicable Hadamard matrices of order 2, we have an orthogonal design of type $(3,3,6,6,6,6,6,6,6)$ in order 48.

We will now show amicable Hadamard matrices exist in the following cases:

AI 2^t t a nonnegative integer.

AII p^r+1 p^r (prime power) $\equiv 3 \pmod 4$.

AIII	$2(q+1)$	$2q+1$ is a prime power, q(prime) $\equiv 1 \pmod 4$.
AIV	$(\|t\|+1)(q+1)$	q(prime power) $\equiv 5 \pmod 8 = s^2+4t^2$, $s \equiv 1 \pmod 4$, and $\|t\|+1$ is the order of amicable orthogonal designs of type $((1, \|t\|);(\frac{1}{2}(\|t\|+1), \frac{1}{2}(\|t\|+1))$.
	$2^r(q+1)$	q(prime power) $\equiv 5 \pmod 8 = s^2+4(2^r-1)^2$, $s \equiv 1 \pmod 4$, r some integer.
AV	S	where S is a product of the above orders.

We first prove AV and AI.

5.77. THEOREM. (Wallis [4]). Suppose there are amicable Hadamard matrices of orders m and n. Then there are amicable Hadamard matrices of order mn. In particular, there exist amicable Hadamard matrices of order 2^t, t a nonnegative integer.

Proof. This is a special case of Corollary 5.49. For the last part, see Corollary 5.50.

5.78. PROPOSITION. There are amicable Hadamard matrices of order p^n+1, $p^n \equiv 3 \pmod 4$, p a prime.

Proof. This is a special case of Theorem 5.52 (the structure is ensured by Lemma 5.43).

This justifies AII.

To prove AIII we introduce a new concept due to Szekeres which arose while he was considering tournaments. Szekeres used supplementary difference sets with one symmetry condition ($a \in M \Rightarrow -a \notin M$) to construct skew-Hadamard matrices. He pointed out to Seidel that there was no skew-Hadamard matrix of order 36 known (at that time). This, in turn, led to Goethals and Seidel publishing their array, which was to prove so significant and useful.

5.79. DEFINITION. Let G be an additive abelian group of order $2m+1$. Then two subsets, M and N, of G, which satisfy

i) M and N are m-sets,

ii) $a \in M \Rightarrow -a \notin M$,

iii) for each $d \in G$, $d \neq 0$, the equations $d = a_1 - a_2$, $d = b_1 - b_2$ have together $m-1$ distinct solution vectors for $a_1, a_2 \in M$, $b_1, b_2 \in N$,

will be called Szekeres difference sets. Alternatively, $2-\{2m+1;m;m-1\}$ supplementary difference sets, M and $N \subset G$, are called Szekeres difference sets if $a \in M \Rightarrow -a \notin M$.

We first show such sets exist.

5.80. THEOREM. (Szekeres). If $q = 4m+3$ is a prime power and G is the cyclic group of order $2m+1$, then there exist Szekeres difference sets M and N in G with $b \in N \Rightarrow -b \in N$.

Proof. Let x be a primitive root of $GF(q)$ and $Q = \{x^{2b} : b = 1, \ldots, 2m+1\}$ the set of quadratic residues in $GF(q)$. Define M and N by the rules

$$a \in M \text{ iff } x^{2a} - 1 \in Q, \tag{7.1}$$

$$b \in N \text{ iff } x^{2b} + 1 \in Q. \tag{7.2}$$

Since

$$-1 = x^{2m+1} \notin Q,$$

$$x^{2a} - 1 \in Q \Rightarrow x^{-2a} - 1 = -x^{-2a}(x^{2a} - 1) \notin Q,$$

$$x^{2b} + 1 \in Q \Rightarrow x^{-2b} + 1 = x^{-2b}(x^{2b} + 1) \in Q,$$

so that $a \in M \Rightarrow -a \notin M$, $b \in N \Rightarrow -b \in N$, and conditions i) and ii) of Definition 5.79 are satisfied. Also, writing N' for the complement of N, we have

$$b' \in N' \text{ if } -(x^{2b'} + 1) \in Q. \tag{7.3}$$

Suppose

$$d = \alpha - a \neq 0, \qquad a, \alpha \in A, \tag{7.4}$$

where

$$x^{2a} = 1+x^{2(i-d)}, \tag{7.5}$$

$$x^{2\alpha} = 1+x^{2j}, \tag{7.6}$$

by (7.1) for suitable $i, j \in G$. Then

$$x^{2\alpha} = x^{2(a+d)} = x^{2d}+x^{2i},$$

by (7.4) and (7.5); hence by (7.6)

$$x^{2d}-1 = x^{2j}-x^{2i}, \tag{7.7}$$

where $x^{2j}+1 \in Q$ by (7.6).

Similarly, if

$$d = b'-\beta' \neq 0, \quad b', \beta' \in N', \tag{7.8}$$

where by (7.8)

$$-x^{2\beta'} = 1+x^{2(i-d)}, \tag{7.9}$$

$$-x^{2b'} = 1+x^{2j}, \tag{7.10}$$

for some $i, j \in G$, we get

$$-x^{2b'} = -x^{2(d+\beta')} = x^{2d}+x^{2i};$$

hence again

$$x^{2d}-1 = x^{2j}-x^{2i},$$

with $-(x^{2j}+1) \in Q$ by (7.10).

Conversely to every solution, $i, j \in G$, of equation (7.7), we can determine uniquely $\alpha \in M'$ or $b \in N'$ from (7.6) or (7.10) depending on whether $1+x^{2j} = x^{2d}+x^{2i}$ is in Q or not; hence a or β from (7.4), (7.8) so that also (7.5) or (7.9) is satisfied, implying $a \in M$, $\beta \in N'$. Thus the total number of solutions of (7.4) and (7.8) is equal to the number of solutions of (7.7) which is m.

5.81. EXAMPLE. Consider $q = 23$ which has primitive root 5 and quadratic residues $Q = \{1,2,3,4,6,8,9,12,13,16,18\}$. Hence

$$M = \{1,2,5,7,8\} \quad \text{and} \quad N = \{2,4,5,6,7,9\}$$

are Szekeres difference sets.

Szekeres and Whiteman (see Wallis [23, p. 323]) have independently shown that there exist Szekeres difference sets of size $(p^t-1)/2$ when $p \equiv 5 \pmod 8$ is a prime power and $t \equiv 2 \pmod 4$. But in this case both sets M, N satisfy the condition, $x \in M, N \Rightarrow -x \notin M, N$, and we have not yet been able to use these sets to construct amicable Hadamard matrices. Nevertheless, the next theorem and corollary indicate the way these Szekeres difference sets may, in some cases, be used:

5.82. THEOREM. Suppose there exist $(1,-1)$ matrices A, B, C, D of order n satisfying

$$C = I+U, \quad U^t = -U, \quad A^t = A, \quad B^t = B, \quad D^t = D,$$

$$AA^t+BB^t = CC^t+DD^t = 2(n+1)I-2J,$$

and with $e = [1, \ldots, 1]$ a $1 \times n$ matrix

$$eA^t = eB^t = eC^t = eD^t = e, \quad AB^t = BA^t, \quad \text{and} \quad CD^t = DC^t.$$

Then if

$$X = \begin{bmatrix} 1 & 1 & e & e \\ 1 & -1 & -e & e \\ e^t & -e^t & A & -B \\ e^t & e^t & -B & -A \end{bmatrix}, \quad Y = \begin{bmatrix} 1 & 1 & e & e \\ -1 & 1 & e & -e \\ -e^t & -e^t & C & D \\ -e^t & -e^t & -D & C \end{bmatrix},$$

X is a symmetric Hadamard matrix and Y is a skew-Hadamard matrix both of order $2(n+1)$. Further, if

$$AC^t-BD^t \quad \text{and} \quad BC^t+AD^t$$

are symmetric, X and Y are amicable Hadamard matrices of order $2(n+1)$.

The next result illustrates how the conditions of the theorem can be satisfied:

5.83. COROLLARY. Let G be an additive abelian group of order $2m+1$. Suppose there exist Szekeres difference sets, M and N, in G.

Further suppose there exist $2-\{2m+1; m+1; m+1\}$ supplementary difference sets P and S in G such that $x \in X \Rightarrow -x \in X$ for

$X \in \{N, P, S\}$. Then there exist amicable Hadamard matrices of order $4(m+1)$.

<u>Proof.</u> Form the type 1 (1, -1) incidence matrix C of M. Form the type 2 (1, -1) incidence matrices, D, A, B of N, P, S, respectively. Now use the properties of type 1 and type 2 matrices in the theorem.

In these theorems, circulant and back circulant can be used to replace type 1 and type 2 incidence matrices, respectively, when the orders are prime.

We now wish to show that sets satisfying the conditions of Corollary 5.83 exist for some orders $n \equiv 1 \pmod 4$. Let $n = 4t+1$ be a prime power, and choose $Q = \{x^{2b} : b = 1, 2, \ldots, 2t\}$ and $R = xQ$, where x is a primitive element of $GF(n)$. Then Q and R are $2\text{-}\{4t+1; 2t; 2t-1\}$ supplementary difference sets. Further, $y \in Q \Rightarrow -y \in Q$, and $y \in R \Rightarrow -y \in R$ since $-1 = x^{2t}$. So Q and R satisfying the conditions of the corollary exist. To find M and N, we use the result of Szekeres in Theorem 5.80. Then we have:

5.84. COROLLARY. There exist amicable Hadamard matrices of order $2(t+1)$ whenever $t \equiv 1 \pmod 4$ is a prime and $2t+1$ is a prime power.

<u>Proof.</u> With $q = 4m+3 = 2t+1$, we form Szekeres difference sets M and N of order $2m+1 = t+1$. Using the notation of Theorem 5.80,

$$b \in N \Rightarrow x^{2b}+1 \in Q \Rightarrow x^{-2b}+1 = x^{-2b}(1+x^{2b}) \in Q \Rightarrow -b \in N,$$

and so M and N are as required by Corollary 5.83.

Choose $P = Q$ and $S = xQ$; then, as observed above, they too satisfy the conditions of the theorem, and we have the result.

This justifies AIII.

5.85. THEOREM. Let $q \equiv 5 \pmod 8$ be a prime power and $q = s^2+4t^2$ be its proper representation with $s \equiv 1 \pmod 4$. Suppose there are amicable orthogonal designs of type $((1, 2r-1); (r, r))$ in order $2r = |t|+1$. Then there exist amicable Hadamard matrices of order $(|t|+1)(q+1)$.

Proof. Using the theory of cyclotomy, we can show that for the q of the enunciation,

$$C_0 \& C_1 \quad \text{and} \quad |t| \text{ copies of } C_0 \& C_2$$

are $(|t|+1) - \{q;(q-1)/2;(|t|+1)(q-3)/4\}$ s.d.s., with the property that

$$x \in C_0 \& C_1 \Rightarrow -x \notin C_0 \& C_1$$

and

$$y \in C_0 \& C_2 \Rightarrow -y \in C_0 \& C_2.$$

Also $(|t|+1)/2$ copies of each of $C_0 \& C_2$ and $C_1 \& C_3$ are $(|t|+1) - \{q;(q-1)/2;(|t|+1)(q-3)/4\}$ s.d.s. with the property that

$$y \in C_0 \& C_2 \Rightarrow -y \in C_0 \& C_2,$$

$$z \in C_1 \& C_3 \Rightarrow -z \in C_1 \& C_3.$$

Let A be the type 1 $(1,-1)$ incidence matrix of $C_0 \& C_1$, and B and C be the type 2 $(1,-1)$ incidence matrices of $C_0 \& C_2$ and $C_1 \& C_3$, respectively.

Then

$$AJ = BJ = CJ = -J, \quad (A+I)^t = -(A+I), \quad B^t = B, \quad C^t = C,$$

$$\begin{aligned} AA^t + |t|BB^t &= (|t|+1)/2(BB^t + CC^t) \\ &= (|t|+1)(q+1)I - (|t|+1)J. \end{aligned}$$

Let $P = x_0U + x_1V$ and $Q = x_3X + x_4Y$ be the amicable orthogonal designs of order $2r$ and type $((1,2r-1);(r,r))$. Further, let e be the $1 \times q$ matrix of all ones. Clearly, we may assume that $U = I$, $V^t = -V$, $X^t = X$, $Y^t = Y$, for if not, we premultiply P and Q by the same matrix W until U, V, X, Y do have the required properties. Now

$$E = \left[\begin{array}{c|c} U+V & (U+V)\times e \\ \hline (-U+V)\times e^t & U\times -A+V\times B \end{array}\right]$$

and

$$F = \left[\begin{array}{c|c} X+Y & (X+Y)\times e \\ \hline (X+Y)\times e^t & X\times C+Y\times D \end{array}\right]$$

are the required amicable Hadamard matrices.

We note that amicable orthogonal designs of type $((1,2r-1);(r,r))$ certainly exist where $2r$ is a power of two (this is proved in Corollary 5.50). Hence we have:

5.86. COROLLARY. Let $q \equiv 5 \pmod 8$ be a prime power and $q = s^2+4t^2$ be its proper representation with $s \equiv 1 \pmod 4$. Further, suppose $|t| = 2^r-1$ for some r. Then there exist amicable Hadamard matrices of order $2^r(q+1)$.

In particular, this leads to two results published elsewhere (Wallis [8],[15]) which now become corollaries:

5.87. COROLLARY. Let $q \equiv 5 \pmod 8$ be a prime power, and suppose $q = s^2+4$ or $q = s^2+36$ with $s \equiv 1 \pmod 4$. Then there exist amicable Hadamard matrices of orders $2(s^2+5)$ or $4(s^2+37)$, respectively.

Corollaries 5.85 and 5.86 justify AIV.

§5.8. CONSTRUCTIONS USING AMICABLE ORTHOGONAL DESIGNS--PRODUCT DESIGNS AND HADAMARD MATRICES

We have already noted the use of amicable orthogonal designs in constructing amicable Hadamard matrices. Here we give a few more general constructions. These constructions arise in a natural way and give some idea that they are simple cases of a far deeper concept.

As we shall see in Chapter 7, when constructing Hadamard matrices, a knowledge of the powers of 2 is crucial. We showed in Chapter 3 that a knowledge of the powers of 2 was also necessary for the solution of the algebraic problem.

Hence we will take the opportunity in the remainder of this section to find out as much as possible about existence in lower powers of 2. We shall, in fact, see that many designs not ruled out by the algebraic theory do not exist in order 16.

Also we can see that some previously known results on Hadamard matrices can be generalized by using amicable orthogonal designs.

PRODUCT DESIGNS

Geramita and Wallis [5] gave the following result.

5.88. CONSTRUCTION. Suppose there exist three matrices R, P and S of order n which give amicable orthogonal designs

$$S, \; x_1R+x_2P,$$

of types $((s_1, \ldots, s_t);(u_1, u_2))$. Then

$$\begin{bmatrix} y_1R+y_2P & y_3R+y_4P & S & y_5R+y_6P \\ -y_3R+y_4P & y_1R-y_2P & -y_5R-y_6P & S \\ -S & y_5R-y_6P & y_1R+y_2P & -y_3R+y_4P \\ -y_5R+y_6P & -S & y_3R+y_4P & y_1R-y_2P \end{bmatrix}$$

is an orthogonal design of order 4n and type

$$(s_1, s_2, \ldots, s_t, u_1, u_1, u_1, u_2, u_2, u_2).$$

We note that the above matrix can be written in the form

$$M_1 \times R + M_2 \times P + N \times S,$$

where M_1, M_2 and N are the 4×4 matrices of the coefficients of R, P and S, respectively.

We are interested in generalizing this idea. With this in mind, we give the following definition.

5.89. DEFINITION. Let M_1, M_2 and N be orthogonal designs of order n and types $(a_1, \ldots, a_r)$, $(b_1, \ldots, b_s)$ and $(c_1, \ldots, c_t)$, respectively. Then $(M_1 ; M_2 ; N)$ are <u>product designs</u> of order n and types $(a_1, \ldots, a_r; b_1, \ldots, b_s; c_1, \ldots, c_t)$ if

i) $M_1 * N = M_2 * N = 0$ (Hadamard product),

ii) M_1+N and M_2+N are orthogonal designs, and

iii) $M_1M_2^t = M_2M_1^t$ (i.e., M_1 and M_2 are amicable orthogonal designs).

NOTE. Construction 5.88 produces product designs of order 4 and types (1, 1, 1;1, 1, 1;1).

CONSTRUCTING PRODUCT DESIGNS

Before we give the main results of this section, we produce two examples of product designs. For convenience we write the designs M_1, M_2 and N in the form M_1+zN and M_2.

5.90. EXAMPLE. Display 5.4 gives product designs of order 8 and types (1, 1, 1;1, 1, 1;5).

$$
\begin{bmatrix}
x_1 & x_2 & z & x_3 & z & z & z & \bar{z} \\
\bar{x}_2 & x_1 & \bar{x}_3 & z & z & \bar{z} & z & z \\
\bar{z} & x_3 & x_1 & \bar{x}_2 & z & \bar{z} & \bar{z} & \bar{z} \\
\bar{x}_3 & \bar{z} & x_2 & x_1 & z & z & \bar{z} & z \\
\bar{z} & \bar{z} & \bar{z} & \bar{z} & x_1 & x_2 & z & \bar{x}_3 \\
\bar{z} & z & z & \bar{z} & \bar{x}_2 & x_1 & x_3 & z \\
\bar{z} & \bar{z} & z & z & \bar{z} & \bar{x}_3 & x_1 & \bar{x}_2 \\
z & \bar{z} & z & \bar{z} & x_3 & \bar{z} & x_2 & x_1
\end{bmatrix}
\qquad
\begin{bmatrix}
y_1 & y_2 & 0 & y_3 & & & & \\
y_2 & \bar{y}_1 & \bar{y}_3 & 0 & & & & \\
0 & \bar{y}_3 & y_1 & y_2 & & 0 & & \\
y_3 & 0 & y_2 & \bar{y}_1 & & & & \\
 & & & & y_2 & y_3 & 0 & \bar{y}_1 \\
 & & & & y_3 & \bar{y}_2 & y_1 & 0 \\
 & 0 & & & 0 & y_1 & y_2 & y_3 \\
 & & & & \bar{y}_1 & 0 & y_3 & \bar{y}_2
\end{bmatrix}
$$

(where $\bar{x} = -x$)

Display 5.4

5.91. EXAMPLE. Display 5.5 gives product designs of order 12 and types (1, 1, 1;1, 1, 1;9).

$$\begin{bmatrix}
x_1 & x_2 & x_3 & \bar{z} & z & z & z & \bar{z} & z & \bar{z} & z & \bar{z} \\
\bar{x}_2 & x_1 & \bar{z} & \bar{x}_3 & z & \bar{z} & \bar{z} & \bar{z} & \bar{z} & \bar{z} & \bar{z} & \bar{z} \\
\bar{x}_3 & z & x_1 & x_2 & z & z & \bar{z} & z & z & z & \bar{z} & \bar{z} \\
z & x_3 & \bar{x}_2 & x_1 & z & \bar{z} & z & z & z & \bar{z} & \bar{z} & z \\
\bar{z} & \bar{z} & \bar{z} & \bar{z} & x_1 & x_2 & x_3 & \bar{z} & z & z & \bar{z} & z \\
\bar{z} & z & \bar{z} & z & \bar{x}_2 & x_1 & \bar{z} & \bar{x}_3 & z & \bar{z} & z & z \\
\bar{z} & z & z & \bar{z} & \bar{x}_3 & z & x_1 & x_2 & \bar{z} & \bar{z} & \bar{z} & z \\
z & z & \bar{z} & \bar{z} & z & x_3 & \bar{x}_2 & x_1 & \bar{z} & z & z & z \\
\bar{z} & z & \bar{z} & \bar{z} & \bar{z} & \bar{z} & z & z & x_1 & x_2 & x_3 & \bar{z} \\
z & z & \bar{z} & z & \bar{z} & z & z & \bar{z} & \bar{x}_2 & x_1 & \bar{z} & \bar{x}_3 \\
\bar{z} & z & z & z & z & \bar{z} & z & \bar{z} & \bar{x}_3 & z & x_1 & x_2 \\
z & z & z & \bar{z} & \bar{z} & \bar{z} & \bar{z} & \bar{z} & z & x_3 & \bar{x}_2 & x_1
\end{bmatrix}$$

$$\begin{bmatrix}
y_1 & y_2 & y_3 & 0 & & & & & & & & \\
y_2 & \bar{y}_1 & 0 & \bar{y}_3 & & 0 & & & & 0 & & \\
y_3 & 0 & \bar{y}_1 & y_2 & & & & & & & & \\
0 & \bar{y}_3 & y_2 & y_1 & & & & & & & & \\
 & & & & y_2 & y_3 & y_1 & 0 & & & & \\
 & 0 & & & y_3 & \bar{y}_2 & 0 & \bar{y}_1 & & 0 & & \\
 & & & & y_1 & 0 & \bar{y}_2 & y_3 & & & & \\
 & & & & 0 & \bar{y}_1 & y_3 & y_2 & & & & \\
 & & & & & & & & y_3 & \bar{y}_1 & \bar{y}_2 & 0 \\
 & 0 & & & & 0 & & & \bar{y}_1 & \bar{y}_3 & 0 & y_2 \\
 & & & & & & & & \bar{y}_2 & 0 & \bar{y}_3 & \bar{y}_1 \\
 & & & & & & & & 0 & y_2 & \bar{y}_1 & y_3
\end{bmatrix}$$

Display 5.5

The next theorem gives us a way of obtaining product designs from product designs of smaller orders.

5.92. THEOREM. (P. J. Robinson). Let $(M_1; y_1M_3+y_2M_4; N)$ be product designs of order n and types $(a_1, \ldots, a_r; b_1, b_2; c_1, \ldots, c_t)$, and let S and x_1R+x_2P be amicable orthogonal designs of order m and types $((u);(v,w))$. Then

$$(x_1P \times M_3 + R \times M_1;\ y_1S \times M_3 + y_2R \times M_4;\ z_1P \times M_4 + S \times N)$$

are product designs of order mn and types

$$(wb_1, va_1, \ldots, va_r;\ ub_1, vb_2;\ wb_2, uc_1, \ldots, uc_t).$$

<u>Proof.</u> By straightforward verification.

A very useful form of this theorem is obtained by using the amicable orthogonal designs of order 2 and types $((1,1);(2))$ (see §5.1). We state this particular case in the following corollary.

5.93. COROLLARY. Let $(M_1;\ y_1M_3+y_2M_4;\ N)$ be product designs of order n and types $(a_1, \ldots, a_r;\ b_1, b_2;\ c_1, \ldots, c_t)$. Then there are product designs of order $2n$ and types $(b_1, a_1, \ldots, a_r;\ 2b_1, b_2;\ b_2, 2c_1, \ldots, 2c_t)$.

In Theorem 5.92 and Corollary 5.93 we may have M_3 or M_4 equal to zero. In this case, however, the next theorem gives a better result.

5.94. THEOREM. (P. J. Robinson). If $(M_1; M_2; N)$ are product designs of order n and types $(a_1, \ldots, a_r; b; c_1, \ldots, c_t)$ and if S and y_1R+P are amicable orthogonal designs of order m and types $((u_1, \ldots, u_j);$ $(v, w_1, \ldots, w_k))$, then there are product designs of order mn and the following types:

i) $(va_1, \ldots, va_r; vb; cu_1, \ldots, cu_j, bw_1, \ldots, bw_k)$ and

ii) $(va_1, \ldots, va_r; vb; uc_1, \ldots, uc_t, bw_1, \ldots, bw_k)$,

where u and c are the sums of the u_i's and c_i's, respectively.

Proof. We consider

$$(R \times M_1;\ R \times M_2;\ S \times N + P \times M_2),$$

with the appropriate variables equated.

The next result gives us a way of obtaining product designs from amicable orthogonal designs.

5.95. THEOREM. (P. J. Robinson). If S and y_1R+y_2P are amicable orthogonal designs of order n and types $((u_1, \dots, u_j);(v, w))$, then

$$\left(\begin{bmatrix} S & x_1R \\ -x_1R & S \end{bmatrix} ; \begin{bmatrix} y_1R+y_3P & y_2R \\ y_2R & -y_1R+y_3P \end{bmatrix} ; \begin{bmatrix} 0 & P \\ -P & 0 \end{bmatrix} \right)$$

are product designs of order $2n$ and types

$$(v, u_1, \dots, u_j;v, v, w;w).$$

Proof. By straightforward verification.

In the following lemma we give examples of product designs which will be used later to produce a very useful orthogonal design.

5.96. LEMMA. (P. J. Robinson). There are product designs of order 2^t, $t \geq 4$, and types

$$(1,1,1,1,2,4,\dots,2^{t-3};2,2^{t-3};2,4,\dots,2^{t-3},2^{t-2},2^{t-2}).$$

Proof. The product designs of order 4 and types $(1,1,1;1,2;1)$ produce product designs of order 8 and types $(1,1,1,1;2,2;2,2)$ (Corollary 5.93). This design in turn produces product designs of order 16 and types $(1,1,1,1,2;2,4;2,4,4)$.

By repeated use of Corollary 5.93 we obtain the required result.

In Table 5.6 we give a list of product designs of orders 4 and 8 which are obtained by using the results given here.

PRODUCT DESIGNS OF ORDER 4 AND 8

(Product designs in higher powers of 2 are given in Appendix I.)

Product Designs	Construction	
Order 4		
(1,1,1;1,1,1;1)	Construction 5.88	
Order 8		
(1,1,1,2;1,1,3;3)	((1,1,2);(1,3))	((1,1);(1,1))
(1,1,2,3;1,3,3;1)	((1,1,2);(3,1))	((1,1);(1,1))
(1,1,2,2;2,2,2;2)	((1,1,2);(2,2))	((1,1);(1,1))
(1,1,1,1;2,2;2,2)	(1,1,1;1,2;1)	
(1,1,1,2;1,4;1,2)	(1,1,1;2,1,1)	
(1,1,1;1,1,1;5)	Example 5.90	

Table 5.6

CONSTRUCTING ORTHOGONAL DESIGNS FROM PRODUCT DESIGNS

We now produce a generalization of Construction 5.88.

5.97. THEOREM. (P. J. Robinson). Let S and y_1R+P be amicable orthogonal designs of order m and types $((u_1,\ldots,u_j);(v,w_1,\ldots,w_k))$, and let $(M_1;M_2;N)$ be product designs of order n and types $(a_1,\ldots,a_r;b_1,\ldots,b_s;c_1,\ldots,c_t)$. Then there exist orthogonal designs of order mn and types

i) $(va_1,\ldots,va_r,wb_1,\ldots,wb_s,uc_1,\ldots,uc_t)$,

ii) $(va_1,\ldots,va_r,wb_1,\ldots,wb_s,u_1c,\ldots,u_jc)$,

iii) $(va_1,\ldots,va_r,w_1b,\ldots,w_kb,uc_1,\ldots,uc_t)$,

iv) $(va_1,\ldots,va_r,w_1b,\ldots,w_kb,u_1c,\ldots,u_jc)$,

where b, c, u and w are the sums of the b_i's, c_i's, u_i's and w_i's, respectively.

Proof. We consider

$$M_1 \times R + M_2 \times P + N \times S,$$

with the appropriate variables equated.

As an example of the use of this theorem, we give the following lemma.

5.98. LEMMA. (P. J. Robinson). There is an orthogonal design of order 2^t, $t \geq 2$, and type $(1,1,1,1,2,2,4,4,\ldots,2^{t-2},2^{t-2})$.

Proof. If $t \geq 5$, we apply the above theorem with the product designs of order 2^{t-1} and types

$$(1,1,1,1,2,4,\ldots,2^{t-4};2,2^{t-3};2,4,\ldots,2^{t-4},2^{t-3},2^{t-3})$$

(Lemma 5.96) and amicable orthogonal designs of order 2 and types $((1,1);(2))$.

The orthogonal design of order 16 and type $(1,1,1,1,2,2,4,4)$ may be obtained in a similar manner by using the product designs of order 8 and type $(1,1,1,1;2,2;2,2)$ (see proof of Lemma 5.96).

The design of order 4 and type $(1,1,1,1)$ is given in §4.1, and the design of type $(1,1,1,1,2,2)$ is obtained from the orthogonal design of order 8 and type $(1,1,1,1,1,1,1,1)$ (see §4.1).

We note that the above orthogonal designs have $2t$ variables. If $t = 4k+1$, $\rho(2^t) = 8k+2 = 2t$, and if $t = 4k+2$, $\rho(2^t) = 8k+4 = 2t$. Therefore, if $t = 4k+1$ or $4k+2$, the above design has the maximum number of variables allowed. We also note that the above design is full. That is, the design contains no zeros.

By equating variables in the above design, we obtain:

5.99. COROLLARY. (P. J. Robinson). All orthogonal designs of type $(1,1,a,b,c)$, $a+b+c = 2^t-2$, exist in order 2^t, $t \geq 3$.

Proof. Noting that any number $< 2^{t-1}-1$ can be formed from $1,2,4,\ldots,2^{t-2}$, we have the result.

In Appendix D we give the types of some orthogonal designs in order 32 obtained from product designs and by doubling orthogonal designs in order 16. We further illustrate Theorem 5.97 by obtaining designs in order 16.

5.100. COROLLARY. (Geramita-Wallis). Suppose there exist amicable orthogonal designs of order n and types $((u_1, \dots, u_p); (v_1, \dots, v_q))$. Then, since there are product designs of type (1,1,1;1,1,1;1) in order 4, there exist orthogonal designs of type

i) $(u_1, u_1, u_1, 3u_2, \dots, 3u_p, v_1, \dots, v_q)$ and

ii) $(u_1, u_1, u_1, w, w, w, v_1, \dots, v_q)$

in order 4n where $w = u_2 + u_3 + \dots + u_p$.

5.101. EXAMPLE. There exist orthogonal designs of types

a) (1,1,1,1,1,1,1,1,2),

b) (1,1,1,1,1,2,2,2,2),

c) (1,1,1,1,1,2,3,3,3),

d) (1,1,2,2,2,2,2,2,2),

e) (1,1,2,2,2,2,3,3)

in order 16.

Proof. Use the amicable orthogonal designs of types ((1,1);(1,1,2)), ((1,2);(1,1,2)), ((1,1,2);(1,1,2)) and ((2,2);(1,1,2)) in order 4 in ii) to obtain a), b), c) and d). For e) we use the amicable orthogonal designs of types ((2,1,1);(1,1,2)) in order 4 in i).

We now give two results to show how product designs may be used to obtain orthogonal designs in orders other than powers of 2.

5.102. LEMMA. (P. J. Robinson). There are product designs of order 12 and types (1,1,1;1,1,4;4), (1,1,4;1,1,1;1), (1,1,4;1,4,4;1), (1,1,4;1,1,4;4), (1,4,4;1,4,4;1) and (1,1,1;1,1,1;9).

Proof. Wolfe [1] gives amicable orthogonal designs of types ((1,1);(1,4)) and ((1,4);(1,4)) in order 6. By using these designs in Theorem 5.95, we obtain the first five designs. The last is given in Example 5.91.

By using Theorem 5.97 with the above designs, we obtain:

5.103. COROLLARY. (P. J. Robinson). There are orthogonal designs of order 24 and types

$(1,1,1,1,1,1,9,9)$, $(1,1,1,1,4,4,4,4)$,
$(1,1,1,1,1,4,4,4)$, $(1,1,1,1,1,1,1,4)$.

APPLICATIONS

By equating variables in the orthogonal designs given in Appendix D, we obtain the following lemmas.

5.104. LEMMA. (P. J. Robinson). All 6-tuples of the form $(a, b, c, d, e, 32-a-b-c-d-e)$, $0 < a+b+c+d+e < 32$, are the types of orthogonal designs in order 32, except possibly

$(1,1,1,1,1,27)$.

5.105. COROLLARY. All 5-tuples, except possibly $(1,1,1,1,27)$, are the types of orthogonal designs in order 32.

5.106. COROLLARY. All n-tuples, $n = 1, 2, 3, 4$, are the types of orthogonal designs in order 32.

HADAMARD MATRICES

We restate a theorem of Wallis [25, p. 337, Theorem 4.20] in terms of our new notation:

5.107. THEOREM. (Wallis). Suppose there exist amicable orthogonal designs of types $((1, m-1);(m))$ and $((1, n-1);(n))$ in orders m and n, respectively. Further suppose there exists an orthogonal design of type $(1, (m-1)n-1)$ in order $(m-1)n$. Then there exists an orthogonal design of type $(1, mn(mn-1)(m-1)-1)$ in order $mn(mn-1)(m-1)$.

<u>Proof.</u> We use Lemma 5.48 with the designs of types $((1, m-1);(m))$ and $(1, (m-1)n-1)$ to see that there is an orthogonal design of type $(1, m-1, mn(m-1)-m)$ in order $mn(m-1)$. We use the designs of types $((1, m-1);(m))$ and $((1, n-1);(n))$ to see that there are amicable orthogonal designs of types

((1, mn-1); (mn)) in order mn.

We now write the amicable orthogonal designs of types ((1, mn-1); (mn)) (after suitable pre- and post-multiplication by monomial matrices) as

$$M = \begin{bmatrix} x & ye \\ -ye^t & P \end{bmatrix} \quad \text{and} \quad N = \begin{bmatrix} 1 & e \\ e^t & D \end{bmatrix},$$

where $e = (1, \ldots, 1)$ is of order $1 \times (mn-1)$, and x, y are commuting variables. Then

$$JP^t = xJ, \quad (P-xI)^t = -(P-xI), \quad D^t = D, \quad JD^t = -J,$$

$$PP^t = (x^2+(mn-1)y^2)I-y^2J,$$

Since

$$MN^t = \begin{bmatrix} x & ye \\ -ye^t & P \end{bmatrix}\begin{bmatrix} 1 & e \\ e^t & D^t \end{bmatrix} = \begin{bmatrix} x+(mn-1)y & (x-y)e \\ (x-y)e^t & -yJ+PD^t \end{bmatrix}$$

$$= \begin{bmatrix} x+(mn-1)y & (-y+x)e \\ (x-y)e^t & -yJ+DP^t \end{bmatrix} = \begin{bmatrix} 1 & e \\ e^t & D \end{bmatrix}\begin{bmatrix} x & -ye \\ ye^t & P^t \end{bmatrix} = NM^t,$$

$$PD^t = DP^t.$$

We now replace the variables of the $(1, m-1, mn(m-1)-m)$ design A by the matrices P, yJ, yD to obtain a matrix B. Now

$$BB^t = (PP^t+(m-1)y^2J^2+(m^2n-mn-m)y^2DD^t) \times I$$

$$= [(x^2+(mn-1)y^2)I-y^2J+(m-1)(mn-1)y^2J$$

$$+ mn(m^2n-m-mn)y^2I-(m^2n-mn-m)y^2J] \times I$$

$$= (x^2+(mn(mn-1)(m-1)-1)y^2)I_{mn(mn-1)(m-1)}.$$

Hence B is the required orthogonal design.

5.108. COROLLARY. Suppose there exist amicable orthogonal designs of type ((1, n-1); (n)) in order n. Then, since there exist amicable orthogonal designs of types ((1, 1); (2)) in order 2, there is an orthogonal design of type (1, 2n(2n-1)-1) in order 2n(2n-1).

Before going to our next result, we first note:

5.109. LEMMA. If m is the order of amicable orthogonal designs of types $((1, m-1);(m))$ and h the order of an orthogonal design of type $(1, h-1)$, then by Wolfe's theorem (5.48) there is an orthogonal design of type $(1, m-1, m(h-1))$ in order mh.

Then Theorem 8.7 of Wallis [25, p. 368] can be restated as:

5.110. THEOREM. (Wallis). Suppose there exists an orthogonal design of type $(1, m-1, m(h-1))$ in order mh. Suppose there exist "suitable" matrices of order n to replace the variables of this design. Then there exists an Hadamard matrix of order mhn.

Proof. Obvious.

5.111. COROLLARY. Let n be the order of any Hadamard matrix H. Suppose there exists an orthogonal design D of type $(1, m-1, nm-n-m)$ in order $n(m-1)$. Then there exists an Hadamard matrix of order $n(n-1)(m-1)$.

Proof. We write H as

$$\begin{bmatrix} 1 & e \\ -e^t & P \end{bmatrix},$$

where e is the $1 \times (n-1)$ matrix of 1's. Then

$$PJ = J, \quad PP^t = nI - J.$$

The result is obtained by replacing the variables of D by P, J, P, respectively.

Many corollaries can be made by finding "suitable" matrices, but we will not proceed further with this here.

Now we will show in Chapter 7 that orthogonal designs of types $(1, m-1, nm-n-m)$ exist in every power of 2, $2^t = (m-1)n$. Hence we have a new result.

5.112. COROLLARY. With t, s any nonnegative integers, there exists an Hadamard matrix of order $2^s(2^s-1)(2^t-1)$.

We note the following result:

5.113. THEOREM. Let $k > 1$ be the order of an Hadamard matrix H, and n be the order of a symmetric conference matrix C. Further, suppose there exist amicable orthogonal designs M, N of order m and types $((1, m-1); (m/2, m/2))$. Then there exists an orthogonal design of type $(k, (m-1)k, (n-1)mk/2, (n-1)nk/2)$ in order nmk.

Proof. Let $P = \begin{bmatrix} 0 & - \\ 1 & 0 \end{bmatrix} \times I_{k/2}$. Then

$$R = C \times H \times N + I \times PH \times M$$

is the required orthogonal design.

Hence we have generalized a theorem of Wallis [25, p. 375, Theorem 8.24]:

5.114. COROLLARY. Suppose H, C, M, N are as in the theorem, and suppose there are "suitable" matrices of order p. Then there exists an Hadamard matrix of order $nmkp$.

Now we note that if m is of the form $2 \; 2^t(p_i^{r_i}+1)$, where $p_i^{r_i} \equiv 3 \pmod 4$ is a prime power, then amicable orthogonal designs of types $((1, m-1); (m/2, m/2))$ exist in order m. Thus we have:

5.115. COROLLARY. Suppose $k > 1$ is the order of an Hadamard matrix and n the order of a symmetric conference matrix. Then there exists an orthogonal design of type $(k, (2m-1)k, (n-1)mk, (n-1)mk)$ where $m = 2^t \prod_i (p_i^{r_i}+1)$, $p_i^{r_i} \equiv 3 \pmod 4$ is a prime power, and $t > 0$ is an integer.

§5.9. USING FAMILIES OF MATRICES--REPEAT DESIGNS

We shall now examine one of the most powerful methods known for constructing orthogonal designs.

The motivation for the constructions of this chapter arises from the following observations:

5.116. CONSTRUCTION. Suppose $(M_1; M_2; N)$ is a product design of type

$(u_1, u_2, \ldots; v_1, v_2, \ldots; w)$ and order n. Then, with x_1, x_2 commuting variables,

$$\begin{bmatrix} M_1+x_1N & M_2+x_2N \\ M_2-x_2N & -M_1+x_1N \end{bmatrix}$$

is an orthogonal design of order $2n$ and type $(w, w, u_1, u_2, \ldots, v_1, v_2, \ldots)$.

(This construction will be discussed further in §5.11.)

5.117. CONSTRUCTION. (Geramita-Wallis [5]). Let P_1, P_2, P_3 be skew-symmetric orthogonal designs of types $(p_{i1}, p_{i2}, \ldots)$, $i = 1, 2, 3$, in order n, and H a symmetric orthogonal design of type $(h_1, h_2, \ldots)$ in order n. Further, suppose $P_iP_j^t = P_jP_i^t$ and $P_kH^t = HP_k^t$. Then

$$\begin{bmatrix} x_1I_n+P_1 & x_3I_n+P_2 & x_5I_n+P_3 & H \\ -x_3I_n+P_2 & x_1I_n-P_1 & H & -x_5I_n-P_3 \\ -x_5I_n+P_3 & -H & x_1I_n-P_1 & x_3I_n+P_2 \\ -H & x_5I_n-P_3 & -x_3I_n+P_2 & x_1I_n+P_1 \end{bmatrix}$$

is an orthogonal design of order $4n$ and type $(1, p_{11}, p_{12}, \ldots, 1, p_{21}, p_{22}, \ldots, 1, p_{31}, p_{32}, \ldots, h_1, h_2, \ldots)$.

Proof. By straightforward verification.

To generalize this design, we introduce the following definition:

5.118. DEFINITION. Let $R, P_1, P_2, \ldots, H$ be orthogonal designs of order n and types $(r_1, r_2, \ldots)$, $(p_{i1}, p_{i2}, \ldots)$, $i = 1, 2, \ldots$, $(h_1, h_2, \ldots)$, respectively. Then $(R; (P_1; P_2; \ldots); H)$ are repeat designs of order n and types $(r_1, r_2, \ldots; (p_{11}, p_{12}, \ldots; p_{21}, p_{22}, \ldots; \ldots); h_1, h_2, \ldots)$ if

i) $R*P_i = 0$, $i = 1, 2, \ldots$,

ii) $R+P_i$, $i = 1, 2, \ldots$ are orthogonal designs,

iii) $R+P_i$ and H, $i = 1, 2, \ldots$, are amicable orthogonal designs,

iv) $P_iP_j^t = P_jP_i^t$, $i \neq j$.

Then we have

5.119. CONSTRUCTION. Let $(L; M_1+M_2+\ldots+M_s; N)$ be product designs of order n and types $(a_1,\ldots,a_p;b_{11},\ldots,b_{1q_1},b_{21},\ldots,b_{2q_2},\ldots,b_{s1},\ldots,b_{sq_s};c_1,\ldots,c_t)$, where M_i is of type $(b_{i1},\ldots,b_{iq_i})$. Further, let $(R;(P_1;P_2;\ldots;P_u);H)$ be repeat designs of order m and types $(r_1,\ldots,r_w;(p_{11},\ldots,p_{1v_1};p_{21},\ldots,p_{2v_2};\ldots;p_{u1},\ldots,p_{uv_u});h_1,\ldots,h_x)$. Then

$$L\times R + M_1\times P_{j1} + \ldots + M_k\times P_{jk} + N\times H$$

is an orthogonal design of order mn and type 1 of

i) $(a_1r,\ldots,a_pr,b_1p_{11},\ldots,b_1p_{1v_1},\ldots,b_sp_{s1},\ldots,b_sp_{sq_s},$
$ch_1,\ldots,ch_x)$,

ii) $(a_1r,\ldots,a_pr,b_1p_{11},\ldots,b_1p_{1v_1},\ldots,b_sp_{s1},\ldots,b_sp_{sq_s},$
$c_1h,\ldots,c_th)$,

iii) $(ar_1,\ldots,ar_w,b_1p_{11},\ldots,b_1p_{1v_1},\ldots,b_sp_{s1},\ldots,b_sp_{sq_s},$
$ch_1,\ldots,ch_x)$,

iv) $(ar_1,\ldots,ar_w,b_1p_{11},\ldots,b_1p_{1v_1},\ldots,b_sp_{s1},\ldots,b_sp_{sq_s},$
$c_1h,\ldots,c_th)$,

where a, c, r, h are the sum of some or all of the a_i, c_i, r_i, h_i, respectively, and $b_i = b_{i1}+\ldots+b_{iq_i}$.

This construction is at first sight quite formidable, but as we shall see, it does in fact lead to new orthogonal designs.

We have previously mentioned product designs, so we need to find some repeat designs to see if any new orthogonal designs can be obtained. First we see that they do lead to new designs:

5.120. EXAMPLE. There are repeat designs of order 4 and types $(1;(1;3);1,3)$, $(1;(2;3);1,3)$, $(1;(1;2);1,1,2)$ and $(1;(2;1,2);1,2)$. They are $(I;(T_1;T_4);T_0)$, $(I;(T_3;T_4);T_0)$, $(I;(T_1;T_3);T_5)$ and $(I;(T_2;T_6);T_7)$, where

$$T_0 = \begin{bmatrix} x & y & y & y \\ y & -x & -y & y \\ y & -y & y & -x \\ y & y & -x & -y \end{bmatrix}, \quad T_1 = \begin{bmatrix} 0 & + & 0 & 0 \\ - & 0 & 0 & 0 \\ 0 & 0 & 0 & - \\ 0 & 0 & + & 0 \end{bmatrix},$$

$$T_2 = \begin{bmatrix} 0 & 0 & + & + \\ 0 & 0 & + & - \\ - & - & 0 & 0 \\ - & + & 0 & 0 \end{bmatrix}, \quad T_3 = \begin{bmatrix} 0 & 0 & + & + \\ 0 & 0 & - & + \\ - & + & 0 & 0 \\ - & - & 0 & 0 \end{bmatrix},$$

$$T_4 \begin{bmatrix} 0 & + & + & + \\ - & 0 & + & - \\ - & - & 0 & + \\ - & + & - & 0 \end{bmatrix}, \quad T_5 = \begin{bmatrix} u & v & w & w \\ v & -u & -w & w \\ w & -w & v & -u \\ w & w & -u & -v \end{bmatrix},$$

$$T_6 = \begin{bmatrix} 0 & a & b & b \\ -a & 0 & -b & b \\ -b & b & 0 & -a \\ -b & -b & a & 0 \end{bmatrix}, \quad T_7 = \begin{bmatrix} u & 0 & w & w \\ 0 & -u & -w & w \\ w & -w & 0 & -u \\ w & w & -u & 0 \end{bmatrix}.$$

Repeat designs in order 4 and types (1;(1,1;1,1);1), (1;(1,1;1,2);2), (1;(1,1;2);1,2), (1;(1;1,2);2,2) and (1;(1,2;1,2);4) can be constructed using Lemma 5.127.

5.121. EXAMPLE. There are product designs of types (1,1,2,3;1,3,3;1), (2,2;1,1,1,1;4) and (1,1,1;1,1,1;5) in order 8. Then using the repeat design of types (1;(2;3);1,3) in order 4 with the matrix of weight 2 used once only, we have orthogonal designs of types (1,1,2,3,2,9,9,1,3), (2,2,2,3,3,3,4,12) and (1,1,1,2,3,3,5,15) in order 32. Since all of these have weight 31, we use the Geramita-Verner theorem to obtain the following orthogonal designs in order 32: (1,1,1,1,2,2,3,3,9,9), (1,2,2,2,3,3,3,4,12) and (1,1,1,1,2,3,3,5,15). These last two designs are new.

The product designs of types (1,1,1;1,1,1;1) in order 4 can be used with the repeat designs of types (1;(p;3);1,3), $p = 1,2$, to obtain the orthogonal designs of types (1,1,1,1,p,p,3,3), $p = 1,2$, in order 16. These have not been constructed previously.

REMARK. In the preceding example we have concentrated on constructing orthogonal designs with no zero. There is considerable scope to exploit these constructions to look for other orthogonal designs in order 32 and higher powers of 2.

We can collect the results from Example 5.120 in the following statement:

5.122. STATEMENT. In order 4 there exist repeat designs of types $(1;(r;s);h)$ for $0 \le r, s \le 3$, $0 \le h \le 4$.

Noting that the repeat designs $(R;(P);H)$ are just amicable orthogonal designs $R+P$ and H, we see that:

5.123. COROLLARY. There exist amicable orthogonal designs of types $((1, r);(h))$ in order 4 for $0 \le r \le 3$, $0 \le h \le 4$.

REMARK. The nonexistence of amicable orthogonal designs of types $((1,7);(5))$ in order 8 and $((1,15);(1))$ in order 16 means there are <u>no</u> repeat designs of types $(1;(r;7);5)$ in order 8 and $(1;(r;15);1)$ in order 16 (see Robinson [4]).

The construction and replication lemmas given later allow us to say:

5.124. COMMENT. In order 8 there, in fact, exist repeat designs $(1;(r);h)$ for all $0 \le r \le 7$ and $0 \le h \le 8$, except $r = 7$, $h = 5$ (which cannot exist).

In order 16 there exist repeat designs $(1;(r);h)$ for all $r = 1, 2, 3, \ldots, 15$, $h = 1, 2, \ldots, 16$, except possibly the following pairs (r, h): $(13,1)$, $(13,5)$, $(13,9)$, $(15,7)$, $(15,9)$, $(15,15)$ which are undecided and $(15,1)$ which does not exist.

CONSTRUCTION AND REPLICATION OF REPEAT DESIGNS

We now show that many repeat designs can be constructed.

5.125. LEMMA. Suppose $((a);(b_1, b_2))$ and $((c);(d_1, d_2))$ are the types of amicable orthogonal designs in orders n_1 and n_2. Then there is a repeat design in order n_1n_2 of type $(b_1d_1;(ad_2, b_2d_1;b_2c, b_1d_2);ac)$.

<u>Proof.</u> Let $A, x_1B_1+x_2B_2$ and $C, y_1D_1+y_2D_2$ be the amicable orthogonal designs. Then $(B_1 \times D_1;(xA \times D_2 + yB_2 \times D_1;uB_2 \times C + wB_1 \times D_2);A \times C)$ are

the required repeat designs.

5.126. EXAMPLE. Let $A = C = \begin{bmatrix} 1 & 1 \\ 1 & - \end{bmatrix}$, $B_1 = D_1 = \begin{bmatrix} 1 & 0 \\ 0 & 1 \end{bmatrix}$, and $B_2 = D_2 = \begin{bmatrix} 0 & 1 \\ - & 0 \end{bmatrix}$. Then the repeat design in order 4 and type $(1;(1,2;1,2);4)$ is

$$\left(I_4; \left(\left[\begin{array}{cc|cc} 0 & y & x & x \\ \bar{y} & 0 & x & \bar{x} \\ \hline \bar{x} & \bar{x} & 0 & y \\ \bar{x} & x & \bar{y} & 0 \end{array}\right]; \left[\begin{array}{cc|cc} 0 & u & w & u \\ \bar{u} & 0 & \bar{u} & w \\ \hline \bar{w} & u & 0 & \bar{u} \\ \bar{u} & \bar{w} & u & 0 \end{array}\right] \right); z \left[\begin{array}{cc|cc} 1 & 1 & 1 & 1 \\ 1 & - & 1 & - \\ \hline 1 & 1 & - & - \\ 1 & - & - & 1 \end{array}\right] \right).$$

Before we proceed to our uses of repeat designs, we first note some replication results.

5.127. LEMMA. Suppose there are repeat designs of type $(r;(p_1,\ldots,p_i;q_1,\ldots,q_j);h_1,\ldots,h_k)$ in order n where $h_1+h_2+\ldots+h_k = h$ and $p_1+\ldots+p_i = p$. Further suppose $A+B$ and $C+D$ are amicable orthogonal designs of types $((a,b);(c,d))$ and order m. Then there are repeat designs of order mn and types

i) $(ar;(cp_1,cp_2,\ldots,br;aq_1,aq_2,\ldots,bh);ch)$,

ii) $(ar;(ap_1,ap_2,\ldots;cq_1,cq_2,\ldots);ah_1,ah_2,\ldots,ch_i,\ldots)$,

iii) $(ar;(ap_1,ap_2,\ldots,bh_1;cq_1,cq_2,\ldots);ch_1,ah_2,ah_3,\ldots)$,

iv) $(ar;(bh_1,bh_2,\ldots;rb+pd,cq_1,cq_2,\ldots);rd+bp)$, where $d = b$,

v) $(ar;(cq_1,cq_2,\ldots;cp);ah_1,ah_2,\ldots,bp)$,

vi) $(ar;(br,dp_1,dp_2,\ldots;aq_1,aq_2,\ldots,bh);dh)$,

vii) $(ar;(cp_1,\ldots;cq_1,\ldots);ch_1,ch_2,\ldots,dr)$,

viii) $(ar;(cp_1,\ldots,dq_1,\ldots);ah_1,ah_2,\ldots,bp_1,bp_2,\ldots)$,

ix) $(ar;(ap_1,\ldots;aq_1,\ldots);ch,dh)$,

x) $(cr;(br;bh_1,bh_2,\ldots);ar)$,

xi) $(cr;(br;bh);ar,abrh)$.

Proof. Use the following constructions:

i) $(A \times R;(C \times P + xB \times R;yA \times Q + zB \times H);\ C \times H)$,

ii) $(A \times R;\ (A \times P;\ C \times Q);\ xA \times V + C \times W)$,

iii) $(A \times R;\ (A \times P + xB \times V;\ C \times Q);\ C \times V + yA \times W)$,

iv) $(A \times R;\ (B \times H;\ xB \times R + yC \times Q - xD \times P);\ D \times R + B \times P)$,

v) $(A \times R;\ (C \times Q;\ C \times P);\ xA \times H + yB \times P)$,

vi) $(A \times R;\ (B \times R + wD \times P;\ xA \times Q + yB \times H);\ D \times H)$,

vii) $(A \times R;\ (C \times P;\ C \times Q);\ C \times H + yD \times R)$,

viii) $(A \times R;\ (C \times P + xD \times Q);\ A \times H + yB \times P)$,

ix) $(A \times R;\ (A \times P;\ A \times Q);\ C \times H + yD \times H)$,

x) $(C \times R;\ (B \times R;\ B \times H);\ A \times R)$,

xi) use Lemma on the result x).

5.128. COROLLARY. There are repeat designs of type $(1;(1,2,\ldots,2^{t-1};1,2,\ldots,2^{t-1});2^t)$ in order 2^t.

Proof. Use part i) of the lemma repeatedly with repeat designs of type $(1;(1,2;1,2);4)$ in order 4 and the amicable orthogonal designs of type $((1,1);(2))$ in order 2.

CONSTRUCTION OF ORTHOGONAL DESIGNS

The use of repeat designs with product designs is so powerful a source of orthogonal designs that it is quite impossible to indicate all the designs constructed. Hence we give only those that are used to give Corollary 5.131. We use Robinson's Ph.D. thesis and Appendix I as a source for product designs.

5.129. COROLLARY. The following types of orthogonal designs exist in order 2^t:

i) $(1,1,1,1,2,2,4,4,\ldots,2^{t-2},2^{t-2})$,

ii) $(1,1,2,1,2,4,8,\ldots,2^{t-3},3,6,12,\ldots,3.2^{t-3})$,

iii) $(1,1,2,4,8,\ldots,2^{t-3},2^{t-3},2^{t-2},3,3,6,\ldots,3.2^{t-4})$,

iv) $(1,1,2,4,8,\ldots,2^{t-3},3,6,9,18,\ldots,9.2^{t-5},3.2^{t-4})$,

v) $(1,2,3,2^{t-4},3.2^{t-4},3.2^{t-3},3,3,6,6,12,12,\ldots,3.2^{t-5},3.2^{t-5})$,

vi) $(2,1,2,\ldots,2^{t-s-4},(2^{t-s-3}-1)(\text{bin}(2^{s+1}-1)),2^{t-s-2}(\text{bin}(2^{s+1}-1)),$
$2^{s+1},2^{t-2},2^{t-s-3}-1,2^{t-s-3}-1)$,

where $x\,\text{bin}(2^y-1)$ means x times the binary expansion of 2^y-1, i.e., $x,x,2x,4x,\ldots,2^{y-1}x$.

Proof. i). Proved by Robinson (Lemma 7.98).

ii). Use the product design $(1,1,2,3,6,12,\ldots,3.2^{t-4};1,3.2^{t-3};1,2,\ldots,2^{t-4})$ with the amicable orthogonal designs of types $((1,1);(1,1))$. For the remainder we use product designs with repeat designs as indicated.

iii). Use $(1,1,1,2,\ldots,2^{t-4};1,2^{t-3};1,2,\ldots,2^{t-4})$ with $(1;(1,2;3);4)$.

iv). Use $(1,1,2,3,6,\ldots,3.2^{t-5};1,3.2^{t-4};1,2,\ldots,2^{t-5})$ with $(1;(1,2;3);4)$.

v). Use $(1,1,2,1,2,\ldots,2^{t-5};3,2^{t-4};3,6,\ldots,3.2^{t-5})$ with $(1;(1,2;3);4)$.

vi). $(1,1,1,1,2,\ldots,2^s;2,2^{s+1};2,\ldots,2^{s+1},2^{s+1})$ with $(1;(\text{bin}(2^{t-s-3}-1);2^{t-s-3}-1);2^{t-s-3})$.

REMARK. This corollary allows us to find all four variable designs of type $(a,b,c,2^t-a-b-c)$ for $t = 5,6,7$, all but the design of type $(13,13,15,215)$ for $t = 8$, and all but the designs of types $(13,13,15,471)$, $(27,29,29,427)$, $(29,29,29,425)$, $(29,29,31,423)$, $(31,45,45,391)$ for $t = 9$.

To eliminate those remaining, we observe:

5.130. COROLLARY. All orthogonal designs of order 2^{t+k} exist of types

i) $(a,b,ma,2^{t+k}-a-ma-b)$, where $a = 2^t-1$, $m < 2^{k-1}$,
$0 \le b \le 2^{t+k-1}$;

ii) $(a,b,ma,2^{t+k}-a-ma-b)$, where $a = 2^t-2^s-1$, $m < 2^{k-1}$,
$0 \le b \le 2^{t+k-1}+1$;

iii) $(a,a,a,2^{t+k}-3a)$, where $0 \le a \le 2^{t+k-2}$;

iv) $(a, a, b, 2^{t+k}-2a-b)$, where $2^{t-1} \le a < 2^t$, $0 \le b \le 2^{k-1}-1$;

v) $(a, a, 2^t-a, \text{bin}\, a, 2^{t+k}-2^t-2a)$, where $2^{t-1} \le a < 2^t$ (note: 2^t-a, bin a can always be used to give 2^t-1 or $2^{t-1}-1$).

Proof. Call the product designs $(1, \text{bin}\, x, 2^j-x, 2^j, \ldots, 2^{k-2}; 1, 2^{k-1}; 1, 2, \ldots, 2^{k-2})$ the product designs A. We now use product designs with repeat designs as indicated:

i). Use A with $(1; (1, 2, \ldots, 2^{t-1}; 2^t-1); 2^t)$.

ii). Use A with $(1, 2^s; (1, 2, \ldots, 2^{s-1}, 2^{s+1}, \ldots, 2^{t-1}; 2^t-2^s-1); 2^t)$.

iv) and v). Use A with $(2^t-a; ((\text{bin}\, a); a); 2^t)$ to get the orthogonal design $(a, a, a\, \text{bin}(2^{k-1}-b-1), b(\text{bin}\, a), 2^t-a, 2^{k-1}(2^t-a), 2^t, 2^{t+1}, \ldots, 2^{t+k-2})$.

iii). We use the amicable orthogonal designs of type $((a, 2^{t+k-2}-a); (2^{t+k}))$ in order 2^{t+k} with product designs $(1,1,1;1,1,1;1)$ in order 4.

So we have:

5.131. COROLLARY. All orthogonal designs of type $(a, b, c, 2^t-a-b-c)$ and of type (a, b, c), $0 \le a+b+c \le 2^t$, exist for $t = 2, 3, 4, 5, 6, 7, 8, 9$.

REMARK. We believe these results do, in fact, allow the construction of all full orthogonal designs (that is, with no zero) with four variables in every power of 2, but we have not been able to prove this result.

§5.10. STRUCTURE OF AMICABLE WEIGHING MATRICES

Lemmas 5.43, 5.44 and 5.67 have already indicated that the amicability condition might force strong constraints on the structure of the component weighing matrices. In this section we study this idea a bit more.

A combinatorial argument lets us obtain the following result:

5.132. THEOREM. Let R be a monomial matrix of order $n \equiv 0 \pmod 4$. Let A be a symmetric weighing matrix of order n. Suppose $RA^t = -AR^t$. Then $A*R = 0$. Further, if A has weight n-1, then R is symmetric if A has any diagonal element zero, and R is skew-symmetric otherwise.

5.133. THEOREM. The existence of amicable orthogonal designs of order $n \equiv 0 \pmod 4$ and types $((1, n-1); (1, n-1))$ is equivalent to the existence of a symmetric weighing matrix of order n and weight $n-1$ with at least one zero on the diagonal.

Proof. Let A be the symmetric weighing matrix. We use the theorem of Delsarte-Goethals-Seidel (2.19) to see that we can find monomial matrices P and Q so that $B = PAQ$ is skew-symmetric. Let $R = P^{-1}Q^{-1}$. Then R is a monomial matrix, and $B^t = (PAQ)^t = Q^t A P^t = Q^{-1} A P^{-1} = -PAQ$, so $RA^t = -AR^t$. Hence by the previous theorem, $A*R = 0$, and R is symmetric. So $uR+vA$ and $xI+yAR$ are the required amicable orthogonal designs.

Now if $xU+yV$ and $uN+vM$ are amicable orthogonal designs of order n and types $((1,n-1);(1,n-1))$, we pre- and post-multiply both matrices by monomial matrices P and Q, where $I = PUQ$. Then the amicable matrices can be written in the form $xI+yPVQ = xI+yS$ and $uPNQ+vPMQ = uR+vA$. Now the amicability and orthogonality gives us R and A are symmetric and $AR^t = -RA^t$. We now assume A has weight $n-1$ and no zero on the diagonal. Then considering the orthogonality conditions on the rows of $A+R$ leads to a contradiction, and we have the result.

REMARK. We note that this proof also shows that the existence of a symmetric weighing matrix C of order n and weight $n-1$ with a zero on the diagonal is equivalent to the existence of a best pair of weights $(n-1, 1)$ in order n. (See Definition 5.136.)

We recall (Theorem 5.52) that amicable orthogonal designs of types $((1,p);(1,p))$ do exist in orders $p+1$ when $p \equiv 3 \pmod 4$ is a prime power. Hence we have:

5.134. COROLLARY. There is a symmetric weighing matrix with a zero on the diagonal of order $p+1$ and weight p where $p \equiv 3 \pmod 4$ is a prime power.

We use Theorem 5.44 to show that symmetric weighing matrices of order n and weight $n-1$ with a zero on the diagonal do not always exist since:

5.135. THEOREM. There is no best pair of weights (15,1) in order 16. Equivalently, there is no symmetric weighing matrix of order 16 and weight 15 with a zero on the diagonal.

Proof. The proof is long and combinatorial. It is given in detail in Robinson [7].

§5.11. A GENERALIZATION OF AMICABILITY--FAMILIES

In the algebraic theory for both orthogonal designs and amicable orthogonal designs, we were concerned with families whose members satisfied given conditions. In Lemma 5.12 we saw a useful family of weighing matrices. Product designs and repeat designs are further kinds of families. Furthermore, constructions such as 5.116 and 5.117 only become powerful if matrices such as the $\{M_1, M_2, N\}$ and $\{P_1, P_2, P_3, H\}$, respectively, mentioned there exist. We shall show later how the results of this section can be used in these constructions.

Let

$$K = \begin{bmatrix} 0 & 1 \\ -1 & 0 \end{bmatrix}, \quad L = \begin{bmatrix} 1 & 0 \\ 0 & -1 \end{bmatrix}, \quad M = \begin{bmatrix} 0 & 1 \\ 1 & 0 \end{bmatrix}, \quad I = \begin{bmatrix} 1 & 0 \\ 0 & 1 \end{bmatrix} \quad \text{and}$$

$$H = \begin{bmatrix} 1 & 1 \\ 1 & -1 \end{bmatrix}.$$

We use + for +1 and - for -1. Also we use I for the identity matrix.

5.136. DEFINITION. Matrices A, B which satisfy $AB^t = BA^t$ will be said to be amicable. A best pair is a pair of amicable weighing matrices (A, B) of weights i, j, respectively, satisfying

$$A^t = -A, \quad B^t = B, \quad AB^t = BA^t.$$

A best pair family of order n is a set of best pairs (A_i, B_j) of order n and weights i and j where

$$i = 1, 2, 3, \ldots, n-1, \quad j = 1, 2, 3, \ldots, n \quad \text{and}$$

$$A_i^t = -A_i, \quad B_j^t = B_j, \quad A_iB_j^t = B_jA_i^t.$$

5.137. DEFINITION. For convenience we will call a repeat design $(P;(R;S);H)$ of type $(1;(r;s);h)$ in order n a triplet when $P = I$. Alternatively, a triplet is three weighing matrices (R, S, H) of order n and weights (r, s, h), respectively, which are pairwise amicable; R, S are skew-symmetric, and H is symmetric.

We note that:

5.138. LEMMA. There are triplets in orders $n = 2$ and 4 for weights (i, j, k) where $i, j = 1, 2, \ldots, n-1$ and $k = 1, 2, \ldots, n$. Hence, there is a best pair family of orders 2 and 4.

Proof. For order 2 consider the pairs (K, M) and (K, H). The required matrices for order 4 are (the weights are given in brackets):

1) $(1,2,1)$ $\{K\times I,\ K\times H;\ L\times I\}$,
2) $(1,2,2)$ $\{K\times I,\ K\times H;\ L\times H\}$,
3) $(1,2,3)$ $\{K\times I,\ K\times H;\ L\times H+M\times I\}$,
4) $(1,2,4)$ $\{K\times I,\ K\times H;\ H\times H\}$,
5) $(1,3,1)$ $\{K\times L,\ K\times I+H\times K;\ M\oplus -L\}$,
6) $(1,3,2)$ $\{K\times L,\ L\times I+H\times K;\ H\times L\}$,
7) $(1,3,3)$ $\{K\times L,\ K\times I+H\times K;\ M\oplus K+I\times M+K\times K\}$,
8) $(1,3,4)$ $\{K\times L,\ L\times I+H\times K;\ L\times I+M\times L+I\times M+K\times K\}$,
9) $(2,3,1)$ $\{I\times L+L\times M,\ L\times I+H\times L;\ L\oplus -M\}$,
10) $(2,3,2)$ $\{K\times H,\ L\times K+M\times K+K\times I;\ M\times I-L\times I\}$,
11) $(2,3,3)$ $\{I\times L+L\times M,\ L\times I+H\times L;\ M\oplus L+I\times M+L\times L\}$,
12) $(2,3,4)$ $\{I\times K+K\times M,\ K\times I+H\times K;\ L\times I+M\times L+I\times M+K\times K\}$.

Because of the extremely powerful constructions that arise from repeat designs, we wished to extend this lemma to higher powers of two. This effort led to the results that follow:

5.139. CONSTRUCTION. If A, B, C is a triplet of weights $(a, b, 1)$ in order n, then (A, B, AC) and $(A, B, AC+C)$ are triplets of weights (a, b, a) and $(a, b, a+1)$.

Since by Lemma 5.65 there is no best pair of weights $(7,5)$ in order 8, we have, regarding a best pair (A, B) as a triplet (X, A, B):

5.140. COROLLARY. There are no triplets of weight (x, 7, 5), (4, 7, 1) or (5, 7, 1) in order 8.

We are grateful to Amnon Neeman for the proof of another result using Lemma 5.44.

5.141. THEOREM. Let $n \equiv 0 \pmod 8$. Then it is not possible to have triplets of weights $(a, n-1, 1)$ in order n, where $a = n-4, n-3, n-2$, or of weights (3, 7, 1) in order 8.

Proof. Lemma 5.44 allows us to consider the triplet (X, Y, Z) where $Z = \bigoplus_{\frac{1}{2}n-1} \begin{bmatrix} 0 & 1 \\ 1 & 0 \end{bmatrix} \oplus \begin{bmatrix} -1 & 0 \\ 0 & 1 \end{bmatrix}$. The result follows by a careful combinatorial argument.

In fact, in order 8 we can say:

5.142. LEMMA. In order 8 all triplets (R, S, H) of weights (r, s, h), $0 \le r, s \le 7$, $0 \le k \le 8$, exist except

i) (3, 7, 1), (4, 7, 1), (5, 7, 1), (6, 7, 1) and (r, 7, 5), $1 \le r \le 6$, which do not exist, and possibly,

ii) (5, 7, 2), (6, 7, 2), (6, 7, 4), (3, 7, 3), (4, 7, 3), (5, 6, 3), (5, 7, 3), (6, 7, 3), (1, 5, 7), (3, 5, 7), (3, 7, 7), (4, 7, 7), (5, 7, 7), (6, 7, 7), which are undecided.

Proof. Part i) follows from the previous corollary and Theorem 5.141 which shows that (3, 7, 1) and (6, 7, 1) do not exist. Lemma 5.127, i), iii), iv), v) and the above construction give all those that exist except (3, 5, 1), (3, 5, 5), (1, 5, 1), (1, 7, 1), (2, 7, 1) and (1, 7, 3). Now we note that a repeat design of type $(r;(p_1, \ldots; q); w_1, \ldots)$ in order n gives a repeat design of type $(r;(q; p_1, \ldots, q); w_1, \ldots)$ in order $2n$ by Lemma 5.127. Hence the repeat design of type (1;(2;3);1) in order 4 gives the (1;(3;5);1) in order 8 and hence the triplet (3, 5, 1) and by Construction 5.139 the (3, 5, 5) in 8. We now give specific constructions for (1, 5, 1), (1, 7, 1), (2, 7, 1) and (1, 7, 3).

$$\text{Let } S = \begin{bmatrix} 0 & 1 & 0 & 0 \\ 0 & 0 & 1 & 0 \\ 0 & 0 & 0 & 1 \\ - & 0 & 0 & 0 \end{bmatrix}, \quad R = \begin{bmatrix} 0 & 0 & 0 & 1 \\ 0 & 0 & 1 & 0 \\ 0 & 1 & 0 & 0 \\ 1 & 0 & 0 & 0 \end{bmatrix}, \quad R^2 = I,$$

and K, L, M be as defined above. Write $E = -S+S^2+S^3$ and $G = S+S^2+S^3$. Then $(L\times S^2,\ I_2\times(S+S^3) + K\times ER,\ L\times S^2)$ and $(L\times S^2,\ L\times G+K\times(E+I)^t R\ (S^3+S-I)^t R \oplus (-S^2-S+I)^t R)$ are (1,5,1) and (1,7,3), respectively. Amnon Neeman found the following (1,7,1):

$$\left[\begin{array}{cccc|cccc} 0 & 1 & & & & & & \\ - & 0 & & & & & & \\ & & 0 & 1 & & & & \\ & & - & 0 & & & & \\ \hline & & & & 0 & 1 & & \\ & & & & - & 0 & & \\ & & & & & & 0 & 1 \\ & & & & & & - & 0 \end{array}\right], \quad \left[\begin{array}{cccc|cccc} 0 & 1 & 1 & - & - & 1 & 1 & 1 \\ - & 0 & 1 & 1 & - & - & - & 1 \\ - & - & 0 & 1 & 1 & 1 & 1 & 1 \\ 1 & - & - & 0 & - & 1 & - & 1 \\ \hline 1 & 1 & - & 1 & 0 & - & 1 & 1 \\ - & 1 & - & - & 1 & 0 & - & 1 \\ - & 1 & - & 1 & - & 1 & 0 & - \\ - & - & - & - & - & - & 1 & 0 \end{array}\right],$$

$$\left[\begin{array}{cccc|cccc} 0 & 0 & 0 & 1 & & & & \\ 0 & 0 & 1 & 0 & & & & \\ 0 & 1 & 0 & 0 & & & & \\ 1 & 0 & 0 & 0 & & & & \\ \hline & & & & 0 & 1 & 0 & 0 \\ & & & & 1 & 0 & 0 & 0 \\ & & & & 0 & 0 & 1 & 0 \\ & & & & 0 & 0 & 0 & - \end{array}\right].$$

The following three matrices give a (2,7,1):

$$\left[\begin{array}{cccc|cccc} 0 & 1 & 1 & 0 & & & & \\ - & 0 & 0 & - & & & & \\ - & 0 & 0 & 1 & & & & \\ 0 & 1 & - & 0 & & & & \\ \hline & & & & 0 & 0 & 1 & 1 \\ & & & & 0 & 0 & 1 & - \\ & & & & - & - & 0 & 0 \\ & & & & - & 1 & 0 & 0 \end{array}\right], \quad \left[\begin{array}{cccc|cccc} 0 & 1 & 1 & - & 1 & - & - & 1 \\ - & 0 & 1 & - & - & 1 & 1 & 1 \\ - & - & 0 & 1 & 1 & 1 & - & 1 \\ 1 & 1 & - & 0 & 1 & 1 & 1 & 1 \\ \hline - & 1 & - & - & 0 & 1 & - & - \\ 1 & - & - & - & - & 0 & - & 1 \\ 1 & - & 1 & - & 1 & 1 & 0 & - \\ - & - & - & - & 1 & - & 1 & 0 \end{array}\right],$$

$$\left[\begin{array}{cccc|cccc} 0 & 1 & 0 & 0 & & & & \\ 1 & 0 & 0 & 0 & & & & \\ 0 & 0 & 0 & 1 & & & & \\ 0 & 0 & 1 & 0 & & & & \\ \hline & & & & 0 & - & 0 & 0 \\ & & & & - & 0 & 0 & 0 \\ & & & & 0 & 0 & 1 & 0 \\ & & & & 0 & 0 & 0 & - \end{array}\right].$$

This gives the results of the enunciation after using Construction 5.139 to obtain (1,5,5), (1,7,7) and (2,7,7).

REMARK. This lemma indicates that the existence problem for triplets (R,S,H) which are repeat designs (I;(R;S);H) is very difficult and far from resolved.

But this lemma does allow us to say:

5.143. COROLLARY. There are best pairs for all (a,b), $0 \leq a \leq 7$, $0 \leq b \leq 8$, in order 8 except (7,5). There are amicable weighing matrices for all (a,b), $0 \leq a, b \leq 8$, in order 8.

The results on repeat designs §5.10 allow us to say:

5.144. THEOREM. There are amicable weighing matrices for all (a,b), $0 \leq a, b \leq 2^t$, in order 2^t.

Now Theorem 5.60, together with the results quoted above, let us say:

5.145. LEMMA. In order 16 there exist best pairs (a,b) (= repeat designs (1;(a);b)) for all a = 1,2,...,15 and b = 1,2,...,16 except possibly the pairs (a,b): (13,1), (13,5), (13,9), (15,7), (15,9), (15,15), which are undecided, and (15,1), which does not exist.

Now, as promised, we apply these results to Constructions 5.116 and 5.117 to obtain:

5.146. CONSTRUCTION. Suppose there is a triplet of weights (a,b,c) in order n. Then there is an orthogonal design of type

i) (1,1,a,b) in order 2n,

and when c = 1, of types,

ii) (1,1,1,a,a,a,c), (1,1,1,a,a,b,c) in order 4n.

5.147. CONSTRUCTION. Suppose there is a best pair of weights (a,b) in order n. Further suppose there is a product design of type

$(a_1, \ldots, a_p; b_1, \ldots, b_q; c_1, \ldots, c_r)$ in order m. Then there is an orthogonal design of type $(a_1, \ldots, a_p, ab_1, \ldots, ab_q, bc_1, \ldots, bc_r)$ in order mn.

5.148. EXAMPLE. There is a product design of type $(1,1,1,1,2,4,\ldots,2^{t-4}; 2, 2^{t-3}; 2, 4, \ldots, 2^{t-4}, 2^{t-3}, 2^{t-3})$ in order 2^t. So using a best pair of weights (a, b) in order n gives an orthogonal design of type $(1,1,1,1,2,4,\ldots,2^{t-4}, 2a, 2^{t-3}a, 2b, 4b, \ldots, 2^{t-4}b, 2^{t-3}b, 2^{t-3}b)$ in order $2^t n$.

CHAPTER 6

ROBINSON'S THEOREM

In trying to decide which orthogonal designs to look for, it would be useful to formulate, and hopefully prove valid, some general principles of the sort, "All orthogonal designs of a certain type exist in certain orders." The Hadamard conjecture, the skew-Hadamard conjecture, the weighing matrix conjecture, and other conjectures that have been made, and extensively verified, provide some solid information which must be dealt with in order to state such principles. We have been singularly unsuccessful in formulating correct principles of a general nature; some conjectures that we have made in the light of those principles have proved to be false.

One principle we had bandied about for awhile was: "In order n, if k is much smaller than $\rho(n)$, then all orthogonal designs on k variables exist in order n." Of course, the key to focusing on this principle is to decide what "much smaller" should mean.

In orders n where n is odd or $n = 2t$, t odd, algebraic conditions appear immediately in deciding if orthogonal designs exist, and so, in terms of deciding what "much smaller" should mean, we put those cases aside. When $n = 4t$, t odd, the situation is different. The algebraic theory says nothing about one-variable designs, i.e., weighing matrices. This fact, coupled with a fair bit of evidence for the weighing matrix conjecture in orders $4t$, t odd, led us to formulate a "sub-principle" for the phrase "much smaller": to wit, we proposed the following: "If, in order n, the algebraic theory imposes <u>no</u> restrictions on any possible k-variable design in order n, then all k-variable designs exist in order n."

If this principle were a sound one, it would say, for example, that whenever $n = 16t$, $t \geq 1$, any orthogonal design on ≤ 7 variables exists. (See Proposition 3.80 and what follows it.) The principle, unfortunately (depending on your point of view), is far from correct. Peter J. Robinson decisively settled that issue and many other alternative ones with the following remarkable theorem.

6.1. THEOREM. (Robinson). If $n > 40$, there is no orthogonal design of type $(1,1,1,1,1,n-5)$ in order n.

PROOF. The proof is a very careful analysis of what such a design would have to look like, and we have expanded Robinson's proof so as to make the verifications a bit easier for the reader.

With no loss of generality we may assume the first 4×4 diagonal block of the orthogonal design is

$$\begin{bmatrix} x_1 & x_2 & x_3 & a_1 \\ -x_2 & x_1 & a_1 & -x_3 \\ -x_3 & -a_1 & x_1 & x_2 \\ -a_1 & x_3 & -x_2 & x_1 \end{bmatrix}.$$

Either $a_1 = \pm x_4$, or not.

If $a_1 = \pm x_4$, we proceed to obtain the following 8×8 diagonal block:

$$\begin{bmatrix} x_1 & x_2 & x_3 & a_1 & x_5 & & & \\ -x_2 & x_1 & a_1 & -x_3 & & -x_5 & & * \\ -x_3 & -a_1 & x_1 & x_2 & & & -x_5 & \\ -a_1 & x_3 & -x_2 & x_1 & & * & & x_5 \\ -x_5 & & & & x_1 & x_2 & x_3 & b_1 \\ & x_5 & * & & -x_2 & x_1 & b_1 & -x_3 \\ * & & x_5 & & -x_3 & -b_1 & x_1 & x_2 \\ & & & -x_5 & -b_1 & x_3 & -x_2 & x_1 \end{bmatrix} \qquad (+)$$

and $b_1 = -a_1$.

In case $a_1 \neq \pm x_4$, we may proceed to make the following 8×8 diagonal block.

$$\begin{bmatrix} x_1 & x_2 & x_3 & a_1 & x_4 & & & \\ -x_2 & x_1 & a_1 & -x_3 & & -x_4 & * & \\ -x_3 & -a_1 & x_1 & x_2 & * & & -x_4 & \\ -a_1 & x_3 & -x_2 & x_1 & & & & x_4 \\ -x_4 & & & & x_1 & x_2 & x_3 & b_1 \\ & x_4 & * & & -x_2 & x_1 & b_1 & -x_3 \\ & * & x_4 & & -x_3 & -b_1 & x_1 & x_2 \\ & & & -x_4 & -b_1 & x_3 & -x_2 & x_1 \end{bmatrix}, \qquad (\dagger\dagger)$$

and $b_1 = -a_1$.

We continue the process and end up with 8×8 blocks of type $(\dagger)$ or $(\dagger\dagger)$ on the diagonal.

CLAIM. There is at most one diagonal block of type $(\dagger)$.

<u>Proof of Claim.</u> Let X denote the orthogonal design of type $(1,1,1,1,1,n-5)$ in order n, and write $X = A_1x_1 + A_2x_2 + A_3x_3 + \dots + A_6x_6$. As we have already seen, we may assume $A_1 = I_n$, $A_2 = \bigoplus_{n/2} \begin{bmatrix} 0 & 1 \\ -1 & 0 \end{bmatrix}$ and

$$A_3 = \bigoplus_{n/4} \begin{bmatrix} 0 & 0 & 1 & 0 \\ 0 & 0 & 0 & -1 \\ -1 & 0 & 0 & 0 \\ 0 & 1 & 0 & 0 \end{bmatrix}.$$

Since X is an orthogonal design, we must have $A_iA_j^t + A_jA_i^t = 0$, $1 \leq i \neq j \leq 6$. The patient reader will then discover that these conditions force A_4, A_5, A_6 to each be skew-symmetric and partitioned into 4×4 blocks, where if $(pqrs)$ is the first row of a block, then the block looks like:

$$\begin{bmatrix} p & q & r & s \\ q & -p & s & -r \\ r & -s & -p & q \\ -s & -r & q & p \end{bmatrix}.$$

Now, assume that there are two 8×8 diagonal blocks like $(\dagger)$, with $a_1 = \pm x_4$, $b_1 = -a_1$, $a_2 = \pm x_4$, $b_2 = -a_2$; we shall obtain a contradiction (we shall just consider the case $a_1 = a_2 = x_4$ and leave the remaining three possibilities for the reader to check). The contradiction shall be obtained by looking at the 8×8 off-diagonal blocks which are at the juncture of the two diagonal positions.

$$\begin{bmatrix} x_1 & x_2 & x_3 & x_4 & & & & \\ -x_2 & x_1 & x_4 & -x_3 & & & * & \\ -x_3 & -x_4 & x_1 & x_2 & & & & \\ -x_4 & x_3 & -x_2 & x_1 & & & & \\ & & & & x_1 & x_2 & x_3 & -x_4 \\ & * & & & -x_2 & x_1 & -x_4 & -x_3 \\ & & & & -x_3 & x_4 & x_1 & x_2 \\ & & & & x_4 & x_3 & -x_3 & x_1 \end{bmatrix} \sim \begin{bmatrix} p & q & r & s & b & c & d & e \\ q & -p & s & -r & c & -b & e & d \\ r & -s & -p & q & d & -e & -b & c \\ -s & -r & q & p & -e & -d & c & b \\ f & g & h & j & k & l & m & v \\ g & -f & j & -h & l & -k & v & -m \\ h & -j & -f & g & m & -v & -k & l \\ -j & -h & g & f & -v & -m & l & k \end{bmatrix}$$

$$\sim \begin{bmatrix} x_1 & x_2 & x_3 & x_4 & & & & \\ -x_2 & x_1 & x_4 & -x_3 & & & * & \\ -x_3 & -x_4 & x_1 & x_2 & & & & \\ -x_4 & x_3 & -x_2 & x_1 & & & & \\ & & & & x_1 & x_2 & x_3 & -x_4 \\ & * & & & -x_2 & x_1 & -x_4 & -x_3 \\ & & & & -x_3 & x_4 & x_1 & x_2 \\ & & & & x_4 & x_3 & -x_2 & x_1 \end{bmatrix} \sim \begin{bmatrix} -p & -q & -r & s & -f & -g & -h & j \\ -q & p & s & r & -g & f & j & h \\ -r & -s & p & -q & -h & -j & f & -g \\ -s & r & -q & p & -j & h & -g & -f \\ -b & -c & -d & e & -k & -l & -m & v \\ -c & b & e & d & -l & k & v & m \\ -d & -e & b & -c & -m & -v & k & -l \\ -e & d & -c & -b & -v & m & -l & -k \end{bmatrix}$$

Display 6.1

If we number the rows of Display 6.1 from 1 to 16, then we obtain (by taking the inner product of rows 1 and 9) a summand $2sx_4$ which we cannot eliminate since x_4 appears only once in each row and column.

(Note: We have used here the fact that $s \neq 0$, but we would get the same result if <u>any</u> of p, q, r or s were $\neq 0$ or if any of k, l, m, v were $\neq 0$. Thus, so far, we could be discussing orthogonal designs of type $(1,1,1,1,1,n-k)$, $5 \leq k \leq 8$, and the conclusions that are drawn would still hold; i.e., no two diagonal blocks of type (+).)

We have now seen that there can only be one diagonal block of type (+) and that all other diagonal blocks are of type (++).

We now seek to discover where in the orthogonal design the x_5's are located. If there is a diagonal block of type (+), then we know where the x_5's in the rows and columns controlled by that diagonal block are, namely, in the diagonal block.

CLAIM. The x_5's are always in the diagonal blocks.

<u>Proof.</u> To prove this, it would be enough to show that there is no x_5 in an off-diagonal block which is above <u>and</u> across from a diagonal block of type (++). Thus, we have the following:

$$\begin{bmatrix} x_1 & x_2 & x_3 & \alpha & x_4 & & & \\ -x_2 & x_1 & \alpha & -x_3 & & -x_4 & & \\ -x_3 & -\alpha & x_1 & x_2 & & & -x_4 & \\ -\alpha & x_3 & -x_2 & x_1 & & & & x_4 \\ -x_4 & & & & x_1 & x_2 & x_3 & -\alpha \\ & x_4 & & & -x_2 & x_1 & -\alpha & -x_3 \\ & & x_4 & & -x_3 & \alpha & x_1 & x_2 \\ & & & -x_4 & \alpha & x_3 & -x_2 & x_1 \end{bmatrix} \sim \begin{bmatrix} p & q & r & s & b & c & d & e \\ q & -p & s & -r & c & -b & e & -d \\ r & -s & -p & q & d & -e & -b & c \\ -s & -r & q & p & -e & -d & c & b \\ f & g & h & j & k & l & m & v \\ g & -f & j & -h & l & -k & v & -m \\ h & -j & -f & g & m & -v & -k & l \\ -j & -h & g & f & -v & -m & l & k \end{bmatrix} = Y$$

$$\wr \qquad\qquad\qquad \wr$$

$$-Y^t = \begin{bmatrix} -p & -q & -r & s & -f & -g & -h & j \\ -q & p & s & r & -g & f & j & h \\ -r & -s & p & -q & -h & -j & f & -g \\ -s & r & -q & -p & -j & h & -g & -f \\ -b & -c & -d & e & -k & -l & -m & v \\ -c & b & e & d & -l & k & v & m \\ -d & -e & b & -c & -m & -v & k & -l \\ -e & d & -c & -b & -v & m & -l & -k \end{bmatrix} \sim \begin{bmatrix} x_1 & x_2 & x_3 & \beta & x_4 & & & \\ -x_2 & x_1 & \beta & -x_3 & & -x_4 & & \\ -x_3 & -\beta & x_1 & x_2 & & & -x_4 & \\ -\beta & x_3 & -x_2 & x_1 & & & & x_4 \\ -x_4 & & & & x_1 & x_2 & x_3 & -\beta \\ & x_4 & & & -x_2 & x_1 & -\beta & -x_3 \\ & & x_4 & & -x_3 & \beta & x_1 & x_2 \\ & & & -x_4 & \beta & x_3 & -x_2 & x_1 \end{bmatrix}$$

$$A_\alpha \sim A_\beta$$

Display 6.2

Now, by checking inner products (just using x_4's), we find that

$$b x_4 - f x_4 = 0, \quad \text{i.e.,} \quad b = f;$$
$$-c x_4 - g x_4 = 0, \quad \text{i.e.,} \quad c = -g;$$
$$-d x_4 - h x_4 = 0, \quad \text{i.e.,} \quad d = -h;$$
$$e x_4 - j x_4 = 0, \quad \text{i.e.,} \quad e = j.$$

Similarly,

$$k = -p,$$
$$l = q,$$
$$m = r,$$
$$v = -s;$$

i.e., we have

$$Y = \left[\begin{array}{cccc|cccc} p & q & r & s & b & c & d & e \\ & \text{etc.} & & & & \text{etc.} & & \\ \hline & \text{etc.} & & & & \text{etc.} & & \\ -e & d & -c & b & s & -r & q & -p \end{array}\right].$$

If we consider the inner product between the two rows and recall that none of p, q, r, s, b, c, d, or e is $= 0$, we find that $\pm x_5 \notin \{p, q, r, s, b, c, d, e\}$. Thus x_5 may only appear in diagonal blocks.

Now, the i-th diagonal block of type $(++)$ has four (as yet) undetermined entries, and we have seen that one of them must be $\pm x_5$. If there are four (or more) blocks of type $(++)$, then in two of them x_5 (up to sign) must occupy the same position. We shall assume that occurs in the i-th and j-th diagonal blocks and write them (along with the (i, j)- and (j, i)-off-diagonal blocks) below.

$$\begin{bmatrix} x_1 & x_2 & x_3 & \alpha_1 & x_4 & \beta_1 & \gamma_1 & \Delta_1 \\ -x_2 & x_1 & \alpha_1 & -x_3 & \beta_1 & -x_4 & \Delta_1 & -\gamma_1 \\ -x_3 & -\alpha_1 & x_1 & x_2 & \gamma_1 & -\Delta_1 & -x_4 & \beta_1 \\ -\alpha_1 & x_3 & -x_2 & x_1 & -\Delta_1 & -\gamma_1 & \beta_1 & x_4 \\ -x_4 & -\beta_1 & -\gamma_1 & \Delta_1 & x_1 & x_2 & x_3 & -\alpha_1 \\ -\beta_1 & x_4 & \Delta_1 & \gamma_1 & -x_2 & x_1 & -\alpha_1 & -x_3 \\ -\gamma_1 & -\Delta_1 & x_4 & -\beta_1 & -x_3 & \alpha_1 & x_1 & x_2 \\ -\Delta_1 & \gamma_1 & -\beta_1 & -x_4 & \alpha_1 & x_3 & -x_2 & x_1 \end{bmatrix} \sim \begin{bmatrix} p & q & r & s & b & c & d & e \\ q & -p & s & -r & c & -b & e & -d \\ r & -s & -p & q & d & -e & -b & c \\ -s & -r & q & p & -e & -d & c & b \\ b & -c & -d & e & -p & q & r & -s \\ -c & -b & e & d & q & p & -s & -r \\ -d & -e & -b & -c & r & s & p & q \\ -e & d & -c & b & s & -r & q & -p \end{bmatrix}$$

$$\begin{bmatrix} -p & -q & -r & s & -b & c & d & e \\ -q & p & s & r & c & b & e & -d \\ -r & -s & p & -q & d & -e & b & c \\ -s & r & -q & -p & -e & -d & c & -b \\ -b & -c & -d & e & p & -q & -r & -s \\ -c & b & e & d & -q & -p & -s & r \\ -d & -e & b & -c & -r & s & -p & -q \\ -e & d & -c & -b & s & r & -q & p \end{bmatrix} \sim \begin{bmatrix} x_1 & x_2 & x_3 & \alpha_2 & x_4 & \beta_2 & \gamma_2 & \Delta_2 \\ -x_2 & x_1 & \alpha_2 & -x_3 & \beta_2 & -x_4 & \Delta_2 & -\gamma_2 \\ -x_3 & -\alpha_2 & x_1 & x_2 & \gamma_2 & -\Delta_2 & -x_4 & \beta_2 \\ -\alpha_2 & x_3 & -x_2 & x_1 & -\Delta_2 & -\gamma_2 & \beta_2 & x_4 \\ -x_4 & -\beta_2 & -\gamma_2 & \Delta_2 & x_1 & x_2 & x_3 & -\alpha_2 \\ -\beta_2 & x_4 & \Delta_2 & \gamma_2 & -x_2 & x_1 & -\alpha_2 & -x_3 \\ -\gamma_2 & -\Delta_2 & x_4 & -\beta_2 & -x_3 & \alpha_2 & x_1 & x_2 \\ -\Delta_2 & \gamma_2 & -\beta_2 & -x_4 & \alpha_2 & x_3 & -x_2 & x_1 \end{bmatrix}$$

Display 6.3

Now, suppose β_1 and β_2 are each $\pm x_5$. Considering rows 1 and 9 of Display 6.3 we find $\beta_1 c+\beta_2 c = 0$, i.e., $\beta_1 = -\beta_2$, but considering rows 1 and 11 we find $-\beta_1 e+\beta_2 e = 0$, i.e., $\beta_1 = \beta_2$. This contradiction establishes that β_1, β_2 cannot both have absolute value x_5. A similar argument gives the same conclusion for γ_1 and γ_2, Δ_1 and Δ_2, and α_1 and α_2. This then completes the proof.

It is possible to use Robinson's Theorem to obtain many other non-existence results. We give just two illustrations.

6.2. COROLLARY. There do not exist amicable orthogonal designs of type $((1,1,m-2);(1,m-1))$ in any order $m > 10$.

Proof. If there did, we could use Construction 5.88 (i.e., a product design of type $(1,1,1;1,1,1;1)$ in order 4) to obtain an orthogonal design in order $4m > 40$ of type $(1,1,m-2,1,1,1,m-1,m-1,m-1)$, contradicting Theorem 6.1.

NOTE. This shows how difficult it is to obtain "full" amicable orthogonal designs which have several 1's in their types.

6.3. COROLLARY. There is no product design of type $(1,1,n-3;1,n-2;1)$ in any order $n > 20$.

Proof. If there were, Construction 5.116 would contradict Robinson's Theorem.

REMARK. This corollary shows how special the product designs constructed in Examples 5.90, 5.91 and Construction 5.88 really are.

CHAPTER 7

THE EXISTENCE OF HADAMARD MATRICES and ASYMPTOTIC EXISTENCE RESULTS FOR ORTHOGONAL DESIGNS

Hadamard matrices were first studied by Sylvester [1] in 1867. In 1893 Hadamard discovered that if $X = (x_{ij})$ is a matrix of order n, then

$$|\det X|^2 \leq \prod_{i=1}^{n} \sum_{j=1}^{n} |x_{ij}|^2 .$$

Hadamard showed that matrices satisfying the equality and with entries in the unit disc (i.e., $|x_{ij}| \leq 1$) have order 1, 2 or $\equiv 0 \pmod 4$ and entries $\{1, -1\}$. He produced examples for orders up to 20. Subsequently, matrices which satisfy the equality of Hadamard's inequality came to be known as Hadamard matrices. Such matrices were further studied by Scarpis in 1898.

The next major work was in 1933 by R. E. A. C. Paley, a brilliant young mathematician who was killed soon afterwards in a skiing accident in Canada. Paley was led to this subject by Todd and Coxeter (whose papers follow his in the journal) in connection with a problem in polytopes.

In 1944 and 1947 Williamson obtained further results of considerable interest.

Many disciplines have an interest in Hadamard matrices: statistics--because they give symmetric balanced incomplete block designs and optimal weighing designs; engineering (in information theory and signal processing)--because they give codes which correct the maximum possible number of errors and some sequences of use in digital communications;

optics--to improve the performance of spectrometers and image scanners.

Thus, in the post-war period, Hadamard matrices have been studied considerably, and many contributions made towards proving the Hadamard conjecture:

> "There exists an Hadamard matrix of order 1, 2 and 4t for every positive integer t."

The first section of this chapter is devoted to proving the most powerful theorem to date on the existence of Hadamard matrices.

The ideas of the first section are continued in the second and third sections to obtain asymptotic existence theorems for some types of orthogonal designs.

§7.1. EXISTENCE OF HADAMARD MATRICES

As in many other combinatorial design problems, recursive constructions are of very great use in constructing orthogonal designs and Hadamard matrices. In fact, in Theorem 7.13 this method gives the most powerful method for constructing Hadamard matrices now known.

7.1. LEMMA. Suppose all orthogonal designs of type $(a, b, n-a-b)$, $0 \leq a+b \leq n$, exist in order n. Then all orthogonal designs of type $(x, y, 2n-x-y)$, $0 \leq x+y \leq 2n$, exist in order $2n$.

Proof. Recall that from Lemma 4.11(ii) we have that the existence of the design of type $(a, b, n-a-b)$ in order n implies the existence of the design of type $(a, a, 2b, 2(n-a-b))$ in order $2n$.

Let (u, v, w) be a full orthogonal design in order $2n$ (we may assume $0 < u \leq v \leq w$). It follows that $v < n$. We distinguish four cases, depending on whether u or v are even or odd.

Case 1: u even, v even.

Then $u = 2a$, $v = 2b$, and $a+b < n$. Then $(a, b, n-a-b)$ exists in order n, and we may apply Lemma 4.11(ii).

Case 2: u even, v odd.

Write $u = 2a$, $v = 2a+\ell$. Then $a+v = 3a+\ell \leq n$, for in this case $w = 2n-4a-\ell$, and $v \leq w \Rightarrow 3a+\ell \leq n$.

Thus, $(v, a, n-a-v)$ exists in order n, and we may apply

Lemma 4.11(ii) to it.

Case 3: u odd, v even.

Write $u = 2a+1$, $v = 2b$. Then $w = 2t+1$ and $u+v+w = 2n$, and so $a+b+t+1 = n$. Notice that $u+b \leq n$, for if not, $u+b > n$, which implies $a > t$.

Thus, there is an orthogonal design of type $(u, b, n-u-b)$ in order n, and we may proceed as before.

Case 4: u odd, v odd.

In this case, write $v = u+2b$, $b \geq 0$. Clearly, $u+b \leq n$, and so there is an orthogonal design of type $(u, b, n-u-b)$ in order n, and we're done.

7.2. COROLLARY. Since all the orthogonal designs of type $(a, b, 4-a-b)$ exist in order 4 for $0 \leq a+b \leq 4$, we have all orthogonal designs of type $(x, y, n-x-y)$ for $0 \leq x+y \leq n$ whenever n is a power of 2.

7.3. COROLLARY. Since all orthogonal designs of type $(x, y, n-x-y)$, $0 \leq x+y \leq n$, exist in all orders for which n is a power of 2, all designs of type (z, w), $0 \leq z+w \leq n$, exist in all orders n which are powers of 2.

7.4. COROLLARY. All orthogonal designs of type $(a, b, 24-a-b)$, $0 \leq a+b \leq 24$, exist in order 24. Hence all orthogonal designs of type $(x, y, m-x-y)$, $0 \leq x+y \leq m$, exist in orders $2^t.3 = m$, $t \geq 3$ a positive integer.

7.5. COROLLARY. All orthogonal designs of type (a, b), $0 \leq a+b \leq 2^t.3$, exist in orders $2^t.3$, $t \geq 3$ a positive integer.

CONJECTURE. (Seberry). All orthogonal designs of type $(x, y, m-x-y)$, $0 \leq x+y \leq m$, exist in orders $m = 2^t q$ for any natural number q and sufficiently large t.

REMARK. This conjecture has not yet been verified for $q = 5$, but the results which are known are given in Appendix F.

Another recursive lemma, but one which is often helpful, is:

7.6. LEMMA. If all orthogonal designs $(1, i, j)$, $1+i+j \leq n$, exist in order n, then all orthogonal designs (ℓ, m), $\ell+m \leq 2n$, exist in order $2n$.

Proof. The design $(1, \frac{1}{2}(\ell-1), \frac{1}{2}(m-1))$ in order n gives, by Lemma 4.11, the design $(1, 1, \ell-1, m-1)$ and hence (ℓ, m) in order $2n$ if ℓ and m are odd. $(1, \frac{1}{2}(\ell-1), \frac{1}{2}m)$ in order n gives $(1, 1, \ell-1, m)$ and hence (ℓ, m) in order $2n$ if ℓ is odd and m is even, while $(1, \frac{1}{2}\ell, \frac{1}{2}m)$ gives the result if both ℓ and m are even.

THE EXISTENCE OF HADAMARD MATRICES

The following theorem of Sylvester, which he studied because of a problem posed by Frobenius, is well known.

7.7. THEOREM. Given any two relatively prime integers x and y, every integer $N > (x-1)(y-1)$ can be written in the form $ax+by$ for some nonnegative integers a and b.

7.8. COROLLARY. Given $x = (v+1)$ and $y = (v-3)$, where v is odd and $v \geq 9$, there exist nonnegative integers a and b such that $a(v+1) + b(v-3) = n = 2^t$ for some t.

Proof. Let g be the greatest common divisor of $v+1$ and $v-3$. Then $g = 1$, 2 or 4. If $g \neq 1$, let m be the smallest power of 2 greater than $N = [((v+1)/g)-1][((v-3)/g)-1]$. Then by the theorem, there exist integers a and b such that $a(v+1)/g + b(v-3)/g = m$, and hence we have the corollary, since g is a power of 2.

7.9. LEMMA. Let $v \equiv 3 \pmod 4$ be a prime ≥ 9. Then there exists a t such that an Hadamard matrix exists in every order $2^s.v$ for $s \geq t$.

Proof. Let $x = v+1$ and $y = v-3$; then by the previous corollary there exists an a and b such that $ax+by = n = 2^t$ for some t. Now we know the orthogonal design D of type $(a, b, n-a-b)$ exists with order 2^t on the variables x_1, x_2, x_3.

Then replace each variable x_1 by the matrix J, each variable x_2 by $J-2I$ and each variable x_3 by the back-circulant $(1, -1)$ matrix

$B = (Q+I)R$, where Q is defined in the proof of Lemma 4.34 and R is the back-diagonal matrix, to form a matrix E. Now

$$B^T = B,\quad BJ = J,\quad B(J-2I) = (J-2I)B,\quad BB^T = (v+1)I-J,$$

$$DD^T = (ax_1^2 + bx_2^2 + (n-a-b)x_3^2)I_n,$$

and

$$\begin{aligned} EE^T &= (aJ^2 + b(J-2I)^2 + (n-a-b)BB^T) \times I_n, \\ &= [avJ + 4bI + b(v-4)J + (n-a-b)(v+1)I - (n-a-b)J] \times I_n, \\ &= ([n(v+1) - a(v+1) - b(v-3)]I + [a(v+1) + b(v-3) - n]J) \times I_n, \\ &= nvI_{nv}. \end{aligned}$$

7.10. LEMMA. Let $v \equiv 1 \pmod 4$ be a prime ≥ 9. Then there exists a t such that an Hadamard matrix exists in every order $2^s.v$ for $s \geq t+1$.

Proof. Choose x, y, n, t, a, b and D as in the previous lemma. Now note there exists an orthogonal design F of type $(2a, 2b, n-a-b, n-a-b)$ in order $2n = 2^{t+1}$ on the variables x_1, x_2, x_3, x_4.

Form the matrix E by replacing each variable x_1 of F by J, each variable x_2 of F by $J-2I$, and the variables x_3 and x_4 by the two circulant $(1,-1)$ incidence matrices $X = I+Q$ and $Y = I-Q$, where Q is defined in the proof of Lemma 4.34. Now

$$X^T = X,\quad Y^T = Y,\quad XY^T = YX^T,\quad XX^T+YY^T = 2(v+1)I-2J,$$

$$XJ = YJ = J,\quad X(J-2I) = (J-2I)X,\quad Y(J-2I) = (J-2I)Y,$$

$$FF^T = (2ax_1^2 + 2bx_2^2 + (n-a-b)x_3^2 + (n-a-b)x_4^2)I_{2n},$$

and

$$\begin{aligned} EE^T &= ((2aJ^2 + 2b(J-2I)^2 + (n-a-b)(XX^T+YY^T)) \times I_{2n}, \\ &= [2avJ + 8bI + 2b(v-4)J + (n-a-b)(2(v+1)I - 2J)] \times I_{2n}, \\ &= [2n(v+1) - 2a(v+1) - 2b(v-3)I_{2nv} + [2a(v+1) + 2b(v-3) - 2n]J_v \times I_{2n}, \\ &= 2nvI_{2nv}. \end{aligned}$$

7.11. LEMMA. There exist Hadamard matrices of orders 2, 4, $2^t.2$, $2^t.3$, $2^t.5$, and $2^t.7$ for all positive integers $t \geq 2$.

Proof. From Corollary 4.36.

7.12. THEOREM. Given any integer q, there exists t dependent on q such that an Hadamard matrix exists of every order $2^s q$ for $s \geq t$.

Proof. Decompose q into its prime factors, and apply the previous lemma to each factor. The result follows because the Kronecker product of any two Hadamard matrices is an Hadamard matrix.

In Corollary 7.8 we chose t so that

$$2^t \geq \left(\frac{v+1}{g}-1\right)\left(\frac{v-3}{g}-1\right).$$

Hence, if $v \equiv 1 \pmod 4$, $g = 2$, and we need $2^t \geq \frac{1}{4}(v-1)(v-5)$. Thus, we may choose $t = [2\log_2(v-3)]-1$ in Lemma 7.10.

For $v \equiv 3 \pmod 4$, $g = 4$, and choosing $t = [2\log_2(v-5)]-3$ ensures the existence of an Hadamard matrix of order $2^t.v$ in Lemma 7.9.

We observe that if $v = p \cdot q$ where p and q are primes $\equiv 1 \pmod 4$, we can ensure the existence of an Hadamard matrix of order $2^r.pq$ where $r = [2\log_2(p-3)] + [2\log_2(q-3)] < [2\log_2(pq-3)]$. Since a v comprising a product of primes $\equiv 1 \pmod 4$ would give the highest theoretical t for which an Hadamard matrix of order $2^t v$ exists, we can say:

7.13. THEOREM. (i) Given any natural number q, there exists an Hadamard matrix of order $2^s q$ for every $s \geq [2\log_2(q-3)]$.

(ii) Given any natural number q, there exists a regular (i.e., constant row sum) symmetric Hadamard matrix with constant diagonal of order $2^{2s}q^2$ for s as before.

Proof. Part (ii) of the theorem follows from a theorem of Goethals and Seidel (see Wallis [25, p. 34]) that if there is an Hadamard matrix of order n, there is a regular symmetric Hadamard matrix with constant diagonal of order n^2.

The result that orthogonal designs $(a, b, n-a-b)$ exist in all orders $n = 2^t.3$, $t \geq 3$, may be used in a similar fashion to that employed in Lemma 7.9 to construct Hadamard matrices of order $2^s.3q$ for sufficiently large s. It may happen that proceeding via this result for appropriate natural numbers $3q$ gives a smaller s than if Theorem 7.13 were used.

This last remark indicates that a knowledge of orthogonal designs $(a, b, n-a-b)$ in orders $n = 2^t p$, p odd, could lead to Hadamard matrices of order $2^s pq$ for smaller s than that given by Theorem 7.13.

Clearly, in general, there will be Hadamard matrices, given by the construction, of order $2^t q$ where $t < [2 \log_2(q-3)]$.

The first twenty-six orders q for which an Hadamard matrix of order $4q$ is not known are now listed by giving the smallest power of 2 for which an Hadamard matrix of order $2^t q$ is known:

$2^5.67$, $2^3.103$, $2^{10}.107$, $2^3.127$, $2^3.151$, $2^3.163$, $2^4.167$, $2^8.179$, $2^{13}.191$,

$2^3.213$, $2^3.219$, $2^3.223$, $2^4.233$, $2^4.239$, $2^3.249$, $2^6.251$, $2^3.267$, $2^8.269$,

$2^3.283$, $2^{14}.311$, $2^6.335$, $2^{15}.347$, $2^4.359$, $2^7.373$, $2^4.419$, $2^6.443$.

A list for $q(\text{odd}) < 10,000$ is given in Appendix K.

§7.2. ASYMPTOTIC EXISTENCE RESULTS FOR ORTHOGONAL DESIGNS

Necessary conditions for the existence of orthogonal designs have been derived from a study of rational matrices. The theorems below show that many of these conditions are also sufficient if the order of the orthogonal design is much larger than the number of nonzero entries in each row.

If n is odd, then $\rho(n) = 1$, so the only orthogonal designs of order n are weighing matrices. We have shown that the weight k of a weighing matrix of odd order n must be a square. If n is much larger than k, this is sufficient for existence; in fact,

7.14. THEOREM. (Geramita-Wallis). Suppose k is a square. Then there is an integer $N = N(k)$ such that for any $n > N$ there is a $W(n, k)$.

This theorem follows from Lemma 7.19 below.

Suppose $n \equiv 2 \pmod 4$; then an orthogonal design of order n has at most two variables. For these orders we derived the following necessary conditions:

(i) If there is an orthogonal design of type (a, b), then there is a 2×2 rational matrix P such that $PP^t = \text{diag}(a, b)$.

(ii) If there is a weighing matrix $W(n, k)$, then k is a sum of at most two squares.

(iii) If there is a skew-symmetric weighing matrix $W(n, k)$, then k is a square.

We can prove an asymptotic converse to (i), (ii) and (iii):

7.15. THEOREM. (Eades). (i) Suppose there is a 2×2 rational matrix P such that $PP^t = \text{diag}(a, b)$ where a and b are positive integers. Then there is an integer $N = N(a, b)$ such that for all $t \geq N$ there is an orthogonal design of type (a, b) and order $2t$.

(ii) Suppose k is a sum of two integer squares. Then there is an integer $N = N(k)$ such that for all $t \geq N$ there is a $W(2t, k)$.

(iii) Suppose k is a square. Then there is an integer $N = N(k)$ such that for all $t \geq N$ there is a skew-symmetric $W(2t, k)$.

The proof of Theorem 7.15 comes later. We merely note here that (ii) and (iii) follow immediately from (i).

Suppose $n \equiv 4 \pmod 8$; then an orthogonal design of order n has at most four variables. The results of Chapter 3 show that for these orders the following necessary conditions apply:

(i) If there is an orthogonal design of type (a, b, c, d) and order $n \equiv 4 \pmod 8$, then there is a 4×4 rational matrix P such that $PP^t = \text{diag}(a, b, c, d)$.

(ii) If there is a skew-symmetric $W(n, k)$ where $n \equiv 4 \pmod 8$, then k is a sum of three integer squares.

There is the following conjecture:

7.16. CONJECTURE. (Seberry). For every positive integer $k \leq n \equiv 0 \pmod 4$ there is a $W(n, k)$.

We can prove an asymptotic result on this conjecture, an asymptotic converse for (ii) and a partial asymptotic converse for (i).

7.17. THEOREM. (Eades). (i) Suppose m, z_1, z_2, z_3 and z_4 are positive integers. Then there is an integer $N = N(m, z_1, z_2, z_3, z_4)$ such that for all $t \geq N$ there is an orthogonal design of type $(mz_1^2, mz_2^2, mz_3^2, mz_4^2)$ and order $4t$.

(ii) Suppose m_1, m_2, z_1, z_2, z_3, z_4 are positive integers and m_1 and m_2 are each sums of two integer squares. Then there is an integer N depending on m_1, m_2, z_1, z_2, z_3, z_4 such that for all $t \geq N$ there is an orthogonal design of type $(m_1z_1^2, m_1z_2^2, m_2z_3^2, m_2z_4^2)$ and order $4t$.

(iii) For any integer k there is an integer $N = N(k)$ such that for all $t \geq N$ there is a $W(4t, k)$.

(iv) If k is a sum of three integer squares, then there is an integer $N = N(k)$ such that for all $t \geq N$ a skew-symmetric $W(4t, k)$ exists.

Again we will leave the proof of this theorem for later and merely note that (iii) and (iv) follow from (i).

For $n \equiv 0 \pmod 8$ it is conjectured that for every $k < n$ there is a skew-symmetric $W(n, k)$. We can prove:

7.18. THEOREM. (Eades). (i) Suppose $u_1, u_2, \ldots, u_8$ are integers. Then there is an integer $N = N(u_1, u_2, \ldots, u_8)$ such that for all $t \geq N$ an orthogonal design of type $(u_1^2, u_2^2, \ldots, u_8^2)$ and order $8t$ exists.

(ii) For any pair (a, b) of integers there is an integer $N = N(a, b)$ such that for all $t \geq N$ there is an orthogonal design of type (a, b) and order $8t$.

(iii) For any integer k there is an integer $N = N(k)$ such that for all $t \geq N$ there is a skew-symmetric $W(8t, k)$.

Since any integer is a sum of four squares of integers, (ii) and (iii) follow from (i).

From Theorem 7.14, Theorem 7.15(ii) and (iii), Theorem 7.17

(iii) and (iv), and Theorem 7.18(iii), we can deduce:

7.19. COROLLARY. (i) For any integer k there is only a finite number, $M_1(k)$, of orders for which the existence of a weighing matrix of weight k is undecided.

(ii) For any integer k there is only a finite number, $M_2(k)$, of orders for which the existence of a skew-symmetric weighing matrix of weight k is undecided.

From Theorem 7.15(i) we can deduce:

7.20. COROLLARY. For any pair (a,b) of integers there are only a finite number, $M_3(a,b)$, of orders $n \equiv 2 \pmod 4$ for which the existence of an orthogonal design of order n and type (a,b) is undecided.

7.21. LEMMA. Suppose $k \neq 0$ is a square, and let $k = \prod_{i=1}^{m} q_i^2$ where each q_i is either 1 or a prime power, for $i = 1, 2, \ldots, m$. Suppose $a_1, a_2, \ldots, a_m$ are any positive integers, and let $n = \prod_{i=1}^{m} a_i(q_i^2+q_i+1)$. Then there is a type-1 $W(n,k)$ and type-2 $W(n,k)$ on the group $\prod_{i=1}^{m} Z_{v_i}$, $v_i = a_i(q_i^2+q_i+1)$.

Proof. If q is either 1 or a prime power, then there is a circulant $W(q^2+q+1, q^2)$. Suppose the first row of this matrix is (w_i), $1 \leq i \leq q^2+q+1$. If a is a positive integer and $1 \leq i \leq a(q^2+q+1)$, then define

$$W'_i = \begin{cases} w_i, & \text{if } i \equiv 0 \pmod a, \\ 0, & \text{otherwise.} \end{cases}$$

Then (W'_i), $1 \leq i \leq a(q^2+q+1)$, is the first row of a circulant $W(a(q^2+q+1), q^2)$.

Proceed now by induction. If there is a type-1 $W(n_i, k_i)$ on G_i, for $i = 1, 2$, then the tensor product of $W(n_1, k_1)$ with $W(n_2, k_2)$ is a type-1 $W(n_1n_2, k_1k_2)$ on $G_1 \times G_2$.

We recall that the existence of a type-1 $W(n,k)$ on G implies the existence of a type-2 $W(n,k)$. (See Corollary 4.23.)

The next lemma is a consequence of Sylvester's Theorem (Theorem 7.7)

7.22. LEMMA. Suppose orthogonal designs of type $(u_1, u_2, \ldots, u_m)$ exist in orders n_1 and n_2. Let $h = \text{g.c.d.}(n_1, n_2)$. Then there is an integer N such that for all $t > N$ there is an orthogonal design of type $(u_1, u_2, \ldots, u_m)$ and order ht.

Proof of Theorem 7.14. By Lemma 7.21, for every square k there is an odd number n such that $W(n,k)$ exists. By Corollary 7.3 there is a $W(2^t, k)$ for some t. Now since n is odd, $\text{g.c.d.}(n, 2^t) = 1$, so Lemma 7.22 gives Theorem 7.14.

7.23. LEMMA. Suppose there is an orthogonal design of type $(u_1, u_2, \ldots, u_k)$ and order n; suppose $z_1, z_2, \ldots, z_k$ are integers. Then there is an odd number, y, such that an orthogonal design of type $(z_1^2 u_1, z_2^2 u_2, \ldots, z_k^2 u_k)$ exists in order yn.

Proof. Let $z = \prod_{i=1}^{b} q_i$ be a decomposition of z into prime powers. Let $y_1 = \prod_{i=1}^{b} (q_i^2 + q_i + 1)$, and let W be the type-2 $W(y_1, z_1^2)$ assured by Lemma 7.21. Type-2 matrices are always symmetric, so the k-tuple $(W, I, I, \ldots, I)$ comprises pairwise amicable matrices. Hence, using Lemma 4.54, there is an orthogonal design of type $(z_1^2 u_1, u_2, \ldots, u_k)$ and order $y_1 n$. Note that y_1 is odd.

We can clearly continue this process to prove the lemma.

In fact, all the weighing matrix results in Theorems 7.17, 7.18 and 7.19 (7.15(ii) and (iii), 7.17(iii) and (iv), 7.18(iii)) follow from Corollary 7.3, Lemmas 7.21, 7.22 and 7.23. However, we will prove the more general results first.

7.24. LEMMA. Suppose m is a sum of two integer squares. Then there is an odd number t such that an orthogonal design of type (m, m) exists in order $2t$.

Proof. Let $m = m_1^2 + m_2^2$. If $m_j \neq 0$, let $m_j^2 = \prod_{i=1}^{k} q_{ij}^2$ where q_{ij} is either 1 or a prime power. For each $i = 1, 2, \ldots, k$ let $b_i = \text{LCM}\{q_{ij}^2 + q_{ij} + 1; m_j \neq 0\}$. If $m_j \neq 0$, let W_j be the type-1

$$W\left(\prod_{i=1}^{k-1} b_i, \prod_{i=1}^{k-1} q_{ij}^2\right)$$

on the group $\prod_{i=1}^{k-1} Z_{b_j}$ from Lemma 7.21, and let C_j be the type-1 $W(b_k, q_{kj}^2)$ on Z_{b_k} from Lemma 7.21. Note that C_j is circulant. Suppose the first row of C_j is (c_{rj}), $1 \le r \le b_k$.

Define (d_{rj}), $1 \le r \le 3b_k$,

$$d_{rj} = \begin{cases} c_{rj}, & \text{if } r \equiv j \pmod 3, \\ 0, & \text{otherwise.} \end{cases}$$

Then the circulant D_j with first row (d_{rj}) is a type-1 $W(3b_k, q_{kj}^2)$ on Z_{3b_k}. Moreover, if $m_1 \ne 0$ and $m_2 \ne 0$, then D_1 and D_2 are mutually disjoint. If $m_j \ne 0$, let $F_j = D_j \times W_j$. If $m_j = 0$, let F_j be the zero matrix of order $3\prod_{i=1}^{k} b_i$. Then F_1 and F_2 are disjoint type-1 weighing matrices of weights m_1^2 and m_2^2, respectively, on the group $\prod_{i=1}^{k-1} Z_{b_i} \times Z_{3b_k}$.

Each b_i is odd, so the order of F_1 is an odd number, $t = 3\prod_{i=1}^{k} b_i$. Hence, we can use the matrices $x_1F_1 + x_2F_2$ and $x_2F_1 - x_1F_2$ in the Wallis-Whiteman generalization of the two circulants construction, obtained by using the remark before Lemma 4.16 with Proposition 4.46, to produce an orthogonal design of type (m, m) and order $2t$ on the variables x_1 and x_2.

To prove Theorem 7.15 we need a simple fact about 2×2 rational matrices.

7.25. LEMMA. Suppose a and b are integers such that there exists a rational 2×2 matrix P satisfying $PP^t = \mathrm{diag}(a,b)$. Then there are integers x, y, z, w such that $a = (z^2+w^2)x^2$ and $b = (z^2+w^2)y^2$.

<u>Proof.</u> Let $a = x^2z_1$ and $b = y^2z_2$, where z_1 and z_2 are squarefree

integers. Now $ab = (\det P)(\det P^t)$ is a rational square. Hence $z_1 = z_2$. Now a is a sum of two rational squares; hence z_1 is a sum of two rational squares. The Cassells-Davenport theorem then implies that z_1 is a sum of two integer squares, say, $z_1 = z^2+w^2$. So $a = x^2(z^2+w^2)$, $b = y^2(z^2+w^2)$.

Proof of Theorem 7.15. Suppose a and b are integers such that there exists a rational 2×2 matrix P satisfying $PP^t = \text{diag}(a, b)$. Then by Lemma 7.25 there are integers x, y and m such that $a = x^2 m$, $b = y^2 m$, and m is a sum of at most two integer squares. Lemma 7.24 gives an orthogonal design of type (m, m) and order $2t$ for some odd t. Lemma 7.23 gives an orthogonal design of type $(x^2 m, y^2 m) = (a, b)$ and order $2ty$, for some odd y. Corollary 7.3 gives an orthogonal design of type (a, b) and order 2^d, for some integer d. Now $\text{g.c.d.}(2^d, 2ty) = 2$, so we can use Lemma 7.22 to obtain Theorem 7.15.

The proof of the next lemma is very similar to the proof of Lemma 7.24 and so is omitted.

7.26. LEMMA. For any positive integer m there exists an odd number t such that an orthogonal design of type (m, m, m, m) and order $4t$ exists.

We have exhibited an orthogonal design of type $(1, 1, \ldots, 1)$ and order n on $\rho(n)$ variables. Since $\rho(2^d)$ is a strictly increasing function of d, equating variables in the Geramita-Pullman orthogonal design gives the following lemma.

7.27. LEMMA. Let $(u_1, u_2, \ldots, u_k)$ be a sequence of positive integers. Then there is an integer d such that an orthogonal design of type $(u_1, u_2, \ldots, u_k)$ and order 2^d exists.

Proof of Theorem 7.17. By Lemma 7.26 there is an orthogonal design of type (m, m, m, m) and order $4t$ for some odd t for any integer m. If z_1, z_2, z_3 and z_4 are integers, then Lemma 7.23 implies the existence of an orthogonal design of type $(z_1^2 m, z_2^2 m, z_3^2 m, z_4^2 m)$ and order $4ty$ for some odd y. Lemma 7.27 gives an orthogonal design of type

$(z_1^2 m, z_2^2 m, z_3^2 m, z_4^2 m)$ and order 2^d for some d. Since g.c.d.$(4ty, 2^d) = 4$, we have Theorem 7.17(i) and (iv) following immediately. Theorem 7.17(ii) follows from Theorem 7.15(i).

Proof of Theorem 7.18. There is an orthogonal design of type $(1,1,1,1,1,1,1,1)$ in order 8. Hence by Lemma 7.23 there is an orthogonal design of type $(z_1^2, z_2^2, \ldots, z_8^2)$ and order $8y$ for some odd y. Lemma 7.27 ensures the existence of an orthogonal design of order 2^d, $d > 3$, and type $(z_1^2, z_2^2, \ldots, z_8^2)$. Hence Lemma 7.22 gives Theorem 7.18(i); Theorem 7.18 (ii) and (iii) follow immediately.

§7.3. N-TUPLES OF THE FORM $(2^p b_1, 2^p b_2, \ldots, 2^p b_n)$

The above discussion led us to believe that while we might not yet know all about orthogonal designs in "small" orders, we might be able to say far more about their existence in "large" orders. The remainder of this section is another case of hindsight: Eades, Robinson, Wallis and Williams first saw the results in powers of 2, and then Eades showed that the same argument could always be used.

This section also shows a distinction in approach between an algebraist and a combinatorialist. The algebraist can ensure the existence of any n-tuple as the type of an orthogonal design in some order $2^t q$ by allowing most of the entries of the design to be zero. The combinatorialist is interested in establishing existence with as few zeros as possible, as this case is more useful in applications.

7.28. DEFINITION. A binary expansion of a positive integer s is a nondecreasing sequence $B = (b_1, b_2, \ldots, b_k)$ of powers of 2 such that $b_1 + b_2 + \ldots + b_k = s$. If B is strictly increasing, we say B is the binary decomposition of s. The binary expansion of $A = (s_1, \ldots, s_n)$, an n-tuple of positive integers, is $(B_1, \ldots, B_n)$, where B_i is the binary expansion of s_i.

From a binary expansion $B = (b_1, b_2, \ldots, b_k)$ of s, we can obtain new binary expansions by combining:

$$(b_1, b_2, \ldots, 2^j, 2^j, \ldots, b_k) \rightarrow (b_1, b_2, \ldots, 2^{j+1}, \ldots, b_k)$$

or by splitting:

$$(b_1, b_2, \ldots, 2^{j+1}, \ldots, b_k) \rightarrow (b_1, b_2, \ldots, 2^j, 2^j, \ldots, b_k)$$

with suitable reordering. If a binary expansion C is obtained from a binary expansion B by repeating these operations, we say C is equivalent to B. Clearly, this relation is transitive.

7.29. LEMMA. Any two binary expansions of s are equivalent.

Proof. All binary expansions of s are equivalent to the binary decomposition of s by combining.

7.30. LEMMA. Suppose there is an orthogonal design of type B and order t, where B is a binary expansion of s. Then for every binary expansion C of s, there is an orthogonal design of type 2^pC and order 2^pt for some integer p.

Proof. If there is an orthogonal design of type

$$(b_1, b_2, \ldots, 2^j, 2^j, \ldots, b_k)$$

and order t, then there is an orthogonal design of type

$$(b_1, b_2, \ldots, 2^{j+1}, \ldots, b_k)$$

and order t. If there is an orthogonal design of type

$$(b_1, b_2, \ldots, 2^{j+1}, \ldots, b_k)$$

and order t, then there is an orthogonal design of type

$$2(b_1, b_2, \ldots, 2^j, 2^j, \ldots, b_k)$$

and order $2t$ (Lemma 4.11).

Clearly, the integer p in the lemma above is the number of splittings to get C from B.

We can now deduce:

7.31. THEOREM. Let B be a binary expansion of s, and suppose there is an orthogonal design of type B and order t.

Let $(a_1, a_2, \ldots, a_u)$ be a u-tuple of positive integers such that $a_1+a_2+\ldots+a_u = 2^k s$ for some k.

Then there is an integer p such that an orthogonal design of type $(2^p a_1, 2^p a_2, \ldots, 2^p a_u)$ exists in order $2^{p+k} t$.

Proof. We can split B to obtain an orthogonal design of type C and order $2^k t$, for some binary expansion C of $2^k s$. Let D be the binary expansion of $2^k s$ which is obtained from the binary decompositions of the a_i. Then the theorem follows by Lemma 7.30.

This theorem is most interesting when $s = t$. The case $s = t = 1$ can be proved using the orthogonal design of order 1 for B.

The integer p depends on the sum of the lengths of the binary decompositions of the a_i. Since there are only a finite number of u-tuples which add to $2^k s$, we can state:

7.32. THEOREM. Suppose B is a binary expansion of s, and suppose there is an orthogonal design of type B and order t. Let u and k be integers. Then there is an integer q such that every orthogonal design of type $(2^q a_1, 2^q a_2, \ldots, 2^q a_u)$ exists in order $2^{q+k} t$, for all $(a_1, a_2, \ldots, a_u)$ such that $a_1+a_2+\ldots+a_u = 2^k s$.

7.33. EXAMPLE. There is an orthogonal design of type $(1,1,2,8)$ and order 12. This gives a $(1,1,2,4,16)$ in order 24. Hence, for every $(a_1, a_2, \ldots, a_u)$ such that $a_1+a_2+\ldots+a_u = 2^k.12$, there is an integer p such that there is an orthogonal design of type $(2^p a_1, 2^p a_2, \ldots, 2^p a_u)$ in order $2^{p+k}.12$.

We can use the same algorithm described in Figure 7.1 to obtain p. If $(a_1, a_2, \ldots, a_u) = (1,1,2,6,14)$, the algorithm is as follows:

DESCRIPTION OF THE CONSTRUCTION ALGORITHM

Let A be an n-tuple of positive integers, and let B be the binary expansion of A described above. We write $B = (b_1, b_2, \ldots, b_j)$. Then the algorithm proceeds as follows:

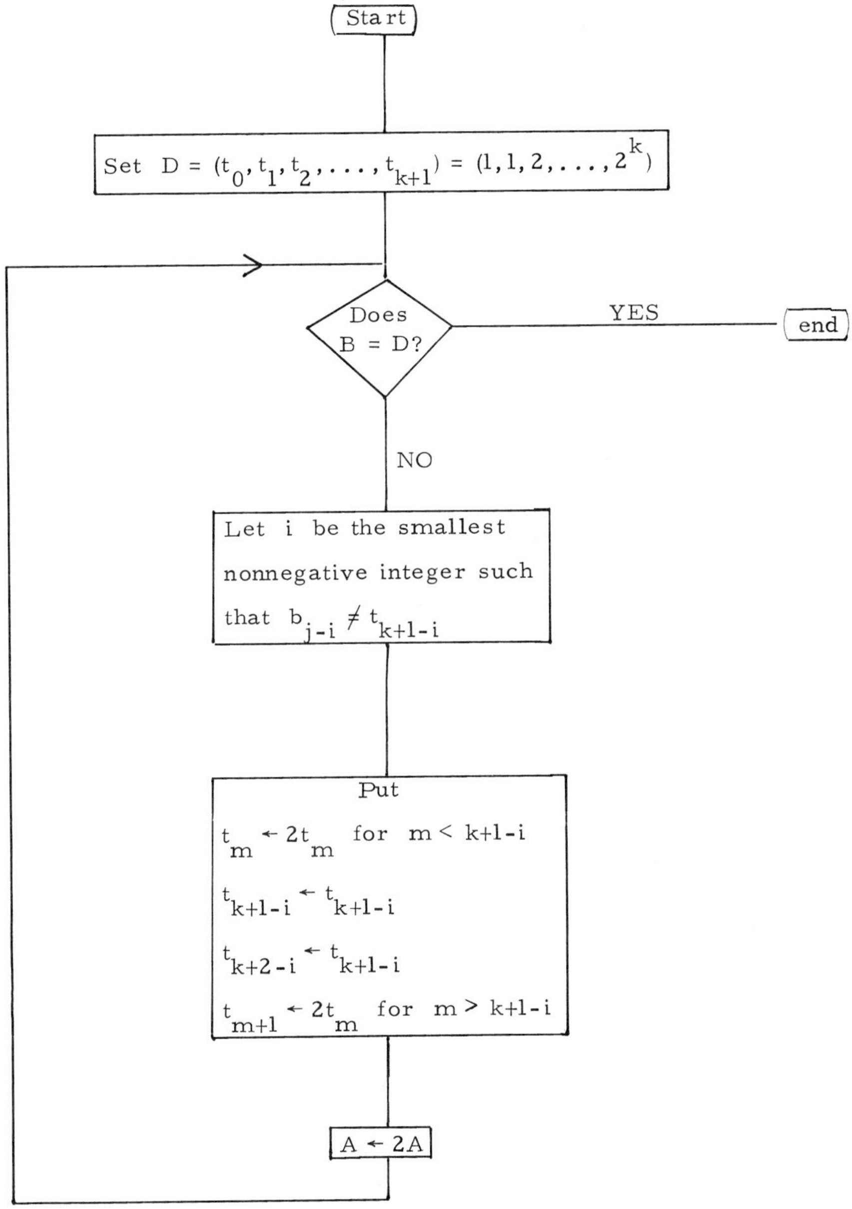

Figure 7.1

```
1 1  2  2  2   4  4   8 | 1 1  2   4         16
2 2  4  4  4   8  8  16 | 2 2  4   8       /    \
4 4  8  8  8  16 16  32 | 4 4  8   16     16     16
8 8 16 16 16  32 32  64 | 8 8 16  /  \   /  \    
                        |        16  16 16  16  32
                        |                  32 32 64 ,
```

and so p = 3, and a design of type $8 \times (1,1,2,6,14)$ exists in order $2^4 \times 12$.

7.34. EXAMPLE. There is an orthogonal design of type (4, 8, 8) and order 20. Let $(a_1, a_2, \ldots, a_u) = (1,1,5,15,18)$. Now from the (4, 8, 8) of order 20, we can find a 4(1,1,2,4,16,16) of order 160. We can use the algorithm as in Figure 7.1:

```
 1,  1,  1,  1,  2,  2,  4,  4,   8,  16 |  1,  1,  2,     4,          16,         16
 2   2   2   2   4   4   8   8   16   32 |  2   2   4      8        16   16        32
 4   4   4   4   8   8  16  16   32   64 |  4   4   8     16     16   16   32      64
 8   8   8   8  16  16  32  32   64  128 |  8   8  16   16  16  32  32   64       128
16  16  16  16  32  32  64  64  128  256 | 16  16 16 16 32  32  64  64  128       256 .
```

Hence, there is an orthogonal design of type

$$4 \times 16 \times (1,1,5,15,18)$$

and order $8 \times 16 \times 20$. Thus, in this case p = 6.

7.35. EXAMPLE. For a third example, consider $(a_1, a_2, \ldots, a_u) =$ (1,1,5,5,5,5,5,5). This 8-tuple expands to (1,1,1,1,1,1,1,1,4,4,4,4,4,4), and so the methods of Figure 7.1 give an orthogonal design of type $(1,1,5,5,5,5,5,5) \times 2^8$. But if we use the orthogonal design of type (1,1,1,1,1,1,1,1) in order 8 for B in Theorem 7.31, we obtain an orthogonal design of type $(1,1,5,5,5,5,5,5) \times 2^4$ and order 32×2^4. The algorithm proceeds as follows:

```
 1,  1,  1,  1,  1,  1,  1,  1,  4,  4,  4,  4,  4,  4 |  1  1  2        4        4  4  4  4  4  4
 2   2   2   2   2   2   2   2   8   8   8   8   8   8 |  2  2  4     4     4      8  8  8  8  8  8
 4   4   4   4   4   4   4   4  16  16  16  16  16  16 |  4  4  8     8    4  4   16 16 16 16 16 16
 8   8   8   8   8   8   8   8  32  32  32  32  32  32 |  8  8 16    8 8   8  8   32 32 32 32 32 32
16  16  16  16  16  16  16  16  64  64  64  64  64  64 | 16 16 16 16 16 16 16 16  64 64 64 64 64 64,
```

and so p = 4.

7.36. EXAMPLE. There is an orthogonal design of type (4, 4, 4, 16) and order 28. The 8-tuple (1,1,1,1,1,2,17,32) illustrates the use of combining as well as splitting. From the (4,4,4,16) we obtain a design of type 4(1,1,2,4,8,8,32) and order 4×56. We use the combining operation to form a design of type 4(1,1,2,4,16,32) and order 4×56 and then proceed with splitting in the usual way:

1	1	1	1	1	1	2	16	32	1	1	2				4	16	32
2	2	2	2	2	2	4	32	64	2	2	4			4	4	32	64
4	4	4	4	4	4	8	64	128	4	4	8		4	4	8	64	128
8	8	8	8	8	8	16	128	256	8	8	8	8	8	8	16	128	256 .

So in this case p = 5.

IMPLEMENTING THE ALGORITHM

We consider the n-tuple $A = (s_1, \ldots, s_n)$, $\Sigma s_i = s$, in order $2^k q$.

Suppose $n \leq 3$. Then if s is a power of 2, by Corollary 7.2, A corresponds to the type of an orthogonal design.

Suppose $n > 3$, and 2^j is the highest power of 2 which divides each s_i. Then we can use Theorem 7.30 with the n-tuple $(s_1/2^j, \ldots, s_n/2^j) = (t_1, \ldots, t_n)$ in order 2^{k-j}.

In fact, for any n and only two odd entries in A, we can usually use the following process to obtain a starting point for the algorithm in a lower power of 2.

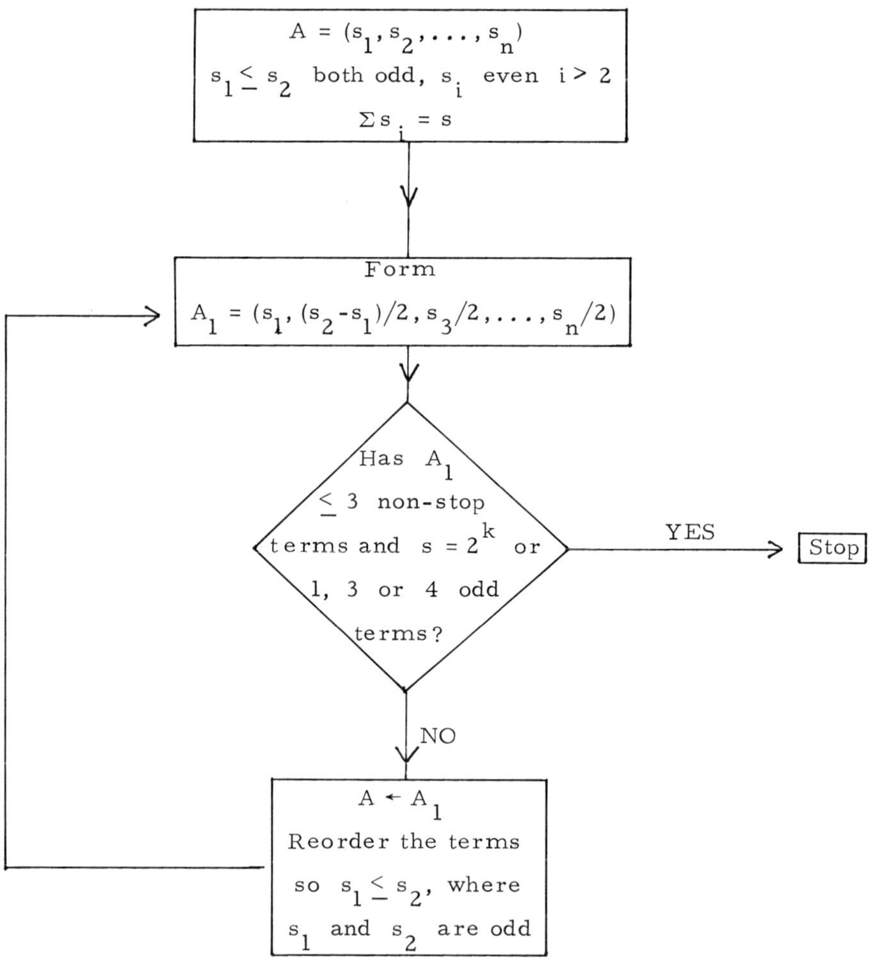

Table 7.2

7.37. EXAMPLE. Consider the binary 5-tuple $A = (3, 3, 6, 20, 96)$ in order 128. Now the binary expansion of A is $(1, 2, 1, 2, 2, 4, 4, 16, 32, 64)$, which is a 10-tuple. So Theorem 7.32 guarantees the existence of a 5-tuple $(2^3.3,\ 2^3.3,\ 2^3.6,\ 2^3.20,\ 2^3.96)$ in order 2^{10}. But if we use the method of Table 7.2, we form $(3, (3-3)/2, 6/2, 20/2, 96/2) = (3, 3, 10, 48)$ in order 64, and then $(3, (3-3)/2, 10/2, 48/2) = (3, 5, 24)$ in order 32. But all 3-tuples $(a, b, 32-a-b)$ exist in order 32, so $(3, 3, 6, 20, 96)$ _is_ the type of an orthogonal design in order 128.

N-TUPLES IN POWERS OF 2 WITH NO ZEROS

In the case where the sum of the entries of A is a power of 2 and A has no zero entries, i.e., the sum of the entries of A is the order of A, it is possible to determine the maximum value of p.

7.38. DEFINITION. Let $A = (s_1, s_2, \ldots, s_n)$ be an n-tuple of positive integers with $s_1+s_2 + \ldots + s_n = s = 2^t$. We write the binary expansion of each s_i, $i = 1, \ldots, n$, and rearrange the order to obtain $B = (b_1, b_2, \ldots, b_k)$, the _binary expansion_ of A, where each b_i is a power of 2 and $b_j \le b_{j+1}$. We say the _binary length_ of A is $k = R_A(n, t)$, the number of entries of B, and use $R(n,t)$ for $\max_A R_A(n,t)$.

7.39. THEOREM. Let $A = \{s_i\}$ be an n-tuple of positive integers such that at least two entries of A are odd and the sum of the entries of A is 2^k. Then there is an orthogonal design of type $2^p A$ in order 2^{p+k}, where p is the binary length of A minus $k+1$.

7.40. EXAMPLE. Robinson's Theorem 6.1 shows that there is no orthogonal design of type $(1, 1, 1, 1, 1, 2^t-5)$ in any order $2^t > 40$. Now 2^t-5 has a binary expansion $1+2+2^3+2^4 + \ldots + 2^{t-1}$, so $B = (1, 1, 1, 1, 1, 1, 2, 8, 16, \ldots, 2^{t-1})$ and $p = t+4-(t+1) = 3$. Hence there is an orthogonal design of type $(2^3.1,\ 2^3.1,\ 2^3.1,\ 2^3.1,\ 2^3.1,\ 2^3.(2^t-5))$ in every order 2^{t+3}.

Proof. Let $A = (a_1, a_2, \ldots, a_n)$ be a sequence such that $\sum_{i=1}^{n} a_i = 2^t$. Let $a_i = b_{i0} + b_{i1}.2 + \ldots + b_{ik_i} 2^{k_i}$ be the binary expansion of a_i.

Now,

$$R_A(n,t) = \sum_{i=1}^{n} \sum_{j=0}^{k} b_{ij}, \qquad k = \max_i k_i.$$

Now assume there exist a j such that one of the b_{ij}'s is nonzero, say b_{1j}, and two of the $b_{i,j-1}$'s are zero, say $b_{n-1,j-1}$ and $b_{n,j-1}$; then we consider the sequence

$$A_1 = (a_1-2^j, a_2, \ldots, a_{n-2}, a_{n-1}+2^{j-1}, a_n+2^{j-1}).$$

Now,

$$R_{A_1}(n,t) = R_A(n,t)+1,$$

and hence any sequence A such that $R_A(n,t) = R(n,t)$ has the property that the sequences $(b_{1j}, b_{2j}, \ldots, b_{nj})$, $j = 0,\ldots,k-1$, contain as many one's as possible. That is, $(b_{ij}, \ldots, b_{nj})$, $j = 0,\ldots,k-1$, contains at most one zero.

Now, we let the binary expansion of $n-1$ be $c_0+c_1.2+\ldots+c_m 2^m$ and consider the following $n \times (m+2)$ matrix:

$$X = \begin{bmatrix} 1 & 1 & \ldots & 1 & 1 & c_0 \\ 2 & 2 & \ldots & 2 & 2 & c_1.2 \\ \vdots & & & & & \vdots \\ 2^m & 2^m & \ldots & 2^m & 2^m & c_m.2^m \\ 2^{m+1} & \ldots & 2^{m+1} & 0 & \ldots & 0 \end{bmatrix}.$$

Let the number of 2^{m+1}'s in the last row be a. The sum S of all the entries in this matrix is:

$$\begin{aligned} S &= (n-1-a)(2^{m+1}-1) + a(2^{m+2}-1) + n-1 \\ &= (n-1-a)2^{m+1} + a.2^{m+2} \\ &= 2^{m+1}(n-1+a). \end{aligned}$$

Now $m = [\log_2(n-1)]$, and so when $n \neq 2^j+1$ for some j, $n-1 \leq 2^{m+1} < 2(n-1)$. However, since $0 \leq a \leq n-1$, we can find an a such that $n-1+a = 2^{m+1}$;

that is, $a = 2^{m+1}-n+1$, and so

$$S = 2^{2m+2}.$$

We now consider the matrix Y, which is obtained from X by replacing all nonzero terms by 1, and define a sequence $A = \{a_i\}$ by letting row j of Y be $(b_{1j}, b_{2j}, \ldots, b_{nj})$ and choosing

$$a_i = \sum_{j=0}^{m+1} b_{ij} 2^j.$$

It is obvious that the sequences $(b_{1j}, \ldots, b_{nj})$, $j = 0, \ldots, m$, are as full as possible, and since

$$S = 2^{2m+2},$$

$$R_A(n, 2m+2) = R(n, 2m+2). \tag{1}$$

But

$$R_A(n, 2m+2) = a+(n-1)(m+1) + B(n-1),$$

where $B(n-1)$ is the number of nonzero terms in the binary expansion of $n-1$.

Therefore,

$$R(n, 2m+2) = 2^{m+1}+(n-1)m+B(n-1).$$

We now consider

$$R(n, 2m+3).$$

From our choice of a, it can be seen that

$$(n-1-a)2^{m+2} + a.2^{m+3} = 2^{2m+3}.$$

Therefore, to obtain an A such that $R_A(n, 2m+3) = R(n, 2m+3)$, we use the A of (1) and put $b_{a+1,m} = b_{a+2,m} = \ldots = b_{n-1,m} = 1$, $b_{nm} = 0$ and $b_{1,m+1} = b_{2,m+1} = \ldots = b_{a,m+1} = 1$. This produces sequences as full as possible, and therefore,

$R(n, 2m+3) = R(n, 2m+2) + (n-1)$. We continue in this way to obtain the following:

$$R(n, 2m+i) = 2^{m+1} + B(n-1) + (n-1)(m+i-2), \text{ where } i = 2, 3, \dots .$$

We note that if $n = 2^m+1$, then

$$2^{2m} = (n-1)(2^m-1) + n-1.$$

So

$$R(2^m+1, 2m) = 2^m m+1.$$

The maximum number of steps in the algorithm of Figure 7.1 is $R(n, t)-(t+1)$. The actual number of steps for an n-tuple A is $p = R_A(n, t)-(t+1) \leq R(n, t)-(t+1)$. We have shown that $R(n, t)$ is finite and may be evaluated easily.

CHAPTER 8

RESULTS ON THE EXISTENCE CONJECTURES - NUMERICAL RESULTS - UNSOLVED PROBLEMS

In this chapter we give the known results in odd orders, orders $\equiv 2 \pmod 4$, and the current status of the following conjectures:

8.1. CONJECTURE. (Seberry). When $n \equiv 0 \pmod 4$, there exists a $W(n, k)$ for all $k \leq n$.

8.2. CONJECTURE. (Geramita-Wallis). When $n \equiv 4 \pmod 8$, there exists an orthogonal design of order n of type $(1, k)$ when $k \leq n-1$, $k = a^2+b^2+c^2$, a, b, c integers except that $n-2$ must be the sum of two squares.

8.3. CONJECTURE. (Geramita-Wallis). When $n \equiv 0 \pmod 8$, there exists an orthogonal design of order n and type $(1, k)$ for all $k \leq n-1$; equivalently, when $n \equiv 0 \pmod 8$, there exists a skew-symmetric $W(n, k)$ for all $k \leq n-1$.

§8.1. $W(n, k^2)$ FOR n ODD

We have the following results which are tabulated in Table 8.4.

8.4. SUMMARY. If a $W(n, k)$ exists and n is odd, then k is a square and $(n-k)^2-(n-k)+2 > n$.

We know that:

(a) a $W(m^2+m+1, m^2)$ exists whenever m is a prime power and only if m is the order of a projective plane;

(b) given any integer k^2 there exists an n dependent on k^2 so that for every $m \geq n$ a $W(m, k^2)$ exists;

(c) there exists a

(i) $W(m, 4)$, where $m \geq 4$ (except $m = 5, 9$, which do not exist);

(ii) $W(m, 9)$, where $m \geq 22$ ($m = 11$ does not exist and $m = 17, 19, 21$ are undecided; $m = 15$ exists);

(iii) $W(m, 16)$, where $m \geq 36$ ($m = 17, 19$ do not exist and $m = 23, 25, 27, 29, 33, 35$ are undecided; $m = 31$ exists);

(iv) $W(m, 25)$, where $m \geq 82$ ($m = 27, 29$ do not exist--there are many undecided cases);

(v) $W(m, 36)$, where $m \geq 138$;

(vi) $W(m, 49)$, where $m \geq 200$.

Proof of (c). First we recall that if q is a prime power, there exists a $W(q^2+q+1, q^2)$. Hence there exist $W(7, 4)$, $W(13, 9)$, $W(21, 16)$, $W(31, 25)$ and $W(57, 49)$. The results in Lemma 4.143 show that $W(n, 4)$ exist in every order $n \equiv 0 \pmod 2$, $n \geq 4$, and $W(n, 16)$ exist in every order $n \equiv 0 \pmod 2$, $n \geq 16$.

Eades found a circulant $W(31, 16)$ with first row,

-0000-0+0--+0++000-+-++00++0+00.

The results later in this chapter may be combined to show that $W(n, 9)$ exist for $n \equiv 0 \pmod 2$, $n \geq 10$, and $W(n, 25)$ exist for $n \equiv 0 \pmod 2$, $n \geq 52$.

Hence we can conclude the result of the enunciation.

There is a symmetric $W(15,9)$:

$$
\left[\begin{array}{ccccc}
000 & 111 & 000 & 111 & --- \\
000 & 111 & --- & 000 & 111 \\
000 & 111 & 111 & --- & 000 \\
\\
111 & 0-1 & 000 & 01- & 0-1 \\
111 & -10 & 000 & 1-0 & -10 \\
111 & 10- & 000 & -01 & 10- \\
\\
0-1 & 000 & 100 & -11 & -11 \\
0-1 & 000 & 001 & 11- & 11- \\
0-1 & 000 & 010 & 1-1 & 1-1 \\
\\
10- & 01- & -11 & 010 & 001 \\
10- & 1-0 & 11- & 100 & 010 \\
10- & -01 & 1-1 & 001 & 100 \\
\\
-10 & 0-1 & -11 & 001 & 010 \\
-10 & -10 & 11- & 010 & 100 \\
-10 & 10- & 1-1 & 100 & 001
\end{array}\right].
$$

From the $W(13,9)$, $W(15,9)$, $W(6,4)$ and $W(7,4)$, we obtain a $W(78,36)$, $W(90,36)$, $W(91,36)$ and a $W(105,36)$. From later in this chapter we have a $W(m,36)$ for $m \in \{36,40,42,44,48,52,56,60,64\}$. The following two sequences of length 37 may be used to form a $W(74,36)$ constructed from two circulants,

11--0-0-11010011010100-101000-0100000,

0000-0100010-00-0-01-001011-0-011-10.

Hence we have that $W(n,36)$ exist for $n \equiv 0 \pmod 2$, $n \geq 72$. Combining results, we have $W(n,36)$ exist for $n \geq 138$.

From later in this chapter we have the existence of $W(m,49)$ for $m \in \{56,60,64,72,76,80,88,96\}$. Lemma 4.34 gives a $W(50,49)$. So we have $W(n,49)$ exist for $n \equiv 0 \pmod 2$, $n \geq 144$. Combining results, we have $W(n,49)$ exist for $n \geq 200$.

Order	Weights eliminated by summary as nonexistent	Other nonexistent weights	Constructed weights	Unsolved weights
5	nil			
7	nil		4	
9	9	4		
11	9		4	
13			4, 9	
15			4, 9	nil
17	16		4	9
19	16		4	9
21			4, 16	9
23			4, 9	16
25	25		4, 9	16
27	25		4	16
29	25		4, 9	16
31			4, 16, 25	nil
33			4, 9	16, 25
35			4, 9	16, 25
37	36		4, 9, 16	25
> 11			4	
> 21			4, 9	
> 35			4, 9, 16	
> 81			4, 9, 16, 25	
> 161			4, 9, 16, 25, 36	
>199			4, 9, 16, 25, 36, 49	

Table 8.1

§8.2. SUMMARY OF UNSOLVED PROBLEMS AND KNOWN RESULTS ON THE EXISTENCE OF $W(n, k^2+\ell^2)$, ORTHOGONAL DESIGNS FOR $n \equiv 2 \pmod 4$

We recall that for $n \equiv 2 \pmod 4$ there can be orthogonal designs on only one or two variables. Further, we recall that:

8.5. LEMMA. If there exists an orthogonal design of order n and type (a, b), then

(i) a, b and $a+b$ must all be the sum of two integer squares;

(ii) ab is an integer square;

(iii) $a+b \le n-1$.

8.6. COROLLARY. If there exists a $W(n,k)$ for $n \equiv 2 \pmod 4$, then

(i) k is the sum of two integer squares;

(ii) $k \le n-1$.

For convenience we restate some constructive results given in Chapter 4.

4.41. THEOREM. (Goethals and Seidel). Let $q \equiv 1 \pmod 4$ be a prime power. Then there exists a $W(q+1, q)$ constructed from two circulant symmetric matrices.

4.142. LEMMA. Let q be a prime power. Then there exist orthogonal designs of types $(1, q^2)$, (q^2, q^2), $(2, 2q^2)$, and (q^2+1, q^2+1) in order $2(q^2+q+1)$. Further, if $q^2+q+1 \equiv 3 \pmod 4$ is prime, then there exists an orthogonal design of type $(1, (q+1)^2)$ in order $2(q^2+q+1)$.

4.116. PROPOSITION. Let $r = 2^a . 10^b . 26^c . 5^d . 13^c$, and let n be any integer $\ge 2^a 10^b 26^c 6^d 14^e$ with a, b, c nonnegative integers. Then there exists an orthogonal design of type (r, r) constructed from two circulants in every order $2n$ where $n \ge r$. Hence there is also a $W(2n, r)$ and a $W(2n, 2r)$ constructed from two circulants.

4.143. LEMMA. There exist orthogonal designs constructed from two circulants for orders $2n$ with types and orders n as indicated:

(i) $(1,1)$ and $n \ge 1$;

(ii) $(2,2)$ and $n \ge 2$;

(iii) $(1,4)$ and $n \ge 3$;

(iv) $(4,4)$ and $n \ge 4$;

(v) $(2,8)$ and $n \ge 6$;

(vi) $(5,5)$ and $n \ge 6$;

(vii) $(8,8)$ and $n \ge 8$;

(viii) $(4,16)$ and $n \ge 10$;

(ix) $(10,10)$ and $n \ge 10$;

(x) (13) and $n \ge 11$;

(xi) $(13,13)$ and $n \ge 14$.

4.144. LEMMA. There exist $W(2n,k)$ constructed from two circulants for:

(i) $n \ge 3$, $k \in \{0,1,2,4,5\}$;

(ii) $n \ge 5$, $k \in \{0,1,2,4,5,8\}$;

(iii) $n \ge 7$, $k \in \{0,1,2,4,5,8,10\}$;

(iv) $n \geq 9$, $k \in \{0,1,2,4,5,8,10,16\}$;

(v) $n \geq 11$, $k \in \{0,1,2,4,5,8,10,13,16,20\}$;

(vi) $n \geq 15$, $k \in \{0,1,2,4,5,8,10,13,16,20,26\}$.

The following result comes from the proof of (c) in the previous section.

8.7. COROLLARY. There exists a $W(m,k)$ where k is the indicated integer square for every $m \equiv 0 \pmod 2$, as shown:

(i) $k = 4$, $n \geq 4$; (iv) $k = 25$, $n \geq 52$;

(ii) $k = 9$, $n \geq 10$; (v) $k = 36$, $n \geq 72$;

(iii) $k = 16$, $n \geq 16$; (vi) $k = 49$, $n \geq 144$.

We quoted the following results of Eades in Chapter 7. The gap between Eades's Theorem and Lemma 4.143 shows that there is a great deal we do not yet know.

7.15. THEOREM. (Eades). Given any two integers a and b, where a, b and $a+b$ are the sum of two integer squares, there exists an m such that an orthogonal design of type (a, b) exists for every $n \geq m$.

Integers which are the sums of two squares: 0, 1, 2, 4, 5, 8, 9, 10, 13, 16, 17, 18, 20, 25, 26, 29, 32, 34, 36, 37, 40, 41, 45, 49, 50, ...

Order	Unsettled cases for weighing matrices	Unsettled cases for skew weighing matrices	Unsettled cases for orthogonal designs
2	nil	nil	nil
6	nil	nil	nil
10	nil	nil	nil
14	nil	nil	nil
18	nil	9*	(1, 9), (4, 9)
22	nil	9	(1, 9), (2, 18) (4, 9), (9, 9)
26	nil	nil	(4, 9), (5, 20)
30	nil	16, 25	(1, 16), (2, 18), (5, 20), (9, 9)
34	17, 18, 25, 29	16, 25	(1, 16), (2, 18), (5, 20), (9, 9)
38	18, 25, 29, 34 36	16, 25, 36	(1, 16), (2, 18), (5, 20), (9, 9)
42	29, 37	25, 36	(1, 25), (1, 36), (4, 25), (4, 36) (8, 18), (8, 32), (9, 9), (9, 16), (9, 25), (18, 18)
46	25, 29		(5, 20) and others
50	18, 25, 29, 34, 36, 37, 41, 45	25, 36	(1, 25), (1, 36), (2, 18), (2, 32), (4, 25), (4, 36), (5, 20), (8, 18), (8, 32), (9, 9), (9, 16), (9, 25), (9, 36), (16, 25), (17, 17), (18, 18)
54	18, 29, 34, 36, 37, 41, 45, 49, 50	25, 36, 49	(1, 25), (1, 36), (1, 49), (2, 18), (2, 32), (2, 50), (4, 25), (4, 36), (4, 49), (8, 18), (9, 9), (9, 16), (9, 25), (9, 36), (10, 40), (16, 25), (16, 36), (17, 17), (18, 18), (18, 32) (25, 25), (26, 26)

*A design of type (1, 16) does not exist.

Table 8.2

Table 8.2 tabulates some results we will now establish. First we see:

8.8. LEMMA. There exist orthogonal designs in order 2n, n odd, of type:

(i) (1, 9) for $n \geq 7$ (except possibly n = 9, 11);

(ii) (1, 16) for n = 11, 13 and $n \geq 21$ (a (1, 16) does not exist in order 18).

Proof. A (1, 9) in order 14 may be obtained from the skew-symmetric W(14, 9) constructed from two circulant matrices with first rows:

011-1--, 0110100.

The fact that a (1, 9) exists in order 12, 16 and 20 and that every even integer $\geq$ 26 is a positive linear combination of 12, 14, 16 and 20 gives the result.

The following first rows are used to generate circulant matrices which give skew-symmetric W(22,16) and W(26,16):

01-111---1-, -1011100010

0101-00001-0-, 1--00110111-1.

Hence there are (1,16) in orders 22 and 26. The (1,16) exists in orders 20, 24, 28 and 32, which gives the rest of the result.

We use Lemma 4.143 and Lemma 8.8 to establish the following results.

8.9. LEMMA. Orthogonal designs of types (1,1), (1,4) and (2,2) exist in order 6. No cases are undecided.

8.10. LEMMA. Orthogonal designs of types (1,1), (1,4), (2,2), (4,4) exist in order 10. No cases are undecided.

8.11. LEMMA. Orthogonal designs of types (1,1), (1,4), (1,9), (2,2), (2,8), (4,4), (4,9) and (5,5) exist in order 14. No cases are undecided.

Proof. The (4,9) is given in Display 8.1; all the other designs are given by Lemma 4.143 and Lemma 8.8.

$$
\begin{bmatrix}
0 & -y & -y & y & x & -x & -x & x & y & -y & y & y & y & -y \\
 & 0 & -y & y & -x & x & -x & x & -y & y & y & y & -y & y \\
 & & 0 & y & -x & -x & x & x & y & -y & -y & -y & -y & y \\
 & & & 0 & x & x & x & x & y & -y & y & y & -y & -y \\
 & & & & 0 & y & y & -y & y & y & y & -y & -y & -y \\
 & & & & & 0 & y & -y & -y & -y & -y & y & -y & -y \\
 & & & & & & 0 & -y & -y & -y & y & -y & y & y \\
 & & & & & & & 0 & -y & -y & y & -y & -y & -y \\
 & & & & & & & & 0 & y & -x & x & -x & x \\
 & & & * & & & & & & 0 & x & -x & -x & x \\
 & & & & & & & & & & 0 & -y & x & x \\
 & & & & & & & & & & & 0 & x & x \\
 & & & & & & & & & & & & 0 & y \\
 & & & & & & & & & & & & & 0
\end{bmatrix}
$$

Display 8.1

8.12. LEMMA. Orthogonal designs of types (1,1), (1,4), (2,2), (2,8), (4,4), (5,5) and (8,8) exist in order 18. A (1,16) design does not exist, and the existence of (1,9) and (4,9) designs is undecided. (A W(18,9) exists from Lemma 8.16.)

8.13. LEMMA. Orthogonal designs of types (1,1), (1,4), (1,16), (2,2), (2,8), (4,4), (4,16), (5,5) and (8,8) exist in order 22. The existence of (1,9), (2,18), (4,9) and (9,9) designs is undecided.

8.14. LEMMA. Orthogonal designs of types (1,1), (1,4), (1,9), (1,16), (2,2), (2,8), (2,18), (4,4), (4,16), (5,5), (8,8), (9,9) and (10,10) exist in order 26. The existence of (4,9) and (5,20) is undecided.

Proof. Use Lemmas 8.8 and 4.116, 4.142 and 4.143.

The remainder of this section is devoted mainly to proving the next two lemmas.

8.15. LEMMA. There exists a W(2n,k), n odd, constructed from two circulants for every $k = a^2+b^2 < 2n$, when n = 3,5,7,11,13. No cases are undecided.

8.16. LEMMA. There exists a W(2n,k), n odd, for every $k = a^2+b^2 < 2n$, for n = 9,15. (Only W(18,9), W(30,9), W(30,18) and W(30,25) are not constructed from two circulants.) No cases are undecided. A complete computer search shows that the W(18, 9) cannot be constructed from two circulant matrices.

Proof of Lemmas 8.15 and 8.16. All the results for n = 3, 5, 7, except the existence of a W(10, 9), W(14, 9) and W(14, 13), exist from Lemma 4.144.

The W(10, 9) is constructed from circulants with first rows:

01--1, -1111.

The W(14, 9) and W(14, 13) are constructed from circulants with first rows:

0110100, 011-1--,

01-11-1, 111--11.

For n = 9: The existence of a W(18, k) for $k \in \{0,1,2,4,5,8,10,16\}$ constructed from circulant matrices is established by Lemma 4.144.

A W (18, 13) and a W(18, 17) constructed from circulant matrices may be obtained by circulating the following first rows respectively:

1-01-1011, 1--010101,

1--1111--, 0-1----1-.

Display 8.2 gives a W(18, 9): A complete computer search has shown it cannot be constructed from two circulants.

$$\begin{bmatrix}
+ & 0 & 0 & 0 & + & + & + & + & + & + & + & + & 0 & 0 & 0 & 0 & 0 & 0 \\
0 & + & 0 & 0 & + & + & + & + & - & - & - & - & 0 & 0 & 0 & 0 & 0 & 0 \\
0 & 0 & + & 0 & + & + & - & - & + & + & - & - & 0 & 0 & 0 & 0 & 0 & 0 \\
0 & 0 & 0 & + & + & + & - & - & - & - & + & + & 0 & 0 & 0 & 0 & 0 & 0 \\
+ & + & + & + & - & 0 & 0 & 0 & 0 & 0 & 0 & 0 & 0 & 0 & + & + & + & + \\
+ & + & + & + & 0 & - & 0 & 0 & 0 & 0 & 0 & 0 & 0 & 0 & - & - & - & - \\
+ & + & - & - & 0 & 0 & - & 0 & 0 & 0 & 0 & 0 & 0 & 0 & + & + & - & - \\
+ & + & - & - & 0 & 0 & 0 & - & 0 & 0 & 0 & 0 & 0 & 0 & - & - & + & + \\
+ & - & + & - & 0 & 0 & 0 & 0 & - & 0 & 0 & 0 & + & + & 0 & 0 & + & - \\
+ & - & + & - & 0 & 0 & 0 & 0 & 0 & - & 0 & 0 & - & - & 0 & 0 & - & + \\
+ & - & - & + & 0 & 0 & 0 & 0 & 0 & 0 & - & 0 & + & - & + & - & 0 & 0 \\
+ & - & - & + & 0 & 0 & 0 & 0 & 0 & 0 & 0 & - & - & + & - & + & 0 & 0 \\
0 & 0 & 0 & 0 & 0 & 0 & 0 & 0 & + & - & + & - & - & 0 & + & - & + & - \\
0 & 0 & 0 & 0 & 0 & 0 & 0 & 0 & + & - & - & + & 0 & - & - & + & + & - \\
0 & 0 & 0 & 0 & - & + & - & + & 0 & 0 & + & - & + & - & - & 0 & 0 & 0 \\
0 & 0 & 0 & 0 & - & + & - & + & 0 & 0 & - & + & - & + & 0 & - & 0 & 0 \\
0 & 0 & 0 & 0 & - & + & + & - & + & - & 0 & 0 & + & + & 0 & 0 & - & 0 \\
0 & 0 & 0 & 0 & - & + & + & - & - & + & 0 & 0 & - & - & 0 & 0 & 0 & -
\end{bmatrix}$$

Display 8.2

For n = 11: The existence of a W(22, k) for k ∈ {0, 1, 2, 4, 5, 8, 10, 16, 20} constructed from circulant matrices is established by Lemma 4.144.

W(22, 9), W(22, 13), W(22, 17) and W(22, 18) constructed from circulant matrices may be obtained by circulating the following first rows:

1 0 0 0 0 0 1 0 - 0 -,	0 1 0 1 1 0 0 1 0 - 0,
1 1 1 0 - 1 1 0 - 1 -,	1 0 1 0 0 0 - 0 0 0 1,
- 1 0 1 1 1 0 0 0 1 0,	1 1 - 1 1 1 - - - 1 -,
1 1 0 1 1 0 1 1 - - -,	1 1 - 1 - 1 0 1 1 0 -.

For n = 13: The existence of W(26, k) for k ∈ {0, 1, 2, 4, 5, 8, 10, 16, 20} constructed from circulant matrices is established by Lemma 4.144.

There exists a circulant W(13, 9) = B.

Hence there exists a W(26, 9) and a W(26, 18) constructed from circulant matrices.

The W(26, 13), W(26, 17) and W(26, 25) may be obtained by using the circulant matrices with the following first rows:

1 1 1 0 - 1 1 0 - 1 - 0 0,	1 0 1 0 0 0 - 0 0 0 1 0 0,
1 1 1 - 1 1 - 1 0 0 - - -,	1 1 0 0 1 0 1 0 1 0 0 0 -,
0 1 - - - 1 - - 1 - - - 1,	1 - 1 1 - - - - - - 1 1 -.

For n = 15: The existence of a W(30, k) for the k of the enunciation with the exception of k = 9, 13, 25, 26 and 29 is given by Lemma 4.144.

The W(30,13), W(30,26) and W(30,29) can be constructed using the following first rows:

1 1 1 0 - 1 1 0 - 1 - 0 0 0 0,	1 0 1 0 0 0 - 0 0 0 1 0 0 0 0;
1 1 1 - - 1 1 1 - 1 - - 0 - 0,	1 1 1 1 - 1 1 - - 1 - 1 0 1 0;
0 1 1 - - 1 - - - - 1 - - 1 1,	1 - 1 - - - - 1 1 - - - - 1 -.

Let D be a W(10, 9) (see above) constructed from two circulants C and B; then $D \oplus D \oplus D$ is a W(30, 9), and replacing the variables of the (2,2) design in order 6 by C' and B gives a W(30,18).

Let A be a W(6, 5); then $A \times B + I \times J$ is a W(30,25).

We can say a little about the orders 34, 38 and 42.

8.17. LEMMA. There exists a W(34,k) for $k \in \{0,1,2,4,5,8,9,10,13,16,20,26,32\}$. All but the W(34, 9) are constructed from two circulants. The undecided cases are $W(34,\ell)$ for $\ell \in \{17,18,25,29\}$.

Proof. $W(18,9) \oplus W(16,9) = W(34,9)$. The Golay sequences of length 16 give W(34, 32). The remaining results follow from Lemma 4.144.

8.18. LEMMA. There exists a W(38,k) for $k \in \{0,1,2,4,5,8,9,10,13,16,17,20,26,32,37\}$. All but the W(38, 9) and W(38,17) are constructed from two circulants. The undecided cases are $W(38,\ell)$ for $\ell \in \{18,25,29,34,36\}$.

Proof. $W(20,9) \oplus W(18,9) = W(38,9)$ and $W(20,17) \oplus W(18,17) = W(38,17)$. The Golay sequences of length 16 give W(38, 32), and Theorem 4.41 gives a W(38, 37) constructed from two circulants. The remaining results come from Lemma 4.144.

8.19. LEMMA. There exists a W(42,k) for $k \in \{0,1,2,4,5,8,9,10,13,16,17,18,20,26,32,34,36,40,41\}$. All but the W(42, 9), W(42,18), W(42,25) and W(42, 36) are constructed from two circulants. The undecided cases are $W(42,\ell)$, $\ell \in \{25,29,37\}$.

Proof. From Lemma 4.144 we have the result for $k \in \{0,1,2,4,5,8,10,13,16,20,26\}$. From Lemma 4.142 we have a $W(42,34)$ and a $W(42,17)$ constructed from two circulants. From Proposition 4.116 we have the result for $k \in \{2,4,8,10,16,20,32,40\}$. Since 41 is a prime, a construction of Goethals and Seidel (see Theorem 4.41) gives a $W(42,41)$ constructed from two circulants.

A $W(42,\ell)$ exists for $\ell \in \{9,18\}$ because $W(20,\ell)$ and $W(22,\ell)$ exist. Let $J-I$ and B be the circulant matrices with first rows:

$$0\,1\,1\,1\,1\,1\,1 \quad \text{and} \quad 0\,1\,1\,-\,1\,-\,-.$$

Then $B^t = -B$ and $JB^t = BJ = 0$. Let A be a symmetric $W(6,5)$; then $(J-I) \times I + B \times A$ is a $W(42,36)$.

Eades found a GGS array of type $(5,5)$ in orders $2n$, $n \geq 6$. There are 2-complementary sequences of type $(1,4)$ for all lengths $m \geq 3$. Hence there are orthogonal designs of type $(5,20)$ in orders $2mn$. In particular, there is an orthogonal design of type $(5,20)$ in order 42 and a $W(42,25)$.

§8. 3. RESULTS ON ORTHOGONAL DESIGNS (1,k) AND SKEW-WEIGHING MATRICES. THE WEIGHING MATRIX CONJECTURE FOR ORDERS ≡ 0 (mod 4)

We proceed case by case to examine the $(1,k)$ conjecture. Later we use these to consider $W(n,k)$ for $n \equiv 0 \pmod 4$.

8.20. LEMMA. All designs of type $(1,k)$, $0 \leq k \leq n-1$, exist in orders $n = 2^t$, t a positive integer.

Proof. See Example 5. 3.

8.21. COROLLARY. All $W(n,k)$ exist for $k = 1,2,\ldots,n$, where $n = 2^t$, t a positive integer. If $k \neq n$, all the weighing matrices may be chosen to be skew-symmetric.

8.22. LEMMA. There exist designs $(1,k)$ for $k = a^2+b^2+c^2$, a, b, c integers, $0 \leq k \leq n-1$, in orders $n = 12,20,28$. All are constructed from circulant matrices.

Proof. See the appropriate appendix giving all orthogonal designs in order n.

8.23. LEMMA. All designs of type (1, k), $0 \leq k \leq n-1$, exist in orders $n = 24$ and $n = 40$.

Proof. Using the result that if a design of type (1, a) exists in order m a design of type (1, 1, a, a) exists in order 2m, we see, using the previous lemma, that we only have to consider the existence of a (1, 14) and (1, 15) in order 24 and a (1, 15), (1, 30) and (1, 31) in order 40.

From the appendices giving all the orthogonal designs in orders 12 and 20, we see that there are (1, 4, 5) and (1, 5, 5) designs in order 12 and (1, 1, 13), (1, 4, 13) and (1, 3, 14) designs in order 20. Then using the result that a design of type (1, i, k) in order m gives a design of type (1, i, 2k) (or (1, 2i, k)) in order 2m, we obtain the required designs.

8.24. COROLLARY. All W(n, k) exist for $k = 1, 2, \ldots, n$, where $n = 24$ and $n = 40$. There exist skew-symmetric weighing matrices for all $k \neq n$.

8.25. COROLLARY. All designs of type (1, k), $0 \leq k \leq n-1$, exist in orders $n = 2^{t+2}.3$ and $n = 2^{t+2}.5$, t a positive integer.

8.26. COROLLARY. All W(n, k) exist for $k = 1, 2, \ldots, n$, where $n = 2^{t+2}.3$ and $n = 2^{t+2}.5$, t a positive integer. There exist skew-symmetric weighing matrices for all $k \neq n$.

8.27. LEMMA. There exists an orthogonal design of type (1, k) in order 56 for $0 \leq k \leq 55$.

Proof. There exists an orthogonal design of type (1, k) for $k \in \{x : x \neq 7, 15, 23, 0 \leq x \leq 27\}$ in order 28 (for example, see the appendix giving orthogonal designs in order 28). Hence using the result that a (1, k) in order n gives a (1, 1, k, k) in order 2n, we have the result for all (1, k) in 56 except k = 30, 31, 46, 47.

There are orthogonal designs of types (1,6,12) and (1,9,13) in order 28 giving designs of types (1,6,24) and (1,18,13) in order 56. Hence we have the designs (1, 30) and (1, 31). It remains to construct a (1,46) and a (1,47). For these we use the (1,1,1,1,1,1,1,1) orthogonal design in order 8 and replace the variables x_i by the circulant matrices X_i indicated below and the variables x_j by the back-circulant matrices X_j indicated below.

For (1,46) use the matrices with first rows (X_2 back-circulant, the rest circulant):

X_1 : x 0 0 0 0 0 0, X_5 : y y y -y -y y y
X_2 : y y y 0 y 0 0, X_6 : y y -y -y -y -y y,
X_3 : y -y y y y y -y, X_7 : y -y y -y -y y -y,
X_4 : y y -y y y -y y, X_8 : y -y -y y y -y -y,

and for (1,47) use the matrices with first rows (X_1 circulant, the rest back-circulant):

X_1 : x y 0 y -y 0 -y, X_5 : y -y -y y y -y -y,
X_2 : 0 y y y y y y, X_6 : y -y y -y -y y -y,
X_3 : 0 y -y y y -y y, X_7 : y y -y -y -y -y y,
X_4 : 0 y y -y -y y y, X_8 : 0 y 0 -y -y 0 y.

8.28. COROLLARY. There exists a W(56,k) for $0 \le k \le 56$.

8.29. COROLLARY. All designs of type (1,k), $0 \le k \le n-1$, exist in order $n = 2^{t+2}.7$, t a positive integer.

8.30. COROLLARY. All W(n,k) exist for k = 1,2,...,n, where $n = 2^{t+2}.7$, t a positive integer. There exist skew-symmetric weighing matrices for all $k \ne n$.

8.31. LEMMA. There exists an orthogonal design (1,k) in order 36 for $k \in \{x : x$ is the sum of three integer squares, $0 \le x \le 36\}$. All may be constructed using four circulant matrices in the Goethals-Seidel array.

Proof. In Theorem 4.149 it is shown there exists a (1, k) in every order 4n, $n \geq 9$, for $k \in \{1,\dots,6,8,\dots,14,16,\dots,22,24,\dots,27,29,30,32,33\}$.

The following four first rows may be used to construct circulant matrices A_1, A_2, A_3, A_4, which may be used as $aI+cA_1$, $bI+cA_2$, cA_3, cA_4 in the Goethals-Seidel array to give orthogonal designs (1,1,18), (1,1,29) and (1,1, 34) in order 36:

(1,1,18):	0 0 0 0 1 - 0 0 0,	0 0 0 0 1 - 0 0 0,
	1 - - 0 1 1 0 1 1,	1 1 1 0 1 - 0 1 -,
(1,1,29):	0 1 1 - 1 - 1 - -,	0 1 1 1 - 1 - - -,
	1 - 0 1 1 1 1 0 1,	1 - 0 1 0 1 0 - 1,
(1,1,34):	0 1 - - 1 - 1 1 -,	0 1 1 - 1 - 1 - -,
	1 1 - 1 1 1 1 - 1,	1 1 - - - - - - 1.

Hence we have orthogonal designs of types (1, 30), (1, 34) and (1, 35) in order 36.

8.32. LEMMA. There exists an orthogonal design (1, k) in order 72, $0 \leq k \leq 71$.

Proof. We know that the existence of a (1, k) in order n means there is a (1,1,k,k) in order 2n. Hence by the results of the previous lemma we have a design (1, k) in order 72 for $k \in \{1\text{-}14, 16\text{-}30, 32\text{-}45, 48\text{-}55, 58\text{-}61, 64\text{-}71\}$.

We recall (from Lemma 8.31) that if there exists a W(2n, k), there exists an orthogonal design (1, k) in order 4n. Now W(36,15), W(36, 30) exist, so we have a (1,15) and (1, 30) in order 72.

There exists a W(18,17) constructed of two circulant symmetric matrices of order 9. Call these matrices P and Q. Write A = P+I, B = P-I. Let L, M be the circulant matrices with first rows 00001-000 and 000011000.

The following orthogonal design may be constructed by replacing the variables of the orthogonal design (1,1,1,1,1,1,1,1) of order 8 by the matrices specified (R, as defined previously, is the back-diagonal matrix):

(1, 4, 17, 36) and $(x_1I+x_2L)R$, x_2M, x_3P, x_3Q, x_4A, x_4B, x_4Q, x_4Q.

Hence we have a (1, 57) in order 72.

Let I be the identity, J the matrix of all 1's, K = J-I, A = J-2I,

$$B = \begin{bmatrix} 0 & 1 & - \\ - & 0 & 1 \\ 1 & - & 0 \end{bmatrix} \quad \text{and} \quad R = \begin{bmatrix} 0 & 0 & 1 \\ 0 & 1 & 0 \\ 1 & 0 & 0 \end{bmatrix}.$$

Then the remaining required designs may be obtained by replacing the variables of the indicated four variable designs (which do exist) in order 24 by the variable matrices shown.

For (1, 46) use x_1I, x_2I, x_2K, x_2A in the (1, 1, 9, 9) design.

For (1, 47) use x_1I, x_2I, x_2K, x_2A in the (1, 2, 9, 9) design.

For (1, 56) use x_1I, x_2I, x_2K, x_2A in the (1, 1, 11, 11) design.

For (1, 62) use x_1I+x_2B, x_2KR, x_2J, x_2AR in the (1, 3, 4, 14) design.

For (1, 63) use x_1I, x_2I, x_2J, x_2A in the (1, 3, 5, 15) design.

8.33. COROLLARY. There exists a W(72, k) for $0 \le k \le 72$.

Proof. The existence of (1, k) in order 72 as stated in Lemma 8.32 gives us the result for all k.

8.34. COROLLARY. All designs of type (1, k), $0 \le k \le n-1$, exist in order $n = 2^{t+2} \cdot 9$, t a positive integer.

8.35. COROLLARY. All W(n, k) exist for k = 1, 2, . . . , n, where $n = 2^{t+2} \cdot 9$, t a positive integer. There exist skew-symmetric weighing matrices for all $k \ne n$.

8.36. LEMMA. There exists an orthogonal design (1, k) in order 44 for $k \in \{x : x$ is the sum of three integer squares, $x \ne 42$, $0 \le x \le 43\}$. A (1, 42) does not exist in order 44. All may be constructed using four circulant matrices in the Goethals-Seidel array.

Proof. From Theorem 4.149 we have the existence of a (1, k) in order

44 for $k \in \{1,\ldots,6,8,\ldots,14,16,\ldots,22,24,\ldots,27,29,30,32,33,35,36,38,40,41,43\}$.

Let A, B, I, T be the circulant matrices of order 11 with the following first rows:

0 1 0 1 1 1 0 0 0 1 0, 0 1 - 1 1 1 - - - 1 -,

1 0 0 0 0 0 0 0 0 0 0, 0 0 1 0 0 0 0 0 0 0 0.

Now we get the orthogonal designs indicated below by using the specified circulant matrices in the Goethals-Seidel array (0 is the zero matrix of order 11; W(22,k) indicates the two circulant matrices which may be used to form this weighing matrix):

for (1, 34) use x_1I+x_2B, $x_2(A-I+T)$, $x_2(A-I-T)$, x_2B;
for (1, 37) use x_1I+x_2B, $x_2(A+I)$, x_2B, $x_2(B+I)$.

If a (1, 42) existed in order 44, then a (1, 1, 42) design would exist, but this is not possible.

8.37. LEMMA. There exists an orthogonal design (1, k) in order 88 for $k \in \{x : 1 \le x \le 87,\ x \ne 56,57,62,63,79\}$.

Proof. The existence of designs (1, 15) and (1, 31) in orders 40 and 48 gives designs of these types in order 88. There is a (1, 43) design in order 44 constructed with three circulant symmetric matrices. Hence there is a design of type (1, 1, 1, 1, 1, 43) in order 88 which gives the (1, 46) and (1, 47) designs. Lemma 4.147 gives a (1, 2, 2, 36) design in 44 and hence a (1, 2, 4, 72) and (1, 78) design in 88. The remaining designs of the enunciation can be constructed from designs (1, k) that exist in order 44.

8.38. LEMMA. There exists an orthogonal design (1, k) in order 52 for $k \in \{x : x$ is the sum of three integer squares, $x \ne 37, 42, 45, 46, 48, 0 \le x \le 51\}$. All may be constructed using four circulant matrices in the Goethals-Seidel array.

Proof. From Lemma 4.149 we have the result for $k \in \{1,\dots,6,8,\dots,14, 16,\dots,22,24,\dots,27,29,30,32,33,35,36,38,40,41,43,44,50,51\}$.

The circulant matrices with the following first rows give a (1,16) design in order 26,

0 - 0 - 1 0 0 0 0 - 1 0 1   0 0 1 1 0 1 1 1 - 1 1 - -.

Call these matrices A and B, respectively.

$aI+bA$, $cI+bA$, $dI+bB$, $dI-bB$ may be used in the Goethals-Seidel array to give a design of type (1,1,2,32) in order 52. Hence we have a (1,34) and (1,35) design.

A (1,1,1,49) design in order 52 can be constructed by using $aI+dX$, $bI+dY$, $cI+dZ$, dW in the Goethals-Seidel array, where X, Y, Z, W are the circulant matrices with the following first rows:

0 1 1 1 1 - 1 - 1 - - - -,   0 1 - 1 1 1 - 1 - - - 1 -,

0 1 - - 1 1 1 - - - 1 1 -,   1 - - - - 1 - - 1 - - - -.

Hence we have designs of types (1,49), (1,50) and (1,51).

8.39. LEMMA. There exists an orthogonal design (1, k) in order 60 for $k \in \{x : x$ is the sum of three integer squares, $x \neq 34,42,48,53,54, 56,57,\ 0 \le x \le 59\}$.

Proof. Theorem 4.149 gives us the result for $k \in \{1,\dots,6,8,\dots,14, 16,\dots,22,24,\dots,27,29,30,32,33,36,38,40,41,43,44,50,51,52,58,59\}$.

Now the following orthogonal designs exist in order 12: (1,1,2,8), (1,2,3,6), (1,1,4,4), (1,1,5,5).

Let I, L, B, K, Y, J be the circulant symmetric matrices with first rows

1 0 0 0 0,   0 1 1 1 1,   0 1 - - 1,

- 1 1 1 1,   0 0 1 1 0,   1 1 1 1 1,

and let X, Z be the back-circulant matrices with first rows

$x_1\ 0\ x_2\ -x_2\ 0$, $x_1\ x_2\ -x_2\ x_2\ -x_2$.

The following orthogonal designs (1,k) may be constructed by replacing the variables of the orthogonal designs, specified by the matrices indicated (0 is the zero matrix of order 5):

(1,35): use $x_1 I$, $x_2 L$, $x_2 I$, $x_2 B$ in the (1,2,3,6) orthogonal design;

(1,37): use $x_1 I$, $x_2 I$, $x_2 K$, $x_2 B$ in the (1,1,4,4) orthogonal design;

(1,45): use $x_1 I$, 0, $x_2 K$, $x_2 B$ in the (1,1,5,5) orthogonal design;

(1,46): use $x_1 I$, $x_2 I$, $x_2 K$, $x_2 B$ in the (1,1,5,5) orthogonal design;

(1,49): use X, $x_2 Y$, $x_2 K$, $x_2 B$ in the (1,1,5,5) orthogonal design.

We now consider the weighing matrix conjecture, "When $n \equiv 0 \pmod 4$, there exists a $W(n,k)$ for all $k \le n$". The results are summarized in Table 8.3. We have first:

8.40. SUMMARY. The weighing matrix conjecture is true for orders 2^t, $2^{t+1}.3$, $2^{t+1}.5$, $2^{t+1}.7$, $2^{t+1}.9$, t a positive integer.

Proof. Use Corollaries 8.21, 8.25, 8.30 and 8.35, and Lemmas 8.22 and 8.31.

REMARK. We observe that for orders $n = 12,20,28$ and 36 all the $W(n,k)$, $k \le n$, could be constructed using four circulant matrices in the Goethals-Seidel array.

8.41. LEMMA. There exists a $W(44,k)$ for $k \in \{x : 0 \le x \le 44\}$. All are constructed using four circulant matrices in the Goethals-Seidel array. Hence, the weighing matrix conjecture is true for order 44.

Proof. Use Lemma 8.36.

8.42. LEMMA. There exists a $W(52,k)$ for $k \in \{x : x \ne 46,\ 0 \le x \le 52\}$. All may be constructed from four circulant matrices using the Goethals-Seidel array.

Proof. The existence of $W(52,k)$ for $k \in \{0,1,2,\dots,45,49,\dots,52\}$

constructed using four circulant matrices in the Goethals-Seidel array follows from Lemma 8.38.

The $W(52,47)$ and $W(52,48)$ may be obtained, respectively, from the circulant matrices with the following first rows:

47: 1 1 1 - 1 1 - 1 1 1 - - -, 1 1 1 - 1 - 0 - 1 - - 1 1,
1 1 1 - 1 1 1 1 - - 1 0 0, 1 1 1 - 1 - - 1 - 1 1 0 0.

48: 0 1 - - - 1 - - 1 - - - 1, 0 - 1 1 - - - - - - 1 1 -,
0 - - - 1 - 1 1 - 1 - - -, 0 1 - 1 1 - - - - 1 1 - 1.

8.43. LEMMA. There exists a $W(60,k)$ for $k \in \{x : 0 \leq x \leq 60\}$. All but $W(60,57)$ are constructed from circulant matrices. Hence the weighing matrix conjecture is true for order 60.

<u>Proof.</u> The existence of $W(60,k)$ for $k \in \{x : 0 \leq x \leq 53, 55, 58, 59\}$ constructed from circulant matrices using the Goethals-Seidel array follows from Lemmas 8.39, 4.147 (for $W(60,48)$), and 4.134 (use the $W(30,29)$ and $W(30,26)$ to form the $W(60,55)$). $W(60,54)$ and $W(60,56)$ constructed using the Goethals-Seidel array may be obtained by using the $W(30,26)$ with Lemma 4.137 (i), (ii).

We define the circulant symmetric $(0,1,-1)$-matrices C, K, each of order 5, by defining their first rows:

1 - 1 1 -, - 1 1 1 1.

We also define the back-circulant $(0,1,-1)$-matrix D, of order 5, with first row:

0 1 1 - -.

Now

$$XY^T = YX^T \text{ for } X, Y \in \{C, D, K\}.$$

Hence the $W(60,57)$ may be constructed by replacing the variables of the orthogonal design of type $(3,3,6)$ in order 12 by D, C and K.

8.44. LEMMA. There exist W(68,k) for k ∈ {1,...,48,50,...,54,56, 58,...,60,62,64,66,67,68} constructed by using four circulant matrices in the Goethals-Seidel array.

Proof. Lemmas 8.10 and 4.133, 4.147 and 4.149 give the result for k ∈ {1,...,48,50,...,53,58,...,60,64}. Lemma 4.136 gives the result for k ∈ {54,56,62,66,68}. The W(68,67) and W(68,68) are constructed using the following first four rows to give circulant matrices:

0 1 - - 1 1 - 1 - 1 - 1 - - 1 1 -, 1 1 - - 1 - - - - - - - - 1 - - 1,
1 1 1 - - - - 1 - - 1 - - - - 1 1, 1 1 - - - 1 - 1 - - 1 - 1 - - - 1.

8.45. LEMMA. There exist W(76,k) for k ∈ {1,...,54, 56,...,60,62,...,70,72,74,75,76} constructed using four circulant matrices in the Goethals-Seidel array.

Proof. Lemmas 8.10 and 4.147, 4.149 and 4.168 give the result for k ∈ {1,...,53,57,...,60,63,64,66,...,69,74}. Lemma 4.136 gives the result for k ∈ {54,56,62,66,68,72,76}. The W(76,65) and W(76,75) are constructed using the following four rows to give circulant matrices:

65: - 1 0 0 1 1 1 1 0 1 0 1 0 0 0 0 1 1 0,
1 1 - - 1 1 1 1 - 1 - 1 - - - - 1 1 -,
0 1 - - 1 1 1 1 - 1 - 1 - - - - 1 1 -,
0 1 - - 1 1 1 1 - 1 - 1 - - - - 1 1 -.

75: 0 - 1 - - - - - 1 1 - - 1 1 1 1 1 - 1,
1 1 - 1 1 1 1 1 - - - - 1 1 1 1 1 - 1,
1 - 1 - - - - 1 1 - - 1 1 - - - - 1 -,
1 - - 1 - 1 - 1 1 - - 1 1 - 1 - 1 - -.

Lemma 4.147 gives 4-complementary sequences $A = \{X, Y, Z, W\}$ of length 9 with $N_A = 0$ giving a (1, 35) design; hence $B = \{\{X, Y\}, \{X, -Y\}, \{Z, W\}, \{Z, -W\}\}$ are 4-complementary sequences of length 18 with $N_B = 0$ giving a (2, 70) design in orders 4n, $n \geq 18$. Hence we have a W(76,70).

8.46. LEMMA. There exists a W(84,k) for $k \in \{0,1,\ldots,58,60,\ldots,62,$ $64,66,\ldots,70,72,\ldots,76,78,80,\ldots,84\}$. All are constructed by using the Goethals-Seidel array.

Proof. The existence of W(42,k) constructed from two circulants gives the result for

$$k \in \{0,1,\ldots,54,56,\ldots,58,60,61,64,66,\ldots,68,72,\ldots,75,80,\ldots,82\}.$$

The W(42,26) used with Lemma 4.136 gives W(84,62). From Lemma 4.139 there are designs of types (2,2,34,34), (1,6,48) and (19,19,19,19) which give a W(84,70), W(84,55) and W(84,76). A (1,83) can be constructed from the circulant matrices with the following first rows:

0 1 1 - - - - - - - 1 - 1 1 1 1 1 1 - -, 1 - - - 1 - - 1 - 1 1 - 1 - - 1 - - -,
1 - 1 - - - - 1 1 - - 1 1 - - - - 1 -, 1 - - - 1 - 1 1 - - - - 1 1 - 1 - - -.

This gives a W(84,83) and a W(84,84).

Lemma 4.148 gives a (1,1,58) design and hence a W(84,59). Lemma 4.147 gives a (1,2,66) design and hence a W(84,69). Lemma 4.147 gives a (1,2,2,36) constructed from sequences X, Y, Z, W of length 11, 11, 10, 10 with zero nonperiodic autocorrelation function; hence {X, Z}, {X, -Z}, {Y, W}, {Y, -W} are 4-complementary sequences of length 21 with zero nonperiodic autocorrelation function which give a (2,4,4,72) design in orders 4n, $n \geq 21$, and a W(84,78).

We summarize the results in order 8n.

8.47. SUMMARY. There exist orthogonal designs of order 8n and type (1,k) where

(i) n = 1,2,3,4,5,6,7,8,9 or 10, $k \in \{1,\ldots,8n-1\}$;

(ii) $n \geq 11$, $k \in \{1,\ldots,47\}$.

Order	Applicable Conjecture	Unresolved Cases	Applicable Conjecture	Unresolved Cases
4	8.1	true	8.2	true
8	8.1	true	8.3	true
12	8.1	true	8.2	true
16	8.1	true	8.3	true
20	8.1	true	8.2	true
24	8.1	true	8.3	true
28	8.1	true	8.2	true
32	8.1	true	8.3	true
36	8.1	true	8.2	true
40	8.1	true	8.3	true
44	8.1	true	8.2	true*
48	8.1	true	8.3	true
52	8.1	46	8.2	37, 42, 45, 46, 48
56	8.1	true	8.3	true
60	8.1	true	8.2	34, 42, 48, 53, 54, 56, 57
64	8.1	true	8.3	true
68	8.1	49, 55, 57, 63, 65	8.2	
72	8.1	true	8.3	true
76	8.1	55, 61, 71, 73	8.2	
80	8.1	true	8.3	true
84	8.1	59, 63, 65, 71, 77, 79	8.2	
88	8.1	57, 63	8.3	56, 57, 62, 63, 79

Table 8.3

True signifies the conjecture is verified.

*(1, n-2) is not possible as n-2 is not the sum of two squares.

DESCRIPTION OF APPENDICES A, B, AND C.

In Appendices A, B, and C we give the first rows of circulant matrices which can be used in the Goethals-Seidel array (Theorem 4.49) to obtain orthogonal designs of the type described in orders 12, 20 and 28.

In each of these appendices the format is the same; we first list all 4-tuples which might be the type of an orthogonal design in that order (i.e. not eliminated by any known theorem); this is then followed by a list of 3-tuples which might be the type of an orthogonal design in that order and is not already known to exist because it can be obtained from a known 4-tuple by equating or killing variables. We continue in this fashion with 2-typles. The blank spaces indicate that we have not been able to construct or prove non-existence of the orthogonal design in question.

These lists were compiled by computer programmes and the constructions of the first rows were done by hand.

APPENDIX A

Order 12

Design	A_1			A_2			A_3			A_4		
(1, 1, 1, 1)	a	0	0	b	0	0	c	0	0	d	0	0
(1, 1, 1, 4)	a	0	0	b	0	0	c	d	-d	0	d	d
(1, 1, 1, 9)	a	d	-d	b	d	-d	c	d	-d	d	d	d
(1, 1, 2, 2)	a	0	0	b	0	0	c	d	0	c	-d	0
(1, 1, 2, 8)	a	d	-d	b	d	-d	c	d	d	-c	d	d
(1, 1, 4, 4)	a	c	-c	0	c	c	b	d	-d	0	d	d
(1, 1, 5, 5)	a	c	-c	b	d	-d	c	d	d	-d	c	c
(1, 2, 2, 4)	a	d	-d	0	d	d	b	c	0	b	-c	0
(1, 2, 3, 6)	a	d	-d	c	d	d	c	-d	b	c	-b	-d
(2, 2, 2, 2)	a	b	0	a	-b	0	c	d	0	c	-d	0
(2, 2, 4, 4)	a	c	d	a	-d	-c	b	c	-d	b	-c	d
(3, 3, 3, 3)	a	b	c	-b	a	d	-c	-d	a	-d	c	-b

APPENDIX B

Order 20

Design	A_1	A_2	A_3	A_4
(1, 1, 1, 1)	a 0 0 0 0	b 0 0 0 0	c 0 0 0 0	d 0 0 0 0
(1, 1, 1, 4)	a 0 0 0 0	b 0 0 0 0	c 0 d -d 0	0 0 d d 0
(1, 1, 1, 9)	a 0 d -d 0	b 0 d -d 0	c -d 0 0 d	d d d 0 0
(1, 1, 2, 2)	a 0 0 0 0	b 0 0 0 0	c d 0 0 0	d -c 0 0 0
(1, 1, 2, 8)	a 0 d -d 0	b 0 d -d 0	c 0 d d 0	-c 0 d d 0
(1, 1, 4, 4)	a 0 c -c 0	0 0 c c 0	b 0 d -d 0	0 0 d d 0
(1, 1, 4, 9)	a 0 d -d 0	b 0 d -d 0	0 d c c -d	d d c -c d
(1, 1, 5, 5)	a 0 c -c 0	b 0 d -d 0	c 0 d d 0	-d 0 c c 0
(1, 1, 8, 8)	a c d -d -c	b c -d d -c	0 c d d c	0 c -d -d c
(1, 1, 9, 9)	a c d -d -c	b d -c c -d	-c c c c c	-d d d d d
(1, 2, 2, 4)	a 0 d -d 0	0 0 d d 0	b c 0 0 0	c -b 0 0 0
(1, 2, 2, 9)	a d 0 0 -d	d d d 0 0	b c d -d 0	c -b 0 d -d
(1, 2, 3, 6)	a 0 d -d 0	b c 0 d 0	b -c 0 -d 0	c 0 -d -d 0
(1, 2, 4, 8)	a 0 d -d 0	b c d d -c	b 0 -d -d 0	d -d c 0 c
(1, 2, 8, 9)	a d c -c -d	d d d c -c	c b d -d c	b -c -c d -d
(1, 3, 6, 8)	if it exists it cannot be constructed from circulants			
(1, 4, 4, 4)	a 0 b -b 0	0 0 b b 0	0 c d -d c	0 c d d -c
(1, 4, 4, 9)	if it exists it cannot be constructed from circulants			

Order 20 (Continued)

Design	A_1	A_2	A_3	A_4
(1, 4, 5, 5)	a d 0 0 -d	-d c 0 0 c	c d b -b d	0 c b b -c
(1, 5, 5, 9)	a b c -c -b	-c b d -d b	b d c c -d	-d d d d d
(2, 2, 2, 2)	a b 0 0 0	b -a 0 0 0	c d 0 0 0	d -c 0 0 0
(2, 2, 2, 8)	a d b -d 0	a -d -b d 0	c 0 d 0 d	c 0 -d 0 -d
(2, 2, 4, 4)	a b 0 0 0	b -a 0 0 0	0 c d -d c	0 c d d -c
(2, 2, 4, 9)	a b d -d 0	a -b d -d 0	0 d c c -d	d d c -c d
(2, 2, 5, 5)	if it exists it cannot be constructed from circulants			
(2, 2, 8, 8)	a d c d -c	a -d -c -d c	b c -d c d	b -c d -c -d
(2, 3, 4, 6)	b d a 0 0	b d -a 0 0	-b d c -c d	0 d c c -d
(2, 3, 6, 9)	b d c a -d	a d -d -b -c	-b d -d c c	d d -c c d
(2, 4, 4, 8)	c 0 b d d	c 0 b -d -d	b a -c d -d	b -a -c -d d
(2, 5, 5, 8)	c b d -d b	a c d d -c	-a b d d -b	-b c d -d c
(3, 3, 3, 3)	a b c 0 0	a -b 0 d 0	a 0 -c -d 0	b -c d 0 0
(3, 3, 6, 6)	0 d -a c -b	0 d a -c -b	b c d d -c	a c d -d c
(4, 4, 4, 4)	0 a b -b a	0 a b b -a	0 c d -d c	0 c d d -c
(4, 4, 5, 5)	d c a -a c	0 d a a -d	0 c b b -c	-c d b -b d
(5, 5, 5, 5)	a b b d -d	-b a a c -c	d c c -a a	-c d d -b b
(1, 1, 13)	a c -c c -c	0 c -c -c -c	b c 0 0 -c	0 c c c 0
(1, 2, 17)	a c -c c -c	b c -c -c -c	b -c c c c	c -c -c -c -c
(1, 2, 11)	a 0 c -c 0	b 0 c -c -c	b 0 -c c c	0 c 0 c c

(1, 3, 14)	a c -c c -c	-b c c c 0	c c c b -c	c c -c 0 b
(1, 4, 13)	a c -c c -c	0 c -c -c -c	0 c b b -c	b c c c -b
(1, 6, 11)	a c b -b -c	c c b b -c	c -b c c 0	c 0 c -c b
(1, 8, 11)	a c -b b -c	-c c b b -c	c -b c c b	c b c -c b
(2, 5, 7)	c c -c 0 -a	a c b -b 0	b b 0 c 0	c -b 0 c 0
(3, 6, 8)	0 b c -b -c	0 b c a c	a b -c 0 -c	a -b c -b -c
(3, 7, 8)	not yet decided			
(7, 10)	0 a a -b -b	b a a -a a	0 a a -a b	b 0 b -b a

APPENDIX C

Order 28

Design	A_1							A_2							A_3							A_4						
(1,1,1,1)	a	0	0	0	0	0	0	b	0	0	0	0	0	0	c	0	0	0	0	0	0	d	0	0	0	0	0	0
(1,1,1,4)	a	0	0	0	0	0	0	b	0	0	0	0	0	0	d	c	-d	0	0	0	0	d	0	d	0	0	0	0
(1,1,1,9)	d	a	-d	0	0	0	0	d	b	-d	0	0	0	0	d	0	c	0	-d	0	0	d	0	d	0	d	0	0
(1,1,1,16)	a	d	d	-d	d	-d	-d	b	0	d	0	0	-d	0	c	0	d	0	0	-d	0	0	d	d	-d	d	d	d
(1,1,1,25)	a	d	d	-d	d	-d	-d	b	d	d	-d	d	-d	-d	c	d	d	-d	d	-d	-d	-d	d	d	d	d	d	d
(1,1,2,2)	a	0	0	0	0	0	0	b	0	0	0	0	0	0	c	d	0	0	0	0	0	c	-d	0	0	0	0	0
(1,1,2,8)	d	a	-d	0	0	0	0	d	b	-d	0	0	0	0	d	c	d	0	0	0	0	d	-c	d	0	0	0	0
(1,1,2,18)	a	b	b	-b	b	-b	-b	c	b	b	-b	b	-b	-b	d	b	b	0	b	0	0	-d	b	b	0	b	0	0
(1,1,4,4)	a	0	0	0	0	0	0	b	0	0	0	0	0	0	c	c	d	-d	0	0	0	d	d	-c	c	0	0	0
(1,1,4,9)	a	d	d	-d	d	-d	-d	0	d	d	0	d	0	0	b	0	0	c	-c	0	0	0	0	0	c	c	0	0
(1,1,4,16)	b	0	b	0	b	0	b	b	0	b	a	-b	0	-b	b	c	-b	0	-b	c	b	b	c	-b	d	b	-c	-b
(1,1,5,5)	a	0	0	0	0	0	0	b	0	0	0	0	0	0	c	c	-c	d	0	d	0	d	d	-d	-c	0	-c	0
(1,1,8,8)	c	d	a	-d	-c	0	0	-c	d	b	-d	c	0	0	c	d	0	d	c	0	0	c	-d	0	-d	c	0	0
(1,1,8,18)	-a	b	b	a	b	a	a	a	b	b	-a	b	-a	-a	c	b	b	-b	b	-b	-b	d	b	b	-b	b	-b	-b
(1,1,9,9)	a	c	c	-c	c	-c	-c	0	c	c	0	c	0	0	b	d	d	-d	d	-d	-d	0	d	d	0	d	0	0
(1,1,10,10)	d	a	a	-a	a	-a	-a	c	b	b	-b	b	-b	-b	-a	b	b	0	b	0	0	b	a	a	0	a	0	0
(1,1,13,13)	-a	b	b	a	b	a	a	-b	-a	-a	b	-a	b	b	c	a	a	-a	a	-a	-a	d	b	b	-b	b	-b	-b
(1,2,2,4)	b	a	-b	0	0	0	0	b	0	b	0	0	0	0	c	d	0	0	0	0	0	c	-d	0	0	0	0	0

(1, 2, 2, 9)	a d d -d d -d -d	0 d d 0 d 0 0	b c 0 0 0 0 0	b -c 0 0 0 0 0
(1, 2, 2, 16)	a d d -d d -d -d	b 0 d c 0 -d 0	b 0 d -c 0 -d 0	0 d d -d d d d
(1, 2, 3, 6)	a b c 0 0 0 0	a b -c 0 0 0 0	b -a b 0 0 0 0	b d -b 0 0 0 0
(1, 2, 4, 8)	c a -c 0 0 0 0	c 0 c 0 0 0 0	d b -d d 0 d 0	d b -d -d 0 -d 0
(1, 2, 4, 18)				
(1, 2, 6, 12)				
(1, 2, 8, 9)	a d d -d d -d -d	0 d d 0 d 0 0	-b b b c b 0 0	-b b b -c b 0 0
(1, 3, 6, 8)	-c c c 0 c a b	c -c -c 0 -c a b	b -a b 0 0 0 0	b d -b 0 0 0 0
(1, 3, 6, 18)				
(1, 4, 4, 4)	b a -b 0 0 0 0	b 0 b 0 0 0 0	c c d -d 0 0 0	d d -c c 0 0 0
(1, 4, 4, 9)	a d d -d d -d -d	0 d d 0 d 0 0	c c b -b 0 0 0	b b -c c 0 0 0
(1, 4, 4, 16)				
(1, 4, 5, 5)	b a -b 0 0 0 0	b 0 b 0 0 0 0	c c -c d 0 d 0	d d -d -c 0 -c 0
(1, 4, 8, 8)				
(1, 4, 9, 9)				
(1, 4, 10, 10)				
(1, 5, 5, 9)	a d d -d d -d -d	0 d d 0 d 0 0	b b -b c 0 c 0	c c -c -b 0 -b 0
(1, 8, 8, 9)				
(1, 9, 9, 9)				
(2, 2, 2, 2)	a b 0 0 0 0 0	a -b 0 0 0 0 0	c d 0 0 0 0 0	c -d 0 0 0 0 0

Order 28 (Continued)

Design	A_1							A_2							A_3							A_4						
(2, 2, 2, 8)	a	b	0	0	0	0	0	a	-b	0	0	0	0	0	c	d	-c	c	0	c	0	c	d	-c	-c	0	-c	0
(2, 2, 2, 18)	a	d	-d	0	b	d	-d	a	d	-d	0	-b	d	-d	c	d	d	d	d	-d	0	-c	d	d	d	d	-d	0
(2, 2, 4, 4)	a	b	0	0	0	0	0	a	-b	0	0	0	0	0	c	c	d	-d	0	0	0	d	d	-c	c	0	0	0
(2, 2, 4, 9)																												
(2, 2, 4, 16)	a	b	-a	a	c	a	0	a	b	-a	a	-c	a	0	a	b	-a	-a	d	-a	0	a	b	-a	-a	-d	-a	0
(2, 2, 5, 5)	a	b	0	0	0	0	0	a	-b	0	0	0	0	0	c	c	-c	d	0	d	0	d	d	-d	-c	0	-c	0
(2, 2, 8, 8)	a	b	-a	a	0	a	0	a	b	-a	-a	0	-a	0	c	d	-c	c	0	c	0	c	d	-c	-c	0	-c	0
(2, 2, 9, 9)	d	b	-d	c	a	-c	0	d	b	-d	-c	-a	c	0	d	c	0	c	-d	c	0	-c	d	0	d	c	d	0
(2, 2, 10, 10)	c	c	-c	d	a	d	0	c	c	-c	d	-a	d	0	d	d	-d	-c	b	-c	0	d	d	-d	-c	-b	-c	0
(2, 3, 4, 6)	a	d	0	-d	a	0	0	a	d	c	d	-a	0	0	b	c	-d	0	0	0	0	-b	c	-d	0	0	0	0
(2, 3, 6, 9)																												
(2, 4, 4, 8)	a	b	-a	a	0	a	0	a	b	-a	-a	0	-a	0	c	c	d	-d	0	0	0	d	d	-c	c	0	0	0
(2, 4, 4, 18)																												
(2, 4, 6, 12)	a	b	c	a	b	-c	0	a	b	c	-a	-b	c	0	b	-a	b	b	d	-b	0	b	-a	b	-b	-d	b	0
(2, 4, 8, 9)																												
(2, 5, 5, 8)	a	d	-a	a	0	a	0	a	d	-a	-a	0	-a	0	c	c	-c	b	0	b	0	b	b	-b	-c	0	-c	0
(2, 8, 8, 8)	a	a	b	-b	c	d	-c	a	a	b	-b	-c	-d	c	a	-a	b	b	c	0	c	a	-a	b	b	-c	0	-c
(2, 8, 9, 9)																												
(3, 3, 3, 3)	a	b	-c	0	0	0	0	a	-b	0	d	0	0	0	a	0	c	-d	0	0	0	b	c	d	0	0	0	0
(3, 3, 3, 12)	-c	c	c	0	c	a	b	c	-c	-c	d	-d	a	0	c	-c	-c	-d	-c	0	b	0	0	0	-d	0	a	-b

(3, 3, 6, 6)	a d b -d a 0 0	a d c d -a 0 0	c -d a 0 -b 0 0	c -d -a 0 b 0 0
(3, 4, 6, 8)	-c c c 0 c a b	c -c -c 0 -c a b	d b 0 -b d 0 0	-d b -a b d 0 0
(3, 6, 8, 9)				
(4, 4, 4, 4)	a b c d 0 0 0	a -b c -d 0 0 0	a b -c -d 0 0 0	a -b -c d 0 0 0
(4, 4, 4, 9)				
(4, 4, 4, 16)	c d -c c a c b	c d -c c -a c -b	c d -c -c a -c -b	c d -c -c -a -c b
(4, 4, 5, 5)	a a b -b 0 0 0	b b -a a 0 0 0	c 0 c d d -d 0	d 0 d -c -c c 0
(4, 4, 8, 8)	a a b -b c d 0	a a b -b -c -d 0	a -a b b c -d 0	a -a b b -c d 0
(4, 4, 9, 9)	known but not constructed from circulants			
(4, 4, 10, 10)	b c a c d d -d	b -c a -c -d -d d	b d -a d -c -c c	b -d -a -d c c -c
(4, 5, 5, 9)				
(5, 5, 5, 5)	a a -a b 0 b 0	-b -b b a 0 a 0	-c -c c -d 0 -d 0	-d -d d c 0 c 0
(5, 5, 8, 8)				
(5, 5, 9, 9)				
(6, 6, 6, 6)	a a b -b c d 0	-b -b a -a d -c 0	-c -c -d d a b 0	-d -d c -c -b a 0
(7, 7, 7, 7)	a a -a b c b d	-b -b b a d a -c	-c -c c -d a -d b	-d -d d c -b c a
(1, 2, 22)	a b 0 b b b -b	a -b 0 -b -b -b b	0 -b b -b b b b	c b b -b b -b -b
(1, 3, 24)	a c -c c -c c -c	b c -c -c -c -c -c	b -c -c c c c c	b -c c c -c c c
(1, 4, 20)	a b b -b b -b -b	c -c b b b b 0	0 b b -b -b b b	c 0 -b b -b b c
(1, 6, 12)	b -b -b 0 -b a a	-b b b 0 b a 0	-b b b 0 b 0 a	c 0 0 a -a 0 0

Order 28 (Continued)

Design	A_1	A_2	A_3	A_4
(1, 6, 18)	c a a b -b -a -a	-b -b a -a -a a 0	a a a -a a b 0	a a -a a -b a 0
(1, 6, 21)	c b -b b -b b -b	a b b b b -b -b	a -a b b -a b b	a a b -b -b b b
(1, 9, 13)	a b b -b b -b -b	0 b b 0 b 0 0	0 c -c c c -c c	c c c -c -c c c
(1, 10, 14)	c -b -b b -b b b	0 -b a a a -a a	0 b -a b a b a	0 b b a -b -a b
(2, 2, 13)	a b 0 0 0 0 0	a -b 0 0 0 0 0	0 c -c c c -c c	c c c -c -c c c
(2, 7, 19)	c a a a -a b -a	-c a b a -a a -a	b -b a a -b a a	b b a -a -a a a
(2, 8, 13)	-b b b c b 0 0	-b b b -c b 0 0	0 a -a a a -a a	a a a -a -a a a
(4, 4, 18)	a -b -c -b -b b -b	a b -c b b -b b	a -b -b b b c 0	a b b -b -b c 0
(5, 5, 13)	a a -a b 0 b 0	b b -b -a 0 -a 0	0 c -c c c -c c	c c c -c -c c c
(3, 23)	-a 0 b b b -b b	a -b b b b -b b	a b b -b b -b 0	b b b b b -b -b
(4, 19)	b b 0 a 0 -a 0	b -b a -a a -a a	0 a a 0 -a a a	a a a a a -a -a
(5, 21)	b -a a b -b a a	b b a -a -a a -a	a a -a -a -a a 0	a a a a -a a 0
(5, 23)	a a -a -b b b b	-a b -a b -b b -b	b b b b b -b -b	b b b -b b b -b
(6, 17)	b -b a a a a -a	b b a a -a -a 0	b a -a a a -a a	b 0 -a 0 0 0 -a
(6, 20)	b a b a -a 0 a	b a -b a 0 -a -a	b a a -a -a a -a	b -a -a -a -a a -a
(7, 10)	a 0 0 b 0 0 0	b -a a 0 0 a 0	a 0 -b b b 0 b	-b b b 0 b -a -a
(7, 15)	b a b a -b 0 a	b -a 0 b 0 0 0	b -a a -a 0 a a	b -a -a -a a a -a
(8, 17)	a -a a -a -a a 0	a a b b b -b 0	b -a -a -a b 0 a	b a a -a a a b
(9, 16)	b b b a b -a -b	b a a a -b a 0	b -b 0 a a -a -a	a -a a -a -a a 0
(9, 17)	a -a a -a -a a 0	a a -a -b -a -a b	a a -b b a a 0	a b -a b b b -b

(11,12)	a	-a	a	a	0	0	0	-b	a	a	a	-a	-a	a	a	b	a	b	b	-b	-b	b	b	-b	b	0	b	0
(11,15)	b	b	-b	b	0	b	-a	b	b	-b	-b	0	-b	-a	b	a	a	-a	-a	a	a	a	a	a	a	-a	a	-a
(11,17)	b	-b	b	b	b	a	-a	b	-b	-b	-b	b	-a	-a	b	a	-a	-a	-a	-a	a	a	a	a	-a	a	a	-a
(12,14)	b	-a	b	b	-b	a	-a	b	a	b	a	b	-b	a	b	b	-a	-a	a	0	-a	a	a	b	-a	-b	-a	0

APPENDIX D: ORTHOGONAL DESIGNS IN POWERS OF 2, ESPECIALLY ORDERS 16, 32 AND 64

We first recall-from Chapters 7 and 4- the strongest results known in the powers of 2.

D.1. THEOREM (P.J. Robinson-See 5.98 of Chapter 5). There is an orthogonal design of order 2^t, $t \geq 2$, and type $(1, 1, 1, 1, 2, 2, 4, 4, \ldots, 2^{t-2}, 2^{t-2})$.

By equating variables we obtain:

D.2. COROLLARY (P.J. Robinson). All orthogonal designs of type $(1, 1, a, b, c)$, $a + b + c = 2^t - 2$, exist in order 2^t, $t \geq 3$. Hence we have, as corollaries, the following results of Wallis.

D.3. COROLLARY. All orthogonal designs of type $(a, b, 2^t - a - b)$ (i.e. full orthogonal designs) exist in order 2^t.

D.4. COROLLARY. All orthogonal designs on two variables exist in orders 2^t.

We now consider specific powers of two.

ORDER 16

D.5. THEOREM (Robinson). The following orthogonal designs and those that can be obtained from them by equating variables or setting variables equal to zero are the only orthogonal designs in order 16:

Nine Variable

(1, 1, 1, 1, 1, 1, 1, 1, 1)	(1, 1, 1, 1, 1, 2, 2, 2, 2)
(1, 1, 1, 1, 1, 1, 1, 1, 2)	(1, 1, 1, 1, 1, 2, 3, 3, 3)
(1, 1, 1, 1, 1, 1, 1, 1, 4)	(1, 1, 1, 1, 2, 2, 2, 2, 2)
(1, 1, 1, 1, 1, 1, 1, 1, 8)	(1, 1, 2, 2, 2, 2, 2, 2, 2)
(1, 1, 1, 1, 1, 1, 2, 2, 2)	

8 Variable

(1, 1, 1, 1, 1, 1, 3, 3)	(1, 1, 1, 1, 2, 2, 3, 3)
(1, 1, 1, 1, 1, 1, 4, 4)	(1, 1, 1, 1, 2, 2, 4, 4)
(1, 1, 1, 1, 1, 1, 5, 5)	(1, 1, 2, 2, 2, 2, 3, 3)

Proof. Many 9 and 8 tuples are proved non-existent by theorems of Shapiro and Geramita-Verner. The remainder are proved non-existent by combinatorial arguments in the PhD thesis of Peter J. Robinson.

The (1, 1, 1, 1, 1, 1, 1, 1, 1) design is exhibited in section 4.1 and the (1, 1, 2, 2, 2, 2, 2, 2, 2) design is obtained by using lemma 4.11 on the (1, 1, 1, 1, 1, 1, 1, 1) design in order 8 exhibited in section 4.1.

The (1, 1, 1, 1, 1, 1, 3, 3) and (1, 1, 1, 1, 2, 2, 3, 3) designs are given in Example 5.121. The (1, 1, 1, 1, 1, 2, 3, 3, 3), (1, 1, 1, 1, 1, 2, 2, 2, 2), (1, 1, 1, 1, 1, 1, 1, 1, 2) and (1, 1, 2, 2, 2, 2, 3, 3) designs are given in Example 5.101.

We exhibit the remaining designs of the enunciation but we first consider the matrix

$$PR = \left[\begin{array}{cccc|cccc|cccc|cccc}
X_1 & X_2 & X_3 & M & X_4 & N & Q & P & A & B & C & D & E & F & G & H \\
-X_2 & X_1 & M & -X_3 & N & -X_4 & P & -Q & B & -A & D & -C & F & -E & H & -G \\
-X_3 & -M & X_1 & X_2 & Q & -P & -X_4 & N & C & -D & -A & B & G & -H & -E & F \\
-M & X_3 & -X_2 & X_1 & -P & -Q & N & X_4 & -D & -C & B & A & -H & -G & F & E \\
\hline
-X_4 & -N & -Q & P & X_1 & X_2 & X_3 & -M & E & -F & -G & H & -A & B & C & -D \\
-N & X_4 & P & Q & -X_2 & X_1 & -M & -X_3 & -F & -E & H & G & B & A & -D & -C \\
-Q & -P & X_4 & -N & -X_3 & M & X_1 & X_2 & -G & -H & -E & -F & C & D & A & B \\
-P & Q & -N & -X_4 & M & X_3 & -X_2 & X_1 & -H & G & -F & E & D & -C & B & -A \\
\hline
-A & -B & -C & D & -E & F & G & H & X_1 & X_2 & X_3 & R & X_4 & S & T & U \\
-B & A & D & C & F & E & H & -G & -X_2 & X_1 & R & -X_3 & S & -X_4 & U & -T \\
-C & -D & A & -B & G & -H & E & F & -X_3 & -R & X_1 & X_2 & T & -U & -X_4 & S \\
-D & C & -B & -A & -H & -G & F & -E & -R & X_3 & -X_2 & X_1 & -U & -T & S & X_4 \\
\hline
-E & -F & -G & H & A & -B & -C & -D & -X_4 & -S & -T & U & X_1 & X_2 & X_3 & -R \\
-F & E & H & G & -B & -A & -D & C & -S & X_4 & U & T & -X_2 & X_1 & -R & -X_3 \\
-G & -H & E & -F & -C & D & -A & -B & -T & -U & X_4 & -S & -X_3 & R & X_1 & X_2 \\
-H & G & -F & -E & D & C & -B & A & -U & T & -S & -X_4 & R & X_3 & -X_2 & X_1
\end{array}\right]$$

This matrix is an orthogonal design if and only if the following conditions are satisfied:

$$\{|M|, |N|, |Q|, |P|\} = \{|R|, |S|, |T|, |U|\}$$

$$(M+R)D + (N+S)F + (Q+T)G + (P+U)H = 0$$
$$(M+R)C + (N+S)E + (Q-T)H - (P-U)G = 0$$
$$(M+R)B + (N-S)H - (Q+T)E - (P-U)F = 0$$
$$(M+R)A + (N-S)G - (Q-T)F + (P+U)E = 0$$
$$(R-M)H + (N+S)B + (Q+T)C + (P-U)D = 0$$
$$(R-M)G + (N+S)A + (Q-T)D - (P+U)C = 0$$
$$(R-M)F + (N-S)D - (Q+T)A - (P+U)B = 0$$
$$(R-M)E + (N-S)C - (Q-T)B + (P-U)A = 0$$
$$AH - BG + CF - DE = 0$$

The following designs can be obtained by choosing the variables of PR in the manner indicated:

(1, 1, 1, 1, 1, 1, 1, 1, 4) $M = S = x_5$, $N = T = x_6$,
$R = -Q = x_7$, $P = U = x_8$,
$A = D = G = -F = x_9$
all others zero;

(1, 1, 1, 1, 1, 1, 1, 1, 8) $M = S = x_5$, $N = T = x_6$,
$R = P = x_7$, $Q = U = x_8$,
$A = -B = D = C = G = -H = -E = -F = x_9$;

(1, 1, 1, 1, 1, 1, 2, 2, 2) $A = x_5$, $P = U = x_6$, $R = -D = -M = x_7$,
$N = -S = F = x_8$, $Q = -T = G = x_9$,
all others zero;

(1, 1, 1, 1, 1, 1, 4, 4) $M = S = x_5$, $R = -Q = x_6$, $A = G = D = -F = x_7$,
$B = H = C = -E = x_8$, all others zero;

(1, 1, 1, 1, 1, 1, 5, 5) $M = S = x_5, \quad N = U = x_6,$

$A = G = P = R = C = -E = x_7,$

$B = H = D = -F = -Q = -T = x_8;$

(1, 1, 1, 1, 2, 2, 4, 4) $D = F = x_5, \quad G = H = x_6,$

$M = R = E = C = -N = -S = x_7,$

$Q = T = B = A = -P = -U = x_8.$

The orthogonal design of type (1, 1, 1, 1, 2, 2, 2, 2, 2) in order 16 is:

x_5	x_5	x_1	x_3	x_2	x_4	0	0	x_6	x_7	x_8	x_9	x_7	$-x_6$	x_9	$-x_8$
$-x_5$	x_5	x_3	$-x_1$	x_4	$-x_2$	0	0	x_7	$-x_6$	$-x_9$	x_8	$-x_6$	$-x_7$	x_8	x_9
$-x_1$	$-x_3$	x_5	x_5	0	0	$-x_2$	x_4	x_9	x_8	$-x_6$	x_7	$-x_8$	x_9	$-x_7$	$-x_6$
$-x_3$	x_1	$-x_5$	x_5	0	0	x_4	x_2	$-x_8$	x_9	x_7	x_6	$-x_9$	$-x_8$	$-x_6$	x_7
$-x_2$	$-x_4$	0	0	x_5	x_5	x_1	$-x_3$	x_7	x_6	x_8	$-x_9$	$-x_6$	x_7	x_9	x_8
$-x_4$	x_2	0	0	$-x_5$	x_5	$-x_3$	$-x_1$	x_6	$-x_7$	x_9	x_8	x_7	x_6	$-x_8$	x_9
0	0	x_2	$-x_4$	$-x_1$	x_3	x_5	x_5	$-x_9$	x_8	$-x_7$	x_6	x_8	x_9	x_6	x_7
0	0	$-x_4$	$-x_2$	x_3	x_1	$-x_5$	x_5	$-x_8$	$-x_9$	x_6	x_7	$-x_9$	x_8	x_7	$-x_6$
$-x_6$	$-x_7$	$-x_8$	$-x_9$	$-x_7$	$-x_6$	$-x_8$	x_9	x_5	x_5	x_1	0	x_2	$-x_4$	x_3	0
$-x_7$	x_6	x_9	$-x_8$	$-x_6$	x_7	$-x_9$	$-x_8$	$-x_5$	x_5	0	$-x_1$	$-x_4$	$-x_2$	0	$-x_3$
$-x_9$	$-x_8$	x_6	$-x_7$	x_9	$-x_8$	x_7	$-x_6$	$-x_1$	0	x_5	x_5	x_3	0	$-x_2$	$-x_4$
x_8	$-x_9$	$-x_7$	$-x_6$	x_8	x_9	$-x_6$	$-x_7$	0	x_1	$-x_5$	x_5	0	$-x_3$	$-x_4$	x_2
$-x_7$	x_6	$-x_9$	x_8	x_6	$-x_7$	$-x_9$	$-x_8$	$-x_2$	x_4	$-x_3$	0	x_5	x_5	x_1	0
x_6	x_7	$-x_8$	$-x_9$	$-x_7$	$-x_6$	x_8	$-x_9$	x_4	x_2	0	x_3	$-x_5$	x_5	0	$-x_1$
x_8	$-x_9$	x_7	x_6	$-x_8$	$-x_9$	$-x_6$	$-x_7$	$-x_3$	0	x_2	$-x_4$	$-x_1$	0	x_5	x_5
x_9	x_8	x_6	$-x_7$	x_9	$-x_8$	$-x_7$	x_6	0	x_3	x_4	$-x_2$	0	x_1	$-x_5$	x_5

D. 6. COROLLARY (Robinson). All 6-tuples are the types of orthogonal designs in order 16 except (1, 1, 1, 1, 4, 7) and (1, 1, 2, 2, 2, 7) which do not exist.

D.7. COROLLARY (Robinson). All n-tuples $n = 1, 2, 3, 4, 5$ are the types of orthogonal designs in order 16.

ORDER 32

In Tables D.1, D.2, D.3, D.4 we give 10, 9, 8 and 7 variable designs in order 32. These designs were constructed in one of the following two ways:

(a) using product designs and amicable orthogonal designs in Theorem 5.97, or

(b) using a doubling construction (Lemma 4.11) with orthogonal designs in order 16.

These tables plus the use of equating and killing variables enable Table D.5 and the following results to be obtained:

D.8. LEMMA (Robinson). All full 6-tuples, except possibly (1, 1, 1, 1, 1, 27), are the types of orthogonal designs in order 32.

D.9. COROLLARY. All 5-tuples, except possibly (1, 1, 1, 1, 27), and all n-tuples, $n = 1, 2, 3, 4$ are the types of orthogonal design in order 32.

10 Variable Designs

Type	Construction
(1, 1, 1, 1, 2, 2, 3, 3, 9, 9)	(1, 1, 2, 3;1, 3, 3;1) and ((1, 3);(1, 1, 2)) in (a)
(1, 1, 1, 1, 2, 2, 4, 4, 8, 8)	(1, 1, 1, 1, 2;2, 4;2, 4, 4) and ((1, 1);(1, 1)) in (a)
(1, 1, 1, 2, 3, 3, 3, 3, 6, 9)	(1, 1, 1, 2;1, 1, 3;3) and ((1, 3);(1, 1, 2)) in (a)
(1, 1, 3, 3, 3, 3, 3, 3, 6, 6)	(1, 1, 1, 2;1, 1, 3;3) and ((3, 1);(1, 1, 2)) in (a)
(1, 2, 2, 2, 2, 2, 3, 6, 6, 6)	(1, 2, 2, 2, 3;2, 2, 6;6) and ((1, 1);(1, 1)) in (a)
(1, 1, 2, 2, 2, 2, 2, 2, 2, 16)	(1, 1, 1, 1, 1, 1, 1, 1, 8) in (b)
(1, 1, 2, 4, 4, 4, 4, 4, 4, 4)	(1, 1, 2, 2, 2, 2, 2, 2, 2) in (b)
(2, 2, 2, 2, 4, 4, 4, 4, 4, 4)	(2, 1, 1, 2, 2, 2, 2, 2, 2) in (b)
(2, 2, 2, 2, 2, 2, 2, 2, 8, 8)	(8, 1, 1, 1, 1, 1, 1, 1, 1) in (b)
(2, 2, 2, 2, 2, 2, 2, 6, 6, 6)	(2, 1, 1, 1, 1, 1, 3, 3, 3) in (b)
(1, 1, 2, 2, 2, 2, 4, 6, 6, 6)	(1, 1, 1, 1, 1, 2, 3, 3, 3) in (b)
(2, 2, 2, 2, 2, 3, 3, 4, 6, 6)	(3, 1, 1, 1, 1, 1, 2, 3, 3) in (b)

TABLE D.1

9 Variable Designs

Type	Construction
(1, 1, 1, 1, 2, 2, 3, 7, 14)	(1, 1, 2, 3;7;1) and ((1, 1, 2);(1, 1, 2)) in (a)
(1, 1, 1, 1, 2, 2, 4, 10, 10)	(1, 1, 1, 1, 2;2, 4;10) and ((1, 1));(1, 1)) in (a)
(1, 1, 1, 1, 2, 4, 7, 7, 8)	(1, 1, 1, 2, 4;1, 8;7) and ((1, 1);(1, 1)) in (a)
(1, 1, 1, 1, 4, 4, 6, 6, 8)	(1, 1, 1, 1;2, 2;4) and ((1, 3);(1, 1, 2)) in (a)
(1, 1, 1, 2, 2, 3, 4, 6, 12)	(1, 1, 1, 2, 2;3, 4;3, 6) and ((1, 1);(2)) in (a)
(1, 1, 1, 2, 2, 3, 6, 8, 8)	(1, 1, 1, 2, 3;2, 6;8) and ((1, 1);(2)) in (a)
(1, 1, 1, 2, 2, 4, 7, 7, 7)	(1, 1, 1;1, 1, 1;1) and ((1, 7);(2, 2, 4)) in (a)
(1, 1, 1, 3, 3, 3, 5, 5, 10)	(1, 1, 1;1, 1, 1;5) and ((1, 3);(1, 1, 2)) in (a)
(1, 1, 1, 2, 3, 3, 5, 6, 10)	(1, 1, 1, 2;5;3) and ((1, 1, 2);(1, 1, 2)) in (a)
(1, 1, 2, 2, 2, 4, 6, 7, 7)	(1, 1, 2, 3;7;1) and ((2, 1, 1);(1, 1, 2)) in (a)
(1, 1, 2, 3, 3, 4, 4, 6, 8)	(1, 1, 2, 3, 4;3, 8;2, 3) and ((1, 1);(2)) in (a)
(1, 1, 2, 3, 3, 4, 5, 5, 8)	(1, 1, 2, 3, 4;3, 8;5) and ((1, 1);(1, 1)) in (a)
(2, 2, 2, 3, 3, 4, 5, 5, 6)	(1, 1, 1, 2;5;3) and ((2, 1, 1);(1, 1, 2)) in (a)
(2, 2, 3, 3, 3, 3, 4, 4, 8)	(1, 1, 1, 1;2, 2;2, 2) and ((3, 1);(1, 1, 2)) in (a)
(2, 2, 3, 3, 3, 4, 5, 5, 5)	(1, 1, 1;1, 1, 1;1) and ((3, 5);(2, 2, 4)) in (a)
(1, 1, 2, 2, 2, 2, 2, 10, 10)	(1, 1, 1, 1, 1, 1, 5, 5) in (b)
(2, 2, 2, 2, 2, 2, 5, 5, 10)	(5, 1, 1, 1, 1, 1, 1, 5) in (b)
(2, 2, 3, 3, 4, 4, 4, 4, 6)	(3, 1, 1, 2, 2, 2, 2, 3) in (b)
(2, 2, 2, 2, 2, 2, 2, 9, 9)	(9, 1, 1, 1, 1, 1, 1, 1) in (b)
(2, 2, 2, 2, 2, 5, 5, 6, 6)	(5, 1, 1, 1, 1, 1, 3, 3) in (b)
(2, 2, 2, 2, 4, 4, 4, 6, 6)	(4, 1, 1, 1, 1, 2, 3, 3) in (b)
(2, 3, 3, 4, 4, 4, 4, 4, 4)	(3, 1, 2, 2, 2, 2, 2, 2) in (b)
(1, 1, 1, 1, 2, 3, 3, 5, 15)	(Example 5.121)

TABLE D. 2

8 Variable Designs

Type	Construction
(1, 1, 1, 1, 7, 7, 7, 7)	(1, 1, 7;1, 1, 7;7) and ((1, 1);(1, 1)) in (a)
(1, 1, 1, 3, 6, 6, 6, 8)	(1, 1, 1;3;1) and ((1, 1, 2, 2, 2);(8)) in (a)
(1, 2, 3, 3, 4, 4, 5, 10)	(1, 2, 3, 3;4, 5;2, 5) and ((1, 1);(2)) in (a)
(1, 2, 3, 3, 4, 5, 7, 7)	(1, 2, 3, 3;4, 5;7) and ((1, 1);(1, 1)) in (a)
(2, 2, 2, 4, 4, 5, 5, 8)	(5, 1, 1, 1, 2, 2, 4) in (b)
(2, 2, 4, 4, 4, 5, 5, 6)	(5, 1, 1, 2, 2, 2, 3) in (b)

TABLE D. 3

7 Variable Designs

Type	Construction
(1, 2, 2, 2, 3, 11, 11)	(1, 2, 2;2, 3;11) and ((1, 1);(1, 1)) in (a)
(1, 1, 1, 2, 5, 11, 11)	(1, 1, 1, 2;5;11) and ((1, 1);(1, 1)) in (a)

TABLE D. 4

Unknown Full 7 Variable Designs in Order 32

1	1	1	1	1	1	26
1	1	1	1	1	2	25
1	1	1	1	1	3	24
1	1	1	1	1	4	23
1	1	1	1	1	5	22
1	1	1	1	1	6	21
1	1	1	1	1	7	20
1	1	1	1	1	8	19
1	1	1	1	1	9	18
1	1	1	1	1	10	17
1	1	1	1	1	11	16
1	1	1	1	1	12	15
1	1	1	1	1	13	14
1	1	1	1	2	13	13
1	1	1	1	3	6	19
1	1	1	1	4	5	19
1	1	1	1	4	11	13
1	1	1	1	5	10	13
1	1	1	2	2	2	23
1	1	1	2	5	5	17
1	1	1	3	4	5	17
1	1	1	4	5	7	13
1	1	1	5	5	5	14
1	1	1	5	5	7	12
1	1	1	5	5	8	11
1	1	1	5	6	7	11
1	1	2	2	2	5	19
1	1	2	5	5	5	13
1	1	3	4	4	4	15
1	1	4	4	4	5	13
1	1	4	4	4	7	11
1	1	4	4	5	5	12
1	1	4	4	6	7	10
1	1	4	5	5	7	9
1	1	5	5	5	5	10
1	1	5	5	5	6	9
1	1	5	5	6	7	7
1	1	5	6	6	6	7
1	2	2	2	3	5	17
1	2	2	2	5	5	15
1	2	2	2	5	9	11
1	2	2	3	3	8	13
1	2	2	3	5	5	14
1	2	2	3	5	8	11
1	2	2	4	4	4	15
1	2	2	4	4	5	14
1	2	2	4	5	5	13
1	2	2	4	5	7	11
1	2	2	5	5	5	12
1	2	2	5	5	6	11
1	2	2	5	5	7	10
1	2	4	4	4	7	10
1	2	4	4	5	7	9
1	2	4	5	5	5	10
1	2	5	5	5	5	9
1	2	5	5	5	7	7
1	2	5	5	6	6	7
1	3	3	3	3	5	14
1	3	3	3	7	7	8
1	3	3	4	4	4	13
1	3	4	4	4	5	11
1	3	4	4	4	7	9
1	3	4	5	6	6	7
1	3	5	5	5	6	7
1	3	5	5	6	6	6
1	4	4	4	5	5	9
1	4	4	4	5	7	7
1	4	4	5	5	6	7
1	4	5	5	5	5	7
1	4	5	5	5	6	6
1	5	5	5	5	5	6
2	2	2	3	7	7	9
2	2	2	4	4	7	11
2	2	2	5	5	5	11
2	2	4	5	5	7	7
2	2	5	5	5	5	8
2	3	3	3	3	5	13
2	3	3	3	7	7	7
2	4	4	5	5	5	7
2	5	5	5	5	5	5
3	3	3	3	4	5	11
3	3	3	3	5	7	8
3	3	4	4	4	5	9
3	3	4	4	4	7	7
3	4	4	4	5	5	7
4	4	4	5	5	5	5

TABLE D.5

ORDER 64

We use the following corollary to Theorem 5.97.

D.10. COROLLARY (Robinson). Suppose there are amicable orthogonal designs of types $((u_1, \ldots, u_p);(v_1, \ldots, v_q))$ in order n. Then there exist orthogonal designs of type

(i) $(u_1, u_1, u_1, 2u_1, w, w, 3w, 3v_1, \ldots, 3v_q)$

(ii) $(u_1, u_1, u_1, 2u_1, 5u_2, \ldots, 5u_p, 3v_1, \ldots, 3v_q)$

(iii) $(u_1, u_1, u_1, w, w, w, 5v_1, 5v_2, \ldots, 5v_q)$

(iv) $(u_1, u_1, u_1, 3u_2, \ldots, 3u_p, 5v_1, \ldots, 5v_q)$

(v) $(u_1, u_1, 2u_1, 3u_1, 7u_2, \ldots, 7u_p, v_1, \ldots, v_q)$

in order $8n$ where $w = u_2 + u_3 + \ldots + u_p$.

Proof: For (i) and (ii) use the theorem with the product design of type (1, 1, 1, 2;1, 1, 3;3) in order 8. For (iii) and (iv) use the product design of type (1, 1, 1;1, 1, 1;5) in order 8 and for (v) use the product design of type (1, 1, 2, 3;7;1) in order 8.

D.11. THEOREM (Robinson). All full orthogonal designs on five variables exist in order 64.

ORTHOGONAL DESIGNS OF ORDER 64

Orthogonal design	Construction: Product designs	Construction: Amicable designs
(1, 1, 1, 1, 2, 2, 4, 4, 8, 8, 16, 16)	(1, 1, 1, 1, 2, 4;2, 8;2, 4, 8, 8)	((1, 1);(2))
(1, 1, 1, 2, 2, 3, 4, 6, 8, 12, 24)	(1, 1, 1, 2, 2, 4;3, 8;3, 6, 12)	((1, 1);(2))
(1, 1, 1, 2, 2, 9, 9, 9, 12, 18)	(1, 1, 1, 2, 2;3, 4;9)	((1, 3);(1, 1, 2))
(1, 1, 1, 2, 4, 7, 7, 9, 14, 18)	(1, 1, 1, 2,4;9;7)	((1, 1, 2);(1, 1, 2))
(1, 1, 1, 2, 5, 6, 6, 10, 12, 20)	(1, 1, 1, 2;5;3)	((1, 1, 2, 4);(2, 2, 4))
(1, 1, 1, 2, 6, 6, 7, 7, 12, 21)	(1, 1, 1, 2;1, 1, 3;3)	((1, 7);(2, 2, 4))
(1, 1, 2, 2, 2, 3, 4, 7, 14, 28)	(1, 1, 2, 3;7;1)	((1, 1, 2, 4);(2, 2, 4))
(1, 1, 2, 3, 4, 5, 5, 10, 11, 22)	(1, 1, 2, 3, 4;11;5)	((1, 1, 2);(1, 1, 2))
(1, 1, 3, 6, 6, 7, 7, 7, 12, 14)	(1, 1, 1, 2;1, 1, 3;3)	((7, 1);(2, 2, 4))
(1, 2, 3, 3, 4, 5, 8, 8, 10, 20)	(1, 2, 3, 3, 4;5, 8;4, 5, 10)	((1, 1);(2))
(1, 3, 3, 3, 4, 6, 8, 8, 12, 16)	(1, 1, 1, 2, 4;1, 8;1, 2, 4)	((3, 1);(4))
(2, 2, 3, 3, 4, 5, 6, 9, 15, 15)	(1, 1, 2, 3;1, 3, 3;1)	((3, 5);(2, 2, 4))
(2, 3, 3, 3, 3, 4, 6, 8, 16, 16)	(1, 1, 1, 1, 2;2, 4;2, 4, 4)	((3, 1);(4))
(3, 3, 3, 5, 5, 6, 6, 6, 12, 15)	(1, 1, 1, 2;1, 1, 3;3)	((3, 5);(2, 2, 4))
(3, 3, 3, 5, 5, 6, 8, 9, 10, 12)	(1, 1, 2, 3, 4;3, 8;5)	((3, 1);(1, 1, 2))
(3, 3, 5, 5, 5, 6, 6, 9, 10, 12)	(1, 1, 1, 2;1, 1, 3;3)	((5, 3);(2, 2, 4))
(1, 1, 1, 2, 3, 7, 7, 21, 21)	(1, 1, 1, 2;1, 1, 3;3)	((1, 7);(1, 7))
(1, 1, 1, 2, 5, 10, 10, 10, 24)	(1, 1, 1, 2;5;3)	((1, 1, 2, 2, 2);(8))
(1, 1, 1, 2, 6, 10, 10, 15, 18)	(1, 1, 1, 2;5;3)	((1, 2, 2, 3);(2, 6))
(1, 1, 1, 3, 6, 10, 10, 12, 20)	(1, 1, 1;3;5)	((1, 1, 2, 4);(2, 2, 4))
(1, 1, 1, 7, 7, 7, 10, 10, 20)	(1, 1, 1;1, 1, 1;5)	((1, 7);(2, 2, 4))
(1, 1, 2, 2, 3, 6, 14, 14, 21)	(1, 1, 2, 3;7;1)	((1, 2, 2, 3);(2, 6))
(1, 1, 2, 3, 7, 8, 14, 14, 14)	(1, 1, 2, 3;7;1)	((1, 1, 2, 2, 2,);(8))
(1, 1, 3, 3, 7, 7, 7, 14, 21)	(1, 1, 1, 2;1, 1, 3;3)	((7, 1);(1, 7))
(1, 2, 2, 3, 4, 4, 6, 14, 28)	(1, 1, 2, 3;7;1)	((2, 2, 4);(3, 1, 4))
(1, 2, 3, 3, 7, 7, 12, 14, 15)	(1, 2, 3, 3;4, 5;7)	((1, 3);(1, 1, 2))
(2, 2, 3, 3, 4, 6, 7, 9, 28)	(1, 1, 2, 3;7;1)	((3, 1, 4);(2, 2, 4))

Order 64 (Continued)

Orthogonal design	Construction: Product designs	Amicable designs
(2, 2, 4, 5, 5, 7, 10, 14, 15)	(1, 1, 2, 3;7;1)	((5, 1, 2);(2, 2, 4))
(2, 3, 3, 6, 6, 7, 9, 14, 14)	(1, 1, 2, 3;7;1)	((3, 1, 2, 2);(2, 6))
(3, 3, 3, 4, 4, 8, 13, 13, 13)	(1, 1, 1;1, 1, 1;1)	((3, 13);(4, 4, 8))
(3, 3, 3, 5, 5, 5, 10, 10, 20)	(1, 1, 1;3;5)	((3, 5);(2, 2, 4))
(3, 3, 3, 5, 6, 6, 6, 12, 20)	(1, 1, 1, 2;5;3)	((3, 1, 4);(2, 2, 4))
(3, 3, 3, 5, 6, 6, 10, 10, 18)	(1, 1, 1, 2;5;3)	((3, 1, 2, 2);(2, 6))
(3, 4, 4, 6, 8, 9, 9, 9, 12)	(1, 1, 1;3;1)	((9, 1, 2, 4);(4, 4, 8))
(4, 4, 5, 5, 5, 8, 11, 11, 11)	(1, 1, 1;1, 1, 1;1)	((5, 11);(4, 4, 8))
(5, 5, 5, 5, 6, 6, 10, 10, 12)	(1, 1, 1, 2;5; 3)	((5, 1, 2);(2, 2, 4))
(1, 1, 1, 3, 6, 6, 6, 40)	(1, 1, 1;3;5)	((1, 1, 2, 2, 2);(8))
(1, 1, 1, 5, 7, 7, 7, 35)	(1, 1, 1;1, 1, 1;5)	((1, 7);(1, 7))
(1, 1, 1, 6, 6, 9, 10, 30)	(1, 1, 1;1, 1, 1;5)	((1, 2, 2, 3);(2, 6))
(2, 2, 4, 7, 7, 7, 14, 21)	(1, 1, 2, 3;7;1)	((7, 1);(2, 2, 4))
(3, 3, 3, 3, 6, 6, 10, 30)	(1, 1, 1;1, 1, 1;5)	((3, 1, 2, 2);(2, 6))
(3, 3, 3, 3, 10, 10, 12, 20)	(1, 1, 1;1, 1, 1;5)	((3, 1, 4);(2, 2, 4))

Table D. 6

14, 13 AND 12 VARIABLE ORTHOGONAL DESIGNS IN ORDER 128

Orthogonal design	Construction: Product designs	Amicable designs
(1, 1, 1, 1, 2, 2, 4, 4, 8, 8, 16, 16, 32, 32)	(1, 1, 1, 1, 2, 2, 4;4, 8;4, 8, 8, 16, 16)	((1, 1); (2))
(1, 1, 1, 2, 2, 3, 4, 6, 8, 12, 16, 24, 48)	(1, 1, 1, 2, 2, 3, 4;6, 8;6, 8, 12, 24)	((1, 1); (2))
(1, 2, 3, 3, 4, 5, 8, 10, 16, 16, 20, 40)	(1, 2, 3, 3, 4, 5;8, 10;8, 8, 10, 20)	((1, 1);(2))
(1, 3, 3, 3, 4, 6, 8, 12, 16, 16, 24, 32)	(1, 1, 1, 2, 4, 8;1, 16;1, 2, 4, 8)	((3, 1);(4))
(2, 3, 3, 3, 3, 6, 8, 8, 12, 16, 32, 32)	(1, 1, 1, 1, 2, 4;2, 8;2, 4, 8, 8)	((3, 1);(4))

Table D. 7

11 VARIABLE ORTHOGONAL DESIGNS IN ORDER 128

Orthogonal designs	Construction: Product designs	Construction: Amicable designs
(1,1,1,1,2,4,8,16,31,31,32)	(1,1,1,2,4,8,16;1,32;31)	((1, 1);(1, 1))
(1,1,1,2,4,9,14,14,18,28,36)	(1,1,1,2,4;9;7)	((1, 1, 2, 4);(2, 2, 4))
(1,1,2,2,3,4,5,8,10,20,72)	(1,1,2,2,3,4;5,8;5,10,36)	((1, 1);(2))
(1,1,2,3,3,4,8,16,29,29,32)	(1,1,2,3,4,8,16;3,32;29)	((1, 1);(1, 1))
(1,1,2,3,3,6,7,7,12,14,72)	(1,1,2,3,6,12;1,24;7)	((1, 3);(1, 1, 2))
(1,1,2,3,4,10,10,11,20,22,44)	(1,1,2,3,4,11;22;10)	((1, 1, 2);(1, 1, 2))
(1,1,2,3,4,4,7,8,14,28,56)	(1,1,2,3;7;1)	((1, 1, 2, 4, 8);(4, 4, 8))
(1,1,2,3,4,8,12,16,19,24,38)	(1,1,2,3,4,8;19;3,4,6)	((1, 1, 2);(4))
(1,1,2,3,4,8,13,13,19,26,38)	(1,1,2,3,4,8;19;13)	((1, 1, 2);(1, 1, 2))
(1,1,2,3,4,8,9,13,13,26,48)	(1,1,2,3,4,8;3,16;13)	((1, 3);(1, 1, 2))
(1,1,2,3,6,7,7,12,14,25,50)	(1,1,2,3,6,12;25;7)	((1, 1, 2);(1, 1, 2))
(1,1,2,3,6,9,10,12,20,24,40)	(1,1,2,3,6,9;10,12;10,12,20)	((1, 1);(2))
(1,1,2,3,7,7,14,18,18,21,36)	(1,1,2,3;1,1,2,3;9)	((1, 7);(2, 2, 4))
(1,1,2,6,6,7,9,12,21,21,42)	(1,1,2,9;1,3,3,6;3)	((1, 7);(2, 2, 4))
(1,1,2,6,7,9,12,14,24,24,28)	(1,1,2,6,7,9;12,14;12,12,14)	((1, 1);(2))
(1,2,3,3,4,6,8,11,22,24,44)	(1,2,3,3,6,11;4,22;4,12,22)	((1, 1);(2))
(1,3,3,4,6,8,9,16,18,24,36)	(1,3,3,6,9,18;4,36;4,8,12)	((1, 1);(2))
(1,3,3,6,6,6,7,7,12,14,63)	(1,1,2,9;1,3,3,6;3)	((7, 1);(2, 2, 4))
(1,3,3,6,7,7,9,14,18,24,36)	(1,1,2,3,6,12;1,24;7)	((3, 1);(1, 1, 2))
(2,2,2,4,4,8,11,11,21,21,42)	(1,1,1,2,2,4;11;21)	((2, 1, 1);(1, 1, 2))
(2,2,2,4,8,15,15,16,17,17,30)	(1,1,1,2,4,8;17;15)	((2, 1, 1);(1, 1, 2))
(2,2,4,6,7,7,12,14,24,25,25)	(1,1,2,3,6,12;25;7)	((2, 1, 1);(1, 1, 2))
(2,2,4,6,8,13,13,16,19,19,26)	(1,1,2,3,4,8;19;13)	((2, 1, 1);(1, 1, 2))
(2,2,4,9,9,9,15,15,15,18,30)	(3,5,5,10;2,2,4,15;9)	((3, 1);(1, 1, 2))
(3,3,5,5,6,9,10,15,18,18,36)	(1,1,2,3;1,1,2,3;9)	((3, 5);(2, 2, 4))
(3,5,5,6,6,9,9,10,12,18,45)	(3,5,5,10;2,2,4,15;9)	((1, 3);(1, 1, 2))

TABLE D. 8

10 VARIABLE ORTHOGONAL DESIGNS OF ORDER 128

Orthogonal design	Construction: Product designs	Construction: Amicable designs
(1,1,2,3,4,4,8,15,45,45)	(1,1,2,3;1,3,3;1)	((1, 15);(4, 4, 8))
(1,2,2,2,5,10,20,40,23,23)	(2,2,2,5,10,20;1,40;23)	((1, 1);(1, 1))
(1,7,7,7,8,8,14,16,28,32)	(1,1,1,2,4;1,8;1,2,4)	((7, 1);(8))
(2,2,4,5,6,9,18,23,23,36)	(2,2,4,6,9,18;5,36;23)	((1,1);(1,1))
(2,2,4,5,9,9,9,10,18,60)	(2,2,4,5,10;3,20;9)	((1, 3);(1, 1, 2))
(2,2,4,6,7,7,9,11,14,66)	(2,2,4,6,11;3,22;7)	((1, 3);(1, 1, 2))
(2,2,4,7,7,8,16,25,25,32)	(2,2,4,7,8,16;7,32;25)	((1, 1);(1, 1))
(2,2,4,7,8,9,9,18,21,48)	(2,2,4,7,8;7,16;9)	((1, 3);(1, 1, 2))
(2,2,4,9,9,18,18,20,23,23)	(1,1,2,9,10;23;9)	((2, 1, 1);(1, 1, 2))
(2,3,8,8,9,9,16,18,22,33)	(1,3,3,6,11;2,22;8)	((3, 1);(1, 1, 2))
(2,4,7,7,7,7,14,20,20,40)	(1,1,1,1,2;2,4;10)	((7, 1);(2, 2, 4))
(2,6,6,9,9,12,18,20,23,23)	(1,3,3,6,10;23;9)	((2, 1, 1);(1, 1, 2))
(3,3,3,5,6,12,14,14,28,40)	(1,1,1,2,4;1,8;7)	((3, 5);(2, 2, 4))
(3,3,4,4,6,8,9,13,39,39)	(1,1,2,3;1,3,3;1)	((3, 13);(4, 4, 8))
(3,3,9,12,12,13,13,13,24,26)	(1,1,1,2;1,1,3;3)	((13, 3);(4, 4, 8))
(3,7,7,8,10,10,14,20,21,28)	(1,1,2,3,4;3,8;5)	((7, 1);(2, 2, 4))
(4,4,4,8,13,13,18,19,19,26)	(2,2,2,4,9;19;13)	((2, 1, 1);(1, 1, 2))
(4,4,5,5,7,8,10,14,15,56)	(1,1,2,3;7;1)	((5, 1, 2, 8);(4,4, 8))
(4,4,5,5,8,10,11,15,33,33)	(1,1,2,3;1,3,3;1)	((5, 11);(4, 4,8))
(4,4,5,8,11,11,15,15,22,33)	(1,1,2,3;1,3,3;1)	((11, 5);(4, 4, 8))
(4,4,7,7,8,9,14,21,27,27)	(1,1,2,3;1,3,3;1)	((7, 9);(4, 4, 8))
(4,4,7,8,9,9,18,21,21,27)	(1,1,2,3;1,3,3;1)	((9, 7);(4, 4, 8))
(4,4,8,9,9,14,16,18,23,23)	(2,2,4,7,8;23;9)	((2, 1, 1);(1, 1, 2))
(5,5,10,10,10,11,15,20,20,22)	(1,1,2,3,4;11;5)	((5, 1, 2);(2, 2, 4))
(5,5,11,11,11,12,12,15,22,24)	(1,1,1,2;1,1,3;3)	((11, 5);(4,4, 8))
(5,5,5,5,6,10,12,16,32,32)	(1,1,1,1,2;2,4;2,4,4)	((5, 3);(8))
(5,5,5,6,10,15,16,16,18,32)	(1,1,1,2,3;2,6;8)	((5, 3);(2, 2, 4))
(5,5,5,7,10,10,14,18,18,36)	(1,1,1,2,2;7;9)	((5, 1, 2);(2,2, 4))
(5,6,6,11,11,12,15,16,22,24)	(2,2,4,5,8;5,16;11)	((3, 1);(1, 1, 2))

TABLE D. 9

9 AND 8 VARIABLE ORTHOGONAL DESIGNS OF ORDER 128

Orthogonal design	Construction	
	Product designs	Amicable designs
(1,1,1,8,8,16,31,31,31)	(1,1,1;1,1,1;1)	((1,31);(8,8,16))
(1,2,6,6,11,11,13,26,52)	(2,6,6,13,26;1,52;11)	((1,1);(1,1))
(1,4,7,7,8,11,16,37,37)	(1,4,7,7,8;11,16;37)	((1,1);(1,1))
(1,6,7,7,9,12,24,31,31)	(1,6,7,7,12;9,24;31)	((1,1);(1,1))
(2,3,4,5,5,10,21,36,42)	(2,3,5,5,10;4,21;18,21)	((1,1);(2))
(2,4,7,7,7,7,10,14,70)	(1,1,1,1,2;2,4;10)	((7,1);(1,7))
(3,3,3,8,8,16,29,29,29)	(1,1,1;1,1,1;1)	((3,29);(8,8,16))
(4,11,11,11,11,16,16,16,32)	(1,1,1,1;4;4)	((11,1,4);(4,4,8))
(4,4,4,5,7,14,28,31,31)	(4,4,4,7,14;5,28;31)	((1,1);(1,1))
(4,4,7,8,12,14,21,29,29)	(4,4,7,8,12;14,21;29)	((1,1);(1,1))
(4,5,7,14,14,14,21,21,28)	(1,2,3,3;4,5;7)	((7,1);(2,2,4))
(5,5,10,11,15,16,20,22,24)	(1,1,2,3,4;11;2,3)	((5,1,2);(8))
(5,5,5,8,8,16,27,27,27)	(1,1,1;1,1,1;1)	((5,27);(8,8,16))
(5,5,9,10,15,16,20,24,24)	(1,1,2,3,4;3,8;2,3)	((5,3);(8))
(6,9,9,10,10,18,20,23,23)	(3,5,5,10;23;9)	((2,1,1);(1,1,2))
(1,1,1,4,28,31,31,31)	(1,1,1;1,1,1;1)	((1,31);(4,28))
(3,4,8,13,13,25,31,31)	(3,4,13,13;8,25;31)	((1,1);(1,1))
(7,7,9,11,12,24,29,29)	(7,7,9,12;11,24;29)	((1,1);(1,1))

TABLE D.10

APPENDIX E: ORTHOGONAL DESIGNS IN ORDER 24

E.1. LEMMA (Wallis-Robinson). The following orthogonal designs, constructed using circulant matrices, exist in order 24:

1) (1, 1, 1, 1, 1, 5, 6);	12) (1, 3, 5, 6, 9);	23) (1, 2, 2, 4, 13);
2) (1, 1, 2, 2, 5, 8);	13) (1, 2, 3, 5, 13);	24) (1, 2, 2, 5, 14);
3) (1, 2, 6, 6, 9);	14) (1, 2, 3, 4, 12);	25) (1, 1, 2, 2, 9, 9);
4) (1, 2, 4, 5, 10);	15) (1, 3, 4, 5, 9);	26) (1, 2, 2, 2, 8, 9);
5) (1, 2, 5, 6, 10);	16) (1, 2, 2, 3, 16);	27) (1, 1, 2, 3, 4, 9);
6) (1, 1, 1, 4, 6, 6);	17) (1, 5, 6, 6, 6);	28) (1, 2, 3, 6, 10);
7) (1, 4, 5, 6, 6);	18) (1, 2, 2, 8, 11);	29) (2, 2, 3, 7, 8);
8) (1, 2, 5, 5, 9);	19) (1, 1, 4, 4, 5);	30) (1, 2, 3, 7, 9);
9) (2, 2, 5, 5, 8);	20) (1, 2, 5, 5, 8);	31) (1, 1, 2, 3, 10);
10) (1, 1, 1, 2, 4, 10);	21) (1, 2, 2, 8, 8);	32) (1, 1, 1, 1, 1, 4, 5):
11) (1, 1, 1, 3, 4, 9);	22) (1, 1, 1, 1, 2, 2, 8);	

Proof: All the designs are found by using various parts of Lemma 4.54, part (iii) for 1), 2), ..., 10); part (iv) for 11), 12), ..., 17); part (v) for 18); part (vi) for 19); part (vii) for 20) and 21); part (viii) for 22), ..., 32). Table 5.1 contains the first rows of the circulant matrices which should be used in Lemma 4.54 for each design.

E.2. LEMMA. There exist orthogonal designs of types (1, 1, 2, 4, 4, 8) and (3, 3, 3, 3, 3, 3, 3, 3) in order 24.

Proof: We note that there is a (1, 4) design in order 6 and so a (1, 1, 2, 4, 4, 8) design exists in order 24. The other design is in section 4.5 and is the Plotkin array of order 24.

COMPUTER SOLUTIONS

The solutions given in the following lemma were obtained by a long computer search.

E.3. LEMMA (Robinson). There are orthogonal designs of order 24 and types (1, 1, 1, 1, 1, 5, 5, 9) and (1, 1, 1, 1, 1, 2, 8, 9).

Proof: Consider the following matrices, M_1 and M_2:

x_5	$x_4\bar{x}_8x_6\bar{x}_7$	$x_7x_6x_8x_8$	$x_6\bar{x}_7\bar{x}_8x_8$	$\bar{x}_7\bar{x}_6\bar{x}_8\bar{x}_8$	$x_7x_6x_8x_8$
$\bar{x}_4x_8\bar{x}_6\bar{x}_7$	$\bar{x}_5$	$x_6x_7x_8x_8$	$\bar{x}_7x_6x_8\bar{x}_8$	$\bar{x}_7x_6x_8\bar{x}_8$	$x_7\bar{x}_6\bar{x}_8x_8$
$\bar{x}_7\bar{x}_6\bar{x}_8x_8$	$\bar{x}_6\bar{x}_7\bar{x}_8x_8$	x_7	$x_4x_8x_5x_6$	$\bar{x}_8x_8x_7x_7$	$x_8x_8\bar{x}_6x_6$
$\bar{x}_6x_7x_8x_8$	$x_7\bar{x}_6\bar{x}_8\bar{x}_8$	$\bar{x}_4\bar{x}_8\bar{x}_5x_6$	$\bar{x}_7$	$x_8\bar{x}_8x_6x_6$	$x_8x_8x_7\bar{x}_7$
$x_7x_6x_8\bar{x}_8$	$x_7\bar{x}_6\bar{x}_8\bar{x}_8$	$x_8\bar{x}_8\bar{x}_7x_7$	$\bar{x}_8x_8\bar{x}_6x_6$	x_6	$x_4x_8x_7x_5$
$\bar{x}_7\bar{x}_6\bar{x}_8x_8$	$\bar{x}_7x_6x_8x_8$	$\bar{x}_8\bar{x}_8x_6x_6$	$\bar{x}_8\bar{x}_8\bar{x}_7\bar{x}_7$	$\bar{x}_4\bar{x}_8\bar{x}_7x_5$	$\bar{x}_6$

$= M_1$

x_5	$x_4\bar{x}_6x_8\bar{x}_6$	$x_8x_8x_7x_7$	$x_8\bar{x}_8\bar{x}_7x_7$	$\bar{x}_8\bar{x}_7\bar{x}_8\bar{x}_7$	$x_8x_7x_8x_7$
$\bar{x}_4x_6\bar{x}_8\bar{x}_6$	$\bar{x}_5$	$x_8x_8x_7x_7$	$\bar{x}_8x_8x_7\bar{x}_7$	$x_8\bar{x}_7\bar{x}_8x_7$	$x_8\bar{x}_7\bar{x}_8x_7$
$\bar{x}_8\bar{x}_8\bar{x}_7x_7$	$\bar{x}_8\bar{x}_8\bar{x}_7x_7$	x_7	$x_4x_7x_5x_8$	$\bar{x}_7x_7x_8x_8$	$x_8x_8\bar{x}_6x_6$
$\bar{x}_8x_8x_7x_7$	$x_8\bar{x}_8\bar{x}_7\bar{x}_7$	$\bar{x}_4\bar{x}_7\bar{x}_5x_8$	$\bar{x}_7$	$x_8\bar{x}_8x_6x_6$	$x_7x_7x_8\bar{x}_8$
$x_8x_7x_8\bar{x}_7$	$\bar{x}_8x_7x_8x_7$	$x_7\bar{x}_7\bar{x}_8x_8$	$\bar{x}_8x_8\bar{x}_6x_6$	x_7	$x_4x_7x_8x_5$
$\bar{x}_8\bar{x}_7\bar{x}_8x_7$	$\bar{x}_8x_7x_8x_7$	$\bar{x}_8\bar{x}_8x_6x_6$	$\bar{x}_7\bar{x}_7\bar{x}_8\bar{x}_8$	$\bar{x}_4x_7x_8x_5$	$\bar{x}_7$

$= M_2$

Let N_1 and N_2 be the matrices obtained from M_1 and M_2 by replacing the diagonal entries, y, of M_i by

$$\begin{array}{cccc} x_1 & x_2 & x_3 & y \\ x_2 & x_1 & y & \bar{x}_3 \\ \bar{x}_3 & \bar{y} & x_1 & x_2 \\ \bar{y} & x_3 & \bar{x}_2 & x_1 \end{array}$$

and the off diagonal block entries $p\ q\ r\ s$ of M_i by

$$\begin{matrix} p & q & r & s \\ q & \bar{p} & s & \bar{r} \\ r & s & \bar{p} & q \\ s & \bar{r} & q & p. \end{matrix}$$

Then N_1 and N_2 give orthogonal designs of order 24 and types (1, 1, 1, 1, 1, 5, 5, 9) and (1, 1, 1, 1, 1, 2, 8, 9) respectively.

We recall Lemma 5.102 and Corollary 5.103.

E.4. LEMMA (Robinson). There are product designs of order 12 and types; (1, 1, 1;1, 1, 4;4), (1, 1, 4;1, 4, 4;1), (1, 1, 4;1, 1, 4;4), (1, 4, 4;1, 4, 4;1), (1, 1, 4;1, 1, 1;1) and (1, 1, 1;1, 1, 1;9).

E.5. LEMMA (Robinson). There are orthogonal designs of order 24 and types:

(1, 1, 1, 1, 1, 1, 9, 9)	(1, 1, 1, 1, 4, 4, 4, 4)
(1, 1, 1, 1, 1, 4, 4, 4)	(1, 1, 1, 1, 1, 1, 1, 4)

By using the designs in the previous section we obtain:

E.6. LEMMA (Robinson). All 4-tuples (a, b, c, 24-a-b-c), $0 \le a+b+c \le 24$, are the types of orthogonal designs in order 24 except possibly

(1, 4, 4, 15)	(3, 3, 7, 11)	(3, 7, 7, 7)
(1, 7, 8, 8)	(3, 4, 4, 13)	(4, 4, 5, 11)
(2, 2, 7, 13)	(3, 4, 6, 11)	(4, 4, 6, 10)
(2, 4, 4, 14)	(3, 4, 7, 10)	(4, 4, 7, 9)
(2, 4, 7, 11)	(3, 5, 8, 8)	(4, 5, 7, 8)
(2, 7, 7, 8)	(3, 6, 7, 8)	(5, 5, 6, 8).

E.7. LEMMA (Robinson). All 3-tuples of the form (a, b, c), $0 \le a+b+c \le 24$, are the types of orthogonal designs in order 24 except possibly

(4, 4, 15) (7, 7, 7) and (7, 8, 8).

TABLE E.1

Design	X_1	X_2	X_3	X_4	X_5	X_6	X_7	X_8
(1, 1, 1, 1, 1, 4, 5)	x_5 $-x_6$ x_6	x_1 0 0	x_2 0 0	x_3 0 0	x_4 0 0	0 x_6 x_6	0 x_7 x_7	$-x_7$ x_7 x_7
(1, 1, 1, 1, 1, 5, 6)	x_1 $-x_7$ x_7	x_2 0 0	x_3 0 0	x_4 0 0	x_5 0 0	x_6 x_7 x_7	$-x_7$ x_6 x_6	$-x_6$ x_6 x_6
(1, 1, 1, 1, 2, 2, 8)	x_1 0 0	x_2 0 0	x_5 x_6 x_6	x_7 $-x_6$ x_6	x_3 0 0	x_4 0 0	$-x_5$ x_6 x_6	x_7 $-x_6$ x_6
(1, 1, 1, 2, 4, 10)	x_1 $-x_5$ x_5	0 x_5 x_5	x_2 0 0	x_3 0 0	x_4 x_6 x_6	$-x_4$ x_6 x_6	$-x_6$ x_6 x_6	$-x_6$ x_6 x_6
(1, 1, 1, 3, 4, 9)	x_1 $-x_5$ x_5	x_4 $-x_6$ x_6	x_4 $-x_6$ x_6	x_4 $-x_6$ x_6	x_6 x_6 x_6	0 x_5 x_5	x_2 0 0	x_3 0 0
(1, 1, 1, 4, 6, 6)	x_1 $-x_4$ x_4	x_2 0 0	x_3 0 0	0 x_4 x_4	x_5 x_6 x_6	$-x_6$ x_5 x_5	$-x_5$ x_5 x_5	$-x_6$ x_6 x_6
(1, 1, 2, 2, 5, 8)	x_3 x_6 $-x_6$	x_3 x_6 $-x_6$	x_1 0 0	x_2 0 0	x_4 x_6 x_6	$-x_4$ x_6 x_6	0 x_5 x_5	$-x_5$ x_5 x_5
(1, 1, 2, 2, 9, 9)	x_2 $-x_5$ x_5	x_1 $-x_6$ x_6	x_3 $-x_5$ x_5	x_4 $-x_6$ x_6	x_5 x_5 x_5	x_6 x_6 x_6	x_3 $-x_5$ x_5	x_4 $-x_6$ x_6
(1, 1, 2, 3, 4, 9)	x_1 0 0	x_1 $-x_6$ x_6	0 x_4 x_5	x_3 $-x_6$ x_6	$-x_4$ x_5 x_5	x_6 x_6 x_6	0 x_4 x_5	x_3 $-x_5$ x_6
(1, 2, 2, 2, 8, 9)	x_2 $-x_5$ x_5	$-x_3$ x_5 x_5	x_1 $-x_6$ x_6	x_4 $-x_6$ x_6	x_2 $-x_5$ x_5	x_3 x_5 x_5	x_6 x_6 x_6	x_4 $-x_6$ x_6
(1, 1, 2, 3, 10)	x_1 0 0	x_3 $-x_4$ x_4	$-x_5$ x_4 x_4	x_5 x_4 0	x_2 0 0	x_3 $-x_4$ x_4	0 x_4 x_4	x_5 x_4 0
(1, 1, 4, 4, 5)	0 $-x_3$ x_3	0 $-x_4$ x_4	x_1 0 0	x_2 0 0	0 x_3 x_3	0 x_4 x_4	0 x_5 x_5	$-x_5$ x_5 x_5
(1, 2, 2, 3, 16)	x_1 $-x_5$ x_5	x_4 $-x_5$ x_5	x_4 $-x_5$ x_5	x_4 $-x_5$ x_5	x_2 x_5 x_5	$-x_2$ x_5 x_5	x_3 x_5 x_5	$-x_3$ x_5 x_5
(1, 2, 2, 4, 13)	x_2 $-x_5$ x_5	x_1 $-x_4$ x_4	x_3 $-x_5$ x_5	0 x_4 x_4	x_2 $-x_5$ x_5	x_5 x_5 x_5	x_3 $-x_5$ x_5	0 x_5 x_5
(1, 2, 2, 5, 14)	x_2 $-x_5$ x_5	x_1 $-x_4$ x_4	x_3 $-x_5$ x_5	x_5 x_4 x_4	x_2 $-x_5$ x_5	x_5 x_5 x_5	x_3 $-x_5$ x_5	$-x_4$ x_5 x_5
(1, 2, 2, 8, 8)	x_1 $-x_4$ x_4	0 $-x_4$ x_4	0 $-x_5$ x_5	0 $-x_5$ x_5	x_2 x_4 x_4	$-x_2$ x_4 x_4	x_3 x_5 x_5	$-x_3$ x_5 x_5
(1, 2, 2, 8, 11)	x_1 $-x_5$ x_5	x_5 x_5 x_5	x_2 $-x_4$ x_4	x_2 $-x_4$ x_4	x_3 x_4 x_4	$-x_3$ x_4 x_4	$-x_5$ x_5 x_5	$-x_5$ x_5 x_5
(1, 2, 3, 4, 12)	x_1 $-x_4$ x_4	x_3 $-x_5$ x_5	x_3 $-x_5$ x_5	x_3 $-x_5$ x_5	x_2 x_5 x_5	$-x_2$ x_5 x_5	0 x_5 x_5	0 x_4 x_4

(1, 2, 3, 5, 13)	$x_1 - x_4 x_4$	$x_3 - x_5 x_5$	$x_3 - x_5 x_5$	$x_3 - x_5 x_5$	$x_2 x_5 x_5$	$-x_2 x_5 x_5$	$x_4 x_5 x_5$	$-x_5 x_4 x_4$
(1, 2, 3, 6, 10)	$x_1 - x_4 x_4$	$x_2 - x_5 x_5$	$x_3 x_4 x_4$	$0 - x_3 x_4$	$x_5 x_5 x_5$	$x_2 - x_5 x_5$	$-x_5 x_5 x_5$	$0 - x_3 x_4$
(1, 2, 3, 7, 9)	$x_1 - x_5 x_5$	$x_2 - x_5 x_5$	$0 x_3 x_4$	$-x_3 x_4 x_4$	$x_5 x_5 x_5$	$x_2 - x_5 x_5$	$0 x_3 x_4$	$-x_4 x_4 x_4$
(1, 2, 4, 5, 10)	$x_1 - x_3 x_3$	$x_2 x_5 x_5$	$-x_2 x_5 x_5$	$-x_5 x_5 x_5$	$-x_5 x_5 x_5$	$0 x_3 x_3$	$0 x_4 x_4$	$-x_4 x_4 x_4$
(1, 2, 5, 5, 8)	$x_1 - x_4 x_4$	$0 - x_3 x_3$	$0 - x_5 x_5$	$0 - x_5 x_5$	$x_3 x_4 x_4$	$-x_4 x_3 x_3$	$x_2 x_5 x_5$	$-x_2 x_5 x_5$
(1, 2, 5, 5, 9)	$x_1 - x_5 x_5$	$x_2 x_5 x_5$	$-x_2 x_5 x_5$	$-x_5 x_5 x_5$	$0 x_3 x_3$	$-x_3 x_3 x_3$	$0 x_4 x_4$	$-x_4 x_4 x_4$
(1, 2, 5, 6, 10)	$x_1 - x_3 x_3$	$x_2 x_5 x_5$	$-x_2 x_5 x_5$	$-x_5 x_5 x_5$	$-x_5 x_5 x_5$	$x_4 x_3 x_3$	$-x_3 x_4 x_4$	$-x_4 x_4 x_4$
(1, 2, 6, 6, 9)	$x_5 x_3 - x_3$	$x_1 x_2 x_2$	$-x_2 x_1 x_1$	$-x_1 x_1 x_1$	$-x_2 x_2 x_2$	$x_4 x_3 x_3$	$-x_4 x_3 x_3$	$-x_3 x_3 x_3$
(1, 3, 4, 5, 9)	$x_1 - x_4 x_4$	$x_2 - x_5 x_5$	$x_2 - x_5 x_5$	$x_2 - x_5 x_5$	$x_5 x_5 x_5$	$0 x_4 x_4$	$0 x_3 x_3$	$-x_3 x_3 x_3$
(1, 3, 5, 6, 9)	$x_1 - x_4 x_4$	$x_2 - x_5 x_5$	$x_2 - x_5 x_5$	$x_2 - x_5 x_5$	$x_5 x_5 x_5$	$x_3 x_4 x_4$	$-x_4 x_3 x_3$	$-x_3 x_3 x_3$
(1, 4, 5, 6, 6)	$x_1 - x_2 x_2$	$0 x_2 x_2$	$0 x_3 x_3$	$-x_3 x_3 x_3$	$x_4 x_5 x_5$	$-x_5 x_4 x_4$	$-x_4 x_4 x_4$	$-x_5 x_5 x_5$
(1, 5, 6, 6, 6)	$x_1 x_2 - x_2$	$x_3 x_2 x_2$	$-x_2 x_3 x_3$	$-x_3 x_3 x_3$	$-x_4 x_5 x_5$	$-x_5 x_5 x_5$	$x_5 x_4 x_4$	$-x_4 x_4 x_4$
(2, 2, 3, 7, 8)	$x_1 x_5 x_5$	$0 x_3 x_4$	$-x_3 x_4 x_4$	$x_2 - x_5 x_5$	$-x_1 x_5 x_5$	$0 x_3 x_4$	$-x_4 x_4 x_4$	$x_2 - x_5 x_5$
(2, 2, 5, 5, 8)	$x_1 - x_5 x_5$	$x_1 - x_5 x_5$	$x_2 x_5 x_5$	$-x_2 x_5 x_5$	$0 x_3 x_3$	$-x_3 x_3 x_3$	$0 x_4 x_4$	$-x_4 x_4 x_4$

We next list (Table E.2) the 74 8-tuples which are not prohibited from being the type of an orthogonal design in order 24 (the number after the type is the sum of the type numbers). The programme which gave this was devised by Roger Magoon while he was an undergraduate at Queen's University. A check (✓) indicates the design has been constructed.

TABLE E.2

ORTHOGONAL DESIGNS OF ORDER 24 OF TYPE

(1, 1, 1, 1, 1, 1, 1, 1): 8(✓)
(1, 1, 1, 1, 1, 1, 1, 4): 11(✓)
(1, 1, 1, 1, 1, 1, 1, 9): 16
(1, 1, 1, 1, 1, 1, 2, 2): 10
(1, 1, 1, 1, 1, 1, 2, 8): 16
(1, 1, 1, 1, 1, 1, 4, 4): 14
(1, 1, 1, 1, 1, 1, 4, 9): 19
(1, 1, 1, 1, 1, 1, 5, 5): 16
(1, 1, 1, 1, 1, 1, 8, 8): 22
(1, 1, 1, 1, 1, 1, 9, 9): 24(✓)
(1, 1, 1, 1, 1, 2, 2, 4): 13
(1, 1, 1, 1, 1, 2, 2, 9): 18
(1, 1, 1, 1, 1, 2, 3, 6): 16
(1, 1, 1, 1, 1, 2, 4, 8): 19
(1, 1, 1, 1, 1, 2, 8, 9): 24(✓)
(1, 1, 1, 1, 1, 3, 6, 8): 22
(1, 1, 1, 1, 1, 4, 4, 4): 17(✓)
(1, 1, 1, 1, 1, 4, 4, 9): 22
(1, 1, 1, 1, 1, 4, 5, 5): 19
(1, 1, 1, 1, 1, 5, 5, 9): 24(✓)
(1, 1, 1, 1, 2, 2, 2, 2): 12
(1, 1, 1, 1, 2, 2, 2, 8): 18
(1, 1, 1, 1, 2, 2, 4, 4): 16
(1, 1, 1, 1, 2, 2, 4, 9): 21
(1, 1, 1, 1, 2, 2, 5, 5): 18
(1, 1, 1, 1, 2, 2, 8, 8): 24
(1, 1, 1, 1, 2, 3, 4, 6): 19
(1, 1, 1, 1, 2, 3, 6, 9): 24
(1, 1, 1, 1, 2, 4, 4, 8): 22
(1, 1, 1, 1, 2, 5, 5, 8): 24
(1, 1, 1, 1, 3, 3, 3, 3): 16
(1, 1, 1, 1, 3, 3, 6, 6): 22
(1, 1, 1, 1, 4, 4, 4, 4): 20(✓)
(1, 1, 1, 1, 4, 4, 5, 5): 22
(1, 1, 1, 1, 5, 5, 5, 5): 24
(1, 1, 1, 2, 2, 2, 2, 4): 15
(1, 1, 1, 2, 2, 2, 2, 9): 20
(1, 1, 1, 2, 2, 2, 3, 6): 18
(1, 1, 1, 2, 2, 2, 4, 8): 21
(1, 1, 1, 2, 2, 3, 6, 8): 24
(1, 1, 1, 2, 2, 4, 4, 4): 19
(1, 1, 1, 2, 2, 4, 4, 9): 24
(1, 1, 1, 2, 2, 4, 5, 5): 21
(1, 1, 1, 2, 3, 4, 4, 6): 22
(1, 1, 1, 2, 3, 5, 5, 6): 24
(1, 1, 1, 3, 3, 3, 3, 4): 19
(1, 1, 1, 3, 3, 3, 3, 9): 24
(1, 1, 2, 2, 2, 2, 2, 2): 14
(1, 1, 2, 2, 2, 2, 2, 8): 20
(1, 1, 2, 2, 2, 2, 4, 4): 18
(1, 1, 2, 2, 2, 2, 5, 5): 20
(1, 1, 2, 2, 2, 3, 4, 6): 21
(1, 1, 2, 2, 2, 4, 4, 8): 24
(1, 1, 2, 2, 3, 3, 3, 3): 18
(1, 1, 2, 2, 3, 3, 6, 6): 24
(1, 1, 2, 2, 4, 4, 4, 4): 22
(1, 1, 2, 2, 4, 4, 5, 5): 24
(1, 1, 2, 3, 3, 3, 3, 8): 24
(1, 1, 3, 3, 3, 3, 4, 4): 22
(1, 1, 3, 3, 3, 3, 5, 5): 24
(1, 2, 2, 2, 2, 2, 2, 4): 17
(1, 2, 2, 2, 2, 2, 2, 9): 22
(1, 2, 2, 2, 2, 2, 3, 6): 20
(1, 2, 2, 2, 2, 4, 4, 4): 21
(1, 2, 2, 2, 3, 4, 4, 6): 24
(1, 2, 2, 3, 3, 3, 3, 4): 21
(1, 2, 3, 3, 3, 3, 3, 6): 24
(2, 2, 2, 2, 2, 2, 2, 2): 16
(2, 2, 2, 2, 2, 2, 2, 8): 22
(2, 2, 2, 2, 2, 2, 4, 4): 20
(2, 2, 2, 2, 2, 2, 5, 5): 22
(2, 2, 2, 2, 3, 3, 3, 3): 20
(2, 2, 2, 2, 4, 4, 4, 4): 24
(2, 2, 3, 3, 3, 3, 4, 4): 24
(3, 3, 3, 3, 3, 3, 3, 3): 24(✓)

APPENDIX F: ORTHOGONAL DESIGNS IN ORDERS 40 AND 80

We note that another Appendix gives all the orthogonal designs known in order 20 and the first rows that may be used in the Goethals-Seidel array to generate them.

For reference we list the following first rows which may be used to generate circulant symmetric matrices which can then be used in the Goethals-Seidel array to give the orthogonal designs in order 20 indicated:

(9)	-aaaaa, 0a-a-aa;
(1, 11)	a-a00-a, a0-a-a0, -aaaaa, b0000;
(13)	a0aa0, 00aa0, 0a-a-aa, 0a-a-aa;
(15)	a-a00-a, 0a00a, -aaaaa, aa-a-aa;
(16)	0aaaa, 0a-a-aa, 0a-a-aa, 0a-a-aa;
(18)	0aaaa, 0a-a-aa, aa-a-aa, a-aaa-a;
(20)	-aaaaa, -aaaaa, aa-a-aa, a-aaa-a;
(8,12)	-abaab, a-abb-a, ab-b-bb, -aaaaa;
(2,13)	ab-b-bb, a-bbb-b, b0bb0, 00bb0;
(10,10)	ab-b-bb, -ba-a-aa, -aaaaa, -bbbbb;
(2,16)	ab-b-bb, a-bbb-b, 0b-b-bb, 0bbbb;
(2,18)	ab-b-bb, a-bbb-b, -bbbbb, -bbbbb;
(9, 9)	0a-a-aa, -aaaaa, 0b-b-bb, -bbbbb;
(1, 1,9)	a0000, c0000, 0b-b-bb, -bbbbb.

The following first rows may be used to generate one circulant and three circulant symmetric matrices which can then be used in the Goethals-Seidel array to give the orthogonal designs in order 20 indicated:

(1, 1, 1, 4)	a0b-b0, 00bb0, c0000, d0000;
(1, 4, 9)	a0b-b0, 00bb0, 0c-c-cc, -ccccc;
(1, 1, 10)	abb-b-b, b0bb0, b-b00-b, c0000;
(1, 14)	abb-b-b, b-b00-b, 0b00b, -bbbbb;
(1, 17)	abb-b-b, b-bbb-b, 0bbbb, 0b-b-bb;
(1, 19)	abb-b-b, -bbbbb, -bbbbb, b-bbb-b,
(5, 15)	aba-ab, b-baa-b, -bbbbb, bb-b-bb;
(4, 11)	a-ab-b-a, 00bb0, a0aa0, aa-a-aa.

We shall not list the 8-tuples which are not prohibited from being the type of an orthogonal design in order 40 since Magoon's programme yielded 703 possibilities!

F.1. LEMMA. The following full orthogonal designs exist in order 40.

1) (1, 2, 2, 2, 16, 17);
2) (1, 10, 14, 15);
3) (2, 2, 14, 22);
4) (1, 2, 3, 34);
5) (1, 8, 12, 19);
6) (5, 8, 12, 15);
7) (1, 2, 10, 27);
8) (2, 4, 13, 21);
9) (2, 2, 5, 31);
10) (1, 2, 2, 11, 24);
11) (1, 2, 12, 25);
12) (1, 10, 10, 19);
13) (1, 2, 14, 23).

Proof: The designs are constructed by using the matrices whose first rows are given in Table F.1 in the (1, 1, 1, 1, 1, 1, 1, 1) design in order 8; we use the back-circulant matrix constructed from X_1, and the circulant matrices constructed from the other X_i, $i = 2, 3, \ldots, 8$. Note that 4) is constructed using Lemma 4.54 (iii).

F.2. COROLLARY. All 3-tuples (a, b, 40-a-b) are the types of orthogonal designs in order 40 except possibly

(1, 7, 32)	(1, 9, 30)	(3, 5, 32)	(3, 6, 31)	(3, 8, 29)
(3, 9, 28)	(4, 7, 29)	(5, 6, 29)	(5, 13, 22)	(6, 7, 27)
(7, 8, 25)	(7, 9, 24)	(7, 11, 22)	(7, 14, 19)	(8, 9, 23)
(8, 11, 21)	(9, 14, 17)			

which are undecided.

Proof: We use the designs of the theorem and the designs

(1, 1, 9, 9)	(1, 5, 5, 9)	(2, 5, 5, 8)	(5, 5, 5, 5)
(1, 2, 8, 9)	(2, 3, 6, 9)	(2, 2, 16)	(4, 8, 8)

in order 20.

F.3. COROLLARY. All 2-tuples are the types of orthogonal designs in order 40 except possibly

(7, 32)   (9, 30)

which are undecided.

F. 4. LEMMA. The following orthogonal designs exist in order 40:

(i) (2, 10, 10, 13)	(v) (1, 2, 2, 4, 25)	(ix) (2, 4, 11, 16)
(ii) (5, 9, 9, 15)	(vi) (1, 4, 9, 9, 9)	(x) (1, 4, 8, 8, 16)
(iii) (1, 2, 6, 9, 20)	(vii) (2, 8, 25)	(xi) (1, 8, 8, 8, 8)
(iv) (1, 2, 16, 19)	(viii) (1, 10, 10, 17)	(xii) (1, 1, 1, 4, 20)
		(xiii) (1, 1, 8, 8, 9).

Proof: Use the first rows given in Table F. 2 to generate circulant or back-circulant matrices which can be used in the (1, 1, 1, 1, 1, 1, 1, 1) design in order 8 to give the desired result. Note that (xii) and (xiii) are constructed using Lemma 4. 54 (iii) and (viii).

F. 5. LEMMA. Let M = {(1, 1, 1, 4), (4, 11), (1, 19), (1, 1, 10), (1, 4, 9), (1, 17), (1, 14), (5, 15)} and N = {(8, 12), (10, 10), (2, 13), (2, 16), (2, 18), (9, 9), (15), (13), (1, 11), (1, 1, 9)}. Then there exist orthogonal designs of type (a, b, 3c, 3d) in order 80 and (a, b, c, d) in order 40 when (a, b) $\in$ M and (c, d) $\in$ N or (a, b) $\in$ N and (c, d) $\in$ M $\cup$ N. In particular there exist orthogonal designs of types

(1, 6, 14, 48)	(2, 3, 16, 51)	(2, 15, 18, 45)
(1, 6, 14, 54)	(2, 3, 16, 57)	(3, 10, 10, 57)

in order 80.

Proof: Each of the designs in N can be constructed using four circulant symmetric matrices of order 5. Each design in M can be constructed using one back-circulant and three circulant symmetric matrices of order 5. These matrices are used in the (1, 1, 1, 1, 3, 3, 3, 3) design in order 16 found in Example 5.101 to obtain the designs of the enunciation.

F. 6. LEMMA. The following orthogonal designs exist in order 80:

1) (1, 4, 6, 19, 50)	8) (1, 6, 28, 43)	15) (1, 1, 6, 16, 43)
2) (1, 4, 6, 34, 35)	9) (1, 2, 18, 59)	16) (2, 3, 14, 26, 35)
3) (2, 3, 16, 24, 35)	10) (3, 10, 10, 57)	17) (2, 2, 3, 6, 32, 35)
4) (2, 3, 4, 35, 36)	11) (1, 14, 65)	18) (4, 6, 11, 54)
5) (2, 3, 20, 20, 35)	12) (1, 6, 15, 58)	19) (4, 11, 24, 36)
6) (2, 3, 4, 18, 53)	13) (1, 2, 28, 45)	20) (4, 11, 30, 30)
7) (2, 3, 12, 28, 35)	14) (1, 6, 29, 44)	

Proof: We use the designs cited in Table F.3 in designs in order 16 as follows: for 1)-5) use the (1, 1, 2, 2, 2, 2, 3, 3) design found in Section 12; for 6)-20) use the (1, 1, 1, 1, 1, 2, 3, 3, 3) design found in Example 5.101. In all cases the first row of the circulant matrix is given, if the first row does not give a circulant symmetric matrix the back circulant matrix should be used.

F.7. COROLLARY. All 3-tuples (a, b, 80-a-b) are the types of orthogonal designs in order 80 except possibly

(3, 13, 64)	(3, 15, 62)	(3, 21, 56)	(5, 11, 64)
(5, 17, 58)	(5, 31, 44)	(6, 9, 65)	(6, 31, 43)
(7, 8, 65)	(7, 9, 64)	(7, 12, 61)	(7, 16, 57)
(7, 22, 51)	(7, 25, 48)	(7, 28, 45)	(8, 29, 43)
(9, 14, 57)	(9, 15, 56)	(9, 16, 55)	(9, 28, 43)
(10, 29, 41)	(11, 14, 55)	(11, 16, 53)	(11, 25, 44)
(12, 27, 41)	(12, 29, 39)	(13, 23, 44)	(14, 19, 47)
(14, 27, 39)	(16, 21, 43)	(16, 23, 41)	(16, 25, 39)
(18, 23, 39)	(19, 28, 33)	(21, 22, 37)	

which are undecided.

Proof: We use the designs of the theorem together with the designs obtained by using lemmas 4.4 and 4.11 on the known designs in 40 and 80 to obtain these results.

F.8. COROLLARY. All 2-tuples (a, b) are the types of orthogonal designs in order 80 except possibly

(13, 64) (15, 62)

which are undecided.

F.9. COROLLARY. All 3-tuples (a, b, 160-a-b) where a, b are even exist in order 160.

TABLE F.1

Design	X_1	X_2	X_3	X_4	X_5	X_6	X_7	X_8
(1, 2, 2, 2, 16, 17)	faa-a-a	a-aaa-a	eb-b-bb	-eb-b-bb	da-b-ba	-da-b-ba	cbaab	-cbaab
(1, 10, 14, 15)	dbb-b-b	-bb-b-bb	babba	aa-b-ba	b-aaa-a	ca-a-aa	-ac-c-cc	-ccccc
(2, 2, 14, 22)	cb-b-bb	-cb-b-bb	da-a-aa	-da-a-aa	-aaaaa	babba	aa-b-ba	b-aaa-a
(1, 2, 3, 34)	cb-bb-b	abb-b-b	abb-b-b	abb-b-b	b-bbb-b	b-bbb-b	dbbbb	-dbbbb
(1, 8, 12, 19)	abb-b-b	b-bbb-b	-cdccd	d-dcc-d	dc-c-cc	-ddddd	-bbbbb	-bbbbb
(5, 8, 12, 15)	aba-ab	b-baa-b	-cdccd	d-dcc-d	dc-c-cc	-ddddd	-aaaaa	aa-a-aa
(1, 2, 10, 27)	dbb-b-b	ab-b-bb	-ab-b-bb	bc-c-cc	-cb-b-bb	-ccccc	b-bbb-b	bbbbb
(2, 4, 13, 21)	-aabb-a	b-abb-a	-cbaab	cabba	cb-b-bb	c-aaa-a	db-b-bb	-db-b-bb
(2, 2, 5, 31)	aabb-a	ba-b-ba	b-bbb-b	cb-b-bb	-cb-b-bb	eb-b-bb	-eb-b-bb	bbbbb
(1, 2, 2, 11, 24)	add-d-d	be-e-ee	-be-e-ee	ce-e-ee	-ce-e-ee	-deeee	d-dee-d	dedde
(1, 2, 12, 25)	add-d-d	cc-c-cc	-dd-d-dd	-cdddd	c-cdd-c	cdccd	bd-d-dd	-bd-d-dd
(1, 10, 10, 19)	add-d-d	-ddddd	-ddddd	d-ddd-d	bc-c-cc	-cb-b-bb	-bbbbb	-ccccc
(1, 2, 14, 23)	bdd-d-d	-ac-c-cc	ac-c-cc	-dd-d-dd	-ddddd	dd-c-cd	cdccd	-cd-d-dd

TABLE F.2

Design	X_1	X_2	X_3	X_4	X_5	X_6	X_7	X_8
(2, 10, 10, 13)	ad-d-dd	a-ddd-d	d0dd0	00dd0	cb-b-bb	-bc-c-cc	-bbbbb	-ccccc
(5, 9, 9, 15)	ada-ad	d-daa-d	-ddddd	dd-d-dd	0b-b-bb	-bbbbb	0c-c-cc	-ccccc
(1, 2, 6, 9, 20)	a-ec-ce	0-ecc-e	ce-e-ee	ceeee	be-e-ee	-be-e-ee	0d-d-dd	-ddddd
(1, 2, 16, 19)	add-d-d	-ddddd	-ddddd	d-ddd-d	bc-c-cc	b-ccc-c	0c-c-cc	0cccc
(1, 2, 2, 4, 25)	a0d-d0	00dd0	be-e-ee	-be-e-ee	ce-e-ee	-ce-e-ee	0eeee	-eeeee
(1, 4, 9, 9, 9)	a0b-b0	00bb0	0c-c-cc	-ccccc	0d-d-dd	-ddddd	0e-e-ee	-eeeee
(2, 8, 25)	bcbb-c	bcbb-c	-bcbbc	-b-cbb-c	b-b00-b	0b00b	ab-b-bb	-ab-b-bb
(1, 10, 10, 17)	add-d-d	d-ddd-d	0dddd	0d-d-dd	bc-c-cc	-cb-b-bb	-bbbbb	-ccccc
(2, 4, 11, 16)	c-cb-b-c	00bb0	c0cc0	cc-c-cc	ad-d-dd	-ad-d-dd	0d-d-dd	0dddd
(1, 4, 8, 8, 16)	acd-d-c	0c-dd-c	0ce-e-c	0c-ee-c	-bcddc	-bc-d-dc	bceec	bc-e-ec
(1, 8, 8, 8, 8)	abc-c-b	0b-cc-b	0de-e-d	0d-ee-d	0bccb	0b-c-cb	0deed	0d-e-ed
(1, 1, 1, 4, 20)	ae00-e	b0000	c0000	0e00e	-ddddd	-ddddd	-dd-d-dd	dd-d-dd
(1, 1, 8, 8, 9)	a0000	b0000	0cc-cc	0ccc-c	-eeeee	0e-e-ee	0dd-dd	0ddd-d

TABLE F.3

Design	X_1	X_2	X_3	X_4	X_5	X_6	X_7	X_8	X_9
(1,4,6,19,50)	fee-e-e	e-eee-e	-eeeee	db-b-bb	-db-b-bb	bbbbb	ab-b-bb	-ab-b-bb	
(1,4,6,24,35)	ecc-c-c	c-ccc-c	-dc-c-cc	dc-c-cc	-bcccc	cbbbb	ab-b-bb	-ab-b-bb	
(2,3,16,24,35)	ebbbb	-ebbbb	-cdccd	c-cdd-c	cd-d-dd	-ccccc	abb-b-b	b-bbb-b	
(2,3,4,35,36)	ebbbb	-ebbbb	cd-d-dd	-cd-d-dd	-ddddd	-ddddd	abb-b-b	b-bbb-b	
(2,3,20,20,35)	ebbbb	-ebbbb	cd-d-dd	-dc-c-cc	-ccccc	-ddddd	abb-b-b	b-bbb-b	
(2,3,4,18,53)	ab-b-bb	-ab-b-bb	-bbbbb	dc-c-cc	-dc-c-cc	ccccc	ecc-c-c	c-ccc-c	
(2,3,12,28,35)	ebbbb	-ebbbb	-ccccc	-dc-c-cc	-cd-d-dd	cd-d-dd	-ddddd	abb-b-b	b-bbb-b
(1,6,28,43)	caa-a-a	0a-a-aa	0b-b-bb	a-aaa-a	bbbbb	-bbbbb	-aaaaa	db-b-bb	-db-b-bb
(1,2,18,59)	cdd-d-d	-ab-b-bb	ab-b-bb	-bbbbb	-bbbbb	dd-d-dd	d-ddd-d	-ddddd	-ddddd
(3,10,10,57)	ab-b-bb	-ba-a-aa	-bbbbb	-aaaaa	cdd-d-d	cdd-d-d	d-ddd-d	-ddddd	-ddddd
(1,14,65)	cbb-b-b	-aaaaa	b-aaa-a	b-bbb-b	-bbbbb	a-bbb-b	ab-b-bb	-bbbbb	-bbbbb
(1,6,15,58)	cbb-b-b	-bb-b-bb	babba	aa-b-ba	b-aaa-a	-aaaaa	da-a-aa	d-aaa-a	-aaaaa
(1,2,28,45)	caa-a-a	0a-a-aa	a-aaa-a	db-b-bb	-db-b-bb	-bbbbb	0b-b-bb	-aaaaa	-bbbbb
(1,6,29,44)	caa-a-a	-ba-a-aa	-ab-b-bb	a-aaa-a	bbbbb	-bbbbb	-aaaaa	db-b-bb	-db-b-bb
(1,1,6,16,43)	c0000	d0000	0b-b-bb	0aaaa	bbbbb	-bbbbb	0a-a-aa	eb-b-bb	-eb-b-bb
(2,3,14,26,35)	ebbbb	-ebbbb	-dc-c-cc	-ccccc	ddddd	-cd-d-dd	cd-d-dd	abb-b-b	b-bbb-b
(2,2,3,6,32,35)	ebbbb	-ebbbb	-cd-d-dd	fdddd	-fdddd	-cd-d-dd	cd-d-dd	abb-b-b	b-bbb-b
(4,6,11,54)	00aa0	b-ba-a-b	b0bb0	bb-b-bb	cd-d-dd	cd-d-dd	-cd-d-dd	-ddddd	-ddddd
(4,11,24,36)	00aa0	b-ba-a-b	b0bb0	bb-b-bb	-cdccd	-cdccd	c-cdd-c	cd-d-dd	-ccccc
(4,11,30,30)	00aa0	b-ba-a-b	b0bb0	bb-b-bb	cd-d-dd	cd-d-dd	-cd-d-dd	-ccccc	-ddddd

APPENDIX G: RESULTS BY ORDER

ORDER 8: The Radon number is 8. There is an O. D. of type (1,1,1,1,1,1,1,1), so all 1, 2, 3, 4, 5, 6, 7, 8 variable designs exist in order 8.

ORDER 12: The Radon number is 4. All designs not excluded by the non-existence theorems exist.

4 VARIABLES: The four variable designs that exist are listed in Appendix A.

3 VARIABLES: The following 3 variable designs are not excluded and all may be found from the 4 variable designs by equating variables or setting variables zero:

(1, 1, 1)	(1, 1, 10)	(1, 2, 9)	(1, 5, 6)	(2, 3, 6)
(1, 1, 2)	(1, 2, 2)	(1, 3, 6)	(2, 2, 2)	(2, 3, 7)
(1, 1, 4)	(1, 2, 3)	(1, 3, 8)	(2, 2, 4)	(2, 4, 4)
(1, 1, 5)	(1, 2, 4)	(1, 4, 4)	(2, 2, 5)	(2, 4, 6)
(1, 1, 8)	(1, 2, 6)	(1, 4, 5)	(2, 2, 8)	(2, 5, 5)
(1, 1, 9)	(1, 2, 8)	(1, 5, 5)	(2, 3, 4)	(3, 3, 3)
				(3, 3, 6)

2 VARIABLES: All (s, t) where $s + t \leq 12$, $1 \leq s \leq t$, (36 possible but 3 excluded leaves 33 possibilities) exist and may be found from 4 variable designs by equating variables or setting variables zero. Those not possible are:

(1, 7) (4, 7) (3, 5)

1 VARIABLE: All 1 variable designs exist.

ORDER 16: The Radon number is 9.

All designs on n variables, $n \leq 9$, which exist can be obtained by equating and killing variables in the 9 variable designs which exist and the following 8 variable designs which exist:

(1, 1, 1, 1, 1, 1, 3, 3) (1, 1, 1, 1, 2, 2, 3, 3) (1, 1, 2, 2, 2, 2, 3, 3)

(1, 1, 1, 1, 1, 1, 5, 5) (1, 1, 1, 1, 2, 2, 4, 4) (See Theorem D. 5 of Appendix D.)

9 VARIABLES: There are 45 possible 9-tuples that might be the type of orthogonal designs. They are listed in Theorem D. 5 of Appendix D.

The following 9-tuples cannot be the type of an orthogonal design:

(1, 1, 1, 1, 1, 1, 1, 1, 7)	(1, 1, 1, 1, 1, 1, 2, 2, 5)	(1, 1, 1, 1, 1, 2, 2, 2, 5)
(1, 1, 1, 1, 1, 1, 1, 2, 3)	(1, 1, 1, 1, 1, 1, 2, 3, 3)	(1, 1, 1, 1, 1, 2, 2, 3, 3)
(1, 1, 1, 1, 1, 1, 1, 2, 5)	(1, 1, 1, 1, 1, 1, 2, 3, 4)	(1, 1, 1, 1, 2, 2, 2, 2, 3)
(1, 1, 1, 1, 1, 1, 1, 2, 6)	(1, 1, 1, 1, 1, 1, 3, 3, 3)	(1, 1, 1, 1, 2, 2, 2, 3, 3)
(1, 1, 1, 1, 1, 1, 1, 3, 3)	(1, 1, 1, 1, 1, 1, 3, 3, 4)	(1, 1, 1, 2, 2, 2, 2, 2, 2)
(1, 1, 1, 1, 1, 1, 1, 3, 5)	(1, 1, 1, 1, 1, 2, 2, 2, 3)	(1, 1, 1, 2, 2, 2, 2, 2, 3)
(1, 1, 1, 1, 1, 1, 1, 4, 4)	(1, 1, 1, 1, 1, 2, 2, 2, 4)	

This leaves the following 17 cases which are excluded by P. Robinson.

(1, 1, 1, 1, 1, 1, 1, 1, 3)	(1, 1, 1, 1, 1, 1, 1, 3, 4)	(1, 1, 1, 1, 1, 1, 2, 3, 5)
(1, 1, 1, 1, 1, 1, 1, 1, 5)	(1, 1, 1, 1, 1, 1, 1, 3, 6)	(1, 1, 1, 1, 1, 1, 2, 4, 4)
(1, 1, 1, 1, 1, 1, 1, 1, 6)	(1, 1, 1, 1, 1, 1, 1, 4, 5)	(1, 1, 1, 1, 1, 2, 2, 3, 4)
(1, 1, 1, 1, 1, 1, 1, 2, 2)	(1, 1, 1, 1, 1, 1, 2, 2, 3)	
(1, 1, 1, 1, 1, 1, 1, 2, 4)	(1, 1, 1, 1, 1, 1, 2, 2, 4)	(1, 1, 1, 1, 2, 2, 2, 2, 4)
(1, 1, 1, 1, 1, 1, 1, 2, 7)	(1, 1, 1, 1, 1, 1, 2, 2, 6)	

8 VARIABLES: There are 67 8-tuples which might be the type of an orthogonal design in order 16. The following 4 cannot be the types of orthogonal designs by the results of Geramita and Verner:

(1, 1, 1, 1, 1, 3, 3, 4)	(1, 1, 1, 1, 2, 2, 2, 5)	(1, 1, 1, 2, 2, 2, 3, 3)
(1, 1, 2, 2, 2, 2, 2, 3).		

The following 37 are known

(1, 1, 1, 1, 1, 1, 1, 1)	(1, 1, 1, 1, 1, 2, 2, 2)	(1, 1, 1, 1, 2, 3, 3, 3)
(1, 1, 1, 1, 1, 1, 1, 2)	(1, 1, 1, 1, 1, 2, 2, 3)	(1, 1, 1, 1, 2, 3, 3, 4)
(1, 1, 1, 1, 1, 1, 1, 3)	(1, 1, 1, 1, 1, 2, 2, 4)	(1, 1, 1, 1, 3, 3, 3, 3)
(1, 1, 1, 1, 1, 1, 1, 4)	(1, 1, 1, 1, 1, 2, 3, 3)	(1, 1, 1, 2, 2, 2, 2, 2)
(1, 1, 1, 1, 1, 1, 1, 5)	(1, 1, 1, 1, 1, 2, 3, 6)	(1, 1, 1, 2, 2, 2, 2, 3)
(1, 1, 1, 1, 1, 1, 1, 8)	(1, 1, 1, 1, 1, 3, 3, 3)	(1, 1, 1, 2, 2, 3, 3, 3)
(1, 1, 1, 1, 1, 1, 1, 9)	(1, 1, 1, 1, 1, 3, 3, 5)	(1, 1, 2, 2, 2, 2, 2, 2)
(1, 1, 1, 1, 1, 1, 2, 2)	(1, 1, 1, 1, 2, 2, 2, 2)	(1, 1, 2, 2, 2, 2, 2, 4)

(1, 1, 1, 1, 1, 1, 2, 4)	(1, 1, 1, 1, 2, 2, 2, 3)	(1, 1, 2, 2, 2, 2, 3, 3)
(1, 1, 1, 1, 1, 1, 2, 8)	(1, 1, 1, 1, 2, 2, 2, 4)	(1, 2, 2, 2, 2, 2, 2, 2)
(1, 1, 1, 1, 1, 1, 3, 3)	(1, 1, 1, 1, 2, 2, 3, 3)	(1, 2, 2, 2, 2, 2, 2, 3)
(1, 1, 1, 1, 1, 1, 4, 4)	(1, 1, 1, 1, 2, 2, 4, 4)	(2, 2, 2, 2, 2, 2, 2, 2)
(1, 1, 1, 1, 1, 1, 5, 5)		

The following 30 do not exist

(1, 1, 1, 1, 1, 1, 1, 6)	(1, 1, 1, 1, 1, 1, 4, 5)	(1, 1, 1, 1, 1, 3, 4, 4)
(1, 1, 1, 1, 1, 1, 1, 7)	(1, 1, 1, 1, 1, 1, 4, 6)	(1, 1, 1, 1, 2, 2, 2, 5)
(1, 1, 1, 1, 1, 1, 2, 3)	(1, 1, 1, 1, 1, 2, 2, 5)	(1, 1, 1, 1, 2, 2, 2, 6)
(1, 1, 1, 1, 1, 1, 2, 5)	(1, 1, 1, 1, 1, 2, 2, 6)	(1, 1, 1, 1, 2, 2, 3, 4)
(1, 1, 1, 1, 1, 1, 2, 6)	(1, 1, 1, 1, 1, 2, 2, 7)	(1, 1, 1, 1, 2, 2, 3, 5)
(1, 1, 1, 1, 1, 1, 2, 7)	(1, 1, 1, 1, 1, 2, 3, 4)	(1, 1, 1, 2, 2, 2, 2, 4)
(1, 1, 1, 1, 1, 1, 3, 4)	(1, 1, 1, 1, 1, 2, 3, 5)	(1, 1, 1, 2, 2, 2, 3, 3)
(1, 1, 1, 1, 1, 1, 3, 5)	(1, 1, 1, 1, 1, 2, 4, 4)	(1, 1, 1, 2, 2, 2, 2, 5)
(1, 1, 1, 1, 1, 1, 3, 6)	(1, 1, 1, 1, 1, 2, 4, 5)	(1, 1, 1, 2, 2, 2, 3, 4)
(1, 1, 1, 1, 1, 1, 3, 7)	(1, 1, 1, 1, 1, 3, 3, 4)	(1, 1, 2, 2, 2, 2, 2, 3)

7 VARIABLES: There are 94 possible 7-tuples. The following 13 do not exist - all others are known:

(1, 1, 1, 1, 1, 1, 7)	(1, 1, 1, 1, 2, 2, 7)	(1, 1, 1, 2, 2, 2, 7)
(1, 1, 1, 1, 1, 2, 7)	(1, 1, 1, 1, 2, 4, 5)	(1, 1, 1, 2, 2, 3, 5)
(1, 1, 1, 1, 1, 3, 7)	(1, 1, 1, 1, 3, 4, 4)	(1, 1, 2, 2, 2, 2, 5)
(1, 1, 1, 1, 1, 4, 6)	(1, 1, 1, 2, 2, 2, 6)	(1, 1, 2, 2, 2, 3, 4)
(1, 1, 1, 1, 1, 4, 7)		

6 VARIABLES: Of the 125 6-tuples that could be the type of an orthogonal design all are, except the following 2 which do not exist.

(1, 1, 1, 1, 4, 7)   (1, 1, 2, 2, 2, 7).

5 VARIABLES: All possible 5-tuples are the types of orthogonal designs.

4 VARIABLES: All possible 4-tuples are the types of orthogonal designs.

3 VARIABLES: All possible 3-tuples are the types of orthogonal designs.

2 VARIABLES: All possible 2-tuples are the types of orthogonal designs.

1 VARIABLE: All one variable designs exist.

ORDER 20: There are at most 4 variables.

4 VARIABLES: The 4-tuples that correspond to Orthogonal Designs are listed in Appendix B.

For the following 4-tuples it is undecided whether an orthogonal design exists:

(1, 3, 6, 8) (1, 4, 4, 9) (2, 2, 5, 5)

It is not possible to construct designs of these types using the Goethals-Seidel array (Eades).

All other 4-tuples are not the types of orthogonal designs.

3 VARIABLES: Of the possible 3-tuples the following 97 are the types of orthogonal designs (all are constructed using four circulant matrices in the Goethals-Seidel array):

(1, 1, 1)	(1, 3, 8)	(2, 2, 4)	(2, 5, 7)	(4, 4, 4)
(1, 1, 2)	(1, 3, 14)	(2, 2, 5)	(2, 5, 8)	(4, 4, 5)
(1, 1, 4)	(1, 4, 4)	(2, 2, 8)	(2, 5, 13)	(4, 4, 8)
(1, 1, 5)	(1, 4, 5)	(2, 2, 9)	(2, 6, 7)	(4, 4, 9)
(1, 1, 8)	(1, 4, 8)	(2, 2, 10)	(2, 6, 9)	(4, 4, 10)
(1, 1, 9)	(1, 4, 9)	(2, 2, 13)	(2, 6, 12)	(4, 5, 5)
(1, 1, 10)	(1, 4, 10)	(2, 2, 16)	(2, 8, 8)	(4, 5, 6)
(1, 1, 13)	(1, 4, 13)	(2, 3, 4)	(2, 8, 9)	(4, 5, 9)
(1, 1, 16)	(1, 5, 5)	(2, 3, 6)	(2, 8, 10)	(4, 6, 8)
(1, 1, 18)	(1, 5, 6)	(2, 3, 7)	(2, 9, 9)	(4, 8, 8)
(1, 2, 2)	(1, 5, 9)	(2, 3, 9)	(3, 3, 3)	(5, 5, 5)
(1, 2, 3)	(1, 5, 14)	(2, 3, 10)	(3, 3, 6)	(5, 5, 8)
(1, 2, 4)	(1, 6, 8)	(2, 3, 15)	(3, 3, 12)	(5, 5, 9)
(1, 2, 6)	(1, 6, 11)	(2, 4, 4)	(3, 4, 6)	(5, 5, 10)
(1, 2, 8)	(1, 8, 8)	(2, 4, 6)	(3, 4, 8)	(5, 6, 9)
(1, 2, 9)	(1, 8, 9)	(2, 4, 8)	(3, 6, 6)	(5, 7, 8)
(1, 2, 11)	(1, 8, 11)	(2, 4, 9)	(3, 6, 8)	(6, 6, 6)
(1, 2, 12)	(1, 9, 9)	(2, 4, 11)	(3, 6, 9)	
(1, 2, 17)	(1, 9, 10)	(2, 4, 12)	(3, 6, 11)	
(1, 3, 6)	(2, 2, 2)	(2, 5, 5)	(3, 8, 9)	

It is undecided whether an orthogonal design exists for the 3-tuple (3, 7, 8). It cannot be constructed by using four circulant matrices in the Goethals-Seidel array (Eades).

All other 3-tuples are not the types of orthogonal designs.

2 VARIABLES: The following eighty-six (86) 2-tuples are the types of orthogonal designs. All other 2-tuples correspond to non-existent designs. All those that exist are constructed using four circulant matrices in the Goethals-Seidel array.

(1, 1)	(2, 3)	(3, 6)	(4, 12)	(6, 11)
(1, 2)	(2, 4)	(3, 7)	(4, 13)	(6, 12)
(1, 3)	(2, 5)	(3, 8)	(4, 14)	(6, 14)
(1, 4)	(2, 6)	(3, 9)	(4, 16)	(7, 7)
(1, 5)	(2, 7)	(3, 10)	(5, 5)	(7, 8)
(1, 6)	(2, 8)	(3, 11)	(5, 6)	(7, 10)
(1, 8)	(2, 9)	(3, 12)	(5, 7)	(7, 11)
(1, 9)	(2, 10)	(3, 14)	(5, 8)	(7, 13)
(1, 10)	(2, 11)	(3, 15)	(5, 9)	(8, 8)
(1, 11)	(2, 12)	(3, 17)	(5, 10)	(8, 9)
(1, 12)	(2, 13)	(4, 4)	(5, 13)	(8, 10)
(1, 13)	(2, 15)	(4, 5)	(5, 14)	(8, 11)
(1, 14)	(2, 16)	(4, 6)	(5, 15)	(8, 12)
(1, 16)	(2, 17)	(4, 8)	(6, 6)	(9, 9)
(1, 17)	(2, 18)	(4, 9)	(6, 7)	(9, 10)
(1, 18)	(3, 3)	(4, 10)	(6, 8)	(9, 11)
(1, 19)	(3, 4)	(4, 11)	(6, 9)	(10, 10)
(2, 2)				

1 VARIABLE: All one variable designs exist.

ORDER 24: At most 8 variables.

8 VARIABLES: The following 8-tuples are the types of orthogonal designs.

(1, 1, 1, 1, 1, 1, 1, 1)	(1, 1, 1, 1, 1, 4, 4, 4)
(1, 1, 1, 1, 1, 1, 1, 4)	(1, 1, 1, 1, 1, 5, 5, 9)
(1, 1, 1, 1, 1, 1, 9, 9)	(1, 1, 1, 1, 4, 4, 4, 4)
(1, 1, 1, 1, 1, 2, 8, 9)	(3, 3, 3, 3, 3, 3, 3, 3)

7 VARIABLES: The following 7-tuples are the types of orthogonal designs.

(1, 1, 1, 1, 1, 1, 1)	(1, 1, 1, 1, 1, 2, 17)	(1, 1, 1, 1, 1, 8, 11)	(1, 1, 1, 1, 4, 4, 8)
(1, 1, 1, 1, 1, 1, 2)	(1, 1, 1, 1, 1, 4, 4)	(1, 1, 1, 1, 1, 9, 9)	(1, 1, 1, 1, 5, 5, 9)
(1, 1, 1, 1, 1, 1, 4)	(1, 1, 1, 1, 1, 4, 5)	(1, 1, 1, 1, 1, 9, 10)	(1, 1, 1, 1, 5, 5, 10)
(1, 1, 1, 1, 1, 1, 5)	(1, 1, 1, 1, 1, 4, 8)	(1, 1, 1, 1, 2, 2, 8)	(1, 1, 1, 1, 5, 6, 9)
(1, 1, 1, 1, 1, 1, 9)	(1, 1, 1, 1, 1, 5, 5)	(1, 1, 1, 1, 2, 8, 9)	(1, 1, 1, 2, 2, 8, 9)
(1, 1, 1, 1, 1, 1, 18)	(1, 1, 1, 1, 1, 5, 6)	(1, 1, 1, 1, 2, 8, 10)	(1, 1, 1, 2, 4, 4, 4)
(1, 1, 1, 1, 1, 2, 4)	(1, 1, 1, 1, 1, 5, 9)	(1, 1, 1, 1, 2, 9, 9)	(1, 1, 1, 2, 5, 5, 9)

(1, 1, 1, 1, 1, 2, 8)	(1, 1, 1, 1, 1, 5, 14)	(1, 1, 1, 1, 4, 4, 4)	(1, 1, 1, 4, 4, 4, 4)
(1, 1, 1, 1, 1, 2, 9)	(1, 1, 1, 1, 1, 8, 9)	(1, 1, 1, 1, 4, 4, 5)	(1, 1, 1, 4, 4, 4, 5)
			(3, 3, 3, 3, 3, 3, 3)
			(3, 3, 3, 3, 3, 3, 6)

The following 7-tuples cannot be the type of an orthogonal design:

(1, 1, 1, 1, 1, a, 7)	a = 1, 2, . . . , 12	(1, 1, 1, 1, 1, 3, 4)
(1, 1, 1, 1, 1, b, 15)	b = 1, 2, 3, 4	(1, 1, 1, 1, 1, 3, 12)
(1, 1, 1, 1, 1, 1, 14)		(1, 1, 1, 1, 1, 4, 11)
(1, 1, 1, 1, 1, 2, 15)		(1, 1, 1, 1, 1, 5, 10)
(1, 1, 1, 1, 1, 2, 13)		(1, 1, 1, 1, 1, 6, 9)
		(1, 1, 1, 1, 1, 7, 8)

6 VARIABLES: See Lemma E.1 for known 6 variable designs.

The following do not correspond to the type of an orthogonal design.

(1, 1, 1, 1, 1, 7) (1, 1, 1, 1, 1, 15) (2, 2, 2, 2, 2, 14).

5 VARIABLES: Known 5 variable designs are listed in Lemma E.1.

4 VARIABLES: (Robinson) All 4-tuples of the form (a, b, c, 24-a-b-c), $0 \le a + b + c < 24$, are the types of orthogonal designs in order 24 except possibly

(1, 4, 4, 15)	(3, 3, 7, 11)	(3, 7, 7, 7)
(1, 7, 8, 8)	(3, 4, 4, 13)	(4, 4, 5, 11)
(2, 2, 7, 13)	(3, 4, 6, 11)	(4, 4, 6, 10)
(2, 4, 4, 14)	(3, 4, 7, 10)	(4, 4, 7, 9)
(2, 4, 7, 11)	(3, 5, 8, 8)	(4, 5, 7, 8)
(2, 7, 7, 8)	(3, 6, 7, 8)	(5, 5, 6, 8)

3 VARIABLES: All possible 3-tuples are the types of orthogonal designs except the following which are undecided:

(4, 4, 15) (7, 7, 7) (7, 8, 8)

2 VARIABLES: Every possible 2-tuple is the type of an orthogonal design.

1 VARIABLE: All one variable designs exist.

<u>ORDER 28:</u> At most four variables.

4 VARIABLES: The "state of the art" is given in Appendix C.

3 VARIABLES: There are orthogonal designs of types (1, 9), (2, 2), (2, 8), (5, 5) and (13) each constructed from two circulant matrices in order 14. Hence there are orthogonal designs of types

(1, 9, 13)   (2, 2, 13)   (2, 8, 13)   (5, 5, 13)

in order 28.

Using the non-existence results quoted above we have listed all 236 3-tuples in order 28 which could be the types of orthogonal designs. The designs marked ✓ may be constructed using four circulant matrices (from the three or four variable designs quoted above) and those marked ⊗ exist but it is unknown whether they can be constructed using four circulant matrices.

(1, 1, 1) (✓)	(1, 3, 24) (✓)	(1, 9, 10) (✓)	(2, 4, 16) (✓)	(2, 10, 12) (✓)
(1, 1, 2) (✓)	(1, 4, 4) (✓)	(1, 9, 13) (✓)	(2, 4, 17)	(2, 11, 11)
(1, 1, 4) (✓)	(1, 4, 5) (✓)	(1, 9, 16)	(2, 4, 18) (✓)	(2, 11, 13)
(1, 1, 5) (✓)	(1, 4, 8) (✓)	(1, 9, 18) (✓)	(2, 4, 19)	(2, 11, 15)
(1, 1, 8) (✓)	(1, 4, 9) (✓)	(1, 10, 10) (✓)	(2, 4, 22)	(2, 13, 13) (✓)
(1, 1, 9) (✓)	(1, 4, 10) (✓)	(1, 10, 11) (✓)	(2, 5, 5) (✓)	(3, 3, 3) (✓)
(1, 1, 10) (✓)	(1, 4, 13) (✓)	(1, 10, 14) (✓)	(2, 5, 7) (✓)	(3, 3, 6) (✓)
(1, 1, 13) (✓)	(1, 4, 16) (✓)	(1, 13, 13) (✓)	(2, 5, 8) (✓)	(3, 3, 12) (✓)
(1, 1, 16) (✓)	(1, 4, 17) (✓)	(1, 13, 14) (✓)	(2, 5, 13) (✓)	(3, 3, 15) (✓)
(1, 1, 17) (✓)	(1, 4, 18)	(2, 2, 2) (✓)	(2, 5, 15)	(3, 4, 6) (✓)
(1, 1, 18) (✓)	(1, 4, 20) (✓)	(2, 2, 4) (✓)	(2, 5, 18)	(3, 4, 8) (✓)
(1, 1, 20) (✓)	(1, 5, 5) (✓)	(2, 2, 5) (✓)	(2, 6, 7) (✓)	(3, 4, 14) (✓)
(1, 1, 25) (✓)	(1, 5, 6) (✓)	(2, 2, 8) (✓)	(2, 6, 9)	(3, 4, 18)
(1, 1, 26) (✓)	(1, 5, 9) (✓)	(2, 2, 9) (✓)	(2, 6, 11)	(3, 6, 6) (✓)
(1, 2, 2) (✓)	(1, 5, 14) (✓)	(2, 2, 10) (✓)	(2, 6, 12)	(3, 6, 8) (✓)
(1, 2, 3) (✓)	(1, 5, 16) (✓)	(2, 2, 13) (✓)	(2, 6, 13)	(3, 6, 9) (✓)
(1, 2, 4) (✓)	(1, 5, 19)	(2, 2, 16) (✓)	(2, 6, 16) (✓)	(3, 6, 11)
(1, 2, 6) (✓)	(1, 5, 20)	(2, 2, 17) (✓)	(2, 6, 17)	(3, 6, 12) (✓)
(1, 2, 8) (✓)	(1, 6, 8) (✓)	(2, 2, 18) (✓)	(2, 7, 10)	(3, 6, 16)
(1, 2, 9) (✓)	(1, 6, 11) (✓)	(2, 2, 20) (✓)	(2, 7, 12)	(3, 6, 17)
(1, 2, 11) (✓)	(1, 6, 12) (✓)	(2, 3, 4) (✓)	(2, 7, 13)	(3, 6, 18)
(1, 2, 12) (✓)	(1, 6, 14)	(2, 3, 6) (✓)	(2, 7, 19) (✓)	(3, 6, 19)
(1, 2, 16) (✓)	(1, 6, 18) (✓)	(2, 3, 7) (✓)	(2, 8, 8) (✓)	(3, 7, 8) (✓)
(1, 2, 17) (✓)	(1, 6, 21) (✓)	(2, 3, 9) (✓)	(2, 8, 9) (✓)	(3, 7, 10)
(1, 2, 18) (✓)	(1, 8, 8) (✓)	(2, 3, 10) (✓)	(2, 8, 10) (✓)	(3, 7, 11)
(1, 2, 19) (✓)	(1, 8, 9) (✓)	(2, 3, 15)	(2, 8, 13) (✓)	(3, 7, 15)
(1, 2, 22) (✓)	(1, 8, 11) (✓)	(2, 3, 16) (✓)	(2, 8, 16) (✓)	(3, 7, 18)
(1, 2, 25) (✓)	(1, 8, 12)	(2, 4, 4) (✓)	(2, 8, 18)	(3, 8, 9) (✓)
(1, 3, 6) (✓)	(1, 8, 16)	(2, 4, 6) (✓)	(2, 9, 9) (✓)	(3, 8, 10) (✓)
(1, 3, 8) (✓)	(1, 8, 17)	(2, 4, 8) (✓)	(2, 9, 11)	(3, 8, 15)
(1, 3, 14) (✓)	(1, 8, 18) (✓)	(2, 4, 9) (✓)	(2, 9, 12)	(3, 9, 14)
(1, 3, 18) (✓)	(1, 8, 19) (✓)	(2, 4, 11)	(2, 9, 17)	(3, 10, 15)
(1, 3, 22)	(1, 9, 9) (✓)	(2, 4, 12) (✓)	(2, 10, 10) (✓)	(3, 11, 14)

(4, 4, 4) (✓)	(4, 5, 19)	(4, 10, 10) (✓)	(5, 7, 10)	(6, 9, 11)
(4, 4, 5) (✓)	(4, 6, 8) (✓)	(4, 10, 11)	(5, 7, 14)	(7, 7, 7) (✓)
(4, 4, 8) (✓)	(4, 6, 11) (✓)	(4, 10, 14) (✓)	(5, 8, 8)	(7, 7, 14) (✓)
(4, 4, 9) (✓)	(4, 6, 12)	(5, 5, 5) (✓)	(5, 8, 13)	(7, 8, 10)
(4, 4, 10) (✓)	(4, 6, 14)	(5, 5, 8) (✓)	(5, 9, 9)	(7, 8, 13)
(4, 4, 13)	(4, 6, 18)	(5, 5, 9) (✓)	(5, 9, 10)	(7, 10, 11)
(4, 4, 16) (✓)	(4, 8, 8) (✓)	(5, 5, 10) (✓)	(5, 9, 14)	(8, 8, 8) (✓)
(4, 4, 17)	(4, 8, 9) (✓)	(5, 5, 13) (✓)	(5, 10, 10)	(8, 8, 9)
(4, 4, 18) (✓)	(4, 8, 11)	(5, 5, 16)	(6, 6, 6) (✓)	(8, 8, 10) (✓)
(4, 4, 20) (✓)	(4, 8, 12) (✓)	(5, 5, 18)	(6, 6, 12) (✓)	(8, 9, 9) ⊗
(4, 5, 5) (✓)	(4, 8, 16) (✓)	(5, 6, 9) (✓)	(6, 7, 8) (✓)	(8, 9, 11)
(4, 5, 6) (✓)	(4, 9, 9) ⊗	(5, 6, 14)	(6, 8, 9)	(8, 10, 10) (✓)
(4, 5, 9) (✓)	(4, 9, 10)	(5, 6, 15)	(6, 8, 11)	(9, 9, 9)
(4, 5, 14)	(4, 9, 13) ⊗	(5, 7, 8) (✓)	(6, 8, 12)	(9, 9, 10)
(4, 5, 16)				

✓ these matrices are constructed using four circulant matrices

⊗ constructed from the (4, 4, 9, 9) design

2 VARIABLES: There are 196 pairs (j, k) such that $j + k \le 28$. Of these, 27 are eliminated as types of orthogonal designs by Wolfe's Theorem and the Geramita-Verner Theorem. The remaining 169 pairs are the types of an orthogonal design in order 28 and may be constructed using four circulant matrices.

The 27 2-tuples that cannot be the types of orthogonal designs in order 28 are:

(1, 7)	(3, 20)	(5, 12)	(7, 17)	(11, 16)
(1, 15)	(3, 21)	(5, 19)	(7, 20)	(12, 13)
(1, 23)	(4, 7)	(5, 22)	(8, 14)	(12, 15)
(2, 14)	(4, 15)	(6, 10)	(9, 15)	
(3, 5)	(4, 23)	(7, 9)	(10, 17)	
(3, 13)	(5, 11)	(7, 16)	(11, 13)	

1 VARIABLE: All one variable designs exist in order 28.

ORDER 32: At most 10 variables.

10 VARIABLES: The following 10-tuples are the types of orthogonal designs.

(1, 1, 2, 2, 2, 2, 2, 2, 2, 4)   (1, 1, 2, 2, 2, 2, 4, 4, 4, 4);

as are the following full 10-tuples:

(1, 1, 1, 1, 2, 2, 3, 3, 9, 9)
(1, 1, 1, 1, 2, 2, 4, 4, 8, 8)
(1, 1, 1, 2, 3, 3, 3, 3, 6, 9)
(1, 1, 2, 2, 2, 2, 2, 2, 2, 16)
(1, 1, 2, 2, 2, 2, 4, 6, 6, 6)
(1, 1, 2, 4, 4, 4, 4, 4, 4, 4)
(1, 1, 3, 3, 3, 3, 3, 3, 6, 6)
(1, 2, 2, 2, 2, 2, 3, 6, 6, 6)
(2, 2, 2, 2, 2, 2, 2, 2, 8, 8)
(2, 2, 2, 2, 2, 2, 2, 6, 6, 6)
(2, 2, 2, 2, 2, 3, 3, 4, 6, 6)
(2, 2, 2, 2, 4, 4, 4, 4, 4, 4)

9 VARIABLES: The following 9-tuples are the types of orthogonal designs:

(1, 1, 1, 2, 2, 4, a, a, a) $\qquad$ a = 1, 2, 3, 4, 5, 6 or 7;

as are the following full 9-tuples given in Table G.1 (see end of appendix G).

8 VARIABLES: The following 8-tuples are the types of orthogonal designs:

(1, 1, 1, 1, 7, 7, 7, 7)
(1, 1, 1, 2, 6, 6, 6, 9)
(1, 1, 1, 2, 6, 7, 7, 7)
(1, 1, 1, 3, 6, 6, 6, 8)
(1, 2, 3, 3, 4, 4, 5, 10)
(1, 2, 3, 3, 4, 5, 7, 7)
(2, 2, 3, 3, 3, 3, 4, 12)
(2, 2, 3, 4, 5, 5, 5, 6)
(2, 3, 3, 3, 3, 6, 6, 6)
(2, 3, 3, 3, 5, 5, 5, 6)

7 VARIABLES: All full 7-tuples except possibly the 86 given in Appendix D, are the types of orthogonal designs.

6 VARIABLES: All full 6-tuples are the type of an orthogonal design except possibly (1, 1, 1, 1, 1, 27).

5 VARIABLES: All 5-tuples are the types of orthogonal designs except possibly (1, 1, 1, 1, 27).

≤ 4 VARIABLES: All possible n-tuples, n = 1, 2, 3, 4 are the types of orthogonal designs.

ORDER 36: At most 4 variables

2 VARIABLES: All (1, k) exist, $k \in \{x \mid 1 \le x \le 35,\ x = a^2 + b^2 + c^2\}$.

The other two variable designs have not been studied.

1 VARIABLE: All one variable designs exist.

ORDER 40: It is a simple matter to find results in n-variables where $3 \le n \le 8$ but except for those listed below this has not been done.

6 VARIABLES: The following 6-tuple is the type of an orthogonal design which exists: (1, 2, 2, 2, 16, 17).

5 VARIABLES: The following 5-tuples correspond to orthogonal designs which exist:

(1, 1, 1, 4, 20)	(1, 2, 2, 4, 25)	(1, 4, 8, 8, 16)
(1, 1, 8, 8, 9)	(1, 2, 2, 11, 24)	(1, 8, 8, 8, 8)
(1, 2, 6, 9, 20)	(1, 4, 9, 9, 9)	(1, 2, 2, 2, 33)

4 VARIABLES: The following 4-tuples correspond to orthogonal designs which exist:

(1, 1, 15, 23)	(1, 9, 13, 15)	(2, 4, 13, 21)
(1, 2, 3, 34)	(1, 10, 10, 17)	(2, 8, 12, 18)
(1, 2, 10, 27)	(1, 10, 10, 19)	(2, 10, 10, 13)
(1, 2, 12, 25)	(1, 10, 14, 15)	(5, 8, 12, 15)
(1, 2, 14, 23)	(2, 2, 5, 31)	(5, 9, 9, 15)
(1, 2, 16, 19)	(2, 2, 14, 22)	(8, 10, 10, 12)
(1, 8, 12, 19)	(2, 4, 11, 16)	

3 VARIABLES: The following 3-tuples correspond to orthogonal designs which exist:

(2, 8, 25)

and all 3-tuples (a, b, 40-a-b) except possibly

(1, 7, 32)	(4, 7, 29)	(7, 11, 22)
(1, 9, 30)	(5, 6, 29)	(7, 14, 19)
(3, 5, 32)	(5, 13, 22)	(8, 9, 23)
(3, 6, 31)	(6, 7, 27)	(8, 11, 21)
(3, 8, 29)	(7, 8, 25)	(9, 14, 17)
(3, 9, 28)	(7, 9, 24)	

which are undecided.

2 VARIABLES: All possible 2-tuples correspond to orthogonal designs which exist except the following which are in doubt:

(7, 32)   (9, 30)

1 VARIABLE: All one variable designs exist.

<u>ORDER 44</u>: At most 4 variables

2 VARIABLES: The designs (1, k) for $k \in \{x : 0 \leq x \leq 43,\ x = a^2 + b^2 + c^2,\ x \neq 42\}$ exist.

1 VARIABLE: All one variable designs exist.

<u>ORDER 48</u>: At most 9 variables.

3 VARIABLES: Every 3-tuple of the form (a, b, 48-a-b) is the type of an orthogonal design.

2 VARIABLES: Every possible 2-tuple is the type of an orthogonal design.

1 VARIABLE: All one variable designs exist.

<u>ORDER 52</u>: At most 4 variables.

2 VARIABLES: The designs (1, k) for $k \in \{x : 0 \leq x \leq 51,\ x = a^2 + b^2 + c^2,\ x \neq 37, 38, 42, 45, 46, 48\}$ exist. The other two variable designs have not been studied.

1 VARIABLE: These are known except for a W(52, 46).

<u>ORDER 56</u>: At most 8 variables.

2 VARIABLES: All (1, k) for k = 1, 2, . . . , 55 are known.

1 VARIABLE: All one variable designs are known.

<u>ORDER 60</u>: At most 4 variables.

2 VARIABLES: The designs (1, k) are known for $k \in \{x : 0 \leq x \leq 59,\ x = a^2 + b^2 + c^2,\ x \neq 34, 38, 42, 48, 53, 54, 56, 57\}$.

1 VARIABLE: All one variable designs exist.

ORDER 64: At most 12 variables.

3 VARIABLES: Every 3-tuple of the form (a, b, 64-a-b) is the type of an orthogonal design.

2 VARIABLES: All possible 2-tuples are the types of orthogonal designs.

1 VARIABLE: All one variable designs exist.

ORDER 72: At most 8 variables.

2 VARIABLES: All designs (1, k) for $k = 1, 2, \ldots, 71$ are known.

1 VARIABLE: All one variable designs exist.

ORDER 80: At most 9 variables.

6 VARIABLES: The following 6-tuples are the types of orthogonal designs:

(1, 1, 1, 1, 9, 10)	(1, 1, 3, 3, 10, 27)
(2, 2, 3, 6, 32, 35)	(1, 1, 3, 3, 9, 30)

5 VARIABLES: The following 5-tuples are the types of orthogonal designs:

(1, 1, 6, 16, 43)	(1, 4, 6, 19, 50)	(1, 4, 6, 34, 35)
(2, 3, 4, 18, 53)	(2, 3, 4, 35, 36)	(2, 3, 12, 28, 35)
(2, 3, 14, 26, 35)	(2, 3, 16, 24, 35)	(2, 3, 20, 20, 35)

(1, 1, 10, c, d) (1, 1, 10, 3c, 3d) (3, 3, 30, c, d)

$(c, d) \in N = \{(8, 12), (10, 10), (2, 13), (2, 16), (2, 18), (1, 11), (9, 9)\}$

(1, 1, 9, a, b) (1, 1, 9, 3a, 3b) (3, 3, 27, a, b)

$(a, b) \in N$ or $(a, b) \in M = \{(4, 11), (1, 19), (1, 17), (1, 14), (5, 15)\}$.

4 VARIABLES: The following 4-tuples are the types of orthogonal designs:

(1, 2, 18, 59)	(1, 6, 15, 58)	
(1, 2, 28, 45)	(1, 6, 28, 43)	(1, 6, 29, 44)
(a, b, c, d)	(a, b, 3c, 3d)	(3a, 3b, c, d)

where $(a, b) \in M$, $(c, d) \in N$ or $(a, b) \in N$, $(c, d) \in M \cup N$.

(4, 6, 11, 54)	(4, 11, 24, 36)	(4, 11, 30, 30)

3 VARIABLES: All 3-tuples (a, b, 80-a-b) are the types of orthogonal designs except possibly

(3, 13, 64)	(7, 22, 51)	(12, 27, 41)
(3, 15, 62)	(7, 25, 48)	(12, 29, 39)
(3, 21, 56)	(7, 28, 45)	(13, 23, 44)
(5, 11, 64)	(8, 29, 43)	(14, 19, 47)
(5, 17, 58)	(9, 14, 57)	(14, 27, 39)
(5, 31, 44)	(9, 15, 56)	(16, 21, 43)
(6, 9, 65)	(9, 16, 55)	(16, 23, 41)
(6, 31, 43)	(9, 28, 43)	(16, 25, 39)
(7, 8, 65)	(10, 29, 41)	(18, 23, 39)
(7, 9, 64)	(11, 14, 55)	(19, 28, 33)
(7, 12, 61)	(11, 16, 53)	(21, 22, 37)
(7, 16, 57)	(11, 25, 44)	

which are undecided.

2 VARIABLES: All 2-tuples are the types of orthogonal designs except possibly

(13, 64) (15, 62).

1 VARIABLE: All one variable designs exist.

ORDER 2^{4c+d}: At most $8c + 2^d$ variables.

3 VARIABLES: Every 3-tuple of the form $(x, y, 2^{4c+d}-x-y)$ is the type of an orthogonal design.

2 VARIABLES: All two variable designs exist.

1 VARIABLE: All one variable designs exist.

ORDER $2^{4c+d}\cdot 3$: At most $8c + 2^d$ variables.

3 VARIABLES: Every 3-tuple of the form $(x, y, 2^{4c+d}\cdot 3-x-y)$ is the type of an orthogonal design.

2 VARIABLES: Every possible 2-tuple is the type of an orthogonal design.

1 VARIABLE: All one variable designs exist.

Known Full 9 Variable Designs in Order 32

1	1	1	1	2	2	3	3	18
1	1	1	1	2	2	3	7	14
1	1	1	1	2	2	3	9	12
1	1	1	1	2	2	4	4	16
1	1	1	1	2	2	4	8	12
1	1	1	1	2	2	4	10	10
1	1	1	1	2	2	6	9	9
1	1	1	1	2	2	8	8	8
1	1	1	1	2	3	3	5	15
1	1	1	1	2	3	3	9	11
1	1	1	1	2	3	5	9	9
1	1	1	1	2	4	4	8	10
1	1	1	1	2	4	6	8	8
1	1	1	1	2	4	7	7	8
1	1	1	1	3	3	4	9	9
1	1	1	1	4	4	4	8	8
1	1	1	1	4	4	6	6	8
1	1	1	2	2	3	3	9	10
1	1	1	2	2	3	4	6	12
1	1	1	2	2	3	4	9	9
1	1	1	2	2	3	6	8	8
1	1	1	2	2	4	4	8	9
1	1	1	2	2	4	5	8	8
1	1	1	2	2	4	7	7	7
1	1	1	2	3	3	3	3	15
1	1	1	2	3	3	3	6	12
1	1	1	2	3	3	3	9	9
1	1	1	2	3	3	5	6	10
1	1	1	2	3	3	6	6	9
1	1	1	2	3	4	4	8	8
1	1	1	3	3	3	3	6	11
1	1	1	3	3	3	3	8	9
1	1	1	3	3	3	5	5	10
1	1	1	3	3	3	5	6	9
1	1	2	2	2	2	2	2	18
1	1	2	2	2	2	2	4	16
1	1	2	2	2	2	2	10	10
1	1	2	2	2	2	4	6	12
1	1	2	2	2	2	6	6	10
1	1	2	2	2	3	3	9	9
1	1	2	2	2	4	4	8	8
1	1	2	2	2	4	6	6	8
1	1	2	2	2	4	6	7	7
1	1	2	2	2	6	6	6	6
1	1	2	2	4	4	6	6	6
1	1	2	3	3	3	3	6	10
1	1	2	3	3	3	3	7	9
1	1	2	3	3	3	4	6	9

1	1	2	3	3	4	4	6	8
1	1	2	3	3	4	5	5	8
1	1	2	4	4	4	4	4	8
1	1	3	3	3	3	3	3	12
1	1	3	3	3	3	3	6	9
1	1	3	3	3	3	6	6	6
1	1	4	4	4	4	4	4	6
1	2	2	2	2	2	2	2	17
1	2	2	2	2	2	2	3	16
1	2	2	2	2	2	3	6	12
1	2	2	2	2	2	6	6	9
1	2	2	2	2	3	6	6	8
1	2	2	2	2	4	6	6	7
1	2	2	2	2	5	6	6	6
1	2	2	2	3	3	3	4	12
1	2	2	2	3	4	6	6	6
1	2	2	3	3	3	3	6	9
1	2	4	4	4	4	4	4	5
1	3	3	3	3	3	3	6	7
1	3	3	3	3	3	4	6	6
1	3	4	4	4	4	4	4	4
2	2	2	2	2	2	2	2	16
2	2	2	2	2	2	2	6	12
2	2	2	2	2	2	2	8	10
2	2	2	2	2	2	2	9	9
2	2	2	2	2	2	4	8	8
2	2	2	2	2	2	5	5	10
2	2	2	2	2	2	6	6	8
2	2	2	2	2	3	3	4	12
2	2	2	2	2	3	3	6	10
2	2	2	2	2	3	4	6	9
2	2	2	2	2	3	6	6	7
2	2	2	2	2	4	6	6	6
2	2	2	2	2	5	5	6	6
2	2	2	2	3	3	4	6	8
2	2	2	2	3	3	6	6	6
2	2	2	2	3	4	5	6	6
2	2	2	2	4	4	4	4	8
2	2	2	2	4	4	4	6	6
2	2	2	3	3	4	4	6	6
2	2	2	3	3	4	5	5	6
2	2	2	4	4	4	4	4	6
2	2	3	3	3	3	4	4	8
2	2	3	3	3	4	5	5	5
2	2	3	3	4	4	4	4	6
2	2	4	4	4	4	4	4	4
2	3	3	3	3	3	3	6	6
2	3	3	4	4	4	4	4	4

APPENDIX H

The sequences given in Tables H.1-H.4 can be used to form first rows of circulant matrices which can then be used in the Goethals-Seidel Array (Theorem 4.49) if there are four sequences, or, if there are two sequences in the "two-circulant" construction (see Definitions 4.39 and 4.47).

For example: Suppose we wish to form an orthogonal design of type (2,12) in order 20. We use the sequences given in Table H.1 as follows:

$abb\bar{b}0$ $\qquad$ $a\bar{b}\bar{b}b0$, $\qquad$ $bbb00$, $\qquad$ $b\bar{b}b00$

give the first rows of 5×5 circulant matrices which may be used in the Goethals-Seidel array to obtain the design we want.

For Table H.3 the process is similar but is designed for the "two-circulant" construction.

In Table H.4 the positioning of the zeroes is a bit more delicate and we use 0_t to denote t consequetive zeroes.

Note that "$\bar{b}$" means "$-b$".

TABLE H.1

Designs with zero non-periodic auto-correlation function

(1, 1, 1, 1)	a, b, c, d
(1, 1, 1, 4)	a, b, $dc\bar{d}$, d0d
(1, 1, 2, 2)	a, b, cd, $c\bar{d}$
(1, 1, 2, 8)	$da\bar{d}$, $db\bar{d}$, dcd, $d\bar{c}d$
(1, 1, 4, 4)	$ba\bar{b}$, b0b, $dc\bar{d}$, d0d
(1, 2, 2, 4)	$ba\bar{b}$, b0b, cd, $c\bar{d}$
(1, 2, 3, 6)	abc, $ab\bar{c}$, $b\bar{a}b$, $bd\bar{b}$
(2, 2, 2, 2)	ab, $a\bar{b}$, cd, $c\bar{d}$
(1, 1, 1, 9)	$da\bar{d}$, $db\bar{d}$, $d0c0\bar{d}$, d0d0d
(1, 1, 8, 8)	$cda\overline{dc}$, $\bar{c}db\bar{d}c$, cd0dc, $c\bar{d}0\bar{d}c$
(1, 1, 9, 9)	$bca\overline{cb}$, $c\bar{b}db\bar{c}$, $ccc\bar{c}c$, $bbb\bar{b}b$
(1, 4, 4, 4)	$ba\bar{b}$, b0b , $ccd\bar{d}$, $dd\bar{c}c$
(2, 2, 4, 4)	ab, $a\bar{b}$, $ccd\bar{d}$, $dd\bar{c}c$
(2, 3, 4, 6)	$ad0\bar{d}a$, $adcd\bar{a}$, $bc\bar{d}$, $b\bar{c}d$
(3, 3, 3, 3)	abc, $a\bar{b}0d$, $a0c\bar{d}$, bcd
(3, 3, 6, 6)	$adb\bar{d}a$, $adcd\bar{a}$, $c\bar{d}a0\bar{b}$, $c\overline{da}0b$
(4, 4, 4, 4)	abcd, $a\bar{b}c\bar{d}$, $ab\overline{cd}$, $a\overline{bc}d$
(2, 12)	$abb\bar{b}$, $a\overline{bb}b$, bbb, $b\bar{b}b$
(1, 1, 4, 16)	a0a0a0a, $a0ab\bar{a}0\bar{a}$, $ac\bar{a}0\bar{a}ca$, $ac\bar{a}d\overline{aca}$
(1, 1, 5, 5)	a, b, $cc\bar{c}d0d$, $dd\overline{dc}0\bar{c}$
(1, 2, 4, 8)	$ca\bar{c}$, c0c, $db\bar{d}d0d$, $db\overline{dd}0\bar{d}$
(1, 2, 6, 12)	ab0ba, $abd\overline{ba}$, $b\bar{a}bbc\bar{b}$, $b\bar{a}b\overline{bc}b$
(1, 4, 5, 5)	$ba\bar{b}$, b0b, $cc\bar{c}d0d$, $dd\overline{dc}0\bar{c}$
(2, 2, 2, 8)	ab, $a\bar{b}$, $cd\bar{c}c0c$, $cd\overline{cc}0\bar{c}$
(2, 2, 4, 16)	$ab\bar{a}aca$, $ab\bar{a}a\bar{c}a$, $ab\overline{aa}d\bar{a}$, $ab\overline{aada}$
(2, 2, 5, 5)	ab, $a\bar{b}$, $cc\bar{c}d0d$, $dd\overline{dc}0\bar{c}$
(2, 2, 8, 8)	$ab\bar{a}a0a$, $ab\overline{aa}0\bar{a}$, $cd\bar{c}c0c$, $cd\overline{cc}0\bar{c}$
(2, 2, 9, 9)	$db\bar{d}ca\bar{c}$, $db\overline{dcac}$, $dc0c\bar{d}c$, $\bar{c}d0dcd$
(2, 2, 10, 10)	$cc\bar{c}dad$, $cc\bar{c}d\bar{a}d$, $dd\overline{dc}b\bar{c}$, $dd\overline{dcbc}$
(2, 4, 4, 8)	$ab\bar{a}a0a$, $ab\overline{aa}0\bar{a}$, $cd\bar{d}c$, $\bar{c}ddc$

Table H.1 (Continued)

(2, 4, 6, 12)	$ab\bar{c}abc$, $ab\overline{cabc}$, $b\bar{a}bbd\bar{b}$, $b\bar{a}b\overline{bd}b$
(2, 5, 5, 8)	$ad\bar{a}a0a$, $ad\overline{aa}0\bar{a}$, $cc\bar{c}b0b$, $bb\overline{bc}0\bar{c}$
(2, 8, 8, 8)	$aab\bar{b}cd\bar{c}$, $aab\overline{bcd}c$, $a\bar{a}bbc0c$, $a\bar{a}bb\bar{c}0\bar{c}$
(4, 4, 4, 16)	$cd\bar{c}cacb$, $cd\bar{c}c\bar{a}c\bar{b}$, $cd\overline{cc}a\overline{cb}$, $cd\overline{ccac}b$
(4, 4, 5, 5)	$aab\bar{b}$, $bb\bar{a}a$, $c0cdd\bar{d}$, $d0d\overline{cc}c$
(4, 4, 8, 8)	$aab\bar{b}cd$, $aab\overline{bcd}$, $a\bar{a}bbc\bar{d}$, $a\bar{a}bb\bar{c}d$
(4, 4, 10, 10)	$bcacdd\bar{d}$, $b\bar{c}a\overline{cdd}d$, $bd\bar{a}d\overline{cc}c$, $b\overline{dad}cc\bar{c}$
(5, 5, 5, 5)	$aa\bar{a}b0b$, $\overline{bb}ba0a$, $\overline{cc}c\bar{d}0\bar{d}$, $\overline{dd}dc0c$
(6, 6, 6, 6)	$aab\bar{b}cd$, $\overline{bb}a\bar{a}d\bar{c}$, $\overline{ccd}dab$, $\overline{dd}\overline{cc}ba$
(7, 7, 7, 7)	$aa\bar{a}bcbd$, $\overline{bb}bada\bar{c}$, $\overline{cc}c\bar{d}a\bar{d}b$, $\overline{dd}dc\bar{b}ca$

TABLE H.2

Designs with zero non-periodic auto-correlation function

(21)	$aaaa\bar{a}\bar{a}$, $aa\bar{a}a\bar{a}a$, $aa\bar{a}aa$, $aa0\bar{a}a$
(23)	$a\bar{a}aaaa\bar{a}$, $aa\bar{a}\bar{a}\bar{a}a\bar{a}$, $a\bar{a}\bar{a}0\bar{a}\bar{a}a$, $a0a000a$
(2, 26)	$abb\bar{b}b\bar{b}\bar{b}$, $\bar{a}bb\bar{b}b\bar{b}\bar{b}$, $bbb\bar{b}bbb$, $bb\bar{b}\bar{b}\bar{b}b\bar{b}$
(1, 24)	$aa\bar{a}ba\bar{a}\bar{a}$, $aaa0\bar{a}\bar{a}a$, $aa\bar{a}a\bar{a}a$, $aaaa\bar{a}a$
(1, 27)	$aa\bar{a}ba\bar{a}\bar{a}$, $aa\bar{a}a\bar{a}a\bar{a}$, $aa\bar{a}\bar{a}\bar{a}\bar{a}a$, $aa\bar{a}aaaa$
(29)	$aaa\bar{a}\bar{a}a\bar{a}\bar{a}$, $aaa\bar{a}a\bar{a}\bar{a}a$, $aaa\bar{a}a\bar{a}a$, $aaa0\bar{a}aa$
(30)	$aaa\bar{a}aa\bar{a}\bar{a}$, $aaaa\bar{a}a\bar{a}a$, $aa\bar{a}\bar{a}a\bar{a}a$, $aaaa\bar{a}\bar{a}a$
(31)	$aa\bar{a}aaa\bar{a}a\bar{a}$, $aa\bar{a}a0\bar{a}a\bar{a}a$, $aaaaa\bar{a}\bar{a}$, $aaa\bar{a}\bar{a}aa$
(1, 1, 32)	$aa\bar{a}ab\bar{a}a\bar{a}\bar{a}$, $aaa\bar{a}ca\bar{a}\bar{a}\bar{a}$, $aa\bar{a}a0a\bar{a}aa$, $aaa\bar{a}0\bar{a}aaa$
(1, 33)	$aaa\bar{a}ba\bar{a}\bar{a}\bar{a}$, $aaa\bar{a}\bar{a}a\bar{a}\bar{a}a$, $aaa\bar{a}a\bar{a}aa$, $aaa\bar{a}a\bar{a}aa$
(1, 35)	$aa\bar{a}ab\bar{a}a\bar{a}\bar{a}$, $aaa\bar{a}\bar{a}\bar{a}a\bar{a}\bar{a}$, $aa\bar{a}a\bar{a}a\bar{a}aa$, $aaaaa\bar{a}\bar{a}aa$
(1, 36)	$aa\bar{a}aaaa\bar{a}a\bar{a}$, $aa\bar{a}aa\bar{a}\bar{a}a\bar{a}a$, $aaa\bar{a}ba\bar{a}\bar{a}\bar{a}$, $aaa\bar{a}0\bar{a}aaa$
(38)	$aaaaa\bar{a}aa\bar{a}\bar{a}$, $aaa\bar{a}\bar{a}a\bar{a}a\bar{a}a$, $aaa\bar{a}\bar{a}aa\bar{a}a$, $aaaa\bar{a}\bar{a}a\bar{a}a$
(39)	$aaa\bar{a}a\bar{a}aa\bar{a}\bar{a}\bar{a}$, $aaa\bar{a}a0aa\bar{a}aa$, $aaa\bar{a}\bar{a}a\bar{a}\bar{a}a$, $aaa\bar{a}a\bar{a}\bar{a}\bar{a}a$
(1, 2, 2, 36)	$aaa\bar{a}ad\bar{a}a\bar{a}\bar{a}\bar{a}$, $aaa\bar{a}a0a\bar{a}aaa$, $aab\bar{a}\bar{a}a\bar{a}ca\bar{a}$, $aab\bar{a}\bar{a}\bar{a}a\bar{c}\bar{a}a$
(42)	$aaa\bar{a}aaa\bar{a}a\bar{a}\bar{a}$, $aaa\bar{a}\bar{a}a\bar{a}\bar{a}a\bar{a}a$, $aaa\bar{a}\bar{a}aaa\bar{a}a$, $aaaaa\bar{a}\bar{a}a\bar{a}a$
(1, 43)	$aaa\bar{a}ab\bar{a}a\bar{a}\bar{a}\bar{a}$, $aaaa\bar{a}\bar{a}aa\bar{a}a\bar{a}$, $aaaa\bar{a}\bar{a}a\bar{a}a\bar{a}a$, $aaa\bar{a}aa\bar{a}\bar{a}aa$
(1, 44)	$aa\bar{a}\bar{a}a\bar{a}a\bar{a}\bar{a}\bar{a}\bar{a}a$, $aa\bar{a}\bar{a}a\bar{a}\bar{a}aaaa\bar{a}$, $a\bar{a}aaab\bar{a}\bar{a}\bar{a}a\bar{a}$, $a\bar{a}aaa0aaa\bar{a}a$
(1, 1, 40)	$a0aaa\bar{a}ba\bar{a}\bar{a}\bar{a}0\bar{a}$, $a0a\bar{a}\bar{a}ac\bar{a}aa\bar{a}0\bar{a}$, $a0aaa\bar{a}0\bar{a}aaa0a$, $a0a\bar{a}\bar{a}a0a\bar{a}\bar{a}a0a$
(47)	$aaa\bar{a}aa\bar{a}aaa\bar{a}\bar{a}\bar{a}$, $aaa\bar{a}a\bar{a}0\bar{a}a\bar{a}\bar{a}aa$, $aaa\bar{a}aaaa\bar{a}\bar{a}a$, $aaa\bar{a}a\bar{a}\bar{a}a\bar{a}aa$
(48)	$aaa\bar{a}aa\bar{a}aa\bar{a}\bar{a}\bar{a}$, $aaaa\bar{a}a\bar{a}\bar{a}aa\bar{a}\bar{a}$, $aaa\bar{a}\bar{a}a\bar{a}a\bar{a}\bar{a}aa$, $aaaa\bar{a}a\bar{a}a\bar{a}aaa$
(50)	$aaaa\bar{a}a\bar{a}\bar{a}aa\bar{a}\bar{a}\bar{a}$, $aaaa\bar{a}a\bar{a}\bar{a}aa\bar{a}\bar{a}a$, $aaa\bar{a}\bar{a}a\bar{a}a\bar{a}\bar{a}aa$, $aaaaa\bar{a}a\bar{a}a\bar{a}aa$
(1, 51)	$aaa\bar{a}a\bar{a}ba\bar{a}a\bar{a}\bar{a}\bar{a}$, $aaa\bar{a}aa\bar{a}\bar{a}a\bar{a}a\bar{a}\bar{a}$, $aaa\bar{a}a\bar{a}aaa\bar{a}\bar{a}aa$, $aaaaa\bar{a}\bar{a}a\bar{a}\bar{a}aaa$
(1, 1, 17, 17)	$aaa\bar{a}0a\bar{a}\bar{a}bbb\bar{b}b\bar{b}bb$, $bbb\bar{b}0b\bar{b}\bar{b}\bar{a}\bar{a}\bar{a}a\bar{a}a\bar{a}\bar{a}$, $a0c0\bar{a}$, $b0d0\bar{b}$
(1, 2, 66)	$aaa\bar{a}ba\bar{a}\bar{a}\bar{a}aaa\bar{a}\bar{a}a\bar{a}\bar{a}a$, $aaa\bar{a}ba\bar{a}\bar{a}\bar{a}\bar{a}\bar{a}\bar{a}aa\bar{a}aa\bar{a}$, $aaa\bar{a}a\bar{a}aa0aa\bar{a}a\bar{a}aaa$, $aaa\bar{a}a\bar{a}aac\bar{a}aa\bar{a}a\bar{a}\bar{a}\bar{a}$

TABLE H. 3

Sequences with zero non-periodic auto-correlation function

Length	Type	
1	(1, 1)	a, b
2	(2, 2)	ab, $a\bar{b}$
3	(1, 4)	$ba\bar{b}$, b0b
4	(4, 4)	$aab\bar{b}$, $bb\bar{a}a$
6	(2, 8)	$ab\bar{a}a0a$, $ab\bar{a}\bar{a}0\bar{a}$
6	(5, 5)	$aa\bar{a}b0b$, $bb\bar{b}\bar{a}0\bar{a}$
8	(8, 8)	$aaa\bar{a}bb\bar{b}b$, $\bar{b}\bar{b}\bar{b}baa\bar{a}a$
10	(4, 16)	$ab\bar{a}a\bar{a}a\bar{a}\bar{a}ba$, $ab\bar{a}aaaaa\bar{b}\bar{a}$
10	(10, 10)	$a\bar{a}\bar{a}ababbb\bar{b}$, $b\bar{b}\bar{b}b\bar{a}b\bar{a}\bar{a}\bar{a}a$
11	(13)	$aaa0\bar{a}aa0\bar{a}a\bar{a}$, $a0a000\bar{a}000a$
14	(13, 13)	$aaab\bar{a}aa\bar{b}\bar{a}a\bar{a}b0b$, $bbb\bar{a}\bar{b}bba\bar{b}b\bar{b}\bar{a}0\bar{a}$

TABLE H.4

Sequences with zero periodic auto-correlation function

$n \geq 5$	(1, 14)	$a0_{\frac{1}{2}(n-5)}bb\bar{b}\bar{b}0_{\frac{1}{2}(n-5)}$, $bb\bar{b}b0_{n-4}$, $bbb0_{n-3}$, $b\bar{b}b0_{n-3}$
$n \geq 7$	(1, 1, 1, 16)	$a0_{\frac{1}{2}(n-7)}bb\bar{b}b\bar{b}\bar{b}0_{\frac{1}{2}(n-7)}$, $c0_{\frac{1}{2}(n-5)}b00\bar{b}0_{\frac{1}{2}(n-5)}$, $d0_{\frac{1}{2}(n-5)}b00\bar{b}0_{\frac{1}{2}(n-5)}$, $bb\bar{b}bbb0_{n-7}$
	(1, 1, 13, 13)	$b0_{\frac{1}{2}(n-7)}cc\bar{c}c\bar{c}\bar{c}0_{\frac{1}{2}(n-7)}$, $ccd\bar{c}dcd0_{n-7}$, $dd\bar{c}\bar{d}\bar{c}d\bar{c}0_{n-7}$, $a0_{\frac{1}{2}(n-7)}dd\bar{d}d\bar{d}\bar{d}0_{\frac{1}{2}(n-7)}$
	(1, 1, 26)	$a0_{\frac{1}{2}(n-7)}bb\bar{b}b\bar{b}\bar{b}0_{\frac{1}{2}(n-7)}$, $c0_{\frac{1}{2}(n-7)}bb\bar{b}b\bar{b}\bar{b}0_{\frac{1}{2}(n-7)}$, $bbb\bar{b}bbb0_{n-7}$, $bb\bar{b}\bar{b}\bar{b}b\bar{b}0_{n-7}$
	(1, 22)	$a0_{\frac{1}{2}(n-7)}bbb\bar{b}\bar{b}\bar{b}0_{\frac{1}{2}(n-7)}$, $bb\bar{b}b\bar{b}b0_{n-6}$, $bb\bar{b}bb0_{n-5}$, $bb\bar{b}bb0_{n-5}$
$n \geq 9$	(1, 25)	$a0_{\frac{1}{2}(n-7)}b0b\bar{b}0\bar{b}0_{\frac{1}{2}(n-7)}$, $b0bb0bbb\bar{b}0_{n-9}$, $b0b\bar{b}\bar{b}b\bar{b}\bar{b}b0_{n-9}$, $bb\bar{b}\bar{b}\bar{b}b0_{n-6}$
	(1, 30)	$a0_{\frac{1}{2}(n-9)}bb\bar{b}b\bar{b}b\bar{b}\bar{b}0_{\frac{1}{2}(n-9)}$, $bbbb\bar{b}\bar{b}\bar{b}b0_{n-8}$, $bbb\bar{b}bbb0_{n-7}$, $b\bar{b}\bar{b}b\bar{b}\bar{b}b0_{n-7}$
$n \geq 11$	(1, 4, 13)	$ab\bar{a}0_{n-3}$, $a0a0_{n-3}$, $ccc0\bar{c}cc0\bar{c}c\bar{c}0_{n-11}$, $c0c000\bar{c}000c0_{n-11}$
	(1, 1, 40)	$a0_{\frac{1}{2}(n-11)}\bar{b}bbb\bar{b}b\bar{b}\bar{b}\bar{b}b0_{\frac{1}{2}(n-11)}$, $c0_{\frac{1}{2}(n-11)}b\bar{b}\bar{b}b\bar{b}b\bar{b}bb\bar{b}0_{\frac{1}{2}(n-11)}$, $b\bar{b}\bar{b}bbbb\bar{b}\bar{b}b0_{n-10}$, $\bar{b}bbbbbbbb\bar{b}0_{n-10}$
$n \geq 13$	(1, 1, 25)	$a0_{\frac{1}{2}(n-13)}bbb\bar{b}0b\bar{b}0b\bar{b}\bar{b}\bar{b}0_{\frac{1}{2}(n-13)}$, $c0_{\frac{1}{2}(n-5)}b00\bar{b}0_{\frac{1}{2}(n-5)}$, $b00000b00b0_{n-10}$, $bbb0\bar{b}b\bar{b}b\bar{b}0bb0_{n-12}$

Table H. 4 (Continued)

	(1, 1, 50)	$a0_{\frac{1}{2}(n-13)}bbb\bar{b}bb\overline{bb}b\overline{bbb}0_{\frac{1}{2}(n-13)}$, $c0_{\frac{1}{2}(n-13)}bbb\overline{bb}b\bar{b}bb\overline{bbb}0_{\frac{1}{2}(n-13)}$, $bbbb\bar{b}b\bar{b}b\bar{b}bbb0_{n-13}$, $bbb\overline{bb}b\bar{b}b\overline{bb}bb\bar{b}0_{n-13}$
$n \geq 15$	(1, 4, 26)	$ab\bar{a}0_{n-3}$, $a0a_{n-3}$, $ccc\overline{cc}ccc\bar{c}c\overline{cc}0\bar{c}0_{n-14}$, $cccc\bar{c}cc\overline{cc}c\bar{c}c0c0_{n-14}$
	(1, 1, 29)	$a0_{\frac{1}{2}(n-15)}b000\bar{b}0000b000\bar{b}0_{\frac{1}{2}(n-15)}$, $c0_{\frac{1}{2}(n-13)}bbb0\bar{b}b\bar{b}b0\overline{bbb}0_{\frac{1}{2}(n-13)}$, $bb00bb\bar{b}0\bar{b}bb00b0_{n-14}$, $b\bar{b}000\bar{b}000\bar{b}b0\bar{b}0_{n-13}$
	(1, 1, 58)	$a0_{\frac{1}{2}(n-15)}bbbb\overline{bb}b\bar{b}bb\overline{bbbb}0_{\frac{1}{2}(n-15)}$, $c0_{\frac{1}{2}(n-15)}b\overline{bbbb}b\bar{b}b\bar{b}bbbb\bar{b}0_{\frac{1}{2}(n-15)}$, $bb\bar{b}bbb\bar{b}b\bar{b}bbb\bar{b}bb0_{n-15}$, $bbb\bar{b}bb\overline{bbb}bb\bar{b}bb\bar{b}0_{n-15}$
	(1, 52)	$a0_{\frac{1}{2}(n-13)}bb\bar{b}b\overline{bb}bb\bar{b}b\overline{bb}0_{\frac{1}{2}(n-13)}$, $bb\bar{b}b\overline{bbbb}b\bar{b}bb0_{n-12}$, $bbb\bar{b}bbbbb\overline{bbb}b\bar{b}0_{n-14}$, $bbb\bar{b}bbb\overline{bb}bbb\bar{b}b0_{n-14}$

TABLE H. 5

4-complementary sequences A, B, C, D such that (A+B)/2, (A-B)/2, (C+D)/2, (C-D)/2 are also 4-complementary sequences

Length	Weight	A, B, C, D
3	10	11-, 11-, 101, 101
3	12	111, 1-1, 11-, 11-
4	14	111-, 1--1, 111, 1-1
5	18	11-11, 11-1-, 111-, 111-
5	20	1-11-, 111--, 111-1, 111-1
6	10	101--1, 10111-
6	12	101--1, 10111-, 1, 1
6	14	101--1, 10111-, 11, 1-
6	16	11-1--, 11-111, 100-, 100-
6	18	11-1--, 11-111, 110-, 1-0-
6	20	111-1-, 1-1111, 1--1, 11--
6	22	11--1-, 1----1, 1-111, 1-111
6	24	11-1--, 11--11, 111-1-, 1-1111
7	26	111-111, 11---1-, 11-1--, 11-1--
7	28	1--1111, 1--1-1-, 11-111-, 1----1
9	28	1---0-11-, 1--101--1, 10110111, 101-0---
9	30	11-10111-, 11--0---1, 101101--1, 101-0-11-
9	34	111-01---, 111--1--1, 111-1-11, 111-1-11
14	26	111--111-1--0-, 1111-11--1-10

TABLE H. 6

2-complementary disjointable sequences

Length	Type	
1	(1, 1)	a, b
2	(2, 2)	$ab, a\bar{b}$
4	(4, 4)	$aab\bar{b}, a\bar{a}bb$
6	(2, 8)	$ab\bar{a}a0a, ab\bar{a}\bar{a}0\bar{a}$
8	(8, 8)	$aaa\bar{a}bb\bar{b}b, a\bar{a}aab\bar{b}\bar{b}\bar{b}$
10	(4, 16)	$ab\bar{a}a\bar{a}a\bar{a}\bar{a}ba, ab\bar{a}aaaaa\bar{b}\bar{a}$
10	(10, 10)	$a\bar{a}\bar{a}ababbb\bar{b}, a\bar{a}\bar{a}\bar{a}b\bar{a}b\bar{b}\bar{b}b$
14	(26)	$aaaa\bar{a}aa\bar{a}\bar{a}a\bar{a}a0a, aaa\bar{a}\bar{a}aaa\bar{a}a\bar{a}\bar{a}0a$

APPENDIX I: PRODUCT DESIGNS

PRODUCT DESIGNS OF ORDER 16

Table I.1

Product designs	Construction	
(1, 1, 2, 3;1, 1, 2, 3;9)	((1, 1, 2);(1, 3))	((1, 1, 2);(1, 3))
(1, 1;2, 9;1, 3, 3, 6;3)	((1, 1, 2);(3, 1))	((1, 1, 2);(1, 3))
(1, 1, 7;1, 1, 7;7)	((1, 7);(1, 7))	((1, 1);(1, 1))
(1, 2, 2, 2, 3;2, 2, 6;6)	((1, 2, 2, 3);(2, 6))	((1, 1);(1, 1))
(2, 2, 2, 4;1, 1, 2, 6;6)	((1, 1, 2);(1, 3))	((1, 1, 2);(2, 2))
(2, 2, 4, 5;3, 5, 5;3)	((2, 2, 4);(5, 3))	((1, 1);(1, 1))
(2, 2, 4, 6;2, 3, 3, 6;2)	((1, 1, 2);(3, 1))	((1, 1, 2);(2, 2))
(2, 2, 4, 7;1, 7, 7;1)	((2, 2, 4);(7, 1))	((1, 1);(1, 1))
(2, 6, 6;1, 1, 2, 4, 6;2)	((1, 1);(1, 1))	((1, 1, 2, 4);(2, 6))
(4, 4, 4;1, 1, 2, 4, 4;4)	((1, 1);(1, 1))	((1, 1, 2, 4);(4, 4))
(1, 1, 1, 1, 2;2, 4;2, 4, 4)	(1, 1, 1, 1;2, 2;2, 2)	
(1, 1, 1, 2, 2;3, 4;3, 6)	(1, 1, 1, 2;2, 3;3)	
(1, 1, 1, 2, 3;2, 6;2, 6)	(1, 1, 2, 3;1, 6;1)	
(1, 1, 1, 2, 4;1, 8;1, 2, 4)	(1, 1, 1, 2;4, 1;1, 2)	
(1, 1, 1, 2;5;3, 3, 5)	(1, 1, 1, 2;5;3)	
(1, 1, 2, 3, 4;3, 8;2, 3)	(1, 1, 2, 3;4, 3;1)	
(1, 1, 2, 3, 6;1, 12;1, 2)	(1, 1, 2, 3;6, 1;1)	
(1, 2, 2;2, 3;3, 4, 4)	(2, 2;1, 3;2, 2)	
(1, 2, 3, 3;4, 5;2, 5)	(1, 3, 3;2, 5;1)	
(2, 2, 2, 5;1, 10;1, 4)	(2, 2, 2;5, 1;2)	

Table I.2

PRODUCT DESIGNS OF ORDER 32

Product designs	Construction	
(13, 3, 13;4, 4, 8, 13; 3)	((1, 1); (1, 1))	((4, 4, 8); (3, 13))
(3, 5, 5, 10;2, 2, 2, 4, 15;9)	((1, 1, 2); (1, 3))	((2, 2, 4); (5, 3))
(4, 4, 8, 12;3, 3, 4, 6, 12;4)	((1, 1, 2); (3, 1))	((1, 1, 2, 4); (4, 4))
(7, 7, 9;4, 4, 7, 8;9)	((1, 1); (1, 1))	((4, 4, 8); (7, 9))
(1, 1, 1, 1, 2, 4;2, 8;2, 4, 8, 8)	(1, 1, 1, 1, 2;4, 2;2, 4, 4)	
(1, 1, 1, 2, 2, 4;3, 8;3, 6, 12)	(1, 1, 1, 2, 2;4, 3;3, 6)	
(1, 1, 1, 2, 4, 8;1, 16;1, 2, 4, 8)	(1, 1, 1, 2, 4;8, 1;1, 2, 4)	
(1, 1, 2, 2, 3;4, 5;5, 18)	(1, 1, 2, 3;2, 5;9)	
(1, 1, 2, 3, 4, 11;22;4, 6)	(1, 1, 2, 3, 4;11;2, 3)	
(1, 1, 2, 3, 4, 8;3, 16;3, 4, 6)	(1, 1, 2, 3, 4;8, 3;2, 3)	
(1, 1, 2, 3, 6, 12;1, 24;1, 2, 4)	(1, 1, 2, 3, 6;12, 1;1, 2)	
(1, 1, 2, 6, 9;7, 12;6, 7)	(1, 1, 2, 9;6, 7;3)	
(1, 1, 2, 3, 9;6, 10;6, 10)	(1, 1, 2, 9;3, 10;3)	
(1, 1, 2, 9, 10;3, 20;3, 6)	(1, 1, 2, 9;10, 3;3)	
(1, 2, 3, 3, 4;5, 8;4, 5, 10)	(1, 2, 3, 3;4, 5;2, 5)	
(1, 2, 3, 3, 6;4, 11;6, 11)	(1, 3, 3, 6;2, 11;3)	
(1, 3, 3, 6, 10;3, 20;3, 6)	(1, 3, 3, 6;10, 3;3)	
(1, 3, 3, 6, 11;2, 22;2, 6)	(1, 3, 3, 6;11, 2;3)	
(1, 3, 3, 6, 9;4, 18;4, 6)	(1, 3, 3, 6;9, 4;3)	
(1, 4, 7, 7;8, 11;2, 11)	(1, 7, 7;4, 11;1)	
(1, 6, 7, 7;9, 12;2, 9)	(1, 7, 7;6, 9;1)	
(2, 2, 2, 4, 9;1, 18;1, 12)	(2, 2, 2, 4;9, 1;6)	
(2, 2, 2, 5, 10;1, 20;1, 2, 8)	(2, 2, 2, 5;10, 1;1, 4)	
(2, 2, 4, 5, 10;3, 20;3, 6)	(2, 2, 4, 5;10, 3;3)	
(2, 2, 4, 5, 8;5, 16;5, 6)	(2, 2, 4, 5;8, 5;3)	
(2, 2, 4, 6, 11;3, 22;3, 4)	(2, 2, 4, 6;11, 3;2)	
(2, 2, 4, 6, 9;5, 18;4, 5)	(2, 2, 4, 6;9, 5;2)	
(2, 2, 4, 7, 8;7, 16;2, 7)	(2, 2, 4, 7;8, 7;1)	
(2, 6, 6, 13;1, 26;1, 4)	(2, 6, 6;13, 1;2)	
(4, 4, 4, 7;5, 14;5, 8)	(4, 4, 4;7, 5;4)	

Table I. 3

PRODUCT DESIGNS OF ORDER 64

Product designs	Construction
(1, 1, 1, 1, 2, 2, 4;4, 8;4, 8, 8, 16, 16)	(1, 1, 1, 1, 2, 4;2, 8;2, 4, 8, 8)
(1, 1, 1, 2, 2, 3, 4;6, 8:6, 8, 12, 24)	(1, 1, 1, 2, 2, 4;3, 8;3, 6, 12)
(1, 1, 1, 2, 4, 8, 16;1, 32;1, 2, 4, 8, 16)	(1, 1, 1, 2, 4, 8;16, 1;1, 2, 4, 8)
(1, 1, 2, 2, 3, 4;5, 8;5, 10, 36)	(1, 1, 2, 2, 3:4, 5;5, 18)
(1, 1, 2, 3, 4, 8, 16;3, 32;3, 6, 8, 12)	(1, 1, 2, 3, 4, 8;16, 3;3, 4, 6)
(1, 1, 2, 3, 6, 9;10, 12;10, 12, 20)	(1, 1, 2, 3, 9;6, 10;6, 10)
(1, 1, 2, 6, 7, 9;12, 14;12, 12, 14)	(1, 1, 2, 6, 9;7, 12;6, 7)
(1, 2, 3, 3, 4, 5;8, 10;8, 8, 10, 20)	(1, 2, 3, 3, 4;5, 8;4, 5, 10)
(1, 2, 3, 3, 6, 11;4, 22;4, 12, 22)	(1, 2, 3, 3, 6;11, 4;6, 11)
(1, 3, 3, 6, 9, 18;4, 36;4, 8, 12)	(1, 3, 3, 6, 9;18, 4;4, 6)
(1, 4, 7, 7, 8;11, 16;4, 11, 22)	(1, 4, 7, 7;8, 11;2, 11)
(1, 6, 7, 7, 12;9, 24;4, 9, 18)	(1, 6, 7, 7;12, 9;2, 9)
(2, 2, 2, 5, 10, 20;1, 40;1, 2, 4, 16)	(2, 2, 2, 5, 10;20, 1;1, 2, 8)
(2, 2, 4, 6, 9, 18;5, 36;5, 8, 10)	(2, 2, 4, 6, 9;18, 5;4, 5)
(2, 2, 4, 7, 8, 16;7, 32;7, 18)	(2, 2, 4, 7, 8;16, 7;9)
(2, 3, 5, 5, 10:4, 21;18, 21)	(3, 5, 5, 10;2, 21;9)
(2, 6, 6, 13, 26;1, 52:1, 2, 8)	(2, 6, 6, 13;26, 1;1, 4)
(3, 4, 13, 13;8, 25;6, 25)	(3, 13, 13;4, 25;3)
(4, 4, 4, 7, 14;5, 28;5, 10, 16)	(4, 4, 4, 7;14, 5;5, 8)
(4, 4, 7, 8, 12;14, 21;8, 21)	(4, 4, 8, 12;7, 21;4)
(7, 7, 9, 12;11, 24;11, 18)	(7, 7, 9;12, 11;9)

APPENDIX J: DESIGNS WHICH MAY EXIST IN ORDER N = 0 (MOD 4)

Roger Magoon (Kingston, Ontario) and Peter J. Robinson (Canberra, Australia) have written computer programmes which generate the 4-tuples not prohibited from being the type of an orthogonal design in order 4n, n odd. This appendix contains the printout for 36 and 44. The results for 4n, $n \leq 25$, are available upon request.

The printout is from Magoon's programme. The number after the colon is the sum of the type numbers and a check indicates that the design has been constructed.

ORDER 36

(1, 1, 1, 1): 4 (✓)
(1, 1, 1, 4): 7 (✓)
(1, 1, 1, 9): 12 (✓)
(1, 1, 1, 16): 19 (✓)
(1, 1, 1, 25): 28
(1, 1, 2, 2): 6 (✓)
(1, 1, 2, 8): 12 (✓)
(1, 1, 2, 18): 22
(1, 1, 2, 32): 36
(1, 1, 4, 4): 10 (✓)
(1, 1, 4, 9): 15 (✓)
(1, 1, 4, 16): 22 (✓)
(1, 1, 4, 25): 31
(1, 1, 5, 5): 12 (✓)
(1, 1, 5, 20): 27
(1, 1, 8, 8): 18 (✓)
(1, 1, 8, 18): 28
(1, 1, 9, 9): 20 (✓)
(1, 1, 9, 16): 27
(1, 1, 9, 25): 36
(1, 1, 10, 10): 22
(1, 1, 13, 13): 28 (✓)
(1, 1, 16, 16): 34
(1, 1, 17, 17): 36
(1, 2, 2, 4): 9 (✓)
(1, 2, 2, 9): 14 (✓)
(1, 2, 2, 16): 21
(1, 2, 2, 25): 30
(1, 2, 3, 6): 12 (✓)
(1, 2, 3, 24): 30
(1, 2, 4, 8): 15 (✓)
(1, 2, 4, 18): 25
(1, 2, 6, 12): 21 (✓)
(1, 2, 6, 27): 36
(1, 2, 8, 9): 20
(1, 2, 8, 16): 27
(1, 2, 8, 25): 36
(1, 2, 9, 18): 30
(1, 2, 11, 22): 36
(1, 3, 6, 8): 18
(1, 3, 6, 18): 28
(1, 3, 8, 24): 36
(1, 4, 4, 4): 13 (✓)
(1, 4, 4, 9): 18
(1, 4, 4, 16): 25
(1, 4, 4, 25): 34
(1, 4, 5, 5): 15 (✓)
(1, 4, 5, 20): 30
(1, 4, 8, 8): 21 (✓)
(1, 4, 8, 18): 31
(1, 4, 9, 9): 23
(1, 4, 9, 16): 30
(1, 4, 10, 10): 25
(1, 4, 13, 13): 31
(1, 5, 5, 9): 20
(1, 5, 5, 16): 27
(1, 5, 5, 25): 36
(1, 6, 8, 12): 27
(1, 8, 8, 9): 26
(1, 8, 8, 16): 33
(1, 8, 9, 18): 36
(1, 9, 9, 9): 28
(1, 9, 10, 10): 30
(1, 9, 13, 13): 36
(2, 2, 2, 2): 8 (✓)
(2, 2, 2, 8): 14 (✓)
(2, 2, 2, 18): 24
(2, 2, 4, 4): 12 (✓)
(2, 2, 4, 9): 17
(2, 2, 4, 16): 24 (✓)
(2, 2, 4, 25): 33
(2, 2, 5, 5): 14 (✓)
(2, 2, 5, 20): 29
(2, 2, 8, 8): 20 (✓)
(2, 2, 8, 18): 30
(2, 2, 9, 9): 22 (✓)
(2, 2, 9, 16): 29
(2, 2, 10, 10): 24 (✓)
(2, 2, 18, 18): 30
(2, 2, 16, 16): 36
(2, 3, 4, 6): 15 (✓)
(2, 3, 4, 24): 33
(2, 3, 6, 9): 20
(2, 3, 6, 16): 27
(2, 3, 6, 25): 36
(2, 3, 10, 15): 30
(2, 4, 4, 8): 18 (✓)
(2, 4, 4, 18): 28
(2, 4, 6, 12): 24 (✓)
(2, 4, 8, 9): 23
(2, 4, 8, 16): 30
(2, 4, 9, 18): 33
(2, 5, 5, 8): 20 (✓)
(2, 5, 5, 18): 30
(2, 6, 7, 21): 36
(2, 6, 9, 12): 29
(2, 6, 12, 16): 36
(2, 8, 8, 8): 26 (✓)
(2, 8, 8, 18): 36
(2, 8, 9, 9): 28
(2, 8, 10, 10): 30 (✓)
(2, 8, 13, 13): 36

(3, 3, 3, 3): 12 (✓)
(3, 3, 3, 12): 21
(3, 3, 3, 27): 36
(3, 3, 6, 6): 18 (✓)
(3, 3, 6, 24): 36
(3, 3, 12, 12): 30
(3, 3, 15, 15): 36
(3, 4, 6, 8): 21
(3, 4, 6, 18): 31
(3, 6, 6, 12): 27
(3, 6, 8, 9): 26
(3, 6, 8, 16): 33
(3, 6, 9, 18): 36
(3, 8, 10, 15): 36
(4, 4, 4, 4): 16 (✓)
(4, 4, 4, 9): 21
(4, 4, 4, 16): 28 (✓)
(4, 4, 5, 5): 18 (✓)
(4, 4, 5, 20): 33
(4, 4, 8, 8): 24 (✓)
(4, 4, 8, 18): 34
(4, 4, 9, 9): 26
(4, 4, 9, 16): 33
(4, 4, 10, 10): 28 (✓)
(4, 4, 13, 13): 34
(4, 5, 5, 9): 23
(4, 5, 5, 16): 30
(4, 6, 8, 12): 30
(4, 8, 8, 9): 29
(4, 8, 8, 16): 36
(4, 9, 9, 9): 31
(4, 9, 10, 10): 33
(5, 5, 5, 5): 20 (✓)
(5, 5, 8, 8): 26 (✓)
(5, 5, 8, 18): 36
(5, 5, 9, 9): 28
(5, 5, 10, 10): 30
(5, 5, 13, 13): 36
(5, 6, 6, 6): 24 (✓)
(6, 6, 12, 12): 36
(7, 7, 7, 7): 28 (✓)
(8, 8, 8, 8): 32 (✓)
(8, 8, 9, 9): 34
(8, 8, 10, 10): 36
(9, 9, 9, 9): 36 (✓)

ORDER 44

1:	1	1	1	1	4
2:	1	1	1	4	7
3:	1	1	1	9	12
4:	1	1	1	16	19
5:	1	1	1	25	28
6:	1	1	1	36	39
7:	1	1	2	2	6
8:	1	1	2	8	12
9:	1	1	2	18	22
10:	1	1	2	32	36
11:	1	1	4	4	10
12:	1	1	4	9	15
13:	1	1	4	16	22
14:	1	1	4	25	31
15:	1	1	4	36	42
16:	1	1	5	5	12
17:	1	1	5	20	27
18:	1	1	8	8	18
19:	1	1	8	18	28
20:	1	1	8	32	42
21:	1	1	9	9	20
22:	1	1	9	16	27
23:	1	1	9	25	36
24:	1	1	10	10	22
25:	1	1	13	13	28
26:	1	1	16	16	34
27:	1	1	17	17	36
28:	1	1	18	18	38
29:	1	1	20	20	42
30:	1	2	2	4	9
31:	1	2	2	9	14
32:	1	2	2	16	21
33:	1	2	2	25	30
34:	1	2	2	36	41
35:	1	2	3	6	12
36:	1	2	3	24	30
37:	1	2	4	8	15
38:	1	2	4	18	25
39:	1	2	4	32	39
40:	1	2	6	12	21
41:	1	2	6	27	36
42:	1	2	8	9	20
43:	1	2	8	16	27
44:	1	2	8	25	36
45:	1	2	9	18	30
46:	1	2	9	32	44
47:	1	2	11	22	36
48:	1	2	12	24	39
49:	1	2	16	18	37
50:	1	3	6	8	18
51:	1	3	6	18	28
52:	1	3	6	32	42
53:	1	3	8	24	36
54:	1	4	4	4	13
55:	1	4	4	9	18
56:	1	4	4	16	25
57:	1	4	4	25	34
58:	1	4	5	5	15
59:	1	4	5	20	30
60:	1	4	8	8	21
61:	1	4	8	18	31
62:	1	4	9	9	23
63:	1	4	9	16	30
64:	1	4	9	25	39
65:	1	4	10	10	25
66:	1	4	13	13	31
67:	1	4	16	16	37
68:	1	4	17	17	39
69:	1	4	18	18	41
70:	1	5	5	9	20
71:	1	5	5	16	27
72:	1	5	5	25	36
73:	1	5	6	30	42
74:	1	5	9	20	35
75:	1	5	16	20	42
76:	1	6	8	12	27
77:	1	6	8	27	42
78:	1	6	12	18	37
79:	1	6	14	21	42
80:	1	8	8	9	26
81:	1	8	8	16	33
82:	1	8	8	25	42
83:	1	8	9	18	36
84:	1	8	11	22	42
85:	1	9	9	9	28
86:	1	9	9	16	35
87:	1	9	9	25	44
88:	1	9	10	10	30
89:	1	9	13	13	36
90:	1	9	16	16	42
91:	1	9	17	17	44
92:	1	10	10	16	37
93:	2	2	2	2	8
94:	2	2	2	8	14
95:	2	2	2	18	24
96:	2	2	2	32	38
97:	2	2	4	4	12
98:	2	2	4	9	17
99:	2	2	4	16	24
100:	2	2	4	25	33
101:	2	2	4	36	44
102:	2	2	5	5	14
103:	2	2	5	20	29
104:	2	2	8	8	20
105:	2	2	8	18	30
106:	2	2	8	32	44
107:	2	2	9	9	22
108:	2	2	9	16	29
109:	2	2	9	25	38
110:	2	2	10	10	24
111:	2	2	13	13	30
112:	2	2	16	16	36
113:	2	2	17	17	38
114:	2	2	18	18	40
115:	2	2	20	20	44
116:	2	3	4	6	15
117:	2	3	4	24	33
118:	2	3	6	9	20
119:	2	3	6	16	27
120:	2	3	6	25	36

121:	(2, 3, 9, 24):	38
122:	(2, 3, 10, 15):	30
123:	(2, 4, 4, 8):	18
124:	(2, 4, 4, 18):	28
125:	(2, 4, 4, 32):	42
126:	(2, 4, 6, 12):	24
127:	(2, 4, 6, 27):	39
128:	(2, 4, 8, 9):	23
129:	(2, 4, 8, 16):	30
130:	(2, 4, 8, 25):	39
131:	(2, 4, 9, 18):	33
132:	(2, 4, 11, 22):	39
133:	(2, 4, 12, 24):	42
134:	(2, 4, 16, 18):	40
135:	(2, 5, 5, 8):	20
136:	(2, 5, 5, 18):	30
137:	(2, 5, 5, 32):	44
138:	(2, 5, 8, 20):	35
139:	(2, 6, 7, 21):	36
140:	(2, 6, 9, 12):	29
141:	(2, 6, 9, 27):	44
142:	(2, 6, 12, 16):	36
143:	(2, 8, 8, 8):	26
144:	(2, 8, 8, 18):	36
145:	(2, 8, 9, 9):	28
146:	(2, 8, 9, 16):	35
147:	(2, 8, 9, 25):	44
148:	(2, 8, 10, 10):	30
149:	(2, 8, 13, 13):	36
150:	(2, 8, 16, 16):	42
151:	(2, 8, 17, 17):	44
152:	(2, 9, 9, 18):	38
153:	(2, 9, 11, 22):	44
154:	(2, 10, 10, 18):	40
155:	(2, 10, 12, 15):	39
156:	(3, 3, 3, 3):	12
157:	(3, 3, 3, 12):	21
158:	(3, 3, 3, 27):	36
159:	(3, 3, 6, 6):	18
160:	(3, 3, 6, 24):	36
161:	(3, 3, 12, 12):	30
162:	(3, 3, 15, 15):	36
163:	(3, 4, 6, 8):	21
164:	(3, 4, 6, 18):	31
165:	(3, 4, 8, 24):	39
166:	(3, 6, 6, 12):	27
167:	(3, 6, 6, 27):	42
168:	(3, 6, 8, 9):	26
169:	(3, 6, 8, 16):	33
170:	(3, 6, 8, 25):	42
171:	(3, 6, 9, 18):	36
172:	(3, 6, 11, 22):	42
173:	(3, 8, 9, 24):	44
174:	(3, 8, 10, 15):	36
175:	(3, 12, 12, 12):	39
176:	(4, 4, 4, 4):	16
177:	(4, 4, 4, 9):	21
178:	(4, 4, 4, 16):	28
179:	(4, 4, 4, 25):	37
180:	(4, 4, 5, 5):	18
181:	(4, 4, 5, 20):	33
182:	(4, 4, 8, 8):	24
183:	(4, 4, 8, 18):	34
184:	(4, 4, 9, 9):	26
185:	(4, 4, 9, 16):	33
186:	(4, 4, 9, 25):	42
187:	(4, 4, 10, 10):	28
188:	(4, 4, 13, 13):	34
189:	(4, 4, 16, 16):	40
190:	(4, 4, 17, 17):	42
191:	(4, 4, 18, 18):	44
192:	(4, 5, 5, 9):	23
193:	(4, 5, 5, 16):	30
194:	(4, 5, 5, 25):	39
195:	(4, 5, 9, 20):	38
196:	(4, 6, 8, 12):	30
197:	(4, 6, 12, 18):	40
198:	(4, 8, 8, 9):	29
199:	(4, 8, 8, 16):	36
200:	(4, 8, 9, 18):	39
201:	(4, 9, 9, 9):	31
202:	(4, 9, 9, 16):	38
203:	(4, 9, 10, 10):	33
204:	(4, 9, 13, 13):	39
205:	(4, 10, 10, 16):	40
206:	(5, 5, 5, 5):	20
207:	(5, 5, 5, 20):	35
208:	(5, 5, 8, 8):	26
209:	(5, 5, 8, 18):	36
210:	(5, 5, 9, 9):	28
211:	(5, 5, 9, 16):	35
212:	(5, 5, 9, 25):	44
213:	(5, 5, 10, 10):	30
214:	(5, 5, 13, 13):	36
215:	(5, 5, 16, 16):	42
216:	(5, 5, 17, 17):	44
217:	(5, 8, 8, 20):	41
218:	(6, 6, 6, 6):	24
219:	(6, 6, 6, 24):	42
220:	(6, 6, 12, 12):	36
221:	(6, 6, 15, 15):	42
222:	(6, 7, 8, 21):	42
223:	(6, 8, 9, 12):	35
224:	(6, 8, 12, 16):	42
225:	(7, 7, 7, 7):	28
226:	(7, 7, 14, 14):	42
227:	(8, 8, 8, 8):	32
228:	(8, 8, 8, 18):	42
229:	(8, 8, 9, 9):	34
230:	(8, 8, 9, 16):	41
231:	(8, 8, 10, 10):	36
232:	(8, 8, 13, 13):	42
233:	(8, 9, 9, 18):	44
234:	(9, 9, 9, 9):	36
235:	(9, 9, 10, 10):	38
236:	(9, 9, 13, 13):	44
237:	(10, 10, 10, 10):	40
238:	(11, 11, 11, 11):	44

APPENDIX K: THE SMALLEST t SUCH THAT AN HADAMARD MATRIX OF ORDER $2^t q$ EXISTS FOR q(ODD) $< 10,000$

In the pages that follow, if an odd number q does not appear in the list, it means that there is an Hadamard matrix of order $4q$. When q appears in the list, the number t next to q indicates there is an Hadamard matrix of order $2^t q$, and, as yet, none known for order $2^s q$, $s < t$.

67 5

103 3
107 10
127 3
151 3
163 3
167 4
179 8
191 13

213 3
219 3
223 3
233 4
239 4
249 3
251 6
267 3
269 8
283 3

303 4
311 14
335 6
347 15
359 4
373 7

419 4
443 6
445 3
463 3
479 12
487 5
491 15

509 8
515 4
519 3
523 3
537 5
571 3
573 3
579 3
581 3
599 8

613 3
623 3
631 13
643 3
647 10
653 10
655 3
657 3
659 17
669 3
699 3

719 4
721 4
723 3
733 18
737 6
739 16
749 11
751 3
757 5
763 3
773 8
781 3
787 5
789 3

803 3
809 12
823 3
839 8
853 3
857 4
859 3
863 4
865 3
879 3
883 3

907 5
913 3
917 3
919 3
933 10
941 6
947 6
949 3
953 14
955 3
959 3
965 3
971 6
979 3
985 3
991 3

			1301 14						
	1103 17		1303 3				1703 3		1903 3
		1205 3							
					1507 3				1907 10
	1109 19				1509 3				1909 3
		1211 3		1411 3					
		1213 16			1513 3		1713 4		1913 12
	1115 4		1315 3						1915 3
1019 8			1319 12			1619 4	1719 3		
									1921 3
	1123 7	1223 8		1423 3			1723 3	1823 4	
		1227 3	1327 5	1427 15	1527 9				
									1929 4
1031 6		1231 3				1631 5		1831 15	
	1133 4			1433 6		1633 3	1733 14		
			1335 3						
				1437 6					1937 3
1039 3				1439 12		1639 3			
		1241 3	1341 3	1441 3				1841 3	
1043 3		1243 3			1543 3				1943 6
							1745 3		
1047 3				1447 19			1747 5	1847 8	
			1349 3						1949 4
1051 3			1351 3	1451 6			1751 4		1951 3
							1753 11		
		1255 3							
	1157 3	1257 5			1557 3				1957 4
1059 3		1259 4	1359 3		1559 4				
						1661 4			
1063 3						1663 11			1963 4
	1165 3			1465 3					
			1367 4		1567 19	1667 19			
	1169 5			1469 3				1869 3	1969 9
				1471 13	1571 10	1671 3		1871 20	
		1273 6	1373 19	1473 3					1973 4
						1675 3			
	1177 5				1577 3				
1079 3			1379 3		1579 5	1679 3		1879 3	1979 4
			1381 3				1781 3		1981 4
		1283 12		1483 3	1583 8		1783 7	1883 6	
			1385 3						1985 3
1087 18	1187 6		1387 3	1487 12			1787 4		1987 16
					1589 3	1689 3		1889 4	
		1291 3		1491 3					
	1193 4			1493 4		1693 7	1793 4	1893 6	1993 15
							1795 5		
			1397 4						
	1199 3	1299 3		1499 16		1699 3			1999 3

		2103	3	2203	3					2503	3	2603	3			2803	3	2903	4
						2305	3												
				2207	4					2507	3								
										2509	3								
												2611	3	2711	10				
				2213	4			2413	4	2513	5							2913	7
				2215	5	2315	4			2515	3					2815	3		
								2417	4									2917	22
2019	3	2119	4	2219	3											2819	8		
												2621	10					2921	4
		2123	3					2423	4							2823	7		
				2225	3														
2027	8			2227	3	2327	6	2427	6	2527	3	2627	3					2927	4
								2429	16	2529	3	2629	3	2729	16				
										2531	6			2731	3	2831	3		
2033	8			2233	3	2333	20			2533	3			2733	3				
						2335	3												
																2837	10	2937	3
2039	20			2239	3	2339	4									2839	5	2939	8
								2441	20							2841	3		
		2143	3					2443	3	2543	6	2643	3			2843	8		
										2545	3								
		2147	3			2347	20					2647	18						
				2249	3					2549	8					2849	3		
2051	3			2251	5							2651	3					2951	3
2053	7	2153	6															2953	3
		2155	3													2855	4		
						2357	6												
2059	3	2159	4					2459	4	2559	4	2659	3	2759	3			2959	3
								2461	3			2661	3						
2063	8			2263	3					2563	5					2863	3	2963	4
		2165	3													2865	4	2965	3
		2167	3											2767	9				
						2369	4							2769	3				
2071	3	2171	5	2271	5	2371	9			2571	5	2671	9					2971	3
										2573	3			2773	4			2973	4
										2575	3								
												2677	9	2777	8				
										2579	4					2879	21		
						2381	22			2581	3					2881	6		
2083	3			2283	3	2383	3					2683	7						
				2285	3														
2087	4			2287	20							2687	14			2887	5	2987	4
								2489	3	2589	5			2789	4				
						2391	3	2491	3	2591	20								
				2293	22	2393	4	2493	3							2893	3		
2095	3																	2995	9
														2797	21				
2099	8					2399	8					2699	21	2799	3	2899	4	2999	20

	3101 6			3401 9		3601 3			3901 4
	3103 7	3203 20		3403 3				3803 6	3903 3
					3505 3			3805 3	
	3107 4		3307 20						3907 21
		3209 4	3309 8	3409 3				3809 3	
					3511 3				
3013 3				3413 8		3613 23			
		3215 4					3715 3		
3017 3	3117 4				3517 5	3617 20		3817 16	
	3119 21			3419 3			3719 19		3919 3
								3821 22	
3023 10		3223 3				3623 6		3823 3	2923 6
	3127 4			3427 3			3727 9		
3029 4						3629 8			3929 12
3031 3						3631 5		3831 3	3931 3
	3133 3				3533 4	3633 3	3733 3		
				3437 10		3637 17			
3039 3					3539 12	3639 5		3839 3	
	3141 3								3941 3
3043 9	3143 3		3343 22		3543 3	3643 7	3743 3		3943 3
					3545 3			3845 3	3945 3
3047 3	3147 3	3247 5	3347 4		3547 9		3747 3	3847 21	
	3149 4		3349 3				3749 4	3849 4	3949 5
				3451 3	3551 7	3651 3		3851 22	
				3453 3		3653 3			3953 4
							3755 4		
	3157 3								3957 4
		3259 3	3359 16		3559 3	3659 4			3959 8
	3161 3			3461 21					
	3163 3	3263 7		3463 3			3763 3		3963 3
3065 4			3365 3			3665 12		3865 7	
	3167 4			3467 21			3767 4		
		3269 3			3569 3	3669 7		3869 4	
3071 3		3271 19	3371 19		3571 7				
	3173 4	3273 3		3473 4				3873 3	3973 3
	3175 3								
						3677 4			
3079 3		3279 3	3379 3	3479 3	3579 3	3679 4	3779 21		3979 3
								3881 10	
3083 6	3183 3				3583 3			3883 3	3983 3
3085 3									3985 3
			3387 3	3487 3					
	3189 4								
3091 3						3691 3	3791 4		
3093 7	3193 4				3593 4				
		3295 9		3495 4	3595 3				
3097 4				3497 8					
		3299 4					3799 3	3899 3	

	4101 5					4601 10			
4003 3				4403 3		4603 23	4703 16		4903 23
				4405 3					
4007 22			4307 4						4907 3
	4109 3				4509 3	4609 3	4709 3		
4011 3			4311 3					4811 4	4911 3
4013 6		4213 3	4313 3	4413 4			4713 4	4813 24	
	4115 4								
		4217 6	4317 5	4417 9	4517 4				
4019 4		4219 3	4319 3		4519 3	4619 3			4919 12
	4121 3						4721 6	4821 3	
		4223 4		4423 3	4523 4		4723 3		
	4127 6	4227 3	4327 13		4527 3		4727 4		
				4429 4	4529 8				
			4331 3		4531 3				4931 18
4033 3	4133 10			4433 3		4633 3	4733 8		
				4435 3	4535 4		4735 3		
		4237 4			4537 3	4637 6			
	4139 21			4439 3	4539 3				4939 3
					4541 5		4741 3	4841 5	
4043 8	4143 4	4243 3		4443 3		4643 4		4843 5	4943 4
						4645 3			
					4547 6		4747 4		
4049 4			4349 12	4449 3		4649 4			
				4451 10	4551 3	4651 3	4751 6		
4053 3	4153 3		4353 6	4453 3					
								4855 3	
	4157 6						4757 7		
		4259 8	4359 3			4659 3	4759 5	4859 3	
4061 3				4461 13					
4063 3			4363 15			4663 7		4863 3	4963 3
		4265 4					4765 3		
		4267 7				4667 5			
				4469 3			4769 4	4869 3	
		4271 21			4571 11				
4073 4					4573 3			4873 7	
4075 3							4775 4		
4077 3						4677 4	4777 3		4977 3
4079 22		4279 3	4379 4	4479 3		4679 16		4879 3	4979 3
	4181 3					4681 3	4781 3		4981 3
4083 3	4183 4	4283 10	4383 3	4483 3	4583 4		4783 7		4983 3
		4285 5			4585 4				
4087 5			4387 9			4687 3	4787 6		
					4589 3			4889 8	4989 4
			4391 6			4691 10			
	4193 9		4393 11			4693 3	4793 10		
		4295 4						4895 3	
		4297 17						4897 4	4997 3
4099 9	4199 3		4399 4	4499 3			4799 8	4899 3	4999 22

5001 5				5401 3			5701 5		
					5503 3	5603 3			5903 4
	5105 3								5905 3
5007 3	5107 23	5207 4		5407 5					
	5109 3		5309 16		5509 3	5609 3	5709 3		5909 4
5011 3		5211 3	5311 3	5411 6			5711 23		
				5413 7		5613 6	5713 3		5913 3
			5315 4		5515 3	5615 4		5815 3	
5017 3			5317 3			5617 3			
	5119 9				5519 8	5619 3		5819 3	5919 3
5021 24		5221 3					5721 3		5921 14
5023 3	5123 4		5323 20			5623 3			5923 24
								5825 4	
			5327 3		5527 22			5827 18	5927 8
5029 3	5129 4			5429 3					
	5131 3	5231 6							5931 3
5033 4			5333 6	5433 3	5533 3				5933 3
									5935 7
				5437 24				5837 3	
5039 23	5139 4		5339 3		5539 7	5639 12	5739 4		5939 12
			5341 3		5541 9				
		5243 11		5443 3			5743 3		
		5245 3			5545 11	5645 3			
	5147 4		5347 5			5647 9	5747 3	5847 5	
			5349 5		5549 9			5849 23	
5051 18			5351 18			5651 6			5951 3
5053 3			5353 4	5453 3				5853 4	
	5155 7						5755 3		
5057 3			5357 4			5657 4			
5059 3		5259 3	5359 5	5459 5			5759 4		5959 4
		5261 6	5361 5	5461 4	5561 7		5761 4	5861 6	5961 17
5063 3	5163 3	5263 3	5363 3		5563 21			5863 3	5963 6
5065 3	5165 3								5965 5
				5467 3	5567 3	5667 5	5767 3	5867 4	
5069 3		5269 7							5969 5
	5171 6	5271 3	5371 3	5471 14		5671 12			
5073 3	5173 7	5273 12					5773 3	5873 4	
5077 22				5477 4			5777 4		
5079 5		5279 4			5579 3	5679 4	5779 11	5879 4	5979 8
						5681 3			5981 10
5083 3		5283 3	5383 3	5483 12		5683 21			5983 3
				5485 3	5585 3		5785 3		
		5287 15	5387 8				5787 4		5987 4
			5389 3				5789 3		5989 3
	5191 9				5591 6		5791 3		
5093 4		5293 6		5493 3		5693 22	5793 3	5893 3	
5095 3	5195 4								
	5197 5				5597 3			5897 4	5997 4
5099 22	5199 8		5399 22					5899 5	5999 5

6001	3													6701	6	6801	3	6901	3
		6103	3	6203	14					6503	3	6603	3	6703	3				
												6605	3			6805	3	6905	4
6007	13	6107	3					6407	3			6607	9	6707	3			6907	21
6009	3	6109	3	6209	3	6309	3	6409	3	6509	4								
										6511	3							6911	10
6013	4	6113	6			6313	3			6513	3	6613	3	6713	3	6813	3	6913	4
		6115	9							6515	4								
		6117	4			6317	16					6617	4	6717	4				
								6419	3			6619	3	6719	8	6819	3		
						6323	6	6423	3	6523	3	6623	4			6823	3		
																		6925	3
																6827	18	6927	3
6029	20							6429	3			6629	7						
6031	3	6131	6			6331	3	6431	4	6531	3	6631	3	6731	5			6931	5
						6333	3	6433	4					6733	7				
										6535	3								
												6637	5						
						6339	3	6439	3	6539	3	6639	5			6839	3		
6041	5					6341	8												
6043	21					6343	23							6743	4			6943	4
6047	6			6247	9													6947	6
								6449	4					6749	3				
				6251	3	6351	3	6451	3			6651	3			6851	3		
6053	4											6653	6			6853	4	6953	3
		6155	4											6755	4				
		6157	4	6257	12			6457	3										
6059	4	6159	4			6359	23					6659	4	6759	3	6859	3	6959	8
				6261	3			6461	3										
		6163	7	6263	4			6463	3					6763	3	6863	4		
6065	4					6365	4	6465	3										
6067	13	6167	3			6367	21					6667	4						
				6269	23	6369	3	6469	24	6569	12								
6071	3					6371	3			6571	3					6871	3		
6073	25					6373	25	6473	4							6873	3		
				6275	4					6575	3								
6077	4	6177	3			6377	3									6877	3	6977	10
		6179	5									6679	5	6779	4				
		6181	4	6281	4					6581	10					6881	3	6981	6
6083	3			6383	3	6383	8			6583	3					6883	23	6983	6
6685	3											6685	3					6985	3
						6387	3							6787	3	6887	3		
6089	4					6389	8											6989	3
6091	3							6491	24					6791	22			6991	3
		6193	3					6493	4			6693	3			6893	3		
								6495	3	6595	5								
		6197	23	6297	9	6397	25	6497	4					6797	4			6997	5
		6199	3	6299	21			6499	3	6599	8					6899	12		

						7301	3											7901	6
7003	4					7303	7					7603	24	7703	8			7903	3
				7205	3	7305	3	7405	3					7705	3				
				7207	21	7307	4												
7009	4	7109	20					7409	4					7709	3				
										7511	3			7711	3	7811	6		
		7113	8					7413	3	7513	3							7913	3
		7115	4					7415	4			7615	3					7915	9
		7117	5					7417	5	7517	6			7717	17				
7019	4					7319	3	7419	3	7519	5			7719	3	7819	3	7919	4
7021	3													7721	18				
7023	3	7123	3	7223	5	7323	4	7423	4					7723	24	7823	6		
						7325	3												
7027	5	7127	4					7427	3			7627	3	7727	4			7927	21
								7429	3	7529	25	7629	3	7729	4	7829	16	7929	3
		7131	3			7331	24					7631	3						
7033	3	7133	8	7233	3			7433	4			7633	3					7933	3
		7135	7	7235	8											7835	10		
7037	3											7637	3			7837	3		
7039	3					7339	3	7439	3					7739	4				
				7241	3			7441	4					7741	3				
7043	4			7243	3	7343	3	7443	3	7543	3	7643	24			7843	3		
7045	3							7445	3	7545	3			7745	3				
				7247	6							7647	5						
7049	3	7149	3	7249	3											7849	3	7949	25
				7251	5	7351	3					7651	3						
		7153	3	7253	4									7753	23	7853	6	7953	3
				7255	7	7355	4			7555	3					7855	5		
		7159	3					7459	3	7559	4	7659	3	7759	3				
				7261	4											7861	8	7961	5
		7163	3									7663	3			7863	3	7963	3
		7165	7											7765	3				
7067	14	7167	3											7767	5				
		6169	8					7469	3					7769	3				
								7471	3	7571	6					7871	4		
7073	4	7173	3	7273	4	7373	4			7573	7	7673	25						
		7177	9	7277	3			7477	5							7877	4		
7079	12	7179	5							7579	3	7679	3	7779	3	7879	3		
7081	3																	7981	16
		7183	7							7583	4							7983	3
																		7985	3
		7187	12			7387	4												
				7289	3	7389	3							7789	3			7989	3
7091	3			7291	3	7391	3			7591	3								
7093	3							7493	5	7593	4			7793	10				
								7495	9					7795	5	7895	6		
		7197	4									7697	8			7897	4		
								7499	4	7599	3	7699	7	7799	3				

					8501 24				
8003 4		8203 14		8403 3	8503 3	8603 3		8803 3	
	8105 3		8305 3						
	8107 6	8207 4					8707 5		8907 3
			8309 7	8409 3	8509 3	8609 4		8809 3	8909 4
8011 3			3811 13	8411 6		8611 3			
		8213 12		8413 10			8713 15		8913 4
			8315 4			8615 4			8915 8
				8417 7		8617 4			8917 3
	8119 3	8219 23		8419 19	8519 3		8719 5	8819 20	
8023 4		8223 3		8423 12	8523 4	8623 19	8723 3		8923 25
8027 3			8327 6			8627 6	8727 3		8927 3
	8129 16			8429 8	8529 4				
	8131 4	8231 14		8431 3			8731 3		
8033 3						8633 3		8833 3	
			8335 5						8935 3
		8237 25				8637 4		8837 6	
8039 24	8139 3		8339 9	8439 3		8639 5		8839 3	8939 3
	8143 13			8443 24			8743 3	8843 5	
			8345 3		8545 3				8945 3
				8447 4			8747 6		8947 3
	8149 3	8249 4	8349 3		8549 5				
		8251 4			8551 3		8751 5	8851 6	
8053 11						8653 7			
	8155 3			8455 3					
8057 3		8257 5				8657 4			
8059 19		8259 3	8359 4	8459 3			8759 3	8859 4	
8061 9		8261 4			8561 9		8761 5	8861 10	
		8263 3	8363 16		8563 15			8863 3	
				8465 6		8665 5			
8067 3		8267 3	8367 5	8467 9			8767 3	8867 4	
8069 8			8369 8			8669 16	8769 3		
	8171 6			8471 3				8871 3	8971 3
	8173 3				8573 6		8773 3		
			8375 4						
8077 3			8377 5	8477 3					8977 25
8079 3	8179 9	8279 6			8579 8	8679 3	8779 5	8879 3	
							8781 5		
8083 3	8183 4		8383 4		8583 3	8683 3	8783 4		
				8485 3				8885 3	
8087 24	8187 5				8587 3			8887 19	
	8189 3			8489 8	8589 3			8889 4	
		8291 21			8591 3			8891 4	
			8393 4	8493 3		8693 6		8893 3	8993 3
				8495 4					
		8297 4					8797 4		8997 10
8099 3	8199 3	8299 3	8399 3	8499 3	8599 3	8699 8		8899 3	8999 25

		9101	13			9301	4	9401	3										
		9103	3	9203	10			9403	3							9803	6	9903	3
9005	3			9205	4														
		9107	8			9307	3	9407	3					9707	4				
												9609	3						
9011	6			9211	4					9511	7	9611	6			9811	3		
				9213	4	9313	6	9413	4			9613	26					9913	3
										9515	3					9815	3		
												9617	4			9817	25		
9019	7							9419	20					9719	20	9819	3		
		9121	4	9221	23														
9023	3	9123	3			9323	6			9523	12	9623	12					9923	24
						9325	3											9925	3
				9227	8			9427	5	9527	3	9627	5	9727	4	9827	3		
						9329	8			9529	3							9929	4
								9431	24			9631	3						
		9133	3							9533	8			9733	3	9833	4		
				9235	9													9935	4
9037	4													9737	3	9837	3	9937	4
				9239	21			9439	3	9539	4					9839	8	9939	3
								9441	3			9641	15	9741	4				
9043	3					9343	3	9443	3			9643	15					9943	4
								9445	5										
						9347	4			9547	9							9947	3
		9149	3	9249	7			9449	4	9549	3			9749	16				
9051	3									9551	6					9851	18		
9053	4			9253	6					9553	3			9753	3	9853	3	9953	8
9055	3	9155	6			9355	5												
				9257	8					9557	3			9757	3				
9059	16	9159	3			9359	3			9559	3	9659	3	9759	3			9959	3
						9361	3												
								9463	3	9563	4			9763	4			9961	3
				9265	3	9365	3	9465	3	9565	3	9665	3			9865	3	9965	16
		9167	5	9267	3	9367	3	9467	4	9567	4			9767	6			9967	9
9069	3	9169	3	9269	3			9469	3	9569	5	9669	3			9869	3		
9071	4											9671	9			9871	7		
				9273	3					9573	3	9673	3					9973	7
																		9977	4
		9179	7	9279	3			9479	20			9679	3			9879	3	9979	3
		9181	5													9881	3		
9083	3			9283	3	9383	4			9583	3					9883	3	9983	7
																9885	3	9985	3
9087	4	9187	5	9287	6					9587	14			9787	13	9887	10		
9089	3	9189	3	9289	3			9489	4	9589	4	9689	8					9989	12
						9391	5											9991	4
9093	3	9193	3					9493	5							9893	3		
																9895	3		
9097	3																		

APPENDIX L: AMICABLE ORTHOGONAL DESIGNS IN ORDER 12

Roger Magoon has also written a computer programme to generate all types of amicable orthogonal designs in order 4n, n odd, not prohibited from existing by known theorems, where each design involves three variables (the maximum number).

We give the printout only in order 12. The number appearing after the colon is the sum of the type numbers in the parentheses. The fourteen full designs are the most important to find with respect to the Hadamard matrix problem.

(1, 1, 1);(1, 1, 1): 6
(1, 1, 1);(1, 1, 4): 9
(1, 1, 1);(1, 1, 9): 14
(1, 1, 1);(1, 2, 2): 8
(1, 1, 1);(1, 2, 8): 14
(1, 1, 1);(1, 4, 4): 12
(1, 1, 1);(1, 5, 5): 14
(1, 1, 1);(2, 2, 4): 11
(1, 1, 1);(2, 3, 6): 14
(1, 1, 1);(4, 4, 4): 15
(1, 1, 2);(1, 1, 2): 8
(1, 1, 2);(1, 1, 8): 14
(1, 1, 2);(1, 2, 4): 11
(1, 1, 2);(1, 2, 9): 16
(1, 1, 2);(1, 3, 6): 14
(1, 1, 2);(2, 2, 2): 10
(1, 1, 2);(2, 2, 8): 16
(1, 1, 2);(2, 4, 4): 14
(1, 1, 2);(2, 5, 5): 16
(1, 1, 4);(1, 1, 4): 12
(1, 1, 4);(1, 1, 9): 17
(1, 1, 4);(1, 2, 2): 11
(1, 1, 4);(1, 2, 8): 17
(1, 1, 4);(1, 4, 4): 15
(1, 1, 4);(1, 5, 5): 17
(1, 1, 4);(2, 2, 4): 14
(1, 1, 4);(2, 3, 6): 17
(1, 1, 4);(4, 4, 4): 18
(1, 1, 5);(1, 1, 5): 14
(1, 1, 5);(1, 4, 5): 17
(1, 1, 5);(2, 2, 5): 16
(1, 1, 8);(1, 1, 8): 20
(1, 1, 8);(1, 2, 4): 17
(1, 1, 8);(1, 2, 9): 22
(1, 1, 8);(1, 3, 6): 20
(1, 1, 8);(2, 2, 2): 16
(1, 1, 8);(2, 2, 8): 22
(1, 1, 8);(2, 4, 4): 20
(1, 1, 8);(2, 5, 5): 22
(1, 1, 9);(1, 1, 9): 22
(1, 1, 9);(1, 2, 2): 16
(1, 1, 9);(1, 2, 8): 22
(1, 1, 9);(1, 4, 4): 20
(1, 1, 9);(1, 5, 5): 22
(1, 1, 9);(2, 2, 4): 19
(1, 1, 9);(2, 3, 6): 22
(1, 1, 9);(4, 4, 4): 23
(1, 1, 10);(1, 1, 10): 24
(1, 2, 2);(1, 2, 2): 10
(1, 2, 2);(1, 2, 8): 16
(1, 2, 2);(1, 4, 4): 14
(1, 2, 2);(1, 5, 5): 16
(1, 2, 2);(2, 2, 4): 13
(1, 2, 2);(2, 3, 6): 16
(1, 2, 2);(4, 4, 4): 17
(1, 2, 3);(1, 2, 3): 12
(1, 2, 3);(1, 3, 8): 18
(1, 2, 3);(2, 3, 4): 15
(1, 2, 3);(3, 3, 6): 18
(1, 2, 4);(1, 2, 4): 14
(1, 2, 4);(1, 2, 9): 19
(1, 2, 4);(1, 3, 6): 17
(1, 2, 4);(2, 2, 2): 13
(1, 2, 4);(2, 2, 8): 19
(1, 2, 4);(2, 4, 4): 17
(1, 2, 4);(2, 5, 5): 19
(1, 2, 6);(1, 2, 6): 18
(1, 2, 6);(2, 4, 6): 21
(1, 2, 6);(3, 3, 3): 18
(1, 2, 8);(1, 2, 8): 22
(1, 2, 8);(1, 4, 4): 20
(1, 2, 8);(1, 5, 5): 22
(1, 2, 8);(2, 2, 4): 19
(1, 2, 8);(2, 3, 6): 22
(1, 2, 8);(4, 4, 4): 23
(1, 2, 9);(1, 2, 9): 24
(1, 2, 9);(1, 3, 6): 22
(1, 2, 9);(2, 2, 2): 18
(1, 2, 9);(2, 2, 8): 24
(1, 2, 9);(2, 4, 4): 22
(1, 2, 9);(2, 5, 5): 24

(1, 3, 6);(1, 3, 6): 20
(1, 3, 6);(2, 2, 2): 16
(1, 3, 6);(2, 2, 8): 22
(1, 3, 6);(2, 4, 4): 20
(1, 3, 6);(2, 5, 5): 22
(1, 3, 8);(1, 3, 8): 24
(1, 3, 8);(2, 3, 4): 21
(1, 3, 8);(3, 3, 6): 24
(1, 4, 4);(1, 4, 4): 18
(1, 4, 4);(1, 5, 5): 20
(1, 4, 4);(2, 2, 4): 17
(1, 4, 4);(2, 3, 6): 20
(1, 4, 4);(4, 4, 4): 21
(1, 4, 5);(1, 4, 5): 20
(1, 4, 5);(2, 2, 5): 19
(1, 5, 5);(1, 5, 5): 22
(1, 5, 5);(2, 2, 4): 19
(1, 5, 5);(2, 3, 6): 22
(1, 5, 5);(4, 4, 4): 23
(1, 5, 6);(1, 5, 6): 24
(2, 2, 2);(2, 2, 2): 12
(2, 2, 2);(2, 2, 8): 18
(2, 2, 2);(2, 4, 4): 16
(2, 2, 2);(2, 5, 5): 18
(2, 2, 4);(2, 2, 4): 16
(2, 2, 4);(2, 3, 6): 19
(2, 2, 4);(4, 4, 4): 20
(2, 2, 5);(2, 2, 5): 18
(2, 2, 8);(2, 2, 8): 24
(2, 2, 8);(2, 4, 4): 22
(2, 2, 8);(2, 5, 5): 24
(2, 3, 4);(2, 3, 4): 18
(2, 3, 4);(3, 3, 6): 21
(2, 3, 6);(2, 3, 6): 22
(2, 3, 6);(4, 4, 4): 23
(2, 3, 7);(2, 3, 7): 24
(2, 4, 4);(2, 4, 4): 20
(2, 4, 4);(2, 5, 5): 22
(2, 4, 6);(2, 4, 6): 24
(2, 4, 6);(3, 3, 3): 21
(2, 5, 5);(2, 5, 5): 24
(3, 3, 3);(3, 3, 3): 18
(3, 3, 6);(3, 3, 6): 24
(4, 4, 4);(4, 4, 4): 24

PROBLEMS

The following list summarizes what, we feel, are the outstanding problems raised in the text. The order in which these problems appear is not related to their difficulty.

The second list of problems is of a different type; although some may be very difficult. We view them as part of the "spade-work" for the first set of problems.

1) Proposition 2.5 gives non-existence for $W(n, k)$, n odd, when $(n-k)^2 - (n-k) + 1 \geq n$. Also $W(n, k)$, n odd, exists when $n = (n-k)^2 - (n-k) + 1$ only if a finite projective plane of order $n - k - 1$ exists.

A). Is it true, in general, that if a projective plane of order m exists then a $W(m^2+m+1, m^2)$ also exists?
(This is known to be the case for every m for which it is currently known that a projective plane exists.)

B). When $(n-k)^2 - (n-k) + 1 > n$, Example 2.6 shows that a $W(n, k)$, n odd, may still not exist. This is probably not the only such example (or is it?). Fit this example into a more general context.

C). Are weighing matrices of odd order related to other incidence structures when $(n-k)^2 - (n-k) + 1 > n$?
(This may be relevant to B) and D).)

D). Find constructions for infinite families of $W(n, k)$, n odd.
(See 4.123 and 4.124.)

E). If n is odd and there is a $W(n, k)$, is there a symmetric $W(n, k)$?

F). Find any combinatorial reasons why there may not be amicable orthogonal designs $((k^2);(\ell^2))$ in order n, n odd, $k, \ell \in \mathbb{Z}$.

G). Construct some infinite families of amicable orthogonal designs in order n, n odd.

2) When $n = 2t$, t odd, there are few combinatorial theorems which prohibit the existence of a $W(n, k)$ or an orthogonal design of type (a, b) in order n.

A). Let n, k be as above and suppose $k < n$ and k is the sum of two squares. Does this imply the existence of a $W(n, k)$?

B). Try to fit the fact that there is no skew-symmetric $W(18, 16)$ into some general framework. (See Geramita-Verner [1].) E.g. Is it true that when $n = 2t$, t odd, and $n - 2 = s^2$, there is no $W(n, n-2)$?

C). Find some general combinatorial theorem which prohibits the existence of orthogonal designs of type (a, b) in order $n \equiv 2 \pmod 4$, when $a + b < n$.

D). We know that when $n \equiv 2 \pmod 4$ and a $W(n, n-1)$ exists, then there is a symmetric $W(n, n-1)$. Is this also true for values of $k \neq n - 1$?

E). Study symmetric orthogonal designs in orders $n \equiv 2 \pmod 4$. Fit Wolfe's proof of no symmetric orthogonal design of type $(2, 2)$, in such an order into a more general framework.

F). Find infinite families of amicable orthogonal designs in orders $n \equiv 2 \pmod 4$, as in 5.52.

G). In 7.15 we found that if a, b are integers and $\langle a, b\rangle = \langle 1, 1\rangle$ (over $\mathbb{Q}$), then there are orthogonal designs of type (a, b) in every order $2n$ where $n > N(a, b)$.

Find an estimate for $N(a, b)$ of the form: There is some fixed function, $f(x, y)$ and

$$N(a, b) = f(a, b).$$

(Does $f(x, y) = \frac{1}{2}(x+y) + c$ (c to be determined) work?)

3) Define an integer family of order n and type $(s_1, \ldots, s_\ell)$ $(s_i \in \mathbb{Z}^+)$, to be a collection of ℓ integer matrices of order n, $A_1, \ldots, A_\ell$, satisfying:

1). $A_i A_i^t = s_i I_n \qquad 1 \le i \le \ell$

2). $A_i A_j^t + A_j A_i^t = 0 \qquad 1 \le i \neq j \le \ell.$

Clearly $\ell \le \rho(n)$ and $\ell = \rho(n)$ is possible. (See 1.6.)

A). Is there some reduction theorem for integer families as in 3.53?

B). When $n \equiv 1, 2, 3 \pmod 4$ it is easy to see that an integer family exists if and only if a rational family of the same order and type exists. Is this true for every n?

C). More simply, let $\langle s_1, s_2, s_3, s_4 \rangle = \langle 1, 1, 1, 1 \rangle$, where $s_i \in \mathbb{Z}$. Is there an integer family of type (s_1, s_2, s_3, s_4) in order 4? Furthermore, if $\langle s_1, s_2 \rangle < \langle 1, 1, 1, 1 \rangle$ (or $\langle s_1, s_2, s_3 \rangle < \langle 1, 1, 1, 1 \rangle$) is there an integer family of type (s_1, s_2) (respectively, (s_1, s_2, s_3)) in order 4?

D). Is there an interesting theory for integer families? Can one find a non-existence theorem for integer families which will have as a consequence a non-existence theorem for orthogonal designs?

4) Study Hadamard equivalence of orthogonal designs, or weighing matrices, to see if there are any interesting groups that arise. (See Hall's work on this topic for Hadamard matrices [2].)

5) Prove: If $W(n, k)$ exists for all $1 \leq k \leq n$ then all $W(2n, j)$ exist for $1 \leq j \leq 2n$.

6) Find new Baumert-Hall arrays in order $4t$, t odd. (See 4.81 and 4.114.)

7) Find more Baumert-Hall-Welch arrays. (See remark after 4.78.)

8) Find more Plotkin arrays of order $8t$ (t odd). These are only known for $t = 1, 3$! Even one for $t = 5$ would be interesting. (See 4.85.)

9) Decide if orthogonal designs of type $(1, 4, 4, 9)$, $(1, 3, 6, 8)$, $(2, 2, 5, 5)$ or $(3, 7, 8)$ exist in order 20. (See 4.67.)

10) Generalized Goethals-Seidel arrays were defined in 4.164.

A). Is there a full theory for these objects? It should, at least, give the exact number of variables one can have for a GGS array in order n.

B). Find more infinite families of these objects.

C). Find new uses for GGS arrays.

11) If there is a $W(n, k)$, is there a symmetric $W(n, k)$? (See 5.11 and what follows.)

12) Develop a theory for symmetric orthogonal designs. (Some facts about these come from amicable rational families of type $((1);\ (s_1, \ldots, s_\ell))$. Thus $\rho_1(n)$ in Definition 5.22 gives the limit on the number of variables in a symmetric orthogonal design of order n.)

13) By analogy with Problem 3, define and study integer amicable families.

14) Find more Golay complementary sequences (see 4.89) or find orders n for which no Golay complementary sequences of length n exist. (See Table 6.1 and remarks preceding it.) Griffin's work (see 4.94) indicates that there is considerable scope for the imaginative use of algebraic number theory.

15) Most combinatorial non-existence theorems for orthogonal designs and amicable orthogonal designs require that the type numbers include one of more 1's. Find other kinds of non-existence theorems for orthogonal designs and amicable orthogonal designs (see §5.5).

16) We have mentioned the non-existence of amicable orthogonal designs of type $((1);\ (2, 2, 2, 2))$, $((1, 2, 2, 3);\ (4, 4))$ and $((1, 7);\ (5))$ in order 8 (see 5.65) and type $((1, 15),\ (1))$ in order 16. Obtain these results as special cases of a theorem (or theorems) valid in all orders $n \equiv 0 \pmod 8$.

17) Prove that there are no amicable orthogonal designs in order $n \equiv 0 \pmod 8$ of types $((1);\ (1, a, n-a-1))$, $a = 1, 2, \ldots, \frac{1}{2}n$. (See 5.45-5.47, 5.140, 5.141, 5.145).

18) Find all amicable orthogonal designs of types $((1, i);\ (j))$ in the powers of 2. (See 5.136 and 5.143; also 5.145 for the status of this problem in orders 8 and 16.)

19) It is known (see §5.11) that there are amicable orthogonal designs, in order n, of types $((i);\ (j))$ for all $i, j \leq n$, when $n = 2^t$. What about other values of n? Is this same fact true for every $n \equiv 0 \pmod 4$?

20) Find infinite families of amicable orthogonal designs as in Theorem 5.52 and what follows.

21) Find new infinite families of amicable Hadamard matrices (see §5.7).

22) There are obvious restrictions on the existence of product designs of type $(a_1, \ldots, a_r; b_1, \ldots, b_s; c_1, \ldots, c_t)$ in order n. Namely;

a) there must be orthogonal designs, in order n, of types

$(a_1, \ldots, a_r, c_1, \ldots, c_t)$ and $(b_1, \ldots, b_s, c_1, \ldots, c_t)$

b) there must be amicable orthogonal designs, in order n, of type $((a_1, \ldots, a_r); (b_1, \ldots, b_s))$.

A). Do a) and b) give the only restrictions on $r, s,$ and t?

B). What new combinatorial facts, apart from those which prohibit designs as in a) and b) from existing, limit the existence of product designs?

23) In Chapter 5, we studied pairs of orthogonal designs X and Y where $XY^t = YX^t$. Wolfe [1] and [4] studied families $\{X_1, \ldots, X_k\}$ of orthogonal designs which are pairwise amicable and discovered limitations on the number of variables in each design. There has not yet been a systematic algebraic study of the possible types for each orthogonal design when $k > 2$. This is a most important question, especially in view of the obvious usefulness of repeat designs. There are also very few examples of three or more amicable orthogonal designs. More examples would be very useful; find some.

24) Let $(R;(P_1, P_2, \ldots, P_k); H)$ be repeat designs in order n of type $(r_1, r_2, \ldots, r_u; (p_{11}, \ldots, p_{1s_1}; p_{21}, p_{22}, \ldots, p_{2s_2}; \ldots; p_{k1}, \ldots, p_{ks_k}); h_1, \ldots, h_v)$ Again, there are obvious restrictions on $u, s_1, \ldots, s_k, v$ and the r_i, $p_{j\ell}$ and h_m coming from the fact that certain orthogonal designs and amicable orthogonal designs must exist in order n. Find new repeat designs and any new combinatorial reasons prohibiting repeat designs from existing.

25) It is clear that product designs, repeat designs and amicable orthogonal designs give strong replication results. It is not clear exactly how strong they are. For example, the existence of amicable orthogonal designs of type $((1,1);(1,1))$, in order 2, was enough to prove if all orthogonal designs of type $(1,k)$ exist in order n, $1 \le k \le n-1$, then all orthogonal designs of type $(1,\ell)$ exist in order $2n$, $1 \le \ell \le 2n-1$.

A). State and prove other theorems of this type.

B). One goal of A) should be to prove:

If all orthogonal designs on t-variables $(t=1,2,3,4)$ exist in order n, then all orthogonal designs on t-variables exist in order $2n$.

C). We saw, in §7.1, if all orthogonal designs of type $(1,i,j)$ exist in order n then all orthogonal designs (k,ℓ) exist in order n. What does one need to know in orders $n, 2n, \ldots, sn$ to guarantee that all orthogonal designs of type $(1,i,j)$ exist in order $(s+1)n$?

D). What does one need to know to prove that all full 4-tuples are the types of orthogonal designs in orders 2^t? (See 5.131.)

E). More specifically than B), try to prove that every 4-tuple (a,b,c,d), $a+b+c+d \le 2^t$, is the type of an orthogonal design in order 2^t. (See 5.106.)

26) (Seberry Conjecture) All 3-tuples $(a,b,2^t s-a-b)$ are the types of orthogonal designs in order $2^t s$, s odd, $t \ge 3$.

27) Are there other theorems like 6.1 for full orthogonal designs? Is there some principle about full orthogonal designs which requires a more "equitable" distribution of the variables? especially if the variables are appearing an odd number of times?

28) A theorem like 6.1 for non-full orthogonal designs would say a great deal about non-existence of amicable orthogonal designs, repeat designs and product designs (even full ones!). Is there such a theorem?

29) If one knew more about full four-variable orthogonal designs in powers of two, how much of an improvement would one get to Theorem 7.13?

30) Prove a theorem like 7.13 for skew-Hadamard and symmetric Hadamard matrices.

31) Let m_1, m_2, m_3, m_4 be any integers for which $\langle m_1, m_2, m_3, m_4 \rangle = \langle 1, 1, 1, 1 \rangle$ over $\mathbb{Q}$. Does there exist an integer $N(m_1, m_2, m_3, m_4)$ such that for $t > n$ there is an orthogonal design of type (m_1, m_2, m_3, m_4) in order $4t$?

32) Improve the asymptotic results in §7.2.

33) Find more Turyn sequences (Definition 4.103) and T-sequences (Definition 4.104A) and analyze the existence problem for such sequences. (Consult also Lemma 4.104, Theorem 4.107 and its corollaries.)

34) Find other constructions, similar to the Goethals-Seidel array, using 8 circulants.

35) We have found several designs in order n of type $(s_1, \ldots, s_\ell)$ where $\sum_{i=1}^{\ell} s_i = n$. Find matrices to use in these designs to construct new Hadamard matrices.

36) Complex orthogonal designs are defined and discussed in Geramita-Geramita [1].

A). If n is odd, is it true that a $CW(n, k)$ exists if and only if a $W(n, k)$ exists?

B). If n is odd and a $CW(n, k)$ exists, does this imply that $k = a^2$ for some $a \in \mathbb{Z}$?

C). If $n = 2t$, t odd, then a complex orthogonal design in order n can involve at most four variables. Is 4-variables ever possible?

D). Settle decisively the problem of how many variables can appear in a complex orthogonal design of order n.

37) Use the theory of cyclotomy, as in §4.9, to find new infinite families of orthogonal designs, amicable orthogonal designs and weighing matrices. (See also, §5.6 and §5.7.)

38) Can you settle any of the unresolved questions about the existence of balanced weighing matrices? Remember any new BW(v, k) may yield new balanced incomplete block designs.

39) Find new constructions for orthogonal designs using negacyclic matrices. In particular use the W(v, v-1) mentioned in § 4.12.

40) Solve Conjecture 8.3.

41) Solve Conjecture 8.2.

42) Solve the weighing matrix conjecture.

43) Solve the Hadamard matrix conjecture (the grand-daddy of them all.)

1) Decide if there is a W(17, 9), W(19, 9) or W(21, 9). (See Summary 8.4.)

2) Decide if there is a W(m, 16) for m = 23, 25, 27, 29, 33, 35. (See Summary 8.4.)

3) Find W(m, 25) for odd m, $m \le 81$.

4) Find W(m, 36) for odd m, $m \le 137$.

5) Find W(m, 49) for odd m, $m \le 199$.

6) Find more results of the type: there exist orthogonal designs of type (a, b) constructed from two circulants in every order 2n, $n \ge k$. (See 4.143 and 7.15.)

7) Further complete Table 8.2.

8) Find orthogonal designs in orders 34, 38, ..., 2n, ... (See 8.9-8.19.)

9) Can W(30, 9), W(30, 18), W(30, 25), W(34, 9), W(38, 9), W(38, 17) be constructed using two circulants?

10) Find more G.G.S. arrays.

11) Find orthogonal designs of type (1, k) in order 88 for $k \in \{56, 57, 62, 63, 69\}$ and hence prove Conjecture 8.3 for orders $n = 2^t \cdot 11$.

12) Find orthogonal designs of type $(1, k)$ in order 52 for $k \in \{37, 42, 45, 46, 48\}$. (See 8.38.)

13) Find orthogonal designs of type $(1, k)$ in order 104 and try to prove Conjecture 8.3 for orders $n = 2^t \cdot 13$.

14) Find orthogonal designs of type $(1, k)$ in order 60 for $k \in \{34, 42, 48, 53, 54, 56, 57\}$.

15) Find orthogonal designs of type $(1, k)$ in order 120. (Can you exploit the fact that 15 is composite to prove Conjecture 8.3 in orders $2^t \cdot 15$?)

16) Find a $W(52, 46)$ and establish the weighing matrix conjecture for order 52. (See 8.1 and 8.42.)

17) Find a $W(60, 57)$ constructed from four circulant matrices. (See 8.43.)

18) Find $W(68, k)$ for $k \in \{49, 55, 57, 61, 63, 65\}$ and establish the weighing matrix conjecture for order 68. (See 8.1 and 8.44.)

19) Find $W(76, k)$ for $k \in \{55, 61, 71, 73\}$ and so establish the weighing matrix conjecture for order 76. (See 8.1 and 8.45.)

20) Find $W(84, k)$. (See 8.46.)

21) Extend Summary 8.47.

22) In Appendix D there is a list of unknown 7 variable designs in order 32 as well as known 7, 8 and 9 variable designs. Completely study orders 32 and 64.

23) Find amicable orthogonal designs $((1, i);(j))$ in order 16 for $(i, j) = \{(13, 1), (13, 5), (13, 9), (15, 7), (15, 9), (15, 15)\}$. (See 5.145.)

24) Decide whether triplets $(5, 7, 2), \ldots, (6, 7, 7)$ (see 5.142) exist in order 8.

25) Study triplets in order 16.

26) The complete facts about the existence of amicable orthogonal designs in order 8 has not yet been settled. Fully settle this. (See 5.64 and 5.65.)

27) Study the existence of amicable orthogonal designs in order 16.

28) Study the existence of amicable orthogonal designs in order 2n (n odd) $n = 7, 9, 11$ in particular.

29) Obtain product designs in orders 4n, n odd.

30) Study amicable orthogonal designs in order 12.

31) Fully analyze the existence of product designs and repeat designs in orders 4 and 8.

32) Give new constructions for product designs and repeat designs.

33) Establish that all 3-tuples $(a, b, 40-a-b)$ are the types of orthogonal designs in order 40 and hence in orders $2^t \cdot 5$ $(t \geq 3)$. (See Section 7.1.)

34) Completely settle the existence problem for orthogonal designs in order 28. (See Appendix G.)

35) Study orthogonal designs in order 36. (See Appendices G and J.)

36) Find more sequences to put in Tables H.1, H.2, H.3, H.4, H.5, H.6.

37) Remove some numbers from Appendix K.

38) Is there an orthogonal design of type (1, 16) in orders 30, 34, or 38?

39) Is there an orthogonal design of type (1, 25) in orders 34, 38, 42, 46, 50, 54?

40) Find $W(42, k)$ for $k \in \{18, 25, 29, 36, 37\}$. (See 8.15 and 8.16.)

41) Is there an orthogonal design of type (1, 9) in orders 18 and 22?

42) Gather evidence for the conjecture: If $n \equiv 0 \pmod 4$ and $k \leq n$, then there is a symmetric $W(n, k)$.

43) In Lemma 4.123 a $W((q^k-1)/(q-1), q^{k-1})$ is constructed for q the power of an odd prime and $k \geq 3$. The case when $q = 3$ was shown to be related to incidence matrices for projective planes. In fact, for $k = 3$, (in Theorem 4.124) a circulant $W((q^k-1)/(q-1), q^{k-1})$ is constructed. For what other values of k is it possible to get a circulant matrix? Are these particular weighing matrices related to other incidence structures.

REFERENCES

Andres, T.H. and Stanton, R.G.

[1] "Golay sequences", Combinatorial Mathematics V: Proc. of Fifth Australian Conference, In Lecture Notes in Mathematics, Vol. 622, Springer-Verlag, Berlin-Heidelberg-New York, (1977), 44-54.

Artin, E.

[1] Geometric Algebra, 3rd Printing, Interscience Tracts in Pure and Applied Mathematics, # 3, Wiley, New York-London, 1974.

Baumert, L.D.

[1] Cyclic Difference Sets, Lecture Notes in Mathematics, Vol. 182, Springer-Verlag, Berlin-Heidelberg-New York, 1971.

[2] "Hadamard matrices of the Williamson type." Math. of Comp., 19 (1965), 442-447.

[3] "Hadamard matrices of orders 116 and 232." Bull. Amer. Math. Soc., 72 (1966), 237.

Baumert, L.D., Golomb, S.W. and Hall, Marshall Jr.

[1] "Discovery of an Hadamard matrix of order 92", Bull. Amer. Math. Soc., 68 (1962), 237-238.

Baumert, L.D. and Hall, Marshall Jr.

[1] "A new construction for Hadamard matrices", Bull. Amer. Math. Soc., 71 (1965), 169-170.

Belevitch, V.

[1] "Conference networks and Hadamard matrices", Ann. Soc. Scientifique Brux., T82 (1968), 13-32.

[2] "Theory of 2n-terminal networks with applications to conference telephony." Electr. Commun., 27, 3 (1950), 231-244.

Berman, Gerald

[1] "Weighing matrices and group divisible designs determined by $EG(t, p^r)$, $p > 2$", Utilitas Math. 12 (1977), 183-191.

[2] "Families of skew-circulant weighing matrices", Ars Combinatoria, 4 (1977), 293-307.

Blake, Ian

[1] "On a generalization of the Pless symmetry codes", Information and control, 27 (1975), 369-373.

Bose, R. C.

[1] "On the construction of balanced incomplete block designs", Ann. of Eugenice, 9 (1939), 353-399.

Cohn, J. N.

[1] "Hadamard matrices and some generalizations", Amer. Math. Monthly, 72 (1965), 515-518.

Cooper, Joan

[1] "A note on Hadamard arrays", Bull. Austral. Math. Soc., 10 (1974), 15-21.

[2] Some investigations of combinatorial integer matrices using cyclotomy, PhD Thesis, University of Newcastle, N.S.W., 1974. (to appear).

[3] "A binary composition for collections and sets", Proceedings of the First Australian Conference on Combinatorial Mathematics, TUNRA, Newcastle (1972), 145-161.

Cooper, Joan and Wallis, Jennifer Seberry

[1] "A construction for Hadamard arrays", Bull. Austral. Math. Soc., 7 (1972), 269-278.

[2] "Using cyclotomy to construct orthogonal designs", J. Austral. Math. Soc.

[3] "A note on orthogonal designs in order eighty", Ars. Combinatoria.

[4] "Equivalence of Hadamard matrices", Combinatorics: Proceedings of the International Conference (ed. D. A. Holton and J. Seberry), Canberra, 1977.

Curtis, Charles, W.

[1] Linear Algebra - an Introductory Approach, 3rd Edition, Allyn and Bacon, Boston, 1974.

Delsarte, P., Goethals, J.M., and Seidel, J.J.

[1] "Orthogonal matrices with zero diagonal II", Canad. J. Math., 23 (1971), 816-832.

Dembowski, P.

[1] Finite Geometries, Ergebnisse der Mathematik, Band 44, Springer-Verlag, 1968.

Eades, Peter

[1] "Orthogonal designs constructed from circulants", Utilitas Math. 11 (1977), 43-55.

[2] "Some asymptotic existence results for orthogonal designs", Ars. Combinatoria, 1 (1976), 109-118.

[3] "Some new construction for orthogonal designs using circulants", Combinatorial Mathematics V, Proc. Fifth Australian Conference, in Lecture Notes in Mathematics, Vol. 622, Springer-Verlag, Berlin-Heidelberg-New York, (1977), 64-75.

[4] On the Existence of Orthogonal Designs, PhD Thesis, Australian National University, Canberra, 1977.

Eades, Peter and Deza, M.

[1] "On circulant codes with prescribed distances," Bull. Aust. Math. Soc., 16 (1977), 361-369.

Eades, Peter and Hain, Richard M.

[1] "On circulant weighing matrices", Ars. Combinatoria 2 (1976), 265-284.

Eades, Peter and Pall, Gordon

[1] "Integral quadratic forms and orthogonal designs", (preprint).

Eades, Peter, Robinson, Peter J., Wallis, Jennifer Seberry, Williams, Ian S.

[1] "An algorithm for orthogonal designs", Proceedings of the Fifth Manitoba Conference on Numerical Mathematics, Congressus Numerantium XVI, University of Manitoba, Winnipeg, (1975), 279-292.

Eades, Peter and Wallis, Jennifer Seberry

[1] "An infinite family of skew weighing matrices", Combinatorial Mathematics IV: Proceedings of the Fourth Australian Conference, Lecture Notes in Mathematico, Vol. 560, Springer-Verlag, Berlin-Heidelberg-New York, (1976), 27-40.

[2] "Some asymptotic results for orthogonal designs: II." Colloque Sur Problemes Combinatoires et Theorie des Graphes, University of Orsay, Paris, 1976.

Eades, Peter, Wallis, Jennifer Seberry and Wormald, Nicholas

[1] "A note on asymptotic existence results for orthogonal designs", Combinatorial Mathematics V: Proc. Fifth Australian Conference, Lecture Notes in Mathematics, Vol. 622, Springer-Verlag, Berlin-Heidelberg-New York, (1977), 76-90.

Elliott, J. E. H. and Butson, A. T.

[1] "Relative difference sets", Illinois J. Math., 10 (1966), 517-531.

Florek, K.

[1] "On the evaluation from below of extremal determinants", Coll. Math., 10 (1963), 111-131.

Gabel, M. R.

[1] "Generic orthogonal stably-free projectives", J. of Algebra, 29 (1974), 477-488.

Geramita, Anthony and Geramita, Joan Murphy

[1] "Complex orthogonal designs", J. Combinatorial Theory. Ser A (to appear)(=Queen's University Math. Preprint No. 1975-10, Queen's University, Kingston, Ont. 1975).

Geramita, Anthony V., Geramita, Joan Murphy and Wallis, Jennifer Seberry

[1] "Orthogonal designs", Linear and Multilinear Algebra, 3 (1975/76), 281-306.

Geramita, A. V. and Pullman, N. J.

[1] "Radon's function and Hadamard arrays", Linear and Multilinear Algebra, 2 (1974), 147-150. (=Report Algebra Group, Queen's University, Kingston, 1972/73, (1973), 100-107.)

[2] "A theorem of Radon-Hurwitz and orthogonal projective modules", Proc. Amer. Math. Soc., 42 (1974), 51-56. (=Report Algebra Group, Queen's University, Kingston, 1972/73, (1973), 47-60.)

Geramita, Anthony V., Pullman, Norman J. and Wallis, Jennifer Seberry

[1] "Families of weighing matrices", Bull. Austral. Math. Soc., 10 (1974), 119-112. (=Report of Algebra Group, Queen's University, Kingston, 1972/73, (1973), 260-264.)

Geramita, A.V. and Verner, J.H.

[1] "Orthogonal designs with zero diagonal", Canad. J. Math. 28 (1976), 215-225.

Geramita, Anthony V. and Wallis, Jennifer Seberry

[1] "Orthogonal designs: II", Aequationes Math., 13 (1975), 299-313.

[2] "Orthogonal designs III: Weighing Matrices", Utilitas Math., 6 (1974), 209-236.

[3] "Orthogonal designs IV: Existence questions", J. Combinatorial Theory, ser. A. 19 (1975), 66-83.

[4] "A survey of orthogonal designs", Proceedings of the Fourth Manitoba Conference on Numerical Mathematics, Congressus Numerantium XII, University of Manitoba, Winnipeg, (1974), 121-168.

[5] "Some new constructions for orthogonal designs", Combinatorial Mathematics IV: Proceedings of the Fourth Australian Conference, Lecture Notes in Mathematics, Vol. 560, Springer-Verlag, Berlin-Heidelberg-New York, (1976), 46-54.

Gilman, R.E.

[1] "On the Hadamard determinant theorem and orthogonal determinants", Bull. Amer. Math. Soc., 37 (1931), 30-31.

Glynn, David

[1] PhD Thesis, University of Adelaide, (to be published).

Goethals, J.M. and Seidel, J.J.

[1] "Orthogonal matrices with zero diagonal", Canad. J. Math., 19 (1967), 1001-1010.

[2] "A skew-Hadamard matrix of order 36", J. Austral. Math. Soc., 11 (1970), 343-344.

Golay, M. J. E.

[1] "Static multisplit spectrometry and its application to the panoramic display of infrared spectra, J. Opt. Soc. Amer., 41 (1951), 468-472.

[2] "Complementary series", IRE Trans. Information Theory, IT-6 (1960), 400-408.

[3] "Note on complementary series", Proc. IRE, 50 (1962), 84.

[4] "Multislit Spectrometry", J. Opt. Soc. Amer., 39 (1949), 434-444.

Golomb, S. W., et. al.

[1] Digital communications with space applications. Prentice-Hall, Englewood Cliffs, New Jersey, 1964.

Golomb, Solomon W. and Baumert, Leonard D.

[1] "The search for Hadamard matrices", Amer. Math. Monthly, 70 (1963), 12-17.

Gordon, Basil

[1] "A note on unequivalent Hadamard matrices", J. Reine Angew. Math., (1974), 427-433.

Gregory, D.

[1] Private communication, 1973.

Griffin, Malcolm

[1] "There are no Golay sequences of length 2.9^t", Aequationes Math., 15 (1977), 73-77.

Hadamard, Jacques

[1] "Resolution d'une question relative aux determinants:, Bull. des Sciences Math., 17 (1893), 240-246.

Hain, Richard

[1] Private communication, 1976.

[2] Circulant weighing matrices, MSc Thesis, Australian National University, Canberra, 1977.

Hall, Marshall Jr.

[1] Combinatorial Theory, (Blaisdell, [Ginn and Co.], Waltham, Massachusetts, 1967).

[2] "Group properties of Hadamard matrices", J. Austral. Math. Soc., Ser. A, 21 (1976), 247-256.

Herstein, I. N.

[1] Non-commutative rings, Carus Mathematical Monographs # 15, M.A.A., 1968, (distributed by John Wiley and Sons).

Hughes, Daniel R. and Piper, Fred C.

[1] Projective Planes, Graduate Texts in Mathematics Vol. 6, Springer-Verlag, Berlin-Heidelberg-New York, 1970.

Hunt, David C.

[1] Private communication, 1973.

[2] "Skew-Hadamard matrices of order less than 100", Proc. First Aust. Conf. on Combinatorial Math., TUNRA, Newcastle, Aust., (1972), 55-59.

Hunt, David C. and Wallis, Jennifer

[1] "Cyclotomy, Hadamard arrays and supplementary difference sets", Proceedings of the Second Manitoba Conference on Numerical Mathematics, 351-381. Congressus Numerantium 7, University of Manitoba, Winnipeg, 1973.

Hurwitz, A.

[1] "Über die Komposition der Quadratischen Formen", Math. Ann. 88 (1923), 1-25.

Ibbett, R.N., Aspinall, D. and Grainger, J.F.

[1] "Real-time multiplexing of dispersed spectra in any wavelength region", Appl. Optics., 7 (1968), 1089-1093.

Jauregui, S. Jr.

[1] "Complementary sequences of length 26", IRE. Trans. Information Theory, IT-8 (1962), 323.

Kawada, Y. and Iwahori, N.

[1] "On the structure and representations of Clifford algebras", J. Math. Soc. Japan, 2 (1950), 34-43.

Krusemeyer, M.

[1] Private communications, 1975.

Kruskal, J.B.

[1] "Golay's complementary series", IRE. Trans. Information Theory, IT-7 (1961), 273-276.

Lakein, Richard B. and Wallis, Jennifer Seberry

[1] "On the matrices used to construct Baumert-Hall arrays", Combinatorics: III, Lecture Notes in Mathematics, Vol. 452, Springer-Verlag, Berlin-Heidelberg-New York, 1975, 156-170.

Lam, T.Y.

[1] The Algebraic Theory of Quadratic Forms, Mathematics Lecture Note Series, Benjamin, Reading, Mass., 1973.

Levenshtein, V.I.

[1] "The application of Hadamard matrices to a problem in coding", Problems of Cybernetics, 5 (1964), 166-184.

van Lint, J.H. and Seidel, J.J.

[1] "Equilateral point sets in elliptic geometry", Kon. Ned. Akad. Wetensch. Amst. Proceeding, Series A69 and Indag. Math., 28 (1966), 335-348.

Longyear, Judith Q.

[1] "Criteria for a Hadamard matrix to be skew-equivalent", Canad. J. Math., 28 (1976), 1216-1223.

Marcus, M. and Minc, H.

[1] A Survey of Matrix Theory and Matrix Inequalities, Allyn and Bacon, Boston, 1964.

Mullin, R.C.

[1] "A note on balanced weighing matrices", Combinatorial Mathematics III: Proceedings of the Third Australian Conference, in

Lecture Notes in Mathematics, Vol. 452, Springer-Verlag, Berlin-Heidelberg-New York, 1975, 28-41.

[2] "Normal affine resolvable designs and orthogonal matrices", Utilitas Math., 6 (1974), 195-208.

Mullin, R.C. and Stanton, R.G.

[1] "Group matrices and balanced weighing designs", Utilitas Math., 8 (1975), 277-301.

[2] "Balanced weighing matrices and group divisible designs", Utilitas Math., 8 (1975), 303-310.

[3] "Corrigenda to: 'Balanced weighing matrices and group divisible designs' ", Utilitas Math., 9 (1976), 347.

Paley, R.E.A.C.

[1] "On orthogonal matrices", J. Math. Phys., 12 (1933), 311-320.

Payne, Stanley E.

[1] The Construction of Hadamard matrices, ARL73-0117, Aerospace Research Labs, United States Air Force, Wright-Patterson Air Force Base, Dayton, Ohio, 1973.

Plackett, R.L. and Burman, J.P.

[1] "Designs of optimum multifactorial experiments", Biometrika, 33 (1946), 305-325.

Plotkin, M.

[1] "Decomposition of Hadamard matrices", J. Combinatorial Theory, ser. A. 13 (1972), 127-130.

Radon, J.

[1] "Lineare Scharen Orthogonaler Matrizen", Abhandlungen aus dem Mathematischen Seminar der Hamburgishen Universität, 1 (1922), 1-14.

Raghavarao, D.

[1] "Some aspects of weighing designs", Ann. Math. Stat., 31 (1960), 878-884.

[2] Constructions and Combinatorial Problems in Design of Experiments, Wiley Series in Probability and Mathematical

Statistics, John Wiley and Sons, Inc., New York-London-Sydney-Toronto, 1971.

Robinson, Peter J.

[1] "A non-existence theorem for orthogonal designs", Utilitas Math. 10 (1976), 179-184.

[2] "Orthogonal designs in order sixteen", Combinatorial Mathematics IV: Proc. Fourth Australian Combinatorics Conference, Lecture Notes in Mathematics, Vol. 560, Springer-Verlag, Berlin-Heidelberg-New York, (1976), 193-197.

[3] "Amicable orthogonal designs", Bull. Austral. Math. Soc., 14 (1976), 303-314.

[4] "The existence of orthogonal designs in order sixteen", Ars Combinatoria, 3 (1977), 209-218.

[5] "Using product designs to construct orthogonal designs", Bull. Austral. Math. Soc., 16 (1977), 297-305.

[6] "Orthogonal designs in order 24", Combinatorial Mathematics V: Proc. Fifth Australian Combinatorics Conf., Lecture Notes in Mathematics, Vol. 622, Springer-Verlag, Berlin-Heidelberg-New York, (1977), 170-174.

[7] Concerning the existence and construction of orthogonal designs, PhD Thesis, Australian National University, Canberra, 1977.

Robinson, Peter J. and Seberry, Jennifer

[1] "On the structure and existence of some amicable orthogonal designs", J. Austral. Math. Soc., (to appear).

[2] "Orthogonal designs in powers of two", Ars Combinatoria, 4 (1977), 43-57.

Robinson, Peter J. and Wallis, Jennifer Seberry

[1] "A note on using sequences to construct orthogonal designs", Colloquia Mathematica Societatis Janos Bolyai, (to appear).

Ryser, Herbert John

[1] Combinatorial Mathematics, Carus Math. Monograph, No. 14, M.A.A., 1963, (distributed by John Wiley and Sons).

Samuel, P.

[1] Algebraic Theory of Numbers, Hermann, Paris, 1967 (English Translation - Houghton Mifflin, Boston, 1970).

Serre, Jean-Pierre

[1] A Course in Arithmetic, (English Translation), Graduate Texts in Mathematics, Springer-Verlag, Berlin-Heidelberg-New York, 1973.

Scarpis, V.

[1] "Sui determinanti di valore massimo", Rend. R. Inst. Lombardo Sci. e Lett., No. 2, 31 (1898), 1441-1446.

Schafer, Richard D.

[1] An Introduction to Nonassociative Algebras, Pure and Applied Mathematics, # 22, Academic Press, New York-London, 1966.

Scharlau, W.

[1] "Quadratic forms", Queen's Papers on Pure and Applied Mathematics, No. 22, Queen's University, Kingston, Ont., 1969.

Schellenberg, P. J.

[1] Private communication, (1976).

Shapiro, D.

[1] Private communication, (1975).

[2] Similarities, Quadratic Forms and Clifford Algebras, PhD Thesis, University of California, Berkeley, 1970.

[3] "Spaces of similarities I - the Hurwitz problem", J. of Alg., 46 (1977), 148-170.

[4] "Spaces of similarities II - Pfister forms", J. of Algebra, 46 (1977), 171-180.

[5] "Spaces of similarities IV - (s, t)-families", Pac. J. Math., 69 (1977), 223-244.

Singer, J.

[1] "A theorem in finite projective geometry and some applications in number theory", Trans. Amer. Math. Soc., 43 (1938), 377-385.

Sloane, Neil J.A. and Harwit, Martin

[1] "Masks for Hadamard transform optics and weighing designs", Appl. Optics, 15 (1975). 107-114.

Speiser, A.

[1] Theorie der Gruppen von endliches ordnung, Springer-Verlag, Berlin, 1937.

Spence, Edward

[1] "Hadamard matrices from relative difference sets", J. Combinatorial Theory, Ser. A, 19 (1975), 287-300.

[2] "A new class of Hadamard matrices", Glasgow J., 8 (1967), 59-62.

[3] "Skew-Hadamard matrices of Goethals-Seidel type", Canad. J. Math., 27 (1975), 555-560.

[4] "Skew-Hadamard matrices of order 2(q+1)", Discrete Math., 18 (1977), 79-86.

Squire, W.D., Whitehouse, H.J. and Alsup, J.M.

[1] "Linear signal processing and ultrasonic transversal filters", IEEE Trans. Microwave Theory and Technol., MIT-17 (1969), 1020-1039.

Stanton, R.G. and Mullin, R.C.

[1] "On the non-existence of a class of circulant balanced weighing matrices", SIAM, J. Appl. Math., 30 (1976), 98-102.

Storer, Thomas

[1] Cyclotomy and Difference Sets, Lectures in Advanced Mathematics, Markham, Chicago, 1967.

Sylvester, J.J.

[1] "Thoughts on inverse orthogonal matrices, simultaneous sign successions, and tesselated pavements in two or more colours, with applications to Newton's Rule, ornamental tile-work, and the theory of numbers", Phil. Mag., No. 4, 34 (1867), 461-475.

Szekeres, G.

[1] "Cyclotomy and complementary difference sets", Acta. Arith., 18 (1971), 349-353.

[2] "Tournaments and Hadamard matrices", Enseignment Math., 15 (1969), 269-278.

[3] "Determinants of skew type ±1 matrices", Period. Math. Hungar., 3 (1973), 229-234.

[4] "The average value of skew-Hadamard matrices", Proc. First Aust. Conf. on Combinatorial Math., TUNRA, Newcastle, Aust. (1972), 55-59.

Szekeres, G. and Turan, P.

[1] "On an extremal problem in determinant theory", [In Hungarian with German abstract], Matematikai es Termeszettudomanyi Ertesito, (1937), 796-804.

Taki, Y., et. al.

[1] "Even shift orthogonal sequences", IEEE Trans. Information Theory, IT-15 (1969), 295-300.

Taussky, Olga

[1] "(1, 2, 4, 8)-sums of squares and Hadamard matrices", Combinatorics, Proc. Symp. in Pure Mathematics, Amer. Math. Soc., Providence, R.I., (1971), 229-233.

Todd, J.A.

[1] "A combinatorial problem", J. Math. and Physics, 12 (1933), 321-

Turan, P.

[1] "On an extremal problem concerning determinants", [In Hungarian with English abstract], Matematikai es Termeszetta-domanyi Ertesito, (1940), 95-105.

[2] "On a problem in the theory of determinants", Acta. Mathematica Sinica, 5 (1955), 411-423.

Turyn, Richard J.

[1] "Computation of certain Hadamard matrices", Notices Amer. Math. Soc., 20 (1973), A-1.

[2] "Hadamard matrices, Baumert-Hall units, four-symbol sequences, pulse compressions and surface wave encodings", J. Combinatorial Theory, Ser. A, 16 (1974), 313-333.

[3] "An infinite class of Williamson matrices", J. Combinatorial Theory, 12 (1972), 319-321.

[4] "On C-matrices of arbitrary powers", Bull. Canad. Math. Soc., 23 (1971), 531-535.

[5] "Ambiguity functions of complementary sequences", IEEE Trans. Information Theory, IT-9 (1963), 46-47.

[6] "Four-phase Barker codes", IEEE Trans. Information Theory, IT-20 (1974), 366-371.

Tseng, C.C.

[1] "Signal multiplexing in surface-wave delay lines using orthogonal pairs of Golay's complementary sequences", IEEE Trans. Sonics and Ultrasonics, SU-18 (1971), 103-107.

Vanstone, G.A. and Mullin, R.C.

[1] "A note on existence of weighing matrices $W(2^{2n-j}, 2^n)$ and associated combinatorial designs", Utilitas Math., 8 (1975), 371-381.

Wadsworth, A. and Shapiro, D.

[1] "Spaces of similarities III - function fields", J. of Alg. 46 (1977), 182-188.

Wallis, Jennifer

[1] "A class of Hadamard matrices", J. Combinatorial Theory, 6 (1969), 40-44.

[2] "A note of a class of Hadamard matrices", J. Combinatorial Theory, 6 (1969), 222-223.

[3] "Hadamard designs", Bull. Austral. Math. Soc., 2 (1970), 45-54.

[4] "(v, k, λ)-configurations and Hadamard matrices", J. Austral. Math. Soc., 11 (1970), 297-309.

[5] "A resume of some recent results on (v, k, λ) graphs, Hadamard matrices and block designs", Combinatorial Structures and their Application, Gordon and Breach, New York, 1970, 463-466.

[6] "Some (1,-1) matrices", J. Combinatorial Theory, ser. B., 10 (1971), 1-11.

[7] "A skew-Hadamard matrix of order 92", Bull. Austral. Math. Soc., 5 (1971), 203-204.

[8] "Amicable Hadamard matrices", J. Combinatorial Theory, ser. A., 11 (1971), 296-298.

[9] "Combinatorial matrices", Bull. Austral. Math. Soc., 5 (1971), 285-286.

[10] "On integer matrices obeying certain matrix equations", J. Combinatorial Theory, ser. A., 12 (1972), 112-118.

[11] "Orthogonal (0, 1, -1) matrices", Proceedings of the First Australian Conference on Combinatorial Mathematics", (ed. Jennifer Wallis and W. D. Wallis) TUNRA Ltd., Newcastle, Australia, (1972), 61-84.

[12] "On supplementary difference sets", Aequationes Math., 8 (1972), 242-257.

[13] "Some matrices of Williamson type", Utilitas Math., 4 (1973), 147-154.

[14] "Recent advances in the construction of Hadamard matrices", Proceedings of the Fourth Southeastern Conference on Combinatorics, Graph Theory and Computing, 53-89, Congressus Numerantium 8, University of Manitoba, Winnipeg, 1973.

[15] "A note on amicable Hadamard matrices", Utiiitas Mathematica, 3 (1973), 119-125.

[16] "Complex Hadamard matrices", Linear and Multilinear Algebra, 1 (1973), 257-272.

[17] "Hadamard matrices of order 28m, 36m, and 44m", J. Combinatorial Theory, ser. A., 15 (1973), 323-328.

[18] "Williamson matrices of even order", Combinatorial Mathematics: Proceedings of the Second Australian Conference (editor D. A. Houlton), Lecture Notes in Mathematics, Vol. 403, Springer-Verlag, Berlin-Heidelberg-New York, 1974, 132-142.

[19] "Some remarks on supplementary difference sets. Infinite and finite sets", Colloquia Mathematica Societatis Janos Bolyai, 10 (1973), 1503-1526.

[20] "On Hadamard matrices", J. Combinatorial Theory, ser. A., 18 (1975), 149-164.

[21] "Construction of amicable orthogonal designs", Bull. Austral. Math. Soc., 12 (1975), 179-182.

[22] "Orthogonal designs V: Orders divisable by eight", Ultilitas Math., 9 (1976), 263-281.

[23] "Construction of Williamson type matrices", J. Linear and Multilinear Alg., 3 (1975), 197-207.

[24] "On the existence of Hadamard matrices", J. Combinatorial Th., Ser. A., 21 (1976), 444-451.

[25] "Hadamard matrices", Part IV of Combinatorics: Room squares, sum-free sets, Hadamard matrices, Lecture Notes in Mathematics, Vol. 292, Springer-Verlag, Berlin-Heidelberg-New York, 1972.

Wallis, Jennifer Seberry and Whiteman, Albert Leon

[1] "Some classes of Hadamard matrices with constant diagonal", Bull. Austral. Math. Soc., 7 (1972), 233-249.

[2] "Some results on weighing matrices", Bull. Austral. Math. Soc., 12 (1975), 433-447.

Wallis, W.D., Street, Anne Penfold, Wallis, Jennifer Seberry

[1] Combinatorics: Room Squares, sum-free sets, Hadamard Matrices, Lecture Notes in Mathematics, 272 Springer-Verlag, Berlin-Heidelberg-New York, 1972.

Welti, G.R.

[1] "Quanternary codes for pulsed radar", IRE Trans. Information Theory, IT-6 (1960), 82-87.

Whiteman, Albert Leon

[1] "An infinite family of skew-Hadamard matrices", Pacific J. Math., 38 (1971), 817-822.

[2] "Hadamard matrices of order 4(2p+1)", J. Number Th., 8 (1976), 1-11.

[3] "Skew-Hadamard matrices of Goethals-Seidel type", Discrete Math., 2 (1972), 397-405.

[4] "Hadamard matrices of Williamson type", J. Austral. Math. Soc., 21 (1976), 481-486.

[5] "An infinite family of Hadamard matrices of Williamson type", J. Combinatorial Theory, (Series A) 14 (1973), 334-340.

Williamson, John

[1] "Hadamard's determinant theorem and the sum of four squares", Duke Math. J., 11 (1944), 65-81.

[2] "Note on Hadamard's determinant theorem", Bull. Amer. Math. Soc., 53 (1947), 608-613.

[3] "Determinants whose elements are 0 and 1", Amer. Math. Monthly, 53 (1946), 427-434.

Wilson, Richard M.

[1] "Cyclotomy and difference families in elementary abelian groups." J. Number Theory, 4 (1972), 17-47.

Wolfe, Warren W.

[1] Orthogonal designs - amicable orthogonal designs - some algebraic and combinatorial techniques, PhD Dissertation, Queen's University, Kingston, Ontario, 1975.

[2] "Clifford Algebras and Amicable Orthogonal Designs", Queen's Mathematical Preprint No. 1974-22.

[3] "Rational quadratic forms and orthogonal designs", Queen's Mathematical Preprint, No. 1975-22. =J. Number Theory, (to appear).

[4] "Amicable orthogonal designs - existence", Canad. J. Math., 28 (1976), 1006-1020.

Yang, C.Y.

[1] "On Hadamard matrices constructible by circulant submatrices", Math. of Computation, 25 (1971), 181-186.

[2] "Maximal binary matrices and sum of two squares", Math. of Computation, (to appear).

Zierler, Neal

[1] "Linear recurring sequences", J. Soc. Indust. Applied Math., 7 (1959), 31-48.

INDEX